Spectral Theory and Applications of Linear Operators and Block Operator Matrices

Aref Jeribi

Spectral Theory and Applications of Linear Operators and Block Operator Matrices

Springer

Aref Jeribi
Department of Mathematics
University of Sfax
Sfax, Tunisia

ISBN 978-3-319-17565-2 ISBN 978-3-319-17566-9 (eBook)
DOI 10.1007/978-3-319-17566-9

Library of Congress Control Number: 2015102923

Springer Cham Heidelberg New York Dordrecht London

Printed on acid-free paper

Springer International Publishing AG Switzerland is part of Springer Science+Business Media (www.springer.com)

To
my mother Sania,
my father Ali,
my wife Fadoua,
my children Adam and Rahma,
my brothers Sofien and Mohamed Amin,
my sister Elhem,
my mother-in-law Zineb,
my father-in-law Ridha, and
all members of my extended family

Preface

Several books have been devoted to the spectral theory and its applications. However, this volume is very special as compared with the previous ones. For example, the perturbation approach has been, for a long time, extensively studied and considered as one of the most useful methods used in order to study some mathematical and applied problems.

The main idea is that, if we know something about the solution for an easier problem, lying "close" to the one we are studying, then we can say something about our problem, provided that the difference or the perturbation is sufficiently weak.

In view of more advanced applications, especially the ones dealing with complicated evolutional problems in Physics, Chemistry, Technology, Biology, etc., where the natural setting doesn't involve single operators but operator matrices and polynomial operator pencils, the concept of compact perturbations is very often used, and it was shown that they were not sufficient for handling such problems. The main advantage of this book is the detailed description of the ways showing how the compactness condition can be relaxed, in a very general Banach space setting, so that the previously impossible problems become suddenly solvable. The method of extending results is not unique. That is why we have to devote a lot of space in order to describe the different extensions of the classical notions, and to demonstrate how they specifically work in different applications.

More precisely, it is well known that the essential spectrum of an operator A consists of those points of the spectrum which cannot be removed from the spectrum by the addition to A of a compact operator. The most powerful result obtained in my thesis is that, in L_1-spaces, the essential spectrum of an operator A is nothing else but the largest subset of the spectrum of A which remains invariant under weakly compact perturbations of A. This unexpected result has opened many prospects to develop innovative ways leading to a rigorous study of the Fredholm theory and in the whole book, we give an account of the recent research on the spectral theory by presenting a wide panorama of techniques including the weak topology, which contributes to an extra insight to the classical results and enables us to solve concrete problems from transport theory arising in their natural setting (L_1-spaces). The main topics include:

- Riesz theory of polynomially compact operators.
- Time behavior of solutions for an abstract Cauchy problem on Banach spaces.
- Fredholm theory and characterization of essential spectra by means of measure of noncompactness, demicompact operator, measure of weak noncompactness, and graph measures.
- S-essential spectra and essential pseudospectra.
- Spectral theory of block operator matrices.
- Spectral graph theory.
- Applications in mathematical physics and biology.

We do hope that this book will be very useful for researchers, since it represents not only a collection of a previously heterogeneous material, but also an innovation through several extensions.

Of course, it is impossible for a single book to cover such a huge field of research. In making personal choices for inclusion of material, we tried to give useful complementary references in this research area, hence probably neglecting some relevant works. We would be very grateful to receive any comments from readers and researchers, providing us with some information concerning some missing references.

We would like to thank Salma Charfi for the improvement she has made in the introduction of this book. So, we are indebted to her. We would like to thank Nedra Moalla for the improvements she has made concerning the spectral mapping theorem. We would also like to thank Aymen Ammar for the improvements he has made throughout this book. So, we are very grateful to him. Concerning the chapter related to graph theory, we were fortunate to have the help of Hatem Baloudi, who assisted in the preparation of this chapter. So, we are indebted to him. We would like to thank Professor Sylvain Golénia for his generous permission to integrate, in this book, the results of Hatem Baloudi dealing with the graph theory. Moreover, we would like to mention that the thesis work results, performed under my direction, by my former students and presently colleagues Nedra Moalla, Afif Ben Amar, Faiçal Abdmouleh, Boulbeba Abdelmoumen, Salma Charfi, Ines Walha, Bilel Krichen, Omar Jedidi, Sonia Yengui, Aymen Ammar, Naouel Ben Ali, Rihab Moalla, Hatem Baloudi, Mohammed Zerai Dhahri, and Bilel Boukettaya, the obtained results have helped us in writing this book. Last but not least, we would like to thank Ridha Damak for improving the English of all chapters of this book. Finally, we apologize in case we have forgotten to quote any author who has contributed, directly or indirectly, to this work.

Sfax, Tunisia
June 2015

Aref Jeribi

Contents

Chapter 1
Introduction

This book is devoted to some recent mathematical developments which cover several topics including Cauchy problem, Fredholm operators, spectral theory, and block operator matrices, both dealing with linear operators. Of course, these topics play a crucial role in many branches of mathematics and also in numerous applications as they are intimately related to the stability of the underlying physical systems.

One of the objectives of this book is the study of the classical Riesz theory of polynomially compact operators, in order to establish the existence results of the second kind operator equations, hence allowing to describe the spectrum, multiplicities, and localization of the eigenvalues of polynomially compact operators. Fredholm theory and perturbation results are also widely investigated. The description of the large time behavior of solutions to an abstract Cauchy problem on Banach spaces without restriction on the initial data is studied. Further, the essential state of the art of research and essential pseudo-spectra of closed, densely defined, and linear operators subjected to additive perturbations is outlined. The spectral theory of block operator matrices is of major interest, since it describes coupled systems of partial differential equations of mixed order and type. For this reason, an important part of this book is devoted to develop essential spectra of 2×2 and 3×3 block operator matrices. Based on the spectral graph theory (which is an active research area), we are interested in the study of the adjacency matrix and the discrete Laplacian acting on forms. Most of the results of this book are motivated by physical transport problems for which we address our applications at the end of the book.

Now, let us describe its contents.

A. Jeribi, *Spectral Theory and Applications of Linear Operators and Block Operator Matrices*, DOI 10.1007/978-3-319-17566-9_1

1.1 Spectral Theory and Cauchy Problem

As it is well known for the second kind operator equations

$$\lambda\varphi - A\varphi = f \tag{1.1.1}$$

in Banach spaces, the existence and uniqueness of a solution can be established by the Neumann series, provided that $\frac{A}{\lambda}$ is a contraction, i.e., $\|A\| < |\lambda|$. The basic theory for the second kind operator equation (1.1.1) with a compact linear operator A on X was developed by F. Riesz [294] and originated by I. Fredholm's work on the second kind integral equations [113]. In [171, 254], A. Jeribi and N. Moalla extended this analysis to the polynomially compact operator A in the more general setting of normed spaces. Such an extension provided some solutions for several physical problems. In fact, if A is a polynomially compact operator on a normed space X, i.e., there exists a nonzero complex polynomial $P(z) = \sum_{r=0}^{p} a_r z^r$ satisfying $P(A) \in \mathcal{K}(X)$ (the set of compact operators), and if $\lambda \in \mathbb{C}$ with $P(\lambda) \neq 0$, then we have two cases:
if the homogeneous equation

$$\lambda\varphi - A\varphi = 0 \tag{1.1.2}$$

only has the trivial solution $\varphi = 0$ then, for all $f \in X$, the non-homogeneous equation (1.1.1) has a unique solution $\varphi \in X$ which depends continuously on f.

If the homogeneous equation (1.1.2) has a nontrivial solution, then the non-homogeneous equation (1.1.1) is either unsolvable or its general solution is of the following form

$$\varphi = \tilde{\varphi} + \sum_{k=1}^{m} \alpha_k \varphi_k,$$

where $\varphi_1, \ldots, \varphi_m$ are linearly independent solutions of the homogeneous equation, $\alpha_1, \ldots, \alpha_m$ represent arbitrary complex numbers, and $\tilde{\varphi}$ denotes a particular solution of the non-homogeneous equation (1.1.1).

The structure of polynomially compact operators was described by F. Gilfeather [117] and by Y. M. Han et al. [146] in the context of Hilbert spaces. F. Gilfeather showed that every polynomially compact operator on a Banach space is the finite direct sum of translates of operators which have the property that the finite power of the operator is compact. Moreover, the spectrum of these operators can be described. This analysis was widely developed by V. I. Istrateescu in [156].

It is well known that, if X is a complex Banach space, and if $A \in \mathcal{K}(X)$, then A and A^* (the dual of A) are Riesz operators, with $\sigma(A^*) = \sigma(A)$ (see [191]). Furthermore, N. Dunford and J. T. Schwartz showed in [101] that, for any eigenvalue $\lambda \in \sigma(A)\backslash\{0\}$, we have $\text{mult}(A, \lambda) < \infty$ and $\text{mult}(A, \lambda) = \text{mult}(A^*, \lambda)$, where $\text{mult}(.,.)$ represents the algebraic multiplicity.

Let us notice that, if A is a Riesz operator on X, then A is a generalized Riesz operator on X. One of the purposes of the Chap. 2 is to prove that a polynomially compact operator is also a generalized Riesz operator.

Let $A \in \mathcal{PK}(X)$, i.e., there exists a nonzero complex polynomial $P(z) = \sum_{r=0}^{p} a_r z^r$ satisfying $P(A) \in \mathcal{K}(X)$. In [171], A. Jeribi and N. Moalla expressed the multiplicity of a nonzero eigenvalue of $P(A)$ according to the one of the eigenvalues of the operator A, and proved that, if $B \in \mathcal{L}(Y)$ such that A and B are related operators, then B is a generalized Riesz operator and $\text{mult}(A, \lambda) = \text{mult}(B, \lambda)$ for all $\lambda \in \sigma(A)\backslash\{0, z_1, \ldots, z_k\}$, where $z_1, \ldots, z_k$ are the zeroes of the minimal polynomial of A.

In [84], J. R. Cuthbert considered a class of C_0-semigroups $(T(t))_{t\geq 0}$ satisfying the property of being near the identity, which means that, for some values of t, $T(t) - I \in \mathcal{K}(X)$. Cuthbert's result asserts that, if $(T(t))_{t\geq 0}$ is a C_0-semigroup with an infinitesimal generator A, then the following conditions are equivalent:

(i) $\{t > 0 \text{ such that } T(t) - I \text{ is compact}\} =]0, \infty[$,
(ii) A is compact, and
(iii) $\lambda(\lambda - A)^{-1} - I$ is compact for some (and then, for all) $\lambda > w$,

where w denotes the type of $(T(t))_{t\geq 0}$. Cuthbert's result was extended by several authors. Their aim was to study other strongly continuous families of operators such as cosine or resolvent families of operators (see [151, 235, 240]). For example, in the paper [218], the authors have shown that the assertions (i), (ii), and (iii) remain equivalent for strongly continuous semigroups $(T(t))_{t\geq 0}$ near the identity, which explains the existence of $t_0 > 0$, such that $T(t_0) - I \in \mathcal{I}(X)$, where $\mathcal{I}(X)$ represents any arbitrary, closed, and proper two-sided ideal of the algebra $\mathcal{L}(X)$ belonging to $\mathcal{F}(X)$ (the set of Fredholm perturbations). Let us remark that, in all these works, the generator A is either compact or belongs to an ideal of $\mathcal{L}(X)$ contained in $\mathcal{F}(X)$. The general case was considered in [155], where A was a Riesz operator, not necessarily belonging to $\mathcal{F}(X)$. We say that an operator $A \in \mathcal{L}(X)$ belongs to $\mathcal{PI}(X)$, if there exists a nonzero complex polynomial $p(.)$, such that the operator $p(A) \in \mathcal{I}(X)$. In [225], the results obtained in [84, 155], and [218] were extended to semigroups for which there exists a nontrivial polynomial $p(.) \in \mathbb{C}[z]$ such that, for some $t > 0$, $p(T(t)) \in \mathcal{I}(X)$. As opposed to the previous results, in this case, the infinitesimal generator of the semigroup is not necessarily a Riesz operator.

In [229], the authors characterized the class of polynomially Riesz strongly continuous semigroups on a Banach space X. In particular, their main results assert that the generators of such semigroups are either polynomially Riesz (then bounded) or there exist two closed, infinite-dimensional, and invariant subspaces X_0 and X_1 of X with $X = X_0 \oplus X_1$, such that the part of the generator in X_0 is unbounded with a resolvent of Riesz type, whereas its part in X_1 is a polynomially Riesz operator.

In Chap. 2 of his thesis [351], M. Yahdi discussed the topological complexity of some subsets of $\mathcal{L}(X)$, under the assumptions that X is a separable Banach space and $\mathcal{L}(X)$ is endowed with the strong operator topology. In particular, M. Yahdi showed that the families of stable, ergodic, and power-bounded operators constitute some Borel subsets of $\mathcal{L}(X)$, whereas the set of superstable operators is coanalytic. We should notice that other results in this direction may be found in Chap. 4 of Yahdi's work.

In [227], the authors have firstly obtained more results in the spirit of those obtained in [351]. Secondly, they applied these new results in order to derive the topological complexity of some subsets of $\mathcal{L}(X)$ equipped with the strong operator topology and also to discuss the properties of strongly continuous semigroups in Banach spaces under some hypotheses. In fact, they presented a characterization of strongly continuous semigroups $(T(t))_{t\geq 0}$ under the assumption that, for all $t > 0$, $\sigma_{e4}(T(t)) = \{\lambda(t)\}$ on Banach spaces with separable duals, where $\sigma_{e4}(.)$ represents the Wolf essential spectrum. We notice that the semigroups satisfying this condition were already discussed in [84, 218].

In the papers [84, 218, 226, 227], the various conditions ensuring the uniform continuity of strongly continuous semigroups and groups were discussed. The common results of these works deal with groups such that, for all real t, $\sigma_{e4}(T(t))$ (representing the spectrum of $T(t)$ in the Calkin algebra) is a finite set; therefore, $\{t \in \mathbb{R}$ such that $\sigma^1_{e4}(T(t)) \neq \mathbb{T}\}$ is equal to $\mathbb{R}$, where $\sigma^1_{e4}(T(t)) := \left\{\frac{\lambda}{|\lambda|} \text{ such that } \lambda \in \sigma_{e4}(T(t))\right\}$ and $\mathbb{T}$ denote the unit circle of $\mathbb{C}$. In particular, $\{t \in \mathbb{R}$ such that $\sigma^1_{e4}(T(t)) \neq \mathbb{T}\}$ has a nonempty interior. In [228], the authors used a weak form of the last observation in order to get a characterization of the uniform continuity of strongly continuous groups. More precisely, they proved that a strongly continuous group $(T(t))_{t\in\mathbb{R}}$ on a Banach space is uniformly continuous if, and only if, $\{t \in \mathbb{R}$ such that $\sigma^1(T(t)) \neq \mathbb{T}\}$ is non-meager, where $\sigma^1(T(t)) := \left\{\frac{\lambda}{|\lambda|} \text{ such that } \lambda \in \sigma(T(t))\right\}$. In particular, they showed that this result holds true if the spectrum is replaced by smaller versions of the spectrum.

1.2 Time Behavior of Solutions to an Abstract Cauchy Problem on Banach Spaces

Let us first notice that the time-dependent linear transport equations arise in a number of diverse applications in biology, chemistry, and physics [150, 275, 296, 336]. These equations can be formulated, in a Banach space X, as the following Cauchy problem

$$\begin{cases} \dfrac{\partial \psi}{\partial t} = A\psi := T\psi + F\psi \\ \psi(0) = \psi_0, \end{cases} \tag{1.2.3}$$

where $T : \mathcal{D}(T) \subseteq X \longrightarrow X$ denotes the infinitesimal generator of a C_0-semigroup of bounded linear operators $(U(t))_{t\geq 0}$ acting on X. Moreover, F is a bounded linear operator on X and $\psi_0 \in X$. Since $A := T + F$ is a bounded perturbation of T, according to the classical perturbation theory, then it generates a C_0-semigroup

$(V(t))_{t\geq 0}$ which solves the Cauchy problem (1.2.3) (see [35, 276]) and which is given by the Dyson–Phillips expansion

$$V(t) = \sum_{j=0}^{n-1} U_j(t) + R_n(t),$$

where $U_0(t) = U(t)$, $U_j(t) = \int_0^t U(s)FU_{j-1}(t-s)\,ds$, $j = 1, 2, \ldots$ and the nth order remainder term $R_n(t)$ can be expressed by

$$R_n(t) = \int_{s_1+\cdots+s_n\leq t,\ s_i\geq 0} U(s_1)F \ldots U(s_n)FV(t - s_1 - \cdots - s_n)\,ds_1 \ldots ds_n.$$

As it is well known, the solution $\psi(t)$ of the problem (1.2.3) exists and is unique for all $\psi_0 \in \mathcal{D}(A)$. When dealing with the time-asymptotic behavior of $\psi(t)$, a useful technique (called the semigroup approach) consists in studying the asymptotic spectrum of $V(t)$ (for more information and a discussion of the recent results, we may refer to [262, Chap. 2]). It is based on the existence of a compact nth order remainder term, $R_n(t)$, of the Dyson–Phillips expansion. Indeed, if some remainder term $R_n(t)$ is compact, then $\sigma(V(t))\bigcap\{\alpha \in \mathbb{C}$ such that $|\alpha| > e^{\eta t}\}$ consists of, at most, isolated eigenvalues with finite algebraic multiplicities and therefore, according to the spectral mapping theorem for the point spectrum, for any $\nu > \eta$, we have $\sigma(A)\bigcap\{\mathrm{Re}\lambda \geq \nu\}$ consists of several finitely isolated eigenvalues $\{\lambda_1, \ldots, \lambda_n\}$. Let $\beta_1 = \sup\{\mathrm{Re}\lambda,\ \lambda \in \sigma(A),\ \mathrm{Re}\lambda < \nu\}$, and let $\beta_2 = \min\{\mathrm{Re}\lambda_j,\ 1 \leq j \leq n\}$. The solution of the problem (1.2.3) satisfies the following equation

$$\left\| \psi(t) - \sum_{i=1}^{n} e^{\lambda_i t} e^{D_i t} P_i \psi_0 \right\| = o(e^{\beta^* t}) \text{ with } \beta_1 < \beta^* < \beta_2, \tag{1.2.4}$$

where $\psi_0 \in \mathcal{D}(A)$, P_i and D_i denote, respectively, the spectral projection and the nilpotent operator associated with λ_i, $i = 1, 2, \ldots, n$. The success of this method is related to the possibility of computing some remainder terms of the Dyson–Phillips series and also to the possibility of discussing their compactness properties. Unfortunately, in some applications, it may happen that the unperturbed semigroup $(U(t))_{t\geq 0}$ is not explicit (this is the case for the streaming operator for general boundary conditions [210]) and therefore, this approach does not work.

An alternative way to discuss the time structure of $\psi(t)$ (called the resolvent approach) is the one initiated by I. Vidav [328] (in a particular case) and developed by M. Mokhtar–Kharroubi [261] (in an abstract setting). In [232], J. Lehner and M. Wing determined the long-time behavior of $\psi(t)$ and expressed $\psi(t)$ as an inverse Laplace transform of $(\lambda - T - F)^{-1}\psi_0$. This technique was systematized in an abstract setting by M. Mokhtar–Kharroubi [261] who has shown that, under the following conditions

$(\mathcal{A}_1)$: There exists an integer m such that $[(\lambda - T)^{-1}F]^m$ is compact for $\mathrm{Re}\lambda > \eta$, where η is the type of $\{U(t), t \geq 0\}$.

$(\mathcal{A}_2)$: There exists an integer m such that
$$\lim_{|\mathrm{Im}\lambda| \to +\infty} \|[(\lambda - T)^{-1}F]^m\| = 0 \text{ uniformly on } \{\lambda \in \mathbb{C} \text{ such that } \mathrm{Re}\lambda \geq \omega\} \ (\omega > \eta),$$

we deduce that $\sigma(A) \bigcap \{\lambda \in \mathbb{C}$ such that $\mathrm{Re}\lambda > \eta\}$ consists, at most, of discrete eigenvalues with finite algebraic multiplicities and $\sigma(A) \bigcap \{\lambda \in \mathbb{C}$ such that $\mathrm{Re}\lambda \geq \omega,\ \omega > \eta\} = \{\lambda_i,\ i = 1, \ldots, n\}$ is finite. Moreover, for any initial data $\psi_0 \in \mathcal{D}(A^2)$, the solution of the Cauchy problem (1.2.3) satisfies the following equation

$$\left\| \psi(t) - \sum_{i=1}^{n} e^{\lambda_i t} e^{D_i t} P_i \psi_0 \right\| = o(e^{\beta^* t}) \text{ with } \beta_1 < \beta^* < \beta_2, \tag{1.2.5}$$

where $\beta_1 = \sup\{\mathrm{Re}\lambda$ such that $\lambda \in \sigma(A)$ and $\mathrm{Re}\lambda < \omega\}$, $\beta_2 = \min\{\mathrm{Re}\lambda_i,\ 1 \leq i \leq n\}$, P_i and D_i denote, respectively, the spectral projection and the nilpotent operator associated with $\lambda_i,\ i = 1, 2, \ldots, n$. Clearly, the weakness of this approach lies in the fact that, unlike (1.2.4), the quantity

$$\psi(t) - \sum_{i=1}^{n} e^{\lambda_i t} e^{D_i t} P_i \psi_0$$

can be evaluated only if the initial data ψ_0 is in $\mathcal{D}(A^2)$. This analysis (1.2.5) was applied by M. Mokhtar–Kharroubi in [261] in order to study the asymptotic behavior of solutions to a transport equation with vacuum boundary conditions in bounded geometry. K. Latrach also applied this analysis in [213] for the study of solutions to one-dimensional transport equation for a large class of boundary conditions. In [166], A. Jeribi studied the spectral analysis of a class of unbounded, linear operators, originally proposed by M. Rotenberg. He gave a spectral decomposition of solutions into an asymptotic term and a transient one which will be estimated for smooth initial data. In [174], A. Jeribi et al. studied the time-asymptotic of solutions of Rotenberg's model of cell populations with general boundary conditions in L_p-spaces ($p \geq 1$). In [177], A. Jeribi, S. Ould Ahmed Mahmoud, and R. Sfaxi studied the time-asymptotic of solutions to a transport equation with Maxwell boundary condition in L_1-space. In [312], D. Song has pointed out that, if the conditions $(\mathcal{A}_1)$ and $(\mathcal{A}_2)$ are fulfilled and if X is a B-convex (resp. Hilbert) space, then the requirement $\psi_0 \in \mathcal{D}(A^2)$ is unnecessary and (1.2.5) holds true for any initial data ψ_0 in $\mathcal{D}(A)$ (resp. in X). Recently, in [313], D. Song studied the time-dependent neutron transport equations with reflecting boundary conditions and considered two typical geometries (slab geometry and spherical geometry) in the setting of L_p including L_1. His work was mainly devoted to eliminate such conditions imposed on the initial distribution for time-dependent transport equation with reflecting boundary conditions under very general assumptions. In [166, 169, 174, 177, 222], under adequate assumptions, the authors proved that, for one-dimensional transport equations concerning a large class of boundary conditions in L_p spaces ($p \in]1, +\infty[$), (1.2.5) holds true for any initial data ψ_0 belonging to $\mathcal{D}(A)$. In [4], B. Abdelmoumen, A. Jeribi, and M. Mnif followed the investigation started in the works [166, 174, 177, 213, 220, 222, 223, 312, 313] concerning the time-asymptotic of solutions to the Cauchy problem of one-dimensional transport equations for a sizable class of anisotropic scattering operators and a variety of boundary conditions. They showed that the solution $\psi(t)$ possesses a nice behavior on L_p spaces, $p \in [1, +\infty[$, independently of the geometry of such spaces, and also satisfies an asymptotic expansion similar to (1.2.5), without restriction on the initial data, even for $p = 1$. Their strategy consisted in proving that, under adequate assumptions, and for all $r \in [0, 1[$, we have

$$\lim_{|\mathrm{Im}\lambda| \to \infty} |\mathrm{Im}\lambda|^r \|\mathcal{K}(\lambda - T_H)^{-1}\mathcal{K}\| = 0 \text{ uniformly on a half plane,}$$

where T_H (resp. $\mathcal{K}$) is the advection (resp. collision) operator, and in using the inverse Laplace transform.

The analysis in [261] was clarified and refined later on by B. Abdelmoumen, A. Jeribi, and M. Mnif [1, 4] who showed that the result (1.2.5) is satisfied even if $\psi_0 \in \mathcal{D}(A)$. Their strategy consisted in replacing the assumption $(\mathcal{A}_2)$ by the following one:

$$(\mathcal{A}_3) \begin{cases} \text{(i) There exist an integer } m, \text{ and a real } r_0 > 0, \text{ for } \omega > \eta, \text{ there exists } C(\omega) \text{ such} \\ \quad \text{that } |\mathrm{Im}\lambda|^{r_0} \|[(\lambda - T)^{-1}F]^m\| \text{ is bounded on } \{\lambda \in \mathbb{C},\ \mathrm{Re}\lambda \geq \omega,\ |\mathrm{Im}\lambda| \geq C(\omega)\}. \\ \text{(ii) There exists a real } c \text{ such that } \|(\lambda - A)^{-1}\| \text{ is bounded on } \{\lambda \in \mathbb{C},\ \mathrm{Re}\lambda \geq c\}. \end{cases}$$

S. Charfi, A. Jeribi, and N. Moalla applied the last analysis in [77] in order to describe the time-asymptotic behavior of the solution to a one-velocity transport operator with Maxwell boundary conditions considered in a homogeneous medium with a spherical symmetry and an isotropic scattering. Another application to a transport operator with a diffuse reflection boundary condition was given by S. Charfi in [75].

Recently, in [3, 157], B. Abdelmoumen, O. Jedidi, and A. Jeribi performed an improvement of the first condition of assumption $(\mathcal{A}_3)$ which was replaced by the following:

$$\begin{cases} \text{There exists a real } r_0 > 0, \text{ for } \omega > \eta, \text{ there exists } C(\omega) \text{ such that} \\ \||\mathrm{Im}\lambda|^{r_0} \|(\lambda - T)^{-1} B_\lambda^m F(\lambda - A)^{-1}\| \text{ is bounded on} \\ \{\lambda \in \mathbb{C} \text{ such that } \mathrm{Re}\lambda \geq \omega, \text{ and } |\mathrm{Im}\lambda| \geq C(\omega)\}, \end{cases}$$

where $B_\lambda := F(\lambda - T)^{-1}$. Their results were also applied in order to characterize the time-asymptotic behavior of the transport operator introduced by M. Rotenberg [296].

Motivated by problems in transport theory, a perturbation technique for studying the spectrum of $V(t)$ was initiated by I. Vidav [329]. His analysis was clarified and refined later on by J. Voigt [331] who showed that, if

$$(\mathcal{A}_4) \begin{cases} \text{there exist } m, n \in \mathbb{N} \text{ such that } (R_n(t)B)^m \text{ is} \\ \text{compact for large } t \text{ and for all } B \in \mathcal{L}(X), \end{cases}$$

then $\sigma(V(t)) \bigcap \{\alpha \in \mathbb{C} \text{ such that } |\alpha| > e^{wt}\}$ consists of, at most, isolated eigenvalues with finite algebraic multiplicity. Within the framework of positive semigroups, the author in [260] has shown the existence of several connections between the assumptions $(\mathcal{A}_1)$ and $(\mathcal{A}_4)$. In particular, the spectral analysis of the perturbed semigroup $V(t)$ is possible with assumptions of type $(\mathcal{A}_1)$. In [215], the author generalized and extended some of the results obtained in [260] to general Banach space contexts. Indeed, he proved that, if there exists $n \in \mathbb{N}$, $n \neq 0$ such that

$$\begin{cases} F\prod_{i=1}^{n}\big((\lambda - T)^{-1}U(t_i)F\big) \text{ is compact for all } n\text{-tuples } (t_1,\ldots,t_n),\ t_i > 0, \\ (t_1,\ldots,t_n) \longrightarrow F\prod_{i=1}^{n}(U(t_i)F) \text{ and is continuous in the uniform topology,} \end{cases}$$

then the remainder $R_{n+1}(t)$ of order $n+1$ in the Dyson–Phillips expansion $V(t) = \sum_{j=0}^{\infty} U_j(t)$ is compact. This shows that the spectral analysis of the perturbed semigroup $V(t)$ is also possible on general Banach spaces (without the constraint of positivity).

1.3 Fredholm Theory and Essential Spectra

The concept of Fredholm operators is one of the attempts to understand the classical Fredholm theory of integral equations. Special types of these operators were considered by many authors and treated in the works of F. V. Atkinson [39], I. C. Gohberg [119–121, 125], and B. Yood [353]. These papers considered bounded operators. Generalizations to unbounded operators were given by M. G. Krein and M. A. Krasnoselskii [196], B. Sz. Nagy [267], R. Kress [197], and I. C. Gohberg [122]. More complete treatments were given by P. Aiena [17], I. C. Gohberg and G. Krein [123], and T. Kato [185]. A general historical account of the theory of Fredholm operators was done in [123]. Further important contributions were due to M. Schechter [300] who gave a simple and unified treatment of this theory which covered all the basic points while avoiding some of the involved concepts already used by previous authors.

These studies of Fredholm theory and perturbation results are of a great importance in the description of the essential spectrum. This set is not as widely known as other parts of the spectrum, in particular, eigenvalues. Nevertheless, in several applications, information about it is interesting for various reasons. In fact, if the whole spectrum lies in a half-plane, then the stability of a physical system is guaranteed. Moreover, near the essential spectrum, numerical calculations of eigenvalues become difficult. Hence, they have to be treated analytically. If the essential spectrum of a closed linear operator is empty, and if its resolvent set is nonempty, then the spectrum consists only of isolated eigenvalues with a finite algebraic multiplicity which accumulate, at most, at ∞.

It is well known that, if A is a self-adjoint operator on a Hilbert space, then the essential spectrum of A is the set of limit points of the spectrum of A (with eigenvalues counted according to their multiplicities), i.e., all points of the spectrum, except some isolated eigenvalues of finite multiplicity (see, for example, [347, 348]). Irrespective of whether A is bounded or not on a Banach space X, there are many ways to define the essential spectrum, most of them are enlargement of the continuous spectrum. Hence, several definitions of the essential spectrum may be found in the literature (see, for example, [144, 299]) or in the comments in [302, Chap. 11, p. 283] which coincide for self-adjoint operators on Hilbert spaces.

The concept of essential spectra was introduced and studied by several mathematicians. We can cite H. Weyl, T. Kato, M. Schechter, F. E. Browder, and R. Mennicken (see, for instance [66, 110, 158, 159, 185, 248, 302, 342]). Further important contributions concerning essential spectra and their applications to transport operators were due to A. Jeribi and his collaborators (see [1, 9, 12, 20, 55, 57, 61, 74, 162–165, 167, 170, 172, 198, 256, 335, 352]).

When dealing with the essential spectra of closed, densely defined, and linear operators on Banach spaces, one of the main problems consists in studying the invariance of the essential spectra of these operators subjected to various kinds of perturbation. In a Banach space X, the Schechter essential spectrum of the operator $A \in \mathcal{C}(X)$ is defined by

$$\sigma_{e5}(A) = \bigcap_{K \in \mathcal{K}(X)} \sigma(A + K),$$

where $\mathcal{K}(X)$ stands for the ideal of all compact operators on X.

One of the main questions in the study of the Schechter essential spectrum of closed and densely defined linear operators on Banach spaces X consists in showing what are the required conditions for $K \in \mathcal{L}(X)$ such that, for $A \in \mathcal{C}(X), \sigma_{e5}(A + K) = \sigma_{e5}(A)$. If K is a compact operator on Banach spaces, then $\sigma_{e5}(A + K) = \sigma_{e5}(A)$.

In [160, 168, 219], the Jeribi essential spectrum was defined by the set

$$\sigma_j(A) := \bigcap_{K \in \mathcal{W}_*(X)} \sigma(A + K),$$

where $\mathcal{W}_*(X)$ stands for each one of the sets $\mathcal{W}(X)$ and $\mathcal{S}(X)$ and where $\mathcal{W}(X)$ (resp. $\mathcal{S}(X)$) denotes the set of all weakly compact (resp. strictly singular) operators on X which contains the set of compact operators. At first sight, $\sigma_j(A)$ and $\sigma_{e5}(A)$ seem to be not equal. However, in L_1-spaces, it was proved in A. Jeribi's thesis [160] that $\sigma_j(A) = \sigma_{e5}(A)$. This result was entirely unexpected. That is why, we have decided to investigate this important and new characterization, hence leading to several studies dealing with the stability of the essential spectra. In fact, if K is strictly singular on L_p-spaces, then $\sigma_{e5}(A + K) = \sigma_{e5}(A)$ (see [221]). If K is weakly compact on Banach spaces which possess the Dunford-Pettis property, then $\sigma_{e5}(A + K) = \sigma_{e5}(A)$ (see [214]). If $K \in \mathcal{L}(X)$ and $(\lambda - A)^{-1}K$ is strictly singular (resp. weakly compact) on L_p-spaces $p > 1$ (resp. on Banach spaces which possess the Dunford-Pettis property), then $\sigma_{e5}(A + K) = \sigma_{e5}(A)$ (see [161, 162]). In [163], A. Jeribi extended this analysis of the Schechter essential spectrum to the case of general Banach spaces, and he proved that $\sigma_{e5}(A + K) = \sigma_{e5}(A)$ for all $K \in \mathcal{L}(X)$ such that $(\lambda - A)^{-1}K \in \mathcal{I}(X)$, where $\mathcal{I}(X)$ is an arbitrary two-sided ideal of $\mathcal{L}(X)$ satisfying $\mathcal{K}(X) \subset \mathcal{I}(X) \subset \mathcal{F}(X)$. In [164], A. Jeribi made an extension of his work [163], where a detailed treatment of the Schechter essential spectrum of closed, densely defined, and linear operators A subjected to additive perturbations K, such that $(\lambda - A)^{-1}K$ or $K(\lambda - A)^{-1}$ belonging to arbitrary subsets of $\mathcal{L}(X)$ (where X denotes a Banach space) contained in the ideal of Fredholm perturbations. His strategy consisted mainly in considering the class of A-closable operator K (not necessarily bounded) which is contained in the set $A\mathcal{J}(X)$, and in proving that $\sigma_{e5}(A + K) = \sigma_{e5}(A)$ for all K in any subset of operators in $A\mathcal{J}(X)$, where $A\mathcal{J}(X)$ designates the set of A-resolvent Fredholm perturbations which zero index. More precisely, let $A \in \mathcal{C}(X)$, then $\sigma_{e5}(A + K) = \sigma_{e5}(A)$ for all $K \in \mathcal{C}(X)$ such that K is A-bounded and $K(\lambda - A - K)^{-1} \in \mathcal{J}(X)$ for all $\lambda \in \rho(A + K)$, where $\mathcal{J}(X)$ designates the set of bounded operators A such that $I + A$ has a zero index. A detailed treatment of the Schechter essential spectrum of closed, densely defined, and linear operators A subjected to additive perturbations contained in the set of A-Fredholm perturbations is given in this book. In [170], A. Jeribi and M. Mnif expressed the Schechter essential spectrum in terms of n-strictly power-compact operators on X and a spectral mapping theorem for the Schechter essential spectrum is also derived. In fact, these results have extended and improved several known ones in the literature (see [11, 45, 158, 159, 161–164, 167, 168, 214, 216, 221, 224]).

Many spectral and basic properties about essential spectra were given in [5, 12, 14, 24, 60] in order to establish the criteria for both the sum and the product of some essential spectra. Moreover, a characterization of approximate point and defect essential spectra by means of semi-Fredholm perturbation operators was investigated by A. Jeribi and N. Moalla in [172]. Moreover, a treatment of the invariance of such essential spectra by the sets of A-Fredholm, upper A-semi-Fredholm and lower A-semi-Fredholm perturbations was performed by A. Jeribi in [164]. H. Baloudi and A. Jeribi have considered in [45] the sum of two bounded linear operators defined on a Banach space and have presented some new and quite general conditions to investigate their essential spectra.

It is worth noticing that the measures of noncompactness and weak noncompactness were introduced respectively by K. Kuratowski [203] and F. S. De Blasi [90] and have been successfully applied in topology, functional analysis, and operator theory in Banach spaces. They were also used in the studies of functional equations, ordinary and partial differential equations, fractional partial differential equations, integral and integro-differential equations, optimal control theory, and in the characterization of compact operators between Banach spaces (see [18, 37, 38, 46, 47, 97, 104, 132, 207, 253, 289]). In [5, 7, 8], B. Abdelmoumen, A. Dehici, A. Jeribi, and M. Mnif also gave some results concerning a certain class of semi-Fredholm and Fredholm operators via the concept of measures of noncompactness and weak noncompactness. Further, they applied their obtained results in order to prove the invariance of the Schechter essential spectrum on Banach spaces, hence establishing a fine description of the Schechter essential spectrum of closed, densely defined, and linear operators. In the paper [255], N. Moalla has used the notion of measure of non-strict-singularity to give some results on Fredholm operators and she has established a fine description of the Schechter essential spectrum of a closed densely defined linear operator.

New measures called, respectively, graph measure of noncompactness and graph measure of weak noncompactness were defined by B. Abdelmoumen, A. Jeribi, and M. Mnif in [1, 6] in order to discuss the incidence of some perturbation results on the behavior of essential spectra of such closed, densely defined, and linear operators on Banach spaces. These new notions of measure were defined in a Banach space X, in the following way: for $T \in \mathcal{C}(X)$, if ν (resp. μ) is a measure of noncompactness (resp. measure of weak noncompactness) in X, the following notations are considered in [1, 6]:

$\mathcal{H}_\nu$ denotes the set of all $T \in \mathcal{C}(X)$ satisfying : for all A_n being bounded and closed in X_T such that $A_{n+1} \subset A_n$, if $\lim\limits_{n\to+\infty} \nu(A_n) = 0$ and $\lim\limits_{n\to+\infty} \nu(T(A_n)) = 0$, then $\bigcap_{n=0}^{+\infty} A_n \neq \emptyset$.
$\mathcal{H}_\mu$ denotes the set of all $T \in \mathcal{C}(X)$ satisfying : for all A_n being bounded and weakly closed in X_T such that $A_{n+1} \subset A_n$, if $\lim\limits_{n\to+\infty} \mu(A_n) = 0$ and $\lim\limits_{n\to+\infty} \mu(T(A_n)) = 0$, then $\bigcap_{n=0}^{+\infty} A_n \neq \emptyset$.

For $T \in \mathcal{H}_\nu$, new measures were constructed in [1, 6] and defined as follows :

$$\begin{aligned} \nu_T : \mathcal{M}_{X_T} &\longrightarrow \mathbb{R}_+ \\ A &\longrightarrow \nu_T(A) = \nu(A) + \nu(T(A)), \end{aligned}$$

where $\mathcal{M}_{X_T}$ denotes the set of all bounded in $X_T := (\mathcal{D}(T), \|.\|_T)$ (which is a Banach space endowed with the graph norm $\|.\|_T := \|.\| + \|T(.)\|$). It was shown that ν_T is a measure of noncompactness in X_T called graph measure of noncompactness associated with T and ν. If $X^* + X^* \circ T$ is dense in $(X_T)^*$, then for a measure of weak noncompactness μ and $T \in \mathcal{H}_\mu$, the map

$$\begin{aligned} \nu_T : \mathcal{M}_{X_T} &\longrightarrow \mathbb{R}_+ \\ A &\longrightarrow \mu_T(A) = \mu(A) + \mu(T(A)) \end{aligned}$$

is a measure of weak noncompactness in X_T called graph measure of weak noncompactness associated with T and μ.

W. V. Petryshyn has proved in [280] that $I - K$ is a Fredholm operator and $i(I - K) = 0$ for every condensing operator K. By using the concept of demicompact operators, W. Chaker, A. Jeribi, and B. Krichen in [73] extended the result to a class of bounded operators K which ensures that μK is demicompact for every $\mu \in (0, 1]$. This class contains the classes of power-compact operators, n-strictly power-compact operators, condensing operators, and demicompact 1-set-contraction operators. Their obtained results were used in order to establish a fine description of the Schechter essential spectrum of closed, densely defined, and linear operators, and to investigate the essential spectrum of the sum of two bounded linear operators defined on a Banach space by means of the essential spectrum of each of the two operators. Recently, B. Krichen introduced in [199] the concept of S-demicompactness with respect to a closed, densely defined, and linear operator, as a generalization of the class of demicompact operators introduced by W. V. Petryshyn in [280] and established some new results in Fredholm theory. Moreover, he applied the obtained results for discussing the incidence of some perturbation results on the behavior of relative essential spectra of unbounded linear operators acting on Banach spaces. He also gave a characterization of the relative Schechter's and approximate essential spectrum.

In [2], the authors have investigated the stability in the set of Fredholm perturbations under composition with bounded operators. They have introduced the concept of a measure of non-Fredholm perturbations which allowed them to give a general approach to the question of obtaining perturbation theorems for semi-Fredholm operators. Finally, they have proved some localization results about the Wolf, the Schechter, and the Browder essential spectra of bounded operators on a Banach space X.

The concept of semi-regularity, and essentially semi-regularity among the various concepts of regularity, originated from the classical treatment of perturbation theory owing to T. Kato and its development has greatly benefited from the work of many authors in the last years, such as M. Mbekhta and A. Ouahab [246], V. Müller [263], V. Rakocevič [286], M. Berkani, and A. Ouahab [62]. Let us recall that an operator A is said to be semi-regular if $R(A)$ is closed and $N(A^n) \subseteq R(A)$, for all $n \geq 0$ (see [246]), where $R(A)$ and $N(A)$ denote the range and the null space of A, respectively. This concept leads naturally to the semi-regular spectrum $\sigma_{se}(A)$, which is an important subset of the ordinary spectrum, defined as the set of all $\lambda \in \mathbb{C}$ for which $\lambda - A$ is not semi-regular and its essential version $\sigma_{es}(A)$ is the set of all $\lambda \in \mathbb{C}$ for which $\lambda - A$ is not essentially semi-regular. The semi-regular spectrum was first introduced by C. Apostol [31] for operators on Hilbert spaces and successively studied by several authors mentioned above in the more general context of operators acting on Banach spaces. An operator A is called a Kato type operator, if we can write $A = A_1 \oplus A_0$ where A_0 is a nilpotent operator and A_1 is a semi-regular one. In 1958, T. Kato proved that a closed semi-Fredholm operator is of Kato type. J. P. Labrousse [205] studied and characterized a new class of operators (named quasi-Fredholm operators) in the case of Hilbert spaces, and he proved that this class coincides with the set of Kato type operators and the Kato decomposition becomes a characterization of the quasi-Fredholm operators. However, in the case of Banach spaces, the Kato type operator is also quasi-Fredholm, but the converse is not true. The study of such a class of operators leads to a new important part of the ordinary spectrum (called the Kato spectrum $\sigma_k(A)$) which represents the set of all complex λ such that $\lambda - A$ is not a Kato type operator. In [61], M. Benharrat, A. Ammar, A. Jeribi, and B. Messirdi studied some properties of the semi-regular, essentially semi-regular, and the operators of Kato type on a Banach space. They have also shown that the essentially semi-regular spectrum of closed, densely defined linear operator is stable under commuting compact perturbation and its Kato spectrum is stable subjected to additive commuting nilpotent perturbations.

1.4 S-Essential Spectra and Essential Pseudospectra

Several mathematical and physical problems lead to operator pencils, $\lambda S - T$ (operator-valued functions of a complex argument) (see for example, [243, 310]). Recently, the spectral theory of operator pencils attracted the attention of many mathematicians. Mainly, the completeness of the root vectors and the asymptotic distributions of characteristic values were considered. In [40], F. V. Atkinson, H. Langer, R. Mennicken, and A. A. Shkalikov proposed a method for dealing with the spectral theory related to pencils of the form $L_0 - \mu M$, where L_0 and M are 2×2 block operator matrices acting on a Banach space and where M is invertible. Then, the authors applied their techniques to a problem occurring in magnetohydrodynamics (see [233]) where the entries of L_0 were ordinary differential operators and M was the identity operator. These entries were completely able to characterize the essential spectrum of the operator L_0. In the same book [233], the authors also considered a problem wherein the entries of L_0 were partial differential operators acting on a rectangular domain in $\mathbb{R}^2$ with coefficients having singularities at one edge of the boundary, and some properties of the essential spectrum of L_0 were proved. In [176], A. Jeribi, N. Moalla, and S. Yengui gave a characterization of the essential spectrum of the operator pencil in order to extend many known results in the literature.

Inspired by [176], F. Abdmouleh, A. Ammar, and A. Jeribi in [13] pursued the study of the S-essential spectra and investigated the S-Browder, the S-upper semi-Browder, and the S-lower semi-Browder essential spectra of bounded linear operators on a Banach space X and they introduced the S-Riesz projection. Moreover, they extended the results of F. Abdmouleh and A. Jeribi [12] to various types of S-essential spectra. In fact, they gave the characterization of the S-essential spectra of the sum of two bounded linear operators.

Historically, the concept of pseudo-spectra was introduced independently by J. M. Varah [327], H. Landau [208], L. N. Trefethen [322], and E. B. Davies [88]. This concept was especially due to L. N. Trefethen, who developed this idea for matrices and operators, and who applied it to several highly interesting problems. This notion of pseudo-spectra arised as a result of the realization that several pathological properties of highly non-self-adjoint operators were closely related. These include the existence of approximate eigenvalues far from the spectrum, the instability of the spectrum even under small perturbations. The analysis of pseudo-spectra has been performed in order to determine and localize the spectrum of operators, hence leading to many applications of the pseudo-spectra. For example, in aeronautics, eigenvalues may determine whether the flow over a wing is laminar or turbulent. In ecology, eigenvalues may determine whether a food web will settle into a steady equilibrium. In electrical engineering, they may determine the frequency response of an amplifier or the reliability of a national power system. Moreover, in probability theory, eigenvalues may determine the convergence rate of a Markov process and, in other fields, we can find the eigenvalues allowing us to examine their properties.

Inspired by the notion of pseudo-spectra, A. Ammar and A. Jeribi in their works [20–23], thought to extend these results for the essential spectra of closed, densely defined, and linear operators on a Banach space. They declared the new concept of the pseudo-essential spectra of closed, densely defined, and linear operators on a Banach space. Because of the existence of several essential spectra, they were interested to focus their study on the pseudo-Browder essential spectrum. This set was shown to be characterized in the way one would expect by analogy with the essential numerical range. As a consequence, the authors in [21] located the pseudo-Browder essential spectrum between the essential spectra and the essential numerical range.

F. Abdmouleh, A. Ammar and A. Jeribi devoted their study to the Browder essential spectrum and they extended the notion of pseudo-spectra to Browder essential spectrum in a Banach space X (see [10]). For $A \in \mathcal{C}(X)$ and for every $\varepsilon > 0$, they defined the pseudo-Browder essential spectrum in the following way

$$\sigma_{e6,\varepsilon}(A) = \sigma_{e6}(A) \bigcup \left\{\lambda \in \mathbb{C} \text{ such that } \|R_b(A,\lambda)\| > \frac{1}{\varepsilon}\right\},$$

where $\sigma_{e6}(.)$ is the Browder essential spectrum and $R_b(.,.)$ is the resolvent of Browder.

The aim of this concept was to study the existence of eigenvalues far from the Browder essential spectrum and also to search the instability of the Browder essential spectrum under every perturbation. Their study of the pseudo-Browder essential spectrum enabled them to determine and localize the Browder essential spectrum of a closed, densely defined linear operator on a Banach space.

1.5 Spectral Theory of Block Operator Matrices

Many problems in mathematical physics are described by the system of partial or ordinary differential equations or linearizations thereof. In applications, the time evolution of a physical system is governed by block operator matrices. Hence, the spectral theory of these matrices plays a crucial role. During the last years, F. V. Atkinson, H. Langer, R. Mennicken, and A. A. Shkalikov (see [40, 309]) studied the Wolf essential spectrum of operators defined by a block operator matrix

$$\mathcal{A}_0 := \begin{pmatrix} A & B \\ C & D \end{pmatrix} \tag{1.5.6}$$

which acts on the product $X \times Y$ of Banach spaces. An account of the research and a wide panorama of methods to investigate the spectrum of block operator matrices were presented by C. Tretter in [323–325]. In general, the operators occurring in $\mathcal{A}_0$ are unbounded and $\mathcal{A}_0$ doesn't need to be closed or to be a closable operator, even if its entries are closed. However, under some conditions, $\mathcal{A}_0$ is closable and its closure $\mathcal{A}$ can be determined.

In the theory of unbounded block operator matrices, the Frobenius-Schur factorization is a basic tool in order to study the spectrum and various spectral properties. This was first recognized by R. Nagel in [265, 266] and, independently and under slightly different assumptions, later in [40]. In fact, F. V. Atkinson, H. Langer, R. Mennicken, and A. A. Shkalikov in [40] were concerned with the Wolf essential spectrum, and they considered the situation where the domains satisfy the conditions $\mathcal{D}(A) \subset \mathcal{D}(C)$ and $\mathcal{D}(B) \subset \mathcal{D}(D)$. Moreover, the compactness was assumed for the operators $(\lambda - A)^{-1}$ (see [40]) or $C(\lambda - A)^{-1}$ and $((\lambda - A)^{-1}B)^*$ (see [309]) for some (and hence, for all) $\lambda \in \rho(A)$, whereas in [86], it was only assumed that $(\lambda - A)^{-1}$ for $\lambda \in \rho(A)$ belongs to a nonzero two-sided closed ideal $\mathcal{I}(X) \subset \mathcal{F}(X)$ of $\mathcal{L}(X)$. In [20, 55, 56, 74, 172, 173, 335], A. Jeribi et al. extended these results to a large class of operators, described their essential spectra, and applied these results for describing the essential spectra of two-group transport operators with general boundary conditions in L_p-spaces. In [50], A. Bátkai, P. Binding, A. Dijksma, R. Hryniv, and H. Langer considered a 2×2 block operator matrix and described its essential spectrum under the assumption that $\mathcal{D}(A) \subset \mathcal{D}(C)$, that the intersection of the domains of the operators B and D is sufficiently large, and that the domain of the operator matrix is defined by an additional relation of the form $\Gamma_X x = \Gamma_Y y$ between the two components of its elements. Moreover, they supposed that the operator $C(A_1 - \lambda)^{-1}$ is compact for some (and hence for all) $\lambda \in \rho(A_1)$, where $A_1 := A|_{\mathcal{D}(A) \bigcap \mathcal{N}(\Gamma_X)}$. However, in classical transport theory in L_1-spaces, this operator is only weakly compact. Recently, S. Charfi, A. Jeribi, and I. Walha [76, 79, 335] extended the results of A. Bátkai, P. Binding, A. Dijksma, R. Hryniv, and H. Langer [50], and they were concerned with the investigation of various essential spectra. Moreover, the use of the Browder resolvent and the lower-upper factorization given by J. Lutgen in [239] allowed them to formulate and to give some supplements to many results already presented by A. Bátkai, P. Binding, A. Dijksma, R. Hryniv, and H. Langer in [50] and by S. Charfi, A. Jeribi, and R. Moalla [78].

Systems of linear evolution equations as well as linear initial value problems with more than one set of initial data lead, in a natural way, to an abstract Cauchy problem involving an operator matrix defined on a product of n Banach spaces. In [175, 335], A. Jeribi, N. Moalla, and I. Walha treated a 3×3 block operator matrix on a product of Banach spaces. They considered the following block operator matrix

$$\begin{pmatrix} A & B & C \\ D & E & F \\ G & H & L \end{pmatrix}, \tag{1.5.7}$$

where the entries of the matrix are generally unbounded operators. The operator (1.5.7) is defined on $(\mathcal{D}(A) \bigcap (D) \bigcap (G)) \times ((B) \bigcap (E) \bigcap (H)) \times ((C) \bigcap (F) \bigcap (L))$. Notice that this operator doesn't need to be closed. It was shown that, under certain conditions, this block operator matrix defines a closable operator and its essential spectra are determined. In [59, 198], A. Ben Amar, A. Jeribi, and B. Krichen, and in [20, 25], A. Ammar, A. Jeribi, and N. Moalla have studied the spectral properties of a 3×3 block operator matrix (1.5.7) with unbounded entries and with a domain consisting of vectors satisfying certain relations between their components.

1.6 Spectral Graph Theory

The paper written by L. Euler and published in 1736 is regarded as the first paper in the history of graph theory [63]. Euler's formula relating the number of edges, vertices, and faces of a convex polyhedron was studied and generalized by Cauchy (see [70]) and S. Huillier (see [154]) and is at the origin of topology.

More than one century after Euler's paper on the bridges of Konigsberg and while Listing introduced topology, A. Cayley was let, by the study of particular analytical forms arising from differential calculus, to examine a particular class of graphs, namely the tree (see [71]). This study has had many implications in theoretical chemistry. The involved techniques mainly concerned the enumeration of graphs having particular properties. Then, the enumerative graph theory started from both the results of A. Cayley and those published by Polya between 1935 and 1937 and their generalization by De Bruijn in 1959. A. Cayley linked his results on trees with contemporary studies dealing with chemical composition (see [72]). The fusion of the ideas coming from mathematics and chemistry is at the origin of a part of the standard terminology in graph theory. The term graph was first introduced by Sylvester in a paper published in 1878 in Nature, where he made an analogy between quantic invariants and co-variants of algebra and molecular diagram (see [182]).

The first textbook on graph theory was written by D. Konig and published in 1936 (see [149]). Another book written by F. Harary and published in 1969 was considered, all over the world, as the fundamental textbook on the subject (see [114]) which enabled mathematicians, chemists, electrical engineers, and social scientists to get in touch with each other. F. Harary donated all the royalties to set up the Polya Prize. In 1969, H. Heesch published a method for solving the problem using computers (see [317]). A computer assistance proof produced in 1976 by K. Appel and W. Haken made fundamental use of the notion of discharging which was developed by K. Appel and W. Haken (see [32, 33]). The proof involved checking the properties of 1936 configurations by computer, and was not fully accepted by that time due to its complexity. A simple proof considering only 633 configurations was provided 20 years later by N. Robertson, D. Sanders, P. Seymour, and R. Thomas (see [295]).

Graphs constitute some useful tools which can be used to model several types of relations and processes in physical, biological (see [245]), social, and information systems. Several practical problems can be represented by using graphs. For example, in computer science, graphs are used to represent communication networks data organization, etc. Graph theory is also used in physics and in chemistry for the study of molecules. In condensed matter physics, the three-dimensional structure of complicated simulated atomic structures can be studied quantitatively by gathering statistics and graph-theoretic properties related to the topology of the atoms. In chemistry, a graph makes a natural model for a molecule, where the vertices represent the atoms and the edges represent the bonds. This approach is especially used in computer processing of molecular structures, ranging from chemical editors to database searching.

In statistical physics, graphs can represent local connections between the interacting parts of a system, as well as the dynamics of a physical process in such systems. Graphs are also used to represent the micro-scale channels of porous media, in which the vertices represent the pores and the edges represent the smaller channels connecting the pores. Graph theory is also widely used in sociology as a way, for example, to measure the actors prestige or to explore the rumor spread, notably through the use of social network analysis software. Under the umbrella of social networks, there are several types of graphs. Acquaintance and friendship graphs describe whether or not people know each other. Influence graphs model whether or not certain people can influence the behavior of others. Finally, collaboration graphs model whether two people can work together in a particular way, such as acting together in a movie. In mathematics, graphs are useful in geometry and certain parts of topology such as knot theory. Algebraic graph theory has close links with group theory. A graph structure can be extended by assigning a weight to each edge of the graph. Graph weights, or weighted graphs, are used to represent structures in which pairwise connections have some numerical values. For example, if a graph represents a road network, the weights could represent the length of each road.

The spectral graph theory is an active research area. The results on spectral theory of discrete Laplacians can be found e.g. in [80, 82, 129, 257]. The search condition for a Laplacian operator is essentially self-adjoint and is a classic problem of mathematical physics. Several definitions of Laplacians on graphs have been proposed such as Laplacians quantum graphs (see [69, 109, 202]), combinatorial Laplacians (see, [82, 131, 181]), or physical Laplacians (see, [337]). In [44], H. Baloudi, S. Golénia, and A. Jeribi considered the discrete Laplacian acting on forms and we discuss the question of the self-adjoint extension.

The results on spectral theory of the standard discrete Laplacian acting on 0-forms, Δ_0, can be found in [129, 130, 188, 345, 346]. By definition, a symmetric operator is essentially self-adjoint, if it has a unique self-adjoint extension. In [131, 250, 251, 282, 345], the authors have studied the uniqueness of a self-adjoint extension of Δ_0. In particular, it was shown that the operator Δ_0 is essentially self-adjoint on $\mathcal{C}_0^c(G)$, when the graph G is simple. In [29], it was shown that the Laplacian $\Delta = \Delta_0 \oplus \Delta_1$ is essentially self-adjoint (so, Δ_0 and Δ_1 are essentially self-adjoint), when the graph G is χ-complete, where Δ_1 is the discrete Laplacian acting on 1-forms. In contrast, there exists a locally finite tree such that Δ_1 is not essentially self-adjoint, see [44] for a concrete example. In [44], the authors gave the sufficient conditions for the Laplacian Δ_1 to be essentially self-adjoint and also studied the unboundedness of the Friedrichs extension $\Delta_1^{\mathcal{F}}$ of Δ_1.

1.7 Applications in Mathematical Physics and Biology

Spectral theory of transport operators is a classical theme in transport theory since the pioneering papers of J. Lehner and M. Wing [231, 232] and K. Jorgens [179] in the late 1980s.

The literature devoted to this topic is huge (see, e.g., [209] and the references quoted there). Roughly speaking, one can say that some aspects of the general theory are, now, quite well known (see [184, Chap. 12]). We can cite the contributions of K. Jorgens [179], I. Vidav [328, 329], J. Voigt [332, 333], I. Marek [242], and others on spectral properties of perturbed strongly continuous semigroups on Banach spaces (and Banach lattices) (for more information, see [184, Chap. 12]).

New results on the general theory are given in [260, 261], F. Andreu, J. Martinez, J. M. Mazon [26] and A. Ben Amar, A. Jeribi, and M. Mnif [58].

The general theory is based on compactness arguments which are already present in the literature devoted to particular models of the neutron transport equation (see, e.g., the references cited in [209]). However, it seems that more general compactness results are quite recent (see J. Voigt [333], G. Greiner [139], P. Takac [316], L. W. Weis [339], A. Palczewski [274], and the systematic analysis of the compactness by M. Mokhtar–Kharroubi [260, 261]). At the same time, the transport equation was considered in different fields of mathematical physics in order to describe the transport processes of particles. In particular, it was investigated according to different boundary conditions. We may recall the works by N. Angelescu, N. Marinescu, and V. Protopopescu [27, 28], A. Belleni-Morante [53], G. Borgioli and S. Totaro [65], and the work by A. Corciovei and V. Protopopescu [83] on linear transport equations with diffuse reflections. Nevertheless, all these papers only deal with particular models. For a general setting, we may mention the work by R. Beals and V. Protopopescu [51], the Habilitationsschrift by J. Voigt [332], and also the monograph by W. Greenberg, C. Van der Mee, and V. Protopopescu [138, Chaps. 11, 12, and 13].

Inspired by [261], K. Latrach [211] made a systematic compactness analysis for the transport equation in a one-dimensional context where the general boundary conditions showing the relationship between the incoming and the outcoming fluxes are modeled by four boundary operators. The known classical boundary conditions (vacuum boundary, specular reflections, periodic, diffuse reflections, generalized and mixed type boundary conditions [27, 28, 53, 65, 83, 231] are special examples of [211]). In fact, the existence and uniqueness theory are well known in a general context (see [138, Chaps. 11, 12, and 13] or [51]), hence the main goal of [211] is to analyze in detail the compactness of the relevant operators by using some techniques of [261], i.e., density and comparison arguments. Then, the spectral theory of the transport equation is a simple consequence of the compactness results and the general theory (cf. [331, 333] and [184, Chap. 12]).

These results in transport theory received a major attention during the last decades (see, for example, the works [92, 93, 261, 262, 312, 313, 357] and the references therein) and has benefited from many engineering, physics, and applied mathematics works and had strong connections with the spectral theory of non-self-adjoint operators, positive operators, semigroup theory, etc.

Asymptotic behavior in transport theory is of major interest, since it is closely related to the stability of systems. It is related to compactness problems in perturbation theory. When dealing with general boundary conditions (including periodic boundary conditions, specular reflections, diffuse reflections, generalized or mixed type boundary conditions), much progress has been made in the recent years in terms of understanding the spectral features of some transport operators. In this book, we investigate the spectral properties of the semigroup governing these operators. As applications, we treat the time-asymptotic description of the solution of a Cauchy problem given by a transport equation, a one-velocity transport operator with Maxwell boundary condition and a transport operator with a diffuse reflection boundary condition.

We also address our study to the question of the stability of the essential spectra of transport operator with general boundary conditions where an abstract boundary operator relates the incoming and the outgoing fluxes. Sufficient conditions are given in terms of boundary and collision operators, ensuring the stability of the essential spectra. These results will enable us to describe many essential spectra for transport operator arising in growing cell populations, for neutron transport operator, and for singular neutron transport operator.

The authors A. Ben Amar, A. Jeribi, and M. Mnif in [58] have been concerned with the spectral analysis of transport operator with general boundary conditions in L_1-setting. They have investigated this subject using results from the theory of positive linear operators, irreducibility, and regularity of the collision operator. Their basic problems were dealing with the notions of essential spectra, spectral bound, and leading eigenvalues.

These characterizations are extended to two-group transport operators governed by 2×2 block operator matrices in order to also describe various essential spectra as well as systems of ordinary differential operators, delay differential equations, λ-rational Sturn–Liouville problem, multi-dimensional transport equations, the two-group radiative transfer equations in a channel, and finally the three-group transport equation.

1.8 Outline of Contents

Our book consists of 13 chapters.

In Chap. 2, we recall some definitions, notations, and basic informations on both bounded and unbounded Fredholm Operators. We also introduce the concept of quasi-inverse operator, semigroup theory, measure of noncompactness, measure of weak noncompactness, and graph measures.

In Chap. 3, we develop the classical Riesz theory, and establish a characterization of a class of bounded Fredholm operators on a Banach space which is developed in order to present some general existence results of the second kind operator equations. The obtained results are used to describe the Riesz theory of compact operators in the more general setting of polynomially compact operators. For this class of operators, the analysis of the spectrum, multiplicities, and localization of the eigenvalues are given.

Chapter 4 is devoted to study the time-asymptotic behavior of the solution to an abstract Cauchy problem. As the time-asymptotic structure ($t \rightarrow \infty$) of evolution transport problems is related to compactness results, we concentrate ourselves to study this property in the first part of this chapter in order to give a nice characterization of the time-asymptotic behavior of the solution to an abstract Cauchy problem acting on Banach spaces without restriction on the initial data.

Chapter 5 deals with the elegant interaction between Fredholm theory and some measures. In the first section, we develop the classical theory of Fredholm operators and in the following sections of this chapter, we examine this theory by means of measure of noncompactness, demicompact operator, measure of weak noncompactness, and graph measures. We also study the Fredholm theory with finite ascent and descent as well as the stability of semi-Browder operators.

In Chap. 6, we focus on the study of spectral theory and perturbation results originating from Fredholm theory. We will also introduce, in the first section, some definitions and basics related to Fredholm operators. The next part contains some results related to semi-Fredholm perturbations, Fredholm inverse operator, quasi-inverse operator, and finally, some results on perturbation theory of operator matrices are investigated in the last section.

Chapter 7 is devoted to the analysis of the essential spectra of linear operators. First, we give some basic definitions and notations related to this subject and then, we establish the essential spectra of the sum of two bounded linear operators as well as their characterization by means of Fredholm inverse and demicompact operators. We study the essential spectra of unbounded operators and we develop some invariance aspects of Kato spectrum by commuting nilpotent perturbation as well as Schechter's essential spectrum and finally, we study the stability of essential spectra under polynomially compact operator perturbations. We also treat Borel mapping, spectral mapping theorem and we give a characterization of polynomially Riesz continuous semigroups.

The aim of Chap. 8 is to develop the concept of essential pseudospectra of linear operators. We concentrate ourselves exclusively on the characterization of pseudo-Browder essential spectrum and on the conditions of its stability. Properties and invariance of Pseudo-Jeribi and Pseudo-Schechter essential spectra of linear operators by means of both Fredholm operators and measures of noncompactness are also studied.

Chapter 9 focuses on S-essential spectra which extend the notion of the essential spectra. In fact, we try to give some definitions and preliminary results. A characterization of S-essential spectra and S-Browder essential spectrum is presented. We also study the S-essential spectra of the sum of bounded, linear operators. S-essential spectra by means of demicompact operators and characterizations of the relative Schechter's and approximate essential spectra are investigated.

In Chap. 10, we exclusively focus on the study of essential spectra of 2×2 block operator matrices on Banach spaces. The first section gives a characterization of the essential spectra in the case where the resolvent of the operator A, defined in (1.5.6), is a Fredholm perturbation. Then, in the second section, we treat the case where the operator A is closed. The third section examines the case where the operator A is closable. In each section, we give sufficient conditions which guarantee the closedness of the block operator matrix, and we describe various essential spectra. The fourth section contains a study of relative boundedness for block operator matrices. The stability of the Wolf essential spectrum of some matrix operators acting in Friedrichs module is investigated in the fifth section. The last section is devoted to study the M-essential spectra of operator matrices.

Chapter 11 contains an extension of the results of Chap. 10 in order to characterize many essential spectra of 3×3 block operator matrices. In fact, we study their closability and we describe their closure in the case where the operator A is closed and also in the case where the operator A is closable. Block operator matrices using Browder resolvent and also perturbations of unbounded Fredholm linear operators are studied.

Chapter 12 gives some elementary properties of the discrete Laplacian defined on a locally finite graph. Next, we discuss the question of essentially self-adjointness. Finally, we study the Laplacian acting on forms in the case of oriented graph.

Finally, Chap. 13 concentrates on a selection of applications in mathematical physics and biology to which the results of the preceding chapter are applied. In the first section, we give the time-asymptotic description of the solution to a transport equation. Then, in the second section, we treat the time-asymptotic behavior of the solution to a Cauchy problem given by a one-velocity transport operator with Maxwell boundary condition. The third section contains the time asymptotic behavior of the transport operator with a diffuse reflection boundary condition. Next, we are concerned with the description of the essential spectra for transport operator arising in growing cell populations. Some applications of the regularity and irreducibility on transport theory are also investigated as well as essential spectra for singular neutron transport operator, systems of ordinary differential operators, two-group transport operators, elliptic problems with λ-dependant boundary conditions delay differential equations, a λ-rational Sturn–Liouville problem, two-group radiative transfer equations in a channel, and the three-group transport equations.

Chapter 2
Fundamentals

The aim of this chapter is to introduce the basic concepts, notations, and elementary results which are used throughout the book. Moreover, the results in this chapter may be found in most standard books dealing with operator theory and functional analysis (see [101, 126, 264, 302, 354]).

2.1 Basic Properties

2.1.1 Closed and Closable Operators

Let X and Y be two Banach spaces. A mapping A which assigns to each element x of a set $\mathcal{D}(A) \subset X$ a unique element $y \in Y$ is called an operator (or transformation). The set $\mathcal{D}(A)$ on which A acts is called the domain of A. The operator A is called linear, if $\mathcal{D}(A)$ is a subspace of X, and if $A(\lambda x + \beta y) = \lambda Ax + \beta Ay$ for all scalars λ, β and all elements x, y in $\mathcal{D}(A)$. The operator A is called bounded, if there is a constant M such that $\|Ax\| \leq M\|x\|$, $x \in X$. The norm of such an operator is defined by

$$\|A\| = \sup_{x\neq 0} \frac{\|Ax\|}{\|x\|}.$$

The graph $G(A)$ of a linear operator A on $\mathcal{D}(A) \subset X$ into Y is the set $\{(x, Ax) \text{ such that } x \in \mathcal{D}(A)\}$ in the product space $X \times Y$. Then, A is called a closed linear operator when its graph $G(A)$ constitutes a closed linear subspace of $X \times Y$. Hence, the notion of a closed linear operator is an extension of the notion of a bounded linear operator. A sequence $(x_n)_n \subset \mathcal{D}(A)$ will be called A-convergent to $x \in X$, if both $(x_n)_n$ and $(Ax_n)_n$ are Cauchy sequences and $x_n \to x$. A linear operator A on $\mathcal{D}(A) \subset X$ into Y is said to be closable, if A has a closed extension. It

A. Jeribi, *Spectral Theory and Applications of Linear Operators and Block Operator Matrices*, DOI 10.1007/978-3-319-17566-9_2

is equivalent to the condition that the graph $G(A)$ is a submanifold (or subspace) of a closed linear manifold (or space) which is at the same time a graph. It follows that A is closable if, and only if, the closure $\overline{G(A)}$ of $G(A)$ is a graph. We are thus led to the criterion: A is closable if, and only if, no element of the form $(0, x)$, $x \neq 0$ is the limit of elements of the form (x, Ax). In other words, A is closable if, and only if, $(x_n)_n \in \mathcal{D}(A)$, $x_n \to 0$ and $Ax_n \to x$ imply $x = 0$. When A is closable, there is a closed operator $\overline{A}$ with $G(\overline{A}) = \overline{G(A)}$. $\overline{A}$ is called the closure of A. It follows immediately that $\overline{A}$ is the smallest closed extension of A, in the sense that any closed extension of A is also an extension of $\overline{A}$. Since $x \in \mathcal{D}(\overline{A})$ is equivalent to $(x, \overline{A}x) \in \overline{G(A)}$, $x \in X$ belongs to $\mathcal{D}(\overline{A})$ if, and only if, there exists a sequence $(x_n)_n$ that is A-convergent to x. In this case we have $\overline{A}x = \lim_{n\to\infty} Ax_n$. By $\mathcal{C}(X, Y)$, we denote the set of all closed, densely defined linear operators from X into Y, and by $\mathcal{L}(X, Y)$ the Banach space of all bounded linear operators from X into Y. By an operator A from X into Y, we mean a linear operator with a domain $\mathcal{D}(A) \subset X$. We denote by $N(A)$ its null space, and $R(A)$ its range. We define the generalized kernel of a closed operator A by $N^{\infty}(A) := \bigcup_{n\in\mathbb{N}} N(A^n)$. We define the generalized range of a closed operator A by $R^{\infty}(A) := \bigcap_{n\in\mathbb{N}} R(A^n)$.

2.1.2 Adjoint Operator

Let $A \in \mathcal{L}(X, Y)$. For each $y' \in Y^*$ (the adjoint space of Y), the expression $y'(Ax)$ assigns a scalar to each $x \in X$. Thus, it is a functional $F(x)$. Clearly F is linear. It is also bounded since $|F(x)| = |y'(Ax)| \leq \|y'\|\|Ax\| \leq \|y'\|\|A\|\|x\|$. Thus, there is an $x' \in X^*$ (the adjoint space of X) such that

$$y'(Ax) = x'(x), \quad x \in X. \tag{2.1.1}$$

This functional is unique, for any other functional satisfying (2.1.1) would have to coincide with x' on each $x \in X$. Thus, (2.1.1) can be written in the form

$$y'(Ax) = A^*y'(x). \tag{2.1.2}$$

The operator A^* is called the adjoint of A, depending on the mood one is in. If A, B are bounded operators defined everywhere, it is easily checked that $(BA)^* = A^*B^*$. We just follow the definition of adjoint for bounded operator (see (2.1.2)) for defining the adjoint of unbounded operator. By an operator A from X into Y, we mean a linear operator with a domain $\mathcal{D}(A) \subset X$. We want $A^*y'(x) = y'(Ax)$, $x \in \mathcal{D}(A)$. Thus, we say that $y' \in \mathcal{D}(A^*)$ if there is an $x' \in X^*$ such that

$$x'(x) = y'(Ax), \; x \in \mathcal{D}(A). \tag{2.1.3}$$

Then we define A^*y' to be x'. In order that this definition make sense, we need x' to be unique, i.e., that $x'(x) = 0$ for all $x \in \mathcal{D}(A)$ should imply that $x' = 0$. Thus is true if $\mathcal{D}(A)$ is dense in X. To summarize, we can define A^* for any linear operator from X into Y provided $\mathcal{D}(A)$ is dense in X. We take $\mathcal{D}(A^*)$ to be the set of those $y' \in Y^*$ for which there is an $x' \in X^*$ satisfying (2.1.3). This x' is unique, and we set $A^*y' = x'$. If A, B are only densely defined, $\mathcal{D}(AB)$ need not be dense, and consequently, $(BA)^*$ need not exist. If $\mathcal{D}(BA)$ is dense, it follows that $\mathcal{D}(A^*B^*) \subset \mathcal{D}((BA)^*)$ and $(BA)^*z' = A^*B^*z'$, for all $z' \in \mathcal{D}(A^*B^*)$. Let N be a subset of a normed vector space X. A functional $x' \in X^*$, is called an annihilator of N if $x'(x) = 0$, $x \in N$. The set of all annihilators of N is denoted by N°. The subspace N° is closed. For any subset R of X^*, we call $x \in X$ an annihilator of R if $x'(x) = 0$, $x' \in R$. We denote the set of such annihilators of R by ${}^\circ R$. The subspace ${}^\circ R$ is closed. If M is a closed subspace of X, then ${}^\circ(M^\circ) = M$.

Theorem 2.1.1 ([302, Theorem 3.7]). *Let X and Y be two normed vector spaces and $A \in \mathcal{L}(X, Y)$. Then*

(i) $R(A)^\circ = N(A^*)$.
(ii) *A necessary and sufficient condition that $R(A) = {}^\circ N(A^*)$ is that $R(A)$ be closed in Y.* ◇

Definition 2.1.1. (i) A densely defined operator A on a Hilbert space is called symmetric, if $A \subset A^*$, that is, if $\mathcal{D}(A) \subset \mathcal{D}(A^*)$ and $A\varphi = A^*\varphi$ for all $\varphi \in \mathcal{D}(A)$. Equivalently, A is symmetric if, and only if, $\langle A\varphi, \psi\rangle = \langle \varphi, A\psi\rangle$ for all $\varphi, \psi \in \mathcal{D}(A)$.

(ii) A is called self-adjoint if $A^* = A$, that is, if, and only if, A is symmetric and $\mathcal{D}(A) = \mathcal{D}(A^*)$.

(iii) A symmetric operator A is called essentially self-adjoint, if its closure $\overline{A}$ is self-adjoint. If A is closed, a subset $D \subset \mathcal{D}(A)$ is called a core for A, if $\overline{A|_D} = A$. ◇

2.1.3 Elementary Results

Lemma 2.1.1 ([302, Lemma 4.14, p. 93]). *If $x'_1, \ldots, x'_m$ are linearly independent vectors in X^*, then there are vectors $x_1, \ldots, x_m$ in X, such that*

$$x'_j(x_k) = \delta_{jk} := \begin{cases} 1 & \text{if } j = k \\ 0 & \text{if } j \neq k \end{cases} \quad 1 \le j, k \le m. \tag{2.1.4}$$

Moreover, if $x_1, \ldots, x_m$ are linearly independent vectors in X, then there are vectors $x'_1, \ldots, x'_m$ in X^*, such that (2.1.4) *holds.* ◇

Lemma 2.1.2 ([302, Lemma 5.2, p. 102]). *Let X_1 be a closed subspace of a normed vector space X, and let M be a finite-dimensional subspace, such that $M \bigcap X_1 = \{0\}$. Then, $X_2 = X_1 \oplus M$ is a closed subspace of X. Moreover, the operator P defined by $Px = x$ for $x \in M$, $Px = 0$ for $x \in X_1$ is in $\mathcal{L}(X_2)$.* ◇

We will recall a simple lemma due to Riesz [294].

Lemma 2.1.3 ([302, Lemma 4.7]). *Let M be a closed subspace of a normed vector space X. If M is not the whole of X, then for each number θ satisfying $0 < \theta < 1$, there is an element $x_\theta \in X$, such that $\|x_\theta\|=1$ and $dist\,(x_\theta, M) := \inf_{x\in M} \|x_\theta - x\| \geq \theta$.* ◇

We will use a simple consequence of Hahn–Banach theorem.

Lemma 2.1.4 ([302, Theorem 2.9]). *Let M be a subspace of a normed vector space X, and suppose that x_0 is an element of X, satisfying $d = dist(x_0, M) > 0$. Then, there is a bounded linear functional F on X, such that $\|F\| = 1$, $F(x_0) = d$, and $F(x) = 0$ for $x \in M$.* ◇

Lemma 2.1.5 ([123, p. 190]). *If M and N are subspaces of X and $\dim M > \dim N$, then there exists an $m \in M$ such that $1 = \|m\| = dist(m, N)$.* ◇

Lemma 2.1.6 ([302, Lemma 5.1]). *Let N be a finite-dimensional subspace of a normed vector space X. Then, there is a closed subspace X_0 of X, such that*

(i) $X_0 \bigcap N = \{0\}$, and
(ii) for each $x \in X$, there exist an $x_0 \in X_0$ and $x_1 \in N$, such that $x = x_0 + x_1$. This decomposition is unique. ◇

Lemma 2.1.7 ([302, Lemma 5.6]). *Let X be a normed vector space, and suppose that $X = N \oplus X_0$, where X_0 is a closed subspace and N is finite-dimensional. If X_1 is a subspace of X containing X_0, then X_1 is closed.* ◇

Lemma 2.1.8 ([302, Lemma 5.3]). *Let X be a normed vector space, and let R be a closed subspace, such that R° is of finite dimension n. Then, there is an n-dimensional subspace M of X, such that $X = R \oplus M$.* ◇

2.1.4 Fredholm Operators

By an operator A from X into Y, we mean a linear operator with a domain $\mathcal{D}(A) \subset X$. If $A \in \mathcal{C}(X, Y)$, then $\alpha(A)$ denotes the dimension of the kernel $N(A)$, and $\beta(A)$ denotes the codimension of $R(A)$ in Y. The classes of Fredholm, upper semi-Fredholm and lower semi-Fredholm operators from X into Y are, respectively, the following:

$$\Phi(X, Y) := \{A \in \mathcal{C}(X, Y) : \alpha(A) < \infty,\ R(A) \text{ closed in } Y,\ \beta(A) < \infty\},$$
$$\Phi_+(X, Y) := \{A \in \mathcal{C}(X, Y) : \alpha(A) < \infty,\ R(A) \text{ closed in } Y\}, \text{ and}$$
$$\Phi_-(X, Y) := \{A \in \mathcal{C}(X, Y) : \beta(A) < \infty,\ R(A) \text{ closed in } Y\}.$$

$\Phi_{\pm}(X,Y)$ denotes the set $\Phi_{\pm}(X,Y) := \Phi_{+}(X,Y)\bigcup\Phi_{-}(X,Y)$. The set of bounded Fredholm operators from X into Y is defined by $\Phi^{b}(X,Y) = \Phi(X,Y)\bigcap\mathcal{L}(X,Y)$. For an operator $A \in \Phi_{+}(X,Y)$ or $\Phi_{-}(X,Y)$, its index is $i(A) := \alpha(A) - \beta(A)$. If $X = Y$, the sets $\mathcal{L}(X,Y)$, $\mathcal{C}(X,Y)$, $\Phi(X,Y)$, $\Phi_{+}(X,Y)$, $\Phi_{-}(X,Y)$, $\Phi_{\pm}(X,Y)$, and $\Phi^{b}(X,Y)$ are replaced, respectively, by $\mathcal{L}(X)$, $\mathcal{C}(X)$, $\Phi(X)$, $\Phi_{+}(X)$, $\Phi_{-}(X)$, $\Phi_{\pm}(X)$, and $\Phi^{b}(X)$. Let $\Phi_{+}^{-}(X) = \{A \in \Phi_{+}(X)$ such that $i(A) \leq 0\}$, and $\Phi_{-}^{+}(X) = \{A \in \Phi_{-}(X)$ such that $i(A) \geq 0\}$. We denote $\Phi_{0}(X)$ by $\Phi_{0}(X) := \{A \in \Phi^{b}(X)$, such that $i(A) = 0\}$. A complex number λ is in Φ_{A}, Φ_{+A}, Φ_{-A} or $\Phi_{\pm A}$, if $\lambda - A$ is in $\Phi(X)$, $\Phi_{+}(X)$, $\Phi_{-}(X)$ or $\Phi_{\pm}(X)$, respectively. A complex λ is in Φ_{A}^{0}, if $\lambda - A \in \Phi(X)$ and $i(\lambda - A) = 0$. Let $S \in \mathcal{L}(X,Y)$. A complex number λ is in $\Phi_{+A,S}$, $\Phi_{-A,S}$, $\Phi_{\pm A,S}$ or $\Phi_{A,S}$, if $\lambda S - A$ is in $\Phi_{+}(X,Y)$, $\Phi_{-}(X,Y)$, $\Phi_{\pm}(X,Y)$ or $\Phi(X,Y)$, respectively. Let $\Phi_{+}^{b}(X,Y)$ and $\Phi_{-}^{b}(X,Y)$ denote the sets $\Phi_{+}(X,Y)\bigcap\mathcal{L}(X,Y)$, and $\Phi_{-}(X,Y)\bigcap\mathcal{L}(X,Y)$, respectively. If $X = Y$, then $\Phi_{+}^{b}(X,Y)$ and $\Phi_{-}^{b}(X,Y)$ are replaced, respectively, by $\Phi_{+}^{b}(X)$ and $\Phi_{-}^{b}(X)$. The subset of all compact operators of $\mathcal{L}(X,Y)$ is designated by $\mathcal{K}(X,Y)$. If $X = Y$, then $\mathcal{K}(X,Y)$ is replaced by $\mathcal{K}(X)$.

Definition 2.1.2. Let W be a linear subspace of X. We say that a linear operator $A : W \longrightarrow Y$ is an invertible modulo compact operator, if there is a linear operator $L : Y \longrightarrow X$ such that $I - AL$ and $I - LA$ are compact, where I represents the identity operator. L is called an inverse of A modulo compact operator. ◇

Remark 2.1.1. Clearly, if L_1 and L_2 are the inverses of A modulo compact operator, then there exists a compact operator K such that $L_1 = L_2 + K$. ◇

Lemma 2.1.9 ([185, Lemma 332]). *If A is a closed linear operator on a Banach space X with $\beta(A) < \infty$, then A has a closed range.* ◇

Let us recall the following closed range theorem of S. Banach:

Theorem 2.1.2 ([354, Theorem p. 205]). *Let X and Y be two Banach spaces, and A a closed linear operator defined in X into Y such that $\overline{\mathcal{D}(A)} = X$. Then, the following propositions are all equivalent:*

(i) *$R(A)$ is closed in Y.*
(ii) *$R(A^*)$ is closed in X^*.*
(iii) *$R(A)^{\circ} = N(A^*)$.*
(iv) *$R(A^*) = {}^{\circ}N(A)$.* ◇

It is easy to see that a bounded operator defined on the whole Banach space X is closed. The inverse is also true and follows from the closed graph theorem which is the following theorem.

Theorem 2.1.3. *If X, Y are Banach spaces, and A is a closed linear operator from X into Y, with $\mathcal{D}(A) = X$, then $A \in \mathcal{L}(X,Y)$.* ◇

2.1.5 Spectrum

Definition 2.1.3. Let A be a closable linear operator in a Banach space X. The resolvent set and the spectrum of A are, respectively, defined as

$$\rho(A) = \{\lambda \in \mathbb{C} \text{ such that } \lambda - A \text{ is injective and } (\lambda - A)^{-1} \in \mathcal{L}(X)\},$$
$$\sigma(A) = \mathbb{C}\backslash\rho(A)$$

and the point spectrum, continuous, and the residual spectrum are defined as

$$\sigma_p(A) = \{\lambda \in \mathbb{C} \text{ such that } \lambda - A \text{ is not injective}\},$$
$$\sigma_c(A) = \{\lambda \in \mathbb{C} \text{ such that } \lambda - A \text{ is injective, } \overline{R(\lambda - A)} = X,\ R(\lambda - A) \neq X\},$$
$$\sigma_r(A) = \{\lambda \in \mathbb{C} \text{ such that } \lambda - A \text{ is injective, } \overline{R(\lambda - A)} \neq X\}.$$

◇

Remark 2.1.2. Note that, if $\rho(A) \neq \emptyset$, then A is closed. In fact, if $\lambda \in \rho(A)$, then $(\lambda - A)^{-1}$ is closed, which is also valid for $\lambda - A$. Then, according to the closed graph theorem (see Theorem 2.1.3), we deduce that $\rho(A) = \{\lambda \in \mathbb{C}$ such that $\lambda - A$ is bijective$\}$ and hence, $\sigma(A) = \sigma_p(A) \bigcup \sigma_c(A) \bigcup \sigma_r(A)$. ◇

Proposition 2.1.1 ([35, Proposition 2.5, p. 67]). *Let $(A, \mathcal{D}(A))$ be a closed, densely defined, and linear operator with a nonempty resolvent set $\rho(A)$. For each $\lambda_0 \in \rho(A)$, we have $\sigma((\lambda_0 - A)^{-1}) = (\lambda_0 - \sigma(A))^{-1}$.* ◇

Lemma 2.1.10. *Let X be a Banach space, $A \in \mathcal{L}(X)$ and let $X = X_0 \oplus X_1$ be a topological decomposition of X such that $A(X_0) \subset X_0$ and $A(X_1) \subset X_1$. If A_0 and A_1 denote, respectively, the restrictions of A to X_0 and X_1 (so $A_0 \in \mathcal{L}(X_0)$ and $A_1 \in \mathcal{L}(X_1)$), then*

(i) $A \in \Phi^b(X)$ if, and only if, $A_0 \in \Phi^b(X_0)$ and $A_1 \in \Phi^b(X_1)$.
(ii) $\sigma(A) = \sigma(A_0) \bigcup \sigma(A_1)$. ◇

Proof. (*i*) It is easy to notice that $N(A) = N(A_0) \oplus N(A_1)$ and $R(A) = R(A_0) \oplus R(A_1)$. This gives the desired result.
(*ii*) The proof is trivial. Q.E.D.

Definition 2.1.4. Let $A \in \mathcal{C}(X)$. We say that $\lambda_0 \in \sigma_p(A)$ is the leading eigenvalue of A if $\lambda_0 \in \mathbb{R}$ and, for every $\lambda \in \sigma(A)$, $\mathrm{Re}\lambda \leq \lambda_0$. ◇

Definition 2.1.5. A set $\mathbb{K} \subset \mathcal{L}(X)$ is called collectively compact, if the set $\mathbb{K}(\overline{B}_X) := \{Kx$ such that $K \in \mathbb{K},\ x \in \overline{B}_X\}$ has a compact closure, where B_X denotes the open unit ball of X and $\overline{B}_X$ its closure. ◇

Proposition 2.1.2. *Let us assume that $A_n \to A$ and $\mathbb{K} \subset \mathcal{L}(X)$ is collectively compact. Then, $\|(A_n - A)K\| \to 0$ uniformly for every $K \in \mathbb{K}$.* ◇

Proof. Since $\overline{\mathbb{K}(\overline{B}_X)}$ is compact, $(A_n - A)Kx \to 0$ uniformly for $K \in \mathbb{K}$ and $x \in \overline{B}_X$. Q.E.D.

Definition 2.1.6. An operator $A \in \mathcal{L}(X, Y)$ is said to be weakly compact, if $A(B)$ is relatively weakly compact in Y, for every bounded subset $B \subset X$. ◇

The family of weakly compact operators from X into Y is denoted by $\mathcal{W}(X, Y)$. If $X = Y$, the family of weakly compact operators on X, $\mathcal{W}(X) := \mathcal{W}(X, X)$, is a closed two-sided ideal of $\mathcal{L}(X)$ containing $\mathcal{K}(X)$ (cf. [101, 126]).

Definition 2.1.7. Let X and Y be Banach spaces, and let S and A be linear operators from X into Y. S is called relatively weakly compact with respect to A (or A-weakly compact), if $\mathcal{D}(A) \subset \mathcal{D}(S)$ and, for every bounded sequence $x_n \in \mathcal{D}(A)$ such that $(Ax_n)_n$ is bounded, the sequence $(Sx_n)_n$ contains a weakly convergent subsequence. ◇

Definition 2.1.8. Let $A \in \mathcal{L}(X)$. A is called a power-compact operator, if there exists $m \in \mathbb{N}^*$ satisfying $A^m \in \mathcal{K}(X)$. ◇

2.1.6 Relatively Boundedness and Relatively Compactness

Definition 2.1.9. Let X, Y, and Z be Banach spaces, and let A and S be two linear operators from X into Y and from X into Z, respectively.

(i) S is called relatively bounded with respect to A (or A-bounded), if $\mathcal{D}(A) \subset \mathcal{D}(S)$ and there exist two constants $a_S \geq 0$, and $b_S \geq 0$, such that

$$\|Sx\| \leq a_S \|x\| + b_S \|Ax\|, \quad x \in \mathcal{D}(A). \tag{2.1.5}$$

The infimum δ of all b_S that (2.1.5) holds for some $a_S \geq 0$ is called relative bound of S with respect to A (or A-bounded of S).

(ii) S is called relatively compact with respect to A (or A-compact), if $\mathcal{D}(A) \subset \mathcal{D}(S)$ and for every bounded sequence $(x_n)_n \in \mathcal{D}(A)$ such that $(Ax_n)_n \subset Y$ is bounded, the sequence $(Sx_n)_n \subset Z$ contains a convergent subsequence. ◇

Remark 2.1.3. The inequality (2.1.5) is equivalent to $\|Sx\|^2 \leq a^2\|x\|^2 + b^2\|Ax\|^2$ for all $x \in \mathcal{D}(A)$, where $a = \sqrt{a_S^2 + a_S b_S}$ and $b = \sqrt{b_S^2 + a_S b_S}$. ◇

Lemma 2.1.11. *If S is A-bounded with an A-bound $\delta < 1$, then S is $(A + S)$-bounded with an A-bound $\leq \frac{\delta}{1-\delta}$.* ◇

Proof. First of all, it should be mentioned that $A+S$ is well defined as $\mathcal{D}(A+S) = \mathcal{D}(S) \bigcap \mathcal{D}(A) = \mathcal{D}(A) \subset \mathcal{D}(S)$. The fact that S is A-bounded, there exist $a_S \geq 0$, and $\delta \leq b_S < 1$ such that, for all $x \in \mathcal{D}(A)$, we have

$$\begin{aligned}\|Sx\| &\leq a_S \|x\| + b_S \|Ax\| \\ &= a_S \|x\| + b_S \|Ax + Sx - Sx\| \\ &\leq a_S \|x\| + b_S \|Ax + Sx\| + b_S \|Sx\|.\end{aligned}$$

Since $b_S < 1$, it follows that

$$\|Sx\| \leq \frac{a_S}{1-b_S}\|x\| + \frac{b_S}{1-b_S}\|(A+S)x\|, \quad x \in \mathcal{D}(A).$$

Q.E.D.

Let $A \in \mathcal{C}(X,Y)$. The graph norm of A is defined by $\|x\|_A = \|x\| + \|Ax\|$, $x \in \mathcal{D}(A)$. From the closedness of A, it follows that $\mathcal{D}(A)$ endowed with the norm $\|.\|_A$ is a Banach space. Let us denote by X_A, the space $\mathcal{D}(A)$ equipped with the norm $\|.\|_A$. Clearly, the operator A satisfies $\|Ax\| \leq \|x\|_A$ and consequently, $A \in \mathcal{L}(X_A, X)$. Let $J : X \longrightarrow Y$ be a linear operator on X. If $\mathcal{D}(A) \subset \mathcal{D}(J)$, then J will be called A-defined. If J is A-defined, we will denote by $\hat{J}$ its restriction to $\mathcal{D}(A)$. Moreover, if $\hat{J} \in \mathcal{L}(X_A, Y)$, we say that J is A-bounded. We can easily check that, if J is closed (or closable), then J is A-bounded.

Proposition 2.1.3. *If A is closed and B is closable, then $\mathcal{D}(A) \subset \mathcal{D}(B)$ implies that B is A-bounded.* ◇

Proof. If A is a closed operator, then $\mathcal{D}(A)$ equipped with the graph norm is a Banach space. If we suppose that $\mathcal{D}(A) \subset \mathcal{D}(B)$ and $(B, \mathcal{D}(B))$ is closable, then $\mathcal{D}(A) \subset \mathcal{D}(\overline{B})$. Because the graph norm on $\mathcal{D}(A)$ is stronger than the norm induced from X, the operator $\overline{B}$, considered as an operator from $\mathcal{D}(A)$ into X is everywhere defined and closed. On the other hand, $\overline{B}_{|\mathcal{D}(A)} = B$, hence $B : \mathcal{D}(A) \longrightarrow X$ is bounded by the closed graph theorem (see Theorem 2.1.3) and thus B is A-bounded. Q.E.D.

Definition 2.1.10. We say that an operator J is A-closed, if $x_n \to x$, $Ax_n \to y$, $Jx_n \to z$ for $(x_n)_n \subseteq \mathcal{D}(A)$ implies that $x \in \mathcal{D}(J)$ and $Jx = z$. An operator J will be called A-closable, if $x_n \to 0$, $Ax_n \to 0$, $Jx_n \to z$ implies $z = 0$. ◇

Remark 2.1.4.

(i) If J is bounded, then J is A-bounded.
(ii) If J is closed, then J is A-closed.
(iii) If J is closable, then J is A-closable.
(iv) If A is closed, then J is A-closed if, and only if, J is A-closable if, and only if, J is A-bounded. ◇

2.1.7 Sum of Closed Operators

In general, the sum of closable or closed operators is not closable or closed, respectively. However, closability and clossedness are stable under relatively bounded perturbations with relative bound < 1. For the stability of bounded invertibility, an additional condition is required.

Theorem 2.1.4. *Let A and S be linear operators from X into Y such that $\mathcal{D}(S) \supset \mathcal{D}(A)$, and $\|Sx\| \leq a\|x\| + b\|Ax\|$, $x \in \mathcal{D}(A)$, for some constants a and b with $b < 1$. Then, A is closed if, and only if, $A + S$ is closed.* ◇

Proof. Let $x \in \mathcal{D}(A)$. We have $\|Ax\| = \|(A + S - S)x\| \leq \|(A + S)x\| + \|Sx\|$. Then, $(1-b)\|Ax\| \leq \|(A+S)x\| + a\|x\|$, $x \in \mathcal{D}(A)$. Hence, if $(x_n)_n \subset \mathcal{D}(A+S) = \mathcal{D}(A)$ such that $x_n \to x$ and $(A+S)x_n \to \psi$, then there exists $N_0 \in \mathbb{N}$ such that, for all $n,\ m \geq N_0$, we have $(1-b)\|A(x_n - x_m)\| \leq a\|x_n - x_m\| + \|(A+S)(x_n - x_m)\|$. Therefore, $(Ax_n)_n$ is a Cauchy sequence in the Banach space Y. Thus, $(Ax_n)_n \to \varphi$. Since A is closed, $x \in \mathcal{D}(A)$ and $Ax = \varphi$. Moreover, $Sx_n = (A + S)x_n - Ax_n \to \psi - \varphi$. However, we have $\|S(x_n - x)\| \leq (a\|x_n - x\| + b\|A(x_n - x)\|) \to 0$, showing that $Sx = \psi - \varphi$. Hence, $(A + S)x = \psi$ and $x \in \mathcal{D}(A + S)$. Conversely, by using the same above reasoning, we find the result. Q.E.D.

The following result may be found in [321].

Theorem 2.1.5. *Let S, A, and B be three linear operators such that $\mathcal{D}(A) \subset \mathcal{D}(S) \subset \mathcal{D}(B)$, such that*

(i) there exist two constants a_1 and $b_1 > 0$, such that $\|Sx\| \leq a_1\|x\| + b_1\|Ax\|$, $x \in \mathcal{D}(A)$,
(ii) there exist two constants a_2 and $b_2 > 0$, such that $b_1(1 + b_2) < 1$, and

$$\|Bx\| \leq a_2\|x\| + b_2\|Sx\|, \quad x \in \mathcal{D}(S).$$

Then, A is closed if, and only if, $A + S + B$ is closed. ◇

Proof. Let $x \in \mathcal{D}(A)$. We have

$$\begin{aligned}\|(A + S + B)x\| &\leq \|Ax\| + \|Sx\| + \|Bx\| \\ &\leq \|Ax\| + a_1\|x\| + b_1\|Ax\| + a_2\|x\| + b_2\|Sx\| \\ &= (1 + b_1)\|Ax\| + (a_1 + a_2)\|x\| + b_2(a_1\|x\| + b_1\|Ax\|).\end{aligned}$$

So,

$$\|(A + S + B)x\| \leq (a_1 + a_2 + a_1 b_2)\|x\| + (1 + b_1 + b_1 b_2)\|Ax\|. \tag{2.1.6}$$

Similarly, for all $x \in \mathcal{D}(A)$, we have

$$\begin{aligned}\|(S+B)x\| &\le \|Sx\| + \|Bx\| \\ &\le a_1\|x\| + b_1\|Ax\| + a_2\|x\| + b_2\|Sx\| \\ &= b_1\|Ax\| + (a_1+a_2)\|x\| + b_2(a_1\|x\| + b_1\|Ax\|).\end{aligned}$$

Hence,

$$\|(S+B)x\| \le (a_1+a_2+a_1b_2)\|x\| + b_1(1+b_2)\|Ax\|. \tag{2.1.7}$$

Now, by combining Eqs. (2.1.6) and (2.1.7), for all $x \in \mathcal{D}(A)$, we get

$$\begin{aligned}\|(A+S+B)x\| &\ge \|Ax\| - \|(S+B)x\| \\ &\ge \|Ax\| - (a_1+a_2+a_1b_2)\|x\| - b_1(1+b_2)\|Ax\| \\ &= -(a_1+a_2+a_1b_2)\|x\| + [1-b_1(1+b_2)]\|Ax\|.\end{aligned}$$

So,

$$\|Ax\| \le \alpha^{-1}\Big(\|(A+S+B)x\| + \beta\|x\|\Big), \tag{2.1.8}$$

where $\alpha = 1-b_1(1+b_2)$ $(0<\alpha<1)$ and $\beta = a_1+a_2+a_1b_2$ $(\beta>0)$. Let $(x_n)_n$ be a sequence in $\mathcal{D}(A+S+B) = \mathcal{D}(A)$ such that $x_n \to x$ and $(A+S+B)x_n \to \psi$. Then, by using Eq. (2.1.8), there exists $N_1 \in \mathbb{N}$ such that, for all $n,\ m \ge N_1$, we have $\|Ax_n - Ax_m\| \le \alpha^{-1}(\|(A+S+B)(x_n-x_m)\| + \beta\|x_n-x_m\|)$. So, $(Ax_n)_n$ is a Cauchy sequence in the Banach space Y and therefore, there exists $y \in Y$ such that $Ax_n \to y$. Since A is closed, then $x \in \mathcal{D}(A)$ and $Ax = y$. From Eq. (2.1.6), it follows that

$$\|(A+S+B)(x_n-x)\| \le \Big((a_1+a_2+a_1b_2)\|x_n-x\| + (1+b_1+b_1b_2)\|A(x_n-x)\|\Big) \to 0$$

which, by letting $n \to +\infty$, implies that $(A+S+B)x_n \to (A+S+B)x$. So, $\psi = (A+S+B)x$, and $A+S+B$ is closed. Conversely, a same reasoning as before leads to the result. Q.E.D.

2.1.8 Strictly Singular and Strictly Cosingular Operators

Definition 2.1.11. Let X and Y be two Banach spaces. An operator $A \in \mathcal{L}(X,Y)$ is called strictly singular if, for every infinite-dimensional subspace M, the restriction of A to M is not a homeomorphism. ◇

Let $\mathcal{S}(X,Y)$ denote the set of strictly singular operators from X into Y.

The concept of strictly singular operators was introduced by T. Kato in his pioneering paper [185] as a generalization of the notion of compact operators. For a detailed study of the properties of strictly singular operators, we may refer to [126, 185]. For our own use, let us recall the following four facts. The set $\mathcal{S}(X, Y)$ is a closed subspace of $\mathcal{L}(X, Y)$. If $X = Y$, $\mathcal{S}(X) := \mathcal{S}(X, X)$ is a closed two-sided ideal of $\mathcal{L}(X)$ containing $\mathcal{K}(X)$. If X is a Hilbert space, then $\mathcal{K}(X) = \mathcal{S}(X)$ and the class of weakly compact operators on L_1-spaces (resp. $C(K)$-spaces with K being a compact Haussdorff space) is, nothing else but, the family of strictly singular operators on L_1-spaces (resp. $C(K)$-spaces) (see [277, Theorem 1]).

Let X be a Banach space. If N is a closed subspace of X, we denote by π_N^X the quotient map $X \longrightarrow X/N$. The codimension of N, denoted codim(N), is defined as the dimension of the vector space X/N.

Definition 2.1.12. Let X and Y be two Banach spaces and let $S \in \mathcal{L}(X, Y)$. S is called strictly cosingular from X into Y, if there exists no closed subspace N of Y with codim$(N) = \infty$, such that $\pi_N^Y S : X \longrightarrow Y/N$ is surjective. ◇

Let $C\mathcal{S}(X, Y)$ denote the set of strictly cosingular operators from X into Y. This class of operators was first introduced by Pelczynski [277]. It constitutes either a closed subspace of $\mathcal{L}(X, Y)$, which is $C\mathcal{S}(X) := C\mathcal{S}(X, X)$, or a closed two-sided ideal of $\mathcal{L}(X)$, if $X = Y$ (cf. [330]). A Banach space is said to be decomposable, if it is the topological direct sum of two closed infinite-dimensional subspaces. A Banach space is said to be hereditarily indecomposable (in short H.I. space), if it does not contain any decomposable subspace. The class of hereditarily indecomposable Banach spaces was first introduced and investigated by Gowers and Maurey in [135]. One of the main related facts to this class is the following result due to Gowers and Maurey [135].

Lemma 2.1.12. *If X is a complex H.I. Banach space, then every operator in $\mathcal{L}(X)$ can be written in the form $\lambda + S$, where $\lambda \in \mathbb{C}$ and $S \in \mathcal{S}(X)$.* ◇

2.1.9 Fredholm and Semi-Fredholm Perturbations

Definition 2.1.13. Let X and Y be two Banach spaces, and let $F \in \mathcal{L}(X, Y)$. F is called a Fredholm perturbation, if $U + F \in \Phi(X, Y)$ whenever $U \in \Phi(X, Y)$. F is called an upper (resp. lower) Fredholm perturbation, if $U + F \in \Phi_+(X, Y)$ (resp. $U + F \in \Phi_-(X, Y)$) whenever $U \in \Phi_+(X, Y)$ (resp. $U \in \Phi_-(X, Y)$). ◇

The sets of Fredholm, upper semi-Fredholm and lower semi-Fredholm perturbations, are, respectively, denoted by $\mathcal{F}(X, Y)$, $\mathcal{F}_+(X, Y)$, and $\mathcal{F}_-(X, Y)$. In general, we have

$$\mathcal{K}(X, Y) \subseteq \mathcal{F}_+(X, Y) \subseteq \mathcal{F}(X, Y)$$
$$\mathcal{K}(X, Y) \subseteq \mathcal{F}_-(X, Y) \subseteq \mathcal{F}(X, Y).$$

If $X = Y$, we may write $\mathcal{F}(X)$, $\mathcal{F}_+(X)$ and $\mathcal{F}_-(X)$ instead of $\mathcal{F}(X,X)$, $\mathcal{F}_+(X,X)$ and $\mathcal{F}_-(X,X)$, respectively. In Definition 2.1.13, if we replace $\Phi(X,Y)$, $\Phi_+(X,Y)$, and $\Phi_-(X,Y)$ by $\Phi^b(X,Y)$, $\Phi^b_+(X,Y)$, and $\Phi^b_-(X,Y)$, we obtain the sets $\mathcal{F}^b(X,Y)$, $\mathcal{F}^b_+(X,Y)$, and $\mathcal{F}^b_-(X,Y)$. These classes of operators were introduced and investigated in [124]. In particular, it was shown that $\mathcal{F}^b(X,Y)$ is a closed subset of $\mathcal{L}(X,Y)$ and $\mathcal{F}^b(X) := \mathcal{F}^b(X,X)$ is a closed two-sided ideal of $\mathcal{L}(X)$. In general, we have

$$\mathcal{K}(X,Y) \subseteq \mathcal{F}^b_+(X,Y) \subseteq \mathcal{F}^b(X,Y) \tag{2.1.9}$$

$$\mathcal{K}(X,Y) \subseteq \mathcal{F}^b_-(X,Y) \subseteq \mathcal{F}^b(X,Y). \tag{2.1.10}$$

In [123], it was shown that $\mathcal{F}^b(X)$ and $\mathcal{F}^b_+(X) := \mathcal{F}^b_+(X,X)$ are closed two-sided ideals of $\mathcal{L}(X)$. It is worth noticing that, in general, the structure ideal of $\mathcal{L}(X)$ is extremely complicated. Most of the results on ideal structure deal with the well-known closed ideals which have arisen from applied work with operators. We can quote, for example, compact operators, weakly compact operators, strictly singular operators, strictly cosingular operators, upper semi-Fredholm perturbations, lower semi-Fredholm perturbations, and Fredholm perturbations. In general, we have

$$\begin{aligned} &\mathcal{K}(X) \subset \mathcal{S}(X) \subset \mathcal{F}^b_+(X) \subset \mathcal{F}^b(X) \subset \mathcal{J}(X), \text{ and} \\ &\mathcal{K}(X) \subset \mathcal{CS}(X) \subset \mathcal{F}^b_-(X) \subset \mathcal{F}^b(X) \subset \mathcal{J}(X), \end{aligned} \tag{2.1.11}$$

where $\mathcal{F}^b_-(X) := \mathcal{F}^b_-(X,X)$, and where $\mathcal{J}(X)$ denotes the set

$$\mathcal{J}(X) = \left\{F \in \mathcal{L}(X), \text{ such that } 1 \in \Phi^0_F\right\}.$$

Remark 2.1.5. $\mathcal{J}(X)$ is not an ideal of $\mathcal{L}(X)$ (since $I \notin \mathcal{J}(X)$). ◇

If X is isomorphic to an L_p-space with $1 \leq p \leq \infty$ or to $C(\Sigma)$ where Σ is a compact Hausdorff space, then

$$\mathcal{K}(X) \subset \mathcal{S}(X) = \mathcal{F}^b_+(X) = \mathcal{F}^b_-(X) = \mathcal{CS}(X) = \mathcal{F}^b(X). \tag{2.1.12}$$

Definition 2.1.14. A subspace $N \subset X$ is said to be complemented, if there exists a closed subspace $M \subset X$, such that $N \oplus M = X$. ◇

A Banach space X is said to be an h-space if each closed infinite-dimensional subspace of X contains a complemented subspace isomorphic to X. Any Banach space isomorphic to an h-space, c, c_0 and l_p $(1 \leq p < \infty)$ are h-spaces. In [343, Theorem 6.2], R. J. Whitley proved that, if X is an h-space, then $\mathcal{S}(X)$ is the greatest proper ideal of $\mathcal{L}(X)$. This implies that

$$\mathcal{K}(X) \subset \mathcal{F}^b_+(X) = \mathcal{S}(X) = \mathcal{F}^b(X),$$

$$\mathcal{K}(X) \subset \mathcal{F}^b_-(X) \subset \mathcal{S}(X) = \mathcal{F}^b(X).$$

We say that X is weakly compactly generating (w.c.g.), if the linear span of some weakly compact subsets is dense in X. For more details and results, see [94]. In particular, all separable and all reflexive Banach spaces are w.c.g., as well as $L_1(\Omega, d\mu)$, if (Ω, μ) is σ-finite. In [338], it was proved that, if X is a w.c.g. Banach space, then $\mathcal{F}_+(X) = \mathcal{S}(X)$ and $\mathcal{F}_-(X) = C\mathcal{S}(X)$.

Remark 2.1.6. Let (Ω, Σ, μ) be a positive measure space. Since the spaces $L_p(\Omega, d\mu)$ with $1 \leq p < \infty$ are w.c.g., then we can deduce, from what precedes, that

$$\mathcal{K}(L_p(\Omega, d\mu)) \subset \mathcal{F}_+(L_p(\Omega, d\mu)) \bigcap \mathcal{F}_-(L_p(\Omega, d\mu)). \qquad \diamond$$

We say that X is subprojective if, for every given closed infinite-dimensional subspace M of X, there exist a closed and finite-dimensional subspace N contained in M, and a continuous projection from X onto N. Clearly, any Hilbert space is subprojective. The spaces c_0, l_p, (with $1 \leq p < \infty$), and L_p (with $2 \leq p < \infty$) are also subprojective [343]. We say that X is superprojective, if every subspace V having infinite-codimension in X is contained in a closed subspace W, which has an infinite-codimension in X, and such that there exists a bounded projection from X into W. The spaces l_p $(1 < p < \infty)$ and L_p $(1 < p \leq 2)$ are superprojective [343]. Let X be a w.c.g. Banach space. It was proved in [306] that, if X is superprojective (resp. subprojective), then $\mathcal{S}(X) \subset C\mathcal{S}(X)$ (resp. $C\mathcal{S}(X) \subset \mathcal{S}(X)$). Accordingly, we have the following result:

Proposition 2.1.4. *Let X be a w.c.g. Banach space. Then,*

(i) If X is superprojective, then $\mathcal{S}(X) \subset \mathcal{F}_+(X) \bigcap \mathcal{F}_-(X)$.
(ii) If X is subprojective, then $C\mathcal{S}(X) \subset \mathcal{F}_+(X) \bigcap \mathcal{F}_-(X)$. $\diamond$

2.1.10 Dunford–Pettis Property

Definition 2.1.15. Let X be a Banach space. The space X is said to have the Dunford–Pettis property (in short DP property) if, for each Banach space Y, every weakly compact operator $T : X \longrightarrow Y$ takes weakly compact sets in X into norm compact sets of Y. $\diamond$

The Dunford–Pettis property, as defined above, was explicitly defined by Grothendieck [143] who undertook an extensive study about it and also about some related properties. It is well known that any L_1-space has the DP property [100]. Moreover, if Ω is a compact Hausdorff space, then $C(\Omega)$ has the DP property [143]. For further examples, we may refer to [95] or [101, p. 494, 479, 508, and 511]. Let us notice that the DP property is not conserved under conjugation. However, if X is a Banach space whose dual has the DP property, then X has also the DP property (see, e.g., [143]). For more information, we may refer to Diestel's paper [95] which contains a real survey of the Dunford-Pettis property, as well as some related topics.

Remark 2.1.7. It was proved in [214, Proposition 3.1] that, if X is a Banach space with the DP property, then $\mathcal{W}(X) \subset \mathcal{F}_-(X) \bigcap \mathcal{F}_+(X)$. ◇

Lemma 2.1.13.

(i) If X has the DP property, then $\mathcal{W}(X)\mathcal{W}(X) \subset \mathcal{K}(X)$.
(ii) Let (Ω, Σ, μ) be a positive measure space and let $p > 1$. If X is isomorphic to one of the spaces $L_p(\Omega, \Sigma, d\mu)$, then $\mathcal{S}(X)\mathcal{S}(X) \subset \mathcal{K}(X)$. ◇

Proof. (*i*) Let T_1 and $T_2 \in \mathcal{W}(X)$. If U is a bounded subset of X, then $T_1(U)$ is relatively weakly compact. Accordingly, since X has the DP property, then $T_2(T_1(U))$ is a relatively compact subset of X. That is, $T_2T_1 \in \mathcal{K}(X)$.
(*ii*) The proof is given in [252, Theorem 1.b]. Q.E.D.

2.2 Basic Notions

In this section, we give a brief listing of the functional analysis properties in a normed vector space. For the proofs of the statements given below, and for further information, the reader may refer to Schechter [302] and Muller [264].

Proposition 2.2.1 ([116, Proposition 3.2, p. 374]). *Let X, Y, and Z be three Banach spaces, and let $A : X \longrightarrow Y$ be a closed operator, with a closed range and a* $\dim N(A) < \infty$ *and let also $C : Z \longrightarrow X$ be a closed operator. Then, AC is a closed operator.* ◇

Theorem 2.2.1 ([302, Theorem 3.12]). *Let X and Y be two Banach spaces, and let A be a one-to-one closed linear operator from X into Y. Then, a necessary and sufficient condition for $R(A)$ to be closed in Y is that, $\|x\| \leq C\|Ax\|$, $x \in X$ holds.* ◇

Theorem 2.2.2 ([301, Theorem 2.12, p. 9]). *Let X and Y be two Banach spaces. If A is a closed linear operator from X into Y, and if B is A-compact, then*

(i) $\|Bx\| \leq c(\|Ax\| + \|x\|)$, $x \in \mathcal{D}(A)$,
(ii) $\|Ax\| \leq c(\|(A+B)x\| + \|x\|)$, $x \in \mathcal{D}(A)$,
(iii) $A + B$ is a closed operator, and
(iv) B is $(A + B)$-compact. ◇

Theorem 2.2.3 ([324, Theorem 2.2.14]). *Suppose that $\mathcal{D}(A) \subset \mathcal{D}(C)$, $\rho(A) \neq \emptyset$, and that, for some (and hence, for all) $\mu \in \rho(A)$, the operator $(A-\mu)^{-1}B$ is bounded on $\mathcal{D}(B)$. Then,*

$$\mathcal{A} := \begin{pmatrix} A & B \\ C & D \end{pmatrix} \tag{2.2.1}$$

is closable (closed, respectively) if, and only if, $D - \mu - C(A-\mu)^{-1}B$ is closable (closed, respectively) for some (and hence, for all) $\mu \in \rho(A)$. ◇

Theorem 2.2.4 ([324, Theorem 2.2.23]). *Suppose that $\mathcal{D}(C) \subset \mathcal{D}(A)$, C is boundedly invertible, and that $C^{-1}D$ is bounded on $\mathcal{D}(D)$. Then, $\mathcal{A}$, given in (2.2.1), is closable (closed, respectively) if, and only if, $B-(A-\mu)C^{-1}(D-\mu)$ is closable (closed, respectively) for some (and hence, for all) $\mu \in \mathbb{C}$.* ◇

Theorem 2.2.5 ([152, Theorem 4.17.4]). *Let $\chi(\xi)$ be a character of the real line and set $F_\tau := \{\chi(\xi)$ such that $0 \le \xi \le \tau\}$ and $F = \bigcap_\tau \overline{F}_\tau$. There are two alternatives: either $F = \{1\}$, in which cases $\chi(\xi)$ is a continuous character and $\chi(\xi) = e^{i\beta\xi}$, or else F contains the unit circle and $\chi(\xi)$ is non-measurable.* ◇

Let A be a closed linear from X into Y. Then $N(A)$ is a closed subspace of X and hence the quotient space $\tilde{X} := X/N(A)$ is a Banach space with respect to the norm

$$\|\tilde{x}\| = \operatorname{dist}(x, N(A)) := \inf\{\|x-y\| \text{ such that } y \in N(A)\}.$$

Since $N(A) \subset \mathcal{D}(A)$, the quotient space $\mathcal{D}(\tilde{A}) := \mathcal{D}(A)/N(A)$ is contained in $\tilde{X}$. Defining $\tilde{A}x = Ax$ for every $\tilde{x} \in \mathcal{D}(\tilde{A})$, it follows that $\tilde{A}$ is a well-defined closed linear operator with $\mathcal{D}(\tilde{A}) \subset \tilde{X}$ and $R(\tilde{A}) = R(A)$. Since $\tilde{A}$ is one-to-one, the inverse $\tilde{A}^{-1}$ exists on $R(A)$. The reduced minimal modulus of A is defined by

$$\tilde{\gamma}(A) := \begin{cases} \inf\limits_{x \notin N(A)} \dfrac{\|Ax\|}{\operatorname{dist}(x, N(A))} & \text{if } A \neq 0, \\ \infty & \text{if } A = 0. \end{cases} \tag{2.2.2}$$

The reduced minimum modulus measures the closedness of the range of operators, in the following sense.

Lemma 2.2.1 ([186, Theorem IV.5.2]). *Let A be a closed linear operator with a domain $\mathcal{D}(A) \subset X$. Then, $R(A)$ is closed if, and only if, $\tilde{\gamma}(A) = \|\tilde{A}^{-1}\|^{-1} > 0$.* ◇

The function $\tilde{\gamma}(.)$ is not continuous. In fact, let $A_n = \begin{pmatrix} 1 & 0 \\ 0 & \frac{1}{n} \end{pmatrix}$ and let $A = \begin{pmatrix} 1 & 0 \\ 0 & 0 \end{pmatrix}$. Then, $A_n \to A$, $\tilde{\gamma}(A_n) = \frac{1}{n}$, and $\tilde{\gamma}(A) = 1$.

2.2.1 Basics on Bounded Fredholm Operators

Theorem 2.2.6 ([302, Theorem 5.4, p. 103]). *Let X and Y be two Banach spaces, and let $A \in \Phi^b(X,Y)$. Then, there is a closed subspace X_0 of X, such that $X = X_0 \oplus N(A)$ and a subspace Y_0 of Y of dimension $\beta(A)$ such that $Y = R(A) \oplus Y_0$. Moreover, there is an operator $A_0 \in \mathcal{L}(Y,X)$, such that $N(A_0) = Y_0$, $R(A_0) = X_0$, $A_0A = I$ on X_0, and $AA_0 = I$ on $R(A)$. In addition, $A_0A = I - F_1$ on X, and*

$AA_0 = I - F_2$ *on* Y, *where* $F_1 \in \mathcal{L}(X)$ *with* $R(F_1) = N(A)$ *and* $F_2 \in \mathcal{L}(Y)$ *with* $R(F_2) = Y_0$. ◇

Theorem 2.2.7 ([299]). *Let* $A \in \mathcal{L}(X,Y)$ *and* $B \in \mathcal{L}(Y,Z)$, *where* X, Y, *and* Z *are Banach spaces. If* A *and* B *are Fredholm operators, upper semi-Fredholm operators, lower semi-Fredholm operators, then* BA *is a Fredholm operator, an upper semi-Fredholm operator, a lower semi-Fredholm operator, respectively, and* $i(BA) = i(B) + i(A)$. ◇

Theorem 2.2.8 ([264, Theorem 14, p. 160]). *Let* X *and* Y *be two Banach spaces. Let* $A \in \mathcal{L}(X,Y)$. *The following conditions are equivalent:*

(i) $A \in \Phi_+^b(X,Y)$ *and* $R(A)$ *is complemented.*
(ii) There exist $S \in \mathcal{L}(Y,X)$ *and* $F \in \mathcal{F}_0(X)$ *such that* $SA = I + F$, *where* $\mathcal{F}_0(X)$ *stands for the ideal of finite rank operators.*
(iii) There exist $S \in \mathcal{L}(Y,X)$ *and* $K \in \mathcal{K}(X)$ *such that* $SA = I + K$. ◇

Theorem 2.2.9 ([264, Theorem 15, p. 160]). *Let* X *and* Y *be two Banach spaces. Let* $A \in \mathcal{L}(X,Y)$. *The following conditions are equivalent:*

(i) $A \in \Phi_-^b(X,Y)$ *and* $N(A)$ *is complemented.*
(ii) There exist $S \in \mathcal{L}(Y,X)$ *and* $F \in \mathcal{F}_0(Y)$ *such that* $AS = I + F$.
(iii) There exist $S \in \mathcal{L}(Y,X)$ *and* $K \in \mathcal{K}(Y)$ *such that* $AS = I + K$. ◇

Theorem 2.2.10 ([302, Theorem 5.13, p. 110]). *Let* X, Y, *and* Z *be three Banach spaces. Assume that* $A \in \mathcal{L}(X,Y)$ *and* $B \in \mathcal{L}(Y,Z)$ *are such that* $BA \in \Phi^b(X,Z)$. *Then,* $A \in \Phi^b(X,Y)$ *if, and only if,* $B \in \Phi^b(Y,Z)$. ◇

Theorem 2.2.11 ([302, Theorem 5.14, p. 111]). *Let* X, Y, *and* Z *be three Banach spaces. Assume that* $A \in \mathcal{L}(X,Y)$ *and* $B \in \mathcal{L}(Y,Z)$ *are such that* $BA \in \Phi^b(X,Z)$. *If* $\alpha(B) < \infty$, *then* $A \in \Phi^b(X,Y)$ *and* $B \in \Phi^b(Y,Z)$. ◇

Theorem 2.2.12 ([302, Theorem 5.16, p. 114]). *Let* X, Y, *and* Z *be three Banach spaces. Assume that* $A \in \mathcal{L}(X,Y)$ *and* $B \in \mathcal{L}(Y,Z)$ *are such that* $BA \in \Phi^b(X,Z)$. *If* $\beta(A) < \infty$, *then* $A \in \Phi^b(X,Y)$ *and* $B \in \Phi^b(Y,Z)$. ◇

Theorem 2.2.13 ([264, Theorem 5, p. 156]). *Let* X, Y, *and* Z *be three Banach spaces,* $A \in \mathcal{L}(X,Y)$ *and* $B \in \mathcal{L}(Y,Z)$. *Then,*

(i) if $A \in \Phi_-^b(X,Y)$ *and* $B \in \Phi_-^b(Y,Z)$, *then* $BA \in \Phi_-^b(X,Z)$.
(ii) if $A \in \Phi_+^b(X,Y)$ *and* $B \in \Phi_+^b(Y,Z)$, *then* $BA \in \Phi_+^b(X,Z)$.
(iii) if $A \in \Phi^b(X,Y)$ *and* $B \in \Phi^b(Y,Z)$, *then* $BA \in \Phi^b(X,Z)$. ◇

Theorem 2.2.14 ([264, Theorem 6, p. 157]). *Let* X, Y, *and* Z *be three Banach spaces,* $A \in \mathcal{L}(X,Y)$ *and* $B \in \mathcal{L}(Y,Z)$.

(i) If $BA \in \Phi_+^b(X,Z)$, *then* $A \in \Phi_+^b(X,Y)$.
(ii) If $BA \in \Phi_-^b(X,Z)$, *then* $B \in \Phi_-^b(Y,Z)$.
(iii) If $BA \in \Phi^b(X,Z)$, *then* $B \in \Phi_-^b(Y,Z)$ *and* $A \in \Phi_+^b(X,Y)$. ◇

Lemma 2.2.2 ([230, Lemma 4.3]). *Let X and Y be Banach spaces. An operator $A \in \mathcal{L}(X,Y)$ is in $\Phi_+^b(X,Y)$ if, and only if, $\alpha(A-K) < \infty$ for all $K \in \mathcal{K}(X,Y)$.* $\diamondsuit$

Theorem 2.2.15 ([302, Theorem 5.11, p. 109]). *Assume that $A \in \Phi^b(X,Y)$. Then, there is an $\eta > 0$ such that, for any $B \in \mathcal{L}(X,Y)$ satisfying $\|B\| < \eta$, one has $A + B \in \Phi^b(X,Y)$, $i(A+B) = i(A)$, and $\alpha(A+B) \leq \alpha(A)$.* $\diamondsuit$

Theorem 2.2.16 ([302, Theorem 5.5, p. 105]). *Let $A \in \mathcal{L}(X,Y)$. Suppose that there exist $A_1, A_2 \in \mathcal{L}(Y,X)$, $F_1 \in \mathcal{K}(X)$, and $F_2 \in \mathcal{K}(Y)$ such that $A_1A = I - F_1$ on X and $AA_2 = I - F_2$ on Y. Then $A \in \Phi^b(X,Y)$.* $\diamondsuit$

Lemma 2.2.3 ([230, Lemma 4.5]). *Let X and Y be Banach spaces. If $A \in \Phi_+^b(X,Y)$ (resp. $\Phi_-^b(X,Y)$, $\Phi^b(X,Y)$), then there exist an $\eta > 0$ such that $B \in \Phi_+^b(X,Y)$ (resp. $\Phi_-^b(X,Y)$, $\Phi^b(X,Y)$) with $i(B) = i(A)$, for all $B \in \mathcal{L}(X,Y)$ satisfying $\|B - A\| < \eta$.* $\diamondsuit$

Theorem 2.2.17 ([264, Theorem 7]). *Let X be a Banach space, and let $A \in \mathcal{L}(X)$. Then, A can be expressed as $A = S + K$, where S, $K \in \mathcal{L}(X)$, K is compact and S is invertible if, and only if, A is Fredholm with $i(A) = 0$.* $\diamondsuit$

Theorem 2.2.18 ([264, Theorem 4, p. 170]). *If $A \in \mathcal{L}(X,Y)$ is semi-Fredholm and $S \in \mathcal{L}(X,Y)$ satisfies $\|S\| < \tilde{\gamma}(A)$, where $\tilde{\gamma}(A)$ is the reduced minimum modulus of A given in (2.2.2). Then $A + S$ is semi-Fredholm, $i(A+S) = i(A)$, $\alpha(A+S) \leq \alpha(A)$ and $\beta(A+S) \leq \beta(A)$.*

Lemma 2.2.4 ([302, Lemma 5.19]). *Let $A_1, \ldots, A_n$ be n operators in $\mathcal{L}(X)$ which commute, and suppose that their product $A = A_1 \ldots A_n$ is in $\Phi^b(X)$. Then, each A_k is in $\Phi^b(X)$.* $\diamondsuit$

Lemma 2.2.5. *Assume that $A \in \mathcal{L}(X)$ and that there exist operators B_0, $B_1 \in \mathcal{L}(X)$ such that B_0A and AB_1 are in $\Phi^b(X)$. Then, $A \in \Phi^b(X)$.* $\diamondsuit$

Proof. By referring to Theorem 2.2.6, there are operators A_0, $A_1 \in \mathcal{L}(X)$ such that $A_0B_0A - I$ and $AB_1A_1 - I$ are in $\mathcal{K}(X)$. This implies that $A \in \Phi^b(X)$ by using the Theorem 2.2.16. Q.E.D.

Theorem 2.2.19 ([264]). *Let $A \in \mathcal{L}(X)$ and $B \in \mathcal{L}(X)$. If $AB \in \Phi^b(X)$ and $BA \in \Phi^b(X)$, then $A \in \Phi^b(X)$ and $B \in \Phi^b(X)$.* $\diamondsuit$

Lemma 2.2.6 ([217, Lemma 2.2]). *Let $F \in \mathcal{L}(X)$. Then, the following statements hold.*

(i) $F \in \mathcal{F}_+^b(X)$ *if, and only if,* $\alpha(A - F) < \infty$ *for each* $A \in \Phi_+^b(X)$.
(ii) $F \in \mathcal{F}_-^b(X)$ *if, and only if,* $\beta(A - F) < \infty$ *for each* $A \in \Phi_-^b(X)$.
(iii) $F \in \mathcal{F}^b(X)$ *if, and only if,* $\alpha(A - F) < \infty$ *or* $\beta(A - F) < \infty$ *for each* $A \in \Phi^b(X)$. $\diamondsuit$

We will make use of a simple lemma due to Yood [353].

Lemma 2.2.7 ([353, Corollary 6.2]). *If $A \in \mathcal{L}(X)$ has a zero index, then $\alpha(A^n) = \alpha(A^{n+1})$ if, and only if, $\beta(A^n) = \beta(A^{n+1})$.* ◇

If $A \in \mathcal{L}(X)$, we define the ascent of A, asc(A), and the descent of A, desc(A), by

$$\operatorname{asc}(A) := \min\left\{n \in \mathbb{N} \text{ such that } N(A^n) = N(A^{n+1})\right\},$$

and

$$\operatorname{desc}(A) := \min\left\{n \in \mathbb{N} \text{ such that } R(A^n) = R(A^{n+1})\right\}.$$

First, we will recall the following result due to Taylor [319].

Proposition 2.2.2 ([319, Theorem 3.6]). *Let $A \in \mathcal{L}(X)$. If asc(A) and desc(A) are finite, then asc(A) = desc(A).* ◇

Lemma 2.2.8 ([136]). *Suppose that A and B are commuting bounded linear operators on the Banach space X. If $A - B$ is compact and A is onto, then B has finite descent.* ◇

Lemma 2.2.9. *Let A be a bounded linear operator on a Banach space X. If $A \in \Phi^b(X)$, with* asc(A) *and* desc(A) *being finite, then $i(A) = 0$.* ◇

Proof. Since asc(A) and desc(A) are finite, and by applying Proposition 2.2.2, there exists an integer k such that $\operatorname{asc}(A) = \operatorname{desc}(A) = k$. Hence, $N(A^k) = N(A^{k+n})$ and $R(A^k) = R(A^{k+n})$, for all $n \in \mathbb{N}$. Therefore, $i(A^k) = i(A^{k+n})$. However, $A \in \Phi^b(X)$. Then, Theorem 2.2.7 implies that $i(A^k) = ki(A) = i(A^{k+n}) = (k+n)i(A)$, for all $n \in \mathbb{N}$. Hence, $i(A) = 0$. Q.E.D.

Lemma 2.2.10. *Let $A \in \mathcal{L}(X)$. If $\alpha(A) < \infty$, then A has finite ascent if, and only if, $N^\infty(A) \bigcap R^\infty(A) = \{0\}$.* ◇

Proof. Suppose that A has ascent k and let $x \in N^\infty(A) \bigcap R^\infty(A)$. Then, we have $x \in N(A^k) \bigcap R^\infty(A)$, so there is $y \in X$ such that $A^k x = 0$ and $x = A^k y$. Then $A^{2k} y = 0$, hence $A^k y = x = 0$. Thus $N^\infty(A) \bigcap R^\infty(A) \subset N(A^k) \bigcap A^k(X) = \{0\}$. Conversely, suppose that $N^\infty(A) \bigcap R^\infty(A) = \{0\}$. $A^n(X)$ is decreasing and $N(A)$ is finite dimensional, therefore the decreasing sequence $N(A) \bigcap A^n(X)$ terminates. Thus, there is an integer k such that $N(A) \bigcap A^k(X) = N(A) \bigcap R(A) \subset N^\infty(A) \bigcap R^\infty(A) \subset N(A^k) \bigcap A^k(X) = \{0\}$. Now, if $A^{k+1}x = 0$, then $A^k x \in N(A) \bigcap A^k(X) = \{0\}$, so $A^k x = 0$. Thus $N(A^k) = N(A^{k+1})$ and A has finite ascent. Q.E.D.

Theorem 2.2.20 ([128, Theorem 3]). *Let X be a Banach space and let A be a closed linear operator acting in X. Let $B \in \mathcal{L}(X)$ such that $\mathcal{D}(A) \subset \mathcal{D}(AB)$ and $BAx = ABx$ for all $x \in \mathcal{D}(A)$. Then, the space $\overline{N^\infty(A+B)} \bigcap R^\infty(A+B)$ is constant, i.e., $\overline{N^\infty(A+B)} \bigcap R^\infty(A+B) = \overline{N^\infty(A)} \bigcap R^\infty(A)$.* ◇

Two important classes of operators in Fredholm theory are given by the classes of semi-Fredholm operators which possess finite ascent or finite descent. We will distinguish two classes of operators: the class of all upper semi-Browder operators on a Banach space X which is defined by $\mathcal{B}_+^b(X) = \{A \in \Phi_+^b(X)$ such that asc $(A) < \infty\}$ and the class of all lower semi-Browder operators which is defined by $\mathcal{B}_-^b(X) = \{A \in \Phi_-^b(X)$ such that desc$(A) < \infty\}$. The class of all Browder operators (also known, in the literature, as Riesz–Schauder operators) is defined by $\mathcal{B}^b(X) = \mathcal{B}_+^b(X) \bigcap \mathcal{B}_-^b(X)$.

Theorem 2.2.21 ([287, Theorem 1]). *Suppose that $A, K \in \mathcal{L}(X)$ and $AK = KA$. Then*

(i) If $A \in \Phi_+^b(X)$, asc$(A) < \infty$ and $K \in \mathcal{F}_+^b(X)$, then asc$(A+K) < \infty$.
(ii) If $A \in \Phi_-^b(X)$, desc$(A) < \infty$ and $K \in \mathcal{F}_-^b(X)$, then desc$(A+K) < \infty$. ◇

Theorem 2.2.22. *Let $A \in \mathcal{L}(X)$ such that $0 \in \rho(A)$. Then, for all $\lambda \neq \mu$, we have $A - (\lambda - \mu) \in \mathcal{B}^b(X)$ if, and only if, $\frac{1}{\lambda-\mu} - A^{-1} \in \mathcal{B}^b(X)$.* ◇

Proof. We note that

$$A - (\lambda - \mu) = (\mu - \lambda)\Big(A^{-1} - \frac{1}{\lambda - \mu}\Big)A. \tag{2.2.3}$$

Now, let us suppose that $\frac{1}{\lambda-\mu} - A^{-1} \in \mathcal{B}^b(X)$. Since $0 \in \rho(A)$ we have $A \in \mathcal{B}^b(X)$. Now, by applying Eq. (2.2.3), we infer that $A - (\lambda - \mu) \in \mathcal{B}^b(X)$. Conversely, assume that $A - (\lambda - \mu) \in \mathcal{B}^b(X)$. Then, the product on the right-hand side of Eq. (2.2.3) is in $\mathcal{B}^b(X)$. Besides, $0 \in \rho(A)$ implies that $A \in \mathcal{B}^b(X)$. Then, $(A^{-1} - \frac{1}{\lambda-\mu}) \in \mathcal{B}^b(X)$. Q.E.D.

Corollary 2.2.1. *Let $A \in \mathcal{L}(X)$ such that $0 \in \rho(A)$. Then, for all $\lambda \neq 0$, we have $A - \lambda \in \mathcal{B}^b(X)$ if, and only if, $\frac{1}{\lambda} - A^{-1} \in \mathcal{B}^b(X)$.* ◇

Proof. It is an obvious consequence of Theorem 2.2.22. Q.E.D.

Definition 2.2.1. Let $A \in \mathcal{L}(X)$. We say that A is Kato's operator, if $R(A)$ is closed and $N(A) \subset R^{\infty}(A)$. ◇

Theorem 2.2.23 ([264, Theorem 10, p. 180]). *Let X be a Banach space. Let $A \in \mathcal{L}(X)$ be upper semi-Browder (lower semi-Browder, Browder) if, and only if, there exists a decomposition $X = X_1 \oplus X_2$ such that $\dim X_1 < \infty$, $A(X_i) \subset X_i$ $(i = 1, 2)$ $A|_{X_1}$ is nilpotent and $A|_{X_2}$ is bounded below (onto, invertible, respectively).* ◇

Theorem 2.2.24 ([264, Theorem 19, p. 185]). *Let X be a Banach space. Let $A \in \mathcal{L}(X)$ be upper semi-Browder (lower semi-Browder, Browder, respectively), and let $B \in \mathcal{L}(X)$ with $BA = AB$ and*

$$\max\{|z| : z \notin \{z \in \Phi_B \text{ such that } i(z - B) = 0 \text{ and all scalars near } z \text{ are in } \rho(B)\}\} = 0.$$

Then, $A + B$ is upper semi-Browder (lower semi-Browder, Browder, respectively). ◇

Definition 2.2.2. An operator A in $\mathcal{L}(X)$ is called quasi-nilpotent if $\lim_{n\to\infty} \|A^n\|^{\frac{1}{n}} = 0$, this is equivalent to the condition that $\sigma(A) = \{0\}$. ◇

Definition 2.2.3. An operator $A \in \mathcal{L}(X)$ is called a Riesz operator, if it satisfies the following conditions

(i) for all $\lambda \in \mathbb{C}^*$, $(\lambda - A)$ is a Fredholm operator on X,
(ii) for all $\lambda \in \mathbb{C}^*$, $(\lambda - A)$ has a finite ascent and a finite descent, and
(iii) all $\lambda \in \sigma(A)\backslash\{0\}$ are eigenvalues of a finite multiplicity, and have no accumulation points, except possibly zero. ◇

Let $\mathcal{R}(X)$ denote the class of all Riesz operators. It is worth noticing that there are many characterizations of Riesz operators. Ruston has characterized $\mathcal{R}(X)$ as being the class of asymptotically quasi-compact operators, i.e., those $A \in \mathcal{L}(X)$, for which

$$\lim_{n\to\infty} \left[\inf_{K\in\mathcal{K}(X)} \|A^n - K\| \right]^{\frac{1}{n}} = 0, \tag{2.2.4}$$

(cf. [67, 68], and [340]). We recall that the Riesz operators satisfy the Riesz–Schauder theory of compact operators and that $\mathcal{R}(X)$ is not an ideal of $\mathcal{L}(X)$ [68]. Moreover, by using (2.2.4), we get the following result established independently by Caradus [67] and West [340].

Proposition 2.2.3. *Let X be a Banach space, and let A and S be two commuting operators of $\mathcal{L}(X)$. We have*

(i) If $A \in \mathcal{R}(X)$, then $AS \in \mathcal{R}(X)$.
(ii) If A and S are in $\mathcal{R}(X)$, then $A + S \in \mathcal{R}(X)$. ◇

Theorem 2.2.25 ([288, Corollary 2]). *Suppose that $A \in \mathcal{L}(X)$, that B is a Riesz operator and that $AB = BA$. Then,*

(i) If $A \in \mathcal{B}^b_+(X)$, then $A + B \in \mathcal{B}^b_+(X)$.
(ii) If $A \in \mathcal{B}^b_-(X)$, then $A + B \in \mathcal{B}^b_-(X)$. ◇

2.2.2 Gap Topology

The gap between two linear subspaces M and N of a normed space X is defined by the following formula

$$\delta(M, N) = \sup_{x\in M,\ \|x\|=1} \operatorname{dist}(x, N), \tag{2.2.5}$$

in the case where $M \neq \{0\}$. Otherwise we define $\delta(\{0\}, N) = 0$ for any subspace N. Moreover, $\delta(M, \{0\}) = 1$ if $M \neq \{0\}$, as shown from the definition (see (2.2.5)). We can also define $\hat{\delta}(M, N) = \max\{\delta(M, N), \delta(N, M)\}$. Sometimes, the latter is called the symmetric or maximal gap between M and N in order to distinguish it from the former. The gap $\delta(M, N)$ can be characterized as the smallest number δ such that $\text{dist}(x, N) \leq \delta\|x\|$, for all $x \in M$. For the properties of $\delta(\cdot, \cdot)$ and $\hat{\delta}(\cdot, \cdot)$ we may refer to [186, p. 197, 200, 201].

Remark 2.2.1.

(i) The gap measures the distance between two subspaces, and it is easy to see the following:

 (a) $\delta(M, N) = \delta(\overline{M}, \overline{N})$ and $\hat{\delta}(M, N) = \hat{\delta}(\overline{M}, \overline{N})$,
 (b) $\delta(M, N) = 0$ if, and only if, $\overline{M} \subset \overline{N}$,
 (c) $\hat{\delta}(M, N) = 0$ if, and only if, $\overline{M} = \overline{N}$.

(ii) Let us notice that $\hat{\delta}$ is a metric on the set $\mathcal{U}(X)$ of all linear, closed subspaces of X and the convergence $M_n \to N$ in $\mathcal{U}(X)$ is obviously defined by $\hat{\delta}(M_n, N) \to 0$. Moreover, $(\mathcal{U}(X), \hat{\delta})$ is a complete metric space.

(iii) The gap between two closed subspaces M and N is introduced by Krein and Krasnoselskii in [195]. ◇

Definition 2.2.4. Let X and Y be two Banach spaces and let T, S be two closed linear operators acting from X into Y. Let us define $\delta(T, S) = \delta(G(T), G(S))$ and $\hat{\delta}(T, S) = \widehat{\delta}(G(T), G(S))$, where $G(T)$ (resp. $G(S)$) is the graph of T (resp. S). $\hat{\delta}(T, S)$ is called the gap between S and T. ◇

Remark 2.2.2. (i) $$\delta(G(T), G(S)) = \sup_{\substack{x \in \mathcal{D}(T) \\ \|x\|^2 + \|Tx\|^2 = 1}} \left(\inf_{y \in \mathcal{D}(S)} \left(\|x - y\|^2 + \|Tx - Sy\|^2\right)^{\frac{1}{2}} \right).$$

(ii) The function $\hat{\delta}(\cdot, \cdot)$ defines a metric on $\mathcal{C}(X, Y)$ which called the gap metric, and the topology induced by this metric is called the Gap topology or Kato topology. ◇

Theorem 2.2.26 ([186]). *Let T and S be two closed densely defined linear operators acting from X into Y. Then, we have*

(i) $\delta(T, S) = \delta(S^*, T^*)$ and $\hat{\delta}(T, S) = \hat{\delta}(S^*, T^*)$.
(ii) If T and S are invertible, then $\delta(S^{-1}, T^{-1}) = \delta(S, T)$ and $\hat{\delta}(S^{-1}, T^{-1}) = \hat{\delta}(S, T)$.
(iii) Let $A \in \mathcal{L}(X, Y)$. Then, $\widehat{\delta}(S + A, T + A) \leq 2(1 + \|A\|^2)\widehat{\delta}(S, T)$.
(iv) Let T be Fredholm (resp. semi-Fredholm). If $\hat{\delta}(S, T) < \tilde{\gamma}(T)(1 + [\tilde{\gamma}(T)]^2)^{-\frac{1}{2}}$, then S is Fredholm (resp. semi-Fredholm), $\alpha(S) \leq \alpha(T)$, and $\beta(S) \leq \beta(T)$. Furthermore, there exists $b > 0$ such that $\hat{\delta}(S, T) < b$ implies $i(S) = i(T)$.
(v) Let $T \in \mathcal{L}(X, Y)$. If $S \in \mathcal{C}(X, Y)$ and $\hat{\delta}(S, T) < (1 + \|T\|^2)^{-\frac{1}{2}}$, then S is bounded (so that $\mathcal{D}(S)$ is closed). ◇

Remark 2.2.3. Let X be a Hilbert space, and let S, T be two essentially self-adjoint (in particular, self-adjoint) linear operators in X. Then, $\hat{\delta}(T,S) = \delta(S,T) = \delta(T,S)$. ◇

Theorem 2.2.27 ([85, Theorem 2.3]). *If $T \in \mathcal{L}(X,Y)$, then we have $\delta(T,0) = \delta(0,T) = \widehat{\delta}(T,0) = \frac{\|T\|}{\sqrt{1+\|T\|^2}}$.* ◇

Definition 2.2.5. Let S and T be two closable operators. We define the gap between T and S by $\delta(T,S) = \delta(\overline{T},\overline{S})$. We can also define the symmetric gap between S and T by $\hat{\delta}(T,S) = \hat{\delta}(\overline{T},\overline{S}) = \max\{\delta(G(\overline{T}),G(\overline{S})),\delta(G(\overline{S}),G(\overline{T}))\}$. ◇

2.2.3 Semi-Regular and Essentially Semi-Regular Operators

Definition 2.2.6. Let $A \in \mathcal{L}(X)$.

(i) A is said to be semi-regular, if $R(A)$ is closed and $N(A) \subseteq R(A^n)$, for all $n \in \mathbb{N}$.
(ii) A is said to be essentially semi-regular if $R(A)$ is closed and there exists a finite dimensional subspace F, such that $N(A) \subseteq R(A^n) + F$, for all $n \in \mathbb{N}$. ◇

Now, let $\mathcal{V}_0(X) := \{A \in \mathcal{L}(X)$ such that A is semi-regular$\}$ and let $\mathcal{V}(X) := \{A \in \mathcal{L}(X)$ such that A is essentially semi-regular$\}$. It is well known that $\Phi_+^b(X) \bigcup \Phi_-^b(X) \subset \mathcal{V}(X)$, and that $\mathcal{V}_0(X)$ and $\mathcal{V}(X)$ are neither semigroups nor open or closed subsets of $\mathcal{L}(X)$. From Shomoeger paper [303], we get $\text{int}(\mathcal{V}(X)) = \Phi_+^b(X) \bigcup \Phi_-^b(X)$ and $\text{int}(\mathcal{V}_0(X)) := \big\{A \in \Phi_\pm(X) \bigcap \mathcal{L}(X)$ such that $\alpha(A) = 0$ or $\beta(A) = 0\big\}$. Trivial examples of semi-regular operators are surjective operators as well as injective ones with a closed range, Fredholm and semi-Fredholm operators with a jump equal to zero. Some other examples of semi-regular operators may be found in Mbekhta and Ouahab [246] and Labrousse [205].

Definition 2.2.7. An operator $A \in \mathcal{L}(X)$ is called a essentially semi-regular perturbation, if $K + A$ is essentially semi-regular for every essentially semi-regular operator K commuting with A. ◇

We denote $\mathcal{F}_e(X)$ by $\mathcal{F}_e(X) = \{A \in \mathcal{L}(X),\ A + K \in \mathcal{V}(X)$ for all $K \in \mathcal{V}(X),\ AK = KA\}$. Examples of essentially semi-regular perturbation operators are the compact operators, operators with a finite rank, Riesz operators, quasi-nilpotent operators, nilpotent operators, and a sufficiently small perturbation of all semi-regular operators. A semi-regular operator A has a closed range.

Definition 2.2.8. Let X, Y be Banach spaces and let $A \in \mathcal{L}(X,Y)$. We define the injectivity modulus of A (sometimes also called the minimum modulus) by

$$j(A) = \inf\{\|Ax\| \text{ such that } x \in X,\ \|x\| = 1\}.$$

◇

If $A \in \mathcal{L}(X)$ is one-to-one, then clearly $\tilde{\gamma}(A) = j(A)$.

Lemma 2.2.11. *Let $A \in \mathcal{L}(X, Y)$. Then $\tilde{\gamma}(A) = j(A_0)$, where $A_0 : X/N(A) \longrightarrow \overline{R(A)}$, $\tilde{x} := x + N(A) \longrightarrow A_0\tilde{x} = Ax$.* ◇

Proof. We have $j(A_0) = \inf\{\|A_0(x + N(A))\|$ such that $\|x + N(A)\|_{X/N(A)} = 1\} = \inf\{\|Ax\|$ such that $\operatorname{dist}(x, N(A)) = 1\} = \tilde{\gamma}(A)$. Q.E.D.

The notion of the reduced minimum modulus is motivated by the following characterization:

Theorem 2.2.28. *Let $A \in \mathcal{L}(X, Y)$. Then $R(A)$ is closed if, and only if, $\tilde{\gamma}(A) > 0$.* ◇

Proof. The statement is clear if $A = 0$. If $A \neq 0$, then $R(A) = R(A_0)$, and $A_0(x + N(A)) = Ax$ and $R(A_0)$ is closed if, and only if, $j(A_0) > 0$. Q.E.D.

Theorem 2.2.29. *Let $A \in \mathcal{L}(X, Y)$. Then $\tilde{\gamma}(A) = \tilde{\gamma}(A^*)$.* ◇

Proof. By Theorem 2.2.28, $\tilde{\gamma}(A) = 0$ if, and only if, $\tilde{\gamma}(A^*) = 0$. Let $\tilde{\gamma}(A) > 0$, so $R(A)$ is closed. We have $A = JA_0Q$, where $Q : X \longrightarrow X/N(A)$ is the canonical projection, $A_0 : X/N(A) \longrightarrow R(A)$ is one-to-one and onto and $J : R(A) \longrightarrow Y$ is the natural embedding. The corresponding decomposition for A^* is $A^* = Q^*A_0^*J^*$. We have $\tilde{\gamma}(A) = j(A_0) = \|A_0^{-1}\|^{-1} = \|A_0^{*-1}\|^{-1} = j(A_0^*) = \tilde{\gamma}(A^*)$. Q.E.D.

The following theorem gives several equivalent definitions of the semi-regular operators.

Theorem 2.2.30 ([246, Theorem 4.1]). *Let $A \in \mathcal{L}(X, Y)$ and $\lambda_0 \in \mathbb{C}$. The following statements are equivalent:*

(i) $\lambda_0 - A$ *is semi-regular.*
(ii) $\tilde{\gamma}(\lambda_0 - A) > 0$ *and the mapping* $\lambda \longrightarrow \tilde{\gamma}(\lambda - A)$ *is continuous at* λ_0.
(iii) $\tilde{\gamma}(\lambda_0 - A) > 0$ *and the mapping* $\lambda \longrightarrow N(\lambda - A)$ *is continuous at* λ_0 *in the gap topology.*
(iv) $R(\lambda - A)$ *is closed in a neighborhood of* λ_0 *and the mapping* $\lambda \longrightarrow R(\lambda - A)$ *is continuous at* λ_0 *in the gap topology.* ◇

Let (M, N) be a pair of closed subspaces of X. A is said to be decomposed according to $X = M \oplus N$, if $A(M) \subset M$, and $A(N) \subset N$. When A is decomposed as above, the pair A_M, A_N of A in M, N, respectively, can be defined: A_M is an operator in the Banach space M with $\mathcal{D}(A_M) = M$ such that $A_M x = Ax \in M$, and A_N is similarly defined. In this case, we write $A = A_M \oplus A_N$.

Definition 2.2.9. (i) An operator $A \in \mathcal{L}(X, Y)$ is said to be of Kato type of order $d \in \mathbb{N}$ if, there exist a pair of closed subspaces (M, N) of X such that $A = A_M \oplus A_N$, where A_M is semi-regular and A_N is nilpotent of order d (i.e., $(A_N)^d = 0$).

(ii) An operator A is said to be of Kato type if, there exists $d \in \mathbb{N}$ such that A is a Kato type of order d. ◇

Clearly, every semi-regular operator is of Kato type with $M = X$ and $N = \{0\}$ and a nilpotent operator has a decomposition with $M = \{0\}$ and $N = X$. Every essentially semi-regular operator admits a decomposition (M, N) such that N is a finite-dimensional vector space.

Definition 2.2.10. Let X and Y be two Banach spaces and let $A \in \mathcal{L}(X, Y)$.

(i) An operator $B \in \mathcal{L}(Y, X)$ is called g_1-inverse of A, if $Au = ABAu$ for all $u \in X$.
(ii) An operator $B \in \mathcal{L}(Y, X)$ is called g_2-inverse (or generalized inverse) of A, if $Au = ABAu$ for all $u \in X$, and $Bv = BABv$ for all $v \in Y$. ◇

We denote $\mathcal{G}_1(A)$ and $\mathcal{G}_2(A)$ by

$$\mathcal{G}_1(A) := \big\{B \in \mathcal{L}(Y, X) \text{ such that } B \text{ is } g_1\text{-inverse of } A\big\},$$

$$\mathcal{G}_2(A) := \big\{B \in \mathcal{L}(Y, X) \text{ such that } B \text{ is } g_2\text{-inverse of } A\big\}.$$

Remark 2.2.4.

(i) The relation (g_2-inverse) is symmetric.
(ii) If A is a one-sided inverse of B, then B is a generalized inverse of A.
(iii) $\mathcal{G}_2(A) \subset \mathcal{G}_1(A)$. ◇

Lemma 2.2.12 ([206, Lemma 1.3]). *Let $A \in \mathcal{L}(X, Y)$ and let $B \in \mathcal{G}_1(A)$. Then,*

(i) AB is a projection of Y onto $R(A)$, and $N(AB) = N(B)$.
(ii) BA is a projection of X onto $R(B)$, and $N(BA) = N(A)$. ◇

Remark 2.2.5. Let $A \in \mathcal{L}(X, Y)$ and let $B \in \mathcal{G}_2(A)$. Then, $\mathcal{D}(B) = N(B) \oplus R(A)$ and $\mathcal{D}(A) = N(A) \oplus R(B)$. ◇

Corollary 2.2.2 ([206, Corollary 1.7]). *Let $A \in \mathcal{L}(X, Y)$ and $B \in \mathcal{G}_1(A)$. Then $X = N(B) \oplus R(A)$.* ◇

Lemma 2.2.13 ([246, Lemma 2.4]). *Let $A \in \mathcal{L}(X, Y)$. If A is semi-regular, then $R^\infty(A)$ is closed and $A(R^\infty(A)) = R^\infty(A)$.* ◇

Lemma 2.2.14. *Let $A \in \mathcal{L}(X)$. If A is semi-regular, then A^n is semi-regular for every $n \in \mathbb{N}$.* ◇

Proof. Since A is regular, then $\tilde{\gamma}(A^n) \geq \tilde{\gamma}(A)^n > 0$. So, $B = A^n$ has a closed range. Furthermore, $N(B) \subset R^\infty(A) = R^\infty(B)$. We conclude that A^n is semi-regular. Q.E.D.

Theorem 2.2.31 ([263]). *Let A, $S \in \mathcal{L}(X)$, such that $AS = SA$. If AS is semi-regular (resp. essentially semi-regular), then both A and S are semi-regular (resp. essentially semi-regular).* ◇

The product of two commuting semi-regular operators need not be semi-regular in general (see [263]). The following two theorems give some cases where the converse of Theorem 2.2.31 is true.

Theorem 2.2.32 ([263]). *Let A, S, C, $D \in \mathcal{L}(X)$ be mutually commuting operators such that $AC + SD = I$. Then, AS is semi-regular if, and only if, both A and S are semi-regular.* ◇

Theorem 2.2.33 ([263]). *Let A, $S \in \mathcal{L}(X)$ such that $AS = SA$ and S is invertible. If A is semi-regular, then AS is also semi-regular.* ◇

Let us denote by X/V the quotient space induced by a closed subspace V of X. Recall the following interesting characterization of the bounded semi-regular (resp. the essentially semi-regular) operators.

Theorem 2.2.34 ([192]). *$A \in \mathcal{L}(X)$ is a semi-regular (resp. essentially semi-regular) operator if, and only if, there exists a closed subspace V of X, such that $A(V) = V$ and the operator $\hat{A} : X/V \longrightarrow X/V$ induced by A is bounded below (resp. upper semi-Fredholm).* ◇

Theorem 2.2.35. *Let $A \in \mathcal{L}(X)$ and let us assume that A is of Kato type of order d with a pair (M, N) of closed subspaces of X. Then*

(i) $R^{\infty}(A) = A(R^{\infty}(A)) = R^{\infty}(A_M)$. *Moreover, $R^{\infty}(A)$ is closed.*
(ii) *For every non-negative integer $n \geq d$, we have $N(A)\bigcap R(A^n) = N(A)\bigcap M = N(A)\bigcap R(A^d)$.*
(iii) *For every non-negative integer $n \geq d$, we have $R(A) + N(A^n) = A(M) \oplus N$ is closed.* ◇

Proof.

(*i*) Since $A = A_M \oplus A_N$, it is clear that $A^n = A_M^n \oplus A_N^n$ for every $n \in \mathbb{N}$ and hence as A_N is nilpotent of degree d, we deduce that $R(A^n) = R(A_M^n)$ for $n \geq d$ and then $R^{\infty}(A) = R^{\infty}(A_M)$. Moreover, since A_M is semi-regular, we infer from Lemma 2.2.14 that A_M^n is also semi-regular, in particular $R(A_M^n)$ is closed for all $n \in \mathbb{N}$ and hence, $R^{\infty}(A_M)$ is closed.
(*ii*) Let $n \geq d$. Then, $N(A)\bigcap R(A^n) = N(A)\bigcap R(A_M^n) \subseteq N(A)\bigcap R(A_M) \subseteq N(A)\bigcap M = N(A_M)$. Since A_M is semi-regular, then we have $N(A_M) \subseteq N(A)\bigcap R(A_M^n) = N(A)\bigcap R(A^n)$. Hence (*ii*) holds.
(*iii*) Let $n \geq d$. Clearly, $N \oplus N(A_M^n) = N(A^n)$ so that $N \subset N(A^n)$ and hence, $R(A_M) \oplus N \subset R(A) + N(A^n)$. Conversely, $N(A^n) = N(A) = N(A_M^n) \oplus N(A_N^n) = N(A_M^n) \oplus N \subseteq R(A_M) \oplus N$, and, by using the semi-regularity of A_M, it follows that $R(A) = R(A_M) \oplus R(A_N) \subset R(A_M) \oplus N$. Hence, $R(A) + N(A^n) \subseteq R(A_M) \oplus N$. Consequently, $R(A) + N(A^n) = A(M) \oplus N$ if $n \geq d$. Now, let $\Psi : (m,n) \in M \times N \longrightarrow \Psi(m,n) = m + n$. Clearly, Ψ is a topological isomorphism and $\Psi(R(A_M), N) = R(A_M) \oplus N$, where $R(A_M)$ is closed in M and hence, $(R(A_M), N)$ is closed. Q.E.D.

Proposition 2.2.4. *Let $A \in \mathcal{L}(X)$, $B \in \mathcal{G}_2(A)$, and $T \in \mathcal{L}(X)$ commuting with A and B. If $R(T)$ is closed, then $R(TA)$ is also closed.* ◇

Proof. Let $(y_n)_n \subset R(TA)$ such that $y_n \to y$. There exists $x_n \in X$, with $y_n = TAx_n$. Since $A = ABA$, $TABAx_n = AB(TAx_n)$ and since AB is a bounded operator, we obtain $ABy = y$. By using Lemma 2.2.12, we infer that there exists $x \in \mathcal{D}(A)$ such that $y = Ax$. Let $z_n = BAx_n - BABTAx_n$. Then, $Tz_n = BABTAx_n - TBABTAx_n = BABy_n - TBABy_n$. Since AB is bounded, it follows from Lemma 2.2.12 that $(Tz_n)_n$ converges to $By - TBy$. The fact that $R(T)$ is closed, there exists $z \in X$ such that $Tz = By - TBy$, which implies that $AT(z + BAx) = y$. Hence, $y \in R(TA)$. Q.E.D.

Theorem 2.2.36. *Let $A \in \mathcal{L}(X)$, and let $B \in \mathcal{G}_2(A)$ where T is essentially semi-regular commuting with A and B. If $N(TA) \subset N(T)$ and A is surjective, then TA is essentially semi-regular.* ◇

Proof. $R(T)$ is closed. Then, by using Proposition 2.2.4, we deduce that $R(TA)$ is closed. T is essentially semi-regular, which implies that there exists a finite-dimensional subspace F, such that $N(TA) \subset N(T) \subset \bigcap_{n\in\mathbb{N}} R(T^n) + F$. Since A is surjective, then $\bigcap_{n\in\mathbb{N}} R(T^n) \subset \bigcap_{n\in\mathbb{N}} R((TA)^n)$ and hence, $N(TA) \subset \bigcap_{n\in\mathbb{N}} R((TA)^n) + F$. Q.E.D.

Corollary 2.2.3. *Let $A \in \mathcal{L}(X)$, and let $B \in \mathcal{G}_2(A)$, where T is semi-regular commuting with A and B. If $N(TA) \subset N(T)$ and A is surjective, then TA is semi-regular.* ◇

Corollary 2.2.4. *Let $A \in \mathcal{L}(X)$, and let $B \in \mathcal{G}_2(A)$, where T is semi-regular (resp. essentially semi-regular) commuting with A and B. If $0 \in \rho(A)$, then TA is semi-regular (resp. essentially semi-regular).* ◇

Sometimes it happens that the spectrum $\sigma(A)$ of a closed operator A contains a bounded part Σ' separated from the rest Σ'' in such a way that a rectifiable, simple closed curve Γ (or, more generally, a finite number of such curves) can be drawn so as to enclose an open set containing Σ' in its interior and Σ'' in its exterior. Under such a circumstance, we have the following decomposition theorem given by Kato in [186].

Theorem 2.2.37. *Let $\sigma(A)$ be separated into two parts Σ' and Σ'' in the way described above. Then we have a decomposition of A according to a decomposition $X = M' \oplus M''$ of the space in such a way that the spectra of the parts $A_{M'}$, $A_{M''}$ coincide with Σ' and Σ'', respectively, and $A_{M'} \in \mathcal{L}(M')$.* ◇

2.2.4 Basics on Unbounded Fredholm Operators

Theorem 2.2.38 ([302, Theorem 7.1, p. 157]). *Let X and Y be Banach spaces, and $A \in \Phi(X,Y)$. Then, there is an operator $A_0 \in \mathcal{L}(Y,X)$, such that $N(A_0) = Y_0$, $R(A_0) = X_0 \bigcap \mathcal{D}(A)$, $A_0A = I$ on $X_0 \bigcap \mathcal{D}(A)$, and $AA_0 = I$ on $R(A)$. There are operators $F_1 \in \mathcal{L}(X)$, $F_2 \in \mathcal{L}(Y)$, such that $A_0A = I - F_1$ on $\mathcal{D}(A)$, $AA_0 = I - F_2$ on Y, $R(F_1) = N(A)$, $N(F_1) = X_0$, and $R(F_2) = Y_0$, $N(F_2) = R(A)$.* ◇

Theorem 2.2.39 ([302, Corollary 7.6, p. 160]). *Let X and Y be Banach spaces. Assume that $A \in \Phi(X, Y)$, and that W is continuously embedded in X in such a way that $\mathcal{D}(A)$ is dense in W. Then, $A \in \Phi(W, Y)$ with $N(A)$ and $R(A)$ being the same.* ◇

Theorem 2.2.40 ([302, Theorem 7.3, p. 157]). *Let X and Y be Banach spaces. If $A \in \Phi(X, Y)$ and $B \in \Phi(Y, Z)$, then $BA \in \Phi(X, Z)$ and $i(BA) = i(B) + i(A)$.* ◇

Theorem 2.2.41 ([302, Theorem 7.12, p. 162]). *Let X, Y, and Z be Banach spaces. If $A \in \Phi(X, Y)$ and B is a densely defined closed linear operator from Y into Z such that $BA \in \Phi(X, Z)$, then $B \in \Phi(Y, Z)$.* ◇

Theorem 2.2.42 ([302, Theorem 7.14, p. 164]). *Let X, Y, and Z be Banach spaces and let A be a densely defined closed linear operator from X into Y. Suppose that $B \in \mathcal{L}(Y, Z)$ with $\alpha(B) < \infty$ and $BA \in \Phi(X, Z)$. Then, $A \in \Phi(X, Y)$.* ◇

Theorem 2.2.43 ([302, Theorem 7.32, p. 175]). *Let X, Y, and Z be Banach spaces. If $A \in \Phi_+(X, Y)$, $B \in \Phi_+(Y, Z)$ and $\mathcal{D}(BA)$ is dense in X, then $BA \in \Phi_+(X, Z)$.* ◇

Theorem 2.2.44 ([302, Theorem 5.10]). *Let $A \in \mathcal{C}(X, Y)$. If $A \in \Phi(X, Y)$ and $K \in \mathcal{K}(X, Y)$, then $A + K \in \Phi(X, Y)$ and $i(A + K) = i(A)$.* ◇

Theorem 2.2.45 ([302, Theorem 7.9, p. 161]). *Let X and Y be Banach spaces. For $A \in \Phi(X, Y)$, there is an $\eta > 0$ such that, for every $T \in \mathcal{L}(X, Y)$ satisfying $\|T\| < \eta$, one has $A + T \in \Phi(X, Y)$, $i(A + T) = i(A)$ and $\alpha(A + T) \leq \alpha(A)$.* ◇

Theorem 2.2.46 ([302, Theorem 7.35, p. 178]). *Let X, Y, Z be Banach spaces, and assume that A is a densely defined, closed linear operator from X into Y such that $R(A)$ is closed in Y and $\beta(A) < \infty$. Let B be a densely defined linear operator from Y into Z. Then $(BA)^* = A^* B^*$.* ◇

Lemma 2.2.15 ([302, Lemma 7.36, p. 178]). *If A, B satisfy the hypotheses of Theorem 2.2.46 and x is any element in $\mathcal{D}(A)$, then there is a sequence $(x_n)_n \subset \mathcal{D}(A)$ such that $Ax_n \to Ax$ in Y and $x_n \to x$ in X. Consequently, $\mathcal{D}(BA)$ is dense in X.* ◇

The following lemma is well known for bounded upper semi-Fredholm operators. The proof is a straightforward adaptation of the proof of Theorem 3.9 in [353].

Lemma 2.2.16. *Let $A \in \Phi_+(X)$. Then, the following statements are equivalent*

(i) $i(A) \leq 0$.
(ii) *A can be expressed in the form $A = U + K$, where $K \in \mathcal{K}(X)$ and $U \in \mathcal{C}(X)$ is an operator with a closed range and $\alpha(U) = 0$.* ◇

Lemma 2.2.17 ([301, Lemma 4.6, p. 16]). *If $0 \in \rho(A)$, then $\lambda \neq 0$ is in Φ_A if, and only if, $\frac{1}{\lambda} \in \Phi_{A^{-1}}$ and $i(A - \lambda) = i(A^{-1} - \lambda^{-1})$.* ◇

Theorem 2.2.47 ([186, Theorem 5.26, p. 238]). *Let X and Y be Banach spaces. Suppose that A is a semi-Fredholm operator and B is an A-compact operator from X into Y, then $A + B$ is also semi-Fredholm with $i(A + B) = i(A)$.* ◇

A consequence of Theorem 2.2.47 is the following:

Theorem 2.2.48 ([302, Theorem 7.26, p. 172]). *$\Phi_{A+K} = \Phi_A$ for all K which are A-compact, and $i(A + K - \lambda) = i(A - \lambda)$ for all $\lambda \in \Phi_A$.* ◇

Theorem 2.2.49 ([185, Theorem 5.31, p. 241]). *Let X and Y be Banach spaces. Let $A \in \mathcal{C}(X,Y)$ be semi-Fredholm and let S be an A-bounded operator from X into Y. Then, $\lambda S + A$ is semi-Fredholm and $\alpha(\lambda S + A)$, $\beta(\lambda S + A)$ are constant for a sufficiently small $|\lambda| > 0$.* ◇

Proposition 2.2.5. *Let $A \in \mathcal{C}(X,Y)$ and let S be a non-null bounded linear operator from X into Y. Then, we have the following results:*

(i) *$\Phi_{A,S}$ is open.*
(ii) *$i(\lambda S - A)$ is constant on any component of $\Phi_{A,S}$.*
(iii) *$\alpha(\lambda S - A)$ and $\beta(\lambda S - A)$ are constant on any component of $\Phi_{A,S}$, except on a discrete set of points on which they have larger values.* ◇

Proof.

(*i*) Let $\lambda_0 \in \Phi_{A,S}$. Then, according to Theorem 2.2.45, there exists $\eta > 0$ such that, for all $\mu \in \mathbb{C}$ with $|\mu| < \frac{\eta}{\|S\|}$, the operator $\lambda_0 S - \mu S - A$ is a Fredholm operator, $i(\lambda_0 S - \mu S - A) = i(\lambda_0 S - A)$ and $\alpha(\lambda_0 S - \mu S - A) \leq \alpha(\lambda_0 S - A)$. Consider $\lambda \in \mathbb{C}$ such that $|\lambda - \lambda_0| < \frac{\eta}{\|S\|}$. Then, $\lambda S - A$ is a Fredholm operator, $i(\lambda S - A) = i(\lambda_0 S - A)$ and $\alpha(\lambda S - A) \leq \alpha(\lambda_0 S - A)$. In particular, this implies that $\Phi_{A,S}$ is open.

(*ii*) Let λ_1 and λ_2 be any two points in $\Phi_{A,S}$ which are connected by a smooth curve Γ whose points are all in $\Phi_{A,S}$. Since $\Phi_{A,S}$ is an open set, then for each $\lambda \in \Gamma$, there exists an $\varepsilon > 0$ such that, for all $\mu \in \mathbb{C}$, $|\lambda - \mu| < \varepsilon$, $\mu \in \Phi_{A,S}$ and $i(\mu S - A) = i(\lambda S - A)$. By using the Heine–Borel theorem, there exists a finite number of such sets which cover Γ. Since each of these sets overlaps with, at least, another set and since $i(\mu S - A)$ is constant on each one, we see that $i(\lambda_1 S - A) = i(\lambda_2 S - A)$.

(*iii*) Let λ_1 and λ_2 be any two points in $\Phi_{A,S}$ which are connected by a smooth curve Γ whose points are all in $\Phi_{A,S}$. Since $\Phi_{A,S}$ is an open set, then for each $\lambda \in \Gamma$, there is a sufficiently small $\varepsilon > 0$ such that, for all $\mu \in \mathbb{C}$, $|\lambda - \mu| < \varepsilon$, $\mu \in \Phi_{A,S}$ and by using Theorem 2.2.49, $\alpha(A + \mu S)$ and $\beta(A + \mu S)$ are constant for all $\mu \in \mathbb{C}, 0 < |\lambda - \mu| < \varepsilon$. By referring to Heine–Borel's theorem, there is a finite number of such sets which cover Γ. Since each of these sets overlaps with, at least, another set, we see that $\alpha(\mu S + A)$ and $\beta(\mu S + A)$ are constant for all $\mu \in \Gamma$, except for a finite number of points of Γ. Q.E.D.

Proposition 2.2.6. *The index map is locally constant and continuous in norm.* ◇

Proof. Let F be a Fredholm operator. By using Theorem 2.2.38, there exist an operator G and a compact operator K such that $FG = I + K$. It suffices to show that, if A is a Fredholm operator with $\|A - F\| < \frac{1}{\|G\|}$, then $i(F) = i(A)$. Indeed, the operator $(A - F)G + I$ is invertible, since its distance from the identity is less than 1. Thus, $i(A) + i(G) = i(AG) = i((A - F + F)G) = i((A - F)G + I + K) = 0$. Hence, $i(A) = -i(G) = i(F)$. Q.E.D.

2.2.5 Quasi-Inverse Operator

Let A be a closed operator on a Banach space X, with the property that $\Phi_A \neq \emptyset$. If $f(\lambda)$ is a complex-valued analytic function of a complex variable, we denote by $\Delta(f)$ the domain of analyticity of f.

Definition 2.2.11. By $\mathcal{R}'_\infty(A)$ we mean the family of all analytic functions $f(\lambda)$ with the following properties:

(i) $\mathbb{C} \backslash \Phi_A \subset \Delta(f)$,
(ii) $\Delta(f)$ contains a neighborhood of ∞ and f is analytic at ∞. ◇

Definition 2.2.12. A bounded operator B is called a quasi-inverse of the closed operator A if $R(B) \subset \mathcal{D}(A)$, $AB = I + K_1$, and $BA = I + K_2$, where K_1, $K_2 \in \mathcal{K}(X)$. ◇

If A is a closed operator such that Φ_A is not empty, then by using Proposition 2.2.5 (i), Φ_A is open. Hence, it is the union of a disjoint collection of connected open sets. Each of them, $\Phi_i(A)$, will be called a component of Φ_A. In each $\Phi_i(A)$, a fixed point λ_i is chosen in a prescribed manner. Since $\alpha(\lambda_i - A) < \infty$, $R(\lambda_i - A)$ is closed and $\beta(\lambda_i - A) < \infty$, then there exist a closed subspace X_i and a subspace Y_i, such that $\dim Y_i = \beta(\lambda_i - A)$ satisfying $X = N(\lambda_i - A) \oplus X_i$ and $X = Y_i \oplus R(\lambda_i - A)$. Now, let P_{1i} be the projection of X onto $N(\lambda_i - A)$ along X_i and let P_{2i} be the projection of X onto Y_i along $R(\lambda_i - A)$. P_{1i} and P_{2i} are bounded finite rank operators. It is shown in [300] that $(\lambda_i - A)|_{\mathcal{D}(A) \bigcap X_i}$ has a bounded inverse A_i, where $A_i : R(\lambda_i - A) \longrightarrow \mathcal{D}(A) \bigcap X_i$. Let T_i be the bounded operator defined by

$$T_i x := A_i (I - P_{2i}) x \tag{2.2.6}$$

satisfying $T_i(\lambda_i - A) = I - P_{1i}$ on $\mathcal{D}(A)$ and $(\lambda_i - A)T_i = I - P_{2i}$ on X. Hence, T_i is a quasi-inverse of $(\lambda_i - A)$. Moreover, when $\lambda \in \Phi_i(A)$ and $\frac{-1}{\lambda - \lambda_i} \in \rho(T_i)$, the operator

$$R'_\lambda(A) := T_i \, [(\lambda - \lambda_i)T_i + I]^{-1} \tag{2.2.7}$$

is shown in [307] to be a quasi-inverse of $(\lambda - A)$. In fact, $R'_\lambda(A)$ is defined and analytic for all $\lambda \in \Phi_A$ except for, at most, an isolated set, $\Phi^0(A)$, having no accumulation point in Φ_A.

Definition 2.2.13. A set D in the complex plane is called a Cauchy domain, if the following conditions are satisfied:

(i) D is open,
(ii) D has a finite number of components, of which the closures of any two are disjoint,
(iii) the boundary of D is composed of a finite positive number of closed rectifiable Jordan curves, of which any two are unable to intersect. ◇

Theorem 2.2.50 ([318, Theorem 3.3]). *Let F and Δ be point sets in the plane. Let F be closed, Δ be open and $F \subset \Delta$. Suppose that the boundary $B(\Delta)$ of Δ is nonempty and bounded. Then, there exists a Cauchy domain D, such that:*

(i) $F \subset D$,
(ii) $\overline{D} \subset \Delta$,
(iii) the curves forming $B(D)$ are polygons, and
(iv) D is unbounded if Δ is unbounded. ◇

Definition 2.2.14. Let $f \in \mathcal{R}'_\infty(A)$. The class of operators $\mathcal{F}(A)$ will be defined as follows: $B \in \mathcal{F}(A)$ if $B = f(\infty) + \frac{1}{2\pi i}\int_{+B(\mathcal{D})} f(\lambda)R'_\lambda(A)d\lambda$, where $f(\infty) := \lim\limits_{\lambda\to\infty} f(\lambda)$ and D is an unbounded Cauchy domain such that $\mathbb{C}\backslash\Phi_A \subset D$, $\overline{D} \subset \Delta(f)$, and the boundary of D, $B(\mathcal{D})$, does not contain any points of $\Phi^0(A)$. ◇

Theorem 2.2.51 ([307, Theorem 7]). *Let $B_1, B_2 \in \mathcal{F}(A)$. Then $B_1 - B_2 = K$, $K \in \mathcal{K}(X)$.* ◇

Definition 2.2.15. Let $f \in \mathcal{R}'_\infty(A)$. By $f(A)$ we mean an arbitrary operator in the set $\mathcal{F}(A)$. ◇

Theorem 2.2.52 ([307, Theorem 9]). *Let $f(\lambda)$ and $g(\lambda)$ be in $\mathcal{R}'_\infty(A)$. Then $f(A).g(A) = (f.g)(A) + K$, $K \in \mathcal{K}(X)$.* ◇

Definition 2.2.16. Let $A \in \mathcal{L}(X)$. By $\mathcal{R}'(A)$ we mean the family of all analytic functions, $f(\lambda)$, such that $\mathbb{C}\backslash\Phi_A \subset \Delta(f)$. ◇

Definition 2.2.17. Let $f \in \mathcal{R}'(A)$. The class of operators $\mathcal{F}^*(A)$ will be defined as follows: $B \in \mathcal{F}^*(A)$ if $B = \frac{1}{2\pi i}\int_{+B(\mathcal{D})} f(\lambda)R'_\lambda(A)d\lambda$, where D is a bounded Cauchy domain such that $\mathbb{C}\backslash\Phi_A \subset D$, $\overline{D} \subset \Delta(f)$, and $B(\mathcal{D})$ does not contain any points of $\Phi^0(A)$. ◇

Definition 2.2.18. Let $f \in \mathcal{R}'(A)$. By $f^*(A)$ we mean an arbitrary operator in the set $\mathcal{F}^*(A)$. ◇

Theorem 2.2.53 ([307, Theorem 12]). *Let $B_1, B_2 \in \mathcal{F}^*(A)$. Then $B_1 - B_2 = K$, $K \in \mathcal{K}(X)$.* ◇

Theorem 2.2.54 ([307, Theorem 13]). *Let $A \in \mathcal{L}(X)$, and let $f(\lambda) = 1$. Then, $f^*(A) = I + K$, $K \in \mathcal{K}(X)$.* ◇

Theorem 2.2.55 ([307, Theorem 14]). *Let $A \in \mathcal{L}(X)$, and let $f(\lambda) = \lambda$. Then, $f^*(A) = A + K$, $K \in \mathcal{K}(X)$.* ◇

Lemma 2.2.18 ([307, Lemma 7.4]). *Let $\mu_i \in \Phi_i^0(A)$ ($\Phi_i^0(A)$ being the set of all $\lambda \in \Phi_i(A)$ such that $\frac{-1}{\lambda-\lambda_i} \in \sigma(T_i)$ and T_i is defined in Eq. (2.2.6)). Let D be a bounded Cauchy domain with the following: $\overline{D} \subset \Phi_i(A)$, $\mu_i \in D$, and no other points of $\Phi^0(A)$ are contained in $\overline{D}$. Then, $\frac{1}{2\pi i} \int_{+B(\mathcal{D})} R'_\lambda(A)d\lambda = K \in \mathcal{K}(X)$.* ◇

Lemma 2.2.19 ([308, Lemma 1.1]). *Let $A \in \mathcal{C}(X)$, such that Φ_A is not empty, and let n be a positive integer. Then, for each $\lambda \in \Phi_A \backslash \Phi^0(A)$, there exists a subspace V_λ dense in X and depending on λ such that, for all $x \in V_\lambda$, we have $R'_\lambda(A)x \in \mathcal{D}(A^n)$.* ◇

2.2.6 Basics on Unbounded Browder Operators

For $A \in \mathcal{C}(X)$ we denote by $\mathcal{D}(A^n) = \{x \in \mathcal{D}(A)$ such that $Ax, \ldots, A^{n-1}x \in \mathcal{D}(A)\}$ and we define A^n on this domain by the equation $A^n x = A(A^{n-1}x)$, where n is any positive integer and $A^0 = I$. It is simple to verify that $\{N(A^k)\}_k$ forms an ascending sequence of subspaces. Suppose that for some k, $N(A^k) = N(A^{k+1})$; we shall then write $\mathrm{asc}(A)$ for the smallest value of k for which this is true, and call the integer $\mathrm{asc}(A)$, the ascent of A. If no such integer exists, we shall say that A has infinite ascent. In a similar way, $\{R(A^k)\}_k$ forms a descending sequence; the smallest integer for which $R(A^k) = R(A^{k+1})$ is called the descent of A and is denoted by $\mathrm{desc}(A)$. If no such integer exists, we shall say that A has infinite descent.

Lemma 2.2.20 ([341]). *Let X be a Banach space and $A \in \mathcal{C}(X)$. If $\alpha(A) < \infty$, then $asc(A) < \infty$ if, and only if, $N^\infty(A) \bigcap R^\infty(A) = \{0\}$.* ◇

Definition 2.2.19. Let X be a Banach space, A and $B \in \mathcal{C}(X)$. We say that A commutes with B if $\mathcal{D}(B) = \mathcal{D}(A)$, $Ax \in \mathcal{D}(B)$ whenever $x \in \mathcal{D}(B)$, and $BAx = ABx$ for $x \in \mathcal{D}(B^2)$. ◇

Definition 2.2.20. Let X be a Banach space, A and $B \in \mathcal{C}(X)$. We say that A and B are mutually commuting operators if A commutes with B and B commutes with A, i.e., we have $\mathcal{D}(B) = \mathcal{D}(A)$, $Ax \in \mathcal{D}(B)$ whenever $x \in \mathcal{D}(B)$, $Bx \in \mathcal{D}(A)$ whenever $x \in \mathcal{D}(A)$, and $BAx = ABx$ for $x \in \mathcal{D}(A)$. ◇

Definition 2.2.21. Let X be a Banach space. We say that an operator $A \in \mathcal{C}(X)$ commutes with itself if A maps $\mathcal{D}(A)$ into $\mathcal{D}(A)$. ◇

Lemma 2.2.21 ([183]). *Let X be a Banach space, A and $B \in \mathcal{C}(X)$. Suppose that A commutes with B. Then for $n \in \mathbb{N}^*$, we have $\mathcal{D}(B) \subset \mathcal{D}(A)$, $B^n A^m x = A^m B^n x$ for all x in $\mathcal{D}(B^{n+1})$ and $m \in \mathbb{N}^*$.* ◇

We close this section by the following results due to Fakhfakh and Mnif [111].

Lemma 2.2.22. *Let X be a Banach space, A and $B \in \mathcal{C}(X)$. If A and B are mutually commuting operators, then A^p and B^k are mutually commuting operators for every $p, k \in \mathbb{N}^*$.* ◇

Proof. Since A and B are mutually commuting operators, then it follows from Lemma 2.2.21 that B^k commutes with A for every $k \in \mathbb{N}^*$ and A^p commutes with B for every $p \in \mathbb{N}^*$. Thus, $\mathcal{D}(A) = \mathcal{D}(B) = \mathcal{D}(A^p) = \mathcal{D}(B^k)$. It is clear that $A^p x \in \mathcal{D}(B^k)$ for all $x \in \mathcal{D}(B^k)$ and $B^k x \in \mathcal{D}(A^p)$ for all $x \in \mathcal{D}(A^p)$. Finally, according to the hypothesis and using Lemma 2.2.21, we infer that $A^p B^k x = B^k A^p x$ for all $x \in \mathcal{D}(A^{p+1}) = \mathcal{D}(A^p)$. Q.E.D.

Lemma 2.2.23. *Let X be a reflexive Banach space, $A \in \mathcal{C}(X)$ and $B = A + K$, where $K \in \mathcal{K}(X)$. If A and B are mutually commuting operators and A^* commutes with itself, then A^{*p} and B^{*k} are mutually commuting operators for every p, $k \in \mathbb{N}^*$.* ◇

Proof. Since $K \in \mathcal{L}(X)$ it is easy to show that $B^* = (A + K)^* = A^* + K^*$. Then $\mathcal{D}(B^*) = \mathcal{D}(A^*)$. From Theorem 2.2.46 and Lemma 2.2.15, A^* and $B^* \in \mathcal{C}(X^*)$. Let $f \in \mathcal{D}(B^*) = \mathcal{D}(A^*)$, $u \in \mathcal{D}(B) = \mathcal{D}(A)$ and $h = B^* f$. So, $B^* f \circ A(u) = h \circ A(u) = h(A(u)) = f \circ B(A(u)) = f \circ A \circ B(u) = g \circ B(u)$, where $g = f \circ A = A^* f$. Therefore, $B^* f \in \mathcal{D}(A^*)$ since $A^* f \in \mathcal{D}(B^*)$ for all $f \in \mathcal{D}(B^*)$. On the other hand, we will check that $B^* A^* f = A^* B^* f$ for all $f \in \mathcal{D}(A^*)$. Indeed, let $f \in \mathcal{D}(A^*) = \mathcal{D}(B^*)$ and $u \in \mathcal{D}(B) = \mathcal{D}(A)$. So, $B^*(A^* f)(u) = (A^* f) \circ B(u) = f \circ A \circ B(u) = f \circ B \circ A(u) = (B^* f) \circ A(u) = A^*(B^* f)(u)$. This implies that A^* and B^* are mutually commuting operators. Now, by Lemma 2.2.22 we conclude that A^{*p} and B^{*k} are mutually commuting operators for every $p, k \in \mathbb{N}^*$. This ends the proof. Q.E.D.

Remark 2.2.6. Let X be a reflexive Banach space, $A \in \mathcal{C}(X)$ and $B = A + K$, where $K \in \mathcal{K}(X)$ such that A and B are mutually commuting operators and A^* commutes with itself. Then,

(i) by Lemma 2.2.22 we infer that $\mathcal{D}(A) = \mathcal{D}(B) = \mathcal{D}(A^p) = \mathcal{D}(B^k) = \mathcal{D}(A^p B^k) = \mathcal{D}(B^k A^p)$ for every $p, k \in \mathbb{N}^*$,
(ii) and by Lemma 2.2.23 we deduce that $\mathcal{D}(A^*) = \mathcal{D}(B^*) = \mathcal{D}(A^{*p}) = \mathcal{D}(B^{*k}) = \mathcal{D}(A^{*p} B^{*k}) = \mathcal{D}(B^{*k} A^{*p})$ for every $p, k \in \mathbb{N}^*$. ◇

Remark 2.2.7. Let X be a Banach space, A and $B \in \mathcal{C}(X)$. If A and B are mutually commuting operators, then $A - \lambda$ and $B - \lambda$ are mutually commuting operators for every $\lambda \in \mathbb{C}$. ◇

Definition 2.2.22. Let X be a Banach space, A and $B \in \mathcal{C}(X)$. We say that A is an extension of B if $\mathcal{D}(B) \subset \mathcal{D}(A)$, and $Ax = Bx$ for $x \in \mathcal{D}(B)$. ◇

Definition 2.2.23. Let X be a Banach space, A and $B \in \mathcal{C}(X)$. We say that A is a k-dimensional extension of B if A is an extension of B, and $\mathcal{D}(A) = \mathcal{D}(B) \oplus Y$, where Y is a subspace of dimension k. ◇

Lemma 2.2.24 ([67]). *Let X be a Banach space, $A \in \mathcal{C}(X)$. Then, for any non negative integer k, $(A^k)^*$ is an extension of $(A^*)^k$.* ◇

Lemma 2.2.25. *Let X be a reflexive Banach space, $A \in \mathcal{C}(X)$ and $B = A + K$, where $K \in \mathcal{K}(X)$. If A and B are mutually commuting operators and A^* commutes with itself, then*

(i) $\mathcal{D}(A^{*p}B^{*k})$ is dense in X^*, and
(ii) $(B^kA^p)^*$ is an extension of $A^{*p}B^{*k}$. ◇

Proof.

(i) The result follows from Remark 2.2.6, Theorem 2.2.46 and Lemma 2.2.15.
(ii) From Lemma 2.2.24, we have $(A^p)^*$ is an extension of A^{*p} and $(B^k)^*$ is an extension of B^{*k}. Therefore, it is easy to see that $(A^p)^*(B^k)^*$ is an extension of $A^{*p}B^{*k}$. Moreover, by Theorem 2.2.47 and Remark 2.2.6 (i) we infer that $(B^kA^p)^*$ is an extension of $(A^p)^*(B^k)^*$. Then $(B^kA^p)^*$ is an extension of $A^{*p}B^{*k}$. This completes the proof. Q.E.D.

Definition 2.2.24. Let X be a Banach space and $A \in \mathcal{C}(X)$ such that all its iterates are densely defined. We say that A is of finite type if, for each k, $(A^k)^*$ is a finite dimensional extension of A^{*k}. ◇

Definition 2.2.25. Let X be a Banach space, A and $B \in \mathcal{C}(X)$ such that for all k, p, B^kA^p is densely defined. We say that (B, A) is of finite type if, for each k, p, $(B^kA^p)^*$ is a finite dimensional extension of $A^{*p}B^{*k}$. ◇

Lemma 2.2.26. *Let X be a Banach space, $A \in \mathcal{C}(X)$ and $B = A + K$, where $K \in \mathcal{K}(X)$ such that A and B are mutually commuting operators. Let Y be a closed subspace of $\mathcal{D}(A)$ satisfying $A(Y) \subset Y$ and $B(Y) \subset Y$. If $A_{|Y}$ is onto, then $B_{|Y}$ has finite descent.* ◇

Proof. It is clear that $A_{|Y}$ is a closed operator since $A \in \mathcal{C}(X)$. Then, from the closed graph theorem we get $A_{|Y}$ and $B_{|Y}$ are bounded operators. According to the hypothesis A and B are mutually commuting operators, it follows that $A_{|Y}$ and $B_{|Y}$ are commuting bounded operators. Moreover, $A_{|Y} - B_{|Y} \in \mathcal{K}(Y)$. Indeed, let $(x_n)_n$ a bounded sequence of $Y \subset X$, then $(Kx_n)_n$ admits a subsequence which converge in X. Moreover, $(Kx_n)_n \in Y$ and Y is a closed subspace of X. Therefore, $(Kx_n)_n$ admits a subsequence which converge in Y. This implies, by the use of Lemma 2.2.8 that $\text{desc}(B_{|Y}) < \infty$. Q.E.D.

Lemma 2.2.27 ([67]). *Let X be a Banach space, $A \in \mathcal{C}(X)$. Then for any $k \in \mathbb{N}$, $\alpha(A^k) \leq asc(A)\alpha(A)$, and $\beta(A^k) \leq desc(A)\beta(A)$.* ◇

An operator $A \in \mathcal{C}(X)$ is called upper semi-Browder if $A \in \Phi_+(X)$, and $\text{asc}(A) < \infty$. A is called lower semi-Browder if $A \in \Phi_-(X)$, and $\text{desc}(A) < \infty$.

Let $\mathcal{B}_+(X)$ (respectively $\mathcal{B}_-(X)$) denote the set of upper (respectively lower) semi-Browder operators. The class of all Browder operators is defined by $\mathcal{B}(X) = \mathcal{B}_+(X) \bigcap \mathcal{B}_-(X)$.

2.3 Positive Operator

2.3.1 Positive Operator on L_p-Spaces

In this section, we recall some facts about positive operators on L_p-spaces. Let Ω be an open subset of $\mathbb{R}^m$, $m \geq 1$, and let $E_p := L_p(\Omega)$, $1 \leq p < \infty$, be the Banach space of equivalence classes of measurable functions on Ω whose p'th power is integrable. Its dual space is E_q where $q = \frac{p}{p-1}$. The positive cone $E^+_{p,0}$ of E_p is given by $E^+_{p,0} := \{f \in E_p$ such that $f(x) \geq 0 \;\; \mu \; a.e. \; x \in \Omega\}$. The set of strictly positive elements in E_p is denoted by $E^+_p := \{f \in E_p$ such that f(x)> 0 $\mu \; a.e. \; x \in \Omega\}$. Note that E^+_p coincides with the set of quasi-interior points of E_p, i.e., $E^+_p := \{f \in E^+_{p,0}$ such that $\langle f, f' \rangle > 0 \;\; \forall f' \in E^+_{q,0} \backslash \{0\}\}$, where $\langle .,. \rangle$ is the duality pairing.

Definition 2.3.1. We say that $A \in \mathcal{L}(E_p)$ is positive on E_p, if $A(E^+_{p,0}) \subseteq E^+_{p,0}$. A is called strictly positive, if $A(E^+_{p,0} \backslash \{0\}) \subseteq E^+_p$. ◇

Definition 2.3.2. A positive operator $T \in \mathcal{L}(E_p)$ is called σ-order continuous whenever it follows from $f_n \downarrow 0$ in E_p that $Tf_n \downarrow 0$ in E_p. ◇

Definition 2.3.3. A positive operator $T \in \mathcal{L}(E_p)$ is called band irreducible, if T leaves no band in E_p invariant except $\{0\}$ and E_p itself. ◇

Definition 2.3.4. An operator $A \in \mathcal{L}(E_p)$ is called irreducible if, for all $f \in E^+_{p,0} \backslash \{0\}$, there exists $n \in \mathbb{N}^*$ such that $A^n f \in E^+_p$. ◇

For notions not explained in the text, we refer the reader to the books of Zaanen [355], Schaefer [298] and Aliprantis, Burkinshow [19]. Consider two positive operators A and B in $\mathcal{L}(E_p)$. It is well known that, if A and B satisfy $A \leq B$ (i.e., $B - A$ is positive), then $r_\sigma(A) \leq r_\sigma(B)$ where $r_\sigma(A)$ is the spectral radius of A. The next result due to Marek [241, Theorem 4.4] provides sufficient conditions under which the latter inequality is strict. More precisely:

Theorem 2.3.1. *Let A and B be two positive operators in $\mathcal{L}(E_p)$ satisfying $A \leq B$ and $A \neq B$. If A is not quasi-nilpotent, B is irreducible and power-compact (i.e., B^n is compact for some integer $n \geq 1$), then $r_\sigma(A) < r_\sigma(B)$.* ◇

The next two results are also required below. The following one is established in [193, p. 67].

Theorem 2.3.2. *Let $A \in \mathcal{L}(E_p)$ be a positive compact operator satisfying $\exists\varphi \geq 0$, $\varphi \neq 0$ and $\alpha > 0$ such that $A\varphi \geq \alpha\varphi$. Then, A has an eigenvalue $\lambda_0 \geq \alpha$ with a corresponding non-negative eigenfunction.* ◇

Corollary 2.3.1. *Let $A \in \mathcal{L}(E_p)$ be a positive, compact, and non-quasi-nilpotent operator. Then, $r_\sigma(A)$ is an eigenvalue of A with a corresponding non-negative eigenfunction.* ◇

Proof. Let $\lambda \in \mathbb{C}$ be an eigenvalue of A such that $|\lambda| = r_\sigma(A)$. We have $A(\varphi) = \lambda\varphi$ with $\varphi \neq 0$. This implies that $|\lambda|\,|\varphi| \leq A(|\varphi|)$. It follows, from Theorem 2.3.2, that there exists $\lambda_0 \geq |\lambda| = r_\sigma(A)$, which completes the proof. Q.E.D.

For the theory of positive operators on general Banach lattices (resp. L_p-spaces), we may refer to [193] or [249] (resp. [356]).

Proposition 2.3.1. *Let (Ω, Σ, μ) be a σ-finite, positive measure space and let S, T be two bounded linear operators on $L_p(\Omega, d\mu)$ with $p \geq 1$. If $0 \leq S \leq T$, then the following assertions hold:*

(i) the set of all weakly compact operators is a norm-closed subset of $\mathcal{L}(L_1(\Omega, d\mu))$,
(ii) if T is weakly compact on $L_1(\Omega, d\mu)$, then S is also weakly compact, and
(iii) if $p > 1$ and T is compact, then S is compact on $L_p(\Omega, d\mu)$. ◇

Part *(ii)* of Proposition 2.3.1 is a consequence of Dunford–Pettis theorem. The part *(iii)* of Proposition 2.3.1 is due to Dodds and Fremlin [96] in a general Banach lattice E, such that E and E^* have order continuous norms.

Theorem 2.3.3 ([258]). *Let $D \subset \mathbb{C}$ be a domain such that $]-\lambda^*, +\infty[\subset D$. Let $N(\lambda)$ be a holomorphic family with a compact value such that $N(\lambda)$ is positive if $\lambda \in]-\lambda^*, +\infty[$, and $\lim_{\lambda\to\infty} \|N(\lambda)\| = 0$. If $r_\sigma(N(\lambda))$ is a decreasing function in the large sense on $]-\lambda^*, +\infty[$, then it is a strictly decreasing function.* ◇

Corollary 2.3.2 ([258]). *Let T_n be a sequence of positive compact operators in $\mathcal{L}(L_p(\Omega))$ such that T_n converges to T. Then $r_\sigma(T_n)$ converges to $r_\sigma(T)$.* ◇

We begin by the following important result which is contained in [142, Theorems 4.13 and 3.14].

Theorem 2.3.4. *Let K be an irreducible positive kernel operator on a Banach function space L such that the spectral radius $r_\sigma(K)$ of K is a pole of the resolvent $(\lambda - K)^{-1}$. Then, $r_\sigma(K)$ is an eigenvalue of K of algebraic multiplicity one, and the corresponding eigenspace is spanned by a strictly positive function.* ◇

It is well known that for K as in Theorem 2.3.4 the assumption that $r_\sigma(K)$ is a pole of the resolvent $(\lambda - K)^{-1}$ is satisfied if some power of K is a compact operator. In this case Theorem 2.3.4 is known as the theorem of Jentzsh and Perron [142, Theorem 5.2].

2.3.2 Positive Operator on Banach Lattice

In the beginning of this section we give some definitions and results concerning some operators with resolvent positive which will be used in the sequel.

Definition 2.3.5. An operator A on a Banach lattice E is called resolvent positive [34] if there exists $w \in \mathbb{R}$ such that $(w, \infty) \subset \rho(A)$ and $(\lambda - A)^{-1}$ is a positive operator for all $\lambda > w$, i.e., $(\lambda - A)^{-1}\varphi \in E^+$, for all $\varphi \in E^+$ and for all $\lambda > w$, where E^+ is the positive cone of E. ◇

In the sequel we denote by $s(A) := \sup\{\text{Re}\lambda \text{ such that } \lambda \in \sigma(A)\}$ the spectral bound of the operator A.

Definition 2.3.6. A real $\overline{\lambda}$ is called the leading eigenvalue of an operator A on a Banach lattice X, if $\overline{\lambda} = s(A)$, where $\overline{\lambda}$ is an eigenvalue and at least one of the corresponding eigenvectors is positive. ◇

In [141] the following extension of the famous Jentzsch–Perron theorem is proved.

Theorem 2.3.5 ([141]). *Let E be a Banach lattice. Let T be a positive σ-ordered continuous, band irreducible, power-compact operator on E. Then, the spectral radius $r_\sigma(T)$ of T is strictly positive and $r_\sigma(T)$ is an eigenvalue of T of algebraic multiplicity one.* ◇

We will also need the following crucial result which is a particular form of Marek theorem given in [259].

Theorem 2.3.6 ([259, Theorem 0]). *Let O_1 and O_2 be two positive power-compact operators on the Banach lattice E. We suppose that $O_1 \leq O_2$ (i.e., $O_1 - O_2 \geq 0$). If the spectral radius $r_\sigma(O_1)$ of O_1 is strictly positive and there exists $N \in \mathbb{N}$ such that O_2^N is a strictly positive operator, then $r_\sigma(O_2) > r_\sigma(O_1)$ if $O_2 \neq O_1$.* ◇

If A is an unbounded linear operator with domain $\mathcal{D}(A)$, we define

$$\mathcal{D}(A)_+ := \{x \in \mathcal{D}(A), \text{ such that } x \geq 0\}.$$

Lemma 2.3.1 ([105, Lemma 2–3]). *If A is a positive resolvent operator, then the following conditions hold true.*

(i) $s(A) \in \sigma(A)$ *whenever* $\sigma(A) \neq \emptyset$.
(ii) $s(A) = \inf\{\lambda \in \rho(A) \bigcap \mathbb{R}$ *such that* $(\lambda - A)^{-1} \geq 0\}$.
(iii) *The resolvent is monotone decreasing, more precisely,* $0 \leq (\lambda - A)^{-1} \leq (\mu - A)^{-1}$ *for all* $\lambda \geq \mu > s(A)$. ◇

The following perturbation result for resolvent positive operators is due to Voigt, see [334, Theorem 1.1].

Lemma 2.3.2. *Let A be a positive resolvent operator on a Banach lattice E and let B be a linear operator on E satisfying $\mathcal{D}(A) \subset \mathcal{D}(B)$ and $Bx \geq 0$ for all $x \in \mathcal{D}(A)_+$. Then, for $\lambda > s(A)$ the following conditions are equivalent.*

(i) $r_\sigma(B(\lambda - A)^{-1}) < 1$.
(ii) $\lambda \in \rho(A + B)$ *and* $(\lambda - A - B)^{-1} \geq 0$.

If these conditions are satisfied, then $A + B$ *has a positive resolvent, where* $s(A + B) < \lambda$ *and* $(\lambda - A)^{-1} \leq (\lambda - A - B)^{-1}$. ◇

Theorem 2.3.7 ([96, Theorem 4.5]). *Let* E *and* F *be Banach lattices, and let* $0 \leq T : E \longrightarrow F$ *be compact. If* E^* *and* F *have order continuous norms, then every* S *such that* $0 \leq S \leq T$ *is compact.* ◇

2.4 Integral Operator

2.4.1 Integral Operator on L_p-Spaces

The objective of this section is to introduce a class of integral operators F. The scattering kernel $\kappa(.,.,.)$ defines a linear operator F by

$$\begin{cases} F : L_p(D \times V) \longrightarrow L_p(D \times V) \\ \qquad\qquad \psi \longrightarrow \displaystyle\int_V \kappa(x, v, v')\psi(x, v')\,dv', \end{cases} \tag{2.4.1}$$

where $(x, v) \in D \times V$ with D and V are open of $\mathbb{R}^N$, $N \geq 1$. Observe that the operator F acts only on the variables v'. So, x may be viewed merely as a parameter in D. Hence, we may consider F as a function $F(.) : x \in D \longrightarrow F(x) \in \mathcal{L}(L_p(V, dv))$. Notice that, if there exists a compact subset $\mathcal{C} \subset \mathcal{L}(L_p(V, dv)$ such that $F(x) \in \mathcal{C}$ a.e. on D, then

$$F(.) \in L_\infty(D, \mathcal{L}(L_p(V, dv))). \tag{2.4.2}$$

Let $\psi \in L_p(D \times V)$. It is easy to see that $(F\psi)(x, v) = F(x)\psi(x, v)$ and, then by using (2.4.2), we have $\int_V |(F\psi)(x, v)|^p\,dv \leq \|F(.)\|_{L_\infty(D,\mathcal{L}(L_p(V,dv)))} \int_V |\psi(x, v)|^p\,dv$ and therefore, $\int_D \int_V |(F\psi)(x, v)|^p\,dvdx \leq \|F(.)\|_{L_\infty(D,\mathcal{L}(L_p(V,dv)))} \int_D \int_V |\psi(x, v)|^p\,dvdx$. This leads to the estimate

$$\|F\|_{\mathcal{L}(L_p(D\times V))} \leq \|F(.)\|_{L_\infty(D,\mathcal{L}(L_p(V,dv)))}. \tag{2.4.3}$$

Lemma 2.4.1. *We suppose that the operator* F*, defined in* (2.4.1)*, satisfies in* $L_p(D \times V, dxdv)$, $p \in [1, +\infty[$*, the following:*

(i) *the function* $F(.)$ *is strongly measurable,*
(ii) *there exists a compact subset* $\mathcal{C} \subset \mathcal{L}(L_p(V, dv))$ *such that* $F(x) \in \mathcal{C}$ *a.e. on* D*, and*
(iii) $F(x) \in \mathcal{K}(L_p(V, dv))$ *a.e.,*

then F can be approximated in the uniform topology, by a sequence $(F_n)_{n\in\mathbb{N}}$ of linear operators with kernels of the form $\sum_{i=1}^{n} \eta_i(x)\theta_i(v)\beta_i(v')$, where $\eta_i(.) \in L_\infty(D, dx)$, $\theta_i(.) \in L_p(V, dv)$ and $\beta_i(.) \in L_q(V, dv)$, $(\frac{1}{p} + \frac{1}{q} = 1)$. ◇

Proof. Let $\mathcal{C}^* = \mathcal{C} \bigcap \mathcal{K}(L_p(V, dv))$. By using (ii) and (iii), $\mathcal{C}^*$ is a nonempty and closed subset of $\mathcal{C}$. Then, $\mathcal{C}^*$ is a compact set of $\mathcal{L}(L_p(V, dv))$. Let $\varepsilon > 0$, then there exist $F_1, \ldots, F_m$ such that $(F_i)_i \subset \mathcal{C}^*$ and $\mathcal{C} \subset \bigcup_{1\leq i\leq m} B(F_i, \varepsilon)$, where $B(F_i, \varepsilon)$ is the open ball in $\mathcal{L}(L_p(V, dv))$, centered at F_i and with a radius ε. Let $C_1 = B(F_1, \varepsilon)$, $C_2 = B(F_2, \varepsilon)\backslash C_1, \ldots, C_m = B(F_m, \varepsilon)\backslash \bigcup_{1\leq i\leq m-1} C_i$. Clearly, $C_i \bigcap C_j = \emptyset$ if $i \neq j$ and $\mathcal{C}^* \subset \bigcup_{1\leq i\leq m} C_i$. Let $1 \leq i \leq m$ and let us denote by I_i the set $F^{-1}(C_i) = \{x \in D$ such that $F(x) \in C_i\}$. Hence, we have $I_i \bigcap I_j = \emptyset$ if $i \neq j$ and $D = \bigcup_{1\leq i\leq m} I_i$. Now, let us consider the following step function from D into $\mathcal{L}(L_p(V, dv))$, defined by $S(x) = \sum_{i=1}^{m} \chi_{I_i}(x)F_i$, where $\chi_{I_i}(.)$ denotes the characteristic function of I_i. Obviously, $S(.)$ satisfies the hypotheses (i), (ii), and (iii). Then, by using (2.4.2), we get $F - S \in L_\infty(D, \mathcal{L}(L_p(V, dv)))$. Moreover, an easy calculation leads to $\|F - S\|_{L_\infty(D,\mathcal{L}(L_p(V,dv)))} \leq \varepsilon$. Now, by using (2.4.3), we obtain $\|F - S\|_{\mathcal{L}(L_p(D\times V))} \leq \|F - S\|_{L_\infty(D,\mathcal{L}(L_p(V,dv)))} \leq \varepsilon$. Hence, we infer that the operator F may be approximated (for the uniform topology) by operators of the form $U(x) = \sum_{i=1}^{n} \eta_i(x)F_i$, where $\eta_i(.) \in L_\infty(D, dx)$ and $F_i \in \mathcal{K}(L_p(V, dv))$. Moreover, each compact operator $F_i \in \mathcal{K}(L_p(V, dv))$ is a limit (for the norm topology) of a sequence of finite rank operators because $L_p(V, dv)$ $(1 \leq p < \infty)$ admits a Schauder's basis. This ends the proof. Q.E.D.

Remark 2.4.1. The regularity of the collision operator allows us to choose $\eta_i(.)$, $\theta_i(.)$ and $\beta_i(.)$, $i \in \{1, \ldots, n\}$, as measurable simple functions. ◇

Definition 2.4.1. An operator integral in the form (2.4.1) is said to be regular, if it satisfies the assumptions (i), (ii), and (iii) of Lemma 2.4.1. ◇

Definition 2.4.2. Let X be a Banach space. $f(\lambda) \in H_p(\alpha, X)$, p fixed, $1 \leq p < \infty$, if

(i) $f(\lambda)$ is a function on complex numbers into X which is holomorphic for $\sigma > \alpha$.

(ii) $\sup_{\sigma>\alpha} \left(\int_{-\infty}^{+\infty} \|f(\sigma + i\tau)\|^p d\tau\right)^{\frac{1}{p}} < \infty$.

(iii) $\lim_{\sigma\to\alpha} f(\sigma + i\tau)$ exists for almost all values of τ and $\lim_{\sigma\to\alpha} f(\sigma + i\tau) \in B_p[(-\infty, +\infty), X]$, where $B_p[(-\infty, +\infty), X]$ is the set function $x(s) : (-\infty, \infty) \longrightarrow X$ is strongly measurable in $(-\infty, \infty)$ and $\int_{-\infty}^{+\infty} \|x(s)\|^p d\mu < \infty$. ◇

Theorem 2.4.1 ([152, Theorem 6.6.1]). *Let $f(\lambda) \in H_p(\alpha, X)$ where $\alpha \geq 0$. Let $\gamma > \alpha$ and $\beta p' > 1$, where $\frac{1}{p} + \frac{1}{p'} = 1$. Then, $a_\beta(\xi) = \frac{1}{2\pi i}\int_{\gamma-i\infty}^{\gamma+i\infty} e^{\mu\xi}\mu^{-\beta} f(\mu)d\mu$ defines a continuous function on $(0,\infty)$ into X and $f(\lambda) = \lambda^\beta \int_0^\infty e^{-\lambda\xi} a_\beta(\xi)d\xi$, the integral being absolutely convergent for $\sigma > \alpha$.* ◇

Theorem 2.4.2 ([194, Theorem 3.10 page 57]). *Let A be a continuous linear operator acting from L_{α_0} into L_{β_0} and from L_{α_1} into L_{β_1} $(0 \leq \alpha_0, \beta_0, \alpha_1, \beta_1 \leq 1)$. Let A be compact as an operator from L_{α_0} into L_{β_0}. Then, for any $\tau \in (0,1)$, A is compact as an operator from $L_{\alpha(\tau)}$ into $L_{\beta(\tau)}$, where $\alpha(\tau) = (1-\tau)\alpha_0 + \tau\alpha_1$, and $\beta(\tau) = (1-\tau)\beta_0 + \tau\beta_1$.* ◇

Theorem 2.4.3 ([180, Theorem 11.6, p. 275]). *For two measure spaces $(X, \mathcal{A}, \mu)$, $(Y, \mathcal{B}, \nu)$ and $p, q \in [1,\infty]$, we consider linear operators from $L_q(Y, \mathcal{B}, \nu)$ into $L_p(X, \mathcal{A}, \mu)$, which can be represented as an integral in the form $Kf(x) = \int_Y \kappa(x,y)f(y)d\nu(y)$ for all $f \in L_q(Y, \mathcal{B}, \nu)$. For $p \in [1,\infty)$, $q \in (1,\infty]$ and $\frac{1}{p} + \frac{1}{q} = 1$, if $\kappa(.,.)$ satisfies the condition*

$$\left\{\int_X \left[\int_Y |\kappa(x,y)|^q \, d\nu(y)\right]^{\frac{p}{q}} d\mu(x)\right\} < +\infty,$$

then K is a compact operator from $L_q(Y, \mathcal{B}, \nu)$ into $L_p(X, \mathcal{A}, \mu)$. ◇

2.4.2 *Integral Operator on L_1-Spaces*

Let Ω be a bounded, smooth, and open subset of $\mathbb{R}^N$ $(N \geq 1)$, and let $d\mu$ and $d\nu$ be two positive Radon measures on $\mathbb{R}^N$ with a common support V. Let

$$K \in \mathcal{L}(L_1(\Omega \times V, dxd\mu(v)), L_1(\Omega \times V, dxd\nu(v)))$$

be given by

$$\begin{cases} K : L_1(\Omega \times V, dxd\mu(v)) \longrightarrow L_1(\Omega \times V, dxd\nu(v)) \\ \qquad\qquad\qquad\qquad \psi \longrightarrow \displaystyle\int_V \kappa(x,v,v')\psi(x,v')\, d\mu(v'), \end{cases} \tag{2.4.4}$$

where the kernel $\kappa(.,.,.)$ is measurable. Note that

$$dx \otimes d\mu - \operatorname*{ess\ sup}_{(x,v')\in\Omega\times V} \int_V |\kappa(x,v,v')| d\nu(v) = \|K\| < \infty. \tag{2.4.5}$$

We have the following definition.

Definition 2.4.3. Let K be the integral operator defined by (2.4.4). Then, K is said to be regular if $\{\kappa(x,.,v') \text{ such that } (x,v') \in \Omega \times V\}$ is a relatively weak compact subset of $L_1(V, d\nu)$. ◇

Remark 2.4.2. The Definition 2.4.3 asserts that for every $x \in \Omega$

$$f \in L_1(V) \longrightarrow \int_V \kappa(x,v,v') f(v') \, d\nu(v') \in L_1(V)$$

is a weakly compact operator and this weak compactness holds collectively in $x \in \Omega$. ◇

The class of regular operators satisfies the following approximate property given in [236].

Theorem 2.4.4. *Let $K \in \mathcal{L}(L_1(\Omega \times V, dxd\mu(v)), L_1(\Omega \times V, dxd\nu(v)))$, introduced in (2.4.4), be a regular and nonnegative operator. Then, there exist $(K_m)_m \subset \mathcal{L}(L_1(\Omega \times V, dxd\mu(v)), L_1(\Omega \times V, dxd\nu(v)))$ such that*

(i) $0 \leq K_m \leq K$ *defined for any* $m \in \mathbb{N}$.
(ii) *For any* $m \in \mathbb{N}$, K_m *is dominated by a rank-one operator in* $\mathcal{L}(L_1(V, d\mu(v)), L_1(V, d\nu(v)))$.
(iii) $\lim_{m \to +\infty} \|K - K_m\| = 0.$ ◇

Proof. According to Definition 2.4.3, the operator K already defined in (2.4.4) and satisfying (2.4.5) and $\{\kappa(x,.,v') \text{ such that } (x,v') \in \Omega \times V\}$ is a relatively weak compact subset of $L_1(V, d\nu)$. According to Takac's version of Dunford–Pettis criterion, we have

$$\lim_{m \to \infty} \sup_{(x,v') \in \Omega \times V} \int_{S_m(x,v')} \kappa(x,v,v') d\nu(v) = 0, \tag{2.4.6}$$

where $S_m(x,v') := \{v \in V \text{ such that } \|v\| \geq m\} \bigcup \{v \in V \text{ such that } \kappa(x,v,v') \geq m\}$, $(x,v') \in \Omega \times V$. For any $m \in \mathbb{N}$, let us define

$$K_m : \varphi \in L_1(\Omega \times V, dxd\mu(v)) \longrightarrow \int_V \kappa_m(x,v,v') \varphi(x,v') \, d\mu(v') \in L_1(\Omega \times V, dxd\nu(v))$$

with $\kappa_m(x,v,v') = \inf\{\kappa(x,v,v'), m\chi_{B_m}(v)\}$ $(x,v,v') \in \Omega \times V \times V$, where $\chi_{B_m}(.)$ denotes the characteristic function of the set $\{v \in V \text{ such that } \|v\| \leq m\}$. Clearly, we have $0 \leq K_m \leq K$. Moreover, we can easily check that

$$\|K - K_m\| \leq (dx \otimes d\mu) - \text{ess} \sup_{(x,v') \in \Omega \times V} \int_V |\kappa(x,v,v') - \kappa_m(x,v,v')| d\nu(v).$$

Besides, for any $(x, v') \in \Omega \times V$, the construction of $\kappa_m(x, ., v')$ yields the following:

$$\int_V |\kappa(x, v, v') - \kappa_m(x, v, v')| d\nu(v)$$
$$= \int_{\{\kappa(x,v,v') \geq m\chi_{B_m}(v)\}} |\kappa(x, v, v') - \kappa_m(x, v, v')| d\nu(v)$$
$$\leq \int_{\{\kappa(x,v,v') \geq m\chi_{B_m}(v)\}} \kappa(x, v, v') d\nu(v).$$

Hence, according to (2.4.6), we have $\lim_{m\to\infty} \|K - K_m\| = 0$. Finally, it is easy to see that, for any $\varphi \in L_1(\Omega \times V, dxd\mu(v))$, $\varphi \geq 0$, $K_m\varphi(x, v) \leq m\chi_{B_m}(v) \int_V \varphi(x, v')d\mu(v')$ which proves the second assertion and achieves the proof. Q.E.D.

Remark 2.4.3. Let us make precise the point (ii) of Theorem 2.4.4. This asserts that, for any $m \in \mathbb{N}$, there exists a nonnegative $f_m \in L_1(V, d\nu(v))$ such that, for any $\varphi \in L_1(\Omega \times V, dxd\mu(v))$, $\varphi \geq 0$, $K_m\varphi(x, v) \leq f_m(v) \int_V \varphi(x, v')\, d\mu(v')$. ◇

Theorem 2.4.5 ([101, Corollary 11, p. 294]). *Let (S, Σ, μ) be a positive measure space. If a set K in $L_1(S, \Sigma, \mu)$ is weakly sequentially compact, then $\lim_{\mu(E)\to 0} \int_E f(s)\mu(ds) = 0$ uniformly for $f \in K$. If $\mu(S) < \infty$, then conversely this condition is sufficient for a bounded set K to be weakly sequentially compact.* ◇

2.4.3 Cauchy's Type Integral

Let X be a Banach space and let A be a bounded linear operator in $\mathcal{L}(X)$. We will define a function $f(A)$ of A by Cauchy's type integral

$$f(A) = (2\pi i)^{-1} \int_C f(\lambda)(\lambda - A)^{-1} d\lambda,$$

where C is a circumference of sufficiently small radius.

For this purpose, we denote by $F(A)$ the family of all complex-valued functions $f(\lambda)$ which are holomorphic in some neighborhood of the spectrum $\sigma(A)$ of A, the neighborhood doesn't need to be connected, and can depend on the function $f(\lambda)$. Let $f \in F(A)$, and let an open set $U \supset \sigma(A)$ of the complex plane be contained in the domain of holomorphy of f, and suppose that the boundary ∂U of U consists of a finite number of rectifiable Jordan curves, oriented in a positive sense. Then, the bounded linear operator $f(A)$ will be defined by

$$f(A) = (2\pi i)^{-1} \int_{\partial U} f(\lambda)(\lambda - A)^{-1} d\lambda,$$

and the integral on the right may be called a Dunford's integral. By using Cauchy's integral theorem, the value $f(A)$ depends only on the function f and the operator A, but not on the choice of the domain U.

Theorem 2.4.6 (N. Dunford). *If f and g are in $F(A)$, and if α and β are complex numbers, then*

(i) *$\alpha f + \beta g$ is in $F(A)$ and $\alpha f(A) + \beta g(A) = (\alpha f + \beta g)(A)$.*
(ii) *$f.g$ is in $F(A)$ and $f(A).g(A) = (f.g)(A)$.* ◇

Proof.

(*i*) Is clear.
(*ii*) Let U_1 and U_2 represent an open neighborhood of $\sigma(A)$ whose boundaries ∂U_1 and ∂U_2 consist of a finite number of rectifiable Jordan curves, and assume that $U_1 + \partial U_1 \subset U_2$ and that $U_2 + \partial U_2$ is contained in the holomorphic domain of f and g. Then, by virtue of both the resolvent equation and Cauchy's integral theorem, we get

$$\begin{aligned}
f(A)g(A) &= -(4\pi^2)^{-1} \int_{\partial U_1} f(\lambda)(\lambda - A)^{-1} d\lambda . \int_{\partial U_2} g(\mu)(\mu - A)^{-1} d\mu \\
&= -(4\pi^2)^{-1} \int_{\partial U_1} \int_{\partial U_2} f(\lambda) g(\mu)(\mu - \lambda)^{-1} \left[(\lambda - A)^{-1} - (\mu - A)^{-1} \right] d\lambda d\mu \\
&= (2\pi i)^{-1} \int_{\partial U_1} f(\lambda)(\lambda - A)^{-1} . \left[(2\pi i)^{-1} \int_{\partial U_2} (\mu - \lambda)^{-1} g(\mu) d\mu \right] d\lambda \\
&\quad - (2\pi i)^{-1} \int_{\partial U_2} g(\mu)(\mu - A)^{-1} . \left[(2\pi i)^{-1} \int_{\partial U_1} (\mu - \lambda)^{-1} f(\lambda) d\lambda \right] d\mu \\
&= (2\pi i)^{-1} \int_{\partial U_1} f(\lambda) g(\lambda)(\lambda - A)^{-1} d\lambda \\
&= (f.g)(A).
\end{aligned}$$

Q.E.D.

Let us recall the spectral mapping theorem:

Corollary 2.4.1. *If f is in $F(A)$, then $f(\sigma(A)) = \sigma(f(A))$.* ◇

Proof. Let $\lambda \in \sigma(A)$, and let us define the function g by $g(\mu) = (f(\lambda) - f(\mu))/(\lambda - \mu)$. According to Theorem 2.4.6, $f(\lambda) - f(A) = (\lambda - A)g(A)$. Hence, if $(f(\lambda) - f(A))$ has a bounded inverse B, then $g(A)B$ would be the bounded inverse of $(\lambda - A)$. Thus, $\lambda \in \sigma(A)$ implies that $f(\lambda) \in \sigma(f(A))$. Conversely, let $\lambda \in \sigma(f(A))$, and assume that $\lambda \in f(\sigma(A))$. Then the function $g(\mu) = (f(\mu) - \lambda)^{-1}$ must belong to $F(A)$ and so, by using the preceding theorem (Theorem 2.4.6), $g(A)(f(A) - \lambda) = I$, which contradicts the assumption $\lambda \in \sigma(f(A))$. Q.E.D.

2.5 Semigroup Theory

2.5.1 Strongly Continuous Semigroup

Let X be a Banach space. A family $(T(t))_{t\geq 0}$ in $\mathcal{L}(X)$ is called a one-parameter semigroup of bounded linear operators in X, if $T(0) = I$, and $T(s+t) = T(s)T(t)$, $s,t \geq 0$. Here, I stands for the identity operator. If, in addition, the function $t \longrightarrow T(t)$ is continuous with respect to the strong operator topology of $\mathcal{L}(X)$, i.e., $t \longrightarrow T(t)x$ is continuous on $[0,+\infty)$ for every $x \in X$, then $(T(t))_{t\geq 0}$ is called a strongly continuous semigroup, or also C_0-semigroup. For the generator $(A,\mathcal{D}(A))$ of a strongly continuous semigroup $(T(t))_{t\geq 0}$ on a Banach space X, we have the identities (see [106, Eqs (3.16) and (3.17), p. 277])

$$\sigma_p(T(t))\backslash\{0\} = e^{t\sigma_p(A)} \tag{2.5.1}$$

$$\sigma_r(T(t))\backslash\{0\} = e^{t\sigma_r(A)}$$

for all $t \geq 0$. A semigroup $(T(t))_{t\geq 0}$ is said to be of Riesz type, if $T(t)$ is a Riesz operator for all $t \in]0,\infty[$.

Theorem 2.5.1 ([152, Theorem 16.3.6]). *A semigroup $(T(t))_{t\geq 0}$ can be embedded in one-parameter strongly continuous group of bounded linear operators on $(-\infty,+\infty)$ if, and only if, $0 \in \rho(T(t))$ for some $t > 0$.* ◇

Theorem 2.5.2 ([276, Theorem 2.2]). *Let $\big(T(t)\big)_{t\geq 0}$ be a C_0-semigroup. Then, there are constants $w \geq 0$ and $M \geq 1$, such that $\|T(t)\| \leq Me^{wt}$ for $0 \leq t < \infty$.* ◇

Let $w(T)$ denote the type of the semigroup $\big(T(t)\big)_{t\geq 0}$ defined by:

$$w(T) = \inf\{w > 0 \text{ such that } \exists\ M_w \text{ satisfies } ||T(t)|| \leq M_w e^{wt} \quad \forall t \geq 0\}.$$

Let $(T(t))_{t\geq 0}$ be a C_0-semigroup on a Banach space X with a generator A. The growth bound $w(T(t))$ of $(T(t))_{t\geq 0}$ is given by

$$w = \inf\left\{\omega \text{ such that } \sup_{t>0} e^{-\omega t}\|T(t)\| < \infty\right\}.$$

Theorem 2.5.3 ([34, Theorem 2.5]). *Let A be a densely defined resolvent positive operator in X. If there exist $\lambda_0 > s(A)$ and $c > 0$ such that $\|(\lambda_0 - A)^{-1}\varphi\| \geq c\|\varphi\|$, $\forall \varphi \in X^+$, then A is the generator of a positive C_0-semigroup on X, and $s(A) = w(A)$ where $s(A)$ (resp. $w(A)$) denotes the spectral bound of A (resp. the type of the C_0-semigroup generated by A).* ◇

Theorem 2.5.4 ([339]). *Let $(T(t))_{t\geq 0}$ be a positive C_0-semigroup on $L_p(\Omega, d\mu)$, $1 < p < \infty$, with a generator A. Then, the spectral bound $s(A)$ equals the growth bound $w(T(t))$.* ◇

Theorem 2.5.5 ([276, Theorem 3.2]). *Let $T(t)$ be a C_0-semigroup. If $T(t)$ is compact for $t > t_0$, then $T(t)$ is continuous in the uniform operator topology for $t > t_0$.* ◇

Theorem 2.5.6 ([339, Corollary 2.3]). *Let (Ω, μ) be a measure space and let $t \in \Omega \longrightarrow T_t \in \mathcal{L}(X, Y)$ be a strongly integrable function, i.e., $Tx = \int_\Omega T_t x d\mu(t)$ exists for all $x \in X$ as a Bochner integral and $\int_\Omega \|T_t\| d\mu(t) < \infty$. If μ-almost all T_t are compact, then T is compact.* ◇

2.5.2 The Hille-Yosida Theorem

Let $T(t)$ be a C_0-semigroup. From Theorem 2.5.2, it follows that there are constants $w \geq 0$ and $M \geq 1$ such that $\|T(t)\| \leq Me^{wt}$ for $t \geq 0$. If $w = 0$, $T(t)$ is called uniformly bounded, and if, moreover $M = 1$, then it is called a C_0-semigroup of contractions.

Theorem 2.5.7 ([276, Theorem 3.1(Hille-Yosida)]). *A linear (unbounded) operator A is the infinitesimal generator of a C_0-semigroup of contractions $T(t)$, $t \geq 0$ if, and only if,*

(i) A is closed and $\overline{\mathcal{D}(A)} = X$, and
(ii) the resolvent set $\rho(A)$ of A contains $\mathbb{R}^+$ and, for every $\lambda > 0$, we have

$$\|(\lambda - A)^{-1}\| \leq \frac{1}{\lambda}.$$

◇

From Theorem 2.5.7, we can deduce the following corollary.

Corollary 2.5.1 ([276, Corollary 3.8]). *A linear operator A is the infinitesimal generator of a C_0-semigroup satisfying $\|T(t)\| \leq e^{wt}$ if, and only if,*

(i) A is closed and $\overline{\mathcal{D}(A)} = X$, and
(ii) the resolvent set $\rho(A)$ of A contains the ray $\{\lambda$ such that $Im\lambda = 0,\ \lambda > w\}$ and, for such a λ, we have $\|(\lambda - A)^{-1}\| \leq \frac{1}{\lambda - w}$. ◇

Theorem 2.5.8 ([276, Theorem 1.1 p. 76]). *Let X be a Banach space, and let A be the infinitesimal generator of a C_0-semigroup $T(t)$ on X, satisfying $\|T(t)\| \leq Me^{wt}$. If B is a bounded linear operator on X, then $A + B$ is the infinitesimal generator of a C_0-semigroup $(S(t))_{t \geq 0}$ on X, satisfying $\|S(t)\| \leq Me^{(w+M\|B\|)t}$.* ◇

We recall the following result owing to Phillips (cf. [281]).

Proposition 2.5.1. *Let $(T(t))_{t \geq 0}$ be a C_0-semigroup on a Banach space X. Let us assume that there exists an unbounded, open, and connected set $\Omega \subset \mathbb{C}$, such that $0 \in \Omega$ and $\sigma(T(t)) \bigcap \Omega = \emptyset$ for t belonging to some interval $]a, b[$ with $0 \leq a < b$. Then, the infinitesimal generator of $(T(t))_{t \geq 0}$ is bounded.* ◇

Theorem 2.5.9 ([35, Theorem 3.3, p. 70]). *Let $T(t)$ be a strongly continuous semigroup on a Banach space X, and assume that the spectrum $\sigma(A)$ of the* *generator A can be decomposed into the disjoint union of two nonempty, closed subsets σ_1 and σ_2. If σ_1 is compact, then there exists a unique corresponding spectral decomposition $X = X_1 \oplus X_2$ such that the restricted semigroup $T_1(t)$ has a bounded generator.* ◇

Theorem 2.5.10 ([81, p. 164]). *Let $(T(t))_{t\geq 0}$ be a positive C_0-semigroup on X with a generator A. Then,*

$$(\lambda - A)^{-1} = \int_0^\infty e^{-\lambda t} T(t)dt, \ \ \mathrm{Re}\lambda > s(A), \tag{2.5.2}$$

where (2.5.2) exists as a norm convergent improper integral. ◇

Theorem 2.5.11 ([186, Eq. (1.17), p. 480]). *Let $(T(t))_{t\geq 0}$ be a C_0-semigroup on X with a generator A. If $A \in \mathcal{C}(X)$, the negative real axis belongs to the resolvent set of A, and the resolvent $(A + \lambda)^{-1}$ satisfies the inequality $\|(A + \lambda)^{-1}\| \leq \frac{1}{\lambda}$. Then, $(I + \frac{t}{n}A)^{-n}$ converges strongly to $T(t)$, for all $t \geq 0$.* ◇

Theorem 2.5.12 ([106, p. 276]). *Let $(T(t))_{t\geq 0}$ be a strongly continuous semigroup with a generator A on a Banach space X. The following spectral inclusion holds $e^{t\sigma(A)} \subset \sigma(T(t))$ for $t \geq 0$. Moreover, if $e^{\lambda_0 t}$ is an isolated eigenvalue of $T(t)$, then λ_0 is an isolated eigenvalue of A and $m_a(\lambda_0, A) \leq m_a(e^{\lambda_0 t}, T(t))$ where $m_a(\lambda_0, A)$ and $m_a(e^{\lambda_0 t}, T(t))$ denote, respectively, the algebraic multiplicities of λ_0 and $e^{\lambda_0 t}$.* ◇

We recall the Gohberg–Shmul'yan's theorem:

Theorem 2.5.13 ([184, Theorem 11.4]). *Let G be an open connected subset of $\mathbb{C}$, and let $\{C(\lambda)$ such that $\lambda \in G\}$ be a holomorphic operator-valued function from G to the space of compact operators. If $I - C(\lambda)$ is boundedly invertible for some $\lambda \in G$, then $I - C(\lambda)$ is boundedly invertible for all $\lambda \in G$, except at a discrete set of points $\{\lambda_k \ : \ k = 1, 2, \ldots\}$, each λ_k is a pole of $(I - C(\lambda))^{-1}$.* ◇

Let $T \in \mathcal{C}(X)$ and $A \in \mathcal{L}(X)$. We call A resolvent compact relative to $-T$ if, for some positive integer n, $((\lambda + T)^{-1}A)^n$ is compact for all $\lambda \in G$, where G is the component of the resolvent set $\rho(-T)$ that contains a right half-plane.

Corollary 2.5.2 ([184, Corollary 11.6]). *Let T be a Hille–Yosida operator, and let A be a bounded linear operator that is resolvent compact relative to $-T$. Let G be the component of the resolvent set $\rho(-T)$ that contains a right half-plane. Then, $\lambda + T + A$ is boundedly invertible for all $\lambda \in G$, except at a discrete set of points $\{\lambda_k$ such that $k = 1, 2, \ldots\}$, each λ_k is a pole of $(\lambda + T + A)^{-1}$ and also an eigenvalue of $T + A$ with a finite (algebraic) multiplicity.* ◇

We recall the following result due to Kato [185].

Lemma 2.5.1. *If $(T_n)_n$ converges strongly to T, then $(T_n u)_n$ converges strongly to Tu uniformly for all u of a compact subset S of X.* ◇

Proof. We may assume that $T = 0$, otherwise, we only have to consider $T_n - T$ instead of T_n. As for every compact subset of a metric space, S is totally bounded, that is, for any $\varepsilon > 0$, there is a finite number of elements $u_k \in S$ such that each

$u \in S$ lies within a distance ε of some u_k. Since $T_n u_k \to 0$, $n \to \infty$, there are positive numbers n_k such that $\|T_n u_k\| < \varepsilon$ for $n > n_k$. Then, for any $u \in S$, we have $\|T_n u\| \le \|T_n(u - u_k)\| + \|T_n u_k\| \le (M+1)\varepsilon$ if $n > \max n_k$, where u_k is such that $\|u - u_k\| < \varepsilon$ and M is an upper bound of $\|T_n\|$ (M is finite). Q.E.D.

Let Y be an arbitrary real or complex Banach space, and let Y^* denote its dual space. For each $x \in Y$, we define $\mathcal{P}(x) = \{x^* \in Y^*$ such that $\|x^*\|^2 = \|x\|^2 = \langle x, x^*\rangle\}$. Let $x \in Y$, $\mathcal{P}(x)$ be nonempty according to Hahn–Banach theorem. A linear operator A, with both the domain and the range in Y, is called dissipative if, for every $x \in \mathcal{D}(A)$, there exists $x^* \in \mathcal{P}(x)$, such that $\mathrm{Re}\langle Ax, x^*\rangle \le 0$. Let (Ω, Σ, μ) be a measure space, and let $\mathcal{Y} = L_p(\Omega, \Sigma, \mu)$ with $1 \le p < \infty$, $\mathcal{P}(0) = \{0\}$ and, for $0 \ne u \in \mathcal{Y}$, $\mathcal{P}(u)$ has solely one element, that is, $\mathcal{P}(u) = \{\|u\|^{2-p}|u|^{p-2}\overline{u}\}$ for $1 < p < \infty$. For $p = 1$, in order to show the dissipativity of the operator A, it is sufficient to show that $\mathrm{Re}\langle Au, s_0(u)\rangle \le 0$ for all $u \in \mathcal{D}(A)$, where

$$s_0(u)(x) = \begin{cases} \dfrac{\overline{u}(x)}{|u(x)|}\|u\| & \text{when } u(x) \ne 0, \\ 0 & \text{when } u(x) = 0. \end{cases}$$

Theorem 2.5.14 ([152, Theorem 4.17.2]). *If $\nu(\xi)$ satisfies $\nu(\xi + \eta) = \nu(\xi)\nu(\eta)$ and $\nu(0) = 1$ non-trivially, and if $|\nu(\xi)|$ is bounded in some interval $[\tau_1, \tau_2]$, then $|\nu(\xi)| = e^{\alpha\xi}$ for some real number α.* ◇

2.6 The Essential Spectral Radius

Let T be a closed operator on a Banach space X, and let $\Delta \subset \rho(T)$ be open. An operator B on X will be called T-power-compact on Δ if B is T-bounded, and there is $n \in \mathbb{N}$ such that $(B(\lambda - T)^{-1})^n$ is compact for all $\lambda \in \Delta$. If $\Delta' \subset \rho(T)$ is open and connected, and $\Delta' \bigcap \Delta \ne \emptyset$, then B will also be T-power-compact on Δ'. This follows from the facts that $\Delta \ni \lambda \longrightarrow (B(\lambda - T)^{-1})^n$ is holomorphic, that the set of compact operators is closed in $\mathcal{L}(X)$, and the Hahn–Banach theorem. The operator B on X will be called strictly power-compact if $B \in \mathcal{L}(X)$, and there is $n \in \mathbb{N}$ such that $(BA)^n$ is compact for all $A \in \mathcal{L}(X)$. Let T be a closed, densely defined linear operator on X. We recall that $\lambda \in \mathbb{C}$ is an eigenvalue of finite algebraic multiplicity of T if λ is an isolated point of $\sigma(T)$ and is a pole of the resolvent $(\lambda - T)^{-1}$ of T with degenerate associated spectral projection P, $n = \dim R(P)$ is the algebraic multiplicity of λ. Let $B \in \mathcal{L}(X)$. The essential spectral radius of B is defined by

$$r_e(B) := \sup\big\{|\lambda|,\ \lambda \in \sigma(B),\ \text{but } \lambda \text{ is not an eigenvalue of finite algebraic multiplicity}\big\}.$$

Let $(U(t))_{t\ge 0}$ be a strongly continuous semigroup on X. Then, $r_\sigma(U(t)) = e^{tw}$ for every $t > 0$, where $r_\sigma(.)$ denotes the spectral radius and w is the type of $(U(t))_{t\ge 0}$, that is $w = \lim\limits_{t\to\infty} \dfrac{1}{t}\log\|U(t)\| = \inf\limits_{t>0} \dfrac{1}{t}\log\|U(t)\|$. Moreover, there exists $w_e \in [-\infty, w]$ such that $r_e(U(t)) = e^{tw_e}$ $(t > 0)$. The real number w_e is called the essential type of $(U(t))_{t\geq 0}$. Let us recall a known result about the essential type of perturbed C_0-semigroups on Banach spaces. We define the "essential resolvent set"

$$\rho_6(T) = \rho(T)\bigcup\{\lambda \in \sigma(T) \text{ such that } \lambda \text{ is an eigenvalue of finite algebraic multiplicity}\}.$$

We recall some basic tools by Ribaric and Vidav [293].

Theorem 2.6.1. *Let $A(z)$ be an analytic function in the punctured disc $0 < |z| < r$, and let its values be bounded, and linear operators on a Banach space X. Suppose $A(z)$ can be written as the sum $A(z) = A_1(z) + A_2(z)$, where (i) $A_1(z)$ is analytic for $0 < |z| < r$, has at $z = 0$ at most a pole, and is such that the ranges of $A_1(z)$ lie in a finite dimensional subspace $Y \subset X$, and (ii) $A_2(z)$ is analytic for $|z| < r$ and is such that $[A_2(0)]^{-1}$ exists. Then, one of the following two possibilities must hold:*

(a) $\lambda = 0$ is an eigenvalue for all operators $A(z)$ in a neighborhood of $z = 0$, or
(b) the inverse $[A(z)]^{-1}$ exists in a neighborhood of $z = 0$ and has there the same properties as $A(z)$. ◇

Proof. Since $[A_2(0)]^{-1}$ exists and $A_2(z)$ is analytic at $z = 0$, $[A_2(z)]^{-1} = U(z)$ exists in a neighborhood of the point $z = 0$ and is analytic there. Thus, we may write in the following form $A(z) = [I + A_1(z)U(z)]A_2(z)$, where the product $A_1(z)U(z)$ is a degenerate operator with its range in Y. The determinant $w(z) := det[I + A_1(z)U(z)]$ is analytic in some neighborhood of $z = 0$, and has at most a pole at $z = 0$. If $w(z) = 0$, all operators $A_1(z)U(z)$ have the eigenvalue $\lambda = -1$. If $e(z) \neq 0$ is a corresponding eigenvector, then $f(z) = U(z)e(z) \neq 0$ is an eigenvector of $A(z)$ for the eigenvalue $\lambda = 0$. Hence, we have case (a). If $w(z) \neq 0$, then in a neighborhood of $z = 0$, $V(z) = I - [I + A_1(z)U(z)]^{-1}$ exists, and $V(z)X \subset Y$, the point $z = 0$ being, at most, a pole of $V(z)$. Furthermore, $[A(z)]^{-1} = U(z)[I + V(z)]$, and the point $z = 0$ is, at most, a pole for $U(z)V(z)$. If we expand $U(z)V(z)$ in a Laurent series, we have $U(z)V(z) = B_1(z) + B'(z)$, where $B_1(z)$ is the sum of the constant terms and of the terms with negative powers of z, and $B'(z)$ denotes the remainder of the series, $B'(z)$ is regular at $z = 0$, and $B'(0) = 0$. All coefficients in $B_1(z)$ are degenerate operators. Hence, there exists a finite dimensional subspace $Y_1 \subset X$ such that $B_1(z)X \subset Y_1$. If we write $B_2(z) = U(z) + B'(z)$, we have $[A(z)]^{-1} = B_1(z) + B_2(z)$, where $B_2(0) = U(0)$ and $[B_2(0)]^{-1} = [U(0)]^{-1} = A_2(0)$. Hence, $B_1(z)$ and $B_2(z)$ have the same properties as $A_1(z)$ and $A_2(z)$, respectively. Q.E.D.

Corollary 2.6.1. *Let $A(z)$ be an essentially meromorphic function in a region D of the complex z plane. Then, either (i) $\lambda = 1$ is an eigenvalue for all operators $A(z)$, $z \in D$, or (ii) the operator $[I - A(z)]^{-1}$ exists for almost all $z \in D$, the function $I - [I - A(z)]^{-1}$ being essentially meromorphic in D.* ◇

Proof. Let z_0 be any point of D. Without loss of generality, we may assume that $z_0 = 0$. Let us expand $A(z)$ in a Laurent series

$$A(z) = \sum_{n=-m_0}^{\infty} z^n A_n, \quad 0 \leq m_0 < \infty, \tag{2.6.1}$$

where $m_0 = 0$ if $A(z)$ is regular at $z = 0$. By assumption, $A_{-1}, \ldots, A_{-m_0}$ are degenerate operators. Since $A(z)$ is compact for any $z \in D$, one can infer from (2.6.1) that the coefficients A_n are also compact operators. Therefore, $\lambda = 1$ is either an isolated point of the spectrum of A_0 and has a finite algebraic multiplicity, or it belongs to the resolvent set $\rho(A_0)$. Let P be the associated eigenprojection (where $P = 0$ if $\lambda = 1$ and is not an eigenvalue). Now, we may write $A_1(z) = \sum_{n=-m_0}^{n=-1} A_n z^n + PA_0$, $A_2(z) = A_0 - PA_0 - I + \sum_{n=1}^{\infty} A_n z^n$ so that $A_1(z) + A_2(z) = A(z) - I$, and apply Theorem 2.6.1. In case (i) all operators $A(z)$ in a neighborhood of $z = 0$ have the eigenvalue $\lambda = 1$. In case (ii), $[I - A(z)]^{-1}$ exists and $I - [I - A(z)]^{-1}$ is essentially meromorphic in this neighborhood. This holds for any point $z_0 \in D$. Hence, the sets of points where (i), respectively (ii), occur, are both open. Due to the connectedness of D, one of these sets must be empty. Q.E.D.

Theorem 2.6.2 ([314]). *If $T(z)$ is an analytic $\mathcal{L}(X)$-valued function for $z \in \Omega \subset \mathbb{C}$, if $T^n(z)$ is compact for some $n \geq 1$, and if $(I - T^n(z))$ is somewhere invertible, then $(I - T(z))^{-1}$ is a meromorphic $\mathcal{L}(X)$-valued function. Here we assume that Ω is open and connected.* ◇

We are now ready to present the concept of essential resolvent set introduced by Voigt [331].

Theorem 2.6.3. *Let A be a closed operator in X and let Ω be a component of $\rho_6(T)$. Let the operator B be A-power-compact on $\Omega \bigcap \rho(T)$, and let $I - (B(\lambda - A)^{-1})^n$ be invertible in $\mathcal{L}(X)$ for some $\lambda \in \Omega \bigcap \rho(T)$ (n from the definition of A-power-compact). Then, $\Omega \subset \rho_6(T + B)$, and B is $(T + B)$-power-compact on $\Omega \bigcap \rho(T + B)$.* ◇

Proof. By the definition of $\rho_6(T)$, $(\lambda - A)^{-1}$ is a degenerate-meromorphic operator function on Ω (i.e., $(\lambda - A)^{-1}$ is holomorphic, except at isolated points, where $(\lambda - A)^{-1}$ has poles, and the coefficients of the main part are degenerate). It follows that $B_\lambda := B(\lambda - A)^{-1}$ is degenerate-meromorphic on Ω, and therefore so is B_λ^n. Also, B_λ^n is a compact operator of the exceptional set. Now, Corollary 2.6.1 implies that either $I - B_\lambda^n$ is nowhere invertible on Ω, or $I - B_\lambda^n$ is invertible, except at a set of isolated points, and $(I - B_\lambda^n)^{-1}$ is degenerate-meromorphic. The hypothesis of our theorem excludes the first alternative. From $I - B_\lambda^n = (I - B_\lambda)(I + B_\lambda + \ldots + B_\lambda^{n-1})$, we conclude that

$$(I - B_\lambda)^{-1} = (I + B_\lambda + \ldots + B_\lambda^{n-1})(I - B_\lambda^n)^{-1}$$

is degenerate-meromorphic on Ω, and so is

$$(\lambda - A - B)^{-1} = (\lambda - A)^{-1}(I - B_\lambda)^{-1}. \tag{2.6.2}$$

This shows that the points of $\Omega \bigcap \sigma(A + B)$ are eigenvalues of finite algebraic multiplicity, $\Omega \subset \rho_6(A + B)$. Also, Eq. (2.6.2) implies $B'_\lambda := B(\lambda - A - B)^{-1} = B_\lambda(I - B_\lambda)^{-1}$, and therefore $(B'_\lambda)^n = B_\lambda^n(I - B_\lambda)^{-n}$ is compact for those $\lambda \in \Omega \bigcap \rho(A)$ for which $I - B_\lambda$ is invertible. Q.E.D.

We note that Theorem 2.6.3 is not symmetric with respect to A and $A+B$, since there is nothing that would tell us that $I-(B(\lambda-A-B)^{-1})^n$ is invertible for some $\lambda \in \Omega \bigcap \rho(A+B)$. The following corollary is given in [331].

Corollary 2.6.2. *Let A be a closed operator, and let Ω be a component of $\rho_6(A)$. Let B be A-power-compact on $\Omega \bigcap \rho(A)$, and let $\|B(\lambda_j - A)^{-1}\| \to 0$ $(j \to \infty)$ for some sequence $(\lambda_j)_j$ in $\Omega \bigcap \rho(A)$. Then, $\lambda_j \in \rho(A+B)$ for j large enough, and $\|B(\lambda_j - A - B)^{-1}\| \to 0$ $(j \to \infty)$. Furthermore, Ω is a component of $\rho_6(A+B)$.* ◇

Proof. Formula (2.6.2) implies $\lambda_j \in \rho(A+B)$, if $\|B_{\lambda_j}\| < 1$, and $B'_{\lambda_j} = B_{\lambda_j}(I - B_{\lambda_j})^{-1} = \sum_{k=1}^{\infty} B^k_{\lambda_j}$. So, $\|B'_{\lambda_j}\| \le \sum_{k=1}^{\infty} \|B_{\lambda_j}\|^k \to 0$ $(j \to \infty)$. Let us denote by Ω' the component of $\rho_6(A+B)$ containing Ω. Now, Theorem 2.6.3, when applied to $A+B$ and $-B$, implies $\Omega' \subset \rho_6(A)$, and therefore $\Omega' = \Omega$. Q.E.D.

Corollary 2.6.3. *Let A, $B \in \mathcal{L}(X)$, and let B be A-power-compact on the unbounded component of $\rho(A)$. Then, the unbounded components of $\rho_6(A)$ and $\rho_6(A+B)$ are the same, especially $r_e(A) = r_e(A+B)$.* ◇

Proof. This follows immediately from Corollary 2.6.2, since $\|(\lambda - A)^{-1}\| \to 0$ for $|\lambda| \to \infty$. Q.E.D.

2.7 Borel Mappings

Let X be a Banach space. An operator $T \in \mathcal{L}(X)$ is called a left topological divisor of zero (briefly, left TDZ) in $\mathcal{L}(X)$, if there exists a sequence $(x_n)_n$ of vectors such that $\|x_n\| = 1$ and $Tx_n \to 0$ as $n \to \infty$. Let $\mathcal{T}(X)$ denote the set

$$\mathcal{T}(X) := \{\text{operators in } \mathcal{L}(X) \text{ which are left TDZ}\}. \tag{2.7.1}$$

On $\mathcal{L}(X)$, we can consider its strong topology which is the coarsest topology for which the maps $\varphi_x : \mathcal{L}(X) \longrightarrow X$, $\varphi_x(T) = Tx$, are continuous for all $x \in X$. Equipped with this topology, $\mathcal{L}(X)$ is a topological vector space, denoted by $\mathcal{L}_s(X)$. We denote the subset of $\mathcal{L}(X)$ consisting of all invertible operators on X by $\text{Inv}(X)$. Let $\mathbb{C}$ be the field of complex numbers, and let $K(\mathbb{C})$ be the collection of all nonempty compact subsets of $\mathbb{C}$. For K and K' in $K(\mathbb{C})$, we define the Hausdorff distance $d_H(K, K')$ between K and K' by $d_H(K,K') = \max(\delta(K,K'), \delta(K',K))$, where $\delta(K,K') = \sup_{x \in K} \text{dist}(x, K')$. In this book, $K(\mathbb{C})$ will be endowed with this metric. The set $\text{Inv}(X)$ may be written in the following form

$$\text{Inv}(X) = \{T \in \mathcal{L}(X) : 0 \notin \sigma(T)\} = \sigma^{-1}(\{K \in K(\mathbb{C}) \text{ such that } 0 \notin K\}). \tag{2.7.2}$$

A topological space is said to be Polish, if it is separable and there is a complete metric determining its topology. Note that $\mathbb{R}$, $\mathbb{C}$, $\mathbb{R}^n$, and $\mathbb{C}^n$ are Polish spaces. For other examples and the properties of this class of spaces, see, for example, [187]. For a later use, we recall that, if a topological space X is Polish, then $\mathcal{K}(X)$ is also Polish (see [187, Theorem 4.25]). Accordingly, $K(\mathbb{C})$ is a Polish space. Let X be a Polish space. A subset A of X is called analytic if there is a Polish space Y and a continuous function $f : Y \longrightarrow X$ with $f(Y) = A$. A subset C of X is called coanalytic if $X \backslash C$ is analytic. The class of coanalytic subsets of X is denoted by Π_1^1. We say that a topological space is a standard Borel's space, if it is a topological space whose Borel's structure (but not necessarily the topological structure) is similar to the one of a Polish space. We will make use of the next result (cf. [187, p. 80]).

Proposition 2.7.1. *If X is a separable Banach space, then $\mathcal{L}_s(X)$ is a standard Borel's space.* ◇

We recall the following result due to Raymond [290].

Proposition 2.7.2. *Let X be a standard Borel space and let Y be a Polish space. If $\Omega \subset X \times Y$ is a Borel set for the product structure and, for every $x \in X$, $C_x = \{y \in Y$ such that $(x, y) \in \Omega\}$ is a K_σ set (a countable union of compact sets) in Y, then $P_X(\Omega)$ (the range of the first projection of Ω) is a Borel set in X.* ◇

Let Z be a topological space. For any subset A of Z, let

$$d(A) = \{x \in A \text{ such that } x \text{ is an accumulation point of } A\}.$$

The map, which assigns to each subset A of Z the set $d(A)$, is called the Cantor-Bendixson derivative. Obviously, $d(A)$ is closed. The set $\mathcal{R}(X)$ may be written in the following form

$$\mathcal{R}(X) = (\sigma \circ d)^{-1}(\{0\}). \tag{2.7.3}$$

Proposition 2.7.3. *(i) The topology of $K(\mathbb{C})$ is generated by two families of open sets in $K(\mathbb{C})$:*

$$\mathcal{T}^1 = \{K \in K(\mathbb{C}) \text{ such that } K \subset U\} \text{ and } \mathcal{T}^2 = \{K \in K(\mathbb{C}) \text{ such that } K \bigcap U \neq \emptyset\}$$

for any open set U in $\mathbb{C}$.

(ii) The Borel structure of $K(\mathbb{C})$ is generated by each of these two families.

(iii) The Cantor-Bendixson derivative $d : K(\mathbb{C}) \longrightarrow K(\mathbb{C})$ is a Borel map. ◇

The following results may be found in [227].

Lemma 2.7.1. *$\Delta := \{(K, \lambda) \in K(\mathbb{C}) \times \mathbb{C}$ such that $\lambda \in K\}$ is a closed set in $K(\mathbb{C}) \times \mathbb{C}$.* ◇

Proof. Let $(K_0, \lambda_0) \notin \Delta$. Then, there exist two open sets U and V, such that $K_0 \subset U$, $\lambda_0 \in V$ and $U \bigcap V = \emptyset$. Hence, $\{K \in K(\mathbb{C})$ such that $K \subset U\} \times V$ is open in $K(\mathbb{C}) \times \mathbb{C}$, contains (K_0, λ_0) and does not intersect Δ (because $\bigcup\{K \in$

$K(\mathbb{C})$ such that $K \subset U\} \bigcap V = \emptyset$). So, $(K(\mathbb{C}) \times \mathbb{C})\backslash\Delta$ is open and therefore, Δ is closed. Q.E.D.

Let $\psi : \mathcal{L}_s(X) \longrightarrow K(\mathbb{C})$, and let U be an open set in $\mathbb{C}$. Let $P_{\mathcal{L}_s(X)} : \mathcal{L}_s(X) \times U \longrightarrow \mathcal{L}_s(X)$ be the projection and let $\Omega_{U,\psi} := \{(T, \lambda) \in \mathcal{L}(X) \times U$ such that $\lambda \in \psi(T)\}$. Note that

$$\begin{aligned} P_{\mathcal{L}(X)}(\Omega_{U,\psi}) &= \{T \in \mathcal{L}(X) \ : \ \exists\lambda \in U \text{ such that } \lambda \in \psi(T)\} \\ &= \{T \in \mathcal{L}(X) \text{ such that } \psi(T) \bigcap U \neq \emptyset\}. \end{aligned}$$

For every $T \in \mathcal{L}(X)$, we have

$$\begin{aligned} \psi(T) \bigcap U &= \{\lambda \in U \text{ such that } \lambda \in \psi(T)\} \\ &= \{\lambda \in U \text{ such that } (T, \lambda) \in \Omega_{U,\psi}\}. \end{aligned} \tag{2.7.4}$$

Lemma 2.7.2. *Let $\psi : \mathcal{L}_s(X) \longrightarrow K(\mathbb{C})$. Then, ψ is a Borel map if, and only if, $\Delta_\psi := \{(T, \lambda) \in \mathcal{L}(X) \times \mathbb{C}$ such that $\lambda \in \psi(T)\}$ is a Borel set in $\mathcal{L}_s(X) \times \mathbb{C}$.* ◇

Proof. Let $\psi : \mathcal{L}_s(X) \longrightarrow K(\mathbb{C})$ be a Borel map. Then, $\psi \times I : \mathcal{L}_s(X) \times \mathbb{C} \longrightarrow K(\mathbb{C})\times\mathbb{C}$, $(\psi\times I)(T, \lambda) = (\psi(T), \lambda)$, is also a Borel map, and we have $\Delta_\psi = (\psi \times I)^{-1}(\{(K, \lambda) \in K(\mathbb{C}) \times \mathbb{C}$ such that $\lambda \in K\})$. Accordingly, ψ being Borel, implies that Δ_ψ is a Borel set (see Lemma 2.7.1). To prove the converse, it is sufficient, by Proposition 2.7.3 (*ii*), to show that $\psi^{-1}(\{K \in K(\mathbb{C})$ such that $K \bigcap U \neq \emptyset\})$ is a Borel set in $\mathcal{L}_s(X)$, for every open set U in $\mathbb{C}$. Since every open set in $\mathbb{C}$ is a K_σ set, so is $\{\lambda \in U$ such that $(T, \lambda) \in \Omega_{U,\psi}\}$ (see Eq. (2.7.4)). Moreover, for every open set U in $\mathbb{C}$, we have $\Omega_{U,\psi} = \Delta_\psi \bigcap(\mathcal{L}_s(X) \times U)$. Since Δ_ψ is a Borel set (by hypothesis) and $\mathcal{L}_s(X) \times U$ is open, it follows that $\Omega_{U,\psi}$ is Borel. Now, the use of Proposition 2.7.2 achieves the proof. Q.E.D.

Let $n \in \mathbb{Z}$, and set $\Phi_n(X) = \{U \in \Phi^b(X)$ such that $i(U) = n\}$.

Lemma 2.7.3. *Let $A \in \mathcal{L}(X)$ be a fixed operator satisfying $i(A) = -n$, where $n \in \mathbb{Z}$. Then, $\Phi_n(X) = \{T \in \mathcal{L}(X)$ such that $AT \in \Phi_0(X)\}$, where $\Phi_0(X)$ is the set $\Phi_n(X)$ with $n = 0$.* ◇

Proof. To prove this, let us first observe that, according to Atkinson's theorem, we have $\Phi_n(X) \subset \{T \in \mathcal{L}(X)$ such that $AT \in \Phi_0(X)\}$. Now, let $T \in \mathcal{L}(X)$ be such that $AT \in \Phi_0(X)$ and consider $B \in \mathcal{L}(X)$ such that $BA = I + F$, where $F \in \mathcal{F}_0(X)$ with $i(B) = -i(A) = n$ and $BAT = T + FT$ (B exists because $A \in \Phi^b(X)$). Since $AT \in \Phi_0(X)$ and $B \in \Phi^b(X)$, we have $T \in \Phi^b(X)$, and $i(T) = i(BAT) = i(B) + i(AT) = n$. This shows that $T \in \Phi_n(X)$. Q.E.D.

2.8 Baire Measurable Functions

Let X be a topological space, and let $K(X)$ be the family of all nonempty compact subsets of X. The topology on $K(X)$ is generated by the sets of the form $\{K \in K(X)$ such that $K \subset U\}$ and $\{K \in K(X)$ such that $K \bigcap U \neq \emptyset\}$, U being open in X. It is called the Vietoris topology [187]. We recall that, for a metric space X, the Vietoris topology is induced by the Hausdorff metric on $K(X)$ [187]. Let $\mathcal{M}$ be the set of all meager subsets of a topological space X. We say that $A, B \subset X$ are equal modulo $\mathcal{M}$, $A \cong B$, if the symmetric difference $A \triangle B = (A\backslash B) \bigcap (B\backslash A)$ belongs to $\mathcal{M}$. This defines an equivalence relation that respects complementation, countable unions, and intersections. Let X and Y be topological spaces. A function $f : X \longrightarrow Y$ is Baire measurable if, for any open subset B of Y, the inverse image of B, $f^{-1}(B)$, satisfies $f^{-1}(B) \cong A$ for some open set $A \subset X$.

Theorem 2.8.1 ([187, Theorem 8.38]). *Let X and Y be two topological spaces and let $f : X \longrightarrow Y$ be Baire measurable. If Y is second countable, there is a set $G \subset X$ that is a countable intersection of dense and open sets, such that $f_{|G}$ is continuous. In particular, if X is Baire, f is continuous on dense G_δ set.* $\diamond$

Let $\mathbb{T}$ denote the unit circle of $\mathbb{C}$. For $U \subset \mathbb{C}\backslash\{0\}$, we define U^1 (a subset of $\mathbb{T}$) by

$$U^1 = \left\{ \frac{z}{|z|} \text{ such that } z \in U \right\}.$$

We recall some results about Baire measurable functions. These results can be found in [228].

Lemma 2.8.1. *Let φ be a homomorphism from $\mathbb{R}$ into $\mathbb{T}$ such that $\varphi \neq 1$. If V is a nonempty open subset of $\mathbb{T}$, then $\varphi^{-1}(V)$ is non-meager.* $\diamond$

Proof. From Theorem 2.2.5, it follows that $\overline{\varphi(\mathbb{R}) \text{ is dense in } \mathbb{T}}$. Since $\mathbb{T}$ is separable, there is a sequence $(t_n)_n$ of $\mathbb{R}$ such that $\overline{\{\varphi(t_n) \text{ such that } n \in \mathbb{N}\}} = \mathbb{T}$. Thus, we can write $\mathbb{T} = \bigcup_{n\in\mathbb{N}} \varphi(t_n)V$. However, $\mathbb{T}$ is compact. So, there exists $m \in \mathbb{N}$ such that $\mathbb{T} = \bigcup_{1\le j\le m} \varphi(t_j)V$ and consequently, $\mathbb{R} = \varphi^{-1}(\mathbb{T}) = \bigcup_{1\le j\le m} \varphi^{-1}(\varphi(t_j)V)$. This implies that there exists $j \in \mathbb{N}$ with $1 \le j \le m$ such that $\varphi^{-1}(\varphi(t_j)V)$ is non-meager. Nevertheless, $\varphi^{-1}(\varphi(t_j)V) = \{t \in \mathbb{R}$ such that $\varphi(t) \in \varphi(t_j)V\} = \{t \in \mathbb{R}$ such that $\varphi(t - t_j) \in V\} = t_j + \varphi^{-1}(V)$. Accordingly, $\varphi^{-1}(V)$ is non-meager.

Q.E.D.

Lemma 2.8.2. *Let φ be a homomorphism from $\mathbb{R}$ into $\mathbb{T}$, and let V be a nonempty open subset of $\mathbb{T}$. If φ is not continuous, then $\varphi^{-1}(V)$ is dense in $\mathbb{R}$.* $\diamond$

Proof. Since φ is not continuous, by applying Theorem 2.2.5, we deduce that the image of each interval of $\mathbb{R}$ is dense in $\mathbb{T}$. So, if $a \in \mathbb{R}$, then for all $\varepsilon > 0$, $\overline{\varphi(]a - \varepsilon, a + \varepsilon[)} = \mathbb{T}$. Therefore, there exists $t \in]a - \varepsilon, a + \varepsilon[$ such that $\varphi(t) \in V$.

Q.E.D.

Lemma 2.8.3. *Assume that φ is a discontinuous homomorphism from $\mathbb{R}$ into $\mathbb{T}$. If U and V are nonempty open subsets of $\mathbb{R}$ and $\mathbb{T}$, respectively, then $\varphi^{-1}(V)\bigcap U$ is non-meager.* ◇

Proof. Since $\mathbb{T}$ is a topological group with unit 1, there exists a nonempty open subset V' of V and a neighborhood V_1 of 1 in $\mathbb{T}$ such that $V_1^{-1}V' \subset V$. Moreover, Lemma 2.8.2 implies that there exists a sequence $(a_n)_n \subset \varphi^{-1}(V_1)$ such that the set $\{a_0, a_1, \ldots, a_n, \ldots\}$ is dense in $\mathbb{R}$. This leads to $\mathbb{R} = \bigcup_{n\in\mathbb{N}}(a_n + U)$ and therefore, the use of Lemma 2.8.1 shows that $\varphi^{-1}(V') = \bigcup_{n\in\mathbb{N}}[\varphi^{-1}(V')\bigcap(a_n + U)]$ is non-meager. Now, by using the fact that a countable union of meager sets is also meager, we infer that there exists $n \in \mathbb{N}$ such that $\varphi^{-1}(V')\bigcap(a_n + U)$ is non-meager. Note that we have

$$\begin{aligned}\varphi^{-1}(V')\bigcap(a_n + U) &= \{t \in a_n + U \text{ such that } \varphi(t) \in V'\}\\ &= \{t + a_n,\ t \in U \text{ such that } \varphi(a_n)\varphi(t) \in V'\}.\end{aligned}$$

Now, making use of the inclusion $V_1^{-1}V' \subset V$, we can write $\varphi^{-1}(V')\bigcap(a_n + U) \subset a_n + \{t \in U \text{ such that } \varphi(t) \in (\varphi(a_n))^{-1}V'\}$. Hence, the set $\{t \in U \text{ such that } \varphi(t) \in V\} = U\bigcap\varphi^{-1}(V)$ is non-meager. Q.E.D.

Theorem 2.8.2. *Let $\theta : \mathbb{R} \longrightarrow K(\mathbb{T})$ be a Baire measurable function and let φ be a homomorphism from $\mathbb{R}$ into $\mathbb{T}$. Assume that $\varphi(t) \in \theta(t)$ for all $t \in \mathbb{R}$, and that $\{t \in \mathbb{R}$ such that $\theta(t) \neq \mathbb{T}\}$ is non-meager. Then, $\varphi(.)$ is continuous.* ◇

Proof. Since θ is Baire measurable, by using Theorem 2.8.1, there exists a meager subset M of $\mathbb{R}$ such that $\theta_{|\mathbb{R}\setminus M}$ is continuous. Let $t_0 \in \mathbb{R}\setminus M$ such that $\theta(t_0) \neq \mathbb{T}$ (such a real exists by hypothesis). Since $\theta(t_0) \in K(\mathbb{T})$ and since $\mathbb{T}$ is a regular topological space, there exist two open subsets U and $V \neq \emptyset$ of $\mathbb{T}$ such that $\theta(t_0) \subset U$ and $U\bigcap V = \emptyset$. However, $\{K \in K(\mathbb{T})$ such that $K \subset U\}$ is an open subset of $K(\mathbb{T})$, so by the continuity of $\theta_{|\mathbb{R}\setminus M}$, there exists $\alpha > 0$ such that $\theta(t) \subset U$ for all $t \in]t_0 - \alpha, t_0 + \alpha[\bigcap(\mathbb{R}\setminus M)$. Accordingly, for all $t \in]t_0 - \alpha, t_0 + \alpha[\bigcap(\mathbb{R}\setminus M)$, we have $\theta(t)\bigcap V = \emptyset$. Thus, $\varphi(t) \notin V$. Hence, the set $\varphi^{-1}(V)\bigcap]t_0 - \alpha, t_0 + \alpha[$ is contained in M and therefore, it is meager. So, according to Lemma 2.8.3, $\varphi(.)$ is necessarily continuous. Q.E.D.

We end this section by the following elementary lemma of which we omit the proof.

Lemma 2.8.4. *Let X be a Banach space, $A \in \mathcal{L}(X)$, and let Y be a A-invariant closed subspace of X. Then, $\rho_\infty(A) \subset \rho_\infty(A_{|Y})$, where $\rho_\infty(.)$ denotes the unbounded connected component of $\rho(.)$. If $0 \in \rho_\infty(A)$, then $\sigma^1(A_{|Y}) \subset \sigma^1(A)$, where*

$$\sigma^1(A) := \left\{\frac{\lambda}{|\lambda|} \text{ such that } \lambda \in \sigma(A)\right\}.$$

◇

2.9 Banach Module

In this section, we recall some definitions and we give some lemmas which will be needed in the sequel.

Definition 2.9.1.

(i) Let H be a Banach space. We say that Banach subalgebra $\mathcal{M}$ of $\mathcal{L}(H)$ is nondegenerate, if the linear subspace of H generated by the elements Mu, with $M \in \mathcal{M}$ and $u \in H$, is dense in H.
(ii) An approximate unit in a Banach algebra $\mathcal{M} \subset \mathcal{L}(H)$ is a net $\{J_n\}_n$ such that $\|J_n\| \leq C$ for a constant C and all n and $\|MJ_n - M\| = 0$, for all $M \in \mathcal{M}$.
(iii) A Banach module is a couple $(H, \mathcal{M})$ consisting of a Banach space H and a nondegenerate Banach subalgebra $\mathcal{M}$ of $\mathcal{L}(H)$ which has an approximate unit. If H is a Hilbert space, we say that H is a Hilbert module. ◇

The distinguished subalgebra $\mathcal{M}$ will be called multiplier algebra of H and, when required by the clarity of the presentation, we will denote it $\mathcal{M}(H)$ (see, for example, [112, p. 404]). A rigged Hilbert space is a pair (H, K) where H is a Hilbert space, and K is a dense subspace, such that K is given a topological vector space structure, for which the inclusion map i is continuous. The specific triple (K, H, K^*) is often called the "Gelfand triple."

Definition 2.9.2. A couple (K, H), consisting of a Hilbert module H and a Hilbert space K such that $K \subset H$ continuously and densely, will be called a Friedrichs module. If $\mathcal{M}(H) \subset \mathcal{K}(K, H)$, we say that (K, H) is a compact Friedrichs module. ◇

In the context of this definition, we always identify H with its adjoint space, which gives us a Gelfand triple $K \subset H \subset K^*$. If (K, H) is a compact Friedrichs module then, each operator M from $\mathcal{M}(H)$ can be extended to a compact operator $M : H \longrightarrow K^*$, and we will have $\mathcal{M}(H) \subset \mathcal{K}(K, H) \bigcap \mathcal{K}(H, K^*)$.

Definition 2.9.3. Let H and K be Banach spaces.

(i) If K is a Banach module, we define $B_0^l(H, K)$ as the closed linear subspace norm generated by the operators MT with $T \in \mathcal{L}(H, K)$ and $M \in \mathcal{M}(K)$.
(ii) If H is a Banach module, then one can similarly define $B_0^r(H, K)$ as the closed linear subspace norm generated by the operators TM with $T \in \mathcal{L}(H, K)$ and $M \in \mathcal{M}(H)$. We say that an operator in $B_0^l(H, K)$ (resp. $B_0^r(H, K)$) is left (resp. right) vanishes at infinity. ◇

Definition 2.9.4. Let H and K be Banach modules and let $S \in \mathcal{L}(H, K)$. We say that S is left decay preserving if, for each $M \in \mathcal{M}(H)$, we have $SM \in B_0^l(H, K)$. We say that S is right decay preserving if, for each $M \in \mathcal{M}(K)$, we have $MS \in B_0^r(H, K)$. We say that S is decay preserving, if S is left and right decay preserving. ◇

We denote these classes of operators by $B_q^l(H,K)$, $B_q^r(H,K)$, and $B_q(H,K)$. These classes represent closed subspaces of $\mathcal{L}(H,K)$. The following theorem follows from the Cohen–Hewitt theorem [112, V-9.2].

Theorem 2.9.1. *Let H be a Banach space, let K be a Banach module, and let $S \in \mathcal{L}(H,K)$. Then $S \in B_0^l(H,K)$ if, and only if, $S = MT$ for some $M \in \mathcal{M}(K)$ and $T \in \mathcal{L}(H,K)$.* ◇

Let (K,H) be a Friedrichs couple and let $K \subset H \subset K^*$ be the Gelfand triple associated with it. To an operator $S \in \mathcal{L}(K,K^*)$, we associate an operator $\hat{S}$ acting in H according to the rules $\mathcal{D}(\hat{S}) = S^{-1}(H)$ and $\hat{S} = S_{|_{\mathcal{D}(\hat{S})}}$. Due to the identification $K^{**} = K$, the operator S^* is an element of $\mathcal{L}(K,K^*)$. So, $\widehat{S^*}$ makes sense.

Lemma 2.9.1 ([115, Lemma 2.4]). *If $S - z : K \longrightarrow K^*$ is bijective for some $z \in \mathbb{C}$, then $\hat{S}$ is a closed, densely defined operator, and we have $(\hat{S})^* = \widehat{S^*}$ and $z \in \rho(\hat{S})$. Moreover, the domains $\mathcal{D}(\hat{S})$ and $\mathcal{D}(\hat{S}^*)$ are dense subspaces of K.* ◇

Remark 2.9.1. Let (K,H) be a compact Friedrich module. It is easy to prove that, if $R \in B_0^l(H) := B_0^l(H,H)$ and $RH \subset K$, then $R \in \mathcal{K}(H)$. ◇

Proposition 2.9.1 ([115, Proposition 1.2]). *The space B_q^l has the following properties:*

(i) if $S \in B_q^l(H,K)$ and $T \in B_q^l(L,K)$, then $ST \in B_q^l(L,K)$, and
(ii) if H and K are reflexive and $S \in B_q^l(H,K)$, then $S^* \in B_q^l(K^*,H^*)$. ◇

If A and B are closed operators acting in a Banach space H, and if there is $z \in \rho(A) \bigcap \rho(B)$ such that $(A-z)^{-1} - (B-z)^{-1}$ is a compact operator, then we say that B is a compact perturbation of A.

2.10 Measure of Noncompactness

The notion of measure of noncompactness turned out to be a useful tool in some topological problems, in functional analysis, and in operator theory (see [18, 47, 97, 253, 289]).

2.10.1 Measure of Noncompactness of a Bounded Subset

In order to recall the measure of noncompactness, let $(X, \|.\|)$ be an infinite-dimensional Banach space. We denote by $\mathcal{M}_X$ the family of all nonempty and bounded subsets of X, while N_X denotes its subfamily consisting of all relatively compact sets. Moreover, let us denote the convex hull of a set $A \subset X$ by $\mathrm{conv}(A)$. Let us recall the following definition.

Definition 2.10.1 ([47]). A mapping $\mu : \mathcal{M}_X \longrightarrow [0, +\infty[$ is said to be a measure of noncompactness in the space X, if it satisfies the following conditions:

(i) The family $\mathrm{Ker}(\mu) := \{D \in \mathcal{M}_X$ such that $\mu(D) = 0\}$ is nonempty, and $\mathrm{Ker}(\mu) \subset N_X$.
For $A, B \in \mathcal{M}_X$, we have the following
(ii) If $A \subset B$, then $\mu(A) \leq \mu(B)$.
(iii) $\mu(\overline{A}) = \mu(A)$.
(iv) $\mu(\overline{\mathrm{conv}(A)}) = \mu(A)$.
(v) $\mu(\lambda A + (1-\lambda)B) \leq \lambda\mu(A) + (1-\lambda)\mu(B)$, for all $\lambda \in [0, 1]$.
(vi) If $(A_n)_{n\in\mathbb{N}}$ is a sequence of sets from $\mathcal{M}_X$ such that $A_{n+1} \subset A_n$, $\overline{A}_n = A_n$ $(n = 1, 2, \ldots)$ and $\lim\limits_{n\to+\infty} \mu(A_n) = 0$, then $A_\infty := \bigcap_{n=1}^{\infty} A_n$ is nonempty and $A_\infty \in \mathrm{Ker}(\mu)$. ◇

The family $\mathrm{Ker}(\mu)$, described in Definition 2.10.1 (i), is called the kernel of the measure of noncompactness μ.

Definition 2.10.2. A measure of noncompactness μ is said to be sublinear if, for all $A, B \in \mathcal{M}_X$, it satisfies the two following conditions:

(i) $\mu(\lambda A) = |\lambda|\mu(A)$ for $\lambda \in \mathbb{R}$ (μ is said to be homogenous), and
(ii) $\mu(A + B) \leq \mu(A) + \mu(B)$ (μ is said to be subadditive). ◇

Definition 2.10.3. A measure of noncompactness μ is referred to as a measure with maximum property if $\max(\mu(A), \mu(B)) = \mu(A \bigcup B)$. ◇

Definition 2.10.4. A measure of noncompactness μ is said to be regular if $\mathrm{Ker}(\mu) = N_X$, sublinear and has a maximum property. ◇

For $A \in \mathcal{M}_X$, the most important examples of measures of noncompactness [253] are

- Kuratowski measure of noncompactness

$$\gamma(A) = \inf\{\varepsilon > 0 \ : \ A \text{ may be covered by a finite number of sets of diameter} \leq \varepsilon\}.$$

- Hausdorff measure of noncompactness

$$\overline{\gamma}(A) = \inf\{\varepsilon > 0 \ : \ A \text{ may be covered by a finite number of open balls of radius} \leq \varepsilon\}.$$

Note that these measures γ and $\overline{\gamma}$ are regular. The relations between these measures are given by the following inequalities, which were obtained by Danes [87]: $\overline{\gamma}(A) \leq \gamma(A) \leq 2\overline{\gamma}(A)$, for any $A \in \mathcal{M}_X$. The following proposition gives some frequently used properties of the Kuratowski's measure of noncompactness.

Proposition 2.10.1. *Let A and A' be two bounded subsets of X. Then, we have the following properties.*

(i) $\gamma(A) = 0$ *if, and only if, A is relatively compact.*
(ii) *If* $A \subseteq A'$, *then* $\gamma(A) \leq \gamma(A')$.
(iii) $\gamma(A + A') \leq \gamma(A) + \gamma(A')$.
(iv) *For every* $\alpha \in \mathbb{C}$, $\gamma(\alpha A) = |\alpha|\, \gamma(A)$. ◇

Let us notice that, throughout the book, we are working on two spaces, for example X and Y with their respective measures $\gamma_X(.)$ and $\gamma_Y(.)$. However, and in order to simplify our reasoning, $\gamma_X(.)$ and $\gamma_Y(.)$ will be simply called $\gamma(.)$. Of course, the reader will be able to link $\gamma(.)$ either to $\gamma_X(.)$ or to $\gamma_Y(.)$.

Definition 2.10.5. Let X and Y be Banach spaces, let $\gamma(.)$ be a Kuratowski measure of noncompactness, and let S, A be two linear operators from X into Y bounded on its domains such that $\mathcal{D}(A) \subset \mathcal{D}(S)$. The operator S is called γ-relatively bounded with respect to A (or A-γ-bounded), if there exist constants $a_S \geq 0$ and $b_S \geq 0$, such that

$$\gamma(S(\mathfrak{D})) \leq a_S \gamma(\mathfrak{D}) + b_S \gamma(A(\mathfrak{D})), \tag{2.10.1}$$

where $\mathfrak{D}$ is a bounded subset of $\mathcal{D}(A)$. The infimum of the constants b_S which satisfy (2.10.1) for some $a_S \geq 0$ is called the A-γ-bound of S. ◇

In general, the sum of closable or closed operators is not closable or closed, respectively. However, we have the following.

Theorem 2.10.1. *If S is A-γ-bounded with a bound* < 1 *and if S and A are closed, then $A + S$ is closed.* ◇

Proof. S is A-γ-bounded with a bound < 1, implies that there exist $a_S \geq 0$ and $b_S \geq 0$ such that $b_S < 1$ and $\gamma(S(\mathfrak{D})) \leq a_S\gamma(\mathfrak{D}) + b_S\gamma(A(\mathfrak{D}))$, $\mathfrak{D} \subset \mathcal{D}(A)$. Then,

$$\gamma((S + A)(\mathfrak{D})) \geq -a_S \gamma(\mathfrak{D}) + (1 - b_S)\gamma(A(\mathfrak{D})). \tag{2.10.2}$$

Let $(x_n)_n \subset \mathcal{D}(A)$ such that $x_n \to x$ and $(S + A)x_n \to y$. Then, $x \in \mathcal{D}(A)$ and $(S + A)x = y$. Indeed, $x_n \to x$ and $Ax_n \to y$ implies that $\{x_n\}$ and $\{(S + A)(x_n)\}$ are relatively compact and hence, $\gamma(\{x_n\}) = \gamma(\{(S + A)x_n\}) = 0$. Then, by using Eq. (2.10.2), we get $\gamma(A\{x_n\}) = 0$, and there exists a subsequence (x_{n_k}) such that $Ax_{n_k} \to \alpha$. Since A is closed and $x_{n_k} \to x$, it follows that $x \in \mathcal{D}(A)$, and $Ax = \alpha$. Since S is closed and $Sx_{n_k} \to y - \alpha$, then $x \in \mathcal{D}(A)$ and $(S + A)x = y$. Q.E.D.

Lemma 2.10.1. *If S is A-γ-bounded with a bound* $\delta < 1$, *then S is $(A + S)$-γ-bounded with a bound* $< \frac{\delta}{1-\delta}$. ◇

Proof. S is A-γ-bounded with a bound $\delta < 1$, implies that there exist a_S and $b_S \geq 0$ such that $\delta \leq b_S < 1$ and, for $\mathfrak{D} \subset \mathcal{D}(A)$, we get

$$\begin{aligned}
\gamma(S(\mathfrak{D})) &\leq a_S\gamma(\mathfrak{D}) + b_S\gamma(A(\mathfrak{D})), \\
&\leq a_S\gamma(\mathfrak{D}) + b_S\gamma((S + A - S)(\mathfrak{D})), \\
&\leq a_S\gamma(\mathfrak{D}) + b_S\gamma((S + A)(\mathfrak{D})) + b_S\gamma(S(\mathfrak{D})).
\end{aligned}$$

Since $b_S < 1$, we obtain $\gamma(S(\mathfrak{D})) \le \frac{a_S}{1-b_S}\gamma(\mathfrak{D}) + \frac{b_S}{1-b_S}\gamma((S+A)(\mathfrak{D}))$. Hence, S is $(A+S)$-γ-bounded with an $(A+S)$-bound $< \frac{\delta}{1-\delta}$. Q.E.D.

2.10.2 Measure of Noncompactness of an Operator

2.10.2.1 Kuratowski and Hausdorff Measures of Noncompactness

Definition 2.10.6 ([204]).

(i) Let $T : \mathcal{D}(T) \subseteq X \longrightarrow X$ be a continuous operator and let $\gamma(.)$ be the Kuratowski measure of noncompactness in X. Let $k \ge 0$. T is said to be k-set-contraction if, for any bounded subset A of $\mathcal{D}(T)$, $T(A)$ is a bounded subset of X and $\gamma(T(A)) \le k\gamma(A)$. T is said to be condensing if, for any bounded subset A of $\mathcal{D}(T)$ such that $\gamma(A) > 0$, $T(A)$ is a bounded subset of X and $\gamma(T(A)) < \gamma(A)$.

(ii) Let $T : \mathcal{D}(T) \subseteq X \longrightarrow X$ be a continuous operator, $\overline{\gamma}(.)$ being the Hausdorff measure of noncompactness in X, and $k \ge 0$. T is said to be k-ball-contraction if, for any bounded subset A of $\mathcal{D}(T)$, $T(A)$ is a bounded subset of X and $\overline{\gamma}(T(A)) \le k\overline{\gamma}(A)$. ◇

Remark 2.10.1. It is well known that:

(i) If $k < 1$, then every k-set-contraction operator is condensing.
(ii) Every condensing operator is 1-set-contraction. ◇

Let $T \in \mathcal{L}(X)$. We define $\gamma(T)$, by $\gamma(T) := \inf\{k$ such that T is k-set-contraction$\}$, and $\overline{\gamma}(T)$, by $\overline{\gamma}(T) := \inf\{k$ such that T is k-ball-contraction$\}$.

In the following lemma, we give some important properties of $\gamma(T)$ and $\overline{\gamma}(T)$.

Lemma 2.10.2 ([37, 102]). *Let X be a Banach space and $T \in \mathcal{L}(X)$. We have the following:*

(i) $\frac{1}{2}\gamma(T) \le \overline{\gamma}(T) \le 2\gamma(T)$.
(ii) $\gamma(T) = 0$ *if, and only if,* $\overline{\gamma}(T) = 0$ *if, and only if,* T *is compact.*
(iii) *If* $T, S \in \mathcal{L}(X)$, *then* $\gamma(ST) \le \gamma(S)\gamma(T)$ *and* $\overline{\gamma}(ST) \le \overline{\gamma}(S)\overline{\gamma}(T)$.
(iv) *If* $K \in \mathcal{K}(X)$, *then* $\gamma(T+K) = \gamma(T)$ *and* $\overline{\gamma}(T+K) = \overline{\gamma}(T)$.
(v) $\gamma(T^*) \le \overline{\gamma}(T)$ *and* $\gamma(T) \le \overline{\gamma}(T^*)$, *where* T^* *denotes the dual operator of* T.
(vi) *If* B *is a bounded subset of* X, *then* $\gamma(T(B)) \le \gamma(T)\gamma(B)$. ◇

2.10.2.2 Measure of Noncompactness in a Cartesian Product

In [47, Theorem 3.3.2], the authors construct the measures of noncompactness in a cartesian product of a given finite collection of Banach spaces. More precisely, we have:

Lemma 2.10.3 ([47, Theorem 3.3.2]). *Let $E_1, \ldots, E_n$ be a finite collection of Banach spaces, and let $\mu_1, \ldots, \mu_n$ be the measures of noncompactness in*

$E_1, \ldots, E_n$, respectively. Assume that the function $F : ([0,+\infty[)^n \longrightarrow [0,+\infty[$ is convex and that $F(x_1,\ldots,x_n) = 0$ if, and only if, $x_i = 0$ for $i = 1,\ldots,n$. Then, $\mu(x) = F(\mu_1(\pi_1(x)),\ldots,\mu_n(\pi_n(x)))$ defines a measure of noncompactness in $E_1 \times E_2 \times \ldots \times E_n$, where $\pi_i(x)$ denotes the natural projection of x into E_i. ◇

According to the previous lemma (Lemma 2.10.3), for γ being a measure of noncompactness in a Banach space X, for all $A \in \mathcal{M}_{X^n}$, the quantity $\gamma(A) = \max(\gamma(\pi_1(A)),\ldots,\gamma(\pi_n(A)))$ defines a measure of noncompactness in X^n. For $T \in \mathcal{L}(X^n)$, let us consider the measure of noncompactness of T, denoted by $\gamma(T)$,

$$\gamma(T) = \sup\left\{\frac{\gamma(T(A))}{\gamma(A)} \text{ such that } A \in \mathcal{M}_{X^n} \text{ and } \gamma(A) > 0\right\}.$$

We start with the following preliminary result which is fundamental for our purpose.

Proposition 2.10.2. *Let X be a Banach space, and let $L = \left(L_{ij}\right)_{1\le i,j\le n}$ be a matrix operator, where $L_{ij} \in \mathcal{L}(X)$, $\forall\ 1 \le i,j \le n$. Then, $\gamma(L) \le \max_{1\le i\le n}\left(\sum_{j=1}^n \gamma(L_{ij})\right)$.* ◇

Proof. For all $A \in \mathcal{M}_{X^n}$, we have $A \subset \pi_1(A) \times \pi_2(A) \times \ldots \times \pi_n(A)$. Hence, $L(A) \subset L(\pi_1(A) \times \pi_2(A) \times \ldots \times \pi_n(A))$. If we denote by $A_i := \pi_i(A)$, $i = 1,\ldots,n$, then

$$\begin{aligned}
\gamma(L) &= \sup\left\{\frac{\gamma(L(A))}{\gamma(A)} \text{ such that } A \in \mathcal{M}_{X^n} \text{ and } \gamma(A) > 0\right\} \\
&\le \sup\left\{\frac{\gamma(L\left(A_j\right)_{1\le j\le n})}{\max\limits_{1\le j\le n}\gamma(A_j)} \text{ such that } \gamma(A_j) > 0,\ \forall j = 1,\ldots,n\right\} \\
&\le \sup\left\{\frac{\max\limits_{1\le i\le n}\gamma\left(\sum\limits_{j=1}^n L_{ij}A_j\right)}{\max\limits_{1\le j\le n}\gamma(A_j)} \text{ such that } \gamma(A_j) > 0,\ \forall j = 1,\ldots,n\right\} \\
&\le \sup\left\{\max_{1\le i\le n}\sum_{j=1}^n \frac{\gamma(L_{ij}A_j)}{\gamma(A_j)} \text{ such that } \gamma(A_j) > 0,\ \forall j = 1,\ldots,n\right\} \\
&\le \max_{1\le i\le n}\left(\sum_{j=1}^n \gamma(L_{ij})\right).
\end{aligned}$$

Q.E.D.

2.10.2.3 Measure of Non-strict Singularity

We end this part by the following definition concerning the measure of non-strict singularity (see [285]). The measure of non-strict singularity of $A \in \mathcal{L}(X, Y)$ is defined by $q(A) = \overline{\gamma}[A(\overline{B}_X)]$. It is proved in [230] that

(*i*) $q(A) \leq \|A\|$,
(*ii*) $q(A) = 0$ if, and only if, $A \in \mathcal{K}(X, Y)$, and
(*iii*) $q(A + K) = q(A)$, for all $K \in \mathcal{K}(X, Y)$.

Definition 2.10.7. For $A \in \mathcal{L}(X, Y)$, set $g_M(A) = \inf_{N \subset M} q(A|_N)$, and $g(A) = \sup_{M \subset X} g_M(A)$, where M and N represent two infinite-dimensional subspaces of X, and where $A|_N$ denotes the restriction of A to the subspace N. ◇

The semi-norm g is a measure of non-strict singularity which was introduced by Schechter in [299]. We recall the following result established in [289].

Proposition 2.10.3. *For $A \in \mathcal{L}(X, Y)$, we have the following:*

(i) $A \in \mathcal{S}(X, Y)$ if, and only if, $g(A) = 0$.
(ii) $A \in \mathcal{S}(X, Y)$ if, and only if, $g(A + T) = g(T)$, for all $T \in \mathcal{L}(X, Y)$.
(iii) If Z is a Banach space and $B \in \mathcal{L}(Y, Z)$, then $g(BA) \leq g(B)g(A)$. ◇

We can easily notice the following relationship between the measure of non-strict singularity g and the measure of noncompactness q.

Proposition 2.10.4. *For $A \in \mathcal{L}(X, Y)$, we have $g(A) \leq q(A)$.* ◇

Proof. This result follows immediately from [289, Theorem 1], [299, Theorem 2.10] and [230, Theorem 3.1]. Q.E.D.

2.11 Measure of Weak Noncompactness

The measure of weak noncompactness has many applications in topology, functional analysis, the theory of differential equations and integral equations (see, for example, [37, 38, 46, 104, 132, 207]). This notion was introduced by De Blasi [90]. In order to recall this notion, we denote by X a Banach space, and by $\mathcal{W}_X$ the subfamily of $\mathcal{M}_X$ consisting of all relatively weakly compact sets. There exists an axiomatic approach in defining the measure of weak noncompactness [49]. Let us recall the following definition:

Definition 2.11.1 ([48]). Let X be a Banach space. A function $\mu : \mathcal{M}_X \longrightarrow [0, +\infty[$ is said to be a measure of weak noncompactness in X if, for all $A, B \in \mathcal{M}_X$, it satisfies the following conditions:

(i) $\mu(A) = 0 \Longleftrightarrow A \in \mathcal{W}_X$.
(ii) $A \subset B \Longrightarrow \mu(A) \leq \mu(B)$.

(iii) $\mu(\overline{\mathrm{conv}(A)}) = \mu(A)$.
(iv) $\mu(A \bigcup B) = \max\{\mu(A), \mu(B)\}$.
(v) $\mu(A + B) \leq \mu(A) + \mu(B)$.
(vi) $\mu(\lambda A) = |\lambda|\mu(A)$, $\lambda \in \mathbb{R}$. ◇

A measure of weak noncompactness in the above sense is said to be regular. Let us recall that each measure of weak noncompactness also satisfies the Cantor intersection condition (see [49, Theorem 5]): If $X_n \in \mathcal{W}_X$, $X_n = \overline{X}_n$ and $X_{n+1} \subset X_n$ for $n = 1, 2, \ldots$ and, if $\lim_{n\to+\infty} \mu(X_n) = 0$, then $X_\infty = \bigcap_{n=1}^{+\infty} X_n \neq \emptyset$. As an example of a regular measure in a Banach space X, we have the measure of weak noncompactness defined by De Blasi in the following formula:

$$\omega(A) = \inf\{t > 0 : \exists\, C \in \mathcal{W}_X \text{ such that } A \subset C + t\overline{B}_X\},\ \forall A \in \mathcal{M}_X. \tag{2.11.1}$$

This function satisfies several useful properties [90]. For example, $\omega(\overline{B}_X) = 1$ whenever X is nonreflexive, and $\omega(\overline{B}_X) = 0$ otherwise. However, for any measure of weak noncompactness ξ and for every $A \in \mathcal{M}_X$, the following inequality holds (see [49, Theorem 4]):

$$\xi(A) \leq \xi(\overline{B}_X)\omega(A). \tag{2.11.2}$$

In order to define another example of measure of weak noncompactness (cf. [200]), let (x_n) be a sequence in X. We say that (y_n) is a sequence of successive convex combinations (in short scc) for (x_n), if there exists a sequence of integers $0 = p_1 < p_2 < \ldots$ such that $y_n \in conv\{x_i\}_{i=p_n+1}^{p_{n+1}}$ for each n. Similarly, the vectors u_1, u_2 are said to be a pair of scc for (x_n), if $u_1 \in conv\{x_i\}_{i=1}^{p}$ and $u_2 \in conv\{x_i\}_{i=p+1}^{\infty}$ for some integer $p \geq 1$. The notion of scc was used in [201] in order to define a measure of weak noncompactness γ which is a counterpart for the weak topology of the separation measure of noncompactness (see [18, 42]). By the convex separation of (x_n), we mean the number $\mathrm{csep}(x_n) = \inf\{\|u_1 - u_2\| \text{ such that } u_1, u_2 \text{ is a pair of scc for } (x_n)\}$. For each $A \in \mathcal{M}_X$, we put

$$\gamma(A) = \sup\{\mathrm{csep}(x_n) \text{ such that } (x_n) \subset \mathrm{conv}(A)\}. \tag{2.11.3}$$

The proof of the following theorem may be found in [201].

Theorem 2.11.1. *Let X be a Banach space and $A \in \mathcal{M}_X$. Then,*

$$\begin{aligned}\gamma(A) &= \sup\{\lim_n \lim_k F_n(x_k) - \lim_k \lim_n F_n(x_k) : (x_n) \subset conv(A),\\ &\qquad (F_n) \subset \overline{B}_{X^*} \text{ and the limits exist}\}\\ &= \sup dist(x^{**}, conv x_n),\end{aligned}$$

where the second supremum is taken over all sequences $(x_n)_n$ in conv(A) and all w^*-cluster points $x^{**} \in X^{**}$ of $(x_n)_n$. ◇

From Theorem 2.11.1 and [38, Corollary 5], we deduce that γ is not equivalent to ω, even though $\gamma(A) \leq \omega(A)$, for all $A \in \mathcal{M}_X$. Moreover, these measures coincide in the space $c_0(X) := \{(x_i)_{i\in\mathbb{N}^*},\ x_i \in X,\ \forall i \in \mathbb{N}^* \text{ such that } \lim_{i\to+\infty} \|x_i\| = 0\}$ endowed by the norm $\|(x_i)\| = \max\{\|x_i\| \text{ such that } i \in \mathbb{N}^*\}$ (see [201]). In the following theorem, we recall the relationship between ω and γ in the Lebesgue space $L_1(\eta)$.

Theorem 2.11.2 ([201, Theorem 2.5]). *Let η be a finite measure, and $A \in \mathcal{M}_{L_1(\eta)}$. Then*

$$\gamma(A) = 2\omega(A).$$

◇

2.12 Graph Measures

Let X be a Banach space, and $T \in \mathcal{C}(X)$. Similarly to the notion of the graph norm $\|.\|_T$ in $X_T := (\mathcal{D}(T), \|.\|_T)$, we define a new measure of noncompactness and weak noncompactness called, respectively, a graph measure of noncompactness and a graph measure of weak noncompactness. The symbol $\overline{A}$ (resp. $\overline{A}^T$) stands for the closure of A in X (resp. in X_T), while the symbol $\overline{A}^\omega$ (resp. $\overline{A}^{\omega T}$) stands for the weak closure of A in X (resp. in X_T).

Lemma 2.12.1. *Let X be a Banach space, $T \in \mathcal{C}(X)$, and let $A \subset X_T$.*

(i) If $x \in \overline{A}^T$, then $x \in \overline{A}$ and $Tx \in \overline{T(A)}$.
(ii) If A is compact in X_T, then A is compact in X. ◇

Proof.

(i) Let $x \in \overline{A}^T$, then there exists $(x_p)_p \subset A$ such that $\lim_{p\to+\infty} \|x_p - x\|_T = 0$. Hence, $\lim_{p\to+\infty} \|x_p - x\| = 0$, and $\lim_{p\to+\infty} \|Tx_p - Tx\| = 0$, which implies that $x \in \overline{A}$ and $Tx \in \overline{T(A)}$.

(ii) Let $(x_n)_n \subset \overline{A}^T$. Then, there exists $x_{\rho(n)} \subset \overline{A}^T$ such that $x_{\rho(n)} \to x$ in X_T (when $n \to +\infty$). Hence, $x_{\rho(n)}$ converges to x in X. Q.E.D.

Remark 2.12.1. Notice that $\mathcal{M}_{X_T} \subset \mathcal{M}_X$. Indeed, let $A \in \mathcal{M}_{X_T}$. Then, there exists $M \geq 0$ such that, for all $x \in A$, $\|x\|_{X_T} = \|x\| + \|Tx\| \leq M$. Hence, $\|x\| \leq M$. ◇

Lemma 2.12.2. *Let X be a Banach space and $T \in \mathcal{C}(X)$. For A in $\mathcal{M}_X$ (resp. A in $\mathcal{M}_{X_T}$). Then,*

(i) $X^* + X^* \circ T \subset (X_T)^*$.
(ii) If $x \in \overline{A}^{\omega T}$, then $x \in \overline{A}^\omega$ and $Tx \in \overline{T(A)}^\omega$.
(iii) Moreover, if we suppose that $X^* + X^* \circ T$ is dense in $(X_T)^*$, then for any $A \in \mathcal{M}_{X_T}$, we have $A,\ T(A) \in \mathcal{W}_X \Longrightarrow A \in \mathcal{W}_{X_T}$.
(iv) If A is weakly compact in X_T, then A is weakly compact in X. ◇

Proof.

(*i*) Let $f_1, f_2 \in X^*$ and $f = f_1 + f_2 \circ T$. Then, f is linear on $\mathcal{D}(T)$ and, for all $x \in \mathcal{D}(T)$, the estimate

$$\begin{aligned}|f(x)| &\leq |f_1(x)| + |f_2 \circ Tx| \\ &\leq \|f_1\|\|x\| + \|f_2\|\|Tx\| \\ &\leq \max(\|f_1\|, \|f_2\|)\|x\|_T\end{aligned}$$

leads to $f \in (X_T)^*$.

(*ii*) For $x \in \overline{A}^{\omega T}$, there exists a net $(x_\alpha)_\alpha \subset A$ such that, for all $f \in X_T^*$, $f(x_\alpha) \to f(x)$, when $\alpha \to +\infty$. By (*i*), for all $f \in X^*$, $f(x_\alpha) \to f(x)$ and $f \circ T(x_\alpha) \to f \circ Tx$, when $\alpha \to +\infty$. So, $x \in \overline{A}^{\omega}$ and $Tx \in \overline{T(A)}^{\omega}$.

(*iii*) Let $(x_n)_n \subset A$, then by (*ii*), $(x_n)_n \subset \overline{A}^{\omega}$ and $(Tx_n)_n \subset \overline{TA}^{\omega}$. Since $\overline{A}^{\omega}$ and $\overline{TA}^{\omega}$ are weakly compact in X, then there exists a subsequence $(x_{\rho(n)})_n$ such that $x_{\rho(n)} \rightharpoonup x$ in X and $Tx_{\rho(n)} \rightharpoonup y$ in X, when $n \to +\infty$. We claim that $x_{\rho(n)} \rightharpoonup x$ in X_T when $n \to +\infty$. To do this, let $f \in (X_T)^*$. Since $X^* + X^* \circ T$ is dense in $(X_T)^*$, then there exists $(f_m = f_{1,m} + f_{2,m} \circ T)_m \subset X^* + X^* \circ T$ such that $\|f_{1,m} + f_{2,m} \circ T - f\|_{(X_T)^*} \to 0$ when $m \to +\infty$. For all $m \in \mathbb{N}$, we have $f_{1,m}(x_{\rho(n)}) \to f_{1,m}(x)$ and $f_{2,m} \circ Tx_{\rho(n)} \to f_{2,m}(y)$, when $n \to +\infty$. Moreover,

$$\begin{aligned}|f(x_{\rho(n)}) - f(x)| &\leq |f(x_{\rho(n)}) - f_m(x_{\rho(n)})| + |f_m(x_{\rho(n)}) - f_m(x)| + |f_m(x) - f(x)| \\ &\leq \|f - f_m\|_{(X_T)^*}\|x_{\rho(n)}\|_T + |f_m(x_{\rho(n)}) \\ &\quad - f_m(x)| + \|f - f_m\|_{(X_T)^*}\|x\|_T.\end{aligned}$$

Since $x_{\rho(n)} \rightharpoonup x$ and $Tx_{\rho(n)} \rightharpoonup y$, then there exists $M > 0$ such that $\max(\|x_{\rho(n)}\|_T, \|x\|_T) \leq M$. Let $\varepsilon > 0$. There exists $m_0 \in \mathbb{N}$ such that, for $m \geq m_0$, we have $\|f - f_m\|_{(X_T)^*} \leq \frac{1}{3M}\varepsilon$. So, for $m \geq m_0$, we have $|f(x_{\rho(n)}) - f(x)| \leq \frac{2}{3}\varepsilon + |f_m(x_{\rho(n)}) - f_m(x)|$. Since $f_m(x_{\rho(n)}) \to f_m(x)$ when $n \to +\infty$, then there exists $n_0 \in \mathbb{N}$ such that, for $n \geq n_0$, $|f_m(x_{\rho(n)}) - f_m(x)| \leq \frac{\varepsilon}{3}$. Hence, for $m \geq m_0$ and $n \geq n_0$, we have $|f(x_{\rho(n)}) - f(x)| \leq \frac{2}{3}\varepsilon + \frac{1}{3}\varepsilon = \varepsilon$. Thus, $f(x_{\rho(n)}) \to f(x)$ when $n \to +\infty$, which proves our claim.

(*iv*) Let $(x_n)_n \subset \overline{A}^{\omega T}$. Then, there exists $x_{\rho(n)} \rightharpoonup x$ in X_T when $n \to +\infty$. Hence, for all $f \in (X_T)^*$, $f(x_{\rho(n)}) \to f(x)$. The result follows from (*i*). Q.E.D.

Remark 2.12.2. If X_T is reflexive, then $X^* + X^* \circ T$ is dense in $(X_T)^*$. Indeed, let $x^{**} \in (X_T)^{**}$ such that, for all $f_1^*, f_2^* \in X^*$, $x^{**}(f_1^* + f_2^* \circ T) = 0$. Since $x^{**} \in (X_T)^{**}$, then $x^{**} = Jx$ with $x \in X_T$, where J is the canonical injection between X and X^{**}. So, $x^{**}(f_1^* + f_2^* \circ T) = 0$ if, and only if, $f_1^*(x) + f_2^* \circ Tx = 0$. From [302, Theorem 3.1], there exists $f_1^*, f_2^* \in X^*$ such that $f_1^*(x) = \|x\|$ and $f_2^* \circ Tx = \|Tx\|$. This leads to $\|x\|_T = 0$ and so, $x = 0$. By applying [302, Corollary 3.2], we get $X^* + X^* \circ T$ is dense in $(X_T)^*$. ◇

2.12.1 Graph Measure of Noncompactness

Let X be a Banach space, let T be in $\mathcal{C}(X)$, and let ν be a measure of noncompactness in X. We define $\mathcal{H}_\nu$ the set of all $T \in \mathcal{C}(X)$ satisfying: for all A_n be bounded and closed in X_T such that $A_{n+1} \subset A_n$, if $\lim\limits_{n\to+\infty} \nu(A_n) = 0$ and $\lim\limits_{n\to+\infty} \nu(T(A_n)) = 0$. Then, $\bigcap_{n=0}^{+\infty} A_n \neq \emptyset$. The set $\mathcal{H}_\nu$ is nonempty. Indeed, let T be injective in $\mathcal{C}(X)$ such that $T(F)$ is closed in X for all F being closed in X_T. Let A_n be bounded and closed in X_T satisfying $A_{n+1} \subset A_n$. Suppose that $\lim\limits_{n\to+\infty} \nu(A_n) = 0$ and $\lim\limits_{n\to+\infty} \nu(T(A_n)) = 0$. Since $T(A_n)$ is bounded and closed in X, then $\bigcap_{n=0}^{+\infty} T(A_n) \neq \emptyset$. The fact that T is injective, then $\bigcap_{n=0}^{+\infty} T(A_n) = T(\bigcap_{n=0}^{+\infty} A_n)$. Hence, $\bigcap_{n=0}^{+\infty} A_n \neq \emptyset$. If $T \in \mathcal{L}(X)$, then $\|.\|_T$ and $\|.\|$ are equivalent and therefore, $T \in H_\nu$. Indeed, let A_n be bounded and closed in X_T. Then, A_n is bounded and closed in X. Now, if $\lim\limits_{n\to+\infty} \nu(A_n) = 0$ and $\lim\limits_{n\to+\infty} \nu(T(A_n)) = 0$, then $\bigcap_{n=0}^{+\infty} A_n \neq \emptyset$. Now, we are ready to state and prove the following.

Theorem 2.12.1. *Let X be a Banach space, let ν be a measure of noncompactness in X and let T be in $\mathcal{H}_\nu$. We define a function $\nu_T : \mathcal{M}_{X_T} \longrightarrow \mathbb{R}_+$ by $\nu_T(A) = \nu(A)+\nu(T(A))$, $\forall A \in \mathcal{M}_{X_T}$. Assume that ν is homogeneous. Then, ν_T is a measure of noncompactness in X_T called a graph measure of noncompactness associated with ν and T.* ◇

Proof. First, since ν is homogeneous, then $\{0\} \in \mathrm{Ker}(\nu)$. Hence, $\{0\} \in \mathrm{Ker}(\nu_T)$ and therefore, $\mathrm{Ker}(\nu_T) \neq \emptyset$. For $A \in \mathcal{M}_{X_T}$, there exists $k > 0$ such that, for all $x \in A$, $\|x\| + \|Tx\| \leq k$, and $T(A)$ is a bounded set in X. So, ν_T is well defined. Moreover, for all $A, B \in \mathcal{M}_{X_T}$, we have

(*i*) If $\nu_T(A) = 0$, then $\overline{A}$ and $\overline{T(A)}$ are compact in X. Let $(x_n)_n$ be a sequence in $\overline{A}^T$. By using Lemma 2.12.1 (*i*), $(x_n)_n \subset \overline{A}$ and $(Tx_n)_n \subset \overline{T(A)}$. Hence, there exists a subsequence $(x_{\rho(n)})_n$ such that $x_{\rho(n)} \to x$ and $Tx_{\rho(n)} \to y$ (when $n \to +\infty$). Since $T \in \mathcal{C}(X)$, then $Tx = y$. Hence, $\|x_{\rho(n)} - x\|_T = \|x_{\rho(n)} - x\| + \|Tx_{\rho(n)} - Tx\| \to 0$ (when $n \to +\infty$), which implies that $\overline{A}^T$ is compact in X_T.

(*ii*) If $A \subset B$, then $T(A) \subset T(B)$. Hence, $\nu(A) \leq \nu(B)$ and $\nu(T(A)) \leq \nu(T(B))$. Thus, $\nu_T(A) \leq \nu_T(B)$.

(*iii*) Since $\overline{\mathrm{conv}(A)}^T \subset \overline{\mathrm{conv}(A)}$, then $\nu(\overline{\mathrm{conv}(A)}^T) \leq \nu(\overline{\mathrm{conv}(A)}) = \nu(A)$. Moreover, $A \subset \overline{\mathrm{conv}(A)}^T$. Then, $\nu(A) \leq \nu(\overline{\mathrm{conv}(A)}^T)$ and therefore, $\nu(A) = \nu(\overline{\mathrm{conv}(A)}^T)$. Since $T(A) \subset T(\overline{\mathrm{conv}(A)}^T)$, then $\nu(T(A)) \leq \nu(T(\overline{\mathrm{conv}(A)}^T))$. Besides, by using Lemma 2.12.1 (*i*), $T(\overline{\mathrm{conv}(A)}^T) \subset \overline{\mathrm{conv}(T(A))}$ and therefore, $\nu(T(\overline{\mathrm{conv}(A)}^T)) \leq \nu(T(A))$. Thus, $\nu(T(A)) = \nu(T(\overline{\mathrm{conv}(A)}^T))$. Hence,

$$\begin{aligned}\nu_T(\overline{\text{conv}(A)}^T) &= \nu(\overline{\text{conv}(A)}^T) + \nu(T(\overline{\text{conv}(A)}^T))\\ &= \nu(A) + \nu(T(A))\\ &= \nu_T(A).\end{aligned}$$

(iv) Let $\lambda \in [0, 1]$. Then, we have

$$\begin{aligned}\nu_T(\lambda A + (1-\lambda)B) &= \nu(\lambda A + (1-\lambda)B) + \nu(\lambda T(A) + (1-\lambda)T(B))\\ &\leq \lambda\nu(A) + (1-\lambda)\nu(B) + \lambda\nu(T(A)) + (1-\lambda)\nu(T(B))\\ &\leq \lambda\nu_T(A) + (1-\lambda)\nu_T(B).\end{aligned}$$

(v) Let A_n be a bounded closed subset in X_T such that $A_{n+1} \subset A_n$ for $n = 1, 2, \ldots$ and $\lim\limits_{n\to+\infty} \nu_T(A_n) = 0$. Since $T \in \mathcal{H}_\nu$, then $\bigcap_{n=0}^{+\infty} A_n \neq \emptyset$. Q.E.D

2.12.2 Graph Measure of Weak Noncompactness

Let X be a Banach space, let T be in $\mathcal{C}(X)$, and let μ be a measure of weak noncompactness in X. We define $\mathcal{H}_\mu$ as the set of all $T \in \mathcal{C}(X)$ satisfying: all A_n are bounded and weakly closed in X_T such that $A_{n+1} \subset A_n$, if $\lim\limits_{n\to+\infty} \mu(A_n) = 0$ and $\lim\limits_{n\to+\infty} \mu(T(A_n)) = 0$. Then, $\bigcap_{n=0}^{+\infty} A_n \neq \emptyset$. The set $\mathcal{H}_\mu$ is nonempty. Indeed, let T be injective in $\mathcal{C}(X)$ such that $T(F)$ is weakly closed in X for all F being weakly closed in X_T. Let A_n be bounded and weakly closed in X_T such that $A_{n+1} \subset A_n$. Suppose that $\lim\limits_{n\to+\infty} \mu(A_n) = 0$ and $\lim\limits_{n\to+\infty} \mu(T(A_n)) = 0$. Since $T(A_n)$ is bounded and weakly closed in X, then $\bigcap_{n=0}^{+\infty} T(A_n) \neq \emptyset$. Moreover, since T is injective, then $\bigcap_{n=0}^{+\infty} T(A_n) = T(\bigcap_{n=0}^{+\infty} A_n)$ and therefore, $\bigcap_{n=0}^{+\infty} A_n \neq \emptyset$. Now, we are ready to state and prove the following.

Theorem 2.12.2. *Let X be a Banach space, let μ be a measure of weak noncompactness in X, and let $T \in \mathcal{H}_\mu$. We define a function $\mu_T : \mathcal{M}_{X_T} \longrightarrow \mathbb{R}_+$ by $\mu_T(A) = \mu(A) + \mu(T(A))$, $\forall A \in \mathcal{M}_{X_T}$. Assume that μ is homogeneous. If we suppose that $X^* + X^* \circ T$ is dense in $(X_T)^*$, then μ_T is a measure of weak noncompactness in X_T, called a graph measure of weak noncompactness associated with μ and T.* ◇

Proof. For $A \in \mathcal{M}_{X_T}$, there exists $k > 0$ such that, for all $x \in A$, $\|x\| + \|Tx\| \leq k$, $T(A)$ is a bounded set in X. So, μ_T is well defined. Moreover, for all $A, B \in \mathcal{M}_{X_T}$, we have

(i) If $\mu_T(A) = 0$, then $\overline{A}^\omega$ and $\overline{T(A)}^\omega$ are weakly compact in X. By applying Lemma 2.12.2 (iii), we deduce that A is relatively weakly compact in X_T.

(ii) If $A \subset B$, then $T(A) \subset T(B)$. Therefore, $\mu(A) \leq \mu(B)$ and $\mu(T(A)) \leq \mu(T(B))$. Thus, $\mu_T(A) \leq \mu_T(B)$.

(*iii*) Arguing as in the proof of Theorem 2.12.1, we have

$$\begin{aligned}\mu_T(\overline{\text{conv}(A)}^T) &= \mu(\overline{\text{conv}(A)}^T) + \mu(T(\overline{\text{conv}(A))}^T)\\ &= \mu(A) + \mu(T(A))\\ &= \mu_T(A).\end{aligned}$$

(*iv*) Let $\lambda \in [0, 1]$. Then, we have

$$\begin{aligned}\mu_T(\lambda A + (1-\lambda)B) &= \mu(\lambda A + (1-\lambda)B) + \mu(\lambda T(A) + (1-\lambda)T(B))\\ &\leq \lambda\mu(A) + (1-\lambda)\mu(B) + \lambda\mu(T(A)) + (1-\lambda)\mu(T(B))\\ &\leq \lambda\mu_T(A) + (1-\lambda)\mu_T(B).\end{aligned}$$

(*v*) Let A_n be a bounded and weakly closed subset in X_T such that $A_{n+1} \subset A_n$ for $n = 1, 2, \ldots$ and $\lim_{n\to+\infty} \mu_T(A_n) = 0$. Then, $\lim_{n\to+\infty} \mu(A_n) = 0$ and $\lim_{n\to+\infty} \mu(T(A_n)) = 0$ and therefore $\bigcap_{n=0}^{+\infty} A_n \neq \emptyset$, because $T \in \mathcal{H}_\mu$. Q.E.D.

Remark 2.12.3. Notice that μ_T is not necessarily a measure with the maximum property. So, μ_T is not regular. However, if ν is sublinear, then μ_T is also sublinear. ◇

2.12.3 Seminorm Related to Upper Semi-Fredholm Perturbations

Let X be a Banach space. Given $\gamma(.)$ the Kuratowski measure of noncompactness, we define, for $T \in \mathcal{L}(X)$, the nonnegative quantities associated with T by

$$\delta(T) = \inf\left\{\frac{\gamma(T(A))}{\gamma(A)} \text{ such that } A \in \mathcal{M}_X \text{ and } \gamma(A) > 0\right\}. \tag{2.12.1}$$

Definition 2.12.1. For $T \in \mathcal{L}(X)$, we define the nonnegative quantity:

$$\varphi(T) = \sup\{\delta(T+S) \text{ such that } \delta(S) = 0\}. \tag{2.12.2}$$

◇

In what follows, we give some fundamental properties satisfied by $\varphi(.)$ already given in [2].

Proposition 2.12.1.

(*i*) $\varphi(T) = 0$ *if, and only if,* $T \in \mathcal{F}_+^b(X)$.

(ii) $\varphi(T + S) = \varphi(S)$, *for all* $T \in \mathcal{F}_+^b(X)$.
(iii) $\varphi(\lambda T) = |\lambda|\varphi(T)$.
(iv) $\delta(T) \leq \varphi(T) \leq \gamma(T)$.
(v) $\varphi(T) - \gamma(S) \leq \varphi(T + S) \leq \varphi(T) + \gamma(S)$.
(vi) $\varphi(T) \leq \|T\|_{\mathcal{F}_+^b(X)} \leq \|T\|_{\mathcal{K}(X)}$, *where* $\|T\|_{\mathcal{K}(X)} = \inf\{\|T - K\|$ *such that* $K \in \mathcal{K}(X)\}$ *and* $\|T\|_{\mathcal{F}_+^b(X)} = \inf\{\|T - K\|$ *such that* $K \in \mathcal{F}_+^b(X)\}$.
(vii) $\varphi(ST) \geq \varphi(T)\delta(S)$, *for all* $S \in \mathcal{L}(X)$.
(viii) *If* $\varphi(T) = 0$, *then, for all* $S \in \mathcal{L}(X)$, $\varphi(ST) = \varphi(TS) = 0$. ◇

Proof.

(*i*) It follows immediately from the fact that $\delta(T) > 0$ if, and only if, $T \in \Phi_+^b(X)$.
(*ii*) Due to the fact that for $T \in \mathcal{F}_+^b(X)$, $\delta(S) = 0$ if, and only if, $\delta(T + S) = 0$.
(*iii*)–(*v*) Follow from the definition of $\varphi(.)$ and the fact that $\delta(T) \leq \gamma(T)$.
(*vi*) Deduction of (*ii*) and (*iv*).
(*vii*) Let S, $S_1 \in \mathcal{L}(X)$. According to Theorem 2.2.14, we have $\delta(S_1) = 0 \Longrightarrow \delta(SS_1) = 0$. Moreover, we have $\delta(ST + SS_1) = \delta(S(T + S_1)) \geq \delta(T + S_1)\delta(S)$. Hence,

$$\sup_{\delta(SS_1)=0} \delta(ST + SS_1) \geq \sup_{\delta(S_1)=0} \delta(T + S_1)\delta(S)$$

and therefore, $\varphi(ST) \geq \varphi(T)\delta(S)$.
(*viii*) The fact that $\mathcal{F}_+^b(X)$ is a two-sided ideal of $\mathcal{L}(X)$ together with (*i*) gives immediately the assertion (*viii*). Q.E.D.

Remark 2.12.4. The assertion (*viii*) of Proposition 2.12.1 is equivalent to say that $\mathcal{F}_+^b(X)$ is a two-sided ideal of $\mathcal{L}(X)$. Moreover by (*v*) of Proposition 2.12.1, $|\varphi(T) - \varphi(S)| \leq \gamma(T - S) \leq \|T - S\|$. This implies that the measure $\varphi(.)$ is continuous. Hence, it follows from (*i*) of Proposition 2.12.1, that $\mathcal{F}_+^b(X)$ is closed. ◇

2.13 Quadratic Forms

Let H be a Hilbert space. One consequence of the Riesz lemma is that, there is a one-to-one correspondence between bounded quadratic forms and bounded operators, that is, any sesquilinear map $q : H \times H \longrightarrow \mathbb{C}$ which satisfies $|q(\varphi, \psi)| \leq M\|\varphi\|\|\psi\|$ is of the form $q(\varphi, \psi) = \langle\varphi, A\psi\rangle$ for some bounded operator A. As one might expect, the situation is more complicated if one removes the boundedness restriction. It is the relationship between unbounded forms and unbounded operators.

Definition 2.13.1. Let H be a Hilbert space. A quadratic form is a map $q : \mathcal{Q}(q) \times \mathcal{Q}(q) \longrightarrow \mathbb{C}$, where $\mathcal{Q}(q)$ is a dense linear subset of H called the form domain, such that $q(.,\psi)$ is conjugate linear and $q(\varphi,.)$ is linear for $\varphi,\psi \in \mathcal{Q}(q)$. If $q(\varphi,\psi) = \overline{q(\psi,\varphi)}$, then we say that q is symmetric. If $q(\varphi,\varphi) \geq 0$ for all $\varphi \in \mathcal{Q}(q)$, then q is called positive. Moreover, if $q(\varphi,\varphi) \geq -M\|\varphi\|^2$ for some M, then we say that q is semi-bounded. ◇

It is clear that if q is semi-bounded, then it is automatically symmetric, provided that H is complex.

Definition 2.13.2. Let q be a semi-bounded quadratic form. q is called closed, if $\mathcal{Q}(q)$ is complete under the norm $\|\psi\|_{+1} = \sqrt{q(\psi,\psi) + (M+1)\|\psi\|^2}$. If q is closed and if $D \subset \mathcal{Q}(q)$ is dense in $\mathcal{Q}(q)$ in the $\|.\|_{+1}$ norm, then D is called a form core for q. ◇

Definition 2.13.3. Let H be a Hilbert space. An operator $B \in \mathcal{L}(H)$ is called positive if $\langle B\varphi,\varphi\rangle \geq 0$ for all $\varphi \in H$. We write $B \geq 0$ if B is positive and, $B \leq A$ if $A - B \geq 0$. ◇

Theorem 2.13.1 ([292, Theorem VIII.15]). *If q is a closed semi-bounded quadratic form, then q is the quadratic form of a unique self-adjoint operator.* ◇

Let us recall the Friedrichs extension theorem.

Theorem 2.13.2 ([292, Theorem X.23]). *Let A be a positive symmetric operator and let $q(\varphi,\psi) = \langle\varphi, A\psi\rangle$ for all $\varphi,\psi \in \mathcal{D}(A)$. Then, q is a closable quadratic form, and its closure $\hat{q}$ is the quadratic form of a unique self-adjoint operator $\hat{A}$. $\hat{A}$ represents a positive extension of A, and the lower bound of its spectrum is the lower bound of q. Moreover, $\hat{A}$ is the only self-adjoint extension of A whose domain is contained in the form domain of $\hat{q}$.* ◇

Proof. Let $\langle\varphi,\psi\rangle_{+1} := q(\varphi,\psi) + \langle\varphi,\psi\rangle$. Then, $\langle .,.\rangle_{+1}$ is an inner product on $\mathcal{D}(A)$. Hence, we can complete $\mathcal{D}(A)$ under $\langle .,.\rangle_{+1}$ in order to obtain a Hilbert space H_{+1}. Clearly, q can be extended to a closed form $\hat{q}$ on H_{+1}. However, in order to show that $\hat{q}$ is a closed form on H, we must demonstrate that H_{+1} is a subset of H. For this purpose, let $i : \mathcal{D}(A) \longrightarrow H$ be the identity map. Since $\|\varphi\| \leq \|\varphi\|_{+1}$, i is bounded and then, it can be extended to a bounded map $\hat{i} : H_{+1} \longrightarrow H$ of a norm less than or equal to one. In order to verify that $H_{+1} \subset H$, we have to show that $\hat{i}$ is injective. Suppose that $\hat{i}(\varphi) = 0$. Then, there exists $(\varphi_n)_n \in \mathcal{D}(A)$, so that $\|\varphi - \varphi_n\|_{+1} \to 0$ and $\|\hat{i}(\varphi_n)\| = \|\varphi_n\| \to 0$. Hence, $\|\varphi\|_{+1} = \lim_{n,m\to\infty} \langle\varphi_n,\varphi_m\rangle_{+1} = \lim_{n\to\infty}\lim_{m\to\infty} (\langle\varphi_m, A\varphi_n\rangle + \langle\varphi_m,\varphi_n\rangle) = 0$ since $\varphi_n \in \mathcal{D}(A)$ and $\|\varphi_m\| \to 0$, then $\hat{i}$ is injective. Notice that the proof that $\hat{i}$ is well-defined uses only the positiveness of q, whereas the proof that $\hat{i}$ is one-to-one uses the hypothesis that q arises from an operator. Since $\hat{q}$ is closed and symmetric, by using Theorem 2.13.1, then there is a unique self-adjoint operator $\hat{A}$, so that $\mathcal{D}(\hat{A}) \subset \mathcal{Q}(\hat{q})$ and $\hat{q}(\varphi,\psi) = \langle\varphi,\hat{A}\psi\rangle$, if $\varphi \in \mathcal{Q}(\hat{q})$ and $\psi \in \mathcal{D}(\hat{A})$. Now, let us suppose that $\varphi \in \mathcal{D}(A)$. Then, by using the continuity of $\hat{q}$, we deduce that $\langle A\varphi,\psi\rangle = \hat{q}(\varphi,\psi) = \langle\varphi,\hat{A}\psi\rangle$. Since this holds for all $\psi \in \mathcal{D}(\hat{A})$, we conclude that $\varphi \in \mathcal{D}(\hat{A}^*) = \mathcal{D}(\hat{A})$ and $\hat{A}^*\varphi = \hat{A}\varphi = A\varphi$. Thus, $\hat{A}$ is the extension of A. The same proof shows that, if A_e is any symmetric extension of A with $\mathcal{D}(A_e) \subset \mathcal{Q}(\hat{q})$, then $\hat{A}$ is the extension of A_e. Hence, if A_e is self-adjoint, then $\hat{A} = A_e$. The easy proof of the statement about the spectrum of A is left to the reader. Q.E.D.

Theorem 2.13.3 ([292, Theorem VIII]). *Let T be a symmetric operator on a Hilbert space H. Then, the following are equivalent.*

(i) T is essentially self-adjoint.
(ii) $N(T^* \pm i) = \{0\}$, where $i^2 = -1$.
(iii) $R(T \pm i)$ are dense. ◇

Theorem 2.13.4 ([292, Theorem X.37]). *Let N be a self-adjoint operator with $N \geq I$. Let A be a symmetric operator with a domain D which is a core for N. Suppose that*

(i) For some c and all $\varphi \in D$, we have $\|A\varphi\| \leq c\|N\varphi\|$.
(ii) For some d and all $\varphi \in D$, we have $|\langle A\varphi, N\varphi\rangle - \langle N\varphi, A\varphi\rangle| \leq d\,\|N^{\frac{1}{2}}\varphi\|^2$.

Then, A is essentially self-adjoint on D, and its closure is essentially self-adjoint on any other core for N. ◇

2.14 Schur Test

Let X and Y be two measurable spaces (such as $\mathbb{R}^n$). Let T be an integral operator with a nonnegative Schwartz kernel $\kappa(x, y)$, $x \in X$, $y \in Y$ and $Tf(x) = \int_Y \kappa(x, y) f(y) dy$. Suppose that, there exist two functions $p(x) > 0$ and $q(x) > 0$, and two numbers $\alpha, \beta > 0$ such that

$$\int_Y \kappa(x, y) q(y) dy \leq \alpha p(x) \tag{2.14.1}$$

for almost all x, and

$$\int_X \kappa(x, y) p(x) dx \leq \beta q(y) \tag{2.14.2}$$

for almost all y. Such functions, namely $p(x)$ and $q(x)$ are called the Schur test functions. By [145], T can be extended to a continuous operator $T : L_2 \longrightarrow L_2$ with the operator norm, and we have $\|T\| \leq \sqrt{\alpha\beta}$. In fact, by using the Cauchy-Schwartz inequality, as well as the inequality (2.14.1), we get

$$\begin{aligned}|Tf(x)|^2 &= \left|\int_Y \kappa(x, y) f(y) dy\right|^2\\ &\leq \left(\int_Y \kappa(x, y) q(y) dy\right)\left(\int_Y \frac{\kappa(x, y)(f(y))^2}{q(y)} dy\right)\end{aligned}$$

$$\leq \alpha p(x) \int_Y \frac{\kappa(x,y)(f(y))^2}{q(y)} dy.$$

By integrating the above relation in x, by using Fubini's theorem, and by applying the inequality (2.14.2), we get

$$\begin{aligned} \|Tf\|^2 &\leq \alpha \int_Y \left(\int_X p(x)\kappa(x,y)dx \right) \frac{(f(y))^2}{q(y)} dy \\ &\leq \alpha\beta \int_Y (f(y))^2 dy \\ &= \alpha\beta \|f\|^2. \end{aligned}$$

We deduce that $\|Tf\| \leq \sqrt{\alpha\beta}\|f\|$ for any $f \in L_2(Y)$.

Theorem 2.14.1 ([305]). *Suppose that T satisfies (2.14.1) and (2.14.2), for $p(x) = q(x) = 1$. Then*

$$\|T\| \leq \left(\sup_{x\in X} \int_Y |\kappa(x,y)|dy \right)^{\frac{1}{2}} \left(\sup_{y\in Y} \int_X |\kappa(x,y)|dx \right)^{\frac{1}{2}}. \qquad \diamond$$

2.15 Generalities about graphs

2.15.1 Unoriented Graph

Let $\mathcal{V}$ be a countable set and $\mathcal{E} : \mathcal{V} \times \mathcal{V} \longrightarrow [0, +\infty]$. We assume that $\mathcal{E}(x,y) = \mathcal{E}(y,x)$, for all $x, y \in \mathcal{V}$. We say that $G = (\mathcal{E}, \mathcal{V})$ is an unoriented weighted graph with vertices $\mathcal{V}$ and weights $\mathcal{E}$. We say that there is a loop in $x \in \mathcal{V}$, if $\mathcal{E}(x,x) > 0$. A graph G is simple, if it has no loops and $\mathcal{E}$ has its values in $\{0, 1\}$. We shall say that $\mathcal{E}$ is bounded from below, if $\inf\{\mathcal{E}(x,y)$ such that $x, y \in \mathcal{V}$ and $\mathcal{E}(x,y) \neq 0\} > 0$. The vertices $x, y \in \mathcal{V}$, with $\mathcal{E}(x,y) > 0$, are called neighbors and, we denote this relationship by $x \sim y$. We say that $G := (\mathcal{E}, \mathcal{V})$ is connected if, for any $x, y \in \mathcal{V}$, there exists a $x - y$ path i.e., there is a finite sequence $x_0, x_1, \ldots, x_n \in \mathcal{V}$ such that $x = x_0$, $y = x_n$ and $x_j \sim x_{j+1}$, for all $0 \leq j \leq n-1$. The degree of $x \in \mathcal{V}$ is, by definition,

$$d_G(x) := |\mathcal{N}_G(x)|, \tag{2.15.1}$$

where $\mathcal{N}_G(x) := \{y \in \mathcal{E}$ such that $x \sim y\}$, and $|\mathcal{N}_G(x)|$ is the cardinal of $\mathcal{N}_G(x)$. We denote by $\eta(\mathcal{A}) := \dim N(\mathcal{A}^* - i) \in \mathbb{N} \bigcup \{+\infty\}$ the deficiency indices of the symmetric operator $\mathcal{A}$. A graph G is locally finite, if $d_G(x)$ is finite for all $x \in \mathcal{V}$. Let $G = (\mathcal{E}, \mathcal{V})$ be a connected, locally finite graph. We endow $\mathcal{V}$ with the metric

$\rho_{\mathcal{V}}$ defined by

$$\rho_{\mathcal{V}}(x,y) := \inf\{n \in \mathbb{N} \text{ such that there exists an } x - y \text{ path of length } n\}. \tag{2.15.2}$$

In the sequel, we assume that: all graphs are locally finite, connected with no loops.

2.15.2 *Tree*

In this section, we define a certain family of trees. It is convenient to choose a root in the tree. Due to its structure, one can take any point of $\mathcal{V}$. We denote it by x. We define the family of spheres $(S_n)_{n\in\mathbb{N}}$ by $S_0 = \{x\}$ and $S_{n+1} = \mathcal{N}_G(S_n) \setminus S_{n-1}$. Note that, if $n \in \mathbb{N}$, $x \in S_n$ and $y \sim x$, then, $y \in S_{n-1} \bigcup S_{n+1}$. We write $x > y$, if $y \in S_{n-1}$ and we also write $x < y$, if $y \in S_{n+1}$. The offspring of an element x is given by $\mathrm{off}(x) := |\{y : \ y \sim x, \ \ y > x\}|$.

Definition 2.15.1. A simple tree $G = (\mathcal{E}, \mathcal{V})$ with an offspring sequence $(b_n)_n$ is a simple tree with a root, such that $b_n = \mathrm{off}(x)$, for all $x \in S_n$ and $n \in \mathbb{N}$. ◇

Example 2.15.1.

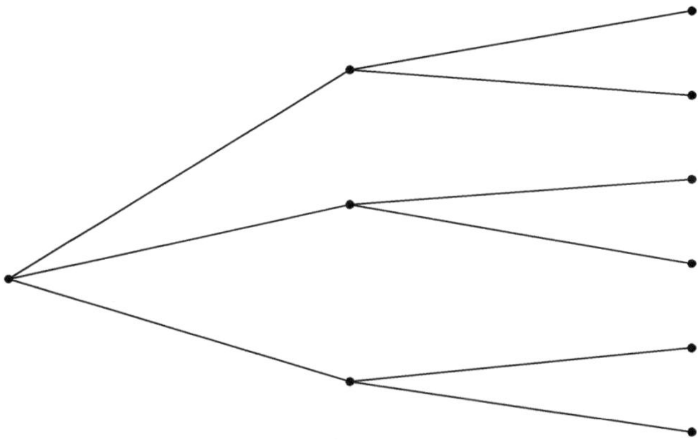

Example of tree with $b_0 = 3$ and $b_1 = 2$

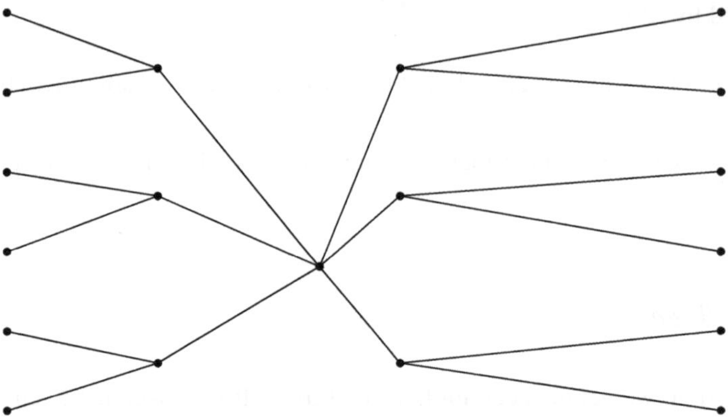

Example of tree with $b_0 = 6$ and $b_1 = 2$

2.15.3 Bipartite Graph

Definition 2.15.2. A simple graph $G = (\mathcal{E}, \mathcal{V})$ is called bipartite, if its vertex set can be partitioned into two disjoint subsets $\mathcal{V} = \mathcal{V}_1 \bigcup \mathcal{V}_2$, such that every edge has the form $e = (a, b)$, where $a \in \mathcal{V}_1$ and $b \in \mathcal{V}_2$. ◇

Example 2.15.2.

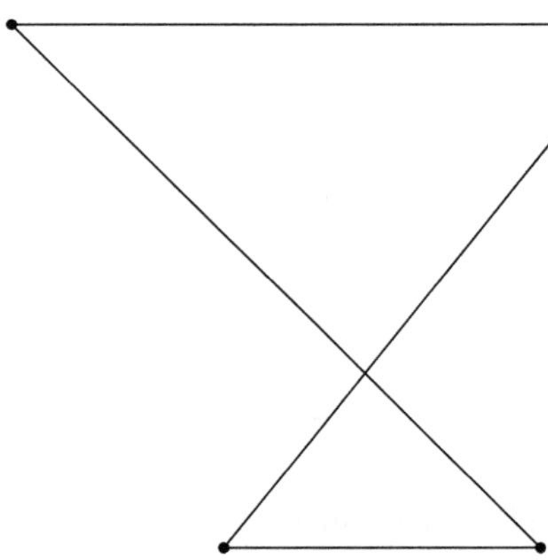

G is bipartite

Example 2.15.3.

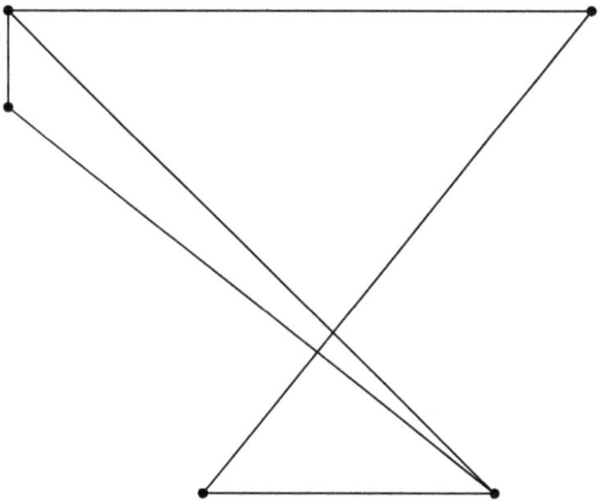

G is not bipartite

Theorem 2.15.1 ([190]). *A graph $G = (\mathcal{E}, \mathcal{V})$ is bipartite if, and only if, it does not have an odd length cycle.* ◇

Proof. Let us fix a vertex $v \in \mathcal{V}$. Let us define two sets of vertices

$$A = \{w \in \mathcal{V} \text{ such that } \exists \text{ odd length path from } v \text{ to } w\}$$

and

$$B = \{w \in \mathcal{V} \text{ such that } \exists \text{ even length path from } v \text{ to } w\}.$$

These sets provide a bipartition. If there is an odd length cycle, a vertex will be present in the path set. Q.E.D.

Definition 2.15.3. A complete bipartite graph $K_{\mathrm{m,n}}$ is a bipartite graph for which each vertex set is adjacent to another one. ◇

Example 2.15.4.

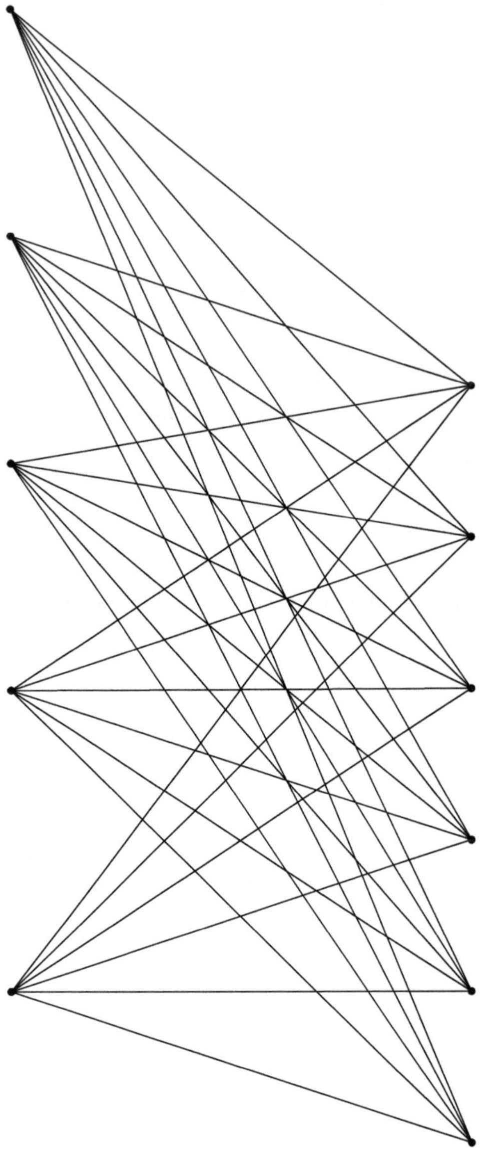

A complete graph $K_{5,6}$

Example 2.15.5.

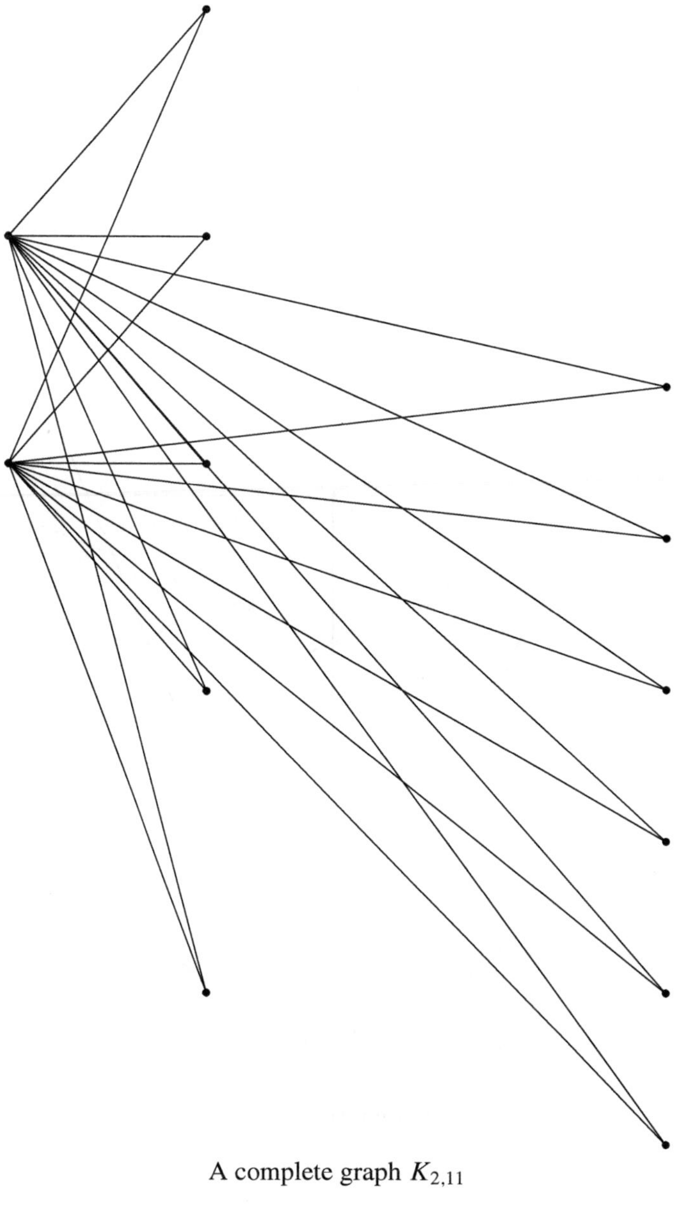

A complete graph $K_{2,11}$

2.15.4 Subgraph

Definition 2.15.4. A subgraph $G' = (\mathcal{E}', \mathcal{V}')$ of a graph $G = (\mathcal{E}, \mathcal{V})$ is a graph whose vertex set is a subset of that of G, and whose adjacency relation is a subset of that of G restricted to this subset, i.e., $\mathcal{V}' \subset \mathcal{V}$ and $\mathcal{E}' = \mathcal{E}|_{\mathcal{V}' \times \mathcal{V}'}$. ◇

Example 2.15.6.

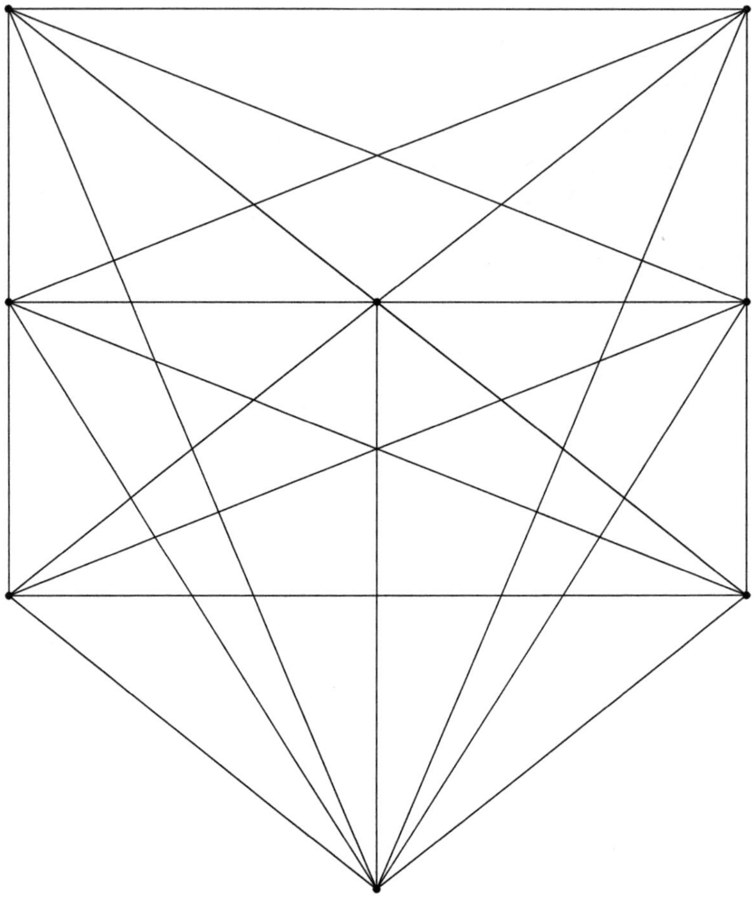

A graph $G = (\mathcal{E}, \mathcal{V})$

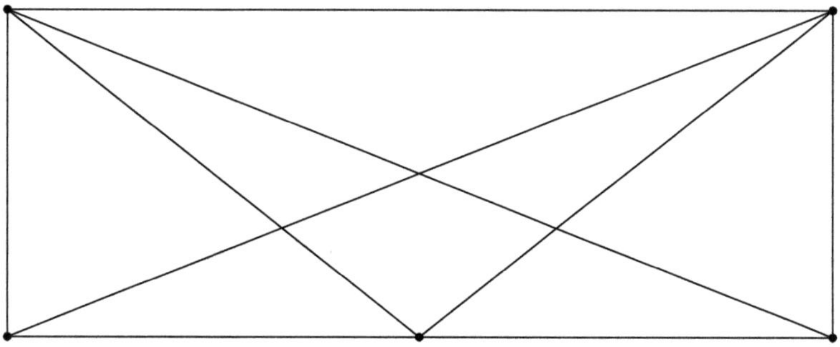

A subgraph $G' = (\mathcal{E}', \mathcal{V}')$ of G

Chapter 3
Fredholm Operators and Riesz Theory for Polynomially Compact Operators

In this chapter, we study the spectral theory of polynomially compact operator on a Banach space. First, we give some preliminary results concerning Fredholm operators which are developed in order to analyze the spectrum of polynomially compact operator on a Banach space. Second, we give some results on multiplicities and localization of the eigenvalues of polynomially compact operators. Finally, we present some results dealing with the polynomially Riesz operators.

3.1 Riesz Theory

3.1.1 Some Results on Polynomially Compact Operators

Let X be a Banach space. We denote by $\mathcal{P}(X)$ the set defined by

$$\mathcal{P}(X) = \Big\{A \in \mathcal{L}(X) \text{ such that there exists a nonzero complex polynomial } P(z) := \sum_{r=0}^{p} a_r z^r \text{ satisfying } P(1) \neq 0,\ P(1) - a_0 \neq 0 \text{ and } P(A) \in \mathcal{K}(X)\Big\}.$$

Definition 3.1.1. We say that an operator $A \in \mathcal{L}(X)$ is polynomially compact if there is a nonzero complex polynomial $P(z)$ such that the operator $P(A)$ is compact. ◇

If $\mathcal{PK}(X)$ denotes the set of polynomially compact operators on X, then $\mathcal{P}(X) \subset \mathcal{PK}(X)$.

Remark 3.1.1. Notice that in the definition of the set $\mathcal{P}(X)$, we are only concerned with polynomials $P(.)$ satisfying the conditions $P(1) \neq 0$ and $P(1) - a_0 \neq 0$. It seems that these hypotheses cannot be dropped. Indeed, let $A = I$ (the identity

A. Jeribi, *Spectral Theory and Applications of Linear Operators and Block Operator Matrices*, DOI 10.1007/978-3-319-17566-9_3

operator) and take $P(z) = 1 - z$. Clearly $P(1) = 0$ and $P(A) \in \mathcal{K}(X)$ ($A \in \mathcal{PK}(X)$), but the results discussed below are false for $F = I - A = 0$ because it is not a Fredholm operator. ◇

We start our investigation with the following lemmas, which constitute a preparation for the proofs of the results of this section.

Lemma 3.1.1. *Let $A \in \mathcal{P}(X)$ and set $F = I - A$. Then,*

(i) $dim[N(F)] < \infty$.
(ii) $R(F)$ *is closed.*
(iii) $codim[R(F)] < \infty$. ◇

Proof.

(*i*) Since $A \in \mathcal{P}(X)$, there exists $P \neq 0$ such that $P(A) \in \mathcal{K}(X)$ with $P(z) = \sum_{r=0}^{p} a_r z^r$, where $a_r \in \mathbb{C}$, $r = 0, 1, \ldots, p$. Let $x \in N(F)$, then $Ax = x$ and therefore $x \in N(I - P(1)^{-1}P(A))$. This shows that $N(F) \subset N(I - P(1)^{-1}P(A))$. Moreover, the identity I restricted to $N(I - P(1)^{-1}P(A))$ is equal to $P(1)^{-1}P(A)$ and consequently is compact. Hence, $N(I - P(1)^{-1}P(A))$ is finite dimensional and therefore $\dim[N(F)] < \infty$.

(*ii*) Since F commutes with I, Newton's binomial formula gives

$$A^r = (I - F)^r = I + \sum_{j=1}^{r} (-1)^j C_r^j F^j$$

and therefore,

$$P(A) = P(1)I + \sum_{r=1}^{p} a_r \left(\sum_{j=1}^{r} (-1)^j C_r^j F^j \right). \tag{3.1.1}$$

Let E be a closed complement for $N(F)$, so that $X = N(F) \oplus E$. Thus, we obtain two linear continuous maps $F_{|E} : E \longrightarrow X$ and $A_{|E} : E \longrightarrow X$, the restrictions of F and A for E. It is clear that the kernel of $F_{|E}$ is $\{0\}$. To conclude, it is sufficient to show that $F_{|E}(E) = F(E) = F(X)$ is closed. For this, it suffices to show that the map $(F_{|E})^{-1} : F(E) \longrightarrow E$ is continuous. By linearity, it even suffices to prove that $(F_{|E})^{-1}$ is continuous at 0. Suppose that this is not the case. Then, we can find a sequence $(x_n)_n$ in E such that $Fx_n \to 0$, but $(x_n)_n$ does not converge to 0. Selecting a suitable subsequence, we can assume, without loss of generality, that $\|x_n\| \geq \eta > 0$ for all n. Then, $\frac{1}{\|x_n\|} \leq \frac{1}{\eta}$ for all n and consequently, $F(\frac{x_n}{\|x_n\|})$ also converges to 0. Furthermore, $\frac{x_n}{\|x_n\|}$ has norm 1 and hence, some subsequence of $P(A)(\frac{x_n}{\|x_n\|})$ converges. It follows from (3.1.1) that $\frac{x_n}{\|x_n\|}$ has a converging subsequence to an element y in E, verifying $\|y\| = 1$ and

$$P(A)(y) = P(1)y. \tag{3.1.2}$$

Moreover, the equation $A^r = A - AF - A^2F - \ldots - A^{r-1}F$ allows us to write in the form $P(A) = a_0 I + (P(1) - a_0)A + \sum_{r=1}^{p} a_r \left(\sum_{j=1}^{r} A^j\right) F$ and so, we get $P(A)(y) = P(1)A(y)$. Since $P(1) - a_0 \neq 0$, the use of (3.1.2) gives $A(y) - y = 0$, which implies that $y \in N(F)$. This contradicts the fact $E \bigcap N(F) = \{0\}$ (because $\|y\| = 1$) and completes the proof of (*ii*).

(*iii*) If $F(X)$ does not have a finite codimension, we can find a sequence of closed subspaces $F(X) = M_0 \subset M_1 \subset M_2 \subset \ldots \subset M_n \subset \ldots$ such that, each M_n is closed and of codimension 1 in M_{n+1} just by adding one-dimensional spaces to $F(X)$ inductively. By Riesz's lemma (Lemma 2.1.3), we can find in each M_n an element x_n such that $\|x_n\| = 1$ and $\|x_n - y\| \geq \frac{1}{2}$ for all y in M_{n-1}. Then, Eq. (3.1.1) together with the fact that $X \supset R(F) \supset R(F^2) \supset \ldots \supset R(F^n) \supset \ldots$ gives, for all $k < n$,

$$\begin{aligned}
&\|P(A)x_n - P(A)x_k\| \\
&= |P(1)| \left\| x_n - x_k + \sum_{r=1}^{p} \frac{a_r}{P(1)} \left(\sum_{j=1}^{r} (-1)^j C_r^j F^j(x_n - x_k) \right) \right\| \\
&\geq \frac{|P(1)|}{2}
\end{aligned}$$

because $x_k - \sum_{r=1}^{p} \frac{a_r}{P(1)} \left(\sum_{j=1}^{r} (-1)^j C_r^j F^j(x_n - x_k)\right) \in M_{n-1}$. This proves that the sequence $(P(A)x_n)_n$ cannot have a convergent subsequence, which contradicts the compactness of $P(A)$. Q.E.D.

Lemma 3.1.2. *Let $A \in \mathcal{P}(X)$ and set $F = I - A$. Then, F is a Fredholm operator and $i(F) = 0$.* ◇

Proof. The first part of the lemma follows from Lemma 3.1.1. To complete the proof, it suffices to show that $\dim[N(F)] = \dim[R(F)^\circ]$. To do this, note first that the operator dual F^* of F is given by $F^* = I^* - A^*$. The use of both Schauder's theorem and Lemma 3.1.1 (*i*) gives $\dim[N(F^*)] < \infty$. Accordingly, using the relation $R(F) = {}^\circ N(F^*)$ (see Theorem 2.1.1), we only need to prove that $\dim[N(F)] = \dim[N(F^*)]$. Now, let us consider the case where $\dim N(F^*) = 0$, i.e., $N(F^*) = \{0\}$. The fact that $R(F)$ is closed, we see, by Theorem 2.1.1, that $R(F) = {}^\circ N(F^*) = X$. We claim that $N(F) = \{0\}$. Indeed, suppose that $N(F) \neq \{0\}$. So, there is an $x_1 \neq 0$ in X such that $Fx_1 = 0$. Since $R(F) = X$, then there is an $x_2 \in X$ such that $Fx_2 = x_1$. So, we can find an element $x_n \in X$ such that $Fx_n = x_{n-1}$. Now, F^n is a bounded operator with $\|F^n\| \leq \|F\|^n$. This implies that $N(F^n)$ is a closed subspace of X. Additionally, we have $N(F) \subset N(F^2) \subset \ldots \subset N(F^n) \subset \ldots$, and what is more, these spaces are actually increasing, because $F^n x_n = F^{n-1}(Fx_n) = F^{n-1}x_{n-1} = \ldots = Fx_1 = 0$, and $F^{n-1}x_n = F^{n-2}(Fx_n) = F^{n-2}x_{n-1} = \ldots = Fx_2 = x_1 \neq 0$. Hence, we can apply the Riesz lemma (Lemma 2.1.3) in order to find a $z_n \in N(F^n)$ such that $\|z_n\| = 1$ and $\text{dist}(z_n, N(F^{n-1})) > \frac{1}{2}$. Since $P(A)$ is compact, then $(P(A)z_n)_n$ has a convergent subsequence. For all $k < n$,

$$\|P(A)z_n - P(A)z_k\| = |P(1)| \left\| z_n - z_k + \sum_{r=1}^{p} \frac{a_r}{P(1)} \left(\sum_{j=1}^{r} (-1)^j C_r^j F^j (z_n - z_k) \right) \right\|$$
$$\geq \frac{|P(1)|}{2}$$

because $F^{n-1}\left(z_k - \sum_{r=1}^{p} \frac{a_r}{P(1)} \left(\sum_{j=1}^{r}(-1)^j C_r^j F^j(z_n - z_k)\right)\right) = 0$. This proves that the sequence $(P(A)z_n)_n$ cannot have a convergent subsequence, which contradicts the compactness of $P(A)$. Hence, $\dim N(F) = 0$. Inversely, if $N(F) = \{0\}$, then $R(F^*) = N(F)^\circ = X^*$ where X^* is the dual of X. Since $F^* = I^* - A^*$, and knowing that A^* is a polynomially compact operator on X^*, the just given argument implies that $N(F^*) = \{0\}$. Hence, we have shown that $\dim N(F) = 0$ if, and only if, $\dim N(F^*) = 0$. Now, suppose that $\dim N(F) = n > 0$ and $\dim N(F^*) = m > 0$. Let $x_1, \ldots, x_n, x_1', \ldots, x_m'$ span $N(F)$ and $N(F^*)$, respectively. We claim the existence of a functional x_0' as well as an element x_0 such that

$$x_0'(x_j) = 0, \quad 1 \leq j < n, \quad x_0'(x_n) \neq 0, \tag{3.1.3}$$

and $x_j'(x_0) = 0, \quad 1 \leq j < m, \quad x_m'(x_0) \neq 0$. Indeed, let M be the subspace spanned by $x_1, \ldots, x_{n-1}$. The element x_n is not in this finite dimensional subspace. This implies that $\text{dist}(x_n, M) > 0$. Then, by Lemma 2.1.4, there is an $x_0' \in X^*$ such that $\|x_0'\| = 1$, $x_0'(x_n) = \text{dist}(x_n, M)$, and $x_0'(x) = 0$ for $x \in M$. This proves the claim. Let $K_0 x = x_0'(x)x_0$, $x \in X$. Then, $K_0^* x' = x'(x_0)x_0'$, $x' \in X'$. Set $F_1 = F - K_0$. We claim that $\dim N(F_1) = n-1$, $\dim N(F_1^*) = m-1$. Suppose that $x \in N(F_1)$. Then, $Fx = K_0 x = x_0'(x)x_0$. Now, x_0 is not in $R(F) = {}^\circ N(F^*)$ (see Theorem 2.1.1), since it does not annihilate x_m'. Hence, we must have $x_0'(x) = 0$, and consequently $Fx = 0$. Since $x \in N(F)$, then $x = \sum_{j=1}^{n} \alpha_j x_j$. Therefore, and using (3.1.3), $x_0'(x) = \sum_{j=1}^{n} \alpha_j x_0'(x_j) = \alpha_n x_0'(x_n) = 0$, showing that $\alpha_n = 0$. Hence, x is of the form

$$x = \sum_{j=1}^{n-1} \alpha_j x_j. \tag{3.1.4}$$

Conversely, every element of the form (3.1.4) is in $N(F_1)$. This follows from the fact that it is in $N(F)$ and satisfies $x_0'(x) = 0$. This shows that $\dim N(F_1) = n-1$. Suppose $x' \in N(F_1^*)$, i.e., $F^* x' = K_0^* x' = x'(x_0)x_0'$. But x_0' is not in $R(F^*) = N(F)^\circ$, since it does not annihilate x_n. Hence, $x'(x_0) = 0$, and consequently, $F^* x' = 0$. Thus, $x' = \sum_{j=1}^{m} \beta_j x_j'$ and $x'(x_0) = \sum_{j=1}^{m} \beta_j x_j'(x_0) = \beta_m x_m'(x_0) = 0$. Showing that $\beta_m = 0$. Hence, x' is of the form

$$x' = \sum_{j=1}^{m-1} \beta_j x_j'. \tag{3.1.5}$$

Conversely, every functional of the form (3.1.5) is in $N(F_1^*)$, since it is in $N(F^*)$ and $N(K_0^*)$. This completes the proof of the claim. Observing that $F_1 = I - (A + K_0)$, and $A + K_0$ is a polynomially compact operator. Thus, we have an operator F_1 of the same form as F with the dimensions of its null space and that of its adjoint exactly one less than those of F, F^*, respectively. If m and n are both greater than one, we can repeat the process and reduce $\dim N(F_1)$ and $\dim N(F_1^*)$ each by one. Continuing in this way, we eventually reach an operator $\tilde{F} = I - \tilde{K}$, where $\tilde{K}$ is polynomially compact and either $\dim N(\tilde{F}) = 0$ or $\dim N(\tilde{F}^*) = 0$. Then, it follows from what we have proved that they must both be equal to zero. Hence, $m = n$, and the proof of lemma is complete. Q.E.D.

Using Lemma 3.1.2 we have the following inclusions $\mathcal{K}(X) \subset \mathcal{P}(X) \subset \mathcal{J}(X)$. If the space X is a Banach space which either satisfies the Dunford–Pettis property or is isomorphic to one of the spaces $L_p(\Omega)$ $p > 1$, then by using Lemmas 2.1.13 and 3.1.2, we have the following inclusions $\mathcal{K}(X) \subset \mathcal{S}(X) \subset \mathcal{J}(X)$.

Theorem 3.1.1. *If $A \in \mathcal{P}(X)$ and $F = I - A$, then there is an integer $n \geq 1$ such that $N(F^n) = N(F^k)$ for all $k \geq n$.* ◇

Proof. If there is an integer k such that $N(F^k) = N(F^{k+1})$. If $j > k$ and $x \in N(F^{j+1})$, then $F^{j-k}x \in N(F^{k+1}) = N(F^k)$, showing that $x \in N(F^j)$. So, $N(F^j) = N(F^{j+1})$ for all $j > k$. If $N(F^k)$ is a proper subspace of $N(F^{k+1})$ for all $k > 1$. Then, using Lemma 2.1.3 for all $k \in \mathbb{N}^*$, there exists $x_k \in N(F^k)$ such that $\|x_k\| = 1$ and $d(x_k, N(F^{k-1})) \geq \frac{1}{2}$. Moreover, since $A \in \mathcal{P}(X)$, then there exists a nonzero complex polynomial $P(z) := \sum_{r=0}^{p} a_r z^r$ satisfying $P(1) \neq 0$, $P(1) - a_0 \neq 0$ and $P(A) \in \mathcal{K}(X)$. Now, using Eq. (3.1.1), we get for all $j < k$,

$$\|P(A)x_k - P(A)x_j\| = |P(1)| \left\| x_k - x_j + \sum_{r=1}^{p} \frac{a_r}{P(1)} \left(\sum_{l=1}^{r} (-1)^l C_r^l F^l (x_k - x_j) \right) \right\|.$$

Since $j \leq k-1$, then $x_j - \sum_{r=1}^{p} \frac{a_r}{P(1)} \left(\sum_{l=1}^{r} (-1)^l C_r^l F^l (x_k - x_j) \right) \in N(F^{k-1})$ and therefore, $\|P(A)x_k - P(A)x_j\| \geq \frac{|P(1)|}{2} > 0$. This contradicts the compactness of $P(A)$. This achieves the proof. Q.E.D.

Definition 3.1.2. Let $\lambda \in \mathbb{C}$ be an eigenvalue of $A \in \mathcal{L}(X)$, the value

$$\dim \bigcup_{n \in \mathbb{N}} N\left[(\lambda - A)^n\right] \in \mathbb{N} \bigcup \{+\infty\}$$

is called the algebraic multiplicity and denoted by $\text{mult}(A, \lambda)$. ◇

Clearly, if $\text{mult}(A, \lambda) < \infty$, then $(\lambda - A)$ has a finite ascent. If $A \in \mathcal{L}(X)$, we consider the quantities $r(A) = \lim_{n \to +\infty} \alpha(A^n)$ and $r'(A) = \lim_{n \to +\infty} \beta(A^n)$. Let $A \in \mathcal{P}(X)$. If $F = I - A$, then for each integer n, the operator $F^n \in \Phi^b(X)$ and $\alpha(F^n)$

are finite. From the inclusion $N(F^n) \subset N(F^{n+1})$, it follows that $\alpha(F^n) \leq \alpha(F^{n+1})$ and consequently, $\alpha(F^n)$ approaches either a finite limit or ∞. The following result excludes the second eventuality.

Proposition 3.1.1. *Let $A \in \mathcal{P}(X)$ and set $F = I - A$. Then,* $\operatorname{asc}(F) =$ *desc*$(F) < \infty$. ◇

Proof. We first prove that $\operatorname{asc}(F) < \infty$. To do this, it suffices to show that there exists an integer k such that $N(F^k) = N(F^{k+1})$. If $N(F^k) \neq N(F^{k+1})$ for all $k \in \mathbb{N}^*$, then we have $N(F) \subsetneq N(F^2) \subsetneq \ldots \subsetneq N(F^k) \subsetneq \ldots$ From Riesz's lemma (Lemma 2.1.3), it follows that, for all $k \in \mathbb{N}^*$ there exists $x_k \in N(F^k)$ such that $\|x_k\| = 1$ and $\operatorname{dist}(x_k, N(F^{k-1})) \geq \frac{1}{2}$. Moreover, since $A \in \mathcal{P}(X)$, then there exists a nonzero complex polynomial $P(z) := \sum_{r=0}^{p} a_r z^r$ satisfying $P(1) \neq 0$, $P(1) - a_0 \neq 0$ and $P(A) \in \mathcal{K}(X)$. Now, arguing as in the proof of Theorem 3.1.1, we show that $(P(A)x_k)_k$ has no convergent subsequence. This contradicts the compactness of $P(A)$. Accordingly, we have $\operatorname{asc}(F) < \infty$. Besides, since F is a Fredholm operator with $i(F) = 0$ (Lemma 3.1.2), the use of Lemma 2.2.7 completes the proof. Q.E.D.

3.1.2 *Generalized Riesz Operator*

It is well known that if X is a complex Banach space and $A \in \mathcal{K}(X)$, then A and A^* are Riesz operators, with $\sigma(A^*) = \sigma(A)$ (see [191]). Furthermore, for any eigenvalue $\lambda \in \sigma(A) \backslash \{0\}$, we have $\operatorname{mult}(A, \lambda) < \infty$ and $\operatorname{mult}(A, \lambda) = \operatorname{mult}(A^*, \lambda)$.

Definition 3.1.3. An operator $A \in \mathcal{L}(X)$ will be called a generalized Riesz operator if there exists E (a finished part of $\mathbb{C}$) such that

(i) For all $\lambda \in \mathbb{C}\backslash E$, $(\lambda - A)$ is a Fredholm operator on X,
(ii) For all $\lambda \in \mathbb{C}\backslash E$, $(\lambda - A)$ has a finite ascent and a finite descent, and
(iii) All $\lambda \in \sigma(A)\backslash E$ are eigenvalues of finite multiplicity, and have no accumulation point except possibly points of E. ◇

Note that, if A is a Riesz operator on X, then A is a generalized Riesz operator on X. It is one of the purposes of this section to prove that a polynomially compact operator is a generalized Riesz one.

Theorem 3.1.2. *Let $A \in \mathcal{PK}(X)$, i.e., there exists a nonzero complex polynomial $P(z) = \sum_{r=0}^{p} a_r z^r$ satisfying $P(A) \in \mathcal{K}(X)$. Let $\lambda \in \mathbb{C}$ with $P(\lambda) \neq 0$ and set $F := \lambda - A$. Then, F is a Fredholm operator on X with both finite ascent and descent.* ◇

Proof. Let $\lambda \in \mathbb{C}$ with $P(\lambda) \neq 0$. We have the following $P(\lambda) - P(A) = \sum_{k=1}^{p} a_k(\lambda^k - A^k)$. Moreover, for any $k \in \{1, \ldots, p\}$, $\lambda^k - A^k = (\lambda - A)\sum_{i=0}^{k-1} \lambda^i A^{k-1-i}$. This allows us to write

$$P(\lambda) - P(A) = (\lambda - A)Q(A) = Q(A)(\lambda - A), \tag{3.1.6}$$

where $Q(A) = \sum_{k=1}^{p} a_k \sum_{i=0}^{k-1} \lambda^i A^{k-1-i}$. Let $n \in \mathbb{N}$. Using Eq. (3.1.6), we have the following $(P(\lambda) - P(A))^n = (\lambda - A)^n Q(A)^n = Q(A)^n(\lambda - A)^n$. Hence, $N[(\lambda - A)^n] \subset N[(P(\lambda) - P(A))^n]$, and $R[(\lambda - A)^n] \supset R[(P(\lambda) - P(A))^n]$, for all $n \in \mathbb{N}$. This implies

$$\bigcup_{n\in\mathbb{N}} N[(\lambda - A)^n] \subset \bigcup_{n\in\mathbb{N}} N[(P(\lambda) - P(A))^n] \tag{3.1.7}$$

and

$$\bigcap_{n\in\mathbb{N}} R[(\lambda - A)^n] \supset \bigcap_{n\in\mathbb{N}} R[(P(\lambda) - P(A))^n]. \tag{3.1.8}$$

Besides, since $P(\lambda) \neq 0$ and $P(A) \in \mathcal{K}(X)$ therefore, from the same reasoning as Lemma 3.1.1 and Proposition 3.1.1, we show that $P(\lambda) - P(A)$ is a Fredholm operator on X with both finite ascent and descent. So, the right-hand sides of (3.1.7) and (3.1.8) are really only finite union and intersections. It follows, from Proposition 2.2.2, that $\operatorname{asc}(P(\lambda) - P(A)) = \operatorname{desc}(P(\lambda) - P(A))$, let n_0 be this quantity. Since $P(\lambda)I$ commutes with $P(A)$, Newton's binomial formula gives $(P(\lambda) - P(A))^{n_0} = \sum_{k=0}^{n_0}(-1)^k C_{n_0}^k (P(\lambda))^{n_0-k} P(A)^k = (P(\lambda))^{n_0} - \mathcal{S}$ where $\mathcal{S} = P(A)\sum_{k=1}^{n_0}(-1)^{k-1} C_{n_0}^k P(\lambda)^{n_0-k} P(A)^{k-1}$ is a compact operator on X. Since $P(\lambda) \neq 0$ and $\mathcal{S}$ is a Riesz operator, we have $\dim \bigcup_{n\in\mathbb{N}} N[(P(\lambda) - P(A))^n] = \dim N[(P(\lambda) - P(A))^{n_0}] < \infty$ and $\operatorname{codim} \bigcap_{n\in\mathbb{N}} R[(P(\lambda) - P(A))^n] = \operatorname{codim} R[(P(\lambda) - P(A))^{n_0}] < \infty$. Using both Eqs. (3.1.7) and (3.1.8), we have the following $\dim \bigcup_{n\in\mathbb{N}} N[(\lambda - A)^n] < \infty$ and $\operatorname{codim} \bigcap_{n\in\mathbb{N}} R[(\lambda - A)^n] < \infty$. This implies that $\operatorname{asc}(\lambda - A) < \infty$ and $\operatorname{desc}(\lambda - A) < \infty$. We also have $\dim N(\lambda - A) < \infty$ and $\operatorname{codim} R(\lambda - A) < \infty$. To prove that $R(\lambda - A)$ is closed, we may assume that $(\lambda - A)$ is injective. Otherwise, the finite dimensional space $N(\lambda - A)$ would have a closed complement M in X and $X = N(\lambda - A) \oplus M$. Define $S : M \longrightarrow X$ by $S := (\lambda - A)|_M$ the restriction of $(\lambda - A)$ to M. Since $R(S) = R(\lambda - A)$ and S is injective, we just replace $(\lambda - A)$ by S in the following. Let $(x_n)_{n\in\mathbb{N}}$ be a sequence in X such that

$$(\lambda - A)(x_n) \to y \text{ as } n \to \infty. \tag{3.1.9}$$

Using Eqs (3.1.6) and (3.1.9), we infer that

$$[P(\lambda) - P(A)](x_n) \to Q(A)y \text{ as } n \to \infty. \tag{3.1.10}$$

We claim that $(x_n)_{n\in\mathbb{N}}$ is bounded. Indeed, if $(x_n)_{n\in\mathbb{N}}$ is unbounded, selecting a suitable subsequence $(x_{\psi(n)}) \in X\backslash\{0\}$ such that

$$||x_{\psi(n)}|| \geq n. \tag{3.1.11}$$

Let $\tilde{x}_n = \frac{x_{\psi(n)}}{\|x_{\psi(n)}\|}$, then by Eqs (3.1.9) and (3.1.11), we have

$$(\lambda - A)\tilde{x}_n \to 0 \text{ as } n \to \infty. \tag{3.1.12}$$

It follows, from Eqs (3.1.6) and (3.1.12), that

$$[P(\lambda) - P(A)]\,\tilde{x}_n \to 0 \text{ as } n \to \infty. \tag{3.1.13}$$

Furthermore, $(\tilde{x}_n)_n$ has a norm 1 and hence, some subsequences of $P(A)\tilde{x}_n$ converge. It follows from Eq. (3.1.13) that $(\tilde{x}_n)$ has a converging subsequence $(\tilde{x}_{n_k})$ to an element $\frac{\tilde{z}}{P(\lambda)}$ in X verifying $\|\frac{\tilde{z}}{P(\lambda)}\| = 1$. So, $(\lambda - A)\tilde{x}_{n_k}$ converges to $(\lambda - A)\frac{\tilde{z}}{P(\lambda)}$. Hence, using Eq. (3.1.12), we infer that $(\lambda - A)\frac{\tilde{z}}{P(\lambda)} = 0$, which implies that $\tilde{z} \in N(\lambda - A)$. This contradicts that $N(\lambda - A) = \{0\}$ (because $\|\tilde{z}\| = |P(\lambda)| \neq 0$) and concludes the proof of the claim. Since $P(A) \in \mathcal{K}(X)$ and (x_n) is bounded, we infer that there is a subsequence $(x_{\varphi(n)})_{n\in\mathbb{N}}$ of (x_n) such that $P(A)(x_{\varphi(n)}) \to z \in X$ as $n \to \infty$. So, it follows from Eq. (3.1.10), that $x_{\varphi(n)} \to \frac{1}{P(\lambda)}(z + Q(A)y)$ as $n \to \infty$ and by Eq. (3.1.9), we have $y = (\lambda - A)\left[\frac{1}{P(\lambda)}(z + Q(A)y)\right]$. This proves that $y \in R(\lambda - A)$, and completes the proof. Q.E.D.

As an immediate consequence of Theorem 3.1.2:

Corollary 3.1.1. *Assume that the hypotheses of Theorem 3.1.2 hold. Then,* $F := \lambda - A$ *is an operator of index zero.* ◇

Remark 3.1.2. The hypotheses of Theorem 3.1.2 are optimal. In fact, let $A = \lambda I$ and take $P(z) = \lambda - z$ with $\lambda \in \mathbb{C}$. Clearly, $P(\lambda) = 0$ and $P(A) \in \mathcal{K}(X)$ (i.e., $A \in \mathcal{PK}(X)$), but the last results are false for $F := \lambda - A = 0$ because it is not a Fredholm operator. ◇

3.2 First and Second Kind Operator Equation

Theorem 3.2.1. *Let* $A \in \mathcal{PK}(X)$*, i.e., there exists a nonzero complex polynomial* $P(z) = \sum_{r=0}^{p} a_r z^r$ *satisfying* $P(A) \in \mathcal{K}(X)$*. Let* $\lambda \in \mathbb{C}$ *with* $P(\lambda) \neq 0$ *and set* $F := \lambda - A$*. If* F *is injective, then the inverse operator* $F^{-1} = (\lambda - A)^{-1} : X \longrightarrow X$ *exists and is bounded.* ◇

By assumption, the operator F is injective, that is, $N(F) = \{0\}$. Therefore, $\text{asc}(F) = 0$ and by Proposition 2.2.2, we conclude that $F(X) = X$, that is, the operator F is surjective. Hence, the inverse operator $F^{-1} = (\lambda - A)^{-1} : X \longrightarrow X$ exists. Assume that F^{-1} is not bounded. Then, there exists a sequence $(f_n)_n$ with $\|f_n\| = 1$ and the sequence $\varphi_n := F^{-1}f_n$ is not bounded. Let us define $g_n := \frac{f_n}{\|\varphi_n\|}$, $\psi_n := \frac{\varphi_n}{\|\varphi_n\|}$, for all $n \in \mathbb{N}$. Then, $g_n \to 0$ as $n \to \infty$, and $\|\psi_n\| = 1$. Since

$$\lambda\psi_n - A\psi_n = g_n, \tag{3.2.1}$$

then we have

$$P(A)\psi_n = P(\lambda)\psi_n - \sum_{k=1}^{p} a_k \left(\sum_{i=0}^{k-1} \lambda^i A^{k-1-i} g_n\right). \tag{3.2.2}$$

Since $P(A)$ is compact, we can choose a subsequence $(\psi_{n(k)})_k$ such that $\left(P(A)\psi_{n(k)}\right)_k \to \psi \in X$ as $k \to \infty$. Using Eq. (3.2.2), we notice that $\psi_{n(k)} \to \frac{\psi}{P(\lambda)}$ as $k \to \infty$. This implies that $\|\psi\| = |P(\lambda)|$. Using Eq. (3.2.1), we show that $\psi \in N(F)$. Hence, $\psi = 0$ which contradicts $P(\lambda) \neq 0$. Q.E.D.

We can rewrite Theorem 3.2.1 in terms of the solvability of the second kind operator equation as follows.

Corollary 3.2.1. *Let $A \in \mathcal{PK}(X)$, i.e., there exists a nonzero complex polynomial $P(z) = \sum_{r=0}^{p} a_r z^r$ satisfying $P(A) \in \mathcal{K}(X)$. Let $\lambda \in \mathbb{C}$ such that $P(\lambda) \neq 0$. If the homogeneous equation $\lambda\varphi - A\varphi = 0$ has only the trivial solution $\varphi = 0$, then for all $f \in X$, the non-homogeneous equation $\lambda\varphi - A\varphi = f$ has a unique solution $\varphi \in X$ which depends continuously on f.* ◇

Theorem 3.2.2. *Let $A \in \mathcal{PK}(X)$, i.e., there exists a nonzero complex polynomial $P(z) = \sum_{r=0}^{p} a_r z^r$ satisfying $P(A) \in \mathcal{K}(X)$. Let $\lambda \in \mathbb{C}$ such that $P(\lambda) \neq 0$ and assume that $\lambda - A$ is not injective. Then, the null space $N(\lambda - A)$ is finite dimensional and the range $(\lambda - A)(X) \neq X$ is a proper closed subspace.* ◇

Proof. Since $\lambda - A$ is not injective, then $N(\lambda - A) \neq \{0\}$. This means that $\text{asc}(\lambda - A) > 0$. Hence, applying Proposition 2.2.2, we conclude that $(\lambda - A)(X) \neq X$. Q.E.D.

Corollary 3.2.2. *Assume that the hypotheses of Corollary 3.2.1 hold. If the homogeneous equation $\lambda\varphi - A\varphi = 0$ has a nontrivial solution, then the non-homogeneous equation $\lambda\varphi - A\varphi = f$ is either unsolvable or its general solution is of the form $\varphi = \tilde{\varphi} + \sum_{k=1}^{m} \alpha_k \varphi_k$, where $\varphi_1, \ldots, \varphi_m$ are linearly independent solutions of the homogeneous equation, $\alpha_1, \ldots, \alpha_m$ are arbitrary complex numbers, and $\tilde{\varphi}$ denotes a particular solution of the non-homogeneous equation.* ◇

Theorem 3.2.3. *Let $A \in \mathcal{PK}(X)$, i.e., there exists a nonzero complex polynomial $Q(z) = \sum_{r=0}^{p} a_r z^r$ satisfying $Q(A) \in \mathcal{K}(X)$. Let $\lambda \in \mathbb{C}$ such that $Q(\lambda) \neq 0$ and set $F := \lambda - A$. Then, the projection $P : X \longrightarrow N\left(F^{\text{asc}(F)}\right)$ defined by the decomposition $X = N\left(F^{\text{asc}(F)}\right) \oplus F^{\text{asc}(F)}(X)$ is compact, and the operator $F - P = \lambda - A - P$ is bijective.* ◇

Proof. Let $n_0 = \text{asc}(F)$, then $F^{n_0} \in \Phi^b(X)$. Since $F \in \Phi^b(X)$, then the null space $N(F^{n_0})$ is of a finite dimension. It is easy to verify that $\|\psi\|_{n_0} := \inf_{\chi \in F^{n_0}(X)} \|\psi + \chi\|$ defines a norm on $N(F^{n_0})$. In particular, the fact that $\|\psi\|_{n_0} = 0$ implies that $\psi = 0$ since the range $F^{n_0}(X)$ is closed by Theorem 2.2.7. Since all the norms are equivalent on a finite dimensional linear space, we infer that there exists a positive number c such that $\|\psi\| \leq c \inf_{\chi \in F^{n_0}(X)} \|\psi + \chi\|$ for all $\psi \in N(F^{n_0})$. Then, for all $\varphi \in X$, we have $P\varphi \in N(F^{n_0})$, and therefore $\|P\varphi\| \le c \inf_{\chi \in F^{n_0}(X)} \|P\varphi + \chi\| \le c\|\varphi\|$ since $\varphi - P\varphi \in F^{n_0}(X)$. Hence, P is bounded. Moreover, P is a compact operator since it has a finite dimensional range $P(X) = N(F^{n_0})$. Besides, the operator $A + P \in \mathcal{PK}(X)$ since $A \in \mathcal{PK}(X)$ and $P \in \mathcal{K}(X)$. Let $\varphi \in N(F - P)$, that is, $F\varphi - P\varphi = 0$. Since $P\varphi \in N(F^{n_0})$, then $F^{n_0+1}\varphi = 0$. Therefore, $\varphi \in N\left(F^{n_0+1}\right) = N(F^{n_0})$ and $P\varphi = \varphi$, which implies $F\varphi = \varphi$. From this, and by iteration, we show that $\varphi = F^{n_0}\varphi = 0$. Hence, $N(F-P) = \{0\}$. Now, applying Theorem 3.2.1 to the operator $A+P$, we conclude that $F - P$ is surjective and the proof is completed. Q.E.D.

3.3 Spectral Analysis

Lemma 3.3.1. *Let $P(.)$ be a nonzero complex polynomial, $P(z) = \sum_{r=0}^{p} a_r z^r$ and $z_1, \dots, z_p$ its zeros. If $(\alpha_n)_n$ is a sequence of complex numbers such that $(P(\alpha_n))_n$ converges to zero. Then, we can choose a subsequence $(\alpha_{n_k})_k$ of $(\alpha_n)_n$ converging to z_i for some $i \in \{1, \dots, p\}$.* ◇

Proof. Let us write $P(z) = a_p(z - z_1)(z - z_2)\dots(z - z_p)$. Then, $P(\alpha_n)$ can be written in the form $P(\alpha_n) = a_p(\alpha_n - z_1)(\alpha_n - z_2)\dots(\alpha_n - z_p)$. Assume that, for all subsequences $(\alpha_{n_k})_k$ of $(\alpha_n)_n$ and for all $i \in \{1, \dots, p\}$, the sequence $(\alpha_{n_k})_k$ does not converge to z_i. Since the sequence $(\alpha_n)_n$ does not converge to z_1, then there is a subsequence $(\alpha_{\varphi_1(n)})_n$ and $\varepsilon_1 > 0$ such that $|\alpha_{\varphi_1(n)} - z_1| > \varepsilon_1$. Similarly, since $(\alpha_{\varphi_1(n)})_n$ does not converge to z_2, then there is a subsequence $(\alpha_{\varphi_2\varphi_1(n)})_n$ and $\varepsilon_2 > 0$ such that $|\alpha_{\varphi_2\varphi_1(n)} - z_2| > \varepsilon_2$. Continuing in the same way, we can find $\varepsilon_p > 0$ and φ_p such that $|\alpha_{\varphi_p\varphi_{p-1}\dots\varphi_1(n)} - z_p| > \varepsilon_p$. Let $\psi(n) = \varphi_p\varphi_{p-1}\dots\varphi_1(n)$. Then, we have $|(\alpha_{\psi(n)} - z_1)(\alpha_{\psi(n)} - z_2)\dots(\alpha_{\psi(n)} - z_p)| > \varepsilon_1\varepsilon_2\dots\varepsilon_p$, $\forall n \in \mathbb{N}$. Hence, $|P(\alpha_{\psi(n)})| > |a_p|\varepsilon_1\varepsilon_2\dots\varepsilon_p > 0$, $\forall n \in \mathbb{N}$. This contradicts the fact that $P(\alpha_{\psi(n)}) \to 0$ as $n \to +\infty$. Q.E.D.

Definition 3.3.1. For a polynomially compact operator A, the nonzero polynomial $P(.)$, of last degree such that $P(A)$ is compact, will be called the minimal polynomial of A. ◇

Now, we may formulate the results of the Riesz theory in terms of spectral analysis.

Theorem 3.3.1. *Let X be an infinite dimensional Banach space, $A \in \mathcal{PK}(X)$ and $P(.)$ denotes the minimal polynomial of A. Let $P(z) = \sum_{r=0}^{p} a_r z^r$ and $z_1, \dots, z_p$ its zeros. Then, $\lambda = z_1, \dots, z_p$ belongs to the spectrum $\sigma(A)$ and $\sigma(A)\backslash\{z_1, \dots, z_p\}$ consists of, at most, a countable set of eigenvalues with no accumulation points, except possibly $\lambda = z_1, \dots, z_p$.* ◇

Proof. Suppose that there is $i \in \{1, \dots, p\}$ such that $z_i \in \rho(A)$, then $(z_i - A)^{-1}$ exists and is bounded. Moreover, $P(A)$ can be written $P(A) = a_p(-1)^p(z_1 - A)\dots(z_p - A)$. Let $Q(.)$ be the following polynomial $Q(z) = a_p(-1)^p$

$(z_1 - z)\dots(z_{i-1} - z)(z_{i+1} - z)\dots(z_p - z)$. Hence, $Q(A) = (z_i - A)^{-1}P(A)$. Since $P(A) \in \mathcal{K}(X)$ and $(z_i - A)^{-1}$ is bounded, then $Q(A) \in \mathcal{K}(X)$. This contradicts the fact that P is the minimal polynomial of A. Therefore, for all $i \in \{1, \dots, p\}$, z_i belongs to the spectrum $\sigma(A)$. If $\lambda \neq \{z_1, \dots, z_p\}$, we will discuss two cases. First, if $N(\lambda - A) = \{0\}$, then by Theorem 3.2.1, the operator $(\lambda - A)^{-1}$ exists and is bounded. Second, if $N(\lambda - A) \neq \{0\}$, then λ is an eigenvalue. Therefore, each $\lambda \neq \{z_1, \dots, z_p\}$ is either in the resolvent set of A or represents an eigenvalue of A. It remains to show that, for each $R > 0$, there exists only a finite number of eigenvalues λ with $\lambda \notin \bigcup_{k=1}^{p} \mathbb{D}(z_k, R)$ where $\mathbb{D}(z_k, R)$ is the disc with the center z_k and the radius R. Assume, in the contrary, that there exists a sequence $(\lambda_n)_n$ of distinct eigenvalues satisfying $\lambda_n \notin \bigcup_{k=1}^{p} \mathbb{D}(z_k, R)$. Let us choose some eigenelements φ_n, such that $A\varphi_n = \lambda_n\varphi_n$, and let us define the finite dimensional subspaces $U_n := \text{span}\{\varphi_1, \dots, \varphi_n\}$. It is easy to verify that the eigenelements corresponding to distinct eigenvalues are linearly independent. Hence, we have $U_{n-1} \subsetneqq U_n$ and, by the Riesz lemma (Lemma 2.1.3), we can choose a sequence $(\psi_n)_n$ of elements $\psi_n \in U_n$ such that $\|\psi_n\| = 1$ and $\|\psi_n - \psi\| \geq \frac{1}{2}$, for all $\psi \in U_{n-1}$. Writing $\psi_n = \sum_{k=1}^{n} \alpha_{n_k}\varphi_k$ we get $P(\lambda_n)\psi_n - P(A)\psi_n = \sum_{k=1}^{n-1}\sum_{r=0}^{p} a_r(\lambda_n^r - \lambda_k^r)\alpha_{n_k}\varphi_k \in U_{n-1}$. Therefore, for $m < n$, we have

$$\begin{aligned} P(A)\psi_n - P(A)\psi_m &= P(\lambda_n)\psi_n - (P(\lambda_n)\psi_n - P(A)\psi_n + P(A)\psi_m) \\ &= P(\lambda_n)(\psi_n - \psi), \end{aligned}$$

where $\psi := \frac{P(\lambda_n)\psi_n - P(A)\psi_n + P(A)\psi_m}{P(\lambda_n)} \in U_{n-1}$. Hence,

$$\begin{aligned} \|P(A)\psi_n - P(A)\psi_m\| &= |P(\lambda_n)|\,\|\psi_n - \psi\| \\ &\geq \frac{|P(\lambda_n)|}{2} \\ &\geq \frac{1}{2}\inf|P(\lambda_n)|. \end{aligned}$$

Combining the fact that, for all $i \in \{1, \dots, p\}$, $|\lambda_n - z_i| > r$, and Lemma 3.3.1, we conclude that $\inf|P(\lambda_n)| > 0$, and the sequence $(P(A)\psi_n)_n$ does not contain a convergent subsequence, which contradicts the compactness of $P(A)$. Q.E.D.

Remark 3.3.1. Let $A \in \mathcal{PK}(X)$ i.e., there exists a nonzero complex polynomial $P(z) = \sum_{r=0}^{p} a_r z^r$ satisfying $P(A) \in \mathcal{K}(X)$ and let $\lambda \in \sigma(A)$ such that $P(\lambda) \neq 0$. Then, using Eq. (3.1.7), we have $\text{mult}(A, \lambda) \leq \text{mult}(P(A), P(\lambda)) < \infty$. ◇

We end this part by the following corollary.

Corollary 3.3.1. *Let X be an infinite dimensional Banach space, $A \in \mathcal{PK}(X)$ and set $P(.)$ as the minimal polynomial of A. Let $P(z) = \sum_{r=0}^{p} a_r z^r$ and $z_1, \dots, z_p$ its zeros. Then, A is a generalized Riesz operator with $E = \{z_1, \dots, z_p\}$.* ◇

3.4 Localization of Eigenvalues of Polynomially Compact Operators

In this section, we give some results about multiplicities and localization of the eigenvalues of polynomially compact operators.

Proposition 3.4.1. *Let* $A \in \mathcal{PK}(X)$*, i.e., there exists a nonzero complex polynomial* $P(z) = \sum_{r=0}^{p} a_r z^r$ *satisfying* $P(A) \in \mathcal{K}(X)$*. The multiplicity of nonzero eigenvalues* μ *of* $P(A)$ *satisfies*

$$mult(P(A), \mu) = \sum_{\substack{\lambda \in \sigma(A) \\ P(\lambda)=\mu}} mult(A, \lambda)$$

and, for all $\lambda \in \sigma(A)$ *with* $P(\lambda) \neq 0$*, we have* $mult(A, \lambda) = mult(A^*, \lambda)$*.* ◇

Proof. Let $\mu \in \sigma(P(A))\backslash\{0\}$. By the spectral mapping theorem, we have $\sigma(P(A)) = P(\sigma(A))$. Then, we can find λ in $\sigma(A)$ such that $P(\lambda) = \mu$. So, using Remark 3.3.1, we have mult$(P(A), \mu) < \infty$ and mult$(A, \lambda) < \infty$. However, the number of values of λ distinct as $P(\lambda) = \mu$ is finite. Then, by Eq. (3.1.7), we obtain

$$\bigcup_{\substack{\lambda \in \sigma(A) \\ P(\lambda)=\mu}} \bigcup_{n \in \mathbb{N}} N[(\lambda - A)^n] \subset \bigcup_{n \in \mathbb{N}} N[(\mu - P(A))^n]. \tag{3.4.1}$$

Combining the fact that $\mu = P(\lambda) \neq 0$ and Theorem 3.1.2, we have asc$(\lambda - A) < \infty$ and

$$\bigcup_{n \in \mathbb{N}} N[(\lambda - A)^n] = N[(\lambda - A)^{n_\lambda}] \tag{3.4.2}$$

where $n_\lambda =$ asc$(\lambda - A)$. Using Eqs (3.4.1) and (3.4.2), we have

$$\bigcup_{\substack{\lambda \in \sigma(A) \\ P(\lambda)=\mu}} N[(\lambda - A)^{n_\lambda}] \subset \bigcup_{n \in \mathbb{N}} N[(\mu - P(A))^n]. \tag{3.4.3}$$

Let $p_1(\lambda) = (\lambda_1 - \lambda)^{n_{\lambda_1}}$ and $p_2(\lambda) = (\lambda_2 - \lambda)^{n_{\lambda_2}}$. Since $\lambda_1 \neq \lambda_2$, the polynomials p_1, p_2 are relatively prime, and hence there exist two polynomials $q_1(\lambda)$ and $q_2(\lambda)$, such that $q_1(\lambda)p_1(\lambda) + q_2(\lambda)p_2(\lambda) = 1$. So, for $x \in X$, we have

$$q_1(A)p_1(A)x + q_2(A)p_2(A)x = x. \tag{3.4.4}$$

In particular, if $x \in N[(\lambda_1 - A)^{n_{\lambda_1}}] \bigcap N[(\lambda_2 - A)^{n_{\lambda_2}}]$, we see that $p_1(A)x = p_2(A)x = 0$. Hence, from Eq. (3.4.4), $x = 0$. So, $N[(\lambda_1 - A)^{n_{\lambda_1}}] \bigcap N[(\lambda_2 - A)^{n_{\lambda_2}}] = \{0\}$. Therefore, Eq. (3.4.3) leads to

$$\sum_{\substack{\lambda \in \sigma(A) \\ P(\lambda)=\mu}} \text{mult}(A, \lambda) \leq \text{mult}(P(A), \mu).$$

Since $\text{mult}(P(A), \mu) < \infty$, then $\text{asc}(\mu - P(A)) < \infty$. Let $n_0 = \text{asc}(\mu - P(A))$ and

$$Y := \bigcup_{n \in \mathbb{N}} N[(\mu - P(A))^n] = N[(\mu - P(A))^{n_0}].$$

Define $A_0 := A_{|Y} : Y \longrightarrow Y$ as the restriction of A to Y. Let $\lambda_0 \in \sigma(A_0)$ be the eigenvalue of A_0 and choose $y \in Y$ as an eigenvector partner to λ_0, i.e., $[(\mu - P(A))^{n_0}]y = 0$ and $Ay = \lambda_0 y$. Since $P(A)$ and $P(\lambda_0)I$ commute, we have

$$\begin{aligned}
[P(\lambda_0) - \mu]^{n_0} y &= [P(\lambda_0) - P(A) + P(A) - \mu]^{n_0} y \\
&= \sum_{k=0}^{n_0} C_{n_0}^k [P(\lambda_0) - P(A)]^k [P(A) - \mu]^{n_0 - k} y \\
&= [P(A) - \mu]^{n_0} y \\
&\quad + \sum_{k=1}^{n_0} C_{n_0}^k [P(A) - \mu]^{n_0 - k} [P(\lambda_0) - P(A)]^k y. \qquad (3.4.5)
\end{aligned}$$

Thus, $[P(\lambda_0) - \mu]^{n_0} y = 0$ with $y \neq 0$. Hence, $P(\lambda_0) = \mu$. Since Y is a finite dimensional space, then by Jordan decomposition theorem, we have

$$Y = \bigoplus_{\lambda \in \sigma(A_0)} N[(\lambda - A_0)^{p_\lambda}] = \bigoplus_{\lambda \in \sigma(A_0)} \bigcup_{n \in \mathbb{N}} N[(\lambda - A_0)^n]$$

where p_λ designates the multiplicity of λ in the characteristic polynomial of A_0. We can deduce

$$Y \subset \bigoplus_{\substack{\lambda \in \sigma(A) \\ P(\lambda)=\mu}} \bigcup_{n \in \mathbb{N}} N[(\lambda - A)^n],$$

since each $\lambda \in \sigma(A_0)$ belongs to $\sigma(A)$ and satisfies $P(\lambda) = \mu$. We conclude

$$\text{mult}(P(A), \mu) \leq \sum_{\substack{\lambda \in \sigma(A) \\ P(\lambda)=\mu}} \text{mult}(A, \lambda).$$

This proves that

$$\operatorname{mult}(P(A),\mu) = \sum_{\substack{\lambda\in\sigma(A)\\ P(\lambda)=\mu}} \operatorname{mult}(A,\lambda).$$

In order to complete the proof, we will verify that $\operatorname{mult}(A,\lambda) = \operatorname{mult}(A^*,\lambda)$ for $\lambda \in \sigma(A)$ and $P(\lambda) \neq 0$. Let $p = \operatorname{asc}(\lambda - A)$. By Theorem 3.1.2, Corollary 3.1.1 and Theorem 2.2.7, the operator $(\lambda - A)^p$ is shown to be a Fredholm operator with index zero. So,

$$\begin{aligned}\operatorname{mult}(A,\lambda) &= \dim N\left[(\lambda - A)^p\right]\\ &= \operatorname{codim} R\left[(\lambda - A)^p\right]\\ &= \dim N\left[(\lambda - A^*)^p\right]\\ &= \operatorname{mult}(A^*,\lambda).\end{aligned}$$

This completes the proof. Q.E.D.

Remark 3.4.1.

(i) Let A be a generalized Riesz operator and set $\lambda \in \sigma(A)\backslash E$ (where E is the set introduced in Definition 3.1.3). Let $Y := \bigcup_{n\in\mathbb{N}} N\left[\lambda - A\right]^n$. In Eq. (3.4.5), if we consider $P(z) = z$, we have $\sigma(A|_Y) = \{\lambda\}$.

(ii) Let A be a generalized Riesz operator. We denote by $(\lambda_n(A))_{n\in\mathbb{N}}$ the sequence of eigenvalues of A. Then, it can be ordered in the following way: the eigenvalues are nonincreasing in absolute values, i.e., $|\lambda_n(A)| \geq |\lambda_{n+1}(A)|$, and each eigenvalue is repeated as often as its multiplicity. For $k \in \mathbb{N}$, if there are less than k eigenvalues in $\sigma(A)\backslash E$, then we let $\lambda_k(A) = \ldots = \lambda_{k+1}(A) = \ldots = 0$. The order may not be unique, we choose a fixed order of this kind. ◇

Proposition 3.4.2. *Let A be a generalized Riesz operator, and set $(\lambda_n(A))_{n\in\mathbb{N}}$ the sequence of eigenvalues of A, ordered following the general rule of Remark 3.4.1. Let $n \in \mathbb{N}$ with $\lambda_n(A) \notin E$ (where E is the set introduced in Definition 3.1.3). Then, there is a n-dimensional subspace X_n of X, invariant under A, such that $A|_{X_n}$, the restriction of A to X_n, has precisely $\lambda_1(A),\ldots,\lambda_n(A)$ as its eigenvalues.* ◇

Proof. Let $\mu_1,\ldots,\mu_l$ represent the different eigenvalues appearing in the sequence $\lambda_1(A),\ldots,\lambda_n(A)$, we have $\mu_l = \lambda_n(A)$ and

$$\sum_{k=1}^{l-1} \operatorname{mult}(A,\mu_k) < n \leq \sum_{k=1}^{l} \operatorname{mult}(A,\mu_k), \tag{3.4.6}$$

since each eigenvalue is repeated as often as its multiplicity. Let $p := n - \sum_{k=1}^{l-1} \operatorname{mult}(A,\mu_k)$, then by Eq. (3.4.6), $0 < p \leq \operatorname{mult}(A,\mu_l)$. For all $j \in \{1,\ldots,l\}$, let $Y_j := \bigcup_{n\in\mathbb{N}} N\left[(\mu_j I - A)^n\right]$. Then, we have $A(Y_j) \subset Y_j$ and $Y_i \bigcap Y_j = \{0\}$ for $i \neq j$, since the eigenvalues $\mu_1,\ldots,\mu_l$ are different. We also have $k := \dim Y_l = \text{mult}(A, \mu_l) < \infty$. Then, by the Jordan decomposition theorem, Y_l has a basis of vectors $e_1, \dots, e_k$ such that $A|_{Y_l} : Y_l \longrightarrow Y_l$ has a matrix representation with basis $e_1, \dots, e_k$ which is block diagonal sum of matrix of the form

$$\begin{pmatrix} \mu_l & 1 & & \\ & & \ddots & \\ & & . & 1 \\ & & & \mu_l \end{pmatrix},$$

since $\sigma(A|_{Y_l}) = \{\mu_l\}$ by Remark 3.4.1. Thus, for all $i \in \{1, \dots, k\}$, $A(e_i) \in \text{span}\{e_1, \dots, e_i\}$. Let $Y_l' := \text{span}\{e_1, \dots, e_p\}$ and define $X_n := \oplus_{j=1}^{l-1} Y_j \oplus Y_l'$. Then, X_n is invariant under A, since Y_l' and Y_j for all $j \in \{1, \dots, l-1\}$ are invariant under A. We have $\dim X_n = \sum_{k=1}^{l-1} \text{mult}(A, \mu_k) + p = n$ and $A|_{X_n}$ has as eigenvalues precisely $\lambda_1(A), \dots, \lambda_n(A)$, thus completes the proof. Q.E.D.

Definition 3.4.1. Let X and Y be two Banach spaces. Two operators $A \in \mathcal{L}(X)$ and $B \in \mathcal{L}(Y)$ are related if there are two operators $S \in \mathcal{L}(X, Y)$ and $T \in \mathcal{L}(Y, X)$ such that $A = TS$ and $B = ST$. ◇

Lemma 3.4.1. *Let $A \in \mathcal{PK}(X)$ and set $P(z) = \sum_{k=0}^{p} a_k z^k$, the minimal polynomial of A, $z_1, \dots, z_p$ its zeros. Let $B \in \mathcal{L}(Y)$ such that A and B are related. Then, $B \in \mathcal{PK}(Y)$ and $\{0, z_1, \dots, z_p\} = \{0, \xi_1, \dots, \xi_q\}$, where $\xi_1, \dots, \xi_q$ are the zeros of $Q(.)$, the minimal polynomial of B.* ◇

Proof. For all $k \in \mathbb{N}$, we have $B^{k+1} = SA^kT$. Then, $BP(B) = SP(A)T$. Thus, $B \in \mathcal{PK}(Y)$ and $Q(.)$ the minimal polynomial of B, divide the polynomial $Q_1(.)$, where $Q_1(z) = zP(z)$. We conclude that $\{\xi_1, \dots, \xi_q\} \subset \{0, z_1, \dots, z_p\}$. Similarly, for all $k \in \mathbb{N}$, $A^{k+1} = TB^kS$ yields $\{z_1, \dots, z_p\} \subset \{0, \xi_1, \dots, \xi_q\}$. It follows that $\{0, z_1, \dots, z_p\} = \{0, \xi_1, \dots, \xi_q\}$. Q.E.D.

We close this section with the following result.

Theorem 3.4.1. *Let X, Y be two infinite dimensional Banach spaces. Let $A \in \mathcal{PK}(X)$ and set $P(z) = \sum_{k=0}^{p} a_k z^k$, the minimal polynomial of A, $z_1, \dots, z_p$ its zeros. Let $B \in \mathcal{L}(Y)$ such that A and B are related. Then, A and B are generalized Riesz operators with $\sigma(A)\backslash\{0, z_1, \dots, z_p\} = \sigma(B)\backslash\{0, z_1, \dots, z_p\}$ and, for all $\lambda \in \sigma(A)\backslash\{0, z_1, \dots, z_p\}$, we have mult$(A, \lambda) =$ mult(B, λ).* ◇

Proof. Let $\lambda \in \rho(A)\backslash\{0, z_1, \dots, z_p\}$ and set $\varphi \in N(\lambda - B)$. Then, $(\lambda - B)\varphi = 0$ and $(\lambda - A)T\varphi = 0$, since $TB = AT$. Thus, $T\varphi = 0$ and $B\varphi = 0$, which implies that $\varphi = 0$, since $\varphi \in N(\lambda - B)$ and $\lambda \neq 0$. It follows that $(\lambda - B)$ is injective. Let $Q(.)$ be the minimal polynomial of B. By Lemma 3.4.1, we have $Q(\lambda) \neq 0$. So, using Theorem 3.2.1, we deduce that $(\lambda - B)^{-1}$ exits and is bounded. Thus, $\lambda \in \rho(B)$. Similarly, $\lambda \in \rho(B)\backslash\{0, z_1, \dots, z_p\}$ which yields $\lambda \in \rho(A)$. Now, let $\lambda \in \sigma(A)\backslash\{0, z_1, \dots, z_p\}$ and set $n := \max(\text{asc}(\lambda - A), \text{asc}(\lambda - B))$. Since $(\lambda - A)^n T = T(\lambda - B)^n$, then we can define the operator H by

$$\begin{cases} H : N\left[(\lambda - B)^n\right] \longrightarrow N\left[(\lambda - A)^n\right] \\ \qquad\qquad\qquad \varphi \longrightarrow \dfrac{1}{\lambda^n} T\varphi. \end{cases}$$

Similarly, since $S(\lambda - A)^n = (\lambda - B)^n S$, then we can define the operator K by

$$\begin{cases} K : N\left[(\lambda - A)^n\right] \longrightarrow N\left[(\lambda - B)^n\right] \\ \qquad\qquad\qquad \varphi \longrightarrow \left(S \displaystyle\sum_{k=1}^{n} C_n^k (-1)^{k-1} \lambda^{n-k} A^{k-1} \right) \varphi. \end{cases}$$

Let $\varphi \in N\left[(\lambda - B)^n\right]$. Since $SA^k T = B^{k+1}$, we have

$$\begin{aligned} KH\varphi &= \frac{1}{\lambda^n} \left(\sum_{k=1}^{n} C_n^k (-1)^{k-1} \lambda^{n-k} S A^{k-1} T \right) \varphi \\ &= \frac{1}{\lambda^n} \left(\sum_{k=1}^{n} C_n^k (-1)^{k-1} \lambda^{n-k} B^k \right) \varphi \\ &= -\frac{1}{\lambda^n} \left[(\lambda - B)^n - \lambda^n I \right] \varphi \\ &= \varphi. \end{aligned}$$

Hence, K is an onto operator. So, $\dim N\left[(\lambda - B)^n\right] \le \dim N\left[(\lambda - A)^n\right]$. Similarly, we consider

$$\begin{cases} \tilde{H} : N\left[(\lambda - A)^n\right] \longrightarrow N\left[(\lambda - B)^n\right] \\ \qquad\qquad\qquad \varphi \longrightarrow \dfrac{1}{\lambda^n} S\varphi. \end{cases}$$

Since $(\lambda - A)^n T = T(\lambda - B)^n$, then we can define

$$\begin{cases} \tilde{K} : N\left[(\lambda - B)^n\right] \longrightarrow N\left[(\lambda - A)^n\right] \\ \qquad\qquad\qquad \varphi \longrightarrow \left(T \displaystyle\sum_{k=1}^{n} C_n^k (-1)^{k-1} \lambda^{n-k} B^{k-1} \right) \varphi. \end{cases}$$

Let $\varphi \in N\left[(\lambda - A)^n\right]$. Since $TB^k S = A^{k+1}$, we have

$$\begin{aligned} \tilde{K}\tilde{H}\varphi &= \frac{1}{\lambda^n} T \left(\sum_{k=1}^{n} C_n^k (-1)^{k-1} \lambda^{n-k} B^{k-1} S \right) \varphi \\ &= \frac{1}{\lambda^n} \left(\sum_{k=1}^{n} C_n^k (-1)^{k-1} \lambda^{n-k} A^k \right) \varphi \\ &= -\frac{1}{\lambda^n} \left[(\lambda - A)^n - \lambda^n I \right] \varphi \\ &= \varphi. \end{aligned}$$

Hence, $\tilde{K}$ is an onto operator. So, $\dim N[(\lambda - A)^n] \leq \dim N[(\lambda - B)^n]$. Hence, we have $\dim N[(\lambda - A)^n] = \dim N[(\lambda - B)^n]$, and then $\text{mult}(A, \lambda) = \text{mult}(B, \lambda)$. Q.E.D.

3.5 Polynomially Riesz Operators

We say that $A \in \mathcal{L}(X)$ is polynomially Riesz operator if there exists a nonzero complex polynomial $p(.)$, such that the operator $p(A) \in \mathcal{R}(X)$. The set of polynomially Riesz operators will be denoted by $P\mathcal{R}(X)$. If $p(.)$ is the nonzero polynomial of the least degree and leading coefficient 1, such that $p(A) \in P\mathcal{R}(X)$, it will be called the minimal polynomial of A. If A belongs to $P\mathcal{R}(X)$, then there exists a nonzero polynomial $p(.)$, such that $p(A) \in \mathcal{R}(X)$. So, $\sigma(p(A))$ must be finite or countable with zero as the only possible accumulation point. Moreover, the nonzero points of $\sigma(p(A))$ are isolated and the corresponding spectral projections are all finite dimensional. According to the spectral mapping theorem, the only possible accumulation points of $\sigma(A)$ are contained in the set of roots of $p(.)$. Let λ_i be a root of $p(.)$ and assume that $\lambda_i \notin \sigma(A)$. Set $q(z) = (z - \lambda_i)^{-1} p(z)$. Obviously, $deg(q) < deg(p)$. Besides, since $q(A) = (A - \lambda_i)^{-1} p(A) = p(A)(A - \lambda_i)^{-1}$, by applying Proposition 2.2.3, we deduce that $q(A) \in P\mathcal{R}(X)$. The following results can be found in [229].

Proposition 3.5.1. *Let $A \in \mathcal{L}(X)$, assume that $\Omega \neq \emptyset$ is a connected open subset of $\mathbb{C}$ such that $\sigma(A) \subset \Omega$, and let $f : \Omega \longrightarrow \mathbb{C}$, $f \neq 0$ be an analytic function. If $f(A) \in \mathcal{R}(X)$, then $A \in P\mathcal{R}(X)$.* ◇

Proof. Obviously, f has only a finite number of zeros on $\sigma(A)$, say $\lambda_1, \ldots, \lambda_m$. Hence, $f(z) = (\prod_{i=1}^{m}(z - \lambda_i)^{\alpha_i})\zeta(z)$, where α_i is the order of the λ_i zero and $\zeta(z) \neq 0$ is an analytic function on a neighborhood of $\sigma(A)$. Set $\vartheta(z) = 1/\zeta(z)$. Clearly, ϑ is analytic on a neighborhood of $\sigma(A)$, and $\prod_{i=1}^{m}(z-\lambda_i)^{\alpha_i} = f(z)\vartheta(z) = \vartheta(z)f(z)$. Therefore, $\prod_{i=1}^{m}(A - \lambda_i)^{\alpha_i} = f(A)\vartheta(A) = \vartheta(A)f(A)$. Since $f(A) \in \mathcal{R}(X)$, and according to Proposition 2.2.3, $\prod_{i=1}^{m}(A - \lambda_i)^{\alpha_i}$ belongs to $\mathcal{R}(X)$ which ends the proof. Q.E.D.

Proposition 3.5.2. *Let $(T(t))_{t\geq 0}$ be a C_0-semigroup on a Banach space X with an infinitesimal generator A. Let n be an integer, and let φ be a function defined from its domain into $\mathbb{C}^n$, that is, $\varphi : \mathcal{D}(\varphi) \subset \mathbb{R} \longrightarrow \mathbb{C}^n$, $t \longrightarrow (\varphi_1(t), \ldots, \varphi_n(t))$. Set $\mathcal{D}^+(\varphi) := \mathcal{D}(\varphi) \bigcap]0, \infty[$. Assume that $\varphi(.)$ is continuous and, for all $t \in \mathcal{D}^+(\varphi)$, $\varphi_i(t) \neq 0$ and $\prod_{i=1}^{n}(T(t) - \varphi_i(t)) \in \mathcal{R}(X)$. If $\mathcal{D}^+(\varphi) \neq \emptyset$, then the set $\mathcal{I} := \{\text{Im}\lambda$ such that $\lambda \in \sigma_p(A)\}$ is bounded.* ◇

Proof. In order to prove this, we will proceed by contradiction. If $\mathcal{I}$ is unbounded, then there exists a sequence $(a_k)_k$ in $\mathcal{I}$ such that $a_{k+1} > a_k$ and $\lim\limits_{k\to\infty} a_k = +\infty$ or $a_{k+1} < a_k$ and $\lim\limits_{k\to\infty} a_k = -\infty$. The treatment of these two cases is the same. Hence, we restrict ourselves to the first one. Thus, there exists a sequence $(\lambda_k)_k$, such that $\lambda_k \in \sigma_p(A)$ and $a_k = \text{Im}\lambda_k$. So, $e^{t\lambda_k} \in \sigma_p(T(t))$ and $\arg(e^{t\lambda_k}) = ta_k + 2m\pi$ for some $m \in \mathbb{Z}$. Since $\varphi_i(t)$, $i = 1, \dots, n$ represent the only possible accumulation points of $\sigma_p(T(t))$, then for all $\alpha > 0$, $\left\{k \in \mathbb{N} \text{ such that } \sup_{1 \leq i \leq n} |e^{t\lambda_k} - \varphi_i(t)| > \alpha\right\}$ is finite. So, for all ε satisfying $0 < \varepsilon < 1/2$, $\left\{k \in \mathbb{N} \text{ such that } \sup_{1 \leq i \leq n,\, m \in \mathbb{Z}} |ta_k - \arg(\varphi_i(t)) + 2m\pi| > \frac{\varepsilon}{2}\right\}$ is finite, i.e., $\left\{k \in \mathbb{N} \text{ such that } ta_k \notin \bigcup_{1 \leq i \leq n,\, m \in \mathbb{Z}} \left[\arg(\varphi_i(t)) + 2m\pi - \frac{\varepsilon}{2}, \arg(\varphi_i(t)) + 2m\pi + \frac{\varepsilon}{2}\right]\right\}$ is finite. Let $t_0 > 0$ be a fixed point, such that the intervals are disjoint or identical in the set $G_\varepsilon = \bigcup_{1 \leq i \leq n,\, m \in \mathbb{Z}} \left[\arg(\varphi_i(t_0)) + 2m\pi - \frac{\varepsilon}{2}, \arg(\varphi_i(t_0)) + 2m\pi + \frac{\varepsilon}{2}\right]$. We also choose a determination of $\arg(\varphi_i(t_0))$ such that none of $\varphi_i(t_0)$ is on the half-axis of discontinuity of arg(.). This is possible, since they are in finite number and also different from zero. Clearly, the complement of G_ε in $\mathbb{R}$ is a reunion of open intervals. Hence, G_ε is closed. Let $\delta > 0$ be such that $|t - t_0| < \delta$ implies $|\arg(\varphi_i(t)) - \arg(\varphi_i(t_0))| < \frac{\varepsilon}{2}$ for $i = 1, 2, \dots, n$ (use the continuity of $\varphi_i(.)$ and arg(.) at $\varphi_i(t_0)$). Then, for all $t \in]t_0 - \delta, t_0 + \delta[$, $\{k \in \mathbb{N} \text{ such that } ta_k \notin G_\varepsilon\}$ is finite and, for all $t \in]t_0 - \delta, t_0 + \delta[$, there is $N_t \in \mathbb{N}$ such that $k \geq N_t$ implies $ta_k \in G_\varepsilon$, (or $t \in \frac{1}{a_k} G_\varepsilon$) and $]t_0 - \delta, t_0 + \delta[\subset \bigcup_{N \in \mathbb{N}} \left(\bigcap_{k \geq N} \frac{1}{a_k} G_\varepsilon\right)$. Using the Baire category theorem allows us to conclude that there exists $N \in \mathbb{N}$, such that $\bigcap_{k \geq N} \frac{1}{a_k} G_\varepsilon$ has a nonempty interior. Accordingly, there are a and b in $]0, \infty[$ with $a < b$ and $]a, b[\subset \bigcap_{k \geq N} \frac{1}{a_k} G_\varepsilon$. However, $]a, b[$ would be contained in one of the connected components of $\bigcap_{k \geq N} \frac{1}{a_k} G_\varepsilon$. Consequently, $b - a \leq \frac{2\varepsilon}{a_k}$ for every $k \in \mathbb{N}$, which leads to a contradiction because $a_k \to \infty$ as $k \to \infty$. This completes the proof. Q.E.D.

3.6 Some Results on Polynomially Fredholm Perturbation

Definition 3.6.1. An operator $A \in \mathcal{L}(X)$ is said to be polynomially Fredholm perturbation if there exists a nonzero complex polynomial P such that $P(A)$ is a Fredholm perturbation. ◇

We denote by $\mathcal{PF}(X)$ the set of polynomially Fredholm perturbation defined by

$$\mathcal{PF}(X) := \Big\{A \in \mathcal{L}(X) \text{ such that there exists a nonzero complex polynomial } P(z) = \sum_{n=0}^{p} a_n z^n \text{ satisfing } P(A) \in \mathcal{F}^b(X)\Big\}.$$

Lemma 3.6.1. *Let P be a complex polynomial and $\lambda \in \mathbb{C}$ such that $P(\lambda) \neq 0$. Then, for all $A \in \mathcal{L}(X)$ satisfying $P(A) \in \mathcal{F}^b(X)$, the operator $P(\lambda) - P(A)$ is a Fredholm operator on X with finite ascent and descent.* ◇

Proof. Put $B = P(\lambda) - P(A) = P(\lambda)\left(I - \frac{P(A)}{P(\lambda)}\right) = P(\lambda)(I - F)$, where $F = \frac{P(A)}{P(\lambda)} \in \mathcal{F}^b(X)$. Let $C = I - F$, then $B = P(\lambda)C$. It is clear that $C + F \in \mathcal{B}^b(X)$ and so we can write $C = C + F - F$, where $C + F \in \mathcal{B}^b(X)$ and $F \in \mathcal{F}^b(X)$. Moreover, we have $(C + F)F = F(C + F)$. By using Theorem 2.2.21, we deduce that $C \in \mathcal{B}^b(X)$ and therefore, $B \in \mathcal{B}^b(X)$. Q.E.D.

Theorem 3.6.1. *Let $A \in \mathcal{PF}(X)$, i.e., there exists a nonzero complex polynomial $P(z) = \sum_{i=0}^{n} a_i z^i$ satisfying $P(A) \in \mathcal{F}^b(X)$. Let $\lambda \in \mathbb{C}$, with $P(\lambda) \neq 0$. Then, $\lambda - A$ is a Fredholm operator on X with index zero and with finite ascent and descent.* ◇

Proof. The proof follows immediately from Theorem 3.1.2. Q.E.D.

Chapter 4
Time-Asymptotic Description of the Solution for an Abstract Cauchy Problem

In this chapter, we give a description of the large time behavior of solutions to an abstract Cauchy problem on Banach spaces without restriction on the initial data. Let X be a Banach space and let $T : \mathcal{D}(T) \subseteq X \longrightarrow X$ be the infinitesimal generator of a C_0-semigroup of bounded linear operators $(U(t))_{t\geq 0}$ acting on X. We consider the Cauchy problem

$$\begin{cases} \dfrac{\partial \psi}{\partial t} = A\psi := T\psi + F\psi \\ \psi(0) = \psi_0, \end{cases} \tag{4.0.1}$$

where F is a bounded linear operator on X and $\psi_0 \in X$.

4.1 Abstract Cauchy Problem

Consider Eq. (4.0.1) in the Banach space X. Since $A := T + F$ is a bounded perturbation of T, by the classical perturbation theory (Theorem 2.5.8), it generates a C_0-semigroup $(V(t))_{t\geq 0}$ which solves the Cauchy problem (4.0.1) and is given by the Dyson–Phillips expansion series (the finite iteration of the Duhamel formula)

$$V(t) = \sum_{j=0}^{n-1} U_j(t) + R_n(t),$$

A. Jeribi, *Spectral Theory and Applications of Linear Operators and Block Operator Matrices*, DOI 10.1007/978-3-319-17566-9_4

where $U_0(t) = U(t)$, $U_j(t) = \int_0^t U(s)FU_{j-1}(t-s)\,ds,\ j = 1, 2, \ldots$ and the nth order remainder term $R_n(t)$ can be expressed by

$$\begin{aligned} R_n(t) &= \sum_{j=n}^{\infty} U_j(t) \\ &= \int_{s_1+\cdots+s_n \le t,\ s_i \ge 0} U(s_1)F \ldots U(s_n)FV(t - s_1 - \ldots - s_n)\,ds_1 \ldots ds_n. \end{aligned}$$

Proposition 4.1.1 ([331]). *With the notation introduced above, suppose that there exist an integer n and a sequence $(t_k)_k$ in $[0, \infty)$, $t_k \to \infty$, such that $R_n(t_k)$ is strictly power-compact for all $k \in \mathbb{N}$. Then, $r_e(V(t)) \le r_\sigma(U(t))$ for all $t \ge 0$, where $(V(t))_{t\ge 0}$ is the C_0-semigroup generated by $T + F$.* ◇

Proof. Let w be the type of $(U(t))_{t\ge 0}$, $r_\sigma(U(t)) = e^{tw}$ $(t > 0)$, and let w'_e be such that $r_e(V(t)) = e^{tw'_e}$. Let $w < w' < w''$, then there exists $M \ge 0$ such that $\|U(t)\| \le Me^{tw'}$ $(t \ge 0)$. It follows that $\|U_j(t)\| \le M^{j+1}e^{tw'}\frac{t^j}{j!}$ $(t \ge 0,\ j \in \mathbb{N}^*)$, and $\left\|\sum_{j=0}^{n-1} U_j(t)\right\| \le e^{tw'}p_{n-1}(t)$, where $p_{n-1}(t) = \sum_{j=0}^{n-1}\frac{M^{j+1}}{j!}t^j$ is a polynomial of degree $n-1$. From $w' < w''$, it follows that $\left\|\sum_{j=0}^{n-1} U_j(t_k)\right\| \le e^{t_k w'}p_{n-1}(t_k) \le e^{t_k w''}$ for a large k, and now, Corollary 2.6.3 implies $e^{t_k w'_e} = r_e(V(t_k)) = r_e\left(\sum_{j=0}^{n-1} U_j(t_k)\right) \le e^{t_k w''}$, $w'_e \le w''$. Since this is true for any $w'' > w$, we obtain $w'_e \le w$, as asserted. Q.E.D.

For more details related to the results of this section, we may refer to [215].

4.1.1 Compactness Results

Lemma 4.1.1. *The following two statements are equivalent:*

(i) The operator $U(t)F$ is compact on X for every $t > 0$.
(ii) The map $(0, +\infty) \ni t \longrightarrow U(t)F$ is continuous in the uniform topology and $(\lambda - T)^{-1}F$ is compact on X for some (every) $\lambda \in \rho(T)$. ◇

Proof. From Theorem 2.5.2, it follows that there are constants $w \ge 0$ and $M \ge 1$ such that $\|U(t)\| \le Me^{wt}$ for $t \ge 0$. Assume that the operator $U(t)F$ is compact on X for every $t > 0$. Since F is bounded and using Theorem 2.5.5, we get the continuity of the map $(0, +\infty) \ni t \longrightarrow U(t)F$ in the uniform topology. Now, we may prove the second part of the assertion. Indeed, from the boundedness of F, it follows from Theorem 2.5.10, that $(\lambda - T)^{-1}F = \int_0^\infty e^{-\lambda t}U(t)F\,dt$, for $\mathrm{Re}\lambda > w$ and the integral exists in the uniform operator topology. Let $\delta > 0$, $\mathrm{Re}\lambda > w$ and

$W(\lambda,\delta) = \int_\delta^\infty e^{-\lambda t}U(t)F dt$. Since $U(t)F$ is compact for every $t > 0$, the use of Theorem 2.5.6 implies the compactness of $W(\lambda,\delta)$. Hence, the estimate

$$\|(\lambda - T)^{-1}F - W(\lambda,\delta)\| = \left\| \int_0^\delta e^{-\lambda t}U(t)F dt \right\|$$
$$\leq \delta M\|F\| \to 0 \text{ as } \delta \to 0$$

shows the compactness of $(\lambda - T)^{-1}F$. Now, the resolvent identity gives the compactness of $(\lambda - T)^{-1}F$ for every $\lambda \in \rho(T)$. Conversely, let $t > 0$. Since $U(t)F$ is bounded, we may write $\lambda(\lambda - T)^{-1}U(t)F = \int_0^\infty \lambda e^{-\lambda s}U(t+s)F ds$ for every $\lambda > w$. Further, the use of the relation $\int_0^\infty \lambda e^{-\lambda s}ds = 1$ leads to $\lambda(\lambda - T)^{-1}U(t)F - U(t)F = \int_0^\infty \lambda e^{-\lambda s}[U(t+s)F - U(t)F]ds$. Let $\lambda > w$. Then, for every $\delta > 0$, we have

$$\begin{aligned}
&\|\lambda(\lambda - T)^{-1}U(t)F - U(t)F\| \\
&\leq \int_0^\infty \lambda e^{-\lambda s}\|U(t+s)F - U(t)F\|ds \\
&= \int_0^\delta \lambda e^{-\lambda s}\|U(t+s)F - U(t)F\|ds \\
&\quad + \int_\delta^\infty \lambda e^{-\lambda s}\|U(t+s)F - U(t)F\|ds \\
&\leq \sup_{0\leq s\leq\delta} \|U(t+s)F - U(t)F\| \\
&\quad + \lambda M\|F\|e^{wt}\left[(\lambda - w)^{-1}e^{-(\lambda-w)\delta} + \lambda^{-1}e^{-\lambda\delta}\right].
\end{aligned}$$

Therefore, for every $\delta > 0$, we get $\lim_{\lambda\to+\infty} \|\lambda(\lambda - T)^{-1}U(t)F - U(t)F\| \leq \sup_{0\leq s\leq\delta} \|U(t+s)F - U(t)F\|$. Since $\delta > 0$ is arbitrary, we have

$$\lim_{\lambda\to+\infty} \|\lambda(\lambda - T)^{-1}U(t)F - U(t)F\| = 0. \tag{4.1.1}$$

Finally, the compactness of $U(t)F$ ($t > 0$) follows from Eq. (4.1.1) and the commutativity of the operators $(\lambda - T)^{-1}$ and $U(t)$. Q.E.D.

Theorem 4.1.1. *Suppose that the map $U(.)F : (0,\infty) \longrightarrow \mathcal{L}(X)$, $t \longrightarrow U(t)F$ is continuous in the uniform topology and that, for some (every) $\lambda \in \rho(T)$, $(\lambda - T)^{-1}F$ is compact. Then, $V(t) - U(t) = \int_0^t U(t-s)FV(s)ds$ is compact on X.* ◇

Proof. The result is immediate if $t = 0$. Now, suppose that $t > 0$. Clearly, Lemma 4.1.1 implies the compactness of $U(t)F$. Accordingly, the operator $U(s)FV(t-s)$ is also compact for every $s \in (0,t]$. Therefore, using Duhamel formula and applying Theorem 2.5.6, one concludes the compactness of $V(t)-U(t)$ for every $t > 0$. Q.E.D.

Corollary 4.1.1. *Suppose that the map* $FU(.) : (0,\infty) \longrightarrow \mathcal{L}(X)$, $t \longrightarrow FU(t)$ *is continuous in the uniform topology and that, for some (every)* $\lambda \in \rho(T)$, $F(\lambda-T)^{-1}$ *is compact. Then,* $V(t) - U(t) = \int_0^t V(t-s)FU(s)ds$ *is compact on* X. ◇

Corollary 4.1.2. *Assume that the hypotheses of Theorem 4.1.1 are satisfied. Then, the map* $(0,+\infty) \ni t \longrightarrow R_1(t)F$ *is continuous in the uniform topology.* ◇

Proof. By virtue of the hypotheses, it is sufficient to show the continuity of the map $(0,+\infty) \ni t \longrightarrow V(t)F$ in the uniform topology. To do this, we first observe that Lemma 4.1.1 and Theorem 4.1.1 imply the compactness of $U(t)F$ and $R_1(t)F$ for all $t > 0$. Therefore, $V(t)F$ is compact for all $t > 0$. Proceeding as in the first part of the proof of Lemma 4.1.1 we obtain the continuity of the map $(0,+\infty) \ni t \longrightarrow V(t)F$ in the uniform topology, which completes the proof. Q.E.D.

4.1.2 The Remainder Term of the Dyson–Philips Expansion

Lemma 4.1.2. *Let* $n \geq 1$ *be a fixed integer and suppose that, for all* $\lambda > w$, *the operator* $F\prod_{i=1}^{n}\left((\lambda-T)^{-1}U(t_i)F\right)$ *is compact for all n-tuples* $(t_1,\dots,t_n)$, $t_i > 0$ *and the map* $(t_1,\dots,t_n) \longrightarrow F\prod_{i=1}^{n}(U(t_i)F)$ *is continuous in the uniform topology. Then, the operator* $F\prod_{i=1}^{n}(U(t_i)F)$ *is compact on* X *for all n-tuples* $(t_1,\dots,t_n)$, $t_i > 0$ *with* $i = 1,\dots,n$. ◇

Proof. For the sake of simplicity, we will only consider the case $n = 2$. The general case can be treated similarly. Let $\lambda > w$ ($\lambda \in \rho(T)$), we can write

$$\begin{aligned}
&\lambda^2 F(\lambda-T)^{-1}U(t_1)F(\lambda-T)^{-1}U(t_2)F - FU(t_1)FU(t_2)F \\
&\quad = \lambda^2 \int_0^\infty Fe^{-\lambda t}U(t+t_1)dt \int_0^\infty Fe^{-\lambda s}U(s+t_2)Fds - FU(t_1)FU(t_2)F \\
&\quad = \lambda^2 \int_0^\infty \int_0^\infty e^{-\lambda(t+s)}FU(t+t_1)FU(s+t_2)Fdtds \\
&\qquad - \lambda^2 \int_0^\infty \int_0^\infty e^{-\lambda(t+s)}FU(t_1)FU(t_2)Fdtds \\
&\quad = \lambda^2 \int_0^\infty \int_0^\infty e^{-\lambda(t+s)}\left[FU(t+t_1)FU(s+t_2)F - FU(t_1)FU(t_2)F\right]dtds.
\end{aligned}$$

Let $\delta > 0$. Writing $J(t,s,t_1,t_2) = \lambda^2 e^{-\lambda(t+s)}\left[FU(t+t_1)FU(s+t_2)F - FU(t_1)\right.$ $\left.FU(t_2)F\right]$, we have

$$\int_0^\infty \int_0^\infty J(t,s,t_1,t_2)dtds$$

$$= \int_0^\delta \int_0^\delta J(t,s,t_1,t_2)dtds + \int_0^\delta dt \int_\delta^\infty J(t,s,t_1,t_2)ds$$
$$+ \int_\delta^\infty dt \int_0^\delta J(t,s,t_1,t_2)ds + \int_\delta^\infty \int_\delta^\infty J(t,s,t_1,t_2)dtds.$$

Putting

$$I_1 = \int_0^\delta \int_0^\delta J(t,s,t_1,t_2)dtds,$$
$$I_2 = \int_0^\delta dt \int_\delta^\infty J(t,s,t_1,t_2)ds,$$
$$I_3 = \int_\delta^\infty dt \int_0^\delta J(t,s,t_1,t_2)ds, \text{ and}$$
$$I_4 = \int_\delta^\infty \int_\delta^\infty J(t,s,t_1,t_2)dtds$$

we get $\int_0^\infty \int_0^\infty J(t,s,t_1,t_2)dtds = I_1 + I_2 + I_3 + I_4$. Using the estimate

$$\|I_1\| \le \int_0^\delta \int_0^\delta \lambda^2 e^{-\lambda(t+s)} \|FU(t+t_1)FU(s+t_2)F - FU(t_1)FU(t_2)F\|dsdt$$
$$\le \sup_{0\le t,s\le\delta} \|FU(t+t_1)FU(s+t_2)F - FU(t_1)FU(t_2)F\| \left(\int_0^\delta \lambda e^{-\lambda t}dt\right)^2$$
$$\le \sup_{0\le t,s\le\delta} \|FU(t+t_1)FU(s+t_2)F - FU(t_1)FU(t_2)F\|.$$

Next,

$$\|I_2\| \le \int_0^\delta \int_\delta^\infty \lambda^2 e^{-\lambda(t+s)} \|FU(t+t_1)FU(s+t_2)F - FU(t_1)FU(t_2)F\|dsdt$$
$$\le \int_\delta^\infty \lambda e^{-\lambda s} \sup_{0\le t\le\delta} \|FU(t+t_1)FU(s+t_2)F$$
$$-FU(t_1)FU(t_2)F\|ds \left(\int_0^\delta \lambda e^{-\lambda t}dt\right).$$

Obviously, the estimate

$$\sup_{0\le s\le\delta} \|FU(t+t_1)FU(s+t_2)F - FU(t_1)FU(t_2)F\|$$
$$\le \|F\|^3 M^2 e^{w(t_1+t_2+t+\delta)} + \|F\|^3 M^2 e^{w(t_1+t_2)}$$
$$= \|F\|^3 M^2 e^{w(t_1+t_2)}\left[1 + e^{w(t+\delta)}\right]$$

and $\int_\delta^\infty e^{-\lambda t} e^{w(t+\delta)} dt = e^{w\delta} \frac{e^{-(\lambda-w)\delta}}{\lambda-w}$ leads to

$$\|I_2\| \le \int_\delta^\infty \lambda e^{-\lambda t} \|F\|^3 M^2 e^{w(t_1+t_2+t+\delta)} dt + \int_\delta^\infty \lambda e^{-\lambda t} \|F\|^3 M^2 e^{w(t_1+t_2)} dt.$$

Since $\int_\delta^\infty \lambda e^{-\lambda t} dt = e^{-\lambda\delta}$, we obtain:

$$\|I_2\| \le \|F\|^3 M^2 e^{w(t_1+t_2)} \left[e^{-\lambda\delta} + \frac{e^{-\delta(\lambda-2w)}}{\lambda - w} \right]. \tag{4.1.2}$$

Note that I_3 is similar to I_2. Then, it satisfies the estimate (4.1.2). Now, let us consider I_4. It is easy to see that

$$\|I_4\| \le \|F\|^3 M^2 \left[\int_\delta^\infty \int_\delta^\infty \lambda^2 e^{-\lambda(t+s)} e^{w(t_1+t_2+t+s)} dt ds + \int_\delta^\infty \int_\delta^\infty \lambda^2 e^{-\lambda(t+s)} e^{w(t_1+t_2)} dt ds \right].$$

Hence, from the relation

$$\begin{aligned} &\int_\delta^\infty \int_\delta^\infty \lambda^2 e^{-\lambda(t+s)} \|F\|^3 M^2 e^{w(t_1+t_2+t+s)} dt ds \\ &= \|F\|^3 M^2 e^{w(t_1+t_2)} \left(\int_\delta^\infty \lambda e^{-(\lambda-w)s} ds \right)^2 \\ &= \|F\|^3 M^2 e^{w(t_1+t_2)} \left(\frac{\lambda}{\lambda - w} \right)^2 e^{-2(\lambda-w)\delta} \end{aligned}$$

and

$$\begin{aligned} \int_\delta^\infty \int_\delta^\infty \lambda^2 e^{-\lambda(t+s)} \|F\|^3 M^2 e^{w(t_1+t_2)} dt ds &= \|F\|^3 M^2 e^{w(t_1+t_2)} \left(\int_\delta^\infty \lambda e^{-\lambda s} ds \right)^2 \\ &= \|F\|^3 M^2 e^{w(t_1+t_2)} e^{-2\lambda\delta} \end{aligned}$$

it follows that

$$\|I_4\| \le \|F\|^3 M^2 e^{w(t_1+t_2)} \left[e^{-2\lambda\delta} + \frac{\lambda^2 e^{-2(\lambda-w)\delta}}{(\lambda - w)^2} \right]. \tag{4.1.3}$$

Since $\delta > 0$ is arbitrary, we have $\|I_1\| = 0$. Moreover, the estimates (4.1.2) and (4.1.3) imply $\lim_{\lambda\to\infty} \|I_i\| = 0$, $i = 2, 3, 4$. Consequently, we have

$$\lim_{\lambda\to\infty} \|\lambda^2 F(\lambda - T)^{-1} U(t_1) F(\lambda - T)^{-1} U(t_2) F - FU(t_1) FU(t_2) F\| = 0. \tag{4.1.4}$$

Now, the use of the compactness of $F(\lambda - T)^{-1}U(t_1)F(\lambda - T)^{-1}U(t_2)F$, together with Eq. (4.1.4), gives the compactness of $FU(t_1)FU(t_2)F$. This completes the proof. Q.E.D.

Theorem 4.1.2. *Under the assumptions of Lemma 4.1.2, the remainder term $R_{n+1}(t)$ of the Dyson–Phillips expansion is compact on X, for all $t \geq 0$.* $\diamond$

Proof. For $t = 0$, the result is obvious. Let $t > 0$. By Lemma 4.1.2, we have the compactness of $FU(t_1)FU(t_2)\ldots FU(t_n)F$ on X. Hence, the integrand of $R_{n+1}(t)$, $t > 0$ is compact on X. Now, the use of Theorem 2.5.6 ends the proof. Q.E.D.

Corollary 4.1.3. *Assume that the hypotheses of Theorem 4.1.2 are satisfied. Then, we have $\sigma(V(t))\bigcap\{\mu \in \mathbb{C}$ such that $|\mu| > e^{tw}\}$ consists of (at most) isolated eigenvalues with finite algebraic multiplicities.* $\diamond$

Proof. Clearly, the hypotheses of Theorem 4.1.2 imply the compactness of $R_{n+1}(t)$ for all $t \geq 0$ and $n \geq 1$. Therefore, for all $B \in \mathcal{L}(X)$ and $m \geq 1$, $(BR_{n+1}(t))^m$ is compact on X. Now, the use of Proposition 4.1.1 gives the desired assertion. Q.E.D.

4.2 Time Behavior of Solutions for an Abstract Cauchy Problem (4.0.1) on Banach Spaces

We suppose the following conditions:

$(\mathcal{A}_1)$: There exists an integer m such that $[(\lambda - T)^{-1}F]^m$ is compact for $Re\lambda > \eta$, where η is the type of $\{U(t),\ t \geq 0\}$

and

$(\mathcal{A}_2)$: There exists an integer m such that $\lim_{|\mathrm{Im}\lambda|\to+\infty} \|[(\lambda - T)^{-1}F]^m\| = 0$ uniformly on $\{\lambda \in \mathbb{C}$ such that $\mathrm{Re}\lambda \geq \omega\}$ $(\omega > \eta)$.

Lemma 4.2.1. *We assume that $(\mathcal{A}_1)$ and $(\mathcal{A}_2)$ are satisfied for some m. Then,*

(i) *$\sigma(A)\bigcap\{\lambda \in \mathbb{C}$ such that $\mathrm{Re}\lambda > \eta\}$ consists of, at most, isolated eigenvalues with finite algebraic multiplicities.*
(ii) *If $\omega > \eta$, then $\sigma(A)\bigcap\{\lambda \in \mathbb{C}$ such that $\mathrm{Re}\lambda \geq \omega\}$ is finite.*
(iii) *If $\omega > \eta$, then $\|(\lambda - A)^{-1}\|$ is uniformly bounded in $\{\lambda \in \mathbb{C}$ such that $\mathrm{Re}\lambda \geq \omega\}$ for large $|\mathrm{Im}\lambda|$.* $\diamond$

Proof.

(*i*) This result was obtained by Vidav [328]. First, let us notice that the function $\lambda \longrightarrow [(\lambda - T)^{-1}F]^m$ is regular analytic in the half-plane $\text{Re}\lambda > \eta$ and its values $[(\lambda - T)^{-1}F]^m$ are, by assumption $(\mathcal{A}_1)$, compact operators. Since, by Corollary 2.5.1, we have $\|[(\lambda - T)^{-1}F]^m\| \leq \|F\|^m \|(\lambda - T)^{-1}\|^m \leq \frac{\|F\|^m}{(\text{Re}\lambda - \eta)^m}$, $[(\lambda - T)^{-1}F]^m$ tends to zero if $\text{Re}\lambda \to \infty$. Therefore, 1 is not an eigenvalue for all operators $[(\lambda - T)^{-1}F]^m$. Hence, applying Shmul'yan theorem we infer that $[I - ((\lambda - T)^{-1}F)^m]^{-1}$ exists as a bounded everywhere defined operator for all λ in the half-plane $\text{Re}\lambda > \eta$, except for a discrete set of values λ_k, where the function $[I - ((\lambda - T)^{-1}F)^m]^{-1}$ has a pole. Since $[I - (\lambda - T)^{-1}F]^{-1} = \sum_{k=0}^{m-1}[(\lambda - T)^{-1}F]^k[I - ((\lambda - T)^{-1}F)^m]^{-1}$, the function $[I - (\lambda - T)^{-1}F]^{-1}$ has a similar behavior as $[I - ((\lambda - T)^{-1}F)^m]^{-1}$ in the half-plane $\text{Re}\lambda > \eta$. Let λ be such that $[I - (\lambda - T)^{-1}F]^{-1}$ exists. Put $R_\lambda = [I - (\lambda - T)^{-1}F]^{-1}(\lambda - T)^{-1}$ and $A_\lambda = \lambda - A = \lambda - T - F$. We remark that

$$(\lambda - T)^{-1}A_\lambda \subset I - (\lambda - T)^{-1}F \tag{4.2.1}$$

since the operator on the right is everywhere defined and on the left is not. Hence, we obtain $R_\lambda A_\lambda \subset I$. Now it is well known that the existence of $[I - (\lambda - T)^{-1}F]^{-1}$ implies the existence of $[I - F(\lambda - T)^{-1}]^{-1}$. Since $[I - (\lambda - T)^{-1}F]^{-1}(\lambda I - T)^{-1} = (\lambda - T)^{-1}[I - F(\lambda - T)^{-1}]^{-1}$, we have $A_\lambda R_\lambda = I$. Hence A_λ^{-1} exists for such a λ as a bounded everywhere defined operator and is equal to R_λ. Consequently, the resolvent $(\lambda - A)^{-1} = R_\lambda = [I - (\lambda - T)^{-1}F]^{-1}(\lambda - T)^{-1}$ is an analytic function of λ in the half-plane $\text{Re}\lambda > \eta$ with the exception of a discrete set of values λ_k, where R_λ has a pole. Any pole λ_k of R_λ is an eigenvalue of A. A corresponding eigenfunction ψ satisfies, according to (4.2.1), the equation $((\lambda_k - T)^{-1}F)\psi = \psi$, or $((\lambda_k - T)^{-1}F)^m\psi = ((\lambda_k - T)^{-1}F)\psi = \psi$. The operator $((\lambda_k - T)^{-1}F)^m$, being compact, the space of solutions of this equation is finite dimensional. This implies that the space of eigenfunctions of A corresponding to the eigenvalue X is finite dimensional too.

(*ii*) Let $a \in]0, 1[$. By $(\mathcal{A}_2)$ there exists $C = C(w)$ such that $\|[(\lambda - T)^{-1}F]^m\| \leq a < 1$ in the region $\mathcal{R}(w) := \{\lambda \in \mathbb{C} \text{ such that } \text{Re}\lambda \geq \omega \text{ and } |\text{Im}(\lambda)| \geq C(w)\}$. It follows that

$$r_\sigma((\lambda - T)^{-1}F) \leq a^{\frac{1}{m}} < 1 \text{ if } \lambda \in \mathcal{R}(w), \tag{4.2.2}$$

where $r_\sigma(.)$ denotes the spectral radius. But λ is an eigenvalue of A if, and only if, 1 is an eigenvalue of $(\lambda - T)^{-1}F$, so $\mathcal{R}(w)$ contains no eigenvalue. Moreover, the general semigroup theory shows that $\|(\lambda - T)^{-1}F\| \to 0$ as $\text{Re}(\lambda) \to 0$ uniformly with respect to $|\text{Im}(\lambda)|$. Thus, the point spectrum of A in the region $\{\lambda \in \mathbb{C} \text{ such that } \text{Re}\lambda \geq \omega\}$ is confined in a compact subregion and, then it is necessarily finite since it is discrete. This proves (*ii*).

(*iii*) Let us consider the problem $\lambda\psi - A\psi = \lambda\psi - T\psi - F\psi = \varphi$ $(\text{Re}\lambda > \eta)$. It is equivalent to $\psi - (\lambda - T)^{-1}F\psi = (\lambda - T)^{-1}\varphi$. If $\lambda \in \mathcal{R}(w)$, then using (4.2.2), we have $(I-(\lambda-T)^{-1}F)^{-1}=\sum_{n=0}^{\infty}((\lambda-T)^{-1}F)^n$ and thus $(\lambda-A)^{-1}=\sum_{n=0}^{\infty}((\lambda-T)^{-1}F)^n(\lambda-T)^{-1}$ for $\lambda\in\mathcal{R}(w)$. Besides, for each n, there are two (unique) integers p and q such that $n=mp+q$ $(q<m)$. Hence,

$$\begin{aligned}\sum_{n=0}^{\infty}\|((\lambda-T)^{-1}F)^n\| &= \sum_{p=0,\ 0\le q<m}^{\infty}\|((\lambda-T)^{-1}F)^{mp+q}\|\\ &\le \sum_{p=0,\ 0\le q<m}^{\infty}\|((\lambda-T)^{-1}F)^{q}\|\|((\lambda-T)^{-1}F)^{m}\|^p\\ &\le \max_{0\le q<m}\|((\lambda-T)^{-1}F)^{q}\|\sum_{p=0}^{\infty}\|((\lambda-T)^{-1}F)^{m}\|^p\\ &= \max_{0\le q<m}\|((\lambda-T)^{-1}F)^{q}\|\frac{1}{1-\|((\lambda-T)^{-1}F)^{m}\|}\\ &\le \max_{0\le q<m}\|((\lambda-T)^{-1}F)^{q}\|\frac{1}{1-a}\end{aligned}$$

because $\|[(\lambda-T)^{-1}F]^m\|\le a<1$ for $\lambda\in\mathcal{R}(w)$. Let $w'>\eta$ be such that $w'<w$. By the general semigroup theory, there exists $M(w')$ such that $\|(\lambda-T)^{-1}\|\le\frac{M(w')}{\mathrm{Re}\lambda-w'}$ for $\mathrm{Re}\lambda>w'$, so, $\|(\lambda-T)^{-1}\|\le\frac{M(w')}{w-w'}$ if $\mathrm{Re}\lambda\ge w$. Therefore, for $\lambda\in\mathcal{R}(w)$, we have the following $\|(I-(\lambda-T)^{-1}F)^{-1}\|\le \max_{0\le q<m}\frac{M(w')^q\|F\|^q}{(w-w')^q}\frac{1}{1-a}$. Finally, for $\lambda\in\mathcal{R}(w)$, we have

$$\begin{aligned}\|(\lambda-A)^{-1}\| &= \|(I-(\lambda-T)^{-1}F)^{-1}(\lambda-T)^{-1}\|\\ &\le \max_{0\le q<m}\frac{M(w')^q\|F\|^q}{(w-w')^q}\frac{M(w')}{(1-a)(w-w')}.\end{aligned}$$

This ends the proof of *(iii)*. Q.E.D.

Now, let us introduce the following hypothesis

$$(\mathcal{A}_3)\begin{cases}(i)\ \text{There exists an integer } m,\ \text{a real } r_0>0,\ \text{for } \omega>\eta,\ \text{there exists } C(\omega)\\ \quad \text{such that}\\ \quad |\mathrm{Im}\lambda|^{r_0}\|[(\lambda-T)^{-1}F]^m\| \text{ is bounded on } \{\lambda\in\mathbb{C}, \mathrm{Re}\lambda\ge\omega, |\mathrm{Im}\lambda|\ge C(\omega)\}.\\ (ii)\ \text{There exists a real } c \text{ such that } \|(\lambda-A)^{-1}\| \text{is bounded on } \{\lambda\in\mathbb{C},\\ \quad \mathrm{Re}\lambda\ge c\}.\end{cases}$$

Proposition 4.2.1. *Assume that* $(\mathcal{A}_1)$ *and* $(\mathcal{A}_3)(i)$ *hold true. Then,*

(i) $\sigma(A)\bigcap\{\lambda\in\mathbb{C}$ *such that* $\mathrm{Re}\lambda>\eta\}$ *consists, at most, of discrete eigenvalues with finite algebraic multiplicities.*
(ii) If $\omega>\eta$, *then the set* $\sigma(A)\bigcap\{\lambda\in\mathbb{C}$ *such that* $\mathrm{Re}\lambda\ge\omega\}$ *is finite.*

(iii) If $\omega > \eta$, *then there exists* $C(\omega) > 0$ *such that* $\|(\lambda - A)^{-1}\|$ *is bounded on*

$$\{\lambda \in \mathbb{C} \text{ such that } \mathrm{Re}\lambda \geq \omega \text{ and } |\mathrm{Im}\lambda| \geq C(\omega)\}.$$

◇

Proof. Using the hypothesis $(\mathcal{A}_3)(i)$, there exists a constant $C > 0$, such that $\|[(\lambda - T)^{-1}F]^m\| \leq \frac{C}{|\mathrm{Im}\lambda|^{r_0}}$. So, $\lim_{|\mathrm{Im}\lambda| \to +\infty} \|[(\lambda - T)^{-1}F]^m\| = 0$. Then, the items (i), (ii) and (iii) follow immediately from Lemma 4.2.1. Q.E.D.

Assume that $(\mathcal{A}_1)$ and $(\mathcal{A}_3)(i)$ hold true. Then, from Proposition 4.2.1, the eigenvalues $\{\lambda_1, \lambda_2, \ldots, \lambda_n, \lambda_{n+1}, \ldots\}$ of A lying in the half plane $\mathrm{Re}\lambda > \eta$, can be ordered in such a way that the real part decreases [186], i.e., $\mathrm{Re}\lambda_1 > \mathrm{Re}\lambda_2 > \cdots > \mathrm{Re}\lambda_{n+1} > \cdots > \eta$ and $\{\lambda \in \mathbb{C} \text{ such that } \mathrm{Re}\lambda > \eta\} \backslash \{\lambda_n, n = 1, 2, \ldots\} \subset \rho(A)$, where $\rho(A)$ is the resolvent set of A. Let P_i and D_i denote respectively, the spectral projection and the nilpotent operator associated with λ_i, $i = 1, 2, \ldots, n$. Then, $P = P_1 + \cdots + P_n$ is the spectral projection of the compact set $\{\lambda_1, \lambda_2, \ldots, \lambda_n\}$. Hence, according to the spectral decomposition theorem corresponding to the sets $\{\lambda_1, \lambda_2, \ldots, \lambda_n\}$ and $\sigma(A) \backslash \{\lambda_1, \lambda_2, \ldots, \lambda_n\}$ (see [35, pp. 68–70]), $V(t)$ splits as $V(t) = \tilde{V}(t) + \sum_{i=1}^{n} e^{\lambda_i t} e^{D_i t} P_i$, where $\tilde{V}(t) := V(t)(I - P)$. $(\tilde{V}(t))_{t \geq 0}$ is a C_0-semigroup on the Banach space $(I - P)X$ with generator $\tilde{A} := A(I - P)$. We first establish the following lemma:

Lemma 4.2.2. *If the hypotheses* $(\mathcal{A}_1)$ *and* $(\mathcal{A}_3)$ *hold true, then*

(i) For any $\varepsilon > 0$, $\|(\lambda - A)^{-1}(I - P)\|$ *is bounded on* $\{\lambda \in \mathbb{C}$ *such that* $\mathrm{Re}\lambda \geq \mathrm{Re}\lambda_{n+1} + \varepsilon\}$.

(ii) For all $q \in \{0, \ldots, m - 1\}$, $\|[(\lambda - T)^{-1}F]^q\|$ *is bounded on* $\{\lambda \in \mathbb{C}$ *such that* $\mathrm{Re}\lambda \geq \mathrm{Re}\lambda_{n+1} + \varepsilon\}$. ◇

Proof.

(i) Let $\varepsilon > 0$. From Proposition 4.2.1(iii), there exists a constant $a > 0$ such that $\|(\lambda - A)^{-1}(I - P)\|$ is bounded on $\{\lambda \in \mathbb{C}$ such that $\mathrm{Re}\lambda \geq \mathrm{Re}\lambda_{n+1} + \varepsilon$ and $|\mathrm{Im}\lambda| \geq a\}$. Since $\sigma(\tilde{A}) = \sigma(A) \backslash \{\lambda_1, \lambda_2, \ldots, \lambda_n\}$ and the function $\lambda \longrightarrow \|(\lambda - \tilde{A})^{-1}\|$ is continuous on $\rho(\tilde{A})$, $\|(\lambda - \tilde{A})^{-1}\|$ is bounded on the compact set $\{\lambda \in \mathbb{C}$ such that $\mathrm{Re}\lambda_{n+1} + \varepsilon \leq \mathrm{Re}\lambda \leq c$ and $|\mathrm{Im}\lambda| \leq a\}$. This holds also true for $\|(\lambda - A)^{-1}(I - P)\|$. Indeed,

$$\begin{aligned} \|(\lambda - A)^{-1}(I - P)\| &= \|(\lambda - A)^{-1}(I - P)(I - P)\| \\ &= \|(\lambda - \tilde{A})^{-1}(I - P)\| \\ &\leq \|(\lambda - \tilde{A})^{-1}\|. \end{aligned}$$

Furthermore, by hypothesis $(\mathcal{A}_3)(ii)$, there exists $M' > 0$ such that $\|(\lambda - A)^{-1}(I - P)\| \leq \|(\lambda - A)^{-1}\| \leq M'$ on $\{\lambda \in \mathbb{C}$ such that $\mathrm{Re}\lambda \geq c\}$.

Consequently, these assertions show that $\|(\lambda - A)^{-1}(I - P)\|$ is bounded on $\{\lambda \in \mathbb{C}$ such that $\mathrm{Re}\lambda \geq \mathrm{Re}\lambda_{n+1} + \varepsilon\}$.

(*ii*) By the general semigroup theory, for $\omega > \eta$, there exists $M(\omega)$ such that $\|(\lambda - T)^{-1}\| \leq \frac{M(\omega)}{\mathrm{Re}\lambda - \omega}$ for $\mathrm{Re}\lambda > \omega$. So, $\|(\lambda - T)^{-1}\| \leq \frac{M(\omega)}{\mathrm{Re}\lambda_{n+1} + \varepsilon - \omega}$ for $\mathrm{Re}\lambda > \mathrm{Re}\lambda_{n+1} + \varepsilon > \omega$. Therefore, for all $q \in \{0, \ldots, m-1\}$, $\|[(\lambda - T)^{-1}F]^q\| \leq \left(\frac{M(\omega)}{\mathrm{Re}\lambda_{n+1} + \varepsilon - \omega}\right)^q \|F\|^q$. Finally, for all $q \in \{0, \ldots, m-1\}$, $\|[(\lambda - T)^{-1}F]^q\|$ is bounded on $\{\lambda \in \mathbb{C}$ such that $\mathrm{Re}\lambda \geq \mathrm{Re}\lambda_{n+1} + \varepsilon\}$. This completes the proof of the lemma. Q.E.D.

Now, we are ready to prove the main result of this section.

Theorem 4.2.1. *If the hypotheses* $(\mathcal{A}_1)$ *and* $(\mathcal{A}_3)$ *are true, then for any* $\varepsilon > 0$, *there exists* $M > 0$ *such that* $\|V(t)(I - P)\| \leq Me^{(\mathrm{Re}\lambda_{n+1} + \varepsilon)t}$, $\forall t > 0$, *where* $P = P_1 + \cdots + P_n$ *is the spectral projection of the compact set* $\{\lambda_1, \lambda_2, \ldots, \lambda_n\}$.◇

Proof. For every $\varepsilon > 0$, there exists $\tilde{M} \geq 1$ such that $\|U(t)\| \leq \tilde{M}e^{(\eta+\varepsilon)t}$, for all $t \geq 0$. By the Dyson–Phillips expansion, the semigroup $(V(t))_{t\geq 0}$ generated by A can be written in the form $V(t) = \sum_{j=0}^{+\infty} U_j(t)$, where $U_0(t) = U(t)$, $U_j(t) = \int_0^t U(s)FU_{j-1}(t-s)\,ds$, $\forall j \in \mathbb{N}^*$. From [276] or Theorem 2.5.10, for any λ such that $\mathrm{Re}\lambda > \eta$, one can write

$$\int_0^{+\infty} e^{-\lambda t} U_k(t)\psi\,dt = (\lambda - T)^{-1}[F(\lambda - T)^{-1}]^k \psi, \quad \psi \in X, k \in \mathbb{N} \tag{4.2.3}$$

and

$$\|U_k(t)\| \leq e^{(\eta+\varepsilon)t}\tilde{M}^{k+1}\|F\|^k \frac{t^k}{k!}, \ k \in \mathbb{N}. \tag{4.2.4}$$

For $r_0 > 0$, there exists an integer $p > 0$, such that $r_0 p > 1$. Let k_0 be an integer such that $2k_0 + 2 = pm + q$, where $q \in \{0, \ldots, m-1\}$. Set $W(t) = V(t)(I-P) - \sum_{k=0}^{2k_0+1} U_k(t)$. It is easy to see that $t \longrightarrow W(t)$ is strongly continuous for $t \geq 0$. For every $\psi \in X$, we have

$$W(t)(I - P)\psi = V(t)(I - P)\psi - \sum_{k=0}^{2k_0+1} U_k(t)(I - P)\psi. \tag{4.2.5}$$

The use of Eqs. (4.2.3) and (4.2.5) gives

$$\int_0^{+\infty} e^{-\lambda t} W(t)(I - P)\psi\,dt = (\lambda - A)^{-1}(I - P)\psi$$
$$- \sum_{k=0}^{2k_0+1} (\lambda - T)^{-1}[F(\lambda - T)^{-1}]^k (I - P)\psi.$$

Moreover, $(\lambda - A)^{-1} = \sum_{k=0}^{+\infty}(\lambda - T)^{-1}[F(\lambda - T)^{-1}]^k$. So,

$$\int_0^{+\infty} e^{-\lambda t} W(t)(I-P)\psi \, dt$$

$$= \sum_{k=2k_0+2}^{+\infty} (\lambda - T)^{-1}[F(\lambda - T)^{-1}]^k (I-P)\psi$$

$$= (\lambda - T)^{-1}[F(\lambda - T)^{-1}]^{2k_0+1} F(\lambda - A)^{-1}(I-P)\psi.$$

Let $\varepsilon > 0$ and set $\beta_{n,\varepsilon} = \mathrm{Re}\lambda_{n+1} + \varepsilon$. For every λ with $\mathrm{Re}\lambda > \beta_{n,\varepsilon} - \frac{\varepsilon}{2}$, define

$$f(\lambda) := (\lambda - T)^{-1}[F(\lambda - T)^{-1}]^{2k_0+1} F(\lambda - A)^{-1}(I-P)\psi. \tag{4.2.6}$$

Hence, $\|f(\lambda)\| \le \|[(\lambda - T)^{-1}F]^{2k_0+2}\| \|(\lambda - A)^{-1}(I-P)\| \|\psi\|$, which implies that $\|f(\lambda)\| \le \|[(\lambda - T)^{-1}F]^m\|^p \|[(\lambda - T)^{-1}F]^q\| \|(\lambda - A)^{-1}(I-P)\| \|\psi\|$. From both Lemma 4.2.2 and hypothesis $(\mathcal{A}_3)(i)$, we deduce that there exists $C > 0$ such that

$$\|f(\lambda)\| \le \frac{C}{|\mathrm{Im}\lambda|^{r_0 p}} \tag{4.2.7}$$

uniformly on $\{\lambda \in \mathbb{C}$ such that $\mathrm{Re}\lambda > \beta_{n,\varepsilon} - \frac{\varepsilon}{2}\}$. From Theorem 2.4.1, it follows that $g(t) = \frac{1}{2i\pi}\int_{\gamma - i\infty}^{\gamma + i\infty} e^{t\lambda} f(\lambda)\, d\lambda$, $\gamma > \max\{0, \beta_{n,\varepsilon}\}$ and $t \ge 0$ is a continuous function, such that

$$\int_0^{+\infty} e^{-t\lambda} g(t)\, dt = f(\lambda). \tag{4.2.8}$$

By virtue of the uniqueness of the Laplace integral, Eqs. (4.2.6) and (4.2.8) imply that $W(t)(I-P)\psi = g(t)$. Since $\lambda \longrightarrow f(\lambda)$ is analytic in the region $\{\lambda \in \mathbb{C}$ such that $\mathrm{Re}\lambda > \beta_{n,\varepsilon} - \frac{\varepsilon}{2}\}$, the integral path in the right-hand side of Eq. (4.2.8) can be shifted to $\mathrm{Re}\lambda = \beta_{n,\varepsilon}$, i.e.,

$$g(t) = \frac{1}{2i\pi} \lim_{y \longrightarrow +\infty} \left[\int_{\beta_{n,\varepsilon} - iy}^{\beta_{n,\varepsilon} + iy} e^{t\lambda} f(\lambda)\, d\lambda + \int_{\beta_{n,\varepsilon}}^{\gamma} e^{t(x+iy)} f(x+iy)\, dx \right.$$
$$\left. + \int_{\gamma}^{\beta_{n,\varepsilon}} e^{t(x-iy)} f(x-iy)\, dx \right].$$

From Eq. (4.2.7), and using the Lebesgue dominated convergence theorem, the second and the third terms of the above equation tend to zero. Then,

$$\begin{aligned} g(t) &= \frac{1}{2i\pi}\int_{\beta_{n,\varepsilon}-i\infty}^{\beta_{n,\varepsilon}+i\infty} e^{t\lambda} f(\lambda)\, d\lambda \\ &= W(t)(I-P)\psi. \end{aligned} \tag{4.2.9}$$

We have

$$\|W(t)(I-P)\psi\| \leq \frac{1}{2\pi} e^{t\beta_{n,\varepsilon}} \int_{-\infty}^{+\infty} \|f(\beta_{n,\varepsilon}+iy)\|\, dy.$$

From Eqs. (4.2.7) and (4.2.9), we deduce that

$$\|W(t)(I-P)\| \leq C_1 e^{t\beta_{n,\varepsilon}}, \tag{4.2.10}$$

where C_1 is a positive constant. From Eqs. (4.2.4), (4.2.5), and (4.2.10), we get

$$\begin{aligned} \|V(t)(I-P)\| &\leq \|W(t)(I-P)\| + \sum_{k=0}^{2k_0+1} \|U_k(t)(I-P)\| \\ &\leq C_1 e^{t\beta_{n,\varepsilon}} + \sum_{k=0}^{2k_0+1} e^{(\eta+\varepsilon)t}\tilde{M}^{k+1}\|F\|^k \frac{t^k}{k!} \\ &\leq M e^{t(\mathrm{Re}\lambda_{n+1}+\varepsilon)}, \end{aligned}$$

where $M = \sup_{t\geq 0}\left(C_1 + e^{(\eta-\mathrm{Re}\lambda_{n+1})t}\sum_{k=0}^{2k_0+1}\tilde{M}^{k+1}\|F\|^k\frac{t^k}{k!}\right)$. This completes the proof. Q.E.D.

We have the following proposition:

Proposition 4.2.2. *Let Q be a complex polynomial satisfying $Q(0) = 0$ and $Q(1) \neq 0$. Assume that $(\lambda - T)^{-1}Q(B_\lambda)$ is compact in $\mathcal{R}_w := \{\lambda \in \mathbb{C}$ such that $\mathrm{Re}\lambda \geq w\}$ ($w > w(U)$), where $B_\lambda := F(\lambda - T)^{-1}$ and $w(U)$ denote the type of the semigroup $(U(t))_{t\geq 0}$. Then, $\sigma(A)\bigcap\{\lambda \in \mathbb{C}$ such that $\mathrm{Re}\lambda > w(U)\}$ consists, at most, of a countable set of isolated points λ_k. Each λ_k is an eigenvalue of finite multiplicity and is a pole for the resolvent $(\lambda - A)^{-1}$.* ◇

Proof. Let $n \in \mathbb{N}^*$. We suppose that the polynomial Q is written as $Q(X) = a_1X + a_2X^2 + a_3X^3 + \cdots + a_nX^n$, where $a_1, a_2, \ldots, a_n \in \mathbb{C}$. The function $\lambda \longrightarrow Q(B_\lambda)$ is regular and analytic in the half-plane $\mathrm{Re}\lambda > w(U)$ and its values $Q(B_\lambda)$ are, by assumption, compact operators. Since, for any $\sigma > \mathrm{Re}\lambda$, we have $||Q(B_\lambda)|| \leq \frac{a_1\|F\|}{\sigma - w(U)} + \frac{a_2\|F\|^2}{(\sigma-w(U))^2} + \cdots + \frac{a_n\|F\|^n}{(\sigma-w(U))^n}$. So, $Q(B_\lambda) \to 0$ if $\mathrm{Re}\lambda \to +\infty$. Therefore, $\mu = \sum_{i=1}^n a_i \neq 0$ is not an eigenvalue for $Q(B_\lambda)$. Hence, Smul'yan's theorem in [311] is applied. Then, except for a discrete set of values $\lambda_k \in \mathcal{R}_w$, the operator $\mu - Q(B_\lambda)$ has a bounded, everywhere defined inverse, while $\left(\mu - Q(B_\lambda)\right)^{-1}$ has a pole at each of the points λ_k. Moreover, we have $\mu - Q(B_\lambda) = a_1(I - B_\lambda) + a_2(I - B_\lambda^2) + \cdots + a_n(I - B_\lambda^n)$. Then,

$$\begin{aligned}
&\left(\mu - Q(B_\lambda)\right)^{-1} \\
&= \left[a_1(I - B_\lambda) + a_2(I - B_\lambda^2) + \cdots + a_n(I - B_\lambda^n)\right]^{-1} \\
&= \left[(I - B_\lambda)\left(a_1 I + a_2(I + B_\lambda) + \cdots + a_n(I + B_\lambda + \cdots + B_\lambda^{n-1})\right)\right]^{-1}.
\end{aligned}$$

So, $\left[a_1 I + a_2(I + B_\lambda) + \cdots + a_n(I + B_\lambda + \cdots + B_\lambda^{n-1})\right]\left(\mu I - Q(B_\lambda)\right)^{-1} = (I - B_\lambda)^{-1}$. Let λ be such that $(I - B_\lambda)^{-1}$ exists. Let $R_\lambda := (\lambda - T)^{-1}(I - B_\lambda)^{-1}$. Let us write $A_\lambda := \lambda - A = \lambda - T - F$. We have $A_\lambda R_\lambda = (\lambda - A)\left(\lambda - T - F\right)^{-1} = I$. In the same way, we have $R_\lambda A_\lambda = I$. Hence, A_λ^{-1} exists for such a λ as a bounded, everywhere defined operator, and is equal to R_λ. Consequently, the resolvent $(\lambda - A)^{-1} = R_\lambda$ is an analytic function of λ in the half-plane $\text{Re}\lambda > w(U)$ with the exception of a discrete set of values λ_k, where R_λ has a pole. Any pole λ_k of R_λ is an eigenvalue of A. A corresponding eigenfunction φ satisfies the equation $B_{\lambda_k}\varphi = \varphi$. The equation $\left(\mu - Q(B_{\lambda_k})\right)\varphi = 0$ implies $\frac{Q(B_{\lambda_k})}{\mu}\varphi = \varphi$. The operator $Q(B_{\lambda_k})$ being compact, the space of solutions of this equation is finite dimensional. This implies that the space of eigenfunctions of A corresponding to the eigenvalue λ_k is finite dimensional too. Q.E.D.

From Proposition 4.2.2, we deduce that the eigenvalues $\lambda_1, \lambda_2, \ldots, \lambda_n, \lambda_{n+1}, \ldots$ of A lying in the half plane $\text{Re}\lambda > w(U)$, can be ordered in such a way that the real part decreases [186, p. 109], i.e., $\text{Re}\lambda_1 > \text{Re}\lambda_2 > \cdots > \text{Re}\lambda_{n+1} > \cdots > w(U)$ and $\{\lambda \in \mathbb{C} \text{ such that } \text{Re}\lambda > w(U)\} \backslash \{\lambda_n, n = 1, 2 \ldots\} \subset \rho(A)$, where $\rho(A)$ is the resolvent set of A. Let P_i and D_i denote respectively, the spectral projection and the nilpotent operator associated with $\lambda_i, i = 1, 2, \ldots, n$. We have the following theorem:

Theorem 4.2.2. *Assume that there exists $m \in \mathbb{N}$ such that $\left[(\lambda - T)^{-1}F\right]^m$ is compact for all λ with $\text{Re}\lambda > w(U)$ and there exists a real $r_0 > 0$, for $w > w(U)$, there exists $C(w)$ such that $|\text{Im}\lambda|^{r_0}\left\|(\lambda - T)^{-1}B_\lambda^m F(\lambda - A)^{-1}\right\|$ is bounded on $\{\lambda \in \mathbb{C}$ such that $\text{Re}\lambda \geq w, |\text{Im}\lambda| \geq C(w)\}$, where $B_\lambda := F(\lambda - T)^{-1}$, and assume that the conditions of Proposition 4.2.2 are satisfied. Then, for any $\varepsilon > 0$, there exists $M > 0$ such that $\|V(t)(I - P)\| \leq M\, e^{(\text{Re}\lambda_{n+1}+\varepsilon)t}$ for all $t > 0$, where $P = P_1 + \cdots + P_n$ is the spectral projection of the compact set $\{\lambda_1, \lambda_2, \ldots, \lambda_n\}$.* ◇

Proof. Let $\varepsilon > 0$ and set $\beta_{n,\varepsilon} = \text{Re}\lambda_{n+1} + \varepsilon$. For every λ with $\text{Re}\lambda > \beta_{n,\varepsilon} - \frac{\varepsilon}{2}$, let us define $f(\lambda) := (\lambda - T)^{-1}B_\lambda^m F(\lambda - A)^{-1}(I - P)\psi$, where $B_\lambda = F(\lambda - T)^{-1}$. From the hypotheses, it follows that there exists $\eta > 0$ such that:

$$\|f(\lambda)\| \leq \frac{\eta}{|\text{Im}\lambda|^{r_0}} \tag{4.2.11}$$

uniformly on $\{\lambda \in \mathbb{C}$ such that $\mathrm{Re}\lambda > \beta_{n,\varepsilon} - \frac{\varepsilon}{2}\}$. According to Theorem 2.4.1, the function $g(t) = \frac{1}{2i\pi}\int_{\gamma-i\infty}^{\gamma+i\infty} e^{\lambda t} f(\lambda)d\lambda$, $\gamma > \max(0, \beta_{n,\varepsilon})$, $t \geq 0$ is continuous and

$$\int_0^{+\infty} e^{-\lambda t} g(t)dt = f(\lambda). \tag{4.2.12}$$

Moreover, let $W(t) = V(t)(I - P) - \sum_{k=0}^{m} U_k(t)$. It is easy to see that $t \longrightarrow W(t)$ is strongly continuous for $t \geq 0$. For every $\psi \in X$, we have:

$$W(t)(I - P)\psi = V(t)(I - P)\psi - \sum_{k=0}^{m} U_k(t)(I - P)\psi. \tag{4.2.13}$$

From Theorem 2.5.10 [152, 276], for any λ such that $\mathrm{Re}\lambda > \omega(U)$, one can write

$$\int_0^{+\infty} e^{-\lambda t} U_k(t)\psi\, dt = (\lambda - T)^{-1} B_\lambda^k \psi,\ \psi \in X,\ k \in \mathbb{N} \tag{4.2.14}$$

and

$$\|U_k(t)\| \leq e^{(\omega(U)+\varepsilon)t} \tilde{M}^{k+1} \|F\|^k \frac{t^k}{k!},\ k \in \mathbb{N}, \tag{4.2.15}$$

where $\tilde{M} \geq 1$ such that $\|U(t)\| \leq \tilde{M} e^{(\omega(U)+\varepsilon)t}$, for all $t \geq 0$. Hence, the use of Eqs. (4.2.13) and (4.2.14) leads to $\int_0^{+\infty} e^{-\lambda t} W(t)(I - P)\psi\, dt = (\lambda - A)^{-1}(I - P)\psi - \sum_{k=0}^{m}(\lambda - T)^{-1} B_\lambda^k (I - P)\psi$. Since $(\lambda - A)^{-1} = \sum_{k=0}^{+\infty}(\lambda - T)^{-1} B_\lambda^k$, then

$$\begin{aligned}\int_0^{+\infty} e^{-\lambda t} W(t)(I - P)\psi\, dt &= \sum_{k=m+1}^{+\infty} (\lambda - T)^{-1} B_\lambda^k (I - P)\psi \\ &= (\lambda - T)^{-1} B_\lambda^m F(\lambda - A)^{-1}(I - P)\psi.\end{aligned}$$

Hence,

$$f(\lambda) = \int_0^{+\infty} e^{-\lambda t} W(t)(I - P)\psi\, dt. \tag{4.2.16}$$

By virtue of the uniqueness of the Laplace integral, Eqs. (4.2.12) and (4.2.16) imply that $W(t)(I - P)\psi = g(t)$. Since $\lambda \longrightarrow f(\lambda)$ is analytic in the region $\{\lambda \in \mathbb{C}$ such that $\mathrm{Re}\lambda > \beta_{n,\varepsilon} - \frac{\varepsilon}{2}\}$, the integral path in the right-hand side of Eq. (4.2.12) can be shifted to $\mathrm{Re}\lambda = \beta_{n,\varepsilon}$, i.e.,

$$g(t) = \frac{1}{2i\pi} \lim_{y \to +\infty} \left[\int_{\beta_{n,\varepsilon}-iy}^{\beta_{n,\varepsilon}+iy} e^{t\lambda} f(\lambda)\, d\lambda + \int_{\beta_{n,\varepsilon}}^{\gamma} e^{t(x+iy)} f(x+iy)\, dx \right.$$

$$\left. + \int_{\gamma}^{\beta_{n,\varepsilon}} e^{t(x-iy)} f(x-iy)\, dx \right].$$

From Eq. (4.2.11), and using the Lebesgue dominated convergence theorem, the second and the third terms of the above equation tend to zero. Then,

$$g(t) = \frac{1}{2i\pi} \int_{\beta_{n,\varepsilon}-i\infty}^{\beta_{n,\varepsilon}+i\infty} e^{t\lambda} f(\lambda)\, d\lambda. \tag{4.2.17}$$

We have $\|g(t)\| \leq \frac{1}{2\pi} e^{t\beta_{n,\varepsilon}} \int_{-\infty}^{+\infty} \|f(\beta_{n,\varepsilon}+iy)\|\, dy$. From Eqs. (4.2.11) and (4.2.17), we deduce that

$$\|g(t)\| \leq \frac{1}{2\pi} \frac{\eta}{|\mathrm{Im}\lambda_{n+1}|^{r_0}}\, e^{\beta_{n,\varepsilon} t}. \tag{4.2.18}$$

Finally, from Eqs. (4.2.13), (4.2.15) and (4.2.18), we get

$$\begin{aligned} \|V(t)(I-P)\| &\leq \|W(t)(I-P)\| + \sum_{k=0}^{m} \|U_k(t)(I-P)\| \\ &\leq \frac{1}{2\pi} \frac{\eta}{|\mathrm{Im}\lambda_{n+1}|^{r_0}}\, e^{t\beta_{n,\varepsilon}} + \sum_{k=0}^{m} e^{(\omega(U)+\varepsilon)t}\, \tilde{M}^{k+1} \|F\|^k \frac{t^k}{k!} \\ &\leq M e^{\beta_{n,\varepsilon} t}, \end{aligned}$$

where $M = \sup_{t \geq 0} \left(\frac{1}{2\pi} \frac{\eta}{|\mathrm{Im}\lambda_{n+1}|^{r_0}} + e^{(\omega(U)-\mathrm{Re}\lambda_{n+1})t} \sum_{k=0}^{m} \tilde{M}^{k+1} \|F\|^k \frac{t^k}{k!} \right)$. This completes the proof. Q.E.D.

4.3 Time Behavior of Solutions for an Abstract Cauchy Problem (4.0.1) on L_p-Spaces $(1 < p < \infty)$

Let (Ω, Σ, μ) be a positive measure space. Consider Eq. (4.0.1) in the setting of $L_p(\Omega)$ $(1 < p < \infty)$. In addition to $(\mathcal{A}_1)$ and $(\mathcal{A}_3)$, it is further assumed that

$(\mathcal{A}_4)$: $\sigma(A) \bigcap \{\lambda \in \mathbb{C}$ such that $\mathrm{Re}\lambda > \eta\}$ and the algebraic multiplicity of every point $\lambda_i \in \sigma(A) \bigcap \{\lambda \in \mathbb{C}$ such that $\mathrm{Re}\lambda > \eta\}$ as well as the projection subspace corresponding to λ_i, do not change with respect to $p \in (1, \infty)$,

$(\mathcal{A}_5)$: $(V(t))_{t\geq 0}$, the C_0-semigroup generated by A in L_p, does not change with respect to $p \in (1, \infty)$. Furthermore, the growth bound of $V(t)$ is equal to $s(A)$, the spectral bound of A in L_p, and $s(A)$ keeps the same for every $p \in (1, \infty)$.

From Proposition 4.2.1, $d := s(A) - \sup\{\text{Re}\lambda$ such that $\lambda \in \sigma(A)$ and $\lambda \neq s(A)\} > 0$. Denoting by P_0 the projection operator corresponding to $\{\lambda$ such that $\text{Re}\lambda = s(A)$ and $\lambda \in \sigma(A)\}$, then for any $\varepsilon \in (0, d)$, it follows from Theorem 4.2.1 that

$$\|V(t)(I - P_0)\|_2 \leq Me^{(s(A)-\varepsilon)t}, \quad \forall t > 0, \tag{4.3.1}$$

where $\|.\|_2$ is the norm in the space L_2. On the other hand, from $(\mathcal{A}_5)$, it can be shown that

$$\|V(t)(I - P_0)\|_p \leq \|V(t)\|_p + \|V(t)P_0\|_p \leq M'e^{(s(A)+\widetilde{\varepsilon})t}, \tag{4.3.2}$$

for any $p \in (1, \infty)$ and $\widetilde{\varepsilon} > 0$, where $\|.\|_p$ is the norm in the space L_p. By virtue of the Riesz–Thorin interpolation theorem, it is not difficult to obtain the following theorem, given in [91], from Eqs. (4.3.1) and (4.3.2).

Theorem 4.3.1. *In the setting of* L_p $(1 < p < \infty)$*, if* $(\mathcal{A}_1)$, $(\mathcal{A}_3)$–$(\mathcal{A}_5)$ *are satisfied, then for any initial distribution* $\psi_0 \in L_p$*, the "solution"* $\psi(t)$ *of Eq. (4.0.1) is given by* $\psi(t) = V(t)\psi_0$*, and* $\|V(t)(I - P_0)\|_p \leq Me^{(s(A)-\varepsilon)t}\|\psi_0\|_p$*, where* P_0 *is the projection operator corresponding to* $\{\lambda \in \sigma(A)$ *such that* $\text{Re}\lambda = s(A)\}$, $\varepsilon \in (0, 2dp^{-1})$ *(if* $p \geq 2$*) or* $\varepsilon \in (0, 2d(1 - p^{-1})))$ *(if* $p < 2$*),* $d = s(A) - \sup\{\text{Re}\lambda$ *such that* $\lambda \in \sigma(A)$ *and* $\lambda \neq s(A)\}$. ◇

Remark 4.3.1. Theorem 4.3.1 indicates that if $\sigma(A) \bigcap \{\lambda \in \mathbb{C}$ such that $\text{Re}\lambda > \eta\}$ is not empty, then the asymptotic behavior of the solution $\psi(t)$ of Eq. (4.0.1) can be exactly determined by $s(A)$. ◇

Corollary 4.3.1. *For any constant* $\varepsilon \in (0, 2dp^{-1})$ *(if* $p \geq 2$*) or* $\varepsilon \in (0, 2d(1 - p^{-1}))$ *(if* $p < 2$*), the spectrum of* $V(t)$ *outside the disc* $\{\lambda \in \mathbb{C}$ *such that* $|\lambda| \leq e^{(s(A)-\varepsilon)t}\}$ *consists of finite isolated eigenvalue of* $V(t)$ *with finite algebraic multiplicity.* ◇

Chapter 5
Fredholm Theory Related to Some Measures

The theory of measures of noncompactness and measures of weak noncompactness has many applications in topology, functional analysis, and operator theory. In this chapter, we consider one axiomatic approach to this notion which includes the most important classical definitions. We give some results concerning a certain class of semi-Fredholm and Fredholm operators via the concept of measures of noncompactness and measures of weak noncompactness.

5.1 Fredholm Operators

First, let us prove the following theorem.

Theorem 5.1.1. *Let $A \in \mathcal{C}(X, Y)$. Suppose that there exist $A_1, A_2 \in \mathcal{L}(Y, X)$, $F_1 \in \mathcal{J}(X)$ and $F_2 \in \mathcal{J}(Y)$ such that $A_1 A = I - F_1$ on $\mathcal{D}(A)$ and $AA_2 = I - F_2$ on Y. Then $A \in \Phi(X, Y)$.* ◇

Proof. In view of the inclusion $N(A) \subset N(A_1 A)$, we have $\alpha(A) \leq \alpha(I - F_1)$. Moreover, $R(A) \supset R(AA_2) = R(I - F_2)$ and therefore, $R(A)^\circ \subset R(I - F_2)^\circ$. Accordingly, we have $\beta(A) \leq \alpha(I^* - F_2^*)$. When combined with the fact that $\alpha(I^* - F_2^*) = \beta(I - F_2)$, this implies that $\alpha(A) < \infty$ and $\beta(A) < \infty$. To complete the proof, it is sufficient to show that $R(A)$ is closed. In fact, since $\beta(I - F_2) < \infty$, then there is a finite dimensional subspace X_1 of X such that $X = R(I - F_2) \oplus X_1$ (Lemma 2.1.8). Since $R(A) \supset R(I - F_2)$, and from Lemma 2.1.7, we deduce that $R(A)$ is closed in X. Q.E.D.

Corollary 5.1.1. *Let A in $\mathcal{L}(X)$ and suppose that there exist A_1, $A_2 \in \mathcal{L}(X)$, and $F_1, F_2 \in \mathcal{PK}(X)$, i.e., there exist two nonzero complex polynomials $P(z) = \sum_{r=0}^{p} a_r z^r$ and $Q(z) = \sum_{r=0}^{n} b_r z^r$ satisfying the compactness of both $P(F_1)$ and $Q(F_2)$. Let λ and $\mu \in \mathbb{C}$ such that $P(\lambda) \neq 0$, $Q(\mu) \neq 0$ and $A_1 A = \lambda - F_1$ on X and $AA_2 = \mu - F_2$ on X. Then, $A \in \Phi^b(X)$.* ◇

A. Jeribi, *Spectral Theory and Applications of Linear Operators and Block Operator Matrices*, DOI 10.1007/978-3-319-17566-9_5

Proof. Using the fact that $N(A) \subset N(A_1A)$, we infer that $\alpha(A) \leq \alpha(\lambda - F_1)$. Moreover, $R(A) \supset R(AA_2) = R(\mu - F_2)$. Using Theorem 3.1.2, we have $\alpha(A) < \infty$ and $\beta(A) < \infty$. To complete the proof, it is sufficient to show that $R(A)$ is closed. In fact, by combining Theorem 3.1.2 with the compactness of $Q(F_2)$, we deduce that $\beta(\mu - F_2) < \infty$. Hence, there is a finite dimensional subspace X_1 of X such that $X = R(\mu - F_2) \oplus X_1$ (Lemma 2.1.8). Since $R(A) \supset R(\mu - F_2)$, and from Lemma 2.1.7, we deduce that $R(A)$ is closed in X. Q.E.D.

Now, let us denote $\Phi^*(X)$ by $\Phi^*(X) := \{F \in \Phi^b(X)$ such that $r(F) < \infty$ and $r'(F) < \infty\}$, where $r(F) = \lim_{n\to\infty} \alpha(F^n)$ and $r'(F) = \lim_{n\to\infty} \beta(F^n)$. It is easy to verify that, if A satisfies the hypotheses of Theorem 3.1.2, then $F := \lambda - A \in \Phi^*(X)$. The next result provides a characterization of the elements of $\Phi^*(X)$.

Theorem 5.1.2. *Let $F \in \mathcal{L}(X)$. Then, $F \in \Phi^*(X)$ if, and only if, there exist $n \in \mathbb{N}^*$, $U \in \mathcal{L}(X)$, and $A \in \mathcal{PK}(X)$, $\lambda \in \mathbb{C}$ with $P(\lambda) \neq 0$ such that $UF^n = F^nU = \lambda - A$.* ◇

Proof. If $r(F) < \infty$, then there must be an integer $m \geq 1$ for which the conclusion of Theorem 3.1.1 must hold. To see this, note that $\alpha(F^k)$ is a nondecreasing sequence of integers bounded from above. Similarly, if $r'(F) < \infty$ there must be an integer $n > 1$ such that $N(F^{*k}) = N(F^{*n})$ for $k \geq n$. If both $r(F) < \infty$ and $r'(F) < \infty$, then we deduce the existence of two integers m and n such that $N(F^k) = N(F^m)\ \forall k \geq m$ and $N(F^{*k}) = N(F^{*n})\ \forall k \geq n$. Let $j = \max(m, n)$. Then, $\alpha(F^k) = \alpha(F^j)$, and $\beta(F^k) = \beta(F^j)$ for all $k > j$. Hence, $i(F^k) = \alpha(F^k) - \beta(F^k) = \alpha(F^j) - \beta(F^j) = i(F^j)$ for all $k > j$, but $i(F^k) = ki(F)$. Then, $(k - j)i(F) = 0$ for all $k > j$. So, $i(F) = 0$. However, if this is the case, we must also have $m = n$. In order to prove that $N(F^n) \bigcap R(F^n) = \{0\}$, let $x \in N(F^n) \bigcap R(F^n)$. Hence, $F^n x = 0$ and there is an $x_1 \in X$ such that $x = F^n x_1$. Then, $F^{2n}x_1 = 0$ and $x_1 \in N(F^{2n}) = N(F^n)$. Consequently, $x = F^n x_1 = 0$. This proves $N(F^n) \bigcap R(F^n) = \{0\}$. Let $x_1, \ldots, x_s$ constitute a basis for $N(F^n)$. Then, by Lemma 2.1.4, there are bounded linear functionals $x'_1, \ldots, x'_s$, which annihilate $R(F^n)$ and satisfy

$$x'_j(x_k) = \delta_{jk}, \quad 1 \leq j, k \leq s. \tag{5.1.1}$$

Let

$$Vx = \sum_{k=1}^{n} x'_k(x)x_k. \tag{5.1.2}$$

The operator $F^n + V \in \Phi^b(X)$ and $i(F^n + V) = 0$ (since V is of finite rank) (Theorem 2.2.44). Let $x \in N(F^n + V)$. Then, $Vx \in N(F^n) \bigcap R(F^n)$. Thus, $Vx = 0$ and $x \in N(F^n)$. Moreover, $Vx = 0$ implies $x'_k(x) = 0$ for each k [see Eq. (5.1.2)], and this can only happen if $x = 0$ [see Eq. (5.1.1)]. So, $F^n + V$ has a bounded inverse E. Besides, since $R(V) \subset N(F^n)$ and $R(F^n) \subset N(V)$, then we have $F^nV = VF^n = 0$. Hence, $(F^n + V)V = V(F^n + V) = V^2$, showing that $V = EV^2 = V^2E$. Since $VE = EV^2E = EV$, we have $EF^n = F^nE = I - EV$ and EV is compact. This proves the necessary part of the theorem. Conversely, denoting F^n by $\tilde{F}$ and using Corollary 5.1.1, we conclude that $\tilde{F} \in \Phi^b(X)$. Using both Theorem 3.1.2 and Proposition 2.2.2, we infer that there exists an integer $k :=$ asc$(\lambda - A) =$ desc$(\lambda - A)$, such that $N\left[(\lambda - A)^j\right] = N\left[(\lambda - A)^k\right]$ $\forall j \geq k$ and $R\left[(\lambda - A)^j\right] = R\left[(\lambda - A)^k\right]$, $\forall j \geq k$. Thus, $\forall j \geq k$, $N(\tilde{F}^j) \subset N(U\tilde{F}^j) \subset N\left[(\lambda - A)^j\right] = N\left[(\lambda - A)^k\right]$, since U and $\tilde{F}$ commute. Hence, $\alpha(\tilde{F}^j)$ is bounded from above. This shows that $r(F) = r(\tilde{F}) < \infty$. Similarly, $R(\tilde{F}^j) \supset R(U\tilde{F}^j) \supset R\left[(\lambda - A)^j\right] = R\left[(\lambda - A)^k\right]$, $\forall j \geq k$. Hence, $\beta(\tilde{F}^j)$ is bounded. This gives $r'(F) = r'(\tilde{F}) < \infty$. To complete the proof, it is sufficient to show that $F \in \Phi^b(X)$. This follows from both Corollary 5.1.1 and Lemma 2.2.4. Q.E.D.

Corollary 5.1.2. *Let $F \in \Phi^b(X)$. If one of the two conditions is satisfied*

(i) $i(F) \geq 0$ *and* $r(F) < \infty$*, or*
(ii) $i(F) \leq 0$ *and* $r'(F) < \infty$.

Then, there exist $n \geq 1$, $U \in \mathcal{L}(X)$, and $A \in \mathcal{PK}(X)$, such that $UF^n = F^nU = I - A$. ◇

Definition 5.1.1. Let $A_1, \ldots, A_n$ be operators in $\mathcal{L}(X)$. We say that $A_1, \ldots, A_n$ is a symmetric family if $A_{\sigma(1)} \ldots A_{\sigma(n)} = A_1 \ldots A_n$, for every permutation σ. ◇

In the following theorem, we prove that the index of each operator is equal to zero.

Theorem 5.1.3. *Let $A_1, \ldots, A_n$ be a symmetric family in $\mathcal{L}(X)$. Suppose that their product $A = A_1 \ldots A_n \in \Phi^b(X)$, with* asc$(A)$ *and* desc(A) *being finite. Then, each $A_j \in \Phi^b(X)$ and $i(A_j) = 0$ for all j, $1 \leq j \leq n$.* ◇

Proof. By Lemma 2.2.4, we have $A_j \in \Phi^b(X)$ for all j with $1 \leq j \leq n$. Now, we prove that $i(A_j) = 0$ for all j, with $1 \leq j \leq n$. In fact,

$$N(A_j^k) \subseteq N(A_1^k \ldots A_n^k) = N(A^k) \text{ for all } j, \text{ with } 1 \leq j \leq n \text{ and } k \in \mathbb{N}. \quad (5.1.3)$$

Since $A \in \Phi^b(X)$, then $A^k \in \Phi^b(X)$ and $\alpha(A^k) < \infty$ for all $k \in \mathbb{N}$. So, asc$(A) < \infty$, then $\alpha(A^{\text{asc}(A)}) = \alpha(A^{\text{asc}(A)+k})$ for all $k \in \mathbb{N}$. By Eq. (5.1.3) we have $\alpha(A_j^k) \leq \alpha(A^k)$ for all $k \in \mathbb{N}$. Therefore, $\alpha(A_j^{\text{asc}(A)+k}) \leq \alpha(A^{\text{asc}(A)})$ for all j, with $1 \leq j \leq n$ and $k \in \mathbb{N}$. If we suppose that $\alpha(A_j^k) < \alpha(A_j^{k+p})$ for all $k \in \mathbb{N}$, and that $p \in \mathbb{N}\backslash\{0\}$, a contradiction follows from the fact that $\alpha(A_j^k) \leq \alpha(A^{\text{asc}(A)})$ and $\alpha(A_j^{k+p}) \leq \alpha(A^{\text{asc}(A)})$ for all $k \in \mathbb{N}$, and $p \in \mathbb{N}\backslash\{0\}$. Then, there exists an integer k_0 such that $N(A_j^{k_0}) = N(A_j^{k_0+p})$ for all $p \in \mathbb{N}$. Therefore,

$$\text{asc}(A_j) < \infty \text{ for all } j, \text{ with } 1 \leq j \leq n. \quad (5.1.4)$$

By using a similar reasoning as before, we show that

$$\mathrm{desc}(A_j) < \infty \text{ for all } j, \text{ with } 1 \leq j \leq n. \tag{5.1.5}$$

We have $A_j \in \Phi^b(X)$ for all j, with $1 \leq j \leq n$. Combining Eqs. (5.1.4), (5.1.5) and Lemma 2.2.9, we deduce that $i(A_j) = 0$ for all j, with $1 \leq j \leq n$. Q.E.D.

Corollary 5.1.3. *Let $A_1, \ldots, A_n$ be a symmetric family in $\mathcal{L}(X)$. Let $K \in \mathcal{PK}(X)$ i.e., there exists a nonzero complex polynomial $P(z) = \sum_{r=0}^{p} a_r z^r$ satisfying $P(K) \in \mathcal{K}(X)$. Let $\lambda \in \mathbb{C}$ with $P(\lambda) \neq 0$. If their product $A := A_1 \ldots A_n = \lambda - K$, then each A_j $(1 \leq j \leq n)$ is a Fredholm operator on X of index zero.* ◇

Proof. This corollary follows immediately from Theorems 3.1.2 and 5.1.3. Q.E.D.

Remark 5.1.1. If $A \in \Phi^b(X)$, then asc$(A) < \infty$ and desc$(A) < \infty$ if, and only if, $r(A) < \infty$ and $r'(A) < \infty$. ◇

In view of Remark 5.1.1, the following corollary is a trivial consequence of Theorem 5.1.3.

Corollary 5.1.4. *Let $A_1, \ldots, A_n$ be a symmetric family in $\mathcal{L}(X)$. Suppose that their product $A = A_1 \ldots A_n \in \Phi^b(X)$, $r(A)$ and $r'(A)$ being finite. Then, each $A_j \in \Phi^b(X)$ and $i(A_j) = 0$ for all j, with $1 \leq j \leq n$.* ◇

Proposition 5.1.1. *Let $A \in \mathcal{L}(X)$. The operator A is essentially semi-regular if, and only if, there exists a closed subspace $V \subset X$ such that $A(V) = V$ and the operator $\hat{A} : X/V \longrightarrow X/V$ induced by A is an upper semi-Fredholm operator.* ◇

Proof. Let $A \in \mathcal{L}(X)$ be essentially semi-regular and set $V = R^{\infty}(A)$. Then, there exist $d \in \mathbb{N}$ and a pair of closed subspaces (M, N) of X such that $A = A_M \oplus A_N$, with A_M is semi-regular and A_N is nilpotent of order d where $\dim N < \infty$. We deduce that $V = R^{\infty}(A_M) \subset M$ and $A(V) = A_M(V) = V$. If $x = m + n$ satisfies $Ax \in V$, then $A_M m \in V$ so that $m \in V$. Thus, $x \in N + V$ and $N(\hat{A}) \subset N + V$. Hence, $\dim N(\hat{A}) < \infty$. Let $Q : X \longrightarrow X/V$ be the canonical projection. Since $V \subset R(A)$ and $R(\hat{A}) = \{Ax + V \text{ such that } x \in V\} = QR(A)$ is closed. Thus, $\hat{A}$ is an upper semi-Fredholm operator. Conversely, let $V \subset X$ be a closed subspace such that $A(V) = V$ and the operator $\hat{A} : X/V \longrightarrow X/V$ induced by A is upper semi-Fredholm. We first prove that $R(A)$ is closed. Let $Q : X \longrightarrow X/V$ be the canonical projection. If $y \in X$ and $Qy \in R(\hat{A})$, then $y \in R(A) + V \subset R(A) + F$ since $V \subset R(A)$. Thus, $R(A)$ is a subspace of finite codimension of the closed space $Q^{-1}(R(\hat{A}))$, so $R(A)$ is closed. Further, $V \subset R^{\infty}(A)$. If $Ax = 0$, then $\hat{A}(x + V) = 0$, i.e., $Qx \in N(\hat{V})$. Thus, $N(A) \subset Q^{-1}(N(\hat{A})) \subset V + F \subset R^{\infty}(A) + F$. Q.E.D.

Theorem 5.1.4. *Let $A \in \mathcal{L}(X)$ be essentially semi-regular and let $K \in \mathcal{K}(X)$ commute with A. Then, $A + K$ is essentially semi-regular.* ◇

Proof. Let $A \in \mathcal{L}(X)$ be essentially semi-regular and let K be a compact operator commuting with A. Let $V = R^{\infty}(A)$. Since $A(V) = V$, by using Lomonosov's theorem, $K(V) \subset V$ and hence, we can define the operators $\hat{A} : X/V \longrightarrow X/V$ and $\hat{K} : X/V \longrightarrow X/V$ induced by A and K, respectively. Then, both $\hat{K}$ and $\hat{A}$ have the same property and consequently, $\hat{A} + \hat{K}$ is upper semi-Fredholm. Thus, according to Proposition 5.1.1, $A + K$ is essentially semi-regular. Q.E.D.

The results appearing in the remaining part of this section may be found in [218].

Proposition 5.1.2. *Let $t_0 > 0$ and let $(U(t))_{t\geq 0}$ be a C_0-semigroup on X. If $U(t_0) \in \Phi_+^b(X)$, then $\alpha(U(t)) = 0$ for all $t \geq 0$.* ◇

Proof. First, we show that $U(t_0)$ is injective. Since $\alpha(U(t_0)) < \infty$, then 0 is an eigenvalue with a finite multiplicity of $U(t_0)$. Let $x \neq 0$ be an eigenvector associated with 0. Let us put $t_1 = t_0/2$. Then, $U(t_0)x = U(t_1)U(t_1)x = 0$ and hence, 0 is an eigenvalue of $U(t_1)$. Proceeding by induction, we define a sequence $(t_n)_n$ with $t_n \to 0$ as $n \to \infty$ such that 0 is an eigenvalue of $U(t_n)$, for all $n \in \mathbb{N}$. For $n \geq 0$, we define the sets $\wedge_n = N(U(t_n)) \bigcap \{x \in X \text{ such that } \|x\| = 1\}$. Clearly, the inclusion $N(U(s)) \subset N(U(t))$, for $s \leq t$, and the compactness of $\wedge_0$ imply that $(\wedge_n)_n$ is a decreasing sequence (in the sense of the inclusion) of nonempty compact subsets of X. Thus, $\bigcap_{n=0}^{\infty} \wedge_n \neq \emptyset$. If $x \in \bigcap_{n=0}^{\infty} \wedge_n$, then

$$\|U(t_n)x - x\| = \|x\| = 1, \ \forall n \geq 1. \tag{5.1.6}$$

Since $t_n \to 0$ as $n \to \infty$, (5.1.6) contradicts the strong continuity of $(U(t))_{t\geq 0}$. This shows that $N(U(t_0)) = \{0\}$, that is, $\alpha(U(t_0)) = 0$. Let $0 \leq t \leq t_0$. The inclusion $N(U(t)) \subset N(U(t_0))$ implies that $\alpha(U(t)) = 0$. Now, let us assume that $t > t_0$ and $x \in N(U(t))$. Then, there exists an integer n such that $nt_0 > t$. Therefore, $U(nt_0)x = U(nt_0 - t)U(t)x = 0$. As a result, we have $x = 0$ and consequently, $N(U(t)) = \{0\}$ for all $t > t_0$. Q.E.D.

Proposition 5.1.3. *Let $t_0 > 0$ and let $(U(t))_{t\geq 0}$ be a C_0-semigroup on X. If $U(t_0) \in \Phi_-^b(X)$, then $\beta(U(t)) = 0$ for all $t \geq 0$.* ◇

Proof. In order to prove this proposition, we will proceed by duality. Let $(U^*(t))_{t\geq 0}$ be the dual semigroup of $(U(t))_{t\geq 0}$. Since $\beta(U(t)) = \alpha(U^*(t))$, then it is sufficient to show that $\alpha(U^*(t)) = 0$ for all $t \geq 0$. By hypothesis, we have $\alpha(U^*(t_0)) < \infty$. Let x^* be an element of $N(U^*(t_0))$. Arguing as in the proof of Proposition 5.1.2, we construct a sequence $(t_n)_n$ with $t_n \to 0$ as $n \to \infty$, such that 0 is an eigenvalue of $U^*(t_n)$ and then, we obtain, for all $n \in \mathbb{N}$, a decreasing sequence $\Sigma_n = N(U^*(t_n)) \bigcap \{x^* \in X^* \text{ such that } \|x^*\| = 1\}$ of nonempty compact subsets of X^*. We deduce that $\bigcap_{n=0}^{\infty} \Sigma_n \neq \emptyset$. Let $x^* \in \bigcap_{n=0}^{\infty} \Sigma_n$. Then, for all $n \in \mathbb{N}$, we have

$$|\langle U^*(t_n)x^* - x^*, x\rangle| = |\langle x^*, x\rangle| \ \forall x \in X. \tag{5.1.7}$$

By using the fact that $(U^*(t))_{t\geq 0}$ is continuous in the weak* topology at $t = 0$, we may conclude that

$$\lim_{t\to 0} |\langle U^*(t_n)x^* - x^*, x\rangle| = 0 \;\forall x \in X. \tag{5.1.8}$$

By combining (5.1.7) and (5.1.8), we get $\langle x^*, x\rangle = 0$ for all $x \in X$. This shows that $x^* = 0$ and hence $\alpha(U^*(t_0)) = 0$. Arguing as above, we show that $\alpha(U^*(t)) = 0$ for all $t \geq 0$. Q.E.D.

Proposition 5.1.4. *Let $t_0 > 0$ and let $(U(t))_{t\geq 0}$ be a C_0-semigroup on X. If $U(t_0) \in \Phi^b(X)$, then $i(U(t)) = 0$ for all $t \geq 0$.* ◇

Proof. This follows from Propositions 5.1.2 and 5.1.3. Q.E.D.

Proposition 5.1.5. *Let $t_0 > 0$ and let $(U(t))_{t\geq 0}$ be a C_0-semigroup on X. If $U(t_0) \in \Phi^b_+(X)$, then $U(t) \in \Phi^b_+(X)$ for all $t \geq 0$.* ◇

Proof. By Proposition 5.1.2, it is sufficient to show that $R(U(t))$ is closed in X for all $t \geq 0$. Let us assume that $U(t_0) \in \Phi^b_+(X)$. Then, $\alpha(U(t_0)) < \infty$ and $\beta(U(t_0)) = \infty$ (if $\beta(U(t_0)) < \infty$ the proof is contained in Proposition 5.1.3). Let $U^*(t_0)$ be the dual operator of $U(t_0)$. Obviously, $U^*(t_0) \in \Phi^b_-(X)$ and consequently, $\beta(U^*(t_0)) < \infty$. Hence, $\beta(U^*(t)) < \infty$ for all $t \geq 0$. Now, by applying Kato's lemma (Lemma 2.1.9), we deduce that $R(U^*(t))$ is closed in X^* for all $t \geq 0$. This, together with the closed range theorem of Banach (Theorem 2.1.2), implies that $R(U(t))$ is closed in X for all $t \geq 0$. Q.E.D.

Proposition 5.1.6. *Let $t_0 > 0$ and let $(U(t))_{t\geq 0}$ be a C_0-semigroup on X. If $U(t_0) \in \Phi^b_-(X)$, then $U(t) \in \Phi^b_-(X)$ for all $t \geq 0$.* ◇

Proof. Now, let us assume that $U(t_0) \in \Phi^b_-(X)$. Then, $\beta(U(t_0)) < \infty$ and $\alpha(U(t_0)) = \infty$ (if $\alpha(U(t_0)) < \infty$, the proof is contained in Proposition 5.1.2). From Proposition 5.1.3, it follows that $\beta(U(t)) < \infty$ for all $t \geq 0$. Again, by using Kato's lemma (Lemma 2.1.9), we show that $R(U(t))$ is closed in X for all $t \geq 0$, which completes the proof. Q.E.D.

Proposition 5.1.7. *Let $t_0 > 0$ and let $(U(t))_{t\geq 0}$ be a C_0-semigroup on X. If $U(t_0) \in \Phi^b(X)$, then $U(t) \in \Phi^b(X)$ for all $t \geq 0$.* ◇

Proof. Now, if $U(t_0) \in \Phi^b(X)$, then $\alpha(U(t_0)) < \infty$ and $\beta(U(t_0)) < \infty$. From Propositions 5.1.5 and 5.1.6, we deduce that $R(U(t))$ is closed in X for all $t \geq 0$. This ends the proof. Q.E.D.

Theorem 5.1.5. *A C_0-semigroup $(U(t))_{t\geq 0}$ can be embedded in a C_0-group on X if, and only if, there exists $t_0 > 0$ such that $U(t_0) \in \Phi^b(X)$.* ◇

Proof. The proof follows immediately from both Proposition 5.1.7 and Theorem 2.5.1. Q.E.D.

We denote by $\mathcal{O}$ the set $\mathcal{O} = \{t > 0$ such that $U(t) - I \in \mathcal{F}^b(X)\}$. Note that, if $\mathcal{O} \neq \emptyset$, then the C_0-semigroup $(U(t))_{t\geq 0}$ can be embedded in a C_0-group on

X (it suffices to write $U(t_0) = I + [U(t_0) - I]$ for some $t_0 \in \mathcal{O}$ and to apply Theorem 5.1.5). Observe that the relation $(U(t)-I)(U(s)-I) = (U(t+s)-I) - (U(s)-I) - (U(t)-I)$ implies that

$$s \in \mathcal{O},\ t \in \mathcal{O} \Longrightarrow s+t \in \mathcal{O}, \quad s \in \mathcal{O},\ t \notin \mathcal{O} \Longrightarrow s+t \notin \mathcal{O}.$$

It follows from these relations that $\mathcal{O}$ is the intersection of an additive subgroup of real number with the positive real line. Therefore, $\mathcal{O}$ may be in one of the following forms:

(i) $\mathcal{O} =]0, \infty[$,
(ii) $\mathcal{O} = \{nx,$ for some $x > 0$ and $n = 1, 2, \ldots\}$,
(iii) $\mathcal{O}$ is a dense subset of $]0, \infty[$ with empty interior.

Remark 5.1.2. The following examples taken from [84] show that all the three types of sets may occur, the above classification of $\mathcal{O}$-sets is not empty, and sets of type (ii) can arise from semigroups having bounded or unbounded infinitesimal generators. In fact, take $X = l_1$, the Banach space of absolutely convergent sequences. The space $\mathcal{K}(X)$ is the sole closed two-sided proper ideal of $\mathcal{L}(X)$, that is, $\mathcal{K}(X) = \mathcal{F}^b(X)$.

1. Let $(U(t))_{t\geq 0}$ be the C_0-semigroup given by $U(t) = I$ for all $t \geq 0$. Clearly, for all $t > 0$, $U(t) - I \in \mathcal{K}(X)$. Accordingly, $\mathcal{O} =]0, \infty[$ and $A = 0$.
2. (a) Assume that $U(t) = \text{diag}\{e^{it}, e^{-it}, e^{it}, e^{-it}, \ldots\}$ for all $t \geq 0$. In this case, we have $\mathcal{O} = \{2n\pi,\ n = 1, 2, 3, \ldots\}$ and $A = \text{diag}\{i, -i, i, -i, \ldots\}$, the infinitesimal generator of $(U(t))_{t\geq 0}$ is bounded.
   (b) Suppose now that $U(t) = \text{diag}\{e^{it}, e^{2it}, e^{3it}, e^{4it}, \ldots\}$ for all $t \geq 0$. Here, we have also $\mathcal{O} = \{2n\pi,\ n = 1, 2, 3, \ldots\}$ but $A = \text{diag}\{i, 2i, 3i, 4i, \ldots\}$, the infinitesimal generator of $(U(t))_{t\geq 0}$ is unbounded.
3. The C_0-semigroup $(U(t))_{t\geq 0}$ with $U(t) = \text{diag}\{e^{it}, e^{2!it}, e^{3!it}, e^{4!it}, \ldots\}$ for all $t \geq 0$ provides an example of $\mathcal{O}$-set of type (iii). ◇

5.2 Fredholm Theory by Means of Noncompactness Measures

We start our investigation with the following lemma.

Lemma 5.2.1. *Let X be a Banach space, $T \in \mathcal{L}(X)$ and P, Q are two complex polynomials satisfying Q divides $P - 1$. Let $M \subset X$ and let $A = \{x \in \overline{B}_X$ such that $Q(T)(x) \in M\}$, where $\overline{B}_X$ denote the closed unit ball of X. If M is compact and $\gamma(P(T)) < 1$, then A is compact or empty, where $\gamma(.)$ is the Kuratowski measure of noncompactness in X.* ◇

Proof. Let us assume that A is not empty. According to the hypothesis ensuring Q divides $P - 1$, we infer that there exists a complex polynomial H such that

$P = HQ + 1$. Consider $x \in A$ and $z \in M$ such that $Q(T)(x) = z$. Then, we get $H(T)Q(T)(x) + x = H(T)(z) + x$, which implies $x = P(T)x - H(T)(z)$. Since a continuous image of a compact set by a continuous operator is also compact, it follows that $\tilde{A} = \{-H(T)(z)$ such that $z \in M\}$ is compact as well. Obviously, $A \subset P(T)A + \tilde{A}$. Then, using the regularity of γ, we get $\gamma(A) \leq \gamma(P(T)A) + \gamma(\tilde{A}) \leq \gamma(A)\gamma(P(T))$. Since $\gamma(P(T)) < 1$, then $\gamma(A) = 0$. Consequently, by using Proposition 2.10.1 and the fact that A is closed, we infer that A is compact. Q.E.D.

Theorem 5.2.1. *Let X be a Banach space, $T \in \mathcal{L}(X)$ and P, Q are two complex polynomials satisfying the fact that Q divides $P - 1$.*

(i) If $\gamma(P(T)) < 1$ then, $Q(T) \in \Phi^b_+(X)$.
(ii) If $\gamma(P(T)) < \frac{1}{2}$, then $Q(T) \in \Phi^b(X)$. ◇

Proof.

(i) First, let us prove that $\alpha(Q(T)) < \infty$. To do so, it is sufficient to establish that the set $N(Q(T)) \bigcap \overline{B}_X$ is compact. The result follows from Lemma 5.2.1 with $M = \{0\}$. In order to complete the proof of (i), we will check that $R(Q(T))$ (the range of $Q(T)$) is closed. Indeed, since $N(Q(T))$ is finite dimensional, then by Lemma 2.1.6, there exists a closed infinite dimensional subspace Y in X such that $X = N(Q(T)) \oplus Y$. We claim that there exists $\delta > 0$ satisfying $\delta\|Q(T)(x)\| \geq \|x\|$ for every $x \in Y$. Now, let us assume the contrary. For every $n \in \mathbb{N}$, there exists $x_n \in Y$ satisfying $\|x_n\| = 1$ and $\|Q(T)(x_n)\| \leq \frac{1}{n}$. Hence, $Q(T)(x_n) \to 0$ (when $n \to +\infty$). From Lemma 5.2.1 with $M = \{Q(T)(x_n)$ such that $n \in \mathbb{N}\} \bigcup \{0\}$, it follows that the sequence $(x_n)_{n\in\mathbb{N}}$ admits a subsequence $(x_{n_k})_{k\in\mathbb{N}}$ which converges to $x_0 \in Y$. Clearly, $\|x_0\| = 1$ and $Q(T)(x_0) = 0$, which is a contradiction. This proves the claim. Using Theorem 2.2.1, it is easy to conclude that $R(Q(T))$ is closed. This ends the proof of (i).

(ii) Assume that $\gamma(P(T)) < \frac{1}{2}$. Combining the assertion (i) with (v) of Lemma 2.10.2, one has $\gamma(P(T)^*) \leq 2\gamma(P(T)) < 1$, where $P(T)^*$ stands for the dual of the operator $P(T)$. Arguing as in the proof of (i), we get $\alpha(Q(T)^*) = \beta(Q(T)) < \infty$. This completes the proof of the theorem. Q.E.D.

As a consequence of Theorem 5.2.1, we have

Corollary 5.2.1. *Let X be a Banach space, $T \in \mathcal{L}(X)$ and let P be a complex, non-constant polynomial satisfying $P(0) = 1$.*

(i) If $\gamma(P(T)) < 1$, then $T \in \Phi^b_+(X)$.
(ii) If $\gamma(P(T)) < \frac{1}{2}$, then $T \in \Phi^b(X)$.
(iii) If $\gamma(I + T) < 1$, then $T \in \Phi^b(X)$. ◇

Proof. (i)–(ii) Since $P(0) = 1$, then $Q(z) := z$ divides $(P(z) - 1)$ and the results of (i) and (ii) are deduced from Theorem 5.2.1.

(iii) If $\gamma(I+T) < 1$, then $\lim_{k\to+\infty} (\gamma(I+T))^k = 0$. So, there exists $k_0 \in \mathbb{N}^*$ such that $(\gamma(I+T))^{k_0} \leq \frac{1}{2}$. Using Lemma 2.10.2 (iii), we deduce that $\gamma((I+T)^{k_0}) \leq \frac{1}{2}$. So, the result is an immediate consequence of Theorem 5.2.1 (ii) with $P(z) := (1+z)^{k_0}$ and $Q(z) := z$. Q.E.D.

Corollary 5.2.2. *Let X be a Banach space and $T \in \mathcal{L}(X)$.*

(i) If $\gamma(T^m) < 1$, for some $m > 0$, then $(I - T)$ is a Fredholm operator with $i(I - T) = 0$.
(ii) Assume that $\gamma(T) \geq 1$ and there exists a complex polynomial $P(z) := a_0 + a_1 z + \cdots + a_n z^n$ such that $(1 - z)$ divides $(P(z) - 1)$ and $\max_{0\leq k\leq n} |a_k| < \frac{1}{2(n+1)(\gamma(T))^n}$. Then, $(I - T) \in \Phi^b(X)$ and $i(I - T) = 0$. ◇

Proof.

(i) If $\gamma(T^m) < 1$, then $\lim_{k\to+\infty} (\gamma(T^m))^k = 0$. Arguing as in the proof of Corollary 5.2.1 (iii), there exists $k_0 \in \mathbb{N}^*$ such that $\gamma(T^{mk_0}) \leq \frac{1}{2}$. So, applying Theorem 5.2.1 (ii) with $P(z) := z^{mk_0}$ and $Q(z) := 1 - z$, we conclude that $Q(T) := (I - T) \in \Phi^b(X)$. Then, we notice that for $t \in [0, 1]$, we have $\gamma((tT)^{mk_0}) < \frac{1}{2}$ and therefore, $(I - tT)$ is a Fredholm operator on X. Moreover, since the index is constant on any component of $\Phi^b(X)$ (see Proposition 2.2.5 (ii)) and $[0, 1]$ is compact, we have $i(Q(T)) = i(I - tT) = i(I) = 0$.
(ii) Let $t \in [0, 1]$. We have

$$\begin{aligned}\gamma(P(tT)) &\leq \sum_{k=0}^{n} t^k |a_k| \gamma(T)^k \\ &\leq (n+1)(\gamma(T))^n \left[\max_{0\leq k\leq n} |a_k|\right] \\ &< \frac{1}{2}.\end{aligned}$$

Now, using Theorem 5.2.1 (ii), we infer that, for all $t \in [0, 1]$, $(I - tT) \in \Phi^b(X)$. Then, arguing as in the proof of the first assertion (i), one sees that, for all $t \in [0, 1]$, $i(I - T) = i(I - tT) = i(I) = 0$. Q.E.D.

As a consequence of Corollaries 5.1.3 and 5.2.2 (i), we have the following:

Corollary 5.2.3. *Let $A_1, \ldots, A_n$ be operators on a Banach space X which mutually commute. If their product $A := A_1 \ldots A_n = \lambda - K$, where $K \in \mathcal{L}(X)$ such that $\gamma(K^m) < |\lambda|^m$ and $r(A)$ and $r'(A)$ are finite, then each A_j $(1 \leq j \leq n)$ is a Fredholm operator on X of zero index.* ◇

Theorem 5.2.2. *Let X be a Banach space, $T \in \mathcal{L}(X)$, $a \in \mathbb{C}^*$ and let P, Q be two complex polynomials, such that Q divides $P - a$. If $\gamma(P(T)) < |a|$, then $Q(T) \in \Phi^b(X)$.* ◇

Proof. Since Q divides $\frac{P}{a} - 1$ and $\gamma(\frac{P(T)}{a}) < 1$, then by Theorem 5.2.1 (i), we get $Q(T) \in \Phi_+^b(X)$. Furthermore, there exists $k_0 \in \mathbb{N}^*$ such that $(\gamma(\frac{P(T)}{a}))^{k_0} < \frac{1}{2}$. According to Lemma 2.10.2 (iii), we deduce that $\gamma((\frac{P(T)}{a})^{k_0}) < \frac{1}{2}$. Combining the assertions (i) and (v) of Lemma 2.10.2 one has $\gamma((\frac{P(T)}{a}^*)^{k_0}) \leq 2\gamma((\frac{P(T)}{a})^{k_0}) < 1$. Hence, $\gamma((\frac{P(T)}{a}^*)^{k_0}) < 1$. Since Q divides $(\frac{P}{a})^{k_0} - 1$, then $\alpha(Q(T)^*) = \beta(Q(T)) < \infty$ and $Q(T) \in \Phi^b(X)$ which completes the proof of theorem. Q.E.D.

Corollary 5.2.4. *Let X be a Banach space, $T \in \mathcal{L}(X)$, $\lambda \in \mathbb{C}^*$, $P(z) = \sum_{k=0}^{n} a_k z^k$, a nonzero complex polynomial satisfying $P(\lambda) \neq 0$ and let $|P|(z) := \sum_{k=0}^{n} |a_k| z^k$.*

(i) If $\gamma(P(T)) < |P(\lambda)|$, then $(\lambda - T)$ is a Fredholm operator.
(ii) If $|P|(\gamma(T)) < |P(\lambda)|$, then $(\lambda - T)$ is a Fredholm operator with zero index. ◇

Proof.

(i) Let $\lambda \in \mathbb{C}^*$ such that $P(\lambda) \neq 0$. We have

$$\begin{aligned} P(\lambda) - P(z) &= \sum_{k=1}^{n} a_k(\lambda^k - z^k) \\ &= (\lambda - z)\left(\sum_{k=1}^{n} a_k \sum_{r=0}^{k-1} \lambda^r z^{k-r-1}\right). \end{aligned}$$

Then, $(\lambda - z)$ divides $P(\lambda) - P(z)$. Applying Theorem 5.2.2 with $a := P(\lambda)$, we get $(\lambda - T) \in \Phi^b(X)$.

(ii) Let $t \in [0, 1]$. We have the following

$$\begin{aligned} \gamma(P(tT)) &\leq \sum_{k=0}^{n} |a_k| t^k \gamma(T^k) \\ &\leq |P|(\gamma(T)). \end{aligned}$$

If $|P|(\gamma(T)) < |P(\lambda)|$, then $\gamma(P(tT)) < |P(\lambda)|$. Applying (i), we get $(\lambda - tT) \in \Phi^b(X)$ for all $t \in [0, 1]$. Moreover, from Proposition 2.2.5, the index is constant on any component of $\Phi^b(X)$, since $[0, 1]$ is connect, we infer that $i(\lambda - tT) = i(I) = 0$. Q.E.D.

5.3 Fredholm Theory by Means of Non-strict Singularity Measures

Our purpose here is to establish some results required in the sequel given by N. Moalla in [255].

Proposition 5.3.1. *Let $A \in \mathcal{L}(X)$. If $g(A^n) < 1$ for some integer $n \geq 1$, then $I - A \in \Phi^b(X)$ with $i(I - A) = 0$, where $g(.)$ is a measure of non-strict singularity introduced in Definition 2.10.7.* ◇

Proof. We start with the case $n = 1$. For this, let us suppose that $I - A \notin \Phi_+^b(X)$. Then, there is an infinite dimensional subspace M of X such that $(I - A)|_M$, the restriction of $I - A$ to M, is compact (see Lemma 2.2.2). Consequently, it is a strictly singular operator. This result, combined with the use of Proposition 2.10.3 (ii), allows us to conclude that $g(I|_M) = g(A|_M) < 1$. However, this is in contradiction to the fact that $g(I|_M) = 1$, since M is infinite dimensional. Hence we must have $I - A \in \Phi_+^b(X)$. The same reasoning holds for $I - \lambda A$, where $0 \leq \lambda \leq 1$. Using the constancy of the index (Proposition 2.2.5), we have $i(I - A) = i(I) = 0$. Therefore, $I - A \in \Phi^b(X)$ and $i(I - A) = 0$. This completes the proof of the case $n = 1$. For $n > 1$, we have

$$(I - A)B = B(I - A) = I - A^n, \tag{5.3.1}$$

where $B = A^{n-1} + A^{n-2} + \cdots + I$. Using the case $n = 1$, we have $I - A^n \in \Phi^b(X)$ and $i(I - A^n) = 0$. In view of Lemma 2.2.5 or Theorem 2.2.19, and Eq. (5.3.1) implies that $I - A \in \Phi^b(X)$. The remaining part of the proof of this case is the same as in the proof related to the case $n = 1$. Q.E.D.

Theorem 5.3.1. *Let $A \in \mathcal{L}(X)$.*

(i) Let P, Q be two complex polynomials satisfying Q divides $P - 1$. If $g(P(A)) < 1$, then $Q(A) \in \Phi^b(X)$.
(ii) Assume that $g(A) \geq 1$ and there exists a complex polynomial $P(z) = \sum_{k=0}^{n} a_k z^k$ such that $1 - z$ divides $P(z) - 1$, and $\sum_{k=0}^{n} |a_k| < \frac{1}{(g(A))^n}$. Then, $I - A \in \Phi^b(X)$ and $i(I - A) = 0$. ◇

Proof.

(i) According to the hypothesis ensuring Q divides $P - 1$, there exists a complex polynomial H such that $P - 1 = QH$. Since $g(P(A)) < 1$, then it follows from Proposition 5.3.1 that $Q(A)H(A) = H(A)Q(A)$ is a Fredholm operator. Hence, $Q(A) \in \Phi^b(X)$.
(ii) Let $\lambda \in [0, 1]$, we have

$$\begin{aligned} g(P(\lambda A)) &\leq \sum_{k=0}^{n} |a_k| \, \lambda^k (g(A))^k \\ &\leq (g(A))^n \sum_{k=0}^{n} |a_k| \\ &< 1. \end{aligned}$$

Using (i), we infer that, for all $\lambda \in [0,1]$, $I - \lambda A \in \Phi^b(X)$. Since the index is constant on any component of $\Phi^b(X)$ (Proposition 2.2.5), we can deduce that, for all $\lambda \in [0,1]$, $i(I - \lambda A) = i(I - A) = i(I) = 0$. Q.E.D.

Corollary 5.3.1. *Let $A \in \mathcal{L}(X)$.*

(i) Assume that $g(A) < 1$, there exists a complex polynomial Q satisfying $Q(0) \neq 0$ and a complex polynomial $P(z) = \sum_{k=0}^{n} a_k z^k$ such that Q divides $P - 1$. If $\sum_{k=0}^{n} |a_k| < 1$, then $Q(A) \in \Phi^b(X)$ and $i(Q(A)) = 0$.
(ii) Let P be a complex and non-constant polynomial satisfying $P(0) = 1$. If $g(P(A)) < 1$, then $A \in \Phi^b(X)$.
(iii) Let P be a complex polynomial such that $P(1) \neq 0$. If $g(P(A)) < |P(1)|$, then $I - A \in \Phi^b(X)$. ◇

Proof.

(i) It follows immediately from the fact that $\sum_{k=0}^{n} |a_k| < 1$ and $g(A) < 1$ that $g(P(\lambda A)) < 1$, for all $\lambda \in [0,1]$. Hence, and from Theorem 5.3.1 (i), we deduce that $Q(\lambda A) \in \Phi^b(X)$, for all $\lambda \in [0,1]$. The use of Proposition 2.2.5 enables us to conclude that $i(Q(A)) = i(Q(0)I) = 0$.
(ii) Since $P(0) = 1$, the complex polynomial $Q(z) = z$ divides $P - 1$ and the result follows from Theorem 5.3.1 (i).
(iii) The result of this assertion also follows from Theorem 5.3.1 (i), since the polynomial $1 - z$ divides the polynomial $1 - \frac{P}{P(1)}$. Q.E.D.

5.4 Fredholm Theory by Means of Demicompact Operator

The concept of demicompactness was introduced in order to discuss the fixed points. Further examples of demicompact operators were introduced by W. V. Petryshyn [279], in particular, we cite compact operator, or the operator T with closed range and for which T^{-1} exists and is continuous. Fore more details and references on this subject, see [36, 273, 279].

5.4.1 Demicompactness

Definition 5.4.1 ([279]). An operator $T : \mathcal{D}(T) \subseteq X \longrightarrow X$ is said to be demicompact if, for every bounded sequence $(x_n)_n$ in $\mathcal{D}(A)$ such that $x_n - Tx_n \to x \in X$, there is a convergent subsequence of $(x_n)_n$. ◇

Definition 5.4.2. An operator $T : \mathcal{D}(T) \subseteq X \longrightarrow X$ is said to be demicontinuous if $(x_n)_n \to x$ $(x_n, x \in \mathcal{D}(T))$ strongly in X implies $Tx_n \to Tx$ weakly in X. ◇

Remark 5.4.1. Let X be a Banach space.

(i) Every compact operator $T : \mathcal{D}(T) \subset X \longrightarrow X$ is demicompact.
(ii) It is well known that every condensing operator is demicompact.
Let X be a Hilbert space.
(iii) Every operator $T : \mathcal{D}(T) \subset X \longrightarrow X$ which satisfies either the condition

$$\mathrm{Re}\langle Tx - Ty, x - y\rangle \leq a\|x - y\|^2, \quad a < 1 \tag{5.4.1}$$

or the condition $\mathrm{Re}\langle Tx - Ty, x - y\rangle \leq a\|Tx - Ty\|^2, \ a < 1$ is demicompact.
(iv) Every operator $T : \mathcal{D}(T) \subset X \longrightarrow X$ for which $(I - T)^{-1}$ exists and is continuous on its range $R(I - T)$ (and, in particular, demicontinuous operators T for which (5.4.1) is valid with $a < 1$ or for which the inequality $|\langle Tx - Ty, x - y\rangle| \leq b\|x - y\|^2$, $0 < b < 1$ is valid for all x and y in X) is demicompact. ◇

Lemma 5.4.1. *Let $T : \mathcal{D}(T) \subseteq X \longrightarrow X$ be a closed linear operator. If $1 \in \rho(T)$, then T is demicompact.* ◇

Proof. Let $(x_n)_n$ be a bounded sequence of $\mathcal{D}(T)$ such that $(I - T)x_n \to y$. Since $1 \in \rho(T)$, we deduce that $(x_n)_n$ converges to $(I - T)^{-1}y$. So, T is demicompact. Q.E.D.

Remark 5.4.2. $-I$ is a demicompact operator. ◇

Lemma 5.4.2. *Let $T : \mathcal{D}(T) \subseteq X \longrightarrow X$ be a closed linear operator. If T is a 1-set-contraction, then μT is demicompact for each $\mu \in [0, 1)$.* ◇

Proof. Obviously for $\mu = 0$, μT is demicompact. We suppose that $\mu \in (0, 1)$. Let $(x_n)_n$ be a bounded sequence of $\mathcal{D}(T)$ such that $y_n := (I - \mu T)x_n \to y$. Let $\gamma(.)$ be the Kuratowski measure of noncompactness in X, and suppose that $\gamma(\{x_n\}) \neq 0$. Since $\{x_n\} \subseteq \{y_n\} + \{\mu Tx_n\}$, it follows that

$$\begin{aligned}\gamma(\{x_n\}) &\leq \gamma(\{y_n\}) + \mu\gamma(\{Tx_n\})\\ &\leq \mu\gamma(\{x_n\})\\ &< \gamma(\{x_n\}),\end{aligned}$$

which is impossible. It follows that $\gamma(\{x_n\}) = 0$. Hence, $\{x_n\}$ is relatively compact, and there is a convergent subsequence $(x_{n_i})_i$ of $(x_n)_n$. Q.E.D.

Lemma 5.4.3. *Let $T \in \mathcal{L}(X)$ be a power-compact operator. Then, for every $\mu \in (0, 1]$, μT is demicompact.* ◇

Proof. Let $T \in \mathcal{L}(X)$ and $m \in \mathbb{N}^*$ such that T^m is a compact operator. Let $(x_n)_n$ be a bounded sequence of X such that $y_n := (I - \mu T)x_n \to y$. Then, x_n is given by $x_n = \mu^m T^m x_n + \sum_{k=0}^{m-1} \mu^k T^k y_n$. Since T^m is a compact operator and $(y_n)_n$ is a convergent sequence, there exists a convergent subsequence of $(x_n)_n$. Q.E.D.

Lemma 5.4.4. *Let $T \in \mathcal{L}(X)$. If $\gamma(T^m) < 1$, for some $m \in \mathbb{N}^*$, then for every $\mu \in (0,1]$, μT is demicompact, where $\gamma(.)$ is the Kuratowski measure of noncompactness in X.* ◇

Proof. Let $T \in \mathcal{L}(X)$ and $m \in \mathbb{N}^*$ such that $\gamma(T^m) < 1$. Let $(x_n)_n$ be a bounded sequence of X such that $y_n := (I - \mu T)x_n \to y$. Similarly as in Lemma 5.4.3, we have

$$\begin{aligned}\gamma(\{x_n\}) &\leq \mu^m \gamma(\{T^m x_n\})\\ &\leq \mu^m \gamma(T^m)\gamma(\{x_n\}).\end{aligned}$$

Since $\mu \in (0,1]$ and $\gamma(T^m) < 1$, we have $\gamma(\{x_n\}) = 0$. Then, there exists a convergent subsequence of $(x_n)_n$. Q.E.D.

Lemma 5.4.5. *Let $T \in \mathcal{C}(X)$ and $T_0 \in \mathcal{L}(X)$ such that $T_0(I-T) = I - K$, where K is a demicompact operator on X. Then T is demicompact.* ◇

Proof. Let $(x_n)_n$ be a bounded sequence of $\mathcal{D}(T)$ such that $(I - T)x_n \to y$. Since $T_0 \in \mathcal{L}(X)$, we have $T_0(I-T)x_n \to T_0 y$. Moreover, $(I-K)x_n = T_0(I-T)x_n \to T_0 y$. According to the fact that K is demicompact, we deduce the result. Q.E.D.

Lemma 5.4.6. *Let $T \in \mathcal{C}(X)$. If $(I - T)$ is a Fredholm operator on X, then T is demicompact.* ◇

Proof. Since $(I - T) \in \Phi(X)$, then by using Theorem 2.2.38, there exists $T_0 \in \mathcal{L}(X)$ such that $T_0(I - T) = I - K_1$, where $K_1 \in \mathcal{K}(X)$. Let $(x_n)_n$ be a bounded sequence of $\mathcal{D}(T)$ such that $(I-T)x_n \to y$. So, $T_0(I-T)x_n = (I-K_1)x_n \to T_0 y$. Now, arguing as in the proof of Lemma 5.4.5, we show that $(x_n)_n$ has a convergent subsequence. Q.E.D.

Theorem 5.4.1. *Let $T \in \mathcal{C}(X)$. If T is demicompact, then $(I - T)$ is an upper semi-Fredholm operator on X.* ◇

Proof. We first show that $N(I - T)$ is finite dimensional. Let $S := \{x \in \mathcal{D}(T)$ such that $(I - T)x = 0$ and $\|x\| = 1\}$ and let $(x_n)_n$ be a bounded sequence of S. Since T is demicompact, there exists a subsequence $(x_{n_i})_i$ of $(x_n)_n$ which converges to $x \in X$. Hence, and from the continuity of the norm and the closedness of T, it follows that $x \in \mathcal{D}(T)$, $x - Tx = 0$ and $\|x\| = 1$. Consequently, $\alpha(I - T)$ is finite. Now, we claim that $R(I-T)$ is closed. Applying Lemma 2.1.6, we can write $\mathcal{D}(T) = N(I-T) \oplus W$, where $W = \mathcal{D}(T) \bigcap X_0$. W is a closed subspace of $\mathcal{D}(T)$ with respect to the graph norm, then it is a Banach space. In view of Theorem 2.2.1, it is sufficient to prove that there is a constant $\lambda > 0$ such that, for every $x \in W$, $\|Tx\| \geq \lambda \|x\|_T$ where $\|.\|_T$ is the graph norm (i.e., $\|x\|_T = \|x\| + \|Tx\|$). If not, there exists a sequence $(x_n)_n$ of W, such that $\|x_n\|_T = 1$ and $\|(I - T)x_n\| \to 0$. Since T is demicompact, there exists a subsequence $(x_{n_i})_i$ of $(x_n)_n$ which converges to $x \in X$. Moreover, $(I - T)$ is closed. Then, $(I - T)x = 0$, which implies $x = 0$. This contradicts the continuity of the norm. Q.E.D.

Theorem 5.4.2. *Let $T \in \mathcal{C}(X)$. If μT is demicompact for each $\mu \in [0, 1]$, then $(I - T)$ is a Fredholm operator of index zero.* ◇

Proof. Since μT is demicompact for each $\mu \in [0, 1]$, Theorem 5.4.1 implies that $(I - \mu T)$ is an upper semi-Fredholm operator on X. Using the stability results for semi-Fredholm operators, the index $i(I - \mu T)$ is continuous in μ (see Proposition 2.2.6). Since it is an integer, including infinite values, it must be constant for every $\mu \in [0, 1]$. Showing that $i(I - \mu T) = i(I - T) = i(I) = 0$, we conclude that $(I - T)$ is a Fredholm operator of index zero. Q.E.D.

Corollary 5.4.1. *If T satisfies the hypothesis of Theorem 5.4.2, then $(I - \lambda T)$ is a Fredholm operator of index zero for every $\lambda \in (0, 1]$.* ◇

Corollary 5.4.2. *Let $T \in \mathcal{C}(X)$. If T is demicompact 1-set-contraction, then $(I - T)$ is a Fredholm operator of index zero.* ◇

Proof. The proof follows from both Lemma 5.4.2 and Theorem 5.4.2. Q.E.D.

Theorem 5.4.3. *Let $T : \mathcal{D}(T) \subset X \longrightarrow Y$ be a densely defined closed linear operator. Suppose that there are linear bounded operators $T_1 : Y \longrightarrow X$, $T_2 : Y \longrightarrow X$, $A_1 : X \longrightarrow X$ and $A_2 : Y \longrightarrow Y$ with A_1 demicompact and A_2 demicompact 1-set-contractive such that*

(i) $T_1 T = I - A_1$ on $\mathcal{D}(T)$,
(ii) $TT_2 = I - A_2$ on Y.

Then, T is a Fredholm operator. ◇

Proof. Since $N(T) \subset N(T_1T)$, we have $\alpha(T) \leq \alpha(T_1T) = \alpha(I - A_1)$. But, $\alpha(I - A_1) < \infty$. In fact, we shall consider $S = \{x \in N(I - A_1) \text{ such that } \|x\| = 1\}$ and prove that S is a compact set in $N(I - A_1)$. Let $(x_n)_n$ be any sequence in S, then $(x_n)_n \subset N(I - A_1)$ and $x_n - A_1x_n = 0$ for each n. Since A_1 is demicompact, there is a subsequence $(x_{n_j})_j$ of $(x_n)_n$ such that $x_{n_j} \to x_0 \in X$ as $n_j \to \infty$. Clearly, $\|x_0\| = 1$, i.e., $x_0 \in S$. This proves that S is a compact set in $N(I - A_1)$ and, consequently $\alpha(I - A_1) < \infty$. Thus $\alpha(T) < \infty$. First note that $R(T) \supset R(TT_2) = R(I - A_2)$. Now, $(I - A_2)$ is Fredholm. Hence by Lemma 2.1.7, $R(T)$ is closed and of finite codimension. Q.E.D.

Theorem 5.4.4. *Let $T : \mathcal{D}(T) \subset X \longrightarrow Y$ be a Fredholm operator. Suppose that $F : X \longrightarrow Y$ is any bounded linear operator such that $-GF$ and $-FG$ are demicompact 1-set-contractive for some operator G which is an inverse of T modulo compact operator. Then $T + F$ is a Fredholm operator with $i(T + F) = i(T)$.* ◇

Proof. First note that, by Remark 2.1.1 we have that $-T_0F$ and $-FT_0$ are demicompact 1-set-contractive for all T_0 which is an inverse of T modulo compact operator. Since T is a Fredholm operator, by Theorem 2.2.38, there is a bounded linear operator $T_0 : Y \longrightarrow X$ such that $T_0T = I - F_1$ on $\mathcal{D}(T)$ and $TT_0 = I - F_2$ on Y, where F_1 and F_2 are compact operators. This implies that T_0 is an inverse of T modulo compact operator. Now, $T_0(T + F) = I - F_1 + T_0F = I - L_1$, on

$\mathcal{D}(T+F)$ and $(T+F)T_0 = I - F_2 + FT_0 = I - L_2$ on Y where $L_1 = F_1 - T_0F$ and $L_2 = F_2 - FT_0$. Since $-T_0F$ and $-FT_0$ are demicompact 1-set-contractive and F_1 and F_2 are compact, we have that L_1 and L_2 are demicompact 1-set-contractive operators. Clearly, $\mathcal{D}(T+F)$ is dense in X. Therefore, by Theorem 5.4.3, $T+F$ is a Fredholm operator. It remains to prove that $i(T+F) = i(T)$. Since T is closed, one can make $\mathcal{D}(T)$ into a Banach space X_T by equipping it with the graph norm $\|x\|_{X_T} := \|x\| + \|Tx\|$. Moreover, X_T is continuously embedded in X and $\overline{\mathcal{D}(T)} = X_T$. Hence, $T \in \Phi^b(X_T, Y)$ by Theorem 2.2.39. So, there is a bounded linear operator $U : Y \longrightarrow X_T$ such that $UT = I - K_1$ on $\mathcal{D}(T)$ and $TU = I - K_2$ on Y, where K_1 and K_2 are compact with $R(K_1) = N(T)$ (see Theorem 2.2.38). In addition, $U \in \Phi^b(Y, X_T)$. Thus, applying Theorem 2.2.40 we have that $i[U(T+F)] = i(U) + i(T+F)$. Let T_1 be the operator $K_1 - UF$. If we consider T_1 as an operator from X into X, T_1 is demicompact and 1-set-contractive. Then, by Corollary 5.4.2 and Remark 5.4.1 (ii) we conclude that $I - T_1 \in \Phi(X)$ with $i(I - T_1) = 0$. Now,

$$i[U(T+F)] = i(I - T_1). \tag{5.4.2}$$

Assuming this for the moment, we see that $i(T+F) = -i(U)$. Theorem 2.2.40 still yields $i(UT) = i(U) + i(T)$. Since $i(UT) = i(I - K_1) = 0$, we have $i(T) = -i(U)$. Then $i(T+F) = i(T)$. Therefore, it remains only to prove (5.4.2). Since $R(K_1) = N(T) \subset \mathcal{D}(T)$ and UF is an operator from X into $\mathcal{D}(T)$, we have that

$$R(T_1) \subset \mathcal{D}(T). \tag{5.4.3}$$

It is clear that $N[U(T+F)] \subset N(I - T_1)$. Conversely, if $x \in N(I - T_1)$, then $x = T_1x \in \mathcal{D}(T)$ by (5.4.3), and hence,

$$N[U(T+F)] = N(I - T_1). \tag{5.4.4}$$

Since $I - T_1 \in \Phi(X)$ and X_T is dense in X, there is a finite dimensional subspace X_1 of X such that

$$X = R(I - T_1) \oplus X_1, \quad X_1 \subset \mathcal{D}(T). \tag{5.4.5}$$

Hence

$$X_T = [R(I - T_1) \bigcap X_T] \oplus X_1. \tag{5.4.6}$$

It is clear that $R[U(T+F)] \subset R(I - T_1) \bigcap X_T$. Conversely, if $z \in R(I - T_1) \bigcap X_T$, then $z = x - T_1x \in X_T$ for some $x \in X$. In view of (5.4.3), $x \in \mathcal{D}(T)$. Hence $z = U(T+F)x$. Thus, $R(I - T_1) \bigcap X_T = R[U(T+F)]$. Combining this with (5.4.6),

$$X_T = R[U(T+F)] \oplus X_1. \tag{5.4.7}$$

It follows from (5.4.5) and (5.4.7) that

$$\beta[U(T+F)] = \dim X_1 = \beta(I - T_1). \tag{5.4.8}$$

(5.4.4) and (5.4.8) imply (5.4.2). Q.E.D.

Remark 5.4.3.

(i) It is important to observe that, Theorem 5.4.4 is valid when GF and FG are condensing operators because condensing operators are demicompact and 1-set-contractive (see Remark 2.10.1).
(ii) When F is compact, Theorem 5.4.4 is also valid (this is the classic result of perturbation theory of Fredholm operators).
(iii) When T is the identity operator, we need $-F$ to be demicompact 1-set-contractive in order to guarantee the validity of Theorem 5.4.4.
(iv) As a consequence of Theorem 5.4.4 we deduce the following classic result whose perturbation operator is not necessarily compact. ◇

Corollary 5.4.3. *For $T \in \Phi(X, Y)$ there is an $\eta > 0$ such that for every linear operator $A : X \longrightarrow Y$ satisfying $\|A\| < \eta$ one has $T + A \in \Phi(X, Y)$ and $i(T + A) = i(T)$.* ◇

Proof. Since $T \in \Phi(X, Y)$ there is a bounded linear operator $T_0 : Y \longrightarrow X$ such that T_0 is an inverse of T modulo compact operator. We take $\eta = \|T_0\|^{-1}$, then $\|T_0 A\| \leq \|T_0\|\|A\| < 1$ and similarly, $\|AT_0\| < 1$. Since a bounded linear operator L is $\|L\|$-set-contractive, $T_0 A$ and AT_0 are k-set-contractive with $k < 1$ and, consequently, $T_0 A$ and AT_0 are condensing. Then, by Remark 5.4.3, A satisfies the hypothesis of Theorem 5.4.4. Q.E.D.

5.4.2 S-Demicompactness

Let us introduce the following definition introduced by B. Krichen in [199].

Definition 5.4.3. Let X be a Banach space, and let $A : \mathcal{D}(A) \subset X \longrightarrow X$, $S : \mathcal{D}(S) \subset X \longrightarrow X$ be densely defined linear operators with $\mathcal{D}(A) \subset \mathcal{D}(S)$. A is called S-demicompact (or relative demicompact with respect to S) if, for every bounded sequence $(x_n)_n$ in $\mathcal{D}(A)$ such that $Sx_n - Ax_n \to x \in X$, and for some x in X, there exists a convergent subsequence of $(x_n)_n$. ◇

When $\mathcal{D}(A)$ lies in a finite dimensional subspace of X, the condition of the relative demicompactness is automatically satisfied. As examples of S-demicompact operators, we cite operators A such that $(S - A)^{-1}$ exists and is continuous on its range $R(S - A)$. We also notice that if S is invertible and $S^{-1}A$ is compact, then A is an S-demicompact operator.

Remark 5.4.4. Obviously, every compact operator is demicompact. However, a compact operator is not necessarily S-demicompact when S is not invertible. ◇

Let X be a Banach space. The following results of this section give a sufficient condition for an operator to be upper semi-Fredholm using S-demicompact operators given by B. Krichen in [199].

Theorem 5.4.5. *Let $A : \mathcal{D}(A) \subset X \longrightarrow X$, $S : \mathcal{D}(S) \subset X \longrightarrow X$ be densely defined closed linear operators with $\mathcal{D}(A) \subset \mathcal{D}(S)$ such that $(S - A)$ is closed. If A is S-demicompact, then $(S - A)$ is an upper semi-Fredholm operator.* ◇

Proof. First, let us prove that $N(S - A)$ is finite dimensional. Let $\Pi^1 := \{x \in \mathcal{D}(A)$ such that $(S - A)x = 0$ and $\|x\| = 1\}$ and let $(x_n)_n$ be a sequence of Π^1. Since A is S-demicompact, there exists a subsequence $(x_{n_i})_i$ of $(x_n)_n$ which converges to $x \in X$. Hence, and from the continuity of the norm and the closedness of $S - A$, it follows that $x \in \mathcal{D}(A)$, $Sx - Ax = 0$ and $\|x\| = 1$. Then, $\alpha(S - A)$ is finite. Now, we claim that $R(S - A)$ is closed. Applying Lemma 2.1.6, there exists a closed subspace X_0 of X such that $\mathcal{D}(A) = N(S - A) \oplus (\mathcal{D}(A) \bigcap X_0)$. In view of Theorem 2.2.1, it is sufficient to prove that there is a constant $\lambda > 0$ such that, for every $x \in \mathcal{D}(A) \bigcap X_0$, $\|Sx - Ax\| \geq \lambda \|x\|_{S-A}$. If not, there exists a sequence $(x_n)_n$ of $\mathcal{D}(A) \bigcap X_0$ such that $\|x_n\|_{S-A} = 1$ and $\|(S - A)x_n\| \to 0$. Since A is S-demicompact, there exists a subsequence $(x_{n_i})_i$ of $(x_n)_n$ which converges to $x \in X$. Since $(S - A)$ is closed, $(S - A)x = 0$, and then $x = 0$, which contradicts the continuity of the norm. Q.E.D.

Remark 5.4.5.

(i) Note that in the assumption of Theorem 5.4.5, the operators S and A need not to be bounded. However, if S is relatively bounded with respect to A with A-bound < 1, then the operator $S - A$ is closed if, and only if, A is closed (see Theorem 2.1.4).
(ii) It is noted that if S is an A-bounded operator with a relative bound lower than one, the family $\{S - \mu A\}_{0 \leq \mu \leq 1}$ of operators in $\mathcal{C}(X)$ is continuous in the gap topology. ◇

Lemma 5.4.7. *Let $A : \mathcal{D}(A) \subset X \longrightarrow X$, $S : \mathcal{D}(S) \subset X \longrightarrow X$ be densely defined linear operators with $\mathcal{D}(A) \subset \mathcal{D}(S)$ and A is closed, such that S is nonzero and is relatively bounded with respect to A with A-bound < 1. If for each $\mu \in [0, 1]$, the operator $S - \mu A$ is closed and μA is S-demicompact, then $(S - A)$ is a Fredholm operator and $i(S - A) = i(S)$.* ◇

Proof. Since μA is S-demicompact for each $\mu \in [0, 1]$, and using Theorem 5.4.5, we deduce that $(S - \mu A)$ is an upper semi-Fredholm operator on X. Using the fact that S is relatively bounded with respect to A with A-bound < 1, the Remark 5.4.5 combined with the stability results for semi-Fredholm operators allow us to infer that the index $i(S - \mu A)$ is continuous in μ (see Proposition 2.2.6). Since it is an integer, including infinite values, it must be constant for every $\mu \in [0, 1]$. Hence, $i(S - \mu A) = i(S - A) = i(S)$. Q.E.D.

From the definition, it may not be easy to recognize a Fredholm operator when one sees such an operator. A useful tool in this connection is the following result.

Theorem 5.4.6. *Let $A : \mathcal{D}(A) \subset X \longrightarrow X$, $S : \mathcal{D}(S) \subset X \longrightarrow X$ be densely defined closed linear operators. Suppose that there are bounded operators S_1, S_2, A_1 and A_2 with $\mathcal{D}(A) \subset \mathcal{D}(S)$, A_1 is S-demicompact and μA_2 is S-demicompact for any $\mu \in [0,1]$, such that $S_1 A = S - A_1$ on $\mathcal{D}(A)$, and $AS_2 = S - A_2$ on $\mathcal{D}(S)$. Then, A is a Fredholm operator.* ◇

Proof. We first show that $\alpha(A) < +\infty$. Since $N(A) \subset N(S_1 A)$, we have $\alpha(A) \leq \alpha(S - A_1) < +\infty$. In fact, we will consider $\Pi^1 = \{x \in \mathcal{D}(A)$ such that $(S - A_1)x = 0$ and $\|x\| = 1\}$ and prove that Π^1 is a compact set in $N(S - A_1)$. Let $(x_n)_n$ be any sequence in Π^1. Then, $(S - A_1)x_n = 0$ and $\|x_n\| = 1$. Since A_1 is S-demicompact, there exists a subsequence $(x_{n_i})_i$ of $(x_n)_n$ such that $x_{n_i} \to x \in X$ as $n_i \to \infty$. Clearly, $N(S - A_1)$ is closed and so $x \in \Pi^1$. This proves that Π^1 is a compact set in $N(S - A_1)$ and consequently, $\alpha(A) < +\infty$. Now, notice that $R(A) \supset R(AS_2) \supset R(S - A_2)$. From Lemmas 2.1.7 and 5.4.7, it follows that $R(A)$ is closed and finite codimension and so, A is a Fredholm operator. Q.E.D.

It was shown, in Lemma 5.4.3, that if $T \in \mathcal{L}(X)$ is a power-compact operator, then for every $\mu \in [0,1)$, μT is demicompact. The following result provides sufficient conditions to ensure that an operator A satisfies the characteristic that μA is S-demicompact for every $\mu \in [0,1]$.

Theorem 5.4.7. *Let $A : \mathcal{D}(A) \subset X \longrightarrow X$, $S : \mathcal{D}(S) \subset X \longrightarrow X$ be densely defined closed linear operators with $\mathcal{D}(A) \subset \mathcal{D}(S)$ such that $S - A$ is closed and let us assume that, for some $\lambda \in \rho(S)$ (hence for all such λ), the operator is closable and its closure $\overline{(S-\lambda)^{-1}A}$ is 1-set contraction and there exists $m \in \mathbb{N}^*$ such that $\overline{(S-\lambda)^{-1}A}^m$ is compact and $(S-\lambda)^{-1}$ is a $k(\lambda)$-set contraction operator with $m|\lambda|k(\lambda) < 1$. Then, for every $\mu \in [0,1)$, the operator μA is S-demicompact.* ◇

Proof. Obviously for $\mu = 0$, μA is S-demicompact. Let $m \in \mathbb{N}^*$ and $\lambda \in \rho(S)$ such that $\overline{(S-\lambda)^{-1}A}^m$ is a compact operator. Take $\mu \in (0,1)$ and $(x_n)_n$ a bounded sequence of $\mathcal{D}(A)$, such that $y_n := (S - \mu A)x_n \to y$. Note first that x_n can be written as

$$\begin{aligned} x_n &= (S-\lambda)^{-1}y_n + (S-\lambda)^{-1}(\mu A - \lambda)x_n \\ &= (S-\lambda)^{-1}y_n + \mu\overline{(S-\lambda)^{-1}A}x_n - \lambda(S-\lambda)^{-1}x_n. \end{aligned}$$

Then,

$$\begin{aligned} x_n - \mu^m \overline{(S-\lambda)^{-1}A}^m x_n &= \sum_{k=0}^{m-1} \mu^k \overline{(S-\lambda)^{-1}A}^k \left(x_n - \mu\overline{(S-\lambda)^{-1}A}x_n\right) \\ &= \sum_{k=0}^{m-1} \mu^k \overline{(S-\lambda)^{-1}A}^k \left((S-\lambda)^{-1}y_n - \lambda(S-\lambda)^{-1}x_n\right). \end{aligned}$$

It follows that

$$x_n = \mu^m \overline{(S-\lambda)^{-1}A}^m x_n + \Upsilon(\lambda, S)y_n - \lambda \Upsilon(\lambda, S)x_n, \tag{5.4.9}$$

where $\Upsilon(\lambda, S) = \sum_{k=0}^{m-1} \mu^k \overline{(S-\lambda)^{-1}A}^k (S-\lambda)^{-1}$. Since $\overline{(S-\lambda)^{-1}A}^m$ is a compact operator, there exists a subsequence $(x_{\varphi(n)})_n \subset \mathcal{D}(A)$ such that $\overline{(S-\lambda)^{-1}A}^m x_{\varphi(n)}$ converges. Let us suppose that $\gamma(\{x_{\varphi(n)}\}) \neq 0$. Then, using the hypotheses and Eq. (5.4.9), we deduce that

$$\begin{aligned}\gamma(\{x_{\varphi(n)}\}) &\leq |\lambda| k(\lambda)\gamma(\{x_{\varphi(n)}\}) + |\lambda|(m-1)k(\lambda)\gamma(\{x_{\varphi(n)}\}) \\ &< \gamma(\{x_{\varphi(n)}\})\end{aligned}$$

which is impossible. It follows that $\gamma(\{x_{\varphi(n)}\}) = 0$. Hence, $\{x_{\varphi(n)}\}$ is relatively compact and, then there is a convergent subsequence of $(x_n)_n$ and μA is S-demicompact. Q.E.D.

Theorem 5.4.8. *Let $A : \mathcal{D}(A) \subset X \longrightarrow X$, $S : \mathcal{D}(S) \subset X \longrightarrow X$ be densely defined closed linear operators with $\mathcal{D}(A) \subset \mathcal{D}(S)$ such that $S - A$ is closed and, for some $\lambda \in \rho(S)$, there exists $m \in \mathbb{N}^*$ such that $\overline{(S-\lambda)^{-1}A}^m$ is compact and $(S-\lambda)^{-1}$ is a $k(\lambda)$-set contraction operator with $m|\lambda|k(\lambda) < 1$. Then, for any $\mu \in [0, 1)$, the operator $\mu A + \lambda$ is S-demicompact.* ◇

Proof. Obviously for $\mu = 0$, $\mu A + \lambda$ is S-demicompact. We suppose that $\mu \in (0, 1)$. Let $(x_n)_n$ be a bounded sequence of $\mathcal{D}(A)$, such that $y_n := (S - \mu A - \lambda)x_n \to y$. Suppose that $\gamma(\{x_n\}) \neq 0$. Since $\{x_n\} \subseteq \{(S-\lambda)^{-1}y_n\} + \{\mu\overline{(S-\lambda)^{-1}A}x_n\}$, it follows that

$$\begin{aligned}\gamma(\{x_n\}) &\leq \gamma(\{(S-\lambda)^{-1}y_n\}) + \mu\gamma(\{\overline{(S-\lambda)^{-1}A}x_n\}) \\ &\leq \mu\gamma(\{x_n\}) \\ &< \gamma(\{x_n\})\end{aligned}$$

which is impossible. It follows that $\gamma(\{x_n\}) = 0$. Hence, $\{x_n\}$ is relatively compact, and there is a convergent subsequence $(x_{n_i})_i$ of $(x_n)_n$. Q.E.D.

5.5 Fredholm Theory by Means of Weak Noncompactness Measures

In order to give an axiomatic approach to the notion of measure of weak noncompactness of operators, let us recall this definition.

Definition 5.5.1. Let X and Y be two Banach spaces, and let μ be a measure of weak noncompactness in Y. We define the function

$$\begin{aligned}\psi_\mu : \mathcal{L}(X,Y) &\longrightarrow [0,+\infty[\\ T &\longrightarrow \psi_\mu(T) = \mu(T(\overline{B}_X)),\end{aligned}$$

where ψ_μ is called a measure of weak noncompactness of operators associated with μ. ◇

From this definition, we can directly obtain the following proposition:

Proposition 5.5.1. *Let X and Y be two Banach spaces, let μ be a measure of weak noncompactness in Y, and let ψ_μ be a measure of weak noncompactness of operators associated with μ. For all S, $T \in \mathcal{L}(X,Y)$, and we have*

(a) $\psi_\mu(T) = 0$ *if, and only if, T is weakly compact (regularity),*
(b) $\psi_\mu(S+T) \leq \psi_\mu(S) + \psi_\mu(T)$ *(algebraic semi-additivity),*
(c) $\psi_\mu(\lambda S) = |\lambda|\psi_\mu(S)$, $\forall \lambda \in \mathbb{R}$ *(semi-homogeneity), and*
(d) $\psi_\mu(S + K) = \psi_\mu(S)$, $\forall K \in \mathcal{W}(X,Y)$ *(invariance under a weak compactness).* ◇

Proof. Using, respectively, the assertions (i), (v), and (vi) in Definition 2.11.1, we can easily prove, respectively, the properties (a), (b), and (c).

(d) Let K be a weakly compact operator. Using the properties (a) and (b), we get $\psi_\mu(S + K) \leq \psi_\mu(S)$. Furthermore, $\psi_\mu(S) = \psi_\mu(S + K - K) \leq \psi_\mu(S + K)$. Q.E.D.

Definition 5.5.2. Let X be a Banach space, let μ be a measure of weak noncompactness in X, and let ψ_μ be a measure of weak noncompactness of operators associated with μ. ψ_μ is said to be algebraic semi-multiplicative if, for all S, $T \in \mathcal{L}(X)$, we have

$$\psi_\mu(ST) \leq \psi_\mu(S)\psi_\mu(T).$$

◇

Proposition 5.5.2. *Let X be a Banach space, let μ be a measure of weak noncompactness in X and let ψ_μ be a measure of weak noncompactness of operators associated with μ. If $\mu(S(D)) \leq \mu(S(\overline{B}_X))\mu(D)$, for every $S \in \mathcal{L}(X)$ and $D \in \mathcal{M}_X$, then ψ_μ has the algebraic semi-multiplicative property.* ◇

Proof. $\psi_\mu(ST) = \mu(ST(\overline{B}_X)) \leq \psi_\mu(S)\mu(T(\overline{B}_X)) \leq \psi_\mu(S)\psi_\mu(T)$. Q.E.D.

Definition 5.5.3. Let X and Y be two Banach spaces, let μ (resp. μ^*) be a measure of weak noncompactness in Y (resp. in X^*) and let ψ_μ (resp. ψ_{μ^*}) be a measure of weak noncompactness of operators associated with μ (resp. to μ^*). (ψ_μ, ψ_{μ^*}) is said to have the adjoint-equivalent property, if there exist a, $b \in \mathbb{R}_+$, such that $a\psi_{\mu^*}(S^*) \leq \psi_\mu(S) \leq b\psi_{\mu^*}(S^*)$ for all $S \in \mathcal{L}(X,Y)$. ◇

Let us recall some examples of measures of weak noncompactness of operators [37, 133]. For X and Y being Banach spaces and $T \in \mathcal{L}(X, Y)$, we denote by

- $\Theta_\omega(T) = \omega(T(\overline{B}_X))$, where ω is the measure of weak noncompactness of De Blasi in the space Y [see (2.11.1)].
- $\Gamma_\gamma(T) = \gamma(T(\overline{B}_X))$, where for $A \in \mathcal{M}_Y, \gamma(A) = \sup\{\text{csep}(x_n)$ such that $(x_n) \subset \text{conv}(A)\}$ [see (2.11.3)].

Remark 5.5.1.

(i) Θ_ω has the algebraic semi-multiplicative property. Indeed, let A be a bounded set in X. Then, for each $\varepsilon > 0$, there exists $0 < t \leq \omega(A) + \varepsilon$ such that $A \subset C + t\overline{B}_X$ and C is a weakly compact operator. Then, $A \subset C + (\omega(A) + \varepsilon)\overline{B}_X$. Applying the operator T and using the assertions (i), (ii), (v), and (vi) in Definition 2.11.1, we get $\omega(T(A)) \leq \omega(A)\omega(T(\overline{B}_X))$. Finally, from Proposition 5.5.2, Θ_ω has the algebraic semi-multiplicative property.

(ii) For general Banach spaces, $(\Theta_\omega, \Theta_{\omega^*})$ doesn't have the adjoint-equivalent property (cf. [38, Theorem 4, p. 371]). ◇

We have the following proposition.

Proposition 5.5.3. $(\Gamma_\gamma, \Gamma_{\gamma^*})$ *has the adjoint-equivalent property.* ◇

Proof. Let X and Y be two Banach spaces. For each $T \in \mathcal{L}(X, Y)$, we have the corresponding operator $R(T) : X^{**}/X \longrightarrow Y^{**}/Y$ given by the formula $R(T)(x^{**} + X) = T^{**}x^{**} + Y$, for every $x^{**} \in X^{**}$. Then, from [200, Lemma 3.1], we have

$$\begin{aligned} \|R(T)\| &= \sup\{\text{dist}(T^{**}x^{**}, Y) \text{ such that } \text{dist}(x^{**}, X) \leq 1\} \\ &= \sup\{\text{dist}(T^{**}x^{**}, Y) \text{ such that } \|x^{**}\| \leq 1\} \end{aligned}$$

and $\|R(T)\| = 0$ if, and only if, T is weakly compact (see [101, p. 482]). From [133, Proposition 1.3], $\|R(T)\|$ has the adjoint-equivalent property. More precisely, we have

$$\frac{1}{2}\|R(S)\| \leq \|R(S^*)\| \leq 2\|R(S)\|, \text{ for all } S \in \mathcal{L}(X, Y). \tag{5.5.1}$$

Moreover, from [200, Theorem 3.2], we deduce that $\|R(.)\|$ and $\Gamma_\gamma(.)$ are equivalent by the following inequality

$$\frac{1}{2}\Gamma_\gamma(S) \leq \|R(S)\| \leq \Gamma_\gamma(S), \text{ for all } S \in \mathcal{L}(X, Y). \tag{5.5.2}$$

Using the above inequalities (5.5.1) and (5.5.2), we get $\frac{1}{4}\Gamma_{\gamma^*}(S^*) \leq \Gamma_\gamma(S) \leq 4\Gamma_{\gamma^*}(S^*)$, for all $S \in \mathcal{L}(X, Y)$, which implies that $(\Gamma_\gamma, \Gamma_{\gamma^*})$ has the adjoint-equivalent property. Q.E.D.

Definition 5.5.4. Let X and Y be two Banach spaces. For $S \in \mathcal{L}(X,Y)$, we define the $\mathcal{W}$-quotient-norm of T by $\|S\|_{\mathcal{W}} := \inf\{\|S-K\|$ such that $K \in \mathcal{W}(X,Y)\}$. ◇

Proposition 5.5.4. *Let X be a Banach space, let $S \in \mathcal{L}(X)$, let μ be a measure of weak noncompactness in X and let ψ_μ be a measure of weak noncompactness of operators associated with μ. Then,*

(i) $\psi_\mu(S) \leq \|S\|\, \mu(\overline{B}_X)$,
(ii) $\psi_\mu(S) \leq \|S\|_{\mathcal{W}}\, \mu(\overline{B}_X)$, *and*
(iii) $\psi_\mu(S) \leq \Theta_\omega(S)\, \mu(\overline{B}_X)$. ◇

Proof.

(i) Since $S(\overline{B}_X) \subset \|S\|\overline{B}_X$, then $\psi_\mu(S) \leq \|S\|\mu(\overline{B}_X)$.
(ii) For all $K \in \mathcal{W}(X)$, we have $\psi_\mu(S+K) \leq \|S+K\|\mu(\overline{B}_X)$. Since $\psi_\mu(S+K) = \psi_\mu(S)$ (see Proposition 5.5.1 (d)), then $\psi_\mu(S) \leq \|S\|_{\mathcal{W}}\, \mu(\overline{B}_X)$.
(iii) From the inequality (2.11.2), we have $\mu(A) \leq \mu(\overline{B}_X)\omega(A)$, for all $A \in \mathcal{M}_X$. The result follows if $A = S(\overline{B}_X)$. Q.E.D.

Definition 5.5.5. Let X and Y be two Banach spaces. An operator $T : X \longrightarrow Y$ is called a Dunford-Pettis operator (for short property DP operator) if T maps weakly compact sets onto compact sets. ◇

Lemma 5.5.1. *Let P and Q be two complex polynomials satisfying Q divides $(P-1)$ and $\mu(.)$ be a measure of weak noncompactness in X satisfies*

$$\mu(P(T)(A)) \leq \Psi_\mu(P(T))\mu(A), \textit{ for every } A \in \mathcal{M}_X, \tag{5.5.3}$$

where Ψ_μ be a measure of weak noncompactness of operators associated with μ. Let $A = \{x \in \overline{B}_X$ such that $Q(T)x - Kx \in M\}$, where $M \subset X$, and $K \in \mathcal{K}(X)$. If M is weakly compact and $\Psi_\mu(P(T)) < 1$, then A is relatively weakly compact or empty. ◇

Proof. Assume that A is not empty. According to the fact that Q divides $(P-1)$, there exists a complex polynomial R such that $P = RQ + 1$. Consider $x \in A$ and $z \in M$ such that $Q(T)(x) - K(x) = z$. Then, we get $R(T)Q(T)(x) - R(T)K(x) + x = R(T)(z) + x$, which implies $x = P(T)(x) - R(T)(z) - R(T)K(x)$. Since a continuous image of a weakly compact set by a continuous operator is also weakly compact, it follows that $\tilde{A} = \{-R(T)(z)$ such that $z \in M\}$ is weakly compact as well. Obviously, $A \subset P(T)A + \tilde{A} - R(T)K(A)$ and $\mu(\tilde{A}) = \mu(R(T)K(A)) = 0$. So, using both Definition 2.11.1 (ii) and (v) and Eq. (5.5.3), we get $\mu(A) \leq \mu(P(T)A) \leq \mu(A)\Psi_\mu(P(T))$. Since $\Psi_\mu(P(T)) < 1$, then $\mu(A) = 0$. Consequently, by using Definition 2.11.1, we infer that $A \in \mathcal{W}_X$. Q.E.D.

Now, we are ready to state and prove the following theorem.

Theorem 5.5.1. *Let X be a Banach space, $T \in \mathcal{L}(X)$, let μ (resp. μ^*) be a measure of weak noncompactness in X (resp. in X^*), and let Ψ_μ (resp. Ψ_{μ^*}) be a measure of weak noncompactness of operators associated with μ (resp. to μ^*).*

Let P and Q be two complex polynomials satisfying the fact that Q divides $(P-1)$, $P(T)$ being a DP operator, and $\mu(.)$ satisfies the Eq. (5.5.3).

(i) If $\Psi_{\mu}(P(T)) < 1$, then $Q(T) \in \Phi_{+}^{b}(X)$.
(ii) Moreover, let us suppose that $(P(T))^{*}$ is a DP operator, and

$$\mu^{*}(P(T)^{*}(D)) \leq \Psi_{\mu^{*}}(P(T)^{*})\mu^{*}(D), \text{ for every } D \in \mathcal{M}_{X^{*}}. \quad (5.5.4)$$

If $\Psi_{\mu}(P(T)) < 1$ and $\Psi_{\mu^{*}}(P(T)^{*}) < 1$, then $Q(T) \in \Phi^{b}(X)$.
(iii) Suppose that $\mu^{*}(.)$ satisfies the Eq. (5.5.4), that $(P(T))^{*}$ is a DP operator and, $(\psi_{\mu}, \psi_{\mu^{*}})$ has the adjoint-equivalent property. If $\Psi_{\mu}(P(T)) < 1$, then $Q(T) \in \Phi^{b}(X)$. ◇

Proof.

(i) By using Lemma 2.2.2, it is sufficient to prove that, for any compact operator $K \in \mathcal{K}(X)$, $\alpha(Q(T) - K) < \infty$. To do so, we only need to establish that the set $N(Q(T) - K) \bigcap \overline{B}_{X}$ is compact. Applying Lemma 5.5.1 with $M = \{0\}$, we show that $N(Q(T) - K) \bigcap \overline{B}_{X}$ is weakly compact and is included in $P(T)(N(Q(T) - K) \bigcap \overline{B}_{X})$. Using the fact that $P(T)$ is a DP operator, we ensure that $N(Q(T) - K) \bigcap \overline{B}_{X}$ is compact. This ends the proof of the first part of the theorem.
(ii) Assume that $\Psi_{\mu^{*}}(P(T)^{*}) < 1$. Proceeding as in the proof of (i), and according to the hypotheses, we get $\alpha(Q(T)^{*}) = \beta(Q(T)) < \infty$. This completes the proof of (ii).
(iii) Suppose that there exists $a \in \mathbb{R}_{+}^{*}$ such that, for all $S \in \mathcal{L}(X)$, $\psi_{\mu^{*}}(S^{*}) \leq a\psi_{\mu}(S)$. If $\Psi_{\mu}(P(T)) < 1$, then $\lim_{k \to +\infty} [\Psi_{\mu}(P(T))]^{k} = 0$. So, there exists $k_0 \in \mathbb{N}^{*}$ such that $[\Psi_{\mu}(P(T))]^{k_0} < \frac{1}{a}$. From Eq. (5.5.3) and using Proposition 5.5.2, we deduce the following $\psi_{\mu}(P(T)^{k_0}) < \frac{1}{a}$. Hence, $\psi_{\mu^{*}}[(P(T)^{*})^{k_0}] < 1$. Since Q divides $P^{k_0} - 1$, then applying (i), we get $Q(T)^{*} \in \Phi_{+}^{b}(X^{*})$ which implies that $Q(T) \in \Phi_{-}^{b}(X)$. Furthermore, since $\Psi_{\mu}(P(T)) < 1$, then $Q(T) \in \Phi_{+}^{b}(X)$. This completes the proof of the theorem. Q.E.D.

As a consequence of Theorem 5.5.1, we have

Corollary 5.5.1. *Let X be a Banach space and let $T \in \mathcal{L}(X)$. Suppose that $\mu(.)$ and $\mu^{*}(.)$ satisfy Eqs. (5.5.3) and (5.5.4), respectively, with $P = z^{m}$ ($m \in \mathbb{N}^{*}$) and let us suppose that $(\psi_{\mu}, \psi_{\mu^{*}})$ has the adjoint-equivalent property. If $\psi_{\mu}(T^{m}) < 1$ and T^{m} is a DP operator, then $(I - T)$ is a Fredholm operator with $i(I - T) = 0$.* ◇

Proof. Applying Theorem 5.5.1 (i) with $P(z) = z^{m}$ and $Q(z) = 1 - z$, we conclude that $(I - T) \in \Phi_{+}^{b}(X)$. Next, note that for $t \in [0, 1]$, we have $\psi_{\mu}((tT)^{m}) < 1$ and $(tT)^{m}$ is a DP operator. Then, $(I - tT) \in \Phi_{+}^{b}(X)$. Moreover, by the continuity of the index on $\Phi_{+}^{b}(X)$, we get $i(I - T) = i(I - tT) = i(I) = 0$. Hence, $(I - T) \in \Phi^{b}(X)$, which completes the proof. Q.E.D.

As a consequence of Theorem 5.5.1, we have

Corollary 5.5.2. *Let X be a Banach space, $T \in \mathcal{L}(X)$, and let P and Q be two complex polynomials such that Q divides $(P-1)$ and $P(T)$ is a DP operator. Suppose that $\mu(.)$ satisfies Eq. (5.5.3) with $P(z) = z^m$ $(m \in \mathbb{N}^*)$. If $\psi_\mu(T^m) < 1$, then $(I-T)$ is a Fredholm operator and $i(I-T) = 0$.* ◇

Proof. By using Theorem 5.5.1, we deduce that $(I-T) \in \Phi_+^b(X)$. Next, let us notice that for $t \in [0,1]$, we have $(\psi_\mu(tT))^{n_0} < 1$ and $(tT)^m$ is a DP operator. Then, $(I-tT) \in \Phi_+^b(X)$. Now, by the continuity of the index on $\Phi_+^b(X)$, we get $i(I-T) = i(I-tT) = i(I) = 0$. Hence, $(I-T) \in \Phi^b(X)$. Q.E.D.

The following corollary is an immediate consequence of Remark 5.5.1 (i), Proposition 5.5.3, and Theorem 5.5.1 (iii).

Corollary 5.5.3. *Let X be a Banach space and let $T \in \mathcal{L}(X)$. Suppose that there exist two complex polynomials P and Q satisfying Q divides $(P-1)$ and $P(T)$ is a DP operator.*

(i) If $\Theta_\omega(P(T)) < 1$, then $Q(T) \in \Phi_+^b(X)$.
(ii) If $\Gamma_\gamma(P(T)) < 1$, then $Q(T) \in \Phi^b(X)$. ◇

Corollary 5.5.4. *Let X be a Banach space and let $T \in \mathcal{L}(X)$. Suppose that there exist two complex polynomials P and Q satisfying the fact that Q divides $(P-1)$, $P(T)$ and $P(T)^*$ being DP operators.*

(i) Let (Ω, Σ, η) be an arbitrary positive measure space and suppose that $X = L_1(\Omega)$. If $\Theta_\omega(P(T)) < \frac{1}{2}$, then $Q(T) \in \Phi^b(X)$.
(ii) Suppose that, for any index set I, $X = l_I^\infty$. If $\Theta_\omega(P(T)) < 1$ and $\|P(T)\|_{\mathcal{W}} < 1$, then $Q(T) \in \Phi^b(X)$. ◇

Proof.

(i) Since $\Gamma_\gamma(P(T)) = 2\Theta_\omega$, then $\Gamma_\gamma(P(T)) < 1$. The result follows from Corollary 5.5.3.
(ii) From [37, Theorem 5.2], we have $\Theta_{\omega^*}(P(T)^*) = \|P(T)^*\|_{\mathcal{W}} = \|P(T)\|_{\mathcal{W}}$. The result follows from Theorem 5.5.1 (i). Q.E.D.

Remark 5.5.2. The result of Corollary 5.5.4 (ii) holds for any Banach space X having the extension property, i.e., for each $T \in \mathcal{L}(M, X)$ there exists an operator $S \in \mathcal{L}(Y, X)$ such that $T = SJ_M$ and $\|T\| = \|S\|$. Here, M is a closed subspace of an arbitrary Banach space Y and $J_M : M \longrightarrow E$ is the canonical injection (see [37]). ◇

Corollary 5.5.5. *Let (Ω, Σ, η) be an arbitrary positive measure space, let $X = L_1(\Omega)$ and let $T \in \mathcal{L}(X)$. Suppose that there exists a complex polynomial P satisfying $P(0) = 1$. Assume that $P(T)$ and $P(T)^*$ are DP operators. If $\Theta_\omega(P(T)) < \frac{1}{2}$, then $T \in \Phi^b(X)$.* ◇

Proof. Since $P(0) = 1$, then $Q(z) = z$ divides $(P(z) - 1)$ and the result follows from Corollary 5.5.4. Q.E.D.

5.6 Fredholm Theory with Finite Ascent and Descent

The following results may be found in [2].

Theorem 5.6.1. *Let T, S be two bounded operators on X. If $\varphi(T) < \delta(S)$, then $T + S \in \Phi_+^b(X)$ and $i(T + S) = i(S)$, where $\varphi(.)$ [resp. $\delta(.)$] is introduced in (2.12.2) [resp. (2.12.1)].* ◇

Proof. Suppose that $\delta(T + S) = 0$. Then $\delta(S) = \delta(T - (T + S)) < \varphi(T)$. Hence, if $\varphi(T) < \delta(S)$, then $\delta(T + S) > 0$ and therefore $T + S \in \Phi_+^b(X)$. Let $t \in [0, 1]$, then $\varphi(tT) < \delta(S)$, and so, by what we have just proved, $tT + S \in \Phi_+^b(X)$. Thus, by the continuity of the index on $\Phi_+^b(X)$, we get $i(T + S) = i(S)$. Q.E.D.

Theorem 5.6.2. *Let T, S be two commuting bounded operators on X. Let $\varphi(T) < \delta(S)$. If $asc(S) < \infty$, then $asc(T + S) < \infty$.* ◇

Proof. For $t \in [0, 1]$, we have $\varphi(tT) < \delta(S)$, then, by Theorem 5.6.1, $tT + S \in \Phi_+^b(X)$. Since S and T are commuting, then according to Theorem 2.2.20, we have $\overline{N^\infty(tT + S)} \bigcap R^\infty(tT + S) = \overline{N^\infty(sT + S)} \bigcap R^\infty(sT + S)$, for all s in some open disk with center t. Hence, $\overline{N^\infty(tT + S)} \bigcap R^\infty(tT + S)$ is locally constant function of t on the interval $[0, 1]$. This yields that for all $t \in [0, 1]$, we have $\overline{N^\infty(tT + S)} \bigcap R^\infty(tT + S) = \overline{N^\infty(S)} \bigcap R^\infty(S)$. Now, since $\mathrm{asc}(S) < \infty$, then from Lemma 2.2.10, $N^\infty(S) \bigcap R^\infty(S) = \overline{N^\infty(S)} \bigcap R^\infty(S) = \{0\}$. Hence, $\overline{N^\infty(T + S)} \bigcap R^\infty(T + S) = \{0\}$. Thus, we have $N^\infty(T + S) \bigcap R^\infty(T + S) = \{0\}$, and again by Lemma 2.2.10, it follows that $\mathrm{asc}(T + S) < \infty$. Q.E.D.

Theorem 5.6.3. *Let T, S be two commuting bounded operators on X. Suppose that there exists $n \in \mathbb{N}^*$ such that $\varphi(T^n) < \delta(S^n)$, then we get*

(i) If $S \in \mathcal{B}_+^b(X)$, then $T + S \in \mathcal{B}_+^b(X)$.
(ii) If $S \in \mathcal{B}^b(X)$, then $T + S \in \mathcal{B}^b(X)$. ◇

Proof.

(i) Let $t \in [0, 1]$. Since $\varphi((tT)^n) < \delta(S^n)$, by Theorem 5.6.1, $tT + S \in \Phi_+^b(X)$. Arguing as in the proof of Theorem 5.6.2, we get the result.
(ii) Since $S \in \mathcal{B}^b(X)$, then $i(S) = 0$. Arguing as in the proof of Theorem 5.6.1, we get $i(T + S) = 0$. Moreover, Theorem 5.6.2 yields $\mathrm{asc}(T + S) < \infty$. According to Lemma 2.2.7, we get $\mathrm{desc}(T + S) < \infty$. Q.E.D.

5.7 Stability of Semi-Browder Operators

The reader interested in the results of this section may also refer to [111] which constitutes the real basis of our work.

Lemma 5.7.1. *Let X be a Banach space, $A \in \mathcal{C}(X)$ such that $\rho(A) \neq \emptyset$ and $\overline{\mathcal{D}(A^p)} = X$ for all $p \in \mathbb{N}$. If $A \in \Phi_-(X)$ and desc$(A) < \infty$, then $A^p \in \Phi_-(X)$ for every $p \in \mathbb{N}$.* ◇

Proof. Since A is a closed operator with a non-empty resolvent set, then we can deduce that A^p is a closed operator. From Lemma 2.2.27, it follows that $\beta(A^p) \leq \text{desc}(A)\beta(A)$. So, $\beta(A^p) < \infty$ for every p. Finally, the use of Lemma 2.1.9 gives $R(A^p)$ is a closed subspace. Q.E.D.

Lemma 5.7.2. *Let X be a reflexive Banach space, $A \in \mathcal{C}(X)$ and $B = A + K$ where $K \in \mathcal{K}(X)$ such that A and B are mutually commuting operators and A^* commutes with itself. If $A \in \Phi_-(X)$, then $A^{*p}B^{*k} \in \Phi_+(X^*)$ for every p, $k \in \mathbb{N}^*$.* ◇

Proof. Let $A \in \Phi_-(X)$, then by Theorem 2.2.46 and Lemma 2.2.15, $A^* \in \Phi_+(X^*)$. Hence, applying Remark 2.2.6 (ii) and Theorem 2.2.43, we get $A^{*p} \in \Phi_+(X^*)$. On the other hand, by Theorem 2.2.47, we deduce that $B \in \Phi_-(X)$. Then a similar reasoning as above shows that $B^{*k} \in \Phi_+(X^*)$. Finally, using Lemma 2.2.25 (i) together with Theorem 2.2.43 we infer that $A^{*p}B^{*k} \in \Phi_+(X^*)$. Q.E.D.

Lemma 5.7.3. *Let X be a reflexive Banach space, $A \in \mathcal{C}(X)$ and $B = A + K$ where $K \in \mathcal{K}(X)$. If A and B are mutually commuting operators and A^* commutes with itself, then (B, A) is of finite type if, and only if, (A, B) is of finite type.* ◇

Proof. Reasoning in the same way as the proof of Lemma 2.2.25 (ii), we can prove that $(A^pB^k)^*$ is an extension of $B^{*k}A^{*p}$. On the other hand, $\mathcal{D}((B^kA^p)^*) = \mathcal{D}((A^pB^k)^*)$. Indeed, let $f \in \mathcal{D}((B^kA^p)^*)$ then there is $g \in X^*$ such that $g(u) = f \circ B^kA^p(u)$ for all $u \in \mathcal{D}(B^kA^p) = \mathcal{D}(A^pB^k)$. The use of Lemma 2.2.22 makes us conclude that $g(u) = f \circ A^pB^k(u)$ for all $u \in \mathcal{D}(B^kA^p) = \mathcal{D}(A^pB^k)$. So, $f \in \mathcal{D}((A^pB^k)^*)$. A similar reasoning as above gives $\mathcal{D}((A^pB^k)^*) \subset \mathcal{D}((B^kA^p)^*)$. Now, since $\mathcal{D}(A^{*p}B^{*k}) = \mathcal{D}(B^{*k}A^{*p})$ and $\mathcal{D}((B^kA^p)^*) = \mathcal{D}((A^pB^k)^*)$, we get $(A^pB^k)^*$ is a finite dimensional extension of $B^{*k}A^{*p}$. Therefore, (A, B) is of finite type. Proceeding as the proof above, we establish that if (A, B) is of finite type, then (B, A) is of finite type. Q.E.D.

Theorem 5.7.1. *Let X be a reflexive Banach space, $A \in \mathcal{C}(X)$ and $B = A + K$ where $K \in \mathcal{K}(X)$ such that A and B are mutually commuting operators and A^* commutes with itself. Assume that $\rho(A) \neq \emptyset$, $\rho(A + K) \neq \emptyset$, and (B, A) is of finite type. If $A \in \Phi_-(X)$, then desc$(A) < \infty$ if, and only if, desc$(A + K) < \infty$.* ◇

Proof. Let $A \in \Phi_-(X)$ such that $\text{desc}(A) = p < \infty$. Then, by Lemma 5.7.1 and the fact that A commutes with itself we deduce that $R(A^p)$ is a closed subspace of

$\mathcal{D}(A)$. On the other hand, from Lemma 2.2.22 and the fact that $\text{desc}(A) = p < \infty$ we show that $B(R(A^p)) \subset R(A^p)$ and $A(R(A^p)) \subset R(A^p)$. Since the restriction of A to $R(A^p)$ is onto, therefore by Lemma 2.2.26, the restriction of B to $R(A^p)$ has finite descent. So that there is an integer k for which

$$R(B^m) \supset R(B^m A^p) = R(B^k A^p) \tag{5.7.1}$$

for all $m \geq k$. Now, it suffices to prove that for all $m \in \mathbb{N}$ $\beta(B^m) \leq N$, where $N \in \mathbb{N}^*$. Indeed, according to the hypothesis (B, A) is of finite type, we can write

$$N((B^k A^p)^*) = N(A^{*p} B^{*k}) \oplus F, \tag{5.7.2}$$

where F is a subspace of finite dimension. On the other hand, from Lemma 5.7.2 we have $\alpha(A^{*p} B^{*k}) < \infty$. Therefore, by Eq. (5.7.2) together with Theorem 2.1.1 we deduce that $\beta(B^k A^p) < \infty$. Then, by Eq. (5.7.1) $\beta(B^m) \leq \beta(B^m A^p) = \beta(B^k A^p) = N < \infty$ for all $m \in \mathbb{N}$. Conversely, let $A \in \Phi_-(X)$ such that $\text{desc}(A + K) < \infty$. So, $B \in \Phi_-(X)$ and $\text{desc}(B) < \infty$. Set $\tilde{K} = -K$, then clearly B and $B + \tilde{K}$ are mutually commuting operators, B^* commutes with itself, $\rho(B) \neq \emptyset$ and by Lemma 5.7.3 $(B + \tilde{K}, B) = (A, B)$ is of finite type. Hence, applying the reasoning above for the operators B and $B + \tilde{K}$, we get $\text{desc}(B + \tilde{K}) = \text{desc}(A) < \infty$. Q.E.D.

Corollary 5.7.1. *Let X be a reflexive Banach space, $A \in \mathcal{C}(X)$ and $B = A + K$ where $K \in \mathcal{K}(X)$ such that A and B are mutually commuting operators and A^* commutes with itself. Assume that $\rho(A) \neq \emptyset$, $\rho(A + K) \neq \emptyset$, and (B, A) is of finite type. Then $A \in \mathcal{B}_-(X)$ if, and only if, $A + K \in \mathcal{B}_-(X)$.* ◇

Proof. We first claim that if $A \in \mathcal{B}_-(X)$, then $A + K \in \mathcal{B}_-(X)$. Indeed, let $A \in \mathcal{B}_-(X)$ thus $A \in \Phi_-(X)$, and $\text{desc}(A) < \infty$. Using Theorem 2.2.47 together with Theorem 5.7.1, we obtain that $A+K \in \mathcal{B}_-(X)$. A similar reasoning gives if $A+K \in \mathcal{B}_-(X)$, then $A \in \mathcal{B}_-(X)$. Q.E.D.

Theorem 5.7.2. *Let X be a Banach space and $A \in \mathcal{C}(X)$ such that A commutes with itself and $\rho(A) \neq \emptyset$. Let $K \in \mathcal{K}(X)$ such that K commutes with A and $\rho(A + K) \neq \emptyset$. If $A \in \Phi_+(X)$, then $asc(A) < \infty$ if, and only if, $asc(A + K) < \infty$.* ◇

Proof. Let $A \in \Phi_+(X)$ such that $\text{asc}(A) < \infty$. According to the hypothesis A is a closed operator with non-empty resolvent set, we get that A^p is a closed operator. Therefore, $N(A^p)$ is a closed subspace. Thus, $N^\infty(A) = \overline{N^\infty(A)}$ since $\text{asc}(A) = p < \infty$. Using Lemma 2.2.20 we infer that $\overline{N^\infty(A)} \bigcap R^\infty(A) = N^\infty(A) \bigcap R^\infty(A) = \{0\}$. Now, set $A_\lambda = A + \lambda K$, where $\lambda \in [0, 1]$. Then, by Theorem 2.2.47 $A_\lambda \in \Phi_+(X)$ for each $\lambda \in [0, 1]$. From Theorem 2.2.20, there exists $\varepsilon = \varepsilon(\lambda)$ such that

$$\overline{N^\infty(A_\lambda)} \bigcap R^\infty(A_\lambda) = \overline{N^\infty(A_\mu)} \bigcap R^\infty(A_\mu)$$

for all μ in the open disc $S(\lambda)$ with center λ and radius ε. Therefore, $\overline{N^{\infty}(A_{\lambda})\bigcap R^{\infty}(A_{\lambda})}$ is a locally constant function of λ on the interval $[0, 1]$. Or every locally constant function on a connected set like $[0, 1]$ is constant, then we conclude that $\overline{N^{\infty}(A + K)\bigcap R^{\infty}(A + K)} = \{0\}$. Thus, $N^{\infty}(A + K)\bigcap R^{\infty}(A + K) = \{0\}$ and again by Lemma 2.2.20 it follows that $\text{asc}(A + K) < \infty$. Conversely, let $A \in \Phi_{+}(X)$ such that $\text{asc}(A + K) < \infty$. So, $A + K \in \Phi_{+}(X)$ and $\text{asc}(A + K) < \infty$. Set $\tilde{K} = -K$, then clearly $\tilde{K} = -K$ commutes with $A + K$ and $A + K$ commutes with itself. Thus, we can apply the reasoning above and deduce that $\text{asc}(A + K + \tilde{K}) = \text{asc}(A) < \infty$. This completes the proof. Q.E.D.

Corollary 5.7.2. *Let X be a Banach space and $A \in \mathcal{C}(X)$ such that A commutes with itself and $\rho(A) \neq \emptyset$. Let $K \in \mathcal{K}(X)$ such that K commutes with A and $\rho(A + K) \neq \emptyset$. Then, $A \in \mathcal{B}_{+}(X)$ if, and only if, $A + K \in \mathcal{B}_{+}(X)$.* ◇

Proof. Reasoning in the same way as the proof of Corollary 5.7.1. Q.E.D.

5.7.1 Convergence to Zero Compactly

In this section we present properties of the nullity, deficiency, and index of operators of the form $T + K_n$, where T is a semi-Fredholm operator and K_n converge to zero compactly. The results of this section may be found in [127].

Definition 5.7.1. A sequence $(K_n)_n$ of bounded linear operators mapping from X into Y is said converge to zero compactly if $K_n x \to 0$ for all $x \in X$, and $\{K_n x_n\}$ is relatively compact for every bounded sequence $(x_n)_n \subset X$. ◇

Remark 5.7.1.

(i) Clearly, $\|K_n\| \to 0$ implies K_n converge to zero compactly.
(ii) If $\bigcup_{n\geq 1}\{K_n x$ such that $\|x\| \leq 1\}$ is relatively compact, then $\{K_n x_n\}$ is relatively compact for every bounded sequence $(x_n)_n \subset X$. In this case $\{K_n\}$ is called collectively compact.
(iii) The assumption $(K_n)_n$ converge to zero compactly does not even imply that the sequence $(K_n^*)_n$ converges strongly. The following simple example taken from [30] confirms this. Take $X = Y = l_2$. Define $K_n x = x_n e_1$ where $x = \sum_{i=1}^{\infty} x_i e_i$, $(e_i)_i$ the usual set of unit vectors. Then $K_n^* e_1 = e_n$ does not converge in l_2. ◇

We recall the following results due to S. Gohberg [123].

Lemma 5.7.4. *If $(A_n)_n$, a sequence of bounded linear operators on X with range in Y, converges strongly to zero, then $(\|A_n\|)_n$ is bounded and $(A_n)_n$ converges to zero uniformly on totally bounded sets.* ◇

Lemma 5.7.5. *Let T be a closed linear operator with domain $\mathcal{D}(T) \subset X$ and range $R(T)$ a closed subspace of Y. If $(y_n)_n$ is a bounded sequence in $R(T)$, then there exists a bounded sequence $(x_n)_n$ such that $Tx_n = y_n$.* ◇

Proof. This follows readily from $y_n = Tv_n$ and $\|y_n\| = \|Tv_n\| \geq \tilde{\gamma}(T)\text{dist}(v_n, R(T))$. Q.E.D.

Lemma 5.7.6. *Let T be a closed linear operator with domain $\mathcal{D}(T) \subset X$ and range $R(T)$ a closed subspace of Y. If $\alpha(T) < \infty$ and $(x_n)_n$ is a bounded sequence such that $(Tx_n)_n$ converges, then $(x_n)_n$ has a convergent subsequence.* ◇

Proof. Since $R(T)$ is closed, $Tx_n \to Tx$ and therefore $x_n + R(T) = \tilde{T}^{-1}Tx_n \to \tilde{T}^{-1}Tx = x + R(T)$ in $X/R(T)$. Thus there exists $z_n \in R(T)$ such that $x_n + z_n \to x$. Since $(z_n)_n$ is bounded in finite dimensional space $R(T)$, it, and therefore $(x_n)_n$, has a convergent subsequence. Q.E.D.

Lemma 5.7.7. *Let T be a closed linear operator with domain $\mathcal{D}(T) \subset X$ and range $R(T)$ a closed subspace of Y. Let $(K_n)_n$ be a sequence of bounded linear operators such that K_n converge to zero compactly. If $\alpha(T) < \infty$ and $R(T)$ is complemented in X by a closed subspace M, then there exists a p and $c > 0$ such that for $n \geq p$, $T_M + K_n$ is one-to-one and $\tilde{\gamma}(T_M + K_n) \geq c$, where T_M is the restriction of T to $M \bigcap \mathcal{D}(T)$.* ◇

Proof. Suppose $\{\tilde{\gamma}(T_M + K_n)\}$ has a subsequence converging to zero. For simplicity, let $\tilde{\gamma}(T_M + K_n) \to 0$. There exists $\{m_n\} \subset M$ such that $\|m_n\| = 1$ and $(T + K_n)m_n \to 0$. Since K_n converge to zero compactly, $\{K_n m_n\}$, and therefore $\{Tm_n\}$, has a convergent subsequence. Thus by Lemma 5.7.6, $(m_n)_n$ has a convergent subsequence and by Lemma 5.7.4, $\{K_n m_n\}$, and therefore $(Tm_n)_n$ has a subsequence converging to zero. This is impossible since T_M has a bounded inverse. This argument also shows that $T_M + K_n$ is one-to-one for sufficiently large n; otherwise, a sequence $(m_n)_n$ with the above properties would obviously exist which leads to a contradiction. Q.E.D.

Lemma 5.7.8. *Let T be a closed linear operator with domain $\mathcal{D}(T) \subset X$ and range $R(T)$ a closed subspace of Y. Let $\mathcal{D}(T)$ be dense in X. Let $(K_n)_n$ be a sequence of bounded linear operators such that K_n converge to zero compactly. If $R(T)$ is complemented in Y by a closed subspace W, then there exists a p and $c > 0$ such that for $n \geq p$, $T^* + K_n^*$ is one-to-one on $W^\circ = \{y' \in Y^* \text{ such that } y'W = 0\}$ and $\tilde{\gamma}(T_0^* + K_n^*) \geq c$, where T_0^* is the restriction of T^* to $W^\circ \bigcap \mathcal{D}(T^*)$.* ◇

Proof. $Y = R(T) \oplus W$ and $Y' = R(T)^\circ \oplus W^\circ$. Suppose $\tilde{\gamma}(T_0^* + K_n^*)$ has a subsequence converging to zero. For simplicity, let $\tilde{\gamma}(T_0^* + K_n^*) \to 0$. There exists $(y'_n)_n \subset W^\circ$ such that $1 = \|y'_n\|$ and $(T^* + K_n^*)y'_n \to 0$. Choose y_n so that $1 = \|y_n\|$ and $y'_n y_n \geq \frac{1}{2}$. Now $y_n = Tv_n + w_n$, $w_n \in W$ and $(Tv_n)_n$ is bounded since $R(T)$ is closed and complemented by W. Hence by Lemma 5.7.5, there exists a bounded sequence $(x_n)_n$ such that $Tx_n = Tv_n$. Furthermore, $y'_n v \to 0$ for all $v \in Y$. To see this, $y'_n Tx = (T^* + K_n^*)y'_n(x) - y'_n K_n x \to 0$. Since y'_n is in W° and $R(T)$ is complemented by W, $y'_n v \to 0$ for all $v \in Y$. Now

$$\frac{1}{2} \leq y'_n y_n = (T^* + K_n^*)y'_n(x_n) - y'_n K_n x_n. \tag{5.7.3}$$

Since $(x_n)_n$ is bounded, $\{K_n x_n\}$ is totally bounded which, together with the observation that $y_n' v \to 0$ for all $v \in Y$, implies that $\{y_n' K_n x_n\}$ converges to zero. Therefore (5.7.3) cannot hold since $(T^* + K_n^*)y_n' \to 0$. The above argument also shows that $T^* + K_n^*$ is one-to-one on W° for all sufficiently large n; otherwise, a sequence $(y_n')_n$ with the above properties would obviously exist which leads to a contradiction. Q.E.D.

Theorem 5.7.3. *Let T be a closed linear operator with domain $\mathcal{D}(T) \subset X$ and range $R(T)$ a closed subspace of Y. Let $(K_n)_n$ be a sequence of bounded linear operators such that K_n converge to zero compactly. Suppose $\alpha(T) < \infty$. Then, there exists a p such that*

(i) $T + K_n$ has a closed range and $\alpha(T + K_n) \le \alpha(T)$, $n \ge p$.
(ii) $\alpha(T + K_n) = \alpha(T)$, $n \ge p$ if, and only if, $\inf_{n \ge p} \tilde{\gamma}(T + K_n) > 0$. In this case, $X = M \oplus R(T + K_n)$, $n \ge p$, where $R(T)$ is complemented by the closed subspace M. ◊

Proof.

(i) $X = M \oplus R(T)$ for some closed subspace M. Let p and $c > 0$ be as in Lemma 5.7.7 and $n \ge p$. Then $(T + K_n)M$ is closed by Lemma 2.2.1 and the finite dimensionality of $R(T)$ implies $R(T + K_n) = (T + K_n)M + K_n R(T)$ is closed. Moreover, by Lemma 5.7.7, $M \bigcap R(T + K_n) = \{0\}$. Hence $X = M \oplus R(T) \supset M \oplus R(T + K_n)$ which implies $\alpha(T + K_n) \le \alpha(T)$.
(ii) Suppose $\alpha(T + K_n) = \alpha(T)$, $n \ge p$. Then $X = M \oplus R(T + K_n)$ and for $x = m_n + z_n$, $m_n \in M$, $z_n \in R(T + K_n)$, we have by Lemma 5.7.7 that

$$\begin{aligned} \|(T + K_n)x\| &= \|(T + K_n)m_n\| \\ &\ge c\,\|m_n\| \\ &\ge c \operatorname{dist}(m_n, R(T + K_n)) \\ &= c \operatorname{dist}(x, R(T + K_n)). \end{aligned}$$

Thus $\tilde{\gamma}(T + K_n) \ge c > 0$, $n \ge p$. Conversely, suppose $\tilde{\gamma}(T + K_n) \ge c > 0$, $n \ge p$, but that $\alpha(T + K_n) \ne \alpha(T)$. Then, from (i), $\alpha(T + K_n) < \alpha(T)$. By Lemma 2.1.5, there exists $(z_n)_n \subset R(T)$ such that $1 = \|z_n\| = \operatorname{dist}(z_n, R(T + K_n))$. Hence for $n \ge p$,

$$0 < c = c \operatorname{dist}(z_n, R(T + K_n)) \le \|(T + K_n)z_n\| = \|K_n z_n\|. \tag{5.7.4}$$

Since $R(T)$ is finite dimensional, $(z_n)_n$ has a convergent subsequence and therefore by Lemma 5.7.4, $\{K_n z_n\}$ has a subsequence converging to zero, contradicting (5.7.4). Q.E.D.

Theorem 5.7.4. *Let T be a closed linear operator with domain $\mathcal{D}(T) \subset X$ and range $R(T)$ a closed subspace of Y. Let $(K_n)_n$ be a sequence of bounded linear operators such that K_n converge to zero compactly. If $R(T)$ is complemented in*

Y (by a closed subspace) and T is densely defined, there exists a p such that for $n \geq p$, $\alpha(T^* + K_n^*) \leq \alpha(T^*)$. If $\beta(T) < \infty$, then there exists a p such that

(i) $T + K_n$ has closed range with $\beta(T + K_n) \leq \beta(T)$, $n \geq p$.
(ii) $\beta(T + K_n) = \beta(T)$, $n \geq p$, implies $\inf_{n \geq p} \tilde{\gamma}(T + K_n) > 0$. ◇

Proof. Let W and p be chosen as in Lemma 5.7.8 with $n \geq p$. Then $Y^* = R(T)^\circ \oplus W^\circ = R(T^*) \oplus W^\circ$. Since $T^* + K_n^*$ is one-to-one on W°, $Y^* \supset R(T^* + K_n^*) \oplus W^\circ$. Thus $\alpha(T^* + K_n^*) \leq \alpha(T^*)$.

(i) By replacing X by $\overline{\mathcal{D}(T)}$, if necessary, we may assume T is densely defined. Since $\beta(T) < \infty$, there exists a p and W as in Lemma 5.7.8. For $n \geq p$, $(T^* + K_n^*)W^\circ$ is closed by preliminary Lemma 2.2.1. Since $\alpha(T^*) = \beta(T) < \infty$, $(T^* + K_n^*)Y^* = (T^* + K_n^*)W^\circ + K_n^* R(T^*)$ is closed; i.e., $T^* + K_n^*$ has a closed range and therefore $T + K_n$ has a closed range. Thus by what we have already shown, $\beta(T + K_n) = \alpha(T^* + K_n^*) \leq \alpha(T^*) = \beta(T)$, $n \geq p$.
(ii) Suppose $\beta(T^* + K_n^*) = \beta(T) < \infty$ or equivalently $\alpha(T^* + K_n^*) = \alpha(T^*)$, $n \geq p$, with p and c chosen as in Lemma 5.7.8. Then $Y^* = R(T^*) \oplus W^\circ = R(T^* + K_n^*) \oplus W^\circ$. Thus for $y' = z_n' + w_n'$, $z_n' \in R(T^* + K_n^*)$, $w_n' \in W^\circ$, we have $\|(T^* + K_n^*)y'\| = \|(T^* + K_n^*)w_n'\| \geq c\,\|w_n'\| \geq c \operatorname{dist}(y', R(T^* + K_n^*))$. Hence $\tilde{\gamma}(T + K_n) = \tilde{\gamma}(T^* + K_n^*) \geq c$, $n \geq p$. Q.E.D.

Theorem 5.7.5. *Let T be a closed linear operator with domain $\mathcal{D}(T) \subset X$ and range $R(T)$ a closed subspace of Y. Let T be a Fredholm operator. Let $(K_n)_n$ be a sequence of bounded linear operators such that K_n converge to zero compactly. Then $\tilde{\gamma}(T + K_n)$ is bounded away from zero for all sufficiently large n if, and only if, $\alpha(T + K_n) = \alpha(T)$ and $\beta(T + K_n) = \beta(T)$ for all sufficiently large n.* ◇

Proof. By replacing X by $\overline{\mathcal{D}(T)}$, if necessary, we may assume T is densely defined. $Y = R(T) \oplus W$, W finite dimensional. Suppose $\tilde{\gamma}(T + K_n) \geq c > 0$ for all but a finite number of n but that $\beta(T + K_n) \neq \beta(T)$ for infinitely many n. Then by Theorem 5.7.4 (i), $\beta(T + K_n) < \beta(T)$ for infinitely many n. For simplicity, suppose $\beta(T + K_n) < \beta(T)$ and $\tilde{\gamma}(T + K_n) \geq c$ for $n \geq p$, where p is chosen so that Lemma 5.7.8 holds. Thus there exists $y_n \in R(T + K_n) \bigcap W$, $\|y_n\| = 1$. Since $\|y_n\|$ is bounded and $\tilde{\gamma}(T + K_n) \geq c > 0$ it follows that there exists a bounded sequence $(x_n)_n$ such that $y_n = (T + K_n)x_n$. Now $(y_n)_n$ has a convergent subsequence since W is finite dimensional; say $y_{n'} \to y \in W$. Since $(K_{n'}x_{n'})_{n'}$ has a convergent subsequence, so does $(Tx_{n'})_{n'}$. Thus by Lemmas 5.7.6 and 5.7.4, $(x_{n'})_{n'}$ has a convergent subsequence and $(K_{n'}x_{n'})_{n'}$ has subsequence $(K_{n''}x_{n''})_{n''}$ converging to zero. Hence $y = \lim y_{n''} = \lim Tx_{n''} \in R(T)$, which shows that y is in $R(T) \bigcap W = \{0\}$. This is impossible since $\|y\| = 1$. The rest of the theorem follows from Theorem 5.7.3. Q.E.D.

Theorem 5.7.6. *Let T be a closed linear operator with domain $\mathcal{D}(T) \subset X$ and range $R(T)$ a closed subspace of Y. Let T be a semi-Fredholm operator. Let $(K_n)_n$ be a sequence of bounded linear operators such that K_n converge to zero compactly. There exists a p such that for $n \geq p$, $T + K_n$ is semi-Fredholm, $\alpha(T + K_n) \leq \alpha(T)$, $\beta(T + K_n) \leq \beta(T)$, and $i(T + K_n) = i(T)$.* ◇

Proof. The first three conclusions are contained in Theorems 5.7.3 and 5.7.4. There exists a p such that for all $\lambda \in [0, 1]$ and $n \geq p$, $T + \lambda K_n$ is semi-Fredholm. If this is not the case, there exists a subsequence $(K_{n'})_{n'}$ and a sequence $\lambda_n \in [0, 1]$ such that $T + \lambda_n K_{n'}$ is not semi-Fredholm. This is impossible by Theorems 5.7.3 and 5.7.4 since $\lambda_n K_{n'}$ converge to zero compactly. Given $n \geq p$, define Φ on $[0, 1]$ with values in the set of extended integers with the discrete topology by $\Phi(\lambda) = i(T + \lambda K_n)$. By Proposition 2.2.6, $\Phi(.)$ is continuous, and since $[0, 1]$ is connected, $\Phi(.)$ is constant. In particular $i(T) = \Phi(0) = \Phi(1) = i(T + K_n)$. Q.E.D.

Chapter 6
Perturbation Results

In this chapter, we present preliminary notions and results on which this book is based, namely, the known properties of Fredholm operators in a Banach space, and some results on semi-Fredholm perturbations, Fredholm inverse, and quasi-Fredholm operator on Banach spaces.

6.1 Definitions and Notations

Let J be an arbitrary A-bounded operator. Hence, we can regard A and J as operators from X_A into Y. They will be denoted by $\hat{A}$ and $\hat{J}$ respectively, and they belong to $\mathcal{L}(X_A, Y)$. Furthermore, we have $\alpha(\hat{A}) = \alpha(A)$, $\beta(\hat{A}) = \beta(A)$, $R(\hat{A}) = R(A)$, $\alpha(\hat{A}+\hat{J}) = \alpha(A+J)$, $\beta(\hat{A}+\hat{J}) = \beta(A+J)$, and $R(\hat{A}+\hat{J}) = R(A+J)$.

Definition 6.1.1. Let X and Y be two Banach spaces, $A \in \mathcal{C}(X, Y)$ and let J be an A-defined linear operator on X. We say that J is A-compact (resp. A-weakly compact, A-strictly singular, A-strictly cosingular) if $\hat{J} \in \mathcal{K}(X_A, Y)$ (resp. $\hat{J} \in \mathcal{W}(X_A, Y)$, $\hat{J} \in \mathcal{S}(X_A, Y)$, $\hat{J} \in \mathcal{CS}(X_A, Y)$). $\diamond$

Let $A\mathcal{K}(X,Y)$, $A\mathcal{W}(X,Y)$, $A\mathcal{S}(X,Y)$, and $A\mathcal{CS}(X,Y)$ denote, respectively, the sets of A-compact, A-weakly compact, A-strictly singular, and A-strictly cosingular on X. If $X = Y$ we write $A\mathcal{K}(X)$, $A\mathcal{W}(X)$, $A\mathcal{S}(X)$, and $A\mathcal{CS}(X)$ for $A\mathcal{K}(X, X)$, $A\mathcal{W}(X, X)$, $A\mathcal{S}(X, X)$, and $A\mathcal{CS}(X, X)$ respectively.

Remark 6.1.1. If J is A-defined and compact (resp. weakly compact, strictly singular, strictly cosingular), then J is A-compact (resp. A-weakly compact, A-strictly singular, A-strictly cosingular). $\diamond$

Definition 6.1.2. Let X and Y be two Banach spaces, $A \in \mathcal{C}(X, Y)$, and let F be an arbitrary A-defined linear operator on X. We say that F is an A-Fredholm

A. Jeribi, *Spectral Theory and Applications of Linear Operators and Block Operator Matrices*, DOI 10.1007/978-3-319-17566-9_6

perturbation if $\hat{F} \in \mathcal{F}^b(X_A, Y)$. F is called an upper (resp. lower) A-Fredholm perturbation if $\hat{F} \in \mathcal{F}_+^b(X_A, Y)$ (resp. $\hat{F} \in \mathcal{F}_-^b(X_A, Y)$). ◇

The sets of A-Fredholm, upper A-semi-Fredholm, and lower A-semi-Fredholm perturbations are denoted by $A\mathcal{F}(X, Y)$, $A\mathcal{F}_+(X, Y)$, and $A\mathcal{F}_-(X, Y)$ respectively. If $X = Y$ we write $A\mathcal{F}(X)$, $A\mathcal{F}_+(X)$, and $A\mathcal{F}_-(X)$ for $A\mathcal{F}(X, X)$, $A\mathcal{F}_+(X, X)$, and $A\mathcal{F}_-(X, X)$ respectively.

Definition 6.1.3. Let X and Y be two Banach spaces, $A \in \mathcal{C}(X, Y)$, and let $F : X \longrightarrow Y$ be an arbitrary A-defined linear operator. We say that F is an unbounded A-Fredholm perturbation if $\hat{F} \in \mathcal{F}(X_A, Y)$. F is called an unbounded upper (resp. unbounded lower) A-Fredholm perturbation if $\hat{F} \in \mathcal{F}_+(X_A, Y)$ (resp. $\hat{F} \in \mathcal{F}_-(X_A, Y)$). ◇

Let $UA\mathcal{F}(X, Y)$, $UA\mathcal{F}_+(X, Y)$, and $UA\mathcal{F}_-(X, Y)$ designate the sets of unbounded A-Fredholm, unbounded upper A-Fredholm, and unbounded lower A-Fredholm perturbations, respectively. If $X = Y$, we write $UA\mathcal{F}(X)$, $UA\mathcal{F}_+(X)$, and $UA\mathcal{F}_-(X)$ for $UA\mathcal{F}(X, X)$, $UA\mathcal{F}_+(X, X)$, and $UA\mathcal{F}_-(X, X)$ respectively.

Remark 6.1.2. As a consequence of Definition 6.1.3 and the inclusions (2.1.9) and (2.1.10), we have $A\mathcal{K}(X, Y) \subseteq UA\mathcal{F}_+(X, Y) \subseteq UA\mathcal{F}(X, Y)$, and $A\mathcal{K}(X, Y) \subseteq UA\mathcal{F}_-(X, Y) \subseteq UA\mathcal{F}(X, Y)$. ◇

Proposition 6.1.1. *Let X and Y be two Banach spaces, and $A \in \mathcal{C}(X, Y)$. The following statements are satisfied:*

(i) $\mathcal{F}(X, Y) \subset UA\mathcal{F}(X, Y)$,
(ii) $\mathcal{F}_+(X, Y) \subset UA\mathcal{F}_+(X, Y)$, *and*
(iii) $\mathcal{F}_-(X, Y) \subset UA\mathcal{F}_-(X, Y)$. ◇

Proof. (i) Let $F \in \mathcal{F}(X, Y)$ and $B \in \Phi(X_A, Y)$. Since X_A is continuously embedded in X and is dense in X, by using Theorem 2.2.39 we have $B \in \Phi(X, Y)$. So, $F + B \in \Phi(X, Y)$. By using Theorem 2.2.39, we get $\hat{F} + B \in \Phi(X_A, Y)$. Hence, $F \in UA\mathcal{F}(X_A, Y)$.

The proofs of (ii) and (iii) may be achieved in a similar way as (i). Q.E.D.

6.2 Fredholm and Semi-Fredholm Operators

Let T be a bounded linear operator from X into Y. The reduced minimum modulus of T will be characterized by the equation

$$\tilde{\gamma}(T) = \begin{cases} \inf\{\|Tx\| \text{ such that } \operatorname{dist}(x, N(T)) = 1\} & \text{if } T \neq 0, \\ 0 & \text{if } T = 0. \end{cases}$$

Theorem 6.2.1. *Let S, T, and K three operators such that $\mathcal{D}(T) \subset \mathcal{D}(S) \subset \mathcal{D}(K)$. If there exist a constant α_1 such that $\|S\varphi\| \leq \alpha_1\big(\|\varphi\| + \|T\varphi\|\big)$, $\varphi \in \mathcal{D}(T)$, if there exist a constant β_1 such that $\alpha_1(1+\beta_1) < 1$, $(1+\beta_1) < \tilde{\gamma}(\tilde{T})$ and $\|K\varphi\| \leq \beta_1\big(\|\varphi\| + \|S\varphi\|\big)$, $\varphi \in \mathcal{D}(T)$, and, if $T \in \Phi(X,Y)$, then the sum $T+S+K \in \Phi(X,Y)$ which satisfies the following properties $\alpha(T+S+K) \leq \alpha(T)$, $\beta(T+S+K) \leq \beta(T)$, and $i(T+S+K) = i(T)$.* ◇

Proof. First using Theorem 2.1.5 it follows that $T+S+K$ is closed operator. Let T_1, S_1, and K_1 be the restrictions of the operators T, S, and K to X_T. Obviously, T_1 is Fredholm linear operator and $S_1 + K_1 : X_T \longrightarrow Y$ is a bounded linear operator. Consequently, $T_1 + S_1 + K_1$ is a Fredholm operator as $\|S_1 + K_1\|_T \leq (\beta_1 + \alpha_1(1+\beta_1)) \leq \tilde{\gamma}(\tilde{T}) = \tilde{\gamma}(\tilde{T}^*)$ (see Theorem 2.2.29). The rest of the proof is a consequence of Theorem 2.2.18. Q.E.D.

6.3 Semi-Fredholm Perturbations

In the beginning of this section, let us prove some results for semi-Fredholm perturbations.

Proposition 6.3.1. *Let X, Y, and Z be three Banach spaces.*

(i) If the set $\Phi^b(Y,Z)$ is not empty, then

$$E_1 \in \mathcal{F}_+^b(X,Y) \text{ and } A \in \Phi^b(Y,Z) \text{ imply } AE_1 \in \mathcal{F}_+^b(X,Z)$$

$$E_1 \in \mathcal{F}_-^b(X,Y) \text{ and } A \in \Phi^b(Y,Z) \text{ imply } AE_1 \in \mathcal{F}_-^b(X,Z).$$

(ii) If the set $\Phi^b(X,Y)$ is not empty, then

$$E_2 \in \mathcal{F}_+^b(Y,Z) \text{ and } B \in \Phi^b(X,Y) \text{ imply } E_2B \in \mathcal{F}_+^b(X,Z)$$

$$E_2 \in \mathcal{F}_-^b(Y,Z) \text{ and } B \in \Phi^b(X,Y) \text{ imply } E_2B \in \mathcal{F}_-^b(X,Z).$$

◇

Proof.

(i) Since $A \in \Phi^b(Y,Z)$, and using Theorem 2.2.6, it follows that there exist $A_0 \in \mathcal{L}(Z,Y)$ and $K \in \mathcal{K}(Z)$ such that $AA_0 = I - K$. From Lemma 3.1.2, we get $AA_0 \in \Phi^b(Z)$. Using Theorem 2.2.10, we have $A_0 \in \Phi^b(Z,Y)$, and so $A_0 \in \Phi_+^b(Z,Y)$ and $A_0 \in \Phi_-^b(Z,Y)$. Let $J \in \Phi_+^b(X,Z)$ (resp. $\Phi_-^b(X,Z)$), using Theorem 2.2.13, we deduce that $A_0J \in \Phi_+^b(X,Y)$ (resp. $\Phi_-^b(X,Y)$). This implies that $(E_1 + A_0J) \in \Phi_+^b(X,Y)$ (resp. $\Phi_-^b(X,Y)$). So, $A(E_1 + A_0J) \in \Phi_+^b(X,Z)$ (resp. $\Phi_-^b(X,Z)$). Now, using the relation

$AE_1 + J - KJ = A(E_1 + A_0 J)$ together with the compactness of the operator KJ, we get $(AE_1 + J) \in \Phi_+^b(X, Z)$ (resp. $\Phi_-^b(X, Z)$). This implies that $AE_1 \in \mathcal{F}_+^b(X, Z)$ (resp. $\mathcal{F}_-^b(X, Z)$).

(ii) The proof of (ii) is obtained in a similar way as the proof of (i). Q.E.D.

Theorem 6.3.1. *Let X, Y, and Z be three Banach spaces.*

(i) If the set $\Phi^b(Y, Z)$ is not empty, then

$$E_1 \in \mathcal{F}_+^b(X, Y) \text{ and } A \in \mathcal{L}(Y, Z) \text{ imply } AE_1 \in \mathcal{F}_+^b(X, Z)$$

$$E_1 \in \mathcal{F}_-^b(X, Y) \text{ and } A \in \mathcal{L}(Y, Z) \text{ imply } AE_1 \in \mathcal{F}_-^b(X, Z).$$

(ii) If the set $\Phi^b(X, Y)$ is not empty, then

$$E_2 \in \mathcal{F}_+^b(Y, Z) \text{ and } B \in \mathcal{L}(X, Y) \text{ imply } E_2 B \in \mathcal{F}_+^b(X, Z)$$

$$E_2 \in \mathcal{F}_-^b(Y, Z) \text{ and } B \in \mathcal{L}(X, Y) \text{ imply } E_2 B \in \mathcal{F}_-^b(X, Z).$$

◇

Proof.

(i) Let $C \in \Phi^b(Y, Z)$ and $\lambda \in \mathbb{C}$. Let $A_1 = A - \lambda C$ and $A_2 = \lambda C$. For a sufficiently large λ, and using Lemma 2.2.3 or Theorem 2.2.15, we have $A_1 \in \Phi^b(Y, Z)$. From Proposition 6.3.1 (i), it follows that $A_1 E_1 \in \mathcal{F}_+^b(X, Z)$ (resp. $\mathcal{F}_-^b(X, Z)$) and $A_2 E_1 \in \mathcal{F}_+^b(X, Z)$ (resp. $\mathcal{F}_-^b(X, Z)$). This implies that $A_1 E_1 + A_2 E_1 = AE_1$ is an element of $\mathcal{F}_+^b(X, Z)$ (resp. $\mathcal{F}_-^b(X, Z)$).

(ii) The proof may be achieved in a similar way as (i). It is sufficient to replace Proposition 6.3.1 (i) by Proposition 6.3.1 (ii). Q.E.D.

Corollary 6.3.1. *Let X be a Banach space. Then, $\mathcal{F}_-^b(X)$ is a closed two-sided ideal of $\mathcal{L}(X)$.* ◇

Theorem 6.3.2. *Let X and Y be two Banach spaces. Then, $\mathcal{F}^b(X, Y) = \mathcal{F}(X, Y)$.* ◇

Proof. Clearly, $\mathcal{F}(X, Y) \subset \mathcal{F}^b(X, Y)$ (because $\Phi^b(X, Y) \subseteq \Phi(X, Y)$). In order to prove the opposite inclusion, let $F \in \mathcal{F}^b(X, Y)$. If $A \in \Phi(X, Y)$, then by Theorem 2.2.38, there exist $A_0 \in \mathcal{L}(Y, X)$ and $K \in \mathcal{L}(Y)$ of finite rank, such that

$$AA_0 = I - K \text{ on } Y. \tag{6.3.1}$$

Hence,

$$(A + F)A_0 = I - K + FA_0 = I + E. \tag{6.3.2}$$

According to Theorem 2.2.39, the fact that $A \in \Phi(X,Y)$ implies that $\hat{A} \in \Phi^b(X_A,Y)$. Also, (6.3.1) implies that AA_0 is a Fredholm operator. Then, by applying Theorem 2.2.41, we get $A_0 \in \Phi^b(Y,X_A)$. Similarly, since $A_0 \in \mathcal{L}(Y,X)$, and using Theorem 6.3.1 (ii), we conclude that $E \in \mathcal{F}^b(Y)$. This, together with (6.3.2), implies that $(A+F)A_0 \in \Phi^b(Y)$. Since $A_0 \in \Phi^b(Y,X_A)$ it follows, from Theorem 2.2.41, that $\hat{A}+\hat{F} \in \Phi^b(X_A,Y)$. Now, by using Theorem 2.2.39, we notice that $A+F \in \Phi(X,Y)$. This shows that $F \in \mathcal{F}(X,Y)$, which ends the proof. Q.E.D.

Open question: In contrast to the result of Theorem 6.3.2, whether or not $\mathcal{F}_+(X,Y)$ (resp. $\mathcal{F}_-(X,Y)$) is equal to $\mathcal{F}_+^b(X,Y)$ (resp. $\mathcal{F}_-^b(X,Y)$) seems to be unknown? □

Corollary 6.3.2. *Let X be a Banach space and $A \in \mathcal{C}(X)$. Then, $UA\mathcal{F}(X) = A\mathcal{F}(X)$.* ◇

Proposition 6.3.2. *Let X, Y, and Z be three Banach spaces. If $\Phi^b(Y,Z)$ is not empty, then $E_1 \in \mathcal{F}_+(X,Y)$ and $A \in \Phi^b(Y,Z)$ imply $AE_1 \in \mathcal{F}_+(X,Z)$.* ◇

Proof. Since $A \in \Phi^b(Y,Z)$, it follows, by using Theorem 2.2.6, that there exist $A_0 \in \mathcal{L}(Z,Y)$ and a finite rank operator K on Z, such that $AA_0 = I - K$. Using Lemma 3.1.2, we have $AA_0 \in \Phi^b(Z)$. By using Theorem 2.2.10, we get $A_0 \in \Phi^b(Z,Y)$ and so $A_0 \in \Phi_+^b(Z,Y)$. Let $J \in \Phi_+(X,Z)$. Since $\mathcal{D}(A_0J) = \mathcal{D}(J)$ and is dense in X, then by using Theorem 2.2.43, we have $A_0J \in \Phi_+(X,Y)$. This implies that $(E_1 + A_0J) \in \Phi_+(X,Y)$. So, $A(E_1 + A_0J) \in \Phi_+(X,Z)$. We claim that KJ is $(AE_1 + J)$-compact. Indeed, let $x \in \mathcal{D}(J)$. We have

$$\begin{aligned}\|Jx\| &= \|(AE_1+J)x - AE_1x\| \\ &\le \|(AE_1+J)x\| + \|AE_1x\| \\ &\le \|(AE_1+J)x\| + \|A\|\,\|E_1\|\,\|x\| \\ &\le \max(1,\|A\|\,\|E_1\|)\,(\|(AE_1+J)x\| + \|x\|)\,.\end{aligned}$$

Hence, using the last inequality, we have

$$\begin{aligned}\|KJx\| \le&\ \|K\|\,\|Jx\| \\ \le&\ \max(1,\|A\|\,\|E_1\|)\|K\|\,(\|(AE_1+J)x\| + \|x\|)\,.\end{aligned}$$

So, KJ is (AE_1+J)-compact, which proves the claim. Now, using the relation $AE_1 + J - KJ = A(E_1 + A_0J)$, KJ being (AE_1+J)-compact and using Theorems 2.2.47 and 2.2.2, one sees that $(AE_1+J) \in \Phi_+(X,Z)$. This implies that $AE_1 \in \mathcal{F}_+(X,Z)$ and completes the proof. Q.E.D.

Proposition 6.3.3. *Let X, Y, and Z be three Banach spaces, $T \in \mathcal{L}(X,Y)$ and $S \in \mathcal{C}(Z,X)$. If $T \in \Phi^b(X,Y)$ and $S \in \Phi_-(Z,X)$, then $TS \in \Phi_-(Z,Y)$.* ◇

Proof. According to Proposition 2.2.1, the operator TS is closed. Let $N_1 = R(S) \bigcap N(T)$. Since $N(T)$ is finite dimensional, then we have $N(T) = N_1 \oplus N_2$, for some finite dimensional subspace N_2. Obviously, $R(S) \bigcap N_2 = \{0\}$. Furthermore, $R(S) \oplus N_2$ is closed, because $R(S)$ is closed and $\dim N_2 < \infty$ (see Lemma 2.1.2). Next, we prove that there exists a finite dimensional subspace N_3, such that

$$(R(S) \oplus N_2) \oplus N_3 = X. \tag{6.3.3}$$

Put $X_0 = R(S) \oplus N_2$ and let $k = \dim X/X_0$. Note that $k \leq \operatorname{codim} R(S) < \infty$. If $k = 0$, then we take $N_3 = \{0\}$ in (6.3.3). Assume $k > 0$. Since X_0 is closed, X is not entirely contained in X_0. So, there exists a vector $x_1 \in X$ such that $x_1 \notin X_0$. Put $X_1 = X_0 \oplus \text{span}\{x_1\}$. Then, X_1 is closed and $\dim X/X_1 = k-1$. Thus, we can repeat the above reasoning for X_1 in place of X_0. Proceeding in this way, we find k steps vectors $x_1, \ldots, x_k$ in $\mathcal{D}(T)$ such that $X = X_0 \oplus \text{span}\{x_1, \ldots, x_k\}$. Put $N_3 = \text{span}\{x_1, \ldots, x_k\}$ and (6.3.3) is fulfilled. The space N_3 is isomorphic to the quotient space $R(T)/R(TS)$ under the map $u \longrightarrow [Tu]$, by virtue of (6.3.3). Indeed, if $x \in X$, then (6.3.3) implies that $x = Sz + v + u$, where $z \in \mathcal{D}(S)$, $v \in N_2 \subset N(T)$ and $u \in N_3$. It follows that $Sz = x - v - u \in \mathcal{D}(T)$ and $T(Sz) = Tx - Tu$, which shows that $[Tx] = [Tu]$. Furthermore, if $[Tu] = [0]$ for $u \in N_3$, then $u \in R(S) + N(T) = R(S) \oplus N_2$, and hence, $u = 0$. So, $u \longrightarrow [Tu]$ has the desired properties, and then

$$\beta(TS) = \beta(T) + \dim N_3. \tag{6.3.4}$$

Now, (6.3.4) shows that $R(TS)$ is closed (see Lemma 2.1.9). Q.E.D.

Proposition 6.3.4. *Let X, Y, and Z be three Banach spaces. If $\Phi^b(Y,Z)$ is not empty, $E_1 \in \mathcal{F}_-(X,Y)$ and $A \in \Phi^b(Y,Z)$. Then, $AE_1 \in \mathcal{F}_-(X,Z)$.* ◇

Proof. Since $A \in \Phi^b(Y,Z)$, and using Theorem 2.2.6, it follows that there exist $A_0 \in \mathcal{L}(Z,Y)$ and a finite rank operator K on Z such that $AA_0 = I - K$. Using Lemma 3.1.2, we have $AA_0 \in \Phi^b(Z)$. According to Theorem 2.2.10, we get $A_0 \in \Phi^b(Z,Y)$ and so, $A_0 \in \Phi^b_+(Z,Y)$. Let $J \in \Phi_-(X,Z)$. By Proposition 6.3.3, we have $A_0J \in \Phi_-(X,Y)$. This implies that $(E_1 + A_0J) \in \Phi_-(X,Y)$. So, $A(E_1 + A_0J) \in \Phi_-(X,Z)$. Now, arguing as in the proof of Proposition 6.3.2, we prove that KJ is $(AE_1 + J)$-compact. Next, using the relation $AE_1 + J - KJ = A(E_1 + A_0J)$, KJ being $(AE_1 + J)$-compact and according to Theorems 2.2.47 and 2.2.2, one sees that $(AE_1 + J) \in \Phi_-(X,Z)$. This implies that $AE_1 \in \mathcal{F}_-(X,Z)$. Q.E.D.

Theorem 6.3.3. *Let X, Y, and Z be three Banach spaces. If $\Phi^b(Y,Z)$ is not empty, then*

$$A \in \mathcal{L}(Y,Z) \text{ and } E_1 \in \mathcal{F}_+(X,Y) \text{ imply } AE_1 \in \mathcal{F}_+(X,Z)$$
$$A \in \mathcal{L}(Y,Z) \text{ and } E_1 \in \mathcal{F}_-(X,Y) \text{ imply } AE_1 \in \mathcal{F}_-(X,Z).$$

◇

Proof. Let $C \in \Phi^b(Y,Z)$ and $\lambda \in \mathbb{C}$. Setting $A_1 = A - \lambda C$ and $A_2 = \lambda C$. For a sufficiently large λ, and using Theorem 2.2.15, we have $A_1 \in \Phi^b(Y,Z)$. From Proposition 6.3.2 (resp. Proposition 6.3.4), it follows that $A_1E_1 \in \mathcal{F}_+(X,Z)$ (resp. $\mathcal{F}_-(X,Z)$) and $A_2E_1 \in \mathcal{F}_+(X,Z)$ (resp. $\mathcal{F}_-(X,Z)$). This implies that $A_1E_1 + A_2E_1 = AE_1$ is an element of $\mathcal{F}_+(X,Z)$ (resp. $\mathcal{F}_-(X,Z)$). Q.E.D.

Proposition 6.3.5. *Let X, Y, and Z be three Banach spaces. If $\Phi^b(Y,Z)$ is not empty, $E_1 \in \mathcal{F}(X,Y)$ and $A \in \Phi^b(Y,Z)$. Then, $AE_1 \in \mathcal{F}(X,Z)$.* ◇

Proof. Since $A \in \Phi^b(Y,Z)$, and referring to Theorem 2.2.6, it follows that there exist $A_0 \in \mathcal{L}(Z,Y)$ and a finite rank operator K on Z, such that $AA_0 = I - K$. Using Lemma 3.1.2, we have $AA_0 \in \Phi^b(Z)$. By using Theorem 2.2.10, we get $A_0 \in \Phi^b(Z,Y)$. Let $J \in \Phi(X,Z)$. According to Theorem 2.2.40, we have $A_0J \in \Phi(X,Y)$. This implies that $(E_1 + A_0J) \in \Phi(X,Y)$. So, $A(E_1 + A_0J) \in \Phi(X,Z)$. Now, arguing as in the proof of Proposition 6.3.2, we prove that KJ is $(AE_1 + J)$-compact. Then, using the relation $AE_1 + J - KJ = A(E_1 + A_0J)$, KJ being $(AE_1 + J)$-compact and referring to Theorems 2.2.47 and 2.2.2, we notice that $(AE_1 + J) \in \Phi(X,Z)$. This implies that $AE_1 \in \mathcal{F}(X,Z)$. Q.E.D.

Theorem 6.3.4. *Let X, Y, and Z be three Banach spaces. If $\Phi^b(Y,Z)$ is not empty, then*

$$A \in \mathcal{L}(Y,Z) \text{ and } E_1 \in \mathcal{F}(X,Y) \text{ imply } AE_1 \in \mathcal{F}(X,Z).$$

◇

Proof. Let $C \in \Phi^b(Y,Z)$ and $\lambda \in \mathbb{C}$. Setting $A_1 = A - \lambda C$ and $A_2 = \lambda C$. For a sufficiently large λ, and using Theorem 2.2.15, we have $A_1 \in \Phi^b(Y,Z)$. It follows, from Proposition 6.3.5, that $A_1E_1 \in \mathcal{F}(X,Z)$ and $A_2E_1 \in \mathcal{F}(X,Z)$. This implies that $A_1E_1 + A_2E_1 = AE_1$ is an element of $\mathcal{F}(X,Z)$. Q.E.D.

Theorem 6.3.5. *Let X be a Banach space and let $\mathcal{I}(X)$ denote an arbitrary nonzero two-sided ideal of $\mathcal{L}(X)$ contained in $\mathcal{F}(X)$. Let $A \in \mathcal{C}(X)$ be such that $\rho(A) \neq \emptyset$. Then $(\lambda - A)^{-1} \in \mathcal{I}(X)$ for some $\lambda \in \rho(A)$ if, and only if, the embedding of $\mathcal{D}(A)$ into X is in $\mathcal{I}(X)$.* ◇

Proof. Let $\lambda \in \rho(A)$ such that $(\lambda - A)^{-1} \in \mathcal{I}(X)$. The operator $\lambda - A : \mathcal{D}(A) \longrightarrow X$ is an isomorphism, when the domain $\mathcal{D}(A)$ of the operator A is equipped with the graph norm. By using the fact that $(\lambda - A)^{-1} \in \mathcal{I}(X)$ and writing the embedding j of $\mathcal{D}(A)$ into X as $j := (\lambda - A)^{-1}(\lambda - A)$ with $\mathcal{D}(j) := \mathcal{D}(A)$, we deduce that $j \in \mathcal{I}(X)$. Inversely, let $\lambda \in \rho(A)$. We can write $(\lambda - A)^{-1} = j \circ (\lambda - A)^{-1}$, where $j : \mathcal{D}(A) \longrightarrow X$ is in $\mathcal{I}(X)$. Then, $(\lambda - A)^{-1}$ is in $\mathcal{I}(X)$ as the compose of a continuous map $(\lambda - A)^{-1}$ and a map j in $\mathcal{I}(X)$. Q.E.D.

Lemma 6.3.1. *Let $A \in \mathcal{C}(X,Y)$ and let $J : X \longrightarrow Y$ be a linear operator. Assume that $J \in \mathcal{UAF}(X,Y)$. Then,*

(i) if $A \in \Phi(X,Y)$, then $A + J \in \Phi(X,Y)$ and $i(A + J) = i(A)$.
Moreover,

(ii) if $A \in \Phi_+(X,Y)$ and $J \in UA\mathcal{F}_+(X,Y)$, then $A+J \in \Phi_+(X,Y)$ and $i(A+J)=i(A)$,

(iii) if $A \in \Phi_-(X,Y)$ and $J \in UA\mathcal{F}_-(X,Y)$, then $A+J \in \Phi_-(X,Y)$ and $i(A+J)=i(A)$,

(iv) if $A \in \Phi_\pm(X,Y)$ and $J \in UA\mathcal{F}_+(X,Y)\bigcap UA\mathcal{F}_-(X,Y)$, then $A+J \in \Phi_\pm(X,Y)$ and $i(A+J)=i(A)$. ◇

Proof. Since $A \in \mathcal{C}(X,Y)$ and $J \in UA\mathcal{F}(X,Y)$ hence, as mentioned above, we can regard A and J as operators from X_A into Y. They will be denoted by $\hat{A}$ and $\hat{J}$ respectively. These operators belong to $\mathcal{L}(X_A,Y)$, and we have

$$\begin{cases} \alpha(\hat{A})=\alpha(A),\ \beta(\hat{A})=\beta(A),\ R(\hat{A})=R(A),\ \alpha(\hat{A}+\hat{J})=\alpha(A+J),\\ \beta(\hat{A}+\hat{J})=\beta(A+J) \text{ and } R(\hat{A}+\hat{J})=R(A+J). \end{cases} \tag{6.3.5}$$

(i) Assume that $A \in \Phi(X,Y)$. Then, using (6.3.5), we infer that $\hat{A} \in \Phi^b(X_A,Y)$. Hence it follows, from Theorem 2.2.6, that there $A_0 \in \mathcal{L}(Y,X_A)$ and $K \in \mathcal{K}(X_A)$ such that:

$$A_0\hat{A} = I - K. \tag{6.3.6}$$

This leads to

$$\begin{aligned} A_0(\hat{A}+\hat{J}) &= I - K + A_0\hat{J} \\ &= I - Q. \end{aligned} \tag{6.3.7}$$

Now, from (6.3.6), it follows that $A_0\hat{A} \in \Phi^b(X_A)$ and $i(A_0\hat{A})=0$. Hence, the use of Theorem 2.2.10, together with Atkinson's theorem (Theorem 2.2.40), implies that $A_0 \in \Phi^b(Y,X_A)$, and

$$i(\hat{A}) = -i(A_0). \tag{6.3.8}$$

Moreover, since $\hat{J} \in UA\mathcal{F}(X,Y)$ and $A_0 \in \mathcal{L}(Y,X_A)$, then applying Theorem 6.3.4, we get $A_0\hat{J} \in \mathcal{F}^b(X_A)$. Using the fact that $\mathcal{K}(X_A) \subset \mathcal{F}^b(X_A)$, we infer that $Q \in \mathcal{F}^b(X_A)$. Therefore, using (2.1.11) and (6.3.7), we get $A_0(\hat{A}+\hat{J}) \in \Phi^b(X_A)$ and $i[A_0(\hat{A}+\hat{J})]=0$. Since $A_0 \in \Phi^b(Y,X_A)$, and using Theorem 2.2.10 together with Atkinson's theorem (Theorem 2.2.40), it follows that $(\hat{A}+\hat{J}) \in \Phi^b(X_A,Y)$ and

$$i(\hat{A}+\hat{J}) = -i(A_0). \tag{6.3.9}$$

Now, using Eqs. (6.3.5), (6.3.8) and (6.3.9), we have $i(A+J) = i(A)$ which completes the proof of (i). Notice that the first part of the assertion (ii), (iii), and

(iv) is immediate. Let $A \in \Phi_+(X, Y)$ and $J \in UA\mathcal{F}_+(X, Y)$. In order to prove that $i(A + J) = i(A)$, we discuss two cases.

1st *case*. If $A \in \Phi_+(X, Y) \backslash \Phi(X, Y)$, then $i(A) = -\infty$. Hence, $i(A + J) = -\infty$. Otherwise, $A + J \in \Phi(X, Y)$ and therefore, $A \in \Phi_+(X, Y)$ since $J \in UA\mathcal{F}_+(X, Y) \subset \mathcal{F}(X, Y)$, which is contradictory.

2nd *case*. If $A \in \Phi(X, Y)$, then the result of (ii) follows from the assertion (i).

Statements (iii) and (iv) can be checked in the same way as (ii). Q.E.D.

Remark 6.3.1. Note that the result of Lemma 6.3.1 (iii) remains true if we suppose that Y is a reflexive space (because if Y is not a reflexive space, we don't guarantee that A^* is densely defined), and if we consider $J^* \in UA\mathcal{F}_+(Y^*, X^*)$. In fact, let $A \in \Phi_-(X, Y)$. Applying Theorem 2.2.46 and Theorem 2.1.2, we infer that $A^* \in \Phi_+(Y^*, X^*)$. Moreover, $J^* \in UA\mathcal{F}_+(Y^*, X^*)$ implies that $A^* + J^* \in \Phi_+(Y^*, X^*)$. This, together with the fact that $\alpha(A^* + J^*) = \beta(A + J)$ (use again Theorem 2.1.2) leads to the result. ◇

6.4 Fredholm Inverse Operator

Lemma 6.4.1. *Let $A \in \Phi(X), B \in \mathcal{L}(X)$, and $F \in \mathcal{F}(X)$. Suppose that $AB_{|V} = F_{|V}$ where V is a dense subspace of X. Then, $B \in \mathcal{F}^b(X)$.* ◇

Proof. Since $A \in \Phi(X)$, then by using Theorem 2.2.38, there exists $A_0 \in \mathcal{L}(X)$ such that $A_0 A = I - K_1$, where $K_1 \in \mathcal{K}(X)$. Hence, $A_0 AB_{|V} = A_0 F_{|V}$, $(I - K_1)B_{|V} = F_{1|V}$, where $F_1 \in \mathcal{F}(X)$. So, we get $B_{|V} = (K_1 B + F_1)_{|V}$. Now, using the fact that the operators B and $K_1 B + F_1$ are bounded and that the subspace V is dense, we deduce by continuity that $B = K_1 B + F_1$. Hence, it is clear that B is a Fredholm perturbation. Q.E.D.

Lemma 6.4.2. *Let $A \in \mathcal{C}(X), B \in \mathcal{L}(X)$, $\lambda \in \Phi_A \backslash \Phi^0(A)$ and $\mu \in \Phi_B \backslash \Phi^0(B)$. If there exist a positive integer n and a Fredholm perturbation F_1, such that $B : \mathcal{D}(A^n) \longrightarrow \mathcal{D}(A)$ and $ABx = BAx + F_1 x$, for all $x \in \mathcal{D}(A^n)$, then there exists a Fredholm perturbation F depending analytically on λ and μ, such that $R'_\lambda(A) R'_\mu(B) = R'_\mu(B) R'_\lambda(A) + F$.* ◇

Proof. Using Lemma 2.2.19, we infer that there exists a subspace V_λ dense in X such that for all $x \in V_\lambda$, we have $R'_\lambda(A)x \in \mathcal{D}(A^n)$. Now, let $x \in V_\lambda$. Then, we have $(\lambda - A)BR'_\lambda(A)x = [B(\lambda - A) - F_1]R'_\lambda(A)x = [B(I - K_1) - F_1 R'_\lambda(A)]x$, where $K_1 \in \mathcal{K}(X)$. Set $F_2 = -F_1 R'_\lambda(A) \in \mathcal{F}(X)$ and $F_3 = -BK_1 + F_2 \in \mathcal{F}(X)$. Hence, we get $(\lambda - A)BR'_\lambda(A)x = Bx + F_3 x$. Moreover, $(\lambda - A)R'_\lambda(A)Bx = (I - K_1)Bx = Bx - K_2 x$, where $K_2 = K_1 B \in \mathcal{K}(X)$. This enables us to conclude that

$$
\begin{aligned}
(\lambda - A)[BR'_\lambda(A) - R'_\lambda(A)B]x &= (F_3 + K_2)x \\
&= F_4 x, \qquad (6.4.1)
\end{aligned}
$$

where $F_4 \in \mathcal{F}(X)$. In fact, Eq. (6.4.1) holds for all $x \in V_\lambda$. Then, the use of Lemma 6.4.1 allows us to conclude that $BR'_\lambda(A) - R'_\lambda(A)B = F_5$, where $F_5 \in \mathcal{F}(X)$. Moreover,

$$\begin{aligned}(\mu - B)[R'_\mu(B)R'_\lambda(A) - R'_\lambda(A)R'_\mu(B)] &\\ = (I - K_3)R'_\lambda(A) - (\mu - B)R'_\lambda(A)R'_\mu(B) &\\ = R'_\lambda(A) - K_4 - \left[R'_\lambda(A)(\mu - B) + F_5\right] R'_\mu(B) &\\ = R'_\lambda(A) - K_4 - R'_\lambda(A)(I - K_5) + F_6 &\\ = -K_4 + K_6 + F_6 &\\ = F_7, &\end{aligned}$$

where $K_i \in \mathcal{K}(X)$ for $i = 3, 4, 5, 6$ and $F_i \in \mathcal{F}(X)$ for $i = 6, 7$. Hence, $R'_\mu(B)R'_\lambda(A) - R'_\lambda(A)R'_\mu(B) = F$, where $F \in \mathcal{F}(X)$. Therefore, $R'_\mu(B)R'_\lambda(A) = R'_\lambda(A)R'_\mu(B) + F$. Furthermore, the analyticity of F in λ and μ is deduced from the analyticity of $R'_\mu(B)$ and $R'_\lambda(A)$. Q.E.D.

Definition 6.4.1. Let X and Y be two Banach spaces.

1. Let $T \in \mathcal{L}(X, Y)$.
   (i) T is said to have a left Fredholm inverse, if there exist $T_l \in \mathcal{L}(Y, X)$ and $K \in \mathcal{K}(X)$ such that $T_l T = I_X - K$. T_l is called a left Fredholm inverse of T.
   (ii) T is said to have a right Fredholm inverse, if there exists $T_r \in \mathcal{L}(Y, X)$ such that $I_Y - TT_r \in \mathcal{K}(Y)$. T_r is called a right Fredholm inverse of T.
   (iii) T is said to have a Fredholm inverse, if there exists a map which is both a left and a right Fredholm inverse of T.
2. Let $T \in \mathcal{C}(X, Y)$. T is said to have a left Fredholm inverse (resp. right Fredholm inverse, Fredholm inverse), if $\hat{T}$ has a left Fredholm inverse (resp. right Fredholm inverse, Fredholm inverse). ◇

Remark 6.4.1. Let X and Y be two Banach spaces.

(i) If $A \in \mathcal{C}(X, Y)$ has a left Fredholm inverse, then there are maps $R_l \in \mathcal{L}(Y, X)$ and $K \in \mathcal{K}(X)$, such that $I_X + K$ extends $R_l A$.
(ii) If $A \in \mathcal{C}(X, Y)$ has a right Fredholm inverse, then there is a map $R_r \in \mathcal{L}(Y, X)$, such that $R_r(Y) \subset \mathcal{D}(A)$ and $AR_r - I_Y \in \mathcal{K}(Y)$. ◇

We will denote $\Phi_l(X, Y)$ and $\Phi_r(X, Y)$ by:

$$\Phi_l(X, Y) := \{T \in \mathcal{C}(X, Y),\ \text{such that } T \text{ has a left Fredholm inverse}\}$$
$$\Phi_r(X, Y) := \{T \in \mathcal{C}(X, Y),\ \text{such that } T \text{ has a right Fredholm inverse}\}.$$

According to the classical theory of Fredholm operators (see for example [185]), $\Phi(X,Y) = \Phi_r(X,Y)\bigcap\Phi_l(X,Y)$. If $X = Y$, the sets $\Phi_l(X,X)$ and $\Phi_r(X,X)$ are replaced respectively by $\Phi_l(X)$ and $\Phi_r(X)$. We will denote by $\Phi_l^b(X,Y)$ (resp. $\Phi_r^b(X,Y)$) the set of bounded operators which have a left (resp. a right) Fredholm inverse. From Theorems 2.2.8 and 2.2.9, it follows that $\Phi_l^b(X,Y) = \{T \in \Phi_+^b(X,Y)$ such that $R(T)$ is complemented$\}$ and $\Phi_r^b(X,Y) = \{T \in \Phi_-^b(X,Y)$ such that $N(T)$ is complemented$\}$. We may notice the following inclusions $\Phi^b(X,Y) \subset \Phi_l^b(X,Y) \subset \Phi_+^b(X,Y)$ and $\Phi^b(X,Y) \subset \Phi_r^b(X,Y) \subset \Phi_-^b(X,Y)$. An operator $A \in \mathcal{C}(X,Y)$ is a left (resp. right) Weyl if A has a left (resp. right) Fredholm inverse and $i(A) \leq 0$ (resp. $i(A) \geq 0$). We use $\mathcal{W}_l(X,Y)$ (resp. $\mathcal{W}_r(X,Y)$) to denote the set of all left (resp. right) Weyl operators. If $X = Y$, the sets $\mathcal{W}_l(X,X)$ and $\mathcal{W}_r(X,X)$ are replaced respectively by $\mathcal{W}_l(X)$ and $\mathcal{W}_r(X)$. Let $\gamma(.)$ be the Kuratowski measure of noncompactness in X. We denote by $\mathcal{P}_\gamma(.)$ the set defined by

$$\mathcal{P}_\gamma(X) = \{A \in \mathcal{L}(X) \text{ such that } \gamma(A^m) < 1, \text{ for some } m > 0\}. \tag{6.4.2}$$

Definition 6.4.2. Let A and B be two operators in $\mathcal{L}(X,Y)$. We denote by $F_{AB}^{\pm}(Y,X)$ the set of left or right Fredholm inverses $R_\pm$ of A satisfying $BR_\pm \in \mathcal{P}_\gamma(X)$ or $R_\pm B \in \mathcal{P}_\gamma(X)$ depending on whether $A \in \Phi_+^b(X,Y)$ or $A \in \Phi_-^b(X,Y)$. ◇

We have the following theorem.

Theorem 6.4.1. *Let X and Y be two Banach spaces, and let A and B be two operators in $\mathcal{L}(X,Y)$. Then,*

(i) If $A \in \Phi^b(X,Y)$ and $R \in \mathcal{L}(Y,X)$ is a Fredholm inverse of A, such that $RB \in \mathcal{P}_\gamma(X)$, then $A+B \in \Phi^b(X,Y)$ and $i(A+B) = i(A)$.
(ii) If $A \in \Phi_+^b(X,Y)$ and $R_l \in \mathcal{L}(Y,X)$ is a left Fredholm inverse of A, such that $BR_l \in \mathcal{P}_\gamma(Y)$, then $A+B \in \Phi_+^b(X,Y)$ and $i(A+B) = i(A)$.
(iii) If $A \in \Phi_-^b(X,Y)$ and $R_r \in \mathcal{L}(Y,X)$ is a right Fredholm inverse of A, such that $R_rB \in \mathcal{P}_\gamma(X)$, then $A+B \in \Phi_-^b(X,Y)$ and $i(A+B) = i(A)$.
(iv) If $A \in \Phi_\pm^b(X,Y)$ and $F_{AB}^{\pm}(Y,X) \neq \emptyset$, then $A+B \in \Phi_\pm^b(X,Y)$. ◇

Proof.

(i) Since R is a Fredholm inverse of A, there exists $F \in \mathcal{F}_0(Y)$ such that

$$AR = I - F \text{ on } Y. \tag{6.4.3}$$

From Eq. (6.4.3), it follows that the operator $A+B$ can be written in the form

$$A+B = A + (AR+F)B = A(I_X + RB) + FB. \tag{6.4.4}$$

Using the fact that $RB \in \mathcal{P}_\gamma(X)$, together with Corollary 5.2.2 (i), we get

$$I_X + RB \in \Phi^b(X) \text{ and } i(I_X + RB) = 0. \tag{6.4.5}$$

Now, applying Theorem 2.2.40 and Eq. (6.4.5), we obtain $A(I_X + RB) \in \Phi^b(X, Y)$ and $i(A(I_X + RB)) = i(A)$. Next, $FB \in \mathcal{F}_0(X, Y)$, and using Eq. (6.4.4), we deduce that $A + B \in \Phi^b(X, Y)$ and $i(A + B) = i(A)$.

(ii) R_l is a left Fredholm inverse of A, then there exists $F \in \mathcal{F}_0(Y)$, such that

$$R_l A = I - F \text{ on } X. \tag{6.4.6}$$

From Eq. (6.4.6), it follows that the operator $A + B$ can be written in the form

$$A + B = A + B(R_l A + F) = (BR_l + I_Y)A + BF. \tag{6.4.7}$$

Using the fact that $BR_l \in \mathcal{P}_\gamma(Y)$, and applying Corollary 5.2.2 (i), we have $BR_l + I_Y \in \Phi^b(Y)$ and $i(BR_l + I_Y) = 0$. Now, using the fact that $A \in \Phi^b_+(X, Y)$, $BF \in \mathcal{K}(X, Y)$, Theorem 2.2.13 and Lemma 6.3.1, we infer that $A + B \in \Phi^b_+(X, Y)$. Moreover, combining Theorem 2.2.7 and Eq. (6.4.7), we get $i(A + B) = i(A)$.

(iii) If R_r is a right Fredholm inverse of A, then there exists $F \in \mathcal{F}_0(Y)$, such that $AR_r = I - F$ on Y. Consequently, $A+B = A+B(AR_r+F) = A(AR_r+I_X)+FB$. Now, arguing as in (ii), we get $A + B \in \Phi^b_-(X, Y)$ and $i(A + B) = i(A)$.

(iv) The statement (iv) is an immediate consequence of the items (ii) and (iii). Q.E.D.

Remark 6.4.2.

(i) The results of Theorem 6.4.1 remain valid if we suppose that $A \in \mathcal{C}(X)$ and B is an A-bounded operator on X. Clearly, applying Theorem 6.4.1, we prove the statements for $\hat{A} \in \mathcal{L}(X_A, X)$ and $\hat{B} \in \mathcal{L}(X_A, X)$ and applying Eq. (6.3.5), we achieve the desired results.

(ii) If we replace $\mathcal{P}_\gamma(X)$ by $\mathcal{J}(X)$ where $\mathcal{J}(X) = \{A \in \mathcal{L}(X) \text{ such that } 1 \in \Phi^0_A\}$, then we can prove the same results as in Theorem 6.4.1. ◇

For $T \in \mathcal{C}(X, Y)$, we denote by $\mathcal{G}^l(T)$ (resp. $\mathcal{G}^r(T)$) the set of left (resp. right) Fredholm inverses of T. A complex number λ is in Φ_{lT} or Φ_{rT}, if $\lambda - T$ is in $\Phi_l(X)$ or $\Phi_r(X)$, respectively. The following theorem deals with the problem of stability in the class of Fredholm operators.

Theorem 6.4.2. *Let X be a Banach space, $T \in \mathcal{C}(X)$ and let S be a T-bounded operator on X. Consider γ (resp. γ_{X_T}) as a measure of noncompactness in X (resp. in X_T). Then, the following statements hold.*

(i) If $T_l \in \mathcal{G}^l(T)$ and $\gamma_{X_T}((T_l\hat{S})^m) < c$, for some $m \in \mathbb{N}^*$ and $c \in [0, 1]$, then $T + S \in \Phi_l(X)$ and $i(T + S) = i(T)$.

(ii) *If* $T_r \in \mathcal{G}^r(T)$ *and* $\gamma((ST_r)^m) < c$, *for some* $m \in \mathbb{N}^*$ *and* $c \in [0,1]$, *then* $T + S \in \Phi_r(X)$ *and* $i(T+S) = i(T)$.

(iii) *If* $\tilde{T} \in \mathcal{G}^l(T) \bigcap \mathcal{G}^r(T)$, $\gamma_{X_T}((\tilde{T}\hat{S})^m) < c$ *and* $\gamma((S\tilde{T})^n) < c'$, *for some* $m, n \in \mathbb{N}^*$ *and* $c, c' \in [0,1]$, *then* $T + S \in \Phi(X)$ *and* $i(T+S) = i(T)$. ◇

Proof. (i) According to the hypotheses, there exists $K_1 \in \mathcal{K}(X_T)$, such that $T_l(\hat{S} + \hat{T}) = T_l\hat{S} + I_{X_T} - K_1$. Since $\gamma_{X_T}((T_l\hat{S} - K_1)^n) = \gamma_{X_T}((T_l\hat{S})^n) < 1$, then by Corollary 5.2.2 (i), $T_l(\hat{S} + \hat{T}) \in \Phi^b(X_T)$ and $i(T_l\hat{S} + I_{X_T}) = 0$. Now, by using Theorem 2.2.6, there exist $A_0 \in \mathcal{L}(X_T)$ and $K' \in \mathcal{K}(X_T)$ such that $A_0T_l(\hat{S} + \hat{T}) = I_{X_T} - K'$, which implies that $\hat{S} + \hat{T} \in \Phi_l(X_T, X)$. Moreover, we have $i(T_l(\hat{S} + \hat{T})) = i(T_l) + i(\hat{S} + \hat{T}) = 0$. Hence, $i(\hat{S} + \hat{T}) = -i(T_l) = i(\hat{T})$. Finally, the result follows from Eq. (6.3.5).

Arguing as in the proof of (i), we prove (ii). Finally, the proof of (iii) is an obvious deduction from (i) and (ii). Q.E.D.

6.5 Fredholm Perturbations

Definition 6.5.1. Let X and Y be two Banach spaces. We say that Y is essentially stronger than X and write $X \preceq Y$, if there exists $R \in \Phi_l^b(X, Y)$. ◇

Remark 6.5.1.

(i) It is clear that, for X a Banach space, $X \preceq X$.

(ii) Let X_1, X_2, X_3 be three Banach spaces such that $X_1 \preceq X_2 \preceq X_3$. Then, there exists $R_1 \in \Phi_+^b(X_1, X_2)$ and $R_2 \in \Phi_+^b(X_2, X_3)$ with $R(R_1)$ complemented in X_2 and $R(R_2)$ complemented in X_3. By Theorem 2.2.8, there exists S_i such that $S_iR_i = I_{X_i} + K_i$ with $K_i \in \mathcal{K}(X_i)$, $i = 1, 2$. Hence, $(S_1S_2)(R_2R_1) = I_{X_1} + K_1 + S_1K_2R_1$. Thus, by Theorem 2.2.14, $R_2R_1 \in \Phi_+^b(X_1, X_3)$ with $R(R_2R_1)$ complemented in X_3 which implies that $X_1 \preceq X_3$.

(iii) To deduce that "$\preceq$" is not antisymmetric, we notice that W. T. Gowers showed in [134] that there is a Banach space Z that is isomorphic to $Z \oplus Z \oplus Z$ but not isomorphic to $Z \oplus Z$. ◇

The following results of this section may be found in [2].

Lemma 6.5.1. *Let* X *and* Y *be two Banach spaces. If* $X \preceq Y$, *then* $X^* \preceq Y^*$. ◇

Proof. If $X \preceq Y$, then there exist $R \in \Phi_+^b(X, Y)$ and $S \in \Phi_-^b(Y, X)$, such that $SR = I_X + K$, with $K \in \mathcal{K}(X)$. This implies that $R^*S^* = I_{X^*} + K^*$. Since $R^* \in \Phi_-^b(Y^*, X^*)$, $S^* \in \Phi_+^b(X^*, Y^*)$ and $K^* \in \mathcal{K}(X^*)$, then we get $X^* \preceq Y^*$. Q.E.D.

Theorem 6.5.1. *Let* (X_1, X_2) *and* (Y_1, Y_2) *be two couples of Banach spaces satisfying* $X_1 \preceq Y_1$, *and let* $U \in \mathcal{L}(X_2, Y_2)$ *and* $V \in \mathcal{L}(Y_1, X_1)$. *Suppose that* $\Phi^b(X_1, X_2) \neq \emptyset$. *If* $S \in \mathcal{F}^b(X_1, X_2)$, *then* $USV \in \mathcal{F}^b(Y_1, Y_2)$. ◇

Proof. Remark that the result is trivial if $\Phi^b(Y_1, Y_2) = \emptyset$. So, let us assume that $\Phi^b(Y_1, Y_2) \neq \emptyset$. Since $X_1 \leq Y_1$ and $\Phi^b(X_1, X_2) \neq \emptyset$, this yields $X_2 \leq Y_2$. Hence, there exists $R_i \in \Phi^b_+(X_i, Y_i)$ and a closed subspace Z_i such that $Y_i := R(R_i) \oplus Z_i$, $i = 1, 2$. Without loss of generality, we can suppose that R_1 and R_2 are injective. For $i = 1, 2$ denote the following invertible operator $R_{i0} : X_i \longrightarrow R(R_i)$, $x \longrightarrow R_{i0}(x) = R_i(x)$. By Theorem 2.2.8, there exists $R_i' \in \mathcal{L}(Y_i, X_i)$ such that $R_i' R_i = I_{X_i} + K_i$, with $K_i \in \mathcal{K}(X_i)$. We can choose R_1' as follows $R_1' : Y_1 = R(R_1) \oplus Z_1 \longrightarrow X_1$, $y = (y_1, z_1) \longrightarrow R_{10}^{-1}(y_1)$. Hence,

$$\begin{cases} R_2SR_1' : Y_1 = R(R_1) \oplus Z_1 \longrightarrow Y_2 = R(R_2) \oplus Z_2, \\ \qquad y = (y_1, z_1) \longrightarrow (R_{20}SR_{10}^{-1}(y_1), 0) = \begin{pmatrix} R_{20}SR_{10}^{-1} & 0 \\ 0 & 0 \end{pmatrix} \begin{pmatrix} y_1 \\ z_1 \end{pmatrix}. \end{cases}$$

Since $S \in \mathcal{F}^b(X_1, X_2)$, then $R_{20}SR_{10}^{-1} \in \mathcal{F}^b(R(R_1), R(R_2))$. First, we claim that $R_2SR_1' \in \mathcal{F}^b(Y_1, Y_2)$. Consider an arbitrary element $L := \begin{pmatrix} A & B \\ C & D \end{pmatrix} \in \Phi^b(Y_1, Y_2)$. It follows, by Atkinson theorem, that there exists $L_0 := \begin{pmatrix} A_0 & B_0 \\ C_0 & D_0 \end{pmatrix} \in \Phi^b(Y_2, Y_1)$ and $K \in \mathcal{K}(Y_2)$ such that $LL_0 = I + K$ on Y_2. Then

$$\begin{aligned} (L + R_2SR_1')L_0 &= I + K + R_2SR_1'L_0 \\ &= K + \begin{pmatrix} I & R_{20}SR_{10}^{-1}B_0 \\ 0 & I \end{pmatrix} \begin{pmatrix} I + R_{20}SR_{10}^{-1}A_0 & 0 \\ 0 & I \end{pmatrix}. \end{aligned}$$

Notice that $\Phi^b(X_1, X_2) \neq \emptyset$ implies $\Phi^b(R(R_1), R(R_2)) \neq \emptyset$. So, $R_{20}SR_{10}^{-1} \in \mathcal{F}^b(R(R_1), R(R_2))$, yields that $R_{20}SR_{10}^{-1}A_0 \in \mathcal{F}^b(R(R_2))$. Thus, $I + R_{20}SR_{10}^{-1}A_0 \in \Phi^b(R(R_2))$ which implies that $\begin{pmatrix} I + R_{20}SR_{10}^{-1}A_0 & 0 \\ 0 & I \end{pmatrix}$ is a Fredholm operator. Observing that $\begin{pmatrix} I & R_{20}SR_{10}^{-1}B_0 \\ 0 & I \end{pmatrix}$ is invertible, with inverse $\begin{pmatrix} I & -R_{20}SR_{10}^{-1}B_0 \\ 0 & I \end{pmatrix}$, we get $(L + R_2SR_1')L_0 \in \Phi^b(Y_2)$. Hence, $(L + R_2SR_1') \in \Phi^b(Y_1, Y_2)$ and therefore $R_2SR_1' \in \mathcal{F}^b(Y_1, Y_2)$. Our claim is proved. Now, since $R_i'R_i = I_{X_i} + K_i$, then $USV = U(R_2'R_2 - K_2)S(R_1'R_1 - K_1)V = UR_2'(R_2SR_1')R_1V + K'$, with $K' \in \mathcal{K}(Y_1, Y_2)$. Finally, the result follows by Theorem 6.3.1, since $UR_2' \in \mathcal{L}(Y_2)$ and $R_1V \in \mathcal{L}(Y_1)$. Q.E.D.

Definition 6.5.2. Let X and Y be two Banach spaces. We denote $\mathcal{F}_l^b(X, Y)$ and $\mathcal{F}_r^b(X, Y)$ by $\mathcal{F}_l^b(X, Y) = \{F \in \mathcal{L}(X, Y)$ such that $T + F \in \Phi_l^b(X, Y)$ whenever $T \in \Phi_l^b(X, Y)\}$ and $\mathcal{F}_r^b(X, Y) = \{F \in \mathcal{L}(X, Y)$ such that $T + F \in \Phi_r^b(X, Y)$ whenever $T \in \Phi_r^b(X, Y)\}$. ◇

If $X = Y$, then the set $\mathcal{F}_l^b(X, X)$ (resp. $\mathcal{F}_r^b(X, X)$) will be denoted by $\mathcal{F}_l^b(X)$ (resp. $\mathcal{F}_r^b(X)$). In the following equalities, we give the results proved by A. Lebow and M. Schechter [230, Theorem 2.7] $\mathcal{F}_l^b(X) = \mathcal{F}_r^b(X) = \mathcal{F}(X)$.

Theorem 6.5.2. *Let (X_1, X_2) and (Y_1, Y_2) be two couples of Banach spaces satisfying $X_1 \leq Y_1$, and let $U \in \mathcal{L}(X_2, Y_2)$ and $V \in \mathcal{L}(Y_1, X_1)$. Suppose that $\Phi_l^b(X_1, X_2) \neq \emptyset$. If $S \in \mathcal{F}_l^b(X_1, X_2)$, then $USV \in \mathcal{F}_l^b(Y_1, Y_2)$.* ◇

Proof. The proof is analogous to the proof of Theorem 6.5.1. Q.E.D.

Theorem 6.5.3. *Let (X_1, X_2) and (Y_1, Y_2) be two couples of Banach spaces satisfying $X_1 \leq Y_1$, and let $U \in \mathcal{L}(X_2, Y_2)$ and $V \in \mathcal{L}(Y_1, X_1)$. Suppose that $\Phi_r^b(X_1, X_2) \neq \emptyset$. If $S \in \mathcal{F}_r^b(X_1, X_2)$, then $USV \in \mathcal{F}_r^b(Y_1, Y_2)$.* ◇

Proof. The proof is analogous to the proof of Theorem 6.5.1. Q.E.D.

Corollary 6.5.1. *Let (X_1, X_2) and (Y_1, Y_2) be two couples of Banach spaces such that $X_1 \leq Y_1$. Assume that $\Phi^b(X_1, X_2) \neq \emptyset$. If $\mathcal{F}^b(Y_1, Y_2) = \mathcal{K}(Y_1, Y_2)$, then $\mathcal{F}^b(X_1, X_2) = \mathcal{K}(X_1, X_2)$.* ◇

Proof. Since $X_i \leq Y_i$, $i = 1, 2$, then there exists $R_1 \in \Phi_+^b(X_1, Y_1)$ and $R_2 \in \Phi_+^b(X_2, Y_2)$ with $R(R_1)$ complemented in Y_1 and $R(R_2)$ complemented in Y_2. By Theorem 2.2.8, there exists R_i' such that $R_i' R_i = I_{X_i} + K_i$ with $K_i \in \mathcal{K}(X_i)$, $i = 1, 2$. Let $T \in \mathcal{F}^b(X_1, X_2)$. The use of Theorem 6.5.1 leads to $R_2 T R_1' \in \mathcal{F}^b(Y_1, Y_2) = \mathcal{K}(Y_1, Y_2)$. Thus, $R_2' R_2 T R_1' R_1 = T + K_2 T + K_2 T K_1 \in \mathcal{K}(X_1, X_2)$ and therefore $T \in \mathcal{K}(X_1, X_2)$. Q.E.D.

Corollary 6.5.2. *Let (X_1, X_2) and (Y_1, Y_2) be two couples of Banach spaces such that $X_1 \leq Y_1$. Assume that $\Phi^b(X_1, X_2) \neq \emptyset$. If $\mathcal{F}^b(Y_1, Y_2) = \mathcal{S}(Y_1, Y_2)$, then $\mathcal{F}^b(X_1, X_2) = \mathcal{S}(X_1, X_2)$.* ◇

Proof. The proof is similar to the proof of Corollary 6.5.1. Q.E.D.

Corollary 6.5.3. *Let X be a Banach space satisfying $X \leq L_p(\Omega, d\mu)$, for some $p \geq 1$, then $\mathcal{F}^b(X) = \mathcal{S}(X) = \mathcal{CS}(X)$.* ◇

Proof. By Eq. (2.1.12) we have $\mathcal{F}^b(L_p(\Omega, d\mu)) = \mathcal{S}(L_p(\Omega, d\mu))$. Since $X \leq L_p(\Omega, d\mu)$, then by Corollary 6.5.2, we get $\mathcal{F}^b(X) = \mathcal{S}(X)$. Now, consider $F \in \mathcal{F}^b(X)$, then $F^* \in \mathcal{F}^b(X^*)$. Since $X \leq L_p(\Omega, d\mu)$, then by Lemma 6.5.1, $X^* \leq L_p^*(\Omega, d\mu) = L_q(\Omega, d\mu)$ for some $q \geq 1$. Again by Corollary 6.5.2, $\mathcal{F}^b(X^*) = \mathcal{S}(X^*)$. Hence $F^* \in \mathcal{S}(X^*)$, and therefore $F \in \mathcal{CS}(X)$. This yields $\mathcal{F}^b(X) \subset \mathcal{CS}(X)$, and we get the result since we have $\mathcal{CS}(X) \subset \mathcal{F}^b(X)$. Q.E.D.

6.6 Some Perturbation Results for Matrix Operators

In this section, we will establish some perturbation results for matrix operators acting on a product of Banach spaces X_1 and X_2. The following two lemmas are fundamental.

Lemma 6.6.1. *Let $A \in \mathcal{L}(X_1)$, $B \in \mathcal{L}(X_2)$ and consider the 2×2 operator matrix $M_C := \begin{pmatrix} A & C \\ 0 & B \end{pmatrix}$ where $C \in \mathcal{L}(X_2, X_1)$. Then,*

(i) If $A \in \Phi^b(X_1)$ and $B \in \Phi^b(X_2)$, then $M_C \in \Phi^b(X_1 \times X_2)$ for every $C \in \mathcal{L}(X_2, X_1)$.
(ii) If $A \in \Phi^b_+(X_1)$ and $B \in \Phi^b_+(X_2)$, then $M_C \in \Phi^b_+(X_1 \times X_2)$ for every $C \in \mathcal{L}(X_2, X_1)$.
(iii) If $A \in \Phi^b_-(X_1)$ and $B \in \Phi^b_-(X_2)$, then $M_C \in \Phi^b_-(X_1 \times X_2)$ for every $C \in \mathcal{L}(X_2, X_1)$. ◇

Proof.

(i) Let us write M_C in the form

$$M_C = \begin{pmatrix} I & 0 \\ 0 & B \end{pmatrix} \begin{pmatrix} I & C \\ 0 & I \end{pmatrix} \begin{pmatrix} A & 0 \\ 0 & I \end{pmatrix}. \tag{6.6.1}$$

Since $A \in \Phi^b(X_1)$ and $B \in \Phi^b(X_2)$, then $\begin{pmatrix} A & 0 \\ 0 & I \end{pmatrix}$ and $\begin{pmatrix} I & 0 \\ 0 & B \end{pmatrix}$ are both Fredholm operators. So, M_C is a Fredholm operator, since $\begin{pmatrix} I & C \\ 0 & I \end{pmatrix}$ is invertible for every $C \in \mathcal{L}(X_2, X_1)$.

(ii) and (iii) can be checked in the same way as (i). Q.E.D.

Remark 6.6.1. Using the same reasoning as in the proof of Lemma 6.6.1, we can show that:

(i) If $A \in \Phi^b(X_1)$ and $B \in \Phi^b(X_2)$, then

$$M_D := \begin{pmatrix} A & 0 \\ D & B \end{pmatrix}$$

is a Fredholm operator on $X_1 \times X_2$ for every $D \in \mathcal{L}(X_1, X_2)$.

(ii) If $A \in \Phi^b_+(X_1)$ and $B \in \Phi^b_+(X_2)$, then $M_D \in \Phi^b_+(X_1 \times X_2)$ for every $D \in \mathcal{L}(X_1, X_2)$.

(iii) If $A \in \Phi^b_-(X_1)$ and $B \in \Phi^b_-(X_2)$, then $M_D \in \Phi^b_-(X_1 \times X_2)$ for every $D \in \mathcal{L}(X_1, X_2)$. ◇

Lemma 6.6.2. *Let $A \in \mathcal{L}(X_1)$, $B \in \mathcal{L}(X_2)$ and consider the 2×2 operator matrix $M_C := \begin{pmatrix} A & C \\ 0 & B \end{pmatrix}$, where $C \in \mathcal{L}(X_2, X_1)$.*

(i) If $M_C \in \Phi^b_+(X_1 \times X_2)$, then $A \in \Phi^b_+(X_1)$.
(ii) If $M_C \in \Phi^b_-(X_1 \times X_2)$, then $B \in \Phi^b_-(X_2)$. ◇

Proof. The result is immediately deduced from Eq. (6.6.1). Q.E.D.

Remark 6.6.2.

(i) Immediately from the last Lemma 6.6.2, it follows that, if $M_C \in \Phi^b(X_1 \times X_2)$, then $A \in \Phi^b_+(X_1)$ and $B \in \Phi^b_-(X_2)$.

(ii) Using the same reasoning as in the proof of Lemma 6.6.1, we can show that if the operator $\begin{pmatrix} A & 0 \\ D & B \end{pmatrix}$ is in $\Phi^b(X_1 \times X_2)$ for some $D \in \mathcal{L}(X_1, X_2)$, then $A \in \Phi^b_-(X_1)$ and $B \in \Phi^b_+(X_2)$. ◇

Theorem 6.6.1. *Let* $F := \begin{pmatrix} F_{11} & F_{12} \\ F_{21} & F_{22} \end{pmatrix}$, *where* $F_{ij} \in \mathcal{L}(X_j, X_i)$, *with* $i, j = 1, 2$. *Then,* $F \in \mathcal{F}^b(X_1 \times X_2)$ *if, and only if,* $F_{ij} \in \mathcal{F}^b(X_j, X_i)$, *with* $i, j = 1, 2$. ◇

Proof. In order to prove the second implication, let us consider the following decomposition:

$$F = \begin{pmatrix} F_{11} & 0 \\ 0 & 0 \end{pmatrix} + \begin{pmatrix} 0 & F_{12} \\ 0 & 0 \end{pmatrix} + \begin{pmatrix} 0 & 0 \\ F_{21} & 0 \end{pmatrix} + \begin{pmatrix} 0 & 0 \\ 0 & F_{22} \end{pmatrix}.$$

It is sufficient to prove that if $F_{ij} \in \mathcal{F}^b(X_j, X_i)$, with $i, j = 1, 2$, then each operator in the right side of the previous equality is a Fredholm perturbation on $X_1 \times X_2$. For example, we will prove the result for the first operator. The proofs for the other operators will be similarly achieved. Consider $L = \begin{pmatrix} A & B \\ C & D \end{pmatrix} \in \Phi^b(X_1 \times X_2)$ and let us denote $\tilde{F} := \begin{pmatrix} F_{11} & 0 \\ 0 & 0 \end{pmatrix}$. From Theorem 2.2.6, it follows that there exist $L_0 = \begin{pmatrix} A_0 & B_0 \\ C_0 & D_0 \end{pmatrix} \in \mathcal{L}(X_1 \times X_2)$ and $K = \begin{pmatrix} K_{11} & K_{12} \\ K_{21} & K_{22} \end{pmatrix} \in \mathcal{K}(X_1 \times X_2)$, such that $LL_0 = I - K$ on $X_1 \times X_2$. Then, $(L + \tilde{F})L_0 = I - K + \tilde{F}L_0 = \begin{pmatrix} I - K_{11} + F_{11}A_0 & -K_{12} + F_{11}B_0 \\ -K_{21} & I - K_{22} \end{pmatrix}$. Since $F_{11} \in \mathcal{F}^b(X_1)$, and using Theorem 6.3.1, we will have $I - K_{11} + F_{11}A_0 \in \Phi^b(X_1)$. This, together with the fact that $I - K_{22} \in \Phi^b(X_2)$, allows us to deduce, from Lemma 6.6.1 (i), that $(L + \tilde{F})L_0 - \begin{pmatrix} 0 & 0 \\ -K_{21} & 0 \end{pmatrix}$ is a Fredholm operator on $X_1 \times X_2$. The fact that K_{21} is a compact operator and $L_0 \in \Phi^b(X_1 \times X_2)$ leads, by Theorem 2.2.10, to $L + \tilde{F} \in \Phi^b(X_1 \times X_2)$.

Conversely, assume that $F \in \mathcal{F}^b(X_1 \times X_2)$. We will prove that $F_{11} \in \mathcal{F}^b(X_1)$. Let $A \in \Phi^b(X_1)$ and let us define the operator $L_1 := \begin{pmatrix} A & -F_{12} \\ 0 & I \end{pmatrix}$. From Lemma 6.6.1 (i), it follows that $L_1 \in \Phi^b(X_1 \times X_2)$. Hence, $F + L_1 = \begin{pmatrix} A + F_{11} & 0 \\ F_{21} & I + F_{22} \end{pmatrix} \in \Phi^b(X_1 \times X_2)$. The use of Remark 6.6.2 (ii) leads to

$$A + F_{11} \in \Phi^b_-(X_1). \tag{6.6.2}$$

In the same way, we may consider the Fredholm operator $\begin{pmatrix} A & 0 \\ -F_{21} & I \end{pmatrix}$. Using Remarks 6.6.1 (i) and 6.6.2 (i), it is easy to deduce that

$$A + F_{11} \in \Phi_+^b(X_1). \tag{6.6.3}$$

From Eqs. (6.6.2) and (6.6.3), it follows that $F_{11} \in \mathcal{F}^b(X_1)$. In the same way, we can prove that $F_{22} \in \mathcal{F}^b(X_2)$. Now, we have to prove that $F_{12} \in \mathcal{F}^b(X_2, X_1)$ and $F_{21} \in \mathcal{F}^b(X_1, X_2)$. For this, let us consider $A \in \Phi^b(X_2, X_1)$ and $B \in \Phi^b(X_1, X_2)$. Then, $\begin{pmatrix} 0 & A \\ B & 0 \end{pmatrix} \in \Phi^b(X_1 \times X_2)$. Using the facts that $F_{11} \in \mathcal{F}^b(X_1)$, that $F_{22} \in \mathcal{F}^b(X_2)$, as well as the result of the second implication, we can deduce that $F + \begin{pmatrix} -F_{11} & 0 \\ 0 & -F_{22} \end{pmatrix} \in \mathcal{F}^b(X_1 \times X_2)$. Hence, $\begin{pmatrix} 0 & A + F_{12} \\ B + F_{21} & 0 \end{pmatrix} \in \Phi^b(X_1 \times X_2)$. So, $A + F_{12} \in \Phi^b(X_2, X_1)$ and $B + F_{21} \in \Phi^b(X_1, X_2)$. Q.E.D.

Theorem 6.6.2. *Let* $F := \begin{pmatrix} F_{11} & F_{12} \\ F_{21} & F_{22} \end{pmatrix}$, *where* $F_{ij} \in \mathcal{L}(X_j, X_i)$, *with* $i, j = 1, 2$. *Then,*

(i) If $F_{ij} \in \mathcal{F}_l^b(X_j, X_i)$, *for all* $i, j = 1, 2$, *then* $F \in \mathcal{F}_l^b(X_1 \times X_2)$.
(ii) If $F_{ij} \in \mathcal{F}_r^b(X_j, X_i)$, *for all* $i, j = 1, 2$, *then* $F \in \mathcal{F}_r^b(X_1 \times X_2)$. ◇

Proof.

(i) Using the same notations as in the proof of Theorem 6.6.1, we get:

$$L_0(L + \tilde{F}) = I - K + L_0\tilde{F} = \begin{pmatrix} I - K_{11} + A_0F_{11} & -K_{12} \\ -K_{21} + C_0F_{11} & I - K_{22} \end{pmatrix}.$$

Since $F_{11} \in \mathcal{F}_l^b(X_1)$, and using Theorem 6.3.1, we can deduce that $I - K_{11} + A_0F_{11} \in \Phi_l^b(X_1)$. So, there exist a bounded operator $H \in \mathcal{L}(X_1)$ and $K_0 \in \mathcal{K}(X_1)$, such that $H(I - K_{11} + A_0F_{11}) = I - K_0$. Therefore,

$$\begin{pmatrix} H & 0 \\ 0 & I \end{pmatrix} L_0(L + \tilde{F}) = I - \begin{pmatrix} K_0 & HK_{12} \\ K_{21} & K_{22} \end{pmatrix} + \begin{pmatrix} 0 & 0 \\ C_0F_{11} & 0 \end{pmatrix}.$$

Using both Theorems 6.3.1 and 6.6.1, we deduce that $\begin{pmatrix} 0 & 0 \\ C_0F_{11} & 0 \end{pmatrix} \in \mathcal{F}^b(X_1 \times X_2)$ and so, $\begin{pmatrix} H & 0 \\ 0 & I \end{pmatrix} L_0(L + \tilde{F}) \in \Phi^b(X_1 \times X_2)$. Hence, there exist $L_1 \in \mathcal{L}(X_1 \times X_2)$ and $\tilde{K} \in \mathcal{K}(X_1 \times X_2)$, such that $L_1 \begin{pmatrix} H & 0 \\ 0 & I \end{pmatrix} L_0(L + \tilde{F}) = I - \tilde{K}$, which implies that $\tilde{F} \in \mathcal{F}_l^b(X_1 \times X_2)$.

(ii) We prove this assertion in the same way as in (i). Q.E.D.

Open question: The following questions remain open:

(i) $F := \begin{pmatrix} F_{11} & F_{12} \\ F_{21} & F_{22} \end{pmatrix} \in \mathcal{F}_+^b(X_1 \times X_2)$ if, and only if, $F_{ij} \in \mathcal{F}_+^b(X_j, X_i)$, $\forall i, j = 1, 2$?

(ii) $F \in \mathcal{F}_-^b(X_1 \times X_2)$ if, and only if, $F_{ij} \in \mathcal{F}_-^b(X_j, X_i)$, $\forall i, j = 1, 2$? □

6.7 Some Fredholm Theory Results for Matrix Operators

The following result is fundamental for our purpose.

Proposition 6.7.1. *Let* $T = \begin{pmatrix} T_1 & 0 \\ 0 & T_2 \end{pmatrix}$ *be in* $\mathcal{C}(X \times X)$. *Then,*

(i) $T^l \in \mathcal{G}^l(T)$ *if, and only if, there exist* $B \in \mathcal{L}(X, X_{T_1})$ *and* $C \in \mathcal{L}(X, X_{T_2})$ *such that*

$$T^l = \begin{pmatrix} T_1^l & B \\ C & T_2^l \end{pmatrix},$$

where $T_i^l \in \mathcal{G}^l(T_i)$, *with* $i = 1, 2$, $C\hat{T}_1 \in \mathcal{K}(X_{T_1}, X_{T_2})$ *and* $B\hat{T}_2 \in \mathcal{K}(X_{T_2}, X_{T_1})$.

(ii) $T^r \in \mathcal{G}^r(T)$ *if, and only if, there exist* $B \in \mathcal{L}(X, X_{T_1})$ *and* $C \in \mathcal{L}(X, X_{T_2})$ *such that*

$$T^r = \begin{pmatrix} T_1^r & B \\ C & T_2^r \end{pmatrix},$$

where $T_i^r \in \mathcal{G}^r(T_i)$, *with* $i = 1, 2$ *and* $\hat{T}_2 C$, $\hat{T}_1 B \in \mathcal{K}(X)$.

(iii) $\tilde{T} \in \mathcal{G}^l(T) \bigcap \mathcal{G}^r(T)$ *if, and only if,*

$$\tilde{T} = \begin{pmatrix} \tilde{T}_1 & K_1 \\ K_2 & \tilde{T}_2 \end{pmatrix},$$

where $K_i \in \mathcal{K}(X, X_{T_i})$ *and* $\tilde{T}_i \in \mathcal{G}^l(T_i) \bigcap \mathcal{G}^r(T_i)$, *with* $i = 1, 2$. ◇

Proof.

(i) Suppose that $T^l = \begin{pmatrix} T_1^l & B \\ C & T_2^l \end{pmatrix}$, where $T_i^l \in \mathcal{G}^l(T_i)$, with $i = 1, 2$, $C\hat{T}_1 \in \mathcal{K}(X_{T_1}, X_{T_2})$ and $B\hat{T}_2 \in \mathcal{K}(X_{T_2}, X_{T_1})$. Then, it is easy to verify that $T^l \hat{T} = I_{X_{T_1} \times X_{T_2}} - K$, where $K \in \mathcal{K}(X_{T_1} \times X_{T_2})$. Conversely, consider $T^l = \begin{pmatrix} A & B \\ C & D \end{pmatrix}$ such that $T^l \hat{T} = I_{X_T} - K$, where $K = \begin{pmatrix} K_1 & K_3 \\ K_2 & K_4 \end{pmatrix} \in \mathcal{K}(X_{T_1} \times X_{T_2})$. We necessarily have $A\hat{T}_1 = I_{X_{T_1}} - K_1$, $D\hat{T}_2 = I_{X_{T_2}} - K_4$, $C\hat{T}_1 = K_2$ and $B\hat{T}_2 = K_3$. Hence, $A \in \mathcal{G}^l(T_1)$ and $D \in \mathcal{G}^l(T_2)$. Arguing as in the proof of (i), we can prove (ii).

(iii) Suppose that $\tilde{T} = \begin{pmatrix} \tilde{T}_1 & K_1 \\ K_2 & \tilde{T}_2 \end{pmatrix}$, where $\tilde{T}_i \in \mathcal{G}^l(T_i) \bigcap \mathcal{G}^r(T_i)$ and K_i is a compact operator in $\mathcal{K}(X, X_{T_i})$, $i = 1, 2$. Applying (i) and (ii), we get $\tilde{T} \in \mathcal{G}^l(T) \bigcap \mathcal{G}^r(T)$. Conversely, consider $\tilde{T} = \begin{pmatrix} A & B \\ C & D \end{pmatrix}$ such that $\tilde{T}\hat{T} = I_{X_{T_1} \times X_{T_2}} - K$ and $\hat{T}\tilde{T} = I_{X \times X} - K'$, where $K = \begin{pmatrix} K_1 & K_3 \\ K_2 & K_4 \end{pmatrix}$ and $K' = \begin{pmatrix} K_1' & K_3' \\ K_2' & K_4' \end{pmatrix}$ are compact operators. We necessarily have $A\hat{T}_1 = I_{X_{T_1}} - K_1$, $\hat{T}_1 A = I_{X_{T_2}} - K_1'$ and $C\hat{T}_1 = K_2$. Then $A \in \mathcal{G}^l(T_1) \bigcap \mathcal{G}^r(T_1)$. Furthermore, $CT_1A = C(I - K) = K_2A$ is a compact operator and therefore, C is also compact. With the same argument, we show that $D \in \mathcal{G}^l(T_2) \bigcap \mathcal{G}^r(T_2)$ and B is compact in $\mathcal{L}(X, X_{T_1})$. Q.E.D.

Lemma 6.7.1. *Let $A \in \mathcal{C}(X)$. Then, for all S and T being invertible in $\mathcal{L}(X)$, such that $R(T) \subset \mathcal{D}(A)$ and $R(A) \subset \mathcal{D}(S)$, we have $\mathcal{G}^l(SAT) = T^{-1}\mathcal{G}^l(A)S^{-1}$ and $\mathcal{G}^r(SAT) = T^{-1}\mathcal{G}^r(A)S^{-1}$.* ◇

Proof. It is easy to see that, if $A_0 \in \mathcal{G}^l(A)$, then the operator $P := T^{-1}A_0S^{-1}$ belongs to $\mathcal{G}^l(SAT)$, which implies that $T^{-1}\mathcal{G}^l(A)S^{-1} \subset \mathcal{G}^l(SAT)$. Conversely, let $P \in \mathcal{G}^l(SAT)$, then $PSAT = I - K$, where $K \in \mathcal{K}(X)$. Thus, $TPS \in \mathcal{G}^l(A)$ and therefore, $P \in T^{-1}\mathcal{G}^r(A)S^{-1}$. With the same argument, we show that $\mathcal{G}^r(SAT) = T^{-1}\mathcal{G}^r(A)S^{-1}$. Q.E.D.

Proposition 6.7.2. *Let $T = \begin{pmatrix} T_1 & T_3 \\ T_4 & T_2 \end{pmatrix}$ be in $\mathcal{L}(X \times X)$. Then,*

(i) Suppose there exists $T_i^l \in \mathcal{G}^l(T_i)$, with $i = 1, 2$, such that $T_1^lT_3$ and $T_2^lT_4$ are compact operators. Then, $T^l = \begin{pmatrix} T_1^l & B \\ C & T_2^l \end{pmatrix} \in \mathcal{G}^l(T)$, where CT_i and BT_j are compact operators (with respectively $i = 1, 3$ and $j = 2, 4$).

(ii) Suppose there exists $T_i^r \in \mathcal{G}^r(T_i)$, $i = 1, 2$ such that $T_4T_1^r$ and $T_3T_2^r$ are compact operators. Then, $T^r = \begin{pmatrix} T_1^r & B \\ C & T_2^r \end{pmatrix} \in \mathcal{G}^r(T)$, where T_iC and T_jB are compact (with respectively $i = 1, 3$ and $j = 2, 4$).

(iii) Suppose there exists $\tilde{T}_k \in \mathcal{G}^l(T_k) \bigcap \mathcal{G}^r(T_k)$, with $k = 1, 2$ such that, for $i = 1, 2$ and $j = 3, 4$, $\tilde{T}_iT_j$, $T_j\tilde{T}_i$ are compact. Then, $\tilde{T} = \begin{pmatrix} \tilde{T}_1 & K_1 \\ K_2 & \tilde{T}_2 \end{pmatrix} \in \mathcal{G}^l(T) \bigcap \mathcal{G}^r(T)$, where K_i are compact operators (with $i = 1, 2$). ◇

Proof. (i)–(ii) It is easy to verify that $T^l\hat{T} = I_{X_T} - K$ and $\hat{T}T^r = I_{X_T} - K'$, where $K, K' \in \mathcal{K}(X_T)$. The assertion (iii) is a direct consequence of (i) and (ii). Q.E.D.

Chapter 7
Essential Spectra of Linear Operators

In this chapter, we investigate the essential spectra of the closed, densely defined linear operators on Banach space.

7.1 Definitions and Notations

It is well known that, if A is a bounded self-adjoint operator on a Hilbert space, the essential spectrum is the set of all points of the spectrum of A that are not isolated eigenvalues of finite algebraic multiplicity (see, for example, [153, 292, 347]). Irrespective of whether A is bounded or not on a Banach space X, there are several definitions of the essential spectrum, most of which constitute an enlargement of the continuous spectrum. Let us define the following sets

$$\sigma_{e1}(A) := \{\lambda \in \mathbb{C} \text{ such that } \lambda - A \notin \Phi_+(X)\} := \mathbb{C}\backslash\Phi_{+A},$$

$$\sigma_{e1l}(A) := \{\lambda \in \mathbb{C} \text{ such that } \lambda - A \notin \Phi_l(X)\} := \mathbb{C}\backslash\Phi_{lA},$$

$$\sigma_{e2}(A) := \{\lambda \in \mathbb{C} \text{ such that} \lambda - A \notin \Phi_-(X)\} := \mathbb{C}\backslash\Phi_{-A},$$

$$\sigma_{e2r}(A) := \{\lambda \in \mathbb{C} \text{ such that } \lambda - A \notin \Phi_r(X)\} := \mathbb{C}\backslash\Phi_{rA},$$

$$\sigma_{e3}(A) := \{\lambda \in \mathbb{C} \text{ such that } \lambda - A \notin \Phi_\pm(X)\} := \mathbb{C}\backslash\Phi_{\pm A},$$

$$\sigma_{e4}(A) := \{\lambda \in \mathbb{C} \text{ such that } \lambda - A \notin \Phi(X)\} := \mathbb{C}\backslash\Phi_A,$$

$$\sigma_{e5}(A) := \bigcap_{K \in \mathcal{K}(X)} \sigma(A+K),$$

$$\sigma_{e6}(A) := \mathbb{C}\backslash\rho_6(A),$$

$$\sigma_{e7}(A) := \bigcap_{K \in \mathcal{K}(X)} \sigma_{ap}(A+K),$$

A. Jeribi, *Spectral Theory and Applications of Linear Operators and Block Operator Matrices*, DOI 10.1007/978-3-319-17566-9_7

$$\sigma_{e7l}(A) := \{\lambda \in \mathbb{C} \text{ such that } \lambda - A \notin \mathcal{W}_l(X)\},$$

$$\sigma_{e8}(A) := \bigcap_{K \in \mathcal{K}(X)} \sigma_\delta(A + K),$$

$$\sigma_{e8r}(A) := \{\lambda \in \mathbb{C} \text{ such that } \lambda - A \notin \mathcal{W}_r(X)\},$$

$$\sigma_{pa}(A) := \{\lambda \in \mathbb{C} \text{ such that } \lambda - A \text{ is not one-to-one with closed range}\},$$

where $\rho_6(A) := \{\lambda \in \rho_5(A)$ such that all scalars near λ are in $\rho(A)\}$ with $\rho_5(A) := \{\lambda \in \Phi_A$ such that $i(\lambda - A) = 0\}$, $\sigma_{ap}(A) := \left\{\lambda \in \mathbb{C} \text{ such that } \inf_{\|x\|=1,\, x \in \mathcal{D}(A)} \|(\lambda - A)x\| = 0\right\}$ and $\sigma_\delta(A) := \{\lambda \in \mathbb{C}$ such that $\lambda - A$ is not surjective$\}$. $\sigma_{e1}(.)$ and $\sigma_{e2}(.)$ are the Gustafson and Weidman's essential spectra [144]. $\sigma_{e3}(.)$ is the Kato's essential spectrum [185]. $\sigma_{e4}(.)$ is the Wolf's essential spectrum [144, 299, 347]. $\sigma_{e5}(.)$ is the Schechter's essential spectrum [144, 299, 302] and $\sigma_{e6}(.)$ denotes the Browder's essential spectrum [144, 299]. $\sigma_{e7}(.)$ was introduced by Rakočević in [284] and designated the essential approximate point spectrum and $\sigma_{e8}(.)$ is the essential defect spectrum and was introduced by Schmoeger [304], $\sigma_{ap}(.)$ is the approximate point spectrum and $\sigma_{pa}(.)$ is the approximate point essential spectrum. Let us notice that all these sets are closed and, in general, we have $\sigma_{e1}(A) \bigcap \sigma_{e2}(A) = \sigma_{e3}(A) \subseteq \sigma_{e4}(A) \subseteq \sigma_{e5}(A) \subseteq \sigma_{e6}(A)$, $\sigma_{e5}(A) = \sigma_{e7}(A) \bigcup \sigma_{e8}(A)$, $\sigma_{e1}(A) \subset \sigma_{e7}(A)$ and $\sigma_{e2}(A) \subset \sigma_{e8}(A)$. However, if X is a Hilbert space and A is self-adjoint, then all these sets coincide.

Remark 7.1.1.

(i) If $\lambda \in \sigma_c(A)$ (the continuous spectrum of A), then $R(\lambda - A)$ is not closed (otherwise $\lambda \in \rho(A)$ see [302, Lemma 5.1 p. 179]). Therefore, $\lambda \in \sigma_{ei}(A)$, $i = 1, \ldots, 6$. Consequently, we have $\sigma_c(A) \subset \bigcap_{i=1}^{6} \sigma_{ei}(A)$. If the spectrum of A is purely continuous, then $\sigma(A) = \sigma_c(A) = \sigma_{ei}(A)$ $i = 1, \ldots, 6$.
(ii) $\sigma_{e5}(A + K) = \sigma_{e5}(A)$ for all $K \in \mathcal{K}(X)$.
(iii) Let E_0 be a core of A, i.e., a linear subspace of $\mathcal{D}(A)$ such that the closure $\overline{A_{|E_0}}$ of its restriction $A_{|E_0}$ equals A. Then, $\sigma_{ap}(A) := \left\{\lambda \in \mathbb{C} \text{ such that } \inf_{\|x\|=1,\, x \in E_0} \|(\lambda - A)x\| = 0\right\}$.
(iv) Set $\tilde{\alpha}(A) := \inf\{\|Ax\|$ such that $x \in \mathcal{D}(A)$ and $\|x\| = 1\}$. Whenever E_0 is a core of A, then $\tilde{\alpha}(A) := \inf\{\|Ax\|$ such that $x \in E_0$ and $\|x\| = 1\}$ holds. Using this quantity we obtain $\sigma_{ap}(A) = \{\lambda \in \mathbb{C}$ such that $\tilde{\alpha}(\lambda - A) = 0\}$. ◇

Proposition 7.1.1 ([302, Theorem 5.4, p. 180]). *Let X be a Banach space and let $A \in \mathcal{C}(X)$. Then, $\lambda \notin \sigma_{e5}(A)$ if, and only if, $\lambda \in \Phi_A^0$, where $\Phi_A^0 = \{\lambda \in \Phi_A$ such that $i(\lambda - A) = 0\}$.* ◇

It is well known that

$$\sigma_{e6}(A) = \sigma(A) \backslash \sigma_d(A), \tag{7.1.1}$$

where $\sigma_d(A)$ stands for the set of all isolated eigenvalues of A with a finite algebraic multiplicity.

Remark 7.1.2. Let $A \in \mathcal{L}(X)$. If λ is an isolated point of $\sigma(A)$, then either $\lambda \in \sigma_d(A)$ or $\lambda \in \sigma_{e5}(A)$ (it suffices to consider the spectral decomposition of A associated with $\{\lambda\}$ and $\sigma(A)\backslash\{\lambda\}$). ◇

Let $A \in \mathcal{L}(X)$. Let us notice that, by using Remark 7.1.2, we may write $\sigma_{e6}(A) = \sigma_{e4}(A) \bigcup d(\sigma(A))$, where d represents the Cantor–Bendixson derivative. The sets $\Phi^b(X)$ and $\Phi_0(X)$ may be written, respectively, in the following forms

$$\Phi^b(X) = \{A \in \mathcal{L}(X) : 0 \notin \sigma_{e4}(A)\} = \sigma_{e4}^{-1}(\{K \in K(\mathbb{C}) \text{ such that } 0 \notin K\}) \tag{7.1.2}$$

$$\Phi_0(X) = \{A \in \mathcal{L}(X) : 0 \notin \sigma_{e5}(A)\} = \sigma_{e5}^{-1}(\{K \in K(\mathbb{C}) \text{ such that } 0 \notin K\}). \tag{7.1.3}$$

Moreover,

$$\sigma_{e4}(A) = \bigcap_{k\in\mathbb{Z}} \{\lambda \in \mathbb{C} \text{ such that } \lambda - A \notin \Phi_k(X)\}, \tag{7.1.4}$$

where $\Phi_k(X) = \{A \in \Phi^b(X) \text{ such that } i(A) = k\}$. It is well known that the approximate point spectrum and $\mathcal{T}(X)$ introduced in (2.7.1) are connected in the following way $\sigma_{ap}(A) = \{\lambda \in \mathbb{C} \text{ such that } A \in \mathcal{T}(X)\}$. Let $A \in \mathcal{L}(X)$, we define

$$\begin{aligned}
\sigma_k(A) &:= \{\lambda \in \mathbb{C} \text{ such that } \lambda - A \text{ is not of Kato type}\},\\
\sigma_{se}(A) &:= \{\lambda \in \mathbb{C} \text{ such that } \lambda - A \text{ is not semi-regular}\},\\
\sigma_{es}(A) &:= \{\lambda \in \mathbb{C} \text{ such that } \lambda - A \text{ is not essentially semi-regular}\}.
\end{aligned}$$

$\sigma_k(.)$ is the Kato's spectrum, $\sigma_{se}(.)$ is the semi-regular spectrum, and $\sigma_{es}(.)$ is the essentially semi-regular spectrum. Let us notice that all these sets are closed and, in general, we have $\sigma_k(A) \subseteq \sigma_{es}(A) \subseteq \sigma_{e1}(A) \bigcap \sigma_{e2}(A) = \sigma_{e3}(A) \subseteq \sigma_{e4}(A) \subseteq \sigma_{e5}(A) \subseteq \sigma_{e6}(A)$, and $\sigma_k(A) \subseteq \sigma_{es}(A) \subseteq \sigma_{se}(A)$. If X be a separable Banach space, then by definition,

$$\begin{aligned}
\sigma_{ap}(A) &= \left\{\lambda \in \mathbb{C} \text{ such that } \inf_{\|x\|=1,\ x\in X} \|(\lambda - A)x\| = 0\right\}\\
&= \left\{\lambda \in \mathbb{C} \ :\ \text{for all } n \in \mathbb{N}, \text{ there is } x \in S_X \text{ such that } \|(\lambda - A)x\| < \frac{1}{n}\right\},
\end{aligned}$$

where S_X denote the unit sphere of X. Let $\mathcal{D}$ be a countable and dense subset of S_X (which exists by separability). Clearly, there exists $x \in S_X$ such that $\|(\lambda - A)x\| < 1/n$ if, and only if, there exists $x \in \mathcal{D}$ such that $\|(\lambda - A)x\| < 1/n$. This leads to

$$\sigma_{ap}(A) = \left\{\lambda \in \mathbb{C} \ : \ \text{for all } n \in \mathbb{N}, \text{ there is } x \in \mathcal{D} \text{ such that } \|(\lambda - A)x\| < \frac{1}{n}\right\}.$$

Following the notations of Lemma 2.7.2, we may write

$$\begin{aligned}\Delta_{\sigma_{ap}} &= \{(A, \lambda) \in \mathcal{L}(X) \times \mathbb{C} \text{ such that } \lambda \in \sigma_{ap}(A)\} \\ &= \Big\{(A, \lambda) \in \mathcal{L}(X) \times \mathbb{C} \ : \ \text{for all } n \in \mathbb{N}, \text{ there is } x \in \mathcal{D} \text{ such that} \\ &\qquad \|(\lambda - A)x\| < \frac{1}{n}\Big\} \\ &= \bigcap_{n\in\mathbb{N}} \bigcup_{x\in\mathcal{D}} A_{x,n},\end{aligned}$$

where $A_{x,n} = \{(A, \lambda) \in \mathcal{L}(X) \times \mathbb{C}$ such that $\|(\lambda - A)x\| < 1/n\}$. Let us recall that $\sigma(A) = \sigma_{ap}(A) \bigcup \sigma_{\text{com}}(A)$, where the compression spectrum of A is defined by $\sigma_{\text{com}}(A) := \{\lambda \in \mathbb{C}$ such that $(\lambda - A)X$ is not dense in $X\}$. So, $\Delta_\sigma = \Delta_{\sigma_{ap}} \bigcup C$, where

$$C := \{(A, \lambda) \in \mathcal{L}(X) \times \mathbb{C} \text{ such that } \lambda \in \sigma_{\text{com}}(A)\}. \tag{7.1.5}$$

Let us observe that, by definition,

$$\begin{aligned}C &= \{(A, \lambda) \in \mathcal{L}(X) \times \mathbb{C} \text{ such that } (\lambda - A)X \text{ is not dense in } X\} \\ &= \Big\{(A, \lambda) \in \mathcal{L}(X) \times \mathbb{C} \ : \ \exists n \in \mathbb{N} \text{ and } y \in X \text{ with } \|y - (A - \lambda)x\| \\ &\qquad > \frac{1}{n} \text{ for all } x \in X\Big\}.\end{aligned}$$

Let us recall that the compression spectrum of A may also be defined by

$$\sigma_{\text{com}}(A) = \{\lambda \in \mathbb{C} \text{ such that } \lambda - A \text{ is a right divisor of zero}\}.$$

Let $\mathcal{Z}(X)$ denote the set of right divisors of zero in $\mathcal{L}(X)$ and let $\mathcal{Z}^c(X) = \mathcal{L}(X)\backslash\mathcal{Z}(X)$. Clearly, we have $\mathcal{Z}^c(X) = \{A \in \mathcal{L}(X)$ such that $0 \notin \sigma_{\text{com}}(A)\}$. Let $\mathcal{T}^c(X)$, the complement of $\mathcal{T}(X)$ in $\mathcal{L}_s(X)$. Let us observe that

$$\mathcal{T}^c(X) = \{A \in \mathcal{L}(X) \text{ such that } 0 \notin \sigma_{ap}(A)\}. \tag{7.1.6}$$

Let $T \in \mathcal{L}(X)$ and set $\sigma^A(T) = \{\lambda \in \mathbb{C}$ such that $\lambda A - T$ is not invertible$\}$. In general $\sigma^A(T)$ is not compact but only closed. Let $q \in \mathbb{N}$, and set $\sigma_q^A(T) := \sigma^A(T) \bigcap B(0,q)$ (where $B(0,q) = \{\lambda \in \mathbb{C}$ such that $|\lambda| \leq q\}$). Hence, $T \longrightarrow \sigma_q^A(T)$ defines a map from $\mathcal{L}(X)$ into $K(\mathbb{C})$. Note that

$$\begin{aligned}
\Delta_{\sigma_q^A(T)} &= \Big\{(T,\lambda) \in \mathcal{L}(X) \times \mathbb{C} \text{ such that } \lambda \in \sigma_q^A(T)\Big\} \\
&= \Big\{(T,\lambda) \in \mathcal{L}(X) \times \mathbb{C} \, : \, |\lambda| \leq q \text{ and } \exists (x_k)_k \subset S_X \text{ with } \lim_{k\to\infty} \|(\lambda A - T)x_k\| = 0\Big\} \\
&\quad \bigcup \Big\{(T,\lambda) \in \mathcal{L}(X) \times \mathbb{C} \text{ such that } |\lambda| \leq q \text{ and } \overline{(\lambda A - T)(X)} \neq X\Big\} \\
&= \Big\{(T,\lambda) \in \mathcal{L}(X) \times B(0,q) \, : \, \forall r \in \mathbb{N}, \, \exists x \in S_X \text{ such that } \|(\lambda A - T)x\| < \frac{1}{r}\Big\} \\
&\quad \bigcup \Big\{(T,\lambda) \in \mathcal{L}(X) \times B(0,q) \, : \, \exists y \in X, \, \exists r \in \mathbb{N} \text{ such that } \forall x \in X, \\
&\qquad\qquad \|y - (\lambda A - T)x\| \geq \frac{1}{r}\Big\}.
\end{aligned}$$

Let us notice that

$$\begin{aligned}
\Omega_1 &:= \Big\{(T,\lambda) \in \mathcal{L}(X) \times B(0,q) \, : \, \forall r \in \mathbb{N}, \, \exists x \in S_X \text{ such that } \|(\lambda A - T)x\| < \frac{1}{r}\Big\} \\
&= \Big\{(T,\lambda) \in \mathcal{L}(X) \times \mathbb{C} \, : \, \forall r \in \mathbb{N}, \, \exists x \in S_X \text{ such that } \|(\lambda A - T)x\| < \frac{1}{r}\Big\} \bigcap \\
&\qquad (\mathcal{L}(X) \times B(0,q)) \\
&= \left(\bigcap_{r\in\mathbb{N}} \bigcup_{x\in X} O_{A,x,r}\right) \bigcap (\mathcal{L}(X) \times B(0,q)), \qquad (7.1.7)
\end{aligned}$$

where $O_{A,x,r} = \Big\{(T,\lambda) \in \mathcal{L}(X) \times \mathbb{C} \text{ such that } \|(\lambda A - T)x\| < \frac{1}{r}\Big\}$. Similarly,

$$\begin{aligned}
\Omega_2 &:= \Big\{(T,\lambda) \in \mathcal{L}(X) \times B(0,q) \, : \, \forall y \in \mathcal{D}, \, \exists r \in \mathbb{N} \text{ such that } \forall x \in X, \\
&\qquad\qquad \|y - (\lambda A - T)x\| \geq \frac{1}{r}\Big\} \\
&= \left(\bigcap_{y\in\mathcal{D}} \bigcup_{r\in\mathbb{N}} \bigcap_{x\in X} F_{A,x,y,r}\right) \bigcap (\mathcal{L}_s(X) \times B(0,q)), \qquad (7.1.8)
\end{aligned}$$

where $F_{A,x,y,r} = \{(T,\lambda) \in \mathcal{L}(X) \times \mathbb{C}$ such that $\|(\lambda A - T)x - y\| \geq \frac{1}{r}\}$ represents a closed subset of $\mathcal{L}_s(X) \times \mathbb{C}$ and $\mathcal{D}$ is a countable and dense subset of X.

7.2 The Jeribi Essential Spectrum

It is important to notice that the unexpected results obtained in this section are fundamental for the study of the essential spectra described in this book. In fact, these ideas have opened new research areas.

Definition 7.2.1. Let X be a Banach space and let $A \in \mathcal{C}(X)$. The Jeribi essential spectrum is defined by

$$\sigma_j(A) := \bigcap_{K \in \mathcal{W}_*(X)} \sigma(A+K),$$

where $\mathcal{W}_*(X)$ stands for each one of the sets $\mathcal{W}(X)$ and $\mathcal{S}(X)$. ◇

Remark 7.2.1.

(i) Since $\mathcal{K}(X) \subset \mathcal{W}_*(X)$, then

$$\sigma_j(A) \subset \sigma_{e5}(A). \tag{7.2.1}$$

(ii) In general, $\mathcal{K}(X)$ is strictly included in $\mathcal{W}_*(X)$.
(iii) Let us notice that, according to Theorem 1 in [277], we have $\mathcal{W}(L_1(\Omega, d\mu)) = \mathcal{S}(L_1(\Omega, d\mu))$, where (Ω, Σ, μ) be an arbitrary positive measure space. If $1 < p < \infty$, $L_p(\Omega, d\mu)$ is reflexive and then, $\mathcal{L}(L_p(\Omega, d\mu)) = \mathcal{W}(L_p(\Omega, d\mu))$. Moreover, from [124, Theorem 5.2] we deduce that $\mathcal{K}(L_p(\Omega, d\mu)) \subsetneq \mathcal{S}(L_p(\Omega, d\mu)) \subsetneq \mathcal{W}(L_p(\Omega, d\mu))$ with $p \neq 2$. For $p = 2$, we have $\mathcal{K}(L_2(\Omega, d\mu)) = \mathcal{S}(L_2(\Omega, d\mu)) = \mathcal{W}(L_2(\Omega, d\mu))$.
(iv) Let X be a reflexive Banach space. Then, $\mathcal{L}(X) = \mathcal{W}(X)$. So, the Jeribi essential spectrum is the smallest essential spectrum in the sense of the inclusion of the other essential spectra, already defined in the beginning of this chapter, in a particular reflexive Banach space.

◇

Open question: Is the inclusion (7.2.1) strict? □

Open question: Is there any relationship between the Jeribi essential spectrum and the other essential spectra, already defined in the beginning of this chapter? □

In this section, we can only give a partial answer to this question which is related to particular spaces. This partial answer has led to a new characterization which was entirely unexpected. That is why, we have decided to investigate this important and new characterization, hence leading to several studies dealing with the stability of the essential spectra. In fact, our motivation for this book was essentially based on this reasoning framework.

7.2.1 Relationship Between Jeribi and Schechter Essential Spectra on L_1-Spaces

In this section, we are concerned with the study of the essential spectrum on L_1-spaces. In particular, we prove the equality in the sense of the inclusion of the sets $\sigma_j(A)$ and $\sigma_{e5}(A)$, where A is a closed, densely defined, and linear operator. The analysis is essentially based on the results obtained in Chaps. 3, 5, and 6.

Theorem 7.2.1. *Let (Ω, Σ, μ) be an arbitrary positive measure space. Let A be a closed, densely defined, and linear operator on $L_1(\Omega, d\mu)$. Then, we have $\sigma_j(A) = \sigma_{e5}(A)$.* ◇

Proof. From $\mathcal{K}(L_1(\Omega, d\mu)) \subset \mathcal{W}_*(L_1(\Omega, d\mu))$, we infer that $\sigma_j(A) \subset \sigma_{e5}(A)$. In order to conclude, it suffices to show that $\sigma_{e5}(A) \subset \sigma_j(A)$. For this purpose, suppose that $\lambda \notin \sigma_j(A)$. Then, there exists $F \in \mathcal{W}_*(L_1(\Omega, d\mu))$ such that $\lambda \in \rho(A+F)$. This implies that $\lambda \in \Phi_{A+F}$ and $i(\lambda - A - F) = 0$. Since $F \in \mathcal{W}_*(L_1(\Omega, d\mu))$ we have $(\lambda - A - F)^{-1}F \in \mathcal{W}_*(L_1(\Omega, d\mu))$. Hence, by using Lemma 2.1.13(i), we get $\left[(\lambda - A - F)^{-1}F\right]^2 \in \mathcal{K}(L_1(\Omega, d\mu))$. Now, by representing $\lambda - A$ as $\lambda - A = (\lambda - A - F)[I + (\lambda - A - F)^{-1}F]$, and by using Theorem 2.2.40, together with Lemma 3.1.2, we obtain $\lambda \in \Phi_A$ and $i(\lambda - A) = 0$. Now Proposition 7.1.1 gives the wanted inclusion and achieves the proof. Q.E.D.

Remark 7.2.2.

(i) Theorem 7.2.1 provides a unified definition of the Schechter essential spectrum on L_1-spaces.
(ii) At first sight, $\sigma_j(A)$ and $\sigma_{e5}(A)$ seem to be not equal. However, in L_1-spaces, it was proved in Theorem 7.2.1 that $\sigma_j(A) = \sigma_{e5}(A)$. This result was entirely unexpected. That is why, we have decided to investigate this important and new characterization. All results concerning the essential spectra in this book are based on the proof of Theorem 7.2.1. ◇

7.2.2 Relationship Between Jeribi and Schechter Essential Spectra on Banach Space Satisfying the Dunford–Pettis Property

Corollary 7.2.1. *If X satisfies the Dunford–Pettis property, and if A is a closed, densely defined, and linear operator on X, then we have $\sigma_j(A) = \sigma_{e5}(A)$.* ◇

Proof. The proof can be checked in the same way as in the proof of Theorem 7.2.1. Q.E.D.

Remark 7.2.3. If X satisfies the Dunford–Pettis property, and if A is a closed, densely defined, and linear operator on X, then $\sigma_{e5}(A+K) = \sigma_{e5}(A)$ for all $K \in \mathcal{W}(X)$. ◇

7.2.3 Other Characterization of the Schechter Essential Spectrum by the Jeribi Essential Spectrum on L_p-Spaces

Let (Ω, Σ, μ) be an arbitrary positive measure space.

Theorem 7.2.2. *Let A be a closed, densely defined, and linear operator on $L_p(\Omega, d\mu)$, and let $p \in [1, \infty)$. In the case where $\mathcal{W}_*(L_p(\Omega, d\mu)) = \mathcal{S}(L_p(\Omega, d\mu))$, we have*

$$\sigma_{e5}(A) = \sigma_j(A).$$

◇

Proof. From $\mathcal{K}(L_p(\Omega, d\mu)) \subset \mathcal{S}(L_p(\Omega, d\mu))$, we infer that $\sigma_j(A) \subset \sigma_{e5}(A)$. In order to conclude, it suffices to show that $\sigma_{e5}(A) \subset \sigma_j(A)$. For this purpose, let us suppose that $\lambda \notin \sigma_j(A)$. Then, there exists $F \in \mathcal{S}(L_p(\Omega, d\mu))$ such that $\lambda \in \rho(A+F)$. This implies that $\lambda \in \Phi_{A+F}$ and $i(\lambda - A - F) = 0$. Since $\mathcal{S}(L_p(\Omega, d\mu))$ is a two-sided ideal of $L_p(\Omega, d\mu)$, we have $(\lambda - A - F)^{-1}F \in \mathcal{S}(L_p(\Omega, d\mu))$. Hence, by using Lemma 2.1.13(*ii*), we get $\left[(\lambda - A - F)^{-1}F\right]^2 \in \mathcal{K}(L_p(\Omega, d\mu))$. By using the equality $\lambda - A$ as $\lambda - A = (\lambda - A - F)[I + (\lambda - A - F)^{-1}F]$, together with Atkinson's theorem (Theorem 2.2.40), we get $\lambda \in \Phi_A$ and $i(\lambda - A) = 0$. Finally, the use of Proposition 7.1.1 gives the wanted inclusion and achieves the proof. Q.E.D.

Remark 7.2.4.

(i) Theorem 7.2.2 provides a unified definition of the Schechter essential spectrum on L_p-spaces, $p \in [1, \infty)$.
(ii) $\sigma_{e5}(A + K) = \sigma_{e5}(A)$ for all $K \in \mathcal{S}(L_p(\Omega, d\mu))$. ◇

7.3 Auxiliary Results

First, let us prove the following theorem.

Theorem 7.3.1. *Let $A \in \mathcal{C}(X)$ such that $\rho(A)$ is not empty.*

(i) If $\mathbb{C}\backslash\sigma_{e4}(A)$ is connected, then $\sigma_{e4}(A) = \sigma_{e5}(A)$, $\sigma_{e2}(A) = \sigma_{e8}(A)$, $\sigma_{e1l}(A) = \sigma_{e7l}(A)$, and $\sigma_{e2r}(A) = \sigma_{e8r}(A)$.
(ii) If $\mathbb{C}\backslash\sigma_{e5}(A)$ is connected, then $\sigma_{e5}(A) = \sigma_{e6}(A)$. ◇

Proof.

(*i*) Since the inclusion $\sigma_{e4}(A) \subset \sigma_{e5}(A)$ is known, it is sufficient to show that $\sigma_{e5}(A) \subset \sigma_{e4}(A)$, which is equivalent to $[\mathbb{C}\backslash\sigma_{e4}(A)]\bigcap\sigma_{e5}(A) = \emptyset$. Suppose that

$$[\mathbb{C}\backslash\sigma_{e4}(A)]\bigcap\sigma_{e5}(A) \neq \emptyset,$$

and let $\lambda_0 \in [\mathbb{C}\backslash\sigma_{e4}(A)]\bigcap\sigma_{e5}(A)$. Since $\rho(A) \neq \emptyset$, then there exists $\lambda_1 \in \mathbb{C}$ such that $\lambda_1 \in \rho(A)$ and consequently, $\lambda_1 - A \in \Phi(X)$ and $i(\lambda_1 - A) = 0$. Moreover, since $\mathbb{C}\backslash\sigma_{e4}(A)$ is connected, and from Proposition 2.2.5(*ii*), it follows that $i(\lambda - A)$ is constant on any component of Φ_A. Therefore, $i(\lambda_1 - A) = i(\lambda_0 - A) = 0$. In this way, we see that $\lambda_0 \notin \sigma_{e5}(A)$, which is a contradiction. This proves that $[\mathbb{C}\backslash\sigma_{e4}(A)]\bigcap\sigma_{e5}(A) = \emptyset$, and completes the proof of the first assertion of (*i*).

It is easy to check that $\sigma_{e1}(A) \subset \sigma_{e7}(A)$. For the second inclusion, we take $\mu \in \mathbb{C}\backslash\sigma_{e1}(A)$. Then, $\mu \in \Phi_{+A} = \Phi_A\bigcup(\Phi_{+A}\backslash\Phi_A)$. Hence, we will discuss these two cases:

First case If $\mu \in \Phi_A$, then $i(A-\mu) = 0$. Indeed, let $\mu_0 \in \rho(A)$. Then, $\mu_0 \in \Phi_A$ and $i(A - \mu_0) = 0$. From Proposition 2.2.5(*ii*), it follows that $i(A-\mu)$ is constant on any component of Φ_A, hence leading to $\rho(A) \subseteq \Phi_A$. Consequently, $i(A-\mu) = 0$ for all $\mu \in \Phi_A$. This shows that $\mu \in \mathbb{C}\backslash\sigma_{e7}(A)$.

Second case If $\mu \in (\Phi_{+A}\backslash\Phi_A)$, then $\alpha(A-\mu) < \infty$ and $\beta(A-\mu) = +\infty$. Consequently, $i(A-\mu) = -\infty < 0$. Hence, we obtain the second inclusion from the above two cases. By following the same reasoning, we get the third equality.

Since the inclusion $\sigma_{e1l}(A) \subset \sigma_{e7l}(A)$ [resp. $\sigma_{e2r}(A) \subset \sigma_{e8r}(A)$] is known, it suffices to show that $\sigma_{e7l}(A) \subset \sigma_{e1l}(A)$ [resp. $\sigma_{e8r}(A) \subset \sigma_{e2r}(A)$] which is equivalent to $[\mathbb{C}\backslash\sigma_{e1l}(A)]\bigcap\sigma_{e7l}(A) = \emptyset$ (resp. $[\mathbb{C}\backslash\sigma_{e2r}(A)]\bigcap\sigma_{e8r}(A) = \emptyset$). Let $\lambda_0 \in \mathbb{C}\backslash\sigma_{e1l}(A)$ [resp. $\mathbb{C}\backslash\sigma_{e2r}(A)$]. We discuss two cases.

First case If $\lambda_0 \in \Phi_{lA}\backslash\Phi_A$ (resp. $\Phi_{rA}\backslash\Phi_A$), then $i(A-\lambda_0) = -\infty < 0$ (resp. $+\infty > 0$). In this way we see that $\lambda_0 \notin \sigma_{e7l}(A)$ [resp. $\sigma_{e8r}(A)$].

Second case $\lambda_0 \in \Phi_A$. Since $\rho(A)$ is not empty, then there exists $\lambda_1 \in \mathbb{C}$ such that $\lambda_1 \in \rho(A)$ and consequently $A - \lambda_1 \in \Phi(X)$ and $i(A-\lambda_1) = 0$. Moreover, Φ_A is connected, it follows from Proposition 2.2.5 that $i(A-\lambda)$ is constant on any component of Φ_A. Therefore $i(A-\lambda_1) = i(A-\lambda_0)$. In this way we see that $\lambda_0 \notin \sigma_{e7l}(A)$ [resp. $\sigma_{e8r}(A)$].

(*ii*) Since the inclusion $\sigma_{e5}(A) \subset \sigma_{e6}(A)$ is known, it is sufficient to show that $\sigma_{e6}(A) \subset \sigma_{e5}(A)$ which is equivalent to $[\mathbb{C}\backslash\sigma_{e5}(A)]\bigcap\sigma_{e6}(A) = \emptyset$. In fact, let us suppose the following $[\mathbb{C}\backslash\sigma_{e5}(A)]\bigcap\sigma_{e6}(A) \neq \emptyset$. Then, there exists $\lambda_0 \in \mathbb{C}$, such that $\lambda_0 \in \mathbb{C}\backslash\sigma_{e5}(A)$ and $\lambda_0 \in \sigma_{e6}(A)$. Let $\lambda \in \mathbb{C}\backslash\sigma_{e5}(A)$. Since $\mathbb{C}\backslash\sigma_{e5}(A)$ is connected, and using Proposition 2.2.5(*iii*), it follows that $\alpha(\lambda - A) = \alpha(\lambda_0 - A)$ and $\beta(\lambda - A) = \beta(\lambda_0 - A)$. We claim that $\lambda \in \sigma_{e6}(A)$. In fact, if $\lambda \in \rho_6(A)$, then there exists a neighborhood V_λ of λ such that $V_\lambda \backslash \{\lambda\} \subset \rho(A)$. Since $\mathbb{C}\backslash\sigma_{e5}(A)$ is connected, there exists $\lambda_1 \in \mathbb{C}\backslash\sigma_{e5}(A)$ such that $\lambda_1 \in V_\lambda\backslash\{\lambda\} \subset \rho(A)$. Hence, $\alpha(\lambda_1 - A) = \beta(\lambda_1 - A) = 0$ and so, $\alpha(\lambda_0 - A) = \beta(\lambda_0 - A) = 0$. In this way, we see that $\lambda_0 \in \rho(A) \subset \rho_6(A)$, which is contrary to the properties of λ_0. This proves the claim and shows that $\mathbb{C}\backslash\sigma_{e5}(A) \subset \sigma_{e6}(A)$. Hence, $\emptyset \neq \rho(A) \subset \rho_5(A) \subset \sigma_{e6}(A) \subset \sigma(A)$ which is a contradiction. This shows that $[\mathbb{C}\backslash\sigma_{e5}(A)] \bigcap \sigma_{e6}(A) = \emptyset$ and completes the proof of the theorem. Q.E.D.

Remark 7.3.1. Let $A \in \mathcal{L}(X)$.

(i) If $\mathbb{C}\backslash\sigma_{e4}(A)$ is connected, then $\sigma_{e4}(A) = \sigma_{e5}(A)$, $\sigma_{e1}(A) = \sigma_{e7}(A)$, $\sigma_{e2}(A) = \sigma_{e8}(A)$, $\sigma_{e1l}(A) = \sigma_{e7l}(A)$, and $\sigma_{e2r}(A) = \sigma_{e8r}(A)$.
(ii) If $\mathbb{C}\backslash\sigma_{e5}(A)$ is connected, then $\sigma_{e5}(A) = \sigma_{e6}(A)$. ◇

Theorem 7.3.2. *Let $A \in \mathcal{C}(X)$. If $0 \in \rho(A)$, then for all $\lambda \in \mathbb{C}$, $\lambda \neq 0$ we have*

(i) $\lambda \in \sigma_{ei}(A)$ if, and only if, $\frac{1}{\lambda} \in \sigma_{ei}(A^{-1})$, for $i = 1, 2, 3, 4, 5, 7, 8, 1l, 2r, 7l, 8r$.
(ii) If $\mathbb{C}\backslash\sigma_{e5}(A)$ and $\mathbb{C}\backslash\sigma_{e5}(A^{-1})$ are connected, then $\lambda \in \sigma_{e6}(A)$ if, and only if, $\frac{1}{\lambda} \in \sigma_{e6}(A^{-1})$. ◇

Proof.

(i) Using Lemma 2.2.17, we deduce that $\lambda \in \sigma_{ei}(A)$ if, and only if, $\frac{1}{\lambda} \in \sigma_{ei}(A^{-1})$ for $i = 4, 5$. Moreover, for $\lambda \neq 0$, we can write $A - \lambda = -\lambda(A^{-1} - \lambda^{-1})A$. Since A is one-to-one and onto, then $\alpha(A - \lambda) = \alpha(A^{-1} - \lambda^{-1})$ and $R(A - \lambda) = R(A^{-1} - \lambda^{-1})$. This shows that $\lambda \in \Phi_{+A}$ (resp. Φ_{-A}) if, and only if, $\frac{1}{\lambda} \in \Phi_{+A^{-1}}$ (resp. $\Phi_{-A^{-1}}$) and we have $i(A - \lambda) = i(A^{-1} - \lambda^{-1})$. Therefore, we infer that $\lambda \in \sigma_{ei}(A)$ if, and only if, $\frac{1}{\lambda} \in \sigma_{ei}(A^{-1})$ for $i = 1, 2, 3, 4, 5, 7, 8, 1l, 2r, 7l, 8r$.
(ii) The sets $\mathbb{C}\backslash\sigma_{e5}(A)$ and $\mathbb{C}\backslash\sigma_{e5}(A^{-1})$ are connected. $\rho(A)$ and $\rho(A^{-1})$ are not empty sets. So, applying Theorem 7.3.1(ii), we deduce that $\lambda \in \sigma_{e6}(A)$ if, and only if, $\frac{1}{\lambda} \in \sigma_{e6}(A^{-1})$. Q.E.D.

Proposition 7.3.1.

(i) Let $A \in \mathcal{L}(X)$. If $0 \in \rho(A)$, then for all $\lambda \in \mathbb{C}$, $\lambda \neq 0$ we have $\lambda \notin \sigma_k(A)$ if, and only if, $\lambda^{-1} \notin \sigma_k(A^{-1})$.
(ii) Let $A \in \mathcal{L}(X)$ and let $\lambda \in \rho(A)$. Then

$$\mu \in \sigma_{se}(A) \text{ if, and only if, } \mu \neq \lambda \text{ and } (\mu - \lambda)^{-1} \in \sigma_{se}((\lambda - A)^{-1}),$$

$$\mu \in \sigma_{es}(A) \text{ if, and only if, } \mu \neq \lambda \text{ and } (\mu - \lambda)^{-1} \in \sigma_{es}((\lambda - A)^{-1}).$$

◇

Proof.

(i) Let $0 \in \rho(A)$. The resolvent identity implies that

$$\lambda - A = -\lambda(A^{-1} - \lambda^{-1})A. \tag{7.3.1}$$

If $\lambda \notin \sigma_k(A)$, then there exists a pair of closed and $(\lambda - A)$-invariant subspaces (M, N) of X such that $(\lambda - A)_M$ is semi-regular and $(\lambda - A)_N$ is nilpotent. Hence $((A^{-1} - \lambda^{-1})A)_M$ is semi-regular and $((A^{-1} - \lambda^{-1})A)_N$ is nilpotent. This shows that $(A^{-1} - \lambda^{-1})_M$ is semi-regular and $(A^{-1} - \lambda^{-1})_N$ is nilpotent. Conversely, if $\lambda^{-1} \notin \sigma_k(A^{-1})$, then $A^{-1} - \lambda^{-1}$ is of Kato type commuting with A (which is invertible). From Eq. (7.3.1), it follows that $\lambda - A$ is of Kato type.

(*ii*) Let us start from the identity $(\lambda - A)^{-1} - (\mu - \lambda)^{-1} = -(\mu - \lambda)^{-1}(\mu - A)(\lambda - A)^{-1}$. Since $(\lambda - A)^{-1}$ is a bounded invertible operator commuting with A, it follows, from both Theorems 2.2.31 and 2.2.33, that $(\lambda - A)^{-1} - (\mu - \lambda)^{-1}$ is semi-regular if, and only if, $(\mu - A)$ is semi-regular. Q.E.D.

Proposition 7.3.2. *Let $A \in \mathcal{C}(X)$. Then,*

(i) $\lambda \notin \sigma_{e7}(A)$ *if, and only if,* $\lambda - A \in \Phi_+(X)$ *and* $i(\lambda - A) \leq 0$.
(ii) $\lambda \notin \sigma_{e8}(A)$ *if, and only if,* $\lambda - A \in \Phi_-(X)$ *and* $i(\lambda - A) \geq 0$.
(iii) If A is a bounded linear operator, then $\sigma_{e8}(A) = \sigma_{e7}(A^*)$, *where A^* stands for the adjoint operator.* ◇

Proof.

(*i*) Let $\lambda \in \Phi_{+A}$ such that $i(\lambda - A) \leq 0$. Then, by using Lemma 2.2.16, $\lambda - A$ can be expressed in the form $\lambda - A = U + K$, where $K \in \mathcal{K}(X)$ and $U \in \mathcal{C}(X)$ is an operator with a closed range and $\alpha(U) = 0$. Hence, by using Theorem 2.2.1, there exists a constant $c > 0$ such that $\|Ux\| \geq c\|x\|$, for all $x \in \mathcal{D}(A)$. Thus, $\lambda \notin \sigma_{ap}(A + K)$ and therefore, $\lambda \notin \sigma_{e7}(A)$. Conversely, if $\lambda \notin \sigma_{e7}(A)$, then there exists $K \in \mathcal{K}(X)$, such that $\inf_{\|x\|=1,\, x \in \mathcal{D}(A)} \|(\lambda - A - K)x\| > 0$. The use of Theorem 2.2.1 leads to $\lambda - A - K \in \Phi_+(X)$ and $\alpha(\lambda - A - K) = 0$. Hence, and from Lemma 6.3.1, it follows that $\lambda - A \in \Phi_+(X)$ and $i(\lambda - A) \leq 0$. This completes the proof of (*i*).

The proof of (*ii*) is a straightforward adoption of the proof of (*i*).

(*iii*) This assertion is immediately deduced from (*i*) and (*ii*). Q.E.D.

Proposition 7.3.3. *Let $A \in \mathcal{C}(X)$. Then,*

(i) $\sigma_c(A) \subset \sigma_{e7}(A) \bigcap \sigma_{e8}(A)$, *and*
(ii) $\sigma_r(A) \subset \sigma_{e8}(A)$. ◇

Proof.

(*i*) Let $\lambda \in \sigma_c(A)$. Then, $R(\lambda - A)$ is not closed, otherwise $\lambda \in \rho(A)$. Thus, by using Proposition 7.3.2, $\lambda \in \sigma_{e7}(A) \bigcap \sigma_{e8}(A)$, which proves (*i*).

(*ii*) Let us consider $\lambda \in \sigma_r(A)$. Then, $\beta(\lambda - A) \neq 0$ and hence, $i(\lambda - A) < 0$, since $\lambda - A$ is one to one. This implies, by the use of Proposition 7.3.2(*ii*), that $\lambda \in \sigma_{e8}(A)$. Q.E.D.

7.4 Essential Spectra of the Sum of Two Bounded Linear Operators

7.4.1 By Means of Fredholm and Semi-Fredholm Perturbations

The following theorem shows the relation between the essential spectra of the sum of the two bounded linear operators and the essential spectra of each of these operators, where their products are Fredholm or semi-Fredholm perturbations on a Banach space X.

Theorem 7.4.1. *Let A and B be two bounded linear operators on a Banach space X.*

(i) If $AB \in \mathcal{F}^b(X)$, then $\sigma_{ei}(A+B)\backslash\{0\} \subset [\sigma_{ei}(A)\bigcup\sigma_{ei}(B)]\backslash\{0\}$, $i = 4, 5$. Furthermore, if $BA \in \mathcal{F}^b(X)$, then $\sigma_{e4}(A+B)\backslash\{0\} = [\sigma_{e4}(A)\bigcup\sigma_{e4}(B)]\backslash\{0\}$. Moreover, if $\mathbb{C}\backslash\sigma_{e4}(A)$ is connected, then

$$\sigma_{e5}(A+B)\backslash\{0\} = \Big[\sigma_{e5}(A)\bigcup\sigma_{e5}(B)\Big]\backslash\{0\}. \tag{7.4.1}$$

(ii) If the hypotheses of (i) are satisfied, and if $\mathbb{C}\backslash\sigma_{e5}(A+B)$, $\mathbb{C}\backslash\sigma_{e5}(A)$ and $\mathbb{C}\backslash\sigma_{e5}(B)$ are connected, then $\sigma_{e6}(A+B)\backslash\{0\} = [\sigma_{e6}(A)\bigcup\sigma_{e6}(B)]\backslash\{0\}$.

(iii) If $AB \in \mathcal{F}_+^b(X)$, then $\sigma_{ei}(A+B)\backslash\{0\} \subset [\sigma_{ei}(A)\bigcup\sigma_{ei}(B)]\backslash\{0\}$, $i = 1, 7$. Besides, if $BA \in \mathcal{F}_+^b(X)$, then

$$\sigma_{e1}(A+B)\backslash\{0\} = \Big[\sigma_{e1}(A)\bigcup\sigma_{e1}(B)\Big]\backslash\{0\}. \tag{7.4.2}$$

Moreover, if $\mathbb{C}\backslash\sigma_{e4}(A)$ is connected, then

$$\sigma_{e7}(A+B)\backslash\{0\} = \Big[\sigma_{e7}(A)\bigcup\sigma_{e7}(B)\Big]\backslash\{0\}. \tag{7.4.3}$$

(iv) If $AB \in \mathcal{F}_-^b(X)$, then $\sigma_{ei}(A+B)\backslash\{0\} \subset [\sigma_{ei}(A)\bigcup\sigma_{ei}(B)]\backslash\{0\}$, $i = 2, 8$. If, further, $BA \in \mathcal{F}_-^b(X)$, then

$$\sigma_{e2}(A+B)\backslash\{0\} = \Big[\sigma_{e2}(A)\bigcup\sigma_{e2}(B)\Big]\backslash\{0\}. \tag{7.4.4}$$

Moreover, if $\mathbb{C}\backslash\sigma_{e4}(A^*)$ is connected, then

$$\sigma_{e8}(A+B)\backslash\{0\} = \Big[\sigma_{e8}(A)\bigcup\sigma_{e8}(B)\Big]\backslash\{0\}. \tag{7.4.5}$$

(v) If $AB \in \mathcal{F}_+^b(X) \bigcap \mathcal{F}_-^b(X)$, *then*

$$\sigma_{e3}(A+B)\backslash\{0\} \subset \Big[\Big(\sigma_{e3}(A)\bigcup\sigma_{e3}(B)\Big)\bigcup\Big(\sigma_{e1}(A)\bigcap\sigma_{e2}(B)\Big)\bigcup\Big(\sigma_{e2}(A)\bigcap\sigma_{e1}(B)\Big)\Big]\backslash\{0\}.$$

Besides, if $BA \in \mathcal{F}_+^b(X) \bigcap \mathcal{F}_-^b(X)$, *then*

$$\sigma_{e3}(A+B)\backslash\{0\} = \Big[\Big(\sigma_{e3}(A)\bigcup\sigma_{e3}(B)\Big)\bigcup\Big(\sigma_{e1}(A)\bigcap\sigma_{e2}(B)\Big)\bigcup\Big(\sigma_{e2}(A)\bigcap\sigma_{e1}(B)\Big)\Big]\backslash\{0\}.$$

(vi) If $AB \in \mathcal{F}_e(X)$, *then we have* $[\sigma_{es}(A)\bigcup\sigma_{es}(B)]\setminus\{0\} \subset \sigma_{es}(A+B)\setminus\{0\}$, *and* $[\sigma_{se}(A)\bigcup\sigma_{se}(B)]\setminus\{0\} \subset \sigma_{se}(A+B)\setminus\{0\}$.

(vii) Let $A, B, C, D \in \mathcal{L}(X)$ *be mutually commuting operators such that* $AC + BD = I$. *If* $AB \in \mathcal{F}_e(X)$, *then we have* $[\sigma_{es}(A)\bigcup\sigma_{es}(B)]\setminus\{0\} = \sigma_{es}(A+B)\setminus\{0\}$, *and* $[\sigma_{se}(A)\bigcup\sigma_{se}(B)]\setminus\{0\} = \sigma_{se}(A+B)\setminus\{0\}$. ◇

Proof. For $\lambda \in \mathbb{C}$, we can write

$$(A-\lambda)(B-\lambda) = AB - \lambda(A+B-\lambda), \tag{7.4.6}$$

and

$$(B-\lambda)(A-\lambda) = BA - \lambda(A+B-\lambda). \tag{7.4.7}$$

(i) Let $\lambda \notin \sigma_{e4}(A)\bigcup\sigma_{e4}(B)\bigcup\{0\}$. Then, $(A-\lambda) \in \Phi^b(X)$ and $(B-\lambda) \in \Phi^b(X)$. Theorem 2.2.40 ensures that $(A-\lambda)(B-\lambda) \in \Phi^b(X)$. Since $AB \in \mathcal{F}^b(X)$, and applying Eq. (7.4.6), we have $(A+B-\lambda) \in \Phi^b(X)$. Hence, $\lambda \notin \sigma_{e4}(A+B)$, and we obtain

$$\sigma_{e4}(A+B)\backslash\{0\} \subset \Big[\sigma_{e4}(A)\bigcup\sigma_{e4}(B)\Big]\backslash\{0\}. \tag{7.4.8}$$

Let $\lambda \notin \sigma_{e5}(A)\bigcup\sigma_{e5}(B)\bigcup\{0\}$. Then, by using Proposition 7.1.1, we get $(A-\lambda) \in \Phi^b(X)$, $i(A-\lambda) = 0$, $(B-\lambda) \in \Phi^b(X)$ and $i(B-\lambda) = 0$ and therefore, Theorem 2.2.40 gives $(A-\lambda)(B-\lambda) \in \Phi^b(X)$ and $i((A-\lambda)(B-\lambda)) = 0$. Moreover, since $AB \in \mathcal{F}^b(X)$, we can apply both Eq. (7.4.6) and Lemma 6.3.1(i), hence ensuring that $(A+B-\lambda) \in \Phi^b(X)$ and $i(A+B-\lambda) = 0$. Again, by applying Proposition 7.1.1, we infer that $\lambda \notin \sigma_{e5}(A+B)$ and, then

$$\sigma_{e5}(A+B)\backslash\{0\} \subset \Big[\sigma_{e5}(A)\bigcup\sigma_{e5}(B)\Big]\backslash\{0\}. \tag{7.4.9}$$

In order to prove the inverse inclusions of Eqs. (7.4.8) and (7.4.9), let us suppose that $\lambda \notin \sigma_{e4}(A+B)\bigcup\{0\}$. Then, $(A+B-\lambda) \in \Phi^b(X)$. Since $AB \in \mathcal{F}^b(X)$ and $BA \in \mathcal{F}^b(X)$, then by using Eqs. (7.4.6) and (7.4.7), we have

$$(A-\lambda)(B-\lambda) \in \Phi^b(X) \text{ and } (B-\lambda)(A-\lambda) \in \Phi^b(X). \tag{7.4.10}$$

Equation (7.4.10) and Theorem 2.2.19 show clearly that $(A-\lambda) \in \Phi^b(X)$ and $(B-\lambda) \in \Phi^b(X)$. Therefore, $\lambda \notin \sigma_{e4}(A)\bigcup\sigma_{e4}(B)$. This proves that $[\sigma_{e4}(A)\bigcup\sigma_{e4}(B)]\backslash\{0\} \subset \sigma_{e4}(A+B)\backslash\{0\}$. Hence, $\sigma_{e4}(A+B)\backslash\{0\} = [\sigma_{e4}(A)\bigcup\sigma_{e4}(B)]\backslash\{0\}$. It remains to prove the following $[\sigma_{e5}(A)\bigcup\sigma_{e5}(B)]\backslash\{0\} \subset \sigma_{e5}(A+B)\backslash\{0\}$. Let $\lambda \notin \sigma_{e5}(A+B)\bigcup\{0\}$. Then, by using Proposition 7.1.1, we have $(A+B-\lambda) \in \Phi^b(X)$ and $i(A+B-\lambda) = 0$. Since $AB \in \mathcal{F}^b(X)$ and $BA \in \mathcal{F}^b(X)$, it is easy to deduce that $(A-\lambda) \in \Phi^b(X)$ and $(B-\lambda) \in \Phi^b(X)$. Again, the use of Eqs. (7.4.6), (7.4.10), Theorem 2.2.40 and Lemma 6.3.1(i) allows us to have

$$i[(A-\lambda)(B-\lambda)] = i(A-\lambda)+i(B-\lambda) = i(A+B-\lambda) = 0. \tag{7.4.11}$$

Since A is a bounded linear operator, we get $\rho(A) \neq \emptyset$. As $\mathbb{C}\backslash\sigma_{e4}(A)$ is a connected set, and from Remark 7.3.1, we deduce that $\sigma_{e4}(A) = \sigma_{e5}(A)$. Using the last equality and the fact that $(A-\lambda) \in \Phi^b(X)$, we deduce that $i(A-\lambda) = 0$. It follows, from Eq. (7.4.11), that $i(B-\lambda) = 0$. We conclude that $\lambda \notin \sigma_{e5}(A)\bigcup\sigma_{e5}(B)$ and hence, we have $[\sigma_{e5}(A)\bigcup\sigma_{e5}(B)]\backslash\{0\} \subset \sigma_{e5}(A+B)\backslash\{0\}$. So, we prove Eq. (7.4.1).

(ii) The sets $\mathbb{C}\backslash\sigma_{e5}(A+B)$, $\mathbb{C}\backslash\sigma_{e5}(A)$ and $\mathbb{C}\backslash\sigma_{e5}(B)$ are connected. Since A and B are bounded operators, we deduce that $\rho(A)$, $\rho(B)$ and $\rho(A+B)$ are not empty sets. So, using Theorem 7.3.1, we show that $\sigma_{e5}(A+B) = \sigma_{e6}(A+B)$, $\sigma_{e5}(A) = \sigma_{e6}(A)$ and $\sigma_{e5}(B) = \sigma_{e6}(B)$. Hence, Eq. (7.4.1) gives $\sigma_{e6}(A+B)\backslash\{0\} = [\sigma_{e6}(A)\bigcup\sigma_{e6}(B)]\backslash\{0\}$.

(iii) Suppose that $\lambda \notin \sigma_{e1}(A)\bigcup\sigma_{e1}(B)\bigcup\{0\}$. Then, $(A-\lambda) \in \Phi_+^b(X)$ and $(B-\lambda) \in \Phi_+^b(X)$. Using Theorem 2.2.13$(ii)$, we have $(A-\lambda)(B-\lambda) \in \Phi_+^b(X)$. Since $AB \in \mathcal{F}_+^b(X)$, we can apply both Eq. (7.4.6) and Lemma 6.3.1(ii), in order to get $(A+B-\lambda) \in \Phi_+^b(X)$. So, $\lambda \notin \sigma_{e1}(A+B)$. Therefore,

$$\sigma_{e1}(A+B)\backslash\{0\} \subset \left[\sigma_{e1}(A)\bigcup\sigma_{e1}(B)\right]\backslash\{0\}. \tag{7.4.12}$$

Now, suppose that $\lambda \notin \sigma_{e7}(A)\bigcup\sigma_{e7}(B)\bigcup\{0\}$. Then, by using Proposition 7.3.2(i), we have $(A-\lambda) \in \Phi_+^b(X)$, $i(A-\lambda) \leq 0$, $(B-\lambda) \in \Phi_+^b(X)$ and $i(B-\lambda) \leq 0$. Using Theorems 2.2.7 and 2.2.13(ii), we obtain $(A-\lambda)(B-\lambda) \in \Phi_+^b(X)$ and $i[(A-\lambda)(B-\lambda)] \leq 0$. Since $AB \in \mathcal{F}_+^b(X)$, and applying Eq. (7.4.6) and Lemma 6.3.1(ii), we deduce that $(A+B-\lambda) \in \Phi_+^b(X)$ and $i(A+B-\lambda) \leq 0$. Again, the use of Proposition 7.3.2(i) clearly shows that $\lambda \notin \sigma_{e7}(A+B)$. Hence,

$$\sigma_{e7}(A+B)\backslash\{0\} \subset \Big[\sigma_{e7}(A)\bigcup\sigma_{e7}(B)\Big]\backslash\{0\}. \tag{7.4.13}$$

Now, it remains to prove the inverse inclusions of Eqs. (7.4.12) and (7.4.13). For this purpose, let us suppose that $\lambda \notin \sigma_{e1}(A+B)\bigcup\{0\}$. Then, $(A+B-\lambda) \in \Phi^b_+(X)$. Since $AB \in \mathcal{F}^b_+(X)$ and $BA \in \mathcal{F}^b_+(X)$, then by using Eqs. (7.4.6), (7.4.7) and Lemma 6.3.1(ii), we get

$$(A-\lambda)(B-\lambda) \in \Phi^b_+(X),\ (B-\lambda)(A-\lambda) \in \Phi^b_+(X). \tag{7.4.14}$$

Combining Eq. (7.4.14) and Theorem 2.2.14(i), we conclude that $(A-\lambda) \in \Phi^b_+(X)$ and $(B-\lambda) \in \Phi^b_+(X)$. Hence, $\lambda \notin \sigma_{e1}(A)\bigcup\sigma_{e1}(B)$. Therefore, $[\sigma_{e1}(A)\bigcup\sigma_{e1}(B)]\backslash\{0\} \subset \sigma_{e1}(A+B)\backslash\{0\}$. This proves Eq. (7.4.2). Now, it remains to prove that $[\sigma_{e7}(A)\bigcup\sigma_{e7}(B)]\backslash\{0\} \subset \sigma_{e7}(A+B)\backslash\{0\}$. In order to achieve this, let $\lambda \notin \sigma_{e7}(A+B)\bigcup\{0\}$. Then, by using Proposition 7.3.2(i), we deduce that $(A+B-\lambda) \in \Phi^b_+(X)$ and $i(A+B-\lambda) \leq 0$. Since $AB \in \mathcal{F}^b_+(X)$ and $BA \in \mathcal{F}^b_+(X)$, a similar reasoning as before leads to $(A-\lambda) \in \Phi^b_+(X)$ and $(B-\lambda) \in \Phi^b_+(X)$. Again, using Eqs. (7.4.6), (7.4.14) and Lemma 6.3.1(ii), we have the following $i[(A-\lambda)(B-\lambda)] = i(A+B-\lambda) \leq 0$. Let $\lambda_0 \in \rho(A)$. Then, $(A-\lambda_0) \in \Phi^b(X)$ and $i(A-\lambda_0) = 0$. Since $\rho(A) \subset \mathbb{C}\backslash\sigma_{e4}(A)$, then $\lambda_0 \in \mathbb{C}\backslash\sigma_{e4}(A)$ which is connected. By using Proposition 2.2.5(ii), we have $i(A-\lambda)$ is constant on any component of Φ_A. Hence, $i(A-\lambda) = 0$, for all $\lambda \in \mathbb{C}\backslash\sigma_{e4}(A)$. By applying Theorem 2.2.7, it is clear that $i[(A-\lambda)(B-\lambda)] = i[(A-\lambda)] + i[(B-\lambda)] \leq 0$, which leads to $i[(B-\lambda)] \leq 0$. According to Proposition 7.3.2(i), we conclude that $\lambda \notin \sigma_{e7}(A)\bigcup\sigma_{e7}(B)$. Then, $[\sigma_{e7}(A)\bigcup\sigma_{e7}(B)]\backslash\{0\} \subset \sigma_{e7}(A+B)\backslash\{0\}$, hence leading to Eq. (7.4.3).

(iv) The proof of Eq. (7.4.4) may be achieved in the same way as in the proof of (iii) for Eq. (7.4.2). Now, let us prove this equality $\sigma_{e8}(A+B)\backslash\{0\} = [\sigma_{e8}(A)\bigcup\sigma_{e8}(B)]\backslash\{0\}$. Let A and B be two bounded operators. Applying Proposition 7.3.2(iii), we have $\sigma_{e8}(A) = \sigma_{e7}(A^*)$, $\sigma_{e8}(B) = \sigma_{e7}(B^*)$ and $\sigma_{e8}(A+B) = \sigma_{e7}(A^*+B^*)$. Applying ($iii$) for Eq. (7.4.3), we get $\sigma_{e7}(A^*+B^*)\backslash\{0\} = [\sigma_{e7}(A^*)\bigcup\sigma_{e7}(B^*)]\backslash\{0\}$. Therefore, we prove Eq. (7.4.5).

(v) Since the following equalities $\sigma_{e3}(A) = \sigma_{e1}(A)\bigcap\sigma_{e2}(A)$, $\sigma_{e3}(B) = \sigma_{e1}(B)\bigcap\sigma_{e2}(B)$ and $\sigma_{e3}(A+B) = \sigma_{e1}(A+B)\bigcap\sigma_{e2}(A+B)$ are known, $AB \in \mathcal{F}^b_+(X)\bigcap\mathcal{F}^b_-(X)$ and $BA \in \mathcal{F}^b_+(X)\bigcap\mathcal{F}^b_-(X)$, then by using Eqs. (7.4.2) and (7.4.4), we deduce that

$$\sigma_{e3}(A+B)\backslash\{0\} = \Big[\Big(\sigma_{e3}(A)\bigcup\sigma_{e3}(B)\Big)\bigcup\Big(\sigma_{e1}(A)\bigcap\sigma_{e2}(B)\Big) \bigcup\Big(\sigma_{e2}(A)\bigcap\sigma_{e1}(B)\Big)\Big]\backslash\{0\}.$$

(*vi*) If $\lambda \notin \sigma_{es}(A+B) \setminus \{0\}$, then $A+B-\lambda$ is essentially semi-regular. Since $AB \in \mathcal{F}_e(X)$, then using Eq. (7.4.6) $(A-\lambda)(B-\lambda)$ is essentially semi-regular. From Theorem 2.2.31, it follows that $(A-\lambda)$ and $(B-\lambda)$ are both essentially semi-regular operators. Then, $\lambda \notin [\sigma_{se}(A) \bigcup \sigma_{se}(B)] \setminus \{0\}$. For the case of semi-regular operators, we may use the same proof.

(*vii*) The proof is ensured by using Theorem 2.2.32. Q.E.D.

Remark 7.4.1. It is easy to see that the condition "$\setminus(\{0\})$" cannot be dropped in Theorem 7.4.1. Indeed, let H be a Hilbert space of infinite dimension, and let $A = diag\{I, 0\}$, $B = diag\{0, I\}$ be diagonal operators in the space $H \oplus H$. Then $\sigma_{e4}(A) = \sigma_{e4}(B) = \{0, 1\}$ while $\sigma_{e4}(A+B) = \{1\}$.

Recall that an operator $T \in \mathcal{L}(X)$ is called a left (*resp.* right) divisor of zero if $TS = 0$ (resp. $ST = 0$) for some nonzero operators $S \in \mathcal{L}(X)$.

Proposition 7.4.1. *Let $T \in \mathcal{L}(X)$ be a left (resp. right) divisor of zero, i.e., $TS = 0$ (resp. $ST = 0$) for $S \in \mathcal{L}(X)$. Then,*

$$\left[\sigma_{es}(T) \bigcup \sigma_{es}(S)\right] \setminus \{0\} = \sigma_{es}(T+S) \setminus \{0\},$$

$$\left[\sigma_{se}(T) \bigcup \sigma_{se}(S)\right] \setminus \{0\} = \sigma_{se}(T+S) \setminus \{0\},$$

$$\left[\sigma_{k}(T) \bigcup \sigma_{k}(S)\right] \setminus \{0\} = \sigma_{k}(T+S) \setminus \{0\},$$

$$\left[\sigma_{e4}(T) \bigcup \sigma_{e4}(S)\right] \setminus \{0\} = \sigma_{e4}(T+S) \setminus \{0\},$$

and $\left(\left[\sigma_{se}(T) \setminus \sigma_{e4}(S)\right] \bigcup \left[\sigma_{se}(S) \setminus \sigma_{e4}(T)\right]\right) \setminus \{0\}$ *is, at most, countable.* ◇

We recall the following lemma (see [359, Theorem 7 and 8] and [358, Corollary 2]).

Lemma 7.4.1. *Let $A \in \mathcal{L}(X)$ and let $E \in \mathcal{R}(X)$.*

(*i*) *If $A \in \Phi^b_+(X)$ and $AE - EA \in \mathcal{F}^b_+(X)$, then $A + E \in \Phi^b_+(X)$ and $i(A+E) = i(A)$.*

(*ii*) *If $A \in \Phi^b_-(X)$ and $AE - EA \in \mathcal{F}^b_-(X)$, then $A + E \in \Phi^b_-(X)$ and $i(A+E) = i(A)$.*

(*iii*) *If $A \in \Phi^b_l(X)$ [resp. $\mathcal{W}_l(X)$, $\Phi^b_r(X)$, $\mathcal{W}_r(X)$] and $AE - EA \in \mathcal{F}^b(X)$, then $A + E \in \Phi^b_l(X)$ [resp. $\mathcal{W}_l(X)$, $\Phi^b_r(X)$, $\mathcal{W}_r(X)$].* ◇

Remark 7.4.2. Let X be a Banach space. It is easy to see that $\Phi^b(X) = \Phi^b_l(X) \bigcap \Phi^b_r(X)$ and $\Phi_0(X) = \mathcal{W}_l(X) \bigcap \mathcal{W}_r(X) \bigcap \mathcal{L}(X)$. So, it follows, immediately, from Lemma 7.4.1 that, if $A \in \Phi^b(X)$ [*resp.* $\Phi_0(X)$] and $AE - EA \in \mathcal{F}^b(X)$, then $A + E \in \Phi^b(X)$ [*resp.* $\Phi_0(X)$]. ◇

Lemma 7.4.2 ([17, Theorem 1.54, p. 32]). *Let $A \in \mathcal{L}(X)$ and let $B \in \mathcal{L}(X)$.*

(*i*) *If $A \in \Phi^b_l(X)$ and $B \in \Phi^b_l(X)$, then $BA \in \Phi^b_l(X)$.*

(*ii*) *If $A \in \Phi^b_r(X)$ and $B \in \Phi^b_r(X)$, then $BA \in \Phi^b_r(X)$.*

(*iii*) *If $BA \in \Phi^b_l(X)$, then $A \in \Phi^b_l(X)$.*

(*iv*) *If $BA \in \Phi^b_r(X)$, then $B \in \Phi^b_r(X)$.* ◇

Theorem 7.4.2. *Let A and B be two bounded linear operators on a Banach space X such that AB is a Riesz operator.*

(i) If $AB - BA \in \mathcal{F}_+^b(X)$, then

$$\sigma_{e1}(A+B)\backslash\{0\} = \left[\sigma_{e1}(A)\bigcup\sigma_{e1}(B)\right]\backslash\{0\} \tag{7.4.15}$$

and $\sigma_{e7}(A+B)\backslash\{0\} \subset [\sigma_{e7}(A)\bigcup\sigma_{e7}(B)]\backslash\{0\}$. If, further, Φ_A and Φ_B are connected, then

$$\sigma_{e7}(A+B)\backslash\{0\} = \left[\sigma_{e7}(A)\bigcup\sigma_{e7}(B)\right]\backslash\{0\}.$$

(ii) If $AB - BA \in \mathcal{F}_-^b(X)$, then $\sigma_{e2}(A+B)\backslash\{0\} = [\sigma_{e2}(A)\bigcup\sigma_{e2}(B)]\backslash\{0\}$ and

$$\sigma_{e8}(A+B)\backslash\{0\} \subset \left[\sigma_{e8}(A)\bigcup\sigma_{e8}(B)\right]\backslash\{0\}.$$

Moreover, if Φ_A and Φ_B are connected, then $\sigma_{e8}(A+B)\backslash\{0\} = [\sigma_{e8}(A)\bigcup\sigma_{e8}(B)]\backslash\{0\}$.

(iii) If $AB - BA \in \mathcal{F}_+^b(X)\bigcap\mathcal{F}_-^b(X)$, then

$$\sigma_{e3}(A+B)\backslash\{0\} = \Big[\Big(\sigma_{e3}(A)\bigcup\sigma_{e3}(B)\Big)\bigcup\Big(\sigma_{e1}(A)\bigcap\sigma_{e2}(B)\Big) \bigcup\Big(\sigma_{e2}(A)\bigcap\sigma_{e1}(B)\Big)\Big]\backslash\{0\}.$$

(iv) If $AB - BA \in \mathcal{F}^b(X)$, then $\sigma_{ei}(A+B)\backslash\{0\} = [\sigma_{ei}(A)\bigcup\sigma_{ei}(B)]\backslash\{0\}$, $i = 1l,\ 2r,\ 4$. and $\sigma_{ei}(A+B)\backslash\{0\} \subset [\sigma_{ei}(A)\bigcup\sigma_{ei}(B)]\backslash\{0\}$, $i = 5,\ 7l,\ 8r$. If, further, Φ_A is connected, then

$$\sigma_{e5}(A+B)\backslash\{0\} = \left[\sigma_{e5}(A)\bigcup\sigma_{e5}(B)\right]\backslash\{0\}. \tag{7.4.16}$$

Moreover, if Φ_B is connected, then $\sigma_{ei}(A+B)\backslash\{0\} = [\sigma_{ei}(A)\bigcup\sigma_{ei}(B)]\backslash\{0\}$, $i = 7l,\ 8r$.

(v) If the hypotheses of (iv) are satisfied and if $\mathbb{C}\backslash\sigma_{e5}(A+B)$ is connected, then

$$\sigma_{e6}(A+B)\backslash\{0\} = \left[\sigma_{e6}(A)\bigcup\sigma_{e6}(B)\right]\backslash\{0\}.$$

◇

Proof. Let $\lambda \in \mathbb{C}$. Using Eqs. (7.4.6) and (7.4.7), we have

$$AB(A+B-\lambda) - (A+B-\lambda)AB = A(BA-AB) + (AB-BA)B \tag{7.4.17}$$

and

$$BA\,(A+B-\lambda)-(A+B-\lambda)\,BA=(BA-AB)\,A+B\,(AB-BA)\,. \qquad (7.4.18)$$

(i) Let $\lambda \notin \sigma_{e1}(A)\bigcup\sigma_{e1}(B)\bigcup\{0\}$, then $(A-\lambda)\in\Phi^b_+(X)$ and $(B-\lambda)\in\Phi_+(X)$. Using Theorem 2.2.13(ii), we have $(A-\lambda)\,(B-\lambda)\in\Phi^b_+(X)$. Since $AB-BA\in\mathcal{F}^b_+(X)$, we can apply Eq. (7.4.17), we infer that $\lambda AB\,(A+B-\lambda)-\lambda\,(A+B-\lambda)\,AB\in\mathcal{F}^b_+(X)$. Also, since $AB\in\mathcal{R}(X)$, then by Lemma 7.4.1(i) and Eq. (7.4.6), $(A+B-\lambda)\in\Phi^b_+(X)$. So, $\lambda\notin\sigma_{e1}(A+B)$. Therefore

$$\sigma_{e1}(A+B)\backslash\{0\}\subset\left[\sigma_{e1}(A)\bigcup\sigma_{e1}(B)\right]\backslash\{0\}. \qquad (7.4.19)$$

Now, suppose that $\lambda\notin\sigma_{e7}(A)\bigcup\sigma_{e7}(B)\bigcup\{0\}$, then $(A-\lambda)\in\Phi^b_+(X)$, $i(A-\lambda)\le 0$, $(B-\lambda)\in\Phi^b_+(X)$ and $i(B-\lambda)\le 0$. Using Theorem 2.2.13(ii) and Theorem 2.2.7, we have $(A-\lambda)\,(B-\lambda)\in\Phi^b_+(X)$ and $i\,[(A-\lambda)\,(B-\lambda)]\le 0$. Since $AB-BA\in\mathcal{F}^b_+(X)$, then by Eq. (7.4.17), it is clear that $\lambda AB\,(A+B-\lambda)-\lambda\,(A+B-\lambda)\,AB\in\mathcal{F}^b_+(X)$. Also, since $AB\in\mathcal{R}(X)$, then by Lemma 7.4.1(i) and Eq. (7.4.6), $(A+B-\lambda)\in\Phi^b_+(X)$ and $i\,(A+B-\lambda)\le 0$. In this way we see that $\lambda\notin\sigma_{e7}(A+B)$ whence $\sigma_{e7}(A+B)\backslash\{0\}\subset[\sigma_{e7}(A)\bigcup\sigma_{e7}(B)]\,\backslash\{0\}$. To prove the inverse inclusion of Eq. (7.4.19). Suppose $\lambda\notin\sigma_{e1}(A+B)\bigcup\{0\}$ then $(A+B-\lambda)\in\Phi^b_+(X)$. Since $AB-BA\in\mathcal{F}^b_+(X)$, then by Eqs. (7.4.17) and (7.4.18), we have $\lambda AB\,(A+B-\lambda)-\lambda\,(A+B-\lambda)\,AB\in\mathcal{F}^b_+(X)$ and $\lambda BA\,(A+B-\lambda)-\lambda\,(A+B-\lambda)\,BA\in\mathcal{F}^b_+(X)$. Also, since $AB\in\mathcal{R}(X)$ and $BA\in\mathcal{R}(X)$, then by Eqs. (7.4.6), (7.4.7) and Lemma 7.4.1(i), we have $(A-\lambda)\,(B-\lambda)\in\Phi^b_+(X)$ and $(B-\lambda)\,(A-\lambda)\in\Phi^b_+(X)$. Again, using Theorem 2.2.14, we have $(A-\lambda)\in\Phi_+(X)$ and $(B-\lambda)\in\Phi^b_+(X)$. Hence $\lambda\notin\sigma_{e1}(A)\bigcup\sigma_{e1}(B)$. Therefore $[\sigma_{e1}(A)\bigcup\sigma_{e1}(B)]\,\backslash\{0\}\subset\sigma_{e1}(A+B)\backslash\{0\}$. This proves that Eq. (7.4.15). Now, it remains to prove the following $[\sigma_{e7}(A)\bigcup\sigma_{e7}(B)]\,\backslash\{0\}\subset\sigma_{e7}(A+B)\backslash\{0\}$. Let $\lambda\notin\sigma_{e7}(A+B)\bigcup\{0\}$ then $(A+B-\lambda)\in\Phi^b_+(X)$ and $i(A+B-\lambda)\le 0$. Since $AB-BA\in\mathcal{F}^b_+(X)$, $AB\in\mathcal{R}(X)$ and $BA\in\mathcal{R}(X)$, a similar reasoning as before, it is clear that $(A-\lambda)\in\Phi^b_+(X)$ and $(B-\lambda)\in\Phi^b_+(X)$. Let $\lambda_0\in\rho(A)$ and $\lambda_1\in\rho(B)$, then $(A-\lambda_0)\in\Phi^b(X)$, $(B-\lambda_1)\in\Phi^b(X)$, $i(A-\lambda_0)=0$ and $i(B-\lambda_1)=0$. Since Φ_A and Φ_B are connected, by Proposition 2.2.5, we have $i(A-\lambda)$ is constant on any component of Φ_A and $i(B-\lambda)$ is constant on any component of Φ_B, then $i(A-\lambda)=i(A-\lambda_0)=0$ for all $\lambda\in\Phi_A$ and $i(B-\lambda)=i(B-\lambda_1)=0$ for all $\lambda\in\Phi_B$. On the other hand, for $\lambda_2\in\Phi_{+A}\backslash\Phi_A$ and $\lambda_3\in\Phi_{+B}\backslash\Phi_B$, $i(A-\lambda_2)=-\infty$ and $i(B-\lambda_3)=-\infty$. So, $i(A-\lambda)\le 0$ and $i(B-\lambda)\le 0$. This proved that $\lambda\notin\sigma_{e7}(A)\bigcup\sigma_{e7}(B)$ whence $[\sigma_{e7}(A)\bigcup\sigma_{e7}(B)]\,\backslash\{0\}\subset\sigma_{e7}(A+B)\backslash\{0\}$.

(ii) The proof may be checked in the same way as the proof of (i).

(*iii*) Since the equalities $\sigma_{e3}(A) = \sigma_{e1}(A)\bigcap\sigma_{e2}(A)$, $\sigma_{e3}(B) = \sigma_{e1}(B)\bigcap\sigma_{e2}(B)$ and $\sigma_{e3}(A+B) = \sigma_{e1}(A+B)\bigcap\sigma_{e2}(A+B)$ are known, $AB \in \mathcal{R}(X), BA \in \mathcal{R}(X)$ and $AB - BA \in \mathcal{F}_+^b(X)\bigcap\mathcal{F}_-^b(X)$ then, by (*i*) and (*ii*) we deduce that

$$\sigma_{e3}(A+B)\backslash\{0\} = \Big[\Big(\sigma_{e3}(A)\bigcup\sigma_{e3}(B)\Big)\bigcup\Big(\sigma_{e1}(A)\bigcap\sigma_{e2}(B)\Big)$$
$$\bigcup\Big(\sigma_{e2}(A)\bigcap\sigma_{e1}(B)\Big)\Big]\backslash\{0\}.$$

(*iv*) Let $\lambda \notin \sigma_{ei}(A)\bigcup\sigma_{ei}(B)\bigcup\{0\}$, $i = 1l,\ 2r,\ 4$. Then, $(A-\lambda) \in \Phi_l^b(X)$ [resp. $\Phi_r^b(X)$, $\Phi^b(X)$] and $(B-\lambda) \in \Phi_l^b(X)$ [resp. $\Phi_r^b(X)$, $\Phi^b(X)$]. Therefore Lemma 7.4.2 and Theorem 2.2.13 give $(A-\lambda)(B-\lambda) \in \Phi_l^b(X)$ [resp. $\Phi_r^b(X)$, $\Phi^b(X)$]. Since $AB - BA \in \mathcal{F}^b(X)$. Then by Eq. (7.4.17), we have $\lambda AB(A+B-\lambda)-\lambda(A+B-\lambda)AB \in \mathcal{F}^b(X)$. By Eq. (7.4.6), Lemma 7.4.1(*iii*) and Remark 7.4.2, it is clear that $(A+B-\lambda) \in \Phi_l^b(X)$ [resp. $\Phi_r^b(X)$, $\Phi^b(X)$]. So, $\lambda \notin \sigma_{ei}(A+B)$, $i = 1l,\ 2r,\ 4$. This proved that

$$\sigma_{ei}(A+B)\backslash\{0\} \subset \Big[\sigma_{ei}(A)\bigcup\sigma_{ei}(B)\Big]\backslash\{0\},\ i = 1l,\ 2r,\ 4. \qquad (7.4.20)$$

Let $\lambda \notin \sigma_{ei}(A)\bigcup\sigma_{ei}(B)\bigcup\{0\}$, $i = 5,\ 7l,\ 8r$. Then $(A-\lambda) \in \Phi^b(X)$ [resp. $\Phi_l^b(X)$, $\Phi_r^b(X)$], $i(A-\lambda) = 0$ (resp. $\leq 0, \geq 0$), $(B-\lambda) \in \Phi^b(X)$ [resp. $\Phi_l^b(X)$, $\Phi_r^b(X)$] and $i(B-\lambda) = 0$ (resp. $\leq 0, \geq 0$). Using Theorem 2.2.40 and Lemma 7.4.2, we have $(A-\lambda)(B-\lambda) \in \Phi^b(X)$ [resp. $\Phi_l^b(X)$, $\Phi_r^b(X)$] and $i\,[(A-\lambda)(B-\lambda)] = 0$ (resp. $\leq 0,\ \geq 0$). Moreover, since $AB-BA \in \mathcal{F}^b(X)$. Then by Eq. (7.4.17), we have $\lambda AB(A+B-\lambda) - \lambda(A+B-\lambda)AB \in \mathcal{F}^b(X)$. Also, since $AB \in \mathcal{R}(X)$, then by Eq. (7.4.6), Remark 7.4.2 and Lemma 7.4.1, we have $(A+B-\lambda) \in \Phi^b(X)$ [resp. $\Phi_l^b(X)$, $\Phi_r^b(X)$] and $i(A+B-\lambda) = 0$. In this way we see that $\lambda \notin \sigma_{ei}(A+B)$, $i = 5,\ 7l,\ 8r$, whence

$$\sigma_{ei}(A+B)\backslash\{0\} \subset \Big[\sigma_{ei}(A)\bigcup\sigma_{ei}(B)\Big]\backslash\{0\},\ i = 5,\ 7l,\ 8r. \qquad (7.4.21)$$

To prove the inverse inclusions of Eqs. (7.4.20) and (7.4.21). Suppose $\lambda \notin \sigma_{ei}(A+B)\bigcup\{0\}$, $i = 1l,\ 2r,\ 4$, then $(A+B-\lambda) \in \Phi_l^b(X)$ [resp. $\Phi_r^b(X)$, $\Phi^b(X)$]. Since $AB-BA \in \mathcal{F}^b(X)$, then by Eqs. (7.4.17) and (7.4.18), it is clear that $\lambda AB(A+B-\lambda)-\lambda(A+B-\lambda)AB \in \mathcal{F}^b(X)$ and $\lambda BA(A+B-\lambda)-\lambda(A+B-\lambda)BA \in \mathcal{F}^b(X)$. Also, since $AB \in \mathcal{R}(X)$ and $BA \in \mathcal{R}(X)$, then by Eqs. (7.4.6), (7.4.7), Lemma 7.4.1 and Remark 7.4.2, we have $(A-\lambda)(B-\lambda) \in \Phi_l^b(X)$ [resp. $\Phi_r^b(X)$, $\Phi^b(X)$] and $(B-\lambda)(A-\lambda) \in \Phi_l^b(X)$ [resp. $\Phi_r^b(X)$, $\Phi^b(X)$]. So, by Lemma 7.4.2 and Theorem 2.2.14, it is clear that $(A-\lambda) \in \Phi_l^b(X)$ [resp. $\Phi_r^b(X)$, $\Phi^b(X)$] and $(B-\lambda) \in \Phi_l^b(X)$ [resp. $\Phi_r^b(X)$, $\Phi^b(X)$]. Therefore $\lambda \notin \sigma_{ei}(A)\bigcup\sigma_{ei}(B)$, $i = 1l, 2r, 4$. This proved that $[\sigma_{ei}(A)\bigcup\sigma_{ei}(B)]\backslash\{0\} \subset \sigma_{ei}(A+B)\backslash\{0\}$, $i = 1l, 2r, 4$.

Now, it remains to prove that $[\sigma_{ei}(A)\bigcup\sigma_{ei}(B)]\backslash\{0\}\subset\sigma_{ei}(A+B)\backslash\{0\}$, $i=5,\ 7l,\ 8r$. Suppose $\lambda\notin\sigma_{e5}(A+B)\bigcup\{0\}$, then $(A+B-\lambda)\in\Phi^b(X)$ and $i(A+B-\lambda)=0$. Since $AB-BA\in\mathcal{F}^b(X)$, $AB\in\mathcal{R}(X)$ and $BA\in\mathcal{R}(X)$, a similar reasoning as before it is clear that $(A-\lambda)\in\Phi^b(X)$ and $(B-\lambda)\in\Phi^b(X)$. Again, we can apply Eq. (7.4.6), Theorem 2.2.40 and Remark 7.4.2 we have $i\ [(A-\lambda)\,(B-\lambda)]=i\ (A-\lambda)+i\ (B-\lambda)=0$. Since A is bounded linear operator, we get $\rho(A)\neq\emptyset$. As, Φ_A we have $i(A-\lambda)=0$. Again, it is clear that $i(B-\lambda)=0$. We conclude $\lambda\notin\sigma_{e5}(A)\bigcup\sigma_{e5}(B)$ whence $[\sigma_{e5}(A)\bigcup\sigma_{e5}(B)]\backslash\{0\}\subset\sigma_{e5}(A+B)\backslash\{0\}$. So, we prove Eq. (7.4.16). Since Φ_A and Φ_B are connected then, by using Theorem 7.3.1(i), we get $\sigma_{e1l}(A)=\sigma_{e7l}(A)$, $\sigma_{e2r}(A)=\sigma_{e8r}(A)$, $\sigma_{e1l}(B)=\sigma_{e7l}(B)$ and $\sigma_{e2r}(B)=\sigma_{e8r}(B)$. Therefore, $[\sigma_{e8r}(A)\bigcup\sigma_{e8r}(B)]\backslash\{0\}\subseteq\sigma_{e2r}(A+B)\backslash\{0\}\subseteq\sigma_{e8r}(A+B)\backslash\{0\}$ and

$$\left[\sigma_{e7l}(A)\bigcup\sigma_{e7l}(B)\right]\backslash\{0\}\subseteq\sigma_{e1l}(A+B)\backslash\{0\}\subseteq\sigma_{e7l}(A+B)\backslash\{0\}.$$

(v) This assertion follows immediately from Theorem 7.3.1 and Eq. (7.4.16). Q.E.D.

Remark 7.4.3. It follows, immediately, from Lemma 7.4.1 and Remark 7.4.2 that, if $AB\in\mathcal{R}(X)$ and $AB-BA\in\mathcal{F}_+^b(X)\bigcup\mathcal{F}_-^b(X)\bigcup\mathcal{F}^b(X)$, then $BA\in\mathcal{R}(X)$. ◇

7.4.2 By Means of Fredholm Inverse

Theorem 7.4.3. *Let X be a Banach space and let A and B be two operators in $\mathcal{L}(X)$. Then, the following statements hold:*

(i) Assume that, for each $\lambda\in\Phi_A$, there exists a Fredholm inverse A_λ of $\lambda-A$ such that $A_\lambda B\in\mathcal{P}_\gamma(X)$ [see (6.4.2)]. Then, $\sigma_{e4}(A+B)\subseteq\sigma_{e4}(A)$ and $\sigma_{e5}(A+B)\subseteq\sigma_{e5}(A)$.

(ii) If the hypothesis of (i) is satisfied, and if $\mathbb{C}\backslash\sigma_{e5}(A)$ and $\mathbb{C}\backslash\sigma_{e5}(A+B)$ are connected, then $\sigma_{e6}(A+B)\subseteq\sigma_{e6}(A)$.

(iii) Assume that, for each $\lambda\in\Phi_{+A}$, there exists a left Fredholm inverse $A_{\lambda l}$ of $\lambda-A$ such that $BA_{\lambda l}\in\mathcal{P}_\gamma(X)$. Then, $\sigma_{e1}(A+B)\subseteq\sigma_{e1}(A)$ and $\sigma_{e7}(A+B)\subseteq\sigma_{e7}(A)$.

(iv) Assume that, for each $\lambda\in\Phi_{-A}$, there exists a right Fredholm inverse $A_{\lambda r}$ of $\lambda-A$ such that $A_{\lambda r}B\in\mathcal{P}_\gamma(X)$. Then, $\sigma_{e2}(A+B)\subseteq\sigma_{e2}(A)$ and $\sigma_{e8}(A+B)\subseteq\sigma_{e8}(A)$.

(v) Assume that, for each $\lambda\in\Phi_{\pm A}$, the set $F^{\pm}_{(\lambda-A)B}(X)\neq\emptyset$. Then, $\sigma_{e3}(A+B)\subseteq\sigma_{e3}(A)$. ◇

Proof.

(*i*) Suppose that $\lambda \notin \sigma_{e4}(A)$ [resp. $\lambda \notin \sigma_{e5}(A)$], then $\lambda \in \Phi_A$ [resp. by Proposition 7.1.1 we have $\lambda \in \Phi_A$ and $i(\lambda - A) = 0$]. Applying Theorem 6.4.1(*i*) to the operators $\lambda - A$ and $\lambda - A - B$, we prove that $\lambda \in \Phi_{A+B}$ and $i(\lambda - A) = i(\lambda - A + B)$. Therefore, $\lambda \notin \sigma_{e4}(A + B)$ [resp. $\lambda \notin \sigma_{e5}(A + B)$]. We obtain $\sigma_{e4}(A + B) \subseteq \sigma_{e4}(A)$ and

$$\sigma_{e5}(A + B) \subseteq \sigma_{e5}(A). \tag{7.4.22}$$

(*ii*) The sets $\mathbb{C}\backslash\sigma_{e5}(A + B)$ and $\mathbb{C}\backslash\sigma_{e5}(A)$ are connected. Since A and B are bounded operators, we deduce that $\rho(A)$ and $\rho(A + B)$ are not empty sets. So, using Theorem 7.3.1 allows us to get $\sigma_{e5}(A + B) = \sigma_{e6}(A + B)$ and $\sigma_{e5}(A) = \sigma_{e6}(A)$. Moreover, Eq. (7.4.22) gives

$$\sigma_{e6}(A + B) \subseteq \sigma_{e6}(A).$$

(*iii*) Suppose that $\lambda \notin \sigma_{e1}(A)$ [resp. $\lambda \notin \sigma_{e7}(A)$], then $\lambda \in \Phi_{+A}$ [resp. by Proposition 7.3.2(*i*), we get $\lambda - A \in \Phi^b_+(X)$ and $i(\lambda - A) \leq 0$]. By applying Theorem 6.4.1(*ii*) to the operators $\lambda - A$ and $\lambda - A - B$, we prove that $\lambda \in \Phi_{+(A+B)}$ and $i(\lambda - A) = i(\lambda - A + B)$. This proves that $\lambda \notin \sigma_{e1}(A + B)$ [resp. $\lambda \notin \sigma_{e7}(A + B)$]. We find $\sigma_{e1}(A + B) \subseteq \sigma_{e1}(A)$ and $\sigma_{e7}(A + B) \subseteq \sigma_{e7}(A)$.

(*iv*) By using a similar proof as in (*iii*), by replacing $\sigma_{e1}(.)$, $\sigma_{e7}(.)$, and $\Phi^b_+(X)$ by $\sigma_{e2}(.)$, $\sigma_{e8}(.)$, and $\Phi^b_-(X)$, respectively, and by combining Proposition 7.3.2(*ii*) and Theorem 6.4.1(*iii*), we get $\sigma_{e2}(A + B) \subseteq \sigma_{e2}(A)$ and $\sigma_{e8}(A + B) \subseteq \sigma_{e8}(A)$.

(*v*) Let $\lambda \notin \sigma_{e3}(A)$. Then, $\lambda \in \Phi_{\pm A}$. Since the set $F^{\pm}_{(\lambda-A)B}(X) \neq \emptyset$, and applying Theorem 6.4.1(*iv*) to the operators $\lambda - A$ and $\lambda - A - B$, we have $\lambda \in \Phi_{\pm(A+B)}$. Therefore,

$$\sigma_{e3}(A + B) \subseteq \sigma_{e3}(A).$$

Q.E.D.

Remark 7.4.4.

(i) The results of Theorem 7.4.3 remain valid if we suppose that $A \in \mathcal{C}(X)$ and B is an A-bounded operator on X.

(ii) If we replace $\mathcal{P}_\gamma(X)$ by $\mathcal{J}(X)$, where $\mathcal{J}(X) = \{A \in \mathcal{L}(X)$ such that $1 \in \Phi^0_A\}$, then we can prove the same results of Theorem 7.4.3. ◇

7.4.3 By Means of Demicompact Operators

Theorem 7.4.4. *Let $A \in \mathcal{L}(X)$ and $B \in \mathcal{L}(X)$. If for every $\lambda \in \Phi_{+A}$, there exists $A_{\lambda l}$ (resp. $A_{\lambda r}$) a left (resp. right) Fredholm inverse of $(\lambda - A)$ such that $BA_{\lambda l}$ (resp. $A_{\lambda r}B$) is demicompact, then $\sigma_{e1}(A + B) \subseteq \sigma_{e1}(A)$.* ◇

Proof. Let $\lambda \in \mathbb{C}$. If $A_{\lambda l}$ is a left Fredholm inverse of $(\lambda - A)$, then there exists a compact operator $K \in \mathcal{K}(X)$, such that $A_{\lambda l}(\lambda - A) = I - K$. Hence, we can write

$$\lambda - A - B = (I - BA_{\lambda l})(\lambda - A) - BK. \tag{7.4.23}$$

In the same way, if there exists $A_{\lambda r}$, a right Fredholm inverse of $(\lambda - A)$, then we can write

$$\lambda - A - B = (\lambda - A)(I - A_{\lambda r}B) - K'B, \tag{7.4.24}$$

where $K' \in \mathcal{K}(X)$. Let $\lambda \notin \sigma_{e1}(A)$. Then, $(\lambda - A) \in \Phi_+^b(X)$. Let $A_{\lambda l}$ (resp. $A_{\lambda r}$) be the left (resp. the right) Fredholm inverse of $(\lambda - A)$. Then, Eq. (7.4.23) [resp. Eq. (7.4.24)] holds. Since $BA_{\lambda l}$ (resp. $A_{\lambda r}B$) is demicompact, Theorem 5.4.1 implies that $(I - BA_{\lambda l}) \in \Phi_+^b(X)$ [resp. $(I - A_{\lambda r}B) \in \Phi_+^b(X)$]. Hence, by applying Theorem 2.2.7 to Eq. (7.4.23) [resp. Eq. (7.4.24)], we get $(I - BA_{\lambda l})(\lambda - A) \in \Phi_+^b(X)$ [resp. $(\lambda - A)(I - A_{\lambda r}B) \in \Phi_+^b(X)$]. Since $BK \in \mathcal{K}(X) \subset \mathcal{F}_+^b(X)$ [resp. $K'B \in \mathcal{K}(X) \subset \mathcal{F}_+^b(X)$], $(\lambda - A - B) \in \Phi_+^b(X)$, hence $\lambda \notin \sigma_{e1}(A + B)$. We conclude that $\sigma_{e1}(A + B) \subseteq \sigma_{e1}(A)$. Q.E.D.

Let X be a Banach space. We define the set Λ_X by

$$\Lambda_X = \{J \in \mathcal{L}(X) \text{ such that } \mu J \text{ is demicompact for every } \mu \in [0, 1]\}. \tag{7.4.25}$$

Theorem 7.4.5. *Let $A \in \mathcal{L}(X)$ and $B \in \mathcal{L}(X)$.*

(i) If for every $\lambda \notin \sigma_{ej}(A)$, where $j \in \{2, 3, 4, 5, 7, 8\}$, there exists $A_{\lambda l}$ (resp. $A_{\lambda r}$) a left (resp. right) Fredholm inverse of $(\lambda - A)$ such that $BA_{\lambda l} \in \Lambda_Y$ (resp. $A_{\lambda r}B \in \Lambda_X$). Then

$$\sigma_{ej}(A + B) \subseteq \sigma_{ej}(A). \tag{7.4.26}$$

(ii) If the hypothesis of (i) is satisfied for the case of $j = 5$, and if Φ_A^0 and Φ_{A+B}^0 are connected, then $\sigma_{e6}(A + B) \subseteq \sigma_{e6}(A)$. ◇

Proof.

(i) We use Eq. (7.4.23) [resp. Eq. (7.4.24)] in the same manner for the different cases. That is why, we will give the proof of only one of them which corresponds to $j = 8$. Let $\lambda \notin \sigma_{e8}(A)$. Then, $(\lambda - A) \in \Phi_-^b(X)$ and $i(\lambda - A) \geq 0$. Let $A_{\lambda l}$ be the left Fredholm inverse of $(\lambda - A)$. Then, Eq. (7.4.23) holds. Since $BA_{\lambda l} \in \Lambda_Y$, $(I - BA_\lambda) \in \Phi^b(X)$ and $i(I - BA_{\lambda l}) = 0$, then using both Theorem 2.2.7 and Eq. (7.4.23), we have $(\lambda - A - B) \in \Phi^b_-(X)$ and $i(\lambda - A - B) \geq 0$. Consequently, $\lambda \notin \sigma_{e8}(A+B)$, which allows us to conclude that $\sigma_{e8}(A + B) \subseteq \sigma_{e8}(A)$.

(ii) The sets Φ^0_A and Φ^0_{A+B} are connected. Since A and B are bounded operators, we deduce that $\rho(A)$ and $\rho(A+B)$ are nonempty sets. Using Theorem 7.3.1(*ii*), we get $\sigma_{e5}(A) = \sigma_{e6}(A)$ and $\sigma_{e5}(A + B) = \sigma_{e6}(A + B)$. So, the inclusion (7.4.26) holds for the case of $j = 5$, which completes the proof. Q.E.D.

Theorem 7.4.6. *Let $A \in \mathcal{L}(X)$ and $B \in \mathcal{L}(X)$. If the following assertions hold:*

(i) For every $\lambda \in \Phi_{+(A+B)}\backslash\{0\}$, there exists $H_{\lambda l}$ (resp. $H_{\lambda r}$) a left (resp. right) Fredholm inverse of $(\lambda - A - B)$, such that $-\lambda^{-1}ABH_{\lambda l}$ (resp. $-\lambda^{-1}H_{\lambda r}AB$) is demicompact.

(ii) For every $\lambda \in \Phi_{+(A+B)}\backslash\{0\}$, there exists $G_{\lambda l}$ (resp. $G_{\lambda r}$) a left (resp. right) Fredholm inverse of $(\lambda - A - B)$, such that $-\lambda^{-1}BAG_{\lambda l}$ (resp. $-\lambda^{-1}G_{\lambda r}BA$) is demicompact. Then, $[\sigma_{e1}(A) \bigcup \sigma_{e1}(B)]\backslash\{0\} \subseteq [\sigma_{e1}(A + B)]\backslash\{0\}$. ◇

Proof. Let $\lambda \in \mathbb{C}\backslash\{0\}$. If there exists $H_{\lambda l}$ a left Fredholm inverse of $(\lambda - A - B)$, then $H_{\lambda l}(\lambda - A - B) = I - K$ where $K \in \mathcal{K}(X)$. Thus, using Eq. (7.4.6) we have $(\lambda - A)(\lambda - B) = \lambda(\lambda - A - B) + ABH_{\lambda l}(\lambda - A - B) + ABK$, and we conclude that

$$(\lambda - A)(\lambda - B) = \lambda(I + \lambda^{-1}ABH_{\lambda l})(\lambda - A - B) + ABK. \tag{7.4.27}$$

In the same manner, we can write

$$(\lambda - B)(\lambda - A) = \lambda(I + \lambda^{-1}BAH_{\lambda l})(\lambda - A - B) + BAK. \tag{7.4.28}$$

If there exists $H_{\lambda r}$ a right Fredholm inverse of $(\lambda - A - B)$, we can write

$$(\lambda - A)(\lambda - B) = \lambda(\lambda - A - B)(I + \lambda^{-1}H_{\lambda r}AB) + K'AB, \tag{7.4.29}$$

and

$$(\lambda - B)(\lambda - A) = \lambda(\lambda - A - B)(I + \lambda^{-1}H_{\lambda r}BA) + K'BA, \tag{7.4.30}$$

where $K' \in \mathcal{K}(X)$. Let $\lambda \in \Phi_{+(A+B)}\backslash\{0\}$, and let $H_{\lambda l}$ and $G_{\lambda l}$ be the left Fredholm inverses of $(\lambda - A - B)$, such that $-\lambda^{-1}ABH_{\lambda l}$ and $-\lambda^{-1}BAG_\lambda$ are demicompact. Applying Theorem 5.4.1, we deduce that $(I + \lambda^{-1}ABH_{\lambda l})$ and $(I + \lambda^{-1}BAG_{\lambda l})$ are upper semi-Fredholm operators on X. Using Theorem 2.2.7, Eqs. (7.4.27) and (7.4.28), we prove that $(\lambda - A)(\lambda - B)$ and $(\lambda - B)(\lambda - A)$ are both upper semi-Fredholm operators. Theorem 2.2.14 completes the proof for the first case. For the other cases, the same arguments have been used, and it is sufficient to replace Eq. (7.4.27) by Eq. (7.4.29) and Eq. (7.4.28) by Eq. (7.4.30). Q.E.D.

Theorem 7.4.7. *Let $A \in \mathcal{L}(X)$ and $B \in \mathcal{L}(X)$. If the following assertions hold:*

(i) For every $\lambda \in \Phi_{A+B}\backslash\{0\}$, there exists $H_{\lambda l}$ (resp. $H_{\lambda r}$) a left (resp. right) Fredholm inverse of $(\lambda - A - B)$, such that $-\lambda^{-1}ABH_{\lambda l} \in \Lambda_X$ (resp. $-\lambda^{-1}H_{\lambda r}AB \in \Lambda_X$).
(ii) For every $\lambda \in \Phi_{A+B}\backslash\{0\}$, there exists $G_{\lambda l}$ (resp. $G_{\lambda r}$) a left (resp. right) Fredholm inverse of $(\lambda - A - B)$ such that $-\lambda^{-1}BAG_{\lambda l} \in \Lambda_X$ (resp. $-\lambda^{-1}G_{\lambda r}BA \in \Lambda_X$).

Then, $[\sigma_{e4}(A)\bigcup\sigma_{e4}(B)]\backslash\{0\} \subseteq [\sigma_{e4}(A+B)]\backslash\{0\}$. ◇

Proof. It is sufficient to replace Theorem 5.4.1 by Theorem 5.4.2 in the proof of Theorem 7.4.6. Q.E.D.

In a similar way, we prove the following theorem.

Theorem 7.4.8. *Let $A \in \mathcal{L}(X)$ and $B \in \mathcal{L}(X)$. If the following assertions hold:*

(i) For every $\lambda \in \Phi_{-(A+B)}\backslash\{0\}$, there exists $H_{\lambda l}$ (resp. $H_{\lambda r}$) a left (resp. right) Fredholm inverse of $(\lambda - A - B)$, such that $-\lambda^{-1}ABH_{\lambda l} \in \Lambda_X$ (resp. $-\lambda^{-1}H_{\lambda r}AB \in \Lambda_X$).
(ii) For every $\lambda \in \Phi_{-(A+B)}\backslash\{0\}$, there exists $G_{\lambda l}$ (resp. $G_{\lambda r}$) a left (resp. right) Fredholm inverse of $(\lambda - A - B)$, such that $-\lambda^{-1}BAG_{\lambda l} \in \Lambda_X$ (resp. $-\lambda^{-1}G_{\lambda r}BA \in \Lambda_X$).

Then, $[\sigma_{e2}(A)\bigcup\sigma_{e2}(B)]\backslash\{0\} \subseteq [\sigma_{e2}(A+B)]\backslash\{0\}$. ◇

7.5 Unbounded Linear Operators

7.5.1 Essential Spectra for the Sum of Closed and Bounded Linear Operators

Theorem 7.5.1. *Let $A \in \mathcal{C}(X)$ and $B \in \mathcal{L}(X)$. Suppose that there exist a positive integer n and $F \in \mathcal{F}(X)$, such that $B : \mathcal{D}(A^n) \longrightarrow \mathcal{D}(A)$ and $ABx = BAx + Fx$, for all $x \in \mathcal{D}(A^n)$. Then,*

(i) $\sigma_{e4}(A+B) \subseteq \sigma_{e4}(A) + \sigma_{e4}(B)$. If $\sigma_{e4}(A)$ is empty, then $\sigma_{e4}(A) + \sigma_{e4}(B)$ is also an empty set.
(ii) If, in addition $\mathbb{C}\backslash\sigma_{e4}(A)$, $\mathbb{C}\backslash\sigma_{e4}(B)$, and $\mathbb{C}\backslash\sigma_{e4}(A+B)$ are connected and if $\rho(A)$ and $\rho(A+B)$ are nonempty sets, then $\sigma_{e5}(A+B) \subseteq \sigma_{e5}(A) + \sigma_{e5}(B)$.
(iii) Moreover, if $\mathbb{C}\backslash\sigma_{e5}(A)$, $\mathbb{C}\backslash\sigma_{e5}(B)$, and $\mathbb{C}\backslash\sigma_{e5}(A+B)$ are connected and if $\rho(A)$ and $\rho(A+B)$ are nonempty sets, then $\sigma_{e6}(A+B) \subseteq \sigma_{e6}(A)+\sigma_{e6}(B)$. ◇

Proof.

(i) First, it is clear that the theorem is trivially true if we suppose that $\sigma_{e4}(A) + \sigma_{e4}(B)$ constitutes the entire complex plane. Hence, let us assume that $\sigma_{e4}(A) + \sigma_{e4}(B)$ is not the entire plane. Second, we fix a point γ such that $\gamma \notin \sigma_{e4}(A) + \sigma_{e4}(B)$ and we define the operator A_1 as $A_1 := \gamma - A$. Hence, it is easy to verify that if $\lambda \in \sigma_{e4}(B)$, the element $\gamma - \lambda$ will be in Φ_A which is equivalent to say that $\lambda \in \Phi_{A_1}$. In the following, we will find a Cauchy domain $\mathcal{D}$ such that $R'_\lambda(A_1)$ and $R'_\lambda(B)$ are analytic on $B(\mathcal{D})$, the boundary of $\mathcal{D}$ where $R'_\lambda(.)$ is defined in (2.2.7). In fact, $\sigma_{e4}(A)$ is closed and $\sigma_{e4}(B)$ is compact. Then, there exists an open set $U \supset \sigma_{e4}(B)$ such that $B(U)$, the boundary of U, is bounded and when $\lambda \in U$, $(\gamma - \lambda) \in \Phi_A$. Therefore, $\sigma_{e4}(B) \subset U \subset \Phi_{A_1}$. Using Theorem 2.2.50, we infer that there exists a bounded Cauchy domain $\mathcal{D}$, such that $\sigma_{e4}(B) \subset \mathcal{D} \subset U$. Note that $\Phi^0(A_1)$ [resp. $\Phi^0(B)$] does not accumulate in Φ_{A_1} [resp. Φ_B]. So, we can choose $\mathcal{D}$ such that $R'_\lambda(A_1)$ and $R'_\lambda(B)$ are analytic on $B(\mathcal{D})$. We also claim that $R'_\lambda(A_1)$ is of the form $TC(\lambda)$ where $C(\lambda)$ is a bounded operator-valued analytic function of λ and T is a fixed bounded operator, such that $T : X \longrightarrow \mathcal{D}(A_1) = \mathcal{D}(A)$. Now, let us define the following operators M_1 and M_2 as follows $M_1 = -\frac{1}{2\pi i}\int_{+B(\mathcal{D})} R'_\lambda(A_1)R'_\lambda(B)d\lambda$ and $M_2 = -\frac{1}{2\pi i}\int_{+B(\mathcal{D})} R'_\lambda(B)R'_\lambda(A_1)d\lambda$. In order to prove this assertion, we will show that $\gamma \in \Phi_{A+B}$. Hence, it is sufficient to find two Fredholm perturbations F_1 and F_2, such that $(\gamma - B - A)M_1 = I + F_1$ and $M_2(\gamma - B - A) = I + F_2$ on $\mathcal{D}(A)$. Now, writing the operator $\gamma - B - A$ as follows $(\gamma - B - A) = (\gamma - \lambda - A) + (\lambda - B) = -(\lambda - A_1) + (\lambda - B)$, we get

$$\begin{aligned}(\gamma - B - A)M_1 = &-\frac{1}{2\pi i}\int_{+B(\mathcal{D})} -(\lambda - A_1)R'_\lambda(A_1)R'_\lambda(B)d\lambda \\ &-\frac{1}{2\pi i}\int_{+B(\mathcal{D})} -(\lambda - B)R'_\lambda(A_1)R'_\lambda(B)d\lambda.\end{aligned} \tag{7.5.1}$$

Obviously, $(\lambda - A_1)R'_\lambda(A_1) = I + \mathfrak{F}$, where $\mathfrak{F}$ is a bounded finite rank operator depending analytically on λ. Then, the first integral of the above equality is of the following form $-\frac{1}{2\pi i}\int_{+B(\mathcal{D})} -(I + \mathfrak{F})R'_\lambda(B)d\lambda$. Using Theorem 2.2.54, we deduce that $\frac{1}{2\pi i}\int_{+B(\mathcal{D})} R'_\lambda(B)d\lambda = I + K_1$, where $K_1 \in \mathcal{K}(X)$. Moreover, we also mention that $\int_{+B(\mathcal{D})} -(I + \mathfrak{F})R'_\lambda(B)d\lambda$ is a compact operator. Hence, we infer that the first integral of (7.5.1) is of the form $I + K_2$, where $K_2 \in \mathcal{K}(X)$. Applying Lemma 6.4.2, we get $R'_\lambda(A_1)R'_\lambda(B) = R'_\lambda(B)R'_\lambda(A_1) + F$, where F is a Fredholm perturbation. Then, the second integral of the same equality is equal to $-\frac{1}{2\pi i}\int_{+B(\mathcal{D})} -(\lambda - B)R'_\lambda(B)R'_\lambda(A_1)d\lambda - \frac{1}{2\pi i}\int_{+B(\mathcal{D})}(\lambda - B)Fd\lambda$. Since $\int_{+B(\mathcal{D})} R'_\lambda(A_1)d\lambda$ is compact (see Lemma 2.2.18), then a same reasoning as in the first part allows us to write $-\frac{1}{2\pi i}\int_{+B(\mathcal{D})} -(\lambda - B)R'_\lambda(B)R'_\lambda(A_1)d\lambda = I + K_3$, where $K_3 \in \mathcal{K}(X)$. Using the fact that $\frac{1}{2\pi i}\int_{+B(\mathcal{D})}(\lambda - B)Fd\lambda$ is also a Fredholm perturbation, we have $(\gamma - B - A)M_1 = I + F_1$, $F_1 \in \mathcal{F}(X)$. By a similar argument, we obtain $M_2(\gamma - B - A) = I + F_2$, where $F_2 \in \mathcal{F}(X)$. Therefore, $(\gamma - B - A) \in \Phi(X)$, and we deduce that $\sigma_{e4}(A + B) \subseteq \sigma_{e4}(A) + \sigma_{e4}(B)$.

(*ii*) This assertion follows immediately from Theorem 7.3.1(*i*).

(*iii*) The proof of this assertion holds from Theorem 7.3.1(*ii*). Q.E.D.

7.5.2 Essential Spectra for the Product of Closed and Bounded Linear Operators

Theorem 7.5.2. *Let $A \in \mathcal{C}(X)$ and $B \in \Phi^b(X)$. Let $B : \mathcal{D}(A) \longrightarrow \mathcal{D}(A)$ and suppose that there exists $F \in \mathcal{F}(X)$ such that $ABx = BAx + Fx$, for all $x \in \mathcal{D}(A)$. Then, BA is closable and*

(*i*) $\sigma_{e4}(\overline{BA}) \subseteq \sigma_{e4}(A)\sigma_{e4}(B)$ *and* $\sigma_{e4}(AB) \subseteq \sigma_{e4}(A)\sigma_{e4}(B)$.

(*ii*) *If, in addition,* $\mathbb{C}\backslash\sigma_{e4}(\overline{BA})$, $\mathbb{C}\backslash\sigma_{e4}(AB)$, $\mathbb{C}\backslash\sigma_{e4}(A)$, $\mathbb{C}\backslash\sigma_{e4}(B)$ *are connected, and if* $\rho(A)$, $\rho(\overline{BA})$, *and* $\rho(AB)$ *are nonempty sets, then* $\sigma_{e5}(\overline{BA}) \subseteq \sigma_{e5}(A)\sigma_{e5}(B)$ *and* $\sigma_{e5}(AB) \subseteq \sigma_{e5}(A)\sigma_{e5}(B)$.

(*iii*) *Moreover, if* $\mathbb{C}\backslash\sigma_{e5}(\overline{BA})$, $\mathbb{C}\backslash\sigma_{e4}(AB)$, $\mathbb{C}\backslash\sigma_{e5}(A)$, $\mathbb{C}\backslash\sigma_{e5}(B)$ *are connected, and if* $\rho(A)$, $\rho(\overline{BA})$, *and* $\rho(AB)$ *are nonempty sets, then* $\sigma_{e6}(\overline{BA}) \subseteq \sigma_{e6}(A)\sigma_{e6}(B)$ *and* $\sigma_{e6}(AB) \subseteq \sigma_{e6}(A)\sigma_{e6}(B)$. ◇

Proof.

(*i*) Since the operator F is bounded and the restriction of the operator AB on $\mathcal{D}(A)$ is closable, then BA is closable. Furthermore, it is clear that $0 \notin \sigma_{e4}(B)$ and $\sigma_{e4}(B)$ is not empty. So, the theorem is trivially true if $\sigma_{e4}(A) = \mathbb{C}$, and we will assume, in the following, that $\sigma_{e4}(A) \neq \mathbb{C}$. Now, let γ be a fixed point not in $\sigma_{e4}(B)\sigma_{e4}(A)$. In what follows, we will show that $\gamma \in \Phi_{\overline{BA}}$. Observing that $\sigma_{e4}(A)$ is closed, $\sigma_{e4}(B)$ is compact and $0 \notin \sigma_{e4}(B)$, we infer that there exists an open set U, with a bounded boundary $B(U)$, containing $\sigma_{e4}(B)$ and satisfying that $0 \notin U$ and $(\gamma - \mu A) \in \Phi(X)$, $\forall \mu \in U$. Let $\mathcal{D}$ be a bounded Cauchy domain, such that $\sigma_{e4}(B) \subset \mathcal{D} \subseteq U$. Writing $(\gamma - \mu A)$ as follows:

$$(\gamma - \mu A) = \mu\gamma\left(\frac{1}{\mu} - \frac{1}{\gamma}A\right) = \frac{\gamma}{\lambda}\left(\lambda - \frac{1}{\gamma}A\right), \quad \lambda = \frac{1}{\mu}$$

and taking $\mathcal{D}'$ as the image of $\mathcal{D}$ under the map $\lambda = \frac{1}{\mu}$, we can assume that $R'_\lambda(A_1)$ is analytic in λ on $B(\mathcal{D}')$, where $A_1 := \frac{1}{\gamma}A$. This assumption holds true, thanks to the fact that $\forall \mu \in \overline{\mathcal{D}}$, $\frac{1}{\mu} \in \Phi_{A_1}$ and that the operator $R'_\lambda(A_1)$ is analytic in λ throughout Φ_{A_1} except for, at most, an isolated set having no accumulation in Φ_{A_1}. Let us define the following operators M_1 and M_2 as follows: $M_1 = -\frac{1}{2\pi i}\int_{+B(\mathcal{D}')} \frac{1}{\gamma\lambda} R'_\lambda(A_1) R'_{\frac{1}{\lambda}}(B) d\lambda$ and $M_2 =$

$-\frac{1}{2\pi i}\int_{+B(\mathcal{D}')}\frac{1}{\gamma\lambda}R'_{\frac{1}{\lambda}}(B)R'_{\lambda}(A_1)d\lambda$. Since $R(M_1)\subset\mathcal{D}(A)$, the operator $(\gamma-\overline{BA})M_1$ is well defined, and we have$(\gamma-\overline{BA})M_1=(\gamma-BA)M_1$. Moreover,

$$\begin{aligned}(\gamma-BA)&=(\gamma-B\gamma A_1)\\&=\gamma B(\lambda-A_1)-\gamma\lambda B+\gamma I\\&=\gamma B(\lambda-A_1)+\gamma(I-\lambda B).\end{aligned}$$

Then,

$$\begin{aligned}(\gamma-\overline{BA})M_1&=-\frac{1}{2\pi i}\int_{+B(\mathcal{D}')}\left(\frac{1}{\lambda}B(\lambda-A_1)R'_{\lambda}(A_1)R'_{\frac{1}{\lambda}}(B)\right.\\&\quad\left.+\left(\frac{1}{\lambda}-B\right)R'_{\lambda}(A_1)R'_{\frac{1}{\lambda}}(B)\right)d\lambda.\end{aligned}$$

On the one hand, the first part of the integrand can be written as follows:

$$\begin{aligned}\int_{+B(\mathcal{D}')}\frac{1}{\lambda}B(\lambda-A_1)R'_{\lambda}(A_1)R'_{\frac{1}{\lambda}}(B)d\lambda&=\int_{+B(\mathcal{D}')}\frac{1}{\lambda}B(I+K_1(\lambda))R'_{\frac{1}{\lambda}}(B)d\lambda\\&=\int_{+B(\mathcal{D}')}\frac{1}{\lambda}BR'_{\frac{1}{\lambda}}(B)d\lambda+K_2\\&=\int_{+B(\mathcal{D})}\frac{1}{\mu}BR'_{\mu}(B)d\mu+K_2,\end{aligned}$$

where $K_i\in\mathcal{K}(X)$, $i=1,2$. On the other hand, since $0\notin D$ hence, using Theorems 2.2.52, 2.2.54, and 2.2.55, we get $\frac{1}{2\pi i}\int_{+B(\mathcal{D})}\frac{1}{\mu}BR'_{\mu}(B)d\mu=I+K_3$, where $K_3\in\mathcal{K}(X)$. Note that the second part of the integrand can also be written as:

$$\begin{aligned}&\int_{+B(\mathcal{D}')}\left(\frac{1}{\lambda}-B\right)R'_{\lambda}(A_1)R'_{\frac{1}{\lambda}}(B)d\lambda\\&=\int_{+B(\mathcal{D}')}\left(\frac{1}{\lambda}-B\right)\left[R'_{\frac{1}{\lambda}}(B)R'_{\lambda}(A_1)+F_1\right]d\lambda\\&=\int_{+B(\mathcal{D}')}[I+K_4(\lambda)]\,R'_{\lambda}(A_1)d\lambda+F_2\\&=\int_{+B(\mathcal{D}')}R'_{\lambda}(A_1)d\lambda+F_3,\end{aligned}$$

where $K_4\in\mathcal{K}(X)$ and $F_i\in\mathcal{F}(X)$, with $i=1,2,3$. We claim that $R'_{\lambda}(A_1)$ is analytic in $\mathcal{D}'$ except for, at most, a finite number of points. Then, by using Lemma 2.2.18, we deduce that $\frac{1}{2\pi i}\int_{+B(\mathcal{D}')}R'_{\lambda}(A_1)d\lambda=K_5\in\mathcal{K}(X)$.

Therefore, $(\gamma - \overline{BA})M_1 = I + F_4$, where $F_4 \in \mathcal{F}(X)$. Now, we can easily check that $\mathcal{D}(\overline{BA}) \subseteq \mathcal{D}(AB)$ and $\overline{BA}x = ABx + Fx \ \forall x \in \mathcal{D}(\overline{BA})$. Hence,

$$\begin{aligned}(\gamma - \overline{BA}) &= \gamma\overline{B(\lambda - A_1)} + \gamma(I - \lambda B)\\ &= \gamma(\lambda - A_1)B + \gamma(I - \lambda B) + F_5,\end{aligned}$$

where $F_5 \in \mathcal{F}(X)$. Then,

$$\begin{aligned}
&M_2(\gamma - \overline{BA})\\
&= \frac{1}{2\pi i}\int_{+B(\mathcal{D}')} \frac{1}{\gamma\lambda} R'_{\frac{1}{\lambda}}(B) R'_{\lambda}(A_1)\left[\gamma(\lambda - A_1)B + \gamma(I - \lambda B) + F_5\right] d\lambda\\
&= -\frac{1}{2\pi i}\int_{+B(\mathcal{D}')} \frac{1}{\lambda} R'_{\frac{1}{\lambda}}(B)\,(I + K_6)\,B d\lambda -\\
&\quad \frac{1}{2\pi i}\int_{+B(\mathcal{D}')} \frac{1}{\gamma\lambda}\left(R'_{\lambda}(A_1)R'_{\frac{1}{\lambda}}(B) - F_1\right)(\gamma(I - \lambda B) + F_5)\, d\lambda\\
&= \left[\frac{1}{2\pi i}\int_{+B(\mathcal{D})} \frac{1}{\mu} R'_{\mu}(B) d\mu\right] B + K_7 -\\
&\quad \frac{1}{2\pi i}\int_{+B(\mathcal{D}')} R'_{\lambda}(A_1)R'_{\frac{1}{\lambda}}(B)\left(\frac{1}{\lambda} - B\right) d\lambda + F_6\\
&= I + F_7 - \frac{1}{2\pi i}\int_{+B(\mathcal{D}')} R'_{\lambda}(A_1)(I + K_8) d\lambda\\
&= I + F_8,
\end{aligned}$$

where $K_i \in \mathcal{K}(X)$, with $i = 6, 7$ and $F_i \in \mathcal{F}(X)$, with $i = 6, 7, 8$. Therefore, we conclude that $(\gamma - \overline{BA}) \in \Phi(X)$, and the proof of the first inclusion is completed. Now, in order to show that $\sigma_{e4}(AB) \subseteq \sigma_{e4}(B)\sigma_{e4}(A)$, we will prove that $(\gamma - AB) \in \Phi(X)$. Since $R(M_1) \subset \mathcal{D}(A)$ and $ABx = BAx - Fx$ for all $x \in \mathcal{D}(A)$, we obtain $(\gamma - AB)M_1 = (\gamma - BA + F)M_1 = (\gamma - BA)M_1 + FM_1 = I + F_4 + F_9 = I + F_{10}$, where $F_i \in \mathcal{F}(X)$, with $i = 9, 10$. Furthermore, we have $M_2(\gamma - \overline{AB}) = M_2\left[\gamma(\lambda - A_1)B + \gamma(I - \lambda B)\right] = I + F_{11}$, where $F_{11} \in \mathcal{F}(X)$. Hence, $(\gamma - \overline{AB}) \in \Phi(X)$ and, we deduce that $\sigma_{e4}(AB) \subseteq \sigma_{e4}(A)\sigma_{e4}(B)$.

(*ii*) The proof of this assertion holds from Theorem 7.3.1(*i*).

(*iii*) This assertion follows immediately from Theorem 7.3.1(*ii*). Q.E.D.

7.5.3 Invariance of the Essential Spectra

Theorem 7.5.3. *Let $A \in \mathcal{C}(X)$ and let J be an operator on X. The following statements are satisfied.*

(i) If $J \in UA\mathcal{F}(X)$, then $\sigma_{ei}(A) = \sigma_{ei}(A+J)$, $i = 4, 5$. Moreover, if $\mathbb{C}\backslash\sigma_{e5}(A)$ is connected and neither $\rho(A)$ nor $\rho(A+J)$ is empty, then $\sigma_{e6}(A) = \sigma_{e6}(A+J)$. Besides,
(ii) if $J \in UA\mathcal{F}_+(X)$, then $\sigma_{e1}(A) = \sigma_{e1}(A+J)$,
(iii) if $J \in UA\mathcal{F}_-(X)$, then $\sigma_{e2}(A) = \sigma_{e2}(A+J)$, and
(iv) if $J \in UA\mathcal{F}_+(X) \bigcap UA\mathcal{F}_-(X)$, then $\sigma_{e3}(A) = \sigma_{e3}(A+J)$. ◇

Proof. The items (ii), (iii), (iv) and the first part of (i) for $i = 4$ are obtained immediately by using Lemma 6.3.1. So, they are omitted. Next, we prove the statement (i) for $i = 5$. If $\lambda \notin \sigma_{e5}(A)$ then, by using Proposition 7.1.1, $\lambda \in \Phi_A$ and $i(\lambda - A) = 0$. Since $J \in UA\mathcal{F}(X)$, and applying Lemma 6.3.1(i), we deduce that $\lambda \in \Phi_{A+J}$ and $i(\lambda - A - J) = 0$ and therefore, $\lambda \notin \sigma_{e5}(A+J)$. Thus, $\sigma_{e5}(A+J) \subset \sigma_{e5}(A)$. Similarly, if $\lambda \notin \sigma_{e5}(A+J)$, then using Lemma 6.3.1$(i)$ and arguing as above, we get $\sigma_{e5}(A) \subset \sigma_{e5}(A+J)$. Q.E.D.

Corollary 7.5.1. *Let X be a Banach space and $A \in \mathcal{C}(X)$. The following statements are satisfied.*

(i) If $\sigma_{e5}(A)$ or $\sigma_{e6}(A)$ is empty, then $\sigma(A+J) = \sigma_p(A+J)$, for every $J \in UA\mathcal{F}(X)$.
(ii) If $\sigma_{e1}(A)$ is empty, then $\sigma(A+J) = \sigma_p(A+J) \bigcup \sigma_r(A+J)$, for every $J \in UA\mathcal{F}_+(X)$.
(iii) If $\sigma_{e2}(A)$ is empty, then $\sigma(A+J) = \sigma_p(A+J) \bigcup \sigma_r(A+J)$, for every $J \in UA\mathcal{F}_-(X)$.
(iv) If $\sigma_{e3}(A)$ is empty, then $\sigma(A+J) = \sigma_p(A+J) \bigcup \sigma_r(A+J)$, for every $J \in UA\mathcal{F}_+(X) \bigcap UA\mathcal{F}_-(X)$.
(v) If $\sigma_{e4}(A)$ is empty, then $\sigma(A+J) = \sigma_p(A+J) \bigcup \sigma_r(A+J)$, for every $J \in UA\mathcal{F}(X)$. ◇

Proof. This corollary follows immediately from Theorem 7.5.3 and from the facts that $\sigma_c(A) \subseteq \bigcap_{i=1}^{6} \sigma_{ei}(A)$, and $\sigma_r(A) \subseteq \sigma_{e5}(A) \subseteq \sigma_{e6}(A)$. Q.E.D.

The following result may be found in [217].

Proposition 7.5.1. *Let $A \in \mathcal{C}(X)$. If $\sigma_{e6}(A) = \sigma_{e5}(A)$, then for each $J \in \mathcal{F}(X)$, there is, at most, a countable set $\mathcal{U}$ of complex numbers, such that $\sigma_{e6}(A + \xi J) = \sigma_{e6}(A)$ for $\xi \notin \mathcal{U}$. If $\mathbb{C}\backslash\sigma_{e6}(A)$ consists of a finite number of components, then $\mathcal{U}$ is discrete.* ◇

Proof. Let ξ be a complex number. Since $\xi J \in \mathcal{F}(X)$, and applying Theorem 7.5.3, we get $\sigma_{e5}(A + \xi J) = \sigma_{e5}(A) = \sigma_{e6}(A)$. Let Σ be an arbitrary component of $\mathbb{C}\backslash\sigma_{e6}(A) = \rho_6(A)$ and let λ_0 be any point of Σ. By definition of $\rho_6(.)$, there is a neighborhood of λ_0, $\mathcal{V}_{\lambda_0}$, such that $\mathcal{V}_{\lambda_0}\backslash\{\lambda_0\} \subset \rho(A)$. Let $\lambda_1 \in \mathcal{V}_{\lambda_0}\backslash\{\lambda_0\} \subset \rho(A)$. Then, by using Lemma 6.3.1(i) for all ξ, $\lambda_1 \in \Phi^0_{A+\xi J}$. Now, by applying Lemma 6.3.1(i), we conclude that, for ξ is not in a discrete set $\mathcal{U}$, $\alpha(\lambda_1 - A - \xi J) = \beta(\lambda_1 - A - \xi J) = 0$, i.e., $\lambda_1 \in \rho(A + \xi J)$. Since $\Sigma \subset \Phi_{A+\xi J}$, it cannot contain any point of the set $\sigma_{e6}(A + \xi J)$. Since $\mathbb{C}\backslash\sigma_{e6}(A)$ consists of, at most, a countable number of components, the proof is complete. Q.E.D.

In general, Theorem 7.5.3 cannot be directly used in applications (see Chap. 13). So, we give, in the following theorem (see [217]), some practical criteria which guarantee the invariance of the various essential spectra by Fredholm and semi-Fredholm perturbations.

Theorem 7.5.4. *Let $A, B \in \mathcal{C}(X)$ and let $\lambda \in \rho(A)\bigcap\rho(B)$. The following statements are satisfied.*

(i) *If $(\lambda - A)^{-1} - (\lambda - B)^{-1} \in \mathcal{F}^b(X)$, then $\sigma_{ei}(A) = \sigma_{ei}(B)$, $i = 4, 5$. Moreover,*
(ii) *if $(\lambda - A)^{-1} - (\lambda - B)^{-1} \in \mathcal{F}^b_+(X)$, then $\sigma_{e7}(A) = \sigma_{e7}(B)$ and $\sigma_{e1}(A) = \sigma_{e1}(B)$,*
(iii) *if $(\lambda - A)^{-1} - (\lambda - B)^{-1} \in \mathcal{F}^b_-(X)$, then $\sigma_{e8}(A) = \sigma_{e8}(B)$ and $\sigma_{e2}(A) = \sigma_{e2}(B)$, and*
(iv) *if $(\lambda - A)^{-1} - (\lambda - B)^{-1} \in \mathcal{F}^b_+(X)\bigcap\mathcal{F}^b_-(X)$, then $\sigma_{e3}(A) = \sigma_{e3}(B)$.*
(v) *If A, $B \in \mathcal{L}(X)$ and $(\lambda - A)^{-1} - (\lambda - B)^{-1}$ is a compact operator commuting with A or B, then, $\sigma_{e6}(A) = \sigma_{e6}(B)$.* ◇

Proof. Without loss of generality, we may assume that $\lambda = 0$. Hence, $0 \in \rho(A)$ and therefore, $\mu - A = -\mu(\mu^{-1} - A^{-1})A$, $\mu \neq 0$. Since A is one-to-one and onto, then $\alpha(\mu - A) = \alpha(\mu^{-1} - A^{-1})$ and $R(\mu - A) = R(\mu^{-1} - A^{-1})$. This shows that $\mu \in \Phi_{+A}$ (resp. Φ_{-A}) if, and only if, $\mu^{-1} \in \Phi_{+A^{-1}}$ (resp. $\Phi_{-A^{-1}}$). Similarly, we have $\mu \in \Phi_A$ if, and only if, $\mu^{-1} \in \Phi_{A^{-1}}$.

(i) If $A^{-1} - B^{-1} \in \mathcal{F}^b(X)$, then using Lemma 6.3.1$(i)$, we conclude that $\Phi_A = \Phi_B$ and $i(\mu - A) = i(\mu - B)$ for all $\mu \in \Phi_A$.
(ii) and (iii) If $A^{-1} - B^{-1} \in \mathcal{F}^b_+(X)$ [resp. $\mathcal{F}^b_-(X)$], then using Lemma 6.3.1(ii) [resp. Lemma 6.3.1(iii)], we conclude that $\Phi_{+A} = \Phi_{+B}$ (resp. $\Phi_{-A} = \Phi_{-B}$) and $i(\mu - A) = i(\mu - B)$, for each $\mu \in \Phi_{+A}$ (resp. Φ_{-A}). This concludes the proof of (ii) [resp. (iii)].
(iv) Using (ii), (iii) and Lemma 6.3.1(iv), we have $\Phi_{-A}\bigcup\Phi_{+A} = \Phi_{-B}\bigcup\Phi_{+B}$. This ends the proof of (iv).
(v) Without loss of generality, we suppose that $\lambda = 0$. Hence, $0 \in \rho(A)\bigcap\rho(B)$ from the fact that $A^{-1} - B^{-1} = K$, where $K \in \mathcal{K}(X)$. Since $KB = BK$, then $\sigma_{e6}(A^{-1}) = \sigma_{e6}(B^{-1} + K) = \sigma_{e6}(B^{-1})$. Now, if we apply Corollary 2.2.1 to both A and B, we notice that $\sigma_{e6}(A) = \sigma_{e6}(B)$. Q.E.D.

Corollary 7.5.2. *Let $A \in \mathcal{C}(X)$ and let J be an A-bounded operator on X, and also assume that there is a complex number $\lambda \in \rho(A)$, such that $r_\sigma(J(\lambda - A)^{-1}) < 1$ where $r_\sigma(A)$ is the spectral radius of A. Then,*

(i) *If $J(\lambda - A)^{-1} \in \mathcal{F}^b(X)$, then $\sigma_{ei}(A) = \sigma_{ei}(A + J)$, $i = 4, 5$.*

(ii) *If* $J(\lambda - A)^{-1} \in \mathcal{F}_+^b(X)$, *then* $\sigma_{e1}(A) = \sigma_{e1}(A + J)$.
(iii) *If* $J(\lambda - A)^{-1} \in \mathcal{F}_-^b(X)$, *then* $\sigma_{e2}(A) = \sigma_{e2}(A + J)$.
(iv) *If* $J(\lambda - A)^{-1} \in \mathcal{F}_+^b(X) \bigcap \mathcal{F}_-^b(X)$, *then* $\sigma_{e3}(A) = \sigma_{e3}(A + J)$. ◇

Proof. Let $\lambda \in \rho(A)$. Since J is A-bounded, and according to Remark 2.1.4(*iv*), $J(\lambda - A)^{-1}$ is a closed linear operator defined on all X and therefore, bounded by the closed graph theorem (see Theorem 2.1.3). Moreover, the assumption $r_\sigma(J(\lambda - A)^{-1}) < 1$ implies that $\lambda \in \rho(A + J)$ and $(\lambda - A - J)^{-1} - (\lambda - A)^{-1} = \sum_{n\geq 1}(\lambda - A)^{-1}[J(\lambda - A)^{-1}]^n$. Clearly, if $J(\lambda - A)^{-1} \in \mathcal{F}^b(X)$ [resp. $\mathcal{F}_+^b(X)$, $\mathcal{F}_-^b(X)$, $\mathcal{F}_+^b(X) \bigcap \mathcal{F}_-^b(X)$], then the closedness of $\mathcal{F}^b(X)$ [resp. $\mathcal{F}_+^b(X)$, $\mathcal{F}_-^b(X)$, $\mathcal{F}_+^b(X) \bigcap \mathcal{F}_-^b(X)$] and the use of Theorem 6.3.4 (resp. Theorem 6.3.3) imply that $(\lambda - A - J)^{-1} - (\lambda - A)^{-1} \in \mathcal{F}^b(X)$ [resp. $\mathcal{F}_+^b(X)$, $\mathcal{F}_-^b(X)$, $\mathcal{F}_+^b(X) \bigcap \mathcal{F}_-^b(X)$]. Now, the items (*i*)–(*iv*) follow immediately from Theorem 7.5.4. Q.E.D.

Let $A \in \mathcal{C}(X)$. A complex number λ is in Φ_{+A}^-, Φ_{-A}^+, Φ_{lA}, Φ_{rA}, $\mathcal{W}_{lA}$ or $\mathcal{W}_{rA}$ if $\lambda - A$ is in $\Phi_+^-(X)$, $\Phi_-^+(X)$, $\Phi_l(X)$, $\Phi_r(X)$, $\mathcal{W}_l(X)$ or $\mathcal{W}_r(X)$, respectively. A is said to be a Weyl operator if A is Fredholm operator having index 0. We first prove the following useful stability result.

Theorem 7.5.5. *Let* $A \in \mathcal{C}(X)$ *and* $B \in \mathcal{C}(X)$. *If there exists* $\lambda_0 \in \rho(A) \bigcap \rho(B)$ *such that* $(A - \lambda_0)^{-1} - (B - \lambda_0)^{-1} \in \mathcal{R}(X)$, *then the following statements are satisfied.*

(i) *If* $(A - \lambda_0)^{-1}(B - \lambda_0)^{-1} - (B - \lambda_0)^{-1}(A - \lambda_0)^{-1} \in \mathcal{F}_+^b(X)$, *then* $\sigma_{ei}(A) = \sigma_{ei}(B)$, $i = 1, 7$.
(ii) *If* $(A - \lambda_0)^{-1}(B - \lambda_0)^{-1} - (B - \lambda_0)^{-1}(A - \lambda_0)^{-1} \in \mathcal{F}_-^b(X)$, *then* $\sigma_{ei}(A) = \sigma_{ei}(B)$, $i = 2, 8$.
(iii) *If* $(A - \lambda_0)^{-1}(B - \lambda_0)^{-1} - (B - \lambda_0)^{-1}(A - \lambda_0)^{-1} \in \mathcal{F}_+^b(X) \bigcap \mathcal{F}_-^b(X)$, *then* $\sigma_{e3}(A) = \sigma_{e3}(B)$.
(iv) *If* $(A - \lambda_0)^{-1}(B - \lambda_0)^{-1} - (B - \lambda_0)^{-1}(A - \lambda_0)^{-1} \in \mathcal{F}^b(X)$, *then* $\sigma_{ei}(A) = \sigma_{ei}(B)$, $i = 1l, 2r, 4, 5, 7l, 8r$.
(v) *If the hypothesis of (iv) is satisfied and if* $\mathbb{C}\backslash\sigma_{e5}(A)$ *and* $\mathbb{C}\backslash\sigma_{e5}(B)$ *are connected, then* $\sigma_{e6}(A) = \sigma_{e6}(B)$. ◇

Proof. Let $R \in \mathcal{R}(X)$ such that $(A - \lambda_0)^{-1} = (B - \lambda_0)^{-1} + R$. Then,

$$R(B - \lambda_0)^{-1} - (B - \lambda_0)^{-1}R = (A - \lambda_0)^{-1}(B - \lambda_0)^{-1} - (B - \lambda_0)^{-1}(A - \lambda_0)^{-1}.$$

(*i*) Since $(A - \lambda_0)^{-1}(B - \lambda_0)^{-1} - (B - \lambda_0)^{-1}(A - \lambda_0)^{-1} \in \mathcal{F}_+^b(X)$, applying Lemma 7.4.1(*i*), we find that $\Phi_{+(A-\lambda_0)^{-1}} = \Phi_{+(B-\lambda_0)^{-1}}$ and $\Phi_{+(A-\lambda_0)^{-1}}^- = \Phi_{+(B-\lambda_0)^{-1}}^-$. Again, applying Theorem 7.3.2 we infer that $\Phi_{+A} = \Phi_{+B}$ and $\Phi_{+A}^- = \Phi_{+B}^-$.
(*ii*) A similar reasoning as before.
(*iii*) Since the equalities $\sigma_{e3}(A) = \sigma_{e1}(A) \bigcap \sigma_{e2}(A)$ and $\sigma_{e3}(B) = \sigma_{e1}(B) \bigcap \sigma_{e2}(B)$ are known and $(A - \lambda_0)^{-1}(B - \lambda_0)^{-1} - (B - \lambda_0)^{-1}(A - \lambda_0)^{-1} \in \mathcal{F}_+^b(X) \bigcap \mathcal{F}_-^b(X)$, then by (*i*) and (*ii*) we deduce that $\sigma_{e3}(A) = \sigma_{e3}(B)$.

(*iv*) Since $(A-\lambda_0)^{-1}(B-\lambda_0)^{-1}-(B-\lambda_0)^{-1}(A-\lambda_0)^{-1} \in \mathcal{F}^b(X)$, then by Lemma 7.4.1(*iii*), Remark 7.4.2 and Theorem 7.3.2(*i*), we have $\sigma_{ei}(A) = \sigma_{ei}(B)$, $i = 1l, 2r, 4, 5, 7l, 8r$.

(*v*) The sets $\mathbb{C}\backslash\sigma_{e5}(A)$ and $\mathbb{C}\backslash\sigma_{e5}(B)$ are connected. So, using Theorem 7.3.1, we deduce that $\sigma_{e5}(A) = \sigma_{e6}(A)$ and $\sigma_{e5}(B) = \sigma_{e6}(B)$. So, (*iv*) gives $\sigma_{e6}(A) = \sigma_{e6}(B)$. Q.E.D.

Theorem 7.5.6. *Let $A, B \in \mathcal{L}(X)$ and let $\lambda \in \rho(A)\bigcap\rho(B)$. If $(\lambda-A)^{-1}-(\lambda-B)^{-1}$ is a compact operator commuting with A or B, then $\sigma_{es}(A) = \sigma_{es}(B)$.* ◇

Proof. By using Theorem 5.1.4, we infer that $\sigma_{es}((\lambda-A)^{-1}) = \sigma_{es}((\lambda-B)^{-1})$ and, according to Proposition 7.3.1, we have $\sigma_{es}(A) = \sigma_{es}(B)$. Q.E.D.

By virtue of the Proposition 7.3.1, we have

Theorem 7.5.7. *Let $T, S \in \mathcal{L}(X)$, and let $\lambda \in \rho(T)\bigcap\rho(S)$. Suppose that one of the following conditions holds*

(i) $(\lambda-T)^{-1}-(\lambda-S)^{-1}$ *is a quasi-nilpotent operator commuting with T or S.*
(ii) If there exists $\varepsilon > 0$, such that $\left\|(\lambda-T)^{-1}-(\lambda-S)^{-1}\right\| < \varepsilon$.

Then $\sigma_i(T) = \sigma_i(S)$, $i = se, es$. ◇

Theorem 7.5.8. *Let X be a Banach space satisfying the Dunford–Pettis property. Let $A \in \mathcal{L}(X)$ and let B be a positive bounded operator on X. If for some $\lambda \in \rho(A)$, $r_\sigma[(\lambda-A)^{-1}B] < 1$, and the operators $(\lambda-A)^{-1}B^{\frac{1}{2}}$ and $B^{\frac{1}{2}}(\lambda-A)^{-1}$ are weakly compact on X, then $\sigma_{es}(A+B) = \sigma_{es}(A)$.* ◇

Proof. Let $\lambda \in \rho(A)$ such that $r_\sigma[(\lambda-A)^{-1}B] < 1$, then $\lambda \in \rho(A+B)$ and we have $(\lambda-A-B)^{-1}-(\lambda-A)^{-1} = (\lambda-A)^{-1}\sum_{n=1}^{+\infty}[B(\lambda-A)^{-1}]^n$. All terms of this series contain the term $(\lambda-A)^{-1}B(\lambda-A)^{-1}$. Besides, $(\lambda-A)^{-1}B(\lambda-A)^{-1} = (\lambda-A)^{-1}B^{\frac{1}{2}}B^{\frac{1}{2}}(\lambda-A)^{-1}$ is a composition of two weakly operators on the Banach space X which satisfy the Dunford–Pettis property. From Lemma 2.1.13(*i*), it follows that $(\lambda-A)^{-1}B(\lambda-A)^{-1}$ is a compact operator commuting with $(\lambda-A)^{-1}$. Hence, $(\lambda-A-B)^{-1}-(\lambda-A)^{-1}$ is a compact operator. Theorem 7.5.6 implies that $\sigma_{es}(A+B) = \sigma_{es}(A)$. Q.E.D.

In the following results, we will compare between the essentially semi-regular spectrum of A and $A+B$, where A is the generator of a one-parameter semigroup and B is a small perturbation.

Theorem 7.5.9. *Let $A, B \in \mathcal{L}(X)$ such that A is a generator of a C_0-semigroup $(T(t))_{t\geq 0}$ on X. Then, $\sigma_{es}(A+B) = \sigma_{es}(A)$.* ◇

Proof. There exists a norm $|.|$ on X such that $\|x\| \leq |x| \leq M\|x\|$ for $x \in X$, $|T(t)| \leq e^{wt}$ and $\left\|(A-\lambda)^{-1}\right\| \leq \frac{1}{\lambda-w}$ for Re$\lambda > w$. Thus, for $\lambda > w+|B|$ the bounded operator $B(A-\lambda)^{-1}$ satisfies $|B(A-\lambda)^{-1}| < 1$ and therefore, $|I-B(A-\lambda)^{-1}| < 1$ is invertible for $\lambda > w+|B|$. Now, let $Q = (A-\lambda)^{-1}[I-B(A-\lambda)^{-1}] = (A-\lambda)^{-1}\sum_{n=0}^{+\infty}[B(A-\lambda)^{-1}]^n$. Then, $(\lambda-A-B)Q = [I-B(A-\lambda)^{-1})]^{-1} - B(A-\lambda)^{-1}[I-B(A-\lambda)^{-1})]^{-1} = I$, and

$$\begin{aligned} Q(\lambda - A - B)x &= (A-\lambda)^{-1}(\lambda - A - B)x \\ &\quad + \sum_{n=1}^{+\infty}(A-\lambda)^{-1}[B(A-\lambda)^{-1}]^n(\lambda - A - B)x \\ &= x - (A-\lambda)^{-1}Bx + \sum_{n=1}^{+\infty}(A-\lambda)^{-1}[B(A-\lambda)^{-1}]^n x \\ &\quad - \sum_{n=2}^{+\infty}(A-\lambda)^{-1}[B(A-\lambda)^{-1}]^n x, \end{aligned}$$

Then, $Q(\lambda - A - B)x = x$. Therefore, the resolvent of $A + B$ exists for $\lambda > w + |B|$ and it is given by Q. Moreover, $|(\lambda - A - B)^{-1}| = |(A - \lambda)^{-1}\sum_{n=1}^{+\infty}[B(A-\lambda)^{-1}]^n| \leq \frac{1}{(\lambda - w - |B|)}$. Since $|(A-\lambda)^{-1} - (B-A-\lambda)^{-1}| \leq \frac{1}{(\lambda-w)} + \frac{1}{(\lambda-w-|B|)}$, then we have the following $\lim_{\mathrm{Re}\lambda\to\infty} |(A-\lambda)^{-1} - (B-A-\lambda)^{-1}| = 0$, and hence, from Theorem 7.5.7, we get $\sigma_{es}(A+B) = \sigma_{es}(A)$. Q.E.D.

We also have the following useful stability result for the Gustafson, Weidmann, Rakočević, and Schmoeger essential spectra.

Theorem 7.5.10. *Let A, $B \in \mathcal{C}(X)$. Assume that there are A_0, $B_0 \in \mathcal{L}(X)$ and K_1, $K_2 \in \mathcal{K}(X)$, such that*

$$AA_0 = I - K_1 \tag{7.5.2}$$

$$BB_0 = I - K_2. \tag{7.5.3}$$

(i) If $0 \in \Phi_A \bigcap \Phi_B$ and $A_0 - B_0 \in \mathcal{F}_+^b(X)$ $\big[$resp. $A_0 - B_0 \in \mathcal{F}_-^b(X)\big]$, then

$$\sigma_{e1}(A) = \sigma_{e1}(B) \quad [\text{resp. } \sigma_{e2}(A) = \sigma_{e2}(B)]\,. \tag{7.5.4}$$

Moreover, if $i(A) = i(B)$, then $\sigma_{e7}(A) = \sigma_{e7}(B)$ [resp. $\sigma_{e8}(A) = \sigma_{e8}(B)$].

(ii) If $0 \in \Phi_A \bigcap \Phi_B$ and $A_0 - B_0 \in \mathcal{F}_+^b(X) \bigcap \mathcal{F}_-^b(X)$, then $\sigma_{e3}(A) = \sigma_{e3}(B)$. ◊

Proof. Let λ be a complex number, Eqs. (7.5.2) and (7.5.3) imply

$$(\lambda - A)A_0 - (\lambda - B)B_0 = K_1 - K_2 + \lambda(A_0 - B_0). \tag{7.5.5}$$

(*i*) If $\lambda \notin \sigma_{e1}(B)$ [resp. $\sigma_{e2}(B)$], then $(\lambda - B) \in \Phi_+(X)$ [resp. $\Phi_-(X)$]. Let $X_B = (\mathcal{D}(B), \|.\|_B)$ be a Banach space for the graph norm $\|.\|_B$. It is clear that $B \in \mathcal{L}(X_B, X)$. We can regard B as an operator from X_B into X. This will be denoted by $\hat{B}$. Using Eq. (6.3.5) we can show that $\lambda - \hat{B} \in \Phi_+^b(X_B, X)$ [resp. $\Phi_-^b(X_B, X)$]. Besides, since $K_2 \in \mathcal{K}(X)$, Eq. (7.5.3), Lemma 3.1.2 and

Theorem 2.2.42 imply that $B_0 \in \Phi^b(X, X_B)$ and consequently, $(\lambda - \hat{B})B_0 \in \Phi^b_+(X)$ [resp. $\Phi^b_-(X)$]. Since $A_0 - B_0 \in \mathcal{F}^b_+(X)$ [resp. $\mathcal{F}^b_-(X)$] and $K_1 - K_2$ is compact, then by using Eq. (7.5.5) and Lemma 6.3.1(*ii*), we can prove that $(\lambda - \hat{A})A_0 \in \Phi^b_+(X)$ [resp. $\Phi^b_-(X)$], and

$$i((\lambda - \hat{A})A_0) = i((\lambda - \hat{B})B_0). \tag{7.5.6}$$

Using a similar reasoning as before, the combination of Eqs. (7.5.2), (7.5.3), Lemma 3.1.2 and Theorem 2.2.42 shows that $A_0 \in \Phi^b(X, X_A)$. By referring to Theorem 2.2.6, we can write

$$A_0 S = I - F \text{ on } X_A, \tag{7.5.7}$$

where $S \in \mathcal{L}(X_A, X)$ and $F \in \mathcal{K}(X_A)$. By using Eq. (7.5.7), we have

$$(\lambda - \hat{A})A_0 S = (\lambda - \hat{A}) - (\lambda - \hat{A})F. \tag{7.5.8}$$

Since $S \in \Phi^b(X_A, X)$, and using Eq. (7.5.8) and Theorem 2.2.43, we show that $(\lambda - \hat{A})A_0 S \in \Phi^b_+(X_A, X)$ [resp. using Theorem 2.2.13(*i*), we show that $(\lambda - \hat{A})A_0 S \in \Phi^b_-(X_A, X)$]. Applying Lemma 6.3.1(*ii*), we can prove that $(\lambda - \hat{A}) \in \Phi^b_+(X_A, X)$ [resp. applying Lemma 6.3.1(*iii*) we prove that $(\lambda - \hat{A}) \in \Phi^b_-(X_A, X)$]. By using Eq. (6.3.5), we have $(\lambda - A) \in \Phi_+(X)$ [resp. $\Phi_-(X)$]. Hence, $\lambda \notin \sigma_{e1}(A)$ [resp. $\sigma_{e2}(A)$], and consequently $\sigma_{e1}(A) \subset \sigma_{e1}(B)$ [resp. $\sigma_{e2}(A) \subset \sigma_{e2}(B)$]. The inverse inclusion follows by symmetry and the proof of Eq. (7.5.4) is complete. Now, we prove $\sigma_{e7}(A) = \sigma_{e7}(B)$ [resp. $\sigma_{e8}(A) = \sigma_{e8}(B)$]. If $\lambda \notin \sigma_7(B)$ [resp. $\sigma_{e8}(B)$], then by Proposition 7.3.2, $(\lambda - B) \in \Phi_+(X)$ and $i(\lambda - B) \leq 0$ (resp. $\Phi_-(X)$ and $i(\lambda - B) \geq 0$). By the first result, we have $(\lambda - A) \in \Phi_+(X)$ [resp. $\Phi_-(X)$]. Since K_1, $K_2 \in \mathcal{K}(X)$, then by using Eqs. (7.5.2), (7.5.3) and Theorem 2.2.40, we have $i(A) + i(A_0) = i(I - K_1) = 0$, $i(B) + i(B_0) = i(I - K_2) = 0$, since $i(A) = i(B)$, then $i(A_0) = i(B_0)$. Using Eqs. (6.3.5), (7.5.6), Lemma 6.3.1 and Theorem 2.2.40, we can write $i(\lambda - A) + i(A_0) = i(\lambda - B) + i(B_0)$. Therefore, $i(\lambda - A) \leq 0$ (resp. $i(\lambda - A) \geq 0$). We conclude that $\lambda \notin \sigma_{e7}(A)$ [resp. $\sigma_{e8}(A)$]. Therefore, we prove the inclusion $\sigma_{e7}(A) \subset \sigma_{e7}(B)$ [resp. $\sigma_{e8}(A) \subset \sigma_{e8}(B)$]. The opposite inclusion follows by symmetry. Therefore, $\sigma_{e7}(A) = \sigma_{e7}(B)$ [resp. $\sigma_{e8}(A) = \sigma_{e8}(B)$].

(*ii*) We know that $\sigma_{e3}(A) = \sigma_{e1}(A) \bigcap \sigma_{e2}(A)$, and using (*i*), we have $\sigma_{e1}(A) = \sigma_{e1}(B)$ and $\sigma_{e2}(A) = \sigma_{e2}(B)$. Therefore, $\sigma_{e3}(A) = \sigma_{e3}(B)$. Q.E.D.

7.5.4 Characterization of the Rakočević and Schmoeger Essential Spectra

We give a characterization of the essential approximate point spectrum (resp. the essential defect spectrum) by means of upper (resp. lower) semi-Fredholm operators. Now, we are ready to express the first result of this section.

Theorem 7.5.11. *Let X be a Banach space and $A \in \mathcal{C}(X)$ with a nonempty resolvent set. Then,*

(i) $\sigma_{e7}(A) = \bigcap_{K \in \mathcal{D}_A(X)} \sigma_{ap}(A+K),$

where $\mathcal{D}_A(X) := \{K \in \mathcal{C}(X)$ *such that* K *is* A*-bounded and* $K(\mu - A)^{-1} \in \mathcal{F}_+^b(X)$, *for some* $\mu \in \rho(A)\}$.

(ii) $\sigma_{e8}(A) = \bigcap_{K \in \mathcal{F}_-^b(X)} \sigma_8(A+K).$ ◇

Proof.

(*i*) Since $\mathcal{K}(X) \subset \mathcal{D}_A(X)$, we infer that $\bigcap_{K \in \mathcal{D}_A(X)} \sigma_{ap}(A+K) \subset \sigma_{e7}(A)$. Conversely, let $\lambda \notin \bigcap_{K \in \mathcal{D}_A(X)} \sigma_{ap}(A+K)$, then there exists $K \in \mathcal{D}_A(X)$, such that

$$\inf_{\|x\|=1,\ x \in \mathcal{D}(A)} \|(\lambda - A - K)x\| > 0.$$

The use of Theorem 2.2.1 enables us to conclude that $\lambda - A - K \in \Phi_+(X)$. Since $Y := R(\lambda - A - K)$ is a closed subspace of X, then Y itself is also a Banach space with the same norm. Therefore, $(\lambda - \hat{A} - \hat{K})^{-1} \in \mathcal{L}(Y, X_A)$. Let $\mu \in \rho(A)$ such that $K(\mu - A)^{-1} \in \mathcal{F}_+^b(X)$. Then, we have

$$\hat{K}(\lambda - \hat{A} - \hat{K})^{-1} = \hat{K}(\mu - \hat{A})^{-1}[\mathcal{I} + (\mu - \lambda + \hat{K})(\lambda - \hat{A} - \hat{K})^{-1}], \tag{7.5.9}$$

where $\mathcal{I}$ denotes the embedding operator which maps every $x \in Y$ onto the same element in X. Since $(\mu - \lambda + \hat{K}) \in \mathcal{L}(X_A, X)$, and $\hat{K}(\mu - \hat{A})^{-1} \in \mathcal{F}_+^b(X)$, then from Theorem 6.3.1(*ii*) and Eq. (7.5.9), it follows that

$$\hat{K}(\lambda - \hat{A} - \hat{K})^{-1} \in \mathcal{F}_+^b(Y, X). \tag{7.5.10}$$

Clearly, $N(\mathcal{I}) = \{0\}$ and $R(\mathcal{I}) = Y$. So, $\mathcal{I} \in \Phi_+^b(Y, X)$ and $i(\mathcal{I}) \leq 0$. Therefore, we can deduce, from (7.5.10) and Lemma 6.3.1, that

$$\mathcal{I} + \hat{K}(\lambda - \hat{A} - \hat{K})^{-1} \in \Phi_+^b(Y, X) \text{ and } i(\mathcal{I} + \hat{K}(\lambda - \hat{A} - \hat{K})^{-1}) \leq 0. \tag{7.5.11}$$

This allows us to write $\lambda - \hat{A}$ in the form $\lambda - \hat{A} = (\mathcal{I} + \hat{K}(\lambda - \hat{A} - \hat{K})^{-1})(\lambda - \hat{A} - \hat{K})$ and, by using (7.5.11) and Theorem 2.2.7, we get $\lambda - \hat{A} \in \Phi_+^b(X_A, X)$ and $i(\lambda - \hat{A}) \leq 0$. Now, using Eq. (6.3.5), we infer that $\lambda - A \in \Phi_+(X)$ and $i(\lambda - A) \leq 0$. Finally, the use of Proposition 7.3.2 shows that $\lambda \notin \sigma_{e7}(A)$, which proves the assertion (i).

(ii) Since $\mathcal{K}(X) \subset \mathcal{F}_-^b(X)$, then $\bigcap_{K \in \mathcal{F}_-^b(X)} \sigma_\delta(A + K) \subset \sigma_{e8}(A)$. It remains to demonstrate that $\sigma_{e8}(A) \subset \bigcap_{K \in \mathcal{F}_-(X)} \sigma_\delta(A + K)$. To do this, we consider $\lambda \notin \bigcap_{K \in \mathcal{F}_-^b(X)} \sigma_\delta(A + K)$. Then, there exists $F \in \mathcal{F}_-^b(X)$ such that $\lambda \notin \sigma_\delta(A + F)$. Thus, $\lambda - A - F$ is surjective. Hence, $\lambda - A - F \in \Phi_-(X)$ and $i(\lambda - A - F) = \alpha(\lambda - A - F) \geq 0$. Therefore, by using Lemma 6.3.1(iii), we deduce that $\lambda - A \in \Phi_-(X)$ and $i(\lambda - A) = i(\lambda - A - F) \geq 0$. We conclude the proof by using Proposition 7.3.2. Q.E.D.

Remark 7.5.1.

(i) From Theorem 7.5.11, it follows immediately that $\sigma_{e7}(A + K) = \sigma_{e7}(A)$ for all $K \in \mathcal{D}_A(X)$ and $\sigma_{e8}(A + K) = \sigma_{e8}(A)$, for all $K \in \mathcal{F}_-^b(X)$.

(ii) Let $A \in \mathcal{C}(X)$. If $\sigma_{e8}(A) = \emptyset$, then for all $K \in \mathcal{F}_-(X)$, $\sigma(A + K) = \sigma_p(A + K)$. ◇

Corollary 7.5.3. *Let X be a Banach space and let $\mathcal{M}(X)$ be any subset of $\mathcal{L}(X)$. Then*

(i) If $\mathcal{K}(X) \subset \mathcal{M}(X) \subset \mathcal{D}_A(X)$, then,

$$\sigma_{e7}(A) = \bigcap_{K \in \mathcal{M}(X)} \sigma_{ap}(A + K).$$

(ii) If $\mathcal{K}(X) \subset \mathcal{M}(X) \subset \mathcal{F}_-^b(X)$, then

$$\sigma_{e8}(A) = \bigcap_{K \in \mathcal{M}(X)} \sigma_\delta(A + K).$$

◇

Remark 7.5.2. It follows immediately, from Corollary 7.5.3, that

(i) $\sigma_{e7}(A+K) = \sigma_{e7}(A)$ for all $K \in \mathcal{M}(X)$, such that $\mathcal{K}(X) \subset \mathcal{M}(X) \subset \mathcal{D}_A(X)$.

(ii) $\sigma_{e8}(A + K) = \sigma_{e8}(A)$ for all $K \in \mathcal{M}(X)$, such that $\mathcal{K}(X) \subset \mathcal{M}(X) \subset \mathcal{F}_-^b(X)$. ◇

In the next theorem, we will give a characterization of $\sigma_{e8}(\cdot)$ by means of A-bounded perturbations.

Theorem 7.5.12. *Let X be a Banach space and $A \in \mathcal{C}(X)$ with a nonempty resolvent set. Then,*

$$\sigma_{e8}(A) = \bigcap_{K \in \mathcal{H}_A(X)} \sigma_8(A + K),$$

where $\mathcal{H}_A(X) := \{K \in \mathcal{C}(X)$, *such that* K *is* A*-bounded and* $((\mu - A)^{-1}\hat{K})^* \in \mathcal{F}_+^b((X_A)^*)$, *for some* $\mu \in \rho(A)\}$. ◇

Proof. Let $\mathcal{O} := \bigcap_{K \in \mathcal{H}_A(X)} \sigma_8(A+K)$. We infer that $\mathcal{O} \subset \sigma_{e8}(A)$. Indeed, if K is a compact operator on X, then $\hat{K} \in \mathcal{K}(X_A, X)$. Hence, $((\mu - A)^{-1}\hat{K})^* \in \mathcal{K}((X_A)^*)$, since $(\mu - A)^{-1} \in \mathcal{L}(X, X_A)$. Using the fact that $\mathcal{K}((X_A)^*) \subset \mathcal{F}_+^b((X_A)^*)$, it follows that $\mathcal{K}(X) \subset \mathcal{H}_A(X)$. Conversely, let $\lambda \notin \mathcal{O}$. Then, there exists $K \in \mathcal{H}_A(X)$, such that $\lambda - A - K$ is surjective. Thus, $\lambda - A - K \in \Phi_-(X)$ and $\beta(\lambda - A - K) = 0$. Therefore, $\lambda - \hat{A} - \hat{K} \in \Phi_-^b(X_A, X)$. Hence, $\lambda - (\hat{A})^* - (\hat{K})^* \in \Phi_+^b(X^*, (X_A)^*)$ and $\alpha(\lambda - (\hat{A})^* - (\hat{K})^*) = 0$. Now, by following the same reasoning as in the proof of Theorem 7.5.11(i), we deduce that $\lambda - (\hat{A})^* \in \Phi_+^b(X^*, (X_A)^*)$ and $i(\lambda - (\hat{A})^*) \leq 0$. This, together with Eq. (6.3.5), allows us to conclude that $\lambda - A \in \Phi_-(X)$ and $i(\lambda - A) \geq 0$. Finally, the result follows from Proposition 7.3.2. Q.E.D.

7.6 Invariance of the Kato Spectrum by Commuting Nilpotent Perturbation

We start by gathering some results, which will be used to demonstrate that the Kato spectrum of an operator is stable by a commuting nilpotent perturbation. We begin this section by the following results:

Proposition 7.6.1. *Let* $A \in \mathcal{L}(X)$ *and let* Q *be a nilpotent operator commuting with* A*. Then,* $A + Q$ *is a nilpotent operator if, and only if,* A *is a nilpotent operator.* ◇

Proof. Let us assume that A is a nilpotent operator. Let r, s be the non-negative integers such that $A^r = 0 \neq A^{r-1}$ and $Q^s = 0 \neq Q^{s-1}$. Let $m = \max(r, s)$. Then,

$$(A+Q)^{2m} = C_{2m}^0 A^{2m} + \cdots + C_{2m}^m Q^m A^m + C_{2m}^{m+1} Q^{m+1} A^{m-1} + \cdots + C_{2m}^{2m} Q^{2m} = 0.$$

Hence, $A + Q$ is a nilpotent operator. For the converse statement, we use the relation $A = (A + Q) - Q$. Q.E.D.

Lemma 7.6.1. *Let* $A \in \mathcal{L}(X)$*.* A *is a Kato type operator if, and only if, there exists a closed subspace* V *of* X *such that* $A(V) = V$ *and the operator* $\hat{A} : X/V \longrightarrow X/V$ *induced by* A *is a direct sum of a bounded below operator and a nilpotent operator.* ◇

Proof. If A is semi-regular, we use Theorem 2.2.34 by taking the zero operator as the nilpotent operator. If A is a nilpotent operator, we take $V = \{0\}$. Now, suppose that A is neither semi-regular, nor nilpotent with admits a Kato decomposition (M, N). Then, set $V = R^{\infty}(A)$. It is well known, according to Theorem 2.2.35, that V is closed, $V \subseteq M$ and $A(V) = V$. Furthermore, $X/V = M/V \oplus N/V$, $\hat{A}(M/V) \subseteq M/V$ and $\hat{A}(N/V) \subseteq N/V$. Let us denote by $\hat{A}_1$ (resp. $\hat{A}_2$) the restriction of $\hat{A}$ on M/V (resp. N/V). Then, we have $\hat{A} = \hat{A}_1 \oplus \hat{A}_2$. Since A_N is a nilpotent operator, then $\hat{A}_2$ is also a nilpotent operator and by using Theorem 2.2.34, $\hat{A}_1$ is bounded below because A_M is semi-regular. Conversely, let V be a closed subspace of X with $A(V) = V$ and $\hat{A}$ is decomposed according to $X/V = M/V \oplus N/V$ and the parts $\hat{A}_1$ and $\hat{A}_2$ are bounded below and nilpotent, respectively, where M, N are two closed subspaces of X. The fact that $A(V) = V$, allows us to prove easily that (M, N) is a Kato decomposition of A and hence, A is an operator of Kato type. Q.E.D.

We recall a result due to Kordula and Muller [192].

Theorem 7.6.1. *Let Q, $A \in \mathcal{L}(X)$, $QA = AQ$, and let A be a quasi-nilpotent. Then $\sigma_{se}(A + Q) = \sigma_{se}(A)$ and $\sigma_{es}(A + Q) = \sigma_{es}(A)$.* ◇

Now, we show that the operators of Kato type are stable under commuting nilpotent perturbations.

Theorem 7.6.2. *Let $A \in \mathcal{L}(X)$, and let $AQ = QA$, where Q is a nilpotent operator on X. Then $\sigma_k(A + Q) = \sigma_k(A)$.* ◇

Proof. Let A be an operator of Kato type and let Q be a nilpotent operator commuting with A. If A is semi-regular, we apply the Theorem 7.6.1, and if A is a nilpotent operator we apply Proposition 7.6.1. Now, let us suppose that A is neither semi-regular nor nilpotent. Let us denote $V = R^{\infty}(A)$, $A_1 = A_V$ and $\hat{A} : X/V \longrightarrow X/V$ induced by A. Clearly $Q(V) \subseteq V$, so that we can define the operators $Q_1 = Q_V$ and $\hat{Q} : X/V \longrightarrow X/V$ induced by Q. Obviously, Q_1 and $\hat{Q}$ are nilpotent operators. Further, $A_1 Q_1 = Q_1 A_1$ and $\hat{A}\hat{Q} = \hat{Q}\hat{A}$. By using the stability of $\sigma_{ap}(A)$ and $\sigma_{\delta}(A)$ under nilpotent perturbation, we have $\sigma_{\delta}(A_1 + Q_1) = \sigma_{\delta}(A_1)$, $\sigma_{ap}(\hat{A} + \hat{Q}) = \sigma_{ap}(\hat{A})$ and $\sigma(\hat{A} + \hat{Q}) = \sigma(\hat{A})$. Hence, $0 \notin \sigma_{\delta}(A_1 + Q_1)$ and so, $(A + Q)(V) = V$. By using Lemma 7.6.1, $\hat{A} = \hat{A}_1 \oplus \hat{A}_2$, where $\hat{A}_1$ is bounded below and $\hat{A}_2$ is a nilpotent operator. Hence, $\sigma_{ap}(\hat{A} + \hat{Q}) = \sigma_{ap}(\hat{A}) = \sigma_{ap}(\hat{A}_1) \bigcup \sigma_{ap}(\hat{A}_2)$ and $\sigma(\hat{A} + \hat{Q}) = \sigma(\hat{A}) = \sigma(\hat{A}_1) \bigcup \sigma(\hat{A}_2)$. Moreover, $\sigma_{ap}(\hat{A}_2) = \sigma(\hat{A}_2) = \{0\}$ and $0 \notin \sigma_{ap}(\hat{A}_1) \subseteq \sigma(\hat{A}_1)$. This implies that $\sigma(\hat{A})$ and hence $\sigma(\hat{A} + \hat{Q})$ is separated in two disjoint parts $\sigma(\hat{A}_1)$ and $\sigma(\hat{A}_2)$. By using Theorem 2.2.37, we have a decomposition of $\hat{A}$ (and hence of $\hat{A} + \hat{Q}$) according to the decomposition of X/V in such a way that $\sigma((\hat{A} + \hat{Q})_{M/V}) = \sigma(\hat{A}_1)$ and $\sigma((\hat{A} + \hat{Q})_{N/V}) = \sigma(\hat{A}_2)$, where M, N are two closed subspaces of X. Thus $(\hat{A} + \hat{Q})_{N/V}$ is a nilpotent operator and $\sigma_{ap}((\hat{A} + \hat{Q})_{M/V}) = \sigma_{ap}(\hat{A}_1)$, i.e., $(\hat{A} + \hat{Q})_{M/V}$ is bounded below. This shows that $\hat{A} + \hat{Q}$ is a direct sum of bounded below operator and a nilpotent operator. Then by using Lemma 7.6.1, we deduce that $A + Q$ is an operator of Kato type. Q.E.D.

Theorem 7.6.3. *Let $A, B \in \mathcal{L}(X)$. If $\lambda \in \rho(A) \bigcap \rho(B)$, such that $(\lambda - A)^{-1} - (\lambda - B)^{-1}$ is a nilpotent operator commuting with A and B, then $\sigma_k(A) = \sigma_k(B)$.* ◇

Proof. The assumptions of Theorem 7.6.2 imply that $\sigma_k((\lambda - A)^{-1}) = \sigma_k((\lambda - B)^{-1})$ and by using Theorem 7.3.2, we have $\sigma_k(A) = \sigma_k(B)$. Q.E.D.

7.7 Invariance of Schechter's Essential Spectrum

7.7.1 *Characterization of Schechter's Essential Spectrum*

Definition 7.7.1. Let X be a Banach space, $A \in \mathcal{C}(X)$, $\rho(A) \neq \emptyset$ and let F be an A-defined linear operator on X. We say that F is an A-resolvent Fredholm perturbation if $(\lambda - \hat{A})^{-1}\hat{F} \in \mathcal{F}^b(X_A)$ for some $\lambda \in \rho(A)$. ◇

Let $A\mathcal{RF}(X)$ designate the set of A-resolvent Fredholm perturbations.

Remark 7.7.1. In the definition of the set $A\mathcal{RF}(X)$, we may notice the following: If an operator satisfies the required condition for a fixed $\lambda \in \rho(A)$, then it satisfies it for every $\lambda \in \rho(A)$. ◇

Let $\mathcal{I}$ denote the imbedding operator which maps every $x \in X_A$ onto the same element $x \in X_{A+F}$. Clearly, we have $N(\mathcal{I}) = \{0\}$ and $R(\mathcal{I}) = X_{A+F}$. Let $\mathcal{J}(X_A, X_{A+F})$ the set

$$\mathcal{J}(X_A, X_{A+F}) := \big\{F \in \mathcal{L}(X_A, X_{A+F}) \text{ such that } \mathcal{I} - F \in \Phi^b(X_A, X_{A+F}) \text{ and }$$
$$i(\mathcal{I} - F) = 0\big\}.$$

Definition 7.7.2. Let X be a Banach space, $A \in \mathcal{C}(X)$, and let F be an A-defined linear operator on X and $\rho(A + F) \neq \emptyset$. We say that F is an A-resolvent Fredholm perturbation with zero index, if $(\lambda - \hat{A} - \hat{F})^{-1}\hat{F} \in \mathcal{J}(X_A, X_{A+F})$ for all $\lambda \in \rho(A + F)$. ◇

Let $A\mathcal{J}(X)$ designate the set of A-resolvent Fredholm perturbations with zero index.

Remark 7.7.2.

(i) For all $\lambda \in \rho(A)$, the operator $(\lambda - \hat{A})^{-1} \in \mathcal{L}(X, X_A)$. In fact, let $x \in X$ and let $y = (\lambda - A)^{-1}x$. From the estimate

$$\begin{aligned}\|y\|_A &= \|y\| + \|\hat{A}y\| \\ &= \|y\| + \|\lambda y - x\| \\ &= \|(\lambda - \hat{A})^{-1}x\| + \|\lambda(\lambda - \hat{A})^{-1}x - x\| \\ &\leq \Big(1 + (1 + |\lambda|)\|(\lambda - \hat{A})^{-1}\|\Big)\|x\|,\end{aligned}$$

it follows that $(\lambda - \hat{A})^{-1} \in \mathcal{L}(X, X_A)$.

(ii) If F is an A-Fredholm perturbation, then F is also an A-resolvent Fredholm perturbation. In fact, if $F \in A\mathcal{F}(X)$, then $\hat{F} \in \mathcal{F}^b(X_A, X)$. By using Remark 7.7.2($i$), we have $(\lambda - \hat{A})^{-1} \in \mathcal{L}(X, X_A)$ and Theorem 6.3.4 gives $(\lambda - \hat{A})^{-1}\hat{F} \in \mathcal{F}^b(X_A)$. This proves that $F \in A\mathcal{RF}(X)$.

(iii) For all $\lambda \in \rho(A+F)$, the operator $(\lambda - \hat{A} + \hat{F})^{-1} \in \mathcal{L}(X, X_{A+F})$. In fact, let $x \in X$ and let $y = (\lambda - A + F)^{-1}x$. From the estimate

$$\begin{aligned}\|y\|_{A+F} &= \|y\| + \|(\hat{A} + \hat{F})y\| \\ &= \|y\| + \|\lambda y - x\| \\ &= \|(\lambda - \hat{A} - \hat{F})^{-1}x\| + \|\lambda(\lambda - \hat{A} - \hat{F})^{-1}x - x\| \\ &\leq \left(1 + (1 + |\lambda|)\|(\lambda - \hat{A} - \hat{F})^{-1}\|\right)\|x\|,\end{aligned}$$

it follows that

$$(\lambda - \hat{A} - \hat{F})^{-1} \in \mathcal{L}(X, X_{A+F}). \tag{7.7.1}$$

(iv) If F is an A-resolvent Fredholm perturbation, then F is an A-resolvent Fredholm perturbation with zero index. In fact, the estimate

$$\begin{aligned}\|\mathcal{I}(x)\|_{X_{A+F}} = \|x\|_{A+F} &\leq \|x\| + \|Ax\|_X + \|Fx\|_X \\ &\leq \left(1 + \|F\|_{\mathcal{L}(X_A,X)}\right)\|x\|_{X_A}, \quad \forall x \in X_A\end{aligned}$$

leads to $\mathcal{I} \in \Phi^b(X_A, X_{A+F})$ and $i(\mathcal{I}) = 0$. Let $\mu \in \rho(A)$. Then, we have

$$(\lambda - \hat{A} - \hat{F})^{-1}\hat{F} = [\mathcal{I} + (\lambda - \hat{A} - \hat{F})^{-1}(\mu - \lambda + \hat{F})](\mu - \hat{A})^{-1}\hat{F}. \tag{7.7.2}$$

Since $(\mu - \lambda + \hat{F}) \in \mathcal{L}(X_A, X)$, and applying (7.7.1), we infer that $(\lambda - \hat{A} - \hat{F})^{-1}(\mu - \lambda + \hat{F}) \in \mathcal{L}(X_A, X_{A+F})$ and therefore, $\mathcal{I} + (\lambda - \hat{A} - \hat{F})^{-1}(\mu - \lambda + \hat{F}) \in \mathcal{L}(X_A, X_{A+F})$. By using (7.7.2), we deduce that F is an A-resolvent Fredholm perturbation, and from Proposition 6.3.1(i), we infer that $(\lambda - \hat{A} - \hat{F})^{-1}\hat{F} \in \mathcal{F}^b(X_A, X_{A+F})$. This proves that $F \in A\mathcal{J}(X)$.

(v) As a consequence of Definition 7.7.2, Remark 7.7.2(ii), (iv) and the inclusions in [124, p. 69], we deduce that

$$A\mathcal{K}(X) \subset A\mathcal{S}(X) \subset A\mathcal{F}_+(X) \subset A\mathcal{F}(X) \subset ARF(X) \subset A\mathcal{J}(X), \text{ and}$$

$$A\mathcal{K}(X) \subset AC\mathcal{S}(X) \subset A\mathcal{F}_-(X) \subset A\mathcal{F}(X) \subset ARF(X) \subset A\mathcal{J}(X).$$

The inclusion $A\mathcal{S}(X) \subset A\mathcal{F}_+(X)$ [resp. $AC\mathcal{S}(X) \subset A\mathcal{F}_-(X)$] was established in [186] (resp. [330]). ◇

Let X be a fixed Banach space. Let $A \in \mathcal{C}(X)$, and let J be an A-bounded operator on X. Let $\lambda \in \rho(A)$. Since J is A-bounded, then $J(\lambda - A)^{-1}$ is a closed linear operator defined on all elements of X and therefore, bounded by the closed graph theorem (see Theorem 2.1.3). We have the following result:

Theorem 7.7.1. *Let X be a Banach space and let $A \in \mathcal{C}(X)$. Then,*

$$\sigma_{e5}(A) = \bigcap_{K \in \mathcal{D}_A(X)} \sigma(A+K),$$

where $\mathcal{D}_A(X) = \{J \in \mathcal{C}(X)$ such that J is A-bounded and $J(\lambda - A)^{-1} \in \mathcal{F}^b(X)$ for some $\lambda \in \rho(A)\}$. ◇

Proof. Since $\mathcal{K}(X) \subset \mathcal{D}_A(X)$, we infer that $\bigcap_{K \in \mathcal{D}_A(X)} \sigma(A+K) \subset \sigma_{e5}(A)$. Conversely, let $\lambda \notin \bigcap_{K \in \mathcal{D}_A(X)} \sigma(A+K)$. Then, there exists $J \in \mathcal{D}_A(X)$ such that $\lambda \in \rho(A+J)$. Hence, $\lambda \in \Phi_{(A+J)}$ and $i(\lambda - A - J) = 0$. Let $\mu \in \rho(A)$. We have

$$J(\lambda - A - J)^{-1} = J(\mu - A)^{-1}[I + (\mu - \lambda + J)(\lambda - A - J)^{-1}]. \tag{7.7.3}$$

By using (7.7.3) and the fact that $\mathcal{F}^b(X)$ is a two-sided ideal of $\mathcal{L}(X)$, we infer that $J(\lambda - A - J)^{-1} \in \mathcal{F}^b(X)$. Applying Lemma 6.3.1, we infer that $I + J(\lambda - A - J)^{-1}$ is a Fredholm operator and $i(I + J(\lambda - A - J)^{-1}) = 0$. Using the equality $\lambda - A = (I + J(\lambda - A - J)^{-1})(\lambda - A - J)$, together with Atkinson's theorem (Theorem 2.2.40), we get $\lambda \in \Phi_A$ and $i(\lambda - A) = 0$. Finally, the use of Proposition 7.1.1 shows that $\lambda \notin \sigma_{e5}(A)$, which completes the proof of the theorem. Q.E.D.

Remark 7.7.3. For all $K \in \mathcal{D}_A(X)$, $\sigma_{e5}(A+K) = \sigma_{e5}(A)$. ◇

We close this section by the following result:

Theorem 7.7.2. *Let X be a Banach space, $A \in \mathcal{C}(X)$, and let $\mathcal{I}(X)$ be any subset of operators satisfying the condition $\mathcal{K}(X) \subset \mathcal{I}(X) \subset \mathcal{ARF}(X)$. Then,*

$$\sigma_{e5}(A) = \bigcap_{J \in \mathcal{I}(X)} \sigma(A+J).$$

◇

Proof. Set $\mathcal{O} := \bigcap_{J \in \mathcal{I}(X)} \sigma(A+J)$. We first claim that $\sigma_{e5}(A) \subset \mathcal{O}$. Indeed, if $\lambda \notin \mathcal{O}$, then there exists $J \in \mathcal{I}(X)$ such that $\lambda \in \rho(A+J)$. Let $x \in X$ and $y = (\lambda - A - J)^{-1}x$. From the following estimate

$$\begin{aligned}
\|y\|_{A+J} &= \|y\| + \|(\hat{A} + \hat{J})y\| \\
&= \|y\| + \|x - \lambda y\| \\
&= \|(\lambda - \hat{A} - \hat{J})^{-1}x\| + \|x - \lambda(\lambda - \hat{A} - \hat{J})^{-1}x\| \\
&\leq \left(1 + (1 + |\lambda|)\|(\lambda - \hat{A} - \hat{J})^{-1}\|\right)\|x\|,
\end{aligned}$$

it follows that

$$(\lambda - \hat{A} - \hat{J})^{-1} \in \mathcal{L}(X, X_{A+J}). \quad (7.7.4)$$

Let $\mathcal{I}$ denote the imbedding operator which maps every $x \in X_A$ onto the same element $x \in X_{A+J}$. Clearly, we have $N(\mathcal{I}) = \{0\}$ and $R(\mathcal{I}) = X_{A+J}$. So, the following estimate

$$\begin{aligned} \|\mathcal{I}(x)\|_{X_{A+J}} = \|x\|_{A+J} &\leq \|x\| + \|Ax\|_X + \|Jx\|_X \\ &\leq \left(1 + \|J\|_{\mathcal{L}(X_A,X)}\right) \|x\|_{X_A}, \quad \forall x \in X_A \end{aligned}$$

leads to $\mathcal{I} \in \Phi^b(X_A, X_{A+J})$ and $i(\mathcal{I}) = 0$. Let $\mu \in \rho(A)$. We have

$$(\lambda - \hat{A} - \hat{J})^{-1}\hat{J} = [\mathcal{I} + (\lambda - \hat{A} - \hat{J})^{-1}(\mu - \lambda + \hat{J})](\mu - \hat{A})^{-1}\hat{J}. \quad (7.7.5)$$

Since $(\mu - \lambda + \hat{J}) \in \mathcal{L}(X_A, X)$, and applying (7.7.4), we deduce that $(\lambda - \hat{A} - \hat{J})^{-1}(\mu - \lambda + \hat{J}) \in \mathcal{L}(X_A, X_{A+J})$ and therefore, $\mathcal{I} + (\lambda - \hat{A} - \hat{J})^{-1}(\mu - \lambda + \hat{J}) \in \mathcal{L}(X_A, X_{A+J})$. By using (7.7.5), $J \in \mathcal{I}(X) \subset \mathcal{ARF}(X)$ and Theorem 6.3.4, we infer that $(\lambda - \hat{A} - \hat{J})^{-1}\hat{J} \in \mathcal{F}^b(X_A, X_{A+J})$. By using Lemma 6.3.1, we get

$$\mathcal{I} + (\lambda - \hat{A} - \hat{J})^{-1}\hat{J} \in \Phi^b(X_A, X_{A+J}) \text{ and } i(\mathcal{I} + (\lambda - \hat{A} - \hat{J})^{-1}\hat{J}) = 0. \quad (7.7.6)$$

Moreover, since $\lambda \in \rho(A + J)$, it follows, from Eq. (6.3.5), that

$$(\lambda - \hat{A} - \hat{J}) \in \Phi^b(X_{A+J}, X) \text{ and } i(\lambda - \hat{A} - \hat{J}) = 0. \quad (7.7.7)$$

Hence, writing $\lambda - \hat{A}$ in the form $\lambda - \hat{A} = (\lambda - \hat{A} - \hat{J})(\mathcal{I} + (\lambda - \hat{A} - \hat{J})^{-1}J)$, and using (7.7.6), (7.7.7) as well as Atkinson's theorem (Theorem 2.2.40), we get $\lambda - \hat{A} \in \Phi^b(X_A, X)$ and $i(\lambda - \hat{A}) = 0$. Now, using (6.3.5), we infer that $\lambda \in \Phi_A$ and $i(\lambda - A) = 0$. Finally, the use of Proposition 7.1.1 shows that $\lambda \notin \sigma_{e5}(A)$, which proves the claim. Besides, since $\mathcal{K}(X) \subset \mathcal{I}(X)$, we infer that $\mathcal{O} \subset \sigma_{e5}(A)$, which completes the proof of the theorem. Q.E.D.

7.7.2 *Invariance by Means of Demicompact Operators*

In this section, we will give a refinement of the definition of the Schechter's essential spectrum. For this purpose, let X be a Banach space and $A \in \mathcal{C}(X)$. We define these two sets $\Upsilon_A(X)$ and $\Psi_A(X)$ by:

$\Upsilon_A(X) = \{K \in \mathcal{L}(X)$ such that $\forall \lambda \in \rho(A+K), -(\lambda - A - K)^{-1}K \in \Lambda_X\}$,
$\Psi_A(X) = \{K$ is A-bounded such that $\forall \lambda \in \rho(A+K), -K(\lambda - A - K)^{-1} \in \Lambda_X\}$,

where Λ_X is the set defined in (7.4.25).

Theorem 7.7.3. *For each $A \in \mathcal{C}(X)$, we have*

$$\sigma_{e5}(A) = \bigcap_{K \in \Upsilon_A(X)} \sigma(A+K) = \bigcap_{K \in \Psi_A(X)} \sigma(A+K).$$

◇

Proof. Note that, if $A \in \mathcal{C}(X)$, then K is an A-bounded operator and $\lambda \in \rho(A+K)$. Hence, $A(\lambda - A - K)^{-1}$ is a closed linear operator defined on X, and therefore bounded. We first claim that $\sigma_{e5}(A) \subset \bigcap_{K \in \Upsilon_A(X)} \sigma(A+K)$ [resp. $\sigma_{e5}(A) \subset \bigcap_{K \in \Psi_A(X)} \sigma(A+K)$]. Indeed, if $\lambda \notin \bigcap_{K \in \Upsilon_A(X)} \sigma(A+K)$ [resp. $\lambda \notin \bigcap_{K \in \Psi_A(X)} \sigma(A+K)$], then there exists $K \in \Upsilon_A(X)$ [resp. $K \in \Psi_A(X)$] such that $-(\lambda - A - K)^{-1}K \in \Lambda_X$ (resp. $-K(\lambda - A - K)^{-1} \in \Lambda_X$) whenever $\lambda \in \rho(A+K)$. Hence, by applying Theorem 5.4.2, we get $[I + (\lambda - A - K)^{-1}K] \in \Phi^b(X)$ and $i[I + (\lambda - A - K)^{-1}K] = 0$ (resp. $[I + K(\lambda - A - K)^{-1}] \in \Phi^b(X)$ and $i[I + K(\lambda - A - K)^{-1}] = 0$). Moreover, we have $\lambda - A = (\lambda - A - K)[I + (\lambda - A - K)^{-1}K]$ [resp. $\lambda - A = [I + K(\lambda - A - K)^{-1}](\lambda - A - K)$]. Then, by using Theorem 2.2.40, one gets $(\lambda - A) \in \Phi(X)$ and $i(\lambda - A) = 0$. By applying Proposition 7.1.1, we conclude that $\lambda \notin \sigma_{e5}(A)$, which proves our claim. For the inverse inclusion, since $\mathcal{K}(X) \subset \Upsilon_A(X)$ [resp. $\mathcal{K}(X) \subset \Psi_A(X)$], we infer that $\bigcap_{K \in \Upsilon_A(X)} \sigma(A+K) \subset \sigma_{e5}(A)$ [resp. $\bigcap_{K \in \Psi_A(X)} \sigma(A+K) \subset \sigma_{e5}(A)$]. Q.E.D.

Corollary 7.7.1. *Let $A \in \mathcal{C}(X)$, and let $\mathcal{E}(X)$ be a subset of $\Upsilon_A(X)$ [resp. $\Psi_A(X)$] containing $\mathcal{K}(X)$. Then,*

$$\sigma_{e5}(A) = \bigcap_{K \in \mathcal{E}(X)} \sigma(A+K).$$

◇

Proof. Since $\mathcal{E}(X) \subset \Upsilon_A(X)$, we obtain $\bigcap_{K \in \Upsilon_A(X)} \sigma(A+K) \subset \bigcap_{K \in \mathcal{E}(X)} \sigma(A+K)$. Applying Theorem 7.7.3, we get $\sigma_{e5}(A) \subset \bigcap_{K \in \mathcal{E}(X)} \sigma(A+K)$. Moreover, since $\mathcal{K}(X) \subset \mathcal{E}(X)$, we have $\bigcap_{K \in \mathcal{E}(X)} \sigma(A+K) \subset \sigma_{e5}(A)$, which completes the proof of the first assertion. For the second one, it is sufficient to replace $\Upsilon_A(X)$ by $\Psi_A(X)$. Q.E.D.

Corollary 7.7.2. *Let $A \in \mathcal{C}(X)$, and let $\mathcal{H}_A(X)$ be a subset of $\Upsilon_A(X)$ [resp. $\Psi_A(X)$], containing $\mathcal{K}(X)$. If, for all $K, K' \in \mathcal{H}_A(X)$, $K \pm K' \in \mathcal{H}_A(X)$, then for every $K \in \mathcal{H}_A(X)$, we have $\sigma_{e5}(A) = \sigma_{e5}(A+K)$.* ◇

Proof. We denote $\sigma'(A)$ by $\sigma'(A) = \bigcap_{K\in\mathcal{H}_A(X)} \sigma(A+K)$. From Corollary 7.7.1, we have $\sigma_{e5}(A) = \sigma'(A)$. Furthermore, for each $K \in \mathcal{H}_A(X)$, we have $\mathcal{H}_A(X) + K = \mathcal{H}_A(X)$. Then, $\sigma'(A+K) = \sigma'(A)$. Hence, we get the desired result. Q.E.D.

7.7.3 Invariance by Means of Noncompactness Measure

The purpose of this subsection is to give a refinement of the definition of Schechter's essential spectrum by means of noncompactness measure. In order to do this, let X be a Banach space and let $n \in \mathbb{N}^*$. We have the following:

Theorem 7.7.4. *Let X be a Banach space and let $n \in \mathbb{N}^*$. For each $A \in \mathcal{C}(X)$, we have*

$$\sigma_{e5}(A) = \bigcap_{K\in\mathcal{G}_A^n(X)} \sigma(A+K),$$

where $\mathcal{G}_A^n(X) = \left\{ K \in \mathcal{L}(X) \,:\, \gamma\left(\left[(\lambda - A - K)^{-1}K\right]^n\right) < \frac{1}{2} \text{ for all } \lambda \in \rho(A+K) \right\}.$ ◇

Proof. First, we claim that $\sigma_{e5}(A) \subset \bigcap_{K\in\mathcal{G}_A^n(X)} \sigma(A+K)$. Indeed, if $\lambda \notin \bigcap_{K\in\mathcal{G}_A^n(X)} \sigma(A+K)$, then there exists $K \in \mathcal{G}_A^n(X)$ such that $\lambda \in \rho(A+K)$. So, $\lambda \in \rho(A+K)$ and $\gamma\left(\left[(\lambda - A - K)^{-1}K\right]^n\right) < \frac{1}{2}$. Hence, applying Corollary 5.2.2(i), we get the following $\left[I + (\lambda - A - K)^{-1}K\right] \in \Phi^b(X)$ and $i\left[I + (\lambda - A - K)^{-1}K\right] = 0$. Moreover, we have $\lambda - A = (\lambda - A - K)\left[I + (\lambda - A - K)^{-1}K\right]$. Then, $(\lambda - A) \in \Phi(X)$ and $i(\lambda - A) = 0$. Finally, the use of Proposition 7.1.1 shows that $\lambda \notin \sigma_{e5}(A)$, which proves our claim. Besides, since $\mathcal{K}(X) \subset \mathcal{G}_A^n(X)$, we infer that $\bigcap_{K\in\mathcal{G}_A^n(X)} \sigma(A+K) \subset \sigma_{e5}(A)$ which completes the proof of the theorem. Q.E.D.

Corollary 7.7.3. *Let $n \in \mathbb{N}^*$, $A \in \mathcal{C}(X)$, and let $\mathcal{H}(X)$ be any subset of $\mathcal{L}(X)$ satisfying $\mathcal{K}(X) \subset \mathcal{H}(X) \subset \mathcal{G}_A^n(X)$. Then,*

$$\sigma_{e5}(A) = \bigcap_{K\in\mathcal{H}(X)} \sigma(A+K).$$

◇

Proof. Since $\mathcal{H}(X) \subset \mathcal{G}_A^n(X)$, then $\bigcap_{K\in\mathcal{G}_A^n(X)} \sigma(A+K) \subset \bigcap_{K\in\mathcal{H}(X)} \sigma(A+K)$. Using Theorem 7.7.4, we get $\sigma_{e5}(A) \subset \bigcap_{K\in\mathcal{H}(X)} \sigma(A+K)$. Moreover, the inclusion $\mathcal{K}(X) \subset \mathcal{H}(X)$ leads to $\bigcap_{K\in\mathcal{H}(X)} \sigma(A+K) \subset \sigma_{e5}(A)$, which completes the proof. Q.E.D.

Corollary 7.7.4. *Let $A \in \mathcal{C}(X)$. Let us consider $\mathcal{I}_A(X)$ included in $\mathcal{G}_A^n(X)$, containing the subspace of all compact operators $\mathcal{K}(X)$ and satisfying : $\forall K, K' \in \mathcal{I}_A(X), K \pm K' \in \mathcal{I}_A(X)$. Then, for each $K \in \mathcal{I}_A(X)$, we have $\sigma_{e5}(A) = \sigma_{e5}(A + K)$.* ◇

Proof. Let us define $\sigma'_W(A)$ by $\sigma'_W(A) = \bigcap_{K \in \mathcal{I}_A(X)} \sigma(A + K)$. From Corollary 7.7.3, we have $\sigma_{e5}(A) = \sigma'_W(A)$. Furthermore, for each $K \in \mathcal{I}_A(X)$, we have $\mathcal{I}_A(X) + K = \mathcal{I}_A(X)$. Then, $\sigma'_W(A + K) = \sigma'_W(A)$. Consequently, for each $K \in \mathcal{I}_A(X)$, we get $\sigma_{e5}(A + K) = \sigma'_W(A + K) = \sigma'_W(A) = \sigma_{e5}(A)$, which completes the proof. Q.E.D.

7.7.4 *Invariance of the Schechter's Essential Spectrum in Dunford–Pettis Space*

In this section, we will establish the invariance of the Schechter's essential spectrum in a Banach space X which satisfies the Dunford–Pettis property. In what follows, we will assume that $A \in \mathcal{C}(X)$ and satisfies the hypothesis $(\mathcal{B})$, that is,

$$(\mathcal{B}) \begin{cases} (i) \text{ For all } R \in \mathcal{L}(X), \text{ there exists } \lambda \in \mathbb{R} \text{ such that }]\lambda, +\infty[\subset \rho(A + R). \\ (ii)\ \rho_5(A) \text{ is a connected set of } \mathbb{C}. \end{cases}$$

Remark 7.7.4. Let $A \in \mathcal{C}(X)$. If A generates a C_0-semigroup and $\rho_5(A)$ is a connected set, then A satisfies the hypothesis $(\mathcal{B})$. ◇

Definition 7.7.3. An operator $R \in \mathcal{L}(X)$ is called A-regular if, for all $\lambda \in \rho(A)$, $R(\lambda - A)^{-1}R$ is weakly compact and $\rho_5(A + R)$ is a connected set of $\mathbb{C}$. ◇

The set of all A-regular operators is denoted $\mathcal{R}_A(X)$. We start by giving some lemmas, remarks, and propositions which are useful for the proof of the main result of this section.

Lemma 7.7.1. *Assume that R is A-regular. Then, for all $\lambda \in \rho(A + R) \bigcap \rho(A)$, $R(\lambda - A - R)^{-1}R$ is weakly compact.* ◇

Proof. The result is directly deduced from the resolvent identity:

$$R(\lambda - A - R)^{-1}R - R(\lambda - A)^{-1}R = R(\lambda - A - R)^{-1}R(\lambda - A)^{-1}R.$$

Q.E.D.

Remark 7.7.5.

(i) If R is A-regular, then for all $\lambda \in \rho(A + R) \bigcap \rho(A)$, $\left[(\lambda - A - R)^{-1}R\right]^4$ is compact.
(ii) If $\rho_5(A)$ is a connected set of $\mathbb{C}$, then $\mathcal{K}(X) \subset \mathcal{R}_A(X)$. ◇

Lemma 7.7.2. *Let Ω be an open and connected set of $\mathbb{C}$, let Y be a Banach space and let $f : \Omega \longrightarrow \mathcal{L}(Y)$ be an analytic operator. We define $K(f) = \{\lambda \in \Omega$ such that $f(\lambda)$ is compact$\}$. Then, $K(f) = \Omega$ or $K(f)$ does not have an accumulation point in Ω.* ◇

Proof. Let $E = \{\lambda \in \Omega$ such that λ is an accumulation point of $K(f)$ in $\Omega\}$. If $\lambda \in E$, then there exists $(\lambda_n)_n \in K(f)$, such that λ_n converges to λ. Since f is continuous, then $f(\lambda_n)$ converges to $f(\lambda)$. As $f(\lambda_n)$ is compact, then $f(\lambda)$ will also be compact, which gives $E \subset K(f)$. Let $\lambda \in K(f)$ be fixed and let us choose $r > 0$, such that $B(\lambda, r) \subset \Omega$. Since f is analytic in $B(\lambda, r)$, then $f(z) = \sum_{n\geq 0} A_n(z-\lambda)^n$, where $(A_n)_n$ are bounded operators and independent of z. We have two possibilities:

(a) A_n is compact for all $n \in \mathbb{N}$, then $B(\lambda, r) \subset K(f)$. So, each point $z \in B(\lambda, r)$ is an accumulation point of $K(f)$. We deduce that $B(\lambda, r) \subset E$ and $\lambda \in \overset{\circ}{E}$.
(b) There exists a smaller integer m, such that A_m is not compact. In this case, for $z \in B(\lambda, r)$, we can write $f(z) = \sum_{k=0}^{m-1} A_k(z-\lambda)^k + (z-\lambda)^m g(z)$, where $g(z) = \sum_{k=0}^{+\infty} A_{m+k}(z-\lambda)^k$. Furthermore $g(\lambda)$ is not compact, and using the continuity of g, we get a neighborhood $V(\lambda)$ of λ included in $B(\lambda, r)$, such that $g(\mu)$ is not compact for all $\mu \in V(\lambda)$. Indeed, suppose that for all $n > 0$, there exists $\lambda_n \in B(\lambda, \frac{1}{n})$ such that $g(\lambda_n)$ is compact. Since $\lim_{n\to+\infty} \lambda_n = \lambda$ and since g is continuous, then $g(\lambda)$ is compact, contradicting the fact that $g(\lambda)$ is not compact. So, $f(\mu)$ is not compact for all $\mu \in V(\lambda)$. Hence, λ is an isolated point of $K(f)$. Let $\lambda \in E$, the first possibility holds, and then E is open. Let $F = \Omega \backslash E$. From the definition of E, it follows that F is open. Since $\Omega = E \bigcup F$, with $E \bigcap F = \emptyset$ and Ω is a connected set of $\mathbb{C}$, then $E = \Omega$. In this case, $K(f) = \Omega$, or $E = \emptyset$. Hence, $K(f)$ does not have an accumulation point in Ω. Q.E.D.

Remark 7.7.6. It should be observed that the result of Lemma 7.7.2 remains valid if we replace $K(f)$ by $K_1(f) = \{\lambda \in \Omega$ such that $f(\lambda)$ is weakly compact$\}$. ◇

Lemma 7.7.3. *If O is an open and connected set of $\mathbb{C}$, and if F is a set of isolated points of O, then $O' := O\backslash F$ is a connected set of $\mathbb{C}$.* ◇

Proposition 7.7.1. *Let $R \in \mathcal{L}(X)$.*

(i) If $\rho_5(A+R)$ is a connected set, then for all K compact operators, $\rho(A+R+K)$ is a connected set.
(ii) If R is A-regular, then for all K compact operators, $\rho(A+R+K)\bigcap\rho(A+R)\bigcap\rho(A)$ has an accumulation point.
(iii) If R is A-regular and $\rho(A+R)$ is a connected set of $\mathbb{C}$, then for all $\lambda \in \rho(A+R)$, $\left[(\lambda - A - R)^{-1}R\right]^4$ is compact. ◇

Proof.

(i) For all K compact operators, we have $\rho_5(A+R) = \rho_5(A+R+K)$. Since $\rho_5(A+R)$ is a connected set, then from Theorem 7.3.1(ii), we have $\mathbb{C}\backslash\rho_6(A+$

$R+K) = \sigma_{e5}(A+R+K) = \sigma_{e5}(A+R)$, where $\rho_6(A+R+K)$ denotes the set of those $\lambda \in \rho_5(A+R+K)$, such that all scalars near λ are in $\rho(A+R+K)$. The result follows from both Lemma 7.7.3 and the following identity

$$\rho(A+R+K) = [\mathbb{C}\backslash\sigma_{e6}(A+R+K)]\backslash\{\lambda \in \sigma(A+R+K),\ \lambda \text{ is an isolated eigenvalue of finite algebraic multiplicity}\}.$$

(*ii*) It is sufficient to show that $\rho(A+R+K)\bigcap\rho(A+R)\bigcap\rho(A)$ is nonempty because all points of every open nonempty set are accumulation points. Since A satisfies the hypothesis $(\mathcal{B})$, there exist λ_1, λ_2 and $\lambda_3 \in \mathbb{R}$ such that $]\lambda_1, +\infty[\subset \rho(A)$, $]\lambda_2, +\infty[\subset \rho(A+R)$ and $]\lambda_3, +\infty[\subset \rho(A+R+K)$. If we take $\overline{\lambda} = \max\{\lambda_1, \lambda_2, \lambda_3\}$, we necessarily have $]\overline{\lambda}, +\infty[\subset \rho(A+R+K)\bigcap\rho(A+R)\bigcap\rho(A)$. Then, the set $\rho(A+R+K)\bigcap\rho(A+R)\bigcap\rho(A)$ has an accumulation point.

(*iii*) Let $E = \Big\{\lambda \in \rho(A+R)$ such that $\big[(\lambda - A - R)^{-1}R\big]^4$ is compact$\Big\}$. From Lemma 7.7.1, we have $\rho(A+R)\bigcap\rho(A) \subset E$. Applying the assertion (*ii*), $\rho(A+R)\bigcap\rho(A)$ has an accumulation point. Finally, according to Lemma 7.7.2, $E = \rho(A+R)$. This completes the proof of the proposition. Q.E.D.

Lemma 7.7.4. *Let K be a compact operator and assume that R is A-regular. Then,*

(*i*) *For all $\lambda \in \rho(A+K)\bigcap\rho(A)$, $R(\lambda - A - K)^{-1}R$ is weakly compact.*
(*ii*) *For all $\lambda \in \rho(A+R+K)$, $\big[(\lambda - A - R - K)^{-1}R\big]^4$ is compact.* ◇

Proof.

(*i*) Using the resolvent equation, we get the following identity

$$R(\lambda - A - K)^{-1}R = R(\lambda - A - K)^{-1}K(\lambda - A)^{-1}R + R(\lambda - A)^{-1}R.$$

Since $R(\lambda - A - K)^{-1}K(\lambda - A)^{-1}R$ is compact and $R(\lambda - A)^{-1}R$ is weakly compact, then $R(\lambda - A - K)^{-1}R$ is weakly compact.

(*ii*) For $\lambda \in \rho(A+R+K)\bigcap\rho(A)$, we have

$$\begin{aligned}&(\lambda - A - R - K)^{-1}R\\ &= (\lambda - A)^{-1}R + (\lambda - A - R - K)^{-1}(R+K)(\lambda - A)^{-1}R\\ &= A_1 + A_2 + A_3,\end{aligned}$$

where $A_1 = (\lambda - A)^{-1}R$, $A_2 = (\lambda - A - R - K)^{-1}R(\lambda - A)^{-1}R$, and $A_3 = (\lambda - A - R - K)^{-1}K(\lambda - A)^{-1}R$. Hence, $\big[(\lambda - A - R - K)^{-1}R\big]^4 = (A_1 + A_2 + A_3)^4 = \sum_{j=1}^{3^4} Q_j$. For each $j \in \{1, \ldots, 3^4\}$, the operator Q_j is compact and, then $\big[(\lambda - A - R - K)^{-1}R\big]^4$ is compact. Let $E' =$

$\Big\{\lambda \in \rho(A+R+K)$ such that $\big[(\lambda - A - R - K)^{-1}R\big]^4$ is compact$\Big\}$. We have $\rho(A+R+K)\bigcap\rho(A) \subset E'$. By Proposition 7.7.1(*ii*), E' has an accumulation point in $\rho(A+R+K)$. By using Proposition 7.7.1(*i*), $\rho(A+R+K)$ is a connected set. Finally, by using Lemma 7.7.2, $E' = \rho(A+R+K)$. Q.E.D.

Lemma 7.7.5. *Let us assume that R is A-regular. Let $\mathcal{J}_A(X)$ be a subgroup of $(\mathcal{L}(X), +)$, such that $\mathcal{J}_A(X) \subset \mathcal{R}_A(X)$, and let*

$$\mathcal{I}_A(X) = \{R + K \in \mathcal{L}(X), \text{ such that } K \text{ is compact and } R \in \mathcal{J}_A(X)\}.$$

We have

(i) $\mathcal{K}(X) \subset \mathcal{I}_A(X) \subset \mathcal{G}_A^4(X)$.
(ii) $\forall (R_1 + K_1), (R_2 + K_2) \in \mathcal{I}_A(X)$, *and we have* $(R_1 + K_1) \pm (R_2 + K_2) \in \mathcal{I}_A(X)$. ◇

Proof.

(*i*) Since the null operator $\tilde{o} \in \mathcal{J}_A(X)$, then $\mathcal{K}(X) \subset \mathcal{I}_A(X)$. Let $R + K \in \mathcal{I}_A(X)$ and let $\lambda \in \rho(A+R+K)$. We have

$$\begin{aligned}&\big[(\lambda - A - R - K)^{-1}(R+K)\big]^4\\ &\quad = \big[(\lambda - A - R - K)^{-1}R + (\lambda - A - R - K)^{-1}K\big]^4 = \sum_{j=1}^{2^4} P_j,\end{aligned}$$

where each P_j is a product of four factors constituted from the operators $(\lambda - A - R - K)^{-1}R$ and $(\lambda - A - R - K)^{-1}K$. According to Lemma 7.7.4, $P_1 = \big[(\lambda - A - R - K)^{-1}R\big]^4$ is compact. For $j \in \{2, \ldots, 2^4\}$, the operator K appears, at least, one time in the expression of P_j. So, P_j is compact. Hence, $\big[(\lambda - A - R - K)^{-1}(R+K)\big]^4$ is compact for all $\lambda \in \rho(A+R+K)$.

(*ii*) It is clear that $\forall (R_1 + K_1), (R_2 + K_2) \in \mathcal{I}_A(X)$, and we have

$$(R_1 + K_1) \pm (R_2 + K_2) = (R_1 \pm R_2) + (K_1 \pm K_2) \in \mathcal{I}_A(X).$$

Q.E.D.

A consequence of Lemma 7.7.5 and Corollary 7.7.4, we have the following.

Theorem 7.7.5. *Let $\mathcal{J}_A(X)$ be a subgroup of $(\mathcal{L}(X), +)$ such that $\mathcal{J}_A(X) \subset \mathcal{R}_A(X)$. Then, for all $R \in \mathcal{J}_A(X)$, $\sigma_{e5}(A+R) = \sigma_{e5}(A)$.* ◇

Corollary 7.7.5. *Let $R \in \mathcal{L}(X)$ such that, for all $n \in \mathbb{Z}$, nR is A-regular. Then,*

$$\sigma_{e5}(A+R) = \sigma_{e5}(A).$$ ◇

Proof. Let $\mathcal{J}_A(X) = \{nR,\ n \in \mathbb{Z}\}$. We have $\mathcal{J}_A(X) \subset \mathcal{R}_A(X)$ and, for all $R_1,\ R_2 \in \mathcal{J}_A(X),\ R_1 \pm R_2 \in \mathcal{J}_A(X)$. Then, by using Theorem 7.7.5, we have $\sigma_{e5}(A+R) = \sigma_{e5}(A)$. Q.E.D.

7.7.5 Invariance Under Perturbation of Polynomially Compact Operators

In this section, we will establish some characterization and invariance of the Schechter essential spectrum in a Banach space X. For this purpose, let $A \in \mathcal{C}(X)$, let S be an A-bounded operator on X, and let $\lambda \in \rho(A+S)$. Since S is A-bounded, then $S(\lambda - A - S)^{-1}$ (resp. $S(\lambda - A)^{-1}$) is a closed linear operator defined on all elements of X and hence, bounded by the closed graph theorem (see Theorem 2.1.3). Let $A \in \mathcal{C}(X)$. We define the three following sets:

- $\mathcal{P}_1\mathcal{K}(X) = \Big\{A \in \mathcal{L}(X)$ such that there exists a nonzero complex polynomial $P(z) = \sum_{k=0}^{p} a_k z^k$ satisfying $P(\frac{1}{n}) \neq 0,\ \forall n \in \mathbb{Z}^*$ and $P(A) \in \mathcal{K}(X)\Big\}$.
- $\mathcal{S}_A(X) = \Big\{S \in \mathcal{C}(X)$ such that S is A-bounded and $S(\lambda - A - S)^{-1} \in \mathcal{P}_1\mathcal{K}(X)$, for all $\lambda \in \rho(A+S)\Big\}$.
- $\mathcal{Q}_A(X) = \Big\{S \in \mathcal{C}(X)$ such that S is A-bounded and there exists a nonzero complex polynomial $P(z) = \sum_{k=0}^{p} a_k z^k$ satisfying $P(-1) \neq 0$ and $|P|\left(\gamma\left(S(\lambda - A - S)^{-1}\right)\right) < |P(-1)|$, for all $\lambda \in \rho(A+S)\Big\}$,

where $|P|(z) = \sum_{k=0}^{p} |a_k| z^k$ and $\gamma(.)$ is the Kuratowski's measure of non-compactness.

Remark 7.7.7. Observe that, for $A \in \mathcal{C}(X)$, we have $\mathcal{K}(X) \subset \mathcal{S}_A(X) \bigcap \mathcal{Q}_A(X)$. Indeed, let $K \in \mathcal{K}(X)$. If we take $P(z) = z$, then for all $\lambda \in \rho(A+K)$, $P(K(\lambda - A - K)^{-1}) \in \mathcal{K}(X)$, $|P|\left(\gamma\left(K(\lambda - A - K)^{-1}\right)\right) = 0 < |P(-1)|$. ◇

In what follows, we say that an operator $A \in \mathcal{C}(X)$ satisfies the hypothesis $(\mathcal{C})$, if

$$(\mathcal{C}) \begin{cases} (i) \text{ For all } R \in \mathcal{C}(X), \text{ such that } R \text{ is } A\text{-bounded, there exists } \lambda \in \mathbb{R}, \\ ]\lambda, +\infty[\subset \rho(A+R). \\ (ii)\ \rho_5(A) \text{ is a connected set of } \mathbb{C}. \end{cases}$$

We start with the following proposition, which is fundamental for our purpose and gives a new characterization of the Schechter's essential spectrum in a Banach space X.

Proposition 7.7.2. *Let $A \in \mathcal{C}(X)$. Then,*

$$\sigma_{e5}(A) = \bigcap_{S \in \mathcal{Q}_A(X) \bigcup \mathcal{S}_A(X)} \sigma(A + S).$$

◇

Proof. Since $\mathcal{K}(X) \subset \mathcal{Q}_A(X) \bigcup \mathcal{S}_A(X)$, we infer that $\bigcap_{S \in \mathcal{Q}_A(X) \bigcup \mathcal{S}_A(X)} \sigma(A + S) \subset \sigma_{e5}(A)$. Moreover, we claim the opposite inclusion. Indeed, suppose that $\lambda \notin \bigcap_{S \in \mathcal{Q}_A(X) \bigcup \mathcal{S}_A(X)} \sigma(A + S)$. Then, there exists $S \in \mathcal{Q}_A(X) \bigcup \mathcal{S}_A(X)$, such that $\lambda \in \rho(A + S)$. We have two possibilities:

(*i*) If $S \in \mathcal{Q}_A(X)$, then there exists a nonzero complex polynomial P satisfying $P(-1) \neq 0$ and $|P|\left(\gamma\left(S(\lambda - A - S)^{-1}\right)\right) < |P(-1)|$. Applying Corollary 5.2.4, we get $I + S(\lambda - A - S)^{-1} \in \Phi^b(X)$ and $i\left(I + S(\lambda - A - S)^{-1}\right) = 0$.

(*ii*) If $S \in \mathcal{S}_A(X)$, then there exists a nonzero complex polynomial P satisfying $P(\frac{1}{n}) \neq 0$, $\forall n \in \mathbb{N}^*$ and $P(S(\lambda - A - S)^{-1})$ is a compact operator on X. Since $P(-1) \neq 0$, then from Theorem 3.1.2 and Corollary 3.1.1, we deduce that $I + S(\lambda - A - S)^{-1} \in \Phi^b(X)$ and $i(I + S(\lambda - A - S)^{-1}) = 0$. For each of the above cases, using the equality $\lambda - A = [I + S(\lambda - A - S)^{-1}](\lambda - A - S)$, together with Atkinson's theorem (Theorem 2.2.40), one gets $\lambda - A \in \Phi(X)$ and $i(\lambda - A) = 0$. Finally, the use of Proposition 7.1.1 shows that $\lambda \notin \sigma_{e5}(A)$ and so, $\sigma_{e5}(A) \subset \bigcap_{S \in \mathcal{Q}_A(X) \bigcup \mathcal{S}_A(X)} \sigma(A + S)$. Q.E.D.

Arguing as in the proof of Corollary 7.7.2, we get

Corollary 7.7.6. *Let $A \in \mathcal{C}(X)$ and let us consider $\mathcal{H}$ including in $\mathcal{S}_A(X)$, containing the subspace of all compact operators $\mathcal{K}(X)$ and satisfying : $\forall K, K' \in \mathcal{H}$, $K \pm K' \in \mathcal{H}$. Then, for each $K \in \mathcal{H}$, we have $\sigma_{e5}(A) = \sigma_{e5}(A + K)$.* ◇

In order to give the main result of this section, we will establish a useful lemma.

Lemma 7.7.6. *Let $A \in \mathcal{C}(X)$ satisfying the hypothesis $(\mathcal{C})$, and let S be an A-bounded operator on X, such that $\rho_5(A+S)$ is a connected set. Then, we have*

(i) If $S \in \mathcal{S}_A(X)$, then for all K compact operators, $S + K \in \mathcal{S}_A(X)$.
(ii) If $S \in \mathcal{S}_A(X)$, then for all $p \in \mathbb{Z}^*$, $pS \in \mathcal{S}_A(X)$. ◇

Proof.

(*i*) Let $K \in \mathcal{K}(X)$ and let $\lambda \in \rho(A + S + K) \bigcap \rho(A + S)$. We have the following identity $S(\lambda - A - S - K)^{-1} = S(\lambda - A - S)^{-1} + S(\lambda - A - S - K)^{-1}K(\lambda - A - S)^{-1}$. Since $S(\lambda - A - S - K)^{-1}K(\lambda - A - S)^{-1}$ is a compact operator, then for all $k \in \mathbb{N}^*$, we have

$$\left[S(\lambda - A - S - K)^{-1}\right]^k = \left[S(\lambda - A - S)^{-1}\right]^k + K_k, \tag{7.7.8}$$

where K_k is a compact operator. According to Eq. (7.7.8), and for all nonzero complex polynomials, $P(z) = \sum_{k=1}^{p} a_k z^k$, $P(S(\lambda - A - S - K)^{-1}) = P(S(\lambda - A - S)^{-1}) + K'$, where $K' = \sum_{k=1}^{p} a_k K_k$ is a compact operator. Then, $P\left(S(\lambda - A - S - K)^{-1}\right) \in \mathcal{K}(X)$ if, and only if, $P\left(S(\lambda - A - S)^{-1}\right) \in \mathcal{K}(X)$. Hence, for all $\lambda \in \rho(A + S + K) \bigcap \rho(A + S)$, we have $S(\lambda - A - S - K)^{-1} \in \mathcal{P}_1\mathcal{K}(X)$ if, and only if, $S(\lambda - A - S)^{-1} \in \mathcal{P}_1\mathcal{K}(X)$. Let $S \in \mathcal{S}_A(X)$ and $K \in \mathcal{K}(X)$. We will prove that $S + K \in \mathcal{S}_A(X)$. To do this, let $G = \left\{\lambda \in \rho(A + S + K) \text{ such that } S(\lambda - A - S - K)^{-1} \in \mathcal{P}_1\mathcal{K}(X)\right\}$. We have $\rho(A + S + K) \bigcap \rho(A + S) \subset G$. According to Proposition 7.7.1 (*ii*), G has an accumulation point and $\rho(A + S + K)$ is a connected set of $\mathbb{C}$. Applying Lemma 7.7.2, we get $G = \rho(A + S + K)$. So, $S + K \in \mathcal{S}_A(X)$.

(*ii*) Let $P(z) = \sum_{i=0}^{p} a_i z^i$ be a nonzero complex polynomial satisfying $P(\frac{1}{n}) \neq 0$, for all $n \in \mathbb{Z}^*$, and let $P(S(\lambda - A - S)^{-1}) \in \mathcal{K}(X)$, for all $\lambda \in \rho(A + S)$. We consider $Q(z) := (1 - (1 - p)z)^p P((1 - (1 - p)z)^{-1} z)$. Then, $Q(\frac{1}{n}) \neq 0$, for all $n \in \mathbb{Z}^*$. In what follows, we claim that, for all $\lambda \in \rho(A + pS)$, $Q(S(\lambda - A - pS)^{-1}) \in \mathcal{K}(X)$. From the hypothesis $(\mathcal{C})$, there exist α_1 and $\beta_1 \in \mathbb{R}$, such that $]\alpha_1, +\infty[\subset \rho(A + S)$ and $]\beta_1, +\infty[\subset \rho(A + pS)$. If we take $\gamma_1 = \max\{\alpha_1, \beta_1\}$, we necessarily have $]\gamma_1, +\infty[\subset \rho(A + S) \bigcap \rho(A + pS)$. Let $\omega_1 > \gamma_1$. Then, for all $\lambda \in \mathbb{R}$ and satisfying $\lambda \geq \omega_1$, we have

$$\begin{aligned}
&\left[I - (1 - p)S(\lambda - A - pS)^{-1}\right]\left[I + (1 - p)S(\lambda - A - S)^{-1}\right] \\
&= \left[I + (1 - p)S(\lambda - A - S)^{-1}\right]\left[I - (1 - p)S(\lambda - A - pS)^{-1}\right] = I.
\end{aligned}$$

Hence, $[I - (1 - p)S(\lambda - A - pS)^{-1}]$ is invertible in $\mathcal{L}(X)$. Let $\lambda \in [\omega_1, +\infty[$. We have the following identity $S(\lambda - A - S)^{-1} = S(\lambda - A - pS)^{-1}\Big[I - (1 - p)S(\lambda - A - pS)^{-1}\Big]^{-1}$. Since

$$\begin{aligned}
&\Big[I - (1 - p)S(\lambda - A - pS)^{-1}\Big]S(\lambda - A - pS)^{-1} \\
&= S(\lambda - A - pS)^{-1}\Big[I - (1 - p)S(\lambda - A - pS)^{-1}\Big],
\end{aligned}$$

then for all $k \in \mathbb{N}$, we have

$$\Big[S(\lambda - A - S)^{-1}\Big]^k = \Big[S(\lambda - A - pS)^{-1}\Big]^k\Big[I - (1 - p)S(\lambda - A - pS)^{-1}\Big]^{-k}.$$

Since, for all $\lambda \in \rho(A + S)$, $P(S(\lambda - A - S)^{-1}) \in \mathcal{K}(X)$, then for all $\lambda \in [\omega_1, +\infty[$, $Q(S(\lambda - A - pS)^{-1}) \in \mathcal{K}(X)$. The fact that $\rho(A + pS)$ is a connected set of $\mathbb{C}$ and $[\omega_1, +\infty[\subset \rho(A + pS)$, we infer, from Lemma 7.7.2, that $Q(S(\lambda -$

$A-pS)^{-1}) \in \mathcal{K}(X)$, for all $\lambda \in \rho(A+pS)$. This proves the claim and completes the proof of the lemma. Q.E.D.

The next result proves the invariance of the Schechter's essential spectrum on Banach spaces by means of polynomially compact perturbations.

Theorem 7.7.6. *Let $A \in \mathcal{C}(X)$ satisfying the hypothesis $(\mathcal{C})$, and let $S \in \mathcal{S}_A(X)$ such that, for all $p \in \mathbb{Z}$, $\rho_5(A + pS)$ is a connected set of $\mathbb{C}$. Then, $\sigma_{e5}(A) = \sigma_{e5}(A + S)$.* ◇

Proof. Let $S \in \mathcal{S}_A(X)$ such that, for all $p \in \mathbb{Z}$, $\rho_5(A+pS)$ is a connected set of $\mathbb{C}$. Let us define the following set $\mathcal{I}_{A,S}(X) = \{K + pS$ such that $K \in \mathcal{K}(X)\}$. It is obvious that $\mathcal{K}(X) \subset \mathcal{I}_{A,S}(X)$. According to Lemma 7.7.6, $\mathcal{I}_{A,S}(X) \subset \mathcal{S}_A(X)$. It is easy to show that, for all $K \in \mathcal{I}_{A,S}(X)$, $\mathcal{I}_{A,S}(X) \pm K = \mathcal{I}_{A,S}(X)$. For $S \in \mathcal{I}_{A,S}(X)$, we have

$$\begin{aligned}
\sigma_{e5}(A + S) &= \bigcap_{K \in \mathcal{I}_{A,S}(X)} \sigma(A + S + K) \\
&= \bigcap_{K-S \in \mathcal{I}_{A,S}(X)} \sigma(A + K) \\
&= \bigcap_{K \in \mathcal{I}_{A,S}(X)} \sigma(A + K) \\
&= \sigma_{e5}(A),
\end{aligned}$$

which completes the proof. Q.E.D.

For $A \in \mathcal{C}(X)$ and satisfying the hypothesis $(\mathcal{C})$, we define the following set

$$\begin{aligned}
\mathcal{M}_A(X) =\{S \in \mathcal{C}(X) \ : \ & S \text{ is } A\text{-bounded and } S(\lambda - A)^{-1} \in \mathcal{P}_1\mathcal{K}(X), \\
& \text{for all } \lambda \in \rho(A)\}.
\end{aligned}$$

Remark 7.7.8. We may notice that, if we take $P(z) = z$, then for all $\lambda \in \rho(A)$, $P(K(\lambda - A)^{-1}) \in \mathcal{K}(X)$ and therefore, $\mathcal{K}(X) \subset \mathcal{M}_A(X)$. ◇

Applying Theorem 7.7.6, we have

Corollary 7.7.7. *Let $S \in \mathcal{M}_A(X)$ such that, for all $p \in \mathbb{Z}$, $\rho_5(A + pS)$ is a connected set of $\mathbb{C}$. Then $\sigma_{e5}(A) = \sigma_{e5}(A + S)$.* ◇

Proof. It is sufficient to prove that $\mathcal{M}_A(X) \subset \mathcal{S}_A(X)$. For this, let $S \in \mathcal{M}_A(X)$. Since A satisfies the hypothesis $(\mathcal{C})$, there exist λ_1 and $\lambda_2 \in \mathbb{R}$ such that $]\lambda_1, +\infty[\subset \rho(A)$ and $]\lambda_2, +\infty[\subset \rho(A + S)$. If we take $\overline{\lambda} = \max\{\lambda_1, \lambda_2\}$, then $]\overline{\lambda}, +\infty[\subset \rho(A + S) \bigcap \rho(A)$. Let $\omega > \overline{\lambda}$, then $\forall \lambda \in \mathbb{R}$ and satisfying $\lambda \geq \omega$, we have

$$\left[I + S(\lambda - A - S)^{-1}\right]\left[I - S(\lambda - A)^{-1}\right]$$

$$= \left[I - S(\lambda - A)^{-1}\right]\left[I + S(\lambda - A - S)^{-1}\right] = I.$$

Hence, $[I + S(\lambda - A - S)^{-1}]$ is invertible in the algebra of bounded linear operators on X. Let $\Omega' = \{\lambda \in \rho(A + S) \bigcap \rho(A)$ such that $\lambda \geq \omega\}$ and let $\lambda \in \Omega'$. We have the following identity $S(\lambda - A)^{-1} = [I + S(\lambda - A - S)^{-1}]^{-1} S(\lambda - A - S)^{-1}$. Since

$$\left[I + S(\lambda - A - S)^{-1}\right] S(\lambda - A - S)^{-1} = S(\lambda - A - S)^{-1}\left[I + S(\lambda - A - S)^{-1}\right],$$

then for all $k \in \mathbb{N}$, we have $[S(\lambda - A)^{-1}]^k = [I + S(\lambda - A - S)^{-1}]^{-k}[S(\lambda - A - S)^{-1}]^k$. Let $P(z) = \sum_{k=0}^{p} a_k z^k$ satisfying $P(\frac{1}{n}) \neq 0$, for all $n \in \mathbb{Z}^*$, and $a_p \neq 0$ such that $P(S(\lambda - A)^{-1}) \in \mathcal{K}(X)$, and let $Q(z) := \sum_{k=0}^{p} a_{p-k}(1 + z)^k z^{p-k} = (1 + z)^p P\left(\frac{z}{1+z}\right)$. Then, for all $n \in \mathbb{Z}$, $Q(\frac{1}{n}) \neq 0$ and $Q(S(\lambda - A - S)^{-1})$ is a compact operator. According to the hypothesis $(\mathcal{C})(i)$, Ω' has an accumulation point. Applying Lemma 7.7.2, we get $Q(S(\lambda - A - S)^{-1}) \in \mathcal{K}(X)$, $\forall \lambda \in \rho(A + S)$, which completes the proof. Q.E.D.

7.7.6 Invariance by Means of Weak Noncompactness Measure

The purpose of this subsection is to give a new characterization and to study the invariance of Schechter's essential spectrum on Banach spaces. In order to do this, let μ (resp. μ^*) be a measure of weak noncompactness in a Banach space X (resp. in X^*). Moreover, let Ψ_μ (resp. Ψ_{μ^*}) be a measure of weak noncompactness of operators associated with μ. Besides, let $A \in \mathcal{C}(X)$, let K be an A-bounded operator on X and let $\lambda \in \rho(A + K)$. Since K is A-bounded and according to Remark 2.1.4(*iv*), $K(\lambda - A - K)^{-1}$ is a closed linear operator defined on all elements of X and therefore, bounded by the closed graph theorem (see Theorem 2.1.3). For $n \in \mathbb{N}^*$, we define the left spectrum of A by

$$\sigma_l^n(A) = \bigcap_{K \in \mathcal{D}_A^n(X)} \sigma(A + K),$$

where

$$\mathcal{D}_A^n(X) = \Big\{K \in \mathcal{C}(X) \text{ such that } K \text{ is } A\text{-bounded and, for all } \lambda \in \rho(A + K),$$
$$\left[K(\lambda - A - K)^{-1}\right]^n \text{ is a DP operator and } \psi_\mu\left(\left[K(\lambda - A - K)^{-1}\right]^n\right) < 1\Big\}.$$

In what follows, we suppose that the inequalities (5.5.3) and (5.5.4) hold true with the operator $K(\lambda - A - K)^{-1}$ and the complex polynomial $P(z) = z^n$. Moreover, suppose that (ψ_μ, ψ_{μ^*}) satisfies the adjoint-equivalent property. Now, we are ready to state and prove the following theorem.

Theorem 7.7.7. *Let $n \in \mathbb{N}^*$. For each $A \in \mathcal{C}(X)$, we have*

(i) $\sigma_{e5}(A) = \bigcap\limits_{K \in \mathcal{D}_A^n(X)} \sigma(A+K)$.

(ii) $\sigma_{e5}(A) = \bigcap\limits_{K \in \mathcal{U}_A^n(X)} \sigma(A+K)$, *where $\mathcal{U}_A^n(X)$ is a subset of $\mathcal{C}(X)$ (not necessarily an ideal one) satisfying $\mathcal{K}(X) \subset \mathcal{U}_A^n(X) \subset \mathcal{D}_A^n(X)$.*

(iii) Let $\mathcal{I}_A^n(X)$ be a subset of $\mathcal{C}(X)$, such that $\mathcal{I}_A^n(X) + K = \mathcal{I}_A^n(X)$ for all $K \in \mathcal{I}_A^n(X)$. If $\mathcal{K}(X) \subset \mathcal{I}_A^n(X) \subset \mathcal{D}_A^n(X)$, then $\sigma_{e5}(A) = \sigma_{e5}(A+K)$ for each K in $\mathcal{I}_A^n(X)$. ◇

Proof.

(i) First, we claim that $\sigma_{e5}(A) \subset \bigcap_{K \in \mathcal{D}_A^n(X)} \sigma(A+K)$. Indeed, if $\lambda \notin \bigcap_{K \in \mathcal{D}_A^n(X)} \sigma(A+K)$, then there exists $K \in \mathcal{D}_A^n(X)$ such that $\lambda \in \rho(A+K)$. So, $\lambda \in \rho(A+K)$, $\left[K(\lambda - A - K)^{-1}\right]^n$ is a DP operator and $\psi_\mu\left(\left[K(\lambda - A - K)^{-1}\right]^n\right) < 1$. Hence, by applying Corollary 5.5.1, we get $\left[I + K(\lambda - A - K)^{-1}\right] \in \Phi^b(X)$ and $i\left[I + K(\lambda - A - K)^{-1}\right] = 0$. Moreover, we have $\lambda - A = \left[I + K(\lambda - A - K)^{-1}\right](\lambda - A - K)$. Then, by applying Atkinson's theorem (Theorem 2.2.40), we get $(\lambda - A) \in \Phi(X)$ and $i(\lambda - A) = 0$. Finally, the use of Proposition 7.1.1 shows that $\lambda \notin \sigma_{e5}(A)$, which proves our claim. Besides, since $\mathcal{K}(X) \subset \mathcal{D}_A^n(X)$, we infer that $\bigcap_{K \in \mathcal{D}_A^n(X)} \sigma(A+K) \subset \sigma_{e5}(A)$, which completes the proof.

(ii) The result is immediately deduced from (i).

(iii) Let us denote $\sigma_l'^n(A)$ by $\sigma_l'^n(A) = \bigcap_{K \in \mathcal{I}_A^n(X)} \sigma(A+K)$. By applying (ii), we get $\sigma_{e5}(A) = \sigma_l'^n(A)$. Furthermore, for $K \in \mathcal{I}_A^n(X)$, we have $\mathcal{I}_A^n(X) + K = \mathcal{I}_A^n(X)$. Then,

$$\begin{aligned}\sigma_l'^n(A+K) &= \bigcap_{K' \in \mathcal{I}_A^n(X)} \sigma(A+K+K') \\ &= \bigcap_{K' \in \mathcal{I}_A^n(X)+K} \sigma(A+K') \\ &= \sigma_l'^n(A).\end{aligned}$$

Hence, for each $K \in \mathcal{I}_A^n(X)$, we get $\sigma_{e5}(A+K) = \sigma_l'^n(A+K) = \sigma_l'^n(A) = \sigma_{e5}(A)$, which completes the proof. Q.E.D.

Corollary 7.7.8. *Let $A \in \mathcal{C}(X)$. We consider $\mathcal{I}_A^n(X)$ as a subgroup of $(\mathcal{L}(X), +)$ satisfying $\mathcal{K}(X) \subset \mathcal{I}_A^n(X) \subset \mathcal{D}_A^n(X)$. Assume that $\sigma_{e5}(A) = \emptyset$. Then, for all $K \in \mathcal{I}_A^n(X)$, we have $\sigma(A+K) = \sigma_p(A+K)$, where $\sigma_p(A+K)$ denotes the point spectrum of $A+K$.* ◇

Proof. We have $\sigma(A+K) = \sigma_c(A+K)\bigcup\sigma_r(A+K)\bigcup\sigma_p(A+K)$. If $\sigma_{e5}(A) = \emptyset$, then from Theorem 7.7.7(*iii*), $\sigma_{e5}(A+K) = \emptyset$. Furthermore, by using Proposition 7.3.3, Remark 7.1.1, and Theorem 7.7.7(*iii*), we have

$$\sigma_c(A+K)\bigcup\sigma_r(A+K) \subset \bigcap_{R\in\mathcal{I}^n_A(X)} \sigma(A+K+R) = \sigma_{e5}(A+K).$$

Hence, $\sigma_c(A+K) = \sigma_r(A+K) = \emptyset$. Finally, $\sigma(A+K) = \sigma_p(A+K)$. Q.E.D.

Definition 7.7.4. Let $\Omega \subset \rho(A)$ and let $n \in \mathbb{N}^*$. An A-bounded operator R on X will be called nA-power weakly compact on Ω, if $[R(\lambda - A)^{-1}]^n$ is weakly compact for all $\lambda \in \Omega$. ◇

Proposition 7.7.3. *Let $n \in \mathbb{N}^*$, let $A \in \mathcal{C}(X)$ satisfying the hypothesis $(\mathcal{C})$, and let R be an A-bounded operator on X and nA-power weakly compact on $\rho(A)$. If $\rho(A+R)$ is a connected set of $\mathbb{C}$, then R is $n(A+R)$-power weakly compact on $\rho(A+R)$.* ◇

Proof. According to the hypothesis $(\mathcal{C})(i)$, there exist λ_1 and λ_2 such that $]\lambda_1, +\infty[\subset \rho(A)$ and $]\lambda_2, +\infty[\subset \rho(A+R)$. If we take $\overline{\lambda} = \max\{\lambda_1, \lambda_2\}$, then $]\overline{\lambda}, +\infty[\subset \rho(A+R)\bigcap\rho(A)$. So, there exists $\omega \geq \overline{\lambda}$ such that, for all $\lambda \in \mathbb{R}$ satisfying $\lambda \geq \omega$, $\left(I - [R(\lambda - A)^{-1}]^n\right)$ is invertible in $\mathcal{L}(X)$. Since we have

$$\begin{aligned} &I - [R(\lambda - A)^{-1}]^n \\ &= \left[I - R(\lambda - A)^{-1}\right]\left[I + R(\lambda - A)^{-1} + \cdots + \left(R(\lambda - A)^{-1}\right)^{n-1}\right], \end{aligned}$$

then for all $\lambda \in \mathbb{R}$ satisfying $\lambda \geq \omega$, $\left(I - R(\lambda - A)^{-1}\right)$ is invertible in $\mathcal{L}(X)$. Moreover, for all $\lambda \in \rho(A+R)\bigcap\rho(A)$, we have $\left[R(\lambda - A - R)^{-1}\right]^n = \left[R(\lambda - A)^{-1}\right]^n\left[I - R(\lambda - A)^{-1}\right]^{-n}$. Then, for all $\lambda \in]\omega, +\infty[$, $[R(\lambda - A - R)^{-1}]^n \in \mathcal{W}(X)$. Let

$$E_n = \left\{\lambda \in \rho(A+R) \text{ such that } \left[R(\lambda - A - R)^{-1}\right]^n \in \mathcal{W}(X)\right\}.$$

We have $]\omega, +\infty[\subset E_n$. Applying Lemma 7.7.2, we get $E_n = \rho(A+R)$, which completes the proof. Q.E.D.

In what follows, we define

$$\begin{aligned} \mathcal{H}^n_A(X) = \{&R \in \mathcal{C}(X) \text{ such that } R \text{ is } A\text{-bounded and } nA\text{-power weakly compact} \\ &\text{on } \rho(A), \rho(A+R) \text{ is a connected set and, for all } \lambda \in \rho(A+R), \\ &[R(\lambda - A - R)^{-1}]^n \text{ is a DP operator}\}. \end{aligned}$$

Theorem 7.7.8. *Let $n \in \mathbb{N}^*$ and let $A \in \mathcal{C}(X)$ satisfying the hypothesis $(\mathcal{C})$. Then, we have*

$$\sigma_{e5}(A) = \bigcap_{K \in \mathcal{H}_A^n(X)} \sigma(A+K).$$

◇

Proof. According to the hypothesis $(\mathcal{C})(ii)$ and using Proposition 7.7.1(i), we deduce that $\rho(A+K)$ is a connected set of $\mathbb{C}$ for all $K \in \mathcal{K}(X)$. Moreover, since $\mathcal{K}(X) \subset \mathcal{H}_A^n(X)$, we have $\bigcap_{K \in \mathcal{H}_A^n(X)} \sigma(A+K) \subset \sigma_{e5}(A)$. Furthermore, by applying Proposition 7.7.3, we get $\mathcal{H}_A^n(X) \subset \mathcal{D}_A^n(X)$. Finally, the result follows directly from Theorem 7.7.7(ii). Q.E.D.

Remark 7.7.9. If X has the DP property, then we can replace the hypothesis ensuring that $[R(\lambda - A - R)^{-1}]^n$ is a DP operator by the fact of being weakly compact. ◇

Now, we are ready to state and prove the main result of this section:

Theorem 7.7.9. *Let $n \in \mathbb{N}^*$, $A \in \mathcal{C}(X)$ satisfying the hypothesis $(\mathcal{C})$, and let R be an A-bounded operator on X and nA-power weakly compact on $\rho(A)$ such that, for all $p \in \mathbb{Z}$, $\rho(A+pR)$ is a connected set of $\mathbb{C}$ and, for all $K \in \mathcal{K}(X)$, for all $\lambda \in \rho(A+K+R)$, $[R(\lambda - A - K - pR)^{-1}]^n$ is a DP operator. Then, $\sigma_{e5}(A) = \sigma_{e5}(A+R)$.* ◇

Proof. We define $\mathcal{I}_{A,R}^n(X) = \{K + pR$, such that $K \in \mathcal{K}(X)$ and $p \in \mathbb{Z}\}$. First, we remark that $\mathcal{I}_{A,R}^n(X)$ contains $\mathcal{K}(X)$, and we have $\mathcal{I}_{A,R}^n(X) + K' = \mathcal{I}_{A,R}^n(X)$, for all $K' \in \mathcal{I}_{A,R}^n(X)$. Furthermore, for all $\lambda \in \rho(A)$ and for all $p \in \mathbb{Z}$, we have

$$\begin{aligned}\left[(pR+K)(\lambda - A)^{-1}\right]^n &= \left[pR(\lambda - A)^{-1} + K(\lambda - A)^{-1}\right]^n \\ &= \left[pR(\lambda - A)^{-1}\right]^n + K_1,\end{aligned}$$

where K_1 is a compact operator. Since pR is nA-power weakly compact on $\rho(A)$, then $(pR+K)$ is also weakly compact. By applying Proposition 7.7.1, $\rho(A+K+pR)$ is a connected set. Moreover, since $[R(\lambda - A - K - pR)^{-1}]^n$ is a DP operator for all $\lambda \in \rho(A+K+R)$, we deduce that $\mathcal{I}_{A,R}^n(X) \subset \mathcal{H}_A^n(X)$. Finally, since $\mathcal{H}_A^n(X) \subset \mathcal{D}_A^n(X)$, then the result follows from Theorem 7.7.7(iii). Q.E.D.

Remark 7.7.10. In Theorem 7.7.9, if X has the DP property, then we can replace the hypothesis ensuring that $[R(\lambda - A - pR)^{-1}]^n$ is a DP operator by the fact of being weakly compact. ◇

7.8 Stability of the Essential Spectra

7.8.1 By Means of Measure of Weak Noncompactness

In what follows, let us consider μ (resp. μ^*) as the measure of weak noncompactness in X (resp. in X^*) and ψ_μ (resp. ψ_{μ^*}) as the measure of weak noncompactness of the operators associated with μ (resp. to μ^*). Assume that ψ_μ (resp. ψ_{μ^*}) has the algebraic semi-multiplicative property. We give the following notations:

- If $A \in \Phi_+(X)$ and if J is an A-bounded operator on X, we denote by $G^+_{J,A}(X)$ the set of left Fredholm inverses A_l of A satisfying:

$$\begin{cases} \bullet\ A_l \in \mathcal{L}(X, X_A). \\ \text{For some } n \in \mathbb{N}^*, \\ \bullet\ (JA_l)^n \text{ is a DP operator in } \mathcal{L}(X), \\ \bullet\ \psi_\mu((JA_l)^n) < 1. \end{cases}$$

The first result of this section is the following:

Theorem 7.8.1. *Let X be a Banach space, let $A \in \mathcal{C}(X)$, and let J be an A-bounded operator on X. Then, the following statements hold.*

(i) If $A \in \Phi_+(X)$ and $G^+_{J,A}(X) \neq \emptyset$, then $A + J \in \Phi_+(X)$.
(ii) If, for each $\lambda \in \Phi_{+A}$, we have $G^+_{J,A-\lambda}(X) \neq \emptyset$, then $\sigma_{e1}(A + J) \subset \sigma_{e1}(A)$. ◇

Proof.

(i) Let A_l be a left Fredholm inverse of A. Then, there exists $K \in \mathcal{K}(X)$ such that $I - K$ extends $A_l A$. Moreover, we have

$$\begin{aligned} \hat{A} + \hat{J} &= \hat{A} + \hat{J}(A_l\hat{A} + K_{|X_A}) \\ &= (I + \hat{J}A_l)\hat{A} + \hat{J}K_{|X_A} \\ &= (I + JA_l)\hat{A} + \hat{J}K_{|X_A}. \end{aligned}$$

By applying Theorem 5.5.1(i), we get $I + JA_l \in \Phi_+(X)$. Finally, since $\hat{A} \in \Phi_+(X_A, X)$ and $\hat{J}K_{|X_A} \in \mathcal{K}(X_A, X)$, then using Lemma 6.3.1$(ii)$, we conclude that $\hat{A} + \hat{J} \in \Phi_+(X_A, X)$.

(ii) First, for each $\lambda \in \mathbb{C}$, J is $(A - \lambda)$-bounded. Indeed,

$$\begin{aligned} \forall x \in X,\ \|\hat{J}x\| &\leq \|\hat{J}\|(\|x\| + \|Ax\|) \\ &\leq \|\hat{J}\|(\|x\| + \|(A-\lambda)x\| + |\lambda|\|x\|) \\ &\leq \|\hat{J}\|(1 + |\lambda|)(\|x\| + \|(A-\lambda)x\|). \end{aligned}$$

Then, $\hat{J} \in \mathcal{L}(X_{A-\lambda}, X)$. By applying (i) to the operator $A - \lambda$, we get $\Phi_{+A} \subset \Phi_{+(A+J)}$. Now, the result follows if we translate it in terms of an essential spectrum. Q.E.D.

Theorem 7.8.2. *Let X be a Banach space, $A \in \mathcal{C}(X)$ and let J be an A-bounded operator on X. Assume that (ψ_μ, ψ_{μ^*}) has the adjoint-equivalent property. Then, the following assertions hold.*

(i) If $A \in \Phi(X)$ and $G^+_{J,A}(X) \neq \emptyset$, then $A + J \in \Phi(X)$ and $i(A + J) = i(A)$.
(ii) If, for each $\lambda \in \Phi_A$, we have $G^+_{J,A-\lambda}(X) \neq \emptyset$, then $\sigma_{e4}(A + J) \subset \sigma_{e4}(A)$ and $\sigma_{e5}(A + J) \subset \sigma_{e5}(A)$. Moreover, if $\mathbb{C}\backslash\sigma_{e5}(A)$ is connected and $\rho(A)$ nor $\rho(A + J)$ is empty, then $\sigma_{e6}(A + J) \subset \sigma_{e6}(A)$. ◇

Proof.

(i) Let A_l be a left Fredholm inverse of A. Then, there exists $K \in \mathcal{K}(X)$, such that $I - K$ extends $A_l A$. Since $\psi_\mu((JA_l)^n) < 1$, then by Theorem 5.5.1(i), we get $I + JA_l$ is a Fredholm operator with a zero index. Using Atkinson's theorem (Theorem 2.2.40), we get $(I + JA_l)\hat{A} \in \Phi^b(X_A, X)$ and $i((I + JA_l)\hat{A}) = i(\hat{A})$. The fact that $\hat{A} + \hat{J} = (I + JA_l)\hat{A} + \hat{J}K_{|X_A}$ and $\hat{J}K_{|X_A} \in \mathcal{K}(X_A, X)$, allows us to deduce that $\hat{A} + \hat{J} \in \Phi^b(X_A, X)$ and $i(\hat{A} + \hat{J}) = i(\hat{A})$. Finally, the results follow from (6.3.5).

(ii) Applying (i) to the operator $A - \lambda$, we get $\Phi_A \subset \Phi_{(A+J)}$ and $i(\lambda - A - J) = i(\lambda - A)$. The results follow if we translate them in terms of essential spectra. Finally, the use of Theorem 7.3.1(ii) leads to $\sigma_{e5}(A) = \sigma_{e6}(A)$ and $\sigma_{e5}(A + J) = \sigma_{e6}(A + J)$, which completes the proof. Q.E.D.

7.8.2 By Means of the Graph Measure of Weak Noncompactness

In what follows, for X_A being a Banach space and for $A \in \mathcal{H}_\mu$, we make the following hypotheses

$$(\mathcal{D}) : \begin{cases} \bullet\ X^* + X^* \circ A \text{ is dense in } (X_A)^*. \\ \bullet\ X^{**} + X^{**} \circ A^* \text{ is dense in } (X^*_{A^*})^*. \end{cases}$$

We consider μ_A (resp. $\mu^*_{A^*}$) as the graph measure of weak noncompactness in X_A (resp. in $X^*_{A^*}$) and ψ_{μ_A} (resp. $\psi_{\mu^*_{A^*}}$) as the measure of weak noncompactness of operators associated with μ_A (resp. to $\mu^*_{A^*}$). Assume that ψ_{μ_A} (resp. $\psi_{\mu^*_{A^*}}$) has the algebraic semi-multiplicative property. We give the following notations :

- If $A \in \Phi_-(X)$ and if J is an A-bounded operator on X, we denote by $G^-_{J,A}(X_A)$ the set of right Fredholm inverses A_r of A satisfying:

$$\begin{cases} \bullet\ A_r \in \mathcal{L}(X, X_A). \\ \text{For some } n \in \mathbb{N}^*, \\ \bullet\ (A_r J)^n \text{ is a DP operator in } \mathcal{L}(X_A), \\ \bullet\ \psi_{\mu_A}\left((A_r \hat{J})^n\right) < 1. \end{cases}$$

Theorem 7.8.3. *Let X be a Banach space, $A \in \mathcal{H}_\mu$ satisfying the hypothesis $(\mathcal{D})$ and let J be an A-bounded operator on X. Suppose that $(\psi_{\mu_A}, \psi_{\mu^*_{A^*}})$ has the adjoint-equivalent property. Then, the following statements hold.*

(i) If $A \in \Phi_-(X)$ and $G^-_{J,A}(X_A) \neq \emptyset$, then $A + J \in \Phi_-(X)$.
(ii) If $A \in \Phi(X)$ and $G^-_{J,A}(X_A) \neq \emptyset$, then $A + J \in \Phi(X)$ and $i(A + J) = i(A)$.
(iii) If, for each $\lambda \in \Phi_{-A}$, we have $G^-_{J,A-\lambda}(X_A) \neq \emptyset$, then $\sigma_{e2}(A + J) \subset \sigma_{e2}(A)$.
(iv) If, for each $\lambda \in \Phi_A$, we have $G^-_{J,A-\lambda}(X_A) \neq \emptyset$, then $\sigma_{e4}(A + J) \subset \sigma_{e4}(A)$ and $\sigma_{e5}(A + J) \subset \sigma_{e5}(A)$. Moreover, if $\mathbb{C}\backslash\sigma_{e5}(A)$ is connected and $\rho(A)$ nor $\rho(A + J)$ is empty, then $\sigma_{e6}(A + J) \subset \sigma_{e6}(A)$. ◇

Proof.

(*i*) Let A_r be a right Fredholm inverse of A. Then, there exists $K \in \mathcal{K}(X)$, such that $AA_r = I - K$ on X. Besides, we have $\|A_r x\|_A = \|A_r x\| + \|AA_r x\| \leq (\|A_r\| + \|I - K\|)\,\|x\|$. Then, $A_r \in \mathcal{L}(X, X_A)$ and therefore, $A_r\hat{J} \in \mathcal{L}(X_A)$. Moreover, we have $\hat{A}+\hat{J} = \hat{A}+(\hat{A}A_r+K)\hat{J} = \hat{A}(I_{X_A}+A_r\hat{J})+K\hat{J}$. Since $\psi_{\mu_A}((A_r\hat{J})^n) < 1$ for some $n \in \mathbb{N}^*$, then from Corollary 5.5.1, we deduce that $I_{X_A} + A_r\hat{J}$ is a Fredholm operator with a zero index. Now, if $A \in \Phi_-(X)$, then from Lemma 6.3.1, the fact that $\hat{A} \in \Phi_-(X_A, X)$ and $K\hat{J} \in \mathcal{K}(X_A, X)$, we conclude that $\hat{A} + \hat{J} \in \Phi_-(X_A, X)$ and the results follow from (6.3.5).

(*ii*) Arguing as in the proof of (*i*), we deduce that $I_{X_A} + A_r\hat{J}$ is a Fredholm operator with a zero index. Using Atkinson's theorem (Theorem 2.2.40), we get $\hat{A}(I_{X_A} + A_r\hat{J}) \in \Phi(X_A, X)$ and $i\left(\hat{A}(I_{X_A} + A_r\hat{J})\right) = i(\hat{A})$. Since $K\hat{J} \in \mathcal{K}(X_A, X)$, we deduce that $\hat{A} + \hat{J} \in \Phi(X_A, X)$ and $i(\hat{A} + \hat{J}) = i(\hat{A})$. Finally, the results follow from (6.3.5).

(*iii*) Since $\hat{J} \in \mathcal{L}(X_{A-\lambda}, X)$ and applying (*i*) to the operators $A-\lambda$, we get $\Phi_{-A} \subset \Phi_{-(A+J)}$. The results follow if we translate them in terms of essential spectra.

(*iv*) Applying (*ii*) to the operator $A - \lambda$, we get $\Phi_A \subset \Phi_{(A+J)}$ and $i(\lambda - A - J) = i(\lambda - A)$. The results follow if we translate in terms of essential spectrum. Finally, arguing as in the proof of Theorem 7.8.2, we get $\sigma_{e6}(A + J) \subset \sigma_{e6}(A)$. Q.E.D.

By using Theorem 7.8.1 as well as Theorem 7.8.3, we have the following corollary.

Corollary 7.8.1. *Let X be a Banach space, $A \in \mathcal{H}_\mu$ satisfying the hypothesis $(\mathcal{D})$, and let J be an A-bounded operator on X. Suppose that $(\psi_{\mu_A}, \psi_{\mu^*_{A^*}})$ has the adjoint-equivalent property. Then, the following statements hold.*

(i) If $A \in \Phi_\pm(X)$, $G^+_{J,A}(X) \neq \emptyset$ and $G^-_{J,A}(X_A) \neq \emptyset$, then $A + J \in \Phi_\pm(X)$.
(ii) If, for each $\lambda \in \Phi_{\pm A}$, we have $G^+_{J,A}(X) \neq \emptyset$ and $G^-_{J,A}(X_A) \neq \emptyset$, then

$$\sigma_{e3}(A + J) \subset \sigma_{e3}(A).$$ ◇

7.8.3 By Measure of Non-upper Semi-Fredholm Perturbations

Recall the w-essential spectral radius $r_w(T) := \max\{|\lambda| \text{ such that } \lambda - T \notin \Phi^b(X)\}$, defined for $T \in \mathcal{L}(X)$. We have $r_w(T) = \lim_{n\to\infty} \|T^n\|_{\mathcal{K}(X)}^{\frac{1}{n}}$. The following results may be found in [2]. By Theorem 5.6.1 and Proposition 2.12.1, we can deduce:

Corollary 7.8.2. $r_w(T) = \lim_{n\to\infty} (\varphi(T^n))^{\frac{1}{n}}$. ◇

Proof. From Proposition 2.12.1(vi) it follows $r_w(T) \geq \lim_{n\to\infty} (\varphi(T^n))^{\frac{1}{n}}$. To prove the opposite inequality, let $\lambda \in \mathbb{C}$ be such that $|\lambda| > (\varphi(T^n))^{\frac{1}{n}}$ for some $n \in \mathbb{N}$, then by Theorem 5.6.1, it follows that $\lambda - T \in \Phi^b(X)$. Hence $r_w(T) \leq (\varphi(T^n))^{\frac{1}{n}}$ for every $n \in \mathbb{N}$. Q.E.D.

For $T \in \mathcal{L}(X)$, define $\delta_0(T)$ to be the limit of the sequence $(\delta(T^n))^{\frac{1}{n}}$. As an application of Theorem 5.6.1, we prove some localization results about the essential spectra $\sigma_{e4}(.)$, $\sigma_{e5}(.)$ and $\sigma_{e6}(.)$ of bounded operators on X. We use $\mathbb{D}(0, r)$ for the disc with center 0 and radius r and $\overline{\mathbb{D}}(0, r)$ for the closure of $\mathbb{D}(0, r)$. We write $C[r_1, r_2] = \overline{\mathbb{D}}(0, r_2)\backslash\mathbb{D}(0, r_1)$, for $r_1 \leq r_2$.

Corollary 7.8.3. *Let T be a bounded operator on X, we have $\sigma_{e5}(T) \subset \overline{\mathbb{D}}(0, r_w(T))$.* ◇

Proof. Let $n \in \mathbb{N}^*$ and suppose that $|\lambda|^n > \varphi(T^n)$, then, by Theorem 5.6.1, we have $\lambda - T \in \Phi^b(X)$ and $i(\lambda - T) = 0$. Hence, if $|\lambda| > r_w(T)$, then $\lambda \notin \sigma_{e5}(T)$. Q.E.D.

Corollary 7.8.4. *Let T be a bounded operator on X, we have*

(i) If $T \in \Phi_-^b(X)$, then $\sigma_{e4}(T) \subset C[\delta_0(T), r_w(T)]$.
(ii) If $0 \notin \sigma_{e5}(T)$, then $\sigma_{e5}(T) \subset C[\delta_0(T), r_w(T)]$. ◇

Proof. Notice that if $\delta(T) = 0$, then $\delta_0(T) = 0$ and the results are all trivial. Suppose that $\delta(T) > 0$. For $|\lambda| < \delta_0(T)$, there exists $n \in \mathbb{N}^*$ such that $|\lambda|^n < \delta(T^n)$. Then, by Theorem 5.6.1, we have $\lambda - T \in \Phi_+^b(X)$ and $i(\lambda - T) = i(T)$. Hence, we get easily (i) and (ii). Q.E.D.

Corollary 7.8.5. *Let T be a bounded operator on X, we have $\sigma_{e6}(T) \subset \mathbb{D}(0, r_w(T))$.* ◇

Proof. For $|\lambda| > r_w(T)$, there exists $n \in \mathbb{N}^*$ such that $|\lambda|^n > \varphi(T^n)$. By Theorem 5.6.3, we have $\lambda - T \in \mathcal{B}^b(X)$. The result follows since we can choose n arbitrary large. Q.E.D.

Corollary 7.8.6. *Let T be a bounded operator on X. If $0 \notin \sigma_{e6}(T)$, then $\sigma_{e6}(T) \subset C[\delta_0(T), r_w(T)]$.* ◇

Proof. Since $0 \notin \sigma_{e6}(T)$, then $T \in \Phi^b(X)$ and hence $\delta(T) > 0$. For $|\lambda| < \delta_0(T)$, there exists $n \in \mathbb{N}^*$ such that $|\lambda|^n < \delta(T^n)$. Theorem 5.6.3 implies that $\lambda - T \in \mathcal{B}^b(X)$ since $T \in \mathcal{B}^b(X)$. Q.E.D.

7.8.4 Generalized Convergence

Definition 7.8.1. Let $(T_n)_{n\in\mathbb{N}}$ be a sequence of closable linear operators from X into Y and let T be a closable linear operator from X into Y. $(T_n)_{n\in\mathbb{N}}$ is said to converge in the generalized sense to T, if $\hat{\delta}(T_n, T)$ converges to 0 as $n \to \infty$. $\diamondsuit$

Now, let us study some basic properties of the convergence in the generalized sense.

Theorem 7.8.4. *Let $(T_n)_{n\in\mathbb{N}}$ be a sequence of closable linear operators from X into Y and let T be a closable linear operator from X into Y. Then, we have*

(i) The sequence T_n converges in the generalized sense to T if, and only if, $T_n + S$ converges in the generalized sense to $T + S$, for all $S \in \mathcal{L}(X, Y)$.
(ii) Let $T \in \mathcal{L}(X, Y)$. T_n converges in the generalized sense to T if, and only if, $T_n \in \mathcal{L}(X, Y)$ for a sufficiently larger n and T_n converges to T.
(iii) Let T_n converge in the generalized sense to T. Then, T^{-1} exists and $T^{-1} \in \mathcal{L}(Y, X)$ if, and only if, T_n^{-1} exists and $T_n^{-1} \in \mathcal{L}(Y, X)$ for a sufficiently larger n and T_n^{-1} converges to T^{-1}. $\diamondsuit$

Proof.

(i) Let $S \in \mathcal{L}(X, Y)$. Then,

$$\begin{aligned}\hat{\delta}(T_n + S, T + S) &= \hat{\delta}(\overline{T_n + S}, \overline{T + S}) \\ &= \hat{\delta}(\overline{T_n} + S, \overline{T} + S).\end{aligned}$$

By using Theorem 2.2.26(*iii*), we have $\hat{\delta}(\overline{T_n} + S, \overline{T} + S) \leq 2(1 + \|S\|^2)\hat{\delta}(\overline{T_n}, \overline{T})$. Hence,

$$\hat{\delta}(T_n + S, T + S) \leq 2(1 + \|S\|^2)\hat{\delta}(T_n, T). \tag{7.8.1}$$

In other terms, $\hat{\delta}(\overline{T_n}, \overline{T}) = \hat{\delta}\left(\overline{T_n + S} - S, \overline{T + S} - S\right)$ and so,

$$\hat{\delta}(T_n, T) \leq 2(1 + \|S\|^2)\hat{\delta}(T_n + S, T + S). \tag{7.8.2}$$

If T_n converges in the generalized sense to T, then $\hat{\delta}(T_n, T) \to 0$. So, by using (7.8.1), we deduce that $\hat{\delta}(T_n + S, T + S) \to 0$. Then, $T_n + S$ converges in the generalized sense to $T + S$. Conversely, if $T_n + S$ converges in the generalized sense to $T + S$, and according to (7.8.2), we have $\hat{\delta}(T_n + S, T + S) \to 0$ as $n \to \infty$. Hence, T_n converges in the generalized sense to T.

(*ii*) Let us assume that T_n converges in the generalized sense to T and that $T \in \mathcal{L}(X,Y)$. Let $n_0 \in \mathbb{N}$ be such that $\hat{\delta}(T, T_n) < \frac{1}{\sqrt{1+\|T\|^2}}$ holds for all $n \geq n_0$. Suppose that $x \in \mathcal{D}(\overline{T_n})$, where $n \geq n_0$ is fixed. First, we show that

$$\|(T - \overline{T_n})x\| \leq \delta(\overline{T_n}, T)(1 + \|T\|^2)^{\frac{1}{2}}. \tag{7.8.3}$$

Indeed, $\|(T - \overline{T_n})x\| \leq \|Ty - \overline{T_n}x\| + \|T\|\|x - y\|$, $\forall y \in X$. By using the Cauchy–Schwarz inequality, we deduce that $\|(T - \overline{T_n})x\| \leq (\|x-y\|^2 + \|Ty - \overline{T_n}x\|^2)^{\frac{1}{2}}(1 + \|T\|^2)^{\frac{1}{2}}$. So, we have

$$\|(T - \overline{T_n})x\| \leq \left(\inf_{y\in X}\left(\|x - y\|^2 + \|Ty - \overline{T_n}x\|^2\right)^{\frac{1}{2}}\right)\left(1 + \|T\|^2\right)^{\frac{1}{2}}. \tag{7.8.4}$$

Hence, the inequality (7.8.3) follows immediately by using both (7.8.4) and Remark 2.2.2(*i*). Now, let $x \in \mathcal{D}(\overline{T_n})$, such that $\|x\|^2 + \|\overline{T_n}x\|^2 = 1$. So, it is easy to prove that

$$1 \leq (1 + \|T\|^2)^{\frac{1}{2}}\|x\| + \|(T - \overline{T_n})x\|.$$

Let $x \in \mathcal{D}(\overline{T}_n)$ such that $\|x\| \leq 1$. Then, we have

$$\|(T - \overline{T_n})x\| \leq \left(\frac{1 + \|T\|^2}{1 - \sqrt{1 + \|T\|^2}\delta(\overline{T_n}, T)}\delta(\overline{T_n}, T)\right)\|x\|. \tag{7.8.5}$$

Since this inequality is homogeneous in x, then (7.8.5) is also true for any $x \in \mathcal{D}(\overline{T_n})$. The fact that $\delta(T_n, T) = \delta(\overline{T_n}, \overline{T}) = \delta(\overline{T_n}, T)$ and $\mathcal{D}(T_n) \subset \mathcal{D}(\overline{T_n})$, allows us to conclude that

$$\|(T - T_n)x\| \leq \left(\frac{1 + \|T\|^2}{1 - \sqrt{1 + \|T\|^2}\delta(T_n, T)}\delta(T_n, T)\right)\|x\|, \forall x \in \mathcal{D}(T_n). \tag{7.8.6}$$

By virtue of (7.8.6), the operator T_n is bounded on $\mathcal{D}(T_n)$. This implies that $\mathcal{D}(T_n)$ is closed. Hence, $\overline{\mathcal{D}(T_n)} = \mathcal{D}(T_n) = X$, $T_n \in \mathcal{L}(X,Y)$, and

$$\|T_n - T\| \leq \left(\frac{1 + \|T\|^2}{1 - \sqrt{1 + \|T\|^2}\delta(T_n, T)}\right)\delta(T_n, T), \ \forall n \geq n_0. \tag{7.8.7}$$

So, the relation (7.8.7) implies that T_n converges to T. Conversely, we suppose that T_n converges to T. So, T is bounded. Now, we can write $\hat{\delta}(T_n, T) = \hat{\delta}((T_n - T) + T, 0 + T)$. This implies that

$$\widehat{\delta}(T_n, T) \leq 2(1 + \|T\|^2)\widehat{\delta}(T_n - T, 0).$$

Theorem 2.2.27 leads to the following inequality $\hat{\delta}(T_n, T) \leq 2(1 + \|T\|^2)\frac{\|T_n - T\|}{\sqrt{1+\|T_n - T\|^2}}$. As a result, T_n converges in the generalized sense to T.

(*iii*) Now, let us argue by contradiction. We assume that there exists $x \in \mathcal{D}(T_n)$ such that $\|x\| = 1$ and $T_n x = 0$, for all $n \in \mathbb{N}$. In other words, since T_n converges in the generalized sense to T, then there exists $N \in \mathbb{N}$ such that $\hat{\delta}(T_n, T) < \frac{1}{\sqrt{1+\|T^{-1}\|^2}}$, $\forall n \geq N$. So, there exists $\delta > 0$ such that $\hat{\delta}(T_n, T) < \delta < \frac{1}{\sqrt{1+\|T^{-1}\|^2}}$. Since $(x, 0) \in G(T_n)$, then there exists $y \in \mathcal{D}(T)$ such that $\|x - y\|^2 + \|T_n x - Ty\|^2 < \delta^2$. Hence, we have

$$1 = \|x\|^2 \leq (\|x - y\| + \|y\|)^2 \leq (\|x - y\| + \|T^{-1}\|\|Ty\|)^2, \tag{7.8.8}$$

and, by using both the Schwarz inequality and (7.8.8), we infer that $1 \leq \delta^2(1 + \|T^{-1}\|^2) < 1$ which is a contradiction. So, T_n^{-1} exists for a sufficiently larger n and, by virtue of (7.8.7), we deduce that $T_n^{-1} \in \mathcal{L}(Y, X)$, and

$$\|T^{-1} - T_n^{-1}\| \leq \left(\frac{1 + \|T^{-1}\|^2}{1 - \sqrt{1 + \|T^{-1}\|^2}\delta(T_n^{-1}, T^{-1})}\right)\delta(T_n^{-1}, T^{-1}). \tag{7.8.9}$$

By using Theorem 2.2.26(*ii*), and also the estimation (7.8.9), we infer that

$$\|T^{-1} - T_n^{-1}\| \leq \left(\frac{1 + \|T^{-1}\|^2}{1 - \sqrt{1 + \|T^{-1}\|^2}\delta(T_n, T)}\right)\delta(T_n, T). \tag{7.8.10}$$

According to (7.8.10), we also emphasize that T_n^{-1} converges to T^{-1}. Conversely, the reasoning is analogous to the proof of (*ii*), where it suffices to replace T and T_n by T^{-1} and T_n^{-1}, respectively. Q.E.D.

As a straightforward consequence of Theorem 7.8.4(*iii*), we can easily obtain the following result.

Corollary 7.8.7. *Let $(T_n)_{n\in\mathbb{N}}$ be a sequence of closable linear operators on X and let T be a closable linear operator on X, such that $\rho(T) \neq \emptyset$. Assume that T_n converges in the generalized sense to T, and let $\lambda \in \mathbb{C}$ arbitrarily. Then, (i) $\lambda \in \rho(T)$ if, and only if, (ii) there exists $n_\lambda \in \mathbb{N}$ such that $\lambda \in \rho(T_n)$ for all $n > n_\lambda$ and $\{\|\lambda - T_n\|^{-1} \, : \, n \geq n_\lambda\}$ is bounded.*

Either (i) or (ii) implies the following (iii) $\|(\lambda - T)^{-1} - (\lambda - T_n)^{-1}\| \to 0$ for a sufficiently large n. ◇

Remark 7.8.1.

(i) We should notice that the case $\{T_n\}_{n\in\mathbb{N}}$ is a sequence of bounded operators. Then, by using [118], we have $\rho(\{T_n\}) = \{\lambda \in \mathbb{C},\ \exists n_\lambda \in \mathbb{N},\ m_\lambda > 0 : \lambda \in \rho(T_n)$ and $\|(\lambda - T_n)^{-1}\| \leq m_\lambda,\ \forall n \geq n_\lambda\}$ and $\sigma\big(\{T_n\}\big) = \mathbb{C}\backslash\rho(\{T_n\})$.

Moreover, if T_n converges to T, then by using [148, Lemma 2.1], we have $\sigma(\{T_n\}) = \sigma(T)$.

(ii) Let T have a nonempty resolvent set. If T_n has a compact resolvent according to Corollary 7.8.7, then T has a compact resolvent.

(iii) We mention that the converse of (ii) is not true (see [186, Remark. IV. 2.27]). ◇

Theorem 7.8.5. *Let $\{T_n\}_{n\in\mathbb{N}}$ be a sequence of closed linear operators, and let T be a closed linear operator mapping on X, such that $0 \in \rho(T)$ and T_n converges in the generalized sense to T.*

(i) If $\mathcal{U} \subseteq \mathbb{C}$ is open and $0 \in \mathcal{U}$, then there exists $n_0 \in \mathbb{N}$ such that, for every $n \geq n_0$, we have

$$\sigma_{e5}(T_n) \subseteq \sigma_{e5}(T) + \mathcal{U}. \tag{7.8.11}$$

In particular, $\delta(\sigma_{e5}(T_n), \sigma_{e5}(T)) = 0$, $\forall n \geq n_0$.

(ii) There exist $\varepsilon > 0$ and $n \in \mathbb{N}$ such that, for all $\|S\| < \varepsilon$, we have $\sigma_{e5}(T_n + S) \subseteq \sigma_{e5}(T) + \mathcal{U}$, for all $n \geq n_0$. In particular, $\delta(\sigma_{e5}(T_n + S), \sigma_{e5}(T)) = \delta(\sigma_{e5}(T + S), \sigma_{e5}(T))$. ◇

Proof.

(i) First, by using Theorem 7.8.4(iii), we have $0 \in \rho(T_n)$ for a sufficiently large n and T_n^{-1} converges to T^{-1}. We have to prove the existence of $n_0 \in \mathbb{N}$, such that $\forall n \geq n_0$, we have

$$\sigma_{e5}(T_n^{-1}) \subseteq \sigma_{e5}(T^{-1}) + \mathcal{U}. \tag{7.8.12}$$

In order to prove (7.8.12), we will proceed by contradiction. Assume that the assertion does not hold. Then, by studying a subsequence (if necessary), we may assume that, for each n, there exists $\lambda_n \in \sigma_{e5}(T_n)$ such that $\lambda_n \notin \sigma_{e5}(T^{-1}) + \mathcal{U}$. Since (λ_n) is bounded, we may assume that $\lim\limits_{n\to+\infty} \lambda_n = \lambda$, which implies that $\lambda \notin \sigma_{e5}(T^{-1}) + \mathcal{U}$. By using the fact that $0 \in \mathcal{U}$, we have $\lambda \notin \sigma_{e5}(T^{-1})$. Therefore, $\lambda - T^{-1} \in \Phi^b(X)$ and $i(\lambda - T^{-1}) = 0$. Since $\Phi^b(X)$ is an open multiplicative semigroup, and by using Theorem 7.8.4, we deduce that $\hat{\delta}(\lambda_n - T_n^{-1}, \lambda - T^{-1}) \to 0$ as $n \to \infty$. Let $\delta := \tilde{\gamma}(\lambda - T^{-1}) > 0$. Then, there exists $N \in \mathbb{N}$ such that, for all $n \geq N$, we have $\hat{\delta}(\lambda_n^{-1} - T_n, \lambda - T^{-1}) \leq \frac{\delta}{\sqrt{1+\delta}}$. By using Theorem 2.2.26(iv), we infer that $\lambda_n - T_n^{-1} \in \Phi^b(X)$ and $i(\lambda_n - T_n^{-1}) = i(\lambda - T^{-1}) = 0$. Then, we obtain $\lambda_n \notin \sigma_{e5}(T_n^{-1})$, which is a contradiction. Hence, (7.8.12) holds. Now, if $\lambda \in \sigma_{e5}(T_n)$, then $\frac{1}{\lambda} \in \sigma_{e5}(T_n^{-1})$. By using (7.8.12), we have

$$\frac{1}{\lambda} \in \sigma_{e5}(T^{-1}) + \mathcal{U}. \tag{7.8.13}$$

Since $\mathcal{U}$ is an arbitrary neighborhood of 0, then (7.8.13) implies that $\frac{1}{\lambda} \in \sigma_{e5}(T^{-1})$. Now, we claim that $\lambda \in \sigma_{e5}(T) + \mathcal{U}$. In fact, let us assume that $\lambda \notin \sigma_{e5}(T) + \mathcal{U}$. The fact that $0 \in \mathcal{U}$ implies that $\lambda \notin \sigma_{e5}(T)$ and so, $\frac{1}{\lambda} \notin \sigma_{e5}(T^{-1})$ which is a contradiction. So, $\lambda \in \sigma_{e5}(T) + \mathcal{U}$. This implies that (7.8.11) holds. Since $\mathcal{U}$ is an arbitrary neighborhood of 0 and by using Eq. (7.8.11), we have $\sigma_{e5}(T_n) \subseteq \overline{\sigma_{e5}(T)}$, for all $n \geq n_0$. Hence, Remark 2.2.1(i) gives $\delta(\sigma_{e5}(T_n), \sigma_{e5}(T)) = \delta(\overline{\sigma_{e5}(T_n)}, \overline{\sigma_{e5}(T)}) = 0, \ \forall n \geq n_0$.

(ii) Let $S \in \mathcal{L}(X)$ such that $\|S\| < \frac{1}{\|T^{-1}\|} = \varepsilon_1$. Let $A_n = T_n + S$ and $A = T + S$. By using Theorem 7.8.4(i), $A_n = T_n + S$ converges in the generalized sense to $A = T + S$. We will establish that $0 \in \rho(T + S)$. In fact, since $\|ST^{-1}\| < 1$, the Neumann series $\sum_{k=0}^{\infty}(-ST^{-1})^k$ converges and equals $(I + ST^{-1})^{-1}$. In other words, $\|(I + ST^{-1})^{-1}\| < \frac{1}{1-\|S\|\|T^{-1}\|}$. It follows that $(T + S)^{-1} = T^{-1}(I + ST^{-1})^{-1}$. In view of the above, $0 \in \rho(T+S)$. Now, by applying (i) to A_n and A, we conclude that there exists $n_0 \in \mathbb{N}$ such that $\sigma_{e5}(T_n + S) \subseteq \sigma_{e5}(T + S) + \mathcal{U}$, for all $n \geq n_0$. Let $\lambda \notin \sigma_{e5}(T)$. Then, $\lambda - T \in \Phi(X)$ and $i(\lambda - T) = 0$. By applying Theorem 2.2.45, there exists $\varepsilon_2 > 0$ such that, for $\|S\| < \varepsilon_2$, one has $\lambda - T - S \in \Phi(X)$ and $i(\lambda - T - S) = i(\lambda - T) = 0$. This implies that $\lambda \notin \sigma_{e5}(T + S)$. In view of the above, and if we take $\varepsilon = \min(\varepsilon_1, \varepsilon_2)$, we have for all $\|S\| < \varepsilon$, there exist $n_0 \in \mathbb{N}$ such that $\sigma_{e5}(T_n + S) \subseteq \sigma_{e5}(T) + \mathcal{U}$, for all $n \geq n_0$. Since $\mathcal{U}$ is an arbitrary neighborhood of the origin, then we have $\delta(\sigma_{e5}(T_n + S), \sigma_{e5}(T)) = 0 = \delta(\sigma_{e5}(T + S), \sigma_{e5}(T))$. Q.E.D.

Remark 7.8.2. Let T, $S \in \mathcal{L}(X)$ and let $T_n \in \mathcal{F}^b(X)$. If T_n converges in the generalized sense to T, then there exists $n_0 \in \mathbb{N}$ such that $\forall n \geq n_0$, $\hat{\delta}(\sigma_{e5}(T_n + S), \sigma_{e5}(T + S)) = 0$. Indeed, according to Theorem 7.8.4(ii), T_n converges to T and so, $T \in \mathcal{F}^b(X)$. Hence, $\hat{\delta}(\sigma_{e5}(T_n+S), \sigma_{e5}(T+S)) = \hat{\delta}(\sigma_{e5}(S), \sigma_{e5}(S)) = 0$. ◇

7.8.5 Convergence to Zero Compactly

Definition 7.8.2. A sequence $(T_n)_{n\in\mathbb{N}}$ of bounded linear operators mapping on X is said to converge to zero compactly if for all $x \in X$, $T_n x \to 0$ and $(T_n x_n)$ is relatively compact for every bounded sequence $(x_n)_n \subset X$. ◇

Remark 7.8.3. Clearly, T_n converges to 0 implies T_n converges to zero compactly. ◇

Proposition 7.8.1. *Let $(T_n)_{n\in\mathbb{N}}$ be a sequence of bounded linear operators and let $T \in \mathcal{L}(X)$ such that $T_n - T$ converges to zero compactly. Then,*

(i) *if $T_n \in \mathcal{F}^b(X)$, then $T \in \mathcal{F}^b(X)$,*
(ii) *if $T_n \in \mathcal{F}_+^b(X)$, then $T \in \mathcal{F}_+^b(X)$, and*
(iii) *if $T_n \in \mathcal{F}_-^b(X)$, then $T \in \mathcal{F}_-^b(X)$.* ◇

Proof.

(*i*) Let $A \in \Phi^b(X)$. Using the fact that $T_n - T$ converges to zero compactly, and by virtue of Theorems 5.7.3 and 5.7.4, there exists $n_0 \in \mathbb{N}$ such that $A-(T_n-T) \in \Phi^b(X)$ for all $n \geq n_0$. Since $T_n \in \mathcal{F}^b(X)$, we have $A-(T_n-T)+T_n \in \Phi^b(X)$ for all $n \geq n_0$. This shows that $T \in \mathcal{F}^b(X)$.

The proofs of (*ii*) and (*iii*) are similar to the previous one. Q.E.D.

Theorem 7.8.6. *Let $(T_n)_{n\in\mathbb{N}}$ be a sequence of bounded linear operators on X and let T be a bounded linear operator on X.*

(*i*) *If $T_n - T$ converges to zero compactly, $\mathcal{U} \subseteq \mathbb{C}$ is open and $0 \in \mathcal{U}$, then there exists $n_0 \in \mathbb{N}$ such that $\sigma_{e5}(T_n) \subseteq \sigma_{e5}(T) + \mathcal{U}$, for all $n \geq n_0$. In particular $\delta\big(\sigma_{e5}(T_n), \sigma_{e5}(T)\big) = 0$, for all $n \geq n_0$.*
(*ii*) *If T_n converges to zero compactly, then there exists $n_0 \in \mathbb{N}$ such that $\sigma_{e5}(T + T_n) \subseteq \sigma_{e5}(T)$ for all $n \geq n_0$. In particular $\delta\big(\sigma_{e5}(T+T_n), \sigma_{e5}(T)\big) = 0$, for all $n \geq n_0$.* ◇

Proof.

(*i*) Assume that the assertion fails. Then by passing to a subsequence, it may be assumed that, for each n, there exists $\lambda_n \in \sigma_{e5}(T_n)$ such that $\lambda_n \notin \sigma_{e5}(T)+\mathcal{U}$. Since $(\lambda_n)_n$ is bounded, we may assume that $\lim_{n\to+\infty} \lambda_n = \lambda$ which implies that $\lambda \notin \sigma_{e5}(T)+\mathcal{U}$. Using the fact that $0 \in \mathcal{U}$, we have $\lambda \notin \sigma_{e5}(T)$ and therefore, $\lambda-T \in \Phi^b(X)$ and $i(\lambda-T) = 0$. Let $F_n = \lambda_n-\lambda+T-T_n$. Since F_n converges to zero compactly, writing $\lambda_n - T_n = \lambda - T + F_n$, and using Theorem 5.7.6, we infer that, there exists $n_0 \in \mathbb{N}$ such that for all $n \geq n_0$, we have $\lambda_n - T_n \in \Phi(X)$ and $i(\lambda_n - T_n) = i(\lambda - T + F_n) = i(\lambda - T) = 0$. So, $\lambda_n \notin \sigma_{e5}(T_n)$, which is a contradiction. So, $\sigma_{e5}(T_n) \subseteq \sigma_{e5}(T) + \mathcal{U}$, for all $n \geq n_0$. Since $\mathcal{U}$ is an arbitrary neighborhood of the origin we have $\sigma_{e5}(T_n) \subseteq \sigma_{e5}(T)$ for all $n \geq n_0$. Now, applying Remark 2.2.1(*i*)–(*b*) we have $\delta\big(\sigma_{e5}(T_n), \sigma_{e5}(T)\big) = 0$, for all $n \geq n_0$.
(*ii*) Let $\lambda \notin \sigma_{e5}(T)$. Then, $\lambda - T \in \Phi^b(X)$ and $i(\lambda - T) = 0$. Since T_n converges to zero compactly and by applying Theorem 5.7.6, there exists $n_0 \in \mathbb{N}$, such that $\lambda - (T + T_n) \in \Phi^b(X)$ for all $n \geq n_0$. Hence, $\lambda \notin \sigma_{e5}(T + T_n)$. Since $\mathcal{U}$ is an arbitrary neighborhood of the origin we have $\sigma_{e5}(T_n) \subseteq \sigma_{e5}(T)$ for all $n \geq n_0$. Now, applying Remark 2.2.1(*i*) we have $\delta\big(\sigma_{e5}(T+T_n), \sigma_{e5}(T)\big) = 0$, for all $n \geq n_0$. Q.E.D.

Corollary 7.8.8. *Let T be a closed linear operator on X, and let $(T_n)_{n\in\mathbb{N}}$ be a sequence of closed linear operators on X. If, for some $\lambda \in \rho(T_n)\bigcap\rho(T)$, $(\lambda - T_n)^{-1} - (\lambda - T)^{-1}$ converges to zero compactly, then there exists $n_0 \in \mathbb{N}$ such that $\sigma_{e5}(T_n) \subseteq \sigma_{e5}(T)$, for all $n \geq n_0$.* ◇

Proof. Without loss of generality, we suppose that $\lambda = 0$. Then, $T_n^{-1} = T^{-1} + F_n$ where F_n converges to zero compactly. By applying Theorem 7.8.6(*ii*), we have

$$\sigma_{e5}(T_n^{-1}) = \sigma_{e5}(T^{-1} + F_n) \subseteq \sigma_{e5}(T^{-1}) \text{ for all } n \geq n_0. \quad (7.8.14)$$

Let $\lambda \in \sigma_{e5}(T_n)$, then $\frac{1}{\lambda} \in \sigma_{e5}(T_n^{-1})$. By (7.8.14), we have $\frac{1}{\lambda} \in \sigma_{e5}(T^{-1})$ and hence, $\lambda \in \sigma_{e5}(T)$. So, $\sigma_{e5}(T_n) \subseteq \sigma_{e5}(T)$. Q.E.D.

We close this section by the following example. Let

$$l_2 = \left\{ (x_j)_{j\geq 1} \text{ such that } x_j \in \mathbb{C} \text{ and } \sum_{j=1}^{+\infty} |x_j|^2 < \infty \right\}.$$

(*i*) Let A_n be a sequence defined by

$$\begin{cases} A_n : \mathcal{D}(A_n) \subset l_2 \longrightarrow l_2 \\ \mathcal{D}(A_n) = \left\{ (x_j)_{j\geq 1} \text{ such that } \sum_{j=1}^{+\infty} j^2 |x_j|^2 < \infty \right\} \\ A_n e_j = \begin{cases} j e_j, \ j \neq n, \\ -n e_j, \ j = n. \end{cases} \end{cases}$$

Then, A_n converges in the generalized sense to A_0. In fact, by using [269, Remark 1.6], we have $\|(i \pm A_n)^{-1} - (i \pm A_0)^{-1}\| = \left| \frac{1}{i+n} - \frac{1}{i-n} \right| = \frac{2n}{1+n^2} \to 0$ (where $i = \sqrt{-1}$). Now, since A_n is closed, we can recall Proposition 7.8.4(*i*) in order to conclude that $\hat{\delta}(A_n, A_0) \to 0$. Consequently, A_n converges in the generalized sense to A_0.

(*ii*) We consider that the operators $\mathcal{T}_n$ and $\mathcal{T}$ in $\mathcal{L}(l_2 \oplus l_2)$ are defined by

$$\mathcal{T}_n := \begin{pmatrix} (A_n + i)^{-1} & 0 \\ 0 & \frac{1}{n+1}(I - \frac{U}{2}) \end{pmatrix} \text{ and } \mathcal{T} := \begin{pmatrix} (A_0 + i)^{-1} & 0 \\ 0 & 0 \end{pmatrix},$$

where $U \in \mathcal{L}(l_2)$ is the forward unilateral shift defined by

$$\begin{cases} U : l_2 \longrightarrow l_2 \\ (x_0, x_1, x_2, \ldots) \longrightarrow (0, x_0, x_1, x_2, \ldots), \end{cases}$$

in terms of the standard basis in l_2 that is $Ue_j = e_{j+1}$ and $\|U\| = 1$. By referring to [103, pp. 70–71], we know that

$$\begin{aligned} \|\mathcal{T}_n - \mathcal{T}\| &= \max\left\{ \|(i + A_n)^{-1} - (i + A_0)^{-1}\|, \left\| \frac{I - \frac{U}{2}}{n+1} \right\| \right\} \\ &\leq \max\left\{ \frac{2n}{1+n^2}, \frac{3}{2(n+1)} \right\} \to 0, \end{aligned} \quad (7.8.15)$$

$$\hat{\delta}(\mathcal{T}_n, \mathcal{T}) \to 0. \tag{7.8.16}$$

Hence, $\mathcal{T}_n$ converges in the generalized sense to $\mathcal{T}$ and by using both Corollary 7.8.7 and Remark 7.8.1, we have $\sigma(\{\mathcal{T}_n\}) = \sigma((A_0+i)^{-1}) \bigcup \{0\} = \{\frac{1}{\lambda}$ such that $\lambda - i \in \sigma(A_0)\} \bigcup \{0\}$. In other words, by using (7.8.15), we have $\mathcal{T}_n - \mathcal{T}$ converges to zero compactly. Hence, according to Theorem 7.8.6, there exists $n_0 \in \mathbb{N}$ such that $\sigma_{e5}(\mathcal{T}_n) \subseteq \sigma_{e5}((A_0+i)^{-1}) \bigcup \{0\}$ for all $n \geq n_0$.

(*iii*) $\mathcal{T}$ is not invertible, $\mathcal{T}_n^{-1}$ exists and

$$\mathcal{T}_n^{-1} := \begin{pmatrix} A_n + i & 0 \\ 0 & (n+1)(I - \frac{U}{2})^{-1} \end{pmatrix}.$$

However, $\mathcal{T}_n^{-1}$ is not convergent. In fact, let us suppose, by contradiction, that there exists $\mathcal{A}$ such that $\mathcal{T}_n^{-1}$ converges in the generalized sense to $\mathcal{A}$. Hence, $\delta(\mathcal{T}_n^{-1}, \mathcal{A}) \to 0$ and there exists $n_0 \in \mathbb{N}$ such that, for all $n \geq n_0$, we have $\hat{\delta}(\mathcal{A}, \mathcal{T}_n^{-1}) < \frac{1}{\sqrt{1+\|\mathcal{T}_n\|^2}}$. Using the same reasoning as in the proof of Theorem 7.8.4(*iii*), we have $\mathcal{A}$ as an invertible operator. According to Theorem 2.2.26(*ii*), we infer that $\hat{\delta}(\mathcal{A}, \mathcal{T}_n^{-1}) = \hat{\delta}(\mathcal{A}^{-1}, \mathcal{T}_n) \to 0$. Together with (7.8.16), $\mathcal{A}^{-1} = \mathcal{T}$. This implies that $\mathcal{T}^{-1} = \mathcal{A}$ constitutes is a contradiction to the fact that $\mathcal{T}$ is not invertible. Q.E.D.

7.9 Borel Mappings

We recall some results about Borel mapping which can also be found in [227].

Theorem 7.9.1. *Let X be a Banach space. If X is separable, then $\sigma_{ap} : \mathcal{L}_s(X) \longrightarrow K(\mathbb{C})$, $T \longrightarrow \sigma_{ap}(T)$, is a Borel map.* ◇

Proof. Let $\mathcal{D}$ be a countable and dense subset of S_X. By using the definition of the strong operator topology, $(T, \lambda) \longrightarrow (\lambda - T)x$ is continuous from $\mathcal{L}_s(X) \times \mathbb{C}$ into X. Therefore, $A_{x,n}$ is an open set in $\mathcal{L}_s(X) \times \mathbb{C}$ (for every $x \in \mathcal{D}$, $n \in \mathbb{N}$) and hence, $\Delta_{\sigma_{ap}}$ is a Borel set in $\mathcal{L}_s(X) \times \mathbb{C}$. Now, the result follows from Lemma 2.7.2. Q.E.D.

Theorem 7.9.2. *Let X be a Banach space. If X is separable, then $\sigma : \mathcal{L}_s(X) \longrightarrow K(\mathbb{C})$, $T \longrightarrow \sigma(T)$, is a Borel map.* ◇

Proof. In order to prove that the map $T \longrightarrow \sigma(T)$ is a Borel one, it suffices to establish that C, given in (7.1.5), is a Borel set. Let $\mathcal{D}$ be a countable and dense subset of X. We can see that

$$C = \Big\{ (T, \lambda) \in \mathcal{L}(X) \times \mathbb{C} \ : \ \exists n \in \mathbb{N} \text{ and } y \in \mathcal{D} \text{ with } \|y - (T - \lambda)x\| > \frac{1}{n} \text{ for all } x \in \mathcal{D} \Big\}.$$

Hence, we can write

$$C = \bigcup_{n\in\mathbb{N}} \bigcup_{y\in\mathcal{D}} \bigcap_{x\in\mathcal{D}} \left\{ (T,\lambda) \in \mathcal{L}(X) \times \mathbb{C} \text{ such that } \|y - (T-\lambda)x\| > \frac{1}{n} \right\}.$$

As in Theorem 7.9.1, by using the continuity for the strong operator topology, we conclude that C is a Borel set. The result follows from Lemma 2.7.2. Q.E.D.

Theorem 7.9.3. *Let X be a Banach space. If X and X^* (the dual space of X) are separable, then $\sigma_{e4} : \mathcal{L}_s(X) \longrightarrow K(\mathbb{C})$, $T \longrightarrow \sigma_{e4}(T)$, is a Borel map.* $\diamond$

Proof. Using Eq. (7.1.7), Ω_1 is an open subset of $\mathcal{L}_s(X)\times\mathbb{C}$. Then, Ω_1 is a Borel set. Using Eq. (7.1.8), Ω_2 is a Borel set of $\mathcal{L}_s(X) \times \mathbb{C}$. Accordingly, $\Delta_{\sigma_q^A} = \Omega_1 \bigcup \Omega_2$ is a Borel set. Then, for all $q \in \mathbb{N}$ and $A \in \mathcal{L}_s(X)$ fixed, $T \longrightarrow \sigma_q^A(T)$ is a Borel map from $\mathcal{L}_s(X)$ into $K(\mathbb{C})$. Let us set $\sigma_{0,q}^A(T) = \{\lambda \in \mathbb{C} \text{ such that } \lambda A - T \notin \Phi_0(X)\} \bigcap B(0,q)$ ($\sigma_{0,q}^A(T)$ being a compact subset of $\mathbb{C}$). Hence, $\sigma_{0,q}^A(T) = \bigcap_{F\in\mathcal{F}_0(X)} \sigma_q^A(T+F) = \bigcap_{F\in\mathcal{F}_0^d(X)} \sigma_q^A(T+F)$, where $\mathcal{F}_0^d(X)$ is a countable and dense subset of $\mathcal{F}_0(X)$ (the set of finite rank operators on X with its norm topology). So, $T \longrightarrow \sigma_{0,q}^A(T)$ is a Borel map. By using Lemma 2.7.3 and Eq. (7.1.4), we may write $\sigma_{e4}(T) = \bigcap_{k\in\mathbb{Z}} \{\lambda \in \mathbb{C} \text{ such that } A_k(\lambda - T) \notin \Phi_0(X)\}$, where, for all k, A_k is a fixed operator with $i(A_k) = -k$. Hence,

$$\begin{aligned} \sigma_{e4}(T) \bigcap B(0,q) &= \bigcap_{k\in\mathbb{Z}} (\{\lambda \in \mathbb{C} \text{ such that } A_k(\lambda - T) \notin \Phi_0(X)\} \bigcap B(0,q)) \\ &= \bigcap_{k\in\mathbb{Z}} \sigma_{0,q}^{A_k}(A_k T). \end{aligned}$$

Since the map $T \longrightarrow A_k T$ is continuous from $\mathcal{L}_s(X)$ into $\mathcal{L}_s(X)$ and since $T \longrightarrow \sigma_{0,q}^{A_k}(T)$ defines a Borel map for all q and k in $\mathbb{N}$, we deduce that $T \longrightarrow \sigma_{e4}(T) \bigcap B(0,q)$ is a Borel map. This shows that $\Delta_{\sigma_{e4}} = \bigcup_{q\in\mathbb{N}} \Delta_{\sigma_{e4}(T)\bigcap B(0,q)}$, which is Borel. Now, using Lemma 2.7.2 gives the result. Q.E.D.

Let $\mathcal{D}$ be a countable and dense subset of $\mathcal{F}_0(X)$. We can easily show that

$$\sigma_{e5}(T) = \bigcap_{F\in\mathcal{F}_0(X)} \sigma(T+F) = \bigcap_{F\in\mathcal{D}} \sigma(T+F)$$

and

$$\begin{aligned} \Delta_{\sigma_{e5}} &= \{(T,\lambda) \in \mathcal{L}(X) \times \mathbb{C} \text{ such that } \lambda \in \sigma_{e5}(T)\} \\ &= \left\{ (T,\lambda) \in \mathcal{L}(X) \times \mathbb{C} \text{ such that } \lambda \in \bigcap_{F\in\mathcal{D}} \sigma(T+F) \right\} \end{aligned}$$

$$= \bigcap_{F \in \mathcal{D}} \{(T, \lambda) \in \mathcal{L}(X) \times \mathbb{C} \text{ such that } \lambda \in \sigma(T + F)\}$$

$$= \bigcap_{F \in \mathcal{D}} \{(T' - F, \lambda) \in \mathcal{L}(X) \times \mathbb{C} \text{ such that } \lambda \in \sigma(T')\}, \qquad (7.9.1)$$

where $T' = T + F$.

Theorem 7.9.4. *Let X be a Banach space. If X and X^* are separable, then $\sigma_{e5} : \mathcal{L}_s(X) \longrightarrow K(\mathbb{C})$, $T \longrightarrow \sigma_{e5}(T)$, is a Borel map.* ◇

Proof. Let $\mathcal{D}$ be a countable and dense subset of $\mathcal{F}_0(X)$. Using Eq. (7.9.1), we have $\Delta_{\sigma_{e5}} = \bigcap_{F \in \mathcal{D}}(\Delta_\sigma - (F, 0))$. Clearly, for every $F \in \mathcal{D}$, the translation by $(-F, 0)$ of the Borel set Δ_σ is a Borel set in $\mathcal{L}_s(X) \times \mathbb{C}$. As a result, $\Delta_{\sigma_{e5}}$ is a Borel set and therefore, $\sigma_{e5}(.)$ is a Borel map. Q.E.D.

Theorem 7.9.5. *Let X be a Banach space. If X and X^* are separable, then $\sigma_{e6} : \mathcal{L}_s(X) \longrightarrow K(\mathbb{C})$, $T \longrightarrow \sigma_{e6}(T)$, is a Borel map.* ◇

Proof. First, let us notice that, $\Delta_{\sigma_{e6}} = \Delta_{\sigma_{e4}} \bigcup \Delta_{d \circ \sigma}$, where d represents the Cantor–Bendixson derivative. By using Theorem 7.9.3, we know that $\Delta_{\sigma_{e4}}$ is a Borel set. Moreover, by applying Lemma 2.7.2 and Proposition 2.7.3(*iii*), we notice that $\Delta_{d \circ \sigma}$ is a Borel set in $\mathcal{L}_s(X) \times \mathbb{C}$. Then, $\Delta_{\sigma_{e6}}$ is also Borel. This gives the desired result. Q.E.D.

Remark 7.9.1. Obviously, in the case where $\mathcal{L}(X)$ is endowed with the operator norm topology, the spectrum and essential spectrum are known to be upper semicontinuous (hence Borel) but, in general, not continuous. ◇

As an immediate consequence of Theorems 7.9.1–7.9.5, we have the following corollary.

Corollary 7.9.1. *Let X be a Banach space. If X is separable, then $\mathcal{T}(X)$,* Inv(X), *and $\mathcal{R}(X)$ are Borel subsets of $\mathcal{L}_s(X)$. Moreover, if X^* is separable, then $\Phi^b(X)$ and $\Phi_0(X)$ are also Borel subsets of $\mathcal{L}_s(X)$.* ◇

Proof. Let us consider the set $\mathcal{T}(X)$. Since the set of Borel sets is closed under complementation, it is sufficient to prove that $\mathcal{T}^c(X)$ is Borel. For this purpose, using Eq. (7.1.6), $\mathcal{T}^c(X)$ may be written as follows: $\mathcal{T}^c(X) = \sigma_{ap}^{-1}(\{K \in K(\mathbb{C}) \text{ such that } 0 \notin K\})$. However, the set $\{K \in K(\mathbb{C}) \text{ such that } 0 \notin K\}$ is open in $K(\mathbb{C})$. Then, by applying Theorem 7.9.1 we conclude that $\mathcal{T}^c(X)$ is Borel. Arguing as above, and using Eqs. (2.7.2), (7.1.2), (7.1.3), and Theorems 7.9.2–7.9.4, we deduce that Inv(X), $\Phi^b(X)$, and $\Phi_0(X)$ are Borel subsets of $\mathcal{L}_s(X)$. Now, in order to prove that $\mathcal{R}(X)$ is a Borel subset of $\mathcal{L}_s(X)$, it suffices to use Eq. (2.7.3) and invokes the Borel character of the functions $\sigma(.)$ and $d(.)$ [cf. Theorem 7.9.2 and Proposition 2.7.3(*iii*)], where $d(.)$ is the Cantor–Bendixson derivative. Q.E.D.

If X^* is separable, then $\mathcal{F}^b(X)$ is a coanalytic subset of $\mathcal{L}_s(X)$. Evidently, here we use the fact that $\Phi^b(X)$ is a Borel subset of $\mathcal{L}_s(X)$ if X^* is separable cf. Corollary 7.9.1. Making use of this observation, our objective here is to describe the topological complexity of some subsets of $\mathcal{L}_s(X)$ for particular Banach spaces. Since the separability of the dual space X^* is required, from the classes of spaces mentioned below, we use only Banach spaces with this property. If X is a h-space, then the ideal of strictly singular operators $\mathcal{S}(X)$ is the greatest proper ideal of $\mathcal{L}(X)$. So, we have $\mathcal{S}(X) = \mathcal{F}^b(X)$. This shows that $\mathcal{S}(X)$ is a coanalytic subset of $\mathcal{L}_s(X)$ for any h-space X with separable dual. This holds true, in particular, for the spaces c, c_0 and l_p $(1 < p < \infty)$. In [68], it is proved that if X is subprojective, then $\mathcal{S}(X) = \mathcal{F}^b(X)$. Accordingly, for any subprojective Banach space X with separable dual, $\mathcal{S}(X)$ is a coanalytic subset of $\mathcal{L}_s(X)$. This holds true, in particular, for the spaces c_0, l_p $(1 < p < \infty)$ and L_p $(2 \leq p < \infty)$. Let (Ω, Σ, μ) be a positive measure space. It follows from Lemma 2.1.13(*ii*) that $\mathcal{S}(L_p(\Omega, d\mu)) = \mathcal{F}^b(L_p(\Omega, d\mu))$ with $1 < p < \infty$ (in fact, this also holds for $p = 1$ and ∞). This implies that the set of strictly singular operators is a Π^1_1 subset of $\mathcal{L}_s(L_p(\Omega, d\mu))$ for $1 < p < \infty$. Note that the identity $\mathcal{S}(L_p(\Omega, d\mu)) = \mathcal{F}^b(L_p(\Omega, d\mu))$ is not specific to L_p-spaces. In fact, it is also fulfilled for $C(\Xi)$ (the Banach space of continuous scalar-valued functions on Ξ with the supremum norm) provided that Ξ is a compact Hausdorff space. So, the conclusion above is also valid for $\mathcal{S}(C(\Xi))$ provided that Ξ is countable. Recall that there are many separable Banach spaces X for which $\mathcal{L}(X)$ has only one proper nonzero closed two-sided ideal. Indeed, Calkin proved that if X is a separable Hilbert space, then $\mathcal{K}(X)$ is the unique proper closed two-sided ideal of $\mathcal{L}(X)$. Gohberg et al. in [124] established the same result for the spaces l_p, $1 \leq p < \infty$, and c_0. Accordingly, if X is separable Hilbert space or one of the spaces l_p, $1 < p < \infty$, or c_0, then $\mathcal{K}(X) = \mathcal{F}^b(X)$ and therefore $\mathcal{K}(X)$ is a Π^1_1 subset of $\mathcal{L}_s(X)$. Corollary 7.9.1 asserts that, if X is a separable Banach space, then $\mathcal{T}(X)$ and $\mathrm{Inv}(X)$ are Borel subsets of $\mathcal{L}_s(X)$, and if X^* is separable, then $\Phi_0(X)$ and $\Phi^b(X)$ satisfy the same property.

Let D_X denote a countable and dense subset of separable Banach space X. If $T \in \mathcal{L}(X)$, then $0 \notin \sigma_{\mathrm{com}}(T)$ is equivalent to the fact that, for all $n \in \mathbb{N}^*$ and $x \in D_X$, there exists $y \in X$ such that $\|Ty - x\| < 1/n$. So,

$$\{T \in \mathcal{L}(X) \,:\, 0 \notin \sigma_{\mathrm{com}}(T)\} = \bigcap_{n\in\mathbb{N}^*} \bigcap_{x\in D_X} \bigcup_{y\in X} \left\{ T \in \mathcal{L}(X) \text{ such that } \|Ty - x\| < \frac{1}{n} \right\}.$$

This shows that $\mathcal{Z}^c(X)$ is a countable intersection of open subsets of $\mathcal{L}_s(X)$. Therefore, $\mathcal{Z}^c(X)$ is a G_δ subset of $\mathcal{L}_s(X)$. Using Eq. (7.1.6), it is not difficult to observe that $T \in \mathcal{T}^c(X)$ if, and only if, $T \in \bigcup_{n\in\mathbb{N}^*}\{U \in \mathcal{L}(X)$ such that $\forall x \in S_X$, $\|Ux\| \geq 1/n\}$, or equivalently $T \in \bigcup_{n\in\mathbb{N}^*} \bigcap_{x\in S_X} \{U \in \mathcal{L}(X) \text{ such that } \|Ux\| \geq \frac{1}{n}\}$. Accordingly, $\mathcal{T}^c(X)$ is a countable union of closed sets, that is, an F_σ subset of $\mathcal{L}_s(X)$. Now, let us consider the set of Weyl operators. It is not difficult to observe that $T \in \Phi_0(X)$ if, and only if, $0 \notin \sigma_{e5}(T)$ if, and only if, $0 \notin \bigcap_{F\in\mathcal{F}_0(X)} \sigma(T + F)$. So,

$$\Phi_0(X) = \left\{ T \in \mathcal{L}(X) \text{ such that } 0 \notin \bigcap_{F \in \mathcal{F}_0(X)} \sigma(T+F) \right\}.$$

Let $\mathcal{D}$ be a countable and dense subset of $\mathcal{F}_0(X)$. Arguing as in the proof of Theorem 7.9.4, we notice that

$$\begin{aligned}\Phi_0(X) &= \left\{ T \in \mathcal{L}(X) \text{ such that } 0 \notin \bigcap_{F \in \mathcal{D}} \sigma(T+F) \right\} \\ &= \{T \in \mathcal{L}(X) \ : \ \exists F_n \in \mathcal{D} \text{ such that } 0 \in \rho(T+F_n)\} \\ &= \{T \in \mathcal{L}(X) \ : \ \exists F_n \in \mathcal{D} \text{ such that } T+F_n \text{ is invertible}\} \\ &= \bigcup_{F \in \mathcal{D}} \{T \in \mathcal{L}(X) \text{ such that } T+F \in \text{Inv}(X)\}.\end{aligned}$$

Let us denote by $\mathcal{T}_F$ the translation from $\mathcal{L}(X)$ into itself which assigns to each T the operator $T+F$. Then, we may write $\Phi_0(X) = \bigcup_{F \in \mathcal{D}} \mathcal{T}_F^{-1}(\text{Inv}(X))$. The following proposition gives more details about the topological structure of these sets given in [227].

Proposition 7.9.1. *If X is a separable Banach space, then $\mathcal{T}(X)$ is a G_δ subset of $\mathcal{L}_s(X)$, $\mathcal{Z}(X)$ is an F_σ subset of $\mathcal{L}_s(X)$, and* Inv(X) *is a Borel subset of $\mathcal{L}_s(X)$ of the form $F_\sigma \backslash F_\sigma$. Moreover, if X^* is separable, then $\Phi_0(X)$ and $\Phi^b(X)$ are countable unions of Borel subsets of $\mathcal{L}_s(X)$ of the form $F_\sigma \backslash F_\sigma$.* ◇

Proof. Concerning the sets $\mathcal{T}(X)$ and $\mathcal{Z}(X)$ are proved before. Let D_X denote a countable and dense subset of X. Now, let us notice that

$$\text{Inv}(X) = \{T \in \mathcal{L}(X) \text{ such that } 0 \notin \sigma_{ap}(T)\} \bigcap \{T \in \mathcal{L}(X) \text{ such that } 0 \notin \sigma_{\text{com}}(T)\}.$$

Consequently, Inv$(X) = \mathcal{T}^c(X) \bigcap \mathcal{Z}^c(X) = \mathcal{T}^c(X) \backslash \mathcal{Z}(X)$, which proves the statement for Inv(X). Now, by using the continuity of $\mathcal{T}_F$ and the fact that Inv$(X) = \mathcal{T}^c(X) \backslash \mathcal{Z}(X)$, we get $\Phi_0(X) = \bigcup_{F \in \mathcal{D}} \mathcal{T}_F^{-1}(\mathcal{T}^c(X)) \backslash \mathcal{T}_F^{-1}(\mathcal{Z}(X))$, where $\mathcal{D}$ be a countable and dense subset of $\mathcal{F}_0(X)$ which ends the proof for $\Phi_0(X)$. When dealing with the set of Fredholm operators, we can write $\Phi^b(X) = \bigcup_{n \in \mathbb{Z}} \Phi_n(X)$. According to Lemma 2.7.3, $\Phi_n(X) = \{T \in \mathcal{L}(X) \text{ such that } A_n T \in \Phi_0(X)\}$ for each $n \in \mathbb{Z}$, where A_n is a fixed operator satisfying $i(A_n) = -n$. Let $\mathcal{C}_A$ [with $A \in \mathcal{L}(X)$] denote the map from $\mathcal{L}(X)$ into itself defined by $\mathcal{C}_A(T) = AT$. By using the continuity of $\mathcal{C}_A$ and the fact that $\Phi_n(X) = \mathcal{C}_{A_n}^{-1}(\Phi_0(X))$ for each $n \in \mathbb{Z}$, we deduce that $\Phi^b(X) = \bigcup_{n \in \mathbb{Z}} \mathcal{C}_{A_n}^{-1}(\Phi_0(X))$, which completes the proof. Q.E.D.

7.10 Spectral Mapping Theorem

In order to make the spectral mapping theorems true in the case of closed, unbounded, and linear operators, we shall include the point at infinity to the essential spectra. So, we will consider the following extended spectra

$$\forall A \in \mathcal{C}(X),\ \tilde{\sigma}_{ei}(A) = \sigma_{ei}(A) \bigcup \{\infty\}, i = 1,2,4,5,6,7,8.$$

The aim of this section is to discuss a spectral mapping theorem for $\tilde{\sigma}_{e5}(\cdot)$, $\tilde{\sigma}_{e7}(\cdot)$, and $\tilde{\sigma}_{e8}(\cdot)$ in a special case which occurs in some applications.

Theorem 7.10.1 ([137, Theorem 7 (a)]). *Let A be a closed, unbounded, and linear operator with a nonempty resolvent set, and let f be a complex-valued function that is locally holomorphic on the extended spectrum of A, $\sigma(A) \bigcup \{\infty\}$. Then, $\tilde{\sigma}_{ei}(f(A)) = f(\tilde{\sigma}_{ei}(A))$, $i = 1,2,4,6$, and, $\tilde{\sigma}_{e5}(f(A)) \subseteq f(\tilde{\sigma}_{e5}(A))$.* ◇

The following result provides a spectral mapping theorem for the Schechter's essential spectrum in a special case which occurs in some applications. Let us recall that the spectral mapping theorem holds true for the Wolf essential spectrum (see Theorem 7.10.1). However, a counter-example given in [137, p. 23] shows that, in general, it is false for $\tilde{\sigma}_{e5}(.)$. In fact, let A be such that $i(I + A) = -1$ and $i(I - A) = 1$. Then $i((I + A)(I - A)) = 0$, so that $0 \notin \sigma_{e5}(f(A))$, where f is $f(\lambda) = (1 + \lambda)(1 - \lambda)$. However, $\pm 1 \in \sigma_{e5}(A)$ and, then $0 \in f(\sigma_{e5}(A))$.

Theorem 7.10.2. *Let $\mathcal{I}(X)$ be any nonzero two-sided ideal of $\mathcal{L}(X)$ satisfying $\mathcal{K}(X) \subset \mathcal{I}(X) \subset \mathcal{F}^b(X)$, and let A_1 and A_2 be two elements of $\mathcal{C}(X)$ such that $(\lambda - A_1)^{-1} - (\lambda - A_2)^{-1} \in \mathcal{I}(X)$ for some $\lambda \in \rho(A_1) \bigcap \rho(A_2)$. If $\sigma_{e4}(A_1) = \sigma_{e5}(A_1)$, and if f is a complex-valued function that is locally holomorphic on the extended spectrum of A, $\sigma(A) \bigcup \{\infty\}$, then $f(\tilde{\sigma}_{e5}(A_k)) = \tilde{\sigma}_{e5}(f(A_k))$, $k = 1,2$.* ◇

Proof. For $k = 1$, the result follows from both the hypotheses $\sigma_{e4}(A_1) = \sigma_{e5}(A_1)$ and Theorem 7.10.1. Now, let's consider the case $k = 2$. The inclusion $\tilde{\sigma}_{e5}(f(A_2)) \subset f(\tilde{\sigma}_{e5}(A_2))$ follows from Theorem 7.10.1. It remains to show that $f(\tilde{\sigma}_{e5}(A_2)) \subset \tilde{\sigma}_{e5}(f(A_2))$. Let $\lambda \in f(\tilde{\sigma}_{e5}(A_2))$. Then, there exists $\mu \in \tilde{\sigma}_{e5}(A_2)$, such that $\lambda = f(\mu)$. Hence, using the hypothesis $\sigma_{e4}(A_1) = \sigma_{e5}(A_1)$ as well as Theorem 7.5.4(i), for $i = 4$, we show that $\mu \in \tilde{\sigma}_{e4}(A_2)$. Next, applying Theorem 7.10.1 for the Wolf essential spectrum, we obtain $f(\mu) \in \tilde{\sigma}_{e4}(f(A_2))$. Since $\tilde{\sigma}_{e4}(f(A_2)) \subset \tilde{\sigma}_{e5}(f(A_2))$, we infer that $f(\mu) \in \tilde{\sigma}_{e5}(f(A_2))$, which completes the proof. Q.E.D.

Theorem 7.10.3 ([285, Theorem 3.3]). *Let $A \in \mathcal{L}(X)$ and let f be an analytic function defined on a neighborhood of $\sigma(A)$. Then, $\sigma_{e7}(f(A)) \subseteq f(\sigma_{e7}(A))$.* ◇

The results of the following theorem were established, respectively, by V. Rakočević (Theorem 7.10.3) and Schmoeger [304, Theorem 3] for bounded linear operators. Using the same method as the one developed in [137, p. 30], we can express the theorem for closed, unbounded, and linear operators. For a better convenience of the reader, we include a proof.

Theorem 7.10.4. *Let $A \in \mathcal{C}(X)$ with a nonempty resolvent set, and let f be a complex-valued function that is holomorphic on an open set containing $\sigma(A)\bigcup\{\infty\}$. Then, $\tilde{\sigma}_{e7}(f(A)) \subseteq f(\tilde{\sigma}_{e7}(A))$ and, $\tilde{\sigma}_{e8}(f(A)) \subseteq f(\tilde{\sigma}_{e8}(A))$.* ◇

Proof. Let β be a fixed point in $\rho(A)$ and let's define the function ψ by

$$\psi : \mathbb{C}\bigcup\{\infty\} \longrightarrow \mathbb{C}\bigcup\{\infty\}$$
$$\lambda \quad \longrightarrow \psi(\lambda) = \begin{cases} (\lambda-\beta)^{-1} \text{ if } \lambda \neq \beta \\ \psi(\beta) = \infty \\ \psi(\infty) = 0. \end{cases}$$

Let $T = \psi(A)$ and let's choose $\lambda \neq \beta$, $\mu = \psi(\lambda)$. Writing $A-\lambda = A-\beta-(\lambda-\beta)$, we get $(A-\lambda)T = \mu^{-1}(\mu - T)$ on X. Since $R(T) = \mathcal{D}(A-\lambda)$, then

$$R(\mu - T) = R(A-\lambda). \tag{7.10.1}$$

Also, since T is one-to-one map of X onto $\mathcal{D}(A-\lambda)$, then

$$\alpha(A-\lambda) = \alpha(\mu - T). \tag{7.10.2}$$

Note that $0 \in \sigma_{e7}(T)$ because $R(T) = \mathcal{D}(A)$ cannot be closed when A is unbounded. Therefore, using Eqs. (7.10.1) and (7.10.2), it is easy to verify that ψ is one-to-one map of $\tilde{\sigma}_{e7}(A)$ onto $\sigma_{e7}(T)$. Now, let's define the function g by $g(\mu) = f \circ \psi^{-1}(\mu)$. Then, g is holomorphic on a neighborhood of $\sigma(T)$ and $g(T) = f(A)$. Hence, from Theorem 7.10.3, it follows that

$$\begin{aligned} \tilde{\sigma}_{e7}(f(A)) &= \tilde{\sigma}_{e7}(g(T)) \\ &\subseteq g(\tilde{\sigma}_{e7}(T)) \\ &= f \circ \psi^{-1}(\tilde{\sigma}_{e7}(T)) \\ &= f(\tilde{\sigma}_{e7}(A)). \end{aligned}$$

This proves the result for $\tilde{\sigma}_{e7}(\cdot)$. Concerning $\tilde{\sigma}_{e8}(\cdot)$, the result can be deduced in the same way. Q.E.D.

Let us recall that the spectral mapping theorem holds true for $\tilde{\sigma}_{e1}(\cdot)$ and $\tilde{\sigma}_{e2}(\cdot)$ (see Theorem 7.10.1). However, a counter-example given in [289] shows that, in general, it is false for $\tilde{\sigma}_{e7}(\cdot)$. The following result provides a spectral mapping theorem for the essential approximate point spectrum and the essential defect spectrum in a special case which occurs in some applications.

Proposition 7.10.1. *Let A_1 and A_2 be two elements of $\mathcal{C}(X)$, such that $(\lambda - A_1)^{-1} - (\lambda - A_2)^{-1} \in \mathcal{F}_+^b(X)$ [resp. $\in \mathcal{F}_-^b(X)$] for some $\lambda \in \rho(A_1)\bigcap\rho(A_2)$. If $\sigma_{e7}(A_1) = \sigma_{e1}(A_1)$ [resp. $\sigma_{e8}(A_1) = \sigma_{e2}(A_1)$], and if f is a complex-valued* *function holomorphic on an open set containing* $\sigma(A)\bigcup\{\infty\}$, *then* $\tilde{\sigma}_{e7}(f(A_k)) = f(\tilde{\sigma}_{e7}(A_k))$, $k = 1, 2$ *(resp.* $\tilde{\sigma}_{e8}(f(A_k)) = f(\tilde{\sigma}_{e8}(A_k))$, $k = 1, 2$*).* ◇

Proof. For $k = 1$, the result follows from both the hypothesis $\sigma_{e7}(A_1) = \sigma_{e1}(A_1)$ and Theorem 7.10.1. For the case $k = 2$, the inclusion $\tilde{\sigma}_{e7}(f(A_2)) \subset f(\tilde{\sigma}_{e7}(A_2))$ follows from Theorem 7.10.4. It remains to show that $f(\tilde{\sigma}_{e7}(A_2)) \subset \tilde{\sigma}_{e7}(f(A_2))$. To do so, we consider $\lambda \in f(\tilde{\sigma}_{e7}(A_2))$. Then, there exists $\mu \in \tilde{\sigma}_{e7}(A_2)$ such that $\lambda = f(\mu)$. Hence, from Theorem 7.5.4 and the hypothesis $\sigma_{e7}(A_1) = \sigma_{e1}(A_1)$, it follows that $\mu \in \tilde{\sigma}_{e1}(A_2)$. Now, applying the spectral mapping theorem for $\tilde{\sigma}_{e1}(\cdot)$ (Theorem 7.10.1), we obtain $f(\mu) \in \tilde{\sigma}_{e1}(f(A_2)) \subseteq \tilde{\sigma}_{e7}(f(A_2))$. Thus, $\lambda \in \tilde{\sigma}_{e7}(f(A_2))$. This proves the result for $\tilde{\sigma}_{e7}(\cdot)$. Concerning $\tilde{\sigma}_{e8}(\cdot)$, the result can be proved in the same way. Q.E.D.

7.11 A Characterization of Polynomially Riesz Strongly Continuous Semigroups

7.11.1 Polynomially Fredholm Perturbations

Let X be a Banach space. We say that an operator $A \in \mathcal{L}(X)$ belongs to $P\mathcal{F}^b(X)$, if there is a nonzero complex polynomial $p(z)$, such that the operator $p(A) \in \mathcal{F}^b(X)$. In view of the good ordered of the positive integers, the minimal polynomial of A can be made unique by specifying the value of the coefficient related to the term with a large degree. Let $A \in P\mathcal{F}^b(X)$. The nonzero polynomial $p(z)$ of the least degree and leading coefficient 1 such that $p(A) \in \mathcal{F}^b(X)$ will be called the minimal polynomial of A.

Remark 7.11.1. Let $A \in P\mathcal{F}^b(X)$ and let $p(z)$ be the minimal polynomial of A. If λ_i is a root of $p(z)$ and $\lambda_i \notin \sigma(A)$, then $q(A) \in P\mathcal{F}^b(X)$, where $q(z) = (z - \lambda_i)^{-1}p(z)$. Let deg($p$) [resp. deg($q$)] denote the degree of $p(z)$ [resp. $q(z)$]. Obviously, deg(q) <deg(p). This contradicts the minimality of $p(z)$. Hence, $\lambda_i \in \sigma(A)$ and therefore, all roots of $p(z)$ belong to $\sigma(A)$. ◇

Let us notice that, if $A \in P\mathcal{F}^b(X)$, then there exists a polynomial $p(.) \neq 0$ such that $p(A) \in \mathcal{F}^b(X)$. Hence, $p(A)$ is a Riesz operator and, then $\sigma_{e6}(p(A)) = \{0\}$. Moreover, if $\dim(X) = \infty$, then we also have $\emptyset \neq \sigma_{e4}(p(A)) \subset \sigma_{e6}(p(A)) = \{0\}$. Consequently, $\sigma_{e4}(p(A)) = \sigma_{e6}(p(A)) = \{0\}$. Moreover, according to the spectral mapping theorem, we have

$$\sigma_{e4}(p(A)) = p(\sigma_{e4}(A)) = \{0\} \text{ and } \sigma_{e6}(p(A)) = p(\sigma_{e6}(A)) = \{0\}. \tag{7.11.1}$$

The following result can be found in [229].

Proposition 7.11.1. *Let* $A \in \mathcal{L}(X)$*. If* $A \in P\mathcal{F}^b(X)$*, then there exists a polynomial* $p(.) \neq 0$ *such that* $\sigma_{e6}(A) = \sigma_{e4}(A) \subset \{\lambda \in \mathbb{C}$ *such that* $p(\lambda) = 0\}$*.* ◇

Proof. Since $A \in P\mathcal{F}^b(X)$, there exists a polynomial $p(.) \neq 0$ such that $p(A) \in \mathcal{F}^b(X)$. So, by using Eq. (7.11.1), we deduce that $\sigma_{e4}(A) \subset \sigma_{e6}(A) \subset \{\lambda \text{ such that } p(\lambda) = 0\}$. Therefore, $\sigma_{e6}(A)$ is a finite set, and all elements in $\sigma(A)\backslash\sigma_{e6}(A)$ are isolated points in $\sigma(A)$. We still have to show that $\sigma_{e6}(A) \subset \sigma_{e4}(A)$. Indeed, let $\lambda_0 \in \sigma_{e6}(A)$. We can write $\sigma(A) = \sigma_0 \bigcup \sigma_1$ where σ_0 and σ_1 are clopen subsets of $\sigma(A)$ and $\sigma_{e6}(A) \bigcap \sigma_0 = \{\lambda_0\}$. So, we have a decomposition of A according to the decomposition $X = X_0 \oplus X_1$ of the space in such a way that the spectra of the parts of A in X_0 and X_1, i.e., A_0 and A_1 coincide with σ_0 and σ_1, respectively. Consequently, for any $\lambda \in \sigma_0$, $\lambda - A_1 \in \Phi^b(X_1)$, and by using Lemma 2.1.10(i), we have $\sigma_{e4}(A_0) = \sigma_0 \bigcap \sigma_{e4}(A) \subset \sigma_{e6}(A) \bigcap \sigma_0 = \{\lambda_0\}$. Since X_0 is not finite-dimensional, then $\sigma_{e4}(A_0) \neq \emptyset$ and therefore, $\sigma_{e4}(A_0) = \sigma_{e4}(A) \bigcap \sigma_0 = \{\lambda_0\}$. Hence, $\lambda_0 \in \sigma_{e4}(A)$ and consequently, $\sigma_{e4}(A) = \sigma_{e6}(A)$. Q.E.D.

7.11.2 Polynomially Riesz Operator

If $A \in P\mathcal{R}(X)$, then $\sigma_{e4}(A)$ is necessarily finite, say $\{\lambda_1, \ldots, \lambda_n\}$, and $p(z) = (z - \lambda_1)\ldots(z - \lambda_n)$ is the nonzero polynomial of the least degree and leading coefficient 1 such that $p(A) \in \mathcal{R}(X)$. It will be called the minimal polynomial of A. Conversely, let $p(z) = (z - \lambda_1)\ldots(z - \lambda_n)$ be the minimal polynomial (in the sense defined above) of A. Arguing as above, we notice that each $\lambda_i \in \sigma_{e4}(A)$ and so, $\sigma_{e4}(A) = \{\lambda_1, \ldots, \lambda_n\}$. This leads to the following characterization of the set of polynomially Riesz operators.

Proposition 7.11.2. *Let X be a Banach space. An operator $A \in \mathcal{L}(X)$ belongs to $P\mathcal{R}(X)$ if, and only if, $\sigma_{e4}(A)$ is finite, say, $\sigma_{e4}(A) = \{\lambda_1, \ldots, \lambda_n\}$. Moreover, the minimal polynomial of A can be written in the form $p(z) = (z - \lambda_1)\ldots(z - \lambda_n)$.* ◇

Let us notice that, if $A \in P\mathcal{R}(X)$, then $p(A) \in \mathcal{R}(X)$, where $p(.)$ is the minimal polynomial of A, and therefore, $\sigma_{e6}(p(A)) = \{0\}$. Moreover, if $\dim(X) = \infty$, we also have $\emptyset \neq \sigma_{e4}(p(A)) \subset \sigma_{e6}(p(A)) = \{0\}$. Consequently, $\sigma_{e4}(p(A)) = \sigma_{e6}(p(A)) = \{0\}$. Besides, and according to the spectral mapping theorem, we have

$$\sigma_{e4}(p(A)) = p(\sigma_{e4}(A)) = \{0\} \text{ and } \sigma_{e6}(p(A)) = p(\sigma_{e6}(A)) = \{0\}. \qquad (7.11.2)$$

All results of this section belong to [229].

Proposition 7.11.3. *Let $A \in \mathcal{L}(X)$. If $A \in P\mathcal{R}(X)$, then except for a finite set, the spectrum of A consists of isolated points representing some eigenvalues with a finite algebraic multiplicity.* ◇

Proof. Since $A \in P\mathcal{R}(X)$, there exists a polynomial $p(.) \neq 0$ such that $p(A) \in \mathcal{R}(X)$. So, it is sufficient to prove that $\sigma_{e6}(A) = \sigma_{e4}(A) \subset \{\lambda \text{ such that } p(\lambda) = 0\}$. For this purpose, let us first observe that (7.1.1) and (7.11.2) imply $\sigma_{e4}(A) \subset \sigma_{e6}(A) \subset \{\lambda$ such that $p(\lambda) = 0\}$. Consequently, $\sigma_{e6}(A)$ is a finite set, and all elements in $\sigma(A)\backslash\sigma_{e6}(A)$ are isolated points in $\sigma(A)$. It remains to demonstrate that $\sigma_{e6}(A) \subset \sigma_{e4}(A)$. Indeed, let $\lambda_0 \in \sigma_{e6}(A)$. We can write $\sigma(A) = \sigma_0 \bigcup \sigma_1$, where σ_0 and σ_1 represent two clopen subsets of $\sigma(A)$ and $\sigma_{e6}(A) \bigcap \sigma_0 = \{\lambda_0\}$. Then, we have a decomposition of A according to the decomposition $X = X_0 \oplus X_1$ of the space, in such a way that the spectra of the parts of A in X_0 and X_1, i.e., A_0 and A_1, coincide with σ_0 and σ_1, respectively. Consequently, for any $\lambda \in \sigma_0$, $\lambda - A_1 \in \Phi^b(X_1)$, and by using Lemma 2.1.10(i), we have $\sigma_{e4}(A_0) = \sigma_0 \bigcap \sigma_{e4}(A) \subset \sigma_{e6}(A) \bigcap \sigma_0 = \{\lambda_0\}$. Since X_0 is not finite-dimensional, then $\sigma_{e4}(A_0) \neq \emptyset$ and therefore, $\sigma_{e4}(A_0) = \sigma_{e4}(A) \bigcap \sigma_0 = \{\lambda_0\}$. Accordingly, $\lambda_0 \in \sigma_{e4}(A)$ and, then $\sigma_{e4}(A) = \sigma_{e6}(A)$ which ends the proof. Q.E.D.

Proposition 7.11.4. *Let $A \in \mathcal{PR}(X)$, such that the minimal polynomial $p(.)$ of A satisfies $p(-1) \neq 0$. Then, $I + A \in \Phi^b(X)$ and $i(I + A) = 0$.* ◇

Proof. Since $p(A) \in \mathcal{R}(X)$, $\sigma_{e6}(P(A)) = \{0\}$. By using the hypothesis that $p(-1) \neq 0$, we infer that $p(-1) \notin \sigma_{e6}(p(F))$. Next, by applying the spectral mapping theorem for the Browder essential spectrum (Theorem 7.10.1), we conclude that $-1 \notin \sigma_{e6}(A)$, i.e., $-1 \in \rho_6(A)$. This completes the proof. Q.E.D.

Proposition 7.11.5. *Let $(T(t))_{t\geq 0}$ be a C_0-semigroup on a Banach space X with an infinitesimal generator A, and let n be an integer. Let φ be a function defined from its domain into $\mathbb{C}^n$, that is, $\varphi : \mathcal{D}(\varphi) \subset \mathbb{R} \longrightarrow \mathbb{C}^n$, $t \longrightarrow (\varphi_1(t), \ldots, \varphi_n(t))$. Set $\mathcal{D}^+(\varphi) := \mathcal{D}(\varphi) \bigcap]0, \infty[$. Let us assume that $\varphi(.)$ is continuous and, for all $t \in \mathcal{D}^+(\varphi)$, $\varphi_i(t) \neq 0$ and $\prod_{i=1}^n (T(t) - \varphi_i(t)) \in \mathcal{R}(X)$. If $\mathcal{D}^+(\varphi) \neq \emptyset$, then $(T(t))_{t\geq 0}$ can be embedded in a C_0-group on X.* ◇

Proof. By hypothesis, there exists $t_0 > 0$ such that $T(t_0) \in \mathcal{PR}(X)$. Let $p_{t_0}(z) = \prod_{i=1}^n (z - \varphi_i(t_0))$ be the minimal polynomial of $T(t_0)$. Then, $p_{t_0}(T(t_0)) = \prod_{i=1}^n (T(t_0) - \varphi_i(t_0)) \in \mathcal{R}(X)$. By writing $T(t_0)$ in the form $T(t_0) = I + (T(t_0) - I)$, we notice that $p_{t_0}(T(t_0)) = \prod_{i=1}^n ((T(t_0) - I) - (\varphi_i(t_0) - 1)I) = \bar{p}_{t_0}(T(t_0) - I)$, where $\bar{p}_{t_0}(z) = \prod_{i=1}^n (z - (\varphi_i(t_0) - 1))$. Clearly, $\bar{p}_{t_0}(-1) = \prod_{i=1}^n (-\varphi_i(t_0)) \neq 0$. Therefore, $T(t_0) - I \in \mathcal{PR}(X)$ such that, the minimal polynomial $\bar{p}_{t_0}(.)$ of $T(t_0) - I$ satisfies $\bar{p}_{t_0}(-1) \neq 0$. From Proposition 7.11.4, it follows that $T(t_0) = I + (T(t_0) - I)$ is a Fredholm operator and $i(T(t_0)) = 0$. Now, the use of Theorem 5.1.5 leads to the desired result. Q.E.D.

Proposition 7.11.6. *Let $(T(t))_{t\geq 0}$ be a C_0-semigroup on a Banach space X with an infinitesimal generator A, and let n be an integer. Let φ be a function defined from its domain into $\mathbb{C}^n$, that is, $\varphi : \mathcal{D}(\varphi) \subset \mathbb{R} \longrightarrow \mathbb{C}^n$, $t \longrightarrow (\varphi_1(t), \ldots, \varphi_n(t))$. Set $\mathcal{D}^+(\varphi) := \mathcal{D}(\varphi) \bigcap]0, \infty[$. Let us assume that $\varphi(.)$ is continuous and, for all $t \in \mathcal{D}^+(\varphi)$, $\varphi_i(t) \neq 0$ and $\prod_{i=1}^n (T(t) - \varphi_i(t)) \in \mathcal{R}(X)$. If $\mathcal{D}^+(\varphi)$ contains a set with a nonempty interior, then A is bounded on X.* ◇

Proof. By using the various hypotheses, there exists a continuous function $\varphi : \mathcal{D}(\varphi) \subset \mathbb{R} \longrightarrow \mathbb{C}^n$, $t \longrightarrow (\varphi_1(t), \ldots, \varphi_n(t))$ (with $n \geq 2$) such that, for all $t \in \mathcal{D}^+(\varphi)$, we have $\varphi_i(t) \neq 0$ and $\prod_{i=1}^n (T(t) - \varphi_i(t)) \in \mathcal{R}(X)$, i.e., for all $t \in \mathcal{D}^+(\varphi)$, we have $T(t) \in \mathcal{PR}(X)$. According to Proposition 7.11.3, the spectrum of $T(t)$ consists of eigenvalues with a finite algebraic multiplicity, possibly accumulating at the points $\varphi_i(t)$, $i = 1,\ldots,n$. Therefore, by using the spectral mapping theorem for the point spectrum [see Eq. (2.5.1)], we deduce that $\sigma(T(t))\backslash\{\varphi_i(t),\ i = 1,\ldots,n\} = \{e^{\eta t}$ such that $\eta \in \sigma_p(A)\}$. According to Proposition 3.5.2, we know that $\mathcal{I} := \{\text{Im}\lambda$ such that $\lambda \in \sigma_p(A)\}$ is bounded. Let $M := \sup\{|\text{Im}\lambda|$ such that $\lambda \in \sigma_p(A)\}$. We will construct an open set satisfying the hypotheses of Proposition 2.5.1. For this purpose, let us first observe that, if $\lambda \in \sigma_p(A)$ and $t \in [0, \frac{\pi}{2M}]$, then $|\arg(e^{\lambda t})| \leq \frac{\pi}{2}$. So, for $t \in [0, \frac{\pi}{2M}]$, we have $e^{t\sigma_p(A)} = \sigma_p(T(t)) \subset \{z$ such that $\text{Re}(z) \geq 0\}$. Moreover, according to Proposition 7.11.5, the semigroup $(T(t))_{t\geq 0}$ can be embedded in a C_0-group, that is,

$$\tilde{T}(t) = \begin{cases} T(t) & \text{if } t \geq 0, \\ T(-t) & \text{if } t \leq 0. \end{cases}$$

Hence, for each $x \in X$, the map $t \in [0, \frac{\pi}{2M}] \longrightarrow T(-t)x$ is continuous. Then, there exists $M_x \geq 0$ such that $\|T(-t)x\| \leq M_x$, for all $t \in [0, \frac{\pi}{2M}]$. Therefore, by using the Banach–Steinhaus theorem, there exists $M' \geq 0$, such that $\|T(-t)\| \leq M'$ for all $t \in \left[0, \frac{\pi}{2M}\right]$. Since $T(-t) = T(t)^{-1}$, we deduce that $\|T(t)^{-1}x\| \leq M'\|x\|$ and, then

$$\|T(t)x\| \geq \frac{1}{M'}\|x\| \text{ for all } x \in X. \tag{7.11.3}$$

Let λ be such that $|\lambda| < \frac{1}{M'}$. From (7.11.3), we infer that, if $t \in [0, \frac{\pi}{2M}]$, then $\|(T(t) - \lambda)x\| \geq \left(\frac{1}{M'} - |\lambda|\right)\|x\|$ and consequently, $\lambda \notin \sigma_p(T(t))$. So, for $t \in [0, \frac{\pi}{2M}]$, we have $\sigma_p(T(t)) \subset \left\{z \in \mathbb{C} \text{ such that } \text{Re}(z) \geq 0 \text{ and } |z| \geq \frac{1}{M'}\right\}$. Recall that, by hypothesis, and for $t > 0$ (use Proposition 7.11.3), we have the following $\sigma(T(t)) = \sigma_p(T(t)) \bigcup \sigma_{e4}(T(t)) \subset \sigma_p(T(t)) \bigcup \{\varphi_1(t), \ldots, \varphi_n(t)\}$. Let $t_0 \in]0, \frac{\pi}{2M}[$ and let $\varepsilon > 0$ be given. Then, there exists $\delta > 0$ such that $]t_0 - \delta, t_0 + \delta[\subset]0, \frac{\pi}{2M}[$ and $|t - t_0| < \delta$ imply that $|\varphi_i(t) - \varphi_i(t_0)| \leq \varepsilon$ for all $i \in \{1, \ldots, n\}$. So, $|t - t_0| < \delta$ implies that $\sigma(T(t)) \subset \mathcal{S}(t_0, \varepsilon)$, where

$$\mathcal{S}(t_0, \varepsilon) := \left\{z \in \mathbb{C} \text{ such that } Re(z) \geq 0 \text{ and } |z| \geq \frac{1}{M'}\right\} \bigcup \left(\bigcup_{1 \leq i \leq n} \{z \in \mathbb{C} \text{ such that } |z - \varphi_i(t_0)| \leq \varepsilon\}\right).$$

This shows that, for a small enough $\varepsilon > 0$, the complement of the set $\mathcal{S}(t_0, \varepsilon)$ in $\mathbb{C}$ is an unbounded, open, and connected set Ω with $0 \in \Omega$. Now, by applying Proposition 2.5.1, we conclude that $A \in \mathcal{L}(X)$, which ends the proof. Q.E.D.

Proposition 7.11.7. *Let $(T(t))_{t\geq 0}$ be a C_0-semigroup on a Banach space X and let $t_0 > 0$. If $0 \in \sigma_{e4}(T(t_0))$, then $0 \in \sigma_{e4}(T(t))$ for all $t > 0$.* ◇

Proof. Let $t_0 > 0$ be such that $0 \in \sigma_{e4}(T(t_0))$. If, for some $t > 0$, $0 \notin \sigma_{e4}(T(t))$, then $T(t)$ is a Fredholm operator. Hence, by using Propositions 5.1.2 and 5.1.3, we deduce that $T(t)$ is invertible for all $t \geq 0$. This contradicts the fact that $0 \in \sigma_{e4}(T(t_0))$. Therefore, for all $t > 0$, $0 \in \sigma_{e4}(T(t))$. Q.E.D.

Proposition 7.11.8. *Let $(T(t))_{t\geq 0}$ be a C_0-semigroup on a Banach space X and let $t_0 > 0$. If $T(t_0) \in \mathcal{R}(X)$, then $(T(t))_{t\geq 0}$ is a C_0-semigroup of Riesz type.* ◇

Proof. Let $t > 0$. If $t > t_0$, then $T(t) = T(t_0)T(t - t_0)$. Since $T(t_0)$ and $T(t - t_0)$ commute, the use of Proposition 2.2.3 implies that $T(t) \in \mathcal{R}(X)$. Now, let us assume that $t < t_0$. There exists $n \in \mathbb{N}^*$ such that $\frac{t_0}{n} < t$. Hence, $T(t) = T(\frac{t_0}{n})T(t - \frac{t_0}{n})$. Since $T(t_0) \in \mathcal{R}(X)$ and $T(t_0) = [T(\frac{t_0}{n})]^n$, then the spectral mapping theorem shows that $T(\frac{t_0}{n})$ is also a Riesz operator. Now, using the fact that $T(\frac{t_0}{n})$ and $T(t-\frac{t_0}{n})$ commute, together with Proposition 2.2.3, we deduce that $T(t) \in \mathcal{R}(X)$. Q.E.D.

Proposition 7.11.9. *Let X be a Banach space and assume that $X = X_1 \oplus X_2$, $A \in \mathcal{L}(X)$ with $A(X_1) \subset X_1$, $A(X_2) \subset X_2$. Let $A_1 = A_{|X_1} \in \mathcal{L}(X_1)$ and $A_2 = A_{|X_2} \in \mathcal{L}(X_2)$. Then, $\sigma_{e4}(A) = \sigma_{e4}(A_1) \bigcup \sigma_{e4}(A_2)$.* ◇

Proof. The proof is trivial. Q.E.D.

Proposition 7.11.10. *Let X be a Banach space and assume that $X = X_1 \oplus X_2$, $A \in \mathcal{L}(X)$ with $A(X_1) \subset X_1$, $A(X_2) \subset X_2$. Let $A_1 = A_{|X_1} \in \mathcal{L}(X_1)$ and $A_2 = A_{|X_2} \in \mathcal{L}(X_2)$. If λ is an isolated point of $\sigma(A_1)$ such that $\lambda \notin \sigma(A_2)$, then $P_\lambda(A)(X) = P_\lambda(A_1)(X_1)$, where $P_\lambda(A)$ [resp. $P_\lambda(A_1)$] denotes the spectral projection associated with $\{\lambda\}$ for A (resp. A_1) in $\mathcal{L}(X)$ [resp. $\mathcal{L}(X_1)$].* ◇

Proof. Let $\lambda \neq 0$ be an isolated point of $\sigma(A_1)$. Then, there exists a spectral decomposition of the space X_1, that is, $X_1 = Y_1 \oplus Y_2$ and $Y_1 = P_\lambda(A_1)X_1$ (where $P_\lambda(A_1)$ denotes the spectral projection associated with the spectral set $\{\lambda\}$). Accordingly, $\sigma(A_{1|Y_1}) = \{\lambda\}$, and $\sigma(A_{1|Y_2}) = \sigma(A_1)\backslash\{\lambda\}$. Let us also notice that $X = Y_1 \oplus (Y_2 \oplus X_2)$, $AY_1 = A_1Y_1 \subset Y_1$ and $A(Y_2 \oplus X_2) \subset Y_2 \oplus X_2$. Now, by applying the first assertion to the operators $A_{|Y_2\oplus X_2}$ and $A_{|Y_2}$, we get $\sigma(A_{|Y_2\oplus X_2}) = \sigma(A_2) \bigcup \sigma(A_{|Y_2})$. However, $\sigma(A_{|Y_2}) = \sigma(A_{1|Y_2}) = \sigma(A_1)\backslash\{\lambda\}$ and $\sigma(A) = \sigma(A_2) \bigcup \sigma(A_1)$, so $\sigma(A_{|Y_2\oplus X_2}) = \sigma(A)\backslash\{\lambda\}$. Accordingly, $\sigma(A_{|Y_1}) = \sigma(A_{1|Y_1}) = \{\lambda\}$. This leads to $Y_1 = R(P_\lambda(A))$ [the range of the spectral projection associated with the spectral set $\{\lambda\}$ of $\sigma(A)$], which completes the proof. Q.E.D.

Theorem 7.11.1. *Every Riesz operator is demicompact operator.* ◇

Proof. Let A be a Riesz operator. We have $\sigma_{e4}(A) = \{0\}$. So, $I - A$ is a Fredholm operator. Now, the result follows from Lemma 5.4.6. Q.E.D.

Lemma 7.11.1. *Let $(T(t))_{t\geq 0}$ be a C_0-semigroup on a Banach space X with an infinitesimal generator A. If $A \in \mathcal{R}(X)$, then $T(t) - I \in \mathcal{R}(X)$.* ◇

Proof. Let $t > 0$ and assume that $A \in \mathcal{R}(X)$. Clearly, the operator $T(t) - I$ may be written in the following form $T(t) - I = e^{tA} - I = \sum_{k=1}^{\infty} \frac{t^k A^k}{k!} = Ag(A)$, where $g(.)$ is the entire function $g(z) = \sum_{k=0}^{\infty} \frac{t^{k+1} z^k}{(k+1)!}$. Since A and $g(A)$ commute, then

$T(t) - I = Ag(A) = g(A)A$. Since A is of Riesz type, using Proposition 2.2.3 implies that $T(t) - I \in \mathcal{R}(X)$. Q.E.D.

Lemma 7.11.2. *Let $(T(t))_{t\geq 0}$ be a C_0-semigroup on a Banach space X with an infinitesimal generator A. If $(A - \lambda) \in \mathcal{R}(X)$ for some $\lambda \in \mathbb{C}$, then $T(t) - e^{\lambda t} \in \mathcal{R}(X)$.* ◇

Proof. Let us suppose that $(A - \lambda) \in \mathcal{R}(X)$ for some $\lambda \in \mathbb{C}$. Notice that the operator $T(t) - e^{\lambda t}$ can be written as follows $T(t) - e^{\lambda t} = e^{\lambda t}(e^{-\lambda t}T(t) - I) = e^{\lambda t}(e^{-\lambda t}e^{tA} - I) = e^{\lambda t}(e^{t(A-\lambda)} - I)$. Since $(A - \lambda) \in \mathcal{R}(X)$, and using the result of Lemma 7.11.1, we get $(e^{t(A-\lambda)} - I) \in \mathcal{R}(X)$ and therefore, $T(t) - e^{\lambda t} \in \mathcal{R}(X)$. Q.E.D.

Lemma 7.11.3. *Let $(T(t))_{t\geq 0}$ be a C_0-semigroup on a Banach space X with an infinitesimal generator A and let n be an integer. Let φ be a function defined from its domain into $\mathbb{C}^n$, that is, $\varphi : \mathcal{D}(\varphi) \subset \mathbb{R} \longrightarrow \mathbb{C}^n$, $t \longrightarrow (\varphi_1(t), \ldots, \varphi_n(t))$. Let us assume that $\varphi(.)$ is continuous and, for all $t \in \mathcal{D}^+(\varphi)$, we have $\varphi_i(t) \neq 0$ and $\prod_{i=1}^n (T(t) - \varphi_i(t)) \in \mathcal{R}(X)$. If the operator A belongs to $P\mathcal{R}(X)$, then $\mathcal{D}^+(\varphi) =]0, \infty[$.* ◇

Proof. Let us suppose that $p(z) = (z - \lambda_1)\ldots(z - \lambda_n)$ is the minimal polynomial of A. Then, $p(A) = (A - \lambda_1)\ldots(A - \lambda_n) \in \mathcal{R}(X)$. This implies that $\sigma_{e6}(A) = \{\lambda_1, \ldots, \lambda_n\}$. Hence, we can write $\sigma(A) = \sigma_1 \bigcup \sigma_2 \bigcup \ldots \bigcup \sigma_n$ where σ_i for $1 \leq i \leq n$ are clopen sets in $\sigma(A)$ such that $\lambda_i \in \sigma_i$ for $i = 1, \ldots, n$ and $\sigma_i \bigcap \sigma_j = \emptyset$ if $i \neq j$. Let $(X_i)_{1\leq i\leq n}$ and $(A_i)_{1\leq i\leq n}$ be the spectral subspaces and the restrictions of A associated with this decomposition, respectively. The fact that the sets X_i, $i = 1, \ldots, n$, are stable by A implies that they are also invariant by e^{tA}. Let $e^{tA}_{|X_i}$ be the restriction of e^{tA} to X_i. Obviously, $e^{tA}_{|X_i} \in \mathcal{L}(X_i)$ and $e^{tA}_{|X_i} = e^{tA_i}$. Now, let us consider the problem separately for each subspace X_i, $i = 1, \ldots, n$. On X_i, we can write $p(A) = (A_i - \lambda_i)\prod_{i\neq j}(A_i - \lambda_j)$. For each $j \neq i$, $\lambda_j \notin \sigma(A_i)$ and so, the operator $\prod_{i\neq j}(A_i - \lambda_j)$ is invertible in $\mathcal{L}(X_i)$. Accordingly, $(A_i - \lambda_i) = p(A_i)\left(\prod_{i\neq j}(A - \lambda_j)\right)^{-1} \in \mathcal{L}(X_i)$. Moreover, the spectral mapping theorem implies that $\sigma(p(A_i)) = p(\sigma(A_i)) = \{p(\tau) \text{ such that } \tau \in \sigma(A_i)\}$. Hence, $\sigma(p(A_i))$ consists of eigenvalues with a finite algebraic multiplicity, accumulating to $p(\lambda_i) = 0$. So, $p(A_i) \in \mathcal{R}(X_i)$. This, combined with the fact that $p(A_i)$ and $\left(\prod_{i\neq j}(A - \lambda_j)\right)^{-1}$ commute, implies, thanks to Proposition 2.2.3, that $(A_i - \lambda_i) \in \mathcal{R}(X_i)$. Moreover, for each $i \in \{1, \ldots, n\}$, we have $\prod_{j=1}^n (e^{tA_i} - e^{\lambda_j t}) = (e^{tA_i} - e^{\lambda_i t})\prod_{j\neq i}(e^{tA_i} - e^{\lambda_j t})$. Since $\prod_{j\neq i}(e^{tA_i} - e^{\lambda_j t})$ is invertible on X_i and $e^{tA_i} - e^{\lambda_i t} \in \mathcal{R}(X_i)$ (use the fact that $(A_i - \lambda_i) \in \mathcal{R}(X_i)$ and Lemma 7.11.2), again by using Proposition 2.2.3, we deduce that $\prod_{j=1}^n (e^{tA}_{|X_i} - e^{\lambda_j t}) = \prod_{j=1}^n (e^{tA} - e^{\lambda_j t})_{|X_i} \in \mathcal{R}(X_i)$. Next, notice that the operator $\prod_{i=1}^n (e^{tA} - e^{\lambda_i t})$ can be written in the form $\sum_{i=1}^n \mathcal{O}_i$ where $\mathcal{O}_i = Ji\left[\prod_{j=1}^n (e^{tA} - e^{\lambda_j t})_{|X_i}\right] P_i$, with $J_i : X_i \longrightarrow X$ representing the canonical embedding and $P_i : X \longrightarrow X_i$ denoting the spectral projection associated with the clopen subset σ_i. Clearly, $\mathcal{O}_i\mathcal{O}_j = \mathcal{O}_j\mathcal{O}_i = 0$

for $i \neq j$. Moreover, by using Proposition 7.11.10, we notice that each $\mathcal{O}_i$, $i = 1, \ldots, n$, belongs to $\mathcal{R}(X)$. Now, by applying Proposition 2.2.3, we get $]0, \infty[\subset \mathcal{D}(\varphi)$. Q.E.D.

Theorem 7.11.2. *Under the assumptions of Lemma 7.11.3, the following conditions are equivalent.*

(i) *There are two constants* a, $b \in]0, \infty[$, $a < b$ *such that* $]a, b[\subset \mathcal{D}^+(\varphi)$.
(ii) $\mathcal{D}^+(\varphi) =]0, \infty[$.
(iii) *The operator* A *belongs to* $\mathcal{PR}(X)$.
(iv) $(\lambda(\lambda - A)^{-1} - I)$ *belongs to* $\mathcal{PR}(X)$ *for every* $\lambda \in \rho(A)$.
(v) $(\lambda(\lambda - A)^{-1} - I)$ *belongs to* $\mathcal{PR}(X)$ *for some* $\lambda \in \rho(A)$. ◇

Proof.

(i) ⇒ *(iii)* From Proposition 7.11.6, we deduce that $A \in \mathcal{L}(X)$. So, it remains to verify that A belongs to $\mathcal{PR}(X)$. For this purpose, let $t \in]a, b[$ and write $T(t) = e^{tA}$. Let $p_t(z) = \prod_{i=1}^{n}(z - \varphi_i(t))$. By using the various hypotheses, we know that $p_t(e^{tA}) = \prod_{i=1}^{n}(e^{tA} - \varphi_i(t)) \in \mathcal{R}(X)$. Accordingly, $f_t(A) \in \mathcal{R}(X)$ where $f_t(.)$ represents the entire function $z \longrightarrow p_t(e^{tz})$. Then, Proposition 3.5.1 implies that $A \in \mathcal{PR}(X)$.
(iii) ⇒ *(ii)* see Lemma 7.11.3.
(ii) ⇒ *(i)* It is trivial.
(iii) ⇒ *(iv)* First, let us notice that, by using the spectral mapping theorem, and for any $\lambda \in \rho(A)$, we have $-1 \in \rho(\lambda(\lambda - A)^{-1} - I)$. Now, let us consider the function f_λ defined by $f_\lambda : \mathbb{C}\backslash\{-1\} \longrightarrow \mathbb{C}$, $z \longrightarrow \lambda - \frac{\lambda}{z+1}$. Clearly, $A = f_\lambda(\lambda(\lambda - A)^{-1} - I)$. By hypothesis, there exists $p(.) \in \mathbb{C}[z]\backslash\{0\}$ such that $p(A) \in \mathcal{R}(X)$. Hence, $(p \circ f_\lambda)(\lambda(\lambda - A)^{-1} - I) \in \mathcal{R}(X)$. Next, by applying Proposition 3.5.1, we conclude that $(\lambda(\lambda - A)^{-1} - I) \in \mathcal{PR}(X)$ for every $\lambda \in \rho(A)$.
(iv) ⇒ *(v)* It is trivial.
(v) ⇒ *(iii)* Let $\lambda \in \rho(A)$ be such that $(\lambda(\lambda - A)^{-1} - I) \in \mathcal{R}(X)$ and let us denote by g_λ the function defined by $g_\lambda : \mathbb{C}\backslash\{\lambda\} \longrightarrow \mathbb{C}$, $z \longrightarrow \frac{\lambda}{\lambda - z} - 1$. Since $g_\lambda(A) = \lambda(\lambda - A)^{-1} - I$, using Proposition 3.5.1 implies that $A \in \mathcal{PR}(X)$. Q.E.D.

Theorem 7.11.3. *Let* $(T(t))_{t\geq 0}$ *be a* C_0*-semigroup on a Banach space* X *with an infinitesimal generator* A *and let* $n \geq 2$ *be an integer. Let* φ *be a function defined by its domain into* $\mathbb{C}^n$, *that is,* $\varphi : \mathcal{D}(\varphi) \subset \mathbb{R} \longrightarrow \mathbb{C}^n$, $t \longrightarrow (\varphi_1(t), \ldots, \varphi_n(t))$. *Suppose that* $\varphi(.)$ *is continuous, there exist* $i_0 \in \{1, \ldots, n\}$ *and* $t_0 \in \mathcal{D}^+(\varphi)$ *such that* $\varphi_{i_0}(t_0) = 0$ *and, for all* $t \in \mathcal{D}^+(\varphi)$, *we have* $\prod_{i=1}^{n}(T(t) - \varphi_i(t)) \in \mathcal{R}(X)$. *Then, there exist two closed subspaces of* X, *namely* X_0 *and* X_1, *such that* $X = X_0 \oplus X_1$ *and, for all* $t \geq 0$, $T(t)X_i \subset X_i$, $i = 0, 1$. *Moreover,* $(T(t)_{|X_1})_{t\geq 0}$ *is a* C_0*-semigroup on* X_1 *which can be embedded in a* C_0*-group. Besides,* $(T(t)_{|X_1})_{t\geq 0}$ *and its generator* $A_{|X_1}$ *satisfy the assertions (i)–(v) of Theorem 7.11.2.* ◇

Proof. The hypotheses say that there is a function $\varphi : \mathcal{D}(\varphi) \subset \mathbb{R} \longrightarrow \mathbb{C}^n$, $t \longrightarrow (\varphi_1(t), \ldots, \varphi_n(t))$ such that, for all $t \in \mathcal{D}^+(\varphi)$, we have $\prod_{i=1}^{n}(T(t) - \varphi_i(t)) \in$

$\mathcal{R}(X)$. By recalling the characterization of polynomially Riesz operators (Proposition 7.11.2) and Proposition 7.11.3, we have $\sigma_{e4}(T(t)) = \{\varphi_i(t),\ i = 1,\ldots,n\}$ and $\sigma(T(t))\backslash\sigma_{e4}(T(t))$ consist of eigenvalues with a finite algebraic multiplicity. Since $\varphi_{i_0}(t_0) = 0$ for some $i_0 \in \{1,\ldots,n\}$, say $i_0 = n$, and from Lemma 2.1.10(*ii*), we deduce that $\varphi_n(t) = 0$ for all $t > 0$. This implies that $0 \in \sigma_{e4}(T(t))$ for all $t > 0$. Consequently, $\sigma_{e4}(T(t)) = \{0, \varphi_1(t),\ldots,\varphi_{n-1}(t)\}$ with $\varphi_i(t) \neq 0$ for all $t \in \mathcal{D}^+(\varphi)$ and $i = 1,\ldots,n-1$. Let τ_0 and τ_1 constitute a partition of $\sigma(T(t_0))$, such that $\tau_0 \bigcap \sigma_{e4}(T(t_0)) = \{0\}$. Therefore, there exist two closed subspaces, namely X_0 and X_1, which reduce $T(t_0)$, that is, $X = X_0 \oplus X_1$. Clearly, for all $t > 0$, $T(t)$ commutes with $T(t_0)$ and hence, with also the associated spectral projectors. So, $T(t)X_k \subset X_k, k = 0, 1$ and, then $(T(t)_{|X_k})_{t\geq 0}$ is a C_0-semigroup on X_k, $A(\mathcal{D}(A) \bigcap X_k) \subset X_k$ and so $A_{|X_k}$ is the generator of $(T(t)_{|X_k})_{t\geq 0}$. Since $0 \notin \sigma_{e4}(T(t_0)_{|X_1})$, and from Proposition 7.11.7, we infer that $0 \notin \sigma_{e4}(T(t)_{|X_1})$ for all $t > 0$. Besides, and using Proposition 7.11.9, we notice that $\sigma_{e4}(T(t)) = \sigma_{e4}(T(t)_{|X_0}) \bigcup \sigma_{e4}(T(t)_{|X_1})$ and therefore, $\sigma_{e4}(T(t)_{|X_1}) = \{\varphi_1(t),\ldots,\varphi_{n-1}(t)\}$. This shows that $(T(t)_{|X_1})_{t\geq 0}$ satisfies the hypotheses of Theorem 7.11.2 and $]0,\infty[\subset \mathcal{D}(\varphi)$. As a result, the statement (*ii*) follows from Propositions 7.11.5, 7.11.6 and Theorem 7.11.2. Q.E.D.

Theorem 7.11.4. *Assume that the hypotheses of Theorem 7.11.3 hold. Then, $(T(t)_{|X_0})_{t\geq 0}$ is a C_0-semigroup of Riesz type on X_0, i.e., $T(t)_{|X_0} \in \mathcal{R}(X_0)$ for all $t > 0$. Its generator $A_{|X_0}$ is unbounded on X_0 and, for any $\lambda \in \rho(A_{|X_0})$, we have $(\lambda - A_{|X_0})^{-1} \in \mathcal{R}(X_0)$.* ◇

Proof. Let us notice that since $0 \in \sigma_{e4}(T(t_0)_{|X_0})$, it follows, from Proposition 7.11.7, that $0 \in \sigma_{e4}(T(t)_{|X_0})$ for all $t > 0$. In order to demonstrate that $(T(t)_{|X_0})_{t\geq 0}$ is a Riesz C_0-semigroup, it is sufficient to prove that $\sigma_{e4}(T(t)_{|X_0}) = \{0\}$ for all $t > 0$. For this purpose, let $\mu \neq 0$ be such that $\mu \in \sigma_{e4}(T(t)_{|X_0})$ for some $t > 0$. So, we can write $X_0 = Z_1 \oplus Z_2$ with $\dim(Z_j) = \infty$, $j = 1, 2$ and $\sigma_{e4}(T(t)_{|Z_2}) = \{\mu\}$. Obviously, $0 \notin \sigma_{e4}(T(t)_{|Z_2})$. So, by using Proposition 7.11.7, we have $0 \notin \sigma_{e4}(T(t_0)_{|Z_2})$. Moreover, by applying Proposition 7.11.9, we show that $\sigma_{e4}(T(t_0)_{|X_0}) = \sigma_{e4}(T(t_0)_{|Z_1}) \bigcup \sigma_{e4}(T(t_0)_{|Z_2}) = \{0\}$. Hence, $\sigma_{e4}(T(t_0)_{|Z_2}) = \emptyset$, which contradicts the fact that $\dim Z_2 = \infty$. Consequently, $\mu \notin \sigma_{e4}(T(t)_{|X_0})$, which proves that $T(t)_{|X_0}$ is a Riesz operator for all $t > 0$. This ends the proof. Q.E.D.

7.12 A Spectral Characterization of the Uniform Continuity of Strongly Continuous Groups

Let $(T(t))_{t\in\mathbb{R}}$ be a strongly continuous group on a Banach space X. Accordingly, the set $\sigma^1(T(t))$ may be defined for each $t \in \mathbb{R}$. In the following, $\chi(T)$ denotes the set

$$\chi(T) := \{t \in \mathbb{R} \text{ such that } \sigma^1(T(t)) \neq \mathbb{T}\}.$$

Remark 7.12.1. Let $(T(t))_{t\in\mathbb{R}}$ be the translation group on the space $L_2(\mathbb{R})$, that is,

$$(T(t)f)(s) = f(t+s),\ t,\ s \in \mathbb{R}$$

for all $f \in L_2(\mathbb{R})$. Its generator is the unbounded operator $Af := f'$ whose domain is given by $\mathcal{D}(A) = \{f \in L_2(\mathbb{R})$ such that f is absolutely continuous and $f' \in L_2(\mathbb{R})\}$. By using the fact that $\sigma(A) = i\mathbb{R}$ and the inclusion $e^{t\sigma(A)} \subset \sigma(T(t))$, we notice that $\sigma(T(t)) = \mathbb{T}$ for all $t \in \mathbb{R}\backslash\{0\}$ and $\chi(T) = \{0\}$. ◇

The following results come from [228].

Lemma 7.12.1. *Let X be a Banach space, and let $(T(t))_{t\in\mathbb{R}}$ be a strongly continuous group on X. If $(T(t))_{t\in\mathbb{R}}$ is uniformly continuous, then $\chi(T)$ has a nonempty interior.* ◇

Proof. According to the hypothesis, there exists $\alpha > 0$ such that $\|T(t) - I\| < 1$ for all $t \in]-\alpha,\alpha[$. This implies that $\sigma(T(t)) \subset B(1,\sqrt{2})$. Hence, $\sigma^1(T(t)) \subset B(1,\sqrt{2})\bigcap\mathbb{T}$. Therefore, $\chi(T)$ contains $(-\alpha,\alpha)$ and, then has a nonempty interior. Q.E.D.

Lemma 7.12.2. *Let X be a Banach space, and let $(T(t))_{t\in\mathbb{R}}$ be a strongly continuous group on X. If $\chi(T)$ has a nonempty interior, then $\chi(T)$ is non-meager.* ◇

Proof. From the Baire category theorem, we deduce the result. Q.E.D.

Let $(T(t))_{t\in\mathbb{R}}$ be a strongly continuous group on X. According to Theorem 7.9.2, we know that the map from $\mathcal{L}(X)$ into $K(\mathbb{C})$, which assigns to each element of $\mathcal{L}(X)$ its spectrum, is Borel where $\mathcal{L}(X)$ and $K(\mathbb{C})$ are equipped, respectively, with the strong operator topology and the Hausdorff topology. The map $t \longrightarrow \sigma(T(t))$ from $\mathbb{R}$ into $K(\mathbb{C})$ is also Borel (thanks to the strong continuity of $(T(t))_{t\in\mathbb{R}}$). Moreover, the map $K \longrightarrow K^1$ from $K(\mathbb{C}\backslash\{0\})$ into $K(\mathbb{T})$ is continuous [187, Exercise 4.29 (*vi*)]. So, by composition, we infer that the function $t \longrightarrow \sigma^1(T(t))$ is Borel from $\mathbb{R}$ into $K(\mathbb{T})$ and therefore, is Baire measurable. Let $\mathcal{A}_T$ be a maximal commutative subalgebra of $\mathcal{L}(X)$ containing the set $\{T(t)$ such that $t \in \mathbb{R}\}$ and let us denote by $\widehat{\mathcal{A}_T}$ the character space (or the spectrum) of $\mathcal{A}_T$. Note that, for each $\psi \in \widehat{\mathcal{A}_T}$, we have $\tilde{\varphi}_\psi(t) := \psi(T(t)) \in \sigma(T(t))$. It is obvious that $\tilde{\varphi}_\psi(t+t') = \tilde{\varphi}_\psi(t)\tilde{\varphi}_\psi(t')$ for all $t,\ t' \in \mathbb{R}$, and $|\tilde{\varphi}_\psi(t)| \leq \|T(t)\|$. Clearly, $|\tilde{\varphi}_\psi(.)|$ is bounded on each compact subset of $\mathbb{R}$ and therefore, according to Theorem 2.5.14, it is continuous. Let us define the function φ_ψ by $\varphi_\psi : \mathbb{R} \longrightarrow \mathbb{T},\ t \longrightarrow \varphi_\psi(t) = \frac{\tilde{\varphi}_\psi(t)}{|\tilde{\varphi}_\psi(t)|}$. Accordingly, $\tilde{\varphi}_\psi(.)$ is continuous if, and only if, $\varphi_\psi(.)$ is continuous. Clearly, for all $t,\ t' \in \mathbb{R}$, $\varphi_\psi(t+t') = \varphi_\psi(t)\varphi_\psi(t')$, $\varphi_\psi(t) \in \sigma^1(T(t))$ and then, it is a character of $\mathbb{R}$.

Lemma 7.12.3. *Let X be a separable Banach space, and let $(T(t))_{t\in\mathbb{R}}$ be a strongly continuous group on X. If $\chi(T)$ is non-meager, then $(T(t))_{t\in\mathbb{R}}$ is uniformly continuous.* ◇

Proof. Since the function $t \longrightarrow \sigma^1(T(t))$ from $\mathbb{R}$ into $K(\mathbb{T})$ is Baire measurable, and since the set $\{t \in \mathbb{R}$ such that $\sigma^1(T(t)) \neq \mathbb{T}\}$ is non-meager, Theorem 2.8.2 implies that, for all $\psi \in \widehat{\mathcal{A}_T}$, $\varphi_\psi(.)$ is continuous and consequently, $\tilde{\varphi}_\psi(.)$ is also continuous. Now, by using [152, Theorem 16.5.1], we conclude that $(T(t))_{t\in\mathbb{R}}$ is uniformly continuous. Q.E.D.

Lemma 7.12.4. *Let X be a Banach space, assume that X is not separable, and let $(T(t))_{t\in\mathbb{R}}$ be a strongly continuous group on X. If $\chi(T)$ is non-meager, then $(T(t))_{t\in\mathbb{R}}$ is uniformly continuous.* ◇

Proof. If $(T(t))_{t\in\mathbb{R}}$ is not uniformly continuous, then there exist $\delta > 0$ and a real sequence $(t_n)_n$ such that $t_n \to 0$ as $n \to \infty$ and $\|T(t_n) - I\| > \delta$. So, there exists a sequence $(x_n)_n$ of vectors in X such that $\|x_n\| = 1$ and $\|T(t_n)x_n - x_n\| > \delta$ for all $n \in \mathbb{N}$. Now, let us set $Y := \overline{\text{span}\left(\bigcup_{n\in\mathbb{N}}\{T(t)x_n,\ t \in \mathbb{R}\}\right)}$. Clearly, Y is $(T(t))_{t\in\mathbb{R}}$-invariant and separable because each orbit is $(T(t))_{t\in\mathbb{R}}$-invariant and separable. So, in view of the vectors x_n, we infer that there exist $\delta > 0$ and a real sequence $(t_n)_n$ such that $t_n \to 0$ as $n \to \infty$ and $\|T_{|Y}(t_n) - I_Y\| > \delta$. This shows that $(T_{|Y}(t))_{t\in\mathbb{R}}$ is not continuous in norm. By using the first part of the proof, we conclude that $\{t \in \mathbb{R}$ such that $\sigma^1(T_{|Y}(t)) \neq \mathbb{T}\}$ is meager. However, Lemma 2.8.4 implies that

$$\chi(T(t)) = \{t \in \mathbb{R} \text{ such that } \sigma^1(T(t)) \neq \mathbb{T}\} \subset \{t \in \mathbb{R} \text{ such that } \sigma^1(T_{|Y}(t)) \neq \mathbb{T}\}.$$

Accordingly, $\chi(T(t))$ is also meager, which ends the proof. Q.E.D.

The following theorem is a consequence of Lemmas 7.12.1–7.12.4:

Theorem 7.12.1. *Let X be a Banach space, and let $(T(t))_{t\in\mathbb{R}}$ be a strongly continuous group on X. The following assertions are equivalent*

(i) $(T(t))_{t\in\mathbb{R}}$ *is uniformly continuous,*
(ii) $\chi(T)$ *has a nonempty interior,*
(iii) $\chi(T)$ *is non-meager.* ◇

Lemma 7.12.5. *Let X be a Banach space and $B \in \mathcal{L}(X)$. If $0 \notin \sigma(B)$, then $\sigma^1(B) \neq \mathbb{T}$ if, and only if, $\sigma^1_*(B) \neq \mathbb{T}$, where $\sigma_*(.)$ stands for each one of the sets $\sigma_{pa}(.)$ and $\sigma_{e3}(.)$.* ◇

Proof. The inclusion $\sigma_{e3}(B) \subset \sigma(B)$ and $\sigma_{pa}(B) \subset \sigma(B)$ imply that $\sigma^1(B) \neq \mathbb{T} \Longrightarrow \sigma^1_{e3}(B) \neq \mathbb{T}$ and $\sigma^1_{pa}(B) \neq \mathbb{T}$. Conversely, let us first assume that $\sigma^1_{e3}(B) \neq \mathbb{T}$ and let S be a sector of the complex plane satisfying $\sigma_{e3}(B) \bigcap S = \emptyset$. Hence, S is contained in the unbounded component of $\mathbb{C}\backslash\sigma_{e3}(B)$ and, then $\sigma(B) \bigcap S$ consists of isolated eigenvalues of a finite algebraic multiplicity with no accumulation point. Hence, $\sigma(B) \bigcap S$ is finite and its projection onto $\mathbb{T}$ cannot fill the whole segment $S \bigcap \mathbb{T}$. For $\sigma_{pa}(.)$, it suffices to observe that the inclusions $\sigma_{pa}(B) \subset \sigma(B)$ and $\partial\sigma(B) \subset \partial\sigma_{pa}(B)$ imply that the unbounded components of $\mathbb{C}\backslash\sigma(B)$ and $\mathbb{C}\backslash\sigma_{pa}(B)$ are the same. This together with the fact that $0 \notin \sigma(B)$ shows that $\sigma^1_{pa}(B) = \sigma^1(B)$ which ends the proof. Q.E.D.

The next corollary is an immediate consequence of Theorem 7.12.1 and Lemma 7.12.5.

Corollary 7.12.1. *Let X be a Banach space, and let $(T(t))_{t\in\mathbb{R}}$ be a strongly continuous group on X. The following assertions are equivalent*

(i) $(T(t))_{t\in\mathbb{R}}$ *is uniformly continuous,*
(ii) $\{t \in \mathbb{R}$ *such that* $\sigma_*^1(T(t)) \neq \mathbb{T}\}$ *has a nonempty interior,*
(iii) $\{t \in \mathbb{R}$ *such that* $\sigma_*^1(T(t)) \neq \mathbb{T}\}$ *is non-meager.* ◇

Remark 7.12.2. It is worth noticing that the results of Lemma 7.12.5 and Corollary 7.12.1 remain valid for all essential spectra containing $\sigma_{e3}(.)$. ◇

Remark 7.12.3. Let $(T(t))_{t\in\mathbb{R}}$ be a strongly continuous group on a Banach space X such that, for all $t \in \mathbb{R}$, $\sigma_{e4}(T(t))$ is finite. Since $\{t \in \mathbb{R}$ such that $\sigma_{e4}^1(T(t)) \neq \mathbb{T}\}$ is non-meager, Remark 7.12.2 implies that $(T(t))_{t\in\mathbb{R}}$ is uniformly continuous. ◇

Corollary 7.12.2. *Let X be a Banach space, and let $(T(t))_{t\geq 0}$ be a strongly continuous semigroup on X. Moreover, let us set $\mathcal{C} := \{t > 0$ such that $T(t) - I \in \mathcal{F}^b(X)\}$. If the interior of $\mathcal{C}$ is nonempty, then $(T(t))_{t\geq 0}$ can be embedded in a uniformly continuous group $(T(t))_{t\in\mathbb{R}}$ on X.* ◇

Proof. For $t_0 \in \mathcal{C}$, we can write $T(t_0) = I + (T(t_0) - I)$, which shows that $T(t_0) \in \Phi^b(X)$. By applying Theorem 5.1.5, we may conclude that $(T(t))_{t\geq 0}$ can be embedded into a strongly continuous group $(T(t))_{t\in\mathbb{R}}$ on X. Moreover, for each $t \in \mathcal{C}$, we have $\sigma_{e4}(T(t)) = \{1\}$. Now, the result follows from Remark 7.12.2. Q.E.D.

Another consequence of Corollary 7.12.1 is the following result established by Rabiger and Ricker in [283] for strongly continuous groups on hereditarily indecomposable Banach spaces.

Corollary 7.12.3. *If X is a H.I. Banach space, then any strongly continuous group $(T(t))_{t\in\mathbb{R}}$ on X is uniformly continuous.* ◇

Proof. Obviously, since $(T(t))_{t\in\mathbb{R}}$ is a group, each $T(t)$ is invertible and, then $T(t) = \lambda_t I + S_t$ where $\lambda_t \neq 0$ and S_t is a strictly singular operator. Accordingly, for each $t \in \mathbb{R}$, $\sigma_{e4}^1(T(t)) = \left\{\frac{\lambda_t}{|\lambda_t|}\right\}$. Now, the use of Remark 7.12.3 ends the proof. Q.E.D.

Let $\mathbb{A}$ be a subset of $\mathbb{R}$. If $\mathbb{A}$ is unbounded, then there exists a sequence $(\lambda_n)_n$ of elements of $\mathbb{A}$ such that $\lim\limits_{n\to\infty} |\lambda_n| = +\infty$. Let us define on l_2 the operator S by

$$\mathcal{D}(S) = \left\{(\alpha_n)_n \in l_2 \text{ such that } \sum_{n=0}^{\infty} |\lambda_n|^2|\alpha_n|^2 < \infty\right\}, \quad S((\alpha_n)_n) = (\lambda_n\alpha_n)_n.$$

It is clear that S is a self-adjoint and unbounded operator on l_2. According to Stone's theorem [276], iS generates a strongly continuous group on l_2, given by $\overline{e^{itS}((\alpha_n)_n)} = \overline{(e^{it\lambda_n}\alpha_n)_n}$. Hence, for each $t \in \mathbb{R}$, we have $\sigma(e^{itS}) = \overline{\{e^{it\lambda_n} : n \in \mathbb{N}\}} \subset \overline{\{e^{itx} : x \in \mathbb{A}\}}$ which implies that $\{t \in \mathbb{R} : \sigma^1(e^{itS}) \neq \mathbb{T}\} = \{t \in \mathbb{R} : \sigma(e^{itS}) \neq \mathbb{T}\} \supset \{t \in \mathbb{R} : \overline{\{e^{itx} : x \in \mathbb{A}\}} \neq \mathbb{T}\}$. Let us recall that the operator iS is unbounded. Therefore, the last inclusion combined with Theorem 7.12.1 implies that $\{t \in \mathbb{R}$ such that $\overline{\{e^{itx} : x \in \mathbb{A}\}} \neq \mathbb{T}\}$ is meager. So, we have the following

Corollary 7.12.4. *Let $\mathbb{A}$ be a subset of $\mathbb{R}$. The following assertions are equivalent*

(i) *$\mathbb{A}$ is not bounded,*
(ii) $\left\{t \in \mathbb{R} \text{ such that } \overline{\{e^{itx} \text{ such that } x \in \mathbb{A}\}} \neq \mathbb{T}\right\}$ *is meager.* ◇

7.13 Some Results on Strongly Continuous Semigroups of Operators

A very interesting discussion (including illustrations examples) about the application to strongly continuous semigroups of operators can be found in [227]. In this section, we are dealing with strongly continuous semigroups $(T(t))_{t\geq 0}$ defined on complex infinite-dimensional Banach spaces X and satisfying the condition

$$(\mathcal{E}) \qquad \sigma_{e4}(T(t)) = \{\lambda(t)\} \text{ for all } t > 0.$$

The following results of this section are given in [227].

7.13.1 Arbitrary Banach Spaces

Let X be a complex infinite-dimensional Banach space and let $(T(t))_{t\geq 0}$ be a strongly continuous semigroup on X and satisfying $(\mathcal{E})$. Clearly, if t, $t' \in [0,\infty)$, and since $T(t)$ and $T(t')$ commute, then we have $\sigma_{e4}(T(t)T(t')) \subset \sigma_{e4}(T(t))\sigma_{e4}(T(t'))$. This implies that

$$\lambda(t + t') = \lambda(t)\lambda(t'). \tag{7.13.1}$$

In the following, we denote by $\lambda(.)$ the map from $[0,\infty)$ into $\mathbb{C}$ which assigns to each $t \in [0,\infty)$ the value $\lambda(t) \in \sigma_{e4}(T(t))$.

Lemma 7.13.1. *If X^* is separable, and if hypothesis $(\mathcal{E})$ holds true, then $\lambda(.)$ is a Borel map.* ◇

Proof. If U is an open set of $\mathbb{C}$, then $\lambda^{-1}(U) = \{t \in [0,\infty)$ such that $\{\lambda(t)\} \subset U\}$. Since $t \longrightarrow T(t)$ is continuous from $[0,\infty)$ into $\mathcal{L}_s(X)$ and since $T(t) \longrightarrow \sigma_{e4}(T(t))$ is a Borel map from $\mathcal{L}_s(X)$ into $K(\mathbb{C})$, by composition the map $t \longrightarrow \{\lambda(t)\}$ is Borel from $[0,\infty)$ into $K(\mathbb{C})$. Let us notice that $\lambda^{-1}(U)$ is the inverse image under $t \longrightarrow \{\lambda(t)\}$ of the set $\{K \in K(\mathbb{C})$ such that $K \subset U\}$. However, this set is open in $K(\mathbb{C})$. Therefore, $\lambda^{-1}(U)$ is a Borel subset of $[0, \infty)$. Q.E.D.

Remark 7.13.1. The separability hypothesis on X^* is required in order to guarantee the Borel character of the function $T \longrightarrow \sigma_{e4}(T)$ from $\mathcal{L}_s(X)$ into $K(\mathbb{C})$ (cf. Theorem 7.9.4). ◇

Lemma 7.13.2. *Let us assume that hypothesis* $(\mathcal{E})$ *is satisfied. Then, only one of the following two statements holds:*

(i) There exists $t_0 > 0$ *such that* $\lambda(t_0) = 0$ *(and therefore* $\lambda(t) = 0$ *for all* $t \in]0, \infty[$*).*
(ii) There exists $\alpha \in \mathbb{C}$ *such that* $\lambda(t) = e^{\alpha t}$ *for all* $t \in [0, \infty)$. ◇

Proof.

(i) Let $t_0 > 0$ be such that $\lambda(t_0) = 0$. By using (7.13.1), we notice that $[\lambda(t_0/n)]^n = \lambda(t_0) = 0$ for every $n \in \mathbb{N}\backslash\{0\}$. Hence, $\lambda(t_0/n) = 0$. Now, let $t \in]0, \infty[$. Then, there exists $n \in \mathbb{N}\backslash\{0\}$ such that $t_0/n \leq t$. Hence, $\lambda(t) = \lambda(t_0/n)\lambda(t - t_0/n) = 0$. Therefore, $\lambda(.) \equiv 0$ on $]0, \infty[$.
(ii) It is well known that any nontrivial Borel solution of the functional Eq. (7.13.1) can be written in the form $\lambda(t) = e^{\alpha t}$ for some $\alpha \in \mathbb{C}$. So, by using the preceding assertion, it is sufficient to show the existence of a real $t_0 > 0$ such that $\lambda(t_0) \neq 0$ in order to get $\lambda(t) \neq 0$ for all $t \in]0, \infty[$. Q.E.D.

Remark 7.13.2. Actually, the second item of Lemma 7.13.2 remains valid if we only assume the measurability of $\lambda(.)$ (see [152, pp. 144–145]). Hence, for such $\lambda(.)$, measurability also implies continuity and hence, $\lambda(.)$ admits an obvious extension to $\mathbb{R}$. ◇

Proposition 7.13.1. *Let* $(T(t))_{t\geq 0}$ *be a strongly continuous semigroup on a Banach space* X *with an infinitesimal generator* A*. Let us assume that* X^* *is separable and that hypothesis* $(\mathcal{E})$ *is satisfied. Then, there are two alternatives: either* $(T(t))_{t\geq 0}$ *is a semigroup of Riesz type and* A *is unbounded with nonempty resolvent set and with a resolvent of Riesz type, or else* $(T(t))_{t\geq 0}$ *is embeddable in a uniformly continuous group and there exists* $\beta \in \mathbb{C}$ *such that* $A - \beta$ *is a Riesz operator.* ◇

Proof. Obviously, if $\lambda(t_0) = 0$ for some $t_0 > 0$, then by using Lemma 7.13.2(i), $\lambda(t) = 0$ for all $t > 0$ and therefore, $\sigma_{e4}(T(t)) = \{\lambda(t)\} = \{0\}$ for all $t \in]0, \infty[$. This proves that $(T(t))_{t\geq 0}$ is a semigroup of Riesz type. Moreover, for all $t \in]0, \infty[$, $\sigma_{e4}(T(t)) = \{0\}$ implies that $T(t)$ is not invertible. Thus, according to Theorem 5.1.5, $(T(t))_{t\geq 0}$ cannot be embedded in a C_0-group and so, A is necessarily unbounded. By using the Hille–Yosida theorem (cf. Theorem 2.5.7), the domain of A, namely $\mathcal{D}(A)$, is dense in X and $\rho(A) \neq \emptyset$. Let $\mu \in \rho(A)$. Then, the range of $(\mu - A)^{-1}$ is the domain of A. So, $(\mu - A)^{-1}$ is closed and has a dense range. If we assume that its range is closed, then we should have $\mathcal{D}(A) = X$ and therefore, A would be bounded. Consequently, $0 \in \sigma_{e4}((\mu - A)^{-1})$.

In order to prove that $(\mu - A)^{-1}$ is a Riesz operator, it remains to show that all other elements of its spectrum are isolated eigenvalues with a finite algebraic multiplicity. Let us notice that every $0 \neq \mu \in \sigma(T(t))$ is an isolated eigenvalue with a finite algebraic multiplicity. So, if $\mu' \in \sigma(A)$ and $e^{t\mu'}$ is an eigenvalue of $T(t)$ with a finite algebraic multiplicity, then by using Theorem 2.5.12, μ' is an eigenvalue of A with a finite algebraic multiplicity. Now, let $\mu'' \neq 0$ be any element of $\sigma((\mu - A)^{-1})$. Then, $\mu - 1/\mu'' \in \sigma(A)$ and represents an isolated eigenvalue by applying the preceding considerations. Hence, μ'' is an eigenvalue of $(\mu - A)^{-1}$ with a finite algebraic multiplicity. So, $(\mu - A)^{-1}$ is a Riesz operator. Now, the resolvent identity, the fact that $(\lambda - A)^{-1}$ and $(\nu - A)^{-1}$ commute for all λ and ν in $\rho(A)$, and also Proposition 2.2.3 show that the resolvent of A is of Riesz type. Now let us suppose that there exists $t > 0$ such that $\lambda(t) \neq 0$. By applying Lemma 7.13.2, (ii) we conclude that $\lambda(t) \neq 0$ for all $t > 0$. By using both Lemma 7.13.2 and Remark 7.13.2, we deduce that the function $t \longrightarrow \lambda(t)$ is continuous. This proves that $(T(t))_{t\geq 0}$ satisfies the hypotheses of Propositions 7.11.5 and 7.11.6 for $n = 1$ and $\lambda(.) = \varphi(.)$. Moreover, since $D(\lambda)$ (the domain of $\lambda(.)$) contains $]0, \infty[$, by applying Propositions 7.11.5 and 7.11.6, we infer that $(T(t))_{t\geq 0}$ can be embedded in a strongly continuous group on X and $A \in \mathcal{L}(X)$. Q.E.D.

7.13.2 *Hereditarily Indecomposable Banach Spaces*

The aim of this subsection is to prove the following result which gives a characterization of strongly continuous semigroups on complex H.I. Banach spaces with separable duals.

Proposition 7.13.2. *Let X be a complex H.I. Banach space, and let $(T(t))_{t\geq 0}$ be a strongly continuous semigroup on X with a generator A. If X^* is separable, then there are two alternatives: either $(T(t))_{t\geq 0}$ is a semigroup of strictly singular operators whose generator A is unbounded with a strictly singular resolvent, or else $(T(t))_{t\geq 0}$ is embeddable in a uniformly continuous group (so, its generator A has the form $\beta + S$ with $\beta \neq 0$ and S being strictly singular).* ◇

Proof. Let $R \in \mathcal{L}(X)$ be a Riesz operator on X. From Lemma 2.1.12, we infer that $R = \beta + S$ for some $\beta \in \mathbb{C}$ and $S \in \mathcal{S}(X)$. Then, according to Theorem 7.5.3(i), we have $\sigma_{e4}(R) = \sigma_{e4}(\beta I) = \{\beta\}$. However, $\sigma_{e4}(R) = \{0\}$ (because $\dim X = \infty$), so $\beta = 0$, that is, $R \in \mathcal{S}(X)$. This shows that $\mathcal{R}(X) = \mathcal{S}(X)$. Hence, in order to prove Proposition 7.13.2, it is sufficient to show that the hypotheses of Proposition 7.13.1 are satisfied. Indeed, by applying Lemma 2.1.12, we get $T(t) = \lambda(t)I + S(t)$, where $\lambda(t) \in \mathbb{C}$ and $S(t) \in \mathcal{S}(X)$ for all $t > 0$. By applying again Theorem 7.5.3(i), we infer that $\sigma_{e4}(T(t)) = \sigma_{e4}(\lambda(t)I) = \{\lambda(t)\}$. This together with the separability of X^* shows that the condition $(\mathcal{E})$ holds, which completes the proof. Q.E.D.

Let us note that if X is a separable H.I. space, it follows from the above proof that $\mathcal{S}(X)$ and $\mathcal{R}(X)$ coincide. This fact does not require the separability of X^*. Now, the use of Corollary 7.9.1 allows us to get the following corollary.

Corollary 7.13.1. *If X is a separable H.I. space, then $\mathcal{S}(X)$ is a Borel subset of $\mathcal{L}_s(X)$.* ◇

Chapter 8
Pseudo-Spectra

In this chapter, we study the essential pseudo-spectra of densely closed, linear operators in the Banach space.

8.1 Pseudo-Spectrum of Linear Operator

Let us start by giving the definition of the pseudo-spectrum of densely closed linear operator A for every $\varepsilon > 0$,

$$\sigma_\varepsilon(A) := \sigma(A) \bigcup \Big\{\lambda \in \mathbb{C} \text{ such that } \|(A-\lambda)^{-1}\| > \frac{1}{\varepsilon}\Big\}.$$

The pseudo-spectrum is the open subset of the complex plane bounded by the ε^{-1} level curve of the norm of the resolvent.

Theorem 8.1.1. *Let $A \in \mathcal{C}(X)$. The following three conditions are equivalent.*

(i) $\lambda \in \sigma_\varepsilon(A)$.
(ii) *There exists a bounded operator D such that $\|D\| < \varepsilon$ and $\lambda \in \sigma(A+D)$.*
(iii) *Either $\lambda \in \sigma(A)$ or $\|(\lambda - A)^{-1}\| < \varepsilon^{-1}$.* ◇

Proof. (*i*) $\Rightarrow$ (*ii*) If $\lambda \in \sigma(A)$, we may put $D = 0$. Otherwise, let $f \in \mathcal{D}(A-\lambda)$, $\|f\| = 1$ and $\|(A-\lambda)f\| < \varepsilon$. Let $\phi \in X^*$ satisfy $\|\phi\| = 1$ and $\phi(f) = 1$. Then, let us define the rank one operator $D : X \longrightarrow X$ by $Dg := -\phi(g)(A-\lambda)f$. We see immediately that $\|D\| < \varepsilon$ and $(A - \lambda + D)f = 0$.

(*ii*) $\Rightarrow$ (*iii*) We derive a contradiction from the assumption that $\lambda \notin \sigma(A)$ and $\|(A-\lambda)^{-1}\| \leq \varepsilon^{-1}$. Let $B : X \longrightarrow X$ be the bounded operator defined by the norm convergent series $B := \sum_{n=0}^{\infty}(A-\lambda)^{-1}\left(-D(A-\lambda)^{-1}\right)^n = (A-\lambda)^{-1}(I + D(A-\lambda)^{-1})^{-1}$. It is immediate from these formulae that B is one-one

A. Jeribi, *Spectral Theory and Applications of Linear Operators and Block Operator Matrices*, DOI 10.1007/978-3-319-17566-9_8

with a range equal to $\mathcal{D}(A-\lambda)$. We also see that $B(I+D(A-\lambda)^{-1})f = (A-\lambda)^{-1}f$, for all $f \in X$. Putting $g = (A-\lambda)^{-1}f$, we conclude that $B(A-\lambda+D)g = g$ for all $g \in \mathcal{D}(A-\lambda)$. The proof that $(A-\lambda+D)Bh = h$ for all $h \in X$ is similar. Hence, $A-\lambda+D$ is invertible, with an inverse B.

$(iii) \Rightarrow (i)$ We assume for nontriviality that $\lambda \notin \sigma(A)$. There exists $g \in X$ such that $\|(A-\lambda)^{-1}g\| > \varepsilon^{-1}\|g\|$. Putting $f := (A-\lambda)^{-1}g$, we see that $\|(A-\lambda)f\| < \varepsilon\|f\|$. Q.E.D.

Remark 8.1.1. From Theorem 8.1.1, it follows immediately that

$$\sigma_\varepsilon(A) = \bigcup_{\|D\|<\varepsilon} \sigma(A+D).$$

◇

Let $A \in \mathcal{C}(X)$ and $\varepsilon > 0$. We define the ε-pseudo-spectrum of A by

$$\Sigma_\varepsilon(A) := \sigma(A) \bigcup \left\{\lambda \in \mathbb{C} \text{ such that } \|(A-\lambda)^{-1}\| \geq \frac{1}{\varepsilon}\right\} := \mathbb{C} \backslash \rho_\varepsilon(A).$$

It is well known that the mapping $T \longrightarrow \sigma(T)$ from the set $\mathcal{L}(X)$ of all bounded linear operators on X into the set of all compact subsets of $\mathbb{C}$ equipped with the Hausdorff metric is not continuous. More precisely, let $(T_n)_n$ be a sequence in $\mathcal{L}(X)$ converging to $T \in \mathcal{L}(X)$ with respect to the operator norm. Then, $\lim_n \operatorname{dist}(\sigma(T_n), \sigma(T)) = 0$ but $\lim_n \operatorname{dist}(\sigma(T), \sigma(T_n)) = 0$ does not hold in general. However the following assertion is true.

Theorem 8.1.2. *Let $(T_n)_n$ be a sequence of bounded linear operators on the Banach space X which converges with respect to the operator norm to the operator T. Then, to every pair $(\varepsilon_1, \varepsilon_2)$ of real numbers with $0 \leq \varepsilon_1 < \varepsilon_2$, there exists $n(\varepsilon_1, \varepsilon_2) \in \mathbb{N}$ such that $\Sigma_{\varepsilon_1}(T) \subset \Sigma_{\varepsilon_2}(T_n)$ holds for all $n \geq n(\varepsilon_1, \varepsilon_2)$.* ◇

Proof. Let $n(\varepsilon_1, \varepsilon_2)$ be such that $\|T - T_n\| < \varepsilon_2 - \varepsilon_1$ holds for all $n \geq n(\varepsilon_1, \varepsilon_2)$. Assume that $\lambda \in \rho_{\varepsilon_2}(T_n)$, where $n \geq n(\varepsilon_1, \varepsilon_2)$ is fixed. Then $\|(T-T_n)(\lambda-T_n)^{-1}\| \leq \|T-T_n\|\|(\lambda-T_n)^{-1}\| < \frac{\varepsilon_2-\varepsilon_1}{\varepsilon_2} = 1-\frac{\varepsilon_1}{\varepsilon_2}$. Therefore the series $\sum_{k=0}^{\infty}[(\lambda-T_n)^{-1}(T-T_n)]^k$ converges to $(I-(\lambda-T_n)^{-1}(T-T_n))^{-1}$. This in turn implies that $\lambda - T = (\lambda-T_n)(I-(\lambda-T_n)^{-1}(T-T_n))$ is invertible and moreover,

$$\begin{aligned}
\|(\lambda-T)^{-1}\| &\leq \|(\lambda-T_n)^{-1}\|\|(I-(\lambda-T_n)^{-1}(T-T_n))^{-1}\| \\
&< \|(\lambda-T_n)^{-1}\| \frac{1}{1-(1-\frac{\varepsilon_1}{\varepsilon_2})} \\
&\leq \frac{1}{\varepsilon_1}
\end{aligned}$$

hence $\lambda \in \rho_{\varepsilon_1}(T)$. So, $\rho_{\varepsilon_2}(T_n) \subset \rho_{\varepsilon_1}(T)$ for all $n \geq n(\varepsilon_1, \varepsilon_2)$ which proves the assertion. Q.E.D.

8.1.1 Approximation of ε-Pseudo-Spectrum

In order to obtain the strongest possible results we have to refine the notion of the ε-pseudo-spectrum of an operator $(A, \mathcal{D}(A))$ defined on the Banach space X. For it turns out that the part $\sigma(A)\backslash\sigma_{ap}(A)$ cannot always be approximated in the general case. For $\varepsilon \geq 0$, we define the ε-approximate spectrum $\sigma_{\varepsilon,ap}(.)$ by $\sigma_{\varepsilon,ap}(A) = \{\lambda \in \mathbb{C}$ such that $\tilde{\alpha}(\lambda - A) \leq \varepsilon\}$, where $\tilde{\alpha}(.)$ is given in Remark 7.1.1 (*iv*). In particular $\sigma_{0,ap}(A) = \sigma_{ap}(A)$ as well as $\sigma_{\varepsilon,ap}(A) \subset \Sigma_\varepsilon(A)$. We denote the norm on F_n by $\|.\|$ as usual. The following results, given in this section, come from [349].

Theorem 8.1.3. *Let $A \in \mathcal{C}(X)$. Then, $\sigma_{\varepsilon,ap}(A)$ is always closed.* ◇

Proof. Let $\lambda \notin \sigma_{\varepsilon,ap}(A)$. Then, $\tilde{\alpha}(\lambda - A) =: \delta > \varepsilon$. Now, let $\mu \in \mathbb{C}$ satisfy $|\lambda - \mu| < \delta - \varepsilon$. Then

$$\begin{aligned}\tilde{\alpha}(\mu - A) &= \inf\{\|(\mu - A)x\| \text{ such that } \|x\| = 1 \text{ and } x \in \mathcal{D}(A)\}\\ &\geq \inf\{|\|(\lambda - A)x\| - |\lambda - \mu|\|x\|| \text{ such that } \|x\| = 1 \text{ and } x \in \mathcal{D}(A)\}\\ &> \varepsilon.\end{aligned}$$

So, the complement of $\sigma_{\varepsilon,ap}(A)$ is open. Q.E.D.

Let X_1 be a dense linear subspace of the Banach space X, let (F_n) be a sequence of Banach spaces and for each n let $P_n : X_1 \longrightarrow F_n$ be a not necessarily bounded linear mapping. If $\lim_n \|P_n x\| = \|x\|$ holds for every x in X_1, then $(X, X_1, (F_n), (P_n))$ is called a discrete approximation scheme. A sequence $(x_n)_n \in \prod_{n\in\mathbb{N}} F_n$ converges discretely to $x \in X_1$ if $\lim_{n\to\infty} \|x_n - P_n x\| = 0$ holds. Then, we write $x = d - \lim x_n$. Let now $(A, \mathcal{D}(A))$ be a closed densely defined linear operator on X and let $X_0 \subset X_1$ be a core of A such that $A(X_0) \subset X_1$ holds. For each n, let $(A_n, \mathcal{D}(A_n))$ be a densely defined operator on F_n such that $P_n(X_1) \subset \mathcal{D}(A_n)$. We say that the sequence (A_n) approximates A discretely if for all $x \in X_0$ the sequence $(A_n P_n x)_n$ converges discretely to Ax, i.e., $\lim_n \|A_n P_n x - P_n Ax\| = 0$ holds for all $x \in X_0$.

Theorem 8.1.4. *Let $(X, X_1, (F_n), (P_n))$ be a fixed discrete approximation scheme. Moreover, let the sequence (A_n) of densely defined linear operators $(A_n, \mathcal{D}(A_n))$ on the Banach space F_n approximate discretely the closed densely defined linear operator $(A, \mathcal{D}(A))$ on X. Then, for every pair $(\varepsilon_1, \varepsilon_2)$ of real numbers with $0 \leq \varepsilon_1 < \varepsilon_2$, $\sigma_{\varepsilon_1,ap}(A) \subset \liminf_n \sigma_{\varepsilon_2,ap}(A_n) = \bigcup_{n=1}^{\infty} \bigcap_{k\geq n} \sigma_{\varepsilon_2,ap}(A_k)$ holds.* ◇

Proof. For the sake of convenience we set $\varepsilon_1 = \eta$, $\varepsilon_2 = \varepsilon$. Let $\lambda \in \sigma_{\eta,ap}(A)$ be arbitrary. Choose $\beta > 0$ such that $\varepsilon - \eta - 2\beta > 0$ and set $\gamma = 1 - \frac{\eta+2\beta}{\varepsilon}$. By hypothesis, there exists a core X_0 of A with $A(X_0) \subset X_1$. Then, there exists $x \in X_0$ of norm 1 such that $\|(\lambda - A)x\| < \eta + \beta/2$. Again by hypothesis, there exists n_0 such that for all $n \geq n_0$, $|\|P_n x\| - \|x\|| < \gamma$, $\|P_n Ax - A_n P_n x\| < \beta$, and

$\| P_n(\lambda - A)x\| < \eta + \beta$. Then for all $n \geq n_0$ we obtain

$$\begin{aligned}\|\lambda P_n x - A_n P_n x\| &\leq \|\lambda P_n x - P_n A x\| + \|P_n A x - A_n P_n x\| \\ &= \|P_n(\lambda - A)x\| + \|P_n A x - A_n P_n x\| \\ &< \eta + 2\beta.\end{aligned}$$

Now $\|x\| = 1$ implies $1 - \gamma < \|P_n x\| < 1 + \gamma$. Dividing the inequalities above by $\|P_n x\|$ we get $\|(\lambda - A_n)(P_n x/\|P_n x\|)\| \leq \varepsilon$ which implies $\lambda \in \sigma_{\varepsilon,ap}(A_n)$ for all $n \geq n_0$. Q.E.D.

Theorem 8.1.5. *Assume in addition to the hypotheses of Theorem 8.1.4 that, there exists $M > 0$ such that $\|(\lambda - A_n)^{-1}\| \leq M\ dist(\lambda, \sigma(A_n))^{-1}$ holds for all $\lambda \in \rho(A_n)$ and for all n. Then, for every compact subset $K \subset \mathbb{C}$, $\lim_{n\to\infty} dist(\sigma_{ap}(A) \bigcap K, \sigma(A_n)) = 0$.* ◇

Proof. Assume that the assertion fails. Then, there exists a $\delta > 0$, a compact set $K \subset \mathbb{C}$ and a sequence $(\lambda_{n_k})_k$ in $\sigma_{ap}(A) \bigcap K$ with $\text{dist}(\lambda_{n_k}, \sigma(A_{n_k})) \geq \delta > 0$ for all k. Since K is compact there exists a subsequence $(\lambda_{n'_k})_k$ converging to a point $z \in \sigma_{ap}(A) \bigcap K$ since $\sigma_{ap}(A)$ is closed. By applying Theorem 8.1.4 with $\varepsilon_1 = 0$ and $\varepsilon_2 = \varepsilon = \frac{\delta}{2(1+2M)}$ we obtain n_0 such that $z \in \Sigma_\varepsilon(A_n)$ for all $n \geq n_0$. Moreover, there exists k_0 with $|z - \lambda_{n'_k}| < \varepsilon$ for all $k \geq k_0$. This implies $z \notin \sigma(A_{n'_k})$. But then, $M\ \text{dist}(z, \sigma(A_{n'_k}))^{-1} \geq \|(z - A_{n'_k})^{-1}\| \geq \frac{1}{\varepsilon}$ yields $M\varepsilon \geq \text{dist}(z, \sigma(A_{n'_k}))$. Hence, there exists $\mu_{n'_k} \in \sigma(A_{n'_k})$ with $|z - \mu_{n'_k}| < 2\varepsilon M$ for all $k \geq k_1 \geq k_0$. This in turn implies $|\lambda_{n'_k} - \mu_{n'_k}| \leq |\lambda_{n'_k} - z| + |z - \mu_{n'_k}| < (1 + 2M)\varepsilon = \frac{\delta}{2}$ for $k \geq k_1$, a contradiction to $\text{dist}(\lambda_{n'_k}, \sigma(A_{n'_k})) \geq \delta$. Q.E.D.

We end this section by the following examples:

(*i*) As for $\varepsilon = 0$ also for $\varepsilon > 0$ it may happen that $\sigma_{\varepsilon,ap}(A) \neq \Sigma_\varepsilon(A)$ holds as the following example shows. In fact, Let $E = l^2(\mathbb{N})$ and let S be the right shift on E given by

$$(Sf)(k) := \begin{cases} 0, & k = 1 \\ f(k-1) & k \geq 2. \end{cases}$$

Since S is an isometry on E, it is easily checked that $\tilde{\alpha}(\lambda - S) \geq 1 - |\lambda|$ holds for all λ with $0 \leq |\lambda| \leq 1$. Moreover, $\sigma_{ap}(S) = \{\lambda \in \mathbb{C}$ such that $|\lambda| = 1\}$ and for $0 < \varepsilon < 1$ we obtain $\Sigma_\varepsilon(S) \backslash \sigma_{\varepsilon,ap}(S) \supset \{\lambda \in \mathbb{C}$ such that $|\lambda| < 1 - \varepsilon\}$, the latter set being a subset of the residual spectrum $\sigma_r(S) = \{\lambda \in \sigma(S)$ such that $\tilde{\alpha}(\lambda - S) > 0\}$ of S. Set $F_n = \mathbb{C}^n$ and $P_n : E \longrightarrow F_n$, $f \longrightarrow (f(1), \ldots, f(n))$. Let $A_n(x) = (x_n, x_1, \ldots, x_{n-1})$. Then, A_n is unitary hence normal, and thus we can apply Theorem 8.1.5 with $M = 1$ to obtain $\lim \text{dist}(\sigma_{ap}(S), \sigma(A_n)) = 0$. However, the residual spectrum $\sigma(S) \backslash \sigma_{ap}(S)$ cannot be approximated by $\sigma(A_n)$.

(*ii*) **Uniform convergence.** Let X be a Banach space, set $E = E_1 = \mathcal{L}(X)$, the Banach algebra of all bounded operators on X, set $F_n = E$ and $P_n = I$. For $T \in \mathcal{L}(X)$, we consider the multiplication operator $A = M_T : U \longrightarrow TU$. Then, the sequence (T_n) converges to T with respect to the operator norm if, and only if, $(A_n) = (M_{T_n})$ approximates A discretely.

(*iii*) **Pointwise (strong) convergence.** Here $E = E_1 = F_n$ and $P_n = I$ for all n. A sequence $(A_n)_n$ of bounded operators A_n converges strongly to A if, and only if, $(A_n)_n$ approximates A discretely.

(*iv*) Let E be a given Banach space and let $(F_n)_n$ be an increasing sequence of closed linear subspaces with $F_\infty := \bigcup_n F_n$ dense in E. Moreover, assume that each F_n is the range of a bounded projection P_n such that $\sup_n(\|P_n\|) < \infty$ as well as $P_{n+k}P_n = P_n$ for $k \geq 0$. Let $(A, \mathcal{D}(A))$ be a closed densely defined operator on E such that for all n $A_{|F_n} =: A_n$ maps $\mathcal{D}(A) \bigcap F_n =: \mathcal{D}(A_n)$ into F_n and moreover that $(A_n, \mathcal{D}(A_n))$ is densely defined and closable on F_n and finally that $\mathcal{D}(A) \bigcap F_\infty$ is a core of A. Then setting $E_1 = F_\infty$, $E_0 = \mathcal{D}(A) \bigcap F_\infty$ we obtain that the sequence $(A_n)_n$ approximates A discretely.

(*v*) Let $E = L_2([0, 1])$, $E_1 = \{f \in E$ such that f continuous, $f(0) = f(1)\}$, $F_n = \mathbb{C}^n$ with the scalar product $\langle x, y\rangle = \frac{1}{n}\sum_{k=1}^{n} \overline{x}_k y_k$, $P_n f = (f(\frac{1}{n}), \ldots, f(\frac{n}{n}))$, $E_0 = \{f \in E_1$ such that $f' \in E_1\}$, and finally let $Af = f'$ with boundary condition $f(0) = f(1)$. For $A_n x = n(x_2 - x_1, \ldots, x_n - x_{n-1}, x_1 - x_n)$ the sequence $(A_n)_n$ approximates A discretely. Then, $\sigma_{ap}(A) = \sigma(A) = 2\pi i\mathbb{Z}$ and the approximating operators A_n are normal with $\sigma(A_n) = \{n(\exp(2\pi ik/n) - 1)$ such that $0 \leq k \leq n-1\}$. We obtain $2\pi i\mathbb{Z} \subset \liminf_n \Sigma_\varepsilon(A_n)$. Moreover, we have $\lim \mathrm{dist}(\sigma(A) \bigcap K, \sigma(A_n)) = 0$ as follows also directly from $2\pi ik = \lim\limits_{n\to\infty} n(\exp(2\pi ik/n) - 1)$ for each fixed k.

(*vi*) Same as (v) up to the A_n. Here we take $A_n x = n(x_2 - x_1, \ldots, x_n - x_{n-1}, 0)$. Then $A_n = n(N_n - Q_n)$, where Q_n is the projection onto the first $(n-1)$ coordinates and $N_n^n = 0$. Hence, $\sigma(A_n) = \{-n, 0\}$. Theorem 8.1.5 does not apply since there is no M satisfying the hypothesis of this theorem. However, $2\pi i\mathbb{Z} \subset \liminf_n \Sigma_\varepsilon(A_n)$ as follows from Theorem 8.1.4. It can also easily be deduced directly by verifying $\|(2\pi ik - A_n)e_{k,n}\| = O(n^{-1/2})$ where $e_{k,n} = (\exp(\frac{2\pi ikl}{n}))_{l=1,\ldots,n}$ for fixed $k \in \mathbb{Z}$.

(*vii*) The spaces are the same as in (v). $Af = f' - f$ with boundary condition $f(0) = f(1)$, $A_n x = n(x_2 - x_1, \ldots, x_n - x_{n-1}, x_1 - x_n) - x$.

(*viii*) Let $X = l^2(\mathbb{N})$ and let T be the left shift given by $(Tf)(k) = f(k+1)$. Let

$$(A_n f)(k) := \begin{cases} f(k+1), & k \leq n-1 \\ 0 & \text{else.} \end{cases}$$

Then, $(A_n)_n$ converges pointwise to T, but $\sigma(A_n) = \{0\}$. Nevertheless $\{\lambda \in \mathbb{C}$ such that $|\lambda| \leq 1\} = \sigma_{ap}(T) = \sigma(T) \subset \liminf_n \sigma_{\varepsilon,ap}(A_n)$ for all $\varepsilon > 0$.

8.1.2 Approximation of the Spectrum

In this section we turn to the problem of when $\bigcap_{\varepsilon>0}\bigcap_{n\geq 1}\bigcup_{k\geq n}\sigma_{\varepsilon,ap}(A_k)\subset\sigma_{ap}(A)$ holds where the sequence $(A_n)_n$ approximates A. Let $(\dot{r}_n)_n$ be an arbitrary sequence of positive real numbers converging to 0. Then

$$\bigcap_{\varepsilon>0}\bigcap_{n\geq 1}\bigcup_{k\geq n}\sigma_{\varepsilon,ap}(A_k)=\bigcap_{n\geq 1}\bigcup_{k\geq n}\sigma_{r_n,ap}(A_k) \tag{8.1.1}$$

is easily seen to hold. So, in the following proofs we prefer the right-hand side description of this set because it contains only two set theoretical operations. Let $\mathcal{E}=(G_n)_n$ be a sequence of Banach spaces. Then, $l_\infty(\mathcal{E})$ is the subspace of all norm bounded sequences of the cartesian product $\prod_n G_n$. Equipped with the norm $\|(x_n)\|:=\sup\{\|x_n\|$ such that $n\in\mathbb{N}\}$, $l_\infty(\mathcal{E})$ is a Banach space. Now, let $\mathcal{U}$ be an arbitrary free ultrafilter on $\mathbb{N}$. Then, $c_{0,\mathcal{U}}(\mathcal{E})=\{(x_n)_n\in l_\infty(\mathcal{E})$ such that $\lim_{\mathcal{U}}\|x_n\|=0\}$ is a closed subspace of $l_\infty(\mathcal{E})$. The quotient space is called the ultraproduct $\mathcal{E}_{\mathcal{U}}$ of $\mathcal{E}$ with respect to $\mathcal{U}$. If $\xi=(x_n)_n$ is in $l_\infty(\mathcal{E})$ then $\|\hat{\xi}\|=\|\xi+c_{0,\mathcal{U}}(\mathcal{E})\|=\lim_{\mathcal{U}}\|x_n\|$ holds. Let $U\in\mathcal{U}$ be arbitrary. Then $\hat{\xi}$ is determined already by the subsequence $(x_{n'})_{n'\in U}$, a fact which we will use tacitely in the sequel. Now, let $(T_n)_n$ be a uniformly bounded sequence of operators $T_n\in\mathcal{L}(G_n)$. Then by $\tilde{T}\xi=(T_nx_n)_n$ there is defined a bounded linear operator $\tilde{T}$ on $l_\infty(\mathcal{E})$ for which $c_{0,\mathcal{U}}(\mathcal{E})$ is an invariant subspace. The operator $\hat{T}$ on the quotient space $\mathcal{E}_{\mathcal{U}}$ is called the ultraproduct of (T_n) with respect to $\mathcal{U}$. Let $U\in\mathcal{U}$ be arbitrary. Similarly, to the fact in the previous paragraph, $\hat{T}$ does not depend on indices not in U. In particular, $\hat{T}$ does not depend on the first n_0 operators. More generally, if the sequence is only defined for $n\in U$, then we may fill up it with arbitrary bounded operators $(T_n)_{n\notin\mathcal{U}}$ obtaining always the same operator $\hat{T}$. Let $(E,E_1,(F_n),(P_n))$ be a given approximation scheme. Let $\mathcal{F}_{\mathcal{U}}$ be the ultraproduct of $(F_n)_n$ with respect to a given free ultrafilter $\mathcal{U}$ on $\mathbb{N}$. Since $\lim\|P_ny\|=\|y\|$ holds by hypothesis for all $y\in E_1$ we obtain an isometry $V_{\mathcal{U}}:E_1\longrightarrow\mathcal{F}_{\mathcal{U}}$ by $V_{\mathcal{U}}(y)=\widehat{(P_ny)_n}$. Since E_1 is dense in E, $V_{\mathcal{U}}$ can be uniquely extended to an equally denoted isometry on E. Let the closed densely defined operator $(A,\mathcal{D}(A))$ on E be discretely approximated by the sequence $(A_n)_n$ of bounded linear operators A_n on F_n. If this sequence is uniformly bounded then we obtain easily $\hat{A}V_{\mathcal{U}|E_0}=V_{\mathcal{U}}A_{|E_0}$. The results, given in this section, come from [349].

Theorem 8.1.6. *Let $(E,E_1,(F_n),(P_n))$ be a given approximation scheme. Let $(A,\mathcal{D}(A))$ be discretely approximated by the sequence $((A_n,\mathcal{D}(A_n)))_n$. Moreover, assume that A as well as all A_n are surjective, and that $\liminf_{n\to\infty}\tilde{\alpha}(A_n)>0$. Then, there exists $n_0\in\mathbb{N}$ such that for all $n\geq n_0$, A_n is bijective. Moreover, A_n^{-1} is bounded, the sequence $(A_n^{-1})_{n\geq n_0}$ is uniformly bounded, and A^{-1} exists and is discretely approximated by $(A_n^{-1})_{n\geq n_0}$. Finally, $V_{\mathcal{U}}A^{-1}=\widehat{(A_n^{-1})}V_{\mathcal{U}}$ holds for every free ultrafilter $\mathcal{U}$ on $\mathbb{N}$.* ◇

Proof. There exists $\eta > 0$ and $n_0 \in \mathbb{N}$ such that $\tilde{\alpha}(A_n) \geq \eta$ for all $n \geq n_0$. Since all A_n are surjective, A_n is bijective for $n \geq n_0$ and $\|A_n^{-1}\| \leq \frac{1}{\eta}$ holds for all these n. Let $\mathcal{U}$ be a free ultrafilter on $\mathbb{N}$. Then, the operator $B = \widehat{(A_n^{-1})_{n\geq n_0}}$ is well defined on $\mathcal{F}_{\mathcal{U}}$ and $\|B\| \leq \frac{1}{\eta}$. Assume now that, there exists $x_0 \in E_0$, $\|x_0\| = 1$ with $\|Ax_0\| < \delta = \min(1,\eta)/2$. By hypothesis the following assertions hold: $\|V_{\mathcal{U}}(x_0)\| = 1$, $\|V_{\mathcal{U}}(Ax_0)\| = \|Ax_0\| < \delta$, $V_{\mathcal{U}}(Ax_0) = \widehat{(A_nP_nx_0)_n}$. But then $\|\widehat{(A_nP_nx_0)_n}\| = \|V_{\mathcal{U}}(Ax_0)\| < \delta$. This in turn implies $\|\widehat{((A_nP_nx_0)/\|P_nx_0\|)_n}\| < 2\delta$, a contradiction to $\tilde{\alpha}(A_n) \geq \eta \geq 2\delta$ for all $n \geq n_0$. So, $\tilde{\alpha}(A) \geq \delta$, since E_0 is a core of A. Since A is surjective, A^{-1} exists and is bounded with norm $\leq 1/\delta$. Moreover, $A(E_0)$ is dense in E. For if $y \in E$ is arbitrary, then $(A^{-1}y, y)$ is contained in the graph $G(A)$ of A. But since E_0 is a core $G(A) = \overline{G(A_{|E_0})}$ and the assertion follows. Now, let $z \in A(E_0)$ be arbitrary and set $y = A^{-1}z$. Then $V_{\mathcal{U}}z = V_{\mathcal{U}}Ay = \widehat{(A_nP_ny)_n}$. This implies $BV_{\mathcal{U}}z = \widehat{(A_n^{-1}(A_nP_ny))_n} = V_{\mathcal{U}}A^{-1}z$ for all $z \in A(E_0)$. Since this latter space is dense in E the final equality follows from the continuity of B, $V_{\mathcal{U}}$ and A^{-1}. Finally, in order to prove that A^{-1} is discretely approximated by the sequence $(A_n^{-1})_n$ we have to specify a core $E_2 \subset E_1$ of A^{-1} with $A^{-1}(E_2) \subset E_1$. But since $A(E_0)$ is dense it may serve as such a core. Q.E.D.

Assume that for every $m < n$, there exists a linear isometric embedding $S_{n,m}$ from F_m into F_n. Moreover, let the sequence $((A_n, \mathcal{D}(A_n)))_n$ approximate the closed densely defined operator $(A, \mathcal{D}(A))$. To every n, let G_n be a core of A_n and as before let E_0 be a core of A with $A(E_0) \subset E_1$.

Theorem 8.1.7. *In addition to the assumptions made in the previous paragraph, let the following condition be satisfied: For all $k \in \mathbb{N}$ and for all $z \in G_k$, there exists $y \in E_0$ and an unbounded sequence $(t_n)_n$ such that $P_{t_n}y = S_{t_n,k}z$ for all t_n as well as $\lim\limits_{n\to\infty} \|A_{t_n}S_{t_n,k}z - S_{t_n,k}A_kz\| = 0$. Then, $\bigcap_{\varepsilon>0}\bigcap_{n\geq 1}\bigcup_{k\geq n}\sigma_{\varepsilon,ap}(A_k) \subset \sigma_{ap}(A)$.* ◇

Proof. We use Eq. (8.1.1) with $r_n = 1/n$. Assume that the assertion fails. Then, there exists $\lambda \in \bigcap_{n\geq 1}\bigcup_{k\geq n}\sigma_{r_n,ap}(A_k)$, with $\tilde{\alpha}(\lambda - A) = \delta > 0$. By hypothesis, there exists a sequence $(k_n)_n$ with $k_n \geq n$ and $\tilde{\alpha}(\lambda - A_{k_n}) \leq r_n$. To each n, there exists $x_{k_n} \in G_{k_n}$ of norm 1 such that $\|(\lambda - A_{k_n})x_{k_n}\| \leq 2r_n$. Choose n_0 such that $r_n < \delta/3$ for all $n \geq n_0$. Fix $n_1 \geq n_0$ and choose an ultrafilter $\mathcal{U}$ with $\{t_n$ such that $n \in \mathbb{N}\} \in \mathcal{U}$ where $(t_n)_n$ is the sequence for the element $z = x_{k_{n_1}}$ as required in the hypothesis. By assumption, there exists an element y in E_0 with $P_{t_n}y = S_{t_n,k_{n_1}}(z)$ for all $n \in \mathbb{N}$. Again, by hypothesis $\lim\limits_{n\to\infty} \|S_{t_n,k_{n_1}}(A_{k_{n_1}}(z)) - A_{t_n}(S_{t_n,k_{n_1}}(z))\| = 0$ holds. Since $S_{t_n,k_{n_1}}(z) = P_{t_n}y$, we obtain $\widehat{(S_{n,k_{n_1}}A_{k_{n_1}}z)} = \widehat{(A_nS_{n,k_{n_1}}z)} = \widehat{(A_nP_ny)} = V_{\mathcal{U}}Ay$. Notice that we have tacitly made use of the fact that elements in the ultraproduct do not depend on values x_n, A_n, etc. for indices n not contained in $\{t_m$ such that $m \in \mathbb{N}\}$. Because $(E, E_1, (F_n), (P_n))$ is an approximation scheme, we have $\|V_{\mathcal{U}}y\| = \|y\| = \|S_{n,k_{n_1}}z\| = \|z\| = 1$. Since $S_{n,k_{n_1}}$ are isometries, we obtain $2r_{n_1} \geq \|(\lambda - A_{k_{n_1}})z\| = \|(\lambda S_{t_n,k_{n_1}}z - S_{t_n,k_{n_1}}A_{k_{n_1}})z\| \geq |\|\lambda P_{t_n}y - A_{t_n}S_{t_n,k_{n_1}}z\| - \|A_{t_n}S_{t_n,k_{n_1}}z - S_{t_n,k_{n_1}}A_{k_{n_1}}z\||$ for all n. This inequality yields $2r_{n_1} \geq \|\lambda V_{\mathcal{U}} y - V_{\mathcal{U}} Ay\| = \|V_{\mathcal{U}}(\lambda - A)y\| \geq \tilde{\alpha}(\lambda - A) = \delta > 3r_{n_1}$, a contradiction. Q.E.D.

A sequence $(x_n)_n \in \prod F_n$ is called discretely compact (or d-compact for short) if, for every $\varepsilon > 0$, there exists a finite set $Y(\varepsilon) \subset E_1$ depending on ε such that $\limsup_n \operatorname{dist}(x_n, P_n(Y(\varepsilon))) < \varepsilon$. Discretely compact sequences can be described as follows:

Lemma 8.1.1. *Let $(x_n)_n$ be a discretely compact sequence. Then, to every free ultrafilter $\mathcal{U}$, there exists $y \in E$ such that $V_{\mathcal{U}} y = \widehat{(x_n)_n}$.* ◇

Proof. Let $\mathcal{U}$ be a fixed free ultrafilter. By hypothesis to every $r \in \mathbb{N}$, there exists a finite set $Y(r)$ in E_1 depending on r such that $\limsup_n \operatorname{dist}(x_n, P_n(Y(r))) < 2^{-r}$. Since $\mathcal{U}$ is an ultrafilter and $Y(r)$ is finite, there exists some $y_r \in Y(r)$ such that $\{n \in \mathbb{N} \text{ such that } \|x_n - P_n y_r\| < 2^{-r}\} \in \mathcal{U}$. This in turn implies $\|(x_n)_n - V_{\mathcal{U}} y_r\| < 2^{-r}$. Let $p \in \mathbb{N}$ be arbitrary. Since $V_{\mathcal{U}}$ is an isometry, we obtain that $\|y_{r+p} - y_r\| \leq \|V_{\mathcal{U}}(y_{r+p}) - \widehat{(x_n)}\| + \|\widehat{(x_n)} - V_{\mathcal{U}}(y_r)\| < 2^{-r+1}$, hence (y_r) is a Cauchy sequence. If $y = \lim y_r$, then obviously $V_{\mathcal{U}} y = \widehat{(x_n)_n}$. Q.E.D.

Let $(A, \mathcal{D}(A))$ be discretely approximated by the sequence $((A_n, \mathcal{D}(A_n)))$. We say that the approximation is discretely compact, if $(A_n)_n$ is uniformly bounded and, for every bounded sequence $(x_n)_n$, the sequence $(A_n x_n)_n$ is d-compact. $(A_n)_n$ is called inverse d-compact, if the sequence $(x_n)_n$ is d-compact whenever $(A_n x_n)_n$ is bounded.

Lemma 8.1.2. *Let $(A_n)_n$ be inverse d-compact. Then $\liminf_{n\to\infty} \tilde{\alpha}(A_n) > 0$. If, moreover, A and all A_n are surjective, then there exists $n_0 \in \mathbb{N}$ such that all A_n are bijective and the approximation of A^{-1} by $(A_n^{-1})_{n \geq n_0}$ is discretely compact.* ◇

Proof. If the assertion fails then to every $k \in \mathbb{N}$, there exists $n_k \geq k$ and x_{n_k} with $\|x_{n_k}\| = 1$ and $\|A_{n_k} x_{n_k}\| < 2^{-k}$. Set

$$y_n = \begin{cases} 0 & n \notin \{n_k \text{ such that } k \in \mathbb{N}\} \\ 2^k x_{n_k} & n = n_k \text{ for some } k \in \mathbb{N}. \end{cases}$$

Then, $(y_n)_n$ is unbounded hence not d-compact though $(A_n y_n)$ is bounded. Now, let all A_n be surjective. By Theorem 8.1.6, there exists n_0 such that for all $n \geq n_0$, A_n is bijective. Moreover, A_n^{-1} is bounded and the sequence $(A_n^{-1})_{n \geq n_0}$ is uniformly bounded and approximates A^{-1} discretely. Finally, let $(x_n)_n$ be bounded and set $y_n = A_n^{-1} x_n$. Then, $(A_n(y_n))_{n \geq n_0}$ is bounded; hence, $(y_n)_n$ is discretely compact by hypothesis. So, the assertion follows. Q.E.D.

Let now $T \in \mathcal{L}(E)$ with $T(E) \subset E_1$ and let $(T_n)_n$ be a uniformly bounded sequence of operators $T_n \in \mathcal{L}(F_n)$. Since E_1 may serve as a core of T the discrete approximation of T by $(T_n)_n$ can be defined unambiguously.

Proposition 8.1.1. *Let T, $(T_n)_n$ be as above. Assume that $(T_n)_n$ approximates T discretely and moreover that (T_n) is discretely compact. Then* $\bigcap_{\varepsilon>0}\bigcap_{n\geq 1}\bigcup_{k\geq n}\sigma_{\varepsilon,ap}(T_k)\subset\sigma_p(T)\bigcup\{0\}$. ◇

Proof. Again we use Eq. (8.1.1) with $r_n = 1/n$. Let $0\neq\lambda\in\bigcap_{n\geq 1}\bigcup_{k\geq n}\sigma_{r_n,ap}(T_k)$ be arbitrary. Then there exists a sequence $(k_n)_n$ with $k_n\geq n$ and $\lambda\in\sigma_{r_n,ap}(T_{k_n})$. This in turn implies the existence of a normalized vector $x_{k_n}\in F_{k_n}$ with $\|(\lambda - T_{k_n})x_{k_n}\|\leq 2r_n$. Set $x_l = 0$ for $l\notin\{k_n$ such that $n\in\mathbb{N}\}$. Let $\mathcal{U}$ be an ultrafilter containing $\{k_n$ such that $n\in\mathbb{N}\}$. Then for $\xi = (x_n)_n$ we obtain $\|\hat{\xi}\| = 1$ as well as $0\neq\lambda\hat{\xi} = \hat{T}\hat{\xi}$. Since by hypothesis the sequence $(T_nx_n)_n$ is discretely compact by Lemma 8.1.1 there exists $y\in E$ with $\hat{T}\hat{\xi} = (T_nx_n)_n = V_{\mathcal{U}}y$ which in turn gives $\lambda\hat{\xi} = V_{\mathcal{U}}y$. All together we obtain $\lambda V_{\mathcal{U}}y = \lambda\hat{T}\hat{\xi} = \hat{T}(\lambda\hat{\xi}) = \hat{T}V_{\mathcal{U}}y = V_{\mathcal{U}}Ty$, where the last equation holds since $(T_n)_n$ approximates T. Because $V_{\mathcal{U}}$ is an isometry this latter equation yields $\lambda y = Ty$. Q.E.D.

In most cases the spaces F_n are finite-dimensional so that $\sigma_{ap}(T_n) = \sigma_p(T_n)$ holds in the following proposition.

Proposition 8.1.2. *Under the assumptions of Proposition 8.1.1*

$$\lim_{n\to\infty}\text{dist}(\sigma_{ap}(T_n),\sigma_p(T)\bigcup\{0\}) = 0$$

holds. ◇

Proof. Assume that the assertion does not hold. Then, there exists $\delta > 0$ and a sequence $(k_n)_n$ with $k_n\geq n$ and moreover to every n a $\lambda_{k_n}\in\sigma_{ap}(T_{k_n})$ such that $\inf\{|\lambda_{k_n}-\nu|$ such that $\nu\in\sigma_p(T)\bigcup\{0\}\}\geq\delta$. Since $(T_n)_n$ is uniformly bounded (λ_{k_n}) is bounded. Set $\lambda_l = 0$ for $l\notin\{k_n$ such that $n\in\mathbb{N}\}$ and let $\mathcal{U}$ be a free ultrafilter containing $\{k_n$ such that $n\in\mathbb{N}\}$. Since $(\lambda_n)_n$ is bounded, it converges along $\mathcal{U}$. Set $\mu = \lim_{\mathcal{U}}\lambda_n$. Then, $|\mu|\geq\delta > 0$. For every n, we choose a normalized vector $x_{k_n}\in F_{k_n}$ with $\|(\lambda_{k_n} - T_{k_n})x_{k_n}\| < 2^{-n}$. We set $x_l = 0$ for $l\notin\{k_n$ such that $n\in\mathbb{N}\}$. Let $\xi = (x_n)_n$. Then, we obtain $\mu\hat{\xi} = \hat{T}\hat{\xi}$. By Lemma 8.1.1 there exists $y\in E$ with $\hat{T}\hat{\xi} = V_{\mathcal{U}}y$. As in the proof of the Proposition 8.1.1, we obtain $\mu\in\sigma_p(T)$, a contradiction to $\text{dist}(\lambda_{k_n},\sigma_p(T)\bigcup\{0\})\geq\delta$. Q.E.D.

Theorem 8.1.8. *Let $((A_n,\mathcal{D}(A_n)))_n$ be an inverse d-compact sequence of operators approximating discretely the closed densely defined operator $(A,\mathcal{D}(A))$ in E. Assume that A as well as all A_n are surjective and that $A(E_0)\subset E_1$ is dense in E. Then, for every compact set $K\neq\emptyset$ in $\mathbb{C}$,* $\lim_n\text{dist}(\sigma_{ap}(A_n)\bigcap K,\sigma_p(A)) = 0$. ◇

Proof. By Lemma 8.1.2, the sequence $(A_n)_n$ satisfies the hypotheses of Theorem 8.1.6 and moreover, the inverse operators A_n^{-1} which exist from some n_0 on form a discretely compact approximation of A^{-1}. Hence, by Proposition 8.1.2

$$\lim_{n\to\infty} \operatorname{dist}(\sigma_{ap}(A_n^{-1}), \sigma_p(A^{-1}) \bigcup \{0\}) = 0 \tag{8.1.2}$$

holds. By the spectral mapping theorem $\sigma_{ap}(A_n) = \{\frac{1}{\lambda}$ such that $\lambda \in \sigma_{ap}(A_n^{-1})\}$ and $\sigma_p(A) = \{\frac{1}{\lambda}$ such that $\lambda \in \sigma_p(A^{-1})\}$. If the assertion does not hold, there exists a sequence $(\lambda_{n_k})_k$ with $\lambda_{n_k} \in \sigma_{ap}(A_{n_k}) \bigcap K$ and

$$\operatorname{dist}(\lambda_{n_k}, \sigma_p(A)) \geq \delta > 0 \tag{8.1.3}$$

for all k. Since K is compact w.l.o.g. we assume that $(\lambda_{n_k})_k$ converges to some $\mu \in K$. Because $\frac{1}{|\lambda_{n_k}|} \leq \|A_{n_k}^{-1}\|$ and this latter sequence is bounded $\mu \neq 0$ holds. Since $(\frac{1}{\lambda_{n_k}})_k$ converges to $\frac{1}{\mu}$, it follows by Eq. (8.1.2) that $\frac{1}{\mu} \in \sigma_p(A^{-1})$ hence, $\mu \in \sigma_p(A)$, a contradiction to the inequality (8.1.3). Q.E.D.

We end this section by the following example. Let H be a separable, infinite dimensional Hilbert space over $\mathbb{R}$ with orthonormal basis $(e_n)_n$. Let E be the space of uniformly continuous complex valued bounded functions on H equipped with the supremum norm. For each n, let Q_n denote the orthogonal projection of H onto the span H_n of $e_1, \ldots, e_n$ and set $F_n = \{f \in E$ such that $f = f \circ Q_n\}$. Then, $P_n : E \longrightarrow F_n$, $f \longrightarrow f \circ Q_n$ is a projection of norm 1, and moreover F_n is isometrically isomorphic to the space of all bounded uniformly continuous functions on $\mathbb{R}^n$. So, we identify these two spaces. Then, the isometry $S_{n,m}$ whose existence is required in Theorem 8.1.7 is nothing else than the inclusion mapping. Let $(\lambda_n)_n$ be a positive summable sequence and set $A_n = \sum_{k=1}^{n} \lambda_k \frac{\partial^2}{\partial x_k^2}$. Then, $(A_n)_n$ approximates the infinite dimensional Laplacian $\sum_{k=1}^{\infty} \lambda_k \frac{\partial^2}{\partial x_k^2}$. In order to apply Theorem 8.1.7, we set $\mathcal{D}(A_n) = G_n$, $E_0 = \bigcup \mathcal{D}(A_n)$ and $E_1 = E$. Moreover, if $z \in G_k$, then we take $(t_n)_n = (n)_{n\geq k}$ and $y = z$. Then, all the assumptions made in Theorem 8.1.7 are satisfied. We show that $\bigcap_{n\geq 1} \bigcup_{k\geq n} \sigma_{1/n,ap}(A_k) = \{\lambda \in \mathbb{C}$ such that $\mathrm{Re}\lambda \leq 0\}$. Let $\lambda \in \mathbb{C}$ with $\mathrm{Re}\lambda := \gamma < 0$ be arbitrary. Then, the function $g_m(x) = \exp\left(\frac{\lambda}{2m} \sum_{j=1}^{m} \frac{x_j^2}{\lambda_j}\right)$ is of norm 1 and a short calculation shows $(\lambda - A_m)g_m(x) = -\frac{\lambda^2}{m^2} g_m(x) \sum_{j=1}^{m} \frac{x_j^2}{\lambda_j}$. The inequality $t \exp\left(\frac{\gamma}{2m} t\right) \leq \frac{2m}{-\gamma e}$ for all $t \geq 0$ which is proved by elementary calculus shows $\alpha(\lambda - A_m) \leq \|(\lambda - A_m)g_m\| \leq \frac{2|\lambda|^2}{|\mathrm{Re}\lambda| em}$ which in turn proves that $\lambda \in \sigma_{1/n,ap}(A_m)$ for $m > \frac{2n|\lambda|^2}{|\mathrm{Re}\lambda| e}$. So, Theorem 8.1.7 implies $\{\lambda \in \mathbb{C}$ such that $\mathrm{Re}\lambda < 0\} \subset \bigcap_{n\geq 1} \bigcup_{k\geq n} \sigma_{1/n,ap}(A_k) \subset \sigma_{ap}(A)$. Finally, we apply Theorem 8.1.4: A_n is known to be the generator of a contraction semigroup. The Hille-Yosida theorem implies $\|(\lambda - A_n)^{-1}\| \leq (\mathrm{Re}\lambda)^{-1}$ for all λ with $\mathrm{Re}\lambda > 0$ independently of n. So, $\lambda \in \rho_\varepsilon(A_n)$ for $\mathrm{Re}\lambda > \varepsilon$. Theorem 8.1.4 then implies $\sigma_{ap}(A) \subset \{\lambda \in \mathbb{C}$ such that $\mathrm{Re}\lambda \leq 0\}$.

8.2 Pseudo-Browder's Essential Spectrum

8.2.1 Extending the Resolvent to a Browder Resolvent $\rho_b(.)$

Let A be a closed operator in a complex Banach space X with a nonempty resolvent set. In what follows, we will denote by $\sigma_d(A)$ the set of isolated points λ of the spectrum, such that the corresponding Riesz projectors

$$P_\lambda = -\frac{1}{2\pi i}\int_{|\lambda-\xi|=\varepsilon}(A-\xi I)^{-1}d\xi$$

(with a sufficiently small ε) are finite dimensional. It is well known that $\sigma_{e6}(A) = \sigma(A)\backslash\sigma_d(A)$ (see [239]). Another part of the spectrum, which is generally larger than $\sigma_{ei}(A)$, $i = 1,\ldots,5$ is $\sigma(A)\backslash\sigma_d(A)$, and was investigated by Browder in [66]. We will also use this terminology here and the notation $\sigma_{e6}(A) := \sigma(A)\backslash\sigma_d(A)$, $\rho_6(A) = \mathbb{C}\backslash\sigma_{e6}(A)$. The largest open set on which the resolvent is finitely meromorphic is precisely $\rho_6(A) = \rho(A)\bigcup\sigma_d(A)$, the points of $\sigma_d(A)$ being poles of finite rank, i.e., around each of these points there is a punctured disk in which the resolvent has a Laurent expansion whose main part has only several finite nonzero terms, the coefficients in these being of finite rank. For $\lambda \in \rho_6(A)$, let P_λ (or $P_\lambda(A)$) denote the corresponding (finite rank) Riesz projector with a range and a kernel denoted by R_λ and K_λ, respectively. Since $\mathcal{D}(A)$ is P_λ-invariant, we may define the operator $A_\lambda := (A-\lambda)(I-P_\lambda)+P_\lambda$ with a domain $\mathcal{D}(A)$ or, with respect to the decomposition $X = K_\lambda \oplus R_\lambda$, $A_\lambda = (A-\lambda)_{|K_\lambda} \oplus I$. We have just cut off the finite dimensional part of $A-\lambda$ in the Riesz decomposition. Since $\sigma((A-\lambda)_{|K_\lambda}) = \sigma(A-\lambda)\backslash\{0\}$, A_λ has a bounded inverse, called the "Browder resolvent", which we denote by $R_b(A,\lambda)$, i.e., $R_b(A,\lambda) := ((A-\lambda)_{|K_\lambda})^{-1}\oplus I$ with respect to $X = K_\lambda\oplus R_\lambda$ or, alternatively, $R_b(A,\lambda) := \left((A-\lambda)_{|K_\lambda}\right)^{-1}(I-P_\lambda)+P_\lambda$ for $\lambda \in \rho_6(A)$. Clearly, this extends the usual resolvent $R(A,\lambda) := (A-\lambda)^{-1}$ from $\rho(A)$ to $\rho_6(A)$ and retains many of its important properties. For example, since $P_\lambda A_\lambda = P_\lambda$ on $\mathcal{D}(A)$ and $A_\lambda P_\lambda = P_\lambda$ on X, it follows that $P_\lambda R_b(A,\lambda) = P_\lambda = R_b(A,\lambda)P_\lambda$, and we also have the "resolvent identity". The following result is due to J. Lutgen [239].

Lemma 8.2.1. *Let A be a closed operator in a complex Banach space X with a nonempty resolvent set. For λ, $\mu \in \rho_6(A)$, we have the resolvent identity*

$$R_b(A,\lambda)-R_b(A,\mu) = (\lambda-\mu)R_b(A,\lambda)R_b(A,\mu)+R_b(A,\lambda)S_A(\lambda,\mu)R_b(A,\mu),$$

where

$$S_A(\lambda,\mu) := (A-(\lambda+1))\,P_\lambda - (A-(\mu+1))\,P_\mu. \tag{8.2.1}$$

The operator defined by $M_A(\lambda,\mu) := R_b(A,\lambda)S_A(\lambda,\mu)R_b(A,\mu)$ *is a finite rank operator with* $\dim R(M_A(\lambda,\mu)) = \dim R(P_\lambda) + \dim R(P_\mu)$ *in the case* $\lambda \neq \mu$. *Moreover, the Browder resolvents commute; hence, the function* $M_A(.,.)$ *is skew-symmetric, i.e.,* $M_A(\lambda,\mu) = -M_A(\mu,\lambda)$. ◇

Proof. The identity follows just by substituting the difference of the equalities for A_μ and A_λ into the equality $R_b(A,\lambda) - R_b(A,\mu) = R_b(A,\lambda)(A_\mu - A_\lambda)R_b(A,\mu)$. Since $S_A(\lambda,\mu) := (A-(\lambda+1))\,P_\lambda - (A-(\mu+1))\,P_\mu$ is a finite rank operator (with $R(S_A(\lambda,\mu)) \subset R_\lambda + R_\mu$), the product $R_b(A,\lambda)S_A(\lambda,\mu)R_b(A,\mu) = M_A(\lambda,\mu)$ is also of finite rank. From $\lambda \neq \mu$, it follows that $P_\lambda P_\mu = P_\mu P_\lambda = 0$ and thus, $S_A(\lambda,\mu) = S_A(\lambda,\mu)(P_\lambda + P_\mu)$. Since $R(R_b(A,\mu)) = \mathcal{D}(A) \supset R_\lambda + R_\mu = R(P_\lambda + P_\mu)$, it follows that $R(S_A(\lambda,\mu)R_b(A,\mu)) \supset R(S_A(\lambda,\mu)(P_\lambda + P_\mu)) = R(S_A(\lambda,\mu))$. Hence, the ranges of $S_A(\lambda,\mu)$ and $S_A(\lambda,\mu)R_b(A,\mu)$ are the same and $S_A(\lambda,\mu)$ and $R_b(A,\lambda)S_A(\lambda,\mu)R_b(A,\mu)$ have also the same rank, since $R_b(A,\lambda)$ is injective, i.e., $\dim R(S_A(\lambda,\mu)) = \dim R(M_A(\lambda,\mu))$. Since $\dim(R_\lambda + R_\mu) = \dim(R_\lambda) + \dim(R_\mu) = \dim R(P_\lambda) + \dim R(P_\mu)$, it is sufficient to show that $R(S_A(\lambda,\mu)) = R_\lambda + R_\mu$. We have seen that $R(S_A(\lambda,\mu)_{|R_\lambda + R_\mu}) = R(S_A(\lambda,\mu)) \subset R_\lambda + R_\mu$. Since $R_\lambda + R_\mu$ is finite dimensional, we only need to show that the restriction of $S_A(\lambda,\mu)$ is injective in order to make the last inclusion an equality. Denoting $A_1 := A_{|R_\lambda}$, $A_2 := A_{|R_\mu}$, we have $\sigma(A_1) = \{\lambda\}$, $\sigma(A_2) = \{\mu\}$. Hence, $A_1 - (\lambda+1)$ and $A_2 - (\mu+1)$ are injective, which implies the injectivity of $S_A(\lambda,\mu)_{|R_\lambda + R_\mu}$. Using the spectral set $\sigma := \{\lambda,\mu\}$ of $\sigma(A)$ and the corresponding decomposition $X = K_\sigma \oplus R_\sigma = K_\sigma \oplus R_\lambda \oplus R_\mu$, induced by the Riesz projector $P_\sigma = P_\lambda + P_\mu$, which completely reduces A as $A = A_0 \oplus A_1 \oplus A_2$ ($A_0 := A_{|K_\sigma}$), we have $R_b(A,\lambda) = (A_0-\lambda)^{-1} \oplus I \oplus (A_2-\lambda)^{-1}$ and $R_b(A,\mu) = (A_0-\mu)^{-1} \oplus (A_1-\mu)^{-1} \oplus I$, and the claimed commutativity follows from that of the usual resolvents. The skew-symmetry is proved simply by exchanging λ and μ in the resolvent identity, adding the result to the original form and using the commutativity. Q.E.D.

The following result is due to J. Lutgen [239].

Lemma 8.2.2. *Let* X *and* Y *be two complex Banach spaces,* A *be a closed operator in* X *with a nonempty resolvent set,* $B : Y \longrightarrow X$, *and* $C : X \longrightarrow Y$ *linear operators. Then,*

(i) $R_b(A,\mu)B$ *is continuous for some* $\mu \in \rho_6(A)$ *if, and only if, it is continuous for all such* μ, *and this is the case if, and only if,* $\mathcal{D}(B^*) \supset \mathcal{D}(A^*)$, *where* A *and* B *are considered as densely defined operators from* $\overline{\mathcal{D}(A)}$ *into* X *and from* $\overline{\mathcal{D}(B)}$ *into* X, *respectively.*

(ii) C *is A-bounded if, and only if,* $CR_b(A,\mu)$ *is bounded for some (or every)* $\mu \in \rho_6(A)$.

(iii) If B *and* C *satisfy the conditions (i) and (ii), respectively, and if* B *is densely defined, then* $\overline{CM_A(\lambda,\mu)}$, $\overline{M_A(\lambda,\mu)B}$ *and* $\overline{CM_A(\lambda,\mu)B}$ *are operators of finite rank for any* $\lambda,\mu \in \rho_6(A)$. ◇

Proof. From the resolvent identity, we have, for any $\lambda,\mu \in \rho_6(A)$,

$$R_b(A,\lambda)B = R_b(A,\mu)B + (\lambda-\mu)R_b(A,\lambda)R_b(A,\mu)B + M_A(\lambda,\mu)B, \tag{8.2.2}$$

$$CR_b(A,\mu) = CR_b(A,\lambda) - (\lambda-\mu)CR_b(A,\lambda)R_b(A,\mu) - CM_A(\lambda,\mu). \tag{8.2.3}$$

Since $S_A(.,.)$ is bounded, the equivalence in (i) is clear from Eq. (8.2.2). Moreover, since $R_b(A,\lambda)S_A(\lambda,\mu)R_b(A,\mu)B$ has a finite-dimensional range, it is clear that $M_A(\lambda,\mu)B$ is of finite rank if B is densely defined, and this is one part of *(iii)*. For the second equivalence in (i) fix $\mu \in \rho_6(A)$, and consider the densely defined operators A, A_μ : $\overline{\mathcal{D}(A)} \longrightarrow X$, B : $\overline{\mathcal{D}(B)} \longrightarrow X$ and their conjugates with domains in X^*. Using standard properties of adjoint operators we obtain the equalities $(R_b(A,\mu)B)^* = B^*R_b(A,\mu)^* = B^*(A_\mu^*)^{-1}$. If the product on the left is bounded, then the product on the right is everywhere defined, i.e., $\mathcal{D}(A_\mu^*) \subset \mathcal{D}(B^*)$. On the other hand, if this inclusion holds, then the adjoint on the left is everywhere defined which implies boundedness of $R_b(A,\mu)B$. Since $\mathcal{D}(A_\mu^*) = \mathcal{D}(A^*)$ due to the boundedness of P_μ, (i) is proved. In *(ii)*, if $CR_b(A,\lambda)$ is bounded for some λ, then clearly $CR_b(A,\lambda)S_A(\lambda,\mu)R_b(A,\mu)$ is also bounded for any μ, and it follows, from Eq. (8.2.3), that $CR_b(A,\mu)$ is bounded for any μ. The well-known fact that C is A-bounded if, and only if, $C(A-\mu)^{-1}$ is bounded for some $\mu \in \rho(A)$, now implies *(ii)* and, in this case, $CR_b(A,\lambda)S_A(\lambda,\mu)R_b(A,\mu)$ is of finite rank. The last part of *(iii)* is also clear, since $CR_b(A,\lambda)S_A(\lambda,\mu)R_b(A,\mu)B$ will again be continuous and densely defined with a finite-dimensional range. Q.E.D.

8.2.2 *Definition of the Pseudo-Browder Essential Spectrum*

The purpose of this section is to give some properties of the pseudo-Browder essential spectrum of closed, densely defined linear operators on a Banach space X. Let us start by defining the pseudo-Browder essential spectrum. Let $A \in \mathcal{C}(X)$ and let $A_\lambda := (A-\lambda)(I-P_\lambda)+P_\lambda$ with a domain $\mathcal{D}(A)$. Then, for $\lambda \in \rho_6(A)$, the operator A_λ is invertible where $A_\lambda^{-1} := R_b(A,\lambda) = \big((A-\lambda)|_{K_\lambda}\big)^{-1}(I-P_\lambda)+P_\lambda$. For $\lambda \in \sigma_d(A)$ and $b \in X$, we consider the operator equation defined by

$$A_\lambda x := \big[(A-\lambda)(I-P_\lambda)+P_\lambda\big]x = b. \tag{8.2.4}$$

The existence of a solution and its uniqueness are guaranteed by $\lambda \in \rho_6(A)$. Then, we have $x = A_\lambda^{-1}b := R_b(A,\lambda)b$. Let us discuss the stability of solutions of the operator Eq. (8.2.4) under perturbations of b or A. We will perturb Eq. (8.2.4) in the following way $\big[(A-\lambda)(I-P_\lambda)+P_\lambda\big]\tilde{x} = b+r$, where $0 < \|r\| < \varepsilon$. Hence, for $\lambda \in \rho_6(A)$, we have $\tilde{x} = R_b(A,\lambda)(b+r)$. We deduce that $\|\tilde{x}-x\| = \|R_b(A,\lambda)r\| \leq \|(R_b(A,\lambda)\|\|r\| \leq \|R_b(A,\lambda)\|\varepsilon$. Then, $\tilde{x}$ is equivalent of x, if

$\| R_b(A,\lambda)r\|$ is small, for example, if $\| R_b(A,\lambda)r\|$ is not too big $\| R_b(A,\lambda)\| \leq \frac{1}{\varepsilon}$. Therefore, the above considerations motivate one to define $\rho_{6,\varepsilon}(A)$ by

$$\rho_{6,\varepsilon}(A) = \rho_6(A) \bigcap \Big\{\lambda \in \mathbb{C} \text{ such that } \| R_b(A,\lambda)\| \leq \frac{1}{\varepsilon}\Big\}.$$

Definition 8.2.1. Let $A \in \mathcal{C}(X)$ and $\varepsilon > 0$. The pseudo-Browder essential spectrum of A is defined by $\sigma_{e6,\varepsilon}(A) = \sigma_{e6}(A) \bigcup \big\{\lambda \in \mathbb{C} \text{ such that } \| R_b(A,\lambda)\| > \frac{1}{\varepsilon}\big\}$. ◇

8.2.3 Pseudo-Browder's Essential Spectrum of Linear Operators

We give the following result.

Theorem 8.2.1. *Let $A \in \mathcal{C}(X)$, ε_1, ε_2 and $\varepsilon > 0$. Then,*

(i) If $\varepsilon_1 < \varepsilon_2$, then $\sigma_{e6}(A) \subset \sigma_{e6,\varepsilon_1}(A) \subset \sigma_{e6,\varepsilon_2}(A)$.
(ii) If $\sigma_d(A) = \emptyset$, then $\sigma_{e6,\varepsilon}(A) = \sigma_\varepsilon(A)$. ◇

Proof. (i) Let $\lambda \in \rho_6(A)$, such that $\|A_\lambda^{-1}\| > \frac{1}{\varepsilon_1}$. Hence, since $\varepsilon_1 < \varepsilon_2$, we have $\|A_\lambda^{-1}\| > \frac{1}{\varepsilon_2}$. This proves that $\sigma_{e6,\varepsilon_1}(A) \subset \sigma_{e6,\varepsilon_2}(A)$.

(ii) Let $A \in \mathcal{C}(X)$, such that $\sigma_d(A) = \emptyset$. So, in the same way, we can deduce, from $\sigma_{e6}(A) = \sigma(A)\backslash\sigma_d(A)$, that $\sigma_{e6}(A) = \sigma(A)$. This implies that $\sigma_{e6,\varepsilon}(A) = \sigma_\varepsilon(A)$. Q.E.D.

Corollary 8.2.1. *Let $A \in \mathcal{C}(X)$, $B \in \mathcal{L}(X)$, $\varepsilon > 0$ and $\varepsilon_1 > 0$, such that $0 \in \rho(B)$. Then*

(i) $\sigma_{e6}(B^{-1}A) \subset \sigma_{e6,\frac{\varepsilon}{\|B\|}}(B^{-1}A) \subset \sigma_{e6,\varepsilon\|B^{-1}\|}(B^{-1}A)$.
(ii) $\sigma_{e6}(B^{-1}A) \subset \sigma_{e6,\varepsilon_1}(B^{-1}A) \bigcap \sigma_{e6,\varepsilon}(B^{-1}A) \subset \sigma_{e6,\varepsilon_1+\varepsilon}(B^{-1}A)$.
(iii) If $0 < \varepsilon < 1$, then $\sigma_{e6}(B^{-1}A) \subset \sigma_{e6,\frac{\varepsilon^2}{\|B\|}}(B^{-1}A) \subset \sigma_{e6,\varepsilon\|B^{-1}\|}(B^{-1}A)$. ◇

Proof. (i) The fact that $\varepsilon\|B^{-1}\| - \frac{\varepsilon}{\|B\|} = \frac{\varepsilon\|B^{-1}\|\|B\|-\varepsilon}{\|B\|} = \frac{\varepsilon(\|B^{-1}\|\|B\|-1)}{\|B\|} \geq 0$ allows us to deduce that $\frac{\varepsilon}{\|B\|} \leq \varepsilon\|B^{-1}\|$. Applying Theorem 8.2.1 (i), we get $\sigma_{e6,\frac{\varepsilon}{\|B\|}}(B^{-1}A) \subset \sigma_{e6,\varepsilon\|B^{-1}\|}(B^{-1}A)$.

(ii) Let $\lambda \in \sigma_{e6,\varepsilon_1}(B^{-1}A) \bigcap \sigma_{e6,\varepsilon}(B^{-1}A)$. Since $\max(\varepsilon_1,\varepsilon) < \varepsilon_1 + \varepsilon$, and applying Theorem 8.2.1 (i), we have $\sigma_{e6,\varepsilon_1}(B^{-1}A) \subset \sigma_{e6,\varepsilon_1+\varepsilon}(B^{-1}A)$ and $\sigma_{e6,\varepsilon}(B^{-1}A) \subset \sigma_{e6,\varepsilon_1+\varepsilon}(B^{-1}A)$.

(iii) If $0 < \varepsilon < 1$, then $\varepsilon^2 < \varepsilon$, which implies that $\frac{\varepsilon^2}{\|B\|} < \varepsilon\|B^{-1}\|$. Therefore, by using Theorem 8.2.1 (i), we have $\sigma_{e6,\frac{\varepsilon^2}{\|B\|}}(B^{-1}A) \subset \sigma_{e6,\varepsilon\|B^{-1}\|}(B^{-1}A)$. Q.E.D.

8.2.4 Characterization of the Pseudo-Browder Essential Spectrum

In this section, we give a characterization of the pseudo-Browder essential spectrum of closed, densely defined linear operators on a Banach space X. Our first result is the following.

Theorem 8.2.2. *Let $A \in \mathcal{C}(X)$ and $\varepsilon > 0$. Then, $\lambda \in \sigma_{e6,\varepsilon}(A)\backslash\sigma_{e6}(A)$ if, and only if, there exists $x \in \mathcal{D}(A)$, such that*

$$\|A_\lambda x\| < \varepsilon\|x\|. \tag{8.2.5}$$

◇

Proof. Let $\lambda \in \sigma_{e6,\varepsilon}(A)\backslash\sigma_{e6}(A)$. Then, $\|A_\lambda^{-1}\| > \frac{1}{\varepsilon}$. Consequently, we infer that $\sup_{y\in X,\ y\neq 0}\frac{\|A_\lambda{}^{-1}y\|}{\|y\|} > \frac{1}{\varepsilon}$. Therefore, there exists a nonzero $y \in X$, such that $\|A_\lambda^{-1}y\| > \frac{1}{\varepsilon}\|y\|$. Putting $x = A_\lambda^{-1}y$. This leads to Eq. (8.2.5). The converse is similar. Q.E.D.

Theorem 8.2.3. *Let $A \in \mathcal{C}(X)$ and $\varepsilon > 0$. Then,*

$$\sigma_{e6,\varepsilon}(A) = \sigma_{e6}(A)\bigcup\Big\{\lambda \in \mathbb{C} \text{ such that } \|A_\lambda x\| < \varepsilon, \text{ for some } x \in \mathcal{D}(A) \text{ and } \|x\| = 1\Big\}.$$

◇

Proof. Let $\lambda \in \sigma_{e6,\varepsilon}(A)$. There are two possible cases:

<u>1st*case*</u> : If $\lambda \in \sigma_{e6}(A)$, then $\lambda \in \sigma_{e6}(A)\bigcup\{\lambda \in \mathbb{C}$ such that $\|A_\lambda x\| < \varepsilon$, for some $x \in \mathcal{D}(A)$ and $\|x\| = 1\}$.

<u>2nd*case*</u> : If $\lambda \in \sigma_{e6,\varepsilon}(A)\backslash\sigma_{e6}(A)$, then $\|A_\lambda^{-1}\| > \dfrac{1}{\varepsilon}$. Moreover,

$$\begin{aligned}
\|A_\lambda{}^{-1}\| &= \sup_{y\in X,\ y\neq 0}\frac{\|A_\lambda{}^{-1}y\|}{\|y\|} \\
&= \sup_{x\in\mathcal{D}(A),\ x\neq 0}\frac{\|x\|}{\|A_\lambda x\|} \\
&= \sup_{x\in\mathcal{D}(A),\ \|x\|=1}\frac{1}{\|A_\lambda x\|} \\
&= \frac{1}{\inf_{x\in\mathcal{D}(A),\ \|x\|=1}\|A_\lambda x\|} \\
&> \frac{1}{\varepsilon}.
\end{aligned}$$

So, $\inf_{x\in\mathcal{D}(A),\ \|x\|=1} \|A_\lambda x\| < \varepsilon$. Conversely, let $\lambda \in \{\lambda \in \mathbb{C}$ such that $\|A_\lambda x\| < \varepsilon$, for some $x \in \mathcal{D}(A)$ and $\|x\| = 1\}$, then there exists $x \in \mathcal{D}(A)$ such that $\|x\| = 1$ and $\|A_\lambda x\| < \varepsilon$, and we get $\|A_\lambda x\| < \varepsilon\|x\|$. Now, by applying Theorem 8.2.2, we infer that $\lambda \in \sigma_{e6,\varepsilon}(A)$. Q.E.D.

Theorem 8.2.4. *Let $A \in \mathcal{C}(X)$ and $\varepsilon > 0$. Then,*

$$\sigma_{e6}(A)\bigcup\Big\{\lambda \in \mathbb{C} : \exists(x_n)_n \in \mathcal{D}(A) \text{ such that } \|x_n\| = 1,\ \lim_{n\to\infty} \|A_\lambda x_n\| < \varepsilon\Big\} \subseteq \sigma_{e6,\varepsilon}(A).$$

◇

Proof. Let $y_n = \frac{A_\lambda x_n}{\|A_\lambda x_n\|}$ implies $\|y_n\| = 1$. We have $\|A_\lambda^{-1}\| \geq \lim_{n\to\infty} \|A_\lambda^{-1} y_n\| = \lim_{n\to\infty} \left(\|A_\lambda x_n\|\right)^{-1}$. Recall that $\lim_{n\to\infty} \|A_\lambda x_n\| < \varepsilon$, so that $\|A_\lambda^{-1}\| > \frac{1}{\varepsilon}$. From Definition 8.2.1, we deduce that $\sigma_{e6}(A)\bigcup\{\lambda \in \mathbb{C} : \exists(x_n)_n \in \mathcal{D}(A)$ such that $\|x_n\| = 1,\ \lim_{n\to\infty} \|A_\lambda x_n\| \leq \varepsilon\} \subseteq \sigma_{e6,\varepsilon}(A)$. Q.E.D.

Theorem 8.2.5. *Let $A \in \mathcal{L}(X)$ and $\varepsilon > 0$. Then, $\sigma_{e6,\varepsilon}(A) = \bigcup_{\|D\|<\varepsilon,\ AD=DA} \sigma_{e6}(A + D)$.* ◇

Proof. Let $\lambda \in \sigma_{e6,\varepsilon}(A)$. There are two possible cases:

1st *case* : If $\lambda \in \sigma_{e6}(A)$, then it is sufficient to take $D = 0$.

2nd *case* : If $\lambda \in \sigma_{e6,\varepsilon}(A)\backslash\sigma_{e6}(A)$, then $\lambda \in \rho_6(A)$ such that $\|A_\lambda^{-1}\| > \frac{1}{\varepsilon}$. First, if $\lambda \in \rho(A)$, then $A_\lambda^{-1} = (A-\lambda)^{-1}$. By applying Theorem 8.2.2, there exists f such that $\|f\| = 1$ and $\|(A-\lambda)f\| < \varepsilon$. Let $\psi \in X^*$ (dual of X) such that $\|\psi\| = 1$ and $\psi(f) = 1$. We suppose that $Dg = -\psi(g)(A-\lambda)f$. It is easy to see that $DA = AD$, $\|D\| < \varepsilon$, and $(A+D-\lambda)f = (A-\lambda)f-(A-\lambda)f = 0$. Hence, $A + D - \lambda$ is not invertible. Therefore, $\lambda \in \sigma(A + D) = \sigma_{e6}(A + D)$. Second, if $\lambda \in \sigma_d(A)$, there exists an $\varepsilon_1 > 0$ such that the disc $\{\mu \in \mathbb{C}$ such that $0 < |\mu - \lambda| < 2\varepsilon_1\}\bigcap\sigma(A) = \{\lambda\}$. Consider the operator $\tilde{A_\lambda}$ introduced in [50] by $\tilde{A_\lambda} = A + \frac{\varepsilon_1}{1+\|P_\lambda\|} P_\lambda$, where P_λ is the Riesz projection corresponding to λ. Then, $\{\mu \in \mathbb{C}$ such that $0 < |\mu - \lambda| < \varepsilon_1\} \subset \rho(A)\bigcap\rho(\tilde{A_\lambda})$. Let $\mu \in \rho(\tilde{A_\lambda})\bigcap\rho(A)$ such that $\|A-\mu\| < \varepsilon$. Then, there exists an $\varepsilon_2 > 0$ such that $\|A-\mu\|+\varepsilon_2 = \varepsilon$. Let us consider the operator $\tilde{A}_{\not\geq} = A + \frac{\varepsilon_3}{1+\|P_\lambda\|} P_\lambda$, where $\varepsilon_3 = \min\left(\frac{\varepsilon_2}{2}, \varepsilon_1\right)$. We have $\|\tilde{A}_{\not\geq} - \mu\| < \|A-\mu\| + \varepsilon_3 < \|A-\mu\| + \varepsilon_1 < \varepsilon$. Moreover, $\sigma_{e6}(A) \subset \sigma_{e6}(\tilde{A}_{\not\geq})$. Then, $\sigma_{e6,\varepsilon}(A) = \sigma_{e6,\varepsilon}(\tilde{A}_{\not\geq})$. Hence, there exists D such that $\|D\| < \varepsilon$ and $D\tilde{A}_{\not\geq} = \tilde{A}_{\not\geq}D$. Therefore, $\sigma_{e6,\varepsilon}(A) \subset \sigma_{e6,\varepsilon}(\tilde{A}_{\not\geq}) = \sigma_{e6}(A+D)$. Conversely, we assume that there exists D such that $\|D\| < \varepsilon$, $AD = DA$ and $\lambda \in \sigma_{e6}(A+D)$. In order to prove that $\lambda \in \sigma_{e6,\varepsilon}(A)$, we suppose that $\lambda \notin \sigma_{e6,\varepsilon}(A)$, hence $\lambda \in \rho_{6,\varepsilon}(A)$. Therefore, $\lambda \in \rho_6(A)\bigcap\{\lambda \in \mathbb{C}$ such that $\|R_b(A,\lambda)\| < \varepsilon\}$. We notice the existence of two cases:

<u>$1^{st}case$</u> : If $\rho_6(A) = \rho(A)$. Let $S : X \longrightarrow X$ be defined by

$$S := \sum_{n=0}^{+\infty} (A-\lambda)^{-1}\Big(- D(A-\lambda)^{-1}\Big)^n.$$

Since $\|D(A-\lambda)^{-1}\| < \varepsilon$, then we can write $S = (A-\lambda)^{-1}\big(I + D(A-\lambda)^{-1}\big)^{-1}$. So, there exists $f \in X$ such that $S\big(I + D(A-\lambda)^{-1}\big)^{-1} f = (A-\lambda)^{-1} f$. Let $g = (A-\lambda)^{-1} f$. We can show that $S(A-\lambda+D)g = g$ for all $g \in \mathcal{D}(A)$. Similarly, we can prove that $(A-\lambda+D)Sh = h$ for all $h \in X$. Therefore $A-\lambda+D$ is invertible, which is absurd.

<u>$2^{nd}case$</u> : If $\lambda \in \sigma(A)\backslash\sigma_d(A)$, then λ is a discrete point of finite multiplicity. Let $\tilde{A} = A + \varepsilon P_\lambda$, then $\rho(\tilde{A}) \neq \emptyset$. Moreover, we have $DA = AD$ and $DP_\lambda = P_\lambda D$. Hence, $\tilde{A}D = D\tilde{A}$. Therefore, $\sigma_{e6}(\tilde{A}+D) = \sigma_{e6}(A+\varepsilon P_\lambda + D) = \sigma_{e6}(A+D)$. Consequently $\lambda \in \sigma_{e6}(\tilde{A}+D)$ and $\|(\tilde{A}-\lambda)^{-1}\| < \varepsilon$. Using a similar reasoning to the first case, we deduce that $\lambda \in \sigma_{e6,\varepsilon}(A)$. Q.E.D.

8.2.5 Stability of the Pseudo-Browder's Essential Spectrum

The aim of this section is the investigation of the stability of the pseudo-Browder's essential spectrum under Riesz operator perturbations satisfying some conditions.

Theorem 8.2.6. *Let $\varepsilon > 0$, $A \in \mathcal{L}(X)$ and $R \in \mathcal{R}(X)$ such that $RA = AR$. If $\|R\| < \frac{\varepsilon}{2}$, then there exists $\varepsilon_1 > 0$, such that $\sigma_{e6,\varepsilon_1}(A+R) = \sigma_{e6,\varepsilon}(A)$.* ◇

Proof. $\sigma_{e6,\varepsilon}(A) = \sigma_{e6}(A) \bigcup \{\lambda \in \mathbb{C} \text{ such that } \|(A-\lambda)^{-1}\| > \frac{1}{\varepsilon}\}$ or as $R \in \mathcal{R}(X)$ and $RA = AR$. So, from Theorem 2.2.25, it follows that $\sigma_{e6}(A+R) = \sigma_{e6}(A)$. Let $\lambda \notin \{\lambda \in \mathbb{C} \text{ such that } \|(A-\lambda)^{-1}\| > \frac{1}{\varepsilon}\}$ implies $\|(A-\lambda)^{-1}\| \leq \varepsilon^{-1}$. Hence, we will discuss these two following cases:

<u>$1^{st}case$</u> : If $A_\lambda := A - \lambda$ (i.e., $\sigma_d(A) = \emptyset$), thus writing $A+R-\lambda$ in the form $A+R-\lambda = \big(A-\lambda\big)\big(I + (A-\lambda)^{-1}R\big)$, together with $\|(A-\lambda)^{-1}R\| \leq \|(A-\lambda)^{-1}\|\|R\| \leq \frac{1}{2}$, we deduce that $I + (A-\lambda)^{-1}R$ is an invertible operator and, we can write that $(A+R-\lambda)^{-1} = \big(I + (A-\lambda)^{-1}R\big)^{-1}(A-\lambda)^{-1}$ and $\big(I+(A-\lambda)^{-1}R\big)^{-1} = \sum_{n=0}^{+\infty}(-1)^n\left[(A-\lambda)^{-1}R\right]^n$. However, $RA = AR$ implies that $\big(I + (A-\lambda)^{-1}R\big)^{-1} = \sum_{n=0}^{+\infty}(-1)^n\left[(A-\lambda)^{-1}\right]^n R^n$. Since $\|\big(I + (A-\lambda)^{-1}R\big)^{-1}\| \leq \sum_{n=0}^{+\infty}\|(A-\lambda)^{-1}\|^n\|R\|^n \leq \sum_{n=0}^{+\infty}\left(\frac{\|R\|}{\varepsilon}\right)^n$, therefore, we have $\|\big(I+(A-\lambda)^{-1}R\big)^{-1}\| \leq \frac{\varepsilon}{\varepsilon-\|R\|}$. This shows that $\|(A+R-\lambda)^{-1}\| \leq \frac{\|(A-\lambda)^{-1}\|\varepsilon}{\varepsilon-\|R\|}$. Consequently, $\|(A+R-\lambda)^{-1}\| \leq \frac{1}{\varepsilon-\|R\|}$, it is sufficient to take $\varepsilon_1 = \varepsilon - \|R\|$.

<u>$2^{nd}case$</u> : If $\lambda \in \sigma_d(A)$, then there exists $\delta > 0$ such that the disc $\{\zeta \in \mathbb{C} \text{ such that } |\zeta - \lambda| \leq 2\delta\}$ does not contain points of $\sigma(A)$ different from λ,

and the Riesz projection P_λ of A corresponding to λ is of finite rank. Let us consider the operator $\tilde{A} := A + \delta P_\lambda$. Then, $\{\mu \in \mathbb{C}$ such that $0 < |\lambda - \mu| < \delta\} \subset \rho(\tilde{A})$. So,

$$\sigma_{e6,\varepsilon}(\tilde{A}) = \bigcup \{\sigma_{e6}(\tilde{A} + D) \text{ such that } \|D\| < \varepsilon \text{ and } \tilde{A}D = D\tilde{A}\},$$

it follows, from $D(A+\delta P_\lambda) = (A+\delta P_\lambda)D$, that $DA = AD$. Then, $(D+A)P_\lambda = P_\lambda(D + A)$. Moreover, P_λ is of finite rank. This shows that $\sigma_{e6}(\tilde{A} + D) = \sigma_{e6}(A + D)$. Obviously,

$$\bigcup \{\sigma_{e6}(\tilde{A} + D) \text{ such that } \|D\| < \varepsilon \text{ and } \tilde{A}D = D\tilde{A}\}$$
$$= \bigcup \{\sigma_{e6}(A + D) \text{ such that } \|D\| < \varepsilon \text{ and } AD = DA\}.$$

This leads us to the conclusion that $\sigma_{e6,\varepsilon}(\tilde{A}) = \sigma_{e6,\varepsilon}(A)$. Let $R \in \mathcal{R}(X)$, such that $RA = AR$ and $\|R\| < \frac{\varepsilon}{2}$, then $\tilde{A}R = R\tilde{A}$. Now, by applying the first part of this proof but for $\tilde{A}$, we deduce that there exists $\varepsilon_1 > 0$ such that $\sigma_{e6,\varepsilon_1}(\tilde{A} + R) \subseteq \sigma_{e6,\varepsilon}(\tilde{A})$. Moreover, $AR = RA$. Then, $RP_\lambda = P_\lambda R$ implies $(D + A + R)P_\lambda = P_\lambda(D + A + R)$. Hence, $\sigma_{e6,\varepsilon}(\tilde{A} + R) = \sigma_{e6,\varepsilon}(A + R)$. This shows that $\sigma_{e6,\varepsilon_1}(A + R) = \sigma_{e6,\varepsilon_1}(\tilde{A} + R) \subseteq \sigma_{e6,\varepsilon_1}(A + R) = \sigma_{e6,\varepsilon}(A)$. This leads us to the conclusion that there exists $\sigma_{e6,\varepsilon_1}(A + R) \subseteq \sigma_{e6,\varepsilon}(A)$. Conversely, let $\lambda \notin \{\lambda \in \mathbb{C}$ such that $\|(A + R)_\lambda^{-1}\| > \frac{1}{\varepsilon_1}\}$ implies $\|(A + R)_\lambda^{-1}\| \leq \varepsilon_1^{-1}$. There are two possible cases:

<u>$1^{st}case$</u> : If $(A + R)_\lambda = A + R - \lambda$. Thus, $A - \lambda = \big(A + R - \lambda\big)\big(I - (A + R - \lambda)^{-1}R\big)$, and writing $\|(A+R-\lambda)^{-1}R\| \leq \|(A+R-\lambda)^{-1}\|\|R\| \leq \frac{\varepsilon}{2(\varepsilon-\|R\|)} < 1$, implies that $I - (A + R - \lambda)^{-1}R$ is an invertible operator and we have $(A - \lambda)^{-1} = \big(I - (A + R - \lambda)^{-1}R\big)^{-1}(A + R - \lambda)^{-1}$. The fact that $\big(I - (A + R - \lambda)^{-1}R\big)^{-1} = \sum_{n=0}^{+\infty} \big((A + R - \lambda)^{-1}R\big)^n$ and knowing that $RA = AR$, implies that $\|\big(I - (A + R - \lambda)^{-1}R\big)^{-1}\| = \left\|\sum_{n=0}^{+\infty} \left[(A + R - \lambda)^{-1}\right]^n R^n\right\| \leq \sum_{n=0}^{+\infty} \|(A + R - \lambda)^{-1}\|^n \|R\|^n$. Then, $\|\big(I - (A + R - \lambda)^{-1}R\big)^{-1}\| < \frac{\varepsilon_1}{\varepsilon_1-\|R\|}$. Consequently, $\|(A - \lambda)^{-1}\| \leq \frac{1}{\varepsilon_1-\|R\|} = \frac{1}{\varepsilon-\|R\|-\|R\|} \leq \frac{1}{\varepsilon}$. Q.E.D.

<u>$2^{nd}case$</u> : Let $\lambda \in \sigma_d(A)$. Writing A_λ in the form $A_\lambda = (A_\lambda + R)(I - (A_\lambda + R)^{-1}R)$ and using the same reasoning as in the proof of $1^{st}case$, we get the desired result.

8.3 Pseudo-Jeribi and Pseudo-Schechter Essential Spectra

Let $\varepsilon > 0$ and $A \in \mathcal{C}(X)$. The pseudo-Jeribi essential spectrum is defined by

$$\sigma_{j,\varepsilon}(A) := \bigcap_{K \in \mathcal{W}_*(X)} \sigma_\varepsilon(A+K).$$

The pseudo-Schechter essential spectrum is defined by

$$\sigma_{e5,\varepsilon}(A) := \bigcap_{K \in \mathcal{K}(X)} \sigma_\varepsilon(A+K).$$

Remark 8.3.1. It follows that

(i) $\sigma_{j,\varepsilon}(A) \subset \sigma_{e5,\varepsilon}(A) \subset \sigma_\varepsilon(A)$.
(ii) $\bigcap_{\varepsilon>0} \sigma_{j,\varepsilon}(A) = \sigma_j(A)$ and $\bigcap_{\varepsilon>0} \sigma_{e5,\varepsilon}(A) = \sigma_{e5}(A)$.
(iii) If $\varepsilon_1 < \varepsilon_2$, then $\sigma_j(A) \subset \sigma_{j,\varepsilon_1}(A) \subset \sigma_{j,\varepsilon_2}(A)$ and $\sigma_{e5}(A) \subset \sigma_{e5,\varepsilon_1}(A) \subset \sigma_{e5,\varepsilon_2}(A)$.
(iv) $\sigma_{j,\varepsilon}(A+K) = \sigma_{j,\varepsilon}(A)$ for all $K \in \mathcal{W}_*(X)$ and $\sigma_{e5,\varepsilon}(A+K) = \sigma_{e5,\varepsilon}(A)$ for all $K \in \mathcal{K}(X)$. ◇

8.3.1 By Means of a Fredholm and Semi-Fredholm Perturbations

The following theorem gives a characterization of the pseudo-Schechter essential spectrum by means of a Fredholm operator.

Theorem 8.3.1. *Let X be a Banach space, $\varepsilon > 0$ and $A \in \mathcal{C}(X)$. Then, $\lambda \notin \sigma_{e5,\varepsilon}(A)$ if, and only if, for all $D \in \mathcal{L}(X)$ such that $\|D\| < \varepsilon$, we have $A + D - \lambda \in \Phi(X)$ and $i(A + D - \lambda) = 0$.* ◇

Proof. Let $\lambda \notin \sigma_{e5,\varepsilon}(A)$. By using Theorem 8.1.1, we infer that there exists a compact operator K on X, such that $\lambda \notin \bigcup_{\|D\|<\varepsilon} \sigma(A+K+D)$. So, for all $D \in \mathcal{L}(X)$ such that $\|D\| < \varepsilon$, we have $\lambda \in \rho(A+D+K)$. Therefore, $A+D+K-\lambda \in \Phi(X)$ and $i(A+D+K-\lambda) = 0$, for all $D \in \mathcal{L}(X)$ such that $\|D\| < \varepsilon$. From Theorem 2.2.44, we deduce that for all $D \in \mathcal{L}(X)$ such that $\|D\| < \varepsilon$, we have $A+D-\lambda \in \Phi(X)$ and $i(A+D-\lambda) = 0$. Conversely, we suppose that, for all $D \in \mathcal{L}(X)$ such that $\|D\| < \varepsilon$, we have $(A+D-\lambda) \in \Phi(X)$ and $i(A+D-\lambda) = 0$. Without loss of generality, we may assume that $\lambda = 0$. Let $n = \alpha(A+D) = \beta(A+D)$, $\{x_1, \ldots, x_n\}$ being the basis for $N(A+D)$ and $\{y'_1, \ldots, y'_n\}$ being the basis for the $N((A+D)^*)$. By using Lemma 2.1.1, there are functionals $x'_1, \ldots, x'_n$ in X^* (the adjoint space of X) and elements $y_1, \ldots, y_n$, such that $x'_j(x_k) = \delta_{jk}$ and $y_j(y_k) = \delta_{jk}$, $1 \leq j, k \leq n$, where $\delta_{jk} = 0$ if $j \neq k$ and $\delta_{jk} = 1$ if $j = k$. The operator K is defined by $Kx = \sum_{k=1}^n x'_k(x) y_k$, $x \in X$. Clearly, K is a linear operator defined everywhere on X. It is bounded, since $\|Kx\| \leq \|x\| \Big(\sum_{k=1}^n \|x'_k\| \|y_k\| \Big)$. Moreover, the range of K is contained in a finite-dimensional subspace of X. Then, K is a finite rank operator in X. So, K is a compact operator in X. We prove that

$$N(A+D)\bigcap N(K)=\{0\} \text{ and } R(A+D)\bigcap R(K)=\{0\}, \qquad (8.3.1)$$

for all $D \in \mathcal{L}(X)$ such that $\|D\| < \varepsilon$. Let $x \in N(A+D)$, then $x = \sum_{k=1}^{n} \alpha_k x_k$, and therefore, $x_j(x) = \alpha_j$, $1 \le j \le n$. Moreover, if $x \in N(K)$, then $x_j'(x) = 0$, with $1 \le j \le n$. This proves the first relation in Eq. (8.3.1). The second inclusion is similar. In fact, if $y \in R(K)$, then $y = \sum_{k=1}^{n} \alpha_k y_k$, and hence, $y_j(y) = \alpha_j$, with $1 \le j \le n$. However, if $y \in R(A+D)$, then $y_j'(y) = 0$, with $1 \le j \le n$. This gives the second relation in Eq. (8.3.1). Besides, K is a compact operator. We deduce, from Theorem 2.2.44, that $0 \in \Phi_{A+K+D}$ and $i(A+D+K) = 0$. If $x \in N(A+D+K)$, then $(A+D)x$ is in $R(A+D)\bigcap R(K)$. This implies that $x \in N(A+D)\bigcap N(K)$ and $x = 0$. Thus, $\alpha(A+D+K) = 0$. In the same way, we can prove that $R(A+D+K) = X$. Hence, $0 \in \rho(A+D+K)$. This implies that, for all $D \in \mathcal{L}(X)$ such that $\|D\| < \varepsilon$, we have $0 \notin \sigma(A+D+K)$. Also, $0 \notin \bigcap_{K \in \mathcal{K}(X)} \sigma_\varepsilon(A+K)$. So, $0 \notin \sigma_{e5,\varepsilon}(A)$. Q.E.D.

Remark 8.3.2. It follows immediately, from Theorem 8.3.1 and Proposition 7.1.1, that

$$\sigma_{e5,\varepsilon}(A) = \bigcup_{\|D\|<\varepsilon} \sigma_{e5}(A+D).$$

◇

Theorem 8.3.2. *If X satisfies the Dunford-Pettis property, $\varepsilon > 0$ and if A is a closed, densely defined, and linear operator on X, then we have $\sigma_{j,\varepsilon}(A) = \sigma_{e5,\varepsilon}(A)$.* ◇

Proof. Since $\mathcal{K}(X) \subset \mathcal{W}_*(X)$, we infer that $\sigma_{j,\varepsilon}(A) \subset \sigma_{e5,\varepsilon}(A)$. Conversely, let $\lambda \notin \sigma_{j,\varepsilon}(A)$. Then, there exists $F \in \mathcal{W}_*(X)$ such that $\lambda \notin \sigma_\varepsilon(A+F)$. Thus, by using Theorem 8.1.1 (*ii*), we notice that $\lambda \in \rho(A+D+F)$, for all $D \in \mathcal{L}(X)$ such that $\|D\| < \varepsilon$. So, $A+D+F-\lambda \in \Phi(X)$ and $i(A+D+F-\lambda) = 0$. Since $F \in \mathcal{W}_*(X)$ we have $(\lambda - A - D - F)^{-1}F \in \mathcal{W}_*(X)$. Hence, by using Lemma 2.1.13 (*i*), we get $\left[(\lambda - A - D - F)^{-1}F\right]^2 \in \mathcal{K}(X)$. Now, by representing $\lambda - A$ as $\lambda - A = (\lambda - A - D - F)[I + (\lambda - A - D - F)^{-1}F]$, and by using Theorem 2.2.40, together with Lemma 3.1.2, we obtain $\lambda \in \Phi_{A+D}$ and $i(\lambda - A - D) = 0$, for all $D \in \mathcal{L}(X)$ such that $\|D\| < \varepsilon$. Finally, Theorem 8.3.1 shows that $\lambda \notin \sigma_{e5,\varepsilon}(A)$. Q.E.D.

Corollary 8.3.1. *Let (Ω, Σ, μ) be an arbitrary positive measure space, and let A be a closed, densely defined, and linear operator on $L_p(\Omega, d\mu)$, $\varepsilon > 0$ and let $p \in [1, \infty)$. In the case where $\mathcal{W}_*(L_p(\Omega, d\mu)) = \mathcal{S}(L_p(\Omega, d\mu))$, we have*

$$\sigma_{e5,\varepsilon}(A) = \sigma_{j,\varepsilon}(A).$$

◇

Theorem 8.3.3. *Let X be a Banach space, $\varepsilon > 0$ and $A \in \mathcal{C}(X)$. Then,*

$$\sigma_{e5,\varepsilon}(A) := \bigcap_{F \in \mathcal{F}^b(X)} \sigma_\varepsilon(A + F).$$

◇

Proof. Let $\mathcal{O} := \bigcap_{F \in \mathcal{F}^b(X)} \sigma_\varepsilon(A + F)$. Since $\mathcal{K}(X) \subset \mathcal{F}^b(X)$, we infer that $\mathcal{O} \subset \sigma_{e5,\varepsilon}(A)$. Conversely, let $\lambda \notin \mathcal{O}$. Then, there exists $F \in \mathcal{F}^b(X)$ such that $\lambda \notin \sigma_\varepsilon(A + F)$. Thus, by using Theorem 8.1.1 (*ii*), we notice that $\lambda \in \rho(A + D + F)$, for all $D \in \mathcal{L}(X)$ such that $\|D\| < \varepsilon$. So, $A + D + F - \lambda \in \Phi(X)$ and $i(A + D + F - \lambda) = 0$. The use of Lemma 6.3.1 (i) allows us to conclude that, for all $D \in \mathcal{L}(X)$ such that $\|D\| < \varepsilon$, $A + D - \lambda \in \Phi(X)$ and $i(A + D - \lambda) = 0$. Finally, Theorem 8.3.1 shows that $\lambda \notin \sigma_{e5,\varepsilon}(A)$. Q.E.D.

Remark 8.3.3. (i) From Theorem 8.3.3, it follows that $\sigma_{e5,\varepsilon}(A + F) = \sigma_{e5,\varepsilon}(A)$ for all $F \in \mathcal{F}^b(X)$.

(ii) If X is a Banach space with the DP property, then $\mathcal{W}(X) \subset \mathcal{F}^b(X)$. Thus, the pseudo-Schechter essential spectrum is invariant under weakly compact perturbations on this class of Banach spaces. ◇

Corollary 8.3.2. *Let X be a Banach space and let $\mathfrak{I}(X)$ be any subset of $\mathcal{L}(X)$. If $\mathcal{K}(X) \subset \mathfrak{I}(X) \subset \mathcal{F}^b(X)$, then*

$$\sigma_{e5,\varepsilon}(A) = \bigcap_{J \in \mathfrak{I}(X)} \sigma_\varepsilon(A + J).$$

◇

Remark 8.3.4. The use of Corollary 8.3.2 implies that $\sigma_{e5,\varepsilon}(A + J) = \sigma_{e5,\varepsilon}(A)$, for all $J \in \mathfrak{I}(X)$ such that $\mathcal{K}(X) \subset \mathfrak{I}(X) \subset \mathcal{F}^b(X)$, and for all $J_1, J_2 \in \mathfrak{I}(X)$, we have $J_1 \pm J_2 \in \mathfrak{I}(X)$. ◇

Theorem 8.3.4. *Let $\varepsilon > 0$, and let A and B be two elements of $\mathcal{C}(X)$, such that $0 \notin \sigma_{e5}(A) \bigcup \sigma_{e5}(B)$. Let us assume the existence of two bounded operators A_0 and B_0 in the Banach space X, such that*

$$AA_0 = I - F_1, \tag{8.3.2}$$

$$BB_0 = I - F_2, \tag{8.3.3}$$

where $F_i \in \mathcal{F}^b(X)$, with $i = 1, 2$. If the difference $A_0 - B_0 \in \mathcal{F}^b(X)$, then

$$\sigma_{e5,\varepsilon}(A) = \sigma_{e5,\varepsilon}(B).$$

◇

Proof. Using Eqs. (8.3.2) and (8.3.3), we infer that, for any scalar λ, we have

$$(A + D - \lambda)A_0 - (B + D - \lambda)B_0 = F_2 - F_1 + (D - \lambda)(A_0 - B_0). \qquad (8.3.4)$$

Let $\lambda \notin \sigma_{e5,\varepsilon}(A)$, then $A + D - \lambda$ is a Fredholm operator and $i(A + D - \lambda) = 0$ for all $D \in \mathcal{L}(X)$, such that $\|D\| < \varepsilon$. Since $A + D$ is closed and $\mathcal{D}(A + D) = \mathcal{D}(A)$ endowed with the graph norm, is a Banach space denoted by X_{A+D} and, using Theorem 2.2.39, we obtain $\widehat{A+D} - \lambda \in \Phi^b(X_{A+D}, X)$. Moreover, $F_1 \in \mathcal{F}^b(X)$. Using Eq. (8.3.2), Lemma 3.1.2, and Theorem 2.2.42, we deduce that $A_0 \in \Phi^b(X, X_{A+D})$. Thus,

$$(\widehat{A+D} - \lambda)A_0 \in \Phi^b(X). \qquad (8.3.5)$$

Now, if the difference $A_0 - B_0 \in \mathcal{F}^b(X)$, and applying Eq. (8.3.4), we get

$$(A + D - \lambda)A_0 - (B + D - \lambda)B_0 \in \mathcal{F}^b(X).$$

Also, from Eq. (8.3.5), it follows that $(\widehat{B+D} - \lambda)B_0 \in \Phi^b(X)$, and

$$i[(\widehat{B+D} - \lambda)B_0] = i[(\widehat{A+D} - \lambda)A_0] = 0. \qquad (8.3.6)$$

Since $B \in \mathcal{C}(X)$, using Eq. (8.3.3) and arguing as in the last part, we conclude that $B_0 \in \Phi^b(X, X_{B+D})$. Since $(B + D - \lambda)B_0$ is a Fredholm operator, the use of Theorem 2.2.41 shows that $\widehat{B+D} - \lambda \in \Phi^b(X_{B+D}, X)$. This implies that $B + D - \lambda$ is a Fredholm operator. Moreover, $0 \notin \sigma_{e5}(A) \bigcup \sigma_{e5}(B)$, then $i(A) = i(B) = 0$. Therefore, using Eqs. (8.3.2), (8.3.3) and also Lemma 3.1.2 and Theorem 2.2.40, we deduce that $i(A_0) = i(B_0) = 0$. This, together with Eq. (8.3.6), shows that $i(A+D-\lambda) = i(B+D-\lambda) = 0$. Thus, $\lambda \notin \sigma_{e5,\varepsilon}(B)$. This proves that $\sigma_{e5,\varepsilon}(B) \subset \sigma_{e5,\varepsilon}(A)$. The opposite inclusion can be obtained by symmetry. Q.E.D.

Theorem 8.3.5. *Let X be a Banach space, $\varepsilon > 0$, A and B being two elements of $\mathcal{C}(X)$. Let us assume that, for all bounded operators D such that $\|D\| < \varepsilon$, the operator B is $(A + D)$-compact. Then, $\sigma_{e5,\varepsilon}(A) = \sigma_{e5,\varepsilon}(A + B)$.* ◇

Proof. Let $\lambda \notin \sigma_{e5,\varepsilon}(A)$. Then, for all $D \in \mathcal{L}(X)$ such that $\|D\| < \varepsilon$, we have $A + D - \lambda$ is a Fredholm operator and $i(A + D - \lambda) = 0$. Since B is $(A + D)$-compact, and applying Theorem 2.2.47, we get $\lambda \in \Phi_{A+B+D}$ and $i(A + B + D - \lambda) = 0$. Therefore, $\lambda \notin \sigma_{e5,\varepsilon}(A + B)$. We conclude that $\sigma_{e5,\varepsilon}(A+B) \subset \sigma_{e5,\varepsilon}(A)$. Conversely, let $\lambda \notin \sigma_{e5,\varepsilon}(A+B)$. Then, for all $D \in \mathcal{L}(X)$ such that $\|D\| < \varepsilon$ we deduce that $A + B + D - \lambda$ is a Fredholm operator and $i(A+B+D-\lambda) = 0$. Moreover, B is $(A+D)$-compact, and using Theorem 2.2.2, we infer that B is $(A + B + D)$-compact. Then, by using Theorem 2.2.48, we have $\lambda \in \Phi_{A+D}$ and $i(A + D - \lambda) = 0$. So, $\lambda \notin \sigma_{e5,\varepsilon}(A)$. This proves that $\sigma_{e5,\varepsilon}(A) \subset \sigma_{e5,\varepsilon}(A + B)$. Q.E.D.

Theorem 8.3.6. *Let X be a Banach space, $\varepsilon > 0$ and $A \in \mathcal{C}(X)$. If $J \in \mathcal{F}(X)$, then*

$$\sigma_{e5,\varepsilon}(A) = \sigma_{e5,\varepsilon}(A + J).$$

◇

Proof. Let $\lambda \notin \sigma_{e5,\varepsilon}(A)$. Then, $\lambda - A - D \in \Phi(X)$ and $i(\lambda - A - D) = 0$ for all $||D|| < \varepsilon$. Hence, by using Lemma 6.3.1 (*ii*), we have $\lambda - A - D - J \in \Phi(X)$ and $i(\lambda - A - D - J) = 0$ for $||D|| < \varepsilon$. Therefore, $\lambda \notin \sigma_{e5,\varepsilon}(A + J)$, i.e., $\sigma_{e5,\varepsilon}(A + J) \subset \sigma_{e5,\varepsilon}(A)$. In order to prove the opposite inclusion, it is sufficient to proceed by symmetry: $\sigma_{e5,\varepsilon}(A) = \sigma_{e5,\varepsilon}(A + J - J) \subset \sigma_{e5,\varepsilon}(A + J)$. Q.E.D.

The following theorem gives a relation between the pseudo-Schechter essential spectrum of the sum of two bounded linear operators and the pseudo-Schechter essential spectrum of each of these operators.

Theorem 8.3.7. *Let A and B be two bounded linear operators on a Banach space and $\varepsilon > 0$. If, for all bounded operators D such that $||D|| < \varepsilon$ and $A(B + D) \in \mathcal{F}^b(X)$, then $\sigma_{e5,\varepsilon}(A + B)\backslash\{0\} \subset [\sigma_{e5}(A) \bigcup \sigma_{e5,\varepsilon}(B)]\backslash\{0\}$. If, further, $(B + D)A \in \mathcal{F}^b(X)$ and $\mathbb{C}\backslash\sigma_{e4}(A)$ is connected, then $\sigma_{e5,\varepsilon}(A + B)\backslash\{0\} = [\sigma_{e5}(A) \bigcup \sigma_{e5,\varepsilon}(B)]\backslash\{0\}$.* ◇

Proof. For $\lambda \in \mathbb{C}$, we can write

$$(\lambda - A)(\lambda - B - D) = A(B + D) + \lambda(\lambda - A - B - D), \tag{8.3.7}$$

and

$$(\lambda - B - D)(\lambda - A) = (B + D)A + \lambda(\lambda - A - B - D). \tag{8.3.8}$$

Let $\lambda \notin \left[\sigma_{e5}(A) \bigcup \sigma_{e5,\varepsilon}(B)\right]\backslash\{0\}$. Then, $(\lambda - A) \in \Phi^b(X)$, $i(\lambda - A) = 0$ and, for all $\|D\| < \varepsilon$, $(\lambda - B - D) \in \Phi^b(X)$ and $i(\lambda - B - D) = 0$. It follows, from Theorem 2.2.13 (*iii*), that $(\lambda - A)(\lambda - B - D) \in \Phi^b(X)$ and $i[(\lambda - A)(\lambda - B - D)] = 0$. Since $A(B + D) \in \mathcal{F}^b(X)$, and applying Eq. (8.3.7), we have $(\lambda - A - B - D) \in \Phi^b(X)$ and $i(\lambda - A - B - D) = 0$. Then, $\lambda \notin \sigma_{e5,\varepsilon}(A + B)$. Therefore,

$$\sigma_{e5,\varepsilon}(A + B)\backslash\{0\} \subseteq \left[\sigma_{e5}(A) \bigcup \sigma_{e5,\varepsilon}(B)\right]\backslash\{0\}. \tag{8.3.9}$$

Now, let us prove the inverse inclusion of Eq. (8.3.9). Suppose $\lambda \notin \sigma_{e5,\varepsilon}(A+B)\backslash\{0\}$. Then, for all $D \in \mathcal{L}(X)$ such that $\|D\| < \varepsilon$, we have $(\lambda - A - B - D) \in \Phi^b(X)$ and $i(\lambda - A - B - D) = 0$. Since $A(B + D) \in \mathcal{F}^b(X)$, $(B + D)A \in \mathcal{F}^b(X)$. Then, by using Eqs. (8.3.7) and (8.3.8), we have $(\lambda - A)(\lambda - B - D) \in \Phi^b(X)$, $(\lambda - B - D)(\lambda - A) \in \Phi^b(X)$ and $i(\lambda - A) + i(\lambda - B - D) = 0$. By applying Theorem 2.2.19, it is clear that $(\lambda - A) \in \Phi^b(X)$ and, for all $\|D\| < \varepsilon$, we have $(\lambda - B - D) \in \Phi^b(X)$. Moreover, $\mathbb{C}\backslash\sigma_{e4}(A)$ is connected, then $i(\lambda - A) = 0$. Therefore, $\lambda \notin \sigma_{e5}(A) \bigcup \sigma_{e5,\varepsilon}(B)$. This proves that $\sigma_{e5,\varepsilon}(A + B)\backslash\{0\} = \left[\sigma_{e5}(A) \bigcup \sigma_{e5,\varepsilon}(B)\right]\backslash\{0\}$. Q.E.D.

8.3.2 By Means of Noncompactness Measure

Let us denote $\mathbf{M}_n^{\varepsilon}(X)$ by

$$\mathbf{M}_n^{\varepsilon}(X) = \Big\{M \in \mathcal{L}(X) : \gamma\big([(\lambda - A - M - D)^{-1}M]^n\big) < 1$$
$$\text{for all } D \in \mathcal{L}(X) \text{ such that } \|D\| < \varepsilon \text{ and } \lambda \in \rho(A + M + D)\Big\},$$

where $\gamma(.)$ is the Kuratowski measure of noncompactness in X and $n \in \mathbb{N}^*$. Let $\mathfrak{I}(X)$ be any subset of $\mathcal{L}(X)$ such that $\mathcal{K}(X) \subset \mathfrak{I}(X) \subset \mathbf{M}_n^{\varepsilon}(X)$. If, for all $K,\ K_1 \in \mathfrak{I}(X)$ such that $K \pm K_1 \in \mathfrak{I}(X)$, then we have $\sigma_{e5,\varepsilon}(A) = \sigma_{e5,\varepsilon}(A + K)$. In what follows, we will give a refinement of the definition of the pseudo-Schechter essential spectrum. Let X be a Banach space and $n \in \mathbb{N}^*$. For each $\varepsilon > 0$ and $A \in \mathcal{C}(X)$, we denote by:

$$\sigma_{\varepsilon}^n(A) = \bigcap_{K \in \mathbf{M}_n^{\varepsilon}(X)} \sigma_{\varepsilon}(A + K).$$

Remark 8.3.5. Let $M \in \mathcal{K}(X)$, then $(\lambda - A - M - D)^{-1}M \in \mathcal{K}(X)$. So, $\gamma\big([(\lambda - A - M - D)^{-1}M]^n\big) = 0$ and therefore, $\mathcal{K}(X) \subset \mathbf{M}_n^{\varepsilon}(X)$. $\diamond$

Theorem 8.3.8. *Let X be a Banach space, $\varepsilon > 0$ and $A \in \mathcal{C}(X)$. Then,*

$$\sigma_{e5,\varepsilon}(A) = \bigcap_{M \in \mathbf{M}_n^{\varepsilon}(X)} \sigma_{\varepsilon}(A + M). \qquad \diamond$$

Proof. We first claim that $\sigma_{e5,\varepsilon}(A) \subset \sigma_{\varepsilon}^n(A)$. Indeed, if $\lambda \notin \sigma_{\varepsilon}^n(A)$, then there exists $M \in \mathbf{M}_n^{\varepsilon}(X)$ such that, for $D \in \mathcal{L}(X)$ verifying $\|D\| < \varepsilon$ and $\lambda \in \rho(A + M + D)$, we have $\gamma\big([(\lambda - A - M - D)^{-1}M]^n\big) < 1$. So, $\lim_{k \to +\infty} \gamma\big([\lambda - A - M - D)^{-1}M]^n\big)^k = 0$. Then, there exists $k_0 \in \mathbb{N}^*$, such that $\gamma\big([\lambda - A - M - D)^{-1}M]^n\big)^{k_0} < \frac{1}{2}$. By using Lemma 2.10.2 *(iii)*, we deduce that$\gamma\big([\lambda - A - M - D)^{-1}M]^{nk_0}\big) < \frac{1}{2}$. Applying Theorem 5.2.1 *(ii)*, $P(z) = z^{nk_0}$ and $Q(z) = 1 - z$, we conclude that $Q(A) = \big[I + (\lambda - A - M - D)^{-1}M\big] \in \Phi^b(X)$. Now, let $t \in [0, 1]$. We have $\gamma\big(t(\lambda - A - M - D)^{-1}M)^{nk_0}\big) < \frac{1}{2}$, which implies that $I + t(\lambda - A - M - D)^{-1}M$ is a Fredholm operator on X. It follows, from Proposition 2.2.5 and the compactness of $[0, 1]$, that $i\big(Q(A)\big) = i\big(I + t(\lambda - A - M - D)^{-1}M\big) = 0$. We can write $(\lambda - A - D) = (\lambda - A - D - M)\big(I + (\lambda - A - D - M)^{-1}M\big)$. Therefore, for all $D \in \mathcal{L}(X)$ such that $\|D\| < \varepsilon$, we have $(\lambda - A - D) \in \Phi(X)$ and $i(\lambda - A - D) = 0$. Finally, the use of Theorem 8.3.1 shows that $\lambda \notin \sigma_{e5,\varepsilon}(A)$. Conversely, since $\mathcal{K}(X) \subset \mathbf{M}_n^{\varepsilon}(X)$ (see Remark 8.3.5), we infer that $\sigma_{\varepsilon}^n(A) \subset \sigma_{e5,\varepsilon}(A)$. Q.E.D.

Theorem 8.3.9. *Let X be a Banach space, $\varepsilon > 0$, $A \in \mathcal{C}(X)$ and let $\mathfrak{I}(X)$ be any subset of $\mathcal{L}(X)$. If $\mathcal{K}(X) \subset \mathfrak{I}(X) \subset \mathbf{M}_n^\varepsilon(X)$, then*

(i) $\sigma_{e5,\varepsilon}(A) = \bigcap\limits_{M \in \mathfrak{I}(X)} \sigma_\varepsilon(A+M).$

(ii) Moreover, if for all $K,\ K_1 \in \mathfrak{I}(X)$, we have $K \pm K_1 \in \mathfrak{I}(X)$. Then, for each $K \in \mathfrak{I}(X)$, we have $\sigma_{e5,\varepsilon}(A) = \sigma_{e5,\varepsilon}(A+K)$. ◇

Proof. (i) Let $\mathcal{O} := \bigcap_{M \in \mathbf{M}_n^\varepsilon(A)} \sigma_\varepsilon(A+M)$. Since $\mathfrak{I}(X) \subset \mathbf{M}_n^\varepsilon(X)$, then we have $\mathcal{O} \subset \bigcap_{M \in \mathfrak{I}(X)} \sigma_\varepsilon(A+M)$. Using Theorem 8.3.8, we get $\sigma_{e5,\varepsilon}(A) \subset \bigcap_{M \in \mathfrak{I}(X)} \sigma_\varepsilon(A+M)$. Moreover, since $\mathcal{K}(X) \subset \mathfrak{I}(X)$, we infer that $\bigcap_{M \in \mathfrak{I}(X)} \sigma_\varepsilon(A+M) \subset \sigma_{e5,\varepsilon}(A)$.

(ii) Let $\sigma_\varepsilon^{\mathfrak{I}}(A) := \bigcap_{M \in \mathfrak{I}(X)} \sigma_\varepsilon(A+M)$. From (i), we deduce that $\sigma_\varepsilon^{\mathfrak{I}}(A) = \sigma_{e5,\varepsilon}(A)$. Furthermore, for $M \in \mathfrak{I}(X)$, we have $K + \mathfrak{I}(X) = \mathfrak{I}(X)$. Then, $\sigma_\varepsilon^{\mathfrak{I}}(A+K) = \sigma_\varepsilon^{\mathfrak{I}}(A)$. Therefore, for all $K \in \mathfrak{I}(X)$, we get $\sigma_{e5,\varepsilon}(A+K) = \sigma_\varepsilon^{\mathfrak{I}}(A+K) = \sigma_\varepsilon^{\mathfrak{I}}(A) = \sigma_{e5,\varepsilon}(A)$. Q.E.D.

Remark 8.3.6. Let $\varepsilon > 0$ and $A \in \mathcal{C}(X)$. Then, $\bigcap_{M \in \mathcal{P}_\gamma(X)} \sigma_\varepsilon(A+M) \subset \sigma_{e5,\varepsilon}(A)$, where $\mathcal{P}_\gamma(X)$ is the set introduced in (6.4.2). ◇

We have the following:

Theorem 8.3.10. *Let $\varepsilon > 0$, and let A and B be two operators in $\mathcal{L}(X)$. Assume that, for all $D \in \mathcal{L}(X)$ such that $\|D\| \leq \varepsilon$ and, for all $\lambda \in \Phi_{A+D}$, there exists a Fredholm inverse A_λ^ε of $A+D-\lambda$ such that $A_\lambda^\varepsilon B \in \mathcal{P}_\gamma(X)$. Then, $\sigma_{e5,\varepsilon}(A+B) \subset \sigma_{e5,\varepsilon}(A)$.* ◇

Proof. Suppose that $\lambda \notin \sigma_{e5,\varepsilon}(A)$. Then, by using Theorem 8.3.1, we deduce that, for all $D \in \mathcal{L}(X)$ such that $\|D\| \leq \varepsilon$, we have $\lambda \in \Phi_{A+D}$ and $i(A+D-\lambda) = 0$. Since A_λ^ε is a Fredholm inverse of $A+D-\lambda$, then there exists $K \in \mathcal{K}(X)$ such that

$$(\lambda - A - D)A_\lambda^\varepsilon = I - K. \tag{8.3.10}$$

From Eq. (8.3.10), it follows that the operator $A+D+B-\lambda$ can be written in the form

$$\lambda - A - D - B = \lambda - A - D - \Big((\lambda - A - D)A_\lambda^\varepsilon + K\Big)B = (\lambda - A - D)(I - A_\lambda^\varepsilon B) - KB. \tag{8.3.11}$$

Using the fact that $A_\lambda^\varepsilon B \in \mathcal{P}_\gamma(X)$, there exists $n \in \mathbb{N}$ such that $\gamma\big((A_\lambda^\varepsilon B)^n\big) < 1$. Then, $\lim\limits_{k \to \infty} \gamma\big((A_\lambda^\varepsilon B)^n\big)^k = 0$. Hence, there exists $k_0 \in \mathbb{N}^*$ such that $\gamma\big((A_\lambda^\varepsilon B)^n\big)^{k_0} < \frac{1}{2}$. By using Lemma 2.10.2 (iii), we deduce that $\gamma\big((A_\lambda^\varepsilon B)^{nk_0}\big) \leq \frac{1}{2}$. Now, by applying Theorem 5.2.1 (ii) with $P(z) = z^{nk_0}$ and $Q(z) = 1 - z$, we conclude that

$$I - A_\lambda^\varepsilon B \in \Phi^b(X). \tag{8.3.12}$$

Arguing as in the proof of Theorem 8.3.8 that $i(I - A_\lambda^\varepsilon B) = 0$, and by using Theorem 2.2.7, we have

$$(\lambda - A - D)(I - A_\lambda^\varepsilon B) \in \Phi^b(X) \text{ and } i\big((\lambda - A - D)(I - A_\lambda^\varepsilon B)\big) = 0. \quad (8.3.13)$$

By applying Eqs. (8.3.11), (8.3.12) and (8.3.13), we get $(\lambda - A - D - B) \in \Phi^b(X)$ and $i(\lambda - A - D - B) = 0$ for all $D \in \mathcal{L}(X)$ such that $\|D\| < \varepsilon$. This proves that $\lambda \notin \sigma_{e5,\varepsilon}(A + B)$. We find that $\sigma_{e5,\varepsilon}(A + B) \subset \sigma_{e5,\varepsilon}(A)$. Q.E.D.

Theorem 8.3.11. *Let $\varepsilon > 0$ and $A \in \mathcal{C}(X)$. Assume that B is $(A + D)$-bounded for all $D \in \mathcal{L}(X)$ such that $\|D\| < \varepsilon$ and, for all $\lambda \in \Phi_{A+D}$, there exists a Fredholm inverse A_λ^ε of $A + D - \lambda$ such that $A_\lambda^\varepsilon B \in \mathcal{P}_\gamma(X)$. Then, $\sigma_{e5,\varepsilon}(A + B) \subset \sigma_{e5,\varepsilon}(A)$.* ◇

Proof. Clearly, $\widehat{A + D} \in \mathcal{L}(X_{A+D}, X)$ and $\hat{B} \in \mathcal{L}(X_{A+D}, X)$. Moreover, it is not difficult to see that

$$\begin{cases} \alpha(\widehat{A+D}) = \alpha(A+D),\ \beta(\widehat{A+D}) = \beta(\widehat{A+D}) \text{ and } R(\widehat{A+D}) = R(A+D), \\ \alpha(\widehat{A + D} + \hat{B}) = \alpha(A + D + B), \text{ and} \\ \beta(\widehat{A + D} + \hat{B}) = \beta(A + D + B) \text{ and } R(\widehat{A + D} + \hat{B}) = R(A + D + B). \end{cases}$$

Hence, if $A + D - \lambda$ belongs to $\Phi(X)$, and using Theorem 2.2.39, we obtain $\widehat{A + D} - \lambda \in \Phi^b(X_{A+D}, X)$. Finally, the use of Theorem 8.3.10 completes the proof. Q.E.D.

Chapter 9
S-Essential Spectra

In this chapter, we give a characterization of **S**-essential spectra of linear operator A on a Banach space X.

9.1 Definitions and Preliminary Results

Let X be a Banach space. Let S be a bounded operator on X, such that $S \neq 0$. For $A \in \mathcal{C}(X)$, we define the S-resolvent set of A by $\rho_S(A) := \{\lambda \in \mathbb{C}$ such that $\lambda S - A$ has a bounded inverse$\}$, and the S-spectrum of A by $\sigma_S(A) = \mathbb{C} \backslash \rho_S(A)$. Note that $\sigma_S(A)$ is not necessarily bounded. In fact, it suffices to see the following examples, where $\sigma_S(A)$ can be discrete or the whole complex plane:

(i) Let $A = \begin{pmatrix} 2 & 2 \\ 0 & 3 \end{pmatrix}$ and $S = \begin{pmatrix} 1 & 0 \\ 0 & 0 \end{pmatrix}$. In this case, $\sigma_S(A) = \{2\}$.

(ii) Let $A = \begin{pmatrix} 1 & 2 \\ 0 & 3 \end{pmatrix}$ and $S = \begin{pmatrix} 0 & 1 \\ 0 & 0 \end{pmatrix}$, then $\sigma_S(A) = \emptyset$ and $\rho_S(A) = \mathbb{C}$.

(iii) Let $A = \begin{pmatrix} 1 & 2 \\ 0 & 0 \end{pmatrix}$ and $S = \begin{pmatrix} 1 & 0 \\ 0 & 0 \end{pmatrix}$, then $\sigma_S(A) = \mathbb{C}$ and $\rho_S(A) = \emptyset$.

In this chapter, we are concerned with the following **S**-essential spectra:

$$\sigma_{e1,S}(A) := \left\{\lambda \in \mathbb{C} \text{ such that } \lambda S - A \notin \Phi_+(X)\right\} := \mathbb{C} \backslash \Phi_{+A,S},$$

$$\sigma_{e2,S}(A) := \left\{\lambda \in \mathbb{C} \text{ such that } \lambda S - A \notin \Phi_-(X)\right\} := \mathbb{C} \backslash \Phi_{-A,S},$$

$$\sigma_{e3,S}(A) := \left\{\lambda \in \mathbb{C} \text{ such that } \lambda S - A \notin \Phi_\pm(X)\right\} := \mathbb{C} \backslash \Phi_{\pm A,S},$$

$$\sigma_{e4,S}(A) := \left\{\lambda \in \mathbb{C} \text{ such that } \lambda S - A \notin \Phi(X)\right\} := \mathbb{C} \backslash \Phi_{A,S},$$

$$\sigma_{e5,S}(A) := \mathbb{C} \backslash \rho_{5,S}(A),$$

A. Jeribi, *Spectral Theory and Applications of Linear Operators and Block Operator Matrices*, DOI 10.1007/978-3-319-17566-9_9

$$\sigma_{e6,S}(A) := \mathbb{C}\backslash\rho_{6,S}(A),$$

$$\sigma_{e7,S}(A) := \bigcap_{K\in\mathcal{K}(X)} \sigma_{ap,S}(A+K),$$

$$\sigma_{e8,S}(A) := \bigcap_{K\in\mathcal{K}(X)} \sigma_{\delta,S}(A+K),$$

where $\rho_{5,S}(A) := \{\lambda \in \Phi_{A,S}$ such that $i(\lambda S - A) = 0\}$, $\rho_{6,S}(A) := \{\lambda \in \rho_{5,S}(A)$ such that all scalars near λ are in $\rho_S(A)\}$,

$$\sigma_{ap,S}(A) := \left\{\lambda \in \mathbb{C} \text{ such that } \inf_{\|x\|=1,\ x\in\mathcal{D}(A)} \|(\lambda S - A)x\| = 0\right\}, \text{ and}$$

$$\sigma_{\delta,S}(A) := \big\{\lambda \in \mathbb{C} \text{ such that } \lambda S - A \text{ is not surjective}\big\}.$$

They can be ordered as $\sigma_{e3,S}(A) := \sigma_{e1,S}(A) \bigcap \sigma_{e2,S}(A) \subset \sigma_{e4,S}(A) \subset \sigma_{e5,S}(A) \subset \sigma_{e6,S}(A)$.

Remark 9.1.1. (*i*) $\sigma_{e5,S}(A) = \sigma_{e4,S}(A) \bigcup \{\lambda \in \mathbb{C}$ such that $i(\lambda S - A) \neq 0\}$.

(*ii*) Note that if $S = I$, we find the usual definition of the essential spectra of a closed densely defined linear operator introduced in Chap. 7.

(*iii*) Note that, even if S is invertible, we don't have $\sigma_{ei}(A) = \sigma_{ei,S}(A)$, $i = 1, \dots, 6$. For example, if X is a finite dimension space and S is an invertible operator, such that $\sigma(S)\backslash\{1\}$ is not empty, then $\sigma_S(S) = \{1\}$, which implies that $\sigma_S(S) \neq \sigma(S)$. ◇

One of the main questions in the study of the S-essential spectra of closed densely defined linear operators consists in showing when the different notions of the essential spectrum coincide, and in studying their invariance by some class of perturbations.

Lemma 9.1.1. *Let $A \in \mathcal{C}(X)$ and $S \in \mathcal{L}(X)$, such that $\rho_S(A)$ is not empty.*

(*i*) *If $\mathbb{C}\backslash\sigma_{e4,S}(A)$ is connected, then $\sigma_{e4,S}(A) = \sigma_{e5,S}(A)$.*

(*ii*) *If $\mathbb{C}\backslash\sigma_{e5,S}(A)$ is connected, then $\sigma_{e5,S}(A) = \sigma_{e6,S}(A)$.*

(*iii*) *If $\Phi_{A,S}$ is connected, then $\sigma_{e1,S}(A) = \sigma_{e7,S}(A)$ and $\sigma_{e2,S}(A) = \sigma_{e8,S}(A)$.* ◇

Proof. (*i*) Since the inclusion $\sigma_{e4,S}(A) \subset \sigma_{e5,S}(A)$ is known, it is sufficient to show that $\sigma_{e5,S}(A) \subset \sigma_{e4,S}(A)$ which is equivalent to $\mathbb{C}\backslash\sigma_{e4,S}(A) \subset \mathbb{C}\backslash\sigma_{e5,S}(A)$. Let $\lambda_0 \in \mathbb{C}\backslash\sigma_{e4,S}(A)$. Since $\rho_S(A) \neq \emptyset$, then there exists $\lambda_1 \in \mathbb{C}$ such that $\lambda_1 \in \rho_S(A)$ and consequently $\lambda_1 S - A \in \Phi(X)$ and $i(\lambda_1 S - A) = 0$. Hence, $\lambda_1 \in \mathbb{C}\backslash\sigma_{e5,S}(A)$ which is a subset of $\mathbb{C}\backslash\sigma_{e4,S}(A)$. Since $\mathbb{C}\backslash\sigma_{e4,S}(A)$ is connected, and from Proposition 2.2.5 (*ii*), it follows that $i(\lambda S - A) = 0$ for all $\lambda \in \mathbb{C}\backslash\sigma_{e4,S}(A)$. In this way, we see that $\lambda_0 \in \mathbb{C}\backslash\sigma_{e5,S}(A)$.

(*ii*) Since the inclusion $\sigma_{e5,S}(A) \subseteq \sigma_{e6,S}(A)$ is known, it is sufficient to show that $\sigma_{e6,S}(A) \subseteq \sigma_{e5,S}(A)$. We have the set $\mathbb{C}\backslash\sigma_{e5,S}(A) \neq \emptyset$, since it contains points of $\rho_S(A)$. Since $\alpha(\lambda S - A)$ and $\beta(\lambda S - A)$ are constant on any component of $\Phi_{S,A}$ except possibly on a discrete set of points where they have large values (see Proposition 2.2.5). Then, $\mathbb{C}\backslash\sigma_{e5,S}(A) \subseteq \mathbb{C}\backslash\sigma_{e6,S}(A)$ which is equivalent to $\sigma_{e6,S}(A) \subseteq \sigma_{e5,S}(A)$ and so, we have the equality.

(*iii*) It is easy to check that $\sigma_{e1,S}(A) \subset \sigma_{e7,S}(A)$. For the second inclusion, we take $\mu \in \mathbb{C}\backslash\sigma_{e1,S}(A)$. Then, $\mu \in \Phi_{+A,S} = \Phi_{A,S} \bigcup(\Phi_{+A,S}\backslash\Phi_{A,S})$. Hence, we will discuss the following two cases:

- If $\mu \in \Phi_{A,S}$, then $i(A - \mu S) = 0$. Indeed, let $\mu_0 \in \rho_S(A)$, then $\mu_0 \in \Phi_{A,S}$ and $i(A - \mu_0 S) = 0$. It follows, from Proposition 2.2.5 (*ii*), that $i(A - \mu S)$ is constant on any component of $\Phi_{A,S}$. Therefore, $\rho_S(A) \subseteq \Phi_{A,S}$ and, then $i(A - \mu S) = 0$ for all $\mu \in \Phi_{A,S}$. This shows that $\mu \in \mathbb{C}\backslash\sigma_{e7,S}(A)$.
- If $\mu \in (\Phi_{+A,S}\backslash\Phi_{A,S})$, then $\alpha(A - \mu S) < \infty$ and $\beta(A - \mu S) = +\infty$. So, $i(A - \mu S) = -\infty < 0$. Thus, from the above reasoning, we can get $\sigma_{e7,S}(A) \subset \sigma_{e1,S}(A)$. Similarly, we obtain the second equality. Q.E.D.

Theorem 9.1.1. *Let A be a closed, densely defined linear operator on X and $J \in \mathcal{L}(X)$. Then,*

(*i*) *if $J \in \mathcal{F}^b(X)$, then $\sigma_{ei,S}(A) = \sigma_{ei,S}(A + J)$, with $i = 4, 5$. Moreover, if $\mathbb{C}\backslash\sigma_{e5,S}(A)$ is connected and neither $\rho_S(A)$ nor $\rho_S(A + J)$ is empty, then $\sigma_{e6,S}(A) = \sigma_{e6,S}(A + J)$.*
(*ii*) *If $J \in \mathcal{F}^b_+(X)$, then $\sigma_{e1,S}(A) = \sigma_{e1,S}(A + J)$.*
(*iii*) *If $J \in \mathcal{F}^b_-(X)$, then $\sigma_{e2,S}(A) = \sigma_{e2,S}(A + J)$.*
(*iv*) *If $J \in \mathcal{F}^b_+(X) \bigcap \mathcal{F}^b_-(X)$, then $\sigma_{e3,S}(A) = \sigma_{e3,S}(A + J)$.* ◇

Proof. (*i*) Let $\lambda \in \mathbb{C}$. By using Lemma 6.3.1 (*i*), we get the equivalence $\lambda S - A \in \Phi(X)$ if, and only if, $\lambda S - A - J \in \Phi(X)$. Moreover, $i(\lambda S - A - J) = i(\lambda S - A)$. Hence, we obtain $\sigma_{ei,S}(A) = \sigma_{ei,S}(A + J)$, with $i = 4, 5$. By using Theorem 7.5.3, the proofs of the items (*ii*), (*iii*), (*iv*) are immediate. So, they are omitted. Q.E.D.

9.2 Characterization of *S*-Essential Spectra

In this section, we give the characterization of different *S*-essential spectra of bounded linear operators on a Banach space X. Let S be a bounded operator on X, such that $S \neq 0$ and let $A \in \mathcal{L}(X)$. A complex number λ is in $\mathcal{B}_{+A,S}$, $\mathcal{B}_{-A,S}$ or $\mathcal{B}_{A,S}$ if $\lambda S - A$ is in $\mathcal{B}^b_+(X)$, $\mathcal{B}^b_-(X)$ or $\mathcal{B}^b(X)$ respectively. An operator $A \in \mathcal{L}(X)$ is called left (resp. right) essentially invertible, if there exist $S \in \mathcal{L}(X)$ and $K \in \mathcal{K}(X)$ such that $SA = I + K$ (resp. $AS = I + K$). We consider the following S-essential spectra

$$\sigma_{eB,S}(A) := \{\lambda \in \mathbb{C} \text{ such that } \lambda S - A \notin \mathcal{B}^b(X)\} := \mathbb{C}\backslash\mathcal{B}_{A,S},$$

$$\sigma_{eB_+,S}(A) := \{\lambda \in \mathbb{C} \text{ such that } \lambda S - A \notin \mathcal{B}^b_+(X)\} := \mathbb{C}\backslash\mathcal{B}_{+A,S},$$

$$\sigma_{eB_-,S}(A) := \{\lambda \in \mathbb{C} \text{ such that } \lambda S - A \notin \mathcal{B}^b_-(X)\} := \mathbb{C}\backslash\mathcal{B}_{-A,S},$$

$$\sigma_{e9,S}(A) := \{\lambda \in \mathbb{C} \text{ such that } \lambda S - A \text{ is not left essentially invertible}\},$$

$$\sigma_{e10,S}(A) := \{\lambda \in \mathbb{C} \text{ such that } \lambda S - A \text{ is not right essentially invertible}\},$$

$$\sigma_{e11,S}(A) := \bigcap_{K \in \mathcal{K}(X)} \sigma_{l,S}(A+K),$$

$$\sigma_{e12,S}(A) := \bigcap_{K \in \mathcal{K}(X)} \sigma_{r,S}(A+K),$$

where $\sigma_{l,S}(A) := \{\lambda \in \mathbb{C}$ such that $\lambda S - A$ is not left invertible$\}$, and $\sigma_{r,S}(A) := \{\lambda \in \mathbb{C}$ such that $\lambda S - A$ is not right invertible$\}$. Our first result is the following theorem.

Theorem 9.2.1. *Let S and A be two bounded linear operators on a Banach space X. Then,*

(*i*) $\sigma_{e5,S}(A) = \bigcap_{K \in \mathcal{K}(X)} \sigma_S(A+K)$.
(*ii*) $\sigma_{e7,S}(A) = \sigma_{e1,S}(A) \bigcup \{\lambda \in \mathbb{C}$ *such that* $i(\lambda S - A) > 0\}$.
(*iii*) $\sigma_{e8,S}(A) = \sigma_{e2,S}(A) \bigcup \{\lambda \in \mathbb{C}$ *such that* $i(\lambda S - A) < 0\}$.
(*iv*) $\sigma_{e11,S}(A) = \sigma_{e9,S}(A) \bigcup \{\lambda \in \mathbb{C}$ *such that* $i(\lambda S - A) > 0\}$.
(*v*) $\sigma_{e12,S}(A) = \sigma_{e10,S}(A) \bigcup \{\lambda \in \mathbb{C}$ *such that* $i(\lambda S - A) < 0\}$. ◇

Proof. (*i*) Let $\lambda \notin \sigma_{e5,S}(A)$. By Remark 9.1.1 (*i*), $\lambda \notin \sigma_{e4,S}(A) \bigcup \{\lambda \in \mathbb{C}$ such that $i(\lambda S - A) \neq 0\}$. Then, $(\lambda S - A) \in \Phi^b(X)$ and $i(\lambda S - A) = 0$. By applying Theorem 2.2.17, there exists $K \in \mathcal{K}(X)$ such that $\lambda S - A - K$ is invertible. Then, $\lambda \in \rho_S(A+K)$. This shows that $\lambda \notin \bigcap_{K \in \mathcal{K}(X)} \sigma_S(A+K)$. Then, we have

$$\bigcap_{K \in \mathcal{K}(X)} \sigma_S(A+K) \subset \sigma_{e5,S}(A). \tag{9.2.1}$$

In order to prove the inverse inclusion of Eq. (9.2.1), suppose that $\lambda \notin \bigcap_{K \in \mathcal{K}(X)} \sigma_S(A+K)$. Then, there exists $K \in \mathcal{K}(X)$ such that $\lambda \in \rho_S(A+K)$. Hence, $\lambda S - A - K \in \Phi^b(X)$ and $i(\lambda S - A - K) = 0$. Now, the operator $\lambda S - A$ can be written in the following form $\lambda S - A = \lambda S - A - K + K$. Since $K \in \mathcal{K}(X)$, and using Lemma 6.3.1 (*i*), we get $\lambda S - A \in \Phi^b(X)$ and $i(\lambda S - A) = i(\lambda S - A - K) = 0$. We conclude that $\lambda \notin \sigma_{e4,S}(A) \bigcup \{\lambda \in \mathbb{C}$ such that $i(\lambda S - A) \neq 0\}$. Hence,

$\sigma_{e5,S}(A) \subset \bigcap_{K\in\mathcal{K}(X)} \sigma_S(A+K)$. The proof of the assertions (*ii*), (*iii*), (*iv*) and (*v*) may be checked in the same way as in the proof of (*i*). Q.E.D.

Corollary 9.2.1. *Let S and A be two bounded linear operators on a Banach space X. Then,*

(*i*) $\lambda \notin \sigma_{e7,S}(A)$ *if, and only if,* $\lambda S - A \in \Phi_+^b(X)$ *and* $i(A - \lambda S) \leq 0$.
(*ii*) $\lambda \notin \sigma_{e8,S}(A)$ *if, and only if,* $\lambda S - A \in \Phi_-^b(X)$ *and* $i(A - \lambda S) \geq 0$. ◇

Proof. The proof of the corollary is immediately deduced from Theorem 9.2.1. Q.E.D.

Proposition 9.2.1. *Let S and $A \in \mathcal{L}(X)$ such that $S \neq A$. Then,*

(*i*) *If* $\alpha(S) < \infty$, *then* $\sigma_{e4}(A) \subset \sigma_{e4,S}(SA)$.
(*ii*) *If* $\beta(S) < \infty$, *then* $\sigma_{e4}(A) \subset \sigma_{e4,S}(AS)$.
(*iii*) *If* $S \in \Phi^b(X)$, *then* $\sigma_{e4}(A) = \sigma_{e4,S}(AS) = \sigma_{e4,S}(SA)$. ◇

Proof. (*i*) Let $\lambda \notin \sigma_{e4,S}(SA)$. Then, $(\lambda S - SA) \in \Phi^b(X)$ which implies that $S(\lambda - A) \in \Phi^b(X)$. Since $\alpha(S) < \infty$, then by using Theorem 2.2.11, we infer that $(\lambda - A) \in \Phi^b(X)$ implies that $\lambda \notin \sigma_{e4}(A)$. Therefore,

$$\sigma_{e4}(A) \subset \sigma_{e4,S}(SA). \tag{9.2.2}$$

(*ii*) Let $\lambda \notin \sigma_{e4,S}(AS)$. Then, $(\lambda S - AS) \in \Phi^b(X)$, which implies that $(\lambda - A)S \in \Phi^b(X)$. Since $\beta(S) < \infty$, then by using Theorem 2.2.12, we conclude that $(\lambda - A) \in \Phi^b(X)$. Hence, $\lambda \notin \sigma_{e4}(A)$. Consequently,

$$\sigma_{e4}(A) \subset \sigma_{e4,S}(AS). \tag{9.2.3}$$

(*iii*) Let $\lambda \notin \sigma_{e4}(A)$. Then, $(\lambda - A) \in \Phi^b(X)$ since $S \in \Phi^b(X)$. Then, by using Theorem 2.2.13, we conclude that $(\lambda - A)S \in \Phi^b(X)$ and $S(\lambda - A) \in \Phi^b(X)$. Then, $\lambda \notin \sigma_{e4,S}(AS)$ and $\lambda \notin \sigma_{e4,S}(SA)$. Therefore $\sigma_{e4,S}(SA) \subset \sigma_{e4}(A)$ and $\sigma_{e4,S}(AS) \subset \sigma_{e4}(A)$. Hence, by using Eqs. (9.2.2) and (9.2.3), we conclude that $\sigma_{e4}(A) = \sigma_{e4,S}(AS) = \sigma_{e4,S}(SA)$. Q.E.D.

9.3 The S-Browder's Essential Spectrum

In this section, we investigate the S-Browder's essential spectrum of a bounded linear operator on a Banach space X. We begin with the following theorem.

Theorem 9.3.1. *Let S and A be two bounded linear operators on a Banach space X. Then,*

$$\sigma_{eB,S}(A) = \sigma_{e4,S}(A) \bigcup \text{acc}\, \sigma_S(A),$$

$$\sigma_{eB_+,S}(A) = \sigma_{e1,S}(A) \bigcup \text{acc } \sigma_S(A),$$

$$\sigma_{eB_-,S}(A) = \sigma_{e2,S}(A) \bigcup \text{acc } \sigma_S(A),$$

where acc $\sigma_S(A)$ *stands for the accumulation* **S**-*spectrum of* A. ◇

Proof. Let $\lambda \notin \sigma_{eB,S}(A)$. Then, $(\lambda S - A) \in \mathcal{B}^b(X)$, and the use of Theorem 2.2.23 proves the existence of a decomposition $X = X_1 \oplus X_2$ such that $\dim X_1 < +\infty$, $\lambda S - A|_{X_1}$ is nilpotent and $\lambda S - A|_{X_2}$ is invertible. So, for all $\mu \neq \lambda$ closed enough to λ, $\mu S - A$ is invertible, then $\mu \in \rho_S(A)$. Thus, λ is not an accumulation point of $\sigma_S(A)$. Moreover, $\lambda S - A|_{X_1}$ is a finite-dimensional nilpotent. So, we get $\lambda \notin \sigma_{e4,S}(A)$ and, then $\lambda \notin \sigma_{e4,S}(A) \bigcup \text{acc } \sigma_S(A)$. Consequently,

$$\sigma_{e4,S}(A) \bigcup \text{acc } \sigma_S(A) \subset \sigma_{eB,S}(A). \tag{9.3.1}$$

In order to prove the inverse inclusion of Eq. (9.3.1), let us suppose $\lambda \notin \sigma_{e4,S}(A) \bigcup \text{acc } \sigma_S(A)$. Then, $(\lambda S - A) \in \Phi^b(X)$. Let $X = X_1 \oplus X_2$ be the Kato decomposition of $\lambda S - A$, so $(\lambda S - A)|_{X_1}$ is a finite-dimensional nilpotent and $(\lambda S - A)|_{X_2}$ is a Kato operator. By hypothesis, $\mu S - A$ is invertible for all μ sufficiently close to λ, $\mu \neq \lambda$ and so $(\mu S - A)|_{X_2}$ is invertible. Then, $(\lambda S - A)|_{X_2}$ is a Kato operator, $(\lambda S - A)|_{X_2}$ is also invertible. Thus, $(\lambda S - A) \in \mathcal{B}^b(X)$, consequently $\lambda \notin \sigma_{eB,S}(A)$. Therefore, $\sigma_{eB,S}(A) \subset \sigma_{e4,S}(A) \bigcup \text{acc } \sigma_S(A)$. The statements for the upper semi-Browder's and lower semi-Browder's spectrum can be proved similarly. Q.E.D.

Proposition 9.3.1. *Let* A, S *and* $B \in \mathcal{L}(X)$ *such that* $r_e(B) := \{|z|$ *such that* $z \notin \sigma_{e6}(B)\} = 0$ *and* B *commute with* S *and* A. *Then,*

(*i*) $\sigma_{eB,S}(A) = \sigma_{eB,S}(A + B)$.
(*ii*) $\sigma_{eB_+,S}(A) = \sigma_{eB_+,S}(A + B)$.
(*iii*) $\sigma_{eB_-,S}(A) = \sigma_{eB_-,S}(A + B)$. ◇

Proof. (*i*) Let $\lambda \notin \sigma_{eB,S}(A)$. Then, $(\lambda S - A) \in \mathcal{B}^b(X)$. Since, $AB = BA$ and $SB = BS$ then $B(\lambda S - A) = (\lambda S - A)B$, and the use of Theorem 2.2.24 gives $\lambda S - A - B \in \mathcal{B}^b(X)$. Hence, $\lambda \notin \sigma_{eB,S}(A + B)$. Therefore,

$$\sigma_{eB,S}(A + B) \subseteq \sigma_{eB,S}(A). \tag{9.3.2}$$

The opposite inclusion of Eq. (9.3.2) follows by symmetry. It is sufficient to replace A and B by $A + B$ and $-B$.

(*ii*) Let $\lambda \notin \sigma_{eB_+,S}(A)$, then $A - \lambda S \in \mathcal{B}^b_+(X)$. Since, $AB = BA$ and $SB = BS$, then $B(\lambda S - A) = (\lambda S - A)B$. The use of Theorem 2.2.24 implies that $\lambda S - A - B \in \mathcal{B}^b_+(X)$, hence $\lambda \notin \sigma_{eB_+,S}(A + B)$, which implies that

$$\sigma_{eB_+,S}(A + B) \subseteq \sigma_{eB_+,S}(A). \tag{9.3.3}$$

The opposite inclusion of Eq. (9.3.3) follows by symmetry. The statement *(iii)* can be checked in the same way from the assertion *(ii)*. Q.E.D.

Remark 9.3.1. If $K \in \mathcal{K}(X)$ or K is quasi-nilpotent and K commutes with S and A, then we have $\sigma_{eB,S}(A) = \sigma_{eB,S}(A+K)$, $\sigma_{eB_-,S}(A) = \sigma_{eB_-,S}(A+K)$, and $\sigma_{eB_+,S}(A) = \sigma_{eB_+,S}(A+K)$. ◇

Theorem 9.3.2. *Let* $A,\ S \in \mathcal{L}(X)$ *and* $\tilde{\mathcal{K}}(X) = \{K \in \mathcal{K}(X)$ *such that* $KA = AK$ *and* $KS = SK\}$. *Then,*

$$\sigma_{eB_-,S}(A) := \bigcap_{K\in\tilde{\mathcal{K}}(X)} \sigma_{\delta,S}(A+K),$$

$$\sigma_{eB_+,S}(A) := \bigcap_{K\in\tilde{\mathcal{K}}(X)} \sigma_{ap,S}(A+K),$$

$$\sigma_{eB,S}(A) := \bigcap_{K\in\tilde{\mathcal{K}}(X)} \sigma_{S}(A+K).$$

◇

Proof. Let $\lambda \in \sigma_{eB_-,S}(A)$. Then, by using Remark 9.3.1, $\lambda \in \sigma_{eB_-,S}(A+K)$ for all $K \in \tilde{\mathcal{K}}(X)$, since $\sigma_{eB_-,S}(A+B) \subseteq \sigma_{\delta,S}(A+K)$ for $K \in \tilde{\mathcal{K}}(X)$. Therefore, we have $\sigma_{eB_-,S}(A) \subset \bigcap_{K\in\tilde{\mathcal{K}}(X)} \sigma_{\delta,S}(A+K)$. Conversely, let $\lambda \notin \sigma_{eB_-,S}(A)$. So, $\lambda S - A$ is lower semi-Browder. Applying Theorem 2.2.23, there exists a decomposition $X = X_1 \oplus X_2$ such that $\dim X_1 < +\infty$, $AX_i \subseteq X_i$ $(i = 1,2)$, $\lambda S - A|_{X_1}$ is nilpotent and $\lambda S - A|_{X_2}$ is onto. We consider the operator K defined by $K = I \oplus 0$ is even a finite rank operator commuting with A and S such that $\lambda S - A - K$ is onto. Also, there exists $K \in \tilde{\mathcal{K}}(X)$ such that $\lambda \notin \sigma_{\delta,S}(A+K)$ which implies that $\lambda \notin \bigcap_{K\in\tilde{\mathcal{K}}(X)} \sigma_{\delta,S}(A+K)$. Therefore, $\bigcap_{K\in\tilde{\mathcal{K}}(X)} \sigma_{\delta,S}(A+K) \subset \sigma_{eB_-,S}(A)$. The statements for the upper semi-Browder and lower semi-Browder's spectrum can be proved similarly. Q.E.D.

Let A be a bounded linear operator on X and let λ_0 be an isolated point of $\sigma_S(A)$. For an admissible contour Γ_{λ_0},

$$P_{\lambda_0,S} = -\frac{S}{2\pi i}\oint_{\Gamma_{\lambda_0}} (A-\lambda S)^{-1}\, d\lambda,$$

is called the **S**-Riesz integral for A, S and λ_0.

Proposition 9.3.2. *Let* $A,\ S \in \mathcal{L}(X)$ *and* λ_0 *is an isolated point of* $\sigma_S(A)$. *Let* $P_{\lambda_0,S}$ *be the* **S**-*Riesz integral for* $A,\ S$, *and* λ_0. *Then,*

(*i*) $P_{\lambda_0,S}$ *is a projection.*
(*ii*) *If* $0 \in \rho(S)$ *and* $AS^{-1} = S^{-1}A$, *then* $N(A-\lambda_0 S) \subset R(P_{\lambda_0,S})$.
(*iii*) *If the hypotheses of* (*ii*) *are satisfied, if* X *is a Hilbert, and if* A *and* S *are self-adjoint, then* $P_{\lambda_0,S}$ *is the orthogonal projection onto* $N(A-\lambda_0 S)$. ◇

Proof. (*i*) Let Γ_{λ_0} and $\tilde{\Gamma}_{\lambda_0}$ be two admissible contours for defining $P_{\lambda_0,S}$. Let us suppose that Γ_{λ_0} is contained in the interior of the region bounded by $\tilde{\Gamma}_{\lambda_0}$. In view of the **S**-resolvent identity, we obtain

$$\begin{aligned}
P^2_{\lambda_0,S} &= \frac{S}{(2i\pi)^2}\oint_{\Gamma_{\lambda_0}\times\tilde{\Gamma}_{\lambda_0}} (A-\lambda S)^{-1}S(A-\mu S)^{-1}d\mu\, d\lambda \\
&= \frac{S}{(2i\pi)^2}\oint_{\Gamma_{\lambda_0}} d\lambda \oint_{\tilde{\Gamma}_{\lambda_0}} \frac{\left[(A-\lambda S)^{-1}-(A-\mu S)^{-1}\right]}{\lambda-\mu}\, d\mu \\
&= \frac{S}{(2i\pi)^2}\oint_{\Gamma_{\lambda_0}} (A-\lambda S)^{-1}\oint_{\tilde{\Gamma}_{\lambda_0}} \frac{1}{\lambda-\mu}\, d\mu\, d\lambda \\
&\quad -\frac{S}{(2i\pi)^2}\oint_{\Gamma_{\lambda_0}}\oint_{\tilde{\Gamma}_{\lambda_0}} \frac{1}{\lambda-\mu}(A-\mu S)^{-1}d\mu\, d\lambda.
\end{aligned} \tag{9.3.4}$$

By applying the residue theorem [244] to the first integral in Eq. (9.3.4), we get

$$\oint_{\Gamma_{\lambda_0}} (A-\lambda S)^{-1}\oint_{\tilde{\Gamma}_{\lambda_0}} \frac{1}{\lambda-\mu} d\mu d\lambda = -2i\pi \oint_{\Gamma_{\lambda_0}} (A-\lambda S)^{-1} d\lambda. \tag{9.3.5}$$

For the second integral in Eq. (9.3.4), let us observe that

$$\oint_{\Gamma_{\lambda_0}}\oint_{\tilde{\Gamma}_{\lambda_0}} \frac{1}{\lambda-\mu}(A-\mu S)^{-1}\, d\mu d\lambda = \oint_{\tilde{\Gamma}_{\lambda_0}} (A-\mu S)^{-1}\oint_{\Gamma_{\lambda_0}} \frac{1}{\lambda-\mu}\, d\lambda\, d\mu = 0,$$

since the integrals are absolutely convergent and $\frac{1}{\lambda-\mu}$ is analytic and inside Γ_{λ_0}. In order to complete the proof, and from Eq. (9.3.5), it follows that $P^2_{\lambda_0,S} = P_{\lambda_0,S}$.

(*ii*) Let $f \in N(A-\lambda_0 S)$, then for $\lambda \neq \lambda_0$, $(A-\lambda S)^{-1}f = (\lambda_0 S-\lambda S)^{-1}f$. We show that

$$\begin{aligned}
P_{\lambda_0,S}f &= -\frac{S}{2\pi i}\oint_{\Gamma_{\lambda_0}} (A-\lambda S)^{-1}f\, d\lambda \\
&= -\frac{S}{2\pi i}\oint_{\Gamma_{\lambda_0}} (\lambda_0 S-\lambda S)^{-1}f\, d\lambda \\
&= f.
\end{aligned}$$

So, $f \in R(P_{\lambda_0,S})$.

(*iii*) Since the operators $A = A^*$ and $S = S^*$, then $((A-\lambda S)^{-1})^* = (A-\overline{\lambda}S)^{-1}$. Let us choose $r > 0$ such that $\Gamma_{\lambda_0} = \{\lambda \in \mathbb{C} \text{ such that } |\lambda-\lambda_0| = r\}$ is an admissible contour and take $\lambda = \lambda_0 + re^{i\theta}$. Then,

$$P_{\lambda_0,S} = -\frac{S}{2\pi}\oint_{-\pi}^{\pi} (A-(\lambda_0+re^{i\theta})S)^{-1}\, rd\theta$$

and

$$\begin{aligned} P^*_{\lambda_0,S} &= -\frac{S}{2\pi}\oint_{-\pi}^{\pi}((A-(\lambda_0+re^{i\theta})S)^{-1})^*\, rd\theta \\ &= -\frac{S}{2\pi}\oint_{-\pi}^{\pi}(A^*-\overline{(\lambda_0+re^{i\theta})}S)^{-1}\, rd\theta \\ &= -\frac{S}{2\pi}\oint_{-\pi}^{\pi}(A-(\lambda_0+re^{-i\theta})S)^{-1}\, rd\theta. \end{aligned}$$

Upon repartitioning with $\theta_1 = -\theta$, we easily find that

$$P^*_{\lambda_0,S} = -\frac{S}{2\pi}\oint_{-\pi}^{\pi}(A-(\lambda_0+re^{i\theta_1})S)^{-1}\, rd\theta_1 = P_{\lambda_0,S}.$$

Finally, we must show that $N(A-\lambda S) = R(P_{\lambda_0,S})$, which, by part (*ii*) requires that we show that $N(A-\lambda S) \supset R(P_{\lambda_0,S})$. We compute the following

$$\begin{aligned} (A-\lambda_0 S)P_{\lambda_0,S} &= -\frac{S}{2\pi i}\oint_{\Gamma_{\lambda_0}}(A-\lambda_0 S)(A-\lambda S)^{-1}\, d\lambda \\ &= -\frac{S}{2\pi i}\oint_{\Gamma_{\lambda_0}}(A-\lambda_0 S+\lambda S-\lambda S)(A-\lambda S)^{-1}\, d\lambda \\ &= -\frac{S}{2\pi i}\oint_{\Gamma_{\lambda_0}}(\lambda S-\lambda_0 S)(A-\lambda S)^{-1}\, d\lambda \\ &= -\frac{S}{2\pi i}\oint_{\Gamma_{\lambda_0}}(\lambda-\lambda_0)(AS^{-1}-\lambda)^{-1}\, d\lambda. \end{aligned}$$

Let U_{λ_0} denote the interior of Γ_{λ_0} on $U_{\lambda_0}\backslash\{\lambda_0\}$, the operator $(\lambda-\lambda_0)$ $(AS^{-1}-\lambda)^{-1}$ is an analytic operator valued function and satisfies the bound

$$|\lambda-\lambda_0|||(AS^{-1}-\lambda)^{-1}|| \leq \frac{|\lambda-\lambda_0|}{\text{dist}(\lambda,\sigma(AS^{-1}))},$$

where $\text{dist}(x,y)$ is the distance between x and y. Now, we take the diameter of Γ_{λ_0} small form standard results that $(\lambda-\lambda_0)(AS^{-1}-\lambda)^{-1}$ extends to analytic function on U_{λ_0} and hence, by Cauchy's theorem, we get

$$-\frac{S}{2\pi i}\oint_{\Gamma_{\lambda_0}}(\lambda-\lambda_0)(AS^{-1}-\lambda)^{-1}\, d\lambda = 0.$$

This establishes that $N(A-\lambda S) \supset R(P_{\lambda_0,S})$. Q.E.D.

The **S**-discrete spectrum of A, denoted $\sigma_{d_S}(A)$, is just the set of isolated points $\lambda \in \mathbb{C}$ of the spectrum such that the corresponding **S**-Riesz projectors $P_{\lambda,S}$ are finite-dimensional. Another part of the spectrum, which is generally larger than $\sigma_{e6,S}(A)$, is $\sigma_S(A)\backslash\sigma_{d_S}(A)$. We will also use this terminology here and the notation $\sigma_{e6,S}(A) = \sigma_S(A)\backslash\sigma_{d_S}(A)$ and $\rho_{6,S}(A) = \mathbb{C}\backslash\sigma_{e6,S}(A)$. The largest open set on which the resolvent is finitely meromorphic is precisely $\rho_{6,S}(A) = \sigma_{d_S}(A)\bigcup\rho_S(A)$. For $\lambda \in \rho_{6,S}(A)$, let $P_{\lambda,S}(A)$ (or $P_{\lambda,S}$) denote the corresponding (finite rank) S-Riesz projector with a range and a kernel denoted by $R_{\lambda,S}$ and $K_{\lambda,S}$. Since $P_{\lambda,S}$ is invariant, we may define the operator $A_{\lambda,S} = (A-\lambda S)(I - P_{\lambda,S}) + P_{\lambda,S}$, with respect to the decomposition $X = K_{\lambda,S} \oplus R_{\lambda,S}$ and $A_{\lambda,S} = ((A-\lambda S)_{|K_{\lambda,S}}) \oplus I$. We have just cut off the finite-dimensional part of $A-\lambda S$ in the S-Riesz decomposition. Since $\sigma\big((A-\lambda S)_{|K_{\lambda,S}}\big) = \sigma(A-\lambda S)\backslash\{0\}$, $A_{\lambda,S}$ has a bounded inverse, denoted by $R_{b,S}(A,\lambda)$ and called the "S-Browder's resolvent", i.e., $R_{b,S}(A,\lambda) = ((A-\lambda)_{|K_{\lambda,S}})^{-1} \oplus I$ with respect to $X = K_{\lambda,S} \oplus R_{\lambda,S}$. Or, alternatively $R_{b,S}(A,\lambda) = \big((A-\lambda S)_{|K_{\lambda,S}}\big)^{-1}(I - P_{\lambda,S}) + P_{\lambda,S}$ for $\lambda \in \rho_{6,S}(A)$. This clearly extends the usual resolvent $R_S(A,\lambda)$ from $\rho_S(A)$ to $\rho_{6,S}(A)$ and retains many of its important properties.

Proposition 9.3.3. *Let A and $S \in \mathcal{L}(X)$. Then, for $\lambda,\ \mu \in \rho_{6,S}(A)$, we have:*

$$\begin{aligned}&R_{b,S}(A,\lambda) - R_{b,S}(A,\mu)\\&\quad = (\lambda-\mu)R_{b,S}(A,\lambda)SR_{b,S}(A,\mu) + R_{b,S}(A,\lambda)\mathbf{N}(\lambda,\mu)R_{b,S}(A,\mu),\end{aligned}$$

where $\mathbf{N}(.,.)$ is a finite rank operator with the following expression

$$\mathbf{N}(\lambda,\mu) = \Big[\big(A-(\lambda S+1)\big)P_{\lambda,S} - \big(A-(\mu S+1)\big)P_{\mu,S}\Big].$$

◇

Proof. We have $R_{b,S}(A,\lambda) - R_{b,S}(A,\mu) = R_{b,S}(A,\lambda)\Big[A_{\mu,S} - A_{\lambda,S}\Big]R_{b,S}(A,\mu)$. So,

$$\begin{aligned}A_{\mu,S} - A_{\lambda,S} &= \big[(A-\mu S)(I-P_{\mu,S}) + P_{\mu,S}\big] - \big[(A-\lambda S)(I-P_{\lambda,S}) + P_{\lambda,S}\big]\\ &= \big[(A-(\lambda S+1))\,P_{\lambda,S} - (A-(\mu S+1))\,P_{\mu,S}\big] + (\lambda-\mu)S.\end{aligned}$$

Therefore,

$$\begin{aligned}R_{b,S}(A,\lambda) - R_{b,S}(A,\mu) &= (\lambda-\mu)R_{b,S}(A,\lambda)SR_{b,S}(A,\mu)\\ &\quad + R_{b,S}(A,\lambda)\Big[(A-(\lambda S+1))\,P_{\lambda,S}\\ &\quad - (A-(\mu S+1))\,P_{\mu,S}\Big]R_{b,S}(A,\mu).\end{aligned}$$

Q.E.D.

9.4 S-Essential Spectra of the Sum of Bounded Linear Operators

The following theorem gives a relation between the S-essential spectra of the sum of two bounded linear operators and the S-essential spectra of each of these operators where their products are Fredholm or semi-Fredholm perturbations in X.

Theorem 9.4.1. *Let S, A, and B be three bounded linear operators on a Banach space X such that $S \neq A$ and $S \neq B$.*

(i) If $AB \in \mathcal{F}^b(X)$, $\alpha(S) < \infty$ and $SA = AS$, then $\sigma_{e4,S}(A+B)\backslash\{0\} \subset [\sigma_{e4,S}(A) \bigcup \sigma_{e4,S}(B)]\backslash\{0\}$. Moreover, if $BA \in \mathcal{F}^b(X)$, $\beta(S) < \infty$ and $SB = BS$, then we have $\sigma_{e4,S}(A+B)\backslash\{0\} = [\sigma_{e4,S}(A) \bigcup \sigma_{e4,S}(B)]\backslash\{0\}$.

(ii) If the hypotheses of (i) are satisfied and $\beta(S) = \alpha(S)$, then $\sigma_{e5,S}(A+B)\backslash\{0\} \subset [\sigma_{e5,S}(A) \bigcup \sigma_{e5,S}(B)]\backslash\{0\}$. Moreover, if $\mathbb{C}\backslash\sigma_{e4,S}(A)$ is connected, then

$$\sigma_{e5,S}(A+B)\backslash\{0\} = \left[\sigma_{e5,S}(A) \bigcup \sigma_{e5,S}(B)\right]\backslash\{0\}. \tag{9.4.1}$$

(iii) If the hypotheses of (ii) are satisfied, and if $\mathbb{C}\backslash\sigma_{e5}(A+B)$, $\mathbb{C}\backslash\sigma_{e5,S}(A)$ and $\mathbb{C}\backslash\sigma_{e5,S}(B)$ are connected, then $\sigma_{e6,S}(A+B)\backslash\{0\} = [\sigma_{e5,S}(A) \bigcup \sigma_{e5,S}(B)]\backslash\{0\}$.

(iv) If $AB \in \mathcal{F}_+^b(X)$, $\alpha(S) < \infty$, $\rho_S(A) \neq \emptyset$ and $SA = AS$, then we have $\sigma_{e1,S}(A+B)\backslash\{0\} \subset [\sigma_{e1,S}(A) \bigcup \sigma_{e1,S}(B)]\backslash\{0\}$. Besides, if $BA \in \mathcal{F}_+^b(X)$ and $R(S)$ is closed in X, then

$$\sigma_{e1,S}(A+B)\backslash\{0\} = \left[\sigma_{e1,S}(A) \bigcup \sigma_{e1,S}(B)\right]\backslash\{0\}. \tag{9.4.2}$$

(v) If the hypotheses of (iv) are satisfied, and $\beta(S) \leq \alpha(S)$, then $\sigma_{e7,S}(A+B)\backslash\{0\} \subset [\sigma_{e7,S}(A) \bigcup \sigma_{e7,S}(B)]\backslash\{0\}$. Moreover, if $\Phi_{A,S}$ is connected, and $BS = SB$, then

$$\sigma_{e7,S}(A+B)\backslash\{0\} = \left[\sigma_{e7,S}(A) \bigcup \sigma_{e7,S}(B)\right]\backslash\{0\}. \tag{9.4.3}$$

(vi) If $AB \in \mathcal{F}_-^b(X)$ and $BS = SB$, then $\sigma_{e2,S}(A+B)\backslash\{0\} \subset [\sigma_{e2,S}(A) \bigcup \sigma_{e2,S}(B)]\backslash\{0\}$. Besides, if $BA \in \mathcal{F}_-^b(X)$ and $S \in \Phi_-^b(X)$, then

$$\sigma_{e2,S}(A+B)\backslash\{0\} = \left[\sigma_{e2,S}(A) \bigcup \sigma_{e2,S}(B)\right]\backslash\{0\}. \tag{9.4.4}$$

(vii) If the hypotheses of (iv) are satisfied and $\beta(S) \geq \alpha(S)$, then $\sigma_{e8,S}(A+B)\backslash\{0\} \subset [\sigma_{e8,S}(A) \bigcup \sigma_{e8,S}(B)]\backslash\{0\}$. If, further, $\Phi_{A,S}$ is connected, then

$$\sigma_{e8,S}(A+B)\backslash\{0\} = \left[\sigma_{e8,S}(A) \bigcup \sigma_{e8,S}(B)\right]\backslash\{0\}.$$

(viii) *If* $AB \in \mathcal{F}_+^b(X) \bigcap \mathcal{F}_-^b(X)$, $\alpha(S) < \infty$ *and* S *commutes with* A *and* B, *then*

$$\sigma_{e3,S}(A+B)\backslash\{0\} \subset \Big(\Big[\sigma_{e3,S}(A) \bigcup \sigma_{e3,S}(B)\Big] \bigcup \Big[\sigma_{e1,S}(A) \bigcap \sigma_{e2,S}(B)\Big] \bigcup \Big[\sigma_{e2,S}(A) \bigcap \sigma_{e1,S}(B)\Big]\Big)\backslash\{0\}.$$

Moreover, if $BA \in \mathcal{F}_+^b(X) \bigcap \mathcal{F}_-^b(X)$, $\beta(S) < \infty$ *and* $R(S)$ *is closed in* X, *then*

$$\sigma_{e3,S}(A+B)\backslash\{0\} = \Big(\Big[\sigma_{e3,S}(A) \bigcup \sigma_{e3,S}(B)\Big] \bigcup \Big[\sigma_{e1,S}(A) \bigcap \sigma_{e2,S}(B)\Big] \bigcup \Big[\sigma_{e2,S}(A) \bigcap \sigma_{e1,S}(B)\Big]\Big)\backslash\{0\}.$$

◇

Proof. For $\lambda \in \mathbb{C}$, we can write

$$(\lambda S - A)(\lambda S - B) = AB + \lambda(\lambda S^2 - AS - SB), \tag{9.4.5}$$

and

$$(\lambda S - B)(\lambda S - A) = BA + \lambda(\lambda S^2 - SA - BS). \tag{9.4.6}$$

(*i*) Let $\lambda \notin \sigma_{e4,S}(A) \bigcup \sigma_{e4,S}(B) \bigcup \{0\}$. Then, $(\lambda S - A) \in \Phi^b(X)$ and $(\lambda S - B) \in \Phi^b(X)$. Theorem 2.2.40 gives $(\lambda S - A)(\lambda S - B) \in \Phi^b(X)$. Since $AB \in \mathcal{F}^b(X)$, by applying Eq. (9.4.5), we have $(\lambda S^2 - AS - SB) \in \Phi^b(X)$. Hence, $S(\lambda S - A - B) \in \Phi^b(X)$. Since $\alpha(S) < \infty$, then by using Theorem 2.2.11, we conclude that $(\lambda S - A - B) \in \Phi^b(X)$. Therefore, $\lambda \notin \sigma_{e4,S}(A+B)$. Consequently,

$$\sigma_{e4,S}(A+B)\backslash\{0\} \subset \Big[\sigma_{e4,S}(A) \bigcup \sigma_{e4,S}(B)\Big]\backslash\{0\}. \tag{9.4.7}$$

In order to prove the inverse inclusion of Eq. (9.4.7), let us suppose $\lambda \notin \sigma_{e4,S}(A+B) \bigcup \{0\}$, then $(\lambda S - A - B) \in \Phi^b(X)$. Since $S \in \Phi^b(X)$ and $SB = BS$, we have $S(\lambda S - A - B) \in \Phi^b(X)$ and $(\lambda S - A - B)S \in \Phi^b(X)$. Since $AB \in \mathcal{F}^b(X)$, $BA \in \mathcal{F}^b(X)$, $\alpha(S) < \infty$ and $\beta(S) < \infty$, then by using Eqs. (9.4.5) and (9.4.6), we have

$$(\lambda S - A)(\lambda S - B) \in \Phi^b(X) \text{ and } (\lambda S - B)(\lambda S - A) \in \Phi^b(X). \tag{9.4.8}$$

By applying Theorem 2.2.19, it is clear that $(\lambda S - A) \in \Phi^b(X)$ and $(\lambda S - B) \in \Phi^b(X)$. Therefore, $\lambda \notin \sigma_{e4,S}(A) \bigcup \sigma_{e4,S}(B)$. This proves that

$$\Big[\sigma_{e4,S}(A) \bigcup \sigma_{e4,S}(B)\Big]\backslash\{0\} \subset \sigma_{e4,S}(A+B)\backslash\{0\}.$$

(*ii*) Let $\lambda \notin [\sigma_{e5,S}(A) \bigcup \sigma_{e5,S}(B)] \backslash\{0\}$. Then, $(\lambda S - A) \in \Phi^b(X)$, $i(\lambda S - A) = 0$, $(\lambda S - B) \in \Phi^b(X)$ and $i(\lambda S - B) = 0$. Using Theorem 2.2.40, we infer that $(\lambda S - A)(\lambda S - B) \in \Phi^b(X)$ and $i((\lambda S - A)(\lambda S - B)) = 0$. Moreover, since $AB \in \mathcal{F}^b(X)$, we can apply Eq. (9.4.5) and Lemma 6.3.1 which ensure that $S(\lambda S - A - B) \in \Phi^b(X)$ and $i(S(\lambda S - A - B)) = 0$. Since $\alpha(S) = \beta(S) < \infty$, then the use of Theorems 2.2.40 and 2.2.42 gives $(\lambda S - A - B) \in \Phi^b(X)$ and $\underbrace{i(S)}_{=0} + i(\lambda S - A - B) = 0$. Therefore, $\lambda \notin \sigma_{e5,S}(A + B)$. Hence,

$$\sigma_{e5,S}(A + B)\backslash\{0\} \subset \left[\sigma_{e5,S}(A) \bigcup \sigma_{e5,S}(B)\right] \backslash\{0\}. \tag{9.4.9}$$

In order to prove the inverse inclusion of Eq. (9.4.9), let $\lambda \notin \sigma_{e5,S}(A + B)\backslash\{0\}$. Then, $(\lambda S - A - B) \in \Phi^b(X)$ and $i(\lambda S - A - B) = 0$. Since $AB \in \mathcal{F}^b(X)$, $BA \in \mathcal{F}^b(X)$ and $\alpha(S) = \beta(S) < \infty$, it is easy to show that $(\lambda S - A) \in \Phi^b(X)$ and $(\lambda S - B) \in \Phi^b(X)$. Moreover, by applying Eqs. (9.4.5) and (9.4.8), Theorem 2.2.40, and Lemma 6.3.1 (i), we have

$$\begin{aligned} i[(\lambda S - A)(\lambda S - B)] &= i(\lambda S - A) + i(\lambda S - B) \\ &= i(S) + i(\lambda S - A - B) \\ &= i(\lambda S - A - B) \\ &= 0. \end{aligned} \tag{9.4.10}$$

Since A is a bounded linear operator, we get $\rho(A) \neq \emptyset$. Besides, $\mathbb{C}\backslash\sigma_{e4,S}(A)$ is connected. This, together with Lemma 9.1.1 (i), allows us to deduce that $\sigma_{e4,S}(A) = \sigma_{e5,S}(A)$. Using the last equality and the fact that $(\lambda S - A) \in \Phi^b(X)$, we deduce that $i(\lambda S - A) = 0$. From Eq. (9.4.10), it follows that $i(\lambda S - B) = 0$. We conclude that $\lambda \notin \sigma_{e5,S}(A) \bigcup \sigma_{e5,S}(B)$. Hence, $[\sigma_{e5,S}(A) \bigcup \sigma_{e5,S}(B)] \backslash\{0\} \subset \sigma_{e5,S}(A + B)\backslash\{0\}$. So, we prove Eq. (9.4.1).

(*iii*) The sets $\mathbb{C}\backslash\sigma_{e5,S}(A + B)$, $\mathbb{C}\backslash\sigma_{e5,S}(A)$ and $\mathbb{C}\backslash\sigma_{e5,S}(B)$ are connected. Since A and B are bounded operators, we deduce that $\rho_S(A)$, $\rho_S(B)$ and $\rho_S(B + A)$ are not empty sets. So, by using Lemma 9.1.1 (*ii*), we obtain $\sigma_{e5,S}(A + B) = \sigma_{e6,S}(A + B)$, $\sigma_{e5,S}(A) = \sigma_{e6,S}(A)$ and $\sigma_{e5,S}(B) = \sigma_{e6,S}(B)$. Therefore, Eq. (9.4.1) gives

$$\sigma_{e6,S}(A + B)\backslash\{0\} = \left[\sigma_{e6,S}(A) \bigcup \sigma_{e6,S}(B)\right] \backslash\{0\}.$$

(*iv*) Suppose that $\lambda \notin \sigma_{e1,S}(A) \bigcup \sigma_{e1,S}(B) \bigcup\{0\}$, then $(\lambda S - A) \in \Phi^b_+(X)$ and $(\lambda S - B) \in \Phi^b_+(X)$. By using Theorem 2.2.13, we have $(\lambda S - A)(\lambda S - B) \in \Phi^b_+(X)$. Since $AB \in \mathcal{F}^b_+(X)$, we can apply Eq. (9.4.5) and Lemma 6.3.1 (*ii*), and we have $S(\lambda S - A - B) \in \Phi^b_+(X)$. Since $\alpha(S) < \infty$, $R(S)$ is closed, and $S \in \mathcal{L}(X)$, then $(\lambda S - A - B) \in \Phi^b_+(X)$. So, $\lambda \notin \sigma_{e1,S}(A + B)$. Therefore, $\sigma_{e1,S}(A + B)\backslash\{0\} \subset \sigma_{e1,S}(A) \bigcup \sigma_{e1,S}(B) \bigcup\{0\}$. Suppose that $\lambda \notin$

$\sigma_{e1,S}(A+B)\bigcup\{0\}$. Then, $(\lambda S - A - B) \in \Phi^b_+(X)$. Since $S \in \Phi^b_+(X)$, then $S(\lambda S - A - B) \in \Phi^b_+(X)$, and $(\lambda S - A - B)S \in \Phi^b_+(X)$. Since $AB \in \mathcal{F}^b_+(X)$ and $BA \in \mathcal{F}^b_+(X)$, and by applying Eqs. (9.4.5), (9.4.6) and Lemma 6.3.1 (*ii*), we have

$$(\lambda S - A)(\lambda S - B) \in \Phi^b_+(X), \ (\lambda S - B)(\lambda S - A) \in \Phi^b_+(X). \qquad (9.4.11)$$

By using Eq. (9.4.11) and Theorem 2.2.14 (i), it is clear that $(\lambda S - A) \in \Phi^b_+(X)$ and $(\lambda S - B) \in \Phi^b_+(X)$. Hence, $\lambda \notin \sigma_{e1,S}(A)\bigcup\sigma_{e1,S}(B)$. Therefore, $[\sigma_{e1,S}(A)\bigcup\sigma_{e1,S}(B)]\backslash\{0\} \subset \sigma_{e1,S}(A+B)\backslash\{0\}$. This proves Eq. (9.4.2).

(v) Now, suppose that $\lambda \notin \sigma_{e7,S}(A)\bigcup\sigma_{e7,S}(B)\bigcup\{0\}$, then by using Corollary 9.2.1, we have $(\lambda S - A) \in \Phi^b_+(X)$, $i(\lambda S - A) \leq 0$, $(\lambda S - B) \in \Phi^b_+(X)$ and $i(\lambda S - B) \leq 0$. By using Theorem 2.2.13 and Theorem 2.2.7, we have $(\lambda S - A)(\lambda S - B) \in \Phi^b_+(X)$ and $i[(\lambda S - A)(\lambda S - B)] \leq 0$. Since $AB \in \mathcal{F}^b_+(X)$, by using Eq. (9.4.5) and Lemma 6.3.1, we deduce that $S(\lambda S - A - B) \in \Phi^b_+(X)$ and $i(S(\lambda S - A - B)) \leq 0$. Then, $i(S) + i(\lambda S - A - B) \leq 0$. Since $\beta(S) \leq \alpha(S)$, then $i(\lambda S - A - B) \leq 0$. Again, by using Corollary 9.2.1 (*ii*), it is clear that $\lambda \notin \sigma_{e7,S}(A)$. Hence,

$$\sigma_{e7,S}(A+B)\backslash\{0\} \subset \sigma_{e7,S}(A)\bigcup\sigma_{e7,S}(B)\bigcup\{0\}. \qquad (9.4.12)$$

Now, it remains to prove the inverse inclusion of Eq. (9.4.12). Let $\lambda \notin \sigma_{e7,S}(A+B)\backslash\{0\}$, then $(\lambda S - A - B) \in \Phi^b_+(X)$ and $i(\lambda S - A - B) \leq 0$. Since $AB \in \mathcal{F}^b_+(X)$, $BA \in \mathcal{F}^b_+(X)$ and $S \in \Phi^b_+(X)$, using a similar reasoning as before leads to the following $(\lambda S - A) \in \Phi^b_+(X)$, $(\lambda S - B) \in \Phi^b_+(X)$, and $i((\lambda S - A)(\lambda S - B)) = i(\lambda S - A) + i(\lambda S - B) \leq 0$. Let $\lambda_0 \in \rho_S(A)$, we have $\lambda_0 S - A \in \Phi^b(X)$ and $i(\lambda_0 S - A) = 0$. Since $\rho_S(A) \subset \mathbb{C}\backslash\sigma_{e4,S}(A)$, we have $\lambda_0 \in \mathbb{C}\backslash\sigma_{e4,S}(A) = \Phi_{A,S}$ which is connected. By using Proposition 2.2.5 (ii), we have $i(\lambda S - A)$ is constant on any component of $\Phi_{A,S}$. Hence, $i(\lambda S - A) = 0$ for all $\lambda \in \sigma_{e4,S}(A)$. So, $i(\lambda S - A) + i(\lambda S - B) \leq 0$. This implies that $i(\lambda S - B) \leq 0$, and we find that $\lambda \notin \sigma_{e7,S}(A)\bigcup\sigma_{e7,S}(B)$. Hence, $[\sigma_{e7,S}(A)\bigcup\sigma_{e7,S}(B)]\backslash\{0\} \subset \sigma_{e7,S}(A+B)\backslash\{0\}$. This proves Eq. (9.4.3).

(*vi*) The proof of (vi) may be checked in the same way as the proof of (iv). It is sufficient to replace $\Phi^b_+(X)$ and $\sigma_{e1,S}(.)$ by $\Phi^b_-(X)$ and $\sigma_{e2,S}(.)$ respectively.

(*vii*) The proof of (*vii*) may be checked in the same way as the proof of (v).

(*viii*) Since the equalities $\sigma_{e3,S}(A) = \sigma_{e1,S}(A)\bigcap\sigma_{e2,S}(A)$, $\sigma_{e3,S}(B) = \sigma_{e1,S}(B)\bigcap\sigma_{e2,S}(B)$, and $\sigma_{e3,S}(A+B) = \sigma_{e1,S}(A+B)\bigcap\sigma_{e2,S}(A+B)$ are known, $AB \in \mathcal{F}^b_+(X)\bigcap\mathcal{F}^b_-(X)$ and $BA \in \mathcal{F}^b_+(X)\bigcap\mathcal{F}^b_-(X)$, then by using Eqs. (9.4.2) and (9.4.4), we deduce that

$$\begin{aligned}\sigma_{e3,S}(A+B)\backslash\{0\} = &\left[\sigma_{e3,S}(A)\bigcup\sigma_{e3,S}(B)\right]\bigcup\left[\sigma_{e3,S}(A)\bigcap\sigma_{e3,S}(B)\right]\\ &\bigcup\left[\sigma_{e2,S}(A)\bigcap\sigma_{e1,S}(B)\right]\backslash\{0\}.\end{aligned}$$

Q.E.D.

Theorem 9.4.2. *Let S, A, and B be three bounded linear operators on a Banach space X such that $S \neq A$, and $S \neq B$.*

(*i*) *If $AB \in \mathcal{F}^b(X)$, then $\sigma_{ei,S^2}(AS+SB)\backslash\{0\} = [\sigma_{ei,S}(A)\bigcup\sigma_{ei,S}(B)]\backslash\{0\}$, with $i = 4,5$. If, further, $BA \in \mathcal{F}^b(X)$, $SA = AS$ and $BS = SB$, then $\sigma_{e4,S^2}(AS+SB)\backslash\{0\} = \sigma_{e4,S^2}(SA+BS)\backslash\{0\} = [\sigma_{e4,S}(A)\bigcup\sigma_{e4,S}(B)]\backslash\{0\}$. Moreover, if $\mathbb{C}\backslash\sigma_{e4,S}(A)$ is connected, then $\sigma_{e5,S^2}(AS+SB)\backslash\{0\} = \sigma_{e5,S^2}(SA+BS) = [\sigma_{e5,S}(A)\bigcup\sigma_{e5,S}(B)]\backslash\{0\}$.*

(*ii*) *If the hypotheses of (i) is satisfied and, if $\mathbb{C}\backslash\sigma_{e5,S^2}(AS+SB)$, $\mathbb{C}\backslash\sigma_{e5,S}(A)$ and $\mathbb{C}\backslash\sigma_{e5,S}(B)$ are connected, then $\sigma_{e6,S^2}(AS+SB)\backslash\{0\} = \sigma_{e6,S^2}(SA+BS) = [\sigma_{e6,S}(A)\bigcup\sigma_{e6,S}(B)]\backslash\{0\}$.*

(*iii*) *If $AB \in \mathcal{F}_+^b(X)$, then $\sigma_{ei,S^2}(AS+SB)\backslash\{0\} \subset [\sigma_{ei,S}(A)\bigcup\sigma_{ei,S}(B)]\backslash\{0\}$, with $i = 1,7$. If, further, $BA \in \mathcal{F}_+^b(X)$, $SA = AS$ and $BS = SB$, then $\sigma_{e1,S^2}(AS+SB)\backslash\{0\} = [\sigma_{e1,S}(A)\bigcup\sigma_{e1,S}(B)]\backslash\{0\}$. Moreover, if $\Phi_{A,S}$ is connected and $\rho_S(A) \neq \emptyset$, then $\sigma_{e7,S^2}(AS+SB)\backslash\{0\} = [\sigma_{e7,S}(A)\bigcup\sigma_{e7,S}(B)]\backslash\{0\}$.*

(*iv*) *If $AB \in \mathcal{F}_-^b(X)$, then $\sigma_{ei,S^2}(AS+SB)\backslash\{0\} \subset [\sigma_{ei,S}(A)\bigcup\sigma_{ei,S}(B)]\backslash\{0\}$, with $i = 2,8$. If, further, $BA \in \mathcal{F}_-^b(X)$, $SA = AS$ and $BS = SB$, then $\sigma_{e2,S^2}(AS+SB)\backslash\{0\} = [\sigma_{e2,S}(A)\bigcup\sigma_{e2,S}(B)]\backslash\{0\}$. Moreover, if $\Phi_{A,S}$ is connected and $\rho_S(A) \neq \emptyset$, then $\sigma_{e8,S^2}(AS+SB)\backslash\{0\} = [\sigma_{e8,S}(A)\bigcup\sigma_{e8,S}(B)]\backslash\{0\}$.*

(*v*) *If $AB \in \mathcal{F}_+^b(X)\bigcap\mathcal{F}_-^b(X)$, then*

$$\sigma_{e3,S^2}(AS+SB)\backslash\{0\} \subset \Big[\sigma_{e3,S}(A)\bigcup\sigma_{e3,S}(B)\Big]\bigcup\Big[\sigma_{e1,S}(A)\bigcap\sigma_{e2,S}(B)\Big]$$
$$\bigcup\Big[\sigma_{e2,S}(A)\bigcap\sigma_{e1,S}(B)\Big]\backslash\{0\}.$$

Moreover, if $BA \in \mathcal{F}_+^b(X)\bigcap\mathcal{F}_-^b(X)$, $SA = AS$ and $BS = SB$, then $\sigma_{e3,S^2}(AS+SB)\backslash\{0\} = [\sigma_{e3,S}(A)\bigcup\sigma_{e3,S}(B)]\bigcup[\sigma_{e1,S}(A)\bigcap\sigma_{e2,S}(B)]\bigcup\big[\sigma_{e2,S}(A)\bigcap\sigma_{e1,S}(B)\big]\backslash\{0\}$. ◇

Proof. The proof of theorem is a straightforward adoption of the proof of Theorem 9.4.1. Q.E.D.

9.5 S-Essential Spectra by Means of Demicompact Operators

Let X be a Banach space and $A \in \mathcal{C}(X)$. We define the sets $\mathcal{F}_A^l(X)$ and $\mathcal{F}_A^r(X)$ by:

$$\mathcal{F}_A^l(X) = \{A_l \in \mathcal{L}(X, X_A) \text{ such that } A_l \text{ is a left Fredholm inverse of } A\},$$
$$\mathcal{F}_A^r(X) = \{A_r \in \mathcal{L}(X, X_A) \text{ such that } A_r \text{ is a right Fredholm inverse of } A\}.$$

Theorem 9.5.1. *Let $A \in \mathcal{C}(X)$ and $S \in \mathcal{L}(X)$ such that $S \neq 0$, and let J be an A-bounded linear operator on X. If, for every $\lambda \in \Phi_{+A,S}$, there exists $A_{\lambda l} \in \mathcal{F}_{\lambda S-A}^l(X)$ such that the operator $JA_{\lambda l}$ is demicompact, then $\sigma_{e1,S}(A+J) \subseteq \sigma_{e1,S}(A)$.* ◇

Proof. First, for each $\lambda \in \mathbb{C}$, J is an $(A - \lambda S)$-bounded linear operator on X. Indeed,

$$\begin{aligned}\forall x \in X, \|\hat{J}x\| &\leq \|\hat{J}\|(\|x\| + \|Ax\|)\\ &\leq \|\hat{J}\|(\|x\| + \|(A - \lambda S)x\| + |\lambda|\|Sx\|)\\ &\leq \|\hat{J}\|(1 + |\lambda|\|S\|)(\|x\| + \|(A - \lambda S)x\|).\end{aligned}$$

Let $\lambda \in \mathbb{C}$ and $A_{\lambda l} \in \mathcal{F}^l_{\lambda S - A}(X)$, then there exists a compact operator $K \in \mathcal{K}(X_A)$, such that $A_{\lambda l}(\lambda \hat{S}_{|X_A} - \hat{A}) + K = I_{X_A}$. Hence, we can write

$$\lambda \hat{S}_{|X_A} - \hat{A} - \hat{J} = (I - \hat{J}A_{\lambda l})(\lambda \hat{S}_{|X_A} - \hat{A}) - \hat{J}K. \tag{9.5.1}$$

Let $\lambda \notin \sigma_{e1,S}(A)$, then $(\lambda S - A) \in \Phi_+(X)$. Since $JA_{\lambda l}$ is demicompact, and from Theorem 5.4.5, it follows that $(I - JA_{\lambda l}) \in \Phi_+(X)$. By applying Theorem 2.2.13 (*ii*), Lemma 6.3.1 (*ii*), Eq. (9.5.1) and using the fact that $\hat{J}K \in \mathcal{K}(X_A, X)$, we conclude that $\lambda \hat{S}_{|X_A} - \hat{A} - \hat{J} \in \Phi^b_+(X_A, X)$. Hence $\lambda \notin \sigma_{e1,S}(A + J)$. Q.E.D.

Theorem 9.5.2. *Let $A \in \mathcal{C}(X)$ and $S \in \mathcal{L}(X)$ such that $S \neq 0$ and let J be an A-bounded linear operator on X. The following statements hold.*

(*i*) *If, for every $\lambda \in \Phi_{A,S}$, there exists $A_{\lambda l} \in \mathcal{F}^l_{\lambda S - A}(X)$ such that the operator $\mu JA_{\lambda l}$ is demicompact for any $\mu \in [0, 1]$, then $\sigma_{e4,S}(A + J) \subseteq \sigma_{e4,S}(A)$ and $\sigma_{e5,S}(A + J) \subseteq \sigma_{e5,S}(A)$.*

(*ii*) *If, for every $\lambda \in \Phi_{A,S}$, there exists $A_{\lambda l} \in \mathcal{F}^l_{\lambda S - A}(X)$ such that the operator $JA_{\lambda l}$ is demicompact, 1-set-contraction such that $\mathbb{C}\backslash\sigma_{e5,S}(A)$ is connected and the resolvent sets $\rho_S(A)$ and $\rho_S(A + J)$ are not empty. Then, $\sigma_{e6,S}(A + J) \subseteq \sigma_{e6,S}(A)$.* ◇

Proof. (*i*) Let $\lambda \in \mathbb{C}$, and $A_{\lambda l} \in \mathcal{F}^l_{\lambda S - A}(X)$, then there exists a compact operator $K \in \mathcal{K}(X_A)$, such that $A_{\lambda l}(\lambda \hat{S}_{|X_A} - \hat{A}) + K = I_{X_A}$. Hence, we can write $\lambda \hat{S}_{|X_A} - \hat{A} - \hat{J} = (I - \hat{J}A_{\lambda l})(\lambda \hat{S}_{|X_A} - \hat{A}) - \hat{J}K$. Let $\lambda \notin \sigma_{e4,S}(A)$, then $(\lambda S - A) \in \Phi(X)$. Since $\mu JA_{\lambda l}$ is demicompact for each $\mu \in [0, 1]$, it follows, from Theorem 5.4.2, that $(I - JA_{\lambda l}) \in \Phi(X)$. By applying Atkinson's theorem (Theorem 2.2.40) and using the fact that $\hat{J}K \in \mathcal{K}(X_A, X)$, we conclude that $\lambda \hat{S}_{|X_A} - \hat{A} - \hat{J} \in \Phi^b(X_A, X)$. Hence, $\lambda \notin \sigma_{e4,S}(A + J)$.

(*ii*) This statement is an application of Theorem 7.3.1 (*ii*). Q.E.D.

Remark 9.5.1. In particular, Theorem 9.5.2 remains true if we replace the assumption $\mu JA_{\lambda l}$ is demicompact for any $\mu \in [0, 1]$ in (*i*) by $JA_{\lambda l}$ is demicompact and 1-set-contraction. ◇

Theorem 9.5.3. *Let A, B, and $S \in \mathcal{L}(X)$, such that $S \neq 0$.*

(*i*) *If, for every $\lambda \in \Phi_{+(A+B),S}\backslash\{0\}$, there exists $A_{\lambda l} \in \mathcal{F}^l_{\lambda S - A - B}(X)$ such that the operator $\mu\frac{1}{\lambda}([A, S] - AB)A_{\lambda l}$ is S-demicompact for any $\mu \in [0, 1)$, then we have $\sigma_{e4,S}(A + B)\backslash\{0\} \subset (\sigma_{e4,S}(A) \bigcup \sigma_{e4,S}(B))\backslash\{0\}$, where $[A, S] := AS - SA$.*

(ii) *Moreover, if there exists* $B_{\lambda l} \in \mathcal{F}^l_{\lambda S-A-B}(X)$, *such that* $\mu\frac{1}{\lambda}([B,S]-BA)B_{\lambda l}$ *is* S*-demicompact for any* $\mu \in [0,1)$, *then* $\sigma_{e4,S}(A+B)\backslash\{0\} = (\sigma_{e4,S}(A)\bigcup\sigma_{e4,S}(B))\backslash\{0\}$. ◇

Proof. (i) Let $\lambda \in \mathbb{C}^*$. If there exists a left inverse modulo compact operator of $(\lambda S - A - B)$, called $A_{\lambda l}$, then $A_{\lambda l}(\lambda S - A - B) = I - K$, where $K \in \mathcal{K}(X)$. Hence, we have

$$\begin{aligned}(\lambda S - A)(\lambda S - B) &= \lambda S(\lambda S - A - B) + \Theta_\lambda(A,B)\\ &= \lambda S(\lambda S - A - B) + \Theta_\lambda(A,B)K\\ &\quad + \Theta_\lambda(A,B)A_{\lambda l}(\lambda S - A - B)\\ &= \lambda(S + \frac{1}{\lambda}\Theta_\lambda(A,B)A_{\lambda l})(\lambda S - A - B) + \Theta_\lambda(A,B)K,\end{aligned}$$

where $\Theta_\lambda(A,B) = AB - \lambda[A,S]$. Let $\lambda \notin (\sigma_{e4,S}(A)\bigcup\sigma_{e4,S}(B))\backslash\{0\}$. Then, $(\lambda S - A) \in \Phi^b(X)$ and $(\lambda S - B) \in \Phi^b(X)$. It follows, from Theorem 2.2.40, that $(\lambda S - A)(\lambda S - B) \in \Phi^b(X)$. Since $\mu\frac{1}{\lambda}([A,S]-AB)A_{\lambda l}$ is S-demicompact, it follows, from Lemma 5.4.7, that $S + \frac{1}{\lambda}\Theta_\lambda(A,B)A_{\lambda l}$ is a Fredholm operator and using the fact that $\Theta_\lambda(A,B)K$ is compact, we conclude, together with Theorem 2.2.41, that $(\lambda S - A - B) \in \Phi^b(X)$. Hence, $\lambda \notin \sigma_{e4,S}(A+B)$.

(ii) Conversely, if $\lambda \notin \sigma_{e4,S}(A+B)$, then arguing as above, we deduce that the operators $(\lambda S - A)(\lambda S - B) \in \Phi^b(X)$ and $(\lambda S - B)(\lambda S - A) \in \Phi^b(X)$. According to Theorem 2.2.19, we conclude that $(\lambda S - A) \in \Phi^b(X)$ and $(\lambda S - B) \in \Phi^b(X)$ and hence, $\lambda \notin (\sigma_{e4,S}(A)\bigcup\sigma_{e4,S}(B))\backslash\{0\}$. Q.E.D.

9.6 Characterization of the Relative Schechter's and Approximate Essential Spectra

Throughout this section, we denote by $\mathcal{DC}(X)$ the class of demicompact, 1-set contraction linear operators. We will give a refinement of the relative Schechter's essential spectrum and the relative approximate essential spectrum definition. For this, let X be a Banach space, $T \in \mathcal{C}(X)$ and $S \in \mathcal{L}(X)$ such that $S \neq 0$. We define these sets $\Omega^L_{S,T}(X)$ and $\Omega^R_{S,T}(X)$ by:

$$\Omega^L_{S,T}(X) = \{K \in \mathcal{L}(X) \text{ such that } \forall\lambda \in \rho_S(T+K), -(\lambda S - T - K)^{-1}K \in \mathcal{DC}(X)\}.$$
$$\Omega^R_{S,T}(X) = \{K \in \mathcal{L}(X) \text{ such that } \forall\lambda \in \rho_S(T+K), -K(\lambda S - T - K)^{-1} \in \mathcal{DC}(X)\}.$$

We have the following theorem.

Theorem 9.6.1. *For any* $T \in \mathcal{C}(X)$ *and* $S \in \mathcal{L}(X)$, *such that* $S \neq 0$, *we have*

$$\sigma_{e5,S}(T) = \bigcap_{K\in\Omega^R_{S,T}(X)} \sigma_S(T+K) = \bigcap_{K\in\Omega^L_{S,T}(X)} \sigma_S(T+K).$$

◇

Proof. It is obvious that $\mathcal{K}(X) \subset \Omega^L_{S,T}(X)$ and $\mathcal{K}(X) \subset \Omega^R_{S,T}(X)$. We infer that $\bigcap_{K \in \Omega^R_{S,T}(X)} \sigma_S(T+K) \subset \sigma_{e5,S}(T)$ and $\bigcap_{K \in \Omega^R_{S,T}(X)} \sigma_S(T+K) \subset \sigma_{e5,S}(T)$. Note that, if $T \in \mathcal{C}(X)$, and if K is a T-bounded operator and $\lambda \in \rho_S(T+K)$, then according to Remark 2.1.4 (*iv*), $T(\lambda S - T - K)^{-1}$ is a closed linear operator defined on X, and therefore bounded. We claim that $\sigma_{e5,S}(T) \subset \bigcap_{K \in \Omega^L_{S,T}(X)} \sigma_S(T+K)$ (resp. $\sigma_{e5,S}(T) \subset \bigcap_{K \in \Omega^R_{S,T}(X)} \sigma_S(T+K)$). Indeed, if $\lambda \notin \bigcap_{K \in \Omega^L_{S,T}(X)} \sigma_S(T+K)$ (resp. $\lambda \notin \bigcap_{K \in \Omega^R_{S,T}(X)} \sigma_S(T+K)$), then there exists $K \in \Omega^L_{S,T}(X)$ (resp. $K \in \Omega^R_{S,T}(X)$) such that $-(\lambda S - T - K)^{-1}K \in \mathcal{DC}(X)$ (resp. $-K(\lambda - T - K)^{-1} \in \mathcal{DC}(X)$) whenever $\lambda \in \rho_S(T+K)$. Hence, by applying Corollary 5.4.2, we get $[I + (\lambda S - T - K)^{-1}K] \in \Phi^b(X)$ and $i[I + (\lambda S - T - K)^{-1}K] = 0$, (resp. $[I + K(\lambda S - T - K)^{-1}] \in \Phi^b(X)$ and $i[I + K(\lambda S - T - K)^{-1}] = 0$). Moreover, we have $\lambda S - T = (\lambda S - T - K)[I + (\lambda S - T - K)^{-1}K]$, (resp. $\lambda S - T = [I + K(\lambda S - T - K)^{-1}](\lambda S - T - K)$). Then, by using Theorem 2.2.40, we get $(\lambda S - T) \in \Phi(X)$ and $i(\lambda S - T) = 0$. By applying Proposition 7.1.1, we conclude that $\lambda \notin \sigma_{e5,S}(T)$, which proves our claim. Q.E.D.

Let $T \in \mathcal{C}(X)$ with $\rho_S(T) \neq \emptyset$. We define the S-upper spectrum of T by

$$\sigma_{+,S}(T) = \bigcap_{K \in \Omega^L_{S,T}(X)} \sigma_{ap,S}(T+K).$$

Theorem 9.6.2. *For any $T \in \mathcal{C}(X)$ and $S \in \mathcal{L}(X)$, such that $S \neq 0$, we have*

$$\sigma_{+,S}(T) = \sigma_{e7,S}(T).$$

◇

Proof. Since $\mathcal{K}(X) \subset \Omega^L_{S,T}(X)$, we infer that $\sigma_{+,S}(T) \subset \sigma_{e7,S}(T)$. Conversely, let $\lambda \notin \sigma_{+,S}(T)$, then there exists $K \in \Omega^L_{S,T}(X)$ such that $\inf_{\|x\|=1, x \in \mathcal{D}(T)} \|(\lambda S - T)x\| > 0$. The use of Theorem 2.2.1 and Lemma 6.3.1 allows us to conclude that $\lambda S - T - K \in \Phi_+(X)$ and $i(\lambda S - T - K) \leq 0$. Since $Y := R(\lambda S - T - K)$ is a closed subset of X, it is itself a Banach space with the same norm. Therefore, $\lambda S - \hat{T} - K \in \mathcal{L}(Y, X_T)$. Now, let us notice that $\lambda S - \hat{T} = \lambda S - \hat{T} - K + K = (I + K(\lambda S - \hat{T} - K)^{-1})(\lambda S - \hat{T} - K)$. Since the operator $-K(\lambda S - \hat{T} - K)^{-1}$ is demicompact, 1-set contraction, we infer, by using Corollary 5.4.2, that $I + K(\lambda S - \hat{T} - K)^{-1}$ is a Fredholm operator and $i(I + K(\lambda S - \hat{T} - K)^{-1}) = 0$. Using Theorem 2.2.7, we conclude that $\lambda S - \hat{T} \in \Phi^b_+(X_T, X)$ and $i(\lambda S - \hat{T}) = i(\lambda S - \hat{T} - K) \leq 0$. Hence, $\lambda \notin \sigma_{e7,S}(T)$. Q.E.D.

Remark 9.6.1. Note that Theorems 9.6.1 and 9.6.2 remain true if we replace $\mathcal{DC}(X)$ by a larger class, that of operators K such that μK is demicompact for any $\mu \in [0, 1]$. ◇

Chapter 10
Essential Spectra of 2 × 2 Block Operator Matrices

Let X and Y be Banach spaces. In the product space $X \times Y$, we consider an operator formally defined by a matrix

$$L_0 := \begin{pmatrix} A & B \\ C & D \end{pmatrix}. \tag{10.0.1}$$

In general, the operators occurring in the representation of L_0 are unbounded. The operator A acts on the Banach space X and has the domain $\mathcal{D}(A)$, D is defined on $\mathcal{D}(D)$ and acts on the Banach space Y, and the intertwining operator B (resp. C) is defined on the domain $\mathcal{D}(B)$ (resp. $\mathcal{D}(C)$) and acts from Y into X (resp. from X into Y). One of the problems in the study of such operators is that in general L_0 is not closed or even closable, even if its entries are closed. The aim of this chapter is to present some hypotheses which should allow the block operator matrix L_0 to be closable. For this purpose, it is interesting to present the suitable conditions for the entries A, B, C, and D. These conditions are mainly based on the Frobenius–Schur decomposition of L_0. Then, our reasoning will be mainly based on the specific properties of the first entry A, hence allowing us to get a closable block matrix operator L_0. It is interesting to notice that this reasoning based on A could also be applied for the other entries, namely B, C, or D. In fact our approach is based on three situations. In the first one, the resolvent of the operator A is a Fredholm perturbation. In the second situation, the operator A is closed, whereas this same operator is closable in the third situation. After that, we will give the essential spectra of the closure of the operator L_0.

A. Jeribi, *Spectral Theory and Applications of Linear Operators and Block Operator Matrices*, DOI 10.1007/978-3-319-17566-9_10

10.1 Case Where the Resolvent of the Operator A Is a Fredholm Perturbation

10.1.1 Frobenius–Schur's Decomposition

First, we will search the Frobenius–Schur's decomposition of the operator L_0 defined in (10.0.1). For this purpose, let $\begin{pmatrix} x \\ y \end{pmatrix} \in \mathcal{D}(L_0)$ and $\lambda \in \mathbb{C}$. Then,

$$(L_0 - \lambda)\begin{pmatrix} x \\ y \end{pmatrix} = \begin{pmatrix} 0 \\ 0 \end{pmatrix} \text{ if, and only if, } \begin{pmatrix} A-\lambda & B \\ C & D-\lambda \end{pmatrix}\begin{pmatrix} x \\ y \end{pmatrix} = \begin{pmatrix} 0 \\ 0 \end{pmatrix}.$$

This leads to the following system:

$$\begin{cases} (A-\lambda)x = -By \\ Cx + (D-\lambda)y = 0. \end{cases} \tag{10.1.1}$$

Suppose that $\rho(A)$ is nonempty and let $\lambda \in \rho(A)$. Then, the first equation of the system (10.1.1) gives $x = -(A-\lambda)^{-1}By$. Consequently, the second equation of (10.1.1) becomes

$$-C(A-\lambda)^{-1}By + (D-\lambda)y = 0. \tag{10.1.2}$$

From Eq. (10.1.2), we must assume that $\mathcal{D}(A) \subset \mathcal{D}(C)$. Then, Eq. (10.1.2) becomes

$$(D - C(A-\lambda)^{-1}B - \lambda)y = 0.$$

Now, let us decompose the operator $L_0 - \lambda$ into the following form

$$L_0 - \lambda I = \begin{pmatrix} I & 0 \\ F(\lambda) & I \end{pmatrix}\begin{pmatrix} A-\lambda & 0 \\ 0 & D - C(A-\lambda)^{-1}B - \lambda \end{pmatrix}\begin{pmatrix} I & \tilde{G}(\lambda) \\ 0 & I \end{pmatrix} \tag{10.1.3}$$

where $F(\lambda)$ and $\tilde{G}(\lambda)$ are unknown. If we suppose that $\mathcal{D}(B) \subset \mathcal{D}(D)$, then $\mathcal{D}(D - C(A-\lambda)^{-1}B) = \mathcal{D}(B)$. As a necessary condition, let us seek $\tilde{G}(\lambda)$ and $F(\lambda)$ satisfying the equality (10.1.3). In fact,

$$\begin{pmatrix} A-\lambda & B \\ C & D-\lambda \end{pmatrix}\begin{pmatrix} x \\ y \end{pmatrix} = \begin{pmatrix} I & 0 \\ F(\lambda) & I \end{pmatrix}\begin{pmatrix} A-\lambda & (A-\lambda)\tilde{G}(\lambda) \\ 0 & D - C(A-\lambda)^{-1}B - \lambda \end{pmatrix}\begin{pmatrix} x \\ y \end{pmatrix}.$$

The above equality leads to the first equation $(A-\lambda)x + (A-\lambda)\tilde{G}(\lambda)y = (A-\lambda)x + By$. Then, $(A-\lambda)\tilde{G}(\lambda)y = By$, and hence $\tilde{G}(\lambda) = (A-\lambda)^{-1}B$. The second equation gives $F(\lambda)(A-\lambda)x + (F(\lambda)(A-\lambda)\tilde{G}(\lambda) + D - C(A-\lambda)^{-1}B - \lambda)y = Cx + (D-\lambda)y$. Under the action of x, let $F(\lambda) = C(A-\lambda)^{-1}$. In our decomposition (10.1.3), we must ensure that the operators

$$\begin{pmatrix} I & 0 \\ F(\lambda) & I \end{pmatrix}$$

and

$$\begin{pmatrix} I & \tilde{G}(\lambda) \\ 0 & I \end{pmatrix}$$

are invertible with bounded inverses on the whole space $X \times Y$. If we assume that C is densely defined and constitutes a closable operator from X into Y, then, for all $\lambda \in \rho(A)$, $C(A-\lambda)^{-1}$ is closable with a domain equal to X. Using the closed graph theorem (see Theorem 2.1.3), we infer that $F(\lambda) = C(A-\lambda)^{-1}$ is bounded. For $\tilde{G}(\lambda)$, if we assume that B is densely defined and constitutes a closable operator from Y into X and, for $\lambda \in \rho(A)$, $(A-\lambda)^{-1}B$ is bounded on its domain $\mathcal{D}(B)$, then $\overline{\mathcal{D}((A-\lambda)^{-1}B)} = \overline{\mathcal{D}(B)} = Y$. Hence, we must choose $G(\lambda) = \overline{(A-\lambda)^{-1}B}$ instead of $\tilde{G}(\lambda)$ in the decomposition (10.1.3), such that

$$\begin{pmatrix} I & G(\lambda) \\ 0 & I \end{pmatrix}$$

is defined on $X \times Y$. Now, let us check the sufficient condition. For $\lambda \in \rho(A)$, let us consider the operator

$$\mathcal{Z}(\lambda) = \begin{pmatrix} I & 0 \\ F(\lambda) & I \end{pmatrix} \begin{pmatrix} A-\lambda & 0 \\ 0 & D - C(A-\lambda)^{-1}B - \lambda \end{pmatrix} \begin{pmatrix} I & G(\lambda) \\ 0 & I \end{pmatrix}.$$

Now, we have to show that $\mathcal{Z}(\lambda) = L_0 - \lambda$. For this purpose, we have to prove that $\mathcal{Z}(\lambda) = L_0 - \lambda$ on $\mathcal{D}(L_0)$ and $\mathcal{D}(\mathcal{Z}(\lambda)) \subset \mathcal{D}(L_0)$.

Let $\begin{pmatrix} x \\ y \end{pmatrix} \in \mathcal{D}(L_0 - \lambda)$, where

$$\mathcal{D}(L_0 - \lambda) := \left\{ \begin{pmatrix} x \\ y \end{pmatrix} \in X \times Y \text{ such that } \mathcal{D}(B) \subset \mathcal{D}(D) \text{ and } \mathcal{D}(A) \subset \mathcal{D}(C) \right\}.$$

We observe that

$$\begin{aligned}
&\begin{pmatrix} I & 0 \\ F(\lambda) & I \end{pmatrix} \begin{pmatrix} A-\lambda & 0 \\ 0 & D - C(A-\lambda)^{-1}B - \lambda \end{pmatrix} \begin{pmatrix} I & G(\lambda) \\ 0 & I \end{pmatrix} \begin{pmatrix} x \\ y \end{pmatrix} \\
&\quad = \begin{pmatrix} (A-\lambda)x + (A-\lambda)G(\lambda)y \\ F(\lambda)(A-\lambda)x + (F(\lambda)(A-\lambda)G(\lambda) + D - C(A-\lambda)^{-1}B - \lambda)y \end{pmatrix} \\
&\quad = \begin{pmatrix} (A-\lambda)x + (A-\lambda)\overline{(A-\lambda)^{-1}B}y \\ Cx + C\overline{(A-\lambda)^{-1}B}y + (D - C(A-\lambda)^{-1}B - \lambda)y \end{pmatrix}.
\end{aligned}$$

Since $y \in \mathcal{D}(B)$, we deduce that $\overline{(A-\lambda)^{-1}B}y = (A-\lambda)^{-1}By$ and we obtain

$$\mathcal{Z}(\lambda)\begin{pmatrix} x \\ y \end{pmatrix} = \begin{pmatrix} (A-\lambda)x + By \\ Cx + (D-\lambda)y \end{pmatrix} = \begin{pmatrix} A-\lambda & B \\ C & D-\lambda \end{pmatrix}\begin{pmatrix} x \\ y \end{pmatrix}.$$

This proves that $\mathcal{Z}(\lambda) = L_0 - \lambda$ on $\mathcal{D}(L_0)$ and $L_0 - \lambda \subset \mathcal{Z}(\lambda)$. Now, we shall verify the inclusion $\mathcal{D}(\mathcal{Z}(\lambda)) \subset \mathcal{D}(L_0)$. In fact, the domain of $\mathcal{D}(\mathcal{Z}(\lambda))$ is given by

$$\begin{aligned}\mathcal{D}(\mathcal{Z}(\lambda)) := \Bigg\{\begin{pmatrix} x \\ y \end{pmatrix} &\in X \times Y \text{ such that } x + G(\lambda)y \in \mathcal{D}(A) \\ &\text{and } y \in \mathcal{D}(D - C(A-\lambda)^{-1}B)\Bigg\} \\ = \Bigg\{\begin{pmatrix} x \\ y \end{pmatrix} &\in X \times Y \text{ such that } \begin{pmatrix} x \\ y \end{pmatrix} \in \begin{pmatrix} I & G(\lambda) \\ 0 & I \end{pmatrix}^{-1} (\mathcal{D}(A) \times \mathcal{D}(B))\Bigg\}.\end{aligned}$$

Since

$$\begin{pmatrix} I & G(\lambda) \\ 0 & I \end{pmatrix}^{-1} = \begin{pmatrix} I & -G(\lambda) \\ 0 & I \end{pmatrix},$$

we infer that

$$\begin{aligned}\mathcal{D}(\mathcal{Z}(\lambda)) := \Bigg\{\begin{pmatrix} x \\ y \end{pmatrix} &\in X \times Y \text{ such that } \begin{matrix} x = x_1 - G(\lambda)y_1 \\ y = y_1 \end{matrix} \text{ with } x_1 \in \mathcal{D}(A) \\ &\text{and } y_1 \in \mathcal{D}(B)\Bigg\}.\end{aligned}$$

observing that $x = x_1 - \tilde{G}(\lambda)y_1 = x_1 - (A-\lambda)^{-1}By_1 \in \mathcal{D}(A)$, we conclude that $\begin{pmatrix} x \\ y \end{pmatrix} \in \mathcal{D}(L_0)$ and $\mathcal{D}(\mathcal{Z}(\lambda)) \subset \mathcal{D}(L_0)$. So, $\mathcal{Z}(\lambda) = L_0 - \lambda$. Since

$$\begin{pmatrix} I & G(\lambda) \\ 0 & I \end{pmatrix} \text{ and } \begin{pmatrix} I & 0 \\ F(\lambda) & I \end{pmatrix}$$

are invertible with bounded inverses on $X \times Y$, we infer that $L_0 - \lambda$ is closable if, and only if,

$$\begin{pmatrix} A-\lambda & 0 \\ 0 & D - C(A-\lambda)^{-1}B - \lambda \end{pmatrix}$$

is closable if, and only if, $A - \lambda$ and $D - C(A-\lambda)^{-1}B - \lambda$ are closable.

Let us collect all the assumptions:

$(H1)$ A is a densely defined operator on X with a nonempty resolvent set $\rho(A)$.
$(H2)$ B and C are densely defined and constitute two closable operators from Y into X and from X into Y, respectively, and $\mathcal{D}(C) \supset \mathcal{D}(A)$.
$(H3)$ For some $\mu \in \rho(A)$, the operator $(A-\mu)^{-1}B$ is bounded on its domain $\mathcal{D}(B)$.
$(H4)$ $\mathcal{D}(B) \subset \mathcal{D}(D)$.
$(H5)$ For some $\mu \in \rho(A)$, the operator $D - C(A-\mu)^{-1}B$ is closable. We denote its closure by $S(\mu)$.

Now, for a better understanding of these hypotheses, let us make some related comments.

Remark 10.1.1.

(i) Using the fact that $\mathcal{D}(C) \supset \mathcal{D}(A)$ and also the closed graph theorem (see Theorem 2.1.3), we deduce that, for each $\mu \in \rho(A)$, the operator $F(\mu) := C(A-\mu)^{-1}$ is defined on X and is bounded.
(ii) If the hypothesis $(H3)$ holds for some $\mu \in \rho(A)$, then it holds for all $\mu \in \rho(A)$. Indeed, let $\mu_0 \in \rho(A)$ be such that the operator $(A-\mu_0)^{-1}B$ is bounded. Then, for an arbitrary $\mu \in \rho(A)$, the relation

$$(A-\mu)^{-1}B = (A-\mu_0)^{-1}B + (\mu-\mu_0)(A-\mu)^{-1}(A-\mu_0)^{-1}B \qquad (10.1.4)$$

shows that $(A-\mu)^{-1}B$ is also bounded.
(iii) We denote the closure of $(A-\mu)^{-1}B$ by $G(\mu)$. Then, the relation (10.1.4) implies

$$G(\mu) = G(\mu_0) + (\mu-\mu_0)(A-\mu)^{-1}G(\mu_0).$$

(iv) The fact that $\mu \in \rho(A)$ implies that the operator $C(A-\mu)^{-1}$ is defined everywhere on X and hence that the operator $C(A-\mu)^{-1}B$ is defined on $\mathcal{D}(B)$.
(v) According to the assumption $(H4)$, the operator $D - C(A-\mu)^{-1}B$ is defined on $\mathcal{D}(B)$.
(vi) If the hypothesis $(H5)$ holds for some $\mu \in \rho(A)$, then it holds for all $\mu \in \rho(A)$.
(vii) The following operator L_0 is defined by

$$\begin{cases} L_0 : \mathcal{D}(L_0) \subset X \times Y \longrightarrow X \times Y \\ \qquad\qquad \begin{pmatrix} x_1 \\ x_2 \end{pmatrix} \longrightarrow L_0 \begin{pmatrix} x_1 \\ x_2 \end{pmatrix} = \begin{pmatrix} Ax_1 + Bx_2 \\ Cx_1 + Dx_2 \end{pmatrix} \\ \mathcal{D}(L_0) = \mathcal{D}(A) \times \mathcal{D}(B). \end{cases}$$

The operator L_0 can be factored in the Frobenius–Schur's sense:

$$L_0 - \mu I = \begin{pmatrix} I & 0 \\ F(\mu) & I \end{pmatrix} \begin{pmatrix} A-\mu & 0 \\ 0 & D - C(A-\mu)^{-1}B - \mu \end{pmatrix} \begin{pmatrix} I & G(\mu) \\ 0 & I \end{pmatrix}.$$

◇

10.1.2 Closability and Closure of the Block Operator Matrix

The closability of the operator L_0 is given by the following theorem.

Theorem 10.1.1. *Let the hypotheses $(H1)$–$(H5)$ be satisfied and let $\mu \in \rho(A)$. Then, L_0 is closable and its closure L is given by the relation*

$$L = \mu I + \begin{pmatrix} I & 0 \\ F(\mu) & I \end{pmatrix} \begin{pmatrix} A-\mu & 0 \\ 0 & S(\mu)-\mu \end{pmatrix} \begin{pmatrix} I & G(\mu) \\ 0 & I \end{pmatrix}$$

or, spelled out

$$\begin{cases} L : \mathcal{D}(L) \subset X \times Y \longrightarrow X \times Y \\ \qquad \begin{pmatrix} x_1 \\ x_2 \end{pmatrix} \longrightarrow L \begin{pmatrix} x_1 \\ x_2 \end{pmatrix} = \begin{pmatrix} A(x_1 + G(\mu)x_2) - \mu G(\mu)x_2 \\ C(x_1 + G(\mu)x_2) + S(\mu)x_2 \end{pmatrix} \\ \mathcal{D}(L) = \left\{ \begin{pmatrix} x_1 \\ x_2 \end{pmatrix} \in X \times Y \text{ such that } x_1 + G(\mu)x_2 \in \mathcal{D}(A) \text{ and } x_2 \in \mathcal{D}(S(\mu)) \right\}. \end{cases}$$

◇

Proof. We observe that for bounded, everywhere defined operators R, T, having bounded, everywhere defined inverses, and a linear operator S, the operator RST is closable if, and only if, S is closable. In this case, the relationship $\overline{RST} = R\overline{S}T$ also holds for the closures. Now, the result follows directly from Remark 10.1.1 (*vii*). Q.E.D

Remark 10.1.2. Using the hypotheses $(H2)$ and $(H3)$, we infer that, for μ, $\mu_0 \in \rho(A)$, the difference

$$\big(D - C(A-\mu)^{-1}B\big) - \big(D - C(A-\mu_0)^{-1}B\big) = (\mu_0 - \mu)C(A-\mu)^{-1}(A-\mu_0)^{-1}B \tag{10.1.5}$$

constitutes a bounded operator. Therefore, if the operator $D - C(A-\mu)^{-1}B$ is closable for some $\mu \in \rho(A)$ then, it is closable for all $\mu \in \rho(A)$. Since the operator L is the closure of L_0, it does not depend on the choice of the point $\mu \in \rho(A)$ in its above description. ◇

From now, $\mathcal{I}(X)$ will denote an arbitrary nonzero two-sided closed ideal of $\mathcal{L}(X)$, satisfying the condition

$$\mathcal{K}(X) \subset \mathcal{I}(X) \subset \mathcal{F}^b(X), \tag{10.1.6}$$

where $\mathcal{K}(X)$ stands for the ideal of compact operators. We conclude this section with the following hypotheses:

$(H6)$ For some $\mu \in \rho(A)$, the resolvent $(A-\mu)^{-1} \in \mathcal{I}(X)$.
$(H7)$ For some $\mu \in \rho(A)$, the operator $F(\mu)G(\mu) := \overline{C(A-\mu)^{-2}B} \in \mathcal{I}(Y)$.

Remark 10.1.3. Let us notice that the six assumptions $(H1)$–$(H6)$ do not imply $(H7)$, even if the operator $C(A-\mu)^{-1}B$ is bounded. In fact, let us consider the Hilbert space $H = H^+ \oplus H^-$, where H^+ and H^- constitute two separable infinite-dimensional Hilbert spaces. Let $(e_k^+)_k$ and $(e_k^-)_k$ be the orthonormal bases in H^+ and H^-, respectively. Now, let us define the operators A, B, and C with the joint domain

$$\begin{aligned}\mathcal{D}(A) &= \mathcal{D}(B) = \mathcal{D}(C)\\ &= \left\{ x \in H \,:\, x = \sum_{k=1}^{\infty}(x_k^+ e_k^+ + x_k^- e_k^-),\ \sum_{k=1}^{\infty} k^2(|x_k^+|^2 + |x_k^-|^2) < \infty \right\}\end{aligned}$$

by the following equations

$$\begin{aligned} Ae_k^+ &= ke_k^-, & Be_k^+ &= ke_k^-, & Ce_k^+ &= e_k^+,\\ Ae_k^- &= ke_k^+, & Be_k^- &= e_k^+, & Ce_k^- &= ke_k^-, \end{aligned}$$

where $k = 1, 2, \ldots$. Then, A^{-1} is compact, B and C are closable, $A^{-1}B$ and $CA^{-1}B$ are bounded, but $CA^{-2}B$ is not compact since $CA^{-2}Be_k^+ = e_k^-$ for all natural numbers k. ◇

10.1.3 Essential Spectra of L

In this section, we present some results dealing with the essential spectra of the operator L. We begin with the following starting result which is fundamental for our purpose. The next result will describe the sufficient condition under which the six assumptions $(H1)$–$(H6)$ imply the assumption $(H7)$.

Theorem 10.1.2. *Under the assumptions $(H1)$–$(H6)$, the condition $(H7)$ is satisfied if, and only if, the operator $S(\mu)$ admits the representation*

$$S(\mu) = S_0 + M(\mu) \qquad (\mu \in \rho(A)) \tag{10.1.7}$$

with a closed operator S_0 *(which is independent of* μ*) and an operator* $M(\mu) \in \mathcal{I}(Y)$*. In this case,* S_0 *can be chosen to be* $S(\mu_0)$ *for any* $\mu_0 \in \rho(A)$*, and* $M(\mu)$ *depends holomorphically on* μ *in* $\rho(A)$*.* ◇

Proof. If $(H7)$ is satisfied, the use of Eq. (10.1.5) allows us to write $S(\mu)$ in the following form

$$\begin{aligned} S(\mu) &= S(\mu_0) + (\mu_0 - \mu)F(\mu)G(\mu_0) \\ &= S(\mu_0) + (\mu_0 - \mu)F(\mu_0)G(\mu_0) + (\mu_0 - \mu)^2 F(\mu_0)(A - \mu)^{-1}G(\mu_0). \end{aligned}$$

By combining the assumptions $(H6)$ and $(H7)$ and Theorem 6.3.1, we can deduce that the representation (10.1.7) follows with $S_0 = S(\mu_0)$. Conversely, Eq. (10.1.7) implies that

$$S(\mu) - S_0 = (M(\mu) - M(\mu_0))_{|\mathcal{D}(S_0)}. \tag{10.1.8}$$

Moreover, we have

$$S(\mu) - S(\mu_0) = (\mu_0 - \mu)F(\mu)G(\mu_0)_{|\mathcal{D}(S_0)}. \tag{10.1.9}$$

Hence, using Eqs. (10.1.8) and (10.1.9), we can deduce that

$$F(\mu)G(\mu_0) = (\mu_0 - \mu)^{-1}(M(\mu) - M(\mu_0)).$$

If $\mu \to \mu_0$, then the operator $F(\mu)G(\mu_0)$ tends to $F(\mu_0)G(\mu_0)$ in the operator norm topology, which means that $(\mu_0 - \mu)^{-1}(M(\mu) - M(\mu_0))$ also converges to $F(\mu)G(\mu_0)$ in the operator norm topology. Furthermore, $(M(\mu) - M(\mu_0)) \in \mathcal{I}(Y)$ and $\mathcal{I}(Y)$ is a closed sided ideal of $\mathcal{L}(Y)$. So, $F(\mu_0)G(\mu_0) \in \mathcal{I}(Y)$, which completes the proof. Q.E.D

For $\mu \in \rho(A)$, we introduce the following matrix operators which will be needed for the sequel:

$$\mathbf{G}(\mu) := \begin{pmatrix} I & 0 \\ F(\mu) & I \end{pmatrix},$$

$$\mathbf{D}(\mu) := \begin{pmatrix} A - \mu & 0 \\ 0 & S_0 + M(\mu) - \mu \end{pmatrix} \text{ and,}$$

$$\mathbf{F}(\mu) := \begin{pmatrix} I & G(\mu) \\ 0 & I \end{pmatrix}.$$

The following remark will be useful for the proof of the next theorem.

Remark 10.1.4. (a) Using Theorems 10.1.1 and 10.1.2, we can write the operator L in the following form:

$$L - \mu I = \mathbf{G}(\mu)\mathbf{D}(\mu)\mathbf{F}(\mu). \tag{10.1.10}$$

(b) If $\mu \in \rho(A)$ then,

(i) $\alpha(A-\mu)=\beta(A-\mu)=0$,
(ii) $\alpha(\mathbf{D}(\mu))=\alpha(S_0+M(\mu)-\mu)$, and
(iii) $\beta(\mathbf{D}(\mu))=\beta(S_0+M(\mu)-\mu)$. ◊

Theorem 10.1.3. *Let us assume that the hypotheses $(H1)$–$(H7)$ hold. Then,*

(i) $\sigma_{ei}(L)=\sigma_{ei}(S_0)$ *with* $i=4,\ 5$. *Moreover, if* $\mathbb{C}\backslash\sigma_{e5}(L)$ *is a connected set and neither* $\rho(S_0)$ *nor* $\rho(S(\mu))$ *is empty, then* $\sigma_{e6}(L)=\sigma_{e6}(S_0)$.
(ii) *If* $\mathcal{I}(Y)\subset\mathcal{F}_+(Y)$, *then* $\sigma_{e1}(L)=\sigma_{e1}(S_0)$.
(iii) *If* $\mathcal{I}(Y)\subset\mathcal{F}_-(Y)$ *or* $[\mathcal{I}(Y)]^*\subset\mathcal{F}_+(Y^*)$, *then* $\sigma_{e2}(L)=\sigma_{e2}(S_0)$.
(iv) *If* $\mathcal{I}(Y)\subset\mathcal{F}_+(Y)\bigcap\mathcal{F}_-(Y)$, *then* $\sigma_{e3}(L)=\sigma_{e3}(S_0)$. ◊

Proof.

(i) First, let us assume that $\lambda\in\rho(A)$. It is clear that $\mathbf{F}(\lambda)$ is a bijection from $\mathcal{D}(L)$ onto $\mathcal{D}(\mathbf{D}(\lambda))=\mathcal{D}(A)\times\mathcal{D}(S_0)$ and $\mathbf{G}(\lambda)$ is also a bijection from $X\times Y$ onto $X\times Y$. Therefore,

$$\alpha(L-\lambda)=\alpha(\mathbf{D}(\lambda)) \tag{10.1.11}$$

and,

$$\beta(L-\lambda)=\beta(\mathbf{D}(\lambda)). \tag{10.1.12}$$

By using Remark 10.1.4 (b) (ii)–(iii), and taking into account (10.1.11) and (10.1.12), we get

$$\alpha(L-\lambda)=\alpha(S_0+M(\lambda)-\lambda) \tag{10.1.13}$$

and,

$$\beta(L-\lambda)=\beta(S_0+M(\lambda)-\lambda). \tag{10.1.14}$$

Since $M(\lambda)\in\mathcal{I}(Y)$, then the numbers $\alpha(L-\lambda)$ and $\beta(L-\lambda)$ are finite if, and only if, $\alpha(S_0-\lambda)$ and $\beta(S_0-\lambda)$ are finite. Consequently, $L-\lambda$ is a Fredholm operator if, and only if, $S_0-\lambda$ is also a Fredholm operator. If this is the case, then $i(L-\lambda)=i(S_0-\lambda)$. Let $\lambda\notin\rho(A)$. By using hypothesis $(H6)$, the spectrum of A is discrete. Therefore, λ is an isolated eigenvalue of A. Let P_λ be the Riesz projection associated with λ. Then, $\lambda\in\rho(A_\lambda)$ where A_λ constitutes the finite-dimensional perturbation of A, given by $A_\lambda:=A(I-P_\lambda)+\delta P_\lambda$, $\delta\neq\lambda$. Now, for $\mu\in\rho(A_\lambda)$, we have $\overline{D-C(A_\lambda-\mu)^{-1}B}=S_0+M_\lambda(\mu)$, where the operator S_0 is introduced in (10.1.7), and the operator $M_\lambda(\mu)$ is an operator in $\mathcal{I}(Y)$. Now, let $L(\lambda)$ be the closure of the operator

$$\begin{pmatrix} A_\lambda & B \\ C & D \end{pmatrix}.$$

Since $L(\lambda)$ is a finite-dimensional perturbation of L, then $L-\lambda I$ is a Fredholm operator on $X\times Y$ if, and only if, $L(\lambda)-\lambda I$ is also a Fredholm operator on $X \times Y$. Now, with the first part of the present proof, we conclude that $\lambda \in \sigma_{ei}(L(\lambda))$ if, and only if, $\lambda \in \sigma_{ei}(S_0)$ with $i = 4,\ 5$. Let us prove the statement for $i = 6$. From the case $i = 5$, it follows that $\mathbb{C}\backslash\sigma_{e5}(L) = \mathbb{C}\backslash\sigma_{e5}(S_0)$. This set contains points belonging to $\rho(S_0)$ and $\rho(L)$. Accordingly, since $\alpha(\lambda - S_0)$ and $\beta(\lambda - S_0)$ (resp. $\alpha(\lambda - L)$ and $\beta(\lambda - L)$) are constant on any component of Φ_{S_0} (resp. Φ_L) except possibly on a discrete set of points where they have larger values (see Proposition 2.2.5 (*iii*)), it cannot contain points of $\sigma_{e6}(S_0)$ or $\sigma_{e6}(L)$. This, together with the inclusions $\sigma_{e5}(S_0) \subset \sigma_{e6}(S_0)$ and $\sigma_{e5}(L) \subset \sigma_{e6}(L)$, leads to $\sigma_{e5}(S_0) = \sigma_{e6}(S_0)$ and $\sigma_{e5}(L) = \sigma_{e6}(L)$, which completes the proof of (i).

(*ii*) Let $\lambda \in \rho(A)$. Using the fact that $\mathcal{I}(Y) \subset \mathcal{F}_+(Y)$ and also Lemma 2.2.6 (i), we deduce that $\alpha(S_0 + M(\lambda) - \lambda I)$ is finite if, and only if, $\alpha(S_0 - \lambda I)$ is finite. Hence, by referring to Eq. (10.1.13), we deduce that $\alpha(L - \lambda I)$ is finite if, and only if, $\alpha(S_0 - \lambda I)$ is finite. Now, let $\lambda \notin \rho(A)$. Since $\mathcal{F}_+(X) \subset \mathcal{F}(X)$ then, by using hypothesis $(H6)$, we infer that λ is an isolated eigenvalue of A. For the remaining part of the proof of (*ii*), it may be done in a similar way as for (i). This completes the proof of (*ii*).

(*iii*) If $\mathcal{I}(Y) \subset \mathcal{F}_-(Y)$, then the proof is the same as in (*ii*). It is sufficient to use both Eq. (10.1.14) and Lemma 2.2.6 (*ii*). If $[\mathcal{I}(Y)]^* \subset \mathcal{F}_+(Y^*)$, the result follows from the fact that $\alpha(S_0^* + M(\lambda)^* - \overline{\lambda} I) = \beta(S_0 + M(\lambda) - \lambda I)$.

(*iv*) The assertion (*iv*) follows directly from combining (*ii*) and (*iii*). Q.E.D

Now, we may deduce the following corollary.

Corollary 10.1.1. *If the operator D is everywhere defined and bounded and if, for some and hence for all $\mu_0 \in \rho(A)$, $C(A-\mu_0)^{-1}B$ is bounded, then S_0 can be chosen as being equal to $S_0 = D - \overline{C(A - \mu_0)^{-1}B}$. In particular, under these assumptions, $\sigma_{ei}(L) \neq \emptyset$ with $i = 1, \ldots, 6$ if $\dim(Y) = \infty$.* ◇

10.1.4 Sufficient Conditions

In the applications (see Chap. 13), the hypotheses $(H3)$, $(H6)$ and the boundedness of the operator $C(A - \mu)^{-1}B$ are not easy to verify. That is why, we will give some sufficient conditions which imply the above assumptions which are easier to check. For this purpose, we start by introducing the following definition.

Definition 10.1.1. The resolvent of the operator A is said to have a ray of minimal growth, if there exist some $\theta \in [0, 2\pi)$ such that $\gamma_\theta := \{\lambda \in \mathbb{C} \text{ such that } \lambda = te^{i\theta},\ t \in \mathbb{R}^+\} \subset \rho(A)$, and there is a positive constant M such that

$$\|(A - \lambda)^{-1}\| \leq \frac{M}{1 + |\lambda|} \text{ holds for all } \lambda \in \gamma_\theta. \tag{10.1.15}$$

◇

Remark 10.1.5. Let $A \in \mathcal{C}(X)$. The domain $\mathcal{D}(A)$ of A is equipped with the graph norm topology, i.e., $\|x\|_1 = \|x\| + \|Ax\|$, hence, $\mathcal{D}(A)$ is a Banach space. Let $X_{1,1}$ denote $(\mathcal{D}(A), \|.\|_1)$. If A is an operator whose resolvent has a ray of minimal growth, then the intermediate spaces $X_{1,\theta} = \mathcal{D}(A^\theta)$, $0 \leq \theta \leq 1$ between X and $X_{1,1} = \mathcal{D}(A)$ with the norm $\|x\|_\theta = \|x\| + \|A^\theta x\|$ are well defined and the same thing holds for the intermediate spaces $X^*_{1,\theta}$ between $X^*_{1,1} = \mathcal{D}(A^*)$ and X^*. ◇

Proposition 10.1.1. *If the operators A and B have the properties $(H1)$ and $(H2)$, then the assumption $(H3)$ holds if, and only if, $\mathcal{D}(B^*) \supset \mathcal{D}(A^*)$.* ◇

Proof. If $(H3)$ holds, then for $\mu \in \rho(A)$, and according to the rules concerning the adjoint operators, we obtain

$$((A-\mu)^{-1}B)^* = B^*(A^* - \overline{\mu})^{-1} \tag{10.1.16}$$

since $(A-\mu)^{-1}$ is a bounded operator on X. As the operator on the left-hand side is bounded on X^*, the same holds for the operator on the right-hand side. Hence, $\mathcal{D}(B^*) \supset \mathcal{D}(A^*)$. Conversely, this inclusion implies the boundedness of the operator on the right-hand side of (10.1.16). Hence also $((A-\mu)^{-1}B)^{**}$ and thus $\overline{(A-\mu)^{-1}B}$ is bounded, and the boundedness of $(A-\mu)^{-1}B$ follows. Q.E.D

Lemma 10.1.1. *Let the conditions $(H1)$, $(H2)$ and $(H6)$ be satisfied and let us assume that the resolvent of A has a ray of minimal growth. Furthermore, assume that for some $\theta \in (0,1)$, the inclusions*

$$\mathcal{D}(B^*) \supset \mathcal{D}((A^*)^\theta) := \mathcal{D}((A^\theta)^*) \tag{10.1.17}$$

and

$$\mathcal{D}(\overline{C}) \supset \mathcal{D}(A^{1-\theta}) \tag{10.1.18}$$

hold. Then, the conditions $(H3)$ and $(H7)$ are also fulfilled and the operator $C(A-\mu)^{-1}B$ is bounded for $\mu \in \rho(A)$. ◇

Proof. From the properties of fractional powers, it follows that $X^*_{1,\theta} \supset X^*_{1,1} = \mathcal{D}(A^*)$ for any $\theta \in [0,1)$. Moreover, the embedding of $X^*_{1,1}$ into $X^*_{1,\theta}$ is a Fredholm perturbation since $0 \in \rho(A)$ by (10.1.15), and $A^{-1} \in \mathcal{I}(X)$ (see Theorem 6.3.5). Using Proposition 10.1.1, we conclude that $(H3)$ holds if, and only if, $\mathcal{D}(B^*) \supset \mathcal{D}(A^*)$. Hence, $(H3)$ is derived from (10.1.17). Now, let us write the operator $CA^{-1}B$ in the following form:

$$CA^{-1}B = \overline{C}A^{-(1-\theta)}A^{-\theta}B. \tag{10.1.19}$$

The operator $\overline{C}A^{-(1-\theta)}$ is bounded on X according to (10.1.18). Furthermore, the operator $A^{-\theta}B$ is bounded on $\mathcal{D}(B)$ by using (10.1.17) and Proposition 10.1.1.

Hence, the boundedness of the operator $CA^{-1}B$ follows from (10.1.19). Finally, by writing $CA^{-2}B = \overline{C}A^{-(1-\theta)}A^{-1}(A^{-\theta}B)$, the hypothesis $(H7)$ follows from both the fact that $A^{-1} \in \mathcal{I}(X)$ and Theorem 6.3.4. Q.E.D

Let X be a Banach space and let T be a closed operator on X. By $\Delta^0(T)$, we denote the maximal open subset of $\mathbb{C}$, where the resolvent $(T-\lambda)^{-1}$ is finitely meromorphic, i.e., it is meromorphic on $\Delta^0(T)$ and all the coefficients in the main parts of the Laurent expansions at the poles are of finite rank.

Remark 10.1.6. $\Delta^0(T)$ is the union of all components w of Φ_T for which $w \bigcap \rho(T) \neq \emptyset$ (see [124, Lemma 2.1]). ◇

Using representation (10.1.10), we can prove the following result:

Corollary 10.1.2. *Under the six assumptions $(H1)$–$(H6)$, the set $\Delta^0(L)$ is the union of all components w of Φ_{S_0} such that, for some $\mu \in w$, the operator $S(\mu) - \mu$ maps Y bijectively onto itself.* ◇

Now, we give a sufficient condition for the fact that $\Delta^0(L)$ contains the unbounded component of $\Delta(S_0)$, denoted by $\Delta^0_{ext}(S_0)$.

Corollary 10.1.3. *Let the conditions $(H1)$, $(H2)$, and $(H6)$ hold. Assume that the resolvent of A has a ray of minimal growth. Moreover, let us assume that, for some $\theta \in (0,1)$, the inclusions (10.1.17) and (10.1.18) hold and the operator D is bounded. Then, the inclusion $\Delta^0(L) \supset \Delta^0_{ext}(S_0)$ holds. In particular, if $\Delta(S_0)$ is simply connected, then the equality $\Delta^0(L) = \Delta^0(S_0)$ holds.* ◇

Proof. Let $\lambda \in \gamma_\theta$, $0 < \theta < 1$, and consider the following identity:

$$S(\lambda) - \lambda = -\lambda + S(0) + \left[\overline{C}A^{-(1-\theta)}\right]\left[\lambda(A-\lambda)^{-1}\right]\left[\overline{A^{-\theta}B}\right].$$

By using Lemma 10.1.1, the operators $S(0)$, $\overline{C}A^{-(1-\theta)}$, and $\overline{A^{-\theta}B}$ are everywhere defined and bounded. Hence, the operator $S(\lambda) - \lambda$ has a bounded inverse for all $\lambda \in \gamma_\theta$ with a sufficiently large $|\lambda|$. By using Corollary 10.1.2, the unbounded component $\Delta^0_{ext}(S_0)$ of $\Delta(S_0)$ is a component of $\Delta^0(L)$. Q.E.D

10.2 Case Where the Operator A Is Closed

In what follows, we will assume that $X = Y$ and the following conditions hold.

$(I1)$ The operator A is a closed, densely defined linear operator on X with a nonempty resolvent set $\rho(A)$.

$(I2)$ The operator B is a densely defined linear operator on X and, for some (hence for all) $\mu \in \rho(A)$, the operator $(A-\mu)^{-1}B$ is closable. (In particular, if B is closable, then $(A-\mu)^{-1}B$ is also closable).

$(I3)$ The operator C satisfies $\mathcal{D}(A) \subset \mathcal{D}(C)$ and, for some (hence for all) $\mu \in \rho(A)$, the operator $C(A-\mu)^{-1}$ is bounded. (In particular, if C is closable, then $C(A-\mu)^{-1}$ is bounded).

$(I4)$ The lineal $\mathcal{D}(B)\bigcap\mathcal{D}(D)$ is dense in X and, for some (hence for all) $\mu \in \rho(A)$, the operator $D - C(A-\mu)^{-1}B$ is closable, and its closure will be denoted by $S(\mu)$.

Remark 10.2.1.

(*i*) From the closed graph theorem (see Theorem 2.1.3), it follows that the operator $G(\mu) := \overline{(A-\mu)^{-1}B}$ is bounded on X.

(*ii*) We emphasize that neither the domain of $S(\mu)$ nor the property of being closable depends on μ. Indeed, from the Hilbert's identity, it follows that

$$S(\lambda) = S(\mu) + (\mu - \lambda)F(\lambda)G(\mu), \tag{10.2.1}$$

where $F(\lambda) := C(A-\lambda)^{-1}$, $\lambda, \mu \in \rho(A)$. Since the operators $F(\lambda)$ and $G(\mu)$ are bounded, then the difference $S(\lambda) - S(\mu)$ is bounded. Therefore, neither the domain of $S(\mu)$ nor the property of being closable depends on μ. ◇

10.2.1 Closability and Closure of the Block Operator Matrix

Arguing as in the proof of Theorem 10.1.1, we have the following result which describes the closure of the operator L_0.

Theorem 10.2.1. *Let the conditions $(I1)$–$(I3)$ be satisfied and the lineal $M := \mathcal{D}(B)\bigcap\mathcal{D}(D)$ be dense in X. Then, the operator L_0 is closable if, and only if, the operator $S(\mu)$, $\mu \in \rho(A)$, is closable in X. Moreover, the closure L of L_0 is given by*

$$\begin{aligned} L &= \mu - U(\mu)V(\mu)W(\mu) \\ &:= \mu - \begin{pmatrix} I & 0 \\ F(\mu) & I \end{pmatrix} \begin{pmatrix} \mu - A & 0 \\ 0 & \mu - S(\mu) \end{pmatrix} \begin{pmatrix} I & G(\mu) \\ 0 & I \end{pmatrix} \end{aligned} \tag{10.2.2}$$

or, spelled out

$$\begin{cases} L : \mathcal{D}(L) \subset X \times X \longrightarrow X \times X \\ \qquad \begin{pmatrix} x \\ y \end{pmatrix} \longrightarrow L\begin{pmatrix} x \\ y \end{pmatrix} = \begin{pmatrix} A(x + G(\mu)y) - \mu G(\mu)y \\ C(x + G(\mu)y) + S(\mu)y \end{pmatrix} \\ \\ \mathcal{D}(L) = \left\{ \begin{pmatrix} x \\ y \end{pmatrix} \in X \times X \text{ such that } x + G(\mu)y \in \mathcal{D}(A),\ y \in \mathcal{D}(S(\mu)) \right\}. \end{cases}$$

◇

Note that, in view of Remark 10.2.1 (*ii*), the description of the operator L does not depend on the choice of the point $\mu \in \rho(A)$. For $\mu \in \rho(A)$, we will denote by $M(\mu)$ the operator

$$M(\mu) := \begin{pmatrix} 0 & G(\mu) \\ F(\mu) & F(\mu)G(\mu) \end{pmatrix}.$$

Lemma 10.2.1. *Let $\mathcal{I}(X)$ be any nonzero two-sided ideal of $\mathcal{L}(X)$ satisfying (10.1.6). If $F(\mu) \in \mathcal{I}(X)$, for some $\mu \in \rho(A)$, then $F(\mu) \in \mathcal{I}(X)$ for all $\mu \in \rho(A)$.* ◇

Proof. Let $\mu_0 \in \rho(A)$, such that $F(\mu_0) \in \mathcal{I}(X)$. We have

$$F(\mu) = F(\mu_0)\left[I + (\mu - \mu_0)(A - \mu)^{-1}\right],$$

for all μ in $\rho(A)$. By using the ideal property of $\mathcal{I}(X)$, this implies that $F(\mu) \in \mathcal{I}(X)$. Q.E.D

10.2.2 Essential Spectra

Lemma 10.2.2. *Let $\mathcal{I}(X)$ be a nonzero two-sided ideal of $\mathcal{L}(X)$ satisfying (10.1.6). If $F(\mu) \in \mathcal{I}(X)$, for some $\mu \in \rho(A)$, then*

(*i*) *$\sigma_{ei}(S(\mu)), i = 4, 5$ does not depend on μ.*
(*ii*) *If $\mathcal{I}(X) \subset \mathcal{F}_+(X)$, then $\sigma_{e1}(S(\mu))$ does not depend on μ.*
(*iii*) *If $\mathcal{I}(X) \subset \mathcal{F}_-(X)$ or $[\mathcal{I}(X)]^* \subset \mathcal{F}_+(X^*)$, then $\sigma_{e2}(S(\mu))$ does not depend on μ.*
(*iv*) *If $\mathcal{I}(X) \subset \mathcal{F}_+(X) \bigcap \mathcal{F}_-(X)$, then $\sigma_{e3}(S(\mu))$ does not depend on μ.* ◇

Proof. The proof of this lemma follows directly from Eq. (10.2.1) and Theorem 7.5.3. Q.E.D

Now, we are ready to express the first result of this section.

Theorem 10.2.2. *Let the matrix operator L_0 satisfy conditions $(I1)$–$(I4)$, and let $\mathcal{I}(X)$ be any nonzero two-sided ideal of $\mathcal{L}(X)$ satisfying (10.1.6). If, for some $\mu \in \rho(A)$, the operator $F(\mu) \in \mathcal{I}(X)$, then*

(*i*) *If $M(\mu) \in \mathcal{F}(X \times X)$, for some $\mu \in \rho(A)$, then $\sigma_{e4}(L) = \sigma_{e4}(A) \bigcup \sigma_{e4}(S(\mu))$ and, $\sigma_{e5}(L) \subseteq \sigma_{e5}(A) \bigcup \sigma_{e5}(S(\mu))$. Moreover, if $\mathbb{C}\backslash\sigma_{e4}(A)$ is a connected set, then $\sigma_{e5}(L) = \sigma_{e5}(A) \bigcup \sigma_{e5}(S(\mu))$. Moreover, if $\mathbb{C}\backslash\sigma_{e5}(L)$ is connected, $\rho(L) \neq \emptyset$, $\mathbb{C}\backslash\sigma_{e5}(S(\mu))$ is connected and $\rho(S(\mu)) \neq \emptyset$, then $\sigma_{e6}(L) = \sigma_{e6}(A) \bigcup \sigma_{e6}(S(\mu))$.*
(*ii*) *If $\mathcal{I}(X) \subseteq \mathcal{F}_+(X)$ and the operator $M(\mu) \in \mathcal{F}_+(X \times X)$, for some $\mu \in \rho(A)$, then*

$$\sigma_{e1}(L) = \sigma_{e1}(A) \bigcup \sigma_{e1}(S(\mu)).$$

(*iii*) *If* $\mathcal{I}(X) \subseteq \mathcal{F}_-(X)$ *and the operator* $M(\mu) \in \mathcal{F}_-(X \times X)$, *then*

$$\sigma_{e2}(L) = \sigma_{e2}(A) \bigcup \sigma_{e2}(S(\mu)).$$

(*iv*) *If* $\mathcal{I}(X) \subseteq \mathcal{F}_+(X) \bigcap \mathcal{F}_-(X)$ *and the operator* $M(\mu) \in \mathcal{F}_+(X \times X) \bigcap \mathcal{F}_-(X \times X)$, *for some* $\mu \in \rho(A)$, *then*

$$\sigma_{e3}(L) = \sigma_{e3}(A) \bigcup \sigma_{e3}(S(\mu)) \bigcup [\sigma_{e2}(A) \bigcap \sigma_{e1}(S(\mu))] \bigcup [\sigma_{e1}(A) \bigcap \sigma_{e2}(S(\mu))].$$

◇

Proof.

(*i*) Let $\mu \in \rho(A)$ be such that $M(\mu) \in \mathcal{F}(X \times X)$ and set $\lambda \in \mathbb{C}$. If we write $\lambda - L = \mu - L + (\lambda - \mu)$ and using relation (10.2.2), we have

$$\begin{aligned} \lambda - L &= \begin{pmatrix} I & 0 \\ F(\mu) & I \end{pmatrix} \begin{pmatrix} \lambda - A & 0 \\ 0 & \lambda - S(\mu) \end{pmatrix} \begin{pmatrix} I & G(\mu) \\ 0 & I \end{pmatrix} - (\lambda - \mu) M(\mu) \\ &:= U(\mu) \tilde{V}(\lambda) W(\mu) - (\lambda - \mu) M(\mu). \end{aligned} \tag{10.2.3}$$

Since $M(\mu) \in \mathcal{F}(X \times X)$, then $\lambda - L$ is a Fredholm operator if, and only if, $U(\mu)\tilde{V}(\lambda)W(\mu)$ is a Fredholm operator. Now, let us notice that the operators $U(\mu)$ and $W(\mu)$ are bounded and have bounded inverses. Hence, the operator $U(\mu)\tilde{V}(\lambda)W(\mu)$ is a Fredholm operator if, and only if, $\tilde{V}(\lambda)$ has this property if, and only if, $\lambda - A$ and $\lambda - S(\mu)$ are Fredholm operators on X. Therefore,

$$\sigma_{e4}(L) = \sigma_{e4}(A) \bigcup \sigma_{e4}(S(\mu)). \tag{10.2.4}$$

The use of Lemma 6.3.1 (*i*) and Eq. (10.2.3) shows that, for $\lambda \in \Phi_L$, we have

$$i(\lambda - L) = i(\lambda - A) + i(\lambda - S(\mu)). \tag{10.2.5}$$

From Eqs. (10.2.4) and (10.2.5), it follows immediately that $\sigma_{e5}(L) \subseteq \sigma_{e5}(A) \bigcup \sigma_{e5}(S(\mu))$. Now, suppose that $\mathbb{C} \backslash \sigma_{e4}(A)$ is connected. By using assumption ($I1$), $\rho(A)$ is nonempty. Let $\mu_0 \in \rho(A)$ then, $\mu_0 - A \in \Phi(X)$ and $i(\mu_0 - A) = 0$. Since $\rho(A) \subseteq \mathbb{C} \backslash \sigma_{e4}(A)$ and $i(\lambda - A)$ is constant on any component of Φ_A, then $i(\lambda - A) = 0$ for all $\lambda \in \mathbb{C} \backslash \sigma_{e4}(A)$. From Eqs. (10.2.4) and (10.2.5), it follows that

$$\sigma_{e5}(L) = \sigma_{e5}(A) \bigcup \sigma_{e5}(S(\mu)). \tag{10.2.6}$$

Moreover, let us assume that $\mathbb{C}\backslash\sigma_{e5}(L)$ is connected. We know that the set $\rho_5(L) := \mathbb{C}\backslash\sigma_{e5}(L)$ contains points of $\rho(L)$, which is a nonempty set. Since $\alpha(\lambda - L)$ and $\beta(\lambda - L)$ are constant on any component of Φ_L, except possibly on a discrete set of points at which they have large values (see Proposition 2.2.5 (*iii*)), then $\rho_5(L) \subset \rho_6(L) := \mathbb{C}\backslash\sigma_{e6}(L)$. This, together with the inclusion $\sigma_{e5}(L) \subset \sigma_{e6}(L)$, leads to $\sigma_{e5}(L) = \sigma_{e6}(L)$. Since $\mathbb{C}\backslash\sigma_{e4}(A)$ is connected, then we deduce that $\sigma_{e5}(A) = \sigma_{e4}(A)$. So, $\mathbb{C}\backslash\sigma_{e5}(A)$ is connected. Using the same reasoning as before, we show that $\sigma_{e5}(A) = \sigma_{e6}(A)$. The condition that $\mathbb{C}\backslash\sigma_{e5}(S(\mu))$ is connected leads to $\sigma_{e5}(S(\mu)) = \sigma_{e6}(S(\mu))$ and, the result of the assertion (i) follows from Eq. (10.2.6).

(*ii*) Let $\mu \in \rho(A)$ be such that $M(\mu)$ is an upper semi-Fredholm perturbation. Then, from Eq. (10.2.3), we have $\lambda - L \in \Phi_+(X \times X)$ if, and only if, $U(\mu)\tilde{V}(\lambda)W(\mu) \in \Phi_+(X \times X)$ if, and only if, $\lambda - A$ and $\lambda - S(\mu)$ are in $\Phi_+(X)$ since the operators $U(\mu)$ and $W(\mu)$ are bounded and have bounded inverses. Then, the result of (*ii*) follows directly.

(*iii*) The proof of this assertion may be checked in the same way as in the proof of (*ii*).

(*iv*) This assertion is an immediate consequence of (*ii*) and (*iii*). Q.E.D

Remark 10.2.2.

(i) If X is a w.c.g. Banach space and superprojective (resp. subprojective), then the ideal $\mathcal{I}(X) = \mathcal{S}(X)$ (resp. $\mathcal{I}(X) = C\mathcal{S}(X)$) satisfies the conditions of Theorem 10.2.2 (see Proposition 2.1.4). Also, if we take X as a Banach space with the DP property and $\mathcal{I}(X) = \mathcal{W}(X)$ (see Remark 2.1.7) or, if we consider the ideal $\mathcal{K}(X_p)$ in the L_p spaces, $1 \leq p \leq \infty$.

(*ii*) The ideal of finite rank operators $\mathcal{F}_0(X)$ is the minimal subset of $\mathcal{L}(X)$ for which the conditions of Theorem 10.2.2 are valid regardless of the Banach spaces.

(*iii*) It is noted that, in Theorem 10.1.3, we suppose that the operator $(A - \mu)^{-1} \in \mathcal{I}(X)$ but, in our case, we only suppose that $C(A - \mu)^{-1} \in \mathcal{I}(X)$, which is a weaker condition, and we usually obtain the same result. So, Theorem 10.2.2 may be regarded as an extension of Theorem 10.1.3 for a larger class of operators.

(*iv*) If $F(\mu)$ and $G(\mu)$ are in $\mathcal{K}(X)$, for some $\mu \in \rho(A)$, then $M(\mu) \in \mathcal{K}(X \times X) \subset \mathcal{F}(X \times X)$.

(v) Let $X = L_1(\Omega, d\mu)$ where (Ω, Σ, μ) is a positive measure space. If $F(\mu)$ and $G(\mu)$ are in $\mathcal{W}(X)$, for some $\mu \in \rho(A)$, then $M(\mu) \in \mathcal{W}(X \times X) \subset \mathcal{F}(X \times X)$.

(*vi*) If the operators A, B, C, and D are everywhere defined and bounded, the hypothesis of Theorem 10.2.2 (*iii*) can be replaced by $[\mathcal{I}(X)]^* \subset \mathcal{F}_+(X^*)$ and $[M(\mu)]^* \in \mathcal{F}_+(X^* \times X^*)$, for some $\mu \in \rho(A)$. Indeed, it is sufficient to write the relation (10.2.3) for the adjoint. Hence, $(\lambda - L)^* = \overline{\lambda} - L^* = W(\mu)^*[\tilde{V}(\lambda)]^*U(\mu)^* - (\overline{\lambda} - \overline{\mu})[M(\mu)]^*$. Now, using the fact that $\alpha(\overline{\lambda} - L^*) = \beta(\lambda - L)$ and $\alpha([\tilde{V}(\lambda)]^*) = \beta(\tilde{V}(\lambda))$ (cf. [126, 185]) and, arguing as in the proof of Theorem 10.2.2 (*ii*), we can easily derive the result.

(*vii*) Assume that the operator L acts on the product of Banach spaces $X \times X$. Using Lemma 10.2.1, we can verify that, if $F(\mu) \in \mathcal{F}^b(X)$, for some $\mu \in \rho(A)$, then $F(\mu) \in \mathcal{F}^b(X)$ for all $\mu \in \rho(A)$ and $\sigma_{ei}(S(\mu)), i = 4, 5$ does not depend on μ. Therefore, it can be shown that the result of Theorem 10.2.2 (i) remains valid if $F(\mu) \in \mathcal{F}^b(X)$ and $M(\mu) \in \mathcal{F}(X \times X)$. ◇

10.2.3 Particular Cases

Let $\gamma(.)$ be the Kuratowski measure of noncompactness in X. Now, we are ready to state and prove the first result of this section.

Theorem 10.2.3. *Let $\mu \in \rho(A)$, and let $\lambda \in \mathbb{D}(\mu, 1) := \{\lambda \in \mathbb{C}$ such that $|\lambda - \mu| < 1\}$.*

(*i*) *Suppose that there exists $A_\lambda^l \in \mathcal{G}^l(\lambda - A)$, $S_\lambda^l(\mu) \in \mathcal{G}^l(\lambda - S(\mu))$ satisfying $\gamma(A_\lambda^l G(\mu)) < \frac{1}{2}$ and $\gamma(S_\lambda^l(\mu)F(\mu)) < 1$. If $\gamma(G(\mu)) < \frac{1}{2}$, then $\lambda - L \in \Phi_l(X \times X)$ and $i(\lambda - L) = i(\tilde{V}(\lambda))$.*

(*ii*) *Suppose that there exists $A_\lambda^r \in \mathcal{G}^r(\lambda - A)$, $S_\lambda^r(\mu) \in \mathcal{G}^r(\lambda - S(\mu))$ satisfying $\gamma(F(\mu)A_\lambda^r) < 1$, $\gamma(S_\lambda^r(\mu)F(\mu)) < 1$ and $\gamma(G(\mu)S_\lambda^r(\mu)) \leq \frac{1}{2}$. If $\gamma(G(\mu)) < \frac{1}{2}$, then $\lambda - L \in \Phi_r(X \times X)$ and $i(\lambda - L) = i(\tilde{V}(\lambda))$.*

(*iii*) *Suppose that the hypotheses of (i) and (ii) hold true. Then, $\lambda - L \in \Phi(X \times X)$ and $i(\lambda - L) = i(\tilde{V}(\lambda))$.* ◇

Proof.

(*i*) Let $T_\lambda := U(\mu)\tilde{V}(\lambda)W(\mu)$ and $V_\lambda^l = \begin{pmatrix} A_\lambda^l & K_1 \\ K_2 & S_\lambda^l(\mu) \end{pmatrix}$ such that K_1 and K_2 are compact operators. According to Proposition 6.7.1 (i), $V_\lambda^l \in \mathcal{G}^l(\tilde{V}(\lambda))$. By using Lemma 6.7.1, we get $T_\lambda^l = W(\mu)^{-1}V_\lambda^l U(\mu)^{-1} \in \mathcal{G}^l(T_\lambda)$. Moreover, we have

$$T_\lambda^l M(\mu) = \begin{pmatrix} K_1 F(\mu) - G(\mu)S_\lambda^l(\mu)F(\mu) & A_\lambda^l G(\mu) - G(\mu)K_2G(\mu) \\ S_\lambda^l(\mu)F(\mu) & K_2G(\mu) \end{pmatrix}.$$

Now, the fact that the measure of noncompactness $\gamma(.)$ is semi-multiplicative and from Proposition 2.10.2, we get $\gamma(T_\lambda^l M(\mu)) \leq \max[\gamma(G(\mu))\gamma(S_\lambda^l(\mu) F(\mu)) + \gamma(A_\lambda^l G(\mu)), \gamma(S_\lambda^l(\mu)F(\mu))]$. According to the hypotheses and the fact that $|\lambda - \mu| < 1$, we deduce that $\gamma((\lambda - \mu)T_\lambda^l M(\mu)) < 1$. Finally, the results follow from Theorem 6.4.2 (i).

(*ii*) Let $V_\lambda^r = \begin{pmatrix} A_\lambda^r & K_1 \\ K_2 & S_\lambda^r(\mu) \end{pmatrix}$ be such that K_1 and K_2 are compact operators. According to Proposition 6.7.1 (*ii*), $V_\lambda^r \in \mathcal{G}^r(\tilde{V}(\lambda))$. By using Lemma 6.7.1, we have $T_\lambda^r = W(\mu)^{-1}V_\lambda^r U(\mu)^{-1} \in \mathcal{G}^r(T_\lambda)$. Moreover, we have

$$M(\mu)T_{\lambda}^{r} = \begin{pmatrix} G(\mu)K_2 - G(\mu)S_{\lambda}^{r}(\mu)F(\mu) & G(\mu)S_{\lambda}^{r}(\mu) \\ F(\mu)A_{\lambda}^{r} - F(\mu)K_1F(\mu) & F(\mu)K_1 \end{pmatrix}.$$

Now, by using Proposition 2.10.2, we get $\gamma(M(\mu)T_{\lambda}^{r}) \leq \max[\gamma(G(\mu))\gamma(S_{\lambda}^{r}(\mu)F(\mu))+\gamma(G(\mu)S_{\lambda}^{r}(\mu)), \gamma(F(\mu)A_{\lambda}^{r})]$. Finally, the results follow from Theorem 6.4.2 (*ii*).

(*iii*) The proof of (*iii*) is an immediate deduction from (*i*) and (*ii*). Q.E.D

Notice that Theorem 10.2.3 can be interpreted in terms of essential spectra as follows:

Corollary 10.2.1. *Let $\mu \in \rho(A)$.*

(*i*) *Suppose that, for each $\lambda \in \mathbb{C}$ such that $\lambda - A \in \Phi_l(X)$ and $\lambda - S(\mu) \in \Phi_l(X)$, we have $\gamma(A_{\lambda}^{l}G(\mu)) < \frac{1}{2}$ and $\gamma(S_{\lambda}^{l}(\mu)F(\mu)) < 1$. If $\gamma(G(\mu)) < \frac{1}{2}$, then*

$$\sigma_{e1l}(L) \subset \sigma_{e1l}(A) \bigcup \sigma_{e1l}(S(\mu)) \bigcup (\mathbb{C}\backslash\mathbb{D}(\mu, 1)).$$

(*ii*) *Suppose that, for each $\lambda \in \mathbb{C}$ such that $\lambda - A \in \Phi_r(X)$ and $\lambda - S(\mu) \in \Phi_r(X)$, we have $\gamma(F(\mu)A_{\lambda}^{r}) < 1$, $\gamma(S_{\lambda}^{r}(\mu)F(\mu)) < 1$ and $\gamma(G(\mu)S_{\lambda}^{r}(\mu)) \leq \frac{1}{2}$. If $\gamma(G(\mu)) < \frac{1}{2}$, then*

$$\sigma_{e2r}(L) \subset \sigma_{e2r}(A) \bigcup \sigma_{e2r}(S(\mu)) \bigcup (\mathbb{C}\backslash\mathbb{D}(\mu, 1)).$$

(*iii*) *Suppose that, for each $\lambda \in \Phi_A \bigcap \Phi_{S(\mu)}$, the hypotheses (i) and (ii) hold true. Then,*

$$\sigma_{e4}(L) \subset \sigma_{e4}(A) \bigcup \sigma_{e4}(S(\mu)) \bigcup (\mathbb{C}\backslash\mathbb{D}(\mu, 1))$$

and

$$\sigma_{e5}(L) \subset \sigma_{e5}(A) \bigcup \sigma_{e5}(S(\mu)) \bigcup (\mathbb{C}\backslash\mathbb{D}(\mu, 1)).$$

◇

Without maintaining the assumption $\gamma(G(\mu)) < \frac{1}{2}$, if we suppose that the operators $A_{\lambda}^{l}G(\mu)$ and $S_{\lambda}^{l}(\mu)F(\mu)$ are compact, then the results of Theorem 10.2.3 remain valid. So, we can deduce the following:

Corollary 10.2.2. *Let $\mu \in \rho(A)$.*

(*i*) *Suppose that, for each $\lambda \in \mathbb{C}$ such that $\lambda - A \in \Phi_l(X)$ and $\lambda - S(\mu) \in \Phi_l(X)$, we have $A_{\lambda}^{l}G(\mu)$ and $S_{\lambda}^{l}(\mu)F(\mu)$ are compact. Then $\sigma_{e1l}(L) \subset \sigma_{e1l}(A) \bigcup \sigma_{e1l}(S(\mu))$.*

(*ii*) *Suppose that, for each $\lambda \in \mathbb{C}$ such that $\lambda - A \in \Phi_r(X)$ and $\lambda - S(\mu) \in \Phi_r(X)$, we have $S_{\lambda}^{r}(\mu)F(\mu)$, $G(\mu)S_{\lambda}^{r}(\mu)$, and $F(\mu)A_{\lambda}^{r}$ are compact. Then, $\sigma_{e2r}(L) \subset \sigma_{e2r}(A) \bigcup \sigma_{e2r}(S(\mu))$.*

(*iii*) *Suppose that, for each* $\lambda \in \Phi_A \bigcap \Phi_{S(\mu)}$*, the hypotheses* ($i$) *and* ($ii$) *hold true. Then,* $\sigma_{e4}(L) \subset \sigma_{e4}(A) \bigcup \sigma_{e4}(S(\mu))$ *and* $\sigma_{e5}(L) \subset \sigma_{e5}(A) \bigcup \sigma_{e5}(S(\mu))$. ◇

Remark 10.2.3. (i) If the hypotheses of statement (iii) of Corollaries 10.2.1 and 10.2.2 hold true for some $\lambda \in \Phi_A \bigcap \Phi_{S(\mu)}$, then they also hold true for each $\alpha \in \Phi_A \bigcap \Phi_{S(\mu)}$. Indeed, suppose, for example, for $\lambda \in \Phi_A \bigcap \Phi_{S(\mu)}$, there exists A'_λ and K_1, K_2 are two compact operators satisfying $A'_\lambda A_\lambda = I + K_1$ on $\mathcal{D}(A_\lambda)$, $A_\lambda A'_\lambda = I + K_2$ on X, and $A'_\lambda G(\mu)$ is a compact operator. Let $\alpha \in \Phi_A \bigcap \Phi_{S(\mu)}$, then there exists A'_α and K_3, K_4 are two compact operators satisfying $A'_\alpha A_\alpha = I + K_3$ on $\mathcal{D}(A_\alpha)$, and $A_\alpha A'_\alpha = I + K_4$ on X. Hence, $A'_\lambda A_\alpha A'_\alpha G(\mu)$ is a compact operator. Now, since $A'_\lambda A_\alpha \in \Phi(X)$, then $A'_\alpha G(\mu)$ is compact.

(ii) If $F(\mu)$ and $G(\mu)$ are compact operators, then we get $\sigma_{e5}(L) = \sigma_{e5}(A) \bigcup \sigma_{e5}(S(\mu))$. Moreover, we obtain $\sigma_{e1l}(L) = \sigma_{e1l}(A) \bigcup \sigma_{e1l}(S(\mu))$ and $\sigma_{e2r}(L) = \sigma_{e2r}(A) \bigcup \sigma_{e2r}(S(\mu))$. Indeed, we have $\lambda - L = U(\mu)\tilde{V}(\lambda)W(\mu) - (\lambda-\mu)M(\mu)$. Since $M(\mu)$ is compact, then $\lambda - L \in \Phi_l(X)$ (resp. $\lambda - L \in \Phi_l(X)$) if, and only if, $U(\mu)\tilde{V}(\lambda)W(\mu) \in \Phi_l(X)$ (resp. $U(\mu)\tilde{V}(\lambda)W(\mu) \in \Phi_r(X)$) if, and only if, $\tilde{V}(\lambda) \in \Phi_l(X)$ (resp. $\tilde{V}(\lambda) \in \Phi_r(X)$). ◇

In the remaining part of this section, we will study the inverse inclusion in Corollary 10.2.2. For this, we suppose that

$$L = \begin{pmatrix} A & B \\ C & B \end{pmatrix} \in \mathcal{L}(X \times X).$$

Theorem 10.2.4. *Let* $\mu \in \rho(A)$*, and let* $\lambda \in \mathbb{D}(\mu, 1)$.

(i) *Suppose that there exist* $A^l_\lambda \in \mathcal{G}^l(\lambda - A)$, $D^l_\lambda \in \mathcal{G}^l(\lambda - D)$ *satisfying* $\gamma(A^l_\lambda G(\mu)) < 1$ *and* $\gamma(D^l_\lambda F(\mu)) < \frac{1}{2}$. *If* $\gamma(G(\mu)) < 1$*, then* $\tilde{V}(\lambda) \in \Phi_l(X)$ *and* $i(\lambda - L) = i(\tilde{V}(\lambda))$.

(ii) *Suppose that there exist* $A^r_\lambda \in \mathcal{G}^r(\lambda - A)$, $D^r_\lambda \in \mathcal{G}^r(\lambda - D)$ *satisfying* $\gamma(G(\mu)D^r_\lambda) < 1$ *and* $\gamma(F(\mu)A^r_\lambda) < \frac{1}{2}$. *If* $\gamma(F(\mu)) < \frac{1}{2}$*, then* $\tilde{V}(\lambda) \in \Phi_r(X)$ *and* $i(\lambda - L) = i(\tilde{V}(\lambda))$.

(iii) *Suppose that the hypotheses of* (i) *and* (ii) *hold true. Then,* $0 \in \Phi_{\tilde{V}(\lambda)}$ *and* $i(\lambda - L) = i(\tilde{V}(\lambda))$. ◇

Proof.

(i) Let us consider $L^l_\lambda = \begin{pmatrix} A^l_\lambda & K_1 \\ K_2 & D^l_\lambda \end{pmatrix}$ such that K_1 and K_2 are compact operators. According to Proposition 6.7.2 (i), $L^l_\lambda \in \mathcal{G}^l(\lambda - L)$. Moreover, we have

$$L^l_\lambda M(\mu) = \begin{pmatrix} K_1 F(\mu) & A^l_\lambda G(\mu) + K_1 F(\mu) G(\mu) \\ D^l_\lambda F(\mu) & K_2 G(\mu) + D^l_\lambda F(\mu) G(\mu) \end{pmatrix}.$$

Now, the fact that the measure of noncompactness $\gamma(.)$ is semi-multiplicative and applying Proposition 2.10.2, we get $\gamma(L_\lambda^l M(\mu)) \leq \max[\gamma(A_\lambda^l G(\mu)), \gamma(D_\lambda^l F(\mu)) + \gamma(D_\lambda^l F(\mu))\gamma(G(\mu))]$. According to the hypotheses and to the fact that $|\lambda - \mu| < 1$, we deduce that $\gamma((\lambda - \mu)L_\lambda^l M(\mu)) < 1$. Finally, the results follow from Theorem 6.4.2 (i).

(ii) Let $L_\lambda^r = \begin{pmatrix} A_\lambda^r & K_1 \\ K_2 & D_\lambda^r \end{pmatrix}$ be such that K_1 and K_2 are compact operators. According to Proposition 6.7.2 (ii), $L_\lambda^r \in \mathcal{G}^r(\lambda - L)$. Besides, we have

$$M(\mu)L_\lambda^r = \begin{pmatrix} G(\mu)K_2 & G(\mu)D_\lambda^r \\ F(\mu)A_\lambda^r + F(\mu)G(\mu)K_2 & F(\mu)K_1 + F(\mu)G(\mu)D_\lambda^r \end{pmatrix}.$$

Now, by using Proposition 2.10.2, we get $\gamma(M(\mu)L_\lambda^r) \leq \max[\gamma(G(\mu)D_\lambda^r), \gamma(F(\mu)A_\lambda^r) + \gamma(F(\mu))\gamma(G(\mu)D_\lambda^r)]$. Finally, the results follow from Theorem 6.4.2 (ii).

(iii) The proof of (iii) is an immediate deduction from (i) and (ii). Q.E.D

Without maintaining the assumption $\gamma(G(\mu)) < 1$ and $\gamma(F(\mu)) < \frac{1}{2}$, if we suppose that the operators $A_\lambda^l G(\mu)$, $D_\lambda^l F(\mu)$ and $S_\lambda^l(\mu)F(\mu)$ are compact, then the results of Theorem 10.2.4 remain valid. So, according to Corollary 10.2.2, we can deduce the following:

Corollary 10.2.3. *Let* $\mu \in \rho(A)$.

(i) *Suppose that, for each* $\lambda \in \mathbb{C}$ *such that* $\lambda - A \in \Phi_l(X)$, $\lambda - S(\mu) \in \Phi_l(X)$ *and* $\lambda - D \in \Phi_l(X)$, *we notice that* $A_\lambda^l G(\mu)$, *and* $D_\lambda^l F(\mu)$ *are compact. Then,* $\sigma_{e1l}(L) = \sigma_{e1l}(A) \bigcup \sigma_{e1l}(S(\mu))$.

(ii) *Suppose that, for each* $\lambda \in \mathbb{C}$ *such that* $\lambda - A \in \Phi_r(X)$, $\lambda - S(\mu) \in \Phi_r(X)$ *and* $\lambda - D \in \Phi_r(X)$, *we notice that* $G(\mu)D_\lambda^r$, *and* $F(\mu)A_\lambda^r$ *are compact. Then,* $\sigma_{e2r}(L) = \sigma_{e2r}(A) \bigcup \sigma_{e2r}(S(\mu))$.

(iii) *Suppose that, for some* $\lambda \in \Phi_A \bigcap \Phi_{S(\mu)} \bigcap \Phi_D$, *the hypotheses* (i) *and* (ii) *hold true. Then* $\sigma_{e4}(L) = \sigma_{e4}(A) \bigcup \sigma_{e4}(S(\mu))$ *and* $\sigma_{e5}(L) = \sigma_{e5}(A) \bigcup \sigma_{e5}(S(\mu))$. *Moreover, if* $\mathbb{C} \backslash \sigma_{e5}(L)$ *is connected,* $\rho(L) \neq \emptyset$ *and* $\rho(S(\mu)) \neq \emptyset$, *then* $\sigma_{e6}(L) = \sigma_{e6}(A) \bigcup \sigma_{e6}(S(\mu))$. ◇

10.3 Case Where the Operator *A* Is Closable

Let Z be a Banach space, we consider the linear operators Γ_X from X into Z and Γ_Y from Y into Z. Therefore, we define the operator $\mathcal{A}_0$ in the Banach space $X \times Y$ as follows:

$$\mathcal{A}_0 := \begin{pmatrix} A & B \\ C & D \end{pmatrix}$$

$$\mathcal{D}(\mathcal{A}_0) := \left\{ \begin{pmatrix} x \\ y \end{pmatrix} \text{ such that } x \in \mathcal{D}(A),\ y \in \mathcal{D}(D) \bigcap \mathcal{D}(B) \text{ and } \Gamma_X x = \Gamma_Y y \right\}.$$

In what follows, we will assume that the following conditions hold:

$(J1)$ The operator A is densely defined and closable.
$\mathcal{D}(\overline{A})$, the domain of closure $\overline{A}$ of A, coincides with the Banach space X_A which is contained in X.
$(J2)$ $\mathcal{D}(A) \subset \mathcal{D}(\Gamma_X) \subset X_A$ and Γ_X is bounded as a mapping from X_A into Z.
$(J3)$ The set $\mathcal{D}(A) \bigcap N(\Gamma_X)$ is dense in X and the resolvent set of the restriction $A_1 := A|_{\mathcal{D}(A) \bigcap N(\Gamma_X)}$ is not empty, i.e., $\rho(A_1) \neq \emptyset$.
$(J4)$ $\mathcal{D}(A) \subset \mathcal{D}(C) \subset X_A$ and C is A_1-bounded.

Remark 10.3.1. From $(J3)$, it follows that A_1 is a closed operator in the Banach space X_A with a nonempty resolvent set. For $\lambda \in \rho_6(A_1)$, let P_λ denote the corresponding finite rank Riesz projector with a range and a kernel denoted by R_λ and N_λ, respectively. Let $A_{1\lambda}$ be the operator defined by $A_{1\lambda} := (A_1 - \lambda)(I - P_\lambda) + P_\lambda$ because $\mathcal{D}(A_1)$ is P_λ-invariant, $A_{1\lambda}$ has the same domain of A_1 with respect to the decomposition $X = R_\lambda \oplus N_\lambda$, we can write $A_{1\lambda} = (A_1 - \lambda|_{N_\lambda}) \oplus I$. Since $\sigma(A_1 - \lambda|_{N_\lambda}) = \sigma(A_1 - \lambda)\backslash\{0\}$, $A_{1\lambda}$ has a bounded inverse denoted by $R_b(A_1, \lambda)$ and called the Browder resolvent. This clearly extends the usual resolvent $(A_1 - \lambda)^{-1}$ from $\rho(A_1)$ to $\rho_6(A_1)$. $\diamond$

10.3.1 Closability and Closure of the Block Operator Matrix

Lemma 10.3.1. *Under the assumptions $(J1)$–$(J3)$, for any $\lambda \in \rho_6(A_1)$, the following decomposition holds:*

$$\mathcal{D}(A) = \mathcal{D}(A_1) \oplus N(A_\lambda), \tag{10.3.1}$$

where A_λ is the operator defined on $\mathcal{D}(A)$ by $A_\lambda := (A - \lambda)(I - P_\lambda) + P_\lambda$. $\diamond$

Proof. Let $\lambda \in \rho_6(A_1)$. It is clear that the sum (10.3.1) is contained in $\mathcal{D}(A)$. Then, we have $\mathcal{D}(A_1) \bigcap N(A_\lambda) = N(A_{1\lambda})$. Since the operator $A_{1\lambda}$ is invertible then, $N(A_{1\lambda}) = \{0\}$, and we get $\mathcal{D}(A_1) \bigcap N(A_\lambda) = \{0\}$. For any $f \in \mathcal{D}(A)$, let $g = R_b(A_1, \lambda) A_\lambda f \in \mathcal{D}(A_1)$. Then, $f - g \in N(A_\lambda)$ and $f = g + f - g \in \mathcal{D}(A_1) + N(A_\lambda)$. Q.E.D

Lemma 10.3.2. *Under the assumptions $(J1)$–$(J3)$, for any $\lambda \in \rho_6(A_1)$, the restriction $\Gamma_\lambda := \Gamma_X|_{N(A_\lambda)}$ is injective, and $R(\Gamma_\lambda) = \Gamma_X(N(A_\lambda)) = \Gamma_X(\mathcal{D}(A)) := Z_1$ does not depend on λ.* $\diamond$

Proof. Let $\lambda \in \rho_6(A_1)$. The injectivity of the operator Γ_λ follows from the fact that $N(\Gamma_\lambda) := N(A_\lambda) \bigcap N(\Gamma_X) = N(A_{1\lambda}) = \{0\}$. From the definition of the operator Γ_λ, we deduce that this range coincides with $\Gamma_X(N(A_\lambda))$. Therefore, from $(J3)$, it follows that $\Gamma_X(\mathcal{D}(A_1)) = \{0\}$. Hence, the use of Lemma 10.3.1 combined with the linearity of the operator Γ_X, allows us to conclude that $\Gamma_X(N(A_\lambda)) = \Gamma_X(\mathcal{D}(A))$. Hence, $R(\Gamma_\lambda)$ does not depend on λ. Q.E.D

In what follows, and for $\lambda \in \rho_6(A_1)$, the inverse of the operator Γ_λ, denoted K_λ, will play an important role $K_\lambda := \left(\Gamma_X|_{N(A_\lambda)}\right)^{-1} : Z_1 \longrightarrow N(A_\lambda) \subset X$. In other words, $K_\lambda z = x$ means that $x \in \mathcal{D}(A)$ and,

$$A_\lambda x = 0, \tag{10.3.2}$$

$$\Gamma_X x = z. \tag{10.3.3}$$

Lemma 10.3.3. *If $\lambda_1, \lambda_2 \in \rho_6(A_1)$, then $K_{\lambda_1} - K_{\lambda_2} = R_b(A_1,\lambda_1)[(\lambda_1 - \lambda_2) + S_{A_1}(\lambda_1,\lambda_2)]K_{\lambda_2}$, where $S_{A_1}(.,.)$ is the finite rank operator defined in (8.2.1). If K_λ is closable for, at least, one $\lambda \in \rho_6(A_1)$, then it is closable for all such λ, and the above relation holds, with K_{λ_j} replaced by the closures $\overline{K}_{\lambda_j}$, $j = 1, 2$.* ◇

Proof. Let $z \in Z_1$ and set $x = x_1 - x_2$ where $x_j = K_{\lambda_j} z$, with $j = 1, 2$. The use of Eq. (10.3.2) shows that

$$\begin{aligned}
A_{\lambda_1} x &= - A_{\lambda_1} x_2 \\
&= - [(A - \lambda_1)(I - P_{\lambda_1}) + P_{\lambda_1}]\, x_2 \\
&= - [(A - \lambda_2)(I - P_{\lambda_1}) + (\lambda_2 - \lambda_1)(I - P_{\lambda_1}) + P_{\lambda_1}]\, x_2 \\
&= [(A - (\lambda_1 + 1)) P_{\lambda_1} - (A - (\lambda_2 + 1)) P_{\lambda_2} + (\lambda_1 - \lambda_2)]\, x_2 \\
&= [(A_1 - (\lambda_1 + 1)) P_{\lambda_1} - (A_1 - (\lambda_2 + 1)) P_{\lambda_2} + (\lambda_1 - \lambda_2)]\, x_2 \\
&= [(\lambda_1 - \lambda_2) + S_{A_1}(\lambda_1, \lambda_2)]\, x_2.
\end{aligned}$$

Therefore, and from Eq. (10.3.3), we deduce that $\Gamma_X x = \Gamma_X x_1 - \Gamma_X x_2 = 0$. Hence, $x \in \mathcal{D}(A_1)$ and $x = R_b(A_1,\lambda_1)\,[(\lambda_1 - \lambda_2) + S_{A_1}(\lambda_1,\lambda_2)]\, x_2$. This allows us to conclude that $K_{\lambda_1} - K_{\lambda_2} = R_b(A_1,\lambda_1)\,[(\lambda_1 - \lambda_2) + S_{A_1}(\lambda_1,\lambda_2)]\, K_{\lambda_2}$. So,

$$K_{\lambda_2} - K_{\lambda_1} = -R_b(A_1,\lambda_2)\,[(\lambda_1 - \lambda_2) + S_{A_1}(\lambda_1,\lambda_2)]\, K_{\lambda_1}.$$

Hence, $[(\lambda_1 - \lambda_2) + S_{A_1}(\lambda_1,\lambda_2)]\, K_{\lambda_1} = A_{1\lambda_2} R_b(A_1,\lambda_1)\,[(\lambda_1 - \lambda_2) + S_{A_1}(\lambda_1,\lambda_2)]$ K_{λ_2}. Since the operator $S_{A_1}(.,.)$ is of finite rank and since $A_{1\lambda_2} R_b(A_1,\lambda_1)$ is bounded and boundedly invertible, K_{λ_1} is closable if K_{λ_2} is similar. In such a case, their closures $\overline{K}_{\lambda_j}$, $j = 1, 2$ satisfy the same relations. Q.E.D

Concerning the operators K_λ, D, Γ_Y, and B, we prescribe the following conditions:

$(J5)$ For some (hence, for all) $\lambda \in \rho_6(A_1)$, the operator K_λ is bounded as a mapping from Z into X.

$(J6)$ The operator D is densely defined and closed.

$(J7)$ $\mathcal{D}(\Gamma_Y) \supset \mathcal{D}(D)\bigcap\mathcal{D}(B)$, the set $Y_1 = \{y$ such that $y \in \mathcal{D}(D)\bigcap\mathcal{D}(B)$ and$\Gamma_Y y \in Z_1\}$ is dense in Y and, the restriction of Γ_Y to this set, is bounded as an operator from Y into Z.

$(J8)$ For some (and hence, for all, see Lemma 8.2.2 (i)) $\lambda \in \rho_6(A_1)$, the operator $R_b(A_1,\lambda)B$ is closable and its closure $\overline{R_b(A_1,\lambda)B}$ is bounded.

Remark 10.3.2. We will denote by

(i) $\overline{\Gamma}_X$ the extension of Γ_X by continuity to $X_A = \mathcal{D}(\overline{A})$. It is a bounded operator from X_A into Z.
(ii) $\overline{\Gamma}^0_Y$ the extension of $\Gamma_Y|_{Y_1}$ by continuity to all elements of Y.
(iii) $\overline{K}_\lambda$ the extension of K_λ to the closure $\overline{Z}_1$ of Z_1, with respect to the norm of Z. Without loss of generality, we assume that $\overline{Z}_1 = Z$. We can easily verify that the operator $\overline{K}_\lambda$ is also bounded as a mapping from $\overline{Z}_1$ to X_A. ◇

In the space Y, and for $\lambda \in \rho_6(A_1)$, let us consider the operator $M_\lambda := D + CK_\lambda\Gamma_Y - C_\lambda B$ where $C_\lambda := CR_b(A_1,\lambda)$. The operator M_λ is defined on the set Y_1, which is dense in Y according to $(J7)$.

Remark 10.3.3. For any λ_1 and $\lambda_2 \in \rho_6(A_1)$, it follows, from the resolvent identity, that $M_{\lambda_1} - M_{\lambda_2} = C_{\lambda_1}\left[(\lambda_2-\lambda_1) - S_{A_1}(\lambda_1,\lambda_2)\right]\left[-K_{\lambda_2}\Gamma_Y + R_b(A_1,\lambda_2)B\right]$. From Lemma 8.2.2 (ii), we immediately deduce that C_λ is bounded. Therefore, we observe that Γ_Y is bounded on this domain by assumption $(J7)$, that K_λ is bounded by assumption $(J5)$, that $R(K_\lambda) \subset \mathcal{D}(A) \subset \mathcal{D}(C)$ and finally $S_{A_1}(.,.)$ is of finite rank. Now, using $(J8)$, we infer that, if M_λ is closable as an operator in Y for some $\lambda \in \rho_6(A_1)$, then it is closable for all $\lambda \in \rho_6(A_1)$. We also emphasize that the domain of $\overline{M}_\lambda$ does not depend on λ. Indeed, the difference

$$\overline{M}_{\lambda_1} - \overline{M}_{\lambda_2} = C_{\lambda_1}\left[(\lambda_2-\lambda_1) - S_{A_1}(\lambda_1,\lambda_2)\right]\left[-\overline{K}_{\lambda_2}\overline{\Gamma}^0_Y + \overline{R_b(A_1,\lambda_2)B}\right] \tag{10.3.4}$$

is a bounded operator. ◇

Lemma 10.3.4. *For $\lambda \in \rho_6(A_1)$ and $x \in \mathcal{D}(A)$, we have $A_\lambda x = A_{1\lambda}(I - K_\lambda\Gamma_X)x$ and, the operator $I - K_\lambda\Gamma_X$ is the projection from $\mathcal{D}(A_1)$ parallel to $N(A_\lambda)$.* ◇

Proof. Let $x \in \mathcal{D}(A)$. Then, we have $x = (I - K_\lambda\Gamma_X)x + K_\lambda\Gamma_X x$. The first summand belongs to $\mathcal{D}(A_1)$ because $x_1 = (I - K_\lambda\Gamma_X)x \in \mathcal{D}(A)$ and $\Gamma_X x_1 = \Gamma_X x - \Gamma_X K_\lambda \Gamma_X x = 0$. Therefore, it is clear that the second summand belongs to $N(A_\lambda)$. Now, we may apply Lemma 10.3.1 in order to get the result. Q.E.D

For each $\lambda \in \rho_6(A_1)$, we define the bounded, lower and upper triangular operator matrices

$$\mathbb{T}_1(\lambda) = \begin{pmatrix} I & 0 \\ C_\lambda & I \end{pmatrix}, \qquad \mathbb{T}_2(\lambda) = \begin{pmatrix} I & -\overline{K}_\lambda\overline{\Gamma}^0_Y + \overline{R_b(A_1,\lambda)B} \\ 0 & I \end{pmatrix},$$

the finite rank operator-matrix

$$\mathbb{N}(\lambda) = \begin{pmatrix} [A_1 - (\lambda + 1)]\, P_\lambda & 0 \\ 0 & 0 \end{pmatrix}$$

and the diagonal operator-matrix

$$\mathbb{D}_0(\lambda) = \begin{pmatrix} A_{1\lambda} & 0 \\ 0 & M_\lambda - \lambda \end{pmatrix}$$

with the domain $\mathcal{D}(A_1) \times Y_1$.

Theorem 10.3.1. *Let us assume that the conditions* $(J1)$–$(J8)$ *are satisfied. Then,* $\mathcal{A}_0$ *is closable in* $X \times Y$ *if, and only if, the operator* $M_\lambda := D + CK_\lambda\Gamma_Y - C_\lambda B$ *is closable for some* $\lambda \in \rho_6(A_1)$*, or equivalently, for all* $\lambda \in \rho_6(A_1)$*. Moreover, the closure* $\mathcal{A}$ *of* $\mathcal{A}_0$ *is given by the relation*

$$\mathcal{A} := \overline{\mathcal{A}_0} = \lambda I + \mathbb{T}_1(\lambda)\mathbb{D}(\lambda)\mathbb{T}_2(\lambda) + \mathbb{N}(\lambda), \tag{10.3.5}$$

where $\mathbb{D}(\lambda) := \overline{\mathbb{D}_0(\lambda)} = \begin{pmatrix} A_{1\lambda} & 0 \\ 0 & \overline{M}_\lambda - \lambda \end{pmatrix}$ *with the domain* $\mathcal{D}(A_1) \times \mathcal{D}(\overline{M_\lambda})$. ◇

Proof. Let $\lambda \in \rho_6(A_1)$. We will show that $\mathcal{A}_0 - \lambda I = \mathcal{G}_\lambda$, where

$$\mathcal{G}_\lambda = \begin{pmatrix} I & 0 \\ C_\lambda & I \end{pmatrix} \begin{pmatrix} A_{1\lambda} & 0 \\ 0 & M_\lambda - \lambda \end{pmatrix} \begin{pmatrix} I & -K_\lambda\Gamma_Y + R_b(A_1, \lambda)B \\ 0 & I \end{pmatrix} + \begin{pmatrix} [A_1 - (\lambda + 1)]\, P_\lambda & 0 \\ 0 & 0 \end{pmatrix}.$$

In order to get this equality, we will prove that $\mathcal{D}(\mathcal{G}_\lambda) \subset \mathcal{D}(\mathcal{A}_0)$ and $\mathcal{A}_0 - \lambda I = \mathcal{G}_\lambda$. First, we notice that $\mathcal{D}(\mathcal{G}_\lambda)$ consists of the elements of the form

$$\begin{pmatrix} x \\ y \end{pmatrix} = \begin{pmatrix} x' - K_\lambda\Gamma_Y y + R_b(A_1, \lambda)By \\ y \end{pmatrix},$$

where x' and y run through $\mathcal{D}(A_1) = \mathcal{D}(A) \bigcap N(\Gamma_X)$ and $\mathcal{D}(M_\lambda)$ respectively. Therefore, $x \in \mathcal{D}(A)$, $y \in \mathcal{D}(D) \bigcap \mathcal{D}(B)$ and $\Gamma_X x = \Gamma_X(K_\lambda\Gamma_Y y) = \Gamma_Y y$. Hence, $\begin{pmatrix} x \\ y \end{pmatrix} \in \mathcal{D}(\mathcal{A}_0)$ and $\mathcal{D}(\mathcal{G}_\lambda) \subset \mathcal{D}(\mathcal{A}_0)$. Second, let $\begin{pmatrix} x \\ y \end{pmatrix} \in \mathcal{D}(\mathcal{A}_0)$, i.e., $x \in \mathcal{D}(A)$, $y \in \mathcal{D}(D) \bigcap \mathcal{D}(B)$ and $\Gamma_X x = \Gamma_Y y$. We have

$$\mathcal{G}_\lambda \begin{pmatrix} x \\ y \end{pmatrix} = \begin{pmatrix} A_{1\lambda} & 0 \\ C & M_\lambda - \lambda \end{pmatrix} \begin{pmatrix} (I - K_\lambda\Gamma_X)x + R_b(A_1, \lambda)By \\ y \end{pmatrix} + \begin{pmatrix} [A_1 - (\lambda + 1)]\, P_\lambda x \\ 0 \end{pmatrix}.$$

Using Lemma 10.3.4, we get

$$\mathcal{G}_\lambda \begin{pmatrix} x \\ y \end{pmatrix} = \begin{pmatrix} A_\lambda x + By \\ C(x - K_\lambda \Gamma_X x + R_b(A_1,\lambda)By) + (M_\lambda - \lambda)y \end{pmatrix}$$
$$+ \begin{pmatrix} [A_1 - (\lambda+1)]\, P_\lambda x \\ 0 \end{pmatrix}$$
$$= \begin{pmatrix} (A-\lambda)x + By \\ Cx + (D-\lambda)y \end{pmatrix} = (\mathcal{A}_0 - \lambda I) \begin{pmatrix} x \\ y \end{pmatrix}.$$

Therefore, $\mathcal{A}_0 - \lambda I = \mathcal{G}_\lambda$. Finally, it is easy to check that $\mathbb{T}_1(\lambda)$ and $\mathbb{T}_2(\lambda)$ are bounded and also have bounded inverses. Then, from the factorization of $\mathcal{A}_0 - \lambda I$, we deduce that $\mathcal{A}_0$ is closable in $X \times Y$ if, and only if, M_λ is closable as a mapping in Y. Moreover, if M_λ is closable and $\overline{M}_\lambda$ denotes its closure, then for the closure $\mathcal{A}$ of $\mathcal{A}_0$, we get

$$\mathcal{A} := \overline{\mathcal{A}_0} = \lambda I + \mathbb{T}_1(\lambda) \begin{pmatrix} A_{1\lambda} & 0 \\ 0 & \overline{M_\lambda} - \lambda \end{pmatrix} \mathbb{T}_2(\lambda) + \mathbb{N}(\lambda).$$

Q.E.D

10.3.2 Essential Spectra of $\mathcal{A}$

Lemma 10.3.5. *Let us assume that, for some (and hence, for all) $\mu \in \rho_6(A_1)$, the operator M_μ is closable and that the set $\Phi^b(Y,X)$ is not empty. Then,*

(i) If $C_\mu \in \mathcal{F}^b(X,Y)$, then $\sigma_{ei}(\overline{M}_\mu)$, $i = 4,5$ does not depend on μ.
(ii) If $C_\mu \in \mathcal{F}_+^b(X,Y)$, then $\sigma_{e1}(\overline{M}_\mu)$ does not depend on μ.
(iii) If $C_\mu \in \mathcal{F}_-^b(X,Y)$, then $\sigma_{e2}(\overline{M}_\mu)$ does not depend on μ.
(iv) If $C_\mu \in \mathcal{F}_+^b(X,Y) \bigcap \mathcal{F}_-^b(X,Y)$, then $\sigma_{e3}(\overline{M}_\mu)$ does not depend on μ.
(v) If $C_\lambda \in \mathcal{F}_+^b(X,Y)$ for some $\lambda \in \rho_6(A_1)$, then $C_\lambda \in \mathcal{F}_+^b(X,Y)$ for all $\lambda \in \rho_6(A_1)$, and $\sigma_{e7}(\overline{M_\lambda})$ does not depend on the choice of λ.
(vi) If $C_\lambda \in \mathcal{F}_-^b(X,Y)$ for some $\lambda \in \rho_6(A_1)$, then $C_\lambda \in \mathcal{F}_-^b(X,Y)$ for all $\lambda \in \rho_6(A_1)$, and $\sigma_{e8}(\overline{M_\lambda})$ does not depend on the choice of λ. ◇

Proof.

(*i*) Since the operator $S_{A_1}(\lambda,\mu)$ is of finite rank, and from both Eq. (10.3.4) and the assumption $(J8)$, it follows that $[(\lambda-\mu) + S_{A_1}(\lambda,\mu)]\left[-\overline{K_\lambda \Gamma_Y^0} + \overline{R_b(A_1,\lambda)B}\right] \in \mathcal{L}(Y,X)$. Using the fact that $C_\mu \in \mathcal{F}^b(X,Y)$, together with Theorem 6.3.1 (*ii*), we infer that $C_\mu[(\lambda-\mu)+S_{A_1}(\lambda,\mu)]\Big[-\overline{K}_\lambda\overline{\Gamma}^0_Y+\overline{R_b(A_1,\lambda)B}\Big]\in\mathcal{F}^b(Y)$. Therefore, from Eq. (10.3.4) and Theorem 7.5.3 (*i*), we can deduce that $\sigma_{ei}(\overline{M}_\mu)=\sigma_{ei}(\overline{M}_\lambda)$, with $i=4,5$. This proves the statement of (*i*).

(*ii*) Since $C_\lambda\in\mathcal{F}^b_+(X,Y)$, (resp. $\mathcal{F}^b_-(X,Y)$), and using both Theorem 6.3.1 (*i*) and Eq. (10.3.4), it follows that $\overline{M}_\mu-\overline{M}_\lambda\in\mathcal{F}^b_+(Y)$ (resp. $\mathcal{F}^b_-(Y)$). The use of Theorem 7.5.3 (*ii*) (resp. (*iii*)) shows that $\sigma_{e1}(\overline{M}_\mu)=\sigma_{e1}(\overline{M}_\lambda)$, (resp. $\sigma_{e2}(\overline{M}_\mu)=\sigma_{e2}(\overline{M}_\lambda)$). Hence, the assertions (*ii*) and (*iii*) hold.

(*iv*) The use of the items (*ii*) and (*iii*) enables us to prove this assertion.

(*v*) Let $\lambda_0\in\rho_6(A_1)$ such that $C_{\lambda_0}\in\mathcal{F}^b_+(X,Y)$. From the resolvent identity, we have $C_\lambda-C_{\lambda_0}=C_{\lambda_0}[(\lambda-\lambda_0)+S_{A_1}(\lambda,\lambda_0)]\,R_b(A_1,\lambda)$ for all $\lambda\in\rho_6(A_1)$. Hence, by writing C_λ in the form $C_\lambda=C_{\lambda_0}[I+((\lambda-\lambda_0)+S_{A_1}(\lambda,\lambda_0))\,R_b(A_1,\lambda)]$ and by using Proposition 6.3.1 (*ii*), we deduce that $C_\lambda\in\mathcal{F}^b_+(X,Y)$ and the difference

$$\overline{M}_\lambda-\overline{M}_{\lambda_0}=C_\lambda\,[(\lambda_0-\lambda)-S_{A_1}(\lambda,\lambda_0)]\Big[-\overline{K}_{\lambda_0}\overline{\Gamma}^0_Y+\overline{R_b(A_1,\lambda_0)B}\Big]$$

belongs to $\mathcal{F}^b_+(Y,Y)$. Now, the use of Theorem 7.5.11 (*i*) and Remark 7.5.1 allows us to conclude that $\sigma_{e7}(\overline{M_\lambda})$ does not depend on the choice of λ.

(*vi*) This assertion can be proved in the same way as for (*i*). Q.E.D

Now, we are ready to express the main results of this section.

Theorem 10.3.2. *Let the assumptions* $(J1)$–$(J8)$ *hold. Assume that the operator* M_μ *is closable for some* $\mu\in\rho_6(A_1)$ *and that the set* $\Phi^b(Y,X)$ *is not empty. Then,*

(*i*) *If for some* $\mu\in\rho_6(A_1)$, *the operator* $C_\mu\in\mathcal{F}^b(X,Y)$, *then* $\sigma_{ei}(\mathcal{A})\bigcap\rho_6(A_1)=\sigma_{ei}(\overline{M}_\mu)\bigcap\rho_6(A_1)$, *with* $i=4,5$. *Moreover, if* $\mathbb{C}\backslash\sigma_{e5}(\mathcal{A})$ *and* $\mathbb{C}\backslash\sigma_{e5}(\overline{M}_\mu)$ *are connected, and if* $\rho(\mathcal{A})$ *and* $\rho(\overline{M}_\mu)$ *are nonempty, then* $\sigma_{e6}(\mathcal{A})\bigcap\rho_6(A_1)=\sigma_{e6}(\overline{M}_\mu)\bigcap\rho_6(A_1)$.

(*ii*) *If for some* $\mu\in\rho_6(A_1)$, *the operator* $C_\mu\in\mathcal{F}^b_+(X,Y)$, *then* $\sigma_{e1}(\mathcal{A})\bigcap\rho_6(A_1)=\sigma_{e1}(\overline{M}_\mu)\bigcap\rho_6(A_1)$, *and* $\sigma_{e7}(\mathcal{A})\bigcap\rho_6(A_1)=\sigma_{e7}(\overline{M}_\mu)\bigcap\rho_6(A_1)$.

(*iii*) *If for some* $\mu\in\rho_6(A_1)$, *the operator* $C_\mu\in\mathcal{F}^b_-(X,Y)$, *then* $\sigma_{e2}(\mathcal{A})\bigcap\rho_6(A_1)=\sigma_{e2}(\overline{M}_\mu)\bigcap\rho_6(A_1)$, *and* $\sigma_{e8}(\mathcal{A})\bigcap\rho_6(A_1)=\sigma_{e8}(\overline{M}_\mu)\bigcap\rho_6(A_1)$.

(*iv*) *If for some* $\mu\in\rho_6(A_1)$, *the operator* $C_\mu\in\mathcal{F}^b_+(X,Y)\bigcap\mathcal{F}^b_-(X,Y)$, *then* $\sigma_{e3}(\mathcal{A})\bigcap\rho_6(A_1)=\sigma_{e3}(\overline{M}_\mu)\bigcap\rho_6(A_1)$. ◊

Proof. (*i*) Let $\lambda\in\rho_6(A_1)$. Then, λ belongs to the union of $\rho(A_1)$ and the discrete spectrum of A_1. So, we will discuss two cases:

First case: If $\lambda\in\rho(A_1)$, then the Schur-Frobenius factorization given by Theorem 10.3.1 can be written as follows:

$$\mathcal{A}-\lambda I=\begin{pmatrix} I & 0\\ C(A_1-\lambda)^{-1} & I\end{pmatrix}\begin{pmatrix} A_1-\lambda & 0\\ 0 & \overline{M}_\lambda-\lambda\end{pmatrix}\begin{pmatrix} I & -\overline{K}_\lambda\overline{\Gamma}^0_Y+\overline{(A_1-\lambda)^{-1}B}\\ 0 & I\end{pmatrix}. \tag{10.3.6}$$

Obviously, the external factors

$$\begin{pmatrix} I & 0 \\ C(A_1-\lambda)^{-1} & I \end{pmatrix} \quad \text{and} \quad \begin{pmatrix} I & -\overline{K_\lambda \Gamma_Y^0 + (A_1-\lambda)^{-1}B} \\ 0 & I \end{pmatrix}$$

are bounded and also have bounded inverses. Hence, by using Eq. (10.3.6), we can conclude that $\mathcal{A}-\lambda$ is a Fredholm operator if, and only if, $\overline{M}_\lambda - \lambda$ is also a Fredholm operator and, in this case, $i(\mathcal{A}-\lambda) = i(\overline{M}_\lambda - \lambda)$. Now, using Lemma 10.3.5, we can deduce that $\lambda \in \sigma_{ei}(\mathcal{A})$, with $i = 4, 5$ if, and only if, $\lambda \in \sigma_{ei}(\overline{M}_\mu)$, with $i = 4, 5$. So, we infer that

$$\sigma_{ei}(\mathcal{A}) \bigcap \rho(A_1) = \sigma_{ei}(\overline{M}_\mu) \bigcap \rho(A_1), \text{ with } i = 4, 5.$$

Second case: If λ belongs to the discrete spectrum of A_1, then there exists $\varepsilon > 0$ such that the disc $\{\xi \in \mathbb{C}$ such that $|\xi - \lambda| \leq 2\varepsilon\}$ does not contain points of $\sigma(A_1)$ different from λ and the Riesz projection P_λ of A_1 corresponding to λ is of finite rank. Now, let us consider the operator $\tilde{A}_1 := A_1 + \varepsilon P_\lambda$. Then, $\{\xi \in \mathbb{C}$ such that $|\xi - \lambda| < \varepsilon\} \subset \rho_6(A_1) \bigcap \rho_6(\tilde{A}_1)$. Until further, we fix $\mu \in \rho_6(A_1) \bigcap \rho_6(\tilde{A}_1)$. Now, we consider the operator matrix $\tilde{\mathcal{A}}_0$ which is the finite rank perturbation of $\mathcal{A}_0$ defined as follows:

$$\tilde{\mathcal{A}}_0 := \begin{pmatrix} \tilde{A} & B \\ C & D \end{pmatrix} := \mathcal{A}_0 + \varepsilon \begin{pmatrix} P_\lambda & 0 \\ 0 & 0 \end{pmatrix},$$

where $\tilde{A} := A + \varepsilon P_\lambda$. In the following, the closure of $\tilde{\mathcal{A}}_0$ will be denoted by $\tilde{\mathcal{A}}$ and satisfies

$$\tilde{\mathcal{A}} := \mathcal{A} + \varepsilon \begin{pmatrix} P_\lambda & 0 \\ 0 & 0 \end{pmatrix}.$$

Clearly, $\tilde{\mathcal{A}}$ is a finite rank perturbation of $\mathcal{A}$. So, $\mathcal{A} - \lambda$ is a Fredholm operator on $X \times Y$ if, and only if, $\tilde{\mathcal{A}} - \lambda$ is a Fredholm operator on $X \times Y$ as well. Therefore, $i(\tilde{\mathcal{A}} - \lambda) = i(\mathcal{A} - \lambda)$. Then, we conclude that $\sigma_{ei}(\tilde{\mathcal{A}}) = \sigma_{ei}(\mathcal{A})$, with $i = 4, 5$. In the following, we are going to apply the result of the first part of this proof for the operator $\tilde{\mathcal{A}}$. For $\mu \in \rho_6(\tilde{A}_1)$, let us consider the operator $\tilde{K}_\mu$, which is defined as K_μ with A replaced by $\tilde{A}$. Hence, $\tilde{K}_\mu z = x$ means that $x \in \mathcal{D}(\tilde{A})$, $\tilde{A}_\mu x = 0$ and $\Gamma_X x = z$. After this, we construct the perturbation of M_μ, defined by $\tilde{M}_\mu := D + C\tilde{K}_\mu \Gamma_Y - CR_b(\tilde{A}_1, \mu)B$. We can easily check that the operator $\tilde{M}_\mu$ is closable and we will denote its closure by $\hat{M}_\mu$. Moreover, it is clear that $CR_b(\tilde{A}_1, \mu) \in \mathcal{F}^b(X, Y)$. So, by applying Lemma 10.3.5, we infer that $\tilde{M}_\mu$ is closable for all $\mu \in \rho_6(\tilde{A}_1)$ and $\sigma_{ei}(\hat{M}_\mu)$ does not depend on μ (with $i = 4, 5$). In what follows, we are going to show that the difference of $\hat{M}_\mu - \overline{M}_\mu$ is of finite rank. This difference $\tilde{M}_\mu - M_\mu$ can be written in the following form

$\tilde{M}_\mu - M_\mu = C[\tilde{K}_\mu - K_\mu]\Gamma_Y - C[R_b(\tilde{A}_1, \mu) - R_b(A_1, \mu)]B$. It remains to prove that $\tilde{K}_\mu - K_\mu$ and that the closure of $C[R_b(\tilde{A}_1, \mu) - R_b(A_1, \mu)]B$ is of finite rank. In fact, on the one hand, if we take $z \in Z_1$ and we put $\tilde{x} = \tilde{K}_\mu z$, $x = K_\mu z$, we obtain $\Gamma_X(\tilde{x} - x) = 0$ and, $A_\mu(\tilde{x} - x) = -\varepsilon P_\lambda \tilde{x}$. From these two latter relations, we infer that $\tilde{x} - x \in \mathcal{D}(A_1)$ and $\tilde{x} - x = -\varepsilon R_b(A_1, \mu) P_\lambda \tilde{x}$. So, $\tilde{K}_\mu - K_\mu = -\varepsilon R_b(A_1, \mu) P_\lambda \tilde{K}_\mu$ which is a finite rank operator. On the other hand, the closure of the right side of $C[R_b(\tilde{A}_1, \mu) - R_b(A_1, \mu)]B = \varepsilon C_\mu P_\lambda R_b(\tilde{A}_1, \mu)B$ is of finite rank. Then, our aim is established, and we deduce that $\sigma_{ei}(\hat{M}_\mu) = \sigma_{ei}(\overline{M}_\mu)$, with $i = 4, 5$. Now, for $\lambda \in \rho_6(\tilde{A}_1)$, we infer, from the first part of this proof, that $\sigma_{ei}(\tilde{\mathcal{A}}) = \sigma_{ei}(\hat{M}_\lambda), i = 4, 5$. Finally, we may deduce that $\sigma_{ei}(\mathcal{A}) = \sigma_{ei}(\overline{M}_\mu), i = 4, 5$. Moreover, if we assume that $\mathbb{C}\backslash\sigma_{e5}(\mathcal{A})$ and $\mathbb{C}\backslash\sigma_{e5}(\overline{M}_\mu)$ are connected, $\rho(\mathcal{A}) \neq \emptyset$ and $\rho(\overline{M}_\mu) \neq \emptyset$, we get $\sigma_{e5}(\mathcal{A}) = \sigma_{e6}(\mathcal{A})$ and $\sigma_{e5}(\overline{M}_\mu) = \sigma_{e6}(\overline{M}_\mu)$. Hence, the proof of this assertion is achieved.

(ii), *(iii)* and *(iv)* can be checked in the same way as for the proof of (i). Q.E.D

We will denote by $\mathbb{Q}(\lambda)$ the operator defined as follows

$$\mathbb{Q}(\lambda) := \begin{pmatrix} 0 & C_\lambda \\ \overline{-K_\lambda \Gamma_Y^0 + R_b(A_1, \lambda)B} & C_\lambda\left[\overline{-K_\lambda \Gamma_Y^0 + R_b(A_1, \lambda)B}\right] \end{pmatrix}.$$

Theorem 10.3.3. *Let the matrix operator $\mathcal{A}_0$ satisfy the assumptions $(J1)$–$(J8)$. If, for some $\mu \in \rho_6(A_1)$, the operator M_μ is closable and the set $\Phi^b(Y, X) \neq \emptyset$, then*

(i) If, for some $\mu \in \rho(A_1)$, the operators $C_\mu \in \mathcal{F}^b(X, Y)$ and $\mathbb{Q}(\mu) \in \mathcal{F}^b(X \times Y)$, then $\sigma_{e4}(\mathcal{A}) = \sigma_{e4}(A_1) \bigcup \sigma_{e4}(\overline{M}_\mu)$, and $\sigma_{e5}(\mathcal{A}) \subseteq \sigma_{e5}(A_1) \bigcup \sigma_{e5}(\overline{M}_\mu)$. Moreover, if $\mathbb{C}\backslash\sigma_{e4}(A_1)$ is connected, then $\sigma_{e5}(\mathcal{A}) = \sigma_{e5}(A_1) \bigcup \sigma_{e5}(\overline{M}_\mu)$. Besides, if $\mathbb{C}\backslash\sigma_{e5}(\mathcal{A})$ and $\mathbb{C}\backslash\sigma_{e5}(\overline{M}_\mu)$ are connected, $\rho(\mathcal{A}) \neq \emptyset$ and $\rho(\overline{M}_\mu) \neq \emptyset$, then $\sigma_{e6}(\mathcal{A}) = \sigma_{e6}(A_1) \bigcup \sigma_{e6}(\overline{M}_\mu)$.

(ii) If, for some $\mu \in \rho_6(A_1)$, the operators $C_\mu \in \mathcal{F}_+^b(X, Y)$ and $\mathbb{Q}(\mu) \in \mathcal{F}_+^b(X \times Y)$, then $\sigma_{e1}(\mathcal{A}) = \sigma_{e1}(A_1) \bigcup \sigma_{e1}(\overline{M}_\mu)$.

(iii) If, for some $\mu \in \rho_6(A_1)$, the operators $C_\mu \in \mathcal{F}_-^b(X, Y)$ and $\mathbb{Q}(\mu) \in \mathcal{F}_-^b(X \times Y)$, then $\sigma_{e2}(\mathcal{A}) = \sigma_{e2}(A_1) \bigcup \sigma_{e2}(\overline{M}_\mu)$.

(iv) If, for some $\mu \in \rho_6(A_1)$, the operators $C_\mu \in \mathcal{F}_+^b(X, Y) \bigcap \mathcal{F}_-^b(X, Y)$ and $\mathbb{Q}(\mu) \in \mathcal{F}_+^b(X \times Y) \bigcap \mathcal{F}_-^b(X \times Y)$, then

$$\sigma_{e3}(\mathcal{A}) = \sigma_{e3}(A_1) \bigcup \sigma_{e3}(\overline{M}_\mu) \bigcup [\sigma_{e1}(A_1) \bigcap \sigma_{e2}(\overline{M}_\mu)]$$
$$\bigcup [\sigma_{e2}(A_1) \bigcap \sigma_{e1}(\overline{M}_\mu)].$$

(v) If, for some $\mu \in \rho_6(A_1)$, the operator $C_\mu \in \mathcal{F}_+^b(X, Y)$ and the operator $\mathbb{Q}(\mu) \in \mathcal{F}_+(X, Y)$, then $\sigma_{e7}(\mathcal{A}) \subseteq \sigma_{e7}(A_1) \bigcup \sigma_{e7}(\overline{M}_\mu)$. In the addition, if we

suppose that the sets $\Phi_{\mathcal{A}}$, Φ_{A_1} and $\Phi_{\overline{M}_\mu}$ are connected and the sets $\rho(\overline{M}_\mu)$ and $\rho(\mathcal{A})$ are not empty, then $\sigma_{e7}(\mathcal{A}) = \sigma_{e7}(A_1)\bigcup\sigma_{e7}(\overline{M}_\mu)$.

(*vi*) *If, for some $\mu \in \rho_6(A_1)$, the operator $C_\mu \in \mathcal{F}_-^b(X,Y)$ and the operator $\mathbb{Q}(\mu) \in \mathcal{F}_-^b(X,Y)$, then $\sigma_{e8}(\mathcal{A}) \subseteq \sigma_{e8}(A_1)\bigcup\sigma_{e8}(\overline{M}_\mu)$. In the addition, if we suppose that the sets $\Phi_{\mathcal{A}}$, Φ_{A_1} and $\Phi_{\overline{M}_\mu}$ are connected and the sets $\rho(\overline{M}_\mu)$ and $\rho(\mathcal{A})$ are not empty, then $\sigma_{e8}(\mathcal{A}) = \sigma_{e8}(A_1)\bigcup\sigma_{e8}(\overline{M}_\mu)$.* ◇

Proof.

(*i*) Let $\mu \in \rho_6(A_1)$ be such that $\mathbb{Q}(\mu) \in \mathcal{F}^b(X \times Y)$ and let $\lambda \in \mathbb{C}$. Using the representation (10.3.5), we can write the following relation:

$$\begin{aligned}
\mathcal{A} - \lambda &= (\mathcal{A} - \mu) + (\mu - \lambda)\\
&= \mathbb{T}_1(\mu)\mathbb{D}(\mu)\mathbb{T}_2(\mu) + \mathbb{N}(\mu) + (\mu - \lambda)\\
&= \mathbb{T}_1(\mu)\mathbb{V}(\lambda)\mathbb{T}_2(\mu) + \mathbb{N}(\mu)\\
&\quad + (\lambda - \mu)\mathbb{Q}(\mu) + \mathbb{T}_1(\mu)\begin{pmatrix} [\mu + 1 - A_1]P_\mu & 0\\ 0 & 0\end{pmatrix}\mathbb{T}_2(\mu)\\
&= \mathbb{T}_1(\mu)\mathbb{V}(\lambda)\mathbb{T}_2(\mu) + (\lambda - \mu)\mathbb{Q}(\mu) + \mathbb{W}(\mu), &(10.3.7)\\
&= \mathbb{T}_1(\mu)\mathbb{V}(\lambda)\mathbb{T}_2(\mu) + (\lambda - \mu)\mathbb{Q}(\mu) - \mathbb{P}(\mu) + \mathbb{N}(\mu). &(10.3.8)
\end{aligned}$$

where

$$\mathbb{V}(\lambda) := \begin{pmatrix} A_1 - \lambda & 0\\ 0 & \overline{M}_\mu - \lambda\end{pmatrix},$$

$$\mathbb{W}(\mu) := \begin{pmatrix} 0 & [\mu + 1 - A_1]P_\mu\left[-\overline{K}_\mu\overline{\Gamma}_Y^0 + \overline{R_b(A_1,\mu)B}\right]\\ C_\mu[\mu + 1 - A_1]P_\mu & C_\mu[\mu + 1 - A_1]P_\mu\left[-\overline{K}_\mu\overline{\Gamma}_Y^0 + \overline{R_b(A_1,\mu)B}\right]\end{pmatrix},$$

and

$$\mathbb{P}(\mu) := \begin{pmatrix} [A_1 - (\mu + 1)]\,P_\mu & [A_1 - (\mu + 1)]\,P_\mu\left[-\overline{K}_\mu\overline{\Gamma}_Y^0 + \overline{R_b(A_1,\mu)B}\right]\\ C_\mu\,[A_1 - (\mu + 1)]\,P_\mu & C_\mu\,[A_1 - (\mu + 1)]\,P_\mu\left[-\overline{K}_\mu\overline{\Gamma}_Y^0 + \overline{R_b(A_1,\mu)B}\right]\end{pmatrix}.$$

Let us notice that the operator P_μ is of finite rank and the operators $\overline{K}_\mu, \overline{\Gamma}_Y^0$ and $\overline{R_b(A_1,\mu)B}$ are bounded. Hence, we deduce that the matrix operator $\mathbb{W}(\mu)$ is of finite rank. Using the fact that $\mathbb{Q}(\mu) \in \mathcal{F}^b(X \times Y)$, we infer that $\mathcal{A} - \lambda$ is a Fredholm operator if, and only if, $\mathbb{T}_1(\mu)\mathbb{V}(\lambda)\mathbb{T}_2(\mu)$ has also this property. Since the operators $\mathbb{T}_1(\mu)$ and $\mathbb{T}_2(\mu)$ are bounded and also have bounded inverses, we claim that $\mathcal{A} - \lambda$ is a Fredholm operator if, and only if, the operators $A_1 - \lambda$ and $\overline{M}_\mu - \lambda$ have also the same characteristic. Therefore,

$$\sigma_{e4}(\mathcal{A}) = \sigma_{e4}(A_1)\bigcup\sigma_{e4}(\overline{M}_\mu). \qquad (10.3.9)$$

For $\lambda \in \Phi_{\mathcal{A}}$, we infer that

$$i(\mathcal{A}-\lambda) = i(A_1-\lambda) + i(\overline{M}_{\mu}-\lambda). \tag{10.3.10}$$

Clearly, the use of Eqs. (10.3.9) and (10.3.10) shows that $\sigma_{e5}(\mathcal{A}) \subseteq \sigma_{e5}(A_1) \bigcup \sigma_{e5}(\overline{M}_{\mu})$. Conversely, let $\lambda \notin \sigma_{e5}(\mathcal{A})$ and suppose that $\mathbb{C}\backslash\sigma_{e4}(A_1)$ is connected. By using the assumption $(J3)$, $\rho(A_1)$ is nonempty. So, let $\mu \in \rho(A_1)$. Then, $A_1-\mu \in \Phi(X)$ and $i(A_1-\mu) = 0$. The fact that $i(A_1-\lambda)$ is constant on any connected component of Φ_{A_1} (see Proposition 2.2.5) and knowing that $\rho(A_1) \subseteq \mathbb{C}\backslash\sigma_{e4}(A_1)$, lead to the following $i(A_1-\lambda) = 0$ for all $\lambda \in \mathbb{C}\backslash\sigma_{e4}(A_1)$. In this case, from Eq. (10.3.10), it follows that $i(\overline{M}_{\mu}-\lambda) = 0$. Now, the use of Eqs. (10.3.9) and (10.3.10) allows us to conclude that

$$\sigma_{e5}(\mathcal{A}) = \sigma_{e5}(A_1) \bigcup \sigma_{e5}(\overline{M}_{\mu}). \tag{10.3.11}$$

By using the assumption $(J3)$ and the fact that $\mathbb{C}\backslash\sigma_{e4}(A_1)$ is connected, we show from Theorem 7.3.1 (i) that $\sigma_{e4}(A_1) = \sigma_{e5}(A_1)$. So, $\mathbb{C}\backslash\sigma_{e5}(A_1)$ is connected. This condition leads to $\sigma_{e5}(A_1) = \sigma_{e6}(A_1)$ (see Theorem 7.3.1 (ii)). Moreover, the fact that $\mathbb{C}\backslash\sigma_{e5}(\mathcal{A})$ (resp. $\mathbb{C}\backslash\sigma_{e5}(\overline{M}_{\mu})$) is connected and $\rho(\mathcal{A}) \neq \emptyset$ (resp. $\rho(\overline{M}_{\mu}) \neq \emptyset$), together with Theorem 7.3.1 (ii), enable us to deduce that $\sigma_{e5}(\mathcal{A}) = \sigma_{e6}(\mathcal{A})$ (resp. $\sigma_{e5}(\overline{M}_{\mu}) = \sigma_{e6}(\overline{M}_{\mu})$). Now, using the relation (10.3.11) and arguing as above, we can easily derive the result for $\sigma_{e6}(.)$.

(ii) Let $\mu \in \rho_6(A_1)$ be such that $\mathbb{Q}(\mu)$ is an upper semi-Fredholm perturbation. Note also that the matrix operator $\mathbb{W}(\mu)$ is of finite rank. Then, the stability theorem for Fredholm operator implies from Eq. (10.3.7) that $\mathcal{A}-\lambda$ is an upper semi-Fredholm operator in $X \times Y$ if, and only if, the product $\mathbb{T}_1(\mu)\mathbb{V}(\lambda)\mathbb{T}_2(\mu)$ has also this property. Since the operators $\mathbb{T}_1(\mu)$ and $\mathbb{T}_2(\mu)$ are bounded and also have bounded inverses, then it follows that the product $\mathbb{T}_1(\mu)\mathbb{V}(\lambda)\mathbb{T}_2(\mu)$ is an upper Fredholm operator if, and only if, $\mathbb{V}(\lambda)$ has the same property. Hence, we infer that $\mathcal{A}-\lambda \in \Phi_+(X \times Y)$ if, and only if, $A_1-\lambda \in \Phi_+(X)$ and $\overline{M}_{\mu}-\lambda \in \Phi_+(Y)$. Then, the result follows for the Gustafson's essential spectrum.

(iii) A similar reasoning to *(ii)* achieves the proof for the Weidmann's essential spectrum. Indeed, it is sufficient to derive easily the result for the lower semi-Fredholm operator.

(iv) This assertion is an immediate consequence of assertions *(ii)* and *(iii)*.

(v) Let $\lambda \in \mathbb{C}$. Since $\mathbb{T}_1(\mu)$ and $\mathbb{T}_2(\mu)$ are bounded and have bounded inverses, $\mathbb{N}(\mu)$ and $\mathbb{P}(\mu)$ are finite rank matrix operators and $\mathbb{Q}(\mu) \in \mathcal{F}_+(X,Y)$, therefore, and for the same reasons as the proof of Theorem 10.3.2, it follows from Eq. (10.3.8) that $(\mathcal{A}-\lambda I)$ is an upper semi-Fredholm operator if, and only if, $\mathbb{V}(\lambda)$ has this property and $i(\mathcal{A}-\lambda I) = i(A_1-\lambda I)+i(\overline{M}_{\lambda}-\lambda)$. This shows that $\sigma_{e7}(\mathcal{A}) \subseteq \sigma_{e7}(A_1) \bigcup \sigma_{e7}(\overline{M}_{\mu})$. Since $\Phi_{\mathcal{A}}$, Φ_{A_1}, and $\Phi_{\overline{M}_{\mu}}$ are connected, and the sets $\rho(\overline{M}_{\mu})$ and $\rho(\mathcal{A})$ are not empty, then, using Theorem 7.3.1 (i),

we get $\sigma_{e7}(\mathcal{A}) = \sigma_{e1}(\mathcal{A})$, $\sigma_{e7}(A_1) = \sigma_{e1}(A_1)$ and $\sigma_{e7}(\overline{M}_{\mu}) = \sigma_{e1}(\overline{M}_{\mu})$. Now, the result follows from Theorem 10.2.2 (ii) and the proof of (v) is completed.

A similar reasoning to (v) allows to reach the result for (vi). Q.E.D

In the following, for an arbitrary fixed $\mu_0 \in \rho_6(A_1)$, let $\overline{\mathcal{A}_{1,\mu_0}}$ be the block diagonal matrix defined as

$$\overline{\mathcal{A}_{1,\mu_0}} := \begin{pmatrix} A_1 & 0 \\ 0 & D + \overline{C\left[K_{\mu_0}\Gamma_Y - R_b(A_1,\mu_0)B\right]} \end{pmatrix} = \begin{pmatrix} A_1 & 0 \\ 0 & \overline{M}_{\mu_0} \end{pmatrix}.$$

Theorem 10.3.4. *Let the assumptions $(J1)$–$(J8)$ hold. Assume that $\rho_6(A_1) \bigcap \rho_6(D) \neq \emptyset$, Y_1 is a core of D and $Y_2 = \{y$ such that $y \in \mathcal{D}(B) \bigcap \mathcal{D}(\Gamma_Y)$ and $\Gamma_Y y \in Z_1\}$ is dense in Y. If for some (and hence for all) $\mu \in \rho_6(A_1) \bigcap \rho_6(D)$, we have:*

(i) The operator $C[-K_\mu\Gamma_Y + R_b(A_1,\mu)B]$ is bounded on Y_2,
(ii) $R_b(D,\mu)CR_b(A_1,\mu) \in \mathcal{F}^b(X,Y)$, and
(iii) $[-\overline{K}_\mu\overline{\Gamma}_Y^0 + \overline{R_b(A_1,\mu)B}]R_b(D,\mu) \in \mathcal{F}^b(Y,X)$

then, for every $\mu_0 \in \rho_6(A_1)$ with $\rho(\mathcal{A}) \bigcap \rho(A_1) \bigcap \rho(\overline{M}_{\mu_0}) \neq \emptyset$, the difference between the resolvents $(\mathcal{A} - \lambda)^{-1} - (\overline{\mathcal{A}_{1,\mu_0}} - \lambda)^{-1} \in \mathcal{F}^b(X \times Y)$ for $\lambda \in \rho(\mathcal{A}) \bigcap \rho(A_1) \bigcap \rho(\overline{M}_{\mu_0})$, in particular $\sigma_{ei}(\mathcal{A}) = \sigma_{ei}(A_1) \bigcup \sigma_{ei}(\overline{M}_{\mu_0})$, for $i = 4,5$. Moreover, if $\mathbb{C}\backslash\sigma_{e5}(\mathcal{A})$, $\mathbb{C}\backslash\sigma_{e5}(A_1)$, and $\mathbb{C}\backslash\sigma_{e5}(\overline{M}_{\mu_0})$ are connected, then $\sigma_{e6}(\mathcal{A}) = \sigma_{e6}(A_1) \bigcup \sigma_{e6}(\overline{M}_{\mu_0})$. $\diamondsuit$

Proof. First, let $\mu_0 \in \rho(A_1)$ and $\lambda \in \rho(\mathcal{A}) \bigcap \rho(A_1) \bigcap \rho(\overline{M_{\mu_0}}) \neq \emptyset$. Using Eq. (10.3.5), we get

$$(\mathcal{A} - \lambda I)^{-1} - (\overline{\mathcal{A}_{1,\mu_0}} - \lambda I)^{-1}$$
$$= \begin{pmatrix} \begin{bmatrix} \left[-\overline{K}_\lambda\overline{\Gamma}_Y^0 + \overline{(A_1-\lambda)^{-1}B}\right] \\ (\overline{M}_\lambda - \lambda)^{-1}C(A_1-\lambda)^{-1} \end{bmatrix} & \left[\overline{K}_\lambda\overline{\Gamma}_Y^0 - \overline{(A_1-\lambda)^{-1}B}\right](\overline{M}_\lambda - \lambda)^{-1} \\ -(\overline{M}_\lambda - \lambda)^{-1}C(A_1-\lambda)^{-1} & (\overline{M}_\lambda - \lambda)^{-1} - (\overline{M}_{\mu_0} - \lambda)^{-1} \end{pmatrix}.$$

It remains to show that all entries of this block matrix operator are Fredholm perturbations. For the left lower corner, we observe that

$$\begin{aligned}(\overline{M}_\lambda - \lambda)^{-1}C(A_1-\lambda)^{-1} &= \left[(\overline{M}_\lambda - \lambda)^{-1} - (D-\lambda)^{-1}\right]C(A_1-\lambda)^{-1} + (D-\lambda)^{-1}C(A_1-\lambda)^{-1} \\ &= (\overline{M}_\lambda - \lambda)^{-1}[D - \overline{M}_\lambda](D-\lambda)^{-1}C(A_1-\lambda)^{-1} + (D-\lambda)^{-1}C(A_1-\lambda)^{-1}\end{aligned}$$

$$= (\overline{M}_\lambda - \lambda)^{-1}\left[\overline{-CK_\lambda\Gamma_Y + C(A_1-\lambda)^{-1}B}\right](D-\lambda)^{-1}C(A_1-\lambda)^{-1}$$
$$+ (D-\lambda)^{-1}C(A_1-\lambda)^{-1}.$$

Hence, the operator $\left(\overline{M_\lambda} - \lambda\right)^{-1} C(A_1-\lambda)^{-1} \in \mathcal{F}^b(X,Y)$. In the same way, we can write the right upper corner as follows:

$$\begin{aligned}
&\left[\overline{K_\lambda\Gamma_Y^0} - \overline{(A_1-\lambda)^{-1}B}\right](\overline{M}_\lambda - \lambda)^{-1}\\
&\quad = \left[\overline{K_\lambda\Gamma_Y^0} - \overline{(A_1-\lambda)^{-1}B}\right](D-\lambda)^{-1}\\
&\quad + \left[\overline{K_\lambda\Gamma_Y^0} - \overline{(A_1-\lambda)^{-1}B}\right]\left[(\overline{M}_\lambda-\lambda)^{-1} - (D-\lambda)^{-1}\right]\\
&\quad = \left[\overline{K_\lambda\Gamma_Y^0} - \overline{(A_1-\lambda)^{-1}B}\right](D-\lambda)^{-1}\\
&\quad + [\overline{K_\lambda\Gamma_Y^0} - \overline{(A_1-\lambda)^{-1}B}](D-\lambda)^{-1}(D-\overline{M}_\lambda)(\overline{M}_\lambda-\lambda)^{-1}\\
&\quad = \left[\overline{K_\lambda\Gamma_Y^0} - \overline{(A_1-\lambda)^{-1}B}\right]\\
&\quad\quad (D-\lambda)^{-1}\left[I - (\overline{CK_\lambda\Gamma_Y + C(A_1-\lambda)^{-1}B})(\overline{M}_\lambda-\lambda)^{-1}\right].
\end{aligned}$$

We conclude that the operator $\left[\overline{K_\lambda\Gamma_Y^0} - \overline{(A_1-\lambda)^{-1}B}\right](\overline{M}_\lambda-\lambda)^{-1} \in \mathcal{F}^b(Y,X)$. This, together with the fact that $C(A_1-\lambda)^{-1}$ is bounded because of the inclusion $\mathcal{D}(A_1) \subset \mathcal{D}(C)$, implies that the left upper corner is also a Fredholm perturbation. Finally, for the right lower corner, the resolvent identity for A_1 and Lemma 10.3.3 shows that

$$\begin{aligned}
&(\overline{M}_\lambda - \lambda)^{-1} - (\overline{M}_{\mu_0} - \lambda)^{-1}\\
&\quad = (\overline{M}_\lambda-\lambda)^{-1}(\overline{M}_{\mu_0} - \overline{M}_\lambda)(\overline{M}_{\mu_0}-\lambda)^{-1}\\
&\quad = (\lambda-\mu_0)(\overline{M}_\lambda-\lambda)^{-1}C(A_1-\lambda)^{-1}\left[\overline{K_{\mu_0}\Gamma_Y^0} - \overline{(A_1-\mu_0)^{-1}B}\right]\\
&\quad \times (\overline{M}_{\mu_0}-\lambda)^{-1}
\end{aligned}$$

This relation, together with Theorem 6.3.1, enables us to conclude that

$$\left(\overline{M}_\lambda - \lambda\right)^{-1} - (\overline{M}_{\mu_0}-\lambda)^{-1} \in \mathcal{F}^b(Y).$$

Hence, by using Theorem 7.5.4, we deduce that $\sigma_{ei}(\mathcal{A}) = \sigma_{ei}(A_1)\bigcup\sigma_{ei}(\overline{M}_{\mu_0})$, for $i = 4,5$. Moreover, if we suppose that $\mathbb{C}\backslash\sigma_{e5}(\mathcal{A})$, $\mathbb{C}\backslash\sigma_{e5}(A_1)$, and $\mathbb{C}\backslash\sigma_{e5}(\overline{M}_{\mu_0})$ are connected, together with the fact that $\rho(\mathcal{A})$, $\rho(A_1)$ and $\rho(\overline{M}_{\mu_0})$ are not empty, we infer from Theorem 7.3.1 (*ii*) that $\sigma_{e6}(\mathcal{A}) = \sigma_{e6}(A_1)\bigcup\sigma_{e6}(\overline{M}_{\mu_0})$. Now, if $\mu_0 \in \sigma_d(A_1)$, then there exists $\varepsilon > 0$ such that the disc $\{\zeta \in \mathbb{C}$ such that $|\zeta - \mu_0| \leq 2\varepsilon\}$

does not contain points of $\sigma(A_1)$ different from μ_0, and the Riesz projection P_{μ_0} of A_1 corresponding to μ_0 is of finite rank. Let us consider the operator $\tilde{A}_1 := A_1 + \varepsilon P_{\mu_0}$. Then, $\{\lambda \in \mathbb{C}$ such that $0 < |\lambda - \mu_0| < \varepsilon\} \subset \rho_6(A_1) \bigcap \rho_6(\tilde{A}_1)$. Until further notice, we fix $\lambda \in \rho_6(A_1) \bigcap \rho_6(\tilde{A}_1)$. We define the operator $\tilde{\mathcal{A}}_0$ like the operator $\mathcal{A}_0$ but with A replaced by $\tilde{A} := A + \varepsilon P_{\mu_0}$. Hence,

$$\tilde{\mathcal{A}}_0 = \begin{pmatrix} \tilde{A} & B \\ C & D \end{pmatrix} = \mathcal{A}_0 + \varepsilon \begin{pmatrix} P_{\mu_0} & 0 \\ 0 & 0 \end{pmatrix},$$

and for the closure of $\tilde{\mathcal{A}}_0$, we obtain

$$\tilde{\mathcal{A}} = \mathcal{A} + \varepsilon \begin{pmatrix} P_{\mu_0} & 0 \\ 0 & 0 \end{pmatrix}.$$

It is clear that $\tilde{\mathcal{A}}$ is a finite rank perturbation of $\mathcal{A}$. Therefore, $\sigma_{ei}(\tilde{\mathcal{A}}) = \sigma_{ei}(\mathcal{A})$ with $i = 4, 5$. In the following, we will apply the obtained result of the first part of this proof for the operator $\tilde{\mathcal{A}}$. Let us consider the operator $\tilde{M}_\lambda := D + C\tilde{K}_\lambda \Gamma_X - CR_b(\tilde{A}_1, \lambda)B$ which is the perturbation of M_λ, for $\lambda \in \rho_6(\tilde{A}_1)$. Here, $\tilde{K}_\lambda, \lambda \in \rho_6(\tilde{A}_1)$, is the operator defined as K_λ with A replaced by $\tilde{A}$. Hence, $\tilde{K}_\lambda z = x$ means that $x \in \mathcal{D}(\tilde{A})$, $\tilde{A}_\lambda x = 0$ and $\Gamma_X x = z$. Then, the operator $\tilde{K}_\lambda$ is well defined for $\lambda \in \rho_6(\tilde{A}_1)$. The difference $\tilde{K}_\lambda - K_\lambda$ is of finite rank. Indeed, take $z \in Z_1$ and put $\tilde{u} = \tilde{K}_\lambda z$, $u = K_\lambda z$, then $\tilde{u} - u$ satisfies the relations $\Gamma_X(\tilde{u} - u) = 0$, and $A_\lambda(\tilde{u} - u) = (\tilde{A}_\lambda - \varepsilon P_{\mu_0})\tilde{u} = -\varepsilon P_{\mu_0}\tilde{u}$. This implies that $\tilde{u} - u \in \mathcal{D}(A_1)$ and $\tilde{u} - u = -\varepsilon R_b(A_1, \lambda) P_{\mu_0} \tilde{u}$, so that

$$\tilde{K}_\lambda - K_\lambda = -\varepsilon P_{\mu_0} R_b(A_1, \lambda) \tilde{K}_\lambda. \tag{10.3.12}$$

We deduce, from this difference and from the fact that K_λ is a closed operator, that $\tilde{K}_\lambda$ is also a closed operator. We denote its closure by $\hat{K}_\lambda$. We can also see that the closure of the difference $CR_b(A_1, \lambda)B - CR_b(\tilde{A}_1, \lambda)B$ is of finite rank. Indeed, $CR_b(A_1, \lambda)B - CR_b(\tilde{A}_1, \lambda)B = \varepsilon CR_b(A_1, \lambda) P_{\mu_0} R_b(\tilde{A}_1, \lambda)B$. Using the last two results, we can easily check that the difference $\tilde{M}_\lambda - M_\lambda$ is of finite rank. Since the operator M_λ is closable in Y, we infer that its perturbation $\tilde{M}_\lambda$ is closable in Y as well, and we will denote its closure by $\hat{M}_\lambda$. Since $\hat{M}_\lambda - \overline{M}_\lambda$ is of finite rank, then $\sigma_{ei}(\hat{M}_\lambda) = \sigma_{ei}(\overline{M}_\lambda)$, with $i = 4, 5$. Now, using the following relations:

$$R_b(D,\lambda)CR_b(A_1,\lambda) - R_b(D,\lambda)CR_b(\tilde{A}_1,\lambda) = \varepsilon R_b(D,\lambda)CR_b(A_1,\lambda)P_{\mu_0}R_b(\tilde{A}_1,\lambda),$$

$$R_b(A_1,\lambda)BR_b(D,\lambda) - R_b(\tilde{A}_1,\lambda)BR_b(D,\lambda) = \varepsilon R_b(A_1,\lambda)P_{\mu_0}R_b(\tilde{A}_1,\lambda)BR_b(D,\lambda)$$

together with Eq. (10.3.12) and the fact that, for some $\lambda \in \rho_6(A_1) \bigcap \rho_6(D)$, the operators $R_b(D, \lambda)CR_b(A_1, \lambda) \in \mathcal{F}^b(X, Y)$ and $\left[-\overline{K_\lambda \Gamma_Y^0} + \overline{R_b(A_1, \lambda)B}\right]$ $R_b(D, \lambda) \in \mathcal{F}^b(Y, X)$, we can easily deduce that, for some $\lambda \in \rho_6(A_1) \bigcap \rho_6(D)$, we have $R_b(D, \lambda)CR_b(\tilde{A}_1, \lambda) \in \mathcal{F}^b(X, Y)$ and, $\left[-\hat{K}_\lambda \overline{\Gamma_Y^0} + \overline{R_b(\tilde{A}_1, \lambda)B}\right]$

$R_b(D, \lambda) \in \mathcal{F}^b(Y, X)$. Hence, we infer the following $\sigma_{ei}(\hat{M}_\lambda)$ with $i = 4, 5$, is independent of $\lambda \in \rho_6(\tilde{A}_1)$. Now, applying the first part of this proof for $\mu_0 \in \rho(\tilde{A}_1)$, we see that $\sigma_{ei}(\tilde{\mathcal{A}}) = \sigma_{ei}(\widetilde{A_1}) \bigcup \sigma_{ei}(\hat{M}_{\mu_0})$, with $i = 4, 5$. Then, we get $\sigma_{ei}(\mathcal{A}) = \sigma_{ei}(\tilde{\mathcal{A}}) = \sigma_{ei}(A_1) \bigcup \sigma_{ei}(\overline{M}_{\mu_0})$, with $i = 4, 5$ for any $\mu_0 \in \rho_6(A_1)$ as required, and the proof of the first part of the theorem is complete. Moreover, if we assume that $\mathbb{C}\backslash\sigma_{e5}(\mathcal{A})$, $\mathbb{C}\backslash\sigma_{e5}(A_1)$, and $\mathbb{C}\backslash\sigma_{e5}(\overline{M}_{\mu_0})$ are connected, and from Theorem 7.3.1 (*ii*), we obtain $\sigma_{e5}(\mathcal{A}) = \sigma_{e6}(\mathcal{A})$, $\sigma_{e5}(A_1) = \sigma_{e6}(A_1)$ and $\sigma_{e5}(\overline{M}_{\mu_0}) = \sigma_{e6}(\overline{M}_{\mu_0})$. Hence, the proof of this assertion is complete. Q.E.D

Open question. In contrast to the result of Theorem 10.3.4, if we replace the second item by $CR_b(A_1, \mu)B$ or $CR_b(D, \mu)B$ are in $\mathcal{F}^b(X, Y)$, can we get the same result? □

10.3.3 The Operator $\mathcal{A}$ as an Infinitesimal Generator of a Holomorphic Semigroup

Definition 10.3.1. An operator T in a Banach space generates a holomorphic semigroup of a semiangle $\eta \in (0, \frac{\pi}{2})$, if there exists $\omega \in \mathbb{R}$, such that the sector

$$S(\omega, \eta) := \left\{\lambda \in \mathbb{C} \text{ such that } |\arg(\lambda - \omega)| < \frac{\pi}{2} + \eta\right\} \tag{10.3.13}$$

belongs to $\rho(T)$ and, for each $\varepsilon \in (0, \eta)$, there is $L_\varepsilon \geq 1$, such that the resolvent of T satisfies the following inequality $\|(T - \lambda)^{-1}\| < \frac{L_\varepsilon}{|\lambda - \omega|}$ in $S(\omega, \eta - \varepsilon)$. ◇

The following results already given in [50].

Theorem 10.3.5. *Let the assumptions $(J1)$–$(J8)$ hold and, for some $\lambda_0 \in \rho(A_1)$, the operator M_{λ_0} is closable. Let us also assume that A_1 and $\overline{M}_{\lambda_0}$ generate holomorphic semigroups in X and Y, respectively, and, for some $\omega > 0$, let $S(\omega, \eta)$ be a sector (10.3.13) corresponding to the operator A_1. If the condition $(J4)$ is strengthened by*

$$\inf_{\lambda \in S(\omega,\eta)} \|C_\lambda\| = 0, \tag{10.3.14}$$

then the operator $\mathcal{A}$ generates a holomorphic semigroup in $X \times Y$. ◇

Proof. From (10.3.6), we may deduce that, for $\lambda \in \rho(A_1)$, the operator $\mathcal{A} - \lambda$ is boundedly invertible if, and only if, $\overline{M}_\lambda - \lambda$ has this same property and, in this case, we have

$$(\mathcal{A}-\lambda I)^{-1} := \begin{pmatrix} I & \overline{K_\lambda \Gamma_Y^0 - (A_1 - \lambda)^{-1} B} \\ 0 & I \end{pmatrix} \begin{pmatrix} (A_1 - \lambda)^{-1} & 0 \\ 0 & (\overline{M}_\lambda - \lambda)^{-1} \end{pmatrix} \begin{pmatrix} I & 0 \\ -C_\lambda & I \end{pmatrix}. \tag{10.3.15}$$

Since the operators A_1 and $\overline{M}_{\lambda_0}$ generate holomorphic semigroups, then there exist constants $L_1 > 0$, $\eta_1 \in (0, \eta)$, and $\omega_1 > 0$, such that the sector $S(\omega_1, \eta_1)$ belongs to the resolvent set of both A_1 and $\overline{M}_{\lambda_0}$ and that, for all $\lambda \in S(\omega_1, \eta_1)$, we have

$$\|(A_1 - \lambda)^{-1}\| \leq \frac{L_1}{|\lambda|}, \qquad \|(\overline{M}_{\lambda_0} - \lambda)^{-1}\| \leq \frac{L_1}{|\lambda|}. \tag{10.3.16}$$

Without loss of generality, we may assume that $\omega = \omega_1$. Let $\lambda \in S(\omega, \eta_1)$. From Eq. (10.3.16), we infer that

$$\|A_1(A_1 - \lambda)^{-1}\| = \|I + \lambda(A_1 - \lambda)^{-1}\| \leq L_1 + 1. \tag{10.3.17}$$

Now, let $\lambda_1 \in \rho(A_1)$. Then, we have $C_\lambda = C_{\lambda_1}[A_1(A_1 - \lambda)^{-1} - \lambda_1(A_1 - \lambda)^{-1}]$. By using the inequalities (10.3.16) and Eq. (10.3.17) and, for $|\lambda| > |\lambda_1|$, we have

$$\|C_\lambda\| \leq \|C_{\lambda_1}\|(2L_1 + 1). \tag{10.3.18}$$

Since $\|C_{\lambda_1}\|$ can be made arbitrarily small in view of condition (10.3.14), we notice that

$$\lim_{|\lambda|\to\infty,\ \lambda\in S(\omega,\eta_1)} \|C_\lambda\| = 0. \tag{10.3.19}$$

By using the first inequality in (10.3.16), Lemma 10.3.3 and also (8.2.2), we conclude that the operators $\overline{(A_1 - \lambda)^{-1}B}$ and $\overline{K}_\lambda\overline{\Gamma}_Y^0$ are uniformly bounded in $\lambda \in S(\omega, \eta_1)$. Therefore, the factors

$$\begin{pmatrix} I & \overline{K}_\lambda\overline{\Gamma}_Y^0 - \overline{(A_1 - \lambda)^{-1}B} \\ 0 & I \end{pmatrix} \text{ and } \begin{pmatrix} I & 0 \\ -C_\lambda & I \end{pmatrix}$$

in (10.3.15) are uniformly bounded in $\lambda \in S(\omega, \eta_1)$. Moreover, by using Eq. (10.3.4), we have

$$\begin{aligned} \overline{M}_\lambda - \lambda =& (\overline{M}_{\lambda_0} - \lambda)\left(I + (\overline{M}_{\lambda_0} - \lambda)^{-1}(\overline{M}_\lambda - \overline{M}_{\lambda_0})\right) \\ =& (\overline{M}_{\lambda_0} - \lambda)\left(I + (\overline{M}_{\lambda_0} - \lambda)^{-1}(\lambda - \lambda_0)C_\lambda(\overline{K}_{\lambda_0}\overline{\Gamma}_Y^0 - \overline{(A_1 - \lambda_0)^{-1}B})\right). \end{aligned}$$

By using the second estimate in (10.3.16), we have $\sup_{\lambda\in S(\omega,\eta_1)} \|(\overline{M}_{\lambda_0} - \lambda)^{-1}(\lambda - \lambda_0)\| < \infty$. Hence, by using Eq. (10.3.19), there exists an element $r > 0$, such that

$$\left\|(\overline{M}_{\lambda_0} - \lambda)^{-1}(\lambda - \lambda_0)C_\lambda\left(\overline{K}_{\lambda_0}\overline{\Gamma}_Y^0 - \overline{(A_1 - \lambda_0)^{-1}B}\right)\right\| < \frac{1}{2} \tag{10.3.20}$$

for all $\lambda \in S(\omega, \eta_1)$, with $|\lambda| > r$. So, the following operator

$$I + (\overline{M}_{\lambda_0} - \lambda)^{-1}(\lambda - \lambda_0)C_\lambda \left(\overline{K}_{\lambda_0}\overline{\Gamma}_Y^0 - \overline{(A_1 - \lambda_0)^{-1}B}\right)$$

is invertible, and the norm of its inverse is not greater than 2. Hence, for all $\lambda \in S(\omega, \eta_1)$, with $|\lambda| > r$, we have $\|\lambda(\overline{M}_\lambda - \lambda)^{-1}\| \leq 2\|\lambda(\overline{M}_{\lambda_0} - \lambda)^{-1}\| \leq 2L_1$. So, $\|\lambda(\mathcal{A} - \lambda I)^{-1}\|$ is uniformly bounded for all $\lambda \in S(\omega + r, \eta_1)$ and therefore, the operator $\mathcal{A}$ generates a holomorphic semigroup in $X \times Y$. Q.E.D

Remark 10.3.4. We can easily notice that the assumption (10.3.14) can be replaced by the weaker condition that the infimum is less than a constant c chosen small enough for the norm in (10.3.20) in order to be less than d, say, where $d < 1$. ◇

Proposition 10.3.1. *Let us suppose that A_1 generates a holomorphic semigroup and let $S(\omega, \eta)$ be the corresponding sector. Then, (10.3.14) holds if the relative A_1-bound of C is zero. This is the case if C is A_1-compact and either X is reflexive or C (as originally defined, i.e., from X into Y) is closable.* ◇

Proof. The relative A_1-boundedness condition implies that, for every $b > 0$, there exists $a > 0$, such that

$$\|Cx\| \leq a\|x\| + b\|A_1x\| \tag{10.3.21}$$

for all $x \in \mathcal{D}(A_1)$. By choosing $x = (A_1 - \lambda)^{-1}u$ in (10.3.21) with being λ in the sector $S(\omega, \eta)$ (as in the proof of Theorem 10.3.5) and by using (10.3.17), we obtain

$$\|C_\lambda u\| \leq a\|(A_1 - \lambda)^{-1}u\| + b\|A_1(A_1 - \lambda)^{-1}u\| \leq \left\{\frac{aL_1}{|\lambda|} + b(L_1 + 1)\right\} \|u\|.$$

The right-hand side can be made arbitrarily small if we choose a sufficiently small b and then a large $|\lambda|$. This leads to the first claim. For the remainder, we notice that if C is A_1-compact and if X is reflexive or C is closable, then the relative A_1-bound of C is zero (see, e.g., [64, Theorem 2] or [106, Lemma III.2.16]). Q.E.D

10.4 Relative Boundedness for Block Operator Matrices

Let $\gamma(.)$ be a measure of noncompactness. In order to ensure that $\mathcal{A}$ is closable or closed, we need more refined assumptions on the strength of the entries with respect to each other, where we consider the γ-diagonally dominant and out γ-diagonally dominant. Let X and Y be Banach spaces. In the product space $X \times Y$, we consider an operator which is formally defined by a matrix

$$\mathcal{A} = \begin{pmatrix} A & B \\ C & D \end{pmatrix}, \tag{10.4.1}$$

where the entries are, in general, unbounded operators. A is acting on a Banach space X, D is acting on a Banach space Y and B, C are acting between these spaces. In general, the operator matrix in (10.4.1) defines a linear operator in $X \times Y$ with a domain $\mathcal{D}(A) \bigcap \mathcal{D}(C) \times \mathcal{D}(B) \bigcap \mathcal{D}(D)$ which is not closed or closable, even if its entries are closed or closable. Now, let us introduce the following definitions.

Definition 10.4.1. Let $\gamma(.)$ be a measure of noncompactness. The block matrix operator $\mathcal{A}$ is called

(*i*) γ-diagonally dominant, if C is A-γ-bounded and B is D-γ-bounded,
(*ii*) off-γ-diagonally dominant, if A is C-γ-bounded and D is B-γ-bounded. ◇

Definition 10.4.2. The block matrix operator $\mathcal{A}$ is called

(*i*) γ-diagonally dominant with bound δ, if C is A-γ-bounded with bound δ_C, and B is D-γ-bounded with bound δ_B, and $\delta = \max\{\delta_B, \delta_C\}$,
(*ii*) off-γ-diagonally dominant with bound δ, if A is C-γ-bounded with bound δ_A, and D is B-γ-bounded with bound δ_D, and $\delta = \max\{\delta_A, \delta_D\}$. ◇

Proposition 10.4.1. *Consider the block operator matrices*

$$\mathcal{T} = \begin{pmatrix} A & 0 \\ 0 & D \end{pmatrix} \quad \textit{and} \quad \mathcal{S} = \begin{pmatrix} 0 & B \\ C & 0 \end{pmatrix}.$$

If $\mathcal{A}$ is γ-diagonally dominant with bound δ, then $\mathcal{S}$ is $\mathcal{T}$-γ-bounded with $\mathcal{T}$-γ-bound δ. ◇

Proof. For $\mathfrak{D} \subset \mathcal{D}(\mathcal{S})$, we get

$$\gamma\left(\begin{pmatrix} 0 & B \\ C & 0 \end{pmatrix}\begin{pmatrix} \pi_1(\mathfrak{D}) \\ \pi_2(\mathfrak{D}) \end{pmatrix}\right) = \max\{\gamma(C(\pi_1(\mathfrak{D}))), \gamma(B(\pi_2(\mathfrak{D})))\}.$$

According to the assumptions, there exist constants a_B, a_C, b_B, $b_C \geq 0$ such that

$$\begin{cases} \gamma(B(\pi_2(\mathfrak{D}))) \leq a_B\gamma(\pi_2(\mathfrak{D})) + b_B\gamma(D(\pi_2(\mathfrak{D}))), \\ \gamma(C(\pi_1(\mathfrak{D}))) \leq a_C\gamma(\pi_1(\mathfrak{D})) + b_C\gamma(A(\pi_1(\mathfrak{D}))). \end{cases}$$

Since $\gamma(\pi_2(\mathfrak{D})) \leq \gamma(\pi_2)\gamma(\mathfrak{D})$ and $\gamma(\pi_1(\mathfrak{D})) \leq \gamma(\pi_1)\gamma(\mathfrak{D})$, then

$$\begin{cases} \gamma(B(\pi_2(\mathfrak{D}))) \leq a_B\gamma(\pi_2)\gamma(\mathfrak{D}) + b_B\gamma(D(\pi_2(\mathfrak{D}))), \\ \gamma(C(\pi_1(\mathfrak{D}))) \leq a_C\gamma(\pi_1)\gamma(\mathfrak{D}) + b_C\gamma(A(\pi_1(\mathfrak{D}))). \end{cases}$$

Hence, we get

$$\begin{aligned}
&\gamma\left(\begin{pmatrix} 0 & B \\ C & 0 \end{pmatrix}\begin{pmatrix} \pi_1(\mathfrak{D}) \\ \pi_2(\mathfrak{D}) \end{pmatrix}\right) \\
&\leq \max\{a_B\gamma(\pi_2), a_C\gamma(\pi_1)\}\gamma(\mathfrak{D}) + \max\{b_B, b_C\}\max\{\gamma(A(\pi_1(\mathfrak{D}))), \gamma(D(\pi_2(\mathfrak{D})))\} \\
&\leq \max\{a_B\gamma(\pi_2), a_C\gamma(\pi_1)\}\gamma(\mathfrak{D}) + \max\{b_B, b_C\}\gamma\left(\begin{pmatrix} A & 0 \\ 0 & D \end{pmatrix}\begin{pmatrix} \pi_1(\mathfrak{D}) \\ \pi_2(\mathfrak{D}) \end{pmatrix}\right),
\end{aligned}$$

which completes the proof. Q.E.D

Remark 10.4.1. If $\mathcal{A}$ is off-γ-diagonally dominant with bound δ, then $\mathcal{T}$ is $\mathcal{S}$-γ-bounded with $\mathcal{S}$-γ-bound δ. ◇

Theorem 10.4.1. *The block matrix operator $\mathcal{A}$ is closed if $\mathcal{A}$ is γ-diagonally dominant, $\rho(A) \neq \emptyset$, C and D are closed, $(A-\lambda)^{-1}B$ is bounded on $\mathcal{D}(B)$, for some $\lambda \in \rho(A)$, and $((a_C + |\lambda| b_C)\gamma((A-\lambda)^{-1}) + b_C)b_B < 1$.* ◇

Proof. The block matrix operator $\mathcal{A}$ is γ-diagonally dominant implies that there exists $a_C,\ a_B,\ b_C,\ b_B \geq 0$, such that

$$\begin{aligned}
\gamma(C(A-\lambda)^{-1}B(\mathfrak{D})) &\leq a_C\gamma((A-\lambda)^{-1}B(\mathfrak{D})) + b_C\gamma(A(A-\lambda)^{-1}B(\mathfrak{D})) \\
&\leq (a_C + b_C|\lambda|)\gamma((A-\lambda)^{-1}) + b_C)\gamma(B(\mathfrak{D})) \\
&\leq (a_C + b_C|\lambda|)\gamma((A-\lambda)^{-1}) + b_C)(a_B\gamma(\mathfrak{D}) + b_B\gamma(D(\mathfrak{D}))).
\end{aligned}$$

So, $C(A-\lambda)^{-1}B$ is D-γ-bounded with D-bound < 1. Since D is closed, then, by using Theorem 2.10.1, $D - C(A-\lambda)^{-1}B$ is also closed. According to Theorem 2.2.3, we conclude that $\mathcal{A}$ is closed. Q.E.D

Theorem 10.4.2. *The block matrix operator $\mathcal{A}$ is closed if $\mathcal{A}$ is γ-diagonally dominant, $\rho(D) \neq \emptyset$, A and B are closed, $(D-\lambda)^{-1}C$ is bounded on $\mathcal{D}(C)$, for some $\lambda \in \rho(D)$, and $((a_B + |\lambda| b_B)\gamma((D-\lambda)^{-1}) + b_B)b_C < 1$.* ◇

Proof. The proof of Theorem 10.4.2 may be checked in a way which is similar to the one in Theorem 10.4.1. Q.E.D

Theorem 10.4.3. *The block matrix operator $\mathcal{A}$ is closed if $\mathcal{A}$ is γ-diagonally dominant with bound < 1.* ◇

Proof. According to Theorem 2.10.1 and Proposition 10.4.1, we deduce that the block matrix operator $\mathcal{A}$ is closed. Q.E.D

Corollary 10.4.1. *The block matrix operator $\mathcal{A}$ is closed if $\mathcal{A}$ is off-γ-diagonally dominant with bound < 1.* ◇

Proof. According to both Theorem 2.10.1 and Remark 10.4.1, we deduce that the block matrix operator $\mathcal{A}$ is closed. Q.E.D

Theorem 10.4.4. *The block matrix operator $\mathcal{A}$ is closed, if $\mathcal{A}$ is off-γ-diagonally dominant, C is boundedly invertible, B, $C^{-1}D$ is bounded on $\mathcal{D}(D)$, A is closed and $(a_A\gamma(C^{-1}) + b_A)b_D < 1$.* ◇

Proof. As $\mathcal{A}$ is off-diagonally dominant, then there exists a_A, a_D, b_A, $b_D \geq 0$ such that

$$\begin{aligned}\gamma(AC^{-1}(D-\lambda)(\mathfrak{D})) &\leq (a_A\gamma(C^{-1}) + b_A)\gamma(D(\mathfrak{D})) \\ &\leq (a_A\gamma(C^{-1}) + b_A)(a_D\gamma(\mathfrak{D}) + b_D\gamma(B(\mathfrak{D})).\end{aligned}$$

Then, $AC^{-1}(D-\lambda)$ is B-γ-bounded with B-bound< 1. Since B is closed and $C^{-1}D$ is bounded on $\mathcal{D}(D)$, then, by using Theorem 2.10.1, $B - AC^{-1}(D-\lambda)$ is closed. According to Theorem 2.2.4, we conclude that $\mathcal{A}$ is closed. Q.E.D

Theorem 10.4.5. *The block matrix operator $\mathcal{A}$ is closed, if $\mathcal{A}$ is off-γ-diagonally dominant, B is boundedly invertible, $B^{-1}A$ is bounded on $\mathcal{D}(A)$, D is closed and $(a_D\gamma(B^{-1}) + b_D)b_A < 1$.* ◇

Proof. The proof may be checked in a way which is similar to that in Theorem 10.4.4. Q.E.D

Let $A_i : \mathcal{D}(A_i) \subset X \longrightarrow Y$, $B_i : \mathcal{D}(B_i) \subset Y \longrightarrow X$, $C_i : \mathcal{D}(C_i) \subset X \longrightarrow Y$ and $D_i : \mathcal{D}(D_i) \subset Y \longrightarrow Y, i = 1, 2$, be unbounded operators, such that $\mathcal{D}(A_1) \subset \mathcal{D}(A_2)$, $\mathcal{D}(B_1) \subset \mathcal{D}(B_2)$, $\mathcal{D}(C_1) \subset \mathcal{D}(C_2)$, $\mathcal{D}(D_1) \subset \mathcal{D}(D_2)$, and we define the block matrix operators by

$$\mathcal{A}_1 = \begin{pmatrix} A_1 & B_1 \\ C_1 & D_1 \end{pmatrix}, \qquad \mathcal{A}_2 = \begin{pmatrix} A_2 & B_2 \\ C_2 & D_2 \end{pmatrix}.$$

Theorem 10.4.6. *Let us decompose the operator $\mathcal{A}_1$ in the following form $\mathcal{A}_1 = \mathcal{S} + \mathcal{T}$, where*

$$\mathcal{T} = \begin{pmatrix} A_1 & 0 \\ 0 & D_1 \end{pmatrix} \quad \text{and} \quad \mathcal{S} = \begin{pmatrix} 0 & B_1 \\ C_1 & 0 \end{pmatrix}.$$

If A_2 and C_2 are A_1-weakly compact, and if B_2 and D_2 are D_1-weakly compact, then $\mathcal{A}_2$ is $\mathcal{T}$-weakly compact. Moreover, if $\mathcal{A}_1$ is diagonally dominant with bound < 1, then $\mathcal{A}_2$ is $\mathcal{A}_1$-weakly compact. ◇

Proof. Assume that $t_n = (x_n, y_n)_n \subset \mathcal{D}(\mathcal{T})$, such that $(t_n)_n$ and $(\mathcal{T}t_n)_n$ are bounded sequences, we have shown that $(\mathcal{A}_2 t_n)_n$ contains a weakly convergent subsequence. Since A_2 and C_2 are A_1-weakly compact and since B_2 and D_2 are D_1-weakly compact, it is sufficient to show that $(A_1x_n)_n$ and $(D_1y_n)_n$ are bounded.

Indeed, since $(\mathcal{A}_1 t_n)_n$ is bounded, then the sequences $(A_1 x_n)_n$ and $(D_1 y_n)_n$ are bounded for $n \in \mathbb{N}$. Then, let us assume that $(h_n)_n \subset \mathcal{D}(\mathcal{A}_1)$ such that $(h_n)_n$ and $(\mathcal{A}_1 h_n)_n$ are bounded sequences. We show that $(\mathcal{A}_2 h_n)_n$ contains a weakly convergent subsequence. Since $\mathcal{A}_2$ is $\mathcal{T}$-weakly compact, it is sufficient to show that $(\mathcal{T} h_n)_n$ is bounded. According to Proposition 10.4.1, the assumption that $\mathcal{A}_1$ is diagonally dominant with bound< 1 implies that $\mathcal{S}$ is $\mathcal{T}$-bounded with $\mathcal{T}$-bound < 1. Thus, there exists $p \in (0, 1]$ and $a_{\mathcal{S}}, b_{\mathcal{S}} \geq 0$, such that

$$\begin{aligned}\|\mathcal{S}h_n\| &\leq a_{\mathcal{S}}\|h_n\| + b_{\mathcal{S}}\|\mathcal{T}h_n\| \\ &\leq a_{\mathcal{S}}\|h_n\| + b_{\mathcal{S}}\big(\|\mathcal{A}_1 h_n\| + \|\mathcal{S}h_n\|\big) \\ &\leq a_{\mathcal{S}}\|h_n\| + b_{\mathcal{S}}\|\mathcal{A}_1 h_n\| + b_{\mathcal{S}}\|\mathcal{S}h_n\|.\end{aligned}$$

Since $b_{\mathcal{S}}< 1$, then $\|\mathcal{S}h_n\| \leq \frac{a_{\mathcal{S}}\|h_n\|+b_{\mathcal{S}}\|\mathcal{A}_1 h_n\|}{1-b_{\mathcal{S}}}$, which implies that $(\mathcal{S}h_n)_n$ is bounded and then, $(\mathcal{T}h_n = (\mathcal{A}_1 - \mathcal{S})h_n)_n$ is also bounded. Consequently, $\mathcal{A}_2$ is $\mathcal{A}_1$-weakly compact. Q.E.D

Remark 10.4.2. The assumptions A_2 and C_2 are A_1-weakly compact and B_2, D_2 are D_1-weakly compact and $\mathcal{A}_1$ which is diagonally dominant with bound < 1 in Theorem *10.4.6*, can be replaced by

(i') A_2, C_2 are C_1-weakly compact and B_2, D_2 are B_1-weakly compact,
(ii') $\mathcal{A}_1$ is off-diagonally dominant with bound < 1

respectively, and we get the same conclusion. ◇

Lemma 10.4.1. *Let S, $T \in \mathcal{C}(X)$, and assume that X has the DP property. If S is T-weakly compact and $\|S(\lambda - T)^{-1}\| < 1$ for some (hence for all) $\lambda \in \rho(T)\bigcap\rho(S+T)$, then $\sigma_{ei}(S+T) = \sigma_{ei}(T)$, with $i = 1,2,3,4,5,7$ and 8. Moreover, if $s(T) < \infty$ and $\mathbb{C}\backslash\sigma_{e5}(T)$ of $\sigma_{e5}(T)$ is connected, then $\sigma_{e6}(S+T) = \sigma_{e6}(T)$.* ◇

Proof. Let $\lambda \in \rho(T)\bigcap\rho(S+T)$. We recall that

$$\begin{aligned}(S+T-\lambda)^{-1} &= \Big[(I + S(T-\lambda)^{-1})(T-\lambda)\Big]^{-1} \\ &= (T-\lambda)^{-1}(I + S(T-\lambda)^{-1}).\end{aligned}$$

Since $\|S(\lambda - T)^{-1}\| < 1$, then $(I + S(T-\lambda)^{-1})^{-1} = \sum_{n\geq 0}(S(T-\lambda)^{-1})^n$. Hence,

$$\begin{aligned}(S+T-\lambda)^{-1} &= (T-\lambda)^{-1}\sum_{n\geq 0}(S(T-\lambda)^{-1})^n \\ &= (T-\lambda)^{-1} + \sum_{n\geq 1}(T-\lambda)^{-1}(S(T-\lambda)^{-1})^n.\end{aligned}$$

Since, $S(T-\lambda)^{-1}$ is weakly compact and $(T-\lambda)^{-1}$ is bounded, for some (hence for all) $\lambda \in \rho(T)\bigcap\rho(S+T)$, then $(S+T-\lambda)^{-1} - (T-\lambda)^{-1}$ is weakly compact.

Consequently, by using Theorem 7.5.4, we get $\sigma_{ei}(S+T) = \sigma_{ei}(T)$, with $i = 1, 2, 3, 4$ and 5. By using Theorem 7.5.4, we have $\sigma_{e6}(S+T) = \sigma_{e6}(T)$. According to Remark 2.1.7, we have $\mathcal{W}(X) \subset \mathcal{F}_+(X) \bigcap \mathcal{F}_-(X)$ since X has the DP property. Consequently, we conclude that $\sigma_{ei}(S+T) = \sigma_{ei}(T)$, with $i = 7, 8$, follows from Theorem 7.5.4 and Lemma 6.3.1. Q.E.D

Theorem 10.4.7. *Suppose that $X \times Y$ has the Dunford-Pettis property and $\mathcal{A}_1$ is closed. If $\mathcal{A}_2$ is $\mathcal{A}_1$-weakly compact and $\|\mathcal{A}_2(\mathcal{A}_1-\lambda)^{-1}\| < 1$, then $\sigma_{ei}(\mathcal{A}_1+\mathcal{A}_2) = \sigma_{ei}(\mathcal{A}_1)$, with $i = 1,2,3,4,5,7$ and 8. Moreover, if $s(\mathcal{A}_1) < \infty$ and $\mathbb{C}\backslash\sigma_{e5}(\mathcal{A}_1)$ is connected, then*

$$\sigma_{e6}(\mathcal{A}_1+\mathcal{A}_2) = \sigma_{e6}(\mathcal{A}_1).$$

◇

Proof. Since $\mathcal{A}_2$ is $\mathcal{A}_1$-weakly compact, $\|\mathcal{A}_2(\mathcal{A}_1-\lambda)^{-1}\| < 1$, $s(\mathcal{A}_1) < \infty$ and $X \times Y$ has the Dunford-Pettis property then, by using Lemma 10.4.1, we get $\sigma_{ei}(\mathcal{A}_1 + \mathcal{A}_2) = \sigma_{ei}(\mathcal{A}_1)$, with $i = 1,2,3,4,5,6,7$ and 8. Q.E.D

10.5 Stability of the Wolf Essential Spectrum of Some Matrix Operators Acting in Friedrichs Module

Let (K, H) be a compact Friedrichs module and let A_1, A_2, B_1, B_2, and C be five closed operators acting in a Banach space H. In the product space $H \times H$, we consider two operators defined by a 2×2 block operator matrix

$$M_C^1 = \begin{pmatrix} A_1 & C \\ 0 & B_1 \end{pmatrix} \text{ and } M_C^2 = \begin{pmatrix} A_2 & C \\ 0 & B_2 \end{pmatrix}.$$

The main purpose of this section is to establish the criteria which ensure that the difference of the resolvent of M_C^1 and M_C^2 is compact. We suppose that $\mathcal{D}(B_i) \subset \mathcal{D}(C)$ and $\rho(A_i) \bigcap \rho(B_i)$ is not empty. It is easy to show that if $z \in \rho(A_i) \bigcap \rho(B_i)$, $z \in \rho(M_C^i)$, with $i = 1,2$, then we get

$$(M_C^i - z)^{-1} = \begin{pmatrix} (A_i - z)^{-1} & -(A_i - z)^{-1}C(B_i - z)^{-1} \\ 0 & (B_i - z)^{-1} \end{pmatrix}.$$

Now, for $E \in \mathcal{C}(H)$, we assume that the following conditions hold:

$(K1)$ $\mathcal{D}(E) \subset K$ densely.
$(K2)$ $\mathcal{D}(E^*) \subset K$.
$(K3)$ E can be extended to a continuous operator $\tilde{E} \in \mathcal{L}(K, K^*)$.

Remark 10.5.1. If $\rho(E)$ is not empty and if E satisfies the assumptions $(K1)$–$(K3)$ then, for $z \in \rho(E)$, $(E-z)^{-1*}H \subset K$. So, we deduce that $(E-z)^{-1}$ extends to a unique continuous $R_z(E)$ operator $K^* \longrightarrow H$, which will be denoted, for the moment, by $(E-z)^{-1}$. ◇

Theorem 10.5.1. *Suppose that A_i, B_i, $i = 1, 2$ satisfy the assumptions $(K1)$–$(K3)$ and let us assume that $\rho(A_1)\bigcap\rho(A_2)\bigcap\rho(B_1)\bigcap\rho(B_2)$ is not empty. If $\mathcal{D}(B_i) \subset \mathcal{D}(C)$, $i = 1, 2$, and if there exists a Banach module L such that the following conditions are satisfied:*

(i) There are an operator $S_1 \in \mathcal{L}(L, K^*)$ and an operator $T_1 \in B_0^l(K, L)$ such that $\widetilde{A_2} - \widetilde{A_1} = S_1T_1$ and $(A_1 - z)^{-1}S_1 \in B_q^l(L, H)$ for some $z \in \rho(A_1)\bigcap\rho(A_2)$.
(ii) There are an operator $S_2 \in \mathcal{L}(L, K^*)$ and an operator $T_2 \in B_0^l(K, L)$ such that $\widetilde{B_2} - \widetilde{B_1} = S_2T_2$ and $(B_1 - z)^{-1}S_2 \in B_q^l(L, H)$ for some $z \in \rho(B_1)\bigcap\rho(B_2)$,

then, M_C^2 is a compact perturbation of the operator M_C^1. In particular, $\sigma_{e4}(M_C^1) = \sigma_{e4}(M_C^2)$. ◇

Proof. Let $z \in \rho(A_1)\bigcap\rho(A_2)\bigcap\rho(B_1)\bigcap\rho(B_2)$. Then, $z \in \rho(M_C^1)\bigcap\rho(M_C^2)$ and

$$(M_C^1 - z)^{-1} - (M_C^2 - z)^{-1} = \begin{pmatrix} R_1(z) & R_3(z) \\ 0 & R_2(z) \end{pmatrix},$$

where $R_1(z) = (A_1-z)^{-1}(A_2-A_1)(A_2-z)^{-1}$, $R_2(z) = (B_1-z)^{-1}(B_2-B_1)(B_2-z)^{-1}$ and $R_3(z) = -R_1(z)C(B_2-z)^{-1} - (A_1-z)^{-1}CR_2(z)$. It remains to demonstrate that all entries of this block operator matrix are compact. Since A_1 satisfies the assumptions $(K1)$–$(K3)$, and by using Remark 10.5.1, we infer that $(A_1 - z)^{-1}$ can be extended to a unique continuous operator $R_z(A_1) : K^* \longrightarrow H$. Now, let us notice that $R_z(A_1)(A_1 - z)x = x$ for $x \in \mathcal{D}(A_1)$. So, by using the density of $\mathcal{D}(A_1)$ in K and the continuity $R_z(A_1)(\widetilde{A_1} - z)x = x$ for $x \in K$, in particular $(A_2 - z)^{-1} = R_z(A_1)(\widetilde{A_1} - z)(A_2 - z)^{-1}$. Moreover, we can write $(A_1 - z)^{-1} = (A_1 - z)^{-1}(A_2 - z)(A_2 - z)^{-1} = R_z(A_1)(\widetilde{A_2} - z)(A_2 - z)^{-1}$. By using the last two relations, we obtain $R_1(z) = R_z(A_1)(\widetilde{A_2} - \tilde{A}_1)(A_2 - z)^{-1}$. It is easy to see that $R_1(z)H \subset K$. So, in order to prove that $R_1(z)$ is a compact operator, we have to show that $R_1(z) \in B_0^l(H)$. Now, according to the factorization assumption, we have $R_1(z) = [R_z(A_1)S_1][T_1(A_2 - z)^{-1}]$. By using Theorem 2.9.1, we have $T_1 = ME_0$ where $M \in \mathcal{M}(L)$ and $E_0 \in \mathcal{L}(K, L)$. Since $R_z(A_1)S_1 \in B_q^l(L, H)$, then $R_z(A_1)S_1M \in B_0^l(L, H)$ and $R_z(A_1)S_1M = NE_1$ where $N \in \mathcal{M}(H)$ and $E_1 \in \mathcal{L}(L, H)$. Finally, we obtain $R_1(z) = N[E_1E_0(A_2 - z)^{-1}]$ where the first factor is in $\mathcal{M}(H)$ and the second is in $B_0^l(H)$. By Remark 2.9.1, we infer that $R_1(z) \in \mathcal{K}(H)$. A similar proof as before, we prove that $R_2(z) = (B_1-z)^{-1}(\widetilde{B_2}-\widetilde{B_1})(B_2-z)^{-1} = [R_z(B_1)S_2][T_2(B_2-z)^{-1}]$, where $R_z(B_1)S_2 \in B_q^l(L, H)$ and $T_2(B_2-z)^{-1} \in B_0^l(H, L)$. Therefore, $R_2(z) \in \mathcal{K}(H)$. Let us observe that $R_3(z) = -R_1(z)C(B_2 - z)^{-1} - (A_1-z)^{-1}CR_2(z)$. Since $R_1(z)$ and $R_2(z)$ are compact operators, then $R_3(z) \in \mathcal{K}(H)$. Q.E.D

We may state the assumptions of the last theorem in a more general form. More precisely, we have the following result:

Corollary 10.5.1. *Suppose that A_i, B_i, $i = 1, 2$ satisfy the assumptions $(K1)$–$(K3)$ and assume that $\rho(A_1)\bigcap\rho(A_2)\bigcap\rho(B_1)\bigcap\rho(B_2)$ is not empty. If $\mathcal{D}(B_i) \subset \mathcal{D}(C)$, with $i = 1, 2$, and if there exists a set of families $(L_k)_{k=1,\ldots,n}$ $(n \in \mathbb{N}^*)$ such that the following conditions are satisfied:*

(i) $\widetilde{A_2} - \widetilde{A_1} = \sum_{k=1}^{n} S_{1,k}T_{1,k}$ *where the operator* $S_{1,k} \in \mathcal{L}(L_k, K^*)$, $T_{1,k} \in B_0^l(K, L_k)$ *and* $R_z(A_1)S_{1,k} \in B_q^l(L_k, H)$ *for some* $z \in \rho(A_1)\bigcap\rho(A_2)$.
(ii) $\widetilde{B_2} - \widetilde{B_1} = \sum_{k=1}^{n} S_{2,k}T_{2,k}$ *where the operator* $S_{2,k} \in \mathcal{L}(L_k, K^*)$, $T_{2,k} \in B_0^l(K, L_k)$ *and* $R_z(B_1)S_{2,k} \in B_q^l(L_k, H)$ *for some* $z \in \rho(B_1)\bigcap\rho(B_2)$.

Then, M_C^2 *is a compact perturbation of the operator* M_C^1 *and* $\sigma_{e4}(M_C^1) = \sigma_{e4}(M_C^2)$. ◇

Proof. The proof follows immediately from Theorem 10.5.1. Q.E.D

10.6 Wolf Essential Spectrum of Block Operator Matrix Acting in Friedrichs Module

Let (J, H) be a compact Friedrichs module. In the product $H \oplus H$, we consider an unbounded block operator matrix

$$\mathcal{A} = \begin{pmatrix} A & B \\ C & D \end{pmatrix},$$

where A, B, C, and D are closable operators with dense domains $\mathcal{D}(A)$, $\mathcal{D}(B)$, $\mathcal{D}(C)$, $\mathcal{D}(D)$ in H obtained as restrictions of some bounded operators $J \longrightarrow J^*$. We always suppose that $\mathcal{A}$ with its natural domain $\mathcal{D}(\mathcal{A}) = (\mathcal{D}(A)\bigcap\mathcal{D}(C)) \oplus (\mathcal{D}(B)\bigcap\mathcal{D}(D))$ is also densely defined. If $\rho(A)$ is not empty, the operator function S defined by $S(\lambda) = D - \lambda - C(A - \lambda)^{-1}B$ is called a Schur complement of $\mathcal{A}$. Now, we assume that the following conditions hold:

$(K4)$ $\mathcal{D}(A) \subset \mathcal{D}(C) \subset J$ densely, $\mathcal{D}(D) \subset \mathcal{D}(B) \subset J$.
$(K5)$ $\mathcal{D}(D^*) \subset J$ and $\mathcal{D}(A^*) \subset J$.

Theorem 10.6.1. *Suppose that A, B, C, and D satisfy the assumptions $(K4)$ and $(K5)$. If for some (and hence for all) $\mu \in \rho(A)\bigcap\rho(D)$, we have*

(i) $(A - \mu)^{-1}B$ *and* $C(A - \mu)^{-1}B$ *are bounded on* $\mathcal{D}(B)$,
(ii) C, B *extend to bounded operators* $\tilde{C}, \tilde{B} \in \mathcal{L}(J, J^*)$,
(iii) there exists $\mu_0 \in \rho(A)$ *with* $\rho(\overline{\mathcal{A}})\bigcap\rho(A)\bigcap\rho(D - \overline{C(A - \mu_0)^{-1}B})$ *is not empty, and*

(*iv*) *there are a Banach module* K *and operators* $S_1,\ S_2 \in \mathcal{L}(K, J^*)$ *and* $T_1,\ T_2 \in B_0^l(J, K)$ *such that* $\tilde{C} = S_1T_1,\ \tilde{B} = S_2T_2,\ (I + \overline{S(\lambda)}^{-1}\overline{C(A-\lambda)^{-1}B})(D-\lambda)^{-1}S_1,\ (A-z)^{-1}S_2 \in B_q^l(K, H)$, *for all* $\lambda \in \rho(\overline{\mathcal{A}}) \bigcap \rho(A) \bigcap \rho(D - \overline{C(A-\mu_0)^{-1}B})$.

Then, $\sigma_{e4}(\overline{\mathcal{A}}) = \sigma_{e4}(A) \bigcup \sigma_{e4}(D - \overline{C(A-\mu_0)^{-1}B})$. ◇

Proof. By using the second condition in (i), $S(\mu)$ is closable for every $\mu \in \rho(A)$ and $\overline{S(\mu)} = D - \mu - \overline{C(A-\mu)^{-1}B}$. According to Theorem 2.2.3, the block operator matrix $\mathcal{A}$ is closable. Let

$$\mathcal{A}_{1,\mu_0} = \begin{pmatrix} A & 0 \\ 0 & S(\mu_0) + \mu_0 \end{pmatrix}.$$

Now, let $\mu_0 \in \rho(A)$ and let $\lambda \in \mathbb{C}$ be such that $\lambda \in \rho(\overline{\mathcal{A}}) \bigcap \rho(A) \bigcap \rho(D - \overline{C(A-\mu_0)^{-1}B})$ is not empty. Using Eq. (10.3.5), we deduce that

$$\begin{aligned} &(\overline{\mathcal{A}} - \lambda)^{-1} - (\mathcal{A}_{1,\mu_0} - \lambda)^{-1} \\ &= \begin{pmatrix} \overline{(A-\lambda)^{-1}B}\ \overline{S(\lambda)}^{-1} C(A-\lambda)^{-1} & -\overline{(A-\lambda)^{-1}B}\ \overline{S(\lambda)}^{-1} \\ -\overline{S(\lambda)}^{-1} C(A-\lambda)^{-1} & \overline{S(\lambda)}^{-1} - (\overline{S(\mu_0)} + \mu_0 - \lambda)^{-1} \end{pmatrix}. \end{aligned}$$

It remains to demonstrate that all entries of this block operator matrix are compact. For this purpose, let $R_1 = \overline{S(\lambda)}^{-1} C(A-\lambda)^{-1}$. Then,

$$\begin{aligned} R_1 &= (D-\lambda)^{-1}C(A-\lambda)^{-1} + (\overline{S(\lambda)}^{-1} - (D-\lambda)^{-1})C(A-\lambda)^{-1} \\ &= (D-\lambda)^{-1}C(A-\lambda)^{-1} + \overline{S(\lambda)}^{-1}(D - \lambda - \overline{S(\lambda)})(D-\lambda)^{-1}C(A-\lambda)^{-1} \\ &= (D-\lambda)^{-1}C(A-\lambda)^{-1} + \overline{S(\lambda)}^{-1}\overline{C(A-\lambda)^{-1}B}(D-\lambda)^{-1}C(A-\lambda)^{-1} \\ &= (I + \overline{S(\lambda)}^{-1}\overline{C(A-\lambda)^{-1}B})(D-\lambda)^{-1}C(A-\lambda)^{-1}. \end{aligned}$$

Note that $z \in \rho(D)$ if, and only if, $z \in \rho(D^*)$ and we have $(D^* - \overline{z})^{-1} = (D - z)^{-1*}$. By using the assumption $(K5)$, $(A-\lambda)^{-1*}H \subset J$. So, $(A-\lambda)^{-1}$ can be extended to a unique continuous operator $R_\lambda(D) : J^* \longrightarrow H$. Since C can be extended to a unique continuous $\tilde{C} = S_1T_1$, then we can write $R_1 = [(I + \overline{S(\lambda)}^{-1}\overline{C(A-\lambda)^{-1}B})R_\lambda(D)S_1][T_1(A-\lambda)^{-1}]$ where the first factor is in $B_q^l(K, H)$ and the second factor is in $B_0^l(H, K)$. So, the product of the two factors is in $B_0^l(H)$. The domain of D is included in J and hence, $R_1H \subset J$. So, $R_1 \in \mathcal{K}(H)$. Let $R_2 = \overline{(A-\lambda)^{-1}B}\ \overline{S(\lambda)}^{-1}C(A-\lambda)^{-1} = \overline{(A-\lambda)^{-1}B}R_1$. Since $R_1 \in \mathcal{K}(H)$, then $R_2 \in \mathcal{K}(H)$. Now, we suppose that $R_3 = \overline{S(\lambda)}^{-1} - (\overline{S(\mu_0)} + \mu_0 - \lambda)^{-1}$. Therefore,

$$\begin{aligned}R_3 &= (\mu_0 - \lambda)\overline{S(\lambda)}^{-1}(\overline{C(A-\lambda)^{-1}B} - \overline{C(A-\mu_0)^{-1}B})(\overline{S(\mu_0)} + \mu_0 - \lambda)^{-1}\\ &= (\lambda - \mu_0)\overline{S(\lambda)}^{-1}C(A-\lambda)^{-1}(A-\mu_0)^{-1}B(\overline{S(\mu_0)} + \mu_0 - \lambda)^{-1}\\ &= (\lambda - \mu_0)R_1(A-\mu_0)^{-1}B(\overline{S(\mu_0)} + \mu_0 - \lambda)^{-1},\end{aligned}$$

where $R_1 \in \mathcal{K}(H)$. So, $R_3 \in \mathcal{K}(H)$.

By using a similar reasoning as before, we prove that $R_4 := \overline{(A-\lambda)^{-1}B}\ \overline{S(\lambda)}^{-1} \in \mathcal{K}(H)$. Q.E.D

Corollary 10.6.1. *Suppose that A, B, C, and D satisfy the assumptions $(K4)$ and $(K5)$. Let $(K_k)_{k=1,\ldots,n}$ be a family of Banach modules ($n \in \mathbb{N}^*$). If, for some (and hence for all) $\mu \in \rho(A)\bigcap\rho(D)$, we have*

(*i*) *$(A-\mu)^{-1}B$ and $C(A-\mu)^{-1}B$ are bounded on $\mathcal{D}(B)$,*
(*ii*) *C, B can extended to be bounded operators $\tilde{C} = \sum_{k=1}^{n} S_{1,k}T_{1,k}$ and $\tilde{B} = \sum_{k=2}^{n} S_{1,k}T_{2,k}$, where $S_{1,k}$, $S_{2,k} \in \mathcal{L}(K_k, J^*)$ and $T_{1,k}$, $T_{2,k} \in B_0^l(J, K_k)$,*
(*iii*) *there exists $\mu_0 \in \rho(A)$ with $\rho(\overline{\mathcal{A}})\bigcap\rho(A)\bigcap\rho(D - \overline{C(A-\mu_0)^{-1}B})$ is not empty, and*
(*iv*) *$(I + \overline{S(\lambda)}^{-1}\overline{C(A-\lambda)^{-1}B})(D-\lambda)^{-1}S_{1,k}$, $(A-z)^{-1}S_{2,k} \in B_q^l(K, H)$, for all $1 \leq k \leq n$, $\lambda \in \rho(\overline{\mathcal{A}})\bigcap\rho(A)\bigcap\rho(D - \overline{C(A-\mu_0)^{-1}B})$.*

Then, $\sigma_{e4}(\overline{\mathcal{A}}) = \sigma_{e4}(A)\bigcup\sigma_{e4}(D - \overline{C(A-\mu_0)^{-1}B})$. ◇

10.7 The M-Essential Spectra of Block Operator Matrices

The purpose of this section is to discuss the M-essential spectra of the 2×2 matrix operator L, the closure of L_0 which acts on the Banach space $X \times Y$ where M is a bounded operator, formally defined on the product space $X \times Y$ by a matrix

$$M = \begin{pmatrix} M_1 & M_2 \\ M_3 & M_4 \end{pmatrix}$$

and L_0 is given by

$$L_0 = \begin{pmatrix} A & B \\ C & D \end{pmatrix}.$$

The operator A acts on the Banach space X and has the domain $\mathcal{D}(A)$, whereas the operator D is defined on $\mathcal{D}(D)$ and acts on the Banach space Y, and the intertwining operator B (resp. C) is defined on the domain $\mathcal{D}(B)$ (resp. $\mathcal{D}(C)$) and acts on the Banach space X (resp. on Y).

10.7.1 Closability and Closure of the Block Operator Matrix

In what follows, we will assume that the following conditions hold:

$(L1)$ The operator A is a closed, densely defined linear operator on X with a nonempty M_1-resolvent set $\rho_{M_1}(A)$.
$(L2)$ The operator B is a densely defined linear operator on X and, for some (hence for all) $\mu \in \rho_{M_1}(A)$, the operator $(A-\mu M_1)^{-1}B$ is closable. (In particular, if B is closable, then $(A-\mu M_1)^{-1}B$ is also closable).
$(L3)$ The operator C satisfies the inclusion $\mathcal{D}(A) \subset \mathcal{D}(C)$ and, for some (hence for all) $\mu \in \rho_{M_1}(A)$, the operator $C(A-\mu M_1)^{-1}$ is bounded.
$(L4)$ The lineal $\mathcal{D}(B)\bigcap\mathcal{D}(D)$ is dense in Y and, for some (hence for all) $\mu \in \rho_{M_1}(A)$, the operator $D - C(A-\mu M_1)^{-1}B$ is closable and, we will denote the closure of the operator $D-(C-\mu M_3)(A-\mu M_1)^{-1}(B-\mu M_2)$ by $S(\mu)$.

Remark 10.7.1.

(*i*) From the closed graph theorem, it follows that the operator $G(\mu) := \overline{(A-\mu M_1)^{-1}(B-\mu M_2)}$ is bounded on Y.
(*ii*) We emphasize that neither the domain of $S(\mu)$ nor the property of being closable depends on μ. Indeed, let us consider $\lambda, \mu \in \rho_{M_1}(A)$. Then, we have:

$$S(\lambda) - S(\mu) = (\mu - \lambda)\left[M_3 G(\mu) + F(\lambda)M_2 + F(\lambda)M_1 G(\mu)\right], \quad (10.7.1)$$

where $F(\lambda) := (C-\lambda M_3)(A-\lambda M_1)^{-1}$. Since the operators $F(\lambda)$ and $G(\mu)$ are bounded, then the difference $S(\lambda) - S(\mu)$ is also bounded. Therefore, neither the domain of $S(\mu)$ nor the property of being closable depends on μ. ◇

Now, let us recall the following result which describes the closure of the operator L_0.

Theorem 10.7.1. *Let the conditions* $(L1)$–$(L3)$ *be satisfied and let the lineal* $\mathcal{D}(B)\bigcap\mathcal{D}(D)$ *be dense in* X. *Then, the operator* L_0 *is closable if, and only if, the operator* $D - C(A-\mu M_1)^{-1}B$ *is closable on* X, *for some* $\mu \in \rho_{M_1}(A)$. *Moreover, the closure* L *of* L_0 *is given by*

$$L = \mu M + \begin{pmatrix} I & 0 \\ F(\mu) & I \end{pmatrix} \begin{pmatrix} A-\mu M_1 & 0 \\ 0 & S(\mu)-\mu M_4 \end{pmatrix} \begin{pmatrix} I & G(\mu) \\ 0 & I \end{pmatrix}. \quad (10.7.2)$$

◇

Proof. Since M is bounded, it is easy to verify that the assumptions $(L1)$–$(L4)$ are also satisfied, if we replace L_0 by $L_0 - \mu M$. Then, the result follows from Theorem 10.1.1 for $(L_0-\mu M) - 0.I$. Q.E.D

10.7.2 Essential Spectra of L

Lemma 10.7.1.

(*i*) *If* $M_3 \in \mathcal{F}^b(X,Y)$ *and* $F(\lambda) \in \mathcal{F}^b(X,Y)$, *for some* $\lambda \in \rho_{M_1}(A)$, *then* $F(\lambda) \in \mathcal{F}^b(X,Y)$, *for all* $\lambda \in \rho_{M_1}(A)$.

(*ii*) *If* $M_2 \in \mathcal{F}^b(Y,X)$ *and, if* $G(\lambda) \in \mathcal{F}^b(Y,X)$, *for some* $\lambda \in \rho_{M_1}(A)$, *then* $G(\lambda) \in \mathcal{F}^b(Y,X)$, *for all* $\lambda \in \rho_{M_1}(A)$.

(*iii*) *If* $F(\lambda)$, $G(\lambda)$, M_2 *and* M_3 *are Fredholm perturbations, for some* $\lambda \in \rho_{M_1}(A)$, *then* $\sigma_{ei,M_4}(S(\lambda))$ *does not depend on* $\lambda \in \rho_{M_1}(A)$, *for* $i = 1,\ldots,6$. ◇

Proof. (*i*) The result follows from the following identity

$$F(\lambda) - F(\mu) = (\lambda-\mu)\big[F(\lambda)M_1 - M_3\big](A-\mu M_1)^{-1}, \text{ for all } \lambda \text{ and } \mu \in \rho_{M_1}(A).$$

(*ii*) The result follows from this identity

$$G(\lambda) - G(\mu) = (\lambda-\mu)(A-\lambda M_1)^{-1}\big[M_1G(\mu) - M_2\big], \text{ for all } \lambda \text{ and } \mu \in \rho_{M_1}(A).$$

(*iii*) The result of this assertion follows directly from Eq. (10.7.1). Q.E.D

Theorem 10.7.2. *Let* L_0 *be the* 2×2 *operator matrix satisfying the conditions* $(L1)$–$(L4)$. *If* M_2 *and* M_3 *are Fredholm perturbations and if, for some (hence for all)* $\mu \in \rho_{M_1}(A)$, $F(\mu)$ *and* $G(\mu)$ *are Fredholm perturbations, then* $\sigma_{e4,M}(L) = \sigma_{e4,M_1}(A)\bigcup\sigma_{e4,M_4}(S(\mu))$ *and,* $\sigma_{e5,M}(L) \subseteq \sigma_{e5,M_1}(A)\bigcup\sigma_{e5,M_4}(S(\mu))$. *Moreover, if* $\mathbb{C}\backslash\sigma_{e4,M_1}(A)$ *is connected, then* $\sigma_{e5,M}(L) = \sigma_{e5,M_1}(A)\bigcup\sigma_{e5,M_4}(S(\mu))$. *Besides, if* $\mathbb{C}\backslash\sigma_{e5,M}(L)$ *is connected,* $\rho_M(L) \neq \emptyset$, $\mathbb{C}\backslash\sigma_{e5,M_4}(S(\mu))$ *is connected and* $\rho_{M_4}(S(\mu)) \neq \emptyset$, *then* $\sigma_{e6,M}(L) = \sigma_{e6,M_1}(A)\bigcup\sigma_{e6,M_4}(S(\mu))$. ◇

Proof. Let $\mu \in \rho_{M_1}(A)$ be such that the operators $F(\mu)$ and $G(\mu)$ are Fredholm perturbations, and let $\lambda \in \mathbb{C}$. Writing $\lambda M - L = \mu M - L + (\lambda-\mu)M$, and using the relation (10.7.2), we have

$$\lambda M - L = UV(\lambda)W - (\lambda-\mu)\begin{pmatrix} 0 & M_1G(\mu) - M_2 \\ F(\mu)M_1 - M_3 & F(\mu)M_1G(\mu) \end{pmatrix}, \tag{10.7.3}$$

where

$$U = \begin{pmatrix} I & 0 \\ F(\mu) & I \end{pmatrix}, W = \begin{pmatrix} I & G(\mu) \\ 0 & I \end{pmatrix} \text{ and } V(\lambda) = \begin{pmatrix} \lambda M_1 - A & 0 \\ 0 & \lambda M_4 - S(\mu) \end{pmatrix}.$$

Since the operators $F(\mu)$, $G(\mu)$, M_2 and M_3 are Fredholm perturbations, then by using Theorem 6.6.1, the second operator in the right-hand side of Eq. (10.7.3) is also a Fredholm perturbation. Hence, $\lambda M - L$ is a Fredholm operator if, and only if, $UV(\lambda)W$ is a Fredholm operator. Now, let us notice that the operators U and W are bounded and also have bounded inverses. Consequently, the operator $UV(\lambda)W$ is a Fredholm operator if, and only if, $V(\lambda)$ has this property if, and only if, $\lambda M_1 - A$ (resp. $\lambda M_4 - S(\mu)$) is a Fredholm operator on X (resp. on Y) and

$$i(\lambda M - L) = i(\lambda M_1 - A) + i(\lambda M_4 - S(\mu)). \tag{10.7.4}$$

Therefore, $\sigma_{e4,M}(L) = \sigma_{e4,M_1}(A) \bigcup \sigma_{e4,M_4}(S(\mu))$, and

$$\sigma_{e5,M}(L) \subseteq \sigma_{e5,M_1}(A) \bigcup \sigma_{e5,M_4}(S(\mu)). \tag{10.7.5}$$

Now, let us now suppose that $\mathbb{C}\backslash\sigma_{e4,M_1}(A)$ is connected. By using the condition $(L1)$, $\rho_{M_1}(A)$ is not empty. Let $\alpha \in \rho_{M_1}(A)$. Then, $\alpha M_1 - A \in \Phi(X)$ and $i(\alpha M_1 - A) = 0$. Since $\rho_{M_1}(A) \subseteq \rho_{4,M_1}(A)$ and by using Proposition 2.2.5, we deduce that $i(\lambda M_1 - A)$ is constant on any component of $\Phi_{M_1,A}$. Then, $i(\lambda M_1 - A) = 0$, for all $\lambda \in \rho_{4,M_1}(A)$. From Eqs. (10.7.4) and (10.7.5), it follows immediately that

$$\sigma_{e5,M}(L) = \sigma_{e5,M_1}(A) \bigcup \sigma_{e5,M_4}(S(\mu)). \tag{10.7.6}$$

Moreover, let us assume that $\mathbb{C}\backslash\sigma_{e5,M_1}(A)$ is connected. Then, by using Lemma 9.1.1 and Eq. (10.7.6), we have $\sigma_{e6,M}(L) = \sigma_{e6,M_1}(A) \bigcup \sigma_{e6,M_4}(S(\mu))$. Q.E.D

In the sequel and, for $\mu \in \rho_{M_1}(A)$, we will denote by $M(\mu)$ the following operator

$$M(\mu) = \begin{pmatrix} 0 & M_1 G(\mu) - M_2 \\ F(\mu)M_1 - M_3 & F(\mu)M_1 G(\mu) \end{pmatrix}.$$

Theorem 10.7.3.

(i) If the operator $M(\mu) \in \mathcal{F}_+(X \times Y)$, for some $\mu \in \rho_{M_1}(A)$, then

$$\sigma_{e1,M}(L) = \sigma_{e1,M_1}(A) \bigcup \sigma_{e1,M_4}(S(\mu)).$$

(ii) If the operator $M(\mu) \in \mathcal{F}_-(X \times Y)$, for some $\mu \in \rho_{M_1}(A)$, then

$$\sigma_{e2,M}(L) = \sigma_{e2,M_1}(A) \bigcup \sigma_{e2,M_4}(S(\mu)).$$

(iii) If $M(\mu) \in \mathcal{F}_+(X \times Y) \bigcap \mathcal{F}_-(X \times Y)$, for some $\mu \in \rho_{M_1}(A)$, then

$$\begin{aligned} &\sigma_{e3,M}(L) \\ &\quad = \sigma_{e3,M_1}(A) \bigcup \sigma_{e3,M_4}(S(\mu)) \bigcup [\sigma_{e2,M_1}(A) \bigcup \sigma_{e1,M_4}(S(\mu))] \\ &\quad \bigcup [\sigma_{e1,M_1}(A) \bigcup \sigma_{e2,M_4}(S(\mu))]. \end{aligned}$$

◇

Proof. The assertions (i) and (ii) follow immediately from Eq. (10.7.3), whereas the assertion (iii) is an immediate consequence of (i) and (ii). Q.E.D

Chapter 11
Essential Spectra of 3 × 3 Block Operator Matrices

In this chapter, we are concerned with the essential spectra of operators defined by a 3×3 block operator matrix

$$\mathrm{L}_0 := \begin{pmatrix} A & B & C \\ D & E & F \\ G & H & L \end{pmatrix}, \tag{11.0.1}$$

where the entries of the matrix are in general unbounded operators. The operator (11.0.1) is defined on $(\mathcal{D}(A) \bigcap \mathcal{D}(D) \bigcap \mathcal{D}(G)) \times (\mathcal{D}(B) \bigcap \mathcal{D}(E) \bigcap \mathcal{D}(H)) \times (\mathcal{D}(C) \bigcap \mathcal{D}(F) \bigcap \mathcal{D}(L))$. Observe that this operator doesn't need to be closed. The aim of this chapter is to present some conditions which should allow the 3×3 block operator matrix L_0 to be closable. For this purpose, our reasoning will be based on the specific properties of the first entry A (always satisfying the Frobenius–Schur decomposition of L_0). So, we will consider two cases. The first one is based on A being closed whereas the second case deals with A being closable.

11.1 Case Where the Operator A Is Closed

11.1.1 The Operator L_0 *and Its Closure*

The essential work in this section is to impose some conditions on the entries of the operator L_0 in order to establish its closedness. In the product of Banach spaces $X \times Y \times Z$, we consider the operator L_0 defined by (11.0.1) where the operator A acts on X and has a domain $\mathcal{D}(A)$, the operator E acts on Y and has a domain $\mathcal{D}(E)$, and the operator L acts on Z and has a domain $\mathcal{D}(L)$. The intertwining operator B is defined on the domain $\mathcal{D}(B) \subset Y$ into X, the operator H is defined on the domain $\mathcal{D}(H) \subset Y$ into Z, the operator C is defined on the domain $\mathcal{D}(C) \subset Z$

A. Jeribi, *Spectral Theory and Applications of Linear Operators and Block Operator Matrices*, DOI 10.1007/978-3-319-17566-9_11

into X, the operator F is defined on the domain $\mathcal{D}(F) \subset Z$ into Y, the operator D is defined on the domain $\mathcal{D}(D) \subset X$ into Y, and the operator G is defined on the domain $\mathcal{D}(G) \subset X$ into Z. In what follows, we will consider the following hypotheses:

$(M1)$ The operator A is a closed, densely defined linear operator on X, with a nonempty resolvent set $\rho(A)$.

$(M2)$ The operator D (resp. G) verifies that $\mathcal{D}(A) \subset \mathcal{D}(D)$ (resp. $\mathcal{D}(A) \subset \mathcal{D}(G)$) and, for some (hence for all) $\mu \in \rho(A)$, the operator $D(A-\mu)^{-1}$ (resp. $G(A-\mu)^{-1}$) is bounded.
Let $F_1(\mu) := D(A-\mu)^{-1}$, and $F_2(\mu) := G(A-\mu)^{-1}$.

- In particular, if D (resp. G) is closable, then from the closed graph theorem (see Theorem 2.1.3), it follows that $F_1(\mu)$ (resp. $F_2(\mu)$) is bounded.

$(M3)$ The operator B (resp. C) is densely defined on Y (resp. Z) and, for some (hence for all) $\mu \in \rho(A)$, the operator $(A-\mu)^{-1}B$ (resp. $(A-\mu)^{-1}C$) is bounded on its domain.
Now, let $G_1(\mu) := \overline{(A-\mu)^{-1}B}$, and $G_2(\mu) := \overline{(A-\mu)^{-1}C}$.

$(M4)$ The lineal $\mathcal{D}(B)\bigcap\mathcal{D}(E)$ is dense in Y and, for some (hence for all) $\mu \in \rho(A)$, the operator $S_1(\mu) := E - D(A-\mu)^{-1}B$ is closed.

$(M5)$ $\mathcal{D}(C) \subset \mathcal{D}(F)$, and the operator $F - D(A-\mu)^{-1}C$ is bounded on its domain, for some $\mu \in \rho(A)$ and therefore, for all $\mu \in \rho(A)$. We will also suppose that there exists μ such that $\mu \in \rho(A)\bigcap\rho(S_1(\mu))$ and we will denote $G_3(\mu)$ by

$$G_3(\mu) := \overline{(S_1(\mu)-\mu)^{-1}(F - D(A-\mu)^{-1}C)}.$$

- To explain this, let $\mu \in \rho(A)$, such that $F - D(A-\mu)^{-1}C$ is bounded on its domain. Then, for an arbitrary $\lambda \in \rho(A)$, we have

$$F - D(A-\lambda)^{-1}C = F - D(A-\mu)^{-1}C + (\mu-\lambda)F_1(\mu)(A-\lambda)^{-1}C.$$

From the assumptions $(M2)$ and $(M3)$, it follows that the operator on the right-hand side is bounded on its domain. Then, the boundedness of the operator $F - D(A-\mu)^{-1}C$ does not depend on $\mu \in \rho(A)$. We will denote $G_4(\mu)$ by $G_4(\mu) := \overline{F - D(A-\mu)^{-1}C}$.

Remark 11.1.1. If the operators A and E generate C_0-semigroups, and if the operators D and B are bounded, then there exists $\mu \in \mathbb{C}$, such that $\mu \in \rho(A)\bigcap\rho(S_1(\mu))$. Indeed, it is well known that if the operators A and E generate C_0-semigroups, then there exist two constants $M > 0$ and $w > 0$, such that $\|(\mu - T)^{-1}\| \leq \frac{M}{\text{Re}\mu - w}$, where $T \in \{A, E\}$ for all μ such that $\text{Re}\mu > w$. For a fixed $\mu \in \mathbb{C}$ chosen in such a way that $\text{Re}\mu > w + \alpha$, where $\alpha > 0$, we consider the following resolvent equation of $S_1(\mu)$

$$(\lambda - E + D(A-\mu)^{-1}B)\varphi = \psi. \tag{11.1.1}$$

Since $\lambda \in \rho(E)$, we deduce that, for $\mathrm{Re}\lambda > w + \alpha$, Eq. (11.1.1) may be transformed into $[I + (\lambda - E)^{-1}D(\mu - A)^{-1}B]\varphi = (\lambda - E)^{-1}\psi$. The fact that $\|(\lambda - E)^{-1}D(\mu - A)^{-1}B\| \leq \frac{M^2\|D\|\|B\|}{\alpha(\mathrm{Re}\lambda - w)}$ allows us to conclude that $\lim_{\mathrm{Re}\lambda \to +\infty} \|(\lambda - E)^{-1}D(\mu - A)^{-1}B\| = 0$. Hence, there exists $\beta > w + \alpha$ such that, for $\mathrm{Re}\lambda > \beta$, we have $r_\sigma((\lambda - E)^{-1}D(\mu - A)^{-1}B) < 1$, where $r_\sigma(.)$ represents the spectral radius. Hence for μ, such that $\mathrm{Re}\mu > \beta$, we have $\mu \in \rho(A)$ and $\mu \in \rho(S_1(\mu))$. Moreover, we can write

$$(\mu - S_1(\mu))^{-1} = \sum_{n \geq 0} [(\mu - E)^{-1}D(\mu - A)^{-1}B]^n(\mu - E)^{-1}.$$

◇

$(M6)$ The operator H satisfies the fact that $\mathcal{D}(B) \subset \mathcal{D}(H)$ and, for some (hence for all) $\mu \in \rho(A) \bigcap \rho(S_1(\mu))$, the operator $(H - G(A - \mu)^{-1}B)(S_1(\mu) - \mu)^{-1}$ is bounded. Set

$$F_3(\mu) := (H - G(A - \mu)^{-1}B)(S_1(\mu) - \mu)^{-1}.$$

$(M7)$ For the operator K, we will assume that $\mathcal{D}(C) \subset \mathcal{D}(K)$ and, for some (hence for all) $\mu \in \rho(A) \bigcap \rho(S_1(\mu))$, the operator

$$L - G(A - \mu)^{-1}C - [H - G(A - \mu)^{-1}B](S_1(\mu) - \mu)^{-1}[F - D(A - \mu)^{-1}C]$$

is closable. Let us denote by $S_2(\mu)$ this operator, and by $\overline{S}_2(\mu)$ its closure.

Remark 11.1.2.

(i) From the Hilbert identity, we get for $\lambda, \mu \in \rho(A)$

$$S_1(\lambda) - S_1(\mu) = (\mu - \lambda)F_1(\mu)(A - \lambda)^{-1}B.$$

Since the operator $F_1(\mu)$ is bounded and $(A - \lambda)^{-1}B$ is bounded on its domain, we deduce that neither the domain of $S_1(\mu)$ nor the property of being closable depends on the choice of μ. Then,

$$S_1(\lambda) - S_1(\mu) = (\mu - \lambda)F_1(\mu)G_1(\lambda). \tag{11.1.2}$$

(ii) Let $\lambda \in \rho(A) \bigcap \rho(S_1(\lambda))$ and $\mu \in \rho(A) \bigcap \rho(S_1(\mu))$. Then,

$$\begin{aligned} S_2(\lambda) - S_2(\mu) & \\ =& (\mu - \lambda)F_2(\mu)(A - \lambda)^{-1}C - F_3(\lambda)[F - D(A - \lambda)^{-1}C] \\ &+ F_3(\mu)[F - D(A - \mu)^{-1}C]. \\ =& (\mu - \lambda)F_2(\mu)(A - \lambda)^{-1}C - F_3(\lambda)[F - D(A - \lambda)^{-1}C] \end{aligned}$$

$$+ F_3(\mu)[F - D(A-\lambda)^{-1}C - (\mu-\lambda)D(A-\mu)^{-1}(A-\lambda)^{-1}C]$$
$$=(\mu-\lambda)F_2(\mu)(A-\lambda)^{-1}C + [F_3(\mu) - F_3(\lambda)][F - D(A-\lambda)^{-1}C]$$
$$+ (\lambda-\mu)F_3(\mu)F_1(\mu)(A-\lambda)^{-1}C.$$

Since the operators $F_i(.)$, with $i = 1, 2, 3$ are bounded everywhere and since the operators $(A-\mu)^{-1}C$ and $F - D(A-\lambda)^{-1}C$ are bounded on their domains, then the closedness of the operator $S_2(\mu)$ does not depend on the choice of μ. Hence,

$$\overline{S}_2(\lambda) - \overline{S}_2(\mu) = (\mu-\lambda)F_2(\mu)G_2(\lambda) + [F_3(\mu) - F_3(\lambda)]G_4(\lambda) + (\lambda-\mu)F_3(\mu)F_1(\mu)G_2(\lambda). \tag{11.1.3}$$

◇

Now, we are able to establish the closedness of the operator L_0.

Theorem 11.1.1. *Let the hypotheses* $(M1)$–$(M6)$ *be satisfied. Then, the operator* L_0 *is closable if, and only if,* $S_2(\mu)$ *is closable on* Z*, for some* $\mu \in \rho(A) \bigcap \rho(S_1(\mu))$. *Moreover, the closure* L *of* L_0 *is given by*

$$\mathrm{L} = \mu - \begin{pmatrix} I & 0 & 0 \\ F_1(\mu) & I & 0 \\ F_2(\mu) & F_3(\mu) & I \end{pmatrix} \begin{pmatrix} \mu - A & 0 & 0 \\ 0 & \mu - S_1(\mu) & 0 \\ 0 & 0 & \mu - \overline{S}_2(\mu) \end{pmatrix} \begin{pmatrix} I & G_1(\mu) & G_2(\mu) \\ 0 & I & G_3(\mu) \\ 0 & 0 & I \end{pmatrix} \tag{11.1.4}$$

or, spelled out,

$$\begin{cases} \mathrm{L} : \mathcal{D}(\mathrm{L}) \subset X \times Y \times Z \longrightarrow X \times Y \times Z \\[2mm] \mathrm{L}\begin{pmatrix} x \\ y \\ z \end{pmatrix} = \begin{pmatrix} A[x + G_1(\mu)y + G_2(\mu)z] - \mu[G_1(\mu)y + G_2(\mu)z] \\ D[x + G_1(\mu)y + G_2(\mu)z] + S_1(\mu)[y + G_3(\mu)z] - \mu G_3(\mu)z \\ G[x + G_1(\mu)y + G_2(\mu)z] + [H - G(A-\mu)^{-1}B][y + G_3(\mu)z] + \overline{S}_2(\mu)z \end{pmatrix} \\[2mm] \mathcal{D}(\mathrm{L}) = \left\{ \begin{pmatrix} x \\ y \\ z \end{pmatrix} \in X \times Y \times Z \text{ such that } \begin{array}{c} x + G_1(\mu)\, y + G_2(\mu)\, z \ \in \ \mathcal{D}(A), \\ y + G_3(\mu)\, z \ \in \ \mathcal{D}(S_1(\mu)) \\ \text{and } \ z \ \in \ \mathcal{D}(\overline{S}_2(\mu)) \end{array} \right\}. \end{cases}$$

◇

Proof. For $\mu \in \rho(A) \bigcap \rho(S_1(\mu))$, the operator L_0 can be factorized in the Frobenius–Schur sense:

$$\mathrm{L}_0 = \mu - \begin{pmatrix} I & 0 & 0 \\ F_1(\mu) & I & 0 \\ F_2(\mu) & F_3(\mu) & I \end{pmatrix} \begin{pmatrix} \mu - A & 0 & 0 \\ 0 & \mu - S_1(\mu) & 0 \\ 0 & 0 & \mu - S_2(\mu) \end{pmatrix} \begin{pmatrix} I & G_1(\mu) & G_2(\mu) \\ 0 & I & G_3(\mu) \\ 0 & 0 & I \end{pmatrix}.$$

Under the assumptions of the theorem, the following operators

$$U := \begin{pmatrix} I & 0 & 0 \\ F_1(\mu) & I & 0 \\ F_2(\mu) & F_3(\mu) & I \end{pmatrix} \quad \text{and} \quad W := \begin{pmatrix} I & G_1(\mu) & G_2(\mu) \\ 0 & I & G_3(\mu) \\ 0 & 0 & I \end{pmatrix}$$

are bounded and also boundedly invertible as mapping from $X \times Y \times Z$ into $X \times Y \times Z$. Hence, we deduce that L_0 is closable in $X \times Y \times Z$ if, and only if, $S_2(\mu)$ is closable as a mapping in Z. In this case, the closure L of L_0 is given by:

$$\mathrm{L} = \mu - \begin{pmatrix} I & 0 & 0 \\ F_1(\mu) & I & 0 \\ F_2(\mu) & F_3(\mu) & I \end{pmatrix} \begin{pmatrix} \mu - A & 0 & 0 \\ 0 & \mu - S_1(\mu) & 0 \\ 0 & 0 & \mu - \overline{S}_2(\mu) \end{pmatrix} \begin{pmatrix} I & G_1(\mu) & G_2(\mu) \\ 0 & I & G_3(\mu) \\ 0 & 0 & I \end{pmatrix}.$$

This achieves the proof of this theorem. Q.E.D.

Remark 11.1.3. From Remark 11.1.2, we notice that the description of the operator L does not depend on the choice of the point $\mu \in \rho(A) \bigcap \rho(S_1(\mu))$. ◇

11.1.2 Essential Spectra of the Operator L

Having obtained the closure L of the operator L_0, we will discuss its essential spectra. As a first step, we will establish the following stability lemma.

Lemma 11.1.1. *Let $\mu \in \rho(A) \bigcap \rho(S_1(\mu))$. If the sets $\Phi^b(Y, X)$, $\Phi^b(Z, X)$, and $\Phi^b(Z, Y)$ are not empty, then*

(i) If $F_1(\mu) \in \mathcal{F}^b(X, Y)$, $F_2(\mu) \in \mathcal{F}^b(X, Z)$, and $F_3(\mu) \in \mathcal{F}^b(Y, Z)$, then $\sigma_{ei}(S_1(\mu))$ and $\sigma_{ei}(\overline{S}_2(\mu))$, $i = 4, 5$ don't depend on μ.
(ii) If $F_1(\mu) \in \mathcal{F}_+^b(X, Y)$, $F_2(\mu) \in \mathcal{F}_+^b(X, Z)$, and $F_3(\mu) \in \mathcal{F}_+^b(Y, Z)$, then $\sigma_{e1}(S_1(\mu))$ and $\sigma_{e1}(\overline{S}_2(\mu))$ don't depend on μ.
(iii) If $F_1(\mu) \in \mathcal{F}_-^b(X, Y)$, $F_2(\mu) \in \mathcal{F}_-^b(X, Z)$, and $F_3(\mu) \in \mathcal{F}_-^b(Y, Z)$, then $\sigma_{e2}(S_1(\mu))$ and $\sigma_{e2}(\overline{S}_2(\mu))$ don't depend on μ.
(iv) If $F_1(\mu) \in \mathcal{F}_+^b(X, Y) \bigcap \mathcal{F}_-^b(X, Y)$, $F_2(\mu) \in \mathcal{F}_+^b(X, Z) \bigcap \mathcal{F}_-^b(X, Z)$, and $F_3(\mu) \in \mathcal{F}_+^b(Y, Z) \bigcap \mathcal{F}_-^b(Y, Z)$, then $\sigma_{e3}(S_1(\mu))$ and $\sigma_{e3}(\overline{S}_2(\mu))$ don't depend on μ. ◇

Proof. (i) Using the fact that $F_1(\mu) \in \mathcal{F}^b(X, Y)$, together with Theorem 6.3.1, we infer that $F_1(\mu)G_1(\mu) \in \mathcal{F}^b(Y)$. Therefore, from Eq. (11.1.2) and Theorem 7.5.3 (i), we can deduce that $\sigma_{ei}(S_1(\mu)) = \sigma_{ei}(S_1(\lambda))$, with $i = 4, 5$. So, $\sigma_{ei}(S_1(\mu))$, with $i = 4, 5$ does not depend on μ. Since $F_2(\mu) \in \mathcal{F}^b(X, Z)$ and $F_3(\mu) \in \mathcal{F}^b(Y, Z)$, it follows, from Theorem 6.3.1, that $F_2(\mu)G_2(\lambda) + [F_3(\mu) - F_3(\lambda)]G_4(\lambda) + (\lambda - \mu)F_3(\mu)F_1(\mu)G_2(\lambda) \in \mathcal{F}^b(Z)$. The use of this result, Eq. (11.1.3) and Theorem 7.5.3 (i) leads to the following equation $\sigma_{ei}(\overline{S}_2(\mu)) = \sigma_{ei}(\overline{S}_2(\lambda))$, with

$i = 4, 5$. A similar reasoning allows us to reach the results for both (ii) and (iii). The assertion (iv) is an immediate consequence of the items (ii) and (iii). Q.E.D.

In the sequel, and for $\mu \in \rho(A) \bigcap \rho(S_1(\mu))$, we will denote by $M(\mu)$ the following operator

$$M(\mu) := \begin{pmatrix} 0 & G_1(\mu) & G_2(\mu) \\ F_1(\mu) & F_1(\mu)G_1(\mu) & F_1(\mu)G_2(\mu) + G_3(\mu) \\ F_2(\mu) & F_2(\mu)G_1(\mu) + F_3(\mu) & F_2(\mu)G_2(\mu) + F_3(\mu)G_3(\mu) \end{pmatrix},$$

and for $\lambda \in \mathbb{C}$, we will denote $V(\lambda)$ the following operator

$$V(\lambda) := \begin{pmatrix} \lambda - A & 0 & 0 \\ 0 & \lambda - S_1(\mu) & 0 \\ 0 & 0 & \lambda - \overline{S}_2(\mu) \end{pmatrix}.$$

Now, we are able to describe the essential spectra of 3×3 block matrix operators.

Theorem 11.1.2. *Let the block matrix operator* L_0 *satisfy the hypotheses* $(M1)$–$(M7)$.

(*i*) *If, for some* $\mu \in \rho(A) \bigcap \rho(S_1(\mu))$, $F_1(\mu) \in \mathcal{F}^b(X, Y)$, $F_2(\mu) \in \mathcal{F}^b(X, Z)$, $F_3(\mu) \in \mathcal{F}^b(Y, Z)$ *and* $M(\mu) \in \mathcal{F}(X \times Y \times Z)$, *then* $\sigma_{e4}(\mathrm{L}) = \sigma_{e4}(A) \bigcup \sigma_{e4}(S_1(\mu)) \bigcup \sigma_{e4}(\overline{S}_2(\mu))$, *and* $\sigma_{e5}(\mathrm{L}) \subseteq \sigma_{e5}(A) \bigcup \sigma_{e5}(S_1(\mu)) \bigcup \sigma_{e5}(\overline{S}_2(\mu))$. *Moreover, if* $\mathbb{C} \backslash \sigma_{e4}(A)$ *and* $\mathbb{C} \backslash \sigma_{e4}(S_1(\mu))$ *are connected, then* $\sigma_{e5}(\mathrm{L}) = \sigma_{e5}(A) \bigcup \sigma_{e5}(S_1(\mu)) \bigcup \sigma_{e5}(\overline{S}_2(\mu))$. *In addition, if* $\mathbb{C} \backslash \sigma_{e5}(\mathrm{L})$ *is connected,* $\rho(\mathrm{L}) \neq \emptyset, \mathbb{C} \backslash \sigma_{e5}(\overline{S}_2(\mu))$ *is connected and* $\rho(\overline{S}_2(\mu)) \neq \emptyset$, *then*
$$\sigma_{e6}(\mathrm{L}) = \sigma_{e6}(A) \bigcup \sigma_{e6}(S_1(\mu)) \bigcup \sigma_{e6}(\overline{S}_2(\mu)).$$

(*ii*) *If* $F_1(\mu) \in \mathcal{F}_+^b(X, Y)$, $F_2(\mu) \in \mathcal{F}_+^b(X, Z)$, $F_3(\mu) \in \mathcal{F}_+^b(Y, Z)$ *and* $M(\mu) \in \mathcal{F}_+(X \times Y \times Z)$, *for some* $\mu \in \rho(A) \bigcap \rho(S_1(\mu))$, *then* $\sigma_{e1}(\mathrm{L}) = \sigma_{e1}(A) \bigcup \sigma_{e1}(S_1(\mu)) \bigcup \sigma_{e1}(\overline{S}_2(\mu))$.

(*iii*) *If* $F_1(\mu) \in \mathcal{F}_-^b(X, Y)$, $F_2(\mu) \in \mathcal{F}_-^b(X, Z)$, $F_3(\mu) \in \mathcal{F}_-^b(Y, Z)$ *and* $M(\mu) \in \mathcal{F}_-(X \times Y \times Z)$, *for some* $\mu \in \rho(A) \bigcap \rho(S_1(\mu))$, *then* $\sigma_{e2}(\mathrm{L}) = \sigma_{e2}(A) \bigcup \sigma_{e2}(S_1(\mu)) \bigcup \sigma_{e2}(\overline{S}_2(\mu))$.

(*iv*) *If* $F_1(\mu) \in \mathcal{F}_+^b(X, Y) \bigcap \mathcal{F}_-^b(X, Y)$, $F_2(\mu) \in \mathcal{F}_+^b(X, Z) \bigcap \mathcal{F}_-^b(X, Z)$, $F_3(\mu) \in \mathcal{F}_+^b(Y, Z) \bigcap \mathcal{F}_-^b(Y, Z)$ *and* $M(\mu) \in \mathcal{F}_+(X \times Y \times Z) \bigcap \mathcal{F}_-(X \times Y \times Z)$, *for some* $\mu \in \rho(A) \bigcap \rho(S_1(\mu))$, *then*
$$\begin{aligned}\sigma_{e3}(\mathrm{L}) = {} & \sigma_{e3}(A) \bigcup \sigma_{e3}(S_1(\mu)) \bigcup \sigma_{e3}(\overline{S}_2(\mu)) \\ & \bigcup [\sigma_{e1}(A) \bigcap \sigma_{e2}(S_1(\mu)) \bigcap \sigma_{e2}(\overline{S}_2(\mu))] \\ & \bigcup [\sigma_{e2}(A) \bigcap \sigma_{e2}(S_1(\mu)) \bigcap \sigma_{e1}(\overline{S}_2(\mu))] \\ & \bigcup [\sigma_{e2}(A) \bigcap \sigma_{e1}(S_1(\mu)) \bigcap \sigma_{e2}(\overline{S}_2(\mu))].\end{aligned}$$

◇

Proof. (i) Let $\mu \in \rho(A) \bigcap \rho(S_1(\mu))$ be such that the operator $M(\mu) \in \mathcal{F}(X \times Y \times Z)$ and set $\lambda \in \mathbb{C}$, then we can rewrite the representation (11.1.4) in the following form:

$$\begin{aligned} \lambda - \mathrm{L} &= (\lambda - \mu) + (\mu - \mathrm{L}) \\ &:= UV(\lambda)W + (\mu - \lambda)M(\mu). \end{aligned} \tag{11.1.5}$$

The remaining part of the proof may be checked in a similar way to that in Theorem 10.2.2. It is sufficient to use Eq. (11.1.5). The details are therefore omitted. Q.E.D.

Remark 11.1.4.

(i) If the operators $F_i(\mu)$ and $G_i(\mu)$, with $i = 1, 2, 3$ are compact operators, for some $\mu \in \rho(A) \bigcap \rho(S_1(\mu))$, then $M(\mu)$ is a compact operator, and so Fredholm perturbation.
(ii) Let $X = Y = Z := L_1(\Omega, d\mu)$ where (Ω, Σ, μ) is a positive measure space. If $F_i(\mu)$ and $G_i(\mu), i = 1, 2, 3$ are in $\mathcal{W}(L_1(\Omega, d\mu))$, for some $\mu \in \rho(A) \bigcap \rho(S_1(\mu))$, then $M(\mu) \in \mathcal{W}(L_1(\Omega, d\mu) \times L_1(\Omega, d\mu) \times L_1(\Omega, d\mu)) \subset \mathcal{F}(L_1(\Omega, d\mu) \times L_1(\Omega, d\mu) \times L_1(\Omega, d\mu))$. ◇

11.2 Case Where the Operator A Is Closable

Let X, Y, Z, and W be four Banach spaces. We consider the following linear operators, A acting in X and has the domain $\mathcal{D}(A)$, E acting in Y and has the domain $\mathcal{D}(E)$, L acting in Z and has the domain $\mathcal{D}(L)$. The intertwining operator B is defined on the domain $\mathcal{D}(B) \subset Y$ into X, the operator C is defined on the domain $\mathcal{D}(C) \subset Z$ into X, the operator D is defined on the domain $\mathcal{D}(D) \subset X$ into Y, the operator F is defined on the domain $\mathcal{D}(F) \subset Z$ into Y, the operator G is defined on the domain $\mathcal{D}(G) \subset X$ into Z, and the operator H is defined on the domain $\mathcal{D}(H) \subset Y$ into Z. We notice that the operators Γ_X go from X into W, Γ_Y go from Y into W and Γ_Z go from Z into W. In the product of the Banach spaces $X \times Y \times Z$, we define the operator $\mathcal{L}_0$ as follows:

$$\mathcal{D}(\mathcal{L}_0) = \left\{ \begin{pmatrix} x \\ y \\ z \end{pmatrix} \text{ such that } \begin{array}{l} x \in \mathcal{D}(A) \\ y \in \mathcal{D}(B) \bigcap \mathcal{D}(E) \\ z \in \mathcal{D}(C) \bigcap \mathcal{D}(F) \bigcap \mathcal{D}(L) \end{array} \text{ and } \Gamma_X x = \Gamma_Y y = \Gamma_Z z \right\}$$

$$\mathcal{L}_0 = \begin{pmatrix} A & B & C \\ D & E & F \\ G & H & L \end{pmatrix}.$$

11.2.1 Closability and Closure of the Operator $\mathcal{L}_0$

In what follows, we will consider the following hypotheses:

$(N1)$ The operator A is densely defined and closable.
It is noted that $\mathcal{D}(A)$ equipped with the graph norm $\|x\|_A = \|x\| + \|Ax\|$ can be completed to a Banach space X_A which coincides with $\mathcal{D}(\overline{A})$, the domain of the closure of A which is contained in X.

$(N2)$ $\mathcal{D}(A) \subset \mathcal{D}(\Gamma_X) \subset X_A$ and $\Gamma_X : X_A \longrightarrow W$ is a bounded mapping. Let us denote by $\overline{\Gamma}_X$ the extension by continuity, which is a bounded operator from X_A into W.

$(N3)$ $\mathcal{D}(A) \bigcap N(\Gamma_X)$ is dense in X and the resolvent set of the restriction of A to this set, $A_1 := A_{|\mathcal{D}(A)\bigcap N(\Gamma_X)}$ is nonempty, i.e., $\rho(A_1) \neq \emptyset$.

Remark 11.2.1. From $(N1)$, $(N3)$, it follows that $\Gamma_X(\mathcal{D}(A_1)) = \{0\}$ and the operator A_1 is closed. Hence, $\mathcal{D}(A_1)$ is a closed subset of X_A. ◇

$(N4)$ The operator B is densely defined and, for some $\mu \in \rho(A_1)$, (hence for all), the operator $(A_1 - \mu)^{-1}B$ is bounded on its domain.
To see this, let us take $\mu, \lambda \in \rho(A_1)$. Then,

$$(A_1 - \lambda)^{-1}B - (A_1 - \mu)^{-1}B = (\lambda - \mu)(A_1 - \lambda)^{-1}(A_1 - \mu)^{-1}B$$

and so, $\overline{(A_1 - \lambda)^{-1}B} - \overline{(A_1 - \mu)^{-1}B} = (\lambda - \mu)(A_1 - \lambda)^{-1}\overline{(A_1 - \mu)^{-1}B}$.

$(N5)$ $\mathcal{D}(A) \subset \mathcal{D}(D) \subset X_A$ and D is a closable operator from X_A into Y.
The closed graph theorem (see Theorem 2.1.3) implies that, for $\lambda \in \rho(A_1)$, the operator $D(A_1 - \lambda)^{-1}$ is bounded from X into Y.

$(N6)$ $\mathcal{D}(A) \subset \mathcal{D}(G) \subset X_A$ and G is a closable operator from X_A into Z.
The closed graph theorem (see Theorem 2.1.3) implies that, for $\lambda \in \rho(A_1)$, the operator $G(A_1-\lambda)^{-1}$ is bounded from X into Z. For every $\mu \in \rho(A_1)$, and under the assumptions $(N1)$–$(N3)$, Lemma 10.3.1 gives the following decomposition $\mathcal{D}(A) = \mathcal{D}(A_1) \oplus N(A - \mu)$. It is easy to see that the restriction of Γ_X to $N(A - \mu)$ is injective. Let us denote the inverse of $\Gamma_{X|N(A-\mu)}$ by $K_\mu := (\Gamma_X|_{N(A-\mu)})^{-1}$. By using Remark 11.2.1, we can write $K_\mu : \Gamma_X(\mathcal{D}(A)) \longrightarrow N(A - \mu) \subset \mathcal{D}(A)$. For $\mu \in \rho(A_1)$, and under the assumptions $(N1)$–$(N3)$, we have

$$(A - \mu)x = (A_1 - \mu)(I - K_\mu\Gamma_X)x. \tag{11.2.1}$$

$(N7)$ For some $\mu \in \rho(A_1)$, K_μ is bounded from $\Gamma_X(\mathcal{D}(A))$ into X and its extension by continuity to $\overline{\Gamma_X(\mathcal{D}(A))}$ is denoted by $\overline{K}_\mu$.
Since, for $x \in N(A - \mu)$, $\|x\|_A = (1 + \mu)\|x\|$, then $\overline{K}_\mu : \overline{\Gamma_X(\mathcal{D}(A))} \longrightarrow X_A$ is bounded and, for $z \in \overline{\Gamma_X(\mathcal{D}(A))}$, we have $\overline{A}\overline{K}_\mu z = \mu\overline{K}_\mu z$ and $\overline{\Gamma}_X\overline{K}_\mu z = z$.
In the following, let us denote $S(\mu)$ by $S(\mu) := E + DK_\mu\Gamma_Y - D(A_1 - \lambda)^{-1}B$.

The operator $S(\mu)$ is defined on the domain

$$Y_1 = \left\{ y \in \mathcal{D}(B) \bigcap \mathcal{D}(E) \text{ such that } \Gamma_Y y \in \Gamma_X(\mathcal{D}(A)) \right\}. \tag{11.2.2}$$

For some $\mu \in \rho(A_1)$, let us denote by $S_1(\mu)$ the restriction of $S(\mu)$ to $Y_1 \bigcap N(\Gamma_Y)$, i.e., $S_1(\mu) := S(\mu)_{|Y_1 \bigcap N(\Gamma_Y)}$.

($N8$) For some $\mu \in \rho(A_1)$, the operator $S_1(\mu)$ is closed.
It is shown in Remark 11.1.1 that, if A_1 and E generate each a C_0-semigroup, and if B and D are bounded then, there exists $\mu \in \mathbb{C}$ such that $\mu \in \rho(A_1) \bigcap \rho(S_1(\mu))$. Hence, for $\mu \in \rho(A_1) \bigcap \rho(S_1(\mu))$, the set Y_1 can be decomposed as follows: $Y_1 = \mathcal{D}(S_1(\mu)) \oplus N(S(\mu) - \mu)$. Indeed, let $y \in \mathcal{D}(S_1(\mu)) \bigcap N(S(\mu) - \mu)$. Then, $y \in N(S_1(\mu) - \mu) = \{0\}$. Now, let us notice that, for $y \in Y_1$, we can write $y = a + y - a$, where $a = (S_1(\mu) - \mu)^{-1}(S(\mu) - \mu)y$. It is easy to check that $a \in \mathcal{D}(S_1(\mu))$ and $y - a \in N(S_1(\mu) - \mu)$. For $\lambda \in \rho(A_1) \bigcap \rho(S_1(\lambda))$, we define the inverse of Γ_Y by $J_\lambda := (\Gamma_Y|_{N(S(\lambda)-\lambda)})^{-1} : \Gamma_Y(Y_1) \longrightarrow N(S(\lambda) - \lambda) \subset Y_1$. Let us assume that, for some $\mu \in \rho(A_1)$, J_μ is bounded from $\Gamma_Y(Y_1)$ into Y, and its extension by continuity to $\overline{\Gamma_Y(Y_1)}$ is denoted by $\overline{J}_\mu$.

($N9$) $\mathcal{D}(B) \bigcap \mathcal{D}(E) \subset \mathcal{D}(\Gamma_Y)$, $\mathcal{D}(B) \bigcap \mathcal{D}(H) \subset \mathcal{D}(\Gamma_Y)$ and the set $Y_1 = \{y \in \mathcal{D}(B) \bigcap \mathcal{D}(E)$ such that $\Gamma_Y y \in \Gamma_X(\mathcal{D}(A))\}$ is dense in Y and the restriction of Γ_Y to Y_1 is bounded as an operator from Y into W. Let us denote the extension by continuity of $\Gamma_{Y|Y_1}$ to Y by $\overline{\Gamma}^0_Y$.

($N10$) L is densely defined and closed with a nonempty resolvent set, i.e., $\rho(L) \neq \emptyset$.

($N11$) For some $\mu \in \rho(A_1)$ (hence for all), the operator $(A_1 - \lambda)^{-1}C$ is bounded on its domain.

($N12$) $\mathcal{D}(C) \bigcap \mathcal{D}(F) \bigcap \mathcal{D}(L) \subset \mathcal{D}(\Gamma_Z)$ and the set:
$Z_1 := \{z \in \mathcal{D}(C) \bigcap \mathcal{D}(F) \bigcap \mathcal{D}(L)$ such that $\Gamma_Z z \in \Gamma_Y(Y_1)\}$ is dense in Z and the restriction of Γ_Z to Z_1 is bounded as an operator from Z into W. Let us denote the extension by continuity of $\Gamma_{Z|Z_1}$ to Z by $\overline{\Gamma}^0_Z$.

($N13$) For some $\mu \in \rho(A_1)$ (hence for all), the operator $F - D(A_1 - \mu)^{-1}C$ is closable and its closure $\overline{F - D(A_1 - \mu)^{-1}C}$ is bounded.

The essential work in this section is to prescribe some conditions on the entries of the operator $\mathcal{L}_0$ in order to establish its closedness. First, we will search the Frobenius–Schur's decomposition of the operator $\mathcal{L}_0$. Let $\begin{pmatrix} x \\ y \\ z \end{pmatrix} \in \mathcal{D}(\mathcal{L}_0)$ and $\mu \in \mathbb{C}$.

$$(\mathcal{L}_0 - \mu)\begin{pmatrix} x \\ y \\ z \end{pmatrix} = 0 \text{ if, and only if, } \begin{pmatrix} A-\mu & B & C \\ D & E-\mu & F \\ G & H & L-\mu \end{pmatrix}\begin{pmatrix} x \\ y \\ z \end{pmatrix} = 0.$$

This is equivalent to the following system

$$\begin{cases} (A-\mu)x + By + Cz = 0 \\ Dx + (E-\mu)y + Fz = 0 \\ Gx + Hy + (L-\mu)z = 0. \end{cases} \tag{11.2.3}$$

From the first equation of (11.2.3) and Eq. (11.2.1), we can deduce that $(A_1-\mu)(I-K_\mu\Gamma_X)x + By + Cz = 0$. Then, $x - K_\mu\Gamma_Y y + (A_1-\mu)^{-1}By + (A_1-\mu)^{-1}Cz = 0$. Hence,

$$x = \left[K_\mu\Gamma_Y - (A_1-\mu)^{-1}B\right]y - (A_1-\mu)^{-1}Cz. \tag{11.2.4}$$

From the second equation of (11.2.3), we have $\left[E + DK_\mu\Gamma_Y - D(A_1-\mu)^{-1}B - \mu\right]$ $y + (F - (A_1-\mu)^{-1}C)z = 0$. Consequently,

$$(S(\mu)-\mu)y + (F - (A_1-\mu)^{-1}C)z = 0. \tag{11.2.5}$$

Lemma 11.2.1. *For some $\mu \in \rho(A_1)\bigcap\rho(S_1(\mu))$ and under assumptions $(N7)$, $(N9)$, we have $(S(\mu)-\mu)y = (S_1(\mu)-\mu)(I-J_\mu\Gamma_Y)y$, where the operator $I-J_\mu\Gamma_Y$ is the projection from Y_1 on $\mathcal{D}(S_1(\mu))$ parallel to $N(S(\mu)-\mu)$.* ◇

Proof. For every $y \in Y_1$, we can write $y = (I-J_\mu\Gamma_Y)y + J_\mu\Gamma_Y y$. Observe that $y \in \mathcal{D}(S_1(\mu))$ because $y_1 = (I-J_\mu\Gamma_Y)y \in \mathcal{D}(S_1(\mu))$ and $y_2 = J_\mu\Gamma_Y y \in N(S(\mu)-\mu)$, then

$$\begin{aligned} (S(\mu)-\mu)y &= (S_1(\mu)-\mu)y_1 \\ &= (S_1(\mu)-\mu)(y-y_2) \\ &= (S_1(\mu)-\mu)(I-J_\mu\Gamma_Y)y. \end{aligned}$$

Q.E.D.

From Eq. (11.2.5), it follows that $(S_1(\mu)-\mu)(I-J_\mu\Gamma_Y)y + (F-D(A-\mu)^{-1}C)z = 0$, and $y - J_\mu\Gamma_Z z + (S_1(\mu)-\mu)^{-1}(F-D(A-\mu)^{-1}C)z = 0$. Hence,

$$y = [J_\mu\Gamma_Z - (S_1(\mu)-\mu)^{-1}(F-D(A-\mu)^{-1}C)]z. \tag{11.2.6}$$

The third equation of the system (11.2.3), together with Eq. (11.2.4), gives:

$$\begin{aligned} Gx + Hy + (L-\mu)z &= 0 \\ (GK_\mu\Gamma_Y - G(A_1-\mu)^{-1}B)y - G(A_1-\mu)^{-1}Cz + Hy + (L-\mu)z &= 0 \\ (H + GK_\mu\Gamma_Y - G(A_1-\mu)^{-1}B)y + [L - G(A_1-\mu)^{-1}C - \mu]z &= 0. \end{aligned}$$

Let us denote by $\Theta(\mu)$ the following operator $\Theta(\mu) := H + GK_\mu\Gamma_Y - G(A_1 - \mu)^{-1}B$. By using Eq. (11.2.6), we deduce that

$$[L - G(A_1-\mu)^{-1}C + \Theta(\mu)(J_\mu\Gamma_Z - (S_1(\mu)-\mu)^{-1}(F - D(A_1-\mu)^{-1}C) - \mu]z = 0.$$

Set

$$S_2(\mu) = [L - G(A_1-\mu)^{-1}C + \Theta(\mu)(J_\mu\Gamma_Z - (S_1(\mu)-\mu)^{-1}(F - D(A_1-\mu)^{-1}C)]. \tag{11.2.7}$$

Now, we can search $F_i(\mu)$ and $G_i(\mu)$ with $i = 1, 2, 3$ such that the operator

$$\begin{pmatrix} I & 0 & 0 \\ F_1(\mu) & I & 0 \\ F_2(\mu) & F_3(\mu) & I \end{pmatrix} \begin{pmatrix} A_1 - \mu & 0 & 0 \\ 0 & S_1(\mu) - \mu & 0 \\ 0 & 0 & S_2(\mu) - \mu \end{pmatrix} \begin{pmatrix} I & G_1(\mu) & G_2(\mu) \\ 0 & I & G_3(\mu) \\ 0 & 0 & I \end{pmatrix}$$

is equal to

$$\begin{pmatrix} A - \mu & B & C \\ D & E - \mu & F \\ G & H & L - \mu \end{pmatrix}.$$

It follows that for $\begin{pmatrix} x \\ y \\ z \end{pmatrix} \in \mathcal{D}(\mathcal{L}_0)$

$$\begin{pmatrix} A_1 - \mu & 0 & 0 \\ F_1(\mu)(A_1 - \mu) & S_1(\mu) - \mu & 0 \\ F_2(\mu)(A_1 - \mu) & F_3(\mu)(S_1(\mu) - \mu) & S_2(\mu) - \mu \end{pmatrix} \begin{pmatrix} I & G_1(\mu) & G_2(\mu) \\ 0 & I & G_3(\mu) \\ 0 & 0 & I \end{pmatrix} \begin{pmatrix} x \\ y \\ z \end{pmatrix}$$

$$= \begin{pmatrix} A - \mu & B & C \\ D & E - \mu & F \\ G & H & E - \mu \end{pmatrix} \begin{pmatrix} x \\ y \\ z \end{pmatrix}. \tag{11.2.8}$$

We can choose $F_i(\mu)$, $i = 1, 2, 3$ and $G_i(\mu)$, $i = 1, 2, 3$ for a necessary condition as follows: $(A_1 - \mu)x + (A_1 - \mu)G_1(\mu)y + (A_1 - \mu)G_2(\mu)z = (A_1 - \mu)(I - K_\mu\Gamma_X)x + By + Cz$. Then, for $\mu \in \rho(A_1)$, we have $x + G_1(\mu)y + G_2(\mu)z = x - K_\mu\Gamma_Y y + (A_1 - \mu)^{-1}By + (A_1 - \mu)^{-1}Cz$. Let us take

$$G_1(\mu) := -K_\mu\Gamma_Y + (A_1 - \mu)^{-1}B \tag{11.2.9}$$

and

$$G_2(\mu) := (A_1 - \mu)^{-1}C. \tag{11.2.10}$$

The second equation of (11.2.8) gives:

$$F_1(\mu)(A_1-\mu)x+(F_1(\mu)(A_1-\mu)G_1(\mu)+S_1(\mu)-\mu)\,y \\ +(F_1(\mu)(A_1-\mu)G_2(\mu)+(S_1(\mu)-\mu)G_3(\mu))z$$

which must be equal to $Dx+(E-\mu)y+Fz$. Let us take

$$F_1(\mu):=D(A_1-\mu)^{-1}. \tag{11.2.11}$$

From the third equation of (11.2.8), we have

$$F_2(\mu)(A_1-\mu)x+(F_2(\mu)(A_1-\mu)G_1(\mu)+F_3(\mu)(S_1(\mu)-\mu))y \\ +(F_2(\mu)(A_1-\mu)G_2(\mu)+F_3(\mu)(S_1(\mu)-\mu)G_3(\mu) \\ +S_2(\mu)-\mu)z=Gx+Hy+(L-\mu)z.$$

Take

$$F_2(\mu):=G(A_1-\mu)^{-1}. \tag{11.2.12}$$

For the action on y, we choose $GG_1(\mu)+F_3(\mu)(S_1(\mu)-\mu)-H=0$. Therefore, for $\mu\in\rho(A_1)\bigcap\rho(S_1(\mu))$, take $F_3(\mu)=[H+GK_\mu\Gamma_Y-G(A_1-\mu)^{-1}B](S_1(\mu)-\mu)^{-1}$, i.e.,

$$F_3(\mu)=\Theta(\mu)(S_1(\mu)-\mu)^{-1}. \tag{11.2.13}$$

Now, for the action on z, take

$$[F_2(\mu)(A_1-\mu)G_2(\mu)+F_3(\mu)(S_1(\mu)-\mu)G_3(\mu)+S_2(\mu)-\mu-L+\mu]=0.$$

Then, $G(A_1-\mu)^{-1}C+\Theta(\mu)G_3(\mu)=L-S_2(\mu)$. From the expression of $S_2(\mu)$ in (11.2.7), we can choose

$$G_3(\mu)=-J_\mu\Gamma_Z+(S_1(\mu)-\mu)^{-1}(F-D(A_1-\mu)^{-1}C). \tag{11.2.14}$$

Now, we can verify the sufficient condition. Let us denote by T_μ the operator defined for every $\mu\in\rho(A_1)\bigcap\rho(S_1(\mu))$ by

$$T_\mu:=\begin{pmatrix} I & 0 & 0\\ F_1(\mu) & I & 0\\ F_2(\mu) & F_3(\mu) & I\end{pmatrix}\begin{pmatrix} A_1-\mu & 0 & 0\\ 0 & S_1(\mu)-\mu & 0\\ 0 & 0 & S_2(\mu)-\mu\end{pmatrix}\begin{pmatrix} I & G_1(\mu) & G_2(\mu)\\ 0 & I & G_3(\mu)\\ 0 & 0 & I\end{pmatrix},$$

where $F_i(\mu), i = 1,2,3$ and $G_i(\mu)$, $i = 1,2,3$ are the operators defined in (11.2.9), (11.2.10), (11.2.11), (11.2.12), (11.2.13), and (11.2.14). Let $\begin{pmatrix} x \\ y \\ z \end{pmatrix} \in \mathcal{D}(\mathcal{L}_0)$. The first row in the product of T_μ gives:

$$\begin{aligned}(A_1 - \mu)x + (A_1 - \mu)G_1(\mu)y + (A_1 - \mu)G_2(\mu)z \\ &= (A_1 - \mu)[I - K_\mu \Gamma_X]x + By + Cz \\ &= (A - \mu)x + By + Cz.\end{aligned}$$

The second row in the product of T_μ gives:

$$\begin{aligned}F_1(\mu)(A_1 - \mu)x + [F_1(\mu)(A_1 - \mu)G_1(\mu) + S_1(\mu) - \mu]y \\ + [F_1(\mu)(A_1 - \mu)G_2(\mu) + (S_1(\mu) - \mu)G_3(\mu)]z\end{aligned}$$

$$\begin{aligned}&= Dx + (E - S(\mu))y + (S_1(\mu) - \mu)(y - J_\mu \Gamma_Y y) + Fz \\ &= Dx + (E - S(\mu) + S_1(\mu) - \mu)y + Fz \\ &= Dx + (E - \mu)y + Fz.\end{aligned}$$

We can also show that the left side of the third row of T_μ, i.e.,

$$\begin{aligned}F_2(\mu)(A_1 - \mu)x + [F_2(\mu)(A_1 - \mu)G_1(\mu) \\ + F_3(\mu)(S_1(\mu) - \mu)]y + [F_2(\mu)(A_1 - \mu)G_2(\mu) \\ + F_3(\mu)(S_1(\mu) - \mu)G_3(\mu) + S_2(\mu) - \mu]z\end{aligned}$$

is equal to $Gx + Hy + (L - \mu)z$. It follows that T_μ is an extension of the operator $\mathcal{L}_0 - \mu$, i.e., $\mathcal{L}_0 - \mu \subset T_\mu$. Now, it remains to prove that $\mathcal{D}(T_\mu) \subset \mathcal{D}(\mathcal{L}_0 - \mu) = \mathcal{D}(\mathcal{L}_0)$. Let us notice that

$$\mathcal{D}(T_\mu) = \left\{ \begin{pmatrix} x' \\ y' \\ z' \end{pmatrix} = \begin{pmatrix} I & -G_1(\mu) & G_1(\mu)G_3(\mu) - G_2(\mu) \\ 0 & I & -G_3(\mu) \\ 0 & 0 & I \end{pmatrix} \begin{pmatrix} x \\ y \\ z \end{pmatrix}, \begin{array}{c} x \in \mathcal{D}(A_1) \\ y \in \mathcal{D}(S_1(\mu)) = Y_1 \bigcap N(\Gamma_Y) \\ z \in \mathcal{D}(S_2(\mu)) = Y_2 \end{array} \right\},$$

where $Y_2 := \{z \in \mathcal{D}(C) \bigcap \mathcal{D}(F) \bigcap \mathcal{D}(L)$ such that $\Gamma_Z z \in \Gamma_Y(Y_1)\}$. Let $\begin{pmatrix} x' \\ y' \\ z' \end{pmatrix} \in \mathcal{D}(T_\mu)$. Then,

$$\begin{cases} x' = x - G_1(\mu)y + [G_1(\mu)G_3(\mu) - G_2(\mu)]\, z \\ y' = y - G_3(\mu)z \\ z' = z. \end{cases}$$

Observe that $z \in Y_2 \subset \mathcal{D}(C) \bigcap \mathcal{D}(F) \bigcap \mathcal{D}(L)$, $y' = y - G_3(\mu)z \in N(S(\mu) - \mu) \subset Y_1$, $Y_1 \subset \mathcal{D}(B) \bigcap \mathcal{D}(E)$ and $x' = x - G_1(\mu)y + (G_1(\mu)G_3(\mu) - G_2(\mu))z \in \mathcal{D}(A)$. Now, let us verify the boundary conditions: $\Gamma_X x = \Gamma_Y y = \Gamma_Z z$

$$\begin{aligned} \Gamma_X x' &= -\Gamma_X \left[G_1(\mu)(y - G_3(\mu)z)\right] \\ &= -\Gamma_X \left[G_1(\mu)y\right] \\ &= -\Gamma_X \left[-K_\mu \Gamma_Y y + (A_1 - \mu)^{-1} B y\right] \\ &= -\Gamma_X \left[-K_\mu \Gamma_Y y\right] \\ &= \Gamma_Y y \end{aligned}$$

and

$$\begin{aligned} \Gamma_Y y' &= \Gamma_Y y - \Gamma_Y (G_1(\mu)z) \\ &= -\Gamma_Y \left[-J_\mu \Gamma_Z z + (S_1(\mu) - \mu)^{-1}(F - D(A - \mu)^{-1} C)z\right] \\ &= -\Gamma_Y \left[-J_\mu \Gamma_Z z\right] \\ &= \Gamma_Z z. \end{aligned}$$

Then, we claim that $\mathcal{L}_0 - \mu \subset T_\mu$. In what follows, we will give some lemmas which will be needed in the sequel.

Lemma 11.2.2. *For every* $\lambda,\ \mu \in \rho(A_1)$

$$S_1(\mu) - S_1(\lambda) = -(\mu - \lambda) D(A_1 - \mu)^{-1}(A_1 - \lambda)^{-1} B. \tag{11.2.15}$$

◇

Proof. Let $\lambda,\ \mu \in \rho(A_1)$. Using Lemma 10.3.3, we have

$$\begin{aligned} &S(\mu) - S(\lambda) \\ &= \left[E + DK_\mu \Gamma_Y - D(A_1 - \mu)^{-1} B\right] - \left[E + DK_\lambda \Gamma_Y - D(A_1 - \lambda)^{-1} B\right] \\ &= D(K_\mu - K_\lambda)\Gamma_Y - D\left[(A_1 - \mu)^{-1} - (A_1 - \lambda)^{-1}\right] B \end{aligned}$$

$$= (\mu - \lambda)D(A_1 - \mu)^{-1}K_\lambda\Gamma_Y - (\mu - \lambda)D(A_1 - \mu)^{-1}(A_1 - \lambda)^{-1}B$$
$$= (\mu - \lambda)D(A_1 - \mu)^{-1}\left[K_\lambda\Gamma_Y - (A_1 - \lambda)^{-1}B\right]$$
$$= -(\mu - \lambda)F_1(\mu)G_1(\lambda).$$

For $y \in \mathcal{D}(S_1(\mu))$, we have $\Gamma_Y y = 0$ and the relation (11.2.15) holds. Q.E.D.

From assumptions $(N4)$ and $(N5)$, the operator $D(A_1 - \mu)^{-1}$ and $(A_1 - \lambda)^{-1}B$ are bounded on their domains, then it follows that if $S_1(\mu)$ is closed for some $\mu \in \rho(A_1)$ then it is closed for all such μ. Assume that for some $\mu \in \rho(A_1) \bigcap \rho(S_1(\lambda))$, J_μ is bounded from $\Gamma_Y(Y_1)$ into Y and its extension by continuity to $\overline{\Gamma_Y(Y_1)}$ is denoted by $\overline{J}_\mu$.

Lemma 11.2.3. *If $\mu \in \rho(A_1) \bigcap \rho(S_1(\mu))$ and $\lambda \in \rho(A_1) \bigcap \rho(S_1(\lambda))$ and under assumptions $(N1)$–$(N3)$, $(N7)$, $(N8)$, then $J_\mu = (S_1(\mu) - \mu)^{-1}(S_1(\lambda) - \lambda)J_\lambda$.* ◇

Proof. Recall that $J_\mu = (\Gamma_{Y|N(S(\mu)-\mu)})^{-1} : W \supset \Gamma_Y(Y_1) \longrightarrow N(S(\mu) - \mu) \subset Y_1$. Let $w \in \Gamma_Y(Y_1)$ and set $y = J_\mu w$ and $y' = J_\lambda w$, then

$$\begin{cases} S(\mu)y = \mu y \\ \Gamma_Y y = w \end{cases} \text{and} \quad \begin{cases} S(\lambda)y' = \lambda y' \\ \Gamma_Y y' = w. \end{cases}$$

Observe that $(y - y') \in N(\Gamma_Y) \bigcap Y_1 = \mathcal{D}(S_1(\mu))$. Now observe the action of $(S_1(\mu) - \mu)$ on $y - y'$. In fact, $(S_1(\mu) - \mu)(y - y') = -(S_1(\mu) - \mu)y' = -S_1(\mu)y' + \mu y'$. Using Lemma 11.2.2, we infer that

$$(S_1(\mu) - \mu)(y - y') = -\left[S_1(\lambda) - (\mu - \lambda)D(A_1 - \mu)^{-1}(A_1 - \lambda)^{-1}B\right]y' + \mu y'$$
$$= (\mu - \lambda)y' + (\mu - \lambda)D(A_1 - \mu)^{-1}(A_1 - \lambda)^{-1}By'.$$

So, by Lemma 11.2.2 we have

$$y - y' = (S_1(\mu) - \mu)^{-1}\left[(\mu - \lambda)I + (\mu - \lambda)D(A_1 - \mu)^{-1}(A_1 - \lambda)^{-1}B\right]y'$$
$$= (S_1(\mu) - \mu)^{-1}\left[(\mu - \lambda)I - S_1(\mu) + S_1(\lambda)\right]y'$$
$$= (S_1(\mu) - \mu)^{-1}\left[(S_1(\lambda) - \lambda) - (S_1(\mu) - \mu)\right]y'$$
$$= \left[(S_1(\mu) - \mu)^{-1}(S_1(\lambda) - \lambda) - I\right]y'.$$

It follows that $J_\mu - J_\lambda = [(S_1(\mu) - \mu)^{-1}(S_1(\lambda) - \lambda) - I]J_\lambda$ and so

$$J_\mu = (S_1(\mu) - \mu)^{-1}(S_1(\lambda) - \lambda)J_\lambda.$$

Q.E.D.

Since for $\mu \in \rho(A_1) \bigcap \rho(S_1(\mu))$ and $\lambda \in \rho(A_1) \bigcap \rho(S_1(\lambda))$, the operator $(S_1(\mu) - \mu)^{-1}(S_1(\lambda) - \lambda)$ is bounded and boundedly invertible, then J_μ is closable if, and only if, J_λ is such. Moreover $\overline{J}_\mu = (S_1(\mu)-\mu)^{-1}(S_1(\lambda)-\lambda)\overline{J}_\lambda$ and so

$$\overline{J}_\mu - \overline{J}_\lambda = \left[(S_1(\mu)-\mu)^{-1}(S_1(\lambda)-\lambda) - I\right]\overline{J}_\lambda.$$

Lemma 11.2.4. *If the operator $\Theta(\mu) := H + GK_\mu\Gamma_Y - G(A_1-\mu)^{-1}B$ is closable for some $\mu \in \rho(A_1)$, then it is closable for all such $\mu \in \rho(A_1)$.* ◇

Proof. Let $\lambda,\ \mu \in \rho(A_1)$. By Lemma 10.3.3, we have

$$\begin{aligned}\Theta(\mu) - \Theta(\lambda) &= G(K_\mu - K_\lambda)\Gamma_Y - G\left[(A_1-\mu)^{-1} - (A_1-\lambda)^{-1}\right]B \\ &= (\mu-\lambda)G(A_1-\mu)^{-1}K_\lambda\Gamma_Y - (\mu-\lambda)G(A_1-\mu)^{-1}(A_1-\lambda)^{-1}B \\ &= (\mu-\lambda)G(A_1-\mu)^{-1}\left[K_\lambda\Gamma_Y - (A_1-\lambda)^{-1}B\right],\end{aligned}$$

where Γ_Y is bounded on Y_1 according to assumption $(N9)$. From the assumptions $(N4)$, $(N6)$ and $(N7)$, it follows that the operators K_λ, $(A_1-\mu)^{-1}B$ and $G(A_1-\mu)^{-1}$ are bounded. Hence the right side of $\Theta(\mu)-\Theta(\lambda)$ is bounded, and the lemma is proved. Q.E.D.

In fact for the closure we have $\overline{\Theta}(\mu) - \overline{\Theta}(\lambda) = (\mu-\lambda)G(A_1-\mu)^{-1}\left[\overline{K_\lambda\Gamma_Y^0} - \overline{(A_1-\lambda)^{-1}B}\right]$. From hypothesis $(N13)$ follows that for every $\mu \in \rho(A_1)\bigcap\rho(S_1(\mu))$,

$$\overline{J}_\mu\overline{\Gamma}_Z^0 - (S_1(\mu)-\mu)^{-1}\left(\overline{F - D(A_1-\mu)^{-1}C}\right)$$

is bounded as an operator from Z into W then

$$\Theta(\mu)(\overline{J}_\mu\overline{\Gamma}_Z^0 - (S_1(\mu)-\mu)^{-1}\left(\overline{F - D(A_1-\mu)^{-1}C}\right))$$

is bounded everywhere. Now by assumption $(N11)$, the operator $\overline{(A_1-\mu)^{-1}C}$ is bounded everywhere, since G is closable then $\overline{G(A_1-\mu)^{-1}C}$ is bounded everywhere. By assumption $(N10)$, L is densely defined closed. Hence $S_2(\mu)$ is closable. In fact the next lemma show that the closedness of $S_2(.)$ does not depend of the choice of $\mu \in \rho(A_1)\bigcap\rho(S_1(\mu))$.

Lemma 11.2.5. *If, for some $\mu \in \rho(A_1)\bigcap\rho(S_1(\mu))$, the operator $S_2(\mu)$ is closable, then it is closable for all such $\mu \in \rho(A_1)\bigcap\rho(S_1(\mu))$.* ◇

Proof. Let $\mu \in \rho(A_1)\bigcap\rho(S_1(\mu))$ and $\lambda \in \rho(A_1)\bigcap\rho(S_1(\lambda))$. Using Lemma 11.2.3, we can write:

$$\begin{aligned}&S_2(\mu) - S_2(\lambda) \\ &\quad= -(\mu-\lambda)F_2(\mu)(A_1-\mu)^{-1}C + \Theta(\mu)[(S_1(\mu)-\mu)^{-1}(S_1(\mu)-\lambda)J_\lambda\Gamma_Z \\ &\qquad - (S_1(\mu)-\mu)^{-1}(F - D(A_1-\mu)^{-1}C)] \\ &\qquad - \Theta(\lambda)[J_\lambda\Gamma_Z - (S_1(\lambda)-\lambda)^{-1}(F - D(A_1-\lambda)^{-1}C)].\end{aligned}$$

Using the resolvent identity, we find

$$\begin{aligned} S_2(\mu) - S_2(\lambda) & \\ = & -(\mu-\lambda)F_2(\mu)(A_1-\mu)^{-1}C + [F_3(\mu)(S_1(\lambda)-\lambda) - \Theta(\mu)]J_\lambda\Gamma_Z \\ & - F_3(\mu)[F - D(A_1-\lambda)^{-1}C - (\mu-\lambda)D(A_1-\mu)^{-1}(A_1-\lambda)^{-1}C] \\ & + F_3(\lambda)(F - D(A_1-\lambda)^{-1}C). \end{aligned}$$

We deduce that

$$\begin{aligned} S_2(\mu) - S_2(\lambda) & \\ = & -(\mu-\lambda)F_2(\mu)(A_1-\mu)^{-1}C + (F_3(\mu)-F_3(\lambda))(S_1(\lambda)-\lambda)[J_\lambda\Gamma_Z \\ & - (S_1(\lambda)-\lambda)^{-1}(F - D(A_1-\lambda)^{-1}C)] + (\mu-\lambda)F_3(\mu)F_1(\mu)(A_1-\lambda)^{-1}C. \end{aligned}$$

Since, on the one hand the operators F_i, $i = 1,2,3$ are bounded everywhere and the operator $(A_1-\lambda)^{-1}C$ is bounded on its domain, on the other hand $(S_1(\lambda)-\lambda)$ is closed, and by assumptions $(N12)$ and $(N13)$, the operator $[J_\lambda\Gamma_Z - (S_1(\lambda) - \lambda)^{-1}(F - D(A_1-\lambda)^{-1}C)]$ is bounded on its domain, then the closedness of the operator $S_2(\mu)$ does not depend of the choice of μ. Q.E.D.

Denote by $\overline{S}_2(\mu)$ its closure, in fact we have

$$\begin{aligned} \overline{S}_2(\mu) - \overline{S}_2(\lambda) = & [F_3(\mu) - F_3(\lambda)]\,(S_1(\lambda)-\lambda) \\ & \times \left[\overline{J_\lambda\Gamma_Z^0} - (S_1(\lambda)-\lambda)^{-1}\overline{(F - D(A_1-\lambda)^{-1}C)}\right] \\ & + (\mu-\lambda)\,[F_3(\mu)F_1(\mu) - F_2(\mu)]\,\overline{(A_1-\lambda)^{-1}C}. \end{aligned} \tag{11.2.16}$$

In what follows, we consider the operators defined by:

$$\begin{aligned} \hat{G}_1(\mu) &:= -\overline{K_\mu\Gamma_Y^0} + \overline{(A_1-\mu)^{-1}B} \\ \hat{G}_2(\mu) &:= \overline{(A_1-\mu)^{-1}C} \\ \hat{G}_3(\mu) &:= -\overline{J_\lambda\Gamma_Z^0} + (S_1(\lambda)-\lambda)^{-1}\overline{(F - D(A_1-\mu)^{-1}C)}. \end{aligned}$$

Now, we are ready to present the first main result of this section.

Theorem 11.2.1. *Under the assumptions $(N1)$–$(N13)$, the operator $\mathcal{L}_0$ is closable if, and only if, $S_2(\mu)$ is closable for some $\mu \in \rho(A_1)\bigcap\rho(S_1(\mu))$. In this case, the closure $\mathcal{L}$ of $\mathcal{L}_0$ is given by*

$$\mathcal{L} = \mu I + \mathcal{G}_1(\mu)\begin{pmatrix} A_1-\mu & 0 & 0 \\ 0 & S_1(\mu)-\mu & 0 \\ 0 & 0 & \overline{S}_2(\mu)-\mu \end{pmatrix}\mathcal{G}_2(\mu),$$

where $\mathcal{G}_1(\mu) = \begin{pmatrix} I & 0 & 0 \\ F_1(\mu) & I & 0 \\ F_2(\mu) & F_3(\mu) & I \end{pmatrix}$ *and* $\mathcal{G}_2(\mu) = \begin{pmatrix} I & \hat{G}_1(\mu) & \hat{G}_2(\mu) \\ 0 & I & \hat{G}_3(\mu) \\ 0 & 0 & I \end{pmatrix}$

or, spelled out,

$$\mathcal{D}(\mathcal{L}) = \left\{ \begin{pmatrix} I & -\hat{G}_1(\mu) & \hat{G}_1(\mu)\hat{G}_3(\mu) - \hat{G}_2(\mu) \\ 0 & I & -\hat{G}_3(\mu) \\ 0 & 0 & I \end{pmatrix} \begin{pmatrix} x \\ y \\ z \end{pmatrix}, \begin{array}{c} x \in \mathcal{D}(A_1) \\ y \in \mathcal{D}(S_1(\mu)) = Y_1 \bigcap N(\Gamma(Y)) \\ z \in \mathcal{D}(\overline{S}_2(\mu))Y_2 \end{array} \right\}$$

$$\mathcal{L}\begin{pmatrix} x - \hat{G}_1(\mu)y + (\hat{G}_1(\mu)\hat{G}_3(\mu) - \hat{G}_2(\mu))z \\ y - \hat{G}_3(\mu)z \\ z \end{pmatrix}$$
$$= \begin{pmatrix} A_1x - \mu\hat{G}_1(\mu)y + \mu(\hat{G}_1(\mu)\hat{G}_3(\mu) - \hat{G}_3(\mu))z \\ F_1(\mu)(A_1 - \mu)x + S_1(\mu)y - \mu\hat{G}_3(\mu)z \\ F_2(\mu)(A_1 - \mu)x + F_3(\mu)(S_1(\mu) - \mu)y + \overline{S}_2(\mu)z \end{pmatrix}.$$

◇

Proof. Let $\mu \in \rho(A_1) \bigcap \rho(S_1(\mu))$. The lower-upper factorization sense is given by

$$\mathcal{L}_0 = \mu I + \begin{pmatrix} I & 0 & 0 \\ F_1(\mu) & I & 0 \\ F_2(\mu) & F_3(\mu) & I \end{pmatrix} \begin{pmatrix} A_1 - \mu & 0 & 0 \\ 0 & S_1(\mu) - \mu & 0 \\ 0 & 0 & S_2(\mu) - \mu \end{pmatrix}$$
$$\times \begin{pmatrix} I & \hat{G}_1(\mu) & \hat{G}_2(\mu) \\ 0 & I & \hat{G}_3(\mu) \\ 0 & 0 & I \end{pmatrix}.$$

The external operators $\mathcal{G}_1(\mu)$ and $\mathcal{G}_2(\mu)$ are boundedly invertible. Hence, $\mathcal{L}_0 - \mu$ is closable if, and only if, $S_2(\mu)$ is closable. Q.E.D.

11.2.2 Gustafson, Weidman, Kato, Wolf, Schechter, Browder, Rakočević, and Schmoeger's Essential Spectra of $\mathcal{L}$

Having obtained the closure of the operator $\mathcal{L}_0$, we will discuss its essential spectra. As a first step, we will establish the following stability lemma.

Lemma 11.2.6. *Let $\mu \in \rho(A_1)\bigcap\rho(S_1(\mu))$. If the sets $\Phi^b(Y,X)$, $\Phi^b(Z,X)$, and $\Phi^b(Z,Y)$ are not empty, then*

(*i*) *If $F_1(\mu) \in \mathcal{F}^b(X,Y)$, $F_2(\mu) \in \mathcal{F}^b(X,Z)$ and $F_3(\mu) \in \mathcal{F}^b(Y,Z)$, then $\sigma_{ei}(S_1(\mu))$ and $\sigma_{ei}(\overline{S}_2(\mu))$, $i = 4,5$ do not depend on the choice of μ.*
(*ii*) *If $F_1(\mu) \in \mathcal{F}_+^b(X,Y)$, $F_2(\mu) \in \mathcal{F}_+^b(X,Z)$ and $F_3(\mu) \in \mathcal{F}_+^b(Y,Z)$, then $\sigma_{e1}(S_1(\mu))$ and $\sigma_{e1}(\overline{S}_2(\mu))$ do not depend on the choice of μ.*
(*iii*) *If $F_1(\mu) \in \mathcal{F}_-^b(X,Y)$, $F_2(\mu) \in \mathcal{F}_-^b(X,Z)$ and $F_3(\mu) \in \mathcal{F}_-^b(Y,Z)$, then $\sigma_{e2}(S_1(\mu))$ and $\sigma_{e2}(\overline{S}_2(\mu))$ do not depend on the choice of μ.*
(*iv*) *If $F_1(\mu) \in \mathcal{F}_+^b(X,Y)\bigcap\mathcal{F}_-^b(X,Y)$, $F_2(\mu) \in \mathcal{F}_+^b(X,Z)\bigcap\mathcal{F}_-^b(X,Z)$ and $F_3(\mu) \in \mathcal{F}_+^b(Y,Z)\bigcap\mathcal{F}_-^b(Y,Z)$, then $\sigma_{e3}(S_1(\mu))$ and $\sigma_{e3}(\overline{S}_2(\mu))$ do not depend on the choice of the scalar μ.*
(*v*) *If $F_1(\mu) \in \mathcal{F}_+^b(X,Y)$, $F_2(\mu) \in \mathcal{F}_+^b(X,Z)$ and $F_3(\mu) \in \mathcal{F}_+^b(Y,Z)$, then $\sigma_{e7}(\overline{S}_2(\mu))$ does not depend on the choice of μ.*
(*vi*) *If $F_1(\mu) \in \mathcal{F}_-^b(X,Y)$, $F_2(\mu) \in \mathcal{F}_-^b(X,Z)$ and $F_3(\mu) \in \mathcal{F}_-^b(Y,Z)$, then $\sigma_{e8}(\overline{S}_2(\mu))$ does not depend on the choice of μ.* ◇

Proof.

(i) Using Eq. (11.2.15), assumption $(N4)$, Lemma 6.3.1 and Theorem 6.3.1, we infer that $\sigma_{ei}(S_1(\mu)) = \sigma_{ei}(S_1(\lambda))$, $i = 4,5$. Hence, $\sigma_{ei}(S_1(\mu))$ does not depend on μ. By using the same above argument, we have $\Big(F_2(\lambda)\overline{G}_2(\mu) + F_2(\lambda)\Big[\overline{K_\lambda\Gamma_Y^0} - \overline{(A_1-\mu)^{-1}}\Big]\overline{G}_3(\mu)\Big) \in \mathcal{F}^b(Z)$ and $\Big(F_3(\lambda)(S_1(\lambda)-\lambda) - F_3(\mu)(S_1(\mu)-\mu)\Big)\overline{G}_3(\mu) \in \mathcal{F}^b(Z)$. From Eq. (11.2.16), Lemma 6.3.1 and Theorem 6.3.1, we can deduce that $\sigma_{ei}(\overline{S}_2(\mu)) = \sigma_{ei}(\overline{S}_2(\lambda))$, with $i = 4,5$. A similar reasoning allows us to reach the results for (ii) and (iii).
(iv) This assertion is an immediate deduction from (ii) and (iii).
(v) Using Theorem 6.3.1, we deduce that the difference $\overline{S}_2(\mu)-\overline{S}_2(\lambda)$ in (11.2.16) is in $\mathcal{F}_+^b(Z)$. Now, from Theorem 7.5.11 and Remark 7.5.1, we conclude that $\sigma_{e7}(\overline{S}_2(\mu))$ does not depend on the choice of μ.
(vi) This assertion can be proved in a similar way as for (v). Q.E.D.

In what follows, and for $\mu \in \rho(A_1)\bigcap\rho(S_1(\mu))$, we will denote by $M(\mu)$ the following operator:

$$M(\mu) := \begin{pmatrix} 0 & \hat{G}_1(\mu) & \hat{G}_2(\mu) \\ F_1(\mu) & F_1(\mu)\hat{G}_1(\mu) & F_1(\mu)\hat{G}_2(\mu) + \hat{G}_3(\mu) \\ F_2(\mu) & F_2(\mu)\hat{G}_1(\mu) + F_3(\mu) & F_2(\mu)\hat{G}_2(\mu) + F_3(\mu)\hat{G}_3(\mu) \end{pmatrix}.$$

Theorem 11.2.2. *Under the hypotheses $(N1)$–$(N13)$, we have*

(*i*) *If, for some $\mu \in \rho(A_1)\bigcap\rho(S_1(\mu))$, $F_1(\mu) \in \mathcal{F}^b(X,Y)$, $F_2(\mu) \in \mathcal{F}^b(X,Z)$, $F_3(\mu) \in \mathcal{F}^b(Y,Z)$ and $M(\mu) \in \mathcal{F}(X\times Y\times Z)$, then $\sigma_{e4}(\mathcal{L}) = \sigma_{e4}(A_1)\bigcup\sigma_{e4}(S_1(\mu))\bigcup\sigma_{e4}(\overline{S}_2(\mu))$ and $\sigma_{e5}(\mathcal{L}) \subseteq \sigma_{e5}(A_1)\bigcup\sigma_{e5}(S_1(\mu))\bigcup\sigma_{e5}(\overline{S}_2(\mu))$. Moreover, if $\mathbb{C}\backslash\sigma_{e4}(A_1)$ and $\mathbb{C}\backslash\sigma_{e4}(S_1(\mu))$ are connected, and*

$\rho(S_1(\mu)) \neq \emptyset$, then $\sigma_{e5}(\mathcal{L}) = \sigma_{e5}(A_1)\bigcup\sigma_{e5}(S_1(\mu))\bigcup\sigma_{e5}(\overline{S}_2(\mu))$. In addition, if $\mathbb{C}\backslash\sigma_{e5}(\mathcal{L})$ and $\mathbb{C}\backslash\sigma_{e5}(\overline{S}_2(\mu))$ are connected, and $\rho(\overline{S}_2(\mu))$, $\rho(\mathcal{L})$ are not empty, then $\sigma_{e6}(\mathcal{L}) = \sigma_{e6}(A_1)\bigcup\sigma_{e6}(S_1(\mu))\bigcup\sigma_{e6}(\overline{S}_2(\mu))$.

(ii) If, for some $\mu \in \rho(A_1)\bigcap\rho(S_1(\mu))$, $F_1(\mu) \in \mathcal{F}_+^b(X,Y)$, $F_2(\mu) \in \mathcal{F}_+^b(X,Z)$, $F_3(\mu) \in \mathcal{F}_+^b(Y,Z)$ and $M(\mu) \in \mathcal{F}_+(X \times Y \times Z)$, then $\sigma_{e1}(\mathcal{L}) = \sigma_{e1}(A_1)\bigcup\sigma_{e1}(S_1(\mu))\bigcup\sigma_{e1}(\overline{S}_2(\mu))$.

(iii) If, for some $\mu \in \rho(A_1)\bigcap\rho(S_1(\mu))$, $F_1(\mu) \in \mathcal{F}_-^b(X,Y)$, $F_2(\mu) \in \mathcal{F}_-^b(X,Z)$, $F_3(\mu) \in \mathcal{F}_-^b(Y,Z)$ and $M(\mu) \in \mathcal{F}_-(X \times Y \times Z)$, then $\sigma_{e2}(\mathcal{L}) = \sigma_{e2}(A_1)\bigcup\sigma_{e2}(S_1(\mu))\bigcup\sigma_{e2}(\overline{S}_2(\mu))$.

(iv) If, for some $\mu \in \rho(A_1)\bigcap\rho(S_1(\mu))$, $F_1(\mu) \in \mathcal{F}_+^b(X,Y)\bigcap\mathcal{F}_-^b(X,Y)$, $F_2(\mu) \in \mathcal{F}_+^b(X,Z)\bigcap\mathcal{F}_-^b(X,Z)$, $F_3(\mu) \in \mathcal{F}_+^b(Y,Z)\bigcap\mathcal{F}_-^b(Y,Z)$ and $M(\mu) \in \mathcal{F}_+(X \times Y \times Z)\bigcap\mathcal{F}_-(X \times Y \times Z)$, then

$$\begin{aligned}\sigma_{e3}(\mathcal{L}) = {} & \sigma_{e3}(A_1)\bigcup\sigma_{e3}(S_1(\mu))\bigcup\sigma_{e3}(\overline{S}_2(\mu)) \\ & \bigcup[\sigma_{e1}(A_1)\bigcap\sigma_{e2}(S_1(\mu))\bigcap\sigma_{e2}(\overline{S}_2(\mu))] \\ & \bigcup[\sigma_{e2}(A_1)\bigcap\sigma_{e2}(S_1(\mu))\bigcap\sigma_{e1}(\overline{S}_2(\mu))] \\ & \bigcup[\sigma_{e2}(A_1)\bigcap\sigma_{e1}(S_1(\mu))\bigcap\sigma_{e2}(\overline{S}_2(\mu))].\end{aligned}$$

*(v) If, for some $\mu \in \rho(A_1)\bigcap\rho(S_1(\mu))$, we have $F_1(\mu) \in \mathcal{F}_+^b(X,Y)$, $F_2(\mu) \in \mathcal{F}_+^b(X,Z)$,
$F_3(\mu) \in \mathcal{F}_+^b(Y,Z)$ and $M(\mu) \in \mathcal{F}_+(X \times Y \times Z)$, then*

$$\sigma_{e7}(\mathcal{L}) \subseteq \sigma_{e7}(A_1)\bigcup\sigma_{e7}(S_1(\mu))\bigcup\sigma_{e7}(\overline{S}_2(\mu)).$$

Moreover, if the sets $\mathbb{C}\backslash\sigma_{e4}(A_1)$, $\mathbb{C}\backslash\sigma_{e4}(S_1(\mu))$, $\mathbb{C}\backslash\sigma_{e4}(\overline{S}_2(\mu))$, and $\mathbb{C}\backslash\sigma_{e4}(\mathcal{L})$ are connected and $\rho(S_1(\mu))$, $\rho(\overline{S}_2(\mu))$ and $\rho(\mathcal{L})$ are not empty, then

$$\sigma_{e7}(\mathcal{L}) = \sigma_{e7}(A_1)\bigcup\sigma_{e7}(S_1(\mu))\bigcup\sigma_{e7}(\overline{S}_2(\mu)).$$

(vi) If, for some $\mu \in \rho(A_1)\bigcap\rho(S_1(\mu))$, we have $F_1(\mu) \in \mathcal{F}_-^b(X,Y)$, $F_2(\mu) \in \mathcal{F}_-^b(X,Z)$, $F_3(\mu) \in \mathcal{F}_-^b(Y,Z)$ and $M(\mu) \in \mathcal{F}_-(X \times Y \times Z)$, then

$$\sigma_{e8}(\mathcal{L}) \subseteq \sigma_{e8}(A_1)\bigcup\sigma_{e8}(S_1(\mu))\bigcup\sigma_{e8}(\overline{S}_2(\mu)).$$

Besides, if the sets $\mathbb{C}\backslash\sigma_{e4}(A_1)$, $\mathbb{C}\backslash\sigma_{e4}(S_1(\mu))$, $\mathbb{C}\backslash\sigma_{e4}(\overline{S}_2(\mu))$, and $\mathbb{C}\backslash\sigma_{e4}(\mathcal{L})$ are connected and $\rho(S_1(\mu))$, $\rho(\overline{S}_2(\mu))$ and $\rho(\mathcal{L})$ are not empty, then

$$\sigma_{e8}(\mathcal{L}) = \sigma_{e8}(A_1)\bigcup\sigma_{e8}(S_1(\mu))\bigcup\sigma_{e8}(\overline{S}_2(\mu)).$$

◇

Proof. Fix $\mu \in \rho(A_1) \bigcap \rho(S_1(\mu))$. Then, for $\lambda \in \mathbb{C}$, we have

$$\begin{aligned}
&\mathcal{L} - \lambda I \\
&= (\mathcal{L} - \mu I) + (\mu - \lambda)I \\
&= \mathcal{G}_1(\mu) \begin{pmatrix} A_1 - \mu & 0 & 0 \\ 0 & S_1(\mu) - \mu & 0 \\ 0 & 0 & \overline{S}_2(\mu) - \mu \end{pmatrix} \mathcal{G}_2(\mu) + (\mu - \lambda)I \\
&= \mathcal{G}_1(\mu) \begin{pmatrix} A_1 - \lambda & 0 & 0 \\ 0 & S_1(\mu) - \lambda & 0 \\ 0 & 0 & \overline{S}_2(\mu) - \lambda \end{pmatrix} \mathcal{G}_2(\mu) - (\mu - \lambda)[\mathcal{G}_1(\mu)\mathcal{G}_2(\mu) - I] \\
&= \mathcal{G}_1(\mu) \begin{pmatrix} A_1 - \lambda & 0 & 0 \\ 0 & S_1(\mu) - \lambda & 0 \\ 0 & 0 & \overline{S}_2(\mu) - \lambda \end{pmatrix} \mathcal{G}_2(\mu) - (\mu - \lambda)M(\mu).
\end{aligned}$$

Since $M(\mu)$ is a Fredholm perturbation, and since $\mathcal{G}_1(\mu)$ and $\mathcal{G}_2(\mu)$ are invertible and boundedly invertible, then $\mathcal{L} - \lambda I$ is a Fredholm operator if, and only if,

$$\begin{pmatrix} A_1 - \lambda & 0 & 0 \\ 0 & S_1(\mu) - \lambda & 0 \\ 0 & 0 & \overline{S}_2(\mu) - \lambda \end{pmatrix}$$

is a Fredholm operator. It follows that $\sigma_{e4}(\mathcal{L}) = \sigma_{e4}(A_1) \bigcup \sigma_{e4}(S_1(\mu)) \bigcup \sigma_{e4}(\overline{S}_2(\mu))$. Moreover, we have

$$i(\mathcal{L} - \lambda I) = i(A_1 - \lambda) + i(S_1(\mu) - \lambda) + i(\overline{S}_2(\mu) - \lambda). \tag{11.2.17}$$

If $i(\mathcal{L} - \lambda I) \neq 0$, then one of the terms of (11.2.17) is nonzero. Hence,

$$\sigma_{e5}(\mathcal{L}) \subseteq \sigma_{e5}(A_1) \bigcup \sigma_{e5}(S_1(\mu)) \bigcup \sigma_{e5}(\overline{S}_2(\mu)).$$

By using the assumption $(N3)$, we have $\rho(A_1) \neq \emptyset$. Since the set $\mathbb{C} \backslash \sigma_{e4}(A_1)$ is connected, then by using Theorem 7.3.1, we can deduce that $\sigma_{e4}(A_1) = \sigma_{e5}(A_1)$. By using the same above argument, we can show that $\sigma_{e4}(S_1(\mu)) = \sigma_{e5}(S_1(\mu))$ and $i(S_1(\mu) - \lambda) = 0$ for each $\lambda \in \mathbb{C} \backslash \sigma_{e4}(S_1(\mu))$. Let $\lambda \in \mathbb{C} \backslash \sigma_{e5}(\mathcal{L})$. Then, $\lambda \in \mathbb{C} \backslash \sigma_{e4}(A_1)$ and $\lambda \in \mathbb{C} \backslash \sigma_{e4}(S_1(\mu))$ and also $\lambda \in \mathbb{C} \backslash \sigma_{e4}(\overline{S}_2(\mu))$, further $i(\mathcal{L} - \lambda I) = i(\overline{S}_2(\mu) - \lambda I)$. Hence, $\lambda \in \mathbb{C} \backslash \sigma_{e5}(\overline{S}_2(\mu))$ and, then $\sigma_{e5}(\mathcal{L}) = \sigma_{e5}(A_1) \bigcup \sigma_{e5}(S_1(\mu)) \bigcup \sigma_{e5}(\overline{S}_2(\mu))$. Moreover, the fact that $\mathbb{C} \backslash \sigma_{e5}(\mathcal{L})$ is connected and $\rho(\mathcal{L}) \neq \emptyset$, together with Theorem 7.3.1 (ii) allow us to deduce that $\sigma_{e6}(\mathcal{L}) = \sigma_{e5}(\mathcal{L})$. To summarize, we have $\sigma_{e4}(A_1) = \sigma_{e5}(A_1)$ and $\sigma_{e4}(S_1(\mu)) =$

$\sigma_{e5}(S_1(\mu))$. Then, $\mathbb{C}\backslash\sigma_{e5}(A_1)$ and $\mathbb{C}\backslash\sigma_{e5}(S_1(\mu))$ are connected. By using the same reasoning as for $\mathcal{L}$, we show that $\sigma_{e5}(A_1) = \sigma_{e6}(A_1)$ and $\sigma_{e5}(S_1(\mu)) = \sigma_{e6}(S_1(\mu))$. Knowing the fact that $\mathbb{C}\backslash\sigma_{e5}(\overline{S}_2(\mu))$ is connected, this allows us to get the same previous result which is $\sigma_{e5}(\overline{S}_2(\mu)) = \sigma_{e6}(\overline{S}_2(\mu))$. The result follows from the first part of the proof.

The proof of the assertions (*ii*), (*iii*), and (*iv*) may be checked in a similar way.

(v) Fix $\mu \in \rho(A_1)\bigcap\rho(S_1(\mu))$. As in the proofs of Theorems 11.2.1 and 11.2.3, we find $\sigma_{e7}(\mathcal{L}) \subseteq \sigma_{e7}(A_1)\bigcup\sigma_{e7}(S_1(\mu))\bigcup\sigma_{e7}(\overline{S}_2(\mu))$. Since $\mathbb{C}\backslash\sigma_{e4}(A_1)$, $\mathbb{C}\backslash\sigma_{e4}(S_1(\mu))$, $\mathbb{C}\backslash\sigma_{e4}(\overline{S}_2(\mu))$, and $\mathbb{C}\backslash\sigma_{e4}(\mathcal{L})$ are connected and $\rho(S_1(\mu))$, $\rho(\overline{S}_2(\mu))$ and $\rho(\mathcal{L})$ are not empty, the result follows from Theorem 7.3.1 (i) together with Theorem 10.2.2.

The proof of (vi) is similar. Q.E.D.

In what follows, we will discuss the Rakočević and the Schmoeger's essential spectra. First, we have to prove again a stability lemma.

Theorem 11.2.3. *Let us assume that* $(N1)$–$(N13)$ *are satisfied.*

(*i*) *If, for some* $\mu \in \rho_6(A_1)\bigcap\rho(S_1(\mu))$, *we have* $F_1(\mu) \in \mathcal{F}_+^b(X,Y)$, $F_2(\mu) \in \mathcal{F}_+^b(X,Z)$ *and* $F_3(\mu) \in \mathcal{F}_+^b(Y,Z)$, *then* $\sigma_{e7}(\mathcal{L})\bigcap\rho_6(A_1)\bigcap\rho(S_1(\mu)) = \sigma_{e7}(\overline{S}_2(\mu))\bigcap\rho_6(A_1)\bigcap\rho(S_1(\mu))$.

(*ii*) *If, for some* $\mu \in \rho_6(A_1)\bigcap\rho(S_1(\mu))$, *we have* $F_1(\mu) \in \mathcal{F}_-^b(X,Y)$, $F_2(\mu) \in \mathcal{F}_-^b(X,Z)$ *and* $F_3(\mu) \in \mathcal{F}_-^b(Y,Z)$, *then* $\sigma_{e8}(\mathcal{L})\bigcap\rho_6(A_1)\bigcap\rho(S_1(\mu)) = \sigma_{e8}(\overline{S}_2(\mu))\bigcap\rho_6(A_1)\bigcap\rho(S_1(\mu))$. ◇

Proof.

(i) We have:

$$\begin{aligned}\rho_6(A_1)\bigcap\rho(S_1(\mu)) &= \Big[\rho(A_1)\bigcup\sigma_d(A_1)\Big]\bigcap\rho(S_1(\mu))\\ &= \Big[\rho(A_1)\bigcap\rho(S_1(\mu))\Big]\bigcup\Big[\sigma_d(A_1)\bigcap\rho(S_1(\mu))\Big].\end{aligned}$$

We notice the existence of two cases.

<u>1*st case*</u>: If $\mu \in \rho(A_1)\bigcap\rho(S_1(\mu))$, it is clear that the external factors $\mathcal{G}_1(\mu)$ and $\mathcal{G}_2(\mu)$ are bounded and also have bounded inverses. Therefore, $\mathcal{L} - \mu I$ is an upper semi-Fredholm operator if, and only if, $\overline{S}_2(\mu) - \mu$ has the same property and, in this case, $i(\mathcal{L}-\mu I) = i(\overline{S}_2(\mu)-\mu)$. Hence, $\sigma_{e7}(\mathcal{L}) = \sigma_{e7}(\overline{S}_2(\mu))$. Now, by using Lemma 11.2.6, we deduce the result.

<u>2*nd case*</u>: If $\mu \in \sigma_d(A_1)\bigcap\rho(S_1(\mu))$, then there exists an $\varepsilon > 0$ such that, for the disc $\mathbb{D}(\mu, 2\varepsilon)$, we have

$$\mathbb{D}(\mu, 2\varepsilon)\backslash\{\mu\} \subset \rho(A_1)\bigcap\rho(S_1(\mu)). \tag{11.2.18}$$

Let us denote $\tilde{A}_1 := A_1 + \varepsilon P_\mu$, where P_μ is the finite rank Riesz projection of A_1 corresponding to μ. We can easily check that $\mathbb{D}(\mu,\varepsilon)\backslash\{\mu\} \subset$

$\rho(A_1)\bigcap\rho(S_1(\mu))\bigcap\rho(\tilde{A}_1)$. Indeed, set $\lambda \in \mathbb{D}(\mu,\varepsilon)\backslash\{\mu\}$ then $|\lambda-\mu| < \varepsilon$. On $N(P_\mu)$, the operator $\tilde{A}_1-\lambda = A_1-\lambda$ is invertible and has bounded inverse. On $R(P_\mu)$, we have $\tilde{A}_1-\lambda = A_1-(\lambda-\varepsilon)$. Observe that $|(\lambda-\varepsilon)-\mu| < 2\varepsilon$ then $\lambda-\varepsilon \in \rho(A_1)\bigcap\rho(S_1(\mu))$, hence the restriction of $\tilde{A}_1-\lambda$ on $R(P_\mu)$ is invertible and has bounded inverse, it follows that $\lambda \in \rho(\tilde{A}_1)$, then by the use of (11.2.18) we deduce the result. Until further notice, we fix $\lambda \in \rho(A_1)\bigcap\rho(S_1(\mu))\bigcap\rho(\tilde{A}_1)$. Let us denote

$$\tilde{\mathcal{L}}_0 := \begin{pmatrix} \tilde{A} & B & C \\ D & E & F \\ G & H & L \end{pmatrix} = \mathcal{L}_0 + \varepsilon \begin{pmatrix} P_\mu & 0 & 0 \\ 0 & 0 & 0 \\ 0 & 0 & 0 \end{pmatrix}.$$

For the closure $\tilde{\mathcal{L}}$ of $\tilde{\mathcal{L}}_0$, we obtain

$$\tilde{\mathcal{L}} = \mathcal{L} + \varepsilon \begin{pmatrix} P_\mu & 0 & 0 \\ 0 & 0 & 0 \\ 0 & 0 & 0 \end{pmatrix}.$$

Clearly, $\tilde{\mathcal{L}}$ is a finite rank perturbation of $\mathcal{L}$. Therefore, $\sigma_{e7}(\tilde{\mathcal{L}}) = \sigma_{e7}(\mathcal{L})$ and $i(\tilde{\mathcal{L}}-\lambda) = i(\mathcal{L}-\lambda)$. Next, we apply the obtained result of the first part of the proof to the operator $\tilde{\mathcal{L}}$. Let us denote $\tilde{S}_2(\lambda)$ by

$$\tilde{S}_2(\lambda) := L - G(\tilde{A}_1-\lambda)^{-1}C + \tilde{\Theta}(\lambda)[\tilde{J}_\lambda\Gamma_Z - (\tilde{S}_1(\lambda)-\lambda)^{-1}(F - D(\tilde{A}_1-\lambda)^{-1}C)].$$

Here, $\tilde{K}_\lambda$, $\tilde{J}_\lambda$, $\tilde{\Theta}(\lambda)$ and $\tilde{S}_1(\lambda)$ are the operators defined as K_λ, J_λ, $\Theta(\lambda)$ and $S_1(\lambda)$ with A replaced by $\tilde{A}$. Hence,

$$\tilde{K}_\lambda w = x \iff x \in N(\tilde{A}-\lambda),\ \Gamma_X x = w$$

and

$$\tilde{J}_\lambda w = y \iff y \in N(\tilde{S}(\lambda)-\lambda),\ \Gamma_Y y = w.$$

If $S_2(\lambda)$ is closable, then its perturbation $\tilde{S}_2(\lambda)$ is also closable. Let us denote its closure by $\hat{S}_2(\lambda)$. We claim the following: if the assumptions $(N1)$–$(N13)$ are satisfied and $S_2(\lambda)$ is closable, then the operator $\hat{S}_2(\lambda) - \tilde{S}_2(\lambda)$ is of finite rank and $\sigma_{e7}(\hat{S}_2(\lambda)) = \sigma_{e7}(\tilde{S}_2(\lambda))$. Indeed, let us first notice that

$$[F - D(\tilde{A}_1-\lambda)^{-1}C] - [F - D(A_1-\lambda)^{-1}C] = -\varepsilon D(\tilde{A}_1-\lambda)^{-1}P_\lambda(A_1-\lambda)^{-1}C.$$

The assumptions $(N5)$ and $(N11)$ imply that this difference is of finite rank. Moreover, we have

$$S_1(\lambda) - \tilde{S}_1(\lambda) = (E + DK_\lambda\Gamma_Y - D(A_1-\lambda)^{-1}B)$$

$$-(E + D\tilde{K}_\lambda \Gamma_Y - D(\tilde{A}_1 - \lambda)^{-1} B)$$
$$= D(K_\lambda - \tilde{K}_\lambda)\Gamma_Y - D((A_1 - \lambda)^{-1} - (\tilde{A}_1 - \lambda)^{-1})B.$$

Obviously, $K_\lambda - \tilde{K}_\lambda = \varepsilon P_\lambda (A_1 - \lambda)^{-1} \tilde{K}_\lambda$. Therefore,

$$S_1(\lambda) - \tilde{S}_1(\lambda) = \varepsilon D(A_1 - \lambda)^{-1} P_\lambda [\tilde{K}_\lambda \Gamma_Y + (\tilde{A}_1 - \lambda)^{-1} B].$$

It follows that $(\tilde{S}_1(\lambda)-\lambda)^{-1}-(S_1(\lambda)-\lambda)^{-1} = (\tilde{S}_1(\lambda)-\lambda)^{-1}[S_1(\lambda)-\tilde{S}_1(\lambda)](S_1(\lambda)-\lambda)^{-1}$. By applying the assumptions $(N4)$ and $(N5)$, we deduce that this difference is of finite rank. Moreover, let us notice that

$$\begin{aligned}\Psi(\lambda) := \tilde{\Theta}(\lambda) - \Theta(\lambda) &= -\varepsilon F_2(\lambda) P_\lambda \tilde{K}_\lambda \Gamma_Y - \varepsilon G(A_1 - \lambda)^{-1} P_\lambda (\tilde{A}_1 - \lambda)^{-1} B \\ &= -\varepsilon F_2(\lambda) P_\lambda [\tilde{K}_\lambda \Gamma_Y + (\tilde{A}_1 - \lambda)^{-1} B].\end{aligned}$$

By using the same argument, $(N6)$ implies that the operator $\Psi(\lambda)$ is of finite rank. The operator $\tilde{\Upsilon}(\lambda) := \tilde{J}_\lambda \Gamma_Z - (\tilde{S}_1(\lambda) - \lambda)^{-1}(F - D(\tilde{A}_1 - \lambda)^{-1} C)$ is bounded on its domain (here $\Upsilon(\lambda) = -G_3(\lambda)$). Hence,

$$\begin{aligned}\tilde{\Theta}(\lambda)\tilde{\Upsilon}(\lambda) - \Theta(\lambda)\Upsilon(\lambda) &= [\Theta(\lambda) + \Psi(\lambda)]\tilde{\Upsilon}(\lambda) - \Theta(\lambda)\Upsilon(\lambda) \\ &= \Theta(\lambda)[\tilde{\Upsilon}(\lambda) - \Upsilon(\lambda)] + \Psi(\lambda)\tilde{\Upsilon}(\lambda).\end{aligned}$$

By hypotheses, $S_2(\lambda)$ is closable in Z. So, its perturbation $\tilde{S}_2(\lambda)$ is closable and we conclude that $\hat{S}_2(\lambda) - \overline{S}_2(\lambda)$ is of finite rank. Therefore, $\sigma_{e7}(\hat{S}_2(\lambda)) = \sigma_{e7}(\overline{S}_2(\lambda))$. Now, by using Lemma 11.2.6, we deduce that $\sigma_{e7}(\hat{S}_2(\lambda))$ is independent of $\rho(\tilde{A}_1)$. Applying the first part of this proof for $\mu \in \rho(\tilde{A}_1) \bigcap \rho(\tilde{S}_1(\mu))$, we find $\sigma_{e7}(\tilde{\mathcal{L}}) = \sigma_{e7}(\hat{S}_2(\lambda))$, and finally $\sigma_{e7}(\mathcal{L}) = \sigma_{e7}(\tilde{\mathcal{L}}) = \sigma_{e7}(\hat{S}_2(\mu)) = \sigma_{e7}(\hat{S}_2(\lambda)) = \sigma_{e7}(\overline{S}_2(\lambda))$.

(ii) The proof is similar as (i). Q.E.D.

11.3 Block Operator Matrices Using Browder Resolvent

11.3.1 The Operator $\mathcal{A}_0$ and Its Closure

Let X, Y, Z and W be Banach spaces. In this section, we consider the linear operators Γ_X, Γ_Y, Γ_Z acting from X, Y, Z into W, respectively. Therefore, we define in the Banach space $X \times Y \times Z$ the operator $\mathcal{A}_0$ as follows:

$$\mathcal{A}_0 := \begin{pmatrix} A & B & C \\ D & E & F \\ G & H & L \end{pmatrix}$$

$$\mathcal{D}(\mathcal{A}_0) = \left\{ \begin{pmatrix} x \\ y \\ z \end{pmatrix} : \begin{array}{l} x \in \mathcal{D}(A) \\ y \in \mathcal{D}(B) \bigcap \mathcal{D}(E) \\ z \in \mathcal{D}(C) \bigcap \mathcal{D}(F) \bigcap \mathcal{D}(L) \end{array} , \; \Gamma_X x = \Gamma_Y y = \Gamma_Z z \right\}.$$

In what follows, we will assume that the following conditions hold:

($\mathcal{O}1$) The operator A is densely defined and closable.
It follows that $\mathcal{D}(A)$, equipped with the graph norm $\|x\|_A = \|x\| + \|Ax\|$ can be completed to a Banach space X_A which coincides with $\mathcal{D}(\overline{A})$, the domain of the closure of A in X.
($\mathcal{O}2$) $\mathcal{D}(A) \subset \mathcal{D}(\Gamma_X) \subset X_A$ and $\Gamma_X : X_A \longrightarrow W$ is a bounded mapping. Denote by $\overline{\Gamma}_X$ the extension by continuity which is a bounded operator from X_A into W.
($\mathcal{O}3$) The set $\mathcal{D}(A) \bigcap N(\Gamma_X)$ is dense in X and the resolvent set of the restriction $A_1 := A|_{\mathcal{D}(A) \bigcap N(\Gamma_X)}$ is not empty, i.e., $\rho(A_1) \neq \emptyset$.

Remark 11.3.1. It follows from ($\mathcal{O}3$) that A_1 is a closed operator in the Banach space X_A with nonempty resolvent set. ◇

($\mathcal{O}4$) $\mathcal{D}(A) \subset \mathcal{D}(D) \subset X_A$ and D is a closable operator from X_A into Y.
($\mathcal{O}5$) $\mathcal{D}(A) \subset \mathcal{D}(G) \subset X_A$ and G is a closable operator from X_A into Z.

The closed graph theorem and the assumptions ($\mathcal{O}3$)–($\mathcal{O}5$) imply that for $\lambda \in \rho_b(A_1)$ the operators $F_1(\lambda) := DR_b(A_1, \lambda)$ and $F_2(\lambda) := GR_b(A_1, \lambda)$ are bounded from X into Y and X into Z, respectively. Let $A_{1\lambda}$ be the operator defined on $\mathcal{D}(A_1)$ by $A_{1\lambda} := (A_1 - \lambda)(I - P_\lambda) + P_\lambda$, where P_λ is the finite rank Riesz projection of A_1 corresponding to λ. Under the assumptions ($\mathcal{O}1$)–($\mathcal{O}3$), and using Lemma 10.3.1, for any $\lambda \in \rho_b(A_1)$, we have the following decomposition $\mathcal{D}(A) = \mathcal{D}(A_1) \oplus N(A_\mu)$, where A_μ is the operator defined on $\mathcal{D}(A)$ by $A_\mu := (A - \mu)(I - P_\mu) + P_\mu$. For $\mu \in \rho_b(A_1)$, we denote the inverse of $\Gamma_X|_{N(A_\mu)}$ by $K_\mu := (\Gamma_X|_{N(A_\mu)})^{-1}$. We can write $K_\mu : \Gamma_X(\mathcal{D}(A)) \longrightarrow N(A_\mu) \subset \mathcal{D}(A)$. For $\mu \in \rho_b(A_1)$, and if assumptions ($\mathcal{O}1$)–($\mathcal{O}3$) are satisfied, then $A_\mu x = A_{1\mu}(I - K_\mu \Gamma_X)x$.

($\mathcal{O}6$) For some (hence for all) $\lambda \in \rho_b(A_1)$, the operator K_λ is bounded on its domain.
($\mathcal{O}7$) E is closable and densely defined linear operator. We denote by Y_E the following Banach space $Y_E := \big(\mathcal{D}(E), \|.\|_E\big)$.
($\mathcal{O}8$) $\mathcal{D}(B) \bigcap \mathcal{D}(E) \subset \mathcal{D}(\Gamma_Y)$, the set $Y_1 = \Big\{ y \in \mathcal{D}(B) \bigcap \mathcal{D}(E)$ such that $\Gamma_Y y \in \Gamma_X(\mathcal{D}(A)) \Big\}$ is dense in Y and the restriction of Γ_Y to Y_1 is bounded as an operator from Y into W. We denote the extension by continuity of $\Gamma_Y|_{Y_1}$ to Y by $\overline{\Gamma}^0_Y$.

In the following, we denote by $S(\lambda)$ the following operator $S(\lambda) := E + DK_\lambda \Gamma_Y - DR_b(A_1, \lambda)B$. For $\lambda \in \rho_b(A_1)$ the operator $S(\lambda)$ is defined on Y_1 and its restriction to $N(\Gamma_Y) \bigcap Y_1$ will be denoted by $S_1(\lambda)$, i.e., $S_1(\lambda) := S(\lambda)_{|N(\Gamma_Y) \bigcap Y_1}$.

Lemma 11.3.1. *Let $\lambda,\ \mu \in \rho_b(A_1)$*

$$S_1(\lambda) - S_1(\mu) = (\mu - \lambda) DR_b(A_1, \lambda) R_b(A_1, \mu) B$$
$$- DR_b(A_1, \lambda) S_{A_1}(\lambda, \mu) R_b(A_1, \mu) B, \qquad (11.3.1)$$

where $S_{A_1}(.,.)$ is the finite rank operator defined in Eq. (8.2.1). ◇

Proof. Let $\lambda,\ \mu \in \rho_b(A_1)$. Using Lemma 8.2.1, we have

$$\begin{aligned} & S(\lambda) - S(\mu) \\ & = D(K_\lambda - K_\mu)\Gamma_Y - D\left[R_b(A_1, \lambda) - R_b(A_1, \mu)\right] B \\ & = DR_b(A_1, \lambda)\Big[(\lambda - \mu) + S_{A_1}(\lambda, \mu)\Big] K_\mu \Gamma_Y \\ & \quad -D\Big[(\lambda - \mu) R_b(A_1, \lambda) R_b(A_1, \mu) + R_b(A_1, \lambda) S_{A_1}(\lambda, \mu) R_b(A_1, \mu)\Big] B \\ & = (\lambda - \mu) DR_b(A_1, \lambda)\Big[K_\mu \Gamma_Y - R_b(A_1, \mu) B\Big] \\ & \quad -\Big[DR_b(A_1, \lambda) S_{A_1}(\lambda, \mu)\Big]\Big[-K_\mu \Gamma_Y + R_b(A_1, \mu) B\Big]. \end{aligned}$$

For $y \in \mathcal{D}(S_1(\mu))$, we have $\Gamma_Y y = 0$ and the relation (11.3.1) holds. Q.E.D.

Remark 11.3.2. By assumptions ($\mathcal{O}4$) and ($\mathcal{O}5$), the operator $F_1(\mu) R_b(A_1, \mu) B$ is bounded on its domain. On the other hand, $S_{A_1}(\lambda, \mu)$ is a finite rank operator, so if $S_1(\mu)$ is closed for some $\mu \in \rho_b(A_1)$ then it is closed for all such μ. ◇

($\mathcal{O}9$) The operator $S_1(\lambda)$ is closed, densely defined in $N(\Gamma_Y) \bigcap Y_1$ with a nonempty resolvent set, i.e., $\rho(S_1(\lambda)) \neq \emptyset$.

($\mathcal{O}10$) The operator H satisfies that $\mathcal{D}(B) \subset \mathcal{D}(H) \subset Y_E$ and for some (hence for all) $\lambda \in \rho_b(A_1) \bigcap \rho_b(S_1(\lambda))$ the operator $\big[H + GK_\lambda \Gamma_Y - GR_b(A_1, \lambda) B\big] R_b(S_1(\lambda), \lambda)$ is bounded.
Set $\Psi(\lambda) := H + GK_\lambda \Gamma_Y - GR_b(A_1, \lambda) B$ and denote by $F_3(\lambda) := \Psi(\lambda) R_b(S_1(\lambda), \lambda)$.

($\mathcal{O}11$) The operator B (resp. C) is densely defined and for some (hence for all) $\lambda \in \rho_b(A_1)$ the operator $R_b(A_1, \lambda) B$ (resp. $R_b(A_1, \lambda) C$) is bounded on its domain. We will denote by $G_1(\lambda) := -K_\lambda \Gamma_Y + R_b(A_1, \lambda) B$ and $G_2(\lambda) := R_b(A_1, \lambda) C$.

Lemma 11.3.2. *If the operator $\Psi(\mu)$ is closable for some $\mu \in \rho_b(A_1)$, then it is closable for all such μ and for all $\lambda \in \rho_b(A_1)$ its closure satisfy:*

$$\overline{\Psi(\lambda)} - \overline{\Psi(\mu)} = (\lambda - \mu) GR_b(A_1, \lambda)\Big[\overline{K_\mu \Gamma_Y^0} - \overline{R_b(A_1, \mu) B}\Big]$$
$$- \Big[GR_b(A_1, \lambda) S_{A_1}(\lambda, \mu)\Big] \overline{G}_1(\mu).$$

◇

Proof. Let $\lambda,\ \mu \in \rho_b(A_1)$.

$$\begin{aligned}
&\Psi(\lambda) - \Psi(\mu)\\
&= G(K_\lambda - K_\mu)\Gamma_Y - G\Big[R_b(A_1,\lambda) - R_b(A_1,\mu)\Big]B\\
&= GR_b(A_1,\lambda)\Big[(\lambda-\mu) + S_{A_1}(\lambda,\mu)\Big]K_\mu\Gamma_Y\\
&\quad -G\Big[(\lambda-\mu)R_b(A_1,\lambda)R_b(A_1,\mu) + R_b(A_1,\lambda)S_{A_1}(\lambda,\mu)R_b(A_1,\mu)\Big]B\\
&= (\lambda-\mu)GR_b(A_1,\lambda)\Big[K_\mu\Gamma_Y - R_b(A_1,\mu)B\Big]\\
&\quad -\Big[GR_b(A_1,\lambda)S_{A_1}(\lambda,\mu)\Big]G_1(\mu),
\end{aligned}$$

here Γ_Y is bounded on Y_1 by assumption $(\mathcal{O}8)$. From $(\mathcal{O}5)$, $(\mathcal{O}6)$ and $(\mathcal{O}11)$ it follows that the operators K_λ, $R_b(A_1,\mu)B$ and $GR_b(A_1,\mu)$ are bounded. Then

$$\begin{aligned}
\overline{\Psi(\lambda)} - \overline{\Psi(\mu)} =&(\lambda-\mu)GR_b(A_1,\lambda)\Big[\overline{K_\mu\Gamma_Y^0} - \overline{R_b(A_1,\mu)B}\Big]\\
&-\Big[GR_b(A_1,\lambda)S_{A_1}(\lambda,\mu)\Big]\overline{G}_1(\mu).
\end{aligned}$$

Q.E.D.

Lemma 11.3.3. *For some $\lambda \in \rho_b(A_1)\bigcap\rho_b(S_1(\lambda))$ and under the assumptions $(\mathcal{O}8)$ and $(\mathcal{O}9)$, we have the following decomposition $Y_1 = \mathcal{D}(S_1(\lambda)) \oplus N(S_\lambda(\lambda))$, where $S_\lambda(\lambda) := (S(\lambda)-\lambda)\,(I-P_\lambda^{'}) + P_\lambda^{'}$ and $P_\lambda^{'}$ is the finite rank Riesz projection of $S(\lambda)$ corresponding to λ.* ◇

Proof. Let $\lambda \in \rho_b(A_1)\bigcap\rho_b(S_1(\lambda))$. The operator $S_{1\lambda}(\lambda)$ is invertible, then $N(S_\lambda(\lambda)) = \{0\}$ and we get $\mathcal{D}(S_1(\lambda))\bigcap N(S_\lambda(\lambda)) = \{0\}$. Now, we set $g = R_b(S_1(\lambda),\lambda)S_\lambda(\lambda)f \in \mathcal{D}(S_1(\lambda))$, for any $f \in Y_1$. We can easily see that $f-g \in N(S_\lambda(\lambda))$ and $f = g + f - g \in \mathcal{D}(S_1(\lambda)) + N(S_\lambda(\lambda))$. Q.E.D.

For $\lambda \in \rho_b(A_1)\bigcap\rho_b(S_1(\lambda))$, we define the inverse of Γ_Y by:

$$J_\lambda := (\Gamma_Y|_{N(S_\lambda(\lambda))})^{-1} : \Gamma_Y(Y_1) \longrightarrow N(S_\lambda(\lambda) \subset Y_1.$$

In other words $J_\lambda w = y$ means that $y \in \mathcal{D}(S_1(\lambda))$, $S_\lambda(\lambda)y = 0$ and $\Gamma_Y y = w$. Assume that for some $\mu \in \rho_b(A_1)\bigcap\rho_b(S_1(\lambda))$, J_μ is bounded from $\Gamma_Y(Y_1)$ into Y and its extension by continuity to $\overline{\Gamma_Y(Y_1)}$ is denoted by $\overline{J}_\mu$.

Lemma 11.3.4. *If $\lambda \in \rho_b(A_1)\bigcap\rho_b(S_1(\lambda))$ and $\mu \in \rho_b(A_1)\bigcap\rho_b(S_1(\mu))$, then*

$$\overline{J}_\lambda - \overline{J}_\mu = R_b(S_1(\lambda),\lambda)\Big[(\lambda-\mu) + \mathbb{U}(\lambda,\mu) + \mathbb{V}(\lambda,\mu)\Big]\overline{J}_\mu,$$

where we define the finite rank operator $\mathbb{U}(.,.)$ *as*

$$\mathbb{U}(\lambda,\mu) := \Big[S_1(\lambda) - (\lambda+1)\Big]P'_\lambda - \Big[S_1(\mu) - (\mu+1)\Big]P'_\mu,$$

and the bounded operator $\mathbb{V}(.,.)$ *as*

$$\mathbb{V}(\lambda,\mu) = (\lambda-\mu)\Big[F_1(\lambda)\overline{R_b(A_1,\mu)B}\Big] + F_1(\lambda)\overline{S_{A_1}(\lambda,\mu)R_b(A_1,\mu)B}.$$

◇

Proof. Let $w \in \Gamma_Y(Y_1)$ and set $y = y_1 - y_2$ such that $y_1 = J_\lambda w$ and $y_2 = J_\mu w$. Then, we have $S_{1\lambda}(\lambda)y = -S_{1\lambda}(\lambda)y_2 = \Big[-S_1(\lambda)+\lambda+\Big(S_1(\lambda)-(\lambda+1)\Big)P'_\lambda\Big]y_2$. Using Lemmas 8.2.1 and 8.2.2, we infer that

$$\begin{aligned} S_{1\lambda}(\lambda)y = \Big[&-S_1(\mu) + \lambda + (\lambda-\mu)F_1(\lambda)R_b(A_1,\mu)B \\ &+F_1(\lambda)S_{A_1}(\lambda,\mu)R_b(A_1,\mu)B + \Big(S_1(\lambda)-(\lambda+1)\Big)P'_\lambda\Big]y_2. \end{aligned}$$

On the other hand, $S_{1\mu}(\mu)y_2 = 0$, then $S_1(\mu)y_2 = \Big[\mu+\Big(S_1(\mu)-(\mu+1)\Big)P'_\mu\Big]y_2$. A short computation, shows that:

$$\begin{aligned} S_{1\lambda}(\lambda)y = \Big[&(\lambda-\mu) + (\lambda-\mu)F_1(\lambda)R_b(A_1,\mu)B + F_1(\lambda)S_{A_1}(\lambda,\mu)R_b(A_1,\mu)B \\ &+\Big(S_1(\lambda)-(\lambda+1)\Big)P'_\lambda - \Big(S_1(\mu)-(\mu+1)\Big)P'_\mu\Big]y_2. \end{aligned}$$

Since $y \in \mathcal{D}(S_1(\lambda))$, then this allow us to conclude that:

$$\begin{aligned} J_\lambda - J_\mu = R_b(S_1(\lambda),\lambda)\Big[&(\lambda-\mu) + (\lambda-\mu)F_1(\lambda)R_b(A_1,\mu)B \\ &+F_1(\lambda)S_{A_1}(\lambda,\mu)R_b(A_1,\mu)B \\ &+\Big(S_1(\lambda)-(\lambda+1)\Big)P'_\lambda - \Big(S_1(\mu)-(\mu+1)\Big)P'_\mu\Big]. \end{aligned}$$

From the above expression of $J_\lambda - J_\mu$, we get $J_\lambda = R_b(S_1(\lambda),\lambda)S_{1\mu}(\mu)J_\mu = S_{1\mu}(\mu)R_b(S_1(\lambda),\lambda)J_\mu$. Since $S_{1\mu}(\mu)R_b(S_1(\lambda),\lambda)$ is bounded and boundedly invertible, then J_λ is closable for each λ if J_μ is too and its closure satisfy $\overline{J}_\lambda - \overline{J}_\mu = R_b(S_1(\lambda),\lambda)\Big[(\lambda-\mu) + \mathbb{U}(\lambda,\mu) + \mathbb{V}(\lambda,\mu)\Big]\overline{J}_\mu$. Q.E.D.

($\mathcal{O}11$) L is densely defined and closed with nonempty resolvent set, i.e., $\rho(L) \neq \emptyset$.

($\mathcal{O}12$) $\mathcal{D}(C)\bigcap\mathcal{D}(F)\bigcap\mathcal{D}(L) \subset \mathcal{D}(\Gamma_Z)$, the set

$$Z_1 := \Big\{z \in \mathcal{D}(C)\bigcap\mathcal{D}(F)\bigcap\mathcal{D}(L) \text{ such that } \Gamma_Z z \in \Gamma_Y(Y_1)\Big\}$$

is dense in Z and the restriction of Γ_Z to Z_1 is bounded as an operator from Z into W. Denote the extension by continuity of $\Gamma_Z|_{Z_1}$ into Z by $\overline{\Gamma}^0_Z$.

($\mathcal{O}13$) For some (and hence for all) $\lambda \in \rho_b(A_1)$, the operator $F - DR_b(A_1,\lambda)C$ is closable and its closure $\overline{F - DR_b(A_1,\lambda)C}$ is bounded and for $\lambda \in \rho_b(A_1)\bigcap\rho_b(S_1(\lambda))$, set

$$G_3(\lambda) := -J_\lambda\Gamma_Z + R_b(S_1(\lambda),\lambda)(F - F_1(\lambda)C).$$

($\mathcal{O}14$) For some (hence for all) $\lambda \in \rho_b(A_1)\bigcap\rho_b(S_1(\lambda))$, the operator

$$S_2(\lambda) = L - F_2(\lambda)C + \Psi(\lambda)\Big[J_\lambda\Gamma_Z - R_b(S_1(\lambda),\lambda)(F - F_1(\lambda)C)\Big]$$

is closable.

Lemma 11.3.5. *If for some $\mu \in \rho_b(A_1)\bigcap\rho_b(S_1(\mu))$ the operator $S_2(\mu)$ is closable, then it is closable for all such μ.* $\diamondsuit$

Proof. Let $\mu \in \rho_b(A_1)\bigcap\rho_b(S_1(\mu))$ and $\lambda \in \rho_b(A_1)\bigcap\rho_b(S_1(\lambda))$. Using the resolvent identity we find

$$\begin{aligned}
S_2(\mu) - S_2(\lambda) &= \big[F_2(\lambda) - F_2(\mu)\big]C + \big[\Psi(\lambda)G_3(\lambda) - \Psi(\mu)G_3(\mu)\big] \\
&= (\lambda-\mu)F_2(\lambda)G_2(\mu) + F_2(\lambda)S_{A_1}(\lambda,\mu)G_2(\mu) \\
&\quad +\Psi(\lambda)G_3(\lambda) - \Psi(\lambda)G_3(\mu) \\
&\quad +(\lambda-\mu)F_2(\lambda)[K_\mu\Gamma_Y - R_b(A_1,\mu)B]G_3(\mu) \\
&\quad -F_2(\lambda)S_{A_1}(\lambda,\mu)G_1(\mu)G_3(\mu) \\
&= (\lambda-\mu)\Big(F_2(\lambda)G_2(\mu) + F_2(\lambda)[K_\mu\Gamma_Y - R_b(A_1,\mu)B]G_3(\mu)\Big) \\
&\quad +F_3(\lambda)S_{1\lambda}(\lambda)(G_3(\lambda) - G_3(\mu)) + F_2(\lambda)S_{A_1}(\lambda,\mu)G_2(\mu) \\
&\quad -F_2(\lambda)S_{A_1}(\lambda,\mu)G_1(\mu)G_3(\mu).
\end{aligned}$$

Since the operators F_i, $i = 1,2,3$ are bounded everywhere and the operators G_i, $i = 1,2,3$ are bounded on its domains and by assumptions ($\mathcal{O}13$) the operator $S_2(\lambda)$ is closable and the closure does not depend on the choice of μ. Q.E.D.

Denote the closure of $S_2(\mu)$ by $\overline{S}_2(\mu)$. Then we have

$$\begin{aligned}
\overline{S}_2(\mu) - \overline{S}_2(\lambda) &= (\lambda-\mu)\Big(F_2(\lambda)\overline{G}_2(\mu) + F_2(\lambda)[\overline{K}_\mu\Gamma_Y^0 - \overline{R_b(A_1,\mu)B}]\overline{G}_3(\mu)\Big) \\
&\quad + F_3(\lambda)S_{1\lambda}(\lambda)(\overline{G}_3(\lambda) - \overline{G}_3(\mu)) + F_2(\lambda)\overline{S_{A_1}(\lambda,\mu)G_2}(\mu) \\
&\quad - F_2(\lambda)\overline{S_{A_1}(\lambda,\mu)G_1}(\mu)\overline{G}_3(\mu).
\end{aligned}$$

Lemma 11.3.6. *For some $\lambda \in \rho_b(A_1)\bigcap\rho_b(S_1(\lambda))$ and $y \in Y_1$, we have the following $S_\lambda(\lambda)y = S_{1\lambda}(\lambda)(I - J_\lambda\Gamma_Y)y$, where the operator $I - J_\lambda\Gamma_Y$ is the projection from Y_1 on $\mathcal{D}(S_{1\lambda}(\lambda))$ parallel to $N(S_\lambda(\lambda))$.* $\diamondsuit$

Proof. Let $y \in Y_1$, then we have $y = (I - J_\mu \Gamma_Y)y + J_\mu \Gamma_Y y$. The first summand belongs to $\mathcal{D}(S_1(\mu))$ because $y_1 = (I - J_\mu \Gamma_Y)y \in \mathcal{D}(S_1(\mu))$ and $y_2 = J_\mu \Gamma_Y y \in N(S_\mu(\mu))$, then

$$\begin{aligned}(S(\mu) - \mu)y &= (S_1(\mu) - \mu)y_1 \\ &= (S_1(\mu) - \mu)(y - y_2) \\ &= (S_1(\mu) - \mu)(I - J_\mu \Gamma_Y)y.\end{aligned}$$

Q.E.D.

In the following we use these assumptions to show the closeness of the operator $\mathcal{A}_0$ and to describe the closure. The main idea is, as in the 2×2 case, a factorization of the 3×3 matrix with a diagonal matrix of Schur complements in the middle and invertible factors to the right and to the left. In the following we consider the operators $\hat{G}_i(\mu) = \overline{G}_i(\mu)$, $i = 1, 2, 3$.

Theorem 11.3.1. *Under assumptions $(\mathcal{O}1)$–$(\mathcal{O}13)$, the operator $\mathcal{A}_0$ is closable if, and only if, $S_2(\mu)$ is closable for some $\mu \in \rho_b(A_1) \bigcap \rho_b(S_1(\mu))$. In this case the closure $\mathcal{A}$ of $\mathcal{A}_0$ is given by*

$$\mathcal{A} = \mu I + \mathcal{G}_1(\mu) \begin{pmatrix} A_{1\mu} & 0 & 0 \\ 0 & S_{1\mu}(\mu) & 0 \\ 0 & 0 & \overline{S}_2(\mu) - \mu \end{pmatrix} \mathcal{G}_2(\mu) + \mathbb{N}(\mu)$$

where $\mathcal{G}_1(\mu) := \begin{pmatrix} I & 0 & 0 \\ F_1(\mu) & I & 0 \\ F_2(\mu) & F_3(\mu) & I \end{pmatrix}$, $\mathcal{G}_2(\mu) = \begin{pmatrix} I & \hat{G}_1(\mu) & \hat{G}_2(\mu) \\ 0 & I & \hat{G}_3(\mu) \\ 0 & 0 & I \end{pmatrix}$

and $\mathbb{N}(\mu) = \begin{pmatrix} [A_1 - (\mu+1)]P_\mu & 0 & 0 \\ 0 & (S_1(\mu) - (\mu+1))P'_\mu & 0 \\ 0 & 0 & 0 \end{pmatrix}$

or, spelled out,

$$\mathcal{D}(\mathcal{A}) = \left\{ \begin{pmatrix} I & -\hat{G}_1(\mu) & \hat{G}_1(\mu)\hat{G}_3(\mu) - \hat{G}_2(\mu) \\ 0 & I & -\hat{G}_3(\mu) \\ 0 & 0 & I \end{pmatrix} \begin{pmatrix} x \\ y \\ z \end{pmatrix}, \begin{matrix} x \in \mathcal{D}(A_1) \\ y \in Y_1 \bigcap N(\Gamma(Y)) \\ z \in Y_2 \end{matrix} \right\},$$

$$\begin{aligned}&\mathcal{A} \begin{pmatrix} x - \hat{G}_1(\mu)y + (\hat{G}_1(\mu)\hat{G}_3(\mu) - \hat{G}_2(\mu))z \\ y - \hat{G}_3(\mu)z \\ z \end{pmatrix} \\ &= \begin{pmatrix} A_{1\mu}x - \mu\hat{G}_1(\mu)y + \mu(\hat{G}_1(\mu)\hat{G}_3(\mu) - \hat{G}_2(\mu))z \\ Dx + S_{1\mu}(\mu)y - \mu\hat{G}_3(\mu)z \\ Gx + \Psi(\mu)y + \overline{S}_2(\mu)z \end{pmatrix}.\end{aligned}$$

◇

Proof. Let $\mu \in \rho_b(A_1) \bigcap \rho_b(S_1(\mu))$ the lower-upper factorization sense

$$\mathcal{A}_0 = \mu I + \begin{pmatrix} I & 0 & 0 \\ F_1(\mu) & I & 0 \\ F_2(\mu) & F_3(\mu) & I \end{pmatrix} \begin{pmatrix} A_{1\mu} & 0 & 0 \\ 0 & S_{1\mu}(\mu) & 0 \\ 0 & 0 & S_2(\mu) - \mu \end{pmatrix} \begin{pmatrix} I & \hat{G}_1(\mu) & \hat{G}_2(\mu) \\ 0 & I & \hat{G}_3(\mu) \\ 0 & 0 & I \end{pmatrix}$$

$$+ \begin{pmatrix} [A_1 - (\mu+1)]P_\mu & 0 & 0 \\ 0 & (S_1(\mu) - (\mu+1))P'_\mu & 0 \\ 0 & 0 & 0 \end{pmatrix}.$$

The external operators $\mathcal{G}_1(\mu)$ and $\mathcal{G}_2(\mu)$ are boundedly invertible and

$$\begin{pmatrix} [A_1 - (\mu+1)]P_\mu & 0 & 0 \\ 0 & (S_1(\mu) - (\mu+1))P'_\mu & 0 \\ 0 & 0 & 0 \end{pmatrix}$$

is a finite rank operator, hence $\mathcal{A}_0 - \mu$ is closable if, and only if, $S_2(\mu)$ is closable. Q.E.D.

11.3.2 Rakočević and Schmoeger Essential Spectra of $\mathcal{A}$

Having obtained the closure $\mathcal{A}$ of the operator $\mathcal{A}_0$, in this section we discuss its essential spectra. As a first step we prove the following stability lemma.

Lemma 11.3.7.

(*i*) *If $F_1(\mu) \in \mathcal{F}_+^b(X,Y)$, $F_2(\mu) \in \mathcal{F}_+^b(X,Z)$ and $F_3(\mu) \in \mathcal{F}_+^b(Y,Z)$ then $\sigma_{e7}(S_1(\mu))$ and $\sigma_{e7}(\overline{S}_2(\mu))$ does not depend on the choice of μ.*

(*ii*) *If $F_1(\mu) \in \mathcal{F}_-^b(X,Y)$, $F_2(\mu) \in \mathcal{F}_-^b(X,Z)$ and $F_3(\mu) \in \mathcal{F}_-^b(Y,Z)$ then $\sigma_{e8}(\overline{S}_2(\mu))$ and $\sigma_{e8}(S_1(\mu))$ does not depend on the choice of μ.* ◇

Proof.

(i) Let $(\lambda,\mu) \in (\rho_b(A_1))^2$, using Eq. (11.3.1) and Theorem 6.3.1 (i), we will have $\sigma_{e7}(S_1(\mu)) = \sigma_{e7}(S_1(\lambda))$. This implies that $\sigma_{e7}(S_1(\mu))$ does not depend on μ. Now, let $\lambda \in \rho_b(A_1) \bigcap \rho_b(S_1(\lambda))$ then from Theorem 6.3.1 (i), we deduce that the difference $\overline{S}_2(\mu) - \overline{S}_2(\lambda) \in \mathcal{F}_+^b(Z)$. Hence by Remark 7.5.1, we infer that $\sigma_{e7}(\overline{S}_2(\mu))$ does not depend on the choice of μ.

(ii) This assertion can be proved in a similar way as (i). Q.E.D.

We will denote for $\mu \in \rho_b(A) \bigcap \rho_b(S_1(\mu))$ by $\mathbb{Q}(\mu)$ the following operator

$$\mathbb{Q}(\mu) := \begin{pmatrix} 0 & \hat{G}_1(\mu) & \hat{G}_2(\mu) \\ F_1(\mu) & F_1(\mu)\hat{G}_1(\mu) & F_1(\mu)\hat{G}_2(\mu) + \hat{G}_3(\mu) \\ F_2(\mu) & F_2(\mu)\hat{G}_1(\mu) + F_3(\mu) & F_2(\mu)\hat{G}_2(\mu) + F_3(\mu)\hat{G}_3(\mu) \end{pmatrix}.$$

Theorem 11.3.2. *Suppose that the assumptions* ($\mathcal{O}1$)–($\mathcal{O}13$) *are satisfied.*

(*i*) *If for some* $\mu \in \rho_b(A_1)\bigcap \rho_b(S_1(\mu))$, *we have* $F_1(\mu) \in \mathcal{F}_+^b(X,Y)$, $F_2(\mu) \in \mathcal{F}_+^b(X,Z)$, $F_3(\mu) \in \mathcal{F}_+^b(Y,Z)$ *and* $\mathbb{Q}(\mu) \in \mathcal{F}_+(X\times Y\times Z)$, *then* $\sigma_{e7}(\mathcal{A}) \subseteq \sigma_{e7}(A_1)\bigcup\sigma_{e7}(S_1(\mu))\bigcup\sigma_{e7}(\overline{S}_2(\mu))$. *If in the addition we suppose that the sets* $\Phi_{\mathcal{A}}$, Φ_{A_1}, $\Phi_{S_1(\mu)}$ *and* $\Phi_{\overline{S}_2(\mu)}$ *are connected and the sets* $\rho(\overline{S}_2(\mu))$ *and* $\rho(\mathcal{L})$ *are not empty, then* $\sigma_{e7}(\mathcal{A}) = \sigma_{e7}(A_1)\bigcup\sigma_{e7}(S_1(\mu))\bigcup\sigma_{e7}(\overline{S}_2(\mu))$.

(*ii*) *If for some* $\mu \in \rho(A_1)\bigcap \rho(S_1(\mu))$, *we have* $F_1(\mu) \in \mathcal{F}_-^b(X,Y)$, $F_2(\mu) \in \mathcal{F}_-^b(X,Z)$, $F_3(\mu) \in \mathcal{F}_-^b(Y,Z)$ *and* $\mathbb{Q}(\mu) \in \mathcal{F}_-(X\times Y\times Z)$, *then* $\sigma_{e8}(\mathcal{A}) \subseteq \sigma_{e8}(A_1)\bigcup\sigma_{e8}(S_1(\mu))\bigcup\sigma_{e8}(\overline{S}_2(\mu))$. *If in the addition we suppose that the sets* $\Phi_{\mathcal{A}}$, Φ_{A_1}, $\Phi_{S_1(\mu)}$ *and* $\Phi_{\overline{S}_2(\mu)}$ *are connected and the sets* $\rho(\overline{S}_2(\mu))$ *and* $\rho(\mathcal{A})$ *are not empty, then* $\sigma_{e8}(\mathcal{A}) = \sigma_{e8}(A_1)\bigcup\sigma_{e8}(S_1(\mu))\bigcup\sigma_{e8}(\overline{S}_2(\mu))$. ◇

Proof.

(i) Fix $\lambda \in \rho_b(A_1)\bigcap\rho_b(S_1(\mu))$. Then, for $\mu \in \mathbb{C}$ we have

$$\mathcal{A}-\lambda I = \mathcal{G}_1(\mu)\mathbb{V}(\lambda)\mathcal{G}_2(\mu) + (\lambda-\mu)\mathbb{Q}(\mu) + \mathbb{P}(\mu) + \mathbb{N}(\mu).$$

The matrices-operators $\mathbb{V}(\lambda)$ and $\mathbb{P}(\lambda)$ are defined by

$$\mathbb{V}(\lambda) = \begin{pmatrix} A-\lambda & 0 & 0 \\ 0 & S_1(\mu)-\lambda & 0 \\ 0 & 0 & \overline{S}_2(\mu)-\lambda \end{pmatrix},$$

$$\mathbb{P}(\mu) = \begin{pmatrix} P_{11} & P_{12} & P_{13} \\ P_{21} & P_{22} & P_{23} \\ P_{31} & P_{32} & P_{33} \end{pmatrix},$$

where

- $P_{11} = [A_1-(\mu+1)]P_\mu$,
- $P_{12} = [A_1-(\mu+1)]P_\mu\hat{G}_1(\mu)$,
- $P_{13} = [A_1-(\mu+1)]P_\mu\hat{G}_2(\mu)$,
- $P_{21} = F_1(\mu)[A_1-(\mu+1)]P_\mu$,
- $P_{22} = F_1(\mu)[A_1-(\mu+1)]P_\mu\hat{G}_1(\mu) + [S_1(\mu)-(\mu+1)]P_\mu^{'}$,
- $F_{23} = F_1(\mu)[A_1-(\mu+1)]P_\mu\hat{G}_2(\mu) + [S_1(\mu)-(\mu+1)]P_\mu^{'}F_1(\mu)\hat{G}_3(\mu)$,
- $P_{31} = F_2(\mu)[A_1-(\mu+1)]P_\mu$,
- $P_{32} = F_2(\mu)[A_1-(\mu+1)]P_\mu\hat{G}_1(\mu) + F_3(\mu)[S_1(\mu)-(\mu+1)]P_\mu^{'}$,
- $P_{33} = F_2(\mu)[A_1-(\mu+1)]P_\mu\hat{G}_2(\mu) + F_3(\mu)[S_1(\mu)-(\mu+1)]P_\mu^{'}\hat{G}_3(\mu)$.

Since $\mathcal{G}_1(\lambda)$ and $\mathcal{G}_2(\lambda)$ are bounded and have bounded inverses, $\mathbb{N}(\lambda)$ and $\mathbb{P}(\lambda)$ are finite rank matrices operators and $\mathbb{Q}(\lambda) \in \mathcal{F}_+(X\times Y\times Z)$, therefore $(\mathcal{A}-\mu I)$ is an upper semi-Fredholm operator if only if $\mathbb{V}(\mu)$ has this property and

$$i(\mathcal{A}-\mu I) = i(A_1-\mu) + i(S_1(\mu)-\mu) + i(\overline{S}_2(\mu)-\mu).$$

This shows that $\sigma_{e7}(\mathcal{A}) = \sigma_{e7}(A_1)\bigcup\sigma_{e7}(S_1(\mu))\bigcup\sigma_{e7}(\overline{S}_2(\mu))$. Since $\Phi_{\mathcal{A}}$, Φ_{A_1}, $\Phi_{S_1(\mu)}$ and $\Phi_{\overline{S}_2(\mu)}$ are connected and the sets $\rho(\overline{S}_2(\mu))$ and $\rho(\mathcal{A})$ are not empty, then using Theorems 7.3.1 and 7.5.11, we completed the proof of (i).

(ii) A same reasoning allows us to reach the result (ii). Q.E.D.

11.4 Perturbations of Unbounded Fredholm Linear Operators

Let us denote by $\mathcal{T}$, $\mathcal{S}$, and $\mathcal{K}$ by

$$\mathcal{T} := \begin{pmatrix} A & 0 & 0 \\ 0 & E & 0 \\ 0 & 0 & K \end{pmatrix}, \quad \mathcal{S} := \begin{pmatrix} 0 & B & 0 \\ 0 & 0 & F \\ G & 0 & 0 \end{pmatrix} \text{ and } \mathcal{K} := \begin{pmatrix} 0 & 0 & C \\ D & 0 & 0 \\ 0 & H & 0 \end{pmatrix}.$$

Then, it is clear that $\mathcal{A} = \mathcal{T} + \mathcal{S} + \mathcal{K}$.

11.4.1 The Operator $\mathcal{A}$ and Its Closure

Lemma 11.4.1.

(*i*) *If B is E-bounded with E-bound δ_1, F is K-bounded with K-bound δ_2, and G is A-bounded with A-bound δ_3, then $\mathcal{S}$ is $\mathcal{T}$-bounded with $\mathcal{T}$-bound $\delta = \max\{\delta_1, \delta_2, \delta_3\}$.*

(*ii*) *If C is F-bounded with F-bound δ_1, D is G-bounded with G-bound δ_2, and H is A-bounded with A-bound δ_3, then $\mathcal{K}$ is $\mathcal{S}$-bounded with $\mathcal{S}$-bound $\delta = \max\{\delta_1, \delta_2, \delta_3\}$.* ◇

Proof.

(i) Let $\varepsilon > 0$. By using the assumptions and Remark 2.1.3, there exist constants $a_1, a_2, a_3, b_1, b_2, b_3 \geq 0$ such that $\delta_1 \leq b_1 < \delta_1 + \varepsilon$, $\delta_2 \leq b_2 < \delta_2 + \varepsilon$, $\delta_3 \leq b_3 < \delta_3 + \varepsilon$ and

$$\|Bg\|^2 \leq a_1^2\|g\|^2 + b_1^2\|Eg\|^2 \text{ for all } g \in \mathcal{D}(E) \subset \mathcal{D}(B),$$

$$\|Fh\|^2 \leq a_2^2\|h\|^2 + b_2^2\|Kh\|^2 \text{ for all } h \in \mathcal{D}(K) \subset \mathcal{D}(F),$$

$$\|Gf\|^2 \leq a_3^2\|f\|^2 + b_3^2\|Af\|^2 \text{ for all } f \in \mathcal{D}(A) \subset \mathcal{D}(G).$$

Hence, for $(f, g, h) \in \mathcal{D}(A) \times \mathcal{D}(E) \times \mathcal{D}(K)$, we get

$$\left\| \begin{pmatrix} 0 & B & 0 \\ 0 & 0 & F \\ G & 0 & 0 \end{pmatrix} \begin{pmatrix} f \\ g \\ h \end{pmatrix} \right\|^2 = \|Bg\|^2 + \|Fh\|^2 + \|Gf\|^2$$

and

$$\left\| \begin{pmatrix} A & 0 & 0 \\ 0 & E & 0 \\ 0 & 0 & K \end{pmatrix} \begin{pmatrix} f \\ g \\ h \end{pmatrix} \right\|^2 = \|Af\|^2 + \|Eg\|^2 + \|Kh\|^2.$$

Then,

$$\begin{aligned} \left\| \begin{pmatrix} 0 & B & 0 \\ 0 & 0 & F \\ G & 0 & 0 \end{pmatrix} \begin{pmatrix} f \\ g \\ h \end{pmatrix} \right\|^2 &= \|Bg\|^2 + \|Fh\|^2 + \|Gf\|^2 \\ &\leq a_3^2\|f\|^2 + b_3^2\|Gf\|^2 + a_1^2\|g\|^2 \\ &\leq \eta \left\| \begin{pmatrix} f \\ g \\ h \end{pmatrix} \right\|^2 + \chi \left\| \begin{pmatrix} A & 0 & 0 \\ 0 & B & 0 \\ 0 & 0 & C \end{pmatrix} \begin{pmatrix} f \\ g \\ h \end{pmatrix} \right\|^2 \end{aligned}$$

where $\eta = \max\{a_1, a_2, a_3\}^2$ and $\chi = \max\{b_1, b_2, b_3\}^2$. Since $\max\{b_1, b_2, b_3\} = \max\{\delta_1 + \varepsilon, \delta_2 + \varepsilon, \delta_3 + \varepsilon\} = \delta + \varepsilon$. This shows that $\mathcal{S}$ is $\mathcal{T}$-bounded with $\mathcal{T}$-bound $< \delta$.

(ii) The proof may be checked in the same way as the proof of (a) Q.E.D.

Theorem 11.4.1. *If $\mathcal{D}(E) \subset \mathcal{D}(B)$, $\mathcal{D}(K) \subset \mathcal{D}(F)$, $\mathcal{D}(A) \subset \mathcal{D}(G)$,$\mathcal{D}(F) \subset \mathcal{D}(C)$, $\mathcal{D}(G) \subset \mathcal{D}(D)$, $\mathcal{D}(B) \subset \mathcal{D}(H)$, and if*

$$\begin{aligned} &\|B\varphi\| \leq a_1\|\varphi\| + b_1\|E\varphi\| \ \textit{for all}\ \varphi \in \mathcal{D}(E), \\ &\|F\varphi\| \leq a_2\|\varphi\| + b_2\|K\varphi\|^2 \ \textit{for all}\ \varphi \in \mathcal{D}(K), \\ &\|G\varphi\| \leq a_3\|\varphi\| + b_3\|A\varphi\| \ \textit{for all}\ \varphi \in \mathcal{D}(A), \\ &\|C\varphi\| \leq a_4\|\varphi\| + b_4\|F\varphi\| \ \textit{for all}\ \varphi \in \mathcal{D}(F), \\ &\|D\varphi\| \leq a_5\|\varphi\| + b_5\|G\varphi\|^2 \ \textit{for all}\ \varphi \in \mathcal{D}(G), \ \textit{and} \\ &\|H\varphi\| \leq a_6\|\varphi\| + b_6\|B\varphi\| \ \textit{for all}\ \varphi \in \mathcal{D}(B), \end{aligned}$$

such that $\max\{b_1, b_2, b_3\}\big(1 + \max\{a_4, a_5, a_6\}\big) < 1$, *then $\mathcal{A}$ is closed if, and only if, A, E, and K are closed.* ◇

Proof. Since $\mathcal{D}(\mathcal{T}) = \mathcal{D}(A) \times \mathcal{D}(E) \times \mathcal{D}(K)$, $\mathcal{D}(\mathcal{S}) = \mathcal{D}(G) \times \mathcal{D}(B) \times \mathcal{D}(F)$, and $\mathcal{D}(\mathcal{K}) = \mathcal{D}(D) \times \mathcal{D}(H) \times \mathcal{D}(C)$. Then, $\mathcal{D}(\mathcal{T}) \subset \mathcal{D}(\mathcal{S}) \subset \mathcal{D}(\mathcal{K})$. Moreover, we have

$$\|\mathcal{S}\varphi\|^2 \leq \eta_1^2 \|\varphi\|^2 + \chi_1^2 \|\mathcal{T}\varphi\|^2, \ \ \varphi \in \mathcal{D}(\mathcal{T})$$

$$\|\mathcal{K}\varphi\|^2 \leq \eta_2^2 \|\varphi\|^2 + \chi_2^2 \|\mathcal{S}\varphi\|^2, \ \ \varphi \in \mathcal{D}(\mathcal{S})$$

where

$$\eta_1 = \max\left\{\sqrt{a_1^2 + a_1 b_1},\ \sqrt{a_2^2 + a_2 b_2},\ \sqrt{a_3^2 + a_3 b_3}\right\},$$

$$\chi_1 = \max\left\{\sqrt{b_1^2 + a_1 b_1},\ \sqrt{b_2^2 + a_2 b_2},\ \sqrt{b_3^2 + a_3 b_3}\right\},$$

$$\eta_2 = \max\left\{\sqrt{a_4^2 + a_4 b_4},\ \sqrt{a_5^2 + a_5 b_5},\ \sqrt{a_6^2 + a_6 b_6}\right\},$$

$$\chi_2 = \max\left\{\sqrt{b_4^2 + a_4 b_4},\ \sqrt{b_5^2 + a_5 b_5},\ \sqrt{b_6^2 + a_6 b_6}\right\}.$$

From Remark 2.1.3, it follows that

$$\|S\varphi\| \leq \max\{a_1, a_2, a_3\}\, \|\varphi\| + \max\{b_1, b_2, b_3\}\, \|T\varphi\| \quad \varphi \in \mathcal{D}(T),$$

$$\|K\varphi\| \leq \max\{a_4, a_5, a_6\}\, \|\varphi\| + \max\{b_4, b_5, b_6\}\, \|S\varphi\| \quad \varphi \in \mathcal{D}(S),$$

where $\max\{b_1, b_2, b_3\}\big(1 + \max\{a_4, a_5, a_6\}\big) < 1$. Now, by using Theorem 2.1.5, we have $\mathcal{A}$ is closed if, and only if, $\mathcal{T}$ is closed if, and only if, A, E and K are closed. Q.E.D.

Now, we introduce the following example. We defined the block operator matrices $\mathcal{A}$ and $\mathcal{B}$ in $L_2(\mathbb{R}^3) \otimes L_2(\mathbb{R}^3) \otimes L_2(\mathbb{R}^3)$ by

$$\mathcal{A} = \begin{pmatrix} \Delta & 0 & 0 \\ 0 & S_0 + 2V & 0 \\ 0 & 0 & \Delta \end{pmatrix} \quad \text{and } \mathcal{B} = \begin{pmatrix} 0 & V & 0 \\ 0 & 0 & -V \\ V & 0 & 0 \end{pmatrix},$$

where S_0 the minimal Schrödinger operator defined in

$$C_0^\infty(\mathbb{R}^3) := \left\{\varphi \in C^\infty(\mathbb{R}^3) : \ \text{supp}(\varphi) \text{ is bounded}\right\},$$

where $\text{supp}(\varphi) := \left\{x \in \mathbb{R}^3 : \varphi(x) \neq 0\right\}$ and the potential $V \in L_{2,\text{loc}}(\mathbb{R}^3)$ is given by $V(x) := -\dfrac{\alpha}{|x|}$ for some constant $\alpha > 0$ (hydrogen atom with Coulomb interaction). Let $\varphi \in L_2(\mathbb{R}^3)$ then

$$\begin{aligned} \left\| V\varphi \right\|^2 &= \alpha^2 \int \frac{|\varphi(x)|^2 dx}{|x|^2} \\ &\leq 4\alpha^2 \int |\nabla\varphi(x)|^2 dx. \end{aligned}$$

It is not difficult to prove that

$$\|V\varphi\|^2 \leq 2\alpha^2\varepsilon\|\Delta\varphi\|^2 + \frac{2\alpha^2}{\varepsilon}\|\varphi\|^2, \text{ for all } \varphi \in C_0^\infty(\mathbb{R}^3),$$

can be formulated as the following inequality:

$$\|V\varphi\|^2 \leq \left(\frac{\theta}{\varepsilon}\right)^2 \|\varphi\|^2 + \theta^2\|\Delta\varphi\|^2, \text{ for all } \varphi \in C_0^\infty(\mathbb{R}^3), \tag{11.4.1}$$

where $\theta = \alpha\sqrt{2\varepsilon}$, $0 < \varepsilon < \frac{1}{4\alpha^2}$ and $\alpha > 0$. Hence, one can check easily that, for all $\varphi \in C_0^\infty(\mathbb{R}^3)$

$$\|V\varphi\|^2 \leq \left(\frac{\theta}{\varepsilon\sqrt{(1-2\theta^2)}}\right)^2 \|\varphi\|^2 + \left(\frac{\sqrt{2}\theta}{\sqrt{1-2\theta^2}}\right)^2 \|(S_0 + 2V)\varphi\|^2. \tag{11.4.2}$$

For each $(\varphi_1, \varphi_2, \varphi_3)^t \in L_2(\mathbb{R}^3) \otimes L_2(\mathbb{R}^3) \otimes L_2(\mathbb{R}^3)$ we have

$$\left\| \begin{pmatrix} 0 & V & 0 \\ 0 & 0 & -V \\ V & 0 & 0 \end{pmatrix} \begin{pmatrix} \varphi_1 \\ \varphi_2 \\ \varphi_3 \end{pmatrix} \right\|^2 = \|V\varphi_2\|^2 + \|-V\varphi_3\|^2 + \|V\varphi_1\|^2.$$

By (11.4.1) and (11.4.2) we observe that

$$\left\| \begin{pmatrix} 0 & V & 0 \\ 0 & 0 & -V \\ V & 0 & 0 \end{pmatrix} \begin{pmatrix} \varphi_1 \\ \varphi_2 \\ \varphi_3 \end{pmatrix} \right\|^2 \leq \gamma^2 \left\| \begin{pmatrix} \varphi_1 \\ \varphi_2 \\ \varphi_3 \end{pmatrix} \right\|^2 + \eta^2 \left(\|\Delta\varphi_1\|^2 + \|(S_0 + V)\varphi_2\|^2 + \|\Delta\varphi_3\|^2\right),$$

where, $\gamma = \max\{\frac{\theta}{\varepsilon}, \frac{\theta}{\varepsilon}, \frac{\theta}{\varepsilon\sqrt{(1-2\theta^2)}}\}$ and $\eta = \max\{\theta, \theta, \frac{\sqrt{2}\theta}{\sqrt{1-2\theta^2}}\}$. Now, applying Lemma 11.4.1, we obtain $\mathcal{B}$ is $\mathcal{A}$-bounded with $\mathcal{A}$-bound $< \delta$, where $\delta < \frac{\sqrt{2}\theta}{\sqrt{1-2\theta^2}}$.

11.4.2 Index of $\mathcal{A}$

Theorem 11.4.2. *Let $\tilde{\mathcal{T}}$ be the bijection associated with $\mathcal{T}$. If we suppose that the conditions of Theorem 11.4.1 are satisfied, and we suppose that $\alpha_1(1 + \beta_1) < 1$ and $(1 + \beta_1) < \gamma(\tilde{\mathcal{T}})$ where the parameters $\alpha_1 = \max\{\max\{a_1, a_2, a_3\}, \max\{b_1, b_2, b_3\}\}$ and $\beta_1 = \max\{\max\{a_4, a_5, a_6\}, \max\{b_4, b_5, b_6\}\}$. Then, if $\mathcal{T}$ is Fredholm operator, then the sum $\mathcal{A}$ is Fredholm which satisfies $\alpha(\mathcal{A}) \leq \alpha(\mathcal{T})$, $\beta(\mathcal{A}) \leq \beta(\mathcal{T})$ and $i(\mathcal{A}) = i(A) + i(E) + i(K)$.* ◇

Proof. From Theorems 6.2.1 and 11.4.1, it follows that $i(\mathcal{A}) = i(\mathcal{T})$. Moreover,

$$\begin{pmatrix} A & 0 & 0 \\ 0 & E & 0 \\ 0 & 0 & K \end{pmatrix} = \begin{pmatrix} A & 0 & 0 \\ 0 & I & 0 \\ 0 & 0 & I \end{pmatrix} \begin{pmatrix} I & 0 & 0 \\ 0 & E & 0 \\ 0 & 0 & I \end{pmatrix} \begin{pmatrix} I & 0 & 0 \\ 0 & I & 0 \\ 0 & 0 & K \end{pmatrix}$$

$i(\mathcal{A}) = i(\mathcal{T}) = i(A) + i(E) + i(K)$. Q.E.D.

Open question:

1. Can we extend the results obtained in Chaps. 10 and 11 for a block operator matrix having the following form

$$\mathcal{A} = \begin{pmatrix} A_{11} & \cdots & A_{1n} \\ \vdots & \ddots & \vdots \\ A_{n1} & \cdots & A_{nn} \end{pmatrix},$$

   where $n \geq 4$?
2. Sufficient conditions are obtained for the Fredholomness of the algebraic conditions of three linear operators (see Theorem 6.2.1). Such a result generalizes some results for the linear operator, see for example in [124, Theorem 4.2, Chap. XVII]. However, if A_1 is a Fredholm linear operator and $A_2,\ A_3, \ldots, A_n$ for $n \geq 4$ are (possible unbounded) linear operators such that

$$\mathcal{D}(A_1) \subset \mathcal{D}(A_2) \subset \ldots \mathcal{D}(A_{n-1}) \subset \mathcal{D}(A_n)$$

$$\|A_{k+1}\varphi\| \leq \alpha_k \Big(\|\varphi\| + \|A_k\varphi\| \Big) \text{ for } k = 1, \ldots n-1 \text{ and } \varphi \in \mathcal{D}(A_1).$$

   It is not clear which additional conditions should be put on the linear operators A_k and the scalars α_k for $k = 1, \ldots, n$, so that the algebraic $\sum_{k=1}^{n} A_k$ is a Fredholm linear operator. This question will be left as an open question. The answer to this question could help in investigating some properties of an $n \times n$ matrix of linear operators $\mathcal{A}$, with $n \geq 4$? □

Chapter 12
Spectral Graph Theory

12.1 Line graph

The concept of the line graph of a given graph is so natural that it has been independently discovered by many authors. Of course, each author gave it a different name: It was called the interchange graph by Ore [272], derivative by H. Sachs [297], derived graph by L. W. Beineke [52], edge-to-vertex dual by M. Reed [291], coverning graph by G. Kirchhoff [189], and adjoint by V. Menon [247]. Various characterizations of line graphs were developed. We also introduce the concept of total graph, which was first studied by M. Belzad [54]. This concept has been discovered only once. That's why, this concept has no other names. The relationships between line graphs and total graphs are usually studied with a particular emphasis on Eulerian and Hamiltonian graphs. Let $\mathcal{V}$ be a countable set and $\mathcal{E} : \mathcal{V} \times \mathcal{V} \longrightarrow [0, +\infty[$. We assume that $\mathcal{E}(x, y) = \mathcal{E}(y, x)$, for all x, $y \in \mathcal{V}$. Let $\widetilde{\mathcal{V}} := \{(x, y) \in \mathcal{V}^2$ such that $x \sim y\}$. The line graph of G is the graph $\widetilde{G} = (\widetilde{\mathcal{E}}, \widetilde{\mathcal{V}})$, where $\widetilde{\mathcal{E}}((x_0, y_0), (x, y)) = \mathcal{E}(x_0, y)1_{\{y \neq y_0\}}(y) + \mathcal{E}(x, y_0)1_{\{x \neq x_0\}}(x)$, where $1_{\{y \neq y_0\}}(.)$ denotes the characteristic function of $\{y \neq y_0\}$.

A. Jeribi, *Spectral Theory and Applications of Linear Operators and Block Operator Matrices*, DOI 10.1007/978-3-319-17566-9_12

Example 12.1.1.

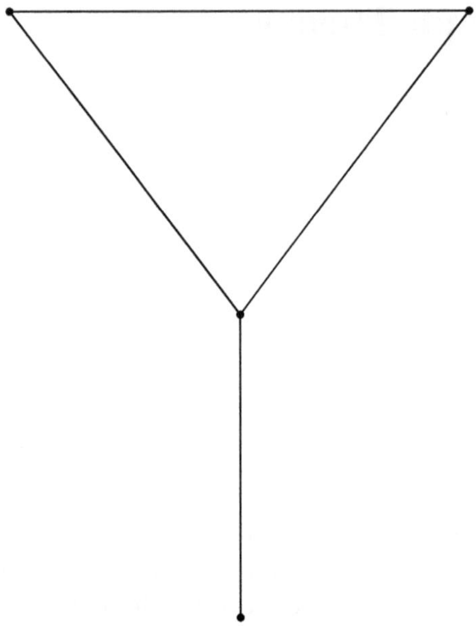

A graph G

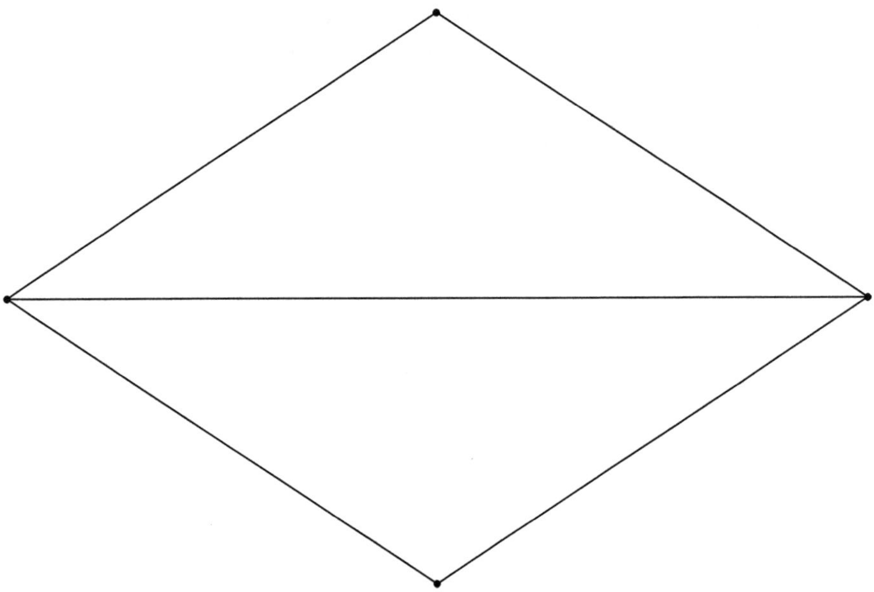

A line graph $\tilde{G}$ of a graph G

If $x = (u, v)$ is a line of G, then the degree of x in $\widetilde{G}$ is clearly $d_{\widetilde{G}}(x) = d_G(u) + d_G(v) - 2$, where $d_G(.)$ is defined in (2.15.1). Given a graph $G = (\mathcal{E}, \mathcal{V})$, the set of 1-cochains (or 1-forms) is given by $\mathcal{C}_1(G) := \{f : \mathcal{V} \times \mathcal{V} \longrightarrow \mathbb{C}$ such that $f(x, y) =$

$-f(y,x)$ for all $x, y \in \mathcal{V}\}$. The subset of $\mathcal{C}_1(G)$ with finite support is denoted by $\mathcal{C}_1^c(G)$. The adjacency matrix $\mathcal{A}_{\widetilde{G}}$ on $\widetilde{G}$ is defined by

$$\mathcal{A}_{\widetilde{G}} f(x_0, y_0) = \sum_{x \sim y_0,\ x \neq x_0} \mathcal{E}(x, y_0) f(x, y_0) + \sum_{y \sim x_0,\ y \neq y_0} \mathcal{E}(x_0, y) f(x_0, y), \ f \in \mathcal{C}_1^c(G).$$

It is complicated to prove that an adjacency matrix is bounded from below. The construction in [129] allows to obtain graphs such that their adjacency matrix is bounded from below but is not essentially self-adjoint. Let $\mathcal{E} := \{(x, y) \in \mathcal{V}^2$ such that $\mathcal{E}(x, y) \neq 0\}$. Choosing an orientation consists in defining a partition of $\mathcal{E} : \mathcal{E} = \mathcal{E}^+ \bigcup \mathcal{E}^-$, $(x, y) \in \mathcal{E}^+$ if, and only if, $(y, x) \in \mathcal{E}^-$. For $z = (x, y) \in \mathcal{E}$, we denote $z^+ = y$ and $z^- = x$. The 0−cochains $\mathcal{C}_0(G)$ are scalar functions on $\mathcal{V}$. The set of 0-cochains with a finite support is denoted by $\mathcal{C}_0^c(G)$. We define the adjacency matrix on G by $\mathcal{A}_G f(x) = \sum_{y \sim x} \mathcal{E}(x, y) f(y)$. We associate to $G = (\mathcal{E}, \mathcal{V})$ the complex Hilbert space

$$\ell^2(\mathcal{E}) := \left\{ f \in \mathcal{C}_1(G) \text{ such that } \|f\|^2 := \frac{1}{2} \sum_{x,y \in \mathcal{V}} \mathcal{E}(x, y) |f(x, y)|^2 < \infty \right\}.$$

The associated scalar product is given by $\langle f, g \rangle := \frac{1}{2} \sum_{x,y \in \mathcal{V}} \mathcal{E}(x, y) \overline{f(x, y)} g(x, y)$, for $f, g \in \ell^2(\mathcal{E})$.

12.2 Operators on Graphs

The difference operator is defined as $d : \mathcal{C}_0^c(G) \longrightarrow \mathcal{C}_1^c(G)$, $d(f)(x, y) = f(y) - f(x)$. The coboundary operator is, by definition, the formal adjoint of d, i.e., $d^* : \mathcal{C}_1^c(G) \longrightarrow \mathcal{C}_0^c(G)$, $d^*(f)(x) = \sum_{y \in \mathcal{V}} \mathcal{E}(y, x) f(y, x)$. The Gauß-Bonnet operator is defined on $\mathcal{C}_0^c(G) \oplus \mathcal{C}_1^c(G)$ by $D := d + d^* \cong \begin{pmatrix} 0 & d^* \\ d & 0 \end{pmatrix}$. This operator is of Dirac type. The associated Laplacian is defined as $\Delta = D^2 = \Delta_0 \oplus \Delta_1$, where Δ_0 is the standard discrete Laplacian acting on 0-forms defined by

$$\Delta_0(f)(x) := \sum_{y \sim x} \mathcal{E}(x, y)(f(y) - f(x)), \text{ with } f \in \mathcal{C}_0^c(G)$$

and the so-called adjacency matrix: $\mathcal{A}_G(f)(x) := \sum_{y \sim x} \mathcal{E}(x, y) f(y)$, with $f \in \mathcal{C}_0^c(G)$. The Δ_1 is the discrete Laplacian acting on 1-forms defined by

$$\Delta_1(f)(x, y) := \sum_{z \in \mathcal{V}} \mathcal{E}(x, z) f(x, z) + \sum_{z \in \mathcal{V}} \mathcal{E}(z, y) f(z, y), \text{ with } f \in \mathcal{C}_1^c(G).$$

This operator is symmetric and thus closable. We denote its closure by Δ_1, its domain by $\mathcal{D}(\Delta_1)$, and its adjoint by Δ_1^*.

12.3 Lower Local Complexity

Now, let us deal with bounded weights $\mathcal{E}$ and will restrict to the case $\mathcal{E}$ bounded from below. Suppose also that d_G is unbounded. Let $\kappa_d(G)$ be the filter generated by $\{x \in \mathcal{V}, \text{ such that } d_G(x) \geq n\}$, with $n \in \mathbb{N}$. We introduce the lower local complexity of a graph G by:

$$\begin{aligned} C_{\text{loc}}(G) &:= \lim \inf_{x \to \kappa_d(G)} \frac{N_G(x)}{d_G^2(x)} \\ &= \inf \bigcap \left\{ \overline{\left\{ \frac{N_G(x)}{d_G^2(x)}, \ x \in \mathcal{V} \text{ and } d_G(x) \geq n \right\}}, \ n \in \mathbb{N} \right\} \end{aligned} \tag{12.3.1}$$

where $N_G(x) = \sharp\{x - \text{triangles}\}$ ($\sharp S$ denotes the numbers of elements in the set S) and the x-triangle given by (x, y, z, x) is different from the one given by (x, z, y, x). The sub-lower local complexity of a graph G is defined by

$$C_{\text{loc}}^{\text{sub}}(G) := \inf_{\{G' \subset G, \ \sup d_{G'} = \infty\}} C_{\text{loc}}(G'). \tag{12.3.2}$$

Let us recall the following result obtained by S. Golénia.

Proposition 12.3.1 ([129, Proposition 3.2]). *Let $G = (\mathcal{E}, \mathcal{V})$ be a locally finite graph and let $\widehat{\mathcal{A}_G}$ be a self-adjoint extension of $\mathcal{A}_G$. Then,*

(i) If $\mathcal{E}$ is not bounded, then the spectrum of $\widehat{\mathcal{A}_G}$ is neither bounded from above nor from below.

(ii) In the sense of inclusion of graphs, we have

$$\sup \sigma(\widehat{\mathcal{A}_G}) \geq \sup_{G' \subset G} \sup_{x \in \mathcal{V}(G')} \left(\frac{1}{\sqrt{d_{G'}(x)}} \sum_{y \sim x, y \in G'} \mathcal{E}(x, y) \right.$$
$$\left. + \frac{1}{2 d_{G'}(x)} \sum_{y \sim x, y \in G'} \sum_{z \sim y, z \sim x, z \in G'} \mathcal{E}(y, z) \right).$$

In particular, if d is not bounded and $\mathcal{E}$ is bounded from below, then the spectrum of $\widehat{\mathcal{A}_G}$ is not bounded from above.

(iii) Suppose that there exists $C > 0$, so that $\inf \sigma(\widehat{\mathcal{A}_G}) \geq -C$. Then, for all $G' \subset G$, we have

$$\frac{1}{C}\left(\sum_{y,x\in V(G')}\mathcal{E}(x,y)\right)^2 \le \sum_{y\sim x,x,y\in V(G')}\ \sum_{z\sim y,z\sim x,z\in V(G')}\mathcal{E}(y,z)+Cd_{G'}(x) \tag{12.3.3}$$

for $x \in G'$. In particular, when $\mathcal{E}$ is with a value in $\{0\}\bigcup[\mathcal{E}_{\min},\mathcal{E}_{\max}]$, with $0<\mathcal{E}_{\min}\le\mathcal{E}_{\max}<\infty$. Recalling Eqs. (12.3.1) and (12.3.2), one obtains

$$\frac{1}{C}\frac{\mathcal{E}_{\min}^2}{\mathcal{E}_{\max}} \le C_{\mathrm{loc}}^{\mathrm{sub}}(G)\le C_{\mathrm{loc}}(G). \tag{12.3.4}$$

◇

Proof. Let G' be a subgraph of G. Fix $x \in \mathcal{V}(G')$ and consider a real-valued function f with support in $\{x\}\bigcup\mathcal{N}_{G'}(x)$. We have

$$\begin{aligned}\langle f,\widehat{\mathcal{A}_G}f\rangle &= f(x)(\mathcal{A}_{G'}f)(x)+\sum_{y\sim x,\ y\in G'} f(y)(\mathcal{A}_{G'}f)(y)\\ &= 2f(x)(\mathcal{A}_{G'}f)(x)+\sum_{y\sim x,\ y\in G'} f(y)\sum_{z\sim y,\ z\sim x\ z\in G'}\mathcal{E}(y,z)f(z).\end{aligned} \tag{12.3.5}$$

We first consider the case. There is a sequence $(x_n,y_n)_n$ of elements of $\mathcal{V}^2$, such that $\mathcal{E}(x_n,y_n)\to\infty$, when n goes to infinity. Take $G'=G$ and $f=f_n$ with support in $\{x_n,y_n\}$ in (12.3.5). We get $\langle f_n,\widehat{\mathcal{A}_G f_n}\rangle = 2\mathcal{E}(x_n,y_n)f(x_n)f(y_n)$. Then, choose $f(y_n)=1$ and $f(x_n)=\pm1$ and let n tend to infinity. For the second case, take $f(x)=1$ and $f(y)=d_{G'}(x)^{-1/2}$ for y neighbor of x in G'. Noting that $\|f\|^2=2$, (12.3.5) establishes the result. Focus finally on the third point. Take $f(x)=1$ and $f(y)=b$ for y neighbor of x in G'. Note that $\|f\|^2=1+d_{G'}(x)b^2$. Now, since $\langle f,\widehat{\mathcal{A}_G f}\rangle\ge -C\|f\|^2$, (12.3.5) entails:

$$2b\sum_{y\sim x,\ y\in G'}\mathcal{E}(x,y)+b^2\sum_{y\sim x,\ y\in G'}\ \sum_{z\sim y,\ z\sim x\ z\in G'}\mathcal{E}(y,z)+C(1+d_{G'}(x)b^2)\ge 0,$$

for all $b\in\mathbb{R}$. Thus, the discriminant of this polynomial in b is non-positive. This gives directly (12.3.3). In turn, this infers:

$$\frac{1}{C}\frac{\mathcal{E}_{\min}^2}{\mathcal{E}_{\max}}\le\frac{N_{G'}(x)}{d_{G'}^2(x)}+\frac{1}{\mathcal{E}_{\max}}\frac{C}{d_{G'}(x)}.$$

The statement (12.3.4) follows right away by taking the limit inferior with respect to the filter $\kappa_d(G')$. Q.E.D.

Consider G a simple graph. If a graph G has a subgraph, being $\bigcup_{n\geq 0} S_{u_n}$ for some sequence $(u_n)_n$ that tends to infinity, then $C_{\text{loc}}^{\text{sub}}(G) = 0$. Here, $S_n = (\mathcal{E}_n, \mathcal{V}_n)$ denotes the star graph of order n, i.e., $|\mathcal{V}_n| = n$ and there is $x_0 \in \mathcal{V}_n$ so that $\mathcal{E}(x, x_0) = 1$ for all $x \neq x_0$ and $\mathcal{E}(x, y) = 0$ for all $x \neq x_0$ and $y \neq x_0$.

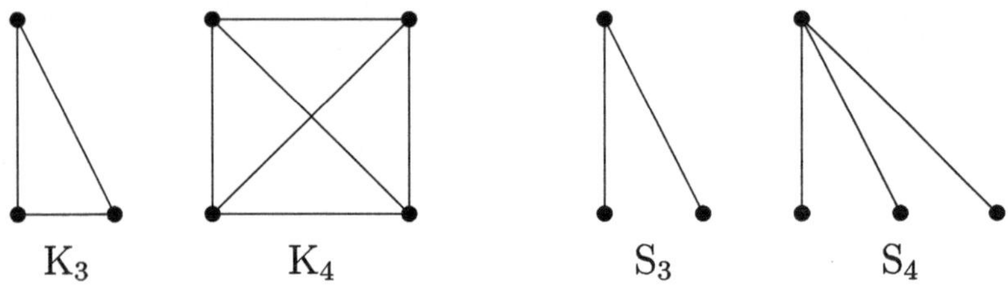

(12.3.6)

We recall the definition of $K_n := (\mathcal{E}_n, \mathcal{V}_n)$ the complete graph of n elements: $\mathcal{V}_n$ is a set of n elements and $\mathcal{E}(a, b) = 1$ for all $a, b \in \mathcal{V}_n$, so that $a \neq b$. One has $N_{K_n}(x)/d^2_{K_n}(x) = (n-1)(n-2)/n^2$, for all $x \in \mathcal{V}_n$. Therefore, one can hope to increase the lower local complexity by having a lot of complete graphs as subgraph. More precisely, it is possible that $C_{\text{loc}}(G)$ is positive, whereas $C_{\text{loc}}^{\text{sub}}(G) = 0$.

Lemma 12.3.1 ([129]). *Let $M \geq 1$. Given a sequence of graphs $G_n = (\mathcal{E}_n, \mathcal{V}_n)$, for $n \in \mathbb{N}$. Choose $x_n \in \mathcal{V}_n$. Let $G^\circ := (\mathcal{E}^\circ, \mathcal{V}^\circ) := \bigcup_{n\in\mathbb{N}} G_n$ be the disjoint union of $\{G_n\}_n$. Set $G := (\mathcal{E}, \mathcal{V})$ with $\mathcal{V} = \mathcal{V}^\circ$ and with $\mathcal{E}(x, y) := \mathcal{E}^\circ(x, y)$, when there is $n \in \mathbb{N}$ so that $x, y \in \mathcal{V}_n$ and where $\sup_{n\in\mathbb{N}} \sum_{m\in\mathbb{N}} \mathcal{E}(x_n, x_m) \leq M$.*

(i) We have $\|(\mathcal{A}_G - \mathcal{A}_{G^\circ})f\| \leq M \sup_{n,m} \mathcal{E}(x_n, x_m)\|f\|$, for all $f \in C_0^c(G) = C_0^c(G^\circ)$.
(ii) The deficiency indices of $\mathcal{A}_G$ are equal to $\eta(\mathcal{A}_G) = \sum_{n\in\mathbb{N}} \eta(\mathcal{A}_{G_n})$.
(iii) In particular, if G_n are all finite graphs, then $\mathcal{A}_G$ is essentially self-adjoint on $C_0^c(G)$. ◇

Proof. We start with the first point. Observe that each x_m has at most M neighbors in $(x_n)_{n\in\mathbb{N}}$. Then,

$$\begin{aligned}
\|(\mathcal{A}_G - \mathcal{A}_{G^\circ})f\|^2 &= \sum_{n\in\mathbb{N}} |((\mathcal{A}_G - \mathcal{A}_{G^\circ})f)(x_n)|^2 \\
&= \sum_{n\in\mathbb{N}} \left| \sum_{m\in\mathbb{N}\setminus\{n\}} \mathcal{E}(x_n, x_m) f(x_m) \right|^2 \\
&\leq M \sum_{n\in\mathbb{N}} \sum_{m\in\mathbb{N}\setminus\{n\}} \mathcal{E}^2(x_n, x_m)|f(x_m)|^2 \\
&\leq M^2 \sup_{n,m} \mathcal{E}^2(x_n, x_m) \sum_{n\in\mathbb{N}} |f(x_n)|^2.
\end{aligned}$$

We turn to the second point. As we have a disjoint union, $\eta(\mathcal{A}_{G^\circ}) = \sum_{n\in\mathbb{N}} \eta(\mathcal{A}_{G_n})$. For a general symmetric operator H, we have the topological direct sum $\mathcal{D}(H^*) = \mathcal{D}(H) \oplus N(H^* + i) \oplus N(H^* - i)$. To conclude, note that $\mathcal{D}(\mathcal{A}_G) = \mathcal{D}(\mathcal{A}_{G^\circ})$ and $\mathcal{D}(\mathcal{A}_G^*) = \mathcal{D}(\mathcal{A}_{G^\circ}^*)$ from the first point. Q.E.D.

Lemma 12.3.2 ([129]). *For each $k, n \in \mathbb{N}^*$, there is a finite graph $K_{k,n}$ and a point $x_{k,n} \in K_{k,n}$ so that:*

(i) We have $\lim_{n\to\infty} N(x_{k,n})/d^2(x_{k,n}) = 1/(2k^2)$.
(ii) The adjacency matrix $\mathcal{A}_{K_{k,n}}$ is bounded from below by $-4k$, in the form sense. ◇

Proof. Consider first the graph given by the disjoint union $K_{k,n}^\circ := \{x_n\} \bigcup (K_n)^k$, where x_n is a point and $K_n := (\mathcal{E}_n, \mathcal{V}_n)$ the complete graph of n elements, i.e., $\mathcal{V}_n$ is a set of n elements and $\mathcal{E}(a, b) = 1$ for all $a, b \in \mathcal{V}_n$, so that $a \neq b$, see (12.3.6). Then connect x_n with each vertices of $(K_n)^k$ to obtain $K_{k,n}$. Note that $K_{1,n-1} = K_n$ and that the first point is fulfilled

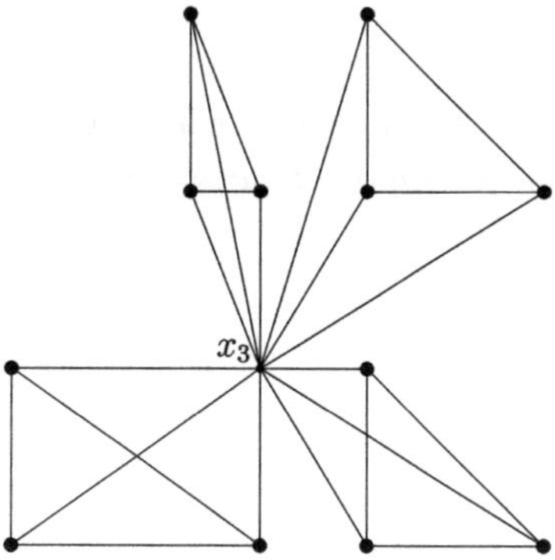

$K_{4,3}$

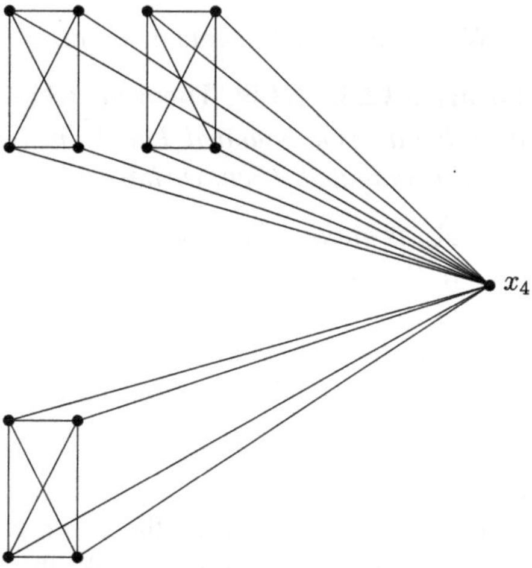

$K_{3,4}$

In a canonical basis, the adjacency matrix of $K_{k,n}$ is represented by the $(n^k + 1)(n^k + 1)$ matrix:

$$M(K_{k,n}) = \begin{pmatrix} 0 & 1 & 1 & \cdots & 1 \\ 1 & M(K_n) & 0 & \cdots & 0 \\ 1 & 0 & M(K_n) & \cdots & 0 \\ \vdots & 0 & 0 & \ddots & 0 \\ 1 & 0 & 0 & \cdots & M(K_n) \end{pmatrix},$$

where

$$M(K_n) = \begin{pmatrix} 0 & 1 & 1 & \cdots & 1 \\ 1 & 0 & 1 & \cdots & 1 \\ 1 & 1 & 0 & \cdots & 1 \\ \vdots & 1 & 1 & \ddots & 1 \\ 1 & 1 & 1 & \cdots & 0 \end{pmatrix}.$$

Easily, the characteristic polynomial of K_n is $\chi_{K_n}(\lambda) = (-\lambda + n - 1)(-\lambda - 1)^{n-1}$. Then we deduce that

$$\chi_{K_{k,n}}(\lambda) = (\lambda^2 - (n-1)\lambda - nk)(-\lambda + n - 1)^{k-1}(-\lambda - 1)^{k(n-1)},$$

by replacing C_1 by $C_1 - (\sum_{i \geq 2} C_i)/(-\lambda + n - 1)$ in the determinant of $M(K_{k,n}) - \lambda$, for instance. Here C_i denotes the i-th column. At last, the second point follows from an elementary computation. Q.E.D.

We recall the S. Golénia theorem.

Theorem 12.3.1 ([129, Theorem 1.1]). *Let $G = (\mathcal{E}, \mathcal{V})$ be a locally finite graph such that d_G is unbounded. Let $\widehat{\mathcal{A}_G}$ be a self-adjoint realization of the $\mathcal{A}_G$. Suppose that $\mathcal{E}$ is bounded. Then, we have*

(i) $\widehat{\mathcal{A}_G}$ is unbounded from above.
(ii) If $C_{\mathrm{loc}}^{\mathrm{sub}}(G) = 0$ and $\mathcal{E}$ is bounded from below, then $\widehat{\mathcal{A}_G}$ is unbounded from below.
(iii) For all $\varepsilon > 0$, there is a connected simple graph G such that $C_{\mathrm{loc}}(G) \in \{0, \varepsilon\}$, $\mathcal{A}_G$ is essentially self-adjoint on $\mathcal{C}_0^c(G)$ and is bounded from below. ◇

Proof. The two first points are proved in Proposition 12.3.1. Consider the last one. Given $\varepsilon > 0$, we choose $k > \sqrt{1/2\varepsilon}$. Given $M = 2$, we apply the Lemma 12.3.1 with $G_n := K_{k,n}$, where the latter is constructed in Lemma 12.3.2 by taking $\mathcal{E}(x_n, x_m) \in \{0, 1\}$, in order to make the graph connected. We obtain a graph G such that $\mathcal{A}_G$ is essentially self-adjoint on $\mathcal{C}_0^c(G)$ and so that $\mathcal{A}_G \geq -4k - M$. Q.E.D.

Let $G = (\mathcal{E}, \mathcal{V})$ be a locally finite graph. Let $\mathcal{E}_{\min} > 0$ such that $\mathcal{E}(x, y) \in \{0\} \bigcup [\mathcal{E}_{\min}, +\infty[$, for all $x, y \in \mathcal{V}$. We denote by $\mathcal{E}_{\max}$ the value such that $\mathcal{E}(x, y) \in \{0\} \bigcup [\mathcal{E}_{\min}, \mathcal{E}_{\max}]$, for all $x, y \in \mathcal{V}$.

Now, let us concentrate on unbounded operators. Since $\Delta_1 \geq 0$, it has a Friedrichs extension (see, Theorem 2.13.2). Let us recall its construction. Consider the quadratic form on $\mathcal{C}_1^c(G)$, $\mathcal{Q}(f, g) := \langle f, \Delta_1 g \rangle + \langle f, g \rangle$. Let $\mathcal{H}_1$ be the completion of $\mathcal{C}_1^c(G)$ under the norm associated with $\mathcal{Q}$, i.e., $\|f\|_{\mathcal{Q}}^2 = \langle f, \Delta_1 f \rangle + \|f\|^2$. The domain of the Friedrichs extension $\Delta_1^{\mathcal{F}}$ of Δ_1 is given by $\mathcal{D}(\Delta_1^{\mathcal{F}}) := \{f \in \mathcal{H}_1$ such that $\mathcal{C}_1^c(G) \ni g \longrightarrow \langle \Delta_1 g, f \rangle + \langle g, f \rangle$ which can be extended to a continuous function on $\ell^2(\mathcal{E})\} = \mathcal{H}_1 \bigcap \mathcal{D}(\Delta_1^*)$. It is a self-adjoint extension of Δ_1, e.g., see Theorem 2.13.2.

Definition 12.3.1. The graph $G = (\mathcal{E}, \mathcal{V})$ is χ-complete, if there exists an increasing sequence of finite sets $(B_n)_n$, such that $\mathcal{V} = \bigcup B_n$ and if there exist some related functions χ_n satisfying the following three conditions:

(*i*) $\chi_n \in \mathcal{C}_0^c(G), 0 \leq \chi_n \leq 1$,
(*ii*) $v \in B_n \Rightarrow \chi_n(v) = 1$, and
(*iii*) $\exists C > 0$ such that for all $n \in \mathbb{N}$ and $x \in \mathcal{V}$, we have $\sum_{e,e^+=x} \mathcal{E}(e)|d\chi_n(e)|^2 \leq C$. ◇

The results obtained in the rest of this chapter are due to H. Baloudi, S. Golénia, and A. Jeribi in [44].

12.4 On the Persson's Lemma

12.4.1 Spectral Measure

Let $C(X)$ be the family of all continuous complex-valued functions on a compact Hausdorff space, X, endowed with the norm $\| f \|_\infty = \sup_{x \in X} |f(x)|$. We recall the Riesz–Markov theorem.

Theorem 12.4.1 ([292, Theorem IV.14]). *Let X be a compact Hausdorff space and $C(X)$ be the family of all continuous complex-valued functions on X. For any positive linear functional ℓ on $C(X)$, there is a unique measure μ on X with $\ell(f) = \int f d\mu$.* ◇

We are now ready to introduce the measure we have anticipated so often before. Let us fix A, be a bounded self-adjoint operator on a Hilbert space H. Let $\psi \in H$, then $f \longrightarrow \langle \psi, f(A)\psi \rangle$ is a positive linear functional on $C(\sigma(A))$. Thus, by the Riesz–Markov Theorem 12.4.1, there is a unique measure μ_ψ on the compact set $\sigma(A)$ with

$$\langle \psi, f(A)\psi \rangle = \int_{\sigma(A)} f(\lambda) d\mu_\psi.$$

The measure μ_ψ is called the spectral measure associated with the vector ψ. Let A be a bounded self-adjoint operator and Ω be a Borel set of $\mathbb{R}$. $P_\Omega \equiv \chi_\Omega(A)$ is called a spectral projection of A, where $\chi_\Omega(.)$ denotes the characteristic function on Ω.

12.4.2 Persson's Lemma

Let A be an unbounded self-adjoint operator on Hilbert space $\mathcal{H}$. We denote by P_Ω the spectral projection of A. For $\psi \in \mathcal{D}(A)$, we denote by $\rho_\psi(.)$ the spectral measure associated with A and ψ. Let f be an unbounded complex valued Borel function and

$$\mathcal{D}_f = \left\{\varphi \text{ such that } \int_{\mathbb{R}} |f(\lambda)|^2 d\rho_\varphi(\lambda) < \infty\right\}.$$

Then, $\mathcal{D}_f$ is dense in $\mathcal{H}$ and the operator $f(A)$ is defined on $\mathcal{D}_f$ by $\langle \varphi, f(A)\varphi\rangle = \int_{\sigma(A)} f(\lambda)d\mu_\varphi(\lambda)$. Let $G = (\mathcal{E}, \mathcal{V})$ be a locally finite graph. Let S be a positive operator defined on $C_1^c(G) \subset \mathcal{D}(S) \subset \ell^2(\mathcal{E})$. We define the quadratic form $Q(f, f) = \langle f, Sf\rangle + \langle f, f\rangle \geq \|f\|^2$, $f \in C_1^c(G)$.

Proposition 12.4.1. *Let $S^{\mathcal{F}}$ be a Friedrichs extension of S associated with Q. Then, $\inf \sigma_{e4}(S^{\mathcal{F}}) = \lim_{K\to\infty} \inf \sigma(S^{\mathcal{F}}_{\mathcal{E}\setminus K})$. If Q is bounded above, then $\sup \sigma_{e4}(S^{\mathcal{F}}) = \lim_{K\to\infty} \sup \sigma(S^{\mathcal{F}}_{\mathcal{E}\setminus K})$, where K is a finite set in $\mathcal{E}$ and $\lim_{K\to\infty}$ we will say K tend to $\mathcal{E}$.* ◇

Proof. Using Theorem 2.13.2, we infer that $S^{\mathcal{F}}$ is a positive self-adjoint operator. So, for all $\lambda_0 < \inf \sigma_{e4}(S^{\mathcal{F}})$ the projection $P_{]-\infty,\lambda_0]}$ is a finite rank operator. Thus imply, for all $\varepsilon > 0$, there exist a finite set K_ε such that for all $K \supset K_\varepsilon \subset \mathcal{E}$ we have $\|P_{]-\infty,\lambda_0]}E_K\| < \varepsilon$, where E_K is the projection of $\ell^2(\mathcal{E})$ onto $\ell^2(\mathcal{E}\setminus K)$. Let $\lambda_1 < \lambda_0 < \inf \sigma_{e4}(S^{\mathcal{F}})$ and $\beta > 0$ such that $\lambda_1 + \beta < \lambda_0$. Therefore, there exists a finite set $K_1 \subset \ell^2(\mathcal{E})$ such that, for all finite site $K \supset K_1$ we have $\|P_{]-\infty,\lambda_0]}\varphi\|^2 \leq \frac{\lambda_0-(\lambda_1+\beta)}{\lambda_0+1} < 1$ for all $\varphi \in \ell^2(\mathcal{E} \setminus K)$ such that $\|\varphi\| = 1$. On the other hand, we have

$$\begin{aligned}
\langle \varphi, S^{\mathcal{F}}\varphi\rangle &= \int_{[0,+\infty]} x d\rho_\varphi(x) \\
&\geq \lambda_0\Big(\int_{-\infty}^{+\infty} d\rho_\varphi(x) - \int_{-\infty}^{\lambda_0} d\rho_\varphi(x)\Big) \\
&\geq \lambda_0 \frac{\lambda_1+\beta}{\lambda_0+1}
\end{aligned}$$

for all $\varphi \in \ell^2(\mathcal{E} \setminus K_1)$ such that $\|\varphi\| = 1$. Now, for all finite set $\mathcal{E} \supset K \supset K_1$ we can choice $\varphi \in \ell^2(\mathcal{E} \setminus K)$ such that $\|\varphi\| = 1$ and $\inf \sigma(S^{\mathcal{F}}_{\mathcal{E}\setminus K}) \geq \langle \varphi, S^{\mathcal{F}}\varphi\rangle - \frac{\lambda_0-(\lambda_1+\beta)}{\lambda_0+1} > \lambda_1$. This shows that $\inf \sigma_{e4}(S^{\mathcal{F}}) \leq \lim_{K\to\infty} \inf \sigma(S^{\mathcal{F}}_{\mathcal{E}\setminus K})$. The opposite inequality is immediately because $\sigma_{e4}(S^{\mathcal{F}}) = \sigma_{e4}(S^{\mathcal{F}}_{\mathcal{E}\setminus K}) \subseteq \sigma(S^{\mathcal{F}}_{\mathcal{E}\setminus K})$ by Weyl theorem. Q.E.D.

12.5 Essential Self-Adjointness

12.5.1 Unbounded Properties

We begin this subsection by the following result which give some elementary unboundedness properties of the Laplacian operator Δ_1 defined on a locally finite graph $G = (\mathcal{E}, \mathcal{V})$. Let $\Delta_1^{\mathcal{F}}$ be a Friedrichs extension of Δ_1.

Proposition 12.5.1. *Let $G = (\mathcal{E}, \mathcal{V})$ be a locally finite graph.*

(i) If $\sup_x \sum_{y\sim x} \mathcal{E}(x, y)$ is finite, then $\Delta_1^{\mathcal{F}}$ is bounded.
(ii) If $\sup_x \sum_{y\sim x} \mathcal{E}(x, y) = \infty$ and $\inf_{x,y} \mathcal{E}(x, y) > 0$, then $\Delta_1^{\mathcal{F}}$ is unbounded. ◇

Proof. (*i*) Take $f \in C_1^c(G)$. We have

$$\begin{aligned}
\langle f, \Delta_1 f\rangle &= \langle \Delta_1 f, f\rangle \\
&= \|d^* f\|^2 \\
&= \sum_x \Big|\sum_y \mathcal{E}(x, y) f(x, y)\Big|^2 \\
&\leq \sum_x \Big(\sum_y \mathcal{E}(x, y)\Big) \sum_z \mathcal{E}(x, z)|f(x, z)|^2 \\
&= \sum_x \sum_z \mathcal{E}(x, z)\overline{f(x, z)} \sum_y \mathcal{E}(x, y) f(x, z) \\
&\leq M\|f\|^2
\end{aligned}$$

where $M = \sup_x \sum_y \mathcal{E}(x, y) < \infty$. So, $\Delta_1^{\mathcal{F}}$ is bounded.

(*ii*) Let $x_0, y_0 \in \mathcal{V}$. It is clear that, $\delta_{x_0,y_0} - \delta_{y_0,x_0} \in C_1^c(G)$ and

$$\begin{aligned}
\left\| \frac{\delta_{x_0,y_0} - \delta_{y_0,x_0}}{\sqrt{\mathcal{E}(x_0, y_0)}} \right\| &= \frac{1}{2\mathcal{E}(x_0, y_0)} \sum_{x,y} \mathcal{E}(x, y)|\delta_{x_0,y_0} - \delta_{y_0,x_0}|^2 \\
&= \frac{1}{2\mathcal{E}(x_0, y_0)} (\mathcal{E}(x_0, y_0) + \mathcal{E}(y_0, x_0)) \\
&= 1.
\end{aligned}$$

Moreover,

$$
\begin{aligned}
&\Delta_1(\delta_{x_0,y_0} - \delta_{y_0,x_0})(a,b) \\
&\quad = \sum_z \mathcal{E}(a,z)(\delta_{x_0,y_0}(a,z) - \delta_{y_0,x_0}(a,z)) + \sum_z \mathcal{E}(z,b)(\delta_{x_0,y_0}(z,b) - \delta_{y_0,x_0}(z,b)) \\
&\quad = \mathcal{E}(a,y_0)(\delta_{x_0,y_0}(a,y_0) - \delta_{y_0,x_0}(a,y_0)) + \mathcal{E}(a,x_0)(\delta_{x_0,y_0}(a,x_0) - \delta_{y_0,x_0}(a,x_0)) \\
&\qquad + \mathcal{E}(x_0,b)(\delta_{x_0,y_0}(x_0,b) - \delta_{y_0,x_0}(x_0,b)) \\
&\qquad + \mathcal{E}(y_0,b)(\delta_{x_0,y_0}(y_0,b) - \delta_{y_0,x_0}(y_0,b)) \\
&\quad = \mathcal{E}(x_0,y_0)\Big(\delta_{y_0}(b) - \delta_{x_0}(b) + \delta_{x_0}(a) - \delta_{y_0}(a)\Big).
\end{aligned}
$$

So, we can write

$$
\begin{aligned}
2\left\|\Delta_1\Big(\frac{\delta_{x_0,y_0} - \delta_{y_0,x_0}}{\sqrt{\mathcal{E}(x_0,y_0)}}\Big)\right\|^2 &= \sum_{x,y}\left|\Delta_1\Big(\frac{\delta_{x_0,y_0} - \delta_{y_0,x_0}}{\sqrt{\mathcal{E}(x_0,y_0)}}\Big)(x,y)\right|^2 \\
&= \sum_{x,y}\mathcal{E}(x,y)\Big|\sqrt{\mathcal{E}(x_0,y_0)}\Big(\delta_{y_0}(y) - \delta_{x_0}(y) \\
&\quad + \delta_{x_0}(x) - \delta_{y_0}(x)\Big)\Big|^2 \\
&= \sum_y \mathcal{E}(x_0,y_0)\Big(\mathcal{E}(x_0,y)|1 + \delta_{y_0}(y)|^2 \\
&\quad + \mathcal{E}(y_0,y)|1 + \delta_{x_0}(y)|^2\Big) \\
&\quad + \sum_{x\neq x_0,\, x\neq y_0}\sum_y \mathcal{E}(x,y)\mathcal{E}(x_0,y_0)|\delta_{y_0}(y) - \delta_{x_0}(y)|^2 \\
&\geq C\Big(8C + \sum_x \mathcal{E}(x,y_0) + \sum_y \mathcal{E}(x_0,y)\Big)
\end{aligned}
$$

where $C = \inf_{x\sim y}\mathcal{E}(x,y) > 0$. Since $\sup_x \sum_y \mathcal{E}(x,y) = +\infty$. So, we can infer that

$$
\sup_{x_0\sim y_0}\left\|\Delta_1\Big(\frac{\delta_{x_0,y_0} - \delta_{y_0,x_0}}{\sqrt{\mathcal{E}(x_0,y_0)}}\Big)\right\| = +\infty.
$$

This proved that $\Delta_1^{\mathcal{F}}$ is unbounded. Q.E.D.

Using the Persson's lemma (Proposition 12.4.1), we have the following result:

Proposition 12.5.2. *Let $G = (\mathcal{E}, \mathcal{V})$ be an infinite and locally finite graph and $\Delta_1^{\mathcal{F}}$ be a Friedrichs extension of Δ_1. Then $\inf \sigma(\Delta_1^{\mathcal{F}}) \le \inf_{x,y} \mathcal{E}(x, y)$ and $\inf \sigma_{e4}(\Delta_1^{\mathcal{F}}) \le \inf_{K \subset \mathcal{E}, K \text{fini}} \inf_{(x,y) \in K^c} \mathcal{E}(x, y)$. In particular, Δ_1 is never with compact resolvent when G is simple.* ◇

Proof. Let $x_0, y_0 \in \mathcal{V}$ such that $\mathcal{E}(x_0, y_0) \neq 0$. Then,

$$\left\langle \frac{\delta_{x_0,y_0} - \delta_{y_0,x_0}}{\sqrt{\mathcal{E}(x_0, y_0)}}, \Delta_1^{\mathcal{F}}\left(\frac{\delta_{x_0,y_0} - \delta_{y_0,x_0}}{\sqrt{\mathcal{E}(x_0, y_0)}}\right)\right\rangle$$

$$= \frac{1}{2\mathcal{E}(x_0, y_0)} \sum_{x,y} \mathcal{E}(x, y) | \left(\delta_{x_0,y_0} - \delta_{y_0,x_0}\right)(x, y) \Delta_1 \left(\delta_{x_0,y_0} - \delta_{y_0,x_0}\right)(x, y)|$$

$$= \frac{1}{2} \sum_{x,y} \mathcal{E}(x, y) \left(\delta_{x_0,y_0} - \delta_{y_0,x_0}\right)(x, y) \left(\delta_{y_0}(y) - \delta_{x_0}(y) + \delta_{x_0}(x) - \delta_{y_0}(x)\right)$$

$$= \frac{1}{2}\mathcal{E}(x_0, y_0)\left((1 + 1) - (-1 - 1)\right)$$

$$= 2\mathcal{E}(x_0, y_0).$$

Using Proposition 12.4.1, we obtain the result. Q.E.D.

12.5.2 A Counter Example

Let $G = (\mathcal{E}, \mathcal{V})$ be a radial simple tree. We denote the origin by v and the spheres by $S_n = \{x \in \mathcal{V}$ such that $\rho_{\mathcal{V}}(v, x) = n\}$. Let $\text{off}(n) := \frac{|S_{n+1}|}{|S_n|}$ be the offspring of the n-th generation.

Theorem 12.5.1. *Let $G = (\mathcal{E}, \mathcal{V})$ be a radial simple tree. Suppose that*

$$n \longrightarrow \frac{\text{off}^2(n)}{\text{off}(n + 1)} \in \ell^1(\mathbb{N}). \tag{12.5.1}$$

Then, Δ_1 does not essentially self-adjoint on $C_1^c(G)$. ◇

Proof. We construct $f \in \ell^2(\mathcal{E})$ such that $f \in N(\Delta_1^* + i)$ and f is constant on $S_n \times S_{n+1}$. We denote the constant value by C_n. Let $\|f\|_{S_n \times S_{n+1}}$ the ℓ^2-norm restricted to $S_n \times S_{n+1}$, i.e., $\|f\|_{S_n \times S_{n+1}} = \frac{1}{2} \sum_{(x,y) \in S_n \times S_{n+1}} \mathcal{E}(x, y)|f(x, y)|^2$. By induction, we prove that $|S_n \times S_{n+1}| = \prod_{i=1}^n \text{off}(i)$. So, we have the following equation

$$(\text{off}(n) + 1 - i)C_n - C_{n+1}\text{off}(n + 1) - C_{n-1} = 0.$$

Therefore,

$$\|f\|_{S_{n+1}\times S_{n+2}} = |C_{n+1}|^2 \prod_0^{n+1} \mathrm{off}(i)$$

$$\leq 2\frac{|\mathrm{off}(n)+1-i|^2}{\mathrm{off}^2(n+1)} \prod_0^{n+1} \mathrm{off}(i)|C_n|^2 + 2\frac{1}{\mathrm{off}^2(n+1)} \prod_0^{n+1} \mathrm{off}(i)|C_{n-1}|^2.$$

Let $U_n = \prod_{i=0}^n \mathrm{off}(i)|C_n|^2$. We infer that

$$U_{n+1} \leq 2U_n \frac{|\mathrm{off}(n)+1-i|^2}{\mathrm{off}^2(n)} \frac{\mathrm{off}^2(n)}{\mathrm{off}(n+1)} + 2U_{n-1}\frac{\mathrm{off}(n)}{\mathrm{off}(n+1)}.$$

Since $n \longrightarrow \frac{\mathrm{off}^2(n)}{\mathrm{off}(n+1)} \in \ell^2(\mathbb{N})$. Then, by induction, we prove that there is $C \in \mathbb{R}_+$ such that $\sup_n \left(\prod_0^n \mathrm{off}(i)|C_n|^2\right) = C < +\infty$. Then, we have

$$\|f\|_{S_{n+1}\times S_{n+2}} \leq 2C \frac{|\mathrm{off}(n)+1-i|^2}{\mathrm{off}^2(n)} \frac{\mathrm{off}^2(n)}{\mathrm{off}(n+1)} + 2C\frac{\mathrm{off}(n)}{\mathrm{off}(n+1)}.$$

By (12.5.1), we conclude that $f \in \ell^2(\mathcal{E})$. Using Theorem 2.13.3 we derive that Δ_1 is not essentially self-adjoint on $\mathcal{C}_1^c(G)$. Q.E.D.

12.5.3 Nelson Commutator Theorem

Let $G = (\mathcal{E}, \mathcal{V})$ be a locally finite graph. For $(x, y) \in \mathcal{E}$ consider

$$\mathcal{M}(x, y) = 1 + \sum_{z\in\mathcal{V}} (\mathcal{E}(x, z) + \mathcal{E}(z, y)).$$

Let $\mathcal{M}(Q)$ be the operator of multiplication by $\mathcal{M}$: i.e.,

$$\mathcal{M}(Q)(f)(x, y) = f(x, y) + \sum_{z\in\mathcal{V}} (\mathcal{E}(x, z) + \mathcal{E}(z, y))f(x, y), \quad f \in \ell^2(\mathcal{E}).$$

Using the Nelson commutator theorem, we prove the criterium a essential self-adjointness for Δ_1.

Theorem 12.5.2. *Let $G = (\mathcal{E}, \mathcal{V})$ be a locally finite graph. Suppose that*

$$\sup_{x,y\in\mathcal{V}} \sum_{z\in\mathcal{V}} \mathcal{E}(x, z)|\mathcal{M}(x, y) - \mathcal{M}(x, z)|^2 < \infty.$$

Then, Δ_1 is essentially self-adjoint on $\mathcal{C}_1^c(G)$. ◇

Proof. Take $f \in \mathcal{C}_1^c(G)$. We denote all constants, which are independent from f, by the same letter C. We have

$$\begin{aligned}2\|\Delta_1 f\|^2 &= \sum_{(x,y)\in\mathcal{V}^2} \mathcal{E}(x,y)\Big|\sum_{z\in\mathcal{V}}\mathcal{E}(x,z)f(x,z) + \sum_{z\in\mathcal{V}}\mathcal{E}(z,y)f(z,y)\Big|^2\\ &\leq 2\big(I_1(f) + I_2(f)\big)\end{aligned}$$

where

$$I_1(f) = \sum_{(x,y)\in\mathcal{V}^2} \mathcal{E}(x,y)\Big(\Big|\sum_{z\in\mathcal{V}}\mathcal{E}(x,z)f(x,z)\Big|^2\Big)$$

and

$$I_2(f) = \sum_{(x,y)\in\mathcal{V}^2} \mathcal{E}(x,y)\Big(\Big|\sum_{z\in\mathcal{V}}\mathcal{E}(z,y)f(z,y)\Big|^2\Big).$$

Then,

$$\begin{aligned}I_1(f) &\leq \sum_{(x,y)\in\mathcal{V}^2} \mathcal{E}(x,y)\Big(\sum_{t\in\mathcal{V}}\mathcal{E}(x,t)\Big)\Big(\sum_{z\in\mathcal{V}}\mathcal{E}(x,z)|f(x,z)|^2\Big)\\ &= \sum_{(x,y)\in\mathcal{V}^2} \Big(\sum_{z\in\mathcal{V}}\mathcal{E}(x,z)\Big)\Big(\sum_{t\in\mathcal{V}}\mathcal{E}(x,t)\Big)\mathcal{E}(x,y)|f(x,y)|^2\\ &= \sum_{(x,y)\in\mathcal{V}^2} \mathcal{E}(x,y)\Big|\sum_{z\in\mathcal{V}}\mathcal{E}(x,z)f(x,y)\Big|^2\\ &\leq \sum_{(x,y)\in\mathcal{V}^2} \mathcal{E}(x,y)\Big(1 + \sum_{z}\mathcal{E}(x,z) + \mathcal{E}(z,y)\Big)^2|f(x,y)|^2\end{aligned}$$

and similarly,

$$\begin{aligned}I_2(f) &\leq \sum_{(x,y)\in\mathcal{V}^2} \mathcal{E}(x,y)\Big(\sum_{t\in\mathcal{V}}\mathcal{E}(t,y)\Big)\Big(\sum_{z\in\mathcal{V}}\mathcal{E}(z,y)|f(z,y)|^2\Big)\\ &= \sum_{(x,y)\in\mathcal{V}^2} \Big(\sum_{z\in\mathcal{V}}\mathcal{E}(z,y)\Big)\Big(\sum_{t\in\mathcal{V}}\mathcal{E}(t,y)\Big)\mathcal{E}(x,y)|f(x,y)|^2\\ &= \sum_{(x,y)\in\mathcal{V}^2} \mathcal{E}(x,y)\Big|\sum_{z\in\mathcal{V}}\mathcal{E}(z,y)f(x,y)\Big|^2\\ &\leq \sum_{(x,y)\in\mathcal{V}^2} \mathcal{E}(x,y)\Big(1 + \sum_{z}\mathcal{E}(x,z) + \mathcal{E}(z,y)\Big)^2|f(x,y)|^2.\end{aligned}$$

Therefore, we obtain

$$\|\Delta_1 f\|^2 \leq 2 \sum_{(x,y)\in\mathcal{V}^2} \mathcal{E}(x,y)\Big|\Big(1+\sum_{z\in\mathcal{V}}(\mathcal{E}(x,z)+\mathcal{E}(z,y))\Big)f(x,y)\Big|^2 = 4\|\mathcal{M}(Q)f\|^2.$$

Moreover,

$$\begin{aligned}
2\langle f, [\Delta_1, \mathcal{M}(Q)]f\rangle &= \sum_{x,y} \overline{f(x,y)}[f, \mathcal{M}(Q)]f \\
&= \sum_{(x,y)\in\mathcal{V}^2} \mathcal{E}(x,y)\overline{f(x,y)} \sum_{z\in\mathcal{V}} \Big(\mathcal{E}(x,z)\big(\mathcal{M}(x,z) \\
&\qquad -\mathcal{M}(x,y)\big)f(x,z) + \mathcal{E}(z,y)\big(\mathcal{M}(z,y) \\
&\qquad -\mathcal{M}(x,y)\big)f(z,y)\Big),
\end{aligned}$$

where $[A, B] := AB - BA$. Therefore,

$$\begin{aligned}
&|2\langle f, [\Delta_1, \mathcal{M}(Q)]f\rangle| \\
&\leq \frac{1}{2}\sum_{(x,y)\in\mathcal{V}^2} \mathcal{E}(x,y)|f(x,y)|^2 + \frac{1}{2}\sum_{(x,y)\in\mathcal{V}^2} \mathcal{E}(x,y) \times \Big(\sum_{z\in\mathcal{V}} \mathcal{E}(x,z)|\mathcal{M}(x,z) \\
&\quad - \mathcal{M}(x,y)||f(x,z)| + \mathcal{E}(z,y)|\mathcal{M}(z,y) - \mathcal{M}(x,y)||f(x,z)|\Big)^2 \\
&\leq J_1(f) + J_2(f) + J_3(f).
\end{aligned}$$

Then,

$$\begin{aligned}
J_1(f) &= \frac{1}{2}\sum_{(x,y)\in\mathcal{V}^2} \mathcal{E}(x,y)|f(x,y)|^2 \\
&\leq \frac{1}{2}\sum_{(x,y)\in\mathcal{V}^2} \mathcal{E}(x,y)\Big(1+\sum_{z\in\mathcal{V}}(\mathcal{E}(x,z)+\mathcal{E}(z,y))\Big)|f(x,y)|^2 = \|\mathcal{M}(Q)^{\frac{1}{2}}f\|^2,
\end{aligned}$$

$$\begin{aligned}
J_2(f) &= \sum_{(x,y)\in\mathcal{V}^2} \mathcal{E}(x,y)\Big(\sum_{z\in\mathcal{V}} \mathcal{E}(x,z)|\mathcal{M}(x,z) - \mathcal{M}(x,y)||f(x,z)|\Big)^2 \\
&\leq \sum_{(x,y)\in\mathcal{V}^2} \mathcal{E}(x,y)\Big(\sum_{t\in\mathcal{V}} \mathcal{E}(x,t)\Big)\Big(\sum_{z\in\mathcal{V}} \mathcal{E}(x,z)|\mathcal{M}(x,z) - \mathcal{M}(x,y)|^2|f(x,z)|^2\Big) \\
&= \sum_{(x,y)\in\mathcal{V}^2}\sum_{z\in\mathcal{V}} \mathcal{E}(x,z)\Big(\sum_{t\in\mathcal{V}} \mathcal{E}(x,t)\Big)\mathcal{E}(x,y)|\mathcal{M}(x,y) - \mathcal{M}(x,z)|^2|f(x,y)|^2 \\
&\leq C\|\mathcal{M}(Q)^{\frac{1}{2}}f\|^2,
\end{aligned}$$

since $\sup_{(x,y)\in\mathcal{V}^2}\sum_{z\in\mathcal{V}}\mathcal{E}(x,z)\,|\mathcal{M}(x,y)-\mathcal{M}(x,z)|^2<\infty$. Similarly, we obtain

$$\begin{aligned}J_3(f) &= \sum_{(x,y)\in\mathcal{V}^2}\mathcal{E}(x,y)\Big(\sum_{z\in\mathcal{V}}\mathcal{E}(x,z)\big|\mathcal{M}(x,z)-\mathcal{M}(x,y)\big|\big|f(x,z)\big|\Big)^2\\ &\le \sum_{(x,y)\in\mathcal{V}^2}\sum_{z\in\mathcal{V}}\mathcal{E}(z,y)\Big(\sum_{t\in\mathcal{V}}\mathcal{E}(t,y)\Big)\mathcal{E}(x,y)\big|\mathcal{M}(x,y)-\mathcal{M}(z,y)\big|^2\big|f(x,y)\big|^2\\ &\le C\|\mathcal{M}(Q)^{\frac{1}{2}}f\|^2.\end{aligned}$$

Finally, we obtain $|\langle f,[\Delta_1,\mathcal{M}(Q)]f\rangle|\le C\|\mathcal{M}(Q)^{\frac{1}{2}}f\|^2$. So, we can apply Theorem 2.13.4, the result follows. Q.E.D.

12.5.4 Application on Schur Test

Proposition 12.5.3. *Let $G=(\mathcal{E},\mathcal{V})$ be a locally finite graph and let $x_0\in\mathcal{V}$. For $x\in\mathcal{V}$, the module of x is defined by $|x|=\rho_{\mathcal{V}}(x_0,x)$, where $\rho_{\mathcal{V}}(.,.)$ is defined in (2.15.2). Let*

$$b_i=\sup\left\{\sum_{x,y}\mathcal{E}(x,y)\text{ such that }|x|=i\text{ and }|y|=i+1\right\}.$$

If $\sum_{i\in\mathbb{N}}\frac{1}{b_i}=+\infty$, then Δ_1 is essentially self-adjoint on $\mathcal{C}_1^c(G)$. ◇

Proof. Let $\mathcal{O}_i=\Big\{(x,y)\in\mathcal{E}\text{ such that }\sup(|x|,|y|)=i\Big\}$. For $n\in\mathbb{N}$, we consider

$$a_n(i):=\begin{cases}1, & \text{if } i\le n,\\ \max\left\{0,1-\dfrac{1}{n}\displaystyle\sum_{j=n+1}^{i}\dfrac{1}{b_j}\right\}, & \text{if } i>n.\end{cases}$$

Set $\chi_n=\sum_{i\in\mathbb{N}}a_n(i)1_{\mathcal{O}_i}$. Since χ_n is with finite support, $\chi_n\mathcal{D}(\Delta_1^*)\subset\mathcal{D}(\Delta_1)\subset\mathcal{D}(\Delta_1^*)$. Let $f\in\mathcal{D}(\Delta_1^*)$ and $f_n=\chi_n f\in\mathcal{C}_1^c(G)$. We have $\|f_m-f_n\|+\|\Delta_1(f_m-f_n)\|\le\|(\chi_m-\chi_n)f\|+\|(\chi_m-\chi_n)\Delta_1^*f\|+\|[\Delta_1^*,\chi_n]f\|+\|[\Delta_1^*,\chi_m]f\|$. Since $\sum_{j\in\mathbb{N}}\frac{1}{b_j}=+\infty$, then a_n has finite support. This prove that $\chi_n f$ and $\chi_n\Delta_1^*f$ tend to f and Δ_1^*f, respectively. The commutator operator $[\Delta_1^*,\chi_n]$ is defined on $\mathcal{D}(\Delta_1^*)$ and extends to a bounded operator, which we denote by $[\Delta_1^*,\chi_n]_o$. Note, that we have

$$[\Delta_1^*,\chi_n]_o f(u,v)=\sum_{(x,y)\in\mathcal{E}}\frac{2}{\mathcal{E}(u,v)}\langle 1_{(u,v)},[\Delta_1^*,\chi_n]_o 1_{(x,y)}\rangle f(x,y)$$

Using the Schur test we have,

$$\begin{aligned}
\|[\Delta_1^*, \chi_n]_o\| &\leq \sup_{(u,v)\in\mathcal{E}} \sum_{(x,y)\in\mathcal{E}} \frac{2}{\mathcal{E}(u,v)} |\langle 1_{(u,v)}, [\Delta_1^*, \chi_n]_o 1_{(x,y)}\rangle| \\
&= \sup_{(u,v)\in\mathcal{E}} \sum_{(x,y)\in\mathcal{E}} |[\Delta_1^*, \chi_n]_o 1_{(x,y)})(u,v)| \\
&= \sup_{(u,v)\in\mathcal{E}} \sum_{(x,y)\in\mathcal{E}} |\mathcal{E}(u,y)\delta_x(u)(\chi_n(u,y) \\
&\quad - \chi_n(u,v)) + \mathcal{E}(x,v)\delta_y(v)(\chi_n(u,v) - \chi_n(x,v))| \\
&\leq \sup_{(u,v)\in\mathcal{E}} \Big(\sum_{y\in\mathcal{V}} \mathcal{E}(u,y)|\chi_n(u,y) - \chi_n(u,v)| \\
&\quad + \sum_{x\in\mathcal{V}} \mathcal{E}(x,v)|\chi_n(x,v) - \chi_n(u,v)| \Big) \\
&= \sup_{(u,v)\in\mathcal{E}} \Big(\sum_{y\in\mathcal{V},\ \rho_{\mathcal{V}}(u,y)=1} \mathcal{E}(u,y)|\chi_n(u,y) - \chi_n(u,v)| \\
&\quad + \sum_{x\in\mathcal{V},\ \rho_{\mathcal{V}}(x,v)=1} \mathcal{E}(x,v)|\chi_n(x,v) - \chi_n(u,v)| \Big) \\
&= \sup_{(u,v)\in\mathcal{E}} \Big(\sum_{y\in\mathcal{V},\ \rho_{\mathcal{V}}(u,y)=1} \mathcal{E}(u,y)|a_n(|u|+1) - a_n(|u|)| \\
&\quad + \sum_{x\in\mathcal{V},\ \rho_{\mathcal{V}}(x,v)=1} \mathcal{E}(x,v)|a_n(|v|+1) - a_n(|v|)| \Big) \\
&\leq \frac{1}{n} + \frac{1}{n} = \frac{2}{n}.
\end{aligned}$$

We conclude that $(f_n)_n$ is a Cauchy sequence in $\mathcal{D}(\Delta_1)$. Let g be its limit. Since Δ_1 is closed, then $g \in \mathcal{D}(\Delta_1)$ and $g = f$. Q.E.D.

12.6 The Adjacency Matrix on Line Graph

12.6.1 Oriented Graph

An orientation of a graph $G = (\mathcal{E}, \mathcal{V})$ is a diagraph obtained from G by giving to edge one of its two possible orientations, i.e., for $x, y \in \mathcal{V}$ such that $\mathcal{E}(x,y) \neq 0$, we write $x \to y$ or $y \to x$.

Example 12.6.1.

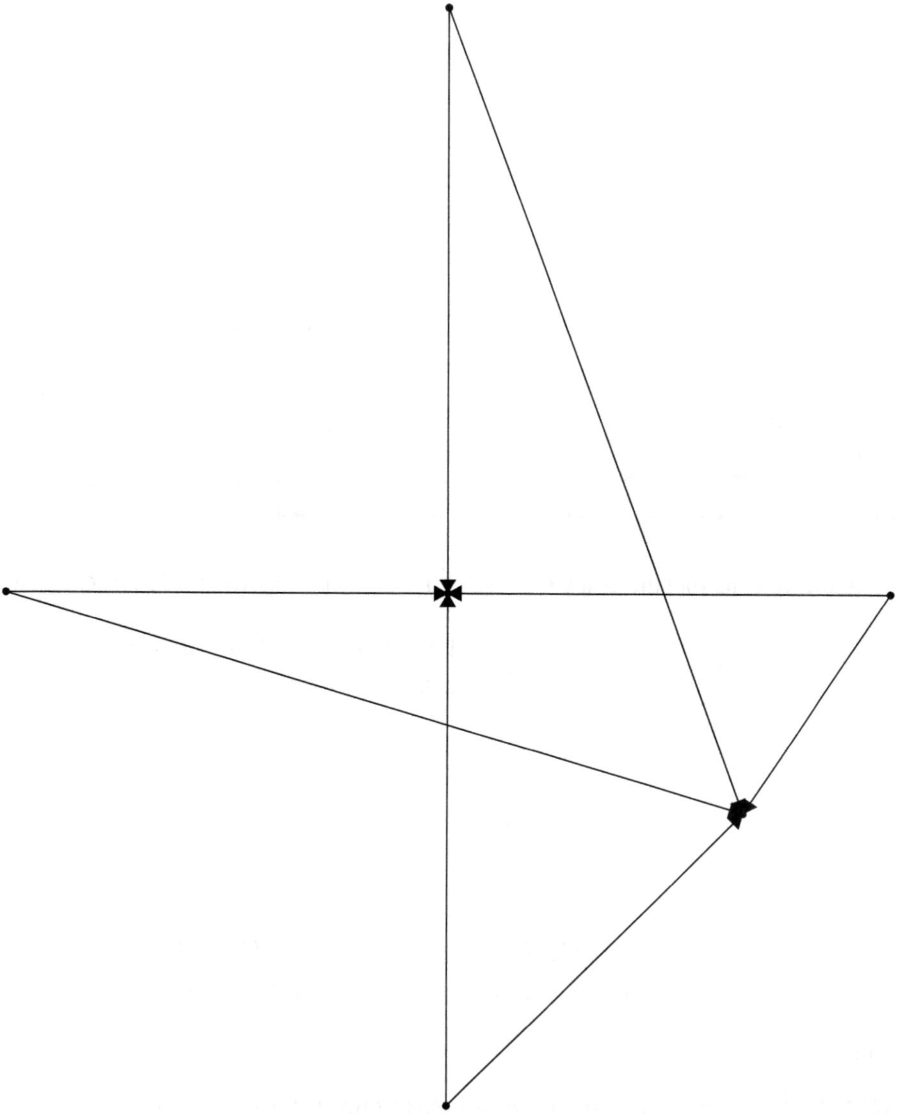

Orientation by $K_{2,4}$

We now change the ℓ^2-space. Above we had $f(x,y) = -f(y,x)$. We set:

$$\ell^2_{\text{sym}}(\mathcal{E}) := \left\{ f : \mathcal{V} \times \mathcal{V} \longrightarrow \mathbb{C} : \ f(x,y) = f(y,x) \text{ and } \sum_{x,y} \mathcal{E}(x,y)|f(x,y)|^2 < \infty \right\},$$

endowed which the scalar product $\langle f, g\rangle := \frac{1}{2}\sum_{x,y}\mathcal{E}(x,y)\overline{f(x,y)}g(x,y)$. We now fix an orientation on G and relate ℓ^2 with ℓ^2_{sym}. Let $U : \ell^2(\mathcal{E}) \longrightarrow \ell^2_{\text{sym}}(\mathcal{E})$ defined by $(Uf)(x,y) = \text{sign}(x,y)f(x,y)$ where

$$\text{sign}(x,y) := \begin{cases} 1 & \text{if } x \to y, \\ -1 & \text{if } y \to x. \end{cases}$$

Then U is a unitary map and $(U^{-1}f)(x,y) = \text{sign}(x,y)f(x,y)$. Therefore, for $x_0,\ y_0 \in \mathcal{V}$ and $f \in \ell^2(\mathcal{E})$ we have

$$U\Delta_1 U^{-1} f(x_0,y_0) =$$
$$\text{sign}(x_0,y_0)\Big(\sum_x \mathcal{E}(x,y_0)\text{sign}(x,y_0)f(x,y_0) + \sum_y \mathcal{E}(x_0,y)\text{sign}(x_0,y)f(x_0,y)\Big).$$

Example 12.6.2. Consider $G = (\mathcal{E}, \mathbb{Z})$ such that $n \to n+1$, $\mathcal{E}(n,n+1) = \mathcal{E}(n+1,n) = 1$.

Now, we construct the graph $\tilde{G} = (\tilde{\mathcal{E}}, \tilde{V})$ such that $\tilde{V} = \{(n,n+1) : n \in \mathbb{Z}\}$ and

$$\tilde{\mathcal{E}}((n,n+1),(m,m+1)) := \begin{cases} 1, & \text{if } m = n+1 \ or \ m = n-1, \\ 0, & \text{otherwise.} \end{cases}$$

Let $f \in \ell^2_{\text{sym}}(G)$ and $n \in \mathbb{Z}$. Then,

$$\begin{aligned} (U\Delta_1 U^{-1})f(n,n+1) &= \text{sign}(n,n+1)\Delta_1(Uf)(n,n+1) \\ &= \sum_{z\sim n}\text{sign}(n,z)f(n,z) + \sum_{z\sim n+1}\text{sign}(z,n+1)f(z,n+1) \\ &= 2f(n,n+1) - f(n,n-1) - f(n+2,n+1) \\ &= 2f(n,n+1) - \mathcal{A}_{\tilde{G}}f(n,n+1), \end{aligned}$$

and

$$\begin{aligned} (U\Delta_1 U^{-1})f(n,n+1) &= \text{sign}(n,n+1)\Delta_1(Uf)(n+1,n) \\ &= -\sum_{z\sim n+1}\text{sign}(n+1,z)f(n+1,z) + \sum_{z\sim n}\text{sign}(z,n)f(z,n) \\ &= 2f(n,n+1) - f(n,n-1) - f(n+2,n+1) \\ &= 2f(n,n+1) - \mathcal{A}_{\tilde{G}}f(n+1,n). \end{aligned}$$

Then, we can infer that Δ_1 and $2I - \mathcal{A}_{\tilde{G}}$ are unitarily equivalent.

Example 12.6.3. Consider $G = (\mathcal{E}, \{k\pi : \ k \in \mathbb{Z}\})$ such that $k\pi \to (k+1)\pi$, $\mathcal{E}(k\pi, (k+1)\pi) = \mathcal{E}((k+1)\pi, k\pi) = 1$.

Now we construct the graph $\tilde{G} = (\tilde{\mathcal{E}}, \tilde{V})$ such that $\tilde{V} = \{(k\pi, (k+1)\pi) : k \in \mathbb{Z}\}$ and

$$\tilde{\mathcal{E}}((n\pi, (n+1)\pi), (m\pi, (m+1)\pi)) := \begin{cases} 1, & \text{if } m = n+1 \ or \ m = n-1, \\ 0, & \text{otherwise.} \end{cases}$$

Let $f \in \ell^2_{\text{sym}}(G)$ and $n \in \mathbb{Z}$. Then,

$$\begin{aligned}(U\Delta_1 U^{-1}) f(n\pi, (n+1)\pi) &= \text{sign}(n\pi, (n+1)\pi)\Delta_1 (Uf)(n\pi, (n+1)\pi) \\ &= \sum_{z \sim n\pi} \text{sign}(n\pi, z) f(n\pi, z) \\ &\quad + \sum_{z \sim (n+1)\pi} \text{sign}(z, (n+1)\pi) f(z, (n+1)\pi) \\ &= 2f(n\pi, (n+1)\pi) - f(n\pi, (n-1)\pi) \\ &\quad - f((n+2)\pi, (n+1)\pi) \\ &= 2f(n\pi, (n+1)\pi) - \mathcal{A}_{\tilde{G}} f(n\pi, (n+1)\pi)\end{aligned}$$

and

$$\begin{aligned}(U\Delta_1 U^{-1}) f((n+1)\pi, n\pi) &= \text{sign}(n\pi, (n+1)\pi)\Delta_1 (Uf)((n+1)\pi, n\pi) \\ &= - \sum_{z \sim (n+1)\pi} \text{sign}((n+1)\pi, z) f((n+1)\pi, z) \\ &\quad + \sum_{z \sim n\pi} \text{sign}(z, n\pi) f(z, n\pi) \\ &= 2f(n\pi, (n+1)\pi) - f(n\pi, (n-1)\pi) \\ &\quad - f((n+2)\pi, (n+1)\pi) \\ &= 2f(n\pi, (n+1)\pi) - \mathcal{A}_{\tilde{G}} f((n+1)\pi, n\pi).\end{aligned}$$

Then, we can infer that Δ_1 and $2I - \mathcal{A}_{\tilde{G}}$ are unitarily equivalent.

Example 12.6.4. Let $G = (\mathcal{E}, \mathbb{Z})$ such that $2k \to 2k+1$ and $2k \to 2k-1$ for all $k \in \mathbb{Z}$. Suppose that $\mathcal{E}(2k, 2k+1) = \mathcal{E}(2k+1, 2k) = 1$ for all $k \in \mathbb{Z}$.

Let $\tilde{G} = (\tilde{\mathcal{E}}, \tilde{V})$ such that $\tilde{V} = \{(2k, 2k+1), (2k, 2k-1) : k \in \mathbb{Z}\}$ and $\tilde{\mathcal{E}}((2k, 2k+1), (2k-1, 2k)) = \tilde{\mathcal{E}}((2k, 2k+1), (2k+1, 2k+2)) = \tilde{\mathcal{E}}((2k-1, 2k), (2k-1, 2k-2)) = 1$. Then, $U\Delta_1 U^{-1} f(2k, 2k+1) = 2f(2k, 2k+1) + \mathcal{A}_{\tilde{G}}(f)(2k, 2k+1)$ and $U\Delta_1 U^{-1} f(2k-1, 2k) = 2f(2k-1, 2k) + \mathcal{A}_{\tilde{G}}(f)(2k-1, 2k)$. This imply that Δ_1 and $2I + \mathcal{A}_{\tilde{G}}$ are unitarily equivalent.

12.6.2 Case of Bipartite Graph

In this section we study a bipartite graph. A bipartite graph is a graph whose vertices can be partitioned into two subsets U and V such that no edge has both endpoint in the same subset, and every possible edge that could connect vertices in different subsets is part of the graph. A bipartite graph with partitions of size $|U| = p$ and $|q| = n$, is denoted by $K_{\mathrm{m,n}}$.

Remark 12.6.1. Let $G = (\mathcal{E}, \mathcal{V})$ be a bipartite graph. Then $\mathcal{A}_{\tilde{G}}$ is independent of the orientation of G. ◇

Example 12.6.5.

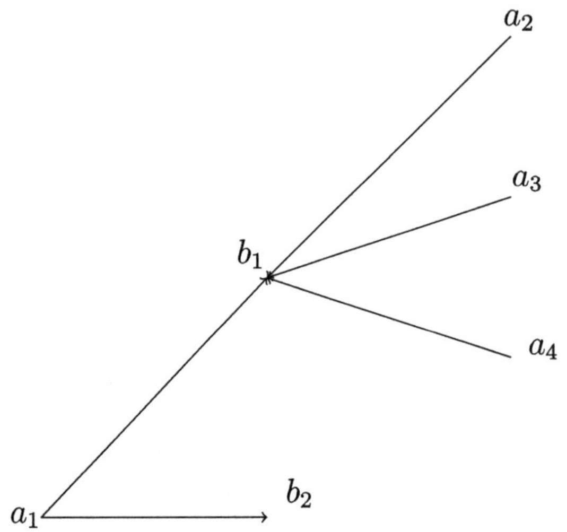

A graph $K_{2,4}$

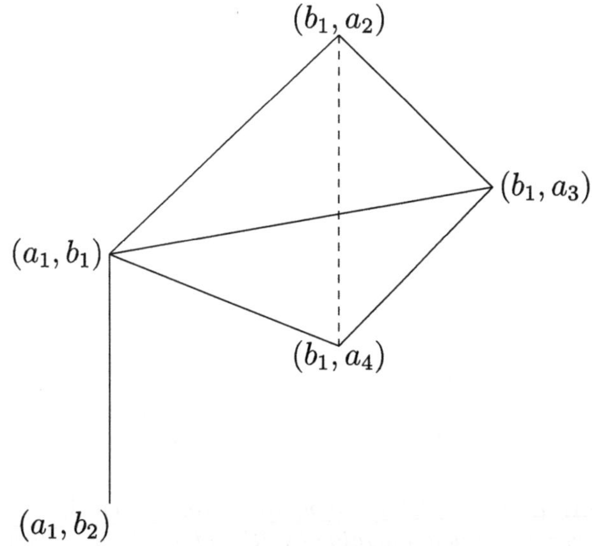

A line graph $\widetilde{K_{2,4}}$

Example 12.6.6.

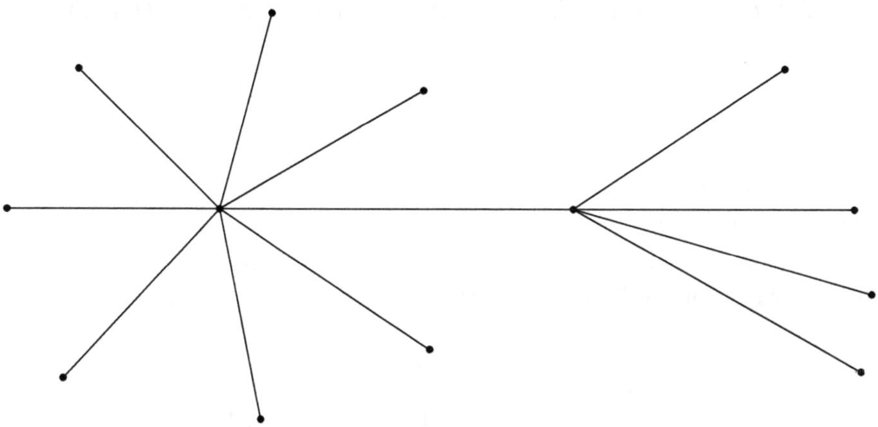

A bipartite graph G

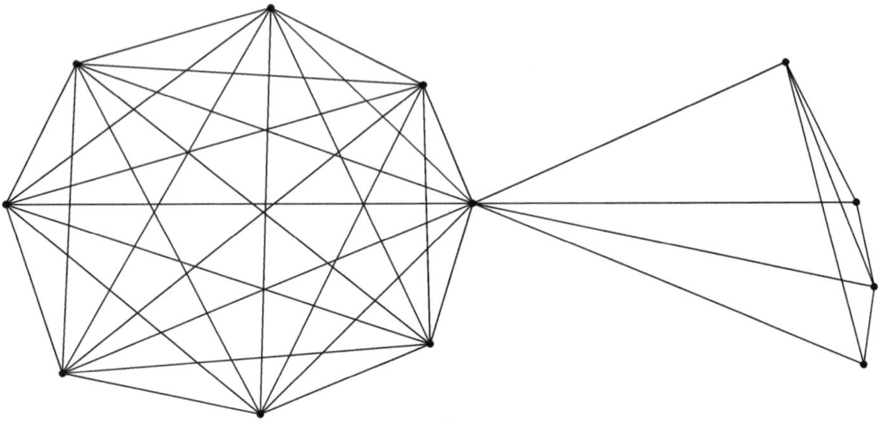

A line graph $\tilde{G} = K_8 \bigcup K_4$

Theorem 12.6.1. *Let $G = (\mathcal{E}, \mathcal{V})$ be a bipartite graph. Then $\mathcal{A}_{\tilde{G}} \geq -2\mathcal{E}(Q)$ where $2\mathcal{E}(Q)$ is the operator of multiplication by $2\mathcal{E}$ and $\mathcal{A}_{\tilde{G}}$ is the adjacency matrix on line graph $\tilde{G}$. In particular, if G is simple, then $\mathcal{A}_{\tilde{G}}$ is bounded from below.* ◇

Proof. Let $(x_0, y_0) \in \tilde{V}$ and $f \in \ell^2_{\text{sym}}(\mathcal{E})$. Then

$$U\Delta_1 U^{-1} f(x_0, y_0) = \sum_{x \sim y_0} \mathcal{E}(x, y_0) f(x, y_0) + \sum_{y \sim x_0} \mathcal{E}(x_0, y) f(x_0, y)$$
$$= 2\mathcal{E}(Q) f(x_0, y_0) + \mathcal{A}_{\widetilde{G}} f(x_0, y_0),$$

where $(Uf)(x, y) = \text{sign}(x, y) f(x, y)$. Therefore, the Laplacian Δ_1 is unitarily equivalent to that $2\mathcal{E}(Q) + \mathcal{A}_{\tilde{G}}$. If G is simple, then $\mathcal{A}_{\tilde{G}} \geq -2I$ since Δ_1 is nonnegative operator. Q.E.D.

Example 12.6.7. Let $n \in \mathbb{N}$ and we consider the bipartite graph $K_{1,n}$. Then Δ_1 and $Q(2\mathcal{E}) + \mathcal{A}_{K_n}$ are unitarily equivalent where K_n is the complete graph of n elements.

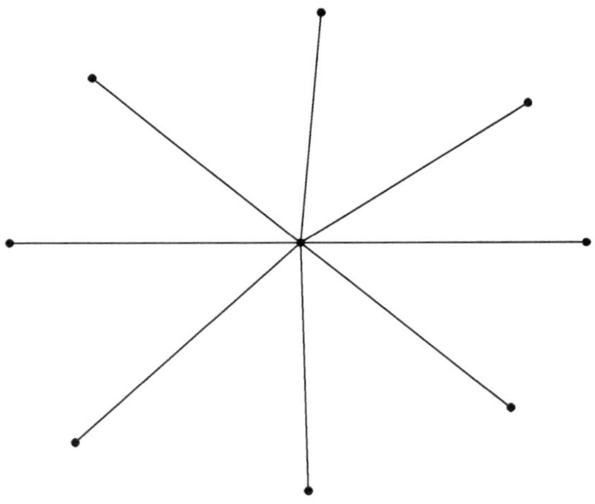

A graph $K_{1,8}$

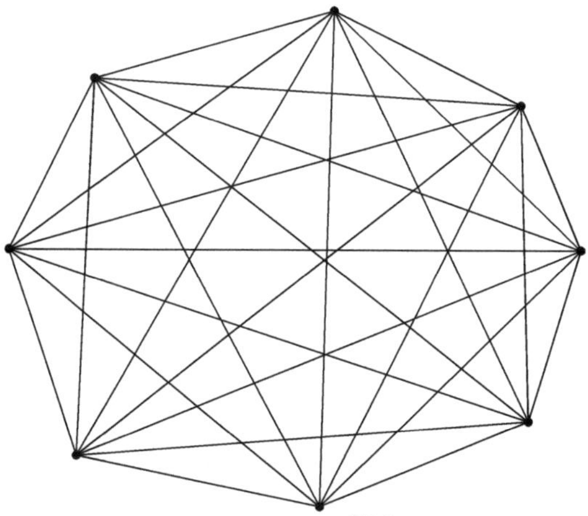

A complete graph $\widetilde{K_{1,8}} = K_8$

We give the main result:

Proposition 12.6.1. *Let G be a locally finite simple bipartite graph and $\widehat{\mathcal{A}_{\tilde{G}}}$ be a self-adjoint realization of $\mathcal{A}_{\tilde{G}}$. Then $C_{\text{loc}}^{\text{sub}}(G) > 0$ or $\tilde{\mathcal{E}}$ is unbounded from below.* $\diamond$

Proof. Since Δ_1 is nonnegative operator. Then $\mathcal{A}_{\tilde{G}} + 2I$ is nonnegative and this show that $\mathcal{A}_{\tilde{G}}$ is bounded from below. Then using Theorem 12.3.1, the result follows. Q.E.D.

We present the relationship between bipartite graphs and line graphs. We recall that C_n (see, [278]) denotes the n-cycle graph i.e., $\mathcal{V} := \mathbb{Z}/_{n\mathbb{Z}}$, where $\mathcal{E}(x, y) = 1$ if, and only if, $|x - y| = 1$.

Proposition 12.6.2. *Let $G = (\mathcal{E}, \mathcal{V})$ be a bipartite graph. Then $G \simeq \tilde{G}$ if, and only if, $G \in \{\mathbb{Z}, \mathbb{N}, C_{2n} : n \in \mathbb{N}\}$.* $\diamond$

Proof. Suppose that $G \simeq \tilde{G}$. Then, $\tilde{G}$ not contain x-triangles. If $d_G(x) \geq 3$, for all $x \in \mathcal{V}$, then there exist a x-triangle of $\tilde{G}$. So, $d_G(x) \leq 2$ for all $x \in \mathcal{V}$. The conversely is trivial. This completes the proof. Q.E.D.

Let $G = (\mathcal{E}, \mathcal{V})$ be a locally finite graph. We recall that if $\mathcal{E}$ is bounded from below, there exist $\mathcal{E}_{\min} > 0$ such that $\mathcal{E}$ is with values in $\{0\} \bigcup [\mathcal{E}_{\min}, \infty)$. We have the following Remark:

Remark 12.6.2. Let $G = (\mathcal{V}, \mathcal{E})$ be a bipartite and $\tilde{G} = (\tilde{\mathcal{V}}, \tilde{\mathcal{E}})$. Using Proposition 12.3.1, we have $\frac{\tilde{\mathcal{E}}_{\min}^2}{2\tilde{\mathcal{E}}_{\max}} \leq \frac{N_{\tilde{G}'}(x)}{d_{\tilde{G}'}(x)} + \frac{2}{\tilde{\mathcal{E}}_{\max} d'_{\tilde{G}}(x)}$, for all $\tilde{G}' \subset \tilde{G}$ and $x \in \tilde{\mathcal{V}}'$. In particular, if $\tilde{G}$ is simple, then $\frac{1}{10} \leq C_{\text{loc}}^{\text{sub}}(\tilde{G})$. $\diamond$

In the next Proposition, we present a characterization of the line graph of simple locally finite tree.

Proposition 12.6.3. *Let H be a locally finite simple connected graph such that $d_H \neq 2$. Then H is the edge graph of a locally finite simple tree if, and only if, there exists a sequence of simple complete subgraph $(K_n = (\mathcal{V}_{K_n}, \mathcal{E}_{K_n}))_n$ such that $|\mathcal{V}_{K_i}| \neq 2$, $|\mathcal{V}_{K_i} \bigcap \mathcal{V}_{K_j}| \leq 1$ for all $i \neq j$, and satisfying the two following assertions:*

(*i*) *$H = \bigcup_i K_i$ and for all $x \in H$ there exist a unique $(i, j) \in \mathbb{N}^2$, $i \neq j$ such that $\{x\} = \mathcal{V}_{K_i} \bigcap \mathcal{V}_{K_j}$.*
(*ii*) *The simple graph $K = (\{K_i$ such that $i \in \mathbb{N}\}, \mathcal{E}_K)$ is a tree, where $\mathcal{E}_K(K_i, K_j) = 1$ if $|K_i \bigcap K_j| = 1$ and 0 otherwise.* ◇

Proof. Assume that (i) and (ii) are satisfied. We construct the locally finite simple graph $G = (\mathcal{E}, \mathcal{V})$ as follows. Take $\mathcal{V} = \{x_i : i \in \mathbb{N}$ such that $x_i \notin \bigcup_j \mathcal{V}_j\}$ and $\mathcal{E}(x_i, x_j) = 1$ if, and only if, $|K_i \bigcap K_j| = 1$. Now, we consider the function $h : \mathcal{V}_{\tilde{G}} \longrightarrow \mathcal{V}_{\tilde{K}}$, $h(x_i, x_j) = (K_i, K_j)$ where $\mathcal{V}_{\tilde{G}}$ is the vertices of the edge graph $\tilde{G}$ and $\mathcal{V}_{\tilde{K}}$ is the vertices of the edge graph $\tilde{K}$. Then h is bijective. Moreover, we prove that $\tilde{\mathcal{E}}((x_i, x_j), (x_n, x_m)) \neq 0$ if, and only if, $\widetilde{\mathcal{E}_K}((K_i, K_j), (K_n, K_m)) \neq 0$. So, $\tilde{G} \simeq \tilde{K}$ and we conclude that G is a tree. Conversely, let $G = (\mathcal{E}, \mathcal{V} = \{x_i : i \in \mathbb{N}\})$ be a simple tree such that $\tilde{G} = H$ and $d_G \neq 2$. Take $B_i = (\{x_i, y \in \mathcal{V} : y \sim x_i\}, \mathcal{E})$ and $K_i := \widetilde{B_i}$. Then K_i is complete for all $i \in \mathbb{N}$. Therefore, we have $\tilde{G} := \bigcup_i K_i$ and $|\mathcal{V}_{K_i}| \neq 2$ since $d_G \neq 2$.

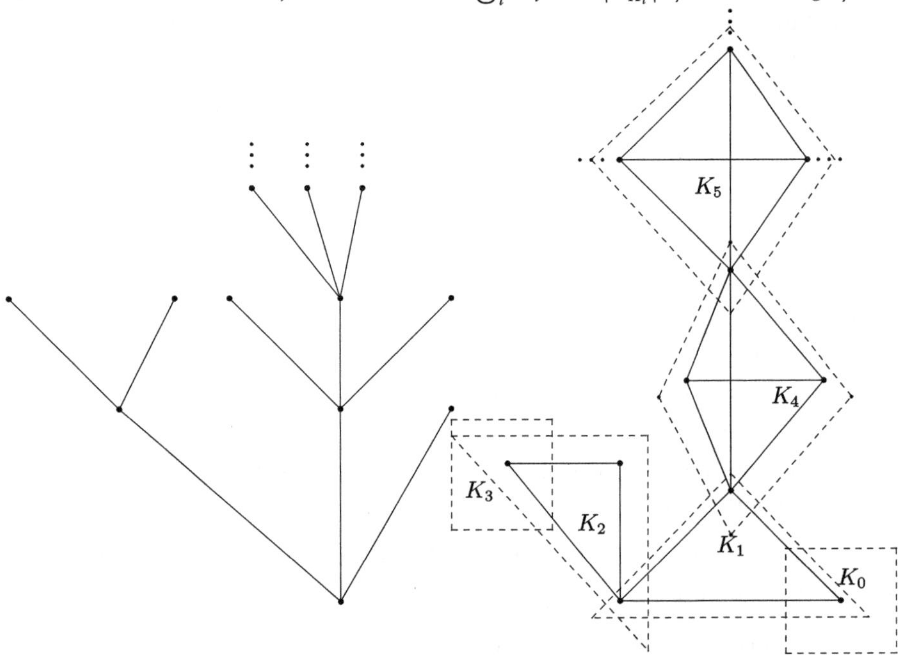

A simple tree $G = (\mathcal{E}, \mathcal{V})$ Edge graph $\widetilde{G} = \bigcup_i K_i$

By construction, $(K_i)_i$ satisfied the two assertion (i) and (ii). Q.E.D.

Open question: Let $\widetilde{G}$ be the line graph of a bipartite graph G. Can we determine the relationship between the adjacency matrix $\mathcal{A}_{\tilde{G}}$ and the discrete Laplacian acting on 1-forms, in the case where the graph G is not bipartite? $\square$

Open question: Let L be the line graph of a bipartite graph G. Can we determine the relationship between the adjacency matrix A_L and the [illegible] Laplacian [illegible] of the [illegible] graph [illegible]? [illegible]

Chapter 13
Applications in Mathematical Physics and Biology

In this chapter, we apply the results of Chaps. 4, 6–11 to five examples: to radiative transfer equations in a channel, to one-velocity transport operator with Maxwell boundary condition, to transport equation in a sphere with a diffuse reflection boundary condition, to transport operator with general boundary conditions, and to a Rotenberg's model in a cell of population. This chapter contains 11 sections. However, it is worth mentioning that each section has its own equations, notations, and symbols. In other words, the reader should remember that the same symbol doesn't have the same meaning or significance from one section to another.

13.1 Time-Asymptotic Description of the Solution for a Transport Equation

On L_p spaces ($p \geq 1$), we study the time-asymptotic behavior of solutions to the initial boundary value problem [43]

$$\begin{cases} \dfrac{\partial \psi}{\partial t}(x,v,t) = -v_3 \dfrac{\partial \psi}{\partial x}(x,v,t) - \sigma(v)\psi(x,v,t) + \displaystyle\int_K \kappa(x,v,v')\psi(x,v',t)\,dv' \\ \qquad\qquad\quad = A_H \psi(x,v,t) \\ \qquad\qquad\quad = T_H \psi(x,v,t) + F\psi(x,v,t) \\ \psi(x,v,0) = \psi_0(x,v), \end{cases} \tag{13.1.1}$$

where K is the unit sphere of $\mathbb{R}^3$, $v = (v_1, v_2, v_3) \in K$, $x \in (0,1)$, H denotes the boundary operator relating the outgoing, ψ^o, and the incoming fluxes, ψ^i, F is the integral part of A_H and $T_H := A_H - F$. The boundary operator H

A. Jeribi, *Spectral Theory and Applications of Linear Operators and Block Operator Matrices*, DOI 10.1007/978-3-319-17566-9_13

describes how the incident energy at the boundary is reflected back inside the domain. The function $\psi(x, v, t)$ represents the energy density having the position x, the velocity v at time $t \geq 0$. $\sigma(.)$ is a nonnegative, measurable, and almost everywhere finite function on K and describes the collision frequency. $\kappa(.,.,.)$ is a measurable function which describes the scattering kernel. The boundary conditions are modeled by $\psi^i = H(\psi^o)$, where

$$H = \begin{pmatrix} H_{11} & H_{12} \\ H_{21} & H_{22} \end{pmatrix}, \quad \psi^i = \begin{pmatrix} \psi(0, v), v_3 > 0 \\ \psi(1, v), v_3 < 0 \end{pmatrix}, \quad \psi^o = \begin{pmatrix} \psi(0, v), v_3 < 0 \\ \psi(1, v), v_3 > 0 \end{pmatrix}$$

and $H_{ij}, i, j = 1, 2$ are bounded operators on appropriate functional spaces.

13.1.1 Preliminaries and Notations

Let $X_p = L_p(D, dxdv)$, where $D = [0, 1] \times K$ and $p \in [1, +\infty[$. Let us define the following sets representing the incoming and the outgoing boundary of the space phase D

$$\begin{aligned} D^i &= D_1^i \bigcup D_2^i = \{0\} \times K^1 \bigcup \{1\} \times K^0 \\ D^o &= D_1^o \bigcup D_2^o = \{0\} \times K^0 \bigcup \{1\} \times K^1, \end{aligned}$$

for $K^0 = K \bigcap \{v_3 < 0\}$ and $K^1 = K \bigcap \{v_3 > 0\}$. We introduce the following boundary spaces

$$\begin{aligned} X_p^- &:= L_p(D^i, |v_3|dv) \sim L_p(D_1^i, |v_3|dv) \oplus L_p(D_2^i, |v_3|dv) \\ &:= X_{1,p}^- \oplus X_{2,p}^- \end{aligned}$$

endowed with the norm

$$\begin{aligned} \|\psi^i\|_{X_p^-} &= \left(\|\psi_1^i\|_{X_{1,p}^-}^p + \|\psi_2^i\|_{X_{2,p}^-}^p\right)^{\frac{1}{p}} \\ &= \left[\int_{K^1} |\psi(0, v)|^p |v_3| \, dv + \int_{K^0} |\psi(1, v)|^p |v_3| \, dv\right]^{\frac{1}{p}} \end{aligned}$$

and

$$\begin{aligned} X_p^+ &:= L_p(D^o, |v_3|dv) \sim L_p(D_1^o, |v_3|dv) \oplus L_p(D_2^o, |v_3|dv) \\ &:= X_{1,p}^+ \oplus X_{2,p}^+ \end{aligned}$$

endowed with the norm

$$\begin{aligned}\|\psi^o\|_{X_p^+} &= \left(\|\psi_1^o\|^p_{X^+_{1,p}} + \|\psi_2^o\|^p_{X^+_{2,p}}\right)^{\frac{1}{p}} \\ &= \left[\int_{K^0} |\psi(0,v)|^p|v_3|\,dv + \int_{K^1} |\psi(1,v)|^p|v_3|\,dv\right]^{\frac{1}{p}},\end{aligned}$$

where $\sim$ means the natural identification of these spaces. Let us introduce the boundary operator H

$$\begin{cases} H : X^+_{1,p} \oplus X^+_{2,p} \longrightarrow X^-_{1,p} \oplus X^-_{2,p} \\ H\begin{pmatrix} u_1 \\ u_2 \end{pmatrix} = \begin{pmatrix} H_{11} & H_{12} \\ H_{21} & H_{22} \end{pmatrix}\begin{pmatrix} u_1 \\ u_2 \end{pmatrix} \end{cases}$$

with $j,k \in \{1,2\}$, $H_{jk} : X^+_{k,p} \longrightarrow X^-_{j,p}$, $H_{jk} \in \mathcal{L}(X^+_{k,p}, X^-_{j,p})$, defined such that, on natural identification, the boundary conditions can be written as $\psi^i = H(\psi^o)$. Now, we define the streaming operator T_H with a domain including the boundary conditions

$$\begin{cases} T_H : \mathcal{D}(T_H) \subseteq X_p \longrightarrow X_p \\ \qquad\qquad \psi \longrightarrow T_H\psi(x,v) = -v_3\dfrac{\partial\psi}{\partial x}(x,v) - \sigma(v)\psi(x,v) \\ \mathcal{D}(T_H) \\ = \left\{\psi \in X_p, v_3\dfrac{\partial\psi}{\partial x} \in X_p, \psi_{|D^i} := \psi^i \in X_p^-, \psi_{|D^o} := \psi^o \in X_p^+ \text{ and } \psi^i = H(\psi^o)\right\}, \end{cases}$$

where $\psi^o = (\psi_1^o, \psi_2^o)^\top$ and $\psi^i = (\psi_1^i, \psi_2^i)^\top$, with ψ_1^o, ψ_2^o, ψ_1^i and ψ_2^i are given by

$$\begin{cases} \psi_1^i(v) = \psi(0,v), & v \in K^1 \\ \psi_2^i(v) = \psi(1,v), & v \in K^0 \\ \psi_1^o(v) = \psi(0,v), & v \in K^0 \\ \psi_2^o(v) = \psi(1,v), & v \in K^1. \end{cases}$$

Remark 13.1.1 ([58]). The derivative of ψ in the definition of T_H is meant in distributional sense. Note that, if $\psi \in \mathcal{D}(T_H)$, then it is absolutely continuous with respect to x. Hence the restrictions of ψ to D^i and D^o are meaningful. Note also that $\mathcal{D}(T_H)$ is dense in X_p because it contains $C_0^\infty(\overset{o}{D})$. ◇

Let $\varphi \in X_p$ and consider the resolvent equation for T_H

$$(\lambda - T_H)\psi = \varphi, \tag{13.1.2}$$

where λ is a complex number and the unknown ψ must be sought in $\mathcal{D}(T_H)$. Let $\underline{\sigma}$ be the real defined by $\underline{\sigma} := \text{ess-}\inf\{\sigma(v),\ v \in K\}$. Thus, for $\mathrm{Re}\lambda > -\underline{\sigma}$, the solution of (13.1.2) is formally given by

$$\psi(x,v) = \psi(0,v)\, e^{-\left(\frac{\sigma(v)+\lambda}{|v_3|}\right)x} + \frac{1}{|v_3|}\int_0^x e^{-\left(\frac{\sigma(v)+\lambda}{|v_3|}\right)(x-x')}\varphi(x',v)\,dx',\quad v \in K^1 \tag{13.1.3}$$

$$\psi(x,v) = \psi(1,v)\, e^{-\left(\frac{\sigma(v)+\lambda}{|v_3|}\right)(1-x)} + \frac{1}{|v_3|}\int_x^1 e^{-\left(\frac{\sigma(v)+\lambda}{|v_3|}\right)(x'-x)}\varphi(x',v)\,dx',\ v \in K^0, \tag{13.1.4}$$

where $\psi(1,v)$ and $\psi(0,v)$ are given by

$$\psi(1,v) = \psi(0,v)\, e^{-\left(\frac{\sigma(v)+\lambda}{|v_3|}\right)} + \frac{1}{|v_3|}\int_0^1 e^{-\left(\frac{\sigma(v)+\lambda}{|v_3|}\right)(1-x')}\varphi(x',v)\,dx',\quad v \in K^1 \tag{13.1.5}$$

$$\psi(0,v) = \psi(1,v)\, e^{-\left(\frac{\sigma(v)+\lambda}{|v_3|}\right)} + \frac{1}{|v_3|}\int_0^1 e^{-\left(\frac{\sigma(v)+\lambda}{|v_3|}\right)x'}\varphi(x',v)\,dx',\quad v \in K^0. \tag{13.1.6}$$

In order to clarify the analysis, let us introduce the following operators depending on the parameter λ

$$\begin{cases} M_\lambda : X_p^- \longrightarrow X_p^+,\ M_\lambda u := (M_\lambda^+ u, M_\lambda^- u),\ \text{with} \\ \\ (M_\lambda^+ u)(0,v) := u(0,v)\, e^{-\left(\frac{\sigma(v)+\lambda}{|v_3|}\right)} \qquad v \in K^1, \\ (M_\lambda^- u)(1,v) := u(1,v)\, e^{-\left(\frac{\sigma(v)+\lambda}{|v_3|}\right)} \qquad v \in K^0, \end{cases}$$

$$\begin{cases} B_\lambda : X_p^- \longrightarrow X_p,\ B_\lambda u := \chi_{K^0}(v) B_\lambda^+ u + \chi_{K^1}(v) B_\lambda^- u,\ \text{with} \\ \\ (B_\lambda^+ u)(x,v) := u(0,v)\, e^{-\left(\frac{\sigma(v)+\lambda}{|v_3|}\right)x} \qquad v \in K^1, \\ (B_\lambda^- u)(x,v) := u(1,v)\, e^{-\left(\frac{\sigma(v)+\lambda}{|v_3|}\right)(1-x)} \qquad v \in K^0, \end{cases}$$

$$\begin{cases} G_\lambda : X_p \longrightarrow X_p^+, G_\lambda u := (G_\lambda^+\varphi, G_\lambda^-\varphi), \text{ with} \\ G_\lambda^+\varphi := \dfrac{1}{|v_3|}\displaystyle\int_0^1 e^{-\left(\frac{\sigma(v)+\lambda}{|v_3|}\right)(1-x)}\varphi(x,v)\,dx, \qquad v \in K^1, \\ G_\lambda^-\varphi := \dfrac{1}{|v_3|}\displaystyle\int_0^1 e^{-\left(\frac{\sigma(v)+\lambda}{|v_3|}\right)x}\varphi(x,v)\,dx, \qquad v \in K^0, \end{cases}$$

and

$$\begin{cases} C_\lambda : X_p \longrightarrow X_p, C_\lambda\varphi := \chi_{K^0}(v)C_\lambda^+\varphi + \chi_{K^1}(v)C_\lambda^-\varphi, \text{ with} \\ C_\lambda^+\varphi := \dfrac{1}{|v_3|}\displaystyle\int_0^x e^{-\left(\frac{\sigma(v)+\lambda}{|v_3|}\right)(x-x')}\varphi(x',v)\,dx', \quad v \in K^1, \\ C_\lambda^-\varphi := \dfrac{1}{|v_3|}\displaystyle\int_x^1 e^{-\left(\frac{\sigma(v)+\lambda}{|v_3|}\right)(x-x')}\varphi(x',v)\,dx', \quad v \in K^0, \end{cases}$$

where $\chi_{K^0}(.)$ and $\chi_{K^1}(.)$ denote, respectively, the characteristic functions of the sets K^0 and K^1. A simple calculation shows that these operators are bounded on their respective spaces. In fact, for $\text{Re}\lambda > -\underline{\sigma}$, the norms of the operators M_λ, B_λ, C_λ, and G_λ are bounded above, respectively, by $e^{-\text{Re}\lambda-\underline{\sigma}}$, $(\frac{1}{p(\underline{\sigma}+\text{Re}\lambda)})^{\frac{1}{p}}$, $\frac{1}{\underline{\sigma}+\text{Re}\lambda}$ and $(\frac{1}{\underline{\sigma}+\text{Re}\lambda})^{\frac{1}{q}}$, where q is the conjugate of p. In what follows, we will assume that the boundary operator H satisfies the assumption

$$(\mathcal{P}) \quad \begin{cases} (i)\ H = \begin{pmatrix} 0 & H_{12} \\ H_{21} & 0 \end{pmatrix}, \\ (ii)\ H_{12} = \alpha J_1 + \beta D_1 \text{ and } H_{21} = \alpha J_2 + \beta D_2 \text{ where } \alpha \text{ and } \beta \in \mathbb{R}^+, \\ J_1 \text{ and } J_2 \text{ are compact while } D_1 \text{ and } D_2 \text{ are respectively given by} \\ u(1,v) \longrightarrow (D_1u)(0,v) = u(0,v),\ u(0,v) \longrightarrow (D_2u)(1,v) = u(1,v), \\ (iii)\ \|H\| \le 1. \end{cases}$$

Note that D_1 and D_2 are given by

$$\begin{aligned} D_1 : X_{2,p}^+ &\longrightarrow X_{1,p}^-,\ u(1,v) \longrightarrow (D_1u)(0,v) = u(0,v), \\ D_2 : X_{1,p}^+ &\longrightarrow X_{2,p}^-,\ u(0,v) \longrightarrow (D_2u)(1,v) = u(1,v). \end{aligned}$$

The boundary conditions are modeled by $\psi_1^i = H_{12}\psi_2^o$, and $\psi_2^i = H_{21}\psi_1^o$. Now, using the above operators and the fact that ψ must satisfy the boundary conditions, Eqs. (13.1.5) and (13.1.6) may be written as $\psi_2^o = M_\lambda^+ H_{12}\psi_2^o + G_\lambda^+\varphi$, and $\psi_1^o = M_\lambda^- H_{21}\psi_1^o + G_\lambda^-\varphi$. Moreover, since $\|M_\lambda^\pm\| \le e^{-\text{Re}\lambda-\underline{\sigma}}$, we infer that, for all λ satisfying $\text{Re}\lambda > -\underline{\sigma}$, the operators $(I - M_\lambda^- H_{21})$ and $(I - M_\lambda^+ H_{12})$ are invertible.

So, we have

$$\psi_2^o = \left(I - M_\lambda^+ H_{12}\right)^{-1} G_\lambda^+ \varphi = \sum_{n\geq 0} (M_\lambda^+ H_{12})^n G_\lambda^+ \varphi \tag{13.1.7}$$

and

$$\psi_1^o = \left(I - M_\lambda^- H_{21}\right)^{-1} G_\lambda^- \varphi = \sum_{n\geq 0} (M_\lambda^- H_{21})^n G_\lambda^- \varphi. \tag{13.1.8}$$

Next, substituting (13.1.7) and (13.1.8) into (13.1.3) and (13.1.4), we get

$$\psi(x,v) = B_\lambda^+ H_{12}(I - M_\lambda^+ H_{12})^{-1} G_\lambda^+ \varphi + C_\lambda^+ \varphi, \quad v \in K^1,$$

$$\psi(x,v) = B_\lambda^- H_{21}(I - M_\lambda^- H_{21})^{-1} G_\lambda^- \varphi + C_\lambda^- \varphi, \quad v \in K^0.$$

Thus, the resolvent $R(\lambda, T_H) := (\lambda - T_H)^{-1}$ of T_H is given by

$$(\lambda - T_H)^{-1} = \chi_{K^1}(v) R^+(\lambda, T_H) + \chi_{K^0}(v) R^-(\lambda, T_H), \tag{13.1.9}$$

where

$$R^+(\lambda, T_H) = \sum_{n\geq 0} B_\lambda^+ H_{12} (M_\lambda^+ H_{12})^n G_\lambda^+ + C_\lambda^+ \tag{13.1.10}$$

and

$$R^-(\lambda, T_H) = \sum_{n\geq 0} B_\lambda^- H_{21} (M_\lambda^- H_{21})^n G_\lambda^- + C_\lambda^-. \tag{13.1.11}$$

It is easy to see that $R(\lambda, T_0) := C_\lambda$ is a positive operator on X_p, where T_0 is the operator T_H with $H = 0$. The positivity of M_λ, B_λ, G_λ, and H implies the positivity of $R(\lambda, T_H)$, i.e.,

$$R(\lambda, T_H) \geq R(\lambda, T_0) \geq 0. \tag{13.1.12}$$

13.1.2 Compactness and Generation Results

The objective of this section is to introduce a class of perturbation operators F which are T_H-compact. This class was already introduced and investigated in [261] for vacuum boundary conditions. The scattering kernel $\kappa(.,.,.)$ defines a linear operator F by

$$\begin{cases} F : X_p \longrightarrow X_p \\ \quad \psi \longrightarrow \displaystyle\int_K \kappa(x, v, v')\psi(x, v')\, dv'. \end{cases} \tag{13.1.13}$$

The operator F satisfies in $X_p := L_p([0, 1] \times K, dxdv)$, $p \in [1, +\infty[$, the assumptions (i)–(iii) of Lemma 2.4.1.

Definition 13.1.1. A collision operator in the form (13.1.13) is said to be regular if it satisfies the assumptions (i)–(iii) of Lemma 2.4.1 for $D := [0, 1]$ and $V := K$.◇

Proposition 13.1.1. *Assume that (i)–(iii) of Lemma 2.4.1 hold true for $D := [0, 1]$ and $V := K$. Then for any $\lambda \in \mathbb{C}$ satisfying $\mathrm{Re}\lambda > -\underline{\sigma}$, the operator $(\lambda - T_H)^{-1}F$ is compact on X_p for $1 < p < \infty$ and weakly compact on X_1.* ◇

Proof. Consider the case where $1 < p < +\infty$. Let λ be such that $\mathrm{Re}\lambda > -\underline{\sigma}$. In view of (13.1.9), we have

$$(\lambda - T_H)^{-1}F = \chi_{K^1}(v)R^+(\lambda, T_H)F + \chi_{K^0}(v)R^-(\lambda, T_H)F. \tag{13.1.14}$$

In order to conclude, it suffices to show that $R^+(\lambda, T_H)F$ and $R^-(\lambda, T_H)F$ are compact on X_p. We claim that $G_\lambda^+ F$, $C_\lambda^+ F$, $G_\lambda^- F$ and $C_\lambda^- F$ are compact on X_p. By using Lemma 2.4.1, it suffices to prove the result for an operator whose kernel is of the form $\eta(x)\theta(v)\beta(v')$, where $\eta(.) \in L_\infty([0, 1], dx)$, $\theta(.) \in L_p(K, dv)$ and $\beta(.) \in L_q(K, dv)$, $(\frac{1}{p} + \frac{1}{q} = 1)$. Consider $\varphi \in X_p$, we have

$$\begin{aligned}(G_\lambda^+ F\varphi)(v) &= \theta(v)\int_K\int_0^1 \frac{1}{v_3}e^{-\frac{(\sigma(v)+\lambda)}{|v_3|}(1-x)}\eta(x)\beta(v')\varphi(x, v')\, dx\, dv', v \in K^1 \\ &= J_\lambda U_\lambda \varphi,\end{aligned}$$

where U_λ and J_λ denote the following bounded operators

$$U_\lambda : \varphi \in X_p \longrightarrow \int_K \beta(v')\varphi(x, v')\, dv' \in L_p((0, 1), dx)$$

and

$$J_\lambda : \psi \in L_p((0, 1), dx) \longrightarrow \int_0^1 \frac{1}{v_3}e^{-\frac{(\sigma(v)+\lambda)}{|v_3|}(1-x)}\eta(x)\theta(v)\psi(x)\, dx \in X_{1,p}^+.$$

Now, it is sufficient to show that J_λ is compact on X_p. This will follow from Theorem 2.4.3 if we show that

$$\int_{K^1}\left(\int_0^1\left|\frac{1}{v_3}e^{-\frac{(\sigma(v)+\lambda)}{|v_3|}(1-x)}\eta(x)\theta(v)\right|^q dx\right)^{\frac{p}{q}}|v_3|\,dv<\infty$$

(J_λ is then a Hille–Tamarkin operator). Indeed, let us first observe that

$$\begin{aligned}\int_0^1\left|\frac{1}{v_3}e^{-\frac{(\sigma(v)+\lambda)}{|v_3|}(1-x)}\eta(x)\theta(v)\right|^q dx &\le \|\eta\|_\infty^q\frac{|\theta(v)|^q}{|v_3|^q}\int_0^1 e^{-q\frac{(\underline{\sigma}+\mathrm{Re}\lambda)}{|v_3|}(1-x)}\,dx\\ &\le \|\eta\|_\infty^q\frac{|\theta(v)|^q}{q|v_3|^{q-1}(\underline{\sigma}+\mathrm{Re}\lambda)}.\end{aligned}$$

Hence,

$$\begin{aligned}\int_{K^1}\left(\int_0^1\left|\frac{1}{v_3}e^{-\frac{(\sigma(v)+\lambda)}{|v_3|}(1-x)}\eta(x)\theta(v)\right|^q dx\right)^{\frac{p}{q}}|v_3|\,dv &\le \frac{\|\eta\|_\infty^{p}}{(q(\underline{\sigma}+\mathrm{Re}\lambda))^{\frac{p}{q}}}\int_{K^1}|\theta(v)|^p\,dv\\ &= \frac{\|\eta\|_\infty^{p}\|\theta\|^p}{(q(\underline{\sigma}+\mathrm{Re}\lambda))^{\frac{p}{q}}}<\infty.\end{aligned}$$

A similar reasoning allows us to reach the same result for the operators C_λ^+F, G_λ^-F and C_λ^-F. This concludes the proof of the claim. By using Eqs. (13.1.10) and (13.1.11), we show that $R^+(\lambda,T_H)F$ and $R^-(\lambda,T_H)F$ are compact. Now, the proof follows from Eq. (13.1.14). For the case $p=1$, let λ be such that $\mathrm{Re}\lambda>-\underline{\sigma}$. It suffices to show that $R^+(\lambda,T_H)F$ and $R^-(\lambda,T_H)F$ are weakly compact on X_1. We claim that $G_\lambda F$ and $C_\lambda F$ are weakly compact on X_1. By using Lemma 2.4.1, it is sufficient to prove the result for an operator whose kernel is of the form $\xi(x,v,v')=\eta(x)\theta(v)\beta(v')$, where $\eta(.)\in L_\infty([0,1],dx)$, $\theta(.)\in L_1(K,dv)$, and $\beta(.)\in L_\infty(K,dv)$. Consider $\varphi\in X_1$. We have

$$\begin{aligned}(G_\lambda^-F\varphi)(v) &= \int_K\int_0^1\frac{1}{|v_3|}\eta(x)\theta(v)e^{-\left(\frac{\sigma(v)+\lambda}{|v_3|}\right)x}\beta(v')\,\varphi(x,v')\,dx\,dv', \quad v\in K^0\\ &= J_\lambda U_\lambda\varphi,\end{aligned}$$

where U_λ and J_λ denote the following bounded operators

$$\begin{cases}U_\lambda : X_1 \longrightarrow L_1([0,1],dx)\\ \qquad\varphi \longrightarrow \displaystyle\int_K \beta(v)\,\varphi(x,v)\,dv\end{cases}$$

and

$$\begin{cases} J_\lambda : L_1([0,1], dx) \longrightarrow X^+_{2,1} \\ \qquad \psi \longrightarrow \displaystyle\int_0^1 \frac{1}{|v_3|}\eta(x)\theta(v)e^{-\left(\frac{\sigma(v)+\lambda}{|v_3|}\right)x}\,\psi(x)\,dx\,. \end{cases}$$

Now, it is sufficient to show that J_λ is weakly compact. To do so, let $\mathcal{O}$ be a bounded set of $L_1([0,1], dx)$, and let $\psi \in \mathcal{O}$. We have $\int_E |J_\lambda\psi(v)|\,|v_3|\,dv \leq \|\eta\|\,\|\psi\| \int_E |\theta(v)|\,dv$, for all measurable subsets E of K^1. Next, by applying Theorem 2.4.5, we infer that the set $J_\lambda(\mathcal{O})$ is weakly compact, since $\lim_{|E|\to 0} \int_E |\theta(v)|\,dv = 0$, $(\theta \in L_1(K, dv))$ where $|E|$ is the measure of E. A similar reasoning allows us to reach the same result for the operators $G^+_\lambda F$ and $C_\lambda F$. Q.E.D.

Lemma 13.1.1. *Assume that* $(\mathcal{P})$ *holds true. Then, for all* λ *such that* $\mathrm{Re}\lambda > -\underline{\sigma}$, *we have* $\|(\lambda - T_H)^{-1}\| \leq \frac{1}{\mathrm{Re}\lambda + \underline{\sigma}}$. ◇

Proof. We define the streaming operator $T'_H : \psi \in \mathcal{D}(T_H) \longrightarrow (T_H + \underline{\sigma})\psi$. Let us first show that T'_H is dissipative on X_p for $p \in [1, \infty)$. For this purpose, we treat separately the case $1 < p < \infty$ and the case $p = 1$. Let $1 < p < \infty$ and consider $\psi \in \mathcal{D}(T_H)$. We have

$$\begin{aligned} \mathrm{Re}\langle T'_H\psi, |\psi|^{p-2}\overline{\psi}\rangle &= \mathrm{Re}\left[\int_K\int_0^1 |\psi|^{p-2}\overline{\psi}\left(-v_3\frac{\partial\psi}{\partial x}(x,v)\right)dx\,dv\right] \\ &\quad - \int_K\int_0^1 (\sigma(v) - \underline{\sigma})|\psi|^p\,dx\,dv \\ &= -\frac{1}{p}\int_K\int_0^1 v_3\frac{\partial|\psi|^p}{\partial x}\,dx\,dv - \int_K\int_0^1 (\sigma(v) - \underline{\sigma})|\psi|^p\,dx\,dv \\ &= \frac{1}{p}\left(\|\psi^i\|^p_{X_p^-} - \|\psi^o\|^p_{X_p^+}\right) - \int_K\int_0^1 (\sigma(v) - \underline{\sigma})|\psi|^p\,dx\,dv \\ &\leq 0 \ \ (\text{because } \|H\| \leq 1). \end{aligned}$$

Now, let us consider the case where $p = 1$. Let $\psi \in \mathcal{D}(T_H)$. Then, we have

$$\begin{aligned} \mathrm{Re}\langle T'_H\psi, s_0(\psi)\rangle &= -\mathrm{Re}\left(\int_K\int_0^1 v_3\frac{\partial\psi}{\partial x}(x,v)s_0(\psi)(x,v)\,dx\,dv\right) \\ &\quad - \left[\int_K\int_0^1 (\sigma(v) - \underline{\sigma})|\psi|\,dx\,dv\right]\|\psi\| \\ &= \left[-\int_K\int_0^1 v_3\frac{\partial|\psi|}{\partial x}\,dx\,dv - \int_K\int_0^1 (\sigma(v) - \underline{\sigma})|\psi|\,dx\,dv\right]\|\psi\| \end{aligned}$$

$$= \left[\left(\|\psi^i\|_{X_1^-} - \|\psi^o\|_{X_1^+} \right) - \int_K \int_0^1 (\sigma(v) - \underline{\sigma})|\psi|\,dx\,dv \right] \|\psi\|$$
$$\leq 0 \ \text{(because } \|H\| \leq 1).$$

We conclude that T'_H is dissipative on X_p $(1 \leq p < \infty)$. Now, for $\mathrm{Re}\lambda + \underline{\sigma} > 0$, consider $\psi \in \mathcal{D}(T_H)$ and let $\varphi = \lambda\psi - T_H\psi$. We have

$$\begin{aligned}
(\mathrm{Re}\lambda + \underline{\sigma})\|\psi\|^2 &= (\mathrm{Re}\lambda + \underline{\sigma})\langle\psi, \psi^*\rangle \\
&= \mathrm{Re}\left[(\lambda + \underline{\sigma})\langle\psi, \psi^*\rangle\right] \\
&\leq \mathrm{Re}\left[\langle(\lambda + \underline{\sigma})\psi, \psi^*\rangle - \langle T'_H\psi, \psi^*\rangle\right] \ \text{(because } T'_H \text{ is dissipative)} \\
&= \mathrm{Re}\langle\varphi, \psi^*\rangle \leq \|\varphi\|\|\psi\|.
\end{aligned}$$

Consequently, $\|\psi\| \leq \frac{\|\varphi\|}{\underline{\sigma}+\mathrm{Re}\lambda}$, which completes the proof. Q.E.D.

Lemma 13.1.2. *Assume that* $(\mathcal{P})$ *holds true. Then,* $\{\lambda \in \mathbb{C}$ *such that* $\mathrm{Re}\lambda > \|F\| - \underline{\sigma} + 1\} \subset \rho(A_H)$ *and, for all* λ *such that* $\mathrm{Re}\lambda > \|F\| - \underline{\sigma} + 1$, *we have* $\|(\lambda - A_H)^{-1}\| \leq 1$. ◇

Proof. By using Lemma 13.1.1, we have $\|(\lambda - T_H)^{-1}F\| \leq \frac{\|F\|}{\mathrm{Re}\lambda+\underline{\sigma}} < 1$. Moreover, $(\lambda - A_H)^{-1} = [I - (\lambda - T_H)^{-1}F]^{-1}(\lambda - T_H)^{-1}$ and therefore,

$$\begin{aligned}
\|(\lambda - A_H)^{-1}\| &\leq \|(\lambda - T_H)^{-1}\| \sum_{k=0}^{+\infty} \|(\lambda - T_H)^{-1}F\|^k \\
&\leq \frac{1}{(\mathrm{Re}\lambda + \underline{\sigma})(1 - \frac{\|F\|}{\mathrm{Re}\lambda+\underline{\sigma}})} \\
&\leq 1.
\end{aligned}$$

Q.E.D.

We end this section by the following result.

Proposition 13.1.2. *Assume that* $(\mathcal{P})$ *holds true. If the boundary operator* H *is nonnegative, then* T_H *generates a strongly continuous semigroup* $(U^H(t))_{t\geq 0}$, *satisfying* $\|U^H(t)\| \leq e^{-\underline{\sigma}t}$. ◇

Proof. The result follows from Lemma 13.1.1 and Corollary 2.5.1. Q.E.D.

13.1.3 Auxiliary Lemmas

The purpose of this section is to establish some lemmas which will be used in the next section. Let $\omega > 0$, and set $R_\omega = \{\lambda \in \mathbb{C}$ such that $\mathrm{Re}\lambda \geq -\underline{\sigma} + \omega\}$. Note that if $\lambda \in R_\omega$, then

$$\underline{\sigma} - \frac{\omega}{2} + \mathrm{Re}\lambda \geq \frac{\omega}{2} > 0 \tag{13.1.15}$$

and, for all $s \in (0, +\infty)$, we have

$$\sigma\left(v_1, v_2, \frac{v_3}{s}\right) - \underline{\sigma} + \frac{\omega}{2} \geq \frac{\omega}{2} > 0. \tag{13.1.16}$$

We define $\mathcal{C}(r)$ by $\mathcal{C}(r) = \{(v_1, v_2) \in \mathbb{R}^2$ such that $v_1^2 + v_2^2 = 1 - r^2\}$, which represents the circle of radius $\sqrt{1-r^2}$ and center O. Let $\varepsilon > 0$, for $x \in (0, 2)$, $(v_1, v_2) \in \mathcal{C}(0)$ and $u > 0$, we consider

$$\begin{cases} \phi_{v_1,v_2,x} : (\varepsilon, +\infty) \longrightarrow \mathbb{R}_+ \\ \qquad\qquad s \longrightarrow \dfrac{1}{s} e^{-s(\sigma(v_1,v_2,\frac{x}{s}) - \underline{\sigma} + \frac{\omega}{2})} \end{cases}$$

and

$$\begin{cases} \qquad \varphi_u : K^1 \longrightarrow \mathbb{R} \\ v = (v_1, v_2, v_3) \longrightarrow e^{-\frac{(\sigma(v) - \underline{\sigma} + \frac{\omega}{2})}{v_3} u}. \end{cases}$$

According to (13.1.16), $0 \leq \phi_{v_1,v_2,x}(.) \in L_1(\varepsilon, +\infty)$ and $0 \leq \varphi_u \in L_1(K^1)$. We denote by $(\varrho_{v_1,v_2,x,n}(.))_{n\in\mathbb{N}}$ (resp. $(L_{u,n}(.))_{n\in\mathbb{N}}$), an increasing sequence of nonnegative step functions with compact support which converges to $\phi_{v_1,v_2,x}(.)$ [resp. $\varphi_u(.)$] almost everywhere. Notice that

$$\phi_{v_1,v_2,x}(s) - \varrho_{v_1,v_2,x,n}(s) \leq \frac{2}{\varepsilon} e^{-\frac{\omega}{2}s}. \tag{13.1.17}$$

Let h be a measurable simple function on K and let A_2 be the operator defined by

$$\begin{cases} A_2 : L_p(0,1) \longrightarrow L_p(0,1) \\ \varphi \longrightarrow \displaystyle\int_0^x \varphi(u) du \int_{\mathcal{C}(\frac{x-u}{s})} dv_1 dv_2 \int_\varepsilon^{+\infty} \frac{h(v_1, v_2, \frac{x-u}{s})}{s} e^{-s(\sigma(v_1,v_2,\frac{x-u}{s}) + \lambda)} \\ \chi_{(x-s,x)}(u) ds. \end{cases}$$

We introduce the sequence $(A_{2,n})_{n\in\mathbb{N}}$ of operators defined, for all $n \in \mathbb{N}$, by

$$\begin{cases} A_{2,n} : L_p(0,1) \longrightarrow L_p(0,1) \\ \varphi \longrightarrow \displaystyle\int_0^x \varphi(u) du \int_{\mathcal{C}(\frac{x-u}{s})} dv_1 dv_2 \int_\varepsilon^{+\infty} h\left(v_1, v_2, \frac{x-u}{s}\right) e^{-s(\underline{\sigma} - \frac{\omega}{2} + \lambda)} \\ \chi_{(x-s,x)}(u) \varrho_{v_1,v_2,x-u,n}(s) ds. \end{cases}$$

Lemma 13.1.3. *We have the following*

(*i*) *The sequence of operators* $(A_{2,n})_{n\in\mathbb{N}}$ *converges uniformly on* R_ω *to* A_2 *in* $\mathcal{L}(L_p(0,1))$.

(*ii*) $\|A_{2,n}\|^p \leq \int_0^1 dx \left[\int_{C(0)} dv_1 dv_2 \left| \int_\varepsilon^{+\infty} h\left(v_1, v_2, \frac{x}{s}\right) e^{-s(\underline{\sigma} - \frac{\omega}{2} + \lambda)} \chi_{(x,+\infty[}(s) \varrho_{v_1,v_2,x,n}(s) ds \right|\right]^p.$

◇

Proof.

(*i*) Let $\varphi \in L_p(0,1)$. Then,

$$\begin{aligned}
&\|(A_{2,n} - A_2)\varphi\|^p \\
&= \int_0^1 dx \Big| \int_0^x \varphi(u) du \int_{C(\frac{x-u}{s})} dv_1 dv_2 \int_\varepsilon^{+\infty} h\left(v_1, v_2, \frac{x-u}{s}\right) e^{-s(\underline{\sigma}-\frac{\omega}{2}+\lambda)} \chi_{(x-s,x)}(u) \\
&\qquad\qquad \times \{\phi_{v_1,v_2,x-u}(s) - \varrho_{v_1,v_2,x-u,n}(s)\} ds \Big|^p \\
&\leq \int_0^1 dx \Bigg(\int_0^1 |\varphi(u)| du \int_{C(\frac{x}{s})} dv_1 dv_2 \int_\varepsilon^{+\infty} \left| h\left(v_1, v_2, \frac{x}{s}\right)\right| e^{-s(\underline{\sigma}-\frac{\omega}{2}+\mathrm{Re}\lambda)} \chi_{(x,+\infty[}(s) \\
&\qquad\qquad \times \{\phi_{v_1,v_2,x}(s) - \varrho_{v_1,v_2,x,n}(s)\} ds \Bigg)^p \\
&\leq \int_0^1 dx \Bigg(\int_0^1 |\varphi(u)| du \int_{C(0)} dv_1 dv_2 \int_\varepsilon^{+\infty} \left| h\left(v_1, v_2, \frac{x}{s}\right)\right| e^{-s(\underline{\sigma}-\frac{\omega}{2}+\mathrm{Re}\lambda)} \\
&\qquad\qquad \times \{\phi_{v_1,v_2,x}(s) - \varrho_{v_1,v_2,x,n}(s)\} ds \Bigg)^p \\
&\leq \|\varphi\|^p \sup |h(.,.,.)|^p \\
&\quad \times \int_0^1 dx \left(\int_{C(0)} dv_1 dv_2 \int_\varepsilon^{+\infty} e^{-s(\underline{\sigma}-\frac{\omega}{2}+\mathrm{Re}\lambda)} \{\phi_{v_1,v_2,x}(s) - \varrho_{v_1,v_2,x,n}(s)\} ds \right)^p.
\end{aligned}$$

Applying Hölder's inequality, we get

$$\begin{aligned}
&\int_{C(0)} dv_1 dv_2 \int_\varepsilon^{+\infty} e^{-s(\underline{\sigma}-\frac{\omega}{2}+\mathrm{Re}\lambda)} \{\phi_{v_1,v_2,x}(s) - \varrho_{v_1,v_2,x,n}(s)\} ds \leq \left(\int_{C(0)} dv_1\, dv_2 \right)^{\frac{1}{q}} \\
&\quad \times \left[\int_{C(0)} dv_1\, dv_2 \left(\int_\varepsilon^{+\infty} e^{-s(\underline{\sigma}-\frac{\omega}{2}+\mathrm{Re}\lambda)} \{\phi_{v_1,v_2,x}(s) - \varrho_{v_1,v_2,x,n}(s)\}\, ds \right)^p \right]^{\frac{1}{p}},
\end{aligned}$$

which implies

$$\|(A_{2,n} - A_2)\varphi\|^p \leq \|\varphi\|^p \sup |h(.,.,.)|^p \pi^{\frac{p}{q}} \int_0^1 dx \int_{C(0)} dv_1 dv_2$$
$$\times \left(\int_\varepsilon^{+\infty} e^{-s(\underline{\sigma} - \frac{\omega}{2} + \mathrm{Re}\lambda)} \{\phi_{v_1,v_2,x}(s) - \varrho_{v_1,v_2,x,n}(s)\} ds \right)^p .$$

Therefore,

$$\|A_{2,n} - A_2\|^p \leq \sup |h(.,.,.)|^p \pi^{\frac{p}{q}} \int_0^1 dx \int_{C(0)} dv_1 dv_2$$
$$\times \left(\int_\varepsilon^{+\infty} e^{-s(\underline{\sigma} - \frac{\omega}{2} + \mathrm{Re}\lambda)} \{\phi_{v_1,v_2,x}(s) - \varrho_{v_1,v_2,x,n}(s)\} ds \right)^p .$$

Using Eqs. (13.1.16) and (13.1.17) and applying two times the Lebesgue dominated convergence theorem, we get $\lim_{n\to+\infty} \|A_{2,n} - A_2\| = 0$ uniformly on R_ω. This completes the proof of (i).

(ii) Let $\varphi \in L_p(0,1)$. We have

$$\|A_{2,n}\varphi\|^p = \int_0^1 dx \Big| \int_0^x \varphi(u) du \int_{C(\frac{x-u}{s})} dv_1\, dv_2$$
$$\times \int_\varepsilon^{+\infty} h\left(v_1, v_2, \frac{x-u}{s}\right) e^{-s(\underline{\sigma} - \frac{\omega}{2} + \lambda)} \chi_{(x-s,x)}(u) \varrho_{v_1,v_2,x-u,n}(s) ds \Big|^p$$
$$\leq \int_0^1 dx \Big| \int_0^{1-x} \varphi(u) du \int_{C(\frac{x}{s})} dv_1 dv_2$$
$$\times \int_\varepsilon^{+\infty} h\left(v_1, v_2, \frac{x}{s}\right) e^{-s(\underline{\sigma} - \frac{\omega}{2} + \lambda)} \chi_{(x,+\infty[}(s) \varrho_{v_1,v_2,x,n}(s)\, ds \Big|^p .$$

We have

$$\Big| \int_0^{1-x} \varphi(u) du \int_{C(\frac{x}{s})} dv_1 dv_2 \int_\varepsilon^{+\infty} h\left(v_1, v_2, \frac{x}{s}\right) e^{-s(\underline{\sigma} - \frac{\omega}{2} + \lambda)}$$
$$\chi_{(x,+\infty[}(s) \varrho_{v_1,v_2,x,n}(s) ds \Big|^p$$
$$\leq \left[\int_0^1 |\varphi(u)| du \int_{C(0)} dv_1 dv_2 \Big| \int_\varepsilon^{+\infty} h\left(v_1, v_2, \frac{x}{s}\right) e^{-s(\underline{\sigma} - \frac{\omega}{2} + \lambda)} \right.$$
$$\left. \chi_{(x,+\infty[}(s) \varrho_{v_1,v_2,x,n}(s) ds \Big| \right]^p$$

$$\leq \|\varphi\|^p \left[\int_{C(0)} dv_1 dv_2 \left| \int_{\varepsilon}^{+\infty} h\left(v_1, v_2, \frac{x}{s}\right) e^{-s(\underline{\sigma}-\frac{\omega}{2}+\lambda)} \right.\right.$$
$$\left.\left. \chi_{(x,+\infty[}(s) \varrho_{v_1,v_2,x,n}(s) ds \right| \right]^p .$$

Therefore,

$$\|A_{2,n}\|^p \leq \int_0^1 dx \left[\int_{C(0)} dv_1 dv_2 \right.$$
$$\left. \left| \int_{\varepsilon}^{+\infty} h\left(v_1, v_2, \frac{x}{s}\right) e^{-s(\underline{\sigma}-\frac{\omega}{2}+\lambda)} \chi_{(x,+\infty[}(s) \varrho_{v_1,v_2,x,n}(s) ds \right| \right]^p .$$

This completes the proof of *(ii)*. Q.E.D.

Lemma 13.1.4. *Let*

$$\mathcal{V}_\lambda : u \in \mathbb{R} \longrightarrow \int_{K^1} \mu(v)\theta(v)\chi_{(0,+\infty[}(u) e^{-\frac{(\sigma(v)+\lambda)}{v_3}u} dv,$$
$$\mathcal{V}_{n,\lambda} : u \in \mathbb{R} \longrightarrow \int_{K^1} \mu(v)\theta(v) e^{-\frac{(\underline{\sigma}+\lambda-\frac{\omega}{2})}{v_3}u} L_{u,n}(v)\chi_{(\varepsilon_n,+\infty[}(u)\, dv,$$
$$H_\lambda : u \in \mathbb{R} \longrightarrow \int_{K^1} \frac{1}{v_3}\theta(v)\beta(v) e^{-\frac{(\sigma(v)+\lambda)}{v_3}u} \chi_{(0,+\infty[}(u)\, dv,$$

and

$$H_{n,\lambda} : u \in \mathbb{R} \longrightarrow \int_{K^1} \frac{1}{v_3}\theta(v)\beta(v) e^{-\frac{(\underline{\sigma}+\lambda-\frac{\omega}{2})}{v_3}u} L_{u,n}(v)\chi_{(\varepsilon_n,+\infty[}(u)\, dv,$$

where $\theta(.) \in L_p(K)$, $\mu(.) \in L_q(K)$ $(\frac{1}{p}+\frac{1}{q}=1)$ *and* $(\varepsilon_n)_{n\in\mathbb{N}}$ *is a sequence of strictly positive numbers converging to zero. Then,*

- *(i)* $\mathcal{V}_\lambda(.) \in L_q(\mathbb{R})$.
- *(ii) The sequence* $(\mathcal{V}_{n,\lambda})_{n\in\mathbb{N}}$ *converges uniformly on* R_ω *to the function* $\mathcal{V}_\lambda$, *in* $L_q(\mathbb{R})$.
- *(iii)* $H_\lambda \in L_q(\mathbb{R})$.
- *(iv)* $(H_{n,\lambda})_{n\in\mathbb{N}}$ *converges uniformly on* R_ω *to* H_λ, *in* $L_q(\mathbb{R})$.

◇

Proof.

(i) Using Hölder's inequality, we get

$$|\mathcal{V}_\lambda(u)| \leq \|\theta\|_{L_p} \left(\int_{K^1} |\mu(v)|^q \chi_{(0,+\infty[}(u) e^{-q\frac{(\underline{\sigma}+\mathrm{Re}\lambda)}{v_3}u} dv \right)^{\frac{1}{q}} .$$

Now, making use of both Fubini's theorem and the change of variable $u = v_3 u'$, we get

$$\int_{\mathbb{R}} |\mathcal{V}_\lambda(u)|^q \, du \leq \|\theta\|_{L_p}^q \int_{K^1} |\mu(v)|^q \, dv \int_0^{+\infty} e^{-q(\underline{\sigma}+\mathrm{Re}\lambda)u'} \, du'$$

and therefore, $\|\mathcal{V}_\lambda\|_{L_q(\mathbb{R})} \leq \frac{\|\theta\|_{L_p}\|\mu\|_{L_q}}{[q(\underline{\sigma}+\mathrm{Re}\lambda)]^{\frac{1}{q}}} < \infty$, which proves that $\mathcal{V}_\lambda(.) \in L_q(\mathbb{R})$.

(ii) For all $n \in \mathbb{N}$, we have

$$\begin{aligned}
&|(\mathcal{V}_{n,\lambda} - \mathcal{V}_\lambda)(u)| \\
&= \left| \int_{K^1} \mu(v)\theta(v) e^{-\frac{(\underline{\sigma}+\lambda-\frac{\omega}{2})}{v_3}u} \{L_{u,n}(v)\chi_{(\varepsilon_n,+\infty[}(u) - \varphi_u(v)\chi_{(0,+\infty[}(u)\} \, dv \right| \\
&\leq \int_{K^1} |\mu(v)||\theta(v)| e^{-\frac{(\underline{\sigma}+\mathrm{Re}\lambda-\frac{\omega}{2})}{v_3}u} \left| L_{u,n}(v)\chi_{(\varepsilon_n,+\infty[}(u) - \varphi_u(v)\chi_{(0,+\infty[}(u) \right| dv.
\end{aligned}$$

According to Eqs. (13.1.15) and (13.1.16), for all $v \in K^1$, we get

$$|(\mathcal{V}_{n,\lambda} - \mathcal{V}_\lambda)(u)|^q \leq (2 \sup|\mu(.)| \sup|\theta(.)|)^q e^{-\frac{q\omega u}{v_3}} \chi_{[0,+\infty[}(u)(\mathrm{vol}(K^1))^q.$$

So, by using the Lebesgue dominated convergence theorem, we have

$$\begin{aligned}
&\lim_{n\to+\infty} \|\mathcal{V}_{n,\lambda} - \mathcal{V}_\lambda\|_{L_q(\mathbb{R})}^q \\
&\leq \int_0^{+\infty} \lim_{n\to+\infty} \left(\int_{K^1} |\mu(v)\theta(v)| e^{-\frac{\omega u}{2v_3}} \{\varphi_u(v) - L_{u,n}(v)\chi_{(\varepsilon_n,+\infty[}(u)\} \, dv \right)^q du.
\end{aligned} \tag{13.1.18}$$

However, for each $n \in \mathbb{N}$ and $u > 0$, we have

$$|\mu(v)\theta(v)| e^{-\frac{\omega u}{2v_3}} \{\varphi_u(v) - L_{u,n}(v)\chi_{(\varepsilon_n,+\infty[}(u)\} \leq 2|\mu(v)\theta(v)| \in L_1(K^1).$$

Then, the convergence of $(L_{u,n}(.))_n$ to $\varphi_u(.)$, $u > 0$, the convergence of $(\varepsilon_n)_n$ to zero, and the inequality (13.1.18), together with the continuity of the power function q, allow us to deduce by a second application of the Lebesgue dominated convergence theorem, that

$$\lim_{n\to+\infty} \|\mathcal{V}_{n,\lambda} - \mathcal{V}_\lambda\| = 0 \text{ uniformly on } R_\omega.$$

(iii) Arguing as in the proof of (i), we show that $H_\lambda \in L_q(\mathbb{R})$.

(*iv*) For all $n \in \mathbb{N}$, we have

$$
\begin{aligned}
&|(H_{n,\lambda} - H_\lambda)(u)| \\
&= \left| \int_{K^1} \frac{1}{v_3} \theta(v)\beta(v) e^{-\frac{(\underline{\sigma}+\lambda-\frac{\omega}{2})}{v_3}u} \{L_{u,n}(v)\chi_{(\varepsilon_n,+\infty[}(u) - \varphi_u(v)\chi_{(0,\infty[}(u)\}\, dv \right| \\
&\leq \int_{K^1} \frac{1}{v_3} |\theta(v)\beta(v)| e^{-\frac{(\underline{\sigma}+\mathrm{Re}\lambda-\frac{\omega}{2})}{v_3}u} \left| L_{u,n}(v)\chi_{(\varepsilon_n,+\infty[}(u) - \varphi_u(v)\chi_{(0,\infty[}(u) \right| dv \\
&\leq 2^3 \sup|\theta(.)\beta(.)| \int_0^1 \frac{1}{v_3} e^{-\omega\frac{u}{v_3}} \chi_{(0,+\infty[}(u)\, dv_3.
\end{aligned}
$$

Let $I(u) = \int_0^1 \frac{1}{v_3} e^{-\omega\frac{u}{v_3}}\, dv_3$. The change of variables $s = \frac{u}{v_3}$ gives $I(u) = \int_u^{+\infty} \frac{1}{s} e^{-\omega s}\, ds$. We claim that $I(.) \in L_1(]0,+\infty[)$. Indeed, we have $I(.)$ is continuous on $]0,+\infty[$, $I(u) \leq \frac{1}{u\omega} e^{-\omega u}$ on $]0,+\infty[$ and $I(u) \sim -\log(u)$ at $u = 0$. So, $I(.) \in L_1(]0,+\infty[)$. However, for each $n \in \mathbb{N}$ and $u > 0$, we have

$$
\begin{aligned}
&\frac{1}{v_3} |\theta(v)\beta(v)| e^{-\frac{(\underline{\sigma}+\mathrm{Re}\lambda-\frac{\omega}{2})}{v_3}u} \left| L_{u,n}(v)\chi_{(\varepsilon_n,+\infty[}(u) - \varphi_u(v)\chi_{(0,\infty[}(u) \right| \\
&\qquad \leq 2 \sup|\theta(.)\beta(.)| \frac{1}{v_3} e^{-\omega\frac{u}{v_3}} \in L_1(K^1).
\end{aligned}
$$

Now, arguing as in the proof of (*ii*), we get the result. Q.E.D.

Lemma 13.1.5. *Let $p \in [1,+\infty[$ and assume that the collision operator F is regular. If the boundary operator H satisfies the hypothesis $(\mathcal{P})$, then $|\mathrm{Im}\lambda| \|FC_\lambda^{\pm}F\|$ is bounded on R_ω.* ◇

Proof. According to Lemma 2.4.1 and Remark 2.4.1, it is sufficient to establish the result for a one rank collision operator F of the form

$$
\begin{cases}
F : X_p \longrightarrow X_p \\
\qquad \varphi \longrightarrow \displaystyle\int_K \eta(x)\theta(v)\beta(v')\varphi(x,v')\, dv',
\end{cases}
$$

where $\eta(.) \in L_\infty(0,1)$, $\theta(.)$ and $\beta(.)$ are measurable simple functions on K. Let $\varphi \in X_p$, we have $C_\lambda^+\varphi(x,v) = \frac{1}{v_3}\int_0^x e^{-\frac{(\sigma(v)+\lambda)}{v_3}(x-x')}\varphi(x',v)\, dx'$. The change of variables $s = \frac{(x-x')}{v_3}$ gives $C_\lambda^+\varphi(x,v) = \int_0^{+\infty} e^{-s(\sigma(v)+\lambda)}\varphi(x - v_3 s, v)\chi_{(0,\frac{x}{v_3})}(s)\, ds$. Now, let us consider the sequence of operators $(C^+_{\lambda,\varepsilon_n})_{n\in\mathbb{N}}$, where $C^+_{\lambda,\varepsilon_n}\varphi(x,v) = \int_{\varepsilon_n}^{+\infty} e^{-s(\sigma(v)+\lambda)}\varphi(x - v_3 s, v)\chi_{(0,\frac{x}{v_3})}(s)\, ds$. Clearly, in the operator topology, the sequence $(C^+_{\lambda,\varepsilon_n})_{n\in\mathbb{N}}$ converges to C_λ^+ and uniformly on R_ω when ε_n goes to zero. Hence, in order to prove the lemma, it suffices to show that, for $\varepsilon > 0$, $|\mathrm{Im}\lambda|\|FC^{+}_{\lambda,\varepsilon}F\|$ is bounded on R_{ω}. An easy calculation shows that

$$\begin{aligned} &FC^{+}_{\lambda,\varepsilon}F\varphi(x,v) \\ &\quad = \eta(x)\theta(v)\int_K h(v'')\,dv''\int_{\varepsilon}^{+\infty}\eta(x-v_3''s)e^{-s(\sigma(v'')+\lambda)}\chi_{(0,\frac{x}{v_3''})}(s)\,ds \\ &\quad\times\int_K \beta(v')\varphi(x-v_3''x'',v')\,dv', \end{aligned}$$

where $h(.) = \beta(.)\theta(.)$. Let $\mathcal{C}(r)$ be the circle of radius $\sqrt{1-r^2}$ and center O. We have

$$\begin{aligned} &FC^{+}_{\lambda,\varepsilon}F\varphi(x,v) \\ &\quad = \eta(x)\theta(v)\int_0^1 dv_3''\int_{\mathcal{C}(v_3'')} h(v'')dv_1''dv_2''\int_{\varepsilon}^{+\infty}\eta(x-v_3''s)e^{-s(\sigma(v'')+\lambda)}\chi_{(0,\frac{x}{v_3''})}(s)ds \\ &\quad\times\int_K \beta(v')\varphi(x-v_3''s,v')\,dv'. \end{aligned}$$

So, putting $u = u(v_3'') = x - v_3''s$, we get

$$\begin{aligned} &FC^{+}_{\lambda,\varepsilon}F\varphi(x,v) \\ &\quad = \eta(x)\theta(v)\int_0^x \eta(u)\,du\int_{\mathcal{C}(\frac{x-u}{s})} dv_1''\,dv_2''\int_{\varepsilon}^{+\infty}\frac{h(v_1'',v_2'',\frac{x-u}{s})}{s}e^{-s(\sigma(v_1'',v_2'',\frac{x-u}{s})+\lambda)} \\ &\quad\times\chi_{(x-s,x)}(u)\,ds\int_K \beta(v')\varphi(u,v')\,dv'. \end{aligned}$$

So, the operator $FC^{+}_{\lambda,\varepsilon}F$ may be decomposed as $FC^{+}_{\lambda,\varepsilon}F = A_1A_2A_3$, where

$$\begin{cases} A_1 : L_p(0,1) \longrightarrow X_p \\ \qquad\qquad \varphi \longrightarrow \eta(x)\theta(v)\varphi(x), \end{cases}$$

$$\begin{cases} A_2 : L_p(0,1) \longrightarrow L_p(0,1) \\ \qquad\qquad \varphi \longrightarrow \displaystyle\int_0^x \varphi(u)du\int_{\mathcal{C}(\frac{x-u}{s})} dv_1dv_2\int_{\varepsilon}^{+\infty}\frac{h(v_1,v_2,\frac{x-u}{s})}{s} \\ \qquad\qquad\qquad e^{-s(\sigma(v_1,v_2,\frac{x-u}{s})+\lambda)}\chi_{(x-s,x)}(u)ds \end{cases}$$

and

$$\begin{cases} A_3 : X_p \longrightarrow L_p(0,1) \\ \qquad \varphi \longrightarrow \eta(x) \displaystyle\int_K \beta(v)\varphi(x,v)\,dv. \end{cases}$$

A_1 and A_3 are uniformly bounded on R_ω. According to Lemma 13.1.3(i), it remains to show that, for all $n \in \mathbb{N}$, we have $|\mathrm{Im}\lambda|^p \|A_{2,n}\|^p$ is bounded on R_ω. From Lemma 13.1.3(ii), we have

$$\|A_{2,n}\|^p \leq \int_0^1 dx \left[\int_{C(0)} dv_1 dv_2 \right.$$
$$\left. \left| \int_\varepsilon^{+\infty} h\left(v_1, v_2, \frac{x}{s}\right) e^{-s(\underline{\sigma}-\frac{\omega}{2}+\lambda)} \chi_{(x,+\infty[}(s) \varrho_{v_1,v_2,x,n}(s) ds \right| \right]^p.$$

Hence, in order to prove Lemma 13.1.5, we have to demonstrate that, for $\varepsilon > 0$, $x \in (0,1)$ and $\varrho_{v_1,v_2,x}(.)$ is a nonnegative step function with a compact support satisfying $\varrho_{v_1,v_2,x}(.) \leq \phi_{v_1,v_2,x}(.) \leq \frac{1}{\varepsilon}$, and

$$|\mathrm{Im}\lambda|^p \int_0^1 dx \left[\int_{C(0)} dv_1 dv_2 \left| \int_\varepsilon^{+\infty} h\left(v_1, v_2, \frac{x}{s}\right) e^{-s(\underline{\sigma}-\frac{\omega}{2}+\lambda)} \chi_{(x,+\infty[}(s) \varrho_{v_1,v_2,x}(s) ds \right| \right]^p$$

is bounded on R_ω. The map

$$\begin{cases} G_{v_1,v_2,x}(.) : (\varepsilon, +\infty) \longrightarrow \mathbb{R} \\ \qquad\qquad s \longrightarrow h(v_1, v_2, \frac{x}{s}) \chi_{(x,+\infty[}(s) \varrho_{v_1,v_2,x}(s), \end{cases}$$

is a simple function. Let $(s_i)_{1 \leq i \leq m}$ denote a subdivision of its support satisfying $G_{v_1,v_2,x}(s) = G_{v_1,v_2,x}(s_i)$, for all $s \in [s_i, s_{i+1}[$, with $i \in \{1, \ldots, m-1\}$. Then,

$$\int_\varepsilon^{+\infty} e^{-s(\underline{\sigma}-\frac{\omega}{2}+\lambda)} G_{v_1,v_2,x}(s)\, ds$$
$$= \left[\sum_{i=1}^{m-1} G_{v_1,v_2,x}(s_i) \int_{s_i}^{s_{i+1}} e^{-s(\underline{\sigma}-\frac{\omega}{2}+\lambda)}\, ds \right]$$
$$= \left[\frac{1}{(\underline{\sigma} - \frac{\omega}{2} + \lambda)} \sum_{i=1}^{m-1} G_{v_1,v_2,x}(s_i) \left(e^{-s_i(\underline{\sigma}-\frac{\omega}{2}+\lambda)} - e^{-s_{i+1}(\underline{\sigma}-\frac{\omega}{2}+\lambda)} \right) \right].$$

We have

$$\begin{cases} \left| e^{-s_i(\underline{\sigma}-\frac{\omega}{2}+\lambda)} - e^{-s_{i+1}(\underline{\sigma}-\frac{\omega}{2}+\lambda)} \right| \le 2 \\ \left| \dfrac{1}{\underline{\sigma}-\frac{\omega}{2}-\lambda} \right| \le \dfrac{1}{|\mathrm{Im}\lambda|} \\ |G_{v_1,v_2,x}(s_i)| \le \dfrac{1}{\varepsilon} \sup |h(.,.,.)|. \end{cases}$$

Therefore,

$$\left| \int_\varepsilon^{+\infty} e^{-s(\underline{\sigma}-\frac{\omega}{2}+\lambda)} G_{v_1,v_2,x}(s)\, ds \right| \le \frac{2(m-1)}{\varepsilon |\mathrm{Im}\lambda|} \sup |h(.,.,.)|.$$

This yields the following

$$\begin{aligned} &|\mathrm{Im}\lambda|^p \int_0^1 dx \left(\int_{C(0)} dv_1\, dv_2 \left| \int_\varepsilon^{+\infty} e^{-s(\underline{\sigma}-\frac{\omega}{2}+\lambda)} G_{v_1,v_2,x}(s)\, ds \right| \right)^p \\ &\quad \le |\mathrm{Im}\lambda|^p \int_0^1 dx \left(\frac{2\pi(m-1)}{\varepsilon |\mathrm{Im}\lambda|} \sup |h(.,.,.)| \right)^p \\ &\quad \le 2^p \left(\frac{\pi(m-1)}{\varepsilon} \sup |h(.,.,.)| \right)^p . \end{aligned}$$

Finally, we have $|\mathrm{Im}\lambda| \|FC_\lambda^+ F\|$ is bounded on R_ω. A similar reasoning allows us to reach the same result for the operator $FC_\lambda^- F$. This completes the proof of lemma.

Q.E.D.

Let $n \in \mathbb{N}$. Since the operators $M_\lambda^+ J_1$ and $M_\lambda^+ D_1$ do not commute, $(M_\lambda^+ H_{12})^n = \sum_{i=1}^{2^n} N_i$, where each N_i is the product of n factors involving both $M_\lambda^+ J_1$ and $M_\lambda^+ D_1$ except the term $N_{2^n} = (M_\lambda^+ D_1)^n$. If $i \in \{1, \ldots, 2^n - 1\}$, then the operator J_1 appears, at least, one time in the expression of N_i. So, there exists $k \in \{1, \ldots, n-1\}$ such that $N_i = Q_i M_\lambda^+ J_1 (M_\lambda^+ D_1)^k$, where Q_i is uniformly bounded on R_ω. Now, we need to prove the following lemma:

Lemma 13.1.6. *Let $p \in [1, +\infty[$ and assume that the collision operator F is regular. If the boundary operator H satisfies the hypothesis ($\mathcal{P}$), then*

(*i*) *for $i \in \{1, \ldots, 2^n - 1\}$, we have $|\mathrm{Im}\lambda| \|FB_\lambda^+ H_{12} N_i G_\lambda^+ F\|$ is bounded on R_ω.*
(*ii*) *$|\mathrm{Im}\lambda| \|FB_\lambda^+ H_{12} P_{2^n} G_\lambda^+ F\|$ is bounded on R_ω.*
(*iii*) *$|\mathrm{Im}\lambda| \|FB_\lambda^+ H_{12} (M_\lambda^+ H_{12})^n G_\lambda^+ F\|$ and $|\mathrm{Im}\lambda| \|FB_\lambda^- H_{21} (M_\lambda^- H_{21})^n G_\lambda^- F\|$ are bounded on R_ω.*

◇

Proof.

(*i*) Let $i \in \{1, \ldots, 2^n - 1\}$. We have $\|FB_\lambda^+ H_{12} N_i G_\lambda^+ F\| \leq \|FB_\lambda^+ H_{12}\| \|N_i G_\lambda^+ F\|$. According to the hypotheses, it is sufficient to prove that $|\text{Im}\lambda| \|J_1 (M_\lambda^+ D_1)^k G_\lambda^+ F\|$ is bounded on R_ω. Since the operator J_1 is compact, we only have to establish the result for an operator of rank one, that is $J_1 : \varphi(1, v) \longrightarrow (J_1\varphi)(0, v) = \alpha(v) \int_{K^1} \mu(v')\varphi(1, v')v_3' \, dv'$, where $\alpha(.)$ and $\mu(.)$ are measurable simple functions. Let $\varphi \in X_p$. Then,

$$\begin{aligned}(J_1(M_\lambda^+ D_1)^k G_\lambda^+ F\varphi)(0, v) &= \alpha(v) \int_{K^1} \mu(v')\theta(v') \, dv' \\ &\quad \int_0^1 e^{-(k+1-x)(\frac{\sigma(v')+\lambda}{v_3'})} \eta(x) \, dx \int_K \beta(v'')\varphi(x, v'') \, dv'' \\ &= (B_1 B_2 B_3 \varphi)(0, v),\end{aligned}$$

where

$$\begin{cases} B_1 : \mathbb{R} \longrightarrow X_{1,p}^- \\ \qquad \delta \longrightarrow \alpha(v)\delta, \end{cases}$$

$$\begin{cases} B_2 : L_p(0, 1) \longrightarrow \mathbb{R} \\ \qquad \psi \longrightarrow \int_{K^1} \mu(v)\theta(v) \, dv \int_0^1 e^{-(k+1-x)(\frac{\sigma(v)+\lambda}{v_3})} \psi(x) \, dx \end{cases}$$

and

$$\begin{cases} B_3 : X_p \longrightarrow L_p(0, 1) \\ \qquad \psi \longrightarrow \eta(x) \int_K \beta(v)\psi(x, v) \, dv. \end{cases}$$

Clearly, we notice that B_1, B_2, and B_3 are three bounded operators, and

$$\|J_1(M_\lambda^+ D_1)^k G_\lambda^+ F\| \leq \|B_1\| \|B_2\| \|B_3\|.$$

Since B_1 and B_3 are independent of λ, it suffices to prove that $|\text{Im}\lambda| \|B_2\|$ is bounded on R_ω. For this purpose, let $\psi \in L_p(0, 1)$ and let us denote by $\tilde{\psi}$ its trivial extension to $\mathbb{R}$. Then,

$$\begin{aligned} B_2\psi &= \int_{\mathbb{R}} \mathcal{V}_\lambda(k + 1 - x)\tilde{\psi}(x) \, dx \quad (\mathcal{V}_\lambda \text{ is defined in Lemma 13.1.4}) \\ &= (\mathcal{V}_\lambda * \tilde{\psi})(k + 1). \end{aligned}$$

So, the use of Young's inequality gives $|B_2\psi| \leq \|\mathcal{V}_\lambda * \tilde{\psi}\|_{L_\infty(\mathbb{R})} \leq \|\mathcal{V}_\lambda\|_{L_q(\mathbb{R})}\|\tilde{\psi}\|_{L_p(\mathbb{R})}$. Since $\|\tilde{\psi}\|_{L_p(\mathbb{R})} = \|\psi\|_{L_p(0,1)}$, then

$$\|B_2\|^q \leq \int_0^{+\infty} \left| \int_{K^1} \mu(v)\theta(v)e^{-\frac{(\sigma(v)+\lambda)}{v_3}u}\, dv\right|^q du.$$

According to Lemma 13.1.4, it is sufficient to prove that, for any $\varepsilon > 0$,

$$|\mathrm{Im}\lambda|^q \int_\varepsilon^{+\infty} \left| \int_{K^1} \mu(v)\theta(v)e^{-\frac{(\underline{\sigma}+\lambda-\frac{\omega}{2})}{v_3}u} L_u(v)\, dv\right|^q du$$

is bounded on R_ω, where $u > 0$ and $L_u(.)$ is a nonnegative step function defined on K^1 such that $L_u(.) \leq \varphi_u(.)$. We have $K^1 \subset P := [-1,1]^2 \times [0,1]$. We denote by $\tilde{\mu}$, $\tilde{\theta}$, and $\tilde{L}_u$ the functions:

$$\tilde{\mu}(v) = \begin{cases} \mu(v) & \text{if } v \in K^1 \\ 0 & \text{if } v \in P\backslash K^1, \end{cases}$$

$$\tilde{\theta}(v) = \begin{cases} \theta(v) & \text{if } v \in K^1 \\ 0 & \text{if } v \in P\backslash K^1 \end{cases}$$

and

$$\tilde{L}_u(v) = \begin{cases} L_u(v) & \text{if } v \in K^1 \\ 0 & \text{if } v \in P\backslash K^1. \end{cases}$$

For all $u \in [\varepsilon, +\infty[$, the function $\tilde{h}_u : v \in P \longrightarrow \tilde{\mu}(v)\tilde{\theta}(v)\tilde{L}_u(v) \in \mathbb{C}$ is a simple function. Let δ be a subdivision of P, associated with $\tilde{h}$, and let $P_1, \ldots, P_N$ be the cells of δ. Then, $\tilde{h}(u) = c_i$ for all $v \in P_i$, $i \in \{1, \ldots, N\}$. We have

$$\begin{aligned}
\left|\int_P \tilde{\mu}(v)\tilde{\theta}(v)e^{-\frac{(\underline{\sigma}+\lambda-\frac{\omega}{2})}{v_3}u}\tilde{L}_u(v)\, dv\right| &= \left|\int_P \tilde{h}_u(v)e^{-\frac{(\underline{\sigma}+\lambda-\frac{\omega}{2})}{v_3}u}\, dv\right| \\
&= \left|\sum_{i=1}^N \int_{P_i} c_i e^{-\frac{(\underline{\sigma}+\lambda-\frac{\omega}{2})}{v_3}u}\, dv\right| \\
&\leq \sup|h_u(.,.,.)| \sum_{i=1}^N \left|\int_{P_i} e^{-\frac{(\underline{\sigma}+\lambda-\frac{\omega}{2})}{v_3}u}\, dv\right|.
\end{aligned}$$

We use the change of variables $a_u : (v_1, v_2, v_3) \longrightarrow (v_1, v_2, \frac{u}{\xi})$, then $a_u(P_i) \subset [-1,1]^2 \times [u, +\infty[$ and we get

$$\begin{aligned}
\left|\int_P \tilde{\mu}(v)\tilde{\theta}(v)e^{-\frac{(\underline{\sigma}+\lambda-\frac{\omega}{2})}{v_3}u}\tilde{L}_u(v)\, dv\right| &\leq \sup|h(.,.,.)| \\
&\quad \times \left(\sum_{i=1}^N \left|\int_{a_u(P_i)} \frac{u}{\xi^2} e^{-(\underline{\sigma}+\mathrm{Re}\lambda-\frac{\omega}{2})\xi} e^{i(\xi\mathrm{Im}\lambda)} dv_1 dv_2 d\xi\right|\right).
\end{aligned}$$

Let

$$F_\lambda(u) = \left| \int_{a_u(P_i)} \frac{u}{\xi^2} e^{-(\underline{\sigma}+\mathrm{Re}\lambda-\frac{\omega}{2})\xi} \cos(\xi \mathrm{Im}\lambda) dv_1 dv_2 d\xi \right|.$$

We have

$$\begin{aligned}
F_\lambda(u) &\leq 4 \left| \int_u^{+\infty} \frac{u}{\xi^2} e^{-(\underline{\sigma}+\mathrm{Re}\lambda-\frac{\omega}{2})\xi} \cos(\xi \mathrm{Im}\lambda) d\xi \right| \\
&\leq 4 \left| \frac{1}{u\, \mathrm{Im}\lambda} e^{-(\underline{\sigma}+\mathrm{Re}\lambda-\frac{\omega}{2})u} \sin(u\, \mathrm{Im}\lambda) \right. \\
&\quad \left. + \frac{1}{\mathrm{Im}\lambda} \int_u^{+\infty} \left(\frac{2u}{\xi^3} + \frac{u}{\xi^2} \left(\underline{\sigma} + \mathrm{Re}\lambda - \frac{\omega}{2} \right) \right) e^{-(\underline{\sigma}+\mathrm{Re}\lambda-\frac{\omega}{2})\xi} \sin(\xi \mathrm{Im}\lambda) d\xi \right| \\
&\leq 4 \frac{(2+e^{-1})}{u|\mathrm{Im}\lambda|} \quad \left(\text{because } \left(\underline{\sigma} + \mathrm{Re}\lambda - \frac{\omega}{2} \right) e^{-(\underline{\sigma}+\mathrm{Re}\lambda-\frac{\omega}{2})\xi} \leq \frac{1}{\xi} e^{-1} \right).
\end{aligned}$$

Let

$$G_\lambda(u) = \left| \int_{a_u(P_i)} \frac{u}{\xi^2} e^{-(\underline{\sigma}+\mathrm{Re}\lambda-\frac{\omega}{2})\xi} \sin(\xi \mathrm{Im}\lambda) d\xi \right|.$$

In the same way, we prove that $G_\lambda(u) \leq 4\frac{(2+e^{-1})}{u|\mathrm{Im}\lambda|}$. So,

$$\left| \int_P \tilde{\mu}(v) \tilde{\theta}(v) e^{-\frac{(\underline{\sigma}+\lambda-\frac{\omega}{2})}{v_3} u} \tilde{L}_u(v)\, dv \right|^q \leq \frac{\left(8N \sup |h(.,.,.)|(2+e^{-1}) \right)^q}{u^q |\mathrm{Im}\lambda|^q}.$$

This yields the following

$$\begin{aligned}
&\int_\varepsilon^{+\infty} \left| \int_{K^1} \tilde{\mu}(v) \tilde{\theta}(v) e^{-\frac{(\underline{\sigma}+\lambda-\frac{\omega}{2})}{v_3} u} \tilde{L}_u(v)\, dv \right|^q du \\
&\quad \leq \frac{\left(8N \sup |h(.,.,.)|(2+e^{-1}) \right)^q}{|\mathrm{Im}\lambda|^q} \int_\varepsilon^{+\infty} \frac{du}{u^q} \\
&\quad \leq \frac{\left(8N \sup |h(.,.,.)|(2+e^{-1}) \right)^q}{|\mathrm{Im}\lambda|^q (q-1) \varepsilon^{q-1}},
\end{aligned}$$

which implies that

$$|\mathrm{Im}\lambda|^q \int_\varepsilon^{+\infty} \left| \int_{K^1} \mu(v) \theta(v) e^{-\frac{(\underline{\sigma}+\lambda-\frac{\omega}{2})}{v_3} u} L_u(v)\, dv \right|^q du \text{ is bounded on } R_\omega.$$

This completes the proof of (i).

(ii) Let $\varphi \in X_p$. Then,

$$(FB_\lambda^+ D_1(M_\lambda^+ D_1)^n G_\lambda^+ F\varphi)(x,v)$$
$$= \eta(x)\theta(v)\int_0^1 dx' \times \int_{K^1} \frac{1}{|v_3'|}\eta(v')\theta(v')\beta(v')e^{-\frac{(\sigma(v')+\lambda)}{|v_3'|}(n+1+x-x')}\,dv'$$
$$\int_K \beta(v'')\varphi(x',v'')\,dv''$$
$$= E_1E_2E_3\varphi(x,v),$$

where

$$\begin{cases} E_1 : L_p(0,1) \longrightarrow X_p \\ \varphi \longrightarrow \eta(x)\theta(v)\varphi(x), \end{cases}$$

$$\begin{cases} E_2 : L_p(0,1) \longrightarrow L_p(0,1) \\ \varphi \longrightarrow \int_0^1 \varphi(x')\,dx' \int_{K^1} \frac{1}{v_3}\theta(v)\beta(v)e^{-\frac{(\sigma(v)+\lambda)}{v_3}(n+1+x-x')}\,dv \end{cases}$$

and

$$\begin{cases} E_3 : X_p \longrightarrow L_p(0,1) \\ \qquad \varphi \longrightarrow \eta(x)\int_K \beta(v)\varphi(x,v)\,dv. \end{cases}$$

Clearly, E_1 and E_3 are bounded and do not depend on λ. Hence, in order to prove the item (ii), we only have to demonstrate that $|\text{Im}\lambda|\|E_2\|$ is bounded on R_ω. The operator E_2 is bounded and the use of Hölder's inequality gives $\|E_2\| \le 2^{\frac{1}{p}}\|H_\lambda\|_{L_q(\mathbb{R})}$. According to Lemma 13.1.4, it suffices to prove that, for any $\varepsilon > 0$,

$$|\text{Im}\lambda|^q \int_\varepsilon^{+\infty} \left|\int_{K^1} \frac{1}{v_3}\theta(v)\beta(v)e^{-\frac{(\underline{\sigma}+\lambda-\frac{\omega}{2})}{v_3}u}L_u(v)\,dv\right|^q du \text{ is bounded on } R_\omega,$$

where $u > 0$ and $L_u(.)$ is a nonnegative measurable simple function on K^1, satisfying $L_u(.) \le \varphi_u(.)$. Arguing as in the proof of Lemma 13.1.6(i), we get the desired result.

(iii) The result follows immediately from (i) and (ii). Q.E.D.

Theorem 13.1.1. *Let $p \in [1,+\infty[$, let $\omega > 0$ and set $R_\omega = \{\lambda \in \mathbb{C}$ such that $\mathrm{Re}\lambda \geq -\underline{\sigma}+\omega\}$. Assume that the collision operator F is regular. If the boundary operator H satisfies the hypothesis $(\mathcal{P})$, then $|\mathrm{Im}\lambda|\|F(\lambda - T_H)^{-1}F\|$ is bounded on R_ω.* ◇

Proof. Let $\lambda \in \mathbb{C}$ such that $\mathrm{Re}\lambda > -\underline{\sigma}$. Using Eqs. (13.1.10) and (13.1.11), we have

$$\|FR^+(\lambda,T_H)F\| \leq \|FC_\lambda^+F\| + \sum_{n\geq 0}\|FB_\lambda^+H_{12}(M_\lambda^+H_{12})^nG_\lambda^+F\| \tag{13.1.19}$$

and

$$\|FR^-(\lambda,T_H)F\| \leq \|FC_\lambda^-F\| + \sum_{n\geq 0}\|FB_\lambda^-H_{21}(M_\lambda^-H_{21})^nG_\lambda^-F\|. \tag{13.1.20}$$

Now, the result follows from (13.1.19), (13.1.20), Lemmas 13.1.5, and 13.1.6(*iii*). Q.E.D.

13.1.4 Solution for the Cauchy Problem (13.1.1)

13.1.5 Generation Results

First, we recall that if the boundary operator H satisfies $\|H\| \leq 1$, then T_H generates a strongly continuous semigroup $(U^H(t))_{t\geq 0}$ and, according to the classical perturbation theory (Theorem 2.5.8), we also recall that A_H generates a C_0-semigroup $(V^H(t))_{t\geq 0}$ on L_p.

Lemma 13.1.7. *The C_0-semigroup $(U^H(t)_{t\geq 0}$ and $(U^0(t)_{t\geq 0}$ satisfy the following inequality*

$$U^H(t) \geq U^0(t) \geq 0, \quad (t \geq 0).$$ ◇

Proof. For $t = 0$, the result is trivial. We fix $t > 0$. By using inequality (13.1.12), it is obvious that, for all integers n such that $\frac{n}{t} > -\underline{\sigma}$ and for all $\psi \geq 0$, we have $\left[R\left(\frac{n}{t},T_H\right)\right]^n \geq \left[R\left(\frac{n}{t},T_0\right)\right]^n \geq 0$. Consequently, $\lim_{n\to\infty}\left[R\left(\frac{n}{t},T_H\right)\right]^n \geq \lim_{n\to\infty}\left[R\left(\frac{n}{t},T_0\right)\right]^n \geq 0$. By using the exponential formula (see Theorem 2.5.11), we deduce that

$$U^H(t) \geq U^0(t) \geq 0, \quad (t \geq 0).$$

Q.E.D.

Lemma 13.1.8. *Using the same notations as previously, if the operator F is positive, then $V^H(t) \geq V^0(t) \geq 0$, $(t \geq 0)$.* ◇

Proof. Let $\lambda \in \rho(A_H)\bigcap \rho(A_0)$ such that $r_\sigma((\lambda - T_H)^{-1}F) < 1$ (spectral radius). Consequently, $(\lambda - A_H)^{-1} - (\lambda - T_H)^{-1} = \sum_{n\geq 1}[(\lambda - T_H)^{-1}F]^n(\lambda - T_H)^{-1}$. The positivity of F and Eq. (13.1.12) imply $[(\lambda - T_H)^{-1}F]^n(\lambda - T_H)^{-1} \geq [(\lambda - T_0)^{-1}F]^n(\lambda - T_0)^{-1} \geq 0$, and therefore, $R(\lambda, A_H) \geq R(\lambda, A_0) \geq 0$. To take into account the last inequality, a similar reasoning to that of Lemma 13.1.7 yields the desired result. Q.E.D.

13.1.6 Time-Asymptotic Behavior

We will prove that the solution $\psi(t)$ has a nice behavior on L_p-spaces, $1 \leq p < +\infty$, independently of the geometry of such spaces and we estimate the norm $\|\psi(t) - \sum_{i=1}^n e^{\lambda_i t}e^{D_i t}P_i\psi_0\|$ without restriction on the initial data ψ_0. We start with discussing the spectrum of the transport operator in the half plane $\{\lambda \in \mathbb{C}$ such that $\mathrm{Re}\lambda > -\underline{\sigma}\}$. Setting $P(A_H) = \sigma(A_H)\bigcap\{\lambda \in \mathbb{C}$ such that $\mathrm{Re}\lambda > -\underline{\sigma}\}$, $\sigma_p(A_H)$ the point spectrum of A_H, $\rho(A_H)$ the resolvent set of the operator A_H, and $s(A_H)$ the spectral bound of A_H. Firstly, we prove the following lemma:

Lemma 13.1.9. *Let $p \in [1, +\infty[$ and assume that the collision operator F is regular on X_p. If the hypothesis $(\mathcal{P})$ is satisfied, then*

(i) $P(A_H)$ consists, at most, of discrete eigenvalues with finite algebraic multiplicities.
(ii) If $\omega > 0$, then the set $\sigma(A_H)\bigcap R_\omega$ is finite.
(iii) If $\omega > 0$, then there exists $C(\omega) > 0$ such that $\|(\lambda - A_H)^{-1}\|$ is bounded on

$$\{\lambda \in \mathbb{C} \text{ such that } \mathrm{Re}\lambda \geq \omega \text{ and } |\mathrm{Im}\lambda| \geq C(\omega)\}.$$ ◇

Proof. Let λ be such that $\mathrm{Re}\lambda > -\underline{\sigma}$. Since F is regular, then using Lemma 2.1.13 and Proposition 13.1.1, we deduce that $[(\lambda - T_H)^{-1}F]^2$ is compact on X_p, for $1 \leq p < +\infty$. Next, using Theorem 13.1.1, we get $\lim_{|\mathrm{Im}\lambda|\to+\infty} \|[(\lambda - T_H)^{-1}F]^2\| = 0$ uniformly on R_ω. Now, the items (i)–(iii) follow immediately from Lemma 4.2.1. Q.E.D.

In the following lemma, we have two classical facts from the spectral theory of transport operators required below (see [184, 261] or [328]).

Lemma 13.1.10. *Let $p \in [1, +\infty[$, let λ be such that $\text{Re}\lambda > \underline{\sigma}$ and assume that the collision operator is regular on X_p. If the hypothesis $(\mathcal{P})$ is satisfied, then*

(*i*) *$\lambda \in \sigma_p(A_H)$ if, and only if, $1 \in \sigma_p((\lambda - T_H)^{-1}F)$ and the corresponding eigen-subspaces are the same.*
(*ii*) *$\lambda \in \rho(A_H)$ if, and only if, $1 \in \rho((\lambda - T_H)^{-1}F)$.* ◇

Proof.

(i) It follows from the equality $(\lambda - T_H)^{-1}(\lambda - A_H) = (I - (\lambda - T_H)^{-1}F)$.
(ii) Let $\lambda \in \sigma(A_H)$ such that $\text{Re}\lambda > -\underline{\sigma}$, then $\lambda \in \sigma_p(A_H)$. Consequently, using ($i$), we get $1 \in \sigma_p((\lambda - T_H)^{-1}F) \subset \sigma((\lambda - T_H)^{-1}F)$. Conversely, since $(\lambda - T_H)^{-1}F$ is compact for $1 < p < +\infty$ and weakly compact for $p = 1$, if $1 \in \sigma((\lambda - T_H)^{-1}F)$, then $1 \in \sigma_p((\lambda - T_H)^{-1}F)$. Now, the result follows from (i). Q.E.D.

Let $p \in [1, +\infty[$, we denote by $T_{H,p}$ and $A_{H,p}$ the streaming operator and the transport operator on the space X_p respectively. Now, we are prepared to establish the following result:

Proposition 13.1.3. *We also assume that the collision operator F is regular on X_p for all $p \in [1, +\infty[$. We also assume that the hypothesis $(\mathcal{P})$ is satisfied. Then,*

(*i*) *$P(A_{H,p})$ is independent of p.*
(*ii*) *For each $\lambda \in P(A_{H,p})$ and $m \in \mathbb{N}^*$, the eigen-subspace $N[(\lambda - A_{H,p})^m]$ is independent of p.*

◇

Proof.

(i) Let $\lambda \in G := \{\lambda \in \mathbb{C}$ such that $\text{Re}\lambda > -\underline{\sigma}\}$. Since $T_{H,p}$ generates a strongly continuous semigroup on X_p and F is bounded, then $\|(\lambda - T_{H,p})^{-1}F\| \to 0$ as $|\text{Re}\lambda| \to +\infty$. So, for $\lambda \in G$ such that $|\text{Re}\lambda|$ is large enough, we have $\left(I - (\lambda - T_{H,p})^{-1}F\right)$ is boundedly invertible. If $p > 1$, then by using Proposition 13.1.1, $(\lambda - T_{H,p})^{-1}F$ is compact. From the Gohberg–Shmul'yan's theorem (see Theorem 2.5.13), $(I - (\lambda - T_{H,p})^{-1}F)$ is boundedly invertible for all $\lambda \in G$, except at a discrete set of points $S_p := \{\lambda_k, k = 1, 2, \ldots\}$. For each $k \in \{1, 2, \ldots\}$, λ_k is a pole with a finite order of $\left(I - (\lambda - T_{H,p})^{-1}F\right)^{-1}$ and 1 is an eigenvalue of $(\lambda_k - T_{H,p})^{-1}F$. We claim that $S_p = P(A_{H,p})$, for all $p > 1$. Indeed, from Lemma 13.1.10, (i) we have $S_p \subset P(A_{H,p})$. Conversely, if $\lambda_0 \in P(A_{H,p})$, then we may use the product formula

$$(\lambda - A_{H,p})^{-1} = (I - [(\lambda - T_{H,p})^{-1}F])^{-1}(\lambda - T_{H,p})^{-1}. \tag{13.1.21}$$

Using Eq. (13.1.21), the non-invertibility of $(\lambda_0 - A_{H,p})$ gives the non-invertibility of$\left(I - (\lambda_0 - T_{H,p})^{-1}F\right)$, which implies that $\lambda_0 \in S_p$. So, $S_p = P(A_{H,p})$, for all $p > 1$. If $p = 1$, then by using Lemma 2.1.13(i), $[(\lambda - T_{H,1})^{-1}F]^2$ is compact for all $\lambda \in G$. From the Gohberg–Shmul'yan's corollary (see Corollary 2.5.2), $(\lambda - A_{H,1})$ is boundedly invertible for all $\lambda \in G$, except at a discrete set of points $S_1 := \{\mu_k, k = 1, 2, \ldots\}$. For each $k \in \{1, 2, \ldots\}$, μ_k is a pole of a finite order of $(\lambda - A_{H,1})^{-1}$, an eigenvalue of $A_{H,1}$ with a finite algebraic multiplicity and 1 is an eigenvalue of $(\mu_k - T_{H,1})^{-1}F$. So, $S_1 = P(A_{H,1})$. It remains to prove that S_p is independent of p. In order to do it, it is sufficient to prove that $S_p = S_1$, for all $p > 1$. It follows, from (13.1.10) and (13.1.11), that

$$(I - (\lambda - T_{H,p})^{-1}F)^{-1} = (I - (\lambda - T_{H,1}^{-1})F)^{-1}_{|X_p}, \quad \text{for any } p \geq 1.$$

So, $S_p \subset S_1$. Conversely, if $\lambda_0 \in S_1$, then it is a pole of $(I - (\lambda - T_{H,1})^{-1}F)^{-1}$. Then, we have the following expansion in the neighborhood of λ_0

$$(I - (\lambda - T_{H,1})^{-1}F)^{-1} = \sum_{k=1}^{n} \frac{A_k}{(\lambda - \lambda_0)^k} + \sum_{k=0}^{+\infty} (\lambda - \lambda_0)^k B_k,$$

where n is the (finite) algebraic multiplicity of λ_0, A_k and B_k are bounded operators and independent of λ. For all $\psi \in X_p$, we have

$$(I - (\lambda - T_{H,1})^{-1}F)^{-1}\psi = \sum_{k=1}^{n} \frac{A_k\psi}{(\lambda - \lambda_0)^k} + \sum_{k=0}^{+\infty} (\lambda - \lambda_0)^k B_k\psi.$$

Suppose that there exists $p > 1$ such that $\lambda_0 \notin S_p$. Then, $A_k\psi = 0$, $\forall \psi \in X_p$, $\forall k \in \{1, 2, \ldots, n\}$. Since X_p is dense in X_1, then $A_k = 0$, $\forall k \in \{1, 2, \ldots, n\}$. This implies that $(I - (\lambda - T_{H,1})^{-1}F)^{-1}$ is analytic at λ_0, which is a contradiction.

(*ii*) Since $(\lambda - A_{H,p})^{-1} = (\lambda - T_{H,p})^{-1}(I - [(\lambda - T_{H,p})^{-1}F])^{-1}$, the use of Eqs. (13.1.9) and (13.1.13) shows that $(\lambda - A_{H,p})^{-1} = (\lambda - A_{H,1})^{-1}_{|X_p}$ for any $\lambda \in G \backslash S$. Let $\lambda \in S$, λ is then an eigenvalue of $A_{H,p}$ with a finite algebraic multiplicity. Let us denote by $P_{\lambda,p}$ the spectral projection associated with $\{\lambda\}$. Let $\delta > 0$ be such that

$$\{\mu \in \mathbb{C} \text{ such that } 0 < |\lambda - \mu| \leq \delta\} \bigcap P(A_{H,p}) = \emptyset.$$

Then,

$$P_{\lambda,p} = \frac{1}{2i\pi} \int_{|\mu - \lambda| = \delta} (\mu - A_{H,p})^{-1} d\mu.$$

Accordingly, for each $\lambda \in P(A_{H,p}) = P(A_{H,1})$, we have

$$P_{\lambda,p} = P_{\lambda,1|X_p}. \tag{13.1.22}$$

Let $\psi \in X_p$. Then, there is a sequence $(\psi_n)_n \in C_0^\infty([0,1] \times K)$ such that $\psi_n \to \psi$ as $n \to \infty$ in X_p. Since $P_{\lambda,p}$ is a bounded operator on X_p, then $P_{\lambda,p}\psi_n \to P_{\lambda,p}\psi$ as $n \to \infty$ in X_p. This shows that $P_{\lambda,p}(C_0^\infty([0,1] \times K))$ is a dense subset of $R(P_{\lambda,p})$. However, $\lambda \in S = P(A_{H,p})$. Then, $R(P_{\lambda,p})$ is a finite-dimensional subset of X_p and then,

$$R(P_{\lambda,p}) = P_{\lambda,p}(C_0^\infty([0,1] \times K)) \text{ for } p \geq 1. \tag{13.1.23}$$

It follows, from Eqs. (13.1.22) and (13.1.23), that $R(P_{\lambda,p}) = R(P_{\lambda,1})$ for $p \geq 1$ and $\lambda \in P(A_{H,p}) = P(A_{H,1})$. Now, observing that, for each $k \geq 1$, $N[(\lambda - A_{H,p})^k] \subset R(P_{\lambda,p}) = R(P_{\lambda,1})$ and $N[(\lambda - A_{H,1})^k] \subset R(P_{\lambda,1}) = R(P_{\lambda,p})$, we conclude that $N[(\lambda - A_{H,p})^k] = N[(\lambda - A_{H,1})^k] \subset X_p \bigcap X_1 = X_p$. This completes the proof.
Q.E.D.

From Proposition 13.1.2, T_H generates a strongly continuous semigroup $(U^H(t))_{t\geq 0}$ on X_p $(p \geq 1)$. Since the operator F is bounded, and from the classical perturbation theory (see Theorem 2.5.8), it follows that $A_H = T_H + F$ also generates a strongly continuous semigroup $(V^H(t))_{t\geq 0}$ on X_p given by the Dyson–Phillips expansion Eq. (4.2.3). Assume that F is regular on X_p and that H satisfies $(\mathcal{P})$. Then, from Lemma 13.1.9, the eigenvalues $\{\lambda_1, \lambda_2, \ldots, \lambda_n, \lambda_{n+1}, \ldots\}$ of A_H lying in the half plane $\text{Re}\lambda > -\underline{\sigma}$, can be ordered in such a way that the real part decreases, i.e., $\text{Re}\lambda_1 > \text{Re}\lambda_2 > \cdots > \text{Re}\lambda_{n+1} > \cdots > -\underline{\sigma}$ and $\{\lambda \in \mathbb{C} \text{ such that } \text{Re}\lambda > -\underline{\sigma}\} \backslash \{\lambda_n, n = 1, 2 \ldots\} \subset \rho(A_H)$. Let P_i and D_i denote, respectively, the spectral projection and the nilpotent operator associated with λ_i, $i = 1, 2, \ldots, n$. Then, $P = P_1 + \cdots + P_n$ is the spectral projection of the compact set $\{\lambda_1, \lambda_2, \ldots, \lambda_n\}$. Hence, according to the spectral decomposition theorem corresponding to the sets $\{\lambda_1, \lambda_2, \ldots, \lambda_n\}$ and $\sigma(A_H) \backslash \{\lambda_1, \lambda_2, \ldots, \lambda_n\}$ (see [35, pp. 68–70]), $V^H(t)$ splits as

$$V^H(t) = \tilde{V}^H(t) + \sum_{i=1}^{n} e^{\lambda_i t} e^{D_i t} P_i,$$

where $\tilde{V}^H(t) := V^H(t)(I - P)$. Furthermore, $(\tilde{V}^H(t))_{t\geq 0}$ is a C_0-semigroup on the Banach space $(I - P)X_p$ with a generator $\tilde{A}_H := A_H(I - P)$. Let

$$\zeta := s(A_H) - \sup\{\text{Re}\lambda \text{ such that } \lambda \in \sigma(A_H) \text{ and } \lambda \neq s(A_H)\}.$$

Clearly, by using Lemma 13.1.9, we can see that $\zeta > 0$. Let $\tilde{P}$ denote the projection operator corresponding to $\{\lambda \in \sigma(A_H) \text{ such that } \text{Re}\lambda = s(A_H)\}$. According to Proposition 13.1.3, this set consists of a finite number of eigenvalues, and then is compact. Then, the spectral decomposition theorem (Theorem 2.5.9) can be applied. Now, we are ready to prove the following.

Theorem 13.1.2. *Let $p \in [1, +\infty[$. Assume that the condition $(\mathcal{P})$ holds and that the collision operator F satisfies the hypotheses (i)–(iii) of Lemma 2.4.1 for $D := [0, 1]$ and $V := K$. Then,*

(i) *For each $\varepsilon > 0$, there exists a positive constant M such that $\|V^H(t)(I-P)\| \leq Me^{(\mathrm{Re}\lambda_{n+1}+\varepsilon)t}$, $\forall t > 0$.*

(ii) *Moreover, if F is positive and $p \in]1, +\infty[$, then there exists a positive constant M' such that $\|V^H(t)(I - \tilde{P})\| \leq M'e^{(s(A_H)-\varepsilon)t}$, for any $\varepsilon \in (0, 2\zeta(1 - p^{-1})))$ [resp. $\varepsilon \in (0, 2\zeta p^{-1})$] if $p \in]1, 2]$ (resp. $p \in]2, +\infty[$), where $\tilde{P}$ denotes the projection operator corresponding to $\{\lambda \in \sigma(A_H)$ such that $\mathrm{Re}\lambda = s(A_H)\}$, and where $s(A_H)$ represents the spectral bound of A_H, and $\zeta := s(A_H) - \sup\{\mathrm{Re}\lambda$ such that $\lambda \in \sigma(A_H)$ and $\lambda \neq s(A_H)\}$.*

◇

Proof.

(i) For $p \geq 1$, the result follows immediately from Proposition 13.1.1, Lemma 13.1.2, Theorems 13.1.1, and 4.2.1.

(ii) According to Proposition 13.1.3, $s(A_H)$ the spectral bound of A_H keeps the same value for every $p \in \mathbb{N}^*$. Moreover, the C_0-semigroup $(V^H(t))_{t\geq 0}$ is positive (see Lemma 13.1.8) and does not change with respect to $p \in [1, +\infty[$. From Weis theorem (Theorem 2.5.4), the types of $(V^H(t))_{t\geq 0}$ and $s(A_H)$ coincide. Now, the result follows from Lemma 13.1.9, Proposition 13.1.3, Weis theorem (Theorem 2.5.4) and Theorem 4.3.1 which holds for $p \in]1, +\infty[$.

Q.E.D

13.2 Time-Asymptotic Behavior of the Solution for a Cauchy Problem Given by a One-Velocity Transport Operator with Maxwell Boundary Condition

This section is devoted to describe the time-asymptotic behavior of the solution of a one-velocity transport operator with Maxwell boundary condition on L_1-spaces. A practical way is given in order to study the behavior of the solution without restriction on the initial data. In a homogeneous medium with spherical symmetry and isotropic scattering, the one-particle distribution function $\Phi(r, \mu, t)$ satisfies the following transport equation

$$\frac{\partial \Phi}{\partial t}(r, \mu, t) = -\mu \frac{\partial \Phi}{\partial r}(r, \mu, t) - \frac{1-\mu^2}{r}\frac{\partial \Phi}{\partial \mu}(r, \mu, t) - \Sigma\Phi(r, \mu, t)$$
$$+\frac{c\Sigma}{2}\int_{-1}^{1} \Phi(r, \mu', t)d\mu'$$

with the following boundary condition of Maxwell type

$$|\mu|\Phi(R,\mu,t)=\int_0^1 \alpha\mu'\Phi(R,\mu',t)d\mu' \quad (-1\le\mu<0, t\ge 0)$$

and with the initial data $\Phi(r,\mu,0)=\Phi_0(r,\mu)$, where $r\in S=[0,R]$, $\mu\in\Omega=[-1,1]$, $t\ge 0$ and R is the radius of the sphere. Both Σ and c are positive constants and $\alpha\ge 0$ is the scattering coefficient on the boundary of the sphere.

13.2.1 Auxiliary Results

Let $x=r\mu$, $y=r\sqrt{1-\mu^2}$, $(r\in S,\mu\in\Omega)$, then this transformation is one to one from $G:=[0,R]\times[-1,1]$ onto $D:=\{(x,y)\in\mathbb{R}^2$ such that $x^2+y^2\le R^2$, $y\ge 0\}$. Hence, under the following isometry J

$$\begin{cases} J: L_1(G,r^2drd\mu)\longrightarrow L_1(D,ydxdy) \\ \qquad f(x,y)=(J\Phi)(x,y)=\Phi\left(\sqrt{x^2+y^2},\dfrac{x}{\sqrt{x^2+y^2}}\right)\end{cases}$$

the transport equation, the Maxwell boundary condition and the initial data are transformed into

$$\begin{cases} \dfrac{\partial f}{\partial t}(x,y,t)=-\dfrac{\partial f}{\partial x}(x,y,t)-\Sigma f(x,y,t)+\dfrac{c\Sigma}{2r}\displaystyle\int_{-r}^{r} f\left(x',\sqrt{r^2-x'^2},t\right)dx' \\ \dfrac{y}{R^2}f\left(-\sqrt{R^2-y^2},y,t\right)=\displaystyle\int_0^R \dfrac{\alpha y}{R\sqrt{R^2-y^2}}\dfrac{y'}{R^2}f\left(\sqrt{R^2-y'^2},y',t\right)dy' \\ f(x,y,0)=f_0(x,y), \end{cases} \tag{13.2.1}$$

where $r=\sqrt{x^2+y^2}$, $(x,y)\in D$ and $t\ge 0$. Let $X:=L_1[D,ydxdy]$ be the Banach space equipped with the norm $\|\psi\|_X=\int_D|\psi(x,y)|ydxdy$ and let $Y:=L_1[S,\frac{y}{R^2}dy]$ be the Banach space equipped with the norm $\|\varphi\|_Y=\int_S|\varphi(y)|\frac{y}{R^2}dy$. The boundary operator H is defined as:

$$\begin{cases} H: Y\longrightarrow Y \\ \qquad f\longrightarrow Hf(y)=\displaystyle\int_0^R \dfrac{\alpha}{\sqrt{R^2-y^2}}f(y')\dfrac{y'}{R}dy'. \end{cases}$$

We introduce the following boundary spaces $\Gamma_e := \left\{(\sqrt{R^2-y^2}, y),\ y \geq 0\right\}$ and $\Gamma_s := \left\{(-\sqrt{R^2-y^2}, y),\ y \geq 0\right\}$. We define the partial Sobolev space W by

$$W = \left\{\psi \in X \text{ such that } \frac{\partial \psi}{\partial x} \in X\right\}.$$

Proposition 13.2.1. *Let $\psi \in W$. If $\psi_{|\Gamma e} \in Y$, then $\psi_{|\Gamma s} \in Y$ and vice versa.* ◇

Proof. Let $\psi \in W$ such that $\psi_{|\Gamma e} \in Y$. We have

$$\psi\left(\sqrt{R^2-y^2}, y\right) - \psi\left(-\sqrt{R^2-y^2}, y\right) = \int_{-\sqrt{R^2-y^2}}^{\sqrt{R^2-y^2}} \frac{\partial \psi}{\partial x}(x,y)dx. \qquad (13.2.2)$$

This implies

$$|\psi\left(-\sqrt{R^2-y^2}, y\right)| \leq |\psi\left(\sqrt{R^2-y^2}, y\right)| + \int_{-\sqrt{R^2-y^2}}^{\sqrt{R^2-y^2}} \left|\frac{\partial \psi}{\partial x}(x,y)\right| dx.$$

Consequently, we have

$$\int_0^R |\psi(-\sqrt{R^2-y^2}, y)| \frac{y}{R^2} dy \leq \int_0^R |\psi(\sqrt{R^2-y^2}, y)| \frac{y}{R^2} dy + \int_D \left|\frac{\partial \psi}{\partial x}(x,y)\right| \frac{y}{R^2} dxdy.$$

So, we have $\left\|\psi_{|\Gamma s}\right\|_Y \leq \left\|\psi_{|\Gamma e}\right\|_Y + \frac{1}{R^2}\left\|\frac{\partial \psi}{\partial x}\right\|_X < \infty$. The converse may be proved in a similar way, starting from a rearrangement of Eq. (13.2.2). This completes the proof. Q.E.D.

Let us define $\mathcal{W}$ by $\mathcal{W} = \{\psi \in W \text{ such that } \psi_{|\Gamma e} \in Y\}$. By using Proposition 13.2.1, we deduce that all functions ψ in $\mathcal{W}$ have traces $\psi_{|\Gamma e}$ and $\psi_{|\Gamma s}$ belonging to the boundary space Y. Next, we introduce the transport operator $A_\alpha := B_\alpha + K$ where B_α is the free streaming operator, namely

$$\begin{cases} B_\alpha : \mathcal{D}(B_\alpha) \subset X \longrightarrow X \\ \qquad \psi \longrightarrow B_\alpha \psi(x,y) = -\dfrac{\partial \psi}{\partial x}(x,y) - \Sigma \psi(x,y) \\ \mathcal{D}(B_\alpha) = \{\psi \in \mathcal{W} \text{ such that} \psi_{|\Gamma s} = H \psi_{|\Gamma e}\} \end{cases}$$

and K is the bounded operator given by

$$\begin{cases} K : X \longrightarrow X \\ \quad \psi \longrightarrow \dfrac{c\Sigma}{2r} \displaystyle\int_{-r}^{r} \psi\left(x', \sqrt{r^2 - x'^2}, t\right) dx'. \end{cases}$$

We will study the time-asymptotic behavior of the solution of the initial boundary value problem (13.2.1) which can be formulated in the following Cauchy problem:

$$\begin{cases} \dfrac{\partial f}{\partial t} = A_\alpha f := B_\alpha f + Kf \\ f(0) = f_0, \end{cases} \tag{13.2.3}$$

where $f_0 \in X := L_1[D, ydxdy]$.

Theorem 13.2.1. *Let $\alpha > 1$. Then, B_α is the generator of a positive C_0-semigroup $(U_\alpha(t))_{t\geq 0}$ on X.* ◇

Proof. Let $\lambda > \lambda_0$ and $\varphi \in X^+$ the positive cone of X. Then, $(\lambda - B_\alpha)^{-1}\varphi = \psi \in \mathcal{D}(B_\alpha) \bigcap X^+$. This implies the following equation $(\lambda + \Sigma)\psi(x, y) + \frac{\partial \psi}{\partial x}(x, y) = \varphi(x, y)$. After integrating this equation with respect to x and y, we obtain

$$(\lambda + \Sigma) \int_0^R \int_{-\sqrt{R^2-y^2}}^{\sqrt{R^2-y^2}} \psi(x, y) ydxdy + \int_0^R \int_{-\sqrt{R^2-y^2}}^{\sqrt{R^2-y^2}} \frac{\partial \psi}{\partial x}(x, y) ydxdy$$
$$= \int_0^R \int_{-\sqrt{R^2-y^2}}^{\sqrt{R^2-y^2}} \varphi(x, y) ydxdy.$$

This implies that $(\lambda+\Sigma)\|\psi\|+\int_0^R \left(\psi(\sqrt{R^2-y^2}, y) - \psi(-\sqrt{R^2-y^2}, y)\right) ydy = \|\varphi\|$ or, equivalently $(\lambda + \Sigma)\|\psi\| + R^2 \left(\|\psi_{|\Gamma s}\| - \|\psi_{|\Gamma e}\|\right) = \|\varphi\|$. Therefore, using the fact that $\psi = (\lambda - B_\alpha)^{-1}\varphi$, we deduce that $\|(\lambda - B_\alpha)^{-1}\varphi\| \geq \frac{1}{\lambda+\Sigma}\|\varphi\|$, $\forall \lambda > \lambda_0$. Now, the result is a consequence of both the last equation and Theorem 2.5.3. Q.E.D.

Theorem 13.2.2. *Let $\alpha > 1$. Then, the operator A_α is the generator of a positive C_0-semigroup $(V_\alpha(t))_{t\geq 0}$ on X.* ◇

Proof. By using Theorem 13.2.1, we deduce that B_α is a generator of a positive C_0-semigroup $(U_\alpha(t))_{t\geq 0}$ on X. According to the perturbation theory, $A_\alpha = B_\alpha + K$ is the generator of a positive C_0-semigroup $(V_\alpha(t))_{t\geq 0}$ (Theorem 2.5.8) and the proof is complete. Q.E.D.

13.2.2 The Resolvent of the Operator B_α

The objective of this section is to determine the solution of the operator equation

$$(\lambda - B_\alpha)\psi = g, \tag{13.2.4}$$

where g is a given function in X, $\lambda \in \mathbb{C}$ and ψ must belong to $\mathcal{D}(B_\alpha)$. For $\text{Re}\lambda > -\Sigma$, a straightforward calculation of (13.2.4) leads to

$$\begin{aligned}\psi(x,y) &= \psi(-\sqrt{R^2-y^2},y)e^{-(\lambda+\Sigma)(x+\sqrt{R^2-y^2})}\\ &\quad + \int_{-\sqrt{R^2-y^2}}^{x} e^{-(\lambda+\Sigma)(x-x')}g(x',y)dx'.\end{aligned} \tag{13.2.5}$$

Accordingly, for $x = \sqrt{R^2-y^2}$, we get

$$\begin{aligned}\psi(\sqrt{R^2-y^2},y) &= \psi(-\sqrt{R^2-y^2},y)e^{-2(\lambda+\Sigma)\sqrt{R^2-y^2}}\\ &\quad + \int_{-\sqrt{R^2-y^2}}^{\sqrt{R^2-y^2}} e^{-(\lambda+\Sigma)(\sqrt{R^2-y^2}-x')}g(x',y)dx'.\end{aligned} \tag{13.2.6}$$

In order to clarify the analysis, let us introduce the following bounded operators:

$$\begin{cases} M_\lambda : Y \longrightarrow Y \\ \qquad f \longrightarrow M_\lambda f(y) = f(y)\, e^{-2(\lambda+\Sigma)\sqrt{R^2-y^2}}, \end{cases}$$

$$\begin{cases} R_\lambda : Y \longrightarrow X \\ \qquad f \longrightarrow (R_\lambda f)(x,y) = f(y)\, e^{-(\lambda+\Sigma)(x+\sqrt{R^2-y^2})}. \end{cases}$$

The operators M_λ and R_λ are bounded and satisfy the following estimates

$$\begin{aligned}\|M_\lambda\| &\le 1,\\ \|R_\lambda\| &\le \frac{R^2}{\text{Re}\lambda + \Sigma}.\end{aligned} \tag{13.2.7}$$

$$\begin{cases} Q_\lambda : X \longrightarrow Y \\ \qquad \psi \longrightarrow (Q_\lambda\psi)(x,y) = \int_{-\sqrt{R^2-y^2}}^{\sqrt{R^2-y^2}} e^{-(\lambda+\Sigma)(\sqrt{R^2-y^2}-x')}\psi(x',y)dx' \end{cases}$$

and finally,

$$\begin{cases} P_\lambda : X \longrightarrow X \\ \qquad \psi \longrightarrow (P_\lambda \psi)(x, y) = \displaystyle\int_{-\sqrt{R^2-y^2}}^{x} e^{-(\lambda+\Sigma)(x-x')} \psi(x', y) dx'. \end{cases}$$

A simple calculation shows that Q_λ and P_λ are bounded and satisfy the following estimates

$$\| Q_\lambda \| \leq \frac{1}{R^2},$$

$$\| P_\lambda \| \leq \frac{1}{Re\lambda + \Sigma}.$$

Proposition 13.2.2. *The operators M_λ, R_λ, Q_λ, and P_λ are bounded for each $\lambda \in \mathbb{C}$ and depend analytically on $\lambda \in \mathbb{C}$.* $\diamond$

Equations (13.2.5) and (13.2.6), in an abstract way, are given as the following equation, in the space Y, by

$$\psi_{|\Gamma_e} = M_\lambda H \psi_{|\Gamma_e} + Q_\lambda g. \tag{13.2.8}$$

In order to pursue our analysis, we need the following proposition:

Proposition 13.2.3. *H is a bounded operator with a norm $\| H \| = \alpha$. Moreover, if $\mathrm{Re}\lambda > -\Sigma$, then $\| M_\lambda H \| = \frac{\alpha}{2(\mathrm{Re}\lambda+\Sigma)R}(1 - e^{-2R(\mathrm{Re}\lambda+\Sigma)})$.* $\diamond$

Proof. For any $f \in Y$, we have $\|Hf\| = \int_0^R \frac{y}{R^2} \left| \int_0^R \frac{\alpha}{\sqrt{R^2-y^2}} f(y') \frac{y'}{R} dy' \right| dy$. So, $\|Hf\| = \alpha \left| \int_0^R f(y') \frac{y'}{R^2} dy' \right|$. Hence, $\|Hf\| \leq \alpha \| f \|$. For $\bar{f} = 2 \in Y$, we have $\| \bar{f} \| = 1$ and $\| H \bar{f} \| = \alpha$. Consequently, $\| H \| = \alpha$. If $\mathrm{Re}\lambda > -\Sigma$, then for any $f \in Y$, we have

$$\|M_\lambda Hf\| = \int_0^R \frac{y}{R^2} \left| \int_0^R \frac{\alpha}{\sqrt{R^2-y^2}} e^{-2(\lambda+\Sigma)\sqrt{R^2-y^2}} f(y') \frac{y'}{R} dy' \right| dy.$$

So,

$$\|M_\lambda Hf\| = \frac{\alpha}{2R(\mathrm{Re}\lambda + \Sigma)} (1 - e^{-2(\mathrm{Re}\lambda+\Sigma)R}) \left| \int_0^R f(y') \frac{y'}{R^2} dy' \right|.$$

This leads to the estimate $\|M_\lambda Hf\| \leq \frac{\alpha}{2R(\mathrm{Re}\lambda+\Sigma)}(1 - e^{-2(\mathrm{Re}\lambda+\Sigma)R})\| f \|$. Similarly, for $\bar{f} = 2 \in Y$, we have $\| \bar{f} \| = 1$ and $\|M_\lambda Hf\| = \frac{\alpha}{2R(\mathrm{Re}\lambda+\Sigma)}(1 - e^{-2(\mathrm{Re}\lambda+\Sigma)R})$.

Hence,

$$\|M_\lambda H\| = \frac{\alpha}{2R(\mathrm{Re}\lambda + \Sigma)}(1 - e^{-2(\mathrm{Re}\lambda+\Sigma)R}).$$

This completes the proof of proposition. Q.E.D.

Let

$$\lambda_0 := \begin{cases} -\Sigma & \text{if} \quad \alpha \leq 1 \\ -\Sigma + \dfrac{\alpha}{2R} & \text{if} \quad \alpha > 1. \end{cases}$$

In view of Eq. (13.2.7) and Proposition 13.2.3, we have $\|M_\lambda H\| < 1$ for all $\lambda \in \mathbb{C}$ such that $\mathrm{Re}\lambda > \lambda_0$ and then using Eq. (13.2.8), $\psi_{|\Gamma_e}$ is given by $\psi_{|\Gamma_e} = (I - M_\lambda H)^{-1} Q_\lambda g$. Moreover, Eq. (13.2.5) can be written as follows $\psi = R_\lambda H \psi_{|\Gamma_e} + P_\lambda g$. Hence, we get $\psi = R_\lambda H(I - M_\lambda H)^{-1} Q_\lambda g + P_\lambda g$. Finally, the resolvent of the operator B_α can be expressed by $R(\lambda, B_\alpha) := (\lambda - B_\alpha)^{-1} = R_\lambda H(I - M_\lambda H)^{-1} Q_\lambda + P_\lambda$.

13.2.3 Generation Result

Lemma 13.2.1. *Assume that* $\alpha \leq 1$. *Then, for all* λ *such that* $\mathrm{Re}\lambda > -\Sigma$, *we have* $\|(\lambda - B_\alpha)^{-1}\| \leq \frac{1}{\mathrm{Re}\lambda + \Sigma}$. ◇

Proof. Let B'_α be the operator defined by:

$$\begin{cases} B'_\alpha : \mathcal{D}(B_\alpha) \subseteq X \longrightarrow X \\ \qquad\qquad \psi \longrightarrow B'_\alpha \psi(x, y) = -\dfrac{\partial \psi}{\partial x}(x, y). \end{cases}$$

First, we will show that the operator B'_α is dissipative. To do this, we take $\psi \in \mathcal{D}(B_\alpha)$ and $\psi^* = S_0 \psi$ where $S_0 \psi$ satisfies

$$S_0 \psi(x) = \begin{cases} 1 & \text{if } \psi(x) > 0 \\ 0 & \text{if } \psi(x) = 0 \\ -1 & \text{if } \psi(x) < 0 \end{cases} \tag{13.2.9}$$

and we will prove, in the following, that $\mathrm{Re}\langle B'_\alpha \psi, S_0 \psi \rangle \leq 0$.

$$\begin{aligned}
\mathrm{Re}\langle B'_\alpha\psi, S_0\psi\rangle &= \mathrm{Re}\left[-\int_0^R\int_{-\sqrt{R^2-y^2}}^{\sqrt{R^2-y^2}}\frac{\partial\psi}{\partial x}(x,y)S_0\psi(x,y)y\,dx\,dy\right]\\
&= -\int_0^R\int_{-\sqrt{R^2-y^2}}^{\sqrt{R^2-y^2}}\frac{\partial|\psi(x,y)|}{\partial x}y\,dx\,dy\\
&= R^2\left(\|\psi_{|\Gamma_s}\| - \|\psi_{|\Gamma_e}\|\right)\\
&\le 0\ \ (\ \|H\|\le 1).
\end{aligned}$$

Now, let $\varphi = (\lambda - B_\alpha)\psi$. Then,

$$\begin{aligned}
\mathrm{Re}(\lambda+\Sigma)\|\psi\|^2 &= \mathrm{Re}(\lambda+\Sigma)\langle\psi,\psi^*\rangle\\
&= \mathrm{Re}\left((\lambda+\Sigma)\langle\psi,\psi^*\rangle\right)\\
&\le \mathrm{Re}[(\lambda+\Sigma)\langle\psi,\psi^*\rangle - \langle B'_\alpha\psi,\psi^*\rangle] \text{ because } B'_\alpha \text{ is dissipative}\\
&= \mathrm{Re}[\lambda\langle\psi,\psi^*\rangle - \langle B_\alpha\psi,\psi^*\rangle]\\
&= \mathrm{Re}\langle\varphi,\psi^*\rangle\\
&\le \|\varphi\|\,\|\psi\|.
\end{aligned}$$

This implies that $\|\psi\| \le \frac{\|\varphi\|}{\mathrm{Re}\lambda+\Sigma}$, and we obtain $\|(\lambda - B_\alpha)^{-1}\| \le \frac{1}{\mathrm{Re}\lambda+\Sigma}$. Q.E.D.

Lemma 13.2.2. *Assume that* $\alpha \le 1$. *Then, we have* $\{\lambda \in \mathbb{C}$ *such that* $\mathrm{Re}\lambda > \|K\| - \Sigma + 1\} \subset \rho(A_\alpha)$ *and, for all* λ *such that* $\mathrm{Re}\lambda > \|K\| - \Sigma + 1$, *we have* $\|(\lambda - A_\alpha)^{-1}\| \le 1$. ◇

Proof. Let $\lambda \in \{\lambda \in \mathbb{C}$ such that $\mathrm{Re}\lambda > \|K\| - \Sigma + 1\}$. Then, $(\lambda - A_\alpha)$ is invertible and we get $(\lambda - A_\alpha)^{-1} = \left[I - (\lambda - B_\alpha)^{-1}K)\right]^{-1}(\lambda - B_\alpha)^{-1}$. Therefore, by using Lemma 13.2.1, we have

$$\begin{aligned}
\|(\lambda - A_\alpha)^{-1}\| &\le \|(\lambda - B_\alpha)^{-1}\|\sum_{k=0}^{\infty}\|(\lambda - B_\alpha)^{-1}K\|^k\\
&\le \frac{1}{(\mathrm{Re}\lambda+\Sigma)(1-\frac{\|K\|}{\mathrm{Re}\lambda+\Sigma})}\\
&\le 1.
\end{aligned}$$

This completes the proof of the lemma. Q.E.D.

We end up this section by the following result.

Proposition 13.2.4. *Assume that* $\alpha \le 1$. *If the boundary operator* H *is nonnegative, then* B_α *generates a strongly continuous semigroup* $(U_\alpha(t))_{t\ge 0}$, *satisfying* $\|U_\alpha(t)\| \le e^{-\Sigma t}$. ◇

Proof. In view of the positivity of the operator H and Lemma 13.2.1, together with Corollary 2.5.1, we can immediately deduce the result. Q.E.D.

13.2.4 Asymptotic Behavior of the Solution

Lemma 13.2.3. *For any complex number λ, both $KR_\lambda HQ_\lambda$ and $(KP_\lambda)^2$ are weakly compact operators on X.* $\diamond$

Proof. By making some computation, we know that $KR_\lambda HQ_\lambda$ is the integral operator

$$(KR_\lambda HQ_\lambda f)(x,y) = \int_0^R dy' \int_{-\sqrt{R^2-y'^2}}^{\sqrt{R^2-y'^2}} K_{\alpha,\lambda}(x,y,x',y') f(x',y') dx',$$

where

$$K_{\alpha,\lambda}(x,y,x',y') = \frac{c\Sigma}{2r}\int_{-r}^{r} e^{-(\lambda+\Sigma)\left(\sqrt{R^2-y^2+z^2}+\sqrt{R^2-y'^2}+z-x'\right)} \frac{\alpha y'}{R\sqrt{R^2-r^2+z^2}} dz,$$

$$r = \sqrt{x^2+y^2}.$$

Let us define

$$\begin{cases} G : D \longrightarrow X \\ (x',y') \longrightarrow K_{\alpha,\lambda}(x,y,x',y'). \end{cases}$$

Then, for $\text{Re}\lambda > -\Sigma$, we have

$$\begin{aligned}
&\max_{(x',y')\in D} \|G(x',y')\| \\
&= \max_{(x',y')\in D} \|K_{\alpha,\lambda}(x,y,x',y')\| \\
&\leq \int_0^R y dy \int_{-\sqrt{R^2-y^2}}^{\sqrt{R^2-y^2}} \left(\frac{c\Sigma}{2r} \int_{-r}^{r} \frac{\alpha y' e^{-(\text{Re}\lambda+\Sigma)\left(\sqrt{R^2-r^2+z^2}+\sqrt{R^2-y'^2}+z-x'\right)}}{R\sqrt{R^2-r^2+z^2}} dz \right) dx \\
&\leq \int_0^R y dy \int_{-\sqrt{R^2-y^2}}^{\sqrt{R^2-y^2}} \left(\frac{c\Sigma}{2\sqrt{x^2+y^2}} \int_{-\sqrt{x^2+y^2}}^{\sqrt{x^2+y^2}} \frac{\alpha R}{R\sqrt{R^2-r^2}} dz \right) dx \\
&\leq \alpha c\Sigma \int_{-R}^{R} dx \int_0^{\sqrt{R^2-x^2}} \frac{y dy}{\sqrt{R^2-x^2-y^2}}.
\end{aligned}$$

For $\mathrm{Re}\lambda < -\Sigma$, and using a similar method, we deduce that

$$\max_{(x',y')\in D} \|G(x',y')\| \leq \alpha c \Sigma e^{-4(\mathrm{Re}\lambda+\Sigma)R} \int_{-R}^{R} dx \int_{0}^{\sqrt{R^2-x^2}} \frac{ydy}{\sqrt{R^2-x^2-y^2}}.$$

So, for any λ, we have

$$\max_{(x',y')\in D} \|G(x',y')\| \leq C(\lambda), \tag{13.2.10}$$

where $C(\lambda)$ is a positive constant. Moreover, let $E \subset D$ be measurable. Then,

$$\begin{aligned} \int_E y|G(x',y')|dxdy &\leq \int_E \left(\frac{c\Sigma}{2\sqrt{x^2+y^2}} \int_{-\sqrt{x^2+y^2}}^{\sqrt{x^2+y^2}} \frac{e^{4|\mathrm{Re}\lambda+\Sigma|R}}{\sqrt{R^2-r^2}} dz \right) ydxdy \\ &\leq \int_E c\Sigma e^{4|\mathrm{Re}\lambda+\Sigma|R} \frac{y}{\sqrt{R^2-x^2-y^2}} dxdy. \end{aligned}$$

Since $\frac{1}{\sqrt{R^2-x^2-y^2}} \in X$, we have

$$\lim_{|E|\to 0} \max_{(x',y')\in D} \int_E y|G(x',y')|dxdy = 0. \tag{13.2.11}$$

From Eqs. (13.2.10) and (13.2.11) and also Theorem 2.4.5, we deduce that $KR_\lambda HQ_\lambda$ is a weakly compact operator on X for each complex number λ. A similar reasoning allows us to reach the same result for the operator $(KP_\lambda)^2$. Q.E.D.

Lemma 13.2.4.

(i) If $\frac{\alpha(1-e^{-2(\lambda+\Sigma)R})}{2(\lambda+\Sigma)R} \neq 0$, *then*

$$(I - M_\lambda H)^{-1} = I + \frac{2(\lambda+\Sigma)}{2(\lambda+\Sigma)R - \alpha(1-e^{-2(\lambda+\Sigma)R})} M_\lambda H.$$

(ii) For any $\lambda \in \mathbb{C}$, *we have*

$$M_\lambda H Q_\lambda = \frac{\alpha(1-e^{-2(\lambda+\Sigma)R})}{2(\lambda+\Sigma)R} Q_\lambda.$$ ◇

Proof.

(*i*) In fact, $(M_\lambda H)^2 = \frac{\alpha(1-e^{-2(\lambda+\Sigma)R})}{2(\lambda+\Sigma)R} M_\lambda H$. Hence, (*i*) is valid.
(*ii*) We can obtain this result by some direct computation.

Q.E.D.

The following theorem is a consequence of Lemmas 13.2.3, 13.2.4 and 2.1.13.

Theorem 13.2.3. *If* $\frac{\alpha(1-e^{-2(\lambda+\Sigma)R})}{2(\lambda+\Sigma)R} \neq 0$, *then the operator* $\left[K(\lambda - B_\alpha)^{-1}\right]^4$ *is compact.* ◇

Let $w > 0$ and set $\mathcal{R}_w = \{\lambda \in \mathbb{C}$ such that $\mathrm{Re}\lambda \geq -\Sigma + w\}$.

Theorem 13.2.4. *If* $\alpha \leq 1$, *then for any* $r \in [0, 1)$, *we have* $\lim_{|\mathrm{Im}\lambda|\to\infty} |\mathrm{Im}\lambda|^r \|[K(\lambda - B_\alpha)^{-1}]^5\| = 0$ *uniformly on* $\mathcal{R}_w$. ◇

Proof. The operator $K(\lambda - B_\alpha)^{-1}$ is given by

$$K(\lambda - B_\alpha)^{-1} = KR_\lambda H(I - M_\lambda H)^{-1} Q_\lambda + KP_\lambda. \tag{13.2.12}$$

By using Eq. (13.2.12) and Lemma 13.2.4, we deduce that

$$\begin{aligned}
&\left[K(\lambda - B_\alpha)^{-1}\right]^5 \\
&= l^5L^5 + l^4L^4A + l^4L^3AL + l^3L^3A^2 + l^4L^2AL^2 + l^3L(LA)^2 \\
&\quad + l^3L^2A^2L + l^2L^2A^3 + l^4LAL^3 + l^3LAL^2A + l^3L(AL)^2 + l^2(LA)^2A \\
&\quad + l^2LA^2LA + l^2LA^3L + lLA^4 + l^4AL^4 + l^3AL^3A + l^3AL^2AL \\
&\quad + l^2AL^2A^2 + l^3(AL)^2L + l^2A(LA)^2 + l^2ALA^2L + lALA^3 + l^3A^2L^3 \\
&\quad + l^2A^2L^2A + l^2A(AL)^2 + lA^2LA^2 + l^2A^3L^2 + lA^3LA + lA^4L \\
&\quad + l^3LA^2L^2 + A^5,
\end{aligned}$$

where $L = KR_\lambda HQ_\lambda$, $A = KP_\lambda$ and $l = \frac{2(\lambda+\Sigma)R}{2(\lambda+\Sigma)R-\alpha(1-e^{-2(\lambda+\Sigma)R})}$. Let us notice that there is an analogy between the components of $\left[K(\lambda - B_\alpha)^{-1}\right]^5$. Then, it suffices to show the result for the operator A^5. By virtue of the compactness of the operator A^4 (see Theorem 13.2.3), we deduce that the family $\{A^4 : X \longrightarrow X$ such that $\lambda \in \mathcal{R}_w\}$ is collectively compact. Hence, by using Proposition 2.1.2, we will show that the family $\{|\mathrm{Im}\lambda|^r A : X \longrightarrow X$ such that $\lambda \in \mathcal{R}_w\}$ converges strongly to zero as $|\mathrm{Im}\lambda|$ goes to infinity. Then, it remains to prove that $\forall r \in [0, 1[$, $\lim_{|\mathrm{Im}\lambda|\to+\infty} |\mathrm{Im}\lambda|^r \|A\psi\| = 0$ uniformly on $\mathcal{R}_w$, where

$$A\psi = \frac{C\Sigma}{2r}\int_{-r}^{r}\left(\int_{-\sqrt{R^2-r^2+x_1^2}}^{x_1} e^{-(\lambda+\Sigma)(x_1-x_2)}\psi\left(x_2, \sqrt{r^2-x_1^2}\right)dx_2\right)dx_1.$$

To do this, we firstly observe that the operator A may be decomposed as follows $A = C\Sigma A_1A_2A_3$, where

$$\begin{cases} A_1 : L_1(S, rdr) \longrightarrow X \\ \qquad\qquad f \longrightarrow A_1f(x, y) = \dfrac{1}{2\sqrt{x^2+y^2}} f\left(\sqrt{x^2+y^2}\right) \end{cases}$$

$$\begin{cases} A_2 : X \longrightarrow L_1(S, rdr) \\ \quad f \longrightarrow A_2 f(r) = \int_{-r}^{r} f\left(z, \sqrt{r^2 - z^2}\right) dz \end{cases}$$

and

$$\begin{cases} A_3 : X \longrightarrow X \\ \quad \psi \longrightarrow A_3\psi(x, y) = \int_{-\sqrt{R^2-y^2}}^{x} e^{-(\lambda+\Sigma)(x-x')}\psi(x', y)dx'. \end{cases}$$

Clearly, the operators A_1 and A_2 are uniformly bounded on $\mathcal{R}_w$. Hence, we will show that $\forall r \in [0, 1[$, $\lim_{|\mathrm{Im}\lambda|\to+\infty} |\mathrm{Im}\lambda|^r \|A_3\psi\| = 0$ uniformly on $\mathcal{R}_w$. Or equivalently, $\forall r \in [0, 1[$, we have

$$\lim_{|\mathrm{Im}\lambda|\to+\infty} |\mathrm{Im}\lambda|^r \int_0^R \int_{-\sqrt{R^2-y^2}}^{\sqrt{R^2-y^2}} \left| \int_{-\sqrt{R^2-y^2}}^{x} e^{-(\lambda+\Sigma)(x-x')}\psi(x', y)dx' \right| ydxdy = 0$$

uniformly on $\mathcal{R}_w$. Since $\psi \in L_1(D, ydxdy)$, it is sufficient to show the result for a measurable simple function ψ in D. Now, using the change of variables $s = x - x'$, it remains to prove that

$$\lim_{|\mathrm{Im}\lambda|\to+\infty} |\mathrm{Im}\lambda|^r \int_0^R y \int_{-\sqrt{R^2-y^2}}^{\sqrt{R^2-y^2}} \left| \int_0^{2R} e^{-(\lambda+\Sigma)s}\psi(x-s, y)\chi_{[0,x+\sqrt{R^2-y^2}]}ds \right| dxdy = 0.$$

To do this, let $y \in [0, R]$, $x \in \left[-\sqrt{R^2 - y^2}, \sqrt{R^2 - y^2}\right]$, and let us consider the map

$$\begin{cases} \varphi_{x,y} : [0, 2R] \longrightarrow \mathbb{R} \\ \quad s \longrightarrow \psi(x - s, y)\chi_{[0,x+\sqrt{R^2-y^2}]}(s). \end{cases}$$

It follows that $\varphi_{x,y}$ is a simple function. Let $(s_i)_{1\le i\le m}$ be the subdivision of its support such that, $\forall 1 \le i \le m - 1$, we have $\varphi_{x,y}(s) = \varphi_{x,y}(s_i)$ $\forall s \in [s_i, s_{i+1}[$. Then, we get

$$\begin{aligned} \int_0^{2R} e^{-(\lambda+\Sigma)s}\varphi_{x,y}(s)ds &= \sum_{i=1}^{m-1} \int_{s_i}^{s_i+1} \varphi_{x,y}(s_i)e^{-(\lambda+\Sigma)s}ds \\ &= \sum_{i=1}^{m-1} \varphi_{x,y}(s_i) \left[\frac{-1}{\lambda+\Sigma}e^{-(\lambda+\Sigma)s}ds\right]_{s_i}^{s_i+1} \\ &= \frac{1}{\lambda+\Sigma} \sum_{i=1}^{m-1} \varphi_{x,y}(s_i)\left(e^{-(\lambda+\Sigma)s_i} - e^{-(\lambda+\Sigma)s_{i+1}}\right). \end{aligned}$$

Hence,

$$\left|\int_0^{2R} e^{-(\lambda+\Sigma)s}\varphi_{x,y}(s)ds\right| \leq \frac{2(m-1)}{|\mathrm{Im}\lambda|}\sup|\psi(.,.)|,$$

and

$$|\mathrm{Im}\lambda|^r \int_0^R y \int_{-\sqrt{R^2-y^2}}^{\sqrt{R^2-y^2}} \left|\int_0^{2R} e^{-(\lambda+\Sigma)s}\psi(x-s,y)\,\chi_{[0,x+\sqrt{R^2-y^2}]}ds\right| dxdy$$

$$\leq \frac{4R^3(m-1)}{3|\mathrm{Im}\lambda|^{1-r}}\sup|\psi(.,.)|.$$

This inequality allows us to reach the desired result. Q.E.D.

Since B_α is the generator of a C_0-semigroup $(U_\alpha(t))_{t\geq 0}$ on X, and since K is a bounded operator, then according to the classical perturbation theory (Theorem 2.5.8), the operator $A_\alpha = B_\alpha + K$ also generates a C_0-semigroup $(V_\alpha(t))_{t\geq 0}$ on X. If the operator H satisfies $\|H\| \leq 1$, then from Proposition 4.2.1, the asymptotic spectrum $P(A_\alpha)$ consists of, at most, discrete eigenvalues with finite algebraic multiplicities $\{\lambda_1, \lambda_2, \ldots, \lambda_n, \lambda_{n+1}, \ldots\}$ which can be ordered in such a way that the real part decreases [186], i.e., $\mathrm{Re}\lambda_1 > \mathrm{Re}\lambda_2 > \cdots > \mathrm{Re}\lambda_n > \mathrm{Re}\lambda_{n+1} > \cdots > \eta$, and $\{\lambda \in \mathbb{C}$ such that $\mathrm{Re}\lambda > \eta\}\backslash\{\lambda_n, n = 1, 2, 3, \ldots\} \subset \rho(A_\alpha)$. By the same way as in the proof of Theorem 4.2.1, we give the description of the asymptotic behavior of the solution of the Cauchy problem (13.2.3) by the following theorem

Theorem 13.2.5. *If $\alpha \leq 1$, then for any $\varepsilon > 0$, there exists $M > 0$ such that*

$$\left\|V_\alpha(t) - \sum_{i=1}^n e^{\lambda_i t} e^{D_i t} P_i\right\| \leq M e^{(\varepsilon + \mathrm{Re}\lambda_{n+1})t}, \quad \forall t > 0.$$

◇

13.3 The Time-Asymptotic Behavior of a Transport Operator with a Diffuse Reflection Boundary Condition

As is well known, the stability of a given transport system is a quite important and interesting topic in transport theory. Various methods have been developed in order to investigate the asymptotic behavior of the time dependent solution (see [138, 179, 184, 209, 232, 275]). In this section, we are concerned with the time dependent transport equation in a sphere with a diffuse reflection boundary condition. For more details concerning the following results, the reader may refer to [75], which constitutes the real basis of our present work. Let us suppose that the region occupied by the reactor media is a sphere of radius $R > 0$, and let $G = [0, R] \times [-1, 1] \times (V_m, V_M)$, $\Omega = [-1, 1] \times (V_m, V_M)$. Then, the weighted L_1-spaces $L_1(G, r^2\, dr\, d\mu\, dv)$ and $L_1(\Omega, |\mu v|\, d\mu\, dv)$ with the following norms

$$\|f\| = \int_G |f(r, \mu, v)| r^2\, dr\, d\mu\, dv, \qquad f \in L_1(G, r^2\, dr\, d\mu\, dv)$$

$$\|g\| = \int_\Omega |g(\mu, v)||v\mu|\, d\mu\, dv, \qquad g \in L_1(\Omega, |v\mu|\, d\mu\, dv)$$

constitute complex Banach spaces, where r is the distance from the center of the sphere, $v \in (V_m, V_M)$ is the speed of migrating particles, and $\mu \in [-1, 1]$ is the cosine of the angle the particle velocity makes with the radius vector. The transport operator A, with a diffuse boundary condition which is the summation of two operators B and K, where B is the streaming operator, and K is a collision operator defined by a scattering fission kernel $\kappa(.,.,.)$, is given by

$$\begin{cases} \bullet \dfrac{\partial f}{\partial t}(r, \mu, v, t) = A\, f(r, \mu, v, t) = B\, f(r, \mu, v, t) + K\, f(r, \mu, v, t) \\ \qquad = -v\mu \dfrac{\partial f}{\partial r}(r, \mu, v, t) - v \dfrac{1-\mu^2}{r} \dfrac{\partial f}{\partial \mu}(r, \mu, v, t) \\ \qquad -v\Sigma(r, \mu, v) f(r, \mu, v, t) + \dfrac{1}{2} \displaystyle\int_{V_m}^{V_M} \int_{-1}^{1} \kappa(r, v, v') \\ \qquad f(r, \mu', v', t) dv' d\mu' \\ \\ \bullet f(R, \mu, v, t) \in L_1(\Omega, |\mu v|\, d\mu\, dv) \\ \bullet |v\mu| f(R, \mu, v, t) = \displaystyle\int_{V_m}^{V_M} dv' \int_0^1 \alpha(\mu, \mu', v, v') v' \mu' f(R, \mu', v', t) d\mu' \\ \quad \text{for } \mu \in [-1, 0), \text{ and} \\ \bullet f(r, \mu, v, 0) = f_0(r, \mu, v). \end{cases} \tag{13.3.1}$$

Here, $f(r, \mu, v, t)$ is the neutron distribution at time t, $\Sigma(r, \mu, v)$ represents the total collision frequency, $\alpha(\mu, \mu', v, v')$ is the diffuse coefficient on the boundary, and $f_0(r, \mu, v)$ represents the initial distribution. Let $D = \{(x, y, v)$ such that $x^2 + y^2 \leq R^2,\ y \geq 0,\ V_m < v < V_M\}$ and $S = [0, R] \times (V_m, V_M)$. By virtue of the transformation [326], we have $x = r\mu,\ y = r\sqrt{1-\mu^2}$, and $\psi(x,y,v,t) = f(r(x,y),\mu(x,y),v,t)$. Equation (13.3.1) can be equivalently written as

$$\begin{cases} \bullet \dfrac{\partial \psi}{\partial t}(x,y,v,t) = A\psi(x,y,v,t) = B\psi(x,y,v,t) + K\psi(x,y,v,t) \\ \qquad = -v\dfrac{\partial \psi}{\partial x}(x,y,v,t) - v\Sigma\left(\sqrt{x^2+y^2}, \dfrac{x}{\sqrt{x^2+y^2}}, v\right)\psi(x,y,v,t) \\ \qquad + \dfrac{1}{2r}\displaystyle\int_{V_m}^{V_M}\int_{-r}^{r} \kappa(r,v,v')\psi\left(z,\sqrt{r^2-z^2},v',t\right)dv'dz \\ \bullet \psi\left(\pm\sqrt{R^2-y^2},y,v,t\right) \in Y \\ \bullet \dfrac{vy}{R^2}\psi\left(-\sqrt{R^2-y^2},y,v,t\right) = \displaystyle\int_S \dfrac{y}{R\sqrt{R^2-y^2}}\alpha\left(-\dfrac{\sqrt{R^2-y^2}}{R}, \dfrac{\sqrt{R^2-y'^2}}{R}, v, v'\right) \\ \qquad \times \dfrac{v'y'}{R^2}\psi\left(\sqrt{R^2-y'^2},y',v',t\right)dy'dv' \\ \bullet \psi(r,\mu,v,0) = \psi_0(r,\mu,v). \end{cases} \tag{13.3.2}$$

Throughout this part, it is assumed that

$(\mathcal{Q}1)$: $\Sigma(r,\mu,v)$ is a bounded, nonnegative, and measurable function defined on G, and let

$$\lambda_0^* := \inf_{(r,\mu,v)\in G}\{v\Sigma(r,\mu,v)\}.$$

$(\mathcal{Q}2)$: $0 < V_m < V_M < +\infty$.

$(\mathcal{Q}3)$: $\alpha(\mu,\mu',v,v')$ is a nonnegative, bounded, and measurable function defined on

$$[-1,0)\times(0,1]\times(V_m,V_M)\times(V_m,V_M), \text{ such that}$$

$$\sup_{(\mu',v')\in(0,1]\times(V_m,V_M)}\left\{\int_{V_m}^{V_M}\int_{-1}^{0}\alpha(\mu,\mu',v,v')d\mu\right\} \le 1.$$

$(\mathcal{Q}4)$: $k(r,v,v')$ is a real bounded and measurable function.

As it is mentioned above, the transport operator (13.3.2) can be formally written as a first order Cauchy problem:

$$\begin{cases} \dfrac{\partial \psi}{\partial t} = A\psi := B\psi + K\psi \\ \psi(0) = \psi_0, \end{cases} \tag{13.3.3}$$

where $\psi_0 \in L_1[D, y\,dx\,dy\,dv]$.

13.3.1 The Resolvent of the Operator B

Let us state precisely the functional setting of the problem. Let $X := L_1(D, y\,dx\,dy\,dv)$ be the Banach space equipped with the norm

$$\|\varphi\|_X = \int_D |\varphi(x, y, v)| y\,dx\,dy\,dv \qquad \varphi \in X,$$

and let $Y := L_1(S, \frac{vy}{R^2}\,dy\,dv)$ be the Banach space equipped with the norm

$$\|\psi\|_Y = \int_S |\psi(y, v)| \frac{vy}{R^2} dy\,dv \qquad \psi \in Y.$$

We denote by Γ_e and Γ_s the following boundary spaces defined by:

$$\Gamma_e := \left\{(\sqrt{R^2 - y^2}, y, v),\ y \geq 0,\ V_m < v < V_M\right\}$$

and

$$\Gamma_s := \left\{(-\sqrt{R^2 - y^2}, y, v),\ y \geq 0,\ V_m < v < V_M\right\}.$$

We define the partial Sobolev space $\mathcal{W}$ by

$$\mathcal{W} = \left\{\psi \in X \text{ such that } v\frac{\partial \psi}{\partial x} \in X\right\}.$$

Proposition 13.3.1. *Let $\psi \in \mathcal{W}$. If $\psi_{|\Gamma_e} \in Y$, then $\psi_{|\Gamma_s} \in Y$ and vice versa.* ◇

Proof. Let $\psi \in \mathcal{W}$ such that $\psi_{|\Gamma_e} \in Y$. We have

$$\psi(\sqrt{R^2 - y^2}, y, v) - \psi(-\sqrt{R^2 - y^2}, y, v) = \int_{-\sqrt{R^2-y^2}}^{\sqrt{R^2-y^2}} \frac{\partial \psi}{\partial x}(x, y, v)\,dx. \tag{13.3.4}$$

This implies that

$$|\psi(-\sqrt{R^2 - y^2}, y, v)| \leq |\psi(\sqrt{R^2 - y^2}, y, v)| + \int_{-\sqrt{R^2-y^2}}^{\sqrt{R^2-y^2}} \left|\frac{\partial \psi}{\partial x}(x, y, v)\right| dx.$$

Consequently, we have

$$\int_S \left|\psi(-\sqrt{R^2-y^2}, y, v)\right| \frac{vy}{R^2} dy\, dv \le \int_S \left|\psi(\sqrt{R^2-y^2}, y, v)\right| \frac{vy}{R^2} dy\, dv + \int_D \left|\frac{\partial \psi}{\partial x}(x, y, v)\right| \frac{vy}{R^2} dx\, dy\, dv.$$

So, we have $\|\psi_{|\Gamma_s}\|_Y \le \|\psi_{|\Gamma_e}\|_Y + \frac{V_M}{R^2}\left\|\frac{\partial \psi}{\partial x}\right\|_X < \infty$. The converse may be proved in a similar way, starting from a rearrangement of Eq. (13.3.4). This achieves the proof. Q.E.D.

Let us define $\tilde{\mathcal{W}} = \{\psi \in \mathcal{W} \text{ such that } \psi_{|\Gamma_e} \in Y\}$. According to Proposition 13.3.1, all functions $\psi \in \tilde{\mathcal{W}}$ have traces $\psi_{|\Gamma_e}$ and $\psi_{|\Gamma_s}$ belonging to the boundary space Y. We define the free streaming operator B by

$$\begin{cases} B : \mathcal{D}(B) \subseteq X \longrightarrow X \\ \qquad \psi \longrightarrow B\psi(x, y) = -v\dfrac{\partial \psi}{\partial x}(x, y, v, t) - v\Sigma \\ \qquad \left(\sqrt{x^2+y^2}, \dfrac{x}{\sqrt{x^2+y^2}}, v\right)\psi(x, y, v, t) \\ \mathcal{D}(B) = \{\psi \in \tilde{\mathcal{W}} \text{ such that } \psi_{|\Gamma_s} = H(\psi_{|\Gamma_e})\}, \end{cases}$$

where H is the boundary operator defined by

$$\begin{cases} H : Y \longrightarrow Y \\ \quad f \longrightarrow Hf(y, v) = \displaystyle\int_S \frac{v'y'}{vR\sqrt{R^2-y^2}} \alpha\left(-\frac{\sqrt{R^2-y^2}}{R}, \frac{\sqrt{R^2-y'^2}}{R}, v, v'\right) \\ \qquad f(y', v')dy'dv' \end{cases}$$

and we suppose that it satisfies the following hypothesis:

$(\mathcal{Q}5)$: $\|H\| \le 1$.

Let φ be a function in X and let us consider the resolvent equation for B, $(\lambda - B)\psi = \varphi$, where λ is a complex number and the unknown ψ must be sought in $\mathcal{D}(B)$. For any $\lambda \in \mathbb{C}$, and in order to clarify our subsequent analysis, we introduce the following operators

$$M(\lambda, y, y', v, v') = \frac{v'y'}{Rv\sqrt{R^2-y^2}} \alpha\left(-\frac{\sqrt{R^2-y^2}}{R}, \frac{\sqrt{R^2-y'^2}}{R}, v, v'\right) \times \exp\left(-\int_{-\sqrt{R^2-y'^2}}^{\sqrt{R^2-y'^2}} \Delta(\lambda, z, y', v')dz\right),$$

$$(y, v) \in S,\ (y', v') \in S,$$

and

$$N(\lambda, y, v, x', y', v')$$
$$= \frac{y'}{Rv\sqrt{R^2 - y^2}}\alpha\left(-\frac{\sqrt{R^2 - y^2}}{R}, \frac{\sqrt{R^2 - y'^2}}{R}, v, v'\right)$$
$$\times \exp\left(-\int_{x'}^{\sqrt{R^2-y'^2}} \Delta(\lambda, z, y', v')dz\right), \ (y, v) \in S, \ (x', y', v') \in D,$$

where

$$\Delta(\lambda, x, y, v) := \frac{1}{v}\left[\lambda + v\Sigma\left(\sqrt{x^2 + y^2}, \frac{x}{\sqrt{x^2 + y^2}}, v\right)\right], \qquad (x, y, v) \in D.$$

Now, let us define the linear operators M_λ, H_λ, L_λ, and P_λ, for $\lambda \in \mathbb{C}$, as follows:

$$\begin{cases} M_\lambda : Y \longrightarrow Y \\ \qquad \psi \longrightarrow M_\lambda\psi(y, v) = \displaystyle\int_S M(\lambda, y, y', v, v')\psi(y', v')dy'dv', \end{cases}$$

$$\begin{cases} H_\lambda : X \longrightarrow Y \\ \qquad \varphi \longrightarrow H_\lambda\psi(y, v) = \displaystyle\int_D N(\lambda, y, v, x', y', v')\psi(x', y', v')dx'dy'dv', \end{cases}$$

$$\begin{cases} L_\lambda : Y \longrightarrow X \\ \qquad \psi \longrightarrow L_\lambda\psi(x, y, v) = \exp\left(-\displaystyle\int_{-\sqrt{R^2-y^2}}^{x} \Delta(\lambda, z, y, v)dz\right)\psi(y, v) \end{cases}$$

and

$$\begin{cases} P_\lambda : X \longrightarrow X \\ \qquad \psi \longrightarrow P_\lambda\varphi(x, y, v) = \displaystyle\int_{-\sqrt{R^2-y^2}}^{x} \frac{1}{v}\exp\left(-\int_{x'}^{x} \Delta(\lambda, z, y, v)dz\right)\varphi(x', y, v)dx'. \end{cases}$$

It is shown in [350] that these operators are bounded on their respective spaces. In fact, for $\text{Re}\lambda > -\lambda_0^*$, the norms of the operators M_λ, H_λ, L_λ, and P_λ are bounded, respectively, by 1, $\frac{1}{R^2}$, $\frac{R^2}{\text{Re}\lambda+\lambda_0^*}$, and $\frac{1}{\text{Re}\lambda+\lambda_0^*}$. Moreover, for λ such that $\text{Re}\lambda > -\lambda_0^*$, the resolvent of the operator B is given by

$$(\lambda - B)^{-1} = L_\lambda(I - M_\lambda)^{-1}H_\lambda + P_\lambda. \qquad (13.3.5)$$

13.3.2 Compactness and Generation Results

Lemma 13.3.1. *For any complex number* λ, *the operators* $(KP_\lambda)^2$, H_λ, *and* M_λ *are weakly compact operators on* X. ◇

Proof. Since the proof of the weak compactness of H_λ and M_λ is similar to that of $(KP_\lambda)^2$, we only have to prove the weak compactness of $(KP_\lambda)^2$. Let us define the operators V and Q as follows:

$$\begin{cases} V : L_1(D, y\,dx\,dy\,dv) \longrightarrow L_1(S, r\,dr\,dv) \\ \qquad\qquad \psi \longrightarrow V\psi(r,v) = \dfrac{1}{2}\displaystyle\int_{-r}^{r} \psi(z, \sqrt{r^2 - z^2}, v)dz \end{cases}$$

and

$$\begin{cases} Q : L_1(S, r\,dr\,dv) \longrightarrow L_1(D, y\,dx\,dy\,dv) \\ \qquad \varphi \longrightarrow Q\varphi(x,y,v) = \dfrac{1}{\sqrt{x^2+y^2}} \displaystyle\int_{V_m}^{V_M} \kappa\left(\sqrt{x^2+y^2}, v, v'\right) \\ \qquad\qquad \varphi\left(\sqrt{x^2+y^2}, v'\right) dv'. \end{cases}$$

Then, V and Q are bounded linear operators. Obviously, $K = QV$, and

$$VP_\lambda Q : L_1(S, r\,dr\,dv) \longrightarrow L_1(S, r\,dr\,dv)$$

is a bounded linear operator for each λ which can be represented by

$$\begin{aligned} &VP_\lambda Q\varphi(r,v) \\ &= \frac{1}{2}\int_{-r}^{r}\int_{-\sqrt{R^2-r^2+z^2}}^{z} \frac{1}{v} e^{-\int_{x'}^{z} \Delta(\lambda, z', \sqrt{r^2-z^2}, v)dz'} \\ &\times \frac{1}{\sqrt{x'^2+r^2-z^2}} \int_{V_m}^{V_M} \kappa(\sqrt{x'^2+r^2-z^2}, v, v')\varphi(\sqrt{x'^2+r^2-z^2}, v')dv'dx'dz. \end{aligned}$$

Taking into account the following change of variables $r' = \sqrt{x'^2 + r^2 - z^2}$, $v' = v$, $s = z - x'$, we can write

$$VP_\lambda Q\varphi(r,v) = \frac{1}{2}\int_{V_m}^{V_M}\int_0^R h(\lambda, v, v', r, r')\psi(r', v')dv'dr',$$

where

$$h(\lambda, v, v', r, r') = \frac{\kappa(r', v, v')}{2v} \int_{|r-r'|}^{r+r'} \frac{1}{s} e^{-\int_{\Delta_1(r,r',s)}^{\Delta_2(r,r',s)} \Delta(\lambda, z', \sqrt{r^2-z^2}, v)dz} ds,$$

$$\Delta_1(r, r', s) = -\frac{r'^2 - r^2}{2s} - \frac{s}{2},$$

and

$$\Delta_2(r, r', s) = -\frac{r'^2 - r^2}{2s} + \frac{s}{2}.$$

In what follows, we will show the weak compactness of $VP_\lambda Q$. To do this, let G_λ be defined by

$$\begin{cases} G_\lambda : S \longrightarrow L_1(S, rdrdv) \\ (r', v') \longrightarrow G_\lambda(r', v') = h(\lambda, v, v', r, r'). \end{cases}$$

Hence,

$$G_\lambda(v', r') = \frac{\kappa(r', v, v')}{2v} \int_{|r-r'|}^{r+r'} \frac{1}{s} e^{-\int_{\Delta_1(r,r',s)}^{\Delta_2(r,r',s)} \Delta(\lambda, z', \sqrt{r^2-z^2}, v)dz} ds.$$

Let $\lambda = \beta + i\tau$. Then, $e^{-\int_{\Delta_1(r,r',s)}^{\Delta_2(r,r',s)} \Delta(\lambda, z', \sqrt{r^2-z^2}, v)dz} = e^{-\int_{\Delta_1(r,r',s)}^{\Delta_2(r,r',s)} \frac{1}{v}(\beta + v\Sigma(.,.,.,.))dz} e^{-\frac{i\tau s}{v}}$. Hence,

$$\begin{aligned} &\max_{(v',r')\in S} \|G_\lambda(v', r')\| \\ &\leq \int_0^R \int_{V_m}^{V_M} \frac{|\kappa(r', v, v')|}{2} \int_{|r-r'|}^{r+r'} \frac{1}{s} e^{-\int_{\Delta_1(r,r',s)}^{\Delta_2(r,r',s)} \frac{1}{v}(\beta + v\Sigma(.,.,.,.))dz} r\, ds\, dr\, dv \\ &\leq \int_0^R \int_{V_m}^{V_M} \frac{|\kappa(r', v, v')|}{2} \int_{|r-r'|}^{r+r'} \frac{1}{s} e^{-\frac{s}{v}(\beta+\lambda_0^*)} r\, ds\, dr\, dv. \end{aligned}$$

For $\beta > -\lambda_0^*$, we get

$$\begin{aligned} \max_{(v',r')\in S} \|G_\lambda(v', r')\| &\leq \int_0^R \int_{V_m}^{V_M} \frac{|\kappa(r', v, v')|}{2} \int_{|r-r'|}^{r+r'} \frac{1}{s} e^{-\frac{s}{V_M}(\beta+\lambda_0^*)} r\, ds\, dr\, dv \\ &\leq \int_0^R \int_{V_m}^{V_M} \frac{|\kappa(r', v, v')|}{2} \int_{|r-r'|}^{r+r'} \frac{1}{s} r\, ds\, dr\, dv \\ &\leq \frac{CR^2}{4} \log(2R)(V_M - V_m), \end{aligned}$$

where C is a positive constant. By using a similar method for $\beta < -\lambda_0^*$, we get

$$\max_{(v',r')\in S} \|G_\lambda(v',r')\| \leq \int_0^R \int_{V_m}^{V_M} \frac{|\kappa(r',v,v')|}{2} \int_{|r-r'|}^{r+r'} \frac{1}{s} e^{-\frac{(r+r')}{V_m}(\beta+\lambda_0^*)} r\, ds\, dr\, dv$$
$$\leq C \int_0^R \int_{V_m}^{V_M} \int_{|r-r'|}^{r+r'} \frac{1}{s} e^{-\frac{2R}{V_m}(\beta+\lambda_0^*)} r\, ds\, dr\, dv$$
$$\leq \frac{CR^2}{4} \log(2R)(V_M - V_m) e^{-\frac{2R}{V_m}(\beta+\lambda_0^*)}.$$

So, for any λ, we have

$$\max_{(v',r')\in S} \|G_\lambda(v',r')\| \leq C(R,\lambda), \tag{13.3.6}$$

where $C(R,\lambda)$ is a positive constant. Moreover, let $E \subset S$ be measurable. Then,

$$\int_E |G_\lambda(v',r')| r\,dr\,dv \leq \int_E \frac{C}{2v_m} \int_{|r-r'|}^{r+r'} \frac{1}{s} e^{-\frac{2R}{v}|\beta+\lambda_0^*|} r\, ds\, dr\, dv.$$

Since $\int_{|r-r'|}^{r+r'} \frac{1}{s} e^{-\frac{2R}{v}|\beta+\lambda_0^*|}\, ds \in L_1(S, r\,dr\,dv)$, we have

$$\lim_{m(E)\to 0} \int_E \|G_\lambda(v',r')\| r\,dr\,dv = 0. \tag{13.3.7}$$

From (13.3.6), (13.3.7), and Theorem 2.4.5, we can deduce that $VP_\lambda Q$ is weakly compact for each λ. This implies that $(KP_\lambda)^2$ is weakly compact. Q.E.D.

Lemma 13.3.2. *Let us assume that the assumptions* ($Q1$)–($Q5$) *hold true. Then, for all* λ *such that* $\mathrm{Re}\lambda > -\lambda_0^*$, *we have* $\|(\lambda - B)^{-1}\| \leq \frac{1}{\mathrm{Re}\lambda + \lambda_0^*}$. ◇

Proof. Let B' be the operator defined by:

$$\begin{cases} B' : \mathcal{D}(B) \subseteq X \longrightarrow X \\ \psi \longrightarrow -v\dfrac{\partial \psi}{\partial x}(x,y,v) - \left(v\Sigma\left(\sqrt{x^2+y^2}, \dfrac{x}{\sqrt{x^2+y^2}}, v\right) - \lambda_0^*\right)\psi(x,y,v). \end{cases}$$

First, we will show that the operator B' is dissipative. To do this, we take $\psi \in \mathcal{D}(B')$ and $\psi^* = S_0\psi$, where $S_0\psi$ satisfies the Eq. (13.2.9) and we will prove, in the following, that $\mathrm{Re}\langle B'\psi, S_0\psi\rangle \leq 0$.

$$
\begin{aligned}
&\operatorname{Re}\langle B'\psi, S_0\psi\rangle \\
&= \operatorname{Re}\Bigg(-\int_{V_m}^{V_M}\int_0^R\int_{-\sqrt{R^2-y^2}}^{\sqrt{R^2-y^2}} v\frac{\partial\psi}{\partial x}(x,y,v)y\,dx\,dy\,dv\Bigg) \\
&\quad + \left(v\Sigma\left(\sqrt{x^2+y^2}, \frac{x}{\sqrt{x^2+y^2}}, v\right) - \lambda_0^*\right)\psi(x,y,v)S_0\psi(x,y,v)y\,dx\,dy\,dv \\
&= \operatorname{Re}\Bigg(-\int_{V_m}^{V_M}\int_0^R\int_{-\sqrt{R^2-y^2}}^{\sqrt{R^2-y^2}} v\frac{\partial}{\partial x}(|\psi|)y\,dx\,dy\,dv\Bigg) \\
&\quad + \left(v\Sigma\left(\sqrt{x^2+y^2}, \frac{x}{\sqrt{x^2+y^2}}, v\right) - \lambda_0^*\psi(x,y,v)|\psi|y\,dx\,dy\,dv\right) \\
&\le R^2\left(\|\psi_{|\Gamma_s}\| - \|\psi_{|\Gamma_e}\|\right) \\
&\le 0 \quad (\|H\| \le 1).
\end{aligned}
$$

Now, let $\varphi = (\lambda - B)\psi$. Then,

$$
\begin{aligned}
\operatorname{Re}(\lambda+\Sigma)\|\psi\|^2 &= \operatorname{Re}(\lambda+\lambda_0^*)\langle\psi,\psi^*\rangle \\
&= \operatorname{Re}\left((\lambda+\Sigma)\langle\psi,\psi^*\rangle\right) \\
&\le \operatorname{Re}[(\lambda+\Sigma)\langle\psi,\psi^*\rangle - \langle B'\psi,\psi^*\rangle] \text{ because } B' \text{ is dissipative} \\
&= \operatorname{Re}[\lambda\langle\psi,\psi^*\rangle - \langle B\psi,\psi^*\rangle] \\
&= \operatorname{Re}\langle\varphi,\psi^*\rangle \\
&\le \|\varphi\|\,\|\psi\|.
\end{aligned}
$$

This implies that $\|\psi\| \le \frac{\|\varphi\|}{\operatorname{Re}\lambda+\lambda_0^*}$, and we obtain $\|(\lambda - B)^{-1}\| \le \frac{1}{\operatorname{Re}\lambda+\lambda_0^*}$. Q.E.D.

Lemma 13.3.3. *Let us assume that the assumptions* $(\mathcal{Q}1)$*–*$(\mathcal{Q}5)$ *hold true. Then, the set* $\{\lambda \in \mathbb{C}$ *such that* $\operatorname{Re}\lambda > \|K\| - \lambda_0^* + 1\} \subset \rho(A)$ *and, for all* λ *such that* $\operatorname{Re}\lambda > \|K\| - \lambda_0^* + 1$, *we have* $\|(\lambda - A)^{-1}\| \le 1$. ◇

Proof. The proof is similar as the proof of Lemma 13.2.2. Q.E.D.

We end this section by the following result.

Proposition 13.3.2. *Let us assume that the assumptions* $(\mathcal{Q}1)$*–*$(\mathcal{Q}5)$ *hold true. If the boundary operator* H *is nonnegative, then* B *generates a strongly continuous semigroup* $(U(t))_{t\ge0}$, *satisfying* $\|U(t)\| \le e^{-\lambda_0^* t}$. ◇

Proof. In view of the positivity of the operator H and also Lemma 13.3.2, together with Corollary 2.5.1, we can immediately deduce the result. Q.E.D.

13.3.3 Asymptotic Behavior of the Solution

In this section, we assume that

$(\mathcal{Q}6)$: The boundary condition is of Maxwell type, that is $\alpha(\mu, \mu', v, v') = \alpha_0 v|\mu|e^{-\frac{v^2}{2\theta}}$, where $\theta = \frac{kT}{m}$, m is the mass of the migrating particles, k represents the Boltzmann constant, T is the absolute temperature, and α_0 represents the adjustment coefficient on the boundary, such that $0 < \alpha_0 \leq \frac{2}{\theta}$.

We also assume that

$(\mathcal{Q}7)$: The transport medium is homogeneous and isotropic, that is $\Sigma(r, \mu, v) = \Sigma(v)$.

It is easy to show that the condition $(\mathcal{Q}6)$ secures the validation of the condition $(\mathcal{Q}3)$. So, all the results obtained above are valid under condition $(\mathcal{Q}6)$. Furthermore, we have the following lemma:

Lemma 13.3.4. *Let $g(\lambda)$ be the entire function defined by*

$$g(\lambda) = \frac{\alpha_0}{R^2} \int_S vye^{-\frac{v^2}{2\theta}} e^{-\frac{2(\lambda + v\Sigma(v))}{v}\sqrt{R^2-y^2}} dy\, dv.$$

(i) *For any $\lambda \in \mathbb{C}$, we have*

$$M_\lambda^2 = g(\lambda) M_\lambda, \qquad M_\lambda H_\lambda = g(\lambda) H_\lambda. \tag{13.3.8}$$

(ii) *If $g(\lambda) \neq 1$, then $(I - M_\lambda)^{-1}$ exists and is bounded. Moreover, we have the following $(I - M_\lambda)^{-1} = I + \frac{1}{1-g(\lambda)} M_\lambda$, and $(\lambda - B)^{-1} = \frac{1}{1-g(\lambda)} L_\lambda H_\lambda + P_\lambda$.*

$\diamond$

Proof.

(i) By making some computation, we can obtain Eq. (13.3.8).
(ii) If $g(\lambda) \neq 1$, then $I + \frac{1}{1-g(\lambda)} M_\lambda$ is a bounded linear operator. From Eq. (13.3.8), it follows that $(I - M_\lambda)\left(I + \frac{1}{1-g(\lambda)} M_\lambda\right) = I + \frac{1}{1-g(\lambda)} M_\lambda - M_\lambda - \frac{g(\lambda)}{1-g(\lambda)} M_\lambda = I$. Similarly, $\left(I + \frac{1}{1-g(\lambda)} M_\lambda\right)(I - M_\lambda) = I$. Hence, $(I - M_\lambda)^{-1}$ exists and is equal to $I + \frac{1}{1-g(\lambda)} M_\lambda$. The remaining part of the proof is a consequence of Eq. (13.3.5).

Q.E.D.

In what follows, we will establish some lemmas that enable us to give the estimation of the resolvent. Let $w > 0$ and set $\mathcal{R}_w = \{\lambda \in \mathbb{C}$ such that $\mathrm{Re}\lambda \geq -\lambda_0^* + w\}$. For $v \in (V_m, V_M)$, we consider

$$\begin{cases} \varphi_v : [0, 2R] \longrightarrow \mathbb{R}_+ \\ \qquad\qquad s \longrightarrow e^{-\frac{1}{v}(v\sigma(v)-\lambda_0^*+\frac{w}{2})s}. \end{cases}$$

Observing that, for all $v \in (V_m, V_M)$, we have $v\sigma(v) - \lambda_0^* + \frac{w}{2} \geq \frac{w}{2} > 0$. We claim that $\varphi_v(.)$ is a bounded, positive, and measurable function. We denote by $(l_{v,n}(.))_{n\in\mathbb{N}}$ an increasing sequence of nonnegative step functions with a compact support which converges to $\varphi_v(.)$ almost everywhere. Now, we introduce the operator P by:

$$\begin{cases} P : L_1(D, y\,dx\,dy\,dv) \longrightarrow L_1(D, y\,dx\,dy\,dv) \\ \qquad\qquad \psi \longrightarrow P\psi(x, y, v) \\ \qquad\qquad\quad = \displaystyle\int_{-\sqrt{R^2-y^2}}^{x} \frac{1}{v} e^{-\frac{1}{v}(\lambda+v\Sigma(v))(x-x')}\psi(x', y, v)dx' \end{cases}$$

and the sequence $(P_n)_{n\in\mathbb{N}}$ of operators defined, for all $n \in \mathbb{N}$, by

$$\begin{cases} P_n : L_1(D, y\,dx\,dy\,dv) \longrightarrow L_1(D, y\,dx\,dy\,dv) \\ \qquad\qquad \psi \longrightarrow P\psi(x, y, v) \\ \qquad\qquad\quad = \int_{-\sqrt{R^2-y^2}}^{x} \frac{1}{v} e^{-\frac{1}{v}(\lambda+\lambda_0^*-\frac{w}{2})(x-x')} l_{v,n}(x - x')\psi(x', y, v)dx'. \end{cases}$$

Lemma 13.3.5. *The sequence of operators $(P_n)_{n\in\mathbb{N}}$ converges uniformly on $\mathcal{R}_w$ to P in $\mathcal{L}(L_1(D, y\,dx\,dy\,dv))$.* ◇

Proof. Let $\psi \in L_1(D, y\,dx\,dy\,dv)$. Then, by using the change of variables $s = x - x'$, we get

$$\begin{aligned} &\|(P_n - P)\psi\| \\ &= \int_D \left| \int_0^{2R} \frac{1}{v} e^{-\frac{1}{v}(\lambda+\lambda_0^*-\frac{w}{2})s}(l_{v,n} - \varphi_v)(s)\chi_{[0,\sqrt{R^2-y^2}-x']}(s)\psi(x', y, v)ds \right| y\,dx'\,dy\,dv \\ &\leq \int_0^R \int_{-\sqrt{R^2-y^2}}^{\sqrt{R^2-y^2}} \int_0^{2R} \frac{1}{v} e^{-\frac{1}{v}(\mathrm{Re}\lambda+\lambda_0^*-\frac{w}{2})s}(\varphi_v - l_{v,n})(s)|\psi(x', y, v)|ds\, y\,dx'\,dy\,dv \\ &\leq \|\psi\| \int_0^{2R} \frac{1}{v} e^{-\frac{1}{v}(\mathrm{Re}\lambda+\lambda_0^*-\frac{w}{2})s}(\varphi_v(s) - l_{v,n}(s))ds. \end{aligned}$$

By applying the Lebesgue dominated convergence theorem, we get $\lim_{n\to+\infty} \|(P_n - P)\psi\| = 0$ uniformly on $\mathcal{R}_w$. This completes the proof of the lemma. Q.E.D.

Lemma 13.3.6. *Let $\psi(.)$ be a measurable simple function in D and let $l(.)$ be a nonnegative step function defined on $[0, 2R]$ and satisfying $l_v(.) \leq \varphi_v(.) \leq \frac{1}{V_m}$. Then, $\forall r \in [0, 1[$, we have*

$$\lim_{|\mathrm{Im}\lambda|\to+\infty} |\mathrm{Im}\lambda|^r \int_D \left| \int_0^{2R} \frac{1}{v} e^{-\frac{1}{v}(\lambda+\lambda_0^*-\frac{w}{2})s} l_v(s)\psi(x-s,y,v)\chi_{[0,x+\sqrt{R^2-y^2}]}(s)ds \right| \times y\,dx'\,dy\,dv = 0$$

uniformly on $\mathcal{R}_w$. ◇

Proof. To do this, let $y \in [0, R]$, $x \in [-\sqrt{R^2-y^2}, \sqrt{R^2-y^2}]$, $v \in (V_m, V_M)$ and let us consider the map

$$\begin{cases} \varphi_{x,y,v} : [0, 2R] \longrightarrow \mathbb{R} \\ \qquad\qquad s \longrightarrow l_v(s)\psi(x-s,y,v)\chi_{[0,x+\sqrt{R^2-y^2}]}(s). \end{cases}$$

It follows that $\varphi_{x,y,v}$ is a simple function. Let $(s_i)_{1\le i\le m}$ be the subdivision of its support such that $\forall 1 \le i \le m-1$. We have $\varphi_{x,y,v}(s) = \varphi_{x,y,v}(s_i)\ \ \forall s \in [s_i, s_{i+1}[$. Then, we get

$$\begin{aligned}
&\int_0^{2R} \frac{1}{v} e^{-\frac{1}{v}(\lambda+\lambda_0^*-\frac{w}{2})s}\varphi_{x,y,v}(s)ds \\
&= \sum_{i=1}^{m-1} \int_{s_i}^{s_i+1} \frac{1}{v}\varphi_{x,y,v}(s_i)e^{-\frac{1}{v}(\lambda+\lambda_0^*-\frac{w}{2})s}ds \\
&= \sum_{i=1}^{m-1} \varphi_{x,y,v}(s_i)\left[\frac{-1}{(\lambda+\lambda_0^*-\frac{w}{2})} e^{-(\lambda+\lambda_0^*-\frac{w}{2})s}ds\right]_{s_i}^{s_i+1} \\
&= \frac{-1}{(\lambda+\lambda_0^*-\frac{w}{2})} \sum_{i=1}^{m-1} \varphi_{x,y,v}(s_i)\left(e^{-(\lambda+\lambda_0^*-\frac{w}{2})s_i} - e^{-(\lambda+\lambda_0^*-\frac{w}{2})s_{i+1}}\right).
\end{aligned}$$

Hence,

$$\left|\int_0^{2R} e^{-\frac{1}{v}(\lambda+\lambda_0^*-\frac{w}{2})s}\varphi_{x,y,v}(s)ds\right| \le \frac{2(m-1)}{V_m|\mathrm{Im}\lambda|}\sup|\psi(.,.,.)|,$$

and

$$\begin{aligned}
&|\mathrm{Im}\lambda|^r \int_D \left|\int_0^{2R} \frac{1}{v}e^{-\frac{1}{v}(\lambda+\lambda_0^*-\frac{w}{2})s} l_v(s)\psi(x-s,y,v)\ \chi_{[0,x+\sqrt{R^2-y^2}]}(s)ds\right| y\,dx'\,dy\,dv \\
&\quad \le \frac{4R^3(m-1)(V_M-V_m)}{3|\mathrm{Im}\lambda|^{1-r}}\sup|\psi(.,.,.)|.
\end{aligned}$$

This inequality allows us to reach the desired result. Q.E.D.

Theorem 13.3.1. *If the boundary operator H satisfies the hypothesis ($\mathcal{Q}5$), then for any $r \in [0, 1)$, we have $\lim_{|\mathrm{Im}\lambda|\to\infty} |\mathrm{Im}\lambda|^r \|[K(\lambda - B)^{-1}]^5\| = 0$ uniformly on $\mathcal{R}_w$.* ◇

Proof. By using Lemma 13.3.4, the operator $K(\lambda - B)^{-1}$ can be written as

$$K(\lambda - B)^{-1} = \frac{1}{1 - g(\lambda)} KL_\lambda H_\lambda + KP_\lambda. \tag{13.3.9}$$

From Eq. (13.3.9), we get

$$\begin{aligned}
&\left[K(\lambda - B)^{-1}\right]^5 \\
&= l^5L^5 + l^4L^4A + l^4L^3AL + l^3L^3A^2 + l^4L^2AL^2 + l^3L(LA)^2 \\
&+ l^3L^2A^2L + l^2L^2A^3 + l^4LAL^3 + l^3LAL^2A + l^3L(AL)^2 + l^2(LA)^2A \\
&+ l^2LA^2LA + l^2LA^3L + lLA^4 + l^4AL^4 + l^3AL^3A + l^3AL^2AL \\
&+ l^2AL^2A^2 + l^3(AL)^2L + l^2A(LA)^2 + l^2ALA^2L + lALA^3 + l^3A^2L^3 \\
&+ l^2A^2L^2A + l^2A(AL)^2 + lA^2LA^2 + l^2A^3L^2 + lA^3LA + lA^4L \\
&+ l^3LA^2L^2 + A^5,
\end{aligned}$$

where $L = KL_\lambda H_\lambda$, $A = KP_\lambda$, and $l = \frac{1}{1-g(\lambda)}$. We claim that it suffices to show the result for the operator A^5, since there is an analogy between the components of $\left[K(\lambda - B)^{-1}\right]^5$. The proof requires the following two steps.

Step 1. By virtue of Lemmas 13.3.1 and 2.1.13, the operator A^4 is compact. Hence, we deduce that the family $\{A^4 : X \longrightarrow X$ such that $\lambda \in \mathcal{R}_w\}$ is bounded in the space X and thus, it is relatively compact by the Heine–Borel's theorem. Hence, this family of operators is collectively compact.

Step 2. By referring to Proposition 2.1.2, we will show that the family

$$\{|\mathrm{Im}\lambda|^r A : X \longrightarrow X \text{ such that } \lambda \in \mathcal{R}_w\}$$

converges strongly to zero as $|\mathrm{Im}\lambda|$ goes to infinity. Then, it remains to prove that $\forall r \in [0, 1[$, $\lim_{|\mathrm{Im}\lambda|\to+\infty} |\mathrm{Im}\lambda|^r \|A\psi\| = 0$ uniformly on $\mathcal{R}_w$, where

$$\begin{aligned}
A\psi &= \frac{1}{2r} \int_{V_M}^{V_m} \int_{-r}^{r} k(r, v, v') \\
&\left(\int_{-\sqrt{R^2-r^2+z^2}}^{z} \frac{1}{v'} e^{-\int_{x'}^{z} \frac{1}{v'}(\lambda + v'\Sigma(v'))dz'} \psi(x', \sqrt{r^2 - z^2}, v')dx' \right) dv'\, dz.
\end{aligned}$$

To do this, we firstly observe that the operator A may be decomposed as follows $A = VQP$, where V, P, and Q are the already defined operators. Since V and Q are bounded uniformly on $\mathcal{R}_w$, and taking into account Lemma 13.3.5, it remains to show that $\forall r \in [0, 1[$, $\lim_{|\text{Im}\lambda| \to +\infty} |\text{Im}\lambda|^r \| P_n \psi \| = 0$ uniformly on $\mathcal{R}_w$. Or equivalently, $\forall r \in [0, 1[$,

$$\lim_{|\text{Im}\lambda| \to +\infty} |\text{Im}\lambda|^r \int_D \left| \int_{-\sqrt{R^2-y^2}}^{x} \frac{1}{v} e^{-\frac{1}{v}(\lambda+\lambda_0^*-\frac{w}{2})(x-x')} l_{v,n}(x-x')\psi(x', y, v)dx' \right| \times y\, dx\, dy\, dv = 0$$

uniformly on $\mathcal{R}_w$. Since $\psi \in L_1(D, y\, dx\, dy\, dv)$, it suffices to show the result for a measurable simple function ψ in D. Now, using the change of variables $s = x - x'$, it remains to prove that

$$\lim_{|\text{Im}\lambda| \to +\infty} |\text{Im}\lambda|^r \int_D \left| \int_0^{2R} \frac{1}{v} e^{-\frac{1}{v}(\lambda+\lambda_0^*-\frac{w}{2})(s)} l_{v,n}(s) \times \chi_{[0, x+\sqrt{R^2-y^2}]}(s)\psi(x-s, y, v)ds \right| y\, dx\, dy\, dv = 0.$$

Now, Lemma 13.3.6 allows us to reach the desired result.

Q.E.D.

In the remaining part of this section, we will give a description of the asymptotic behavior of the solution to the Cauchy problem (13.3.3). Since B is an infinitesimal generator of a C_0-semigroup $(U(t))_{t\geq 0}$ acting on $L_1[D, y\, dx\, dy\, dv]$ and since K is a bounded linear operator, then by using the classical perturbation theory (Theorem 2.5.8), the operator $A = B + K$ also generates a C_0-semigroup $(V(t))_{t\geq 0}$ on X. We suppose that the operator H satisfies $(\mathcal{Q}5)$. So, from Proposition 4.2.1, the asymptotic spectrum $P(A)$ consists of, at most, discrete eigenvalues with finite algebraic multiplicities $\{\lambda_1, \lambda_2, \ldots, \lambda_n, \lambda_{n+1}, \ldots\}$ which can be ordered in such a way that the real part decreases [186], i.e., $\text{Re}\lambda_1 > \text{Re}\lambda_2 > \cdots > \text{Re}\lambda_n > \text{Re}\lambda_{n+1} > \cdots > \eta$, and $\{\lambda \in \mathbb{C}$ such that $\text{Re}\lambda > \eta\} \backslash \{\lambda_n, n = 1, 2, 3, \ldots\} \subset \rho(A)$. By applying Theorem 4.2.1, we obtain the description of the asymptotic behavior of the solution of the transport operator with diffuse reflection boundary conditions.

Theorem 13.3.2. *Let us assume that the hypotheses $(\mathcal{Q}1)$–$(\mathcal{Q}7)$ hold true. Then, for any $\varepsilon > 0$, there exists $M > 0$, such that*

$$\left\| V(t) - \sum_{i=1}^{n} e^{\lambda_i t} e^{D_i t} P_i \right\| \leq M e^{(\varepsilon + \text{Re}\lambda_{n+1})t}, \quad \forall t > 0.$$

◇

13.4 Essential Spectra of Transport Operator Arising in Growing Cell Populations

In [296] Rotenberg proposed the following partial differential equation

$$\frac{\partial \psi}{\partial t}(\mu, v, t) = -v\frac{\partial \psi}{\partial \mu}(\mu, v, t) - \sigma(\mu, v)\psi(\mu, v, t) + \int_0^c r(\mu, v, v')\psi(\mu, v', t)dv'$$
$$= A_K\psi := S_K\psi + B\psi \tag{13.4.1}$$

in order to describe the growth of cells of a population where S_K denotes the streaming operator and B stands for the collision operator (the integral part of A_K). In this model, the cells are distinguished by their degree of maturity $\mu \in [0, a]$, $a > 0$, and their maturation velocity $v \in [0, c]$, $c > 0$. The degree of maturation μ is then defined so that $\mu = 0$ at the birth time and $\mu = a$ at the death time. Equation (13.4.1) describes the number density of cell population as a function of the degree of maturation μ, the maturation velocity v and time t. The function $r(., ., .)$ denotes the transition rate at which cells change their maturation velocity from v to v'.

13.4.1 The Resolvent of the Operator S_K

First, let us state precisely the functional setting of the problem. Let X_p the space defined by $X_p := L_p([0, a]\times[0, c], d\mu dv)$ where $a > 0$, $c > 0$ and $1 \leq p < \infty$. We denote by X_p^0 and X_p^1 the following boundary spaces $X_p^0 := L_p(\{0\} \times [0, c], vdv)$, and $X_p^1 := L_p(\{a\}\times[0, c], vdv)$ endowed with their natural norms. In the sequel, X_p^0 and X_p^1 will often be identified with $L_p([0, c], vdv)$. We define the partial Sobolev space W_p by $W_p = \left\{\psi \in X_p \text{ such that } v\frac{\partial \psi}{\partial \mu} \in X_p\right\}$. It is well known (see [89] or [138]) that any ψ in W_p has traces on the spatial boundary $\{0\}$ and $\{a\}$ which belong to the spaces X_p^0 and X_p^1, respectively. They are denoted, respectively, by ψ^0 and ψ^1. Let K be the following boundary operator

$$\begin{cases} K : X_p^1 \longrightarrow X_p^0, \\ \qquad u \longrightarrow Ku. \end{cases}$$

We define the free streaming operator S_K by

$$\begin{cases} S_K : \mathcal{D}(S_K) \subset X_p \longrightarrow X_p \\ \qquad \psi \longrightarrow S_K\psi(\mu, v) = -v\frac{\partial \psi}{\partial \mu}(\mu, v) - \sigma(\mu, v)\psi(\mu, v) \\ \qquad \mathcal{D}(S_K) = \{\psi \in W_p \text{ such that } \psi^0 = K\psi^1\}, \end{cases}$$

where the function $\sigma(.,.)$ is bounded below and belongs to $L_{1\text{loc}}[(0,a)\times(0,c)]$. Now, let us consider the resolvent equation for the operator S_K,

$$(\lambda - S_K)\psi = \varphi, \tag{13.4.2}$$

where φ is a given function of X_p, $\lambda \in \mathbb{C}$ and the unknown ψ must be sought in $\mathcal{D}(S_K)$. Let $\underline{\sigma}$ be the real defined by $\underline{\sigma} = \text{ess-}\inf\{\sigma(\mu,v),\ (\mu,v)\in[0,a]\times[0,c]\}$. For $\text{Re}\lambda > -\underline{\sigma}$, the solution is formally given by

$$\psi(\mu,v) = \psi(0,v)\, e^{-\frac{1}{v}\int_0^\mu(\lambda+\sigma(\mu',v))d\mu'} + \frac{1}{v}\int_0^\mu e^{-\frac{1}{v}\int_{\mu'}^\mu(\lambda+\sigma(\tau,v))d\tau}\,\varphi(\mu',v)\,d\mu'. \tag{13.4.3}$$

Accordingly, for $\mu = a$, we get

$$\psi(a,v) = \psi(0,v)\, e^{-\frac{1}{v}\int_0^a(\lambda+\sigma(\mu',v))d\mu'} + \frac{1}{v}\int_0^a e^{-\frac{1}{v}\int_{\mu'}^a(\lambda+\sigma(\tau,v))d\tau}\,\varphi(\mu',v)\,d\mu'. \tag{13.4.4}$$

In the sequel, we will need the following operators

$$P_\lambda : X_p^0 \longrightarrow X_p^1,\ u \longrightarrow (P_\lambda u)(0,v) := u(0,v)\, e^{-\frac{1}{v}\int_0^a(\lambda+\sigma(\mu',v))d\mu'},$$

$$Q_\lambda : X_p^0 \longrightarrow X_p, u \longrightarrow (Q_\lambda u)(0,v) := u(0,v)\, e^{-\frac{1}{v}\int_0^\mu(\lambda+\sigma(\mu',v))d\mu'},$$

$$\begin{cases} \Pi_\lambda : X_p \longrightarrow X_p^1, \\ \qquad \varphi \longrightarrow (\Pi_\lambda\varphi)(\mu,v) := \dfrac{1}{v}\displaystyle\int_0^a e^{-\frac{1}{v}\int_{\mu'}^a(\lambda+\sigma(\tau,v))d\tau}\,\varphi(\mu',v)\,d\mu', \end{cases}$$

and

$$\begin{cases} \Xi_\lambda : X_p \longrightarrow X_p, \\ \qquad \varphi \longrightarrow (\Xi_\lambda\varphi)(\mu,v) := \dfrac{1}{v}\displaystyle\int_0^\mu e^{-\frac{1}{v}\int_{\mu'}^\mu(\lambda+\sigma(\tau,v))d\tau}\,\varphi(\mu',v)\,d\mu'. \end{cases}$$

Clearly, for λ satisfying $\text{Re}\lambda > -\underline{\sigma}$, the operators P_λ, Q_λ, Π_λ, and Ξ_λ are bounded. We can easily check that the norms of P_λ and Q_λ satisfy $\|P_\lambda\| \le e^{-\frac{a}{c}(\text{Re}\lambda+\underline{\sigma})}$ and $\|Q_\lambda\| \le (p\ (\text{Re}\lambda + \underline{\sigma}))^{-\frac{1}{p}}$. Moreover, a simple calculation using the Hölder inequality shows that $\|\Pi_\lambda\| \le (\text{Re}\lambda + \underline{\sigma})^{-\frac{1}{q}}$ and $\|\Xi_\lambda\| \le (\text{Re}\lambda + \underline{\sigma})^{-1}$ where q is the conjugate exponent of p, i.e., $q = \frac{p}{p-1}$. Using the above operators and the fact that ψ must satisfy the boundary conditions, (13.4.4) may be written abstractly in the form

$$\psi^1 = P_\lambda K\psi^1 + \Pi_\lambda\varphi. \tag{13.4.5}$$

Similarly, Eq. (13.4.3) becomes

$$\psi = Q_\lambda K \psi^1 + \Xi_\lambda \varphi. \tag{13.4.6}$$

Throughout this section, we denote by λ_K the real

$$\lambda_K := \begin{cases} -\underline{\sigma} & \text{if } r_\sigma(K) \leq 1, \\ -\underline{\sigma} + \dfrac{c}{a}\log(r_\sigma(K)) & \text{if } r_\sigma(K) > 1, \end{cases}$$

where $r_\sigma(K)$ is the spectral radius of K. Clearly, the solution of Eq. (13.4.5) reduces to the invertibility of the operator $\mathcal{U}(\lambda) := I - P_\lambda K$ (which is the case if $\text{Re}\lambda > \lambda_K$). This leads to $\psi^1 = \{\mathcal{U}(\lambda)\}^{-1}\Pi_\lambda\varphi$ where $\{\mathcal{U}(\lambda)\}^{-1} = \sum_{n\geq 0}(P_\lambda K)^n$. Together with (13.4.6), this gives $\psi = Q_\lambda K\{\mathcal{U}(\lambda)\}^{-1}\Pi_\lambda\varphi + \Xi_\lambda\varphi$. Accordingly, for $\text{Re}\lambda > \lambda_K$, the resolvent of the operator S_K may be written in the form

$$(\lambda - S_K)^{-1} = \sum_{n\geq 0} Q_\lambda K (P_\lambda K)^n \Pi_\lambda + \Xi_\lambda. \tag{13.4.7}$$

13.4.2 Spectral Properties of S_K

The purpose of this section is to derive, under reasonable hypotheses about the transition operator K, a precise description of the spectrum of the streaming operator S_K. We will also discuss the influence of the transition operators on the leading eigenvalue (when it exists). To do so, we will first consider the case of smooth transition operators, i.e., K satisfies the assumption:

$$(\mathcal{R}1) \quad \begin{cases} K \text{ is a positive operator (in the lattice sense)} \\ \text{and some power of } K \text{ is compact.} \end{cases}$$

We recall the following result obtained by Vidav [328].

Theorem 13.4.1. *Suppose that $B(\lambda - S_K)^{-1}B$ is a compact operator on X_p for all λ in the open half-plane $\text{Re}(\lambda) > -\underline{\sigma}$, and that the intersection of the set $\sigma(A_K)$ with the strip $-\underline{\sigma} < \text{Re}(\lambda) < \|B\| - \underline{\sigma}$ is not empty. Then, there exists a real eigenvalue λ_0 of A_K with a corresponding nonnegative eigenfunction. The half-plane $\text{Re}(\lambda) > \lambda_0$ belongs to the resolvent set of A_K. If $p > 1$, the nonnegative eigenfunction is uniquely determined up to a positive constant factor.* ◇

We define the sets $\mathbf{U} = \{\lambda \in \mathbb{C} \text{ such that } \text{Re}\lambda > -\underline{\sigma}\}$ and $P(S_K) = \sigma(S_K)\bigcap \mathbf{U}$. Our first result is the following.

Theorem 13.4.2. *Let $p \in [1, +\infty)$ and let us assume that the transition operator K satisfies the hypothesis ($\mathcal{R}1$). Then,*

(*i*) *$P(S_K)$ consists of, at most, isolated eigenvalues with a finite algebraic multiplicity.*
(*ii*) *If $P(S_K) \neq \emptyset$, then S_K has a leading eigenvalue $\lambda(a)$.*
(*iii*) *$P(S_K) \neq \emptyset$ if, and only if, $\lim_{\lambda \to -\underline{\sigma}} r_\sigma(P_\lambda K) > 1$. Furthermore, if $\lambda(a)$ exists, then $-\underline{\sigma} \leq \lambda(a) \leq -\underline{\sigma} + \frac{c}{a} \log(r_\sigma(K))$. In particular, if $\sigma(\mu, v) = \sigma$, then $P(S_K) \neq \emptyset$ if, and only if, $r_\sigma(K) > 1$ (regardless of a).*
(*iv*) *If $r_\sigma(K) \leq 1$, then $P(S_K) = \emptyset$ for all a.*
(*v*) *If $r_\sigma(K) > 1$, then $P(S_K) \neq \emptyset$, at least, for a small a and $\lambda(a) \to +\infty$ as $a \to 0$.*

$\diamond$

Proof. Let us first notice that if $r_\sigma(P_\lambda K) < 1$ for all $\lambda \in \mathbf{U}$, then $I - P_\lambda K$ is boundedly invertible. Hence, the solution of (13.4.5) can be written as $\psi^1 = (I - P_\lambda K)^{-1}\Pi_\lambda \varphi$, $\forall \lambda \in \mathbf{U}$. This shows that $\mathbf{U} \subseteq \rho(S_K)$ and, then $P(S_K) = \emptyset$. Now, we suppose that $r_\sigma(P_\lambda K) > 1$ for some $\lambda \in \mathbf{U}$. Clearly, for all $\lambda > -\underline{\sigma}$, we have $P_\lambda \leq e^{-\frac{a}{c}(\lambda + \underline{\sigma})} I$, where I denotes the identity operator on $L_p([0, c], vdv)$ [here, we make the identification $X_p^1 \sim X_p^0 \sim L_p([0, c], vdv)$]. Consequently,

$$P_\lambda K \leq e^{-\frac{a}{c}(\lambda + \underline{\sigma})} K, \quad \forall \lambda \geq -\underline{\sigma}. \tag{13.4.8}$$

Moreover, by using ($\mathcal{R}1$), there exists $N \in \mathbb{N}^*$ such that $(K)^N$ is compact. Besides, (13.4.8) implies $(P_\lambda K)^N \leq (K)^N \ \forall \lambda \geq -\underline{\sigma}$. So, by applying the Dodds–Fremlin comparison theorem for compact operators (see Theorem 2.3.7), we deduce that $(P_\lambda K)^N$ is compact for $\lambda \geq -\underline{\sigma}$. Next, using the analyticity of the operator valued function $\mathbf{U} \ni \lambda \to (P_\lambda K)^N$ [186, p. 365], we infer the compactness of $(P_\lambda K)^N$ for all λ in $\mathbf{U}$. Moreover, the inequality $(P_\lambda K)^{N+1} \leq P_\lambda K K^N$ implies that $\|(P_\lambda K)^{N+1}\| \leq \|P_\lambda K(K)^N\|$. Since $P_\lambda K \to 0$ strongly as $\lambda \to +\infty$, the use of Lemma 2.5.1, together with the compactness of K^N, implies that $P_\lambda K\,(K)^N \to 0$ in the operator norm as $\lambda \to +\infty$. This shows that $\|(P_\lambda K)^{N+1}\| \to 0$ as $\lambda \to +\infty$ and therefore,

$$r_\sigma((P_\lambda K)^{N+1}) \to 0 \text{ as } \lambda \to \infty. \tag{13.4.9}$$

From (13.4.9), together with Gohberg–Shmul'yan's theorem (see Theorem 2.5.13), it follows that $(I - (P_\lambda K)^{N+1})^{-1}$ is a degenerate-meromorphic operator function on $\mathbf{U}$, (i.e., $(I - (P_\lambda K)^{N+1})^{-1}$ is holomorphic on $\mathbf{U}$ except for a set S of isolated points where $(I - (P_\lambda K)^{N+1})^{-1}$ has poles and the coefficients of the main part have a finite rank). From $I - (P_\lambda K)^{N+1} = (I - P_\lambda K)\,(I + P_\lambda K + \cdots + (P_\lambda K)^N)$, we conclude that $(I - P_\lambda K)^{-1} = (I + P_\lambda K + \cdots + (P_\lambda K)^N)\,(I - (P_\lambda K)^{N+1})^{-1}$ is a degenerate-meromorphic on $\mathbf{U}$. So, if $\lambda \notin S$, Eq. (13.4.5) becomes $\psi^1 = (I - P_\lambda K)^{-1}\Pi_\lambda \varphi$.

By inserting ψ^1 into (13.4.6), we get $\psi = (\lambda - S_K)^{-1}\varphi$ where $(\lambda - S_K)^{-1} = Q_\lambda K(I - P_\lambda K)^{-1}\Pi_\lambda + \Xi_\lambda$. Thus, $(\lambda - S_K)^{-1}$ is degenerate-meromorphic on $\mathbf{U}$ which completes the proof of (i).

(ii) If $\lambda_0 \in P(S_K)$, then there exists $\varphi \neq 0$ such that $P_{\lambda_0}K\varphi = \varphi$. Thus, $(P_{\lambda_0}K)^N\varphi = \varphi$ and therefore, $|\varphi| \leq |(P_{\beta_0}K)^N\varphi| \leq (P_{\beta_0}K)^N|\varphi|$ where $\beta_0 = \text{Re}\lambda_0$. This implies that

$$r_\sigma((P_{\beta_0}K)^N) \geq 1. \qquad (13.4.10)$$

Moreover, according to the analyticity arguments (see Theorem 2.3.3), $r_\sigma((P_\beta K)^N)$ is a continuous and strictly decreasing function of β in $]-\underline{\sigma}, +\infty[$. Besides, by using the spectral mapping theorem (see Corollary 2.4.1), there exists $\alpha(\beta_0) \in \sigma(P_{\beta_0}K)$, such that $(\alpha(\beta_0))^N = r_\sigma((P_{\beta_0}K)^N)$, i.e., $\alpha(\beta_0) = \sqrt[N]{r_\sigma((P_{\beta_0}K)^N)}$. Hence, $\alpha(\beta)$ is also a continuous and strictly decreasing function of β in $]-\underline{\sigma}, +\infty[$. Moreover, (13.4.10) [resp. (13.4.9)] shows that $\alpha(\beta_0) \geq 1$ (resp. $\lim_{\beta\to+\infty} \alpha(\beta) = 0$). Accordingly, there exists (a unique) $\lambda \geq \beta_0$ such that $\alpha(\lambda) = 1$, i.e., $\lambda = \lambda(a)$ which represents the leading eigenvalue of S_K.

(iii) In order to prove this statement, we restrict ourselves to $\sigma(S_K)\bigcap(-\underline{\sigma}, +\infty)$. Hence, proceeding as in the proof of the second assertion, we deduce that the leading eigenvalue $\lambda(a)$ is characterized by

$$r_\sigma(P_{\lambda(a)}K) = 1. \qquad (13.4.11)$$

Hence, $\lambda(a)$ exists if, and only if, $\lim_{\lambda\to-\underline{\sigma}} r_\sigma(M_\lambda H) > 1$. If $\lambda(a)$ exists, using (13.4.8) and (13.4.11), we get $1 \leq e^{\frac{a}{c}(\lambda(a)-\underline{\sigma})}r_\sigma(K)$. Then $-\underline{\sigma} \leq \lambda(a) \leq -\underline{\sigma} + \frac{c}{a}\log(r_\sigma(K))$. Now, let us assume that $\sigma(\mu, v) = \sigma$. Then, $P_{-\underline{\sigma}} \leq I$ and consequently, $P_{-\underline{\sigma}}K \leq K$, which completes the proof of (iii).

(iv) Similarly as in (iii), we notice that $P_{(-\sigma)}K \leq K$. Hence, if $\lim_{\lambda\to-\sigma} r_\sigma(K) \leq 1$, then $\lim_{\lambda\to-\sigma} r_\sigma(P_\lambda K) \leq 1$. Then, the assertion is an immediate consequence of (iii).

(v) Let $\overline{\lambda}$ be an arbitrary real satisfying $\overline{\lambda} > -\underline{\sigma}$. Clearly, $P_{\overline{\lambda}} \to I$ strongly as $a \to 0$. Now, using the compactness of K^N, we deduce that $\lim_{a\to0} \|(P_{\overline{\lambda}}K)^{N+1} - (K)^{N+1}\| = 0$ and consequently, $\lim_{a\to0} r_\sigma(P_{\overline{\lambda}}K) = r_\sigma(K) > 1$. This shows that, for a small enough $r_\sigma(P_{\overline{\lambda}}K) > 1$, $\lambda(a)$ exists and $\lambda(a) > \overline{\lambda}$. Next, using the fact that $\overline{\lambda}$ is an arbitrary real in $]-\underline{\sigma}, +\infty[$, we infer that $\lambda(a) \to \infty$ as $a \to 0$. Q.E.D.

Let K be the operator defined by

$$K. := \langle ., \vartheta \rangle \zeta, \tag{13.4.12}$$

where $\zeta \in X_p^0$ and $\vartheta \in X_q^1$, and let λ be a complex number such that $\mathrm{Re}\lambda > -\underline{\sigma}$. The dual of the operator $P_\lambda K$ is given by $(P_\lambda K)^* = K^* \tilde{P}_\lambda$, where

$$\tilde{P}_\lambda : X_q^1 \longrightarrow X_q^0,\ u \longrightarrow (\tilde{P}_\lambda u)(0, v) := u(a, v)\, e^{-\frac{1}{v}\int_0^a (\lambda + \sigma(\mu', v))d\mu'} \tag{13.4.13}$$

and

$$K^* : X_q^0 \longrightarrow X_q^1,\ \ u \longrightarrow (K^* u)(0, v) := \langle \zeta, u \rangle \vartheta, \tag{13.4.14}$$

where ζ and ϑ represent the functions appearing in the expression of K. Let λ_0 be the real defined by $\lambda_0 := -\underline{\sigma} + \frac{c}{a}\log(r_\sigma(K))$.

Lemma 13.4.1. *Let K be the operator defined in (13.4.12). If λ belongs to the strip $-\underline{\sigma} < \mathrm{Re}\lambda \leq \lambda_0$, then $(K^* \tilde{P}_\lambda)$ converges to 0, for the strong operator topology, as $|\mathrm{Im}\lambda| \to +\infty$.* ◇

Proof. Let $\varphi \in X_q^1$. From (13.4.13) and (13.4.14), it follows that

$$K^* \tilde{P}_\lambda \varphi := \langle \zeta, P_\lambda \varphi \rangle \vartheta = \int_0^c \vartheta(v)\zeta(v') e^{-\frac{1}{v'}\int_0^a (\lambda + \sigma(\mu', v'))d\mu'} \varphi(a, v') v' dv'.$$

Let $(\lambda_n)_n$ be a sequence of complex numbers such that $\lambda_n = \eta + it_n$ where $\eta \in] - \underline{\sigma}, \lambda_0]$ and $t_n \to +\infty$ as $n \to \infty$. Hence,

$$|(K^* \tilde{P}_{\lambda_n} \varphi)(a, v)| = \left| \int_0^c \vartheta(v)\zeta(v') e^{-\frac{1}{v'}\int_0^a (\eta + \sigma(\mu', v'))d\mu'} e^{\frac{ai}{v'} t_n} \varphi(a, v') v' dv' \right|.$$

By applying the Riemann–Lebesgue lemma, we deduce that

$$\lim_{n \to \infty} \left| \int_0^c \vartheta(v)\zeta(v') e^{-\frac{1}{v'}\int_0^a (\eta + \sigma(\mu', v'))d\mu'} e^{\frac{ai}{v'} t_n} \varphi(a, v') v' dv' \right| = 0 \text{ a.e. on } \{a\} \times (0, c).$$

Accordingly, $\lim\limits_{n \to +\infty} |(K^* \tilde{P}_{\lambda_n} \varphi)(a, v)| = 0$ a.e. on $\{a\} \times (0, c)$. Furthermore, for every integer n, we have $|(K^* \tilde{P}_{\lambda_n} \varphi)(a, v)| \leq \int_0^c |\vartheta(v)|\, |\zeta(v')|\, |\varphi(a, v')| v' dv' \in X_q^1$. Then, according to Lebesgue's the dominated convergence theorem, we have $\lim\limits_{n \to +\infty} \|K^* \tilde{P}_{\lambda_n} \varphi\|_{X_q^1} = 0$. This proves the lemma. Q.E.D.

Lemma 13.4.2. *Let K be the operator defined in (13.4.12). The family $\{K^* \widetilde{P_\lambda}$ such that $-\underline{\sigma} < \mathrm{Re}\lambda \leq \lambda_0\}$ is collectively compact.* ◇

Proof. Let $\mathbf{B}_q$ denote the unit ball of the space X_q^1, and let $(\psi_n)_{n\in\mathbb{N}}$ be a sequence in $\bigcup_\lambda (K^* \tilde{P}_\lambda \mathbf{B}_q)$, $\lambda \in \{\lambda \in \mathbb{C}$ such that $-\underline{\sigma} < \mathrm{Re}\lambda \leq \lambda_0\}$. Then, there exists a sequence $(\varphi_n)_{n\in\mathbb{N}}$ in $\mathbf{B}_q$ such that $\psi_n = K^* \tilde{P}_{\lambda_n} \varphi_n$, $n = 1, 2, \ldots$. It is clear that the sequence $(y_n = \tilde{P}_{\lambda_n} \varphi_n)_{n\in\mathbb{N}}$ is bounded in X_q^0. So, from the compactness of K^*, it follows that $(\psi_n = K^* y_n)_{n\in\mathbb{N}}$ has a converging subsequence in $\overline{\bigcup_\lambda (K^* \tilde{P}_\lambda \mathbf{B}_q)}$. This completes the proof of lemma. Q.E.D.

Lemma 13.4.3. *Let K be the operator defined in (13.4.12). Let λ be in the strip $-\underline{\sigma} < \mathrm{Re}\lambda \leq \lambda_0$. Then,* $\lim_{|\mathrm{Im}\lambda| \to +\infty} r_\sigma (P_\lambda K) = 0.$ ◇

Proof. In view of Lemmas 13.4.1, 13.4.2 and Proposition 2.1.2, we have the following $\lim_{|\mathrm{Im}\lambda| \to +\infty} \|(K^* \tilde{P}_\lambda)^2\| = 0$ uniformly on $\{\lambda \in \mathbb{C}$ such that $-\underline{\sigma} < \mathrm{Re}\lambda \leq \lambda_0\}$. Therefore, since $r_\sigma(K^* \tilde{P}_\lambda) \leq \|(K^* \tilde{P}_\lambda)^n\|^{\frac{1}{n}}$ with $n = 1, 2, \ldots$, we conclude that $\lim_{|\mathrm{Im}\lambda| \to +\infty} r_\sigma(K^* \tilde{P}_\lambda) = 0$ uniformly on $\{\lambda \in \mathbb{C}$ such that $-\underline{\sigma} < \mathrm{Re}\lambda \leq \lambda_0\}$. Next, the use of the equality $r_\sigma(K^* \tilde{P}_\lambda) = r_\sigma(P_\lambda K)$ proves the lemma. Q.E.D.

Lemma 13.4.4. *Let K be an arbitrary compact transition operator. Then, $(I - P_\lambda K)^{-1}$ exists for λ in the half-plane $\{\lambda \in \mathbb{C}$ such that $\mathrm{Re}\lambda > -\underline{\sigma}\}$ with a sufficiently large $|\mathrm{Im}\lambda|$.* ◇

Proof. Notice that, if the transition operator K is compact, then there exists a sequence of finite rank operators, which converges in the operator norm, to K. Hence, it is sufficient to establish the result for a finite rank operator, that is, $K = \sum_{k=1}^n K_k$, $K_k = \langle ., \vartheta_k \rangle \zeta_k$ where $n \in \mathbb{N}$, $\vartheta_k \in X_1^q$, $\zeta_k \in X_p^0$ and where q denotes the conjugate exponent of p. Thus, we may restrict ourselves to a transition operator of rank one which we also denote by K, namely, $K. := \langle ., \vartheta \rangle \zeta$ where $\zeta \in X_p^0$ and $\vartheta \in X_q^1$. Clearly, if $\mathrm{Re}\lambda > \lambda_0$, then $\|P_\lambda K\| < 1$ and consequently, the half plane $\mathrm{Re}\lambda > \lambda_0$ is contained in $\rho(S_K)$. So, we only have to establish the lemma in the strip $\{\lambda \in \mathbb{C}$ such that $-\underline{\sigma} < \mathrm{Re}\lambda \leq \lambda_0\}$. Now, according to Lemma 13.4.3, there exists $M > 0$ such that, for any λ in the strip $-\underline{\sigma} < \mathrm{Re}\lambda \leq \lambda_0$ satisfying $|\mathrm{Im}\lambda| > M$, we have $r_\sigma(P_\lambda K) < 1$. This completes the proof of the lemma. Q.E.D.

Theorem 13.4.3. *Let $p \in [1, \infty)$ and let us assume that K is a nonnegative compact transition operator. Then, the following statements hold:*

(*i*) *$P(S_K)$ is bounded and, for every $\eta > 0$, $\sigma(S_K) \bigcap \{\lambda \in \mathbb{C}$ such that $\mathrm{Re}\lambda > -\underline{\sigma} + \eta\}$ is finite.*
(*ii*) *Assume that $\sigma \in L_\infty[(0, a) \times (0, c)]$. If $r_\sigma(K) > 1$, then there exists a positive constant ν such that $\lambda(a) \geq -\|\sigma\|_{L_\infty} + \frac{\nu}{a}$.*
(*iii*) *Let $\sigma(\mu, v) = \sigma$. If $r_\sigma(K) = 1$, then $-\sigma \in \sigma_p(S_K)$.* ◇

Proof.

(i) As mentioned above, if $\mathrm{Re}\lambda > \lambda_0$ then, $r_\sigma(P_\lambda K) < 1$ and therefore, $\sigma(S_K)\bigcap\{\lambda \in \mathbb{C}$ such that $\mathrm{Re}\lambda > \lambda_0\} = \emptyset$. Next, by using Lemma 13.4.4, we conclude that there exists $M > 0$, such that $P(S_K) \subseteq \{\lambda \in \mathbb{C}$ such that $-\underline{\sigma} < \mathrm{Re}\lambda \leq \lambda_0$ and $|\mathrm{Im}\lambda| \leq M\}$. This proves the boundedness of $P(S_K)$. Moreover, for any $\eta > 0$ such that $-\underline{\sigma} + \eta < \lambda_0$, $P(S_K)\bigcap\{\lambda \in \mathbb{C}$ such that $-\underline{\sigma} + \eta < \mathrm{Re}\lambda \leq \lambda_0\}$ is confined in a compact subset of the complex plane and then it is necessarily finite since it is discrete.

(ii) Let $\varepsilon \in (0, c)$ and let us define the operator K_ε by $K_\varepsilon : u \longrightarrow I_\varepsilon Ku$, where I_ε denotes the operator $I_\varepsilon : u \longrightarrow \chi_{(\varepsilon,c)}u$ and $\chi_{(\varepsilon,c)}(.)$ stands for the characteristic function of (ε, c). Obviously, $K_\varepsilon \leq K$ and $\|K_\varepsilon - K\| \to 0$ as $\varepsilon \to 0$ (by using the compactness of K). Let φ_ε be a positive eigenfunction of K_ε associated with the eigenvalue $r_\sigma(K_\varepsilon)$. Let $\lambda > -\underline{\sigma}$. It is clear that $P_\lambda K\varphi_\varepsilon \geq P_\lambda K_\varepsilon\varphi_\varepsilon$. Moreover, the fact that $\varphi_\varepsilon(v) = 0$ if $v \in [0, \varepsilon[$ implies that $P_\lambda\varphi_\varepsilon \geq e^{-a\left(\frac{\lambda+\|\sigma\|_{L_\infty}}{\varepsilon}\right)}\varphi_\varepsilon$. Similarly, $P_\lambda K_\varepsilon\varphi_\varepsilon \geq e^{-a\left(\frac{\lambda+\|\sigma\|_{L_\infty}}{\varepsilon}\right)}K_\varepsilon\varphi_\varepsilon$. Hence, $P_\lambda K \geq e^{-a\left(\frac{\lambda+\|\sigma\|_{L_\infty}}{\varepsilon}\right)}K_\varepsilon$ and consequently,

$$r_\sigma(P_\lambda K) \geq e^{-a\left(\frac{\lambda+\|\sigma\|_{L_\infty}}{\varepsilon}\right)}\, r_\sigma(K_\varepsilon). \tag{13.4.15}$$

Owing to the fact that $r_\sigma(P_{\lambda(a)}K) = 1$, thus for $\lambda = \lambda(a)$, (13.4.15) becomes

$$1 \geq e^{-a\left(\frac{\lambda(a)+\|\sigma\|_{L_\infty}}{\varepsilon}\right)} r_\sigma(K_\varepsilon).$$

Let ε be small enough so that $r_\sigma(K_\varepsilon) > 1$ (note that by using Corollary 2.3.2, $r_\sigma(K_\varepsilon) \to r_\sigma(K) > 1$ as $\varepsilon \to 0$). Then, $\lambda(a) \geq -\|\sigma\|_{L_\infty} + \frac{\varepsilon}{a}\log(r_\sigma(K_\varepsilon))$. This ends the proof. Q.E.D.

In the following, we denote by $\lambda(K)$ the leading eigenvalue of the operator S_K (when it exists). Now, we may discuss the monotonicity properties of $\lambda(K)$. To do so, we consider two transition operators K_1 and K_2 satisfying $K_1 \leq K_2$ and $K_1 \neq K_2$.

Theorem 13.4.4. *Let K_1 and K_2 be two transition operators satisfying $(\mathcal{R}1)$. If $\lambda(K_1)$ exists, then $\lambda(K_2)$ exists and $\lambda(K_1) \leq \lambda(K_2)$. Moreover, if there exists an integer m such that $(P_{\lambda(K_1)}K_2)^m$ is strictly positive, then $\lambda(K_1) < \lambda(K_2)$.* ◇

Proof. By hypothesis, there exist two integers n_1 and n_2, such that $(K_1)^{n_1}$ and $(K_2)^{n_2}$ are compact. Let $n_3 = \max(n_1, n_2)$. From (13.4.8), together with the Dodds–Fremlin's theorem (see Theorem 2.3.7), it follows that $(P_\lambda K_1)^{n_3}$ and $(P_\lambda K_2)^{n_3}$ are compact for all λ belonging to $]-\underline{\sigma}, \infty[$. In particular, $(P_{\lambda(K_1)}K_1)^{n_3}$ and $(P_{\lambda(K_1)}K_2)^{n_3}$ are positive compact operators on X_p^1. As already seen in the proof of Theorem 13.4.2, λ is an eigenvalue of S_K if, and only if, 1 is an eigenvalue of $P_\lambda K$. So, we conclude that

$$r_\sigma(P_{\lambda(K_1)}K_1) \geq 1. \tag{13.4.16}$$

Moreover, since $K_1 \leq K_2$ and $K_1 \neq K_2$, then $P_{\lambda(K_1)}K_1 \leq P_{\lambda(K_1)}K_2$ and $P_{\lambda(K_1)}K_1 \neq P_{\lambda(K_1)}K_2$. This implies that $r_\sigma(P_{\lambda(K_1)}K_1) \geq r_\sigma(P_{\lambda(K_1)}K_2)$. However, $P_{\lambda(K_1)}K_2$ is irreducible and power-compact, then using (13.4.16) and Theorem 2.3.1, we infer that

$$r_\sigma[P_{\lambda(K_1)}K_2]^{n_3} > 1. \tag{13.4.17}$$

Clearly, $[P_\lambda K_2]^{n_3}$ is an analytic operator-valued function whose values are compact for all $\lambda > -\underline{\sigma}$. Moreover, we have $\lim_{\lambda\to\infty} \|[P_\lambda K_2]^{n_3}\| = 0$ (see the proof of Theorem 13.4.2). Hence, the use of the analyticity arguments (see Theorem 2.3.3) implies that the function $]-\underline{\sigma},+\infty) \ni \lambda \to r_\sigma([P_\lambda K_2]^{n_3})$ is strictly decreasing. This, together with (13.4.17), implies that there exists a unique $\overline{\lambda} > \lambda(K_1)$, such that $r_\sigma([P_{\overline{\lambda}}K_2]^{n_3}) = 1$. Now, the spectral mapping theorem yields $\overline{\lambda} = \lambda(K_2)$ and the proof is complete. Q.E.D.

Now, let us consider the case of partly smooth transition operators:

$$(\mathcal{R}2) \qquad \begin{cases} K = K_1 + K_2 \text{ with } K_i \geq 0 \; i = 1,2, \;\; K_2 \text{ is either compact} \\ \text{if } 1 < p < \infty \text{ or weakly compact if } p = 1. \end{cases}$$

Theorem 13.4.5. *Let* $p \in [1, \infty)$ *and suppose that the hypothesis* $(\mathcal{R}2)$ *is satisfied. Then, the following assertions hold:*

(i) $\sigma(S_K)\bigcap\{\lambda \in \mathbb{C}$ *such that* $\mathrm{Re}\lambda > \lambda_{K_1}\}$ *consists of, at most, isolated eigenvalues with a finite algebraic multiplicity.*
(ii) If $\sigma(S_K)\bigcap\{\lambda \in \mathbb{C}$ *such that* $\mathrm{Re}\lambda > \lambda_{K_1}\} \neq \emptyset$, *then* S_K *has a leading eigenvalue* $\lambda(a)$.
(iii) If $\lim_{\lambda\to\lambda_{K_1}} r_\sigma(P_\lambda K_2) > 1$, *then* $\sigma(S_K)\bigcap\{\lambda \in \mathbb{C}$ *such that* $\mathrm{Re}\lambda > \lambda_{K_1}\} \neq \emptyset$.

◇

Proof.

(i) Let us again consider the problem (13.4.2), which is now equivalent to solving in X_p^1 the following one

$$\psi^1 = P_\lambda K_1 \psi^1 + P_\lambda K_2 \psi^1 + \Pi_\lambda \varphi. \tag{13.4.18}$$

Clearly, if $\lambda > \lambda_{K_1}$, then the operator $I - P_\lambda K_1$ is boundedly invertible and (13.4.18) becomes $\psi^1 = F_\lambda \psi^1 + L_\lambda \varphi$ where $F_\lambda := (I - P_\lambda K_1)^{-1} P_\lambda K_2$ and $L_\lambda := (I - P_\lambda K_1)^{-1}\Pi_\lambda$. As already mentioned, $P_\lambda \to 0$ strongly as

$\lambda \to \infty$ for all p in $[1, \infty)$. For $p \in (1, \infty)$, K_2 is compact and therefore, $\|P_\lambda K_2\| \to 0$ as $\lambda \to \infty$ in the operator topology (use Lemma 2.5.1). Now, let $p = 1$, and let $\lambda_2 > \lambda_{K_1}$. From the estimate $(I - P_\lambda K_1)^{-1} \leq (I - P_{\lambda_2} K_1)^{-1}$ (valid for $\lambda > \lambda_2$), it follows that $(F_\lambda)^3 \leq (I - P_{\lambda_2} K_2)^{-1} P_\lambda K_2 (F_{\lambda_2})^2$ for all $\lambda > \lambda_2$. Since K_2 is weakly compact, by applying Lemma 2.1.13, we infer that $(F_{\lambda_2})^2$ is compact. Using again Lemma 2.5.1, we get $\|(F_\lambda)^3\| \leq \|(I - P_{\lambda_2} K_1)^{-1}\| \, \|P_\lambda K_2 (F_{\lambda_1})^2\| \to 0$ as $\lambda \to \infty$. Since $r_\sigma(F_\lambda) \leq \|F_\lambda^n\|^{\frac{1}{n}}$, $n = 1, 2, 3, \ldots$, we have $r_\sigma(F_\lambda) \to 0$ as $\lambda \to +\infty$ for all $p \in [1, \infty)$. Now, by applying the Gohberg–Shmul'yan's theorem (see Theorem 2.5.13), we get the desired result.

(*ii*) This assertion follows from the fact that $(\lambda - S_K)^{-1}$ is positive for a large λ [see Eq. (13.1.12)].

(*iii*) Let $\lambda > \lambda_{K1}$. The estimate $F_\lambda \geq P_\lambda K_2$ implies that $r_\sigma(F_\lambda)) \geq r_\sigma(P_\lambda K_2)$. Hence, if $\lim_{\lambda \to \lambda_{K_1}} r_\sigma(P_\lambda K_2) > 1$, then $\lim_{\lambda \to \lambda_{K_1}} r_\sigma(F_\lambda) \geq \lim_{\lambda \to \lambda_{K_1}} r_\sigma(P_\lambda K_2) > 1$. Moreover, since F_λ^3 is compact on X_p^0, $1 \leq p < \infty$ [see the proof of (*i*)] and satisfies $\lim_{\lambda \to \infty} \|(F_\lambda)^3\| \to 0$, the use of the analyticity arguments (see Theorem 2.3.3) and also the spectral mapping theorem shows that $r_\sigma(F_\lambda)$ is a continuous and strictly decreasing function of λ satisfying $\lim_{\lambda \to +\infty} r_\sigma(F_\lambda) = 0$. Therefore, there exists $\bar{\lambda} > \lambda_{K_1}$ such that $r_\sigma(F_{\bar{\lambda}}) = 1$ which is the leading eigenvalue. Q.E.D.

13.4.3 Generation Results

We are going to start with a well-known generation result in the case of contractive transition operators (see [138]) which is a simple consequence of Hille–Yosida's Theorem.

Theorem 13.4.6. *Assume that* $\|K\| < 1$, *then* S_K *generates a* C_0*-semigroup* $\{U_K(t), t \geq 0\}$ *satisfying* $\|U_K(t)\| \leq e^{-\underline{\sigma} t}$, $t \geq 0$. ◇

For $\|K\| \geq 1$, as in [138, p. 478] we will make a suitable change of unknown, such that the new equivalent problem involves a contractive boundary operator and is governed by Theorem 13.4.6. For $q > 0$, we define the boundary multiplication operator by

$$\begin{cases} M_q : X_p^1 \longrightarrow X_p^1 \\ \qquad \varphi \longrightarrow M_q \varphi(1, v) := q^a \varphi(1, v) \end{cases}$$

and the unbounded operator by

$$\begin{cases} T_{K_q} : \mathcal{D}(T_{K_q}) \subset X_p \longrightarrow X_p \\ \qquad \psi \longrightarrow T_{K_q}\psi(\mu, v) := -v\,\dfrac{\partial\psi}{\partial\mu}(\mu, v) - (\sigma(\mu, v) + v\log q)\psi(\mu, v) \\ \mathcal{D}(T_{K_q}) = \left\{\psi \in W_p \text{ such that } \psi^0 = KM_q\psi^1\right\}. \end{cases}$$

Let

$$\begin{cases} \tilde{M}_q : X_p \longrightarrow X_p \\ \qquad \varphi \longrightarrow \tilde{M}_q\varphi(\mu, v) := q^{\mu}\varphi(\mu, v). \end{cases}$$

Remark 13.4.1. It is not difficult to verify that for $0 < q < 1$, the operator $\tilde{M}_q$ defines a continuous bijection from W_p into itself. Moreover, some easy calculations show that $\|\tilde{M}_q\| \le 1$ and $\|\tilde{M}_q^{-1}\| \le q^{-a}$. ◇

The following lemma shows a relation between the operators T_{K_q} and S_K.

Lemma 13.4.5. *For a fixed* $0 < q < 1$, $\tilde{M}_q^{-1}(\mathcal{D}(S_K)) = \mathcal{D}(T_{K_q})$ *and* $S_K = \tilde{M}_q T_{K_q} \tilde{M}_q^{-1}$. ◇

Proof. Let us take $\varphi \in \mathcal{D}(S_K)$ and $\psi = \tilde{M}_q^{-1}\varphi$, then $\psi \in W_p$, hence verifying

$$\begin{cases} \psi^0 = (q^{-\mu}\varphi)^0 = \varphi^0 \\ \psi^1 = (q^{-\mu}\varphi)^1 = q^{-a}\varphi^1. \end{cases}$$

Then, $\psi^0 = KM_q\psi^1$ and $\psi \in \mathcal{D}(T_{K_q})$. Conversely, we similarly prove that if $\psi \in \mathcal{D}(T_{K_q})$, $\tilde{M}_q\psi \in \mathcal{D}(S_K)$. So, $\tilde{M}_q^{-1}(\mathcal{D}(S_K)) = \mathcal{D}(T_{K_q})$. Next, for any $\varphi \in \mathcal{D}(S_K)$, we have

$$\begin{aligned} \tilde{M}_q T_{K_q} \tilde{M}_q^{-1}\varphi &= q^{\mu} T_{K_q}(q^{-\mu}\varphi) \\ &= q^{\mu}\left[-v\frac{\partial}{\partial\mu}(q^{-\mu}\varphi) - (\sigma(\mu, v) + v\log q)q^{-\mu}\varphi\right] \\ &= -v\frac{\partial\varphi}{\partial\mu}(\mu, v) - \sigma(\mu, v)\varphi \\ &= S_K\varphi. \end{aligned}$$

Q.E.D.

According to Lemma 13.4.5 and Remark 13.4.1, the following proposition holds.

Proposition 13.4.1. *Let* $0 < q < 1$, *then a* C_0*-semigroup* $\{V_K(t), t \ge 0\}$ *in* X_p *is generated by the operator* T_{K_q} *if, and only if,* S_K *is a generator of a* C_0*-semigroup* $\{U_K(t), t \ge 0\}$ *in* X_p *given by* $U_K(t) = \tilde{M}_q V_{K_q}(t)\tilde{M}_q^{-1}$. *Furthermore,* $\|U_K(t)\| \le q^{-a}\|V_K(t)\|$, $t \ge 0$. ◇

Proof. Let us assume that T_{K_q} generates a C_0-semigroup $\{V_K(t), t \geq 0\}$ in X_p. Let $\psi \in \mathcal{D}(T_{K_q})$. Then, by using Lemma 13.4.5, $\psi = \tilde{M}_q^{-1}\varphi$, where $\varphi \in \mathcal{D}(S_K)$. Thus,

$$\begin{aligned} T_{K_q}\psi &= \lim_{t\to 0+} t^{-1}[V_{K_q}(t)\psi - \psi] \\ &= \lim_{t\to 0+} t^{-1}\tilde{M}_q^{-1}\left[\tilde{M}_q V_{K_q}(t)\tilde{M}_q^{-1}\varphi - \varphi\right], \end{aligned}$$

which implies, by using Lemma 13.4.5, that $S_K\varphi = \lim_{t\to 0+} t^{-1}\left[\tilde{M}_q V_{K_q}(t)\tilde{M}_q^{-1}\varphi - \varphi\right]$. Therefore, S_K generates the C_0-semigroup $\{U_K(t),\ t \geq 0\}$ in X_p. Similarly, we prove the converse of our assumption, which ends the proof. Q.E.D.

Consequently, and according to the previous results, we deduce that the following time-dependent problem

$$\begin{cases} \dfrac{\partial\varphi}{\partial t}(\mu, v, t) = -v\dfrac{\partial\varphi}{\partial\mu}(\mu, v, t) - \sigma(\mu, v)\varphi(\mu, v, t) =: S_K\varphi(\mu, v, t) \\ \varphi^0 = K\varphi^1 \\ \varphi(\mu, v, 0) = \varphi_0(\mu, v) \end{cases}$$

is equivalent to

$$\begin{cases} \dfrac{\partial\psi}{\partial t}(\mu, v, t) = -v\dfrac{\partial\psi}{\partial\mu}(\mu, v, t) - (\sigma(\mu, v) + v\log q)\psi(\mu, v, t) =: T_{K_q}\psi(\mu, v, t) \\ \psi^0 = KM_q\psi^1 \\ \psi(\mu, v, 0) = q^{-\mu}\psi_0(\mu, v), \end{cases}$$

where $\varphi(\mu, v, t) = q^{-\mu}\psi(\mu, v, t)$.

Theorem 13.4.7. *Let $K \in \mathcal{L}(X_p^1, X_p^0)$ be an arbitrary bounded boundary operator. Then, the streaming operator S_K generates a C_0-semigroup $\{U_K(t), t \geq 0\}$ in X_p.* ◇

Proof. According to Theorem 13.4.6, it is sufficient to prove the result for $\|K\| \geq 1$. This will be done by taking $q := \frac{1}{2}\|K\|^{-\frac{1}{a}}$, then $\|KM_q\| < 1$, and the result follows from Proposition 13.4.1. Q.E.D.

13.4.4 Compactness Results

Now, let us consider the transport operator $A_K = S_K + B$ where B is the bounded operator given by

$$\begin{cases} B : X_p \longrightarrow X_p \\ \quad \psi \longrightarrow \displaystyle\int_0^c r(\mu, v, v')\psi(\mu, v')dv', \end{cases} \tag{13.4.19}$$

where $r(.,.,.)$ is a measurable function from $[0,a] \times [0,c] \times [0,c]$ into $\mathbb{R}^+$.

Theorem 13.4.8. *Assume that (i)–(iii) of Lemma 2.4.1 hold true for $D := (0,a)$ and $V := (0,c)$. Then, for any $\lambda \in \mathbb{C}$ such that $\mathrm{Re}\lambda > \lambda_K$, the operator $(\lambda - S_K)^{-1}B$ is compact on X_p, $1 < p < \infty$, and weakly compact on X_1.* ◇

Remark 13.4.2. Let λ be such that $\mathrm{Re}\lambda > \lambda_K$. We know from Eq. (13.4.7) that

$$(\lambda - S_K)^{-1}B = \sum_{n\geq 0} Q_\lambda K(P_\lambda K)^n \Pi_\lambda B \; + \; \Xi_\lambda B.$$

In order to prove the compactness (resp. the weak compactness) of $(\lambda - S_K)^{-1}B$ on X_p (resp. X_1), it suffices to show that the operators $\Pi_\lambda B$ and $\Xi_\lambda B$ are compact (resp. weakly compact) on X_p (resp. X_1). ◇

Lemma 13.4.6. *Assume that (i)–(iii) of Lemma 2.4.1 hold true for $D := (0,a)$ and $V := (0,c)$. Then, the operators $\Pi_\lambda B$ and $\Xi_\lambda B$ are compact on X_p and weakly compact on X_1.* ◇

Proof. Since (i)–(iii) of Lemma 2.4.1 are satisfied, then from Lemma 2.4.1, it follows that B can be approximated, in the uniform topology by a sequence B_n of finite rank operators on $L_p([0,c], dv)$ which converges, in the operator norm, to B. Then, it suffices to establish the result for a finite rank operator, that is $\kappa_n(\mu, v, v') = \sum_{j=1}^n \eta_j(\mu)\theta_j(v)\beta_j(v')$ where $\eta_j(.) \in L_\infty([0,a], d\mu)$, $\theta_j(.) \in L_p([0,c], dv)$ and $\beta_j(.) \in L_q([0,c], dv)$ (q denotes the conjugate of p). So, from the linearity and the stability of the compactness by summation, we infer that it suffices to prove the result for an operator B whose kernel is in the form $\kappa(\mu, v, v') = \eta(\mu)\, \theta(v)\beta(v')$ where $\eta(.) \in L_\infty([0,a], d\mu)$, $\theta(.) \in L_p([0,c], dv)$ and $\beta(.) \in L_q([0,c], dv)$. Consider $g \in X_p$,

$$\begin{cases} (\Pi_\lambda B g)(v) = \displaystyle\int_0^c \int_0^a \frac{1}{v}\eta(\mu)\theta(v)\, e^{-\frac{1}{v}\int_\mu^1(\lambda + \sigma(\tau, v))d\tau}\beta(v')g(\mu, v')\, d\mu dv' \\ \qquad\qquad\quad = J_\lambda\, Ug, \end{cases}$$

where U and J_λ denote the following bounded operators

$$\begin{cases} U : X_p \longrightarrow L_p([0,a], d\mu) \\ \qquad \varphi \longrightarrow (U\varphi)(\mu) = \displaystyle\int_0^c \beta(v)\varphi(\mu, v)\, dv \end{cases}$$

and

$$\begin{cases} J_\lambda : L_p([0,a], d\mu) \longrightarrow X_p^1 \\ \qquad \psi \longrightarrow \displaystyle\int_0^a \frac{\eta(\mu)\,\theta(v)}{v}\, e^{-\frac{1}{v}\int_\mu^a(\lambda+\sigma(\tau,v))d\tau}\psi(\mu)\, d\mu. \end{cases}$$

We first consider the case $p \in (1, \infty)$. Then, it is sufficient to check that J_λ is compact. This will follow from Theorem 2.4.3, if we show that

$$\int_0^c \left[\int_0^a \left|\frac{1}{v}\,\eta(\mu)\,\theta(v)\, e^{-\frac{1}{v}\int_\mu^a(\lambda+\sigma(\tau,v))d\tau}\right|^q d\mu\right]^{\frac{p}{q}} v dv < +\infty$$

(J_λ is then a Hille–Tamarkin operator). To do so, let us first observe that we have

$$\begin{aligned}\int_0^a \left|\frac{1}{v}\,\eta(\mu)\,\theta(v)\, e^{-\frac{1}{v}\int_\mu^a(\lambda+\sigma(\tau,v))d\tau}\right|^q d\mu &\leq \|\eta\|_\infty^q \frac{|\theta(v)|^q}{v^q}\int_0^a e^{-q\frac{(\mathrm{Re}\lambda+\underline{\sigma})}{v}(a-\mu)}d\mu \\ &\leq \|\eta\|_\infty^q \frac{|\theta(v)|^q}{q(\mathrm{Re}\lambda+\underline{\sigma})v^{(q-1)}}\end{aligned}$$

which leads to

$$\left[\int_0^a \left|\frac{1}{v}\,\eta(\mu)\,\theta(v)\, e^{-\frac{1}{v}\int_\mu^a(\lambda+\sigma(\tau,v))d\tau}\right|^q d\mu\right]^{\frac{p}{q}} \leq \|\eta\|_\infty^{p} \frac{|\theta(v)|^p}{(q(\mathrm{Re}\lambda+\underline{\sigma}))^{\frac{p}{q}}}\, v^{(\frac{p}{q}-p)}.$$

Integrating in v from 0 to c, we obtain

$$\begin{aligned}\int_0^c \left[\int_0^a \left|\frac{1}{v}\,\eta(\mu)\,\theta(v)\, e^{-\frac{1}{v}\int_\mu^a(\lambda+\sigma(\tau,v))d\tau}\right|^q d\mu\right]^{\frac{p}{q}} vdv \\ \leq \int_0^c \|\eta\|_\infty^{p} \frac{|\theta(v)|^p}{(q(\mathrm{Re}\lambda+\underline{\sigma}))^{\frac{p}{q}}}\, v^{(\frac{p}{q}-p)} vdv \\ \leq \|\eta\|_\infty^{p} \frac{\|\theta\|^p}{(q(\mathrm{Re}\lambda+\underline{\sigma}))^{\frac{p}{q}}}.\end{aligned}$$

Now, we consider the case $p = 1$. Let λ be such that $\mathrm{Re}\lambda > -\underline{\sigma} + \frac{c}{a}\log(r_\sigma(K))$. As above, according to Lemma 2.4.1, it suffices to establish the result for an operator B with a kernel of the form $\kappa(\mu, v, v') = \eta(\mu)\theta(v)\beta(v')$, where $\eta(.) \in L_\infty([0,a], d\mu)$, $\theta(.) \in L_1([0,c], dv)$ and $\beta(.) \in L_\infty([0,c], dv)$. The operator $\Pi_\lambda B$ can be written in the form $\Pi_\lambda B = \Gamma_\lambda R_\beta$ where R_β and Γ_λ are two bounded operators given by $R_\beta : X_1 \longrightarrow L_1([0,a], d\mu)$, $\varphi \longrightarrow (R_\beta\varphi)(\mu) := \int_0^c \beta(v)\varphi(\mu, v)dv$ and

$$\begin{cases} \Lambda_\lambda : L_1([0,a], d\mu) \longrightarrow X_1^1, \\ \qquad\qquad \varphi \longrightarrow \dfrac{1}{v}\displaystyle\int_0^a \eta(\mu')\theta(v)e^{-\frac{1}{v}\int_{\mu'}^a(\lambda+\sigma(\tau,v))d\tau}\,\varphi(\mu')\,d\mu'. \end{cases}$$

It is sufficient to prove that Λ_λ is weakly compact. For this purpose, let $\mathcal{O}$ be a bounded subset of $L_1([0,a], d\mu)$ and let $\varphi \in \mathcal{O}$. We have $\int_E |(\Lambda_\lambda\varphi)(v)|vdv \leq \|\eta\|_\infty\|\varphi\| \int_E |\theta(v)|dv$, for all measurable subsets of $[0,c]$. Next, by applying Theorem 2.4.5 we infer that the set $\Lambda_\lambda(\mathcal{O})$ is weakly compact, since $\lim_{|E|\to 0}\int_E |\theta(v)|dv = 0$, where $|E|$ is the measure of E. A similar reasoning allows us to reach the same results for the operator $\Xi_\lambda B$. This completes the proof. Q.E.D.

Proof of Theorem 13.4.8. This follows from Lemma 13.4.6 and Remark 13.4.2. Q.E.D.

13.4.5 The Irreducibility of the Semigroup e^{tA_k}

In this section, we will study the irreducibility of the C_0-semigroup e^{tA_K} generated by the transport operator $A_K := S_K + B$. In fact, if e^{tA_K} is irreducible, then the leading eigenvalue (if it exists) of A_K is strictly dominant with multiplicity 1 and the associated eigenprojection is strictly positive (=positivity improving). Thus, if this eigenvalue is strictly dominant, we obtain a much easier description of the time-asymptotic behavior ($t \to \infty$) of the solution of the Cauchy problem (13.4.1). Since S_K is a generator of a C_0-semigroup, according to the perturbation theory, $A_K = S_K + B$ generates a C_0-semigroup $\{U_{K,B}(t), t \geq 0\}$ given by Dyson–Phillips expansion.

Definition 13.4.1. Let Q be a positive operator on $L_p(\Omega)$. Q is named strictly positive if $Qf > 0$ a.e. on Ω for all $f \geq 0$, $f \neq 0$. ◇

We will use the following definition of the irreducibility of a C_0-semigroup $\{V(t), t \geq 0\}$.

Definition 13.4.2. Let $\{V(t), t \geq 0\}$ be a positive C_0-semigroup on L_p and let A be its infinitesimal generator. $V(t)$ is irreducible on L_p if, for all $\lambda > s(A)$ (where $s(A)$ denotes the spectral bound of A), the operator $(\lambda - A)^{-1}$ is strictly positive, i.e., for all $f \in X_p$, $f \geq 0$, $f \neq 0$, we have: $(\lambda - A)^{-1}f$ is strictly positive a.e. ◇

Lemma 13.4.7. *If K is positive (in the lattice sense), then the C_0-semigroups $\{U_{K,0}(t), t \geq 0\}$ and $\{U_{0,0}(t), t \geq 0\}$ generated respectively by S_K and S_0 (S_K with $K = 0$), satisfy the following inequality $U_{K,0}(t) \geq U_{0,0}(t) \geq 0, \quad (t \geq 0)$.* $\diamond$

Proof. For $t = 0$, the result is trivial.

We fixed $t > 0$. Let $\lambda > \lambda_K$. Then, $\lambda \in \rho(S_K)$ and the resolvent of S_K is given by

$$(\lambda - S_K)^{-1} = Q_\lambda K(I - P_\lambda K)^{-1}\Pi_\lambda + \Xi_\lambda. \tag{13.4.20}$$

It is noted that the operator Ξ_λ is nothing else but the resolvent of the operator S_0. The positivity of the operators Q_λ, P_λ, Π_λ, Ξ_λ and K implies that

$$(\lambda - S_K)^{-1} \geq (\lambda - S_0)^{-1} \geq 0. \tag{13.4.21}$$

Hence, for each integer n such that $\frac{n}{t} > \lambda_K$, we have

$$\left[\frac{n}{t}R\left(\frac{n}{t}, S_K\right)\right]^n \psi \geq \left[\frac{n}{t}R\left(\frac{n}{t}, S_0\right)\right]^n \psi \geq 0, \quad (\psi \geq 0).$$

Therefore, $\lim\limits_{n\to+\infty}\left[\frac{n}{t}R\left(\frac{n}{t}, S_K\right)\right]^n \psi \geq \lim\limits_{n\to+\infty}\left[\frac{n}{t}R\left(\frac{n}{t}, S_0\right)\right]^n \psi \geq 0, \quad (\psi \geq 0)$. By using the exponential formula (see Theorem 2.5.11), we deduce that $U_{K,0}(t) \geq U_{0,0}(t) \geq 0, \quad (t \geq 0)$. Q.E.D.

Lemma 13.4.8. *If K is positive (in the lattice sense) and the operator B is positive, then $U_{K,B}(t) \geq U_{0,B}(t) \geq 0$, $(t \geq 0)$, and $U_{K,B}(t) \geq U_{K,0}(t) \geq 0$, $(t \geq 0)$.* $\diamond$

Proof. Let $\lambda \in \rho(S_K)$ be such that $r_\sigma[(\lambda - S_K)^{-1}B] < 1$. Then, $\lambda \in \rho(A_K)$ and

$$(\lambda - A_K)^{-1} - (\lambda - S_K)^{-1} = \sum_{n=1}^{+\infty}\left[(\lambda - S_K)^{-1}B\right]^n (\lambda - S_K)^{-1}. \tag{13.4.22}$$

Due to Eq. (13.4.21) and the positivity of B, we have

$$\left[(\lambda - S_K)^{-1}B\right]^n (\lambda - S_K)^{-1} \geq \left[(\lambda - S_0)^{-1}B\right]^n (\lambda - S_0)^{-1} \geq 0.$$

Therefore, the equality (13.4.22) leads to $(\lambda - A_K)^{-1} \geq (\lambda - S_K)^{-1} \geq 0$ and $(\lambda - A_K)^{-1} \geq (\lambda - A_0)^{-1} \geq 0$, where A_0 is the operator A_K with $K = 0$. To deduce the result, we use a similar reasoning to that of Lemma 13.4.7. Q.E.D.

It is well known that if A and B are two positive linear operators on a lattice Banach space X satisfying $0 \leq A \leq B$, then if A is irreducible, B is irreducible, too. So, as an immediate consequence of Lemma 13.4.8, we have the following.

Theorem 13.4.9. *Assume that K is positive. Then, if $\{U_{K,0}(t), t \geq 0\}$ is irreducible, $\{U_{K,B}(t), t \geq 0\}$ is irreducible, too.* $\diamond$

In the following, we give some sufficient conditions in terms of boundary operator guaranteeing the irreducibility of $\{U_{K,B}(t), t \geq 0\}$.

Theorem 13.4.10. *If the boundary operator K is positive and if the operator $K(I - P_\lambda K)^{-1}$ is strictly positive, then the C_0-semigroup $\{U_{K,B}(t), t \geq 0\}$ generated by A_K is irreducible on X_p.* ◇

Proof. By using Eq. (13.4.20) and the fact that Ξ_λ is a positive operator, we have the following

$$(\lambda - S_K)^{-1} \geq Q_\lambda K(I - P_\lambda K)^{-1}\Pi_\lambda. \tag{13.4.23}$$

Knowing that Q_λ is a multiplication operator by a strictly positive function and Π_λ is a strictly positive operator from X_p into X_p^1, we infer that, for all $f \geq 0$, $f \neq 0$, $Q_\lambda K(I - P_\lambda K)^{-1}\Pi_\lambda f > 0$ a.e. Consequently, $(\lambda - S_K)^{-1} f > 0$, $f \geq 0$, $f \neq 0$. Now, the result follows from Definition 13.4.2 and Theorem 13.4.9. Q.E.D.

Corollary 13.4.1. *If there exists $n_0 \in \mathbb{N}$ such that $(P_\lambda K)^{n_0}$ is strictly positive, the C_0-semigroup $\{U_{K,B}(t), t \geq 0\}$ generated by A_K is irreducible on X_p.* ◇

Proof. Knowing that P_λ is a multiplication operator by a strictly positive function, it is easy to verify that $K(I - P_\lambda K)^{-1}$ is strictly positive if, and only if, $P_\lambda K(I - P_\lambda K)^{-1}$ is strictly positive. Since K is positive and $P_\lambda K(I - P_\lambda K)^{-1} = \sum_{n=1}^{+\infty}(P_\lambda K)^n$, we have $P_\lambda K(I - P_\lambda K)^{-1} \geq (P_\lambda K)^{n_0}$. Now, the result follows from Eq. (13.4.23) and Theorem 13.4.9. Q.E.D.

13.4.6 Existence of the Leading Eigenvalues of A_K

Let us denote by $L_p(dv)$ the space of functions $L_p[(0,c), dv]$. Notice that $L_p(dv)$ is a subspace of X_p^0 and the imbedding $L_p(dv) \hookrightarrow X_p^0$ is continuous. By $\overline{B}$, we mean the integral operator on X_p whose kernel is given by $\overline{r}(\mu, v, v') = \frac{r(\mu,v,v')}{v}$.

Theorem 13.4.11. *Suppose that the operator $\overline{B}$ is bounded on X_p and that K is bounded from X_p^0 into $L_p(dv)$ with $\|K\| < 1$. Then, $\sigma(A_K)\bigcap\{\lambda \in \mathbb{C}$ such that $Re\lambda > -\underline{\sigma}\} = \emptyset$ for a small enough a.* ◇

Proof. Let $\psi \in X_p$ and put $\varphi = B\psi$. Then, we have $|\Xi_\lambda\varphi(\mu, v)|^p \leq a^{\frac{p}{q}} \int_0^a \frac{|\varphi(\mu,v)|^p}{v^p}\, d\mu$ and so,

$$\begin{aligned}\int_0^a \int_0^c |\Xi_\lambda\varphi(\mu, v)|^p dv d\mu &\leq a^{(\frac{p}{q}+1)} \int_0^a \int_0^c \frac{|\varphi(\mu, v)|^p}{v^p}\, dv d\mu \\ &= a^p \int_0^a \int_0^c |\overline{B}\psi(\mu, v)|^p dv d\mu,\end{aligned}$$

where q is the conjugate of p. Thus, we can write $\left[\int_0^a \int_0^c |\Xi_\lambda \varphi(\mu, v)|^p dv d\mu\right]^{\frac{1}{p}} \leq a \,\|\overline{B}\|\, \|\psi\|$ which gives the estimate

$$\|\Xi_\lambda B\| \leq a\|\overline{B}\|. \tag{13.4.24}$$

Moreover, the operator Π_λ satisfies the following inequality

$$|\Pi_\lambda \varphi(\mu, v)| \leq \frac{1}{v}\int_0^a e^{-\frac{1}{v}(\mathrm{Re}\lambda + \underline{\sigma})(a-\mu)}\, |\varphi(\mu, v)| d\mu \leq \frac{1}{v}\int_0^a |\varphi(\mu, v)| d\mu.$$

By using Hölder's inequality, we obtain

$$|\Pi_\lambda \varphi(\mu, v)| \leq \frac{a^{1/q}}{v}\Big[\int_0^a |\varphi(\mu, v)|^p d\mu\Big]^{1/p} \leq a^{1/q}\Big[\int_0^a \frac{|\varphi(\mu, v)|^p}{v^p} d\mu\Big]^{1/p}.$$

Finally, we have the estimate

$$\|\Pi_\lambda B\| \leq a^{1/q}\|\overline{B}\|. \tag{13.4.25}$$

Next, the hypothesis about K, together with the estimate $\|P_\lambda\| \leq e^{-\frac{a}{c}\mathrm{Re}(\lambda+\underline{\sigma})}$, gives $\|P_\lambda K\| < 1$ uniformly on $\{\lambda \in \mathbb{C}$ such that $\mathrm{Re}\lambda \geq -\underline{\sigma}\}$, which implies

$$\|(I - P_\lambda K)^{-1}\| \leq \frac{1}{1 - \|K\|}, \quad \text{for } \mathrm{Re}\lambda \geq -\underline{\sigma}. \tag{13.4.26}$$

Moreover, a simple calculation leads to

$$\|Q_\lambda\|_{\mathcal{L}(L_p(dv), X_p)} \leq a^{1/p}. \tag{13.4.27}$$

Now, by combining (13.4.24)–(13.4.27) with the hypothesis on K ($\|Ku\|_{L_p(dv)} \leq \rho\, \|u\|_{X_p^0},\ \rho > 0$), we may write

$$\begin{aligned}
\|(\lambda - S_K)^{-1}B\| &\leq \frac{a^{1/p}\rho\, a^{1/q}\|\overline{B}\|}{1 - \|K\|} + a\,\|\overline{B}\| \\
&= \left[\frac{\rho + 1 - \|K\|}{1 - \|K\|}\right]\|\overline{B}\|\, a \\
&= f(a).
\end{aligned}$$

Clearly, f is a continuously increasing function on $[0, \infty[$ which satisfies $f(0) = 0$ and $\lim_{a\to\infty} f(a) = +\infty$. Hence, there exists $a_0 > 0$, such that $f(a_0) < 1$. This completes the proof. Q.E.D.

In what follows, we turn our attention to the bounded part of the transport operator A_K which we denote by $\mathcal{N}$. We will discuss the relationship between the real eigenvalues of A_K and those of $\mathcal{N}$. For the sake of simplicity, we will deal here with the homogeneous case, i.e., $\sigma(\mu, v) = \sigma(v)$ and $r(\mu, v, v') = r(v, v')$. Hence, the bounded part of A_K is then defined by

$$\begin{cases} \mathcal{N} : L_p([0,c], dv) \longrightarrow L_p([0,c], dv) \\ \qquad\qquad \varphi \longrightarrow (\mathcal{N}\varphi)(v) = -\sigma(v)\,\varphi(v) + \int_0^c r(v, v')\varphi(v')dv'. \end{cases}$$

In the following, we denote by $P(A_K)$ [resp. $P(\mathcal{N})$] the set

$$\sigma(A_K)\bigcap\{\lambda \in \mathbb{C} \text{ such that } \mathrm{Re}\lambda > \lambda_K\} \text{ (resp. } \sigma(\mathcal{N})\bigcap\{\lambda \in \mathbb{C} \text{ such that } \mathrm{Re}\lambda > \lambda_K\}).$$

Theorem 13.4.12. *Suppose that B is a positive regular operator on X_p, and that $K \leq Id$. Hence, if $P(\mathcal{N}) = \emptyset$, then $P(A_K) = \emptyset$ $\forall\, a > 0$ and the leading eigenvalue of A_K is less than or equal to that of $\mathcal{N}$. Moreover, the latter is less than or equal to $-\underline{\sigma} + r_\sigma(B)$.* ◇

Proof. Since B is regular, then according to Theorem 13.4.8, for all λ such that $\mathrm{Re}\lambda > -\underline{\sigma}$, $(\lambda - S_K)^{-1}B$ is power-compact on X_p, with $1 \leq p < +\infty$. By applying Theorem 13.4.1, we conclude that A_K has a leading eigenvalue $\overline{\lambda}$ with a corresponding nonnegative eigenfunction $\overline{\psi}$, i.e., $A_K\overline{\psi} = \overline{\lambda}\,\overline{\psi}$. This equation may be written as

$$-v\frac{\partial\overline{\psi}}{\partial\mu}(\mu, v) - (\overline{\lambda} + \sigma(v))\overline{\psi}(\mu, v) + \int_0^c r(v, v')\overline{\psi}(\mu, v')dv' = 0. \qquad (13.4.28)$$

Set $\overline{\varphi}(v) = \int_0^a \overline{\psi}(\mu, v)d\mu$. It is clear that $\overline{\varphi} \geq 0$ and $\overline{\varphi} \neq 0$. By integrating (13.4.28) with respect to μ, we get $-v\,[\overline{\psi}(a, v) - \overline{\psi}(0, v)] - \sigma(v)\overline{\varphi}(v) + \int_0^c r(v, v')\overline{\varphi}(v')dv' = \overline{\lambda}\,\overline{\varphi}(v)$. Taking into account the hypotheses and the sign of $\overline{\psi}$, we obtain

$$-v\left[\overline{\psi}(a, v) - \overline{\psi}(0, v)\right] = -v\left[\overline{\psi}^1 - \overline{\psi}^0\right] = -v(I - K)\overline{\psi}^1 \leq 0 \;\; \forall v \in [0, c]. \qquad (13.4.29)$$

Now, Eqs. (13.4.28) and (13.4.29) lead to $-\sigma(v)\,\overline{\varphi} + B\overline{\varphi} \geq \overline{\lambda}\overline{\varphi}$ and therefore,

$$\int_0^c \frac{r(v, v')}{\overline{\lambda} + \sigma(v)}\,\overline{\varphi} \geq \overline{\varphi}. \qquad (13.4.30)$$

Let $\lambda \in]-\underline{\sigma}, +\infty[$ and let us define the operator B_λ on $L_p([0,c], dv)$ by

$$\begin{cases} B_\lambda : L_p([0,c], dv) \longrightarrow L_p([0,c], dv) \\ \qquad\qquad \varphi \longrightarrow (B_\lambda\varphi)(v) = \int_0^c \frac{r(v, v')}{\lambda + \sigma(v)}\,\varphi(v')dv'. \end{cases}$$

Since B is a positive regular operator on X_p, then B_λ is positive and compact on $L_p([0,c],dv)$. From Corollary 2.3.1, it follows that $r_\sigma(B_\lambda)$ is an eigenvalue of B_λ depending continuously on λ. Moreover, using both Eq. (13.4.30) and Theorem 2.3.2, we conclude that $r_\sigma(B_{\overline{\lambda}}) \geq 1$. Since $\lim_{\lambda\to+\infty} r_\sigma(B_\lambda) = 0$, then there exists $\lambda_0 \geq \overline{\lambda}$ such that $r_\sigma(B_{\lambda_0}) = 1$. Consequently, there exists $\varphi_0 \neq 0$ and $\varphi_0 \geq 0$ in $L_p([0,c],dv)$ such that

$$B_{\lambda_0}\varphi_0 = \varphi_0. \tag{13.4.31}$$

This leads to $\mathcal{N}\varphi_0 = \lambda_0\varphi_0$ and proves the first part of the theorem. Besides, (13.4.31) may be written in the form $\int_0^c r(v,v')\varphi_0(v')dv' = (\lambda_0 + \sigma(v))\,\varphi_0(v) \geq (\lambda_0 + \underline{\sigma})\varphi_0(v)$. Since $\varphi_0 \neq 0$ and $\varphi_0 \geq 0$, and applying Theorem 2.3.2, we conclude that $r_\sigma(B) \geq \underline{\sigma} + \lambda_0$ which ends the proof. Q.E.D.

Corollary 13.4.2. *Suppose that the hypotheses of Theorem 13.4.12 hold. If the operator $\mathcal{N}$ is subcritical (i.e., $P(\mathcal{N}) \subseteq \{\lambda \in \mathbb{R}$ such that $\lambda < 0\}$), then the transport operator A_K is subcritical for all $a > 0$.* ◇

Remark 13.4.3. Let λ be in $\rho(A_K)\bigcap\rho(A_0)$ such that $r_\sigma((\lambda - S_K)^{-1}B) < 1$. Then, $(\lambda - S_K - B)^{-1} = \sum_{n\geq 0}[(\lambda - S_K)^{-1}B]^n(\lambda - S_K)^{-1}$. The positivity of B and the fact that $(\lambda - S_K)^{-1} \geq (\lambda - S_0)^{-1} \geq 0$ imply that

$$[(\lambda - S_K)^{-1}B]^n(\lambda - S_B)^{-1} \geq [(\lambda - S_0)^{-1}S]^n(\lambda - S_0)^{-1} \geq 0$$

and therefore,

$$R(\lambda, A_K) \geq R(\lambda, A_0) \geq 0. \tag{13.4.32}$$

Next, by using (13.4.32) and Proposition 2.1.1, it follows that if $P(A_0) \neq \emptyset$, then $P(A_K) \neq \emptyset$. ◇

13.4.7 The Strict Monotonicity of the Leading Eigenvalue of A_K

The objective of this section is to study the strict growth properties of the leading eigenvalue with respect to the parameters of the equation. We start our study by discussing the incidence of the boundary operators on the monotony of the leading eigenvalue. For this purpose, we consider two positive boundary operators K_1 and K_2 satisfying $K_1 \leq K_2$ and $K_1 \neq K_2$. We denote by $\lambda(K)$ the leading eigenvalue of A_K (when it exists).

Theorem 13.4.13. *Suppose that the assumptions* (i)–(iii) *of Lemma 2.4.1 are satisfied for* $D := (0,a)$ *and* $V := (0,c)$ *and* $\lambda(K_1)$ *exists, then* $\lambda(K_2)$ *exists and* $\lambda(K_1) \leq \lambda(K_2)$. *Moreover, if one of the following conditions is satisfied*

(i) *there exists an integer* $n \geq 1$, *such that* $[(\Xi_{\lambda(K_1)})B]^n$ *is strictly positive,*

(ii) *there exists an integer* $n \geq 1$, *such that* $\left(Q_{\lambda(K_1)}K_2(I - P_{\lambda(K_1)}K_2)^{-1}\Pi_{\lambda(K_1)}B\right)^n$ *is strictly positive,*

then, $\lambda(K_1) < \lambda(K_2)$. ◇

Proof. Since $K_1 \leq K_2$, then $\lambda_{K_1} \leq \lambda_{K_2}$. The positivity of the operators K_1, K_2, B and the fact that $K_1 \leq K_2$ imply that, for all $\lambda > \lambda_{K_2}$, $(\lambda - S_{K_1})^{-1}B \leq (\lambda - S_{K_2})^{-1}B$ and therefore, $r_\sigma((\lambda - S_{K_1})^{-1}B) \leq r_\sigma((\lambda - S_{K_2})^{-1}B)$. Moreover, according to Theorem 13.4.8, $(\lambda - S_{K_1})^{-1}B$ is power-compact on X_p, $1 \leq p < +\infty$. So, by using Gohberg–Shmul'yan's theorem (see Theorem 2.5.13), and arguing as in the proof of Theorem 13.4.1, we infer that $P(A_{K_1})$ consists of, at most, eigenvalues with a finite algebraic multiplicity. Besides, it is clear that $\lambda \in P(A_{K_1})$ if, and only if, 1 is an eigenvalue of $(\lambda - S_{K_1})^{-1}B$. Accordingly, since $\lambda(K_1) \in P(A_{K_1})$, we have

$$r_\sigma[(\lambda(K_1) - S_{K_1})^{-1}B] \geq 1. \tag{13.4.33}$$

Set $\chi_1 = (\lambda(K_1) - S_{K_1})^{-1}B$ and $\chi_2 = (\lambda(K_1) - S_{K_2})^{-1}B$. According to Theorem 13.4.8, χ_2 is power-compact on X_p. Moreover, if one of the above conditions is satisfied, then χ_2 has a strictly positive power. Now, the fact that $\chi_1 \leq \chi_2$, (13.4.33) and using Theorem 2.3.1, we get $r_\sigma(\chi_2) = r_\sigma[(\lambda(K_1) - S_{K_2})^{-1}B] > 1$. However, the function $]s(S_{K_2}), +\infty[\ni \lambda \to r_\sigma[(\lambda - S_{K_2})^{-1}B]$ is strictly decreasing. Hence, there exists a unique $\lambda' > \lambda(K_1)$, such that $r_\sigma[(\lambda' - S_{K_2})^{-1}B] = 1$. This immediately implies that $\lambda' = \lambda(K_2)$, which completes the proof. Q.E.D.

We deduce the following corollary which provides a practical criterion of monotonicity of $\lambda(K)$.

Corollary 13.4.3. *Suppose that* B *satisfies the items* (i)–(iii) *of Lemma 2.4.1 for* $D := (0,a)$ *and* $V := (0,c)$ *and that* $\lambda(K_1)$ *exists. Then,* $\lambda(K_2)$ *exists and* $\lambda(K_1) \leq \lambda(K_2)$. *Further, if one of the following conditions is satisfied*

(i) K_2 *is strictly positive and* $N(B)\bigcap\{\varphi \in X_p$ *such that* $\varphi \geq 0\} = \{0\}$,

(ii) *there exists an integer* $n \geq 1$, *such that* $(P_{\lambda(K_1)}K_2)^n$ *is strictly positive and* $N(B)\bigcap\{\varphi \in X_p$ *such that* $\varphi \geq 0\} = \{0\}$,

then, $\lambda(K_1) < \lambda(K_2)$. ◇

The proof of this corollary is similar to that of Theorem 13.4.13. It uses the fact that, for $\lambda > -\underline{\sigma}$, the operators P_λ and Q_λ are two multiplication operators by strictly positive functions.

In what follows, we will study the strict monotonicity of the leading eigenvalue of A_K with respect to the collision operators. In fact, let us consider B_1 and B_2 as two operators satisfying the following hypothesis

$$(\mathcal{R}3) \qquad B_1 \leq B_2 \text{ and } B_1 \neq B_2.$$

We denote by $\lambda(B)$ the leading eigenvalue of $A_K = S_K + B$ (when it exists).

Proposition 13.4.2. *Let us assume that B_1 and B_2 satisfy $(\mathcal{R}3)$ and that $\lambda(B_1)$ exists. Then, $\lambda(K_2)$ exists and $\lambda(B_1) \leq \lambda(B_2)$. Further, if one of the following conditions is satisfied*

(i) there exists an integer $n \geq 1$, such that $[\Xi_{\lambda(B_1)} B_2]^n$ is strictly positive,
(ii) there exists an integer $n \geq 1$, such that $[Q_{\lambda(B_1)} K(I - P_{\lambda(B_1)} K)^{-1} \Pi_{\lambda(B_1)} B_2]^n$ is strictly positive,

then, $\lambda(B_1) < \lambda(B_2)$. ◇

Proof. Since B_1 is regular, as in the proof of Theorem 13.4.13, we have $P(S_K + B_1) \neq \emptyset$ and $\lambda(B_1) \in P(S_K + B_1)$. This implies that

$$r_\sigma[(\lambda(B_1) - S_K)^{-1} B_1] \geq 1. \tag{13.4.34}$$

Set $\chi_1 = (\lambda(B_1) - S_K)^{-1} B_1$ and $\chi_2 = (\lambda(B_1) - S_K)^{-1} B_2$. Clearly, $\chi_1 \leq \chi_2$ and, according to Theorem 13.4.8, χ_2 is power-compact on X_p. Moreover, if one of the above conditions is satisfied, then χ_2 has a strictly positive power. By using (13.4.34) and applying Theorem 2.3.1, we conclude that $r_\sigma(\chi_2) = r_\sigma[(\lambda(B_1) - S_K)^{-1} B_2] > 1$. Since the function $]\lambda_K, +\infty[\ni \lambda \to r_\sigma[(\lambda - S_K)^{-1} B_2]$ is strictly decreasing, there exists a unique $\lambda' > \lambda(B_1)$, such that $r_\sigma[(\lambda' - S_K)^{-1} B_2] = 1$. This implies that $\lambda' = \lambda(B_2)$, which completes the proof. Q.E.D.

As an immediate consequence of Proposition 13.4.2, we have:

Corollary 13.4.4. *Let us assume that $\lambda(B_1)$ exists, then $\lambda(B_2)$ exists and $\lambda(B_1) \leq \lambda(B_2)$. Moreover, if one of the following conditions is satisfied*

(i) K is strictly positive and $N(B_2) \bigcap \{\varphi \in X_p$ such that $\varphi \geq 0\} = \{0\}$,
(ii) there exists an integer $n \geq 1$, such that $(P_{\lambda(B_1)} K)^n$ is strictly positive and $N(B_2) \bigcap \{\varphi \in X_p$ such that $\varphi \geq 0\} = \{0\}$,

then, $\lambda(B_1) < \lambda(B_2)$. ◇

13.4.8 Essential Spectra of A_K

The aim of this section is to describe in detail the various essential spectra of the operator A_K for large classes of transition and collision operators. From Eq. (13.4.7), we know that, if $\mathrm{Re}\lambda > \lambda_K$, then $\lambda \in \rho(S_K)$ and $(\lambda - S_K)^{-1}$ is given by $(\lambda - S_K)^{-1} = \sum_{n\geq 0} Q_\lambda K(P_\lambda K)^n \Pi_\lambda + \Xi_\lambda$. Moreover, the operator Ξ_λ is nothing else but $(\lambda - S_0)^{-1}$, i.e., $K = 0$. So, if $\text{Re}\lambda > \lambda_K$, then $\lambda \in \rho(S_K) \bigcap \rho(S_0)$, and

$$(\lambda - S_K)^{-1} - (\lambda - S_0)^{-1} = \mathcal{V}_\lambda, \tag{13.4.35}$$

where $\mathcal{V}_\lambda := \sum_{n\geq 0} Q_\lambda K(P_\lambda K)^n \Pi_\lambda$. Let $\lambda \in \mathbb{C}$ be such that $\text{Re}\lambda \leq -\lambda^*$. The solution of the eigenvalue problem $(\lambda - S_0)\psi = 0$ is formally given by $\psi(\mu, v) = k(v)e^{-\frac{1}{v}(\lambda+\sigma(v))\mu}$. Moreover, ψ must satisfy the boundary conditions, i.e., $\psi^0 = 0$. So, we obtain $k(v) = 0$ and consequently, $\psi = 0$. This shows that the point spectrum of the operator S_0 is empty, i.e., $\sigma_p(S_0) = \emptyset$. Let S_0^* denote the dual operator S_0. It is given by

$$\begin{cases} S_0^* : \mathcal{D}(S_0^*) \subset X_q \longrightarrow X_q \\ \qquad\qquad \psi \longrightarrow S_0^*\psi(\mu, v) = v\dfrac{\partial\psi}{\partial\mu}(\mu, v) - \sigma(\mu, v)\psi(\mu, v) \\ \qquad \mathcal{D}(S_0^*) = \{\psi \in W_q \text{ such that } \psi^1 = 0\}, \end{cases}$$

where q is the conjugate of p. Now, let us consider the eigenvalue problem $(\lambda - S_0^*)\psi = 0$ with $\text{Re}\lambda \leq -\lambda^*$ (because $\sigma(S_0) = \sigma(S_0^*)$). In view of the boundary conditions, a straightforward computation shows that the above problem admits only the trivial solution, i.e., $\sigma_p(S_0^*) = \emptyset$. Now, by using the inclusion $\sigma_r(S_0) \subseteq \sigma_p(S_0^*)$, we conclude that $\sigma_r(S_0) = \emptyset$. This leads to the following lemma.

Lemma 13.4.9. *With the above introduced notations, we have*

$$\sigma(S_0) = \sigma_c(S_0) = \{\lambda \in \mathbb{C} \textit{ such that } \text{Re}\lambda \leq -\lambda^*\}.$$

◇

As an immediate consequence of Lemma 13.4.9, and the fact that all essential spectra constitute some enlargements of the continuous spectrum, we have

$$\sigma_{ei}(S_0) = \{\lambda \in \mathbb{C} \text{ such that } \text{Re}\lambda \leq -\underline{\sigma}\} \text{ for } i = 1, \ldots, 6. \tag{13.4.36}$$

Let us notice that the perturbation of the boundary conditions of the operator S_0 leads to the above Eq. (13.4.35). So, if the transition operator K is strictly singular (in applications, K is compact or weakly compact), and since $\mathcal{S}(X_p)$ is a closed two-sided ideal of $\mathcal{L}(X_p)$, then $\mathcal{V}_\lambda$ is strictly singular too. Hence, Lemma 13.4.9, Theorem 7.5.4, and Eq. (13.4.36) give

$$\sigma_{ei}(S_K) = \{\lambda \in \mathbb{C} \text{ such that } \text{Re}\lambda \leq -\underline{\sigma}\},\ i = 1, 2, 3, 4 \text{ and } 5. \tag{13.4.37}$$

Let us recall that the transport operator A_K is defined as a bounded perturbation of S_K, i.e., $A_K = S_K + B$, where B is the operator defined by (13.4.19). Now, we may introduce the class $\mathcal{G}(X_p)$ of collision operators which is defined by

$$\mathcal{G}(X_p) = \Big\{B \in \mathcal{L}(X_p) \text{ such that } (\lambda - S_K)^{-1}B \in \mathcal{S}(X_p) \text{ for some } \lambda \in \rho(S_K)\Big\}.$$

Clearly, if B is a collision operator on X_p satisfying (i)–(iii) of Lemma 2.4.1 for $D := (0,a)$ and $V := (0,c)$, then from Theorem 13.4.8, it follows that $(\lambda - S_K)^{-1}B$ is compact on X_p for $1 < p < \infty$ (resp. weakly compact on X_1). Hence, by using the inclusion $\mathcal{K}(X_p) \subseteq \mathcal{S}(X_p)$ [resp. the fact that the set of weakly compact operators on X_1 coincides with $\mathcal{S}(X_1)$ (cf. [277]), we infer that $B \in \mathcal{G}(X_p)$. In particular, the set of collision operators with kernels in the form $r(v,v') = f(v)\, g(v')$ with $f \in L_p([0,c],dv)$ and $g \in L_q([0,c],dv)$, $q = \frac{p}{p-1}$, is contained in $\mathcal{G}(X_p)$. This shows that $\mathcal{G}(X_p) \neq \emptyset$. Let $\lambda \in \rho(S_K)$ be such that $r_\sigma((\lambda - S_K)^{-1}B) < 1$, then $\lambda \in \rho(S_K + B)$, and

$$(\lambda - A_K)^{-1} - (\lambda - S_K)^{-1} = \sum_{n\geq 1}[(\lambda - S_K)^{-1}B]^n(\lambda - S_K)^{-1}. \qquad (13.4.38)$$

Theorem 13.4.14. *Let $p \in [1,\infty)$. If the collision operator $B \in \mathcal{G}(X_p)$, then $\sigma_{ei}(A_K) = \sigma_{ei}(S_K)$, for $i = 1,\ldots,5$. Moreover, if K is strictly singular, then $\sigma_{ei}(A_K) = \{\lambda \in \mathbb{C}$ such that $\mathrm{Re}\lambda \leq -\underline{\sigma}\}$, for $i = 1,\ldots,5$.* ◇

Proof. Since $B \in \mathcal{G}(X_p)$, and according to (13.4.37) and Theorem 13.4.8, we infer that $(\lambda - A_H)^{-1} - (\lambda - S_K)^{-1} \in \mathcal{G}(X_p)$. Then, the first claim follows from Theorem 7.5.4. In order to establish the second claim, let us notice that Eqs. (13.4.35) and (13.4.38) give

$$(\lambda - A_K)^{-1} - (\lambda - S_0)^{-1} = \mathcal{V}_\lambda + \sum_{n\geq 1}[(\lambda - S_K)^{-1}B]^n\,(\lambda - S_K)^{-1}.$$

Next, if K is strictly singular, then $\mathcal{V}_\lambda$ is strictly singular too. This, together with Theorem 13.4.8, leads to $(\lambda - A_K)^{-1} - (\lambda - S_0)^{-1} \in \mathcal{S}(X_p)$. Again, the use of Lemma 13.4.9 and Theorem 7.5.4 gives the result. Q.E.D.

13.5 Some Applications of the Regularity and Irreducibility on Transport Theory

This section deals with the spectral analysis of the following integro-differential operator:

$$A_H\psi(x,\xi) = -\xi\frac{\partial\psi}{\partial x}(x,\xi) - \sigma(\xi)\psi(x,\xi) + \int_{-1}^{1}\kappa(x,\xi,\xi')\psi(x,\xi')\,d\xi'$$

with the following boundary conditions:

$$\begin{cases} \psi_1^i = H_{11}\psi_1^o + H_{12}\psi_2^o \\ \psi_2^i = H_{21}\psi_1^o + H_{22}\psi_2^o, \end{cases}$$

where $x \in (-a, a), a > 0, \xi \in (-1, 1)$,

$$\begin{cases} \psi_1^i \;:\; \xi \in (0,1) & \longrightarrow \psi(-a,\xi), \\ \psi_2^i \;:\; \xi \in (-1,0) & \longrightarrow \psi(a,\xi), \\ \psi_1^o \;:\; \xi \in (-1,0) & \longrightarrow \psi(-a,\xi), \\ \psi_2^o \;:\; \xi \in (0,1) & \longrightarrow \psi(a,\xi), \end{cases} \tag{13.5.1}$$

H_{11}, H_{12}, H_{21} and H_{22} are abstract linear operators defined on suitable boundary spaces. Here ψ represents the angular density of particles (for instance, gas molecules, photons, or neutrons) in a homogeneous slab of thickness $2a$. The real function $\sigma(.)$ in $L_\infty(-1,1)$ is called the collision frequency. The function $\kappa(.,.,.)$ is called the scattering kernel and is defined on $(-a,a) \times (-1,1) \times (-1,1)$. Both functions are assumed to be measurable. Let us introduce the functional setting of the problem:

$$D = (-a,a) \times (-1,1), (a > 0),$$

$$\begin{aligned} D^i &= D_1^i \bigcup D_2^i = \{-a\} \times (0,1) \bigcup \{a\} \times (-1,0), \\ D^o &= D_1^o \bigcup D_2^o = \{-a\} \times (-1,0) \bigcup \{a\} \times (0,1). \end{aligned}$$

D^o and D^i represent respectively the outgoing and the incoming boundary of the phase space D ("o" for outgoing and "i" for incoming). Now let

$$X = L_1(D, dxd\xi),$$

$$\begin{aligned} X^i &:= L_1(D^i, |\xi| d\xi) \\ &:= X_1^i \times X_2^i \end{aligned}$$

endowed with the norm:

$$\begin{aligned} \|\psi^i, X^i\| &= \left(\|\psi_1^i, X_1^i\| + \|\psi_2^i, X_2^i\|\right) \\ &= \left[\int_0^1 |\psi(-a,\xi)||\xi|\, d\xi + \int_{-1}^0 |\psi(a,\xi)||\xi|\, d\xi\right] \end{aligned}$$

and

$$\begin{aligned} X^o &:= L_1(D^o, |\xi| d\xi) \\ &:= X_1^o \times X_2^o \end{aligned}$$

endowed with the norm:

$$\|\psi^o, X^o\| = (\|\psi_1^o, X_1^o\| + \|\psi_2^o, X_2^o\|)$$
$$= \left[\int_{-1}^{0} |\psi(-a,\xi)||\xi|\,d\xi + \int_0^1 |\psi(a,\xi)||\xi|\,d\xi\right].$$

We define the partial Sobolev space $\mathcal{W}$ by $\mathcal{W} = \left\{\psi \in X \text{ such that } \xi\frac{\partial\psi}{\partial x} \in X\right\}$. It is well known (see [89, 138]) that any $\psi \in \mathcal{W}$ has traces on the spatial boundary $\{-a\}$ and $\{a\}$ which belong respectively to the spaces X^o and X^i. They are denoted respectively, by ψ^o and ψ^i. We are now in position to define the boundary operator H,

$$\begin{cases} H : X_1^o \times X_2^o \longrightarrow X_1^i \times X_2^i \\ H\begin{pmatrix} u_1 \\ u_2 \end{pmatrix} = \begin{pmatrix} H_{11} & H_{12} \\ H_{21} & H_{22} \end{pmatrix}\begin{pmatrix} u_1 \\ u_2 \end{pmatrix} \end{cases}$$

with for $j,k \in \{1,2\}$, $H_{jk} : X_k^o \longrightarrow X_j^i$, $H_{jk} \in \mathcal{L}(X_k^o, X_j^i)$, defined such that, on natural identification, the boundary conditions can be written as $\psi^i = H(\psi^o)$. We define now the streaming operator T_H with domain including the boundary conditions:

$$\begin{cases} T_H : \mathcal{D}(T_H) \subseteq X \longrightarrow X \\ \qquad\qquad \psi \longrightarrow T_H\psi(x,\xi) = -\xi\dfrac{\partial\psi}{\partial x}(x,\xi) - \sigma(\xi)\psi(x,\xi) \\ \mathcal{D}(T_H) = \{\psi \in \mathcal{W} \text{ such that } \psi^i = H(\psi^o)\}, \end{cases}$$

where $\psi^o = (\psi_1^o, \psi_2^o)^\top$ and $\psi^i = (\psi_1^i, \psi_2^i)^\top$, with $\psi_1^o, \psi_2^o, \psi_1^i$, and ψ_2^i are given by (13.5.1). We define the bounded collision operator K by

$$\begin{cases} K : X \longrightarrow X \\ \qquad \varphi \longrightarrow \displaystyle\int_{-1}^{1} \kappa(x,\xi,\xi')\varphi(x,\xi')\,d\xi', \end{cases}$$

where the kernel $\kappa(.,.,.)$ is measurable. Such an operator brings compactness with respect to the velocity $\xi \in (-1,1)$. To make these compactness properties precise, M. Mokhtar-Kharroubi introduced the class of regular collision operators in a general L_p-spaces setting ($1 \le p < \infty$) (see Definition 2.4.1). Roughly speaking, a bounded operator $K \in \mathcal{L}(L_p(-a,a)\times(-1,1))$ ($1 \le p < \infty$) is said to be regular if K is local with respect to $x \in (-a,a)$ and compact with respect to $\xi \in (-1,1)$. In a L_1-space setting, M. Mokhtar-Kharroubi noticed that the compactness assumption is too restrictive and that a nonnegative collision operator K can be assumed to be only dominated by a compact operator. Unfortunately, this class of operators dominated by a compact operator is not yet optimal. In this section, we deal with the class of collision operators, introduced and studied by Lods in [236], which are weakly compact in velocities (see Definition 2.4.3).

13.5.1 Weak Compactness Results

We begin this section by giving the expression of the resolvent $(\lambda - T_H)^{-1}$. To this end we consider the problem $(\lambda - T_H)\psi = \phi$. Set $\lambda^* = \inf\{\sigma(\xi), -1 \leq \xi \leq 1\}$, a straightforward calculation for $\mathrm{Re}\lambda + \lambda^* > 0$ gives

$$\psi(x,\xi) = \psi(-a,\xi)\, e^{-\frac{(\lambda+\sigma(\xi))|a+x|}{|\xi|}} + \frac{1}{|\xi|}\int_{-a}^{x} e^{-\frac{(\lambda+\sigma(\xi))|x-x'|}{|\xi|}} \phi(x',\xi)\, dx',\ 0 < \xi < 1 \tag{13.5.2}$$

$$\psi(x,\xi) = \psi(a,\xi)\, e^{-\frac{(\lambda+\sigma(\xi))|a-x|}{|\xi|}} + \frac{1}{|\xi|}\int_{x}^{a} e^{-\frac{(\lambda+\sigma(\xi))|x-x'|}{|\xi|}} \phi(x',\xi)\, dx',\ -1 < \xi < 0, \tag{13.5.3}$$

where as $\psi(a,\xi)$ and $\psi(-a,\xi)$ are given by

$$\psi(a,\xi) = \psi(-a,\xi)\, e^{-2a\frac{(\lambda+\sigma(\xi))}{|\xi|}} + \frac{1}{|\xi|}\int_{-a}^{a} e^{-\frac{(\lambda+\sigma(\xi))|a-x|}{|\xi|}} \phi(x,\xi)\, dx,\ 0 < \xi < 1 \tag{13.5.4}$$

$$\psi(-a,\xi) = \psi(a,\xi)\, e^{-2a\frac{(\lambda+\sigma(\xi))}{|\xi|}} + \frac{1}{|\xi|}\int_{-a}^{a} e^{-\frac{(\lambda+\sigma(\xi))|a+x|}{|\xi|}} \phi(x,\xi)\, dx,\ -1 < \xi < 0. \tag{13.5.5}$$

For the lucidity of analysis, we introduce the following bounded operators:

$$\begin{cases} M_\lambda : X^i \longrightarrow X^o, M_\lambda u := (M_\lambda^+ u, M_\lambda^- u) & \text{with} \\ (M_\lambda^+ u)(-a,\xi) := u(a,\xi)\, e^{-2a\frac{(\lambda+\sigma(\xi))}{|\xi|}}, & -1 < \xi < 0, \\ (M_\lambda^- u)(a,\xi) := u(-a,\xi)\, e^{-2a\frac{(\lambda+\sigma(\xi))}{|\xi|}}, & 0 < \xi < 1, \end{cases}$$

$$\begin{cases} B_\lambda : X^i \longrightarrow X, B_\lambda u := \chi_{(-1,0)}(\xi) B_\lambda^+ u + \chi_{(0,1)}(\xi) B_\lambda^- u & \text{with} \\ (B_\lambda^- u)(x,\xi) := u(-a,\xi)\, e^{-\frac{(\lambda+\sigma(\xi))|a+x|}{|\xi|}}, & 0 < \xi < 1, \\ (B_\lambda^+ u)(x,\xi) := u(a,\xi)\, e^{-\frac{(\lambda+\sigma(\xi))|a-x|}{|\xi|}}, & -1 < \xi < 0, \end{cases}$$

$$\begin{cases} G_\lambda : X \longrightarrow X^o, G_\lambda u := (G_\lambda^+\phi, G_\lambda^-\phi) \quad \text{with} \\ G_\lambda^-\phi := \dfrac{1}{|\xi|}\displaystyle\int_{-a}^{a} e^{-\frac{(\lambda+\sigma(\xi))|a-x|}{|\xi|}}\phi(x,\xi)\,dx, \qquad 0<\xi<1, \\ G_\lambda^+\phi := \dfrac{1}{|\xi|}\displaystyle\int_{-a}^{a} e^{-\frac{(\lambda+\sigma(\xi))|a+x|}{|\xi|}}\phi(x,\xi)\,dx, \qquad -1<\xi<0, \end{cases}$$

and

$$\begin{cases} C_\lambda : X \longrightarrow X, C_\lambda\phi := \chi_{(-1,0)}(\xi)C_\lambda^+\phi + \chi_{(0,1)}(\xi)C_\lambda^-\phi \quad \text{with} \\ C_\lambda^-\phi := \dfrac{1}{|\xi|}\displaystyle\int_{-a}^{x} e^{-\frac{(\lambda+\sigma(\xi))|x-x'|}{|\xi|}}\phi(x',\xi)\,dx', \quad 0<\xi<1, \\ C_\lambda^+\phi := \dfrac{1}{|\xi|}\displaystyle\int_{x}^{a} e^{-\frac{(\lambda+\sigma(\xi))|x-x'|}{|\xi|}}\phi(x',\xi)\,dx', \quad -1<\xi<0, \end{cases}$$

where $\chi_{(-1,0)}(.)$ and $\chi_{(0,1)}(.)$ denote, respectively, the characteristic functions of the sets $(-1,0)$ and $(0,1)$. A simple calculation shows for $\mathrm{Re}\lambda + \lambda^* > 0$, the norms of the operators M_λ, B_λ, G_λ and C_λ are bounded above, respectively, by $e^{-2a(\mathrm{Re}\lambda + \lambda^*)}$, $\frac{1}{\mathrm{Re}\lambda+\lambda^*}$, 1 and $\frac{1}{\mathrm{Re}\lambda+\lambda^*}$. Let λ_0 denote the real defined by

$$\lambda_0 := \begin{cases} -\lambda^* & \text{if } \|H\| \le 1 \\ -\lambda^* + \dfrac{1}{2a}\log(\|H\|) & \text{if } \|H\| > 1. \end{cases}$$

The use of preceding operators and spaces allows us to write abstractly equations (13.5.4) and (13.5.5) as an equation in the space X^o, $\psi^o = M_\lambda H\psi^o + G_\lambda\phi$. For $\mathrm{Re}\lambda > \lambda_0$, we have $\|M_\lambda H\| < 1$, then ψ^o is given by

$$\psi^o = \sum_{n\ge 0}(M_\lambda H)^n G_\lambda\phi. \tag{13.5.6}$$

On the other hand, Eqs. (13.5.2) and (13.5.3) can be written as follows $\psi = B_\lambda H\psi^o + C_\lambda\psi$. Hence by (13.5.6), we obtain $\psi = \sum_{n\ge 0} B_\lambda H(M_\lambda H)^n G_\lambda\phi + C_\lambda\phi$. Finally, the resolvent of the operator T_H can be expressed by

$$R(\lambda, T_H) := (\lambda - T_H)^{-1} = \sum_{n\ge 0} B_\lambda H(M_\lambda H)^n G_\lambda + C_\lambda. \tag{13.5.7}$$

Note that C_λ is not either $(\lambda - T_0)^{-1}$. The rest of this section is devoted to establish some results of weak compactness of the operators $(\lambda - T_H)^{-1}K$ and $K(\lambda - T_H)^{-1}$. For this, we have the following.

Theorem 13.5.1. *We assume that the collision operator K is nonnegative, regular in the sense of Definition 2.4.3 and the boundary operator H is positive, then* (*i*) *For any* $\lambda \in \mathbb{C}$ *satisfying* $\mathrm{Re}\lambda > s(T_H)$, *the operator* $(\lambda - T_H)^{-1}K$ *is weakly compact on* X.

(*ii*) $\lim_{\mathrm{Re}\lambda \to +\infty} \|(\lambda - T_H)^{-1}K\| = \lim_{\mathrm{Re}\lambda \to +\infty} \|K(\lambda - T_H)^{-1}\| = 0.$

◇

Proof.

(*i*) Let $\lambda \in \mathbb{C}$ such that $\mathrm{Re}\lambda > \lambda_0$. Using Eq. (13.5.7) We have

$$\begin{aligned}\|(\lambda - T_H)^{-1}\| &\leq \|B_\lambda\|\|H\| \frac{\|G_\lambda\|}{1 - \|M_\lambda H\|} + \|C_\lambda\| \\ &\leq \frac{1}{\mathrm{Re}\lambda + \lambda^*}\left(\frac{\|H\|}{1 - \|M_\lambda\|\|H\|} + 1\right).\end{aligned}$$

Let $\varepsilon > 0$, for $\mathrm{Re}\lambda > \lambda_0 + \varepsilon$, we have $\|(\lambda - T_H)^{-1}\| \leq \frac{1}{\varepsilon}\left(\frac{\|H\|}{1 - e^{-2a\varepsilon}\|H\|} + 1\right)$, then

$$\|(\lambda - T_H)^{-1}K\| \leq \frac{1}{\varepsilon}\left(\frac{\|H\|}{1 - e^{-2a\varepsilon}\|H\|} + 1\right)\|K\|. \tag{13.5.8}$$

Then, $(\lambda - T_H)^{-1}K$ depends continuously on K, uniformly on $\{\mathrm{Re}\lambda > \lambda_0 + \varepsilon\}$. According to Theorem 2.4.4 and Proposition 2.3.1(*i*), it suffices to prove the result when K is dominated by a rank-one operator in $\mathcal{L}(L_1((-1,1), d\xi))$. Moreover, by Remark 2.4.3 and Proposition 2.3.1(*ii*) we may assume that K itself is a rank-one collision operator in $\mathcal{L}(L_1((-1,1), d\xi))$. This asserts that K have kernel $\kappa(\xi, \xi') = \kappa_1(\xi)\kappa_2(\xi')$, $\kappa_1(.) \in L_1(-1,1)$, $\kappa_2(.) \in L_\infty(-1,1)$. To conclude, it suffices to show that $\sum_{n\geq 0} B_\lambda H(M_\lambda H)^n G_\lambda K$ and $C_\lambda K$ are weakly compact on X. We claim that $G_\lambda K$ and $C_\lambda K$ are weakly compact on X. Consider $\varphi \in X$,

$$\begin{aligned}(G_\lambda^- K\varphi)(\xi) &= \frac{1}{|\xi|}\int_{-a}^{a} e^{-\frac{(\lambda+\sigma(\xi))|a-x|}{|\xi|}} K\varphi(x,\xi)\,dx, \\ &= \frac{1}{|\xi|}\int_{-a}^{a}\int_{-1}^{1} e^{-\frac{(\lambda+\sigma(\xi))|a-x|}{|\xi|}} \kappa_1(\xi)\kappa_2(\xi')\varphi(x,\xi')\,dxd\xi', \quad 0 < \xi < 1, \\ &= J_\lambda U_\lambda \varphi,\end{aligned}$$

where U_λ and J_λ denote the following bounded operators

$$\begin{cases} U_\lambda : X \longrightarrow L_1((-a,a), dx) \\ \qquad \varphi \longrightarrow \int_{-1}^{1} \kappa_2(\xi)\,\varphi(x,\xi)\,d\xi \end{cases}$$

and

$$\begin{cases} J_\lambda : L_1((-a,a),\, dx) \longrightarrow X_2^o \\ \qquad\qquad \psi \longrightarrow \dfrac{1}{|\xi|}\displaystyle\int_{-a}^{a} e^{-\frac{(\lambda+\sigma(\xi))|a-x|}{|\xi|}} \kappa_1(\xi)\, \psi(x)\, dx. \end{cases}$$

It is now sufficient to show that J_λ is weakly compact. To do so let $\mathcal{O}$ be a bounded set of $L_1((-a,a),dx)$, and let $\psi \in \mathcal{O}$. We have $\int_E |J_\lambda\psi(\xi)|\,|\xi|\,d\xi \le \|\psi\| \int_E |\kappa_1(\xi)|\,d\xi$, for all measurable subset E of $(0,1)$. Next, applying Theorem 2.4.5 we infer that the set $J_\lambda(\mathcal{O})$ is weakly compact, since $\lim_{|E|\to 0}\int_E |\kappa_1(\xi)|\,d\xi = 0$, $(\kappa_1(.) \in L_1((-1,1),d\xi))$ where $|E|$ is the measure of E. A similar reasoning allows us to reach the same result for the operators $G_\lambda^+ K$ and $C_\lambda K$ for any λ such that $\mathrm{Re}\lambda > \lambda_0$. Using the resolvent identity we have the result for all λ such that $\mathrm{Re}\lambda > s(T_H)$.

(*ii*) Let λ such that $\mathrm{Re}\lambda > \lambda_0$. From (13.5.8) when we set $\varepsilon = \frac{\mathrm{Re}\lambda-\lambda_0}{2} > 0$ we have $\mathrm{Re}\lambda > \lambda_0+\varepsilon$ and so, $\|(\lambda - T_H)^{-1}K\| \le \frac{2}{\mathrm{Re}\lambda-\lambda_0}\left(\frac{\|H\|}{1-e^{-2a\frac{\mathrm{Re}\lambda-\lambda_0}{2}}\|H\|} + 1\right)\|K\|$. This implies $\lim_{\mathrm{Re}\lambda\to+\infty}\|(\lambda - T_H)^{-1}K\| = 0$, and by the same way we get $\lim_{\mathrm{Re}\lambda\to+\infty}\|K(\lambda - T_H)^{-1}\| = 0$. Q.E.D.

Theorem 13.5.2. *We assume that the collision operator K is nonnegative with kernel $\kappa(.,.,.)$ satisfying $\left\{\frac{\kappa(x,.,\xi')}{|\xi'|}, (x,\xi') \in (-a,a)\times(-1,1)\right\}$ is relatively weakly compact on $L_1((-1,1),d\xi)$. Then, for any $\lambda \in \mathbb{C}$ satisfying* $\mathrm{Re}\lambda > \lambda_0$, *the operator $K(\lambda - T_H)^{-1}$ is weakly compact on X.* ◇

Proof. Let $\lambda \in \mathbb{C}$ such that $\mathrm{Re}\lambda > \lambda_0$, We have

$$K(\lambda - T_H)^{-1} = \sum_{n\ge 0} KB_\lambda H(M_\lambda H)^n G_\lambda + KC_\lambda.$$

It suffices to show that KB_λ and KC_λ are weakly compact on X. Let $\varphi \in X_1^i$

$$\begin{aligned}(KB_\lambda^-\varphi)(x,\xi) &= \int_0^1 \kappa(x,\xi,\xi')\, e^{-\frac{(\lambda+\sigma(\xi'))|a-x|}{|\xi'|}} \varphi(-a,\xi')\,d\xi', \\ &= \int_0^1 \frac{\kappa(x,\xi,\xi')}{|\xi'|}\, e^{-\frac{(\lambda+\sigma(\xi'))|a-x|}{|\xi'|}} \varphi(-a,\xi')\,|\xi'|\,d\xi', \\ &= K'\tilde{B}_\lambda\varphi,\end{aligned}$$

where K' and $\tilde{B}_\lambda$ denote the following bounded operators

$$\begin{cases} K' : X \longrightarrow X \\ \qquad \varphi \longrightarrow \displaystyle\int_{-1}^{1} \frac{\kappa(x,\xi,\xi')}{|\xi'|}\varphi(x,\xi')\,d\xi' \end{cases}$$

and

$$\begin{cases} \tilde{B}_\lambda : X_1^i \longrightarrow X \\ \qquad \psi \longrightarrow e^{-\frac{(\lambda+\sigma(\xi'))|a-x|}{|\xi'|}}\psi(-a,\xi')\,|\xi'|\,\chi_{(0,1)}(\xi'). \end{cases}$$

We claim that $K'\tilde{B}_\lambda$ depend continuously on K'. Let $\varphi \in X_1^i$

$$\begin{aligned} \|\tilde{B}_\lambda\varphi\|_1 &= \int_{-a}^{a}\int_0^1 \left| e^{-\frac{(\lambda+\sigma(\xi'))|a-x|}{|\xi'|}}\varphi(-a,\xi')\right| |\xi'|\,dxd\xi', \\ &\le 2a\int_0^1 |\varphi(-a,\xi')|\,|\xi'|\,d\xi', \\ &\le 2a\,\|\varphi, X_1^i\|, \end{aligned}$$

then $\|K'\tilde{B}_\lambda\| \le 2a\|K'\|$. According to Theorem 2.4.4 and Proposition 2.3.1(i), it suffices to prove the result when K' is dominated by a rank-one operator in $\mathcal{L}\big(L_1((-1,1),d\xi)\big)$. Moreover, by Remark 2.4.3 and Proposition 2.3.1(ii) we may assume that K' itself is a rank-one collision operator in $\mathcal{L}\big(L_1((-1,1),d\xi)\big)$. This asserts that K' have kernel $\kappa'(\xi,\xi') = \kappa_1'(\xi)\kappa_2'(\xi')$, $\kappa'_1(.) \in L_1(-1,1)$, $\kappa'_2(.) \in L_\infty(-1,1)$. Let $\mathcal{O}$ be a bounded set of X_1^i, and let $\psi \in \mathcal{O}$. We have

$$\begin{aligned} \int_E |K'\tilde{B}_\lambda\psi(x,\xi)|\,dxd\xi &\le \int_E \left|\int_0^1 \frac{\kappa(x,\xi,\xi')}{|\xi'|} e^{-\frac{(\lambda+\sigma(\xi'))|a-x|}{|\xi'|}}\psi(-a,\xi')\,|\xi'|\,d\xi'\right| dxd\xi, \\ &\le \|\kappa_2'\|_\infty\,\|\psi, X_1^i\| \int_E \kappa_1'(\xi)\,dxd\xi, \end{aligned}$$

for all measurable subset E of D. Next, applying Theorem 2.4.5 we infer that the set $K'\tilde{B}_\lambda(\mathcal{O})$ is weakly compact, since $\lim_{|E|\to 0}\int_E |\kappa_1'(\xi)|\,dxd\xi = 0$, ($\kappa_1' \in L_1((-1,1),d\xi)$) where $|E|$ is the measure of E. A similar reasoning allows us to reach the same result for the operators KB_λ^+ and KC_λ. Q.E.D.

13.5.2 Essential Spectra

It is well known that if H is weakly compact, then $\sigma(T_H) = \{\lambda \in \mathbb{C}$ such that $\mathrm{Re}\lambda \le \lambda_0\}$. In fact, we can easily show that $\sigma(T_H)$ is reduced to $\sigma_c(T_H)$, the continuous spectrum of T_H, that is

$$\sigma(T_H) = \sigma_c(T_H) = \{\lambda \in \mathbb{C} \text{ such that } \operatorname{Re}\lambda \le \lambda_0\}. \tag{13.5.9}$$

On the other hand, if $\lambda \in \sigma_c(T_H)$ then $R(\lambda - T_H)$ is not closed [otherwise $\lambda \in \rho(T_H)$]. So, $\lambda \in \sigma_{ei}(T_H)$, $i = 1, \ldots, 6$. This implies that $\sigma_c(T_H) \subseteq \bigcap_{i=1}^{6} \sigma_{ei}(T_H)$. Thus, according to (13.5.9) we have

$$\sigma_{ei}(T_H) = \{\lambda \in \mathbb{C} \text{ such that } \operatorname{Re}\lambda \le \lambda_0\} \text{ for } i = 1, \ldots, 6. \tag{13.5.10}$$

Theorem 13.5.3. *We assume that the collision operator is positive, regular and H is positive and weakly compact. Then, $\sigma_{ei}(A_H) = \sigma_{ei}(T_H) = \{\lambda \in \mathbb{C}$ such that $\operatorname{Re}\lambda \le \lambda_0\}$, $i = 1, \ldots, 6$.* ◇

Proof. By Theorem 13.5.1(*ii*) we have $\lim_{\operatorname{Re}\lambda \to +\infty} \|(\lambda - T_H)^{-1}K\| = 0$. So, there is a complex number $\lambda \in \rho(T_H)$ such that $r_\sigma((\lambda - T_H)^{-1}K) < 1$. Now, the result follows from Eq. (13.5.10), Theorem 13.5.1, Corollary 7.5.2 and Theorem 7.3.1. Q.E.D.

Let M the multiplication operator $M\varphi = \sigma\varphi$ and let B the bounded part of transport operator A_H defined by

$$\begin{cases} B : L_1(-1,1) \longrightarrow L_1(-1,1) \\ \qquad\qquad \varphi \longrightarrow -\sigma(\xi)\varphi(\xi) + \displaystyle\int_{-1}^{1} \kappa(\xi,\xi')\varphi(\xi')\,d\xi'. \end{cases}$$

Theorem 13.5.4. *We assume that the collision operator K is positive, regular and the function σ is continuous, then $-\lambda^* \in \sigma_{e5}(B)$.* ◇

Proof. The use of the Theorem 13.5.2 and Lemma 2.4 in [270] implies that $-\lambda^* \in \sigma_{e5}(M)$. For $\operatorname{Re}\lambda > -\lambda^*$, we have $(\lambda - M)^{-1}K$ is weakly compact and $\lim_{\operatorname{Re}\lambda \to +\infty} \|(\lambda - M)^{-1}K\| = 0$. Now the result follows from Corollary 7.5.2. Q.E.D.

13.5.3 Existence Results of Eigenvalues

We denote by $P_{s(T_H)}(A_H)$ (resp. $P_{s(T_H)}(B)$) the set $\sigma(A_H) \bigcap \{\lambda \in \mathbb{C}$ such that $\operatorname{Re}\lambda > s(T_H)\}$ (resp. $\sigma(B) \bigcap \{\lambda \in \mathbb{C}$ such that $\operatorname{Re}\lambda > s(T_H)\}$), where $\sigma(A_H)$ [resp. $\sigma(B)$] stands for the spectrum of the operator A_H (resp. B).

Remark 13.5.1. If the boundary operator H is positive in the lattice sense then for any $\lambda > \lambda_0$, we have $(\lambda - T_H)^{-1} \ge 0$, so T_H has a positive resolvent and by Lemma 2.3.1, we conclude that $s(T_H) \le \lambda_0$. ◇

In the remainder of this section we denote by K_λ the operator $(\lambda - T_H)^{-1}K$ for $\lambda \in]s(T_H), +\infty[$.

Lemma 13.5.1. *Suppose that K is regular and positive. If the boundary operator H is positive, then the spectral radius $r_\sigma(K_\lambda)$ as a function of $\lambda \in]s(T_H), +\infty[$ is continuous.* ◇

Proof. Let λ_1 and λ_2 in $]s(T_H), +\infty[$, we have $K_{\lambda_1} - K_{\lambda_2} = (\lambda_1 - \lambda_2)(\lambda_1 - T_H)^{-1}K_{\lambda_2}$, so

$$\|K_{\lambda_1} - K_{\lambda_2}\| \le |\lambda_1 - \lambda_2| \, \|(\lambda_1 - T_H)^{-1}\| \|K_{\lambda_2}\|. \tag{13.5.11}$$

The Eq. (13.5.11) proves that the function $]s(T_H), +\infty[\ni \lambda \longrightarrow K_\lambda$ is continuous. On the other hand for any $\lambda \in]s(T_H), +\infty[$, K_λ^2 is compact, so K_λ has a finite or countable spectrum. Then by a consequence of Newburgh theorem [41, pp. 51–52], the function $\lambda \longrightarrow r_\sigma(K_\lambda)$ is continuous on $]s(T_H), +\infty[$. Q.E.D.

Remark 13.5.2. We remark that the continuity of $r_\sigma(K_\lambda)$ is proved without restrictive conditions on K providing the irreducibility of K_λ. ◇

Lemma 13.5.2. *Suppose that K is regular and positive, the boundary operator H is positive and $P_{s(T_H)}(A_H) \neq \emptyset$. Let $\lambda_1 = \inf\{\lambda \in]s(T_H), +\infty[$ such that $r_\sigma(K_\lambda) = 0\}$, then:*

(*i*) $s(T_H) < \lambda_1 \le +\infty$.
(*ii*) *The spectral radius $r_\sigma(K_\lambda)$ as a function of $\lambda \in]s(T_H), \lambda_1[$ is strictly decreasing and* $\lim_{\mathrm{Re}\lambda \to \lambda_1} r_\sigma(K_\lambda) = 0$.
(*iii*) *λ is an eigenvalue of A_H if, and only if, 1 is an eigenvalue of K_λ and the corresponding eigenspace is the same.*

◇

Proof. (*i*) For $\mathrm{Re}\lambda > s(T_H)$, using the identity

$$\lambda - A_H = (\lambda - T_H)(I - K_\lambda) \tag{13.5.12}$$

we have, $\lambda - A_H$ is invertible if, and only if, $(I - K_\lambda)$ is invertible. Since $P_{s(T_H)}(A_H) \neq \emptyset$, then there exists λ_2 such that $\mathrm{Re}\lambda_2 > s(T_H)$ and $r_\sigma(K_{\lambda_2}) \ge 1$. The positivity of the operator K means that $|R(\lambda_2, T_H)Kf| \le R(\mathrm{Re}\lambda_2, T_H)K\,|f|$, for all f in the Banach lattice X (see [271, Remark 1.1] and [237, Proposition 1.6]). Therefore, we obtain $\|R(\lambda_2, T_H)K\| \le \|R(\mathrm{Re}\lambda_2, T_H)K\|$. We deduce that $1 \le r_\sigma(K_{\lambda_2}) \le r_\sigma(K_{\mathrm{Re}\lambda_2})$ and for $s(T_H) \le \lambda \le \mathrm{Re}\lambda_2$, we have $r_\sigma(K_\lambda) \neq 0$. So, $s(T_H) < \mathrm{Re}\lambda_2 \le \lambda_1$.

(*ii*) By Lemma 2.3.1(*iii*), we have $r_\sigma(K_\lambda)$ as a function of λ is decreasing. We suppose that there exists α and β in $]s(T_H), \lambda_1[$ such that $\alpha < \beta$ and $r_\sigma(K_\alpha) = r_\sigma(K_\beta) = m$, then for all $\lambda \in [\alpha, \beta]$, $r_\sigma(K_\lambda) = m$. From (*i*), $m > 0$ and m is an eigenvalue of K_λ for all $\lambda \in [\alpha, \beta]$. So, there exists $\psi_\lambda \neq 0$

such that $K_\lambda \psi_\lambda = m\psi_\lambda$. Therefore, $(T_H + \frac{K}{m})\psi_\lambda = \lambda\psi_\lambda$ and then $[\alpha, \beta] \subset \sigma(T_H + \frac{K}{m}) \bigcap]s(T_H), \lambda_1[$. But, by Theorem 2.6.2, $\sigma(T_H + \frac{K}{m}) \bigcap]s(T_H), \lambda_1[$ is a discrete set which is a contradiction. Hence $r_\sigma(K_\lambda)$ is a strictly decreasing function.

(iii) Let $\lambda \in]s(T_H), +\infty[$ an eigenvalue of A_H and $\psi_\lambda \in \mathcal{D}(A_H)$ a corresponding eigenfunction. We have $A_H\psi_\lambda = \lambda\psi_\lambda$ if, and only if, $K_\lambda\psi_\lambda = \psi_\lambda$. That is 1 is an eigenvalue of K_λ with eigenfunction ψ_λ. Q.E.D.

Lemma 13.5.3. *Suppose that K is regular and positive, the boundary operator H is positive and $P_{s(T_H)}(A_H) \neq \emptyset$. The following assertions are equivalent:*

(i) *There exists $\lambda_3 > s(T_H)$ with $r_\sigma(K_{\lambda_3}) > 1$.*
(ii) *There exists a unique $\lambda_4 > s(T_H)$ with $r_\sigma(K_{\lambda_4}) = 1$. In particular $\lambda_4 = s(A_H)$.*
(iii) $s(A_H) > s(T_H)$.

◇

Proof. *(i)* $\Rightarrow$ *(ii)* By Lemmas 13.5.1 and 13.5.2, there exists a unique $\lambda_4 > s(T_H)$ such that $r_\sigma(K_{\lambda_4}) = 1$. The fact that $\lambda \in \sigma(A_H)$ with $\text{Re}\lambda > \lambda_4$ would imply $r_\sigma(K_\lambda) \leq r_\sigma(K_{\text{Re}\lambda}) < r_\sigma(K_{\lambda_4}) = 1$, which together with the Eq. (13.5.12) leads to a contradiction. Then $\lambda_4 \geq s(A_H)$. On the other hand, since $K_{\lambda_4} \geq 0$, then $r_\sigma(K_{\lambda_4}) \in \sigma(K_{\lambda_4})$ and so by (13.5.12), we get $\lambda_4 \in \sigma(A_H)$. Hence $\lambda_4 \leq s(A_H)$.

(ii) $\Rightarrow$ *(iii)* This is obvious, since $s(A_H) = \lambda_4 > s(T_H)$.

(iii) $\Rightarrow$ *(i)* Suppose that $r_\sigma(K_\lambda) \leq 1$ for all $\lambda > s(T_H)$, then by Lemma 13.5.2, we have $r_\sigma(K_\lambda) < 1$ for all $\lambda > s(T_H)$. Hence, $R(1, K_\lambda)$ exists for $\lambda \in]s(T_H), +\infty[$. Using the fact that $R(1, K_\lambda)$ exists for all $\lambda \in]s(T_H), +\infty[$ and the Eq. (13.5.12), it follows that $R(\lambda, A_H)$ exists for all $\lambda \in]s(T_H), +\infty[$ and $R(\lambda, A_H) = R(1, K_\lambda)R(\lambda, T_H)$. Now, we know that $R(1, K_\lambda) \geq 0$ and $R(\lambda, T_H) \geq 0$ for all $\lambda \in]s(T_H), +\infty[$, so $R(\lambda, A_H) \geq 0$ for all $\lambda \in]s(T_H), +\infty[$. Using Lemma 2.3.1*(ii)*, we deduce that $s(A_H) \leq s(T_H)$ which is the desired contradiction. Q.E.D.

Lemma 13.5.4. *Suppose that K is regular, positive, and the boundary operator H is positive, then*

(i) *$P_{s(T_H)}(A_H)$ consists of, at most, isolated eigenvalues with finite algebraic multiplicities.*
(ii) *If $P_{s(T_H)}(A_H) \neq \emptyset$, then there exists a leading eigenvalue $\overline{\lambda}$ for the operator A_H.*
(iii) *$P_{s(T_H)}(B)$ consists of, at most, isolated eigenvalues with finite algebraic multiplicities.*
(iv) *If $P_{s(T_H)}(B) \neq \emptyset$, then there exists a leading eigenvalue $\overline{\lambda_1}$ for the operator B.*

◇

Proof.

(*i*) Since X is a Dunford–Pettis space, then by Theorem 13.5.1, $(K_\lambda)^2$ is compact for $\mathrm{Re}\lambda > s(T_H)$. Hence, (*i*) follows from Theorem 2.6.2.

(*ii*) By using Lemma 13.5.1 and Theorem 13.5.1(*ii*) , there exists $\omega > s(T_H)$ such that, $r_\sigma(K(\lambda - T_H)^{-1}) < 1$ for all $\lambda \in (\omega, +\infty)$. So, by Lemma 2.3.2, we get $(\omega, +\infty) \subset \rho(A_H)$ and $R(\lambda, A_H) \geq 0$ for all $\lambda \in (\omega, +\infty)$. Since $P_{s(T_H)}(A_H) \neq \emptyset$, (*i*) and Lemma 2.3.1(*i*) give that $s(A_H)$ exists and strictly greater then $s(T_H)$. By Proposition 13.5.1, $s(A_H)$ is characterized by $r_\sigma(K_{s(A_H)}) = 1$. Since $K_{s(A_H)}$ is power-compact, 1 is a pole of the resolvent $R(\lambda, K_{s(A_H)})$. So, using [298, Ex. 7 p. 352], $K_{s(A_H)}$ possess a positive eigenfunction ψ_λ associated with the eigenvalue 1. By Lemma 13.5.2(*ii*), ψ_λ is a positive eigenfunction of A_H associated with the eigenvalue $s(A_H)$ which imply that $s(A_H)$ is the leading eigenvalue of A_H. The same arguments provide the results of (*iii*) and (*iv*).

Q.E.D.

Theorem 13.5.5. *Suppose that K is homogeneous, regular and irreducible, $H_{11} = H_{22} = 0$, $0 \leq H_{ij} \leq Id, i \neq j, i, j = 1, 2$ and the function σ is continuous. If $P_{s(T_H)}(A_H) \neq \emptyset$, then $P_{s(T_H)}(B) \neq \emptyset$, $(\forall\, a > 0)$ and $\overline{\lambda} \leq \overline{\lambda_1} \leq -\lambda^* + r_\sigma(K)$, where $\overline{\lambda}$ is the leading eigenvalue of A_H and $\overline{\lambda_1}$ is the leading eigenvalue of B.* $\diamondsuit$

Proof. We start the proof by the case when $\overline{\lambda} > -\lambda^*$. Since $(\lambda - T_H)^{-1}K$ is weakly compact for $\mathrm{Re}\lambda > s(T_H)$ (see Theorem 13.5.1) and X is a Dunford–Pettis space, then $[(\lambda - T_H)^{-1}K]^2$ is compact. Denote, by $\overline{\psi}$ an associated positive eigenfunction to $\overline{\lambda}$. Hence $A_H\overline{\psi} = \overline{\lambda}\,\overline{\psi}$. This equation may be written in the form

$$-\xi\frac{\partial\overline{\psi}}{\partial x}(x,\xi) - (\overline{\lambda} + \sigma(\xi))\overline{\psi}(x,\xi) + \int_{-1}^{1}\kappa(\xi,\xi')\overline{\psi}(x,\xi')\,d\xi' = 0. \qquad (13.5.13)$$

Set $\overline{\varphi}(\xi) = \int_{-a}^{a}\overline{\psi}(x,\xi)\,dx$. It is clear that $\overline{\varphi} \geq 0$ and $\overline{\varphi} \neq 0$. By integrating (13.5.13) with respect to x, we get $-\xi[\overline{\psi}(a,\xi) - \overline{\psi}(-a,\xi)] - \sigma(\xi)\overline{\varphi}(\xi) + \int_{-1}^{1}\kappa(\xi,\xi')\overline{\varphi}(\xi')\,d\xi' = \overline{\lambda}\,\overline{\varphi}(\xi)$. Taking into account the hypotheses and the sign of $\overline{\psi}$, we get

$$-\xi[\overline{\psi}(a,\xi) - \overline{\psi}(-a,\xi)] \leq 0 \qquad \forall\xi \in (-1,1). \qquad (13.5.14)$$

In fact, if $\xi < 0$, then $\overline{\psi}(a,\xi) - \overline{\psi}(-a,\xi) = \overline{\psi}_2^{\,i} - \overline{\psi}_1^{\,o} = H_{21}\overline{\psi}_1^{\,o} - \overline{\psi}_1^{\,o} = (H_{21} - I)\overline{\psi}_1^{\,o} \leq 0$, and therefore $-\xi[\overline{\psi}(a,\xi) - \overline{\psi}(-a,\xi)] \leq 0$. If $\xi > 0$, then $\overline{\psi}(a,\xi) - \overline{\psi}(-a,\xi) = \overline{\psi}_2^{\,o} - H_{12}\overline{\psi}_2^{\,o} = (I - H_{12})\overline{\psi}_2^{\,o}$, which implies $-\xi[\overline{\psi}(a,\xi) - \overline{\psi}(-a,\xi)] \leq 0$. Now, (13.5.13) and (13.5.14) lead to $\sigma(\xi)\overline{\varphi} + K\overline{\varphi} \geq \overline{\lambda}\overline{\varphi}$. Therefore, we get

$$\int_{-1}^{1}\frac{\kappa(\xi,\xi')}{\overline{\lambda} + \sigma(\xi)}\overline{\varphi}(\xi')d\xi' \geq \overline{\varphi}(\xi). \qquad (13.5.15)$$

Let $\lambda \geq \overline{\lambda}$ and define the operator K_λ on $L_1((-1,1), d\xi)$ by

$$\begin{cases} \tilde{K_\lambda} : L_1((-1,1), d\xi) \longrightarrow L_1((-1,1), d\xi) \\ \qquad\qquad \varphi \longrightarrow \tilde{K_\lambda}\varphi = \displaystyle\int_{-1}^{1} \frac{\kappa(\xi,\xi')}{\lambda + \sigma(\xi)}\varphi(\xi')d\xi'. \end{cases}$$

Notice that $\tilde{K_\lambda}$ is a weakly compact operator on $L_1((-1,1), d\xi)$, then we have $\tilde{K}_\lambda^2$ is compact. This implies with the fact that $\tilde{K_\lambda}$ is irreducible and by the theorem of Jentzch and Perron (see Theorem 2.3.4), that $r_\sigma(\tilde{K_\lambda})$ is a pole of the resolvent $(\mu - \tilde{K_\lambda})^{-1}$. Then by Theorem 2.3.4, $r_\sigma(\tilde{K_\lambda}) > 0$, $r_\sigma(\tilde{K_\lambda})$ is an eigenvalue of $\tilde{K_\lambda}$ of algebraic multiplicity one and the corresponding eigenspace is spanned by a strictly positive function. On the other hand, by Lemma 13.5.1, $r_\sigma(\tilde{K_\lambda})$ is an eigenvalue of $\tilde{K_\lambda}$, depending continuously on λ. By (13.5.15) we have $\|(\tilde{K_{\overline{\lambda}}})^n\| \geq 1$ for all $n \in \mathbb{N}^*$. So, $r_\sigma(\tilde{K_{\overline{\lambda}}}) \geq 1$. On the other hand, $\lim\limits_{\lambda\to\infty} r_\sigma(\tilde{K_\lambda}) = 0$. Hence, there exists $\lambda' \geq \overline{\lambda}$ such that $r_\sigma(\tilde{K_{\lambda'}}) = 1$. Consequently, there exists $\varphi_0 \neq 0$ in $L_1((-1,1), d\xi)$ satisfying $\tilde{K_{\lambda'}}\varphi_0 = \varphi_0$. This leads to $B\varphi_0 = \lambda'\varphi_0$. Then $P_{s(T_H)}(B) \neq \emptyset$ and $\overline{\lambda} \leq \overline{\lambda_1}$. If $\overline{\lambda} \leq -\lambda^*$ then by Theorem 13.5.4, we have $\overline{\lambda} \leq \overline{\lambda_1}$. To establish the second part of the theorem, we use the fact that for $\overline{\lambda_1}$ there exists $\varphi_1 \neq 0$ in $L_1((-1,1), d\xi)$ such that $B\varphi_1 = \overline{\lambda_1}\varphi_1$. This equality allows us to write $-\sigma(\xi)\varphi_1(\xi) + \int_{-1}^{1} \kappa(\xi,\xi')\varphi_1(\xi')\,d\xi' = \overline{\lambda_1}\varphi_1(\xi)$. Since $\overline{\lambda_1} > -\lambda^*$, we have $(\overline{\lambda_1} + \lambda^*)\,|\varphi_1(\xi)| \leq \int_{-1}^{1} \kappa(\xi,\xi')\,|\varphi_1(\xi')|\,d\xi'$, and therefore $\overline{\lambda_1} + \lambda^* \leq r_\sigma(K)$. So, $\overline{\lambda_1} \leq -\lambda^* + r_\sigma(K)$. Q.E.D.

Proposition 13.5.1. *Suppose that K is regular, strictly positive ($\kappa > 0$) and the boundary operator H is positive. Then, for all $\lambda \in]s(T_H), +\infty[$,*

(*i*) *the operator K_λ is irreducible,*
(*ii*) *the spectral radius $r_\sigma(K_\lambda) > 0$ and $r_\sigma(K_\lambda)$ is an eigenvalue of K_λ of algebraic multiplicity one.* ◇

Proof.

(*i*) We start by the case when $\lambda > \lambda_0$. By (13.5.7), we have the following inequality $((\lambda - T_0)^{-1}K)^2 \leq ((\lambda - T_H)^{-1}K)^2$. By the hypothesis on κ ($\kappa > 0$) we have $((\lambda - T_0)^{-1}K)^2$ is strictly positive. So, $((\lambda - T_H)^{-1}K)^2$ is strictly positive on the L_1-space X, then it is an irreducible operator. Therefore, K_λ is irreducible. Let $\lambda \in]s(T_H), \lambda_0]$ and $\lambda' > \lambda_0$. By Lemma 2.3.1(*iii*), we have $0 \leq K_{\lambda'} \leq K_\lambda$. Since $K_{\lambda'}$ is irreducible, then K_λ is also irreducible.
(*ii*) In our proof, we use some ideas from the paper of Drnovšek [98]. Since every closed ideal of the L_1-space X is a band, every positive operator on X is σ-order continuous (see [19, Theorem 4.8]). The same argument shows that K_λ is band irreducible. The result follows from Theorem 2.3.5. Q.E.D.

Proposition 13.5.2. *We suppose that the assumptions of Proposition 13.5.1 hold true. Then:*

(*i*) *The spectral radius $r_\sigma(K_\lambda)$ is strictly decreasing as a function of $\lambda \in]s(T_H), +\infty[$.*
(*ii*) *If $P_{s(T_H)}(A_H) \neq \emptyset$, then the eigenspace associated with $s(A_H)$ is spanned by a quasi-interior point of $\mathcal{D}(A_H)_+$ (i.e., $\psi > 0$ a.e).* ◇

Proof. The item (*i*) is a consequence of Lemma 13.5.2(*ii*) and Proposition 13.5.1(*ii*).

The assertion (*ii*) follows from Proposition 3.1 and Theorem 5.2 in [298]. Q.E.D.

13.5.4 *Monotonicity of the Spectral Bound*

We are concerned in this section with the monotonicity dependence of the spectral bound with respect to the parameters of the transport operator. We consider two boundary operators H_1 and H_2 such that $0 \leq H_1 \leq H_2$ and $H_1 \neq H_2$.

Theorem 13.5.6. *We consider two boundary operators H_1 and H_2 such that $0 \leq H_1 \leq H_2$ and $H_1 \neq H_2$. Suppose that K is regular, strictly positive ($\kappa > 0$) and the spectrum of A_{H_1} satisfy $\sigma(A_{H_1}) \bigcap \{\lambda \in \mathbb{C}$ such that $\mathrm{Re}\lambda > \lambda_0\} \neq \emptyset$. Then $s(A_{H_1}) < s(A_{H_2})$.* ◇

Proof. Let $\lambda > \lambda_0$. According to the positivity of the operators H_1, H_2, and K, the fact that $H_1 \leq H_2$ and the expression of the resolvent Eq. (13.5.7), we have the following $(\lambda - T_{H_1})^{-1}K \leq (\lambda - T_{H_2})^{-1}K$. Set $\chi_1 := (s(A_{H_1}) - T_{H_1})^{-1}K$ and $\chi_2 := (s(A_{H_1}) - T_{H_2})^{-1}K$. Since K is regular, it follows from Theorem 13.5.1, that χ_2 is weakly compact on X and so χ_2^2 is compact. Moreover, since K is strictly positive, we have by Proposition 13.5.1, χ_2 is irreducible. Next, Theorem 2.3.6 gives $r_\sigma(\chi_2) > r_\sigma(\chi_1)$, which implies by Proposition 13.5.1(*ii*) that $r_\sigma(\chi_2) > 1$. On the other hand, by Theorem 13.5.1 we have $\lim\limits_{\lambda \to \infty} r_\sigma((\lambda - T_{H_2})^{-1}K) = 0$. Hence, by Lemmas 13.5.1 and 13.5.2(*ii*) there exists a unique $\lambda' > s(A_{H_1})$ such that $r_\sigma((\lambda' - T_{H_2})^{-1}K) = 1$. By Lemma 2.3.1 it follows that $s(A_{H_2}) = \lambda' > s(A_{H_1})$, which completes the proof. Q.E.D.

Remark 13.5.3. If K is regular, positive, H_1 is positive and

$$\sigma(A_{H_1}) \bigcap \{\lambda \in \mathbb{C} \text{ such that } \mathrm{Re}\lambda > \lambda_0\} \neq \emptyset$$

then $s(A_{H_1}) > \lambda_0$. ◇

In the following, we shall study the strict monotonicity of the spectral bound of A_H with respect to the collision operators. In fact, consider K_1 and K_2, two regular positive collisions operators satisfying $K_1 \leq K_2$ and $K_1 \neq K_2$. We denote by $s_K(T_H)$ the spectral bound of $A_H = T_H + K$ (when it exists).

Theorem 13.5.7. *Suppose that $s_{K_1}(T_H) > s(T_H)$, K_2 is strictly positive ($\kappa_2 > 0$) and the boundary operator H is positive. Then, $s_{K_2}(T_H)$ exists and $s_{K_1}(T_H) < s_{K_2}(T_H)$.* ◇

Proof. From the positivity of the operators H, K_1, and K_2 and the fact that $K_1 \leq K_2$ we deduce, for any $\lambda > s(T_H)$ the inequality $(\lambda - T_H)^{-1}K_2 \geq (\lambda - T_H)^{-1}K_1$. We have already shown in Proposition 13.5.1(*ii*) that the spectral bound of $T_H + K_1$, $s_{K_1}(T_H)$, is characterized by $r_\sigma(s_{K_1}(T_H) - T_H)^{-1}K_1) = 1$. Set $\chi_1 := (s_{K_1}(T_H) - T_H)^{-1}K_1$ and $\chi_2 := (s_{K_1}(T_H) - T_H)^{-1}K_2$. As in the proof of Theorem 13.5.6, from the regularity and the strict positivity of K_2, we have χ_2 is irreducible and positive. Then the Theorem 2.3.6 gives $r_\sigma(\chi_2) > r_\sigma(\chi_1) = 1$. Since the function $]s(T_H), +\infty[\ni \lambda \longrightarrow r_\sigma((\lambda - T_H)^{-1}K_2)$ is strictly decreasing, there exists a unique $\lambda' > s_{K_1}(T_H)$ such that $r_\sigma((\lambda' - T_H)^{-1}K_2) = 1$. But this equation characterizes $s_{K_2}(T_H)$, so we have $\lambda' = s_{K_2}(T_H)$. This completes the proof of the theorem. Q.E.D.

13.6 Singular Neutron Transport Operator

This section is concerned with the application of Corollary 7.5.2 with the aim to study the essential spectra of the following singular neutron transport operator

$$A\psi(x,v) = -v.\nabla_x\psi(x,v) - \sigma(v)\psi(x,v) + \int_{\mathbb{R}^n} \kappa(v,v')\psi(x,v')\,d\mu(v'),$$

where $(x,v) \in D \times \mathbb{R}^n$, D represents an open bounded subset of $\mathbb{R}^n$, $d\mu(.)$ is a bounded positive Radon measure on $\mathbb{R}^n$ and where K denotes the integral part of A. This operator describes the transport of particles (neutrons, photons, molecules of gas, etc.) in the domain D. The function $\psi(x,v)$ represents the number (or probability) density of gas particles having the position x and the velocity v. The collision frequency and the scattering kernel are denoted, respectively, by the functions $\sigma(.)$ and $\kappa(.,.)$. First, let us state precisely the functional setting of our problem:

$$X_p = L_p(D \times \mathbb{R}^n, dx\,d\mu(v)), p \in (1,\infty),$$

$$X_p^\sigma = L_p(D \times \mathbb{R}^n, \sigma(v)dx\,d\mu(v)),$$

$$L_p^\sigma(\mathbb{R}^n) = L_p(\mathbb{R}^n, \sigma(v)d\mu(v)), \text{ and}$$

$$L_p(\mathbb{R}^n) = L_p(\mathbb{R}^n, d\mu(v)).$$

Let $\mathcal{W}_p$ be the space defined by $\mathcal{W}_p = \{\psi \in X_p \text{ such that } v.\nabla_x\psi \in X_p\}$. Then, we introduce the following subspace of $\mathcal{W}_p$ by $\mathcal{W}_p^0 = \{\psi \in \mathcal{W}_p \text{such that}$

$\psi_{|\Gamma_-} = 0\}$, with Γ_- is defined by $\Gamma_- = \{(x, v) \in \partial D \times \mathbb{R}^n$ such that $v.\nu_x < 0\}$, where ν_x stands for the outer unit normal vector at $x \in \partial D$. Here, the main characteristic is that the collision frequency $\sigma(.)$ and the collision operator K are unbounded. In fact, an unbounded collision frequency $\sigma(.)$ acts as a strong absorption which enables the unboundedness of K. In the following, we will assume the existence of a closed subset $\mathcal{O} \subset \mathbb{R}^n$ with a zero $d\mu$ measure and a constant $\sigma_0 > 0$, such that

$$\sigma(.) \in L^\infty_{\mathrm{loc}}(\mathbb{R}^n \backslash \mathcal{O}), \quad \sigma(v) > \sigma_0 \text{ a.e., and} \tag{13.6.1}$$

$$\left[\int_{\mathbb{R}^n} \left(\frac{\kappa(., v')}{\sigma(v')^{\frac{1}{p}}}\right)^q d\mu(v')\right]^{\frac{1}{q}} \in L_p(\mathbb{R}^n), \tag{13.6.2}$$

where q represents the conjugate exponent of p. K represents the collision operator, which is defined as follows $K : \psi \longrightarrow K\psi(v) := \int_{\mathbb{R}^n} \kappa(v, v')\psi(x, v')\, d\mu(v') \in L_p(\mathbb{R}^n)$, where the scattering kernel $\kappa(., .)$ will be assumed to be unbounded. Now, we are ready to define the streaming operator T by

$$\begin{cases} T : \mathcal{D}(T) \subseteq X_p \longrightarrow X_p \\ \qquad\qquad\quad \psi \longrightarrow T\psi(x, v) = -v.\nabla_x \psi(x, v) - \sigma(v)\psi(x, v) \\ \mathcal{D}(T) = \mathcal{W}^0_p \bigcap X^\sigma_p. \end{cases}$$

Remark 13.6.1.

(i) From the assumption (13.6.2), we deduce that $K \in \mathcal{L}(L^\sigma_p(\mathbb{R}^n), L_p(\mathbb{R}^n))$ and

$$\|K\|_{\mathcal{L}(L^\sigma_p(\mathbb{R}^n), L_p(\mathbb{R}^n))} \leq \left\| \left[\int_{\mathbb{R}^n} \left(\frac{\kappa(., v')}{\sigma(v')^{\frac{1}{p}}}\right)^q d\mu(v')\right]^{\frac{1}{q}} \right\|_{L_p(\mathbb{R}^n)}.$$

Besides, by using the boundedness of D, we infer that $K \in \mathcal{L}(X^\sigma_p, X_p)$, where

$$\|K\|_{\mathcal{L}(X^\sigma_p, X_p)} \leq \left\| \left[\int_{\mathbb{R}^n} \left(\frac{\kappa(., v')}{\sigma(v')^{\frac{1}{p}}}\right)^q d\mu(v')\right]^{\frac{1}{q}} \right\|_{L_p(\mathbb{R}^n)}.$$

By using the assumption (13.6.1), a simple calculation allows us to show that X^σ_p is a subset of X_p and that the embedding $X^\sigma_p \hookrightarrow X_p$ is continuous.

(ii) Let $\varphi \in X_p$ and $\lambda \in \mathbb{C}$, such that $\mathrm{Re}\lambda > -\sigma_0$. We seek $\psi \in \mathcal{D}(T)$ satisfying

$$(\lambda - T)\psi = \varphi. \tag{13.6.3}$$

The formal and intermediate solution of (13.6.3) can be given by

$$\psi(x,v) = \psi(x - t^-(x,v)v, v)e^{-(\lambda+\sigma(v))t^-(x,v)} + \int_0^{t^-(x,v)} e^{(-\lambda+\sigma(v))}\varphi(x - sv, v)ds,$$

where $t^-(x,v) = \sup\{t > 0, x - sv \in D, 0 < s < t\}$. Since ψ must belong to $\mathcal{D}(T)$, then we can infer that $\psi(x - t^-(x,v)v, v) = 0$ for all $(x,v) \in D \times \mathbb{R}^n$. Hence, the final solution of (13.6.3) is given by $\psi(x,v) = \int_0^{t^-(x,v)} e^{(-\lambda+\sigma(v))}\varphi(x - sv, v)ds$. An immediate consequence of these facts is that $\{\lambda \in \mathbb{C}$ such that $\mathrm{Re}\lambda > -\sigma_0\} \subset \rho(T)$. Since $\sigma(.)$ is bounded below by σ_0, a reasoning similar to the one in Lemma 13.4.9 (see also [184, Corollary 12.11, p. 272]) shows that $\sigma(T) = \{\lambda \in \mathbb{C}$ such that $\mathrm{Re}\lambda \leq -\sigma_0\}$. Actually, by using Lemma 13.4.9 (see [184, Chapter 12]), we can easily check that $\sigma(T)$ is reduced to $\sigma_c(T)$. Then, we have

$$\sigma_{ei}(T) = \sigma_c(T) = \{\lambda \in \mathbb{C} \text{ such that } \mathrm{Re}\lambda \leq -\sigma_0\}, i = 1, \ldots, 6. \tag{13.6.4}$$

◇

Lemma 13.6.1 ([262, Theorem 3.2 (*ii*)]). *Let us assume that the measure $d\mu$ satisfies*

$$\begin{cases} \text{the hyperplanes have a zero } d\mu \text{ measure, i.e.,} \\ \text{for each } e \in S^{n-1}, d\mu\{v \in \mathbb{R}^n \text{ such that } v.e = 0\}, \end{cases} \tag{13.6.5}$$

where S^{n-1} represents the unit sphere of $\mathbb{R}^n$, and let M be the averaging operator

$$\varphi \in L_p(\mathbb{R}^n, \mathbb{R}^n) \longrightarrow \int_{\mathbb{R}^n} \varphi(., v')\, d\mu(v') \in L_p(\mathbb{R}^n),$$

where $1 < p < \infty$. Then, $M : \mathcal{W}_p^0 \longrightarrow L_p(D)$ is compact. ◇

We start by recalling a lemma which is given in [224].

Lemma 13.6.2. *Let D be a bounded subset of $\mathbb{R}^n$ and $1 < p < \infty$. If the hypotheses (13.6.1), (13.6.2) and (13.6.5) are all satisfied, and if the collision operator $K : L_p^\sigma(\mathbb{R}^n) \longrightarrow L_p(\mathbb{R}^n)$ is compact, then for any λ satisfying* $\mathrm{Re}\lambda >$ $-\sigma_0$, *the operator $K(\lambda - T)^{-1}$ is compact on X_p.* ◇

Proof. Since K is compact from $L_p^\sigma(\mathbb{R}^n)$ into $L_p(\mathbb{R}^n)$, and by using the linearity and approximation arguments, we may restrict ourselves to the case where the scattering kernel has the form $\kappa(v,v') = f(v)g(v')$, and where $f(.) \in L_p(\mathbb{R}^n)$, $g(.)\sigma^{\frac{1}{q}} \in L_q(\mathbb{R}^n)$ and q represents the conjugate exponent of p. Again, the use of a density argument enables us to assume that $f(.) \in C_c(\mathbb{R}^n)$ (continuous functions with a compact support). In these conditions, the operator $K(\lambda - T)^{-1}$ maps X_η into itself, for all $\eta \in [1, +\infty]$. By applying the interpolation arguments (see Theorem 2.4.2), we can restrict ourselves to the case where $p = 2$. Let $\mathcal{A}_g$ be the averaging operator defined by

$$\mathcal{A}_g : \varphi \in X_2^\sigma \longrightarrow \int g(v')\varphi(x, v')\, d\mu(v') \in L_2(D).$$

It is sufficient to show that $\mathcal{A}_g(\lambda - T)^{-1}$ is a compact operator from X_2 into $L_2(D)$. This leads to $\mathcal{A}_g : \mathcal{D}(T) = \mathcal{W}_2^0 \bigcap X_2^\sigma \longrightarrow L_2(D)$ is compact. We may notice that $\mathcal{D}(T)$ is equipped with the norm $\|\psi\|_{\mathcal{D}(T)} = \|\psi\|_{\mathcal{W}_2^0} + \|\psi\|_{X_2^\sigma}$ for all $\psi \in \mathcal{D}(T)$, where $\|\psi\|_{\mathcal{W}_2^0} = \|\psi\|_{X_2} + \|v.\nabla_x\psi\|_{X_2}$ is a Banach space. If U is a bounded subset of $\mathcal{D}(T)$, then there exists $\rho > 0$, such that $\|\psi\|_{\mathcal{D}(T)} \leq \rho$ for every $\psi \in U$. This implies, in particular, that U is bounded as a set of $\mathcal{W}_2^0$. Now, by using Lemma 13.6.1, we deduce that $\mathcal{A}_g U$ is relatively compact in $L_2(D)$. This shows the compactness of $\mathcal{A}_g$, which ends the proof. Q.E.D.

Theorem 13.6.1. *Let us assume that the hypotheses of Lemma 13.6.2 are satisfied. Then,*

$$\sigma_{ei}(A) = \{\lambda \in \mathbb{C} \text{ such that } \mathrm{Re}\lambda \leq -\sigma_0\}, \text{ with } i = 1, \ldots, 6.$$

◇

Proof. Let $\lambda \in \rho(T)$. Since the collision operator K is T-defined, and by applying the closed graph theorem, we deduce that $K(\lambda - T)^{-1} \in \mathcal{L}(X_p)$. Moreover, the domain $\mathcal{D}(T)$ is continuously embedded in X_p^σ and, for any $\lambda > 0$, we have

$$\|(\lambda - T)^{-1}\|_{\mathcal{L}(X_p, X_p^\sigma)} \leq \left(\frac{1}{\lambda q}\right)^{\frac{1}{q}} \left(\frac{1}{p}\right)^{\frac{1}{p}}.$$

Since X_p^σ is continuously embedded in X_p, we deduce that $\lim_{\lambda\to\infty} \|K(\lambda - T)^{-1}\|_{\mathcal{L}(X_p)} = 0$. Therefore, there exists $\lambda \in \rho(T)$, such that $r_\sigma(K(\lambda - T)^{-1}) < 1$. Then, by using Lemma 13.6.2 we infer that $K(\lambda - T)^{-1} \in \mathcal{K}(X_p)$. The result follows from both (13.6.4) and Corollary 7.5.2. Q.E.D.

Open question. The result of Theorem 13.6.1 is open in $L_1(D \times \mathbb{R}^n, dx\, d\mu(v))$-space? □

Open question. Consider the multidimensional neutron transport operators on L_1-spaces (cf., [99, 138, 184, 344]):

$$\begin{aligned} A_0\psi(x, v) &= -v.\nabla_x\psi(x, v) - \sigma(x, v)\psi(x, v) \\ &\quad + \int_V \kappa(x, v, v')\psi(x, v')\, dv' = T_0\psi + R\psi, \end{aligned} \tag{13.6.6}$$

where $(x, v) \in \mathcal{D}\times V$ (here $\mathcal{D}$ and V are open subsets of $\mathbb{R}^N$ $(N \geq 1)$), $v.\nabla_x\psi(x, v)$ is the inner product of v and $\nabla_x\psi(x, v)$, and R is the integral operator with a kernel $\kappa(x, v, v')$. Let Γ_- denote the set $\Gamma_- = \{(x, v) \in \partial\mathcal{D}\times V$ such that $v.\nu_x \leq 0\}$, where ν_x stands for the outer normal unit vector at $x \in \partial D$. The unbounded operator A_0 is studied in the Banach space $L_1(\mathcal{D} \times V, dxdv)$, and its domain is

$$\mathcal{D}(A_0) = \Big\{\psi \in L_1(\mathcal{D} \times V, dxdv), \text{ such that}$$
$$v.\nabla_x\psi(x, v) \in L_1(\mathcal{D} \times V, dxdv),\ \psi_{|\Gamma_-} = 0\Big\},$$

where $\psi_{|\Gamma_-}$ denotes the trace of ψ on Γ_-. The functions $\sigma(.,.)$ and $\kappa(.,.,.)$ are respectively called the collision frequency and the scattering kernel. This operator describes the transport of particle neutrons in the domain $\mathcal{D}$. The function $\psi(x, v)$ represents the number (or probability) density of gas particles having the position x, and the velocity v. Let R_1 and R_2 be two regular collision operators on $L_1(\mathcal{D} \times V, dxdv)$ and $\text{Re}\lambda > \eta$, where η is the type of the C_0-semigroup generated by T_0. It was proved in [261] that $R_1(\lambda - T_0)^{-1}R_2$ is weakly compact on $L_1(\mathcal{D} \times V, dxdv)$. However, if $\sigma(x, v) = \sigma(v)$ and if $\mathcal{D}$ is convex, then $R_1(\lambda - T_0)^{-1}R_2$ is compact on $L_1(\mathcal{D}\times V, dxdv)$. In both cases, can we determine the essential spectra of A_0? □

13.7 Systems of Ordinary Differential Operators

In this section, we study the case where A, B, and C constitute three ordinary differential operators in spaces of vector functions and where D represents a multiplication operator. Let us consider the operators defined by differential expressions of the form

$$(A\varphi)(x) = a_0(x)\varphi^{(l)}(x) + a_1(x)\varphi^{(l-1)}(x) + \cdots + a_l(x)\varphi(x)$$
$$(B\psi)(x) = b_0(x)\psi^{(s)}(x) + b_1(x)\psi^{(s-1)}(x) + \cdots + b_s(x)\psi(x)$$
$$(C\varphi)(x) = c_0(x)\varphi^{(h)}(x) + c_1(x)\varphi^{(h-1)}(x) + \cdots + c_h(x)\varphi(x),$$

where $l > 0$, $0 \leq s, h \leq l$ and $s + h = l$, $\varphi \in X_1 := (L_p(0, 1))^n$, $p > 1$, $n \in \mathbb{N}$, $\psi \in X_2 := (L_p(0, 1))^m$, $m \in \mathbb{N}$, and with suitable domains to be specified below. Here, a_i $0 \leq i \leq l$ represent $n\times n$-matrix functions with sufficiently smooth entries and $\det a_0(x) \neq 0$ (whence a_0^{-1} is also a smooth matrix function). Moreover, b_i, $0 \leq i \leq s$ are $n \times m$-matrix functions, and c_i, $0 \leq i \leq h$ are $m \times n$-matrix functions. Let us notice that both b_i and c_i have sufficiently smooth entries. Finally, with an $m\times m$-matrix function d, which is assumed to be measurable and essentially bounded, we define

$$\begin{cases} D : X_2 \longrightarrow X_2 \\ \quad \psi \longrightarrow D\psi(x) = d(x)\psi(x), \end{cases}$$

which is a bounded operator in X_2. The domain $\mathcal{D}(A)$ is assumed to be given by general boundary conditions

$$U(\varphi) := U_0 \begin{pmatrix} \varphi(0) \\ \varphi'(0) \\ . \\ . \\ \varphi^{(l-1)}(0) \end{pmatrix} + U_1 \begin{pmatrix} \varphi(1) \\ \varphi'(1) \\ . \\ . \\ \varphi^{(l-1)}(1) \end{pmatrix} = 0, \tag{13.7.1}$$

with $nl \times nl$ matrix U_0 and U_1. We suppose that these boundary conditions are normalized and Birkhoff regular (see [268, Chapter I, III]). Hence, the domain of A can be written as $\mathcal{D}(A) = H^l_{p,U} := \{\varphi \in (H^l_p)^n$ such that $U(\varphi) = 0\}$ where $(H^l_p)^n := (H^l_p(0,1))^n$ is a Sobolev space of n-vector functions. Moreover, let

$$\mathcal{D}(C) = H^h_{p,U}, \tag{13.7.2}$$

where $H^h_{p,U}$ consists of all functions $\varphi \in (H^h_p(0,1))^n$ satisfying all the boundary conditions in (13.7.1) of order $\leq h-1$.

The definition of $\mathcal{D}(B)$ is more complicated. Our aim is, on the one hand, to satisfy the condition $\mathcal{D}(B^*) \supset \mathcal{D}(A^*)$ (see Proposition 10.1.1) and, on the other hand, $\mathcal{D}(B)$ has to be chosen as large as possible, in order to cover examples which are interesting for the applications. The inclusion $\mathcal{D}(B^*) \supset \mathcal{D}(A^*)$ would be satisfied, e.g., with $\mathcal{D}(B) = \Big\{\psi \in (H^s_p(0,1))^m$ such that $\psi^{(j)}(0) = \psi^{(j)}(1) = 0$, $0 \leq j \leq s-1\Big\}$. However, this definition of B is too restrictive for the applications. Let us notice that the construction of $\mathcal{D}(B)$ can be described through the following steps.

(*i*) Note that $\mathcal{D}(A^*) = H^l_{q,U^*} := \{\varphi \in (H^l_q(0,1))^n$ such that $U^*(\varphi) = 0\}$, where $q = \frac{p}{p-1}$ and the system of boundary conditions $U^*(\varphi) = 0$ is adjoint to the system (13.7.1) in the sense of Lagrange identity with respect to the differential operator A (see [268, Chapter I, III]). Moreover, we suppose that the boundary conditions $U^*(\varphi) = 0$ are normalized.

(*ii*) Let us define the formally adjoint differential expression B^*_f from $\mathcal{D}(B^*_f) := (H^s_q(0,1))^n$, a subspace of $X^*_1 := (L_q(0,1))^n$, into $X^*_2 := (L_q(0,1))^m$, $q = \frac{p}{p-1}$, by the expression

$$(B^*_f v_1)(x) = (-1)^s (b^*_0(x)v_1(x))^{(s)} + (-1)^{s-1}(b^*_1(x)v_1(x))^{(s-1)} + \cdots + b^*_s(x)v_1(x).$$

(*iii*) Let us take all the boundary conditions of order $\leq s-1$ in the system of boundary conditions $U^*(v_1) = 0$ and let us denote the corresponding subsystem of linear forms by $\hat{U}^*(v_1)$. Then, we define the operator $B_{\hat{U}}^*$ as the restriction of B_f^* to the space

$$\mathcal{D}(B_{\hat{U}}^*) := \left\{ v_1 \in (H_q^s(0,1))^n \text{ such that } \hat{U}^*(v_1) = 0 \right\} \ (= H_{q,U^*}^s).$$

(*iv*) By using the Lagrange identity for the operator $B_{\hat{U}}^*$, we find the boundary conditions $\hat{U}(\psi) = 0$ which are adjoint to the conditions $\hat{U}^*(v_1) = 0$ with respect to B_f^*. Finally, we define

$$\mathcal{D}(B) = \left\{ \psi \in (H_p^s(0,1))^m \text{ such that } \hat{U}(\psi) = 0 \right\}. \tag{13.7.3}$$

Then, we have the following operators

$$\begin{cases} A : \mathcal{D}(A) \subset X_1 \longrightarrow X_1 \\ \qquad \varphi \longrightarrow A\varphi(x) = \sum_{k=0}^{l} a_k(x)\varphi^{(l-k)}(x) \\ \mathcal{D}(A) = \{\varphi \in (H_p^l)^n \text{ such that } U(\varphi) = 0\}, \end{cases} \tag{13.7.4}$$

$$\begin{cases} B : \mathcal{D}(B) \subset X_2 \longrightarrow X_1 \\ \qquad \psi \longrightarrow B\psi(x) = \sum_{k=0}^{s} b_k(x)\psi^{(s-k)}(x) \\ \mathcal{D}(B) = \{\psi \in (H_p^s)^m \text{ such that } \hat{U}(\psi) = 0\}, \end{cases}$$

$$\begin{cases} C : \mathcal{D}(C) \subset X_1 \longrightarrow X_2 \\ \qquad \varphi \longrightarrow C\varphi(x) = \sum_{k=0}^{h} c_k(x)\varphi^{(h-k)}(x) \\ \mathcal{D}(C) = H_{p,U}^h := \{\varphi \in (H_p^h)^n \text{such that } U\varphi = 0\}, \end{cases}$$

and

$$\begin{cases} D : X_2 \longrightarrow X_2 \\ \qquad \psi \longrightarrow D\psi(x) = d(x)\psi(x). \end{cases}$$

For more details, we may refer to [40]. The next proposition (see [40]) contains the sufficient conditions which will be needed in the sequel.

Proposition 13.7.1. *The operator B with the domain (13.7.3) is closable and the inclusion*

$$\mathcal{D}(B^*) \supset \mathcal{D}((A^*)^{\frac{s}{l}}) \tag{13.7.5}$$

holds. ◇

Proof. The resolvent of the differential operator A has a ray of minimal growth. Hence, the fractional powers of A and A^* are well defined. The results of [107, 108] imply that

$$\mathcal{D}((A^*)^{\frac{s}{l}}) = \mathcal{D}(B_U^*) = H^s_{q,U^*}. \tag{13.7.6}$$

The operators B and B_U^* are adjoint to each other, i.e., $(By_2, v_1) = (y_2, B_U^* v_1)$ for $y_2 \in \mathcal{D}(B)$, $v_1 \in \mathcal{D}(B_U^*)$. Besides, both operators are densely defined. Hence, they are closable. Obviously, $B^* \supset B_U^*$ and the inclusion (13.7.5) follows from (13.7.6). Q.E.D.

Proposition 13.7.2. *The operator C with the domain given by (13.7.2) is closable, and*

$$\mathcal{D}(\overline{C}) \supset \mathcal{D}(A^{\frac{h}{l}}).$$

◇

Proof. First, we note that, as in (13.7.6), we have $\mathcal{D}(A^{\frac{h}{l}}) = H^h_{p,U}$. As before, we construct the formal adjoint $C_f^* : X_2^* \longrightarrow X_1^*$ with the domain $\mathcal{D}(C_f^*) = (H_q^h(0,1))^m$ and then we consider the restriction C_U^* of C_f^* such that C_U^* and C are adjoint to each other in the sense of Lagrange identity. Since C_U^* is densely defined, then we deduce that C is closable and $\mathcal{D}(\overline{C}) \supset \mathcal{D}(C) = \mathcal{D}(A^{\frac{h}{l}})$. Q.E.D.

Lemma 13.7.1 ([40]). *Let A be a differential operator generated by an expression as in (13.7.4). Let us denote the Green's (matrix) function of A by $G(x,\xi)$. Then, the partial derivatives of the Green's function with respect to x and ξ up to the order $l-1$, i.e.,*

$$\frac{\partial^{i+j} G}{\partial x^i \partial \xi^j}(x,\xi) \quad (i+j \le l-1)$$

exist and are continuous on $[0,1] \times [0,1] \backslash \Gamma$, where Γ denotes the diagonal of the square $[0,1] \times [0,1]$. Moreover, if $0 < x < 1$, then the limits

$$\frac{\partial^{i+j} G}{\partial x^i \partial \xi^j}(x, x \pm 0) \quad (i+j \le l-1) \tag{13.7.7}$$

exist and satisfy the following relationships

$$\frac{\partial^{i+j} G}{\partial x^i \partial \xi^j}(x, x+0) - \frac{\partial^{i+j} G}{\partial x^i \partial \xi^j}(x, x-0) = \begin{cases} 0, & \text{if } i+j \leq l-2, \\ (-1)^j a_0^{-1}(x), & \text{if } i+j = l-1. \end{cases} \tag{13.7.8}$$

◇

Proof. For more simplicity, we consider the case $n = 1$, which means that A is a scalar differential operator. Making the standard substitution $z(x) = e^{p(x)} y(x)$, $p(x) = \frac{1}{l} \int_0^x \frac{a_1(\tau)}{a_0(\tau)} d\tau$, we get $Ay_1 = a_0 e^{-p(.)} A_0 e^{p(.)} y_1$, where the operator A_0 is defined by the expression $(A_0 z)(x) = \left(\frac{d^l}{dx^l} + \tilde{a}_2(x) \frac{d^{l-2}}{dx^{l-2}} + \cdots + \tilde{a}_l(x) \right) z(x)$, and the boundary conditions are $\tilde{U}(z) = U(e^{-p(.)} z) = 0$. Hence, $G(x,\xi) = e^{-p(x)} G_0(x,\xi) e^{p(\xi)} a_0^{-1}(\xi)$, where $G_0(x,\xi)$ represents the Green's function of the differential operator A_0. Hence, it is sufficient to prove the assertions of Lemma 13.7.1 for the case $a_0(x) = 1$ and $a_1(x) = 0$. Let $\eta_1, \ldots, \eta_l$ be the fundamental system of the differential equation $Ay_1 = 0$ satisfying the initial conditions $\eta_j^{(i-1)}(0) = \delta_{ij}$ $(i, j = 1, 2, \ldots, l)$. It is known [268, Ch. I, Section 3.8] that the Green's function $G(x,\xi)$ has the following form

$$G(x,\xi) = \frac{(-1)^l}{\Delta} H(x,\xi), \tag{13.7.9}$$

where

$$\Delta = \det \begin{pmatrix} U_1(\eta_1) & \cdots & U_1(\eta_l) \\ \vdots & & \vdots \\ U_l(\eta_1) & \cdots & U_l(\eta_l) \end{pmatrix},$$

$$H(x,\xi) = \det \begin{pmatrix} \eta_1(x) & \cdots & \eta_l(x) & g(x,\xi) \\ U_1(\eta_1) & \cdots & U_1(\eta_l) & U_1(g(.,\xi)) \\ \vdots & & \vdots & \vdots \\ U_l(\eta_1) & \cdots & U_l(\eta_l) & U_l(g(.,\xi)) \end{pmatrix},$$

and

$$g(x,\xi) = \pm \frac{1}{2W(\xi)} \det \begin{pmatrix} \eta_1(x) & \cdots & \eta_l(x) \\ \eta_1^{(l-2)}(\xi) & \cdots & \eta_l^{(l-2)}(\xi) \\ \vdots & & \vdots \\ \eta_1(\xi) & \cdots & \eta_l(\xi) \end{pmatrix},$$

with the sign + if $x > \xi$, and the sign − if $x < \xi$. Here, the function $W(\xi)$ in the denominator denotes the Wronskian

$$\det \begin{pmatrix} \eta_1^{(l-1)}(\xi) \cdots \eta_l^{(l-1)}(\xi) \\ \vdots \qquad\qquad \vdots \\ \eta_1(\xi) \quad \cdots \quad \eta_l(\xi) \end{pmatrix}.$$

Since $a_1(x) = 0$, the Wronskian is identical to 1. The statements dealing with the differentiability of the Green's function follow immediately from the representation (13.7.9). The existence of the limits (13.7.7) as well as the validity of the formulas (13.7.8) depend only on the function $g(x, \xi)$. The calculation of the corresponding derivatives of $g(x, \xi)$ shows that the limits

$$\frac{\partial^{i+j} g}{\partial x^i \partial \xi^j}(x, x \pm 0) \quad (i + j \le l - 1)$$

exist if $0 < x < 1$, and satisfy the relationships

$$\frac{\partial^{i+j} g}{\partial x^i \partial \xi^j}(x, x + 0) - \frac{\partial^{i+j} g}{\partial x^i \partial \xi^j}(x, x - 0) = \begin{cases} 0, & \text{if } i + j \le l - 2, \\ (-1)^j, & \text{if } i + j = l - 1. \end{cases}$$

Consequently, the limits (13.7.7) exist and satisfy the Eq. (13.7.8) with $a_0(x) = 1$. Q.E.D.

Theorem 13.7.1. *For the operators A, B, C, and D as defined in this section, and for $\mu \in \rho(A)$, the operator $D - C(A - \mu)^{-1}B$, which is defined on $\mathcal{D}(B)$, admits a bounded closure $S(\mu) = S_0 + K(\mu)$, where S_0 represents the multiplication operator, by the function*

$$d - c_0 a_0^{-1} b_0, \tag{13.7.10}$$

and where $K(\mu)$ is a compact operator in X_2. ◇

Proof. From Propositions 13.7.1, 13.7.2 and Lemma 10.1.1, it follows that the operator $S(\mu)$ is bounded and defined on X_2. Let $\mu \in \rho(A)$. Then, the differential operator $A - \mu$ satisfies the assumptions of Lemma 13.7.1. Let $G(x, \xi, \mu)$ be its Green's function. Let φ be a C^∞-function on $\mathbb{R}$, with a compact support in the interval $(0, 1)$. Let $j \in \mathbb{N}$ be such that $j \le l$, and let us consider

$$\begin{aligned}&[(A - \mu)^{-1}\varphi^{(j)}](x) \\ &= \int_0^x G(x, \xi, \mu)\varphi^{(j)}(\xi)\, d\xi + \int_x^1 G(x, \xi, \mu)\varphi^{(j)}(\xi)\, d\xi\end{aligned}$$

$$= \sum_{k=0}^{j-1}(-1)^{k+1}\left(\frac{\partial^k G}{\partial \xi^k}(x,x+0,\mu) - \frac{\partial^k G}{\partial \xi^k}(x,x-0,\mu)\right)\varphi^{(j-k-1)}(x)$$

$$+(-1)^j\left(\int_0^x + \int_x^1\right)\frac{\partial^j G}{\partial \xi^j}(x,\xi,\mu)\varphi(\xi)\,d\xi$$

which is obtained using an integration by parts, where special attention has to be paid to the cases $x = 0$ and $x = 1$. Let $i + j \leq l$. By using Eq. (13.7.8), we deduce that

$$[(A-\mu)^{-1}\varphi^{(j)}]^{(i)}(x) = \delta_{i+j,l}a_0^{-1}(x)\varphi(x) + (-1)^j\int_0^1 \frac{\partial^{i+j}G}{\partial x^i\partial \xi^j}(x,\xi,\mu)\varphi(\xi)\,d\xi.$$

Since the differential operator B can be written in the form

$$By_2 = (b_0y_2)^{(s)} + (\tilde{b}_1y_2)^{(s-1)} + \cdots + \tilde{b}_sy_2$$

with appropriate functions $\tilde{b}_1,\ldots,\tilde{b}_s$, and since the space of C^∞-functions with a compact support in $(0,1)$ is dense in $L_p(0,1)$, then the representation (13.7.10) is proved. Q.E.D.

Theorem 13.7.2. *Let L_0 be the operator defined in* (10.0.1), *and let L denote the closure of L_0. Then, for $i = 1,\ldots,6$*

$$\sigma_{ei}(L) = \left\{\lambda \in \mathbb{C} \text{ such that } ess - \inf\left|det\left[d(x) - c_0(x)a_0^{-1}(x)b_0(x) - \lambda I\right]\right| = 0\right\}. \tag{13.7.11}$$

Moreover, if the complement of this set is connected, then this complement coincides with the domain of a finite meromorphy of the operator function $(L-\lambda I)^{-1}$. ◇

Proof. By using Theorems 10.1.3 and 13.7.1, we infer that $\sigma_{ei}(L) = \sigma_{ei}(S_0)$, with $i = 1,\ldots,6$ and where S_0 represents the multiplication operator by the matrix function $d - c_0a_0^{-1}b_0$. Moreover, it is shown in [147] that the spectrum of S_0 is purely continuous and is given by the expression on the right-hand side of (13.7.11). Now, the result follows from Remark 7.1.1. Q.E.D.

Remark 13.7.1. By virtue of the Frobenius–Schur factorization, we have

$$\det\left[d(x) - c_0(x)a_0^{-1}(x)b_0(x) - \lambda I\right] = (\det a_0(x))^{-1}\det\begin{pmatrix} a_0(x) & b_0(x) \\ c_0(x) & d(x) - \lambda I\end{pmatrix}.$$

◇

13.8 Essential Spectra of Two-Group Transport Operators

Let $X_p := L_p((-a,a)\times(-1,1), dxd\xi)$, with $a>0$, and $1 \leq p < \infty$. Let us consider the following two-group transport operators with abstract boundary conditions $A_H = T_H + K$, where

$$T_H\psi = \begin{pmatrix} -\xi\dfrac{\partial\psi_1}{\partial x} - \sigma_1(\xi)\psi_1 & 0 \\ 0 & -\xi\dfrac{\partial\psi_2}{\partial x} - \sigma_2(\xi)\psi_2 \end{pmatrix} = \begin{pmatrix} T_{H_1} & 0 \\ 0 & T_{H_2} \end{pmatrix}\begin{pmatrix} \psi_1 \\ \psi_2 \end{pmatrix},$$

and

$$K = \begin{pmatrix} K_{11} & K_{12} \\ K_{21} & K_{22} \end{pmatrix}$$

with $K_{ij}, i,j = 1,2$ as bounded linear operators defined on X_p by

$$\begin{cases} K_{ij} : X_p \longrightarrow X_p \\ \qquad u \longrightarrow K_{ij}u(x,\xi) = \displaystyle\int_{-1}^{1} \kappa_{ij}(x,\xi,\xi')\, u(x,\xi')\, d\xi', \end{cases}$$

and where the kernels $\kappa_{ij} : (-a,a)\times(-1,1)\times(-1,1) \longrightarrow \mathbb{R}$ are assumed to be measurable. Each operator $T_{H_j}, j = 1,2$ is defined by

$$\begin{cases} T_{H_j} : \mathcal{D}(T_{H_j}) \subset X_p \longrightarrow X_p \\ \qquad \varphi \longrightarrow (T_{H_j}\varphi)(x,\xi) = -\xi\,\dfrac{\partial\varphi}{\partial x}(x,\xi) - \sigma_j(\xi)\varphi(x,\xi) \\ \mathcal{D}(T_{H_j}) = \Big\{\varphi \in W_p \text{ such that } \varphi^i = H_j\varphi^o\}, \end{cases}$$

where W_p is the space defined by $W_p = \Big\{\varphi \in X_p \text{ such that } \xi\frac{\partial\varphi}{\partial x} \in X_p\Big\}$ and where $\sigma_j(.) \in L_\infty(-1,1)$. φ^o, φ^i represent the outgoing and the incoming fluxes related by the boundary operator H_j ("o" for the outgoing and "i" for the incoming) and given by

$$\begin{cases} \varphi^i(\xi) = \varphi(-a,\xi) \ \xi \in (0,1) \\ \varphi^i(\xi) = \varphi(a,\xi) \ \xi \in (-1,0) \\ \varphi^o(\xi) = \varphi(-a,\xi) \ \xi \in (-1,0) \\ \varphi^o(\xi) = \varphi(a,\xi) \ \xi \in (0,1). \end{cases}$$

We denote by X_p^o and X_p^i the following boundary spaces

$$X_p^o := L_p[\{-a\}\times(-1,0), |\xi|d\xi] \times L_p[\{a\}\times(0,1), |\xi|d\xi] := X_{1,p}^o \times X_{2,p}^o$$

equipped with the norm

$$\|u^o, X_p^o\| := \left(\|u_1^o, X_{1,p}^o\|^p + \|u_2^o, X_{2,p}^o\|^p\right)^{\frac{1}{p}}$$
$$= \left[\int_{-1}^{0} |u(-a,\xi)|^p \, |\xi| d\xi + \int_{0}^{1} |u(a,\xi)|^p \, |\xi| d\xi\right]^{\frac{1}{p}},$$

and $X_p^i := L_p[\{-a\} \times (0,1), |\xi| d\xi] \times L_p[\{a\} \times (-1,0), |\xi| d\xi] := X_{1,p}^i \times X_{2,p}^i$ equipped with the norm

$$\|u^i, X_p^i\| := \left(\|u_1^i, X_{1,p}^i\|^p + \|u_2^i, X_{2,p}^i\|^p\right)^{\frac{1}{p}}$$
$$= \left[\int_{0}^{1} |u(-a,\xi)|^p \, |\xi| d\xi + \int_{-1}^{0} |u(a,\xi)|^p \, |\xi| d\xi\right]^{\frac{1}{p}}.$$

It is well known that any function u in W_p possesses some traces on the spatial boundary $\{-a\} \times (-1,0)$ and $\{a\} \times (0,1)$ which respectively belong to the spaces X_p^o and X_p^i (see, for instance, [89] or [138]). These traces are denoted, respectively, by u^o and u^i. It is clear that the operator A_H is defined on $\mathcal{D}(T_{H_1}) \times \mathcal{D}(T_{H_2})$. We will denote the operator A_H by

$$A_H := \begin{pmatrix} A_{11} & A_{12} \\ A_{21} & A_{22} \end{pmatrix},$$

where

$$\begin{cases} A_{11} = T_{H_1} + K_{11} \\ A_{12} = K_{12} \\ A_{21} = K_{21} \\ A_{22} = T_{H_2} + K_{22}. \end{cases}$$

13.8.1 The Expression of the Resolvent of T_{H_1}

We will determine the expression of the resolvent of the operator T_{H_1}. Let $\varphi \in X_p$, $\lambda \in \mathbb{C}$ and consider the following resolvent equation for T_{H_1}

$$(\lambda - T_{H_1})\psi_1 = \varphi, \tag{13.8.1}$$

where the unknown ψ_1 must be in $\mathcal{D}(T_{H_1})$. Let $\lambda_j^* = \liminf_{|\xi|\to 0} \sigma_j(\xi)$, with $j = 1, 2$ and

$$\lambda_0^j := \begin{cases} -\lambda_j^* & \text{if } \|H_j\| \leq 1 \\ -\lambda_j^* + \dfrac{1}{2a}\log(\|H_j\|) & \text{if } \|H_j\| > 1. \end{cases}$$

Therefore, for $\lambda \in \mathbb{C}$ such that $\mathrm{Re}\lambda > -\lambda_1^*$, the solution of (13.8.1) is formally given by

$$\psi_1(x,\xi) = \begin{cases} \psi_1(-a,\xi)e^{-\frac{(\lambda+\sigma_1(\xi))|a+x|}{|\xi|}} + \dfrac{1}{|\xi|}\displaystyle\int_{-a}^{x} e^{-\frac{(\lambda+\sigma_1(\xi))|x-x'|}{|\xi|}}\varphi(x',\xi)dx', & \text{if } 0 < \xi < 1 \\ \psi_1(a,\xi)e^{-\frac{(\lambda+\sigma_1(\xi))|a-x|}{|\xi|}} + \dfrac{1}{|\xi|}\displaystyle\int_{x}^{a} e^{-\frac{(\lambda+\sigma_1(\xi))|x-x'|}{|\xi|}})\varphi(x',\xi)dx', & \text{if } -1 < \xi < 0. \end{cases} \tag{13.8.2}$$

Accordingly, $\psi_1(a,\xi)$ and $\psi_1(-a,\xi)$ are given by

$$\psi_1(a,\xi) = \psi_1(-a,\xi)e^{-2a\frac{(\lambda+\sigma_1(\xi))}{|\xi|}} + \frac{1}{|\xi|}\int_{-a}^{a} e^{-\frac{(\lambda+\sigma_1(\xi))|a-x|}{|\xi|}}\varphi(x,\xi)dx, \text{ if } 0 < \xi < 1 \tag{13.8.3}$$

$$\psi_1(-a,\xi) = \psi_1(a,\xi)e^{-2a\frac{(\lambda+\sigma_1(\xi))}{|\xi|}} + \frac{1}{|\xi|}\int_{-a}^{a} e^{-\frac{(\lambda+\sigma_1(\xi))|a+x|}{|\xi|}}\varphi(x,\xi)dx, \text{ if } -1 < \xi < 0. \tag{13.8.4}$$

In order to clarify our subsequent analysis, we introduce the following bounded operators:

$$\begin{cases} M_\lambda : X_p^i \longrightarrow X_p^o, \quad M_\lambda u := (M_\lambda^+ u, M_\lambda{}^- u) \text{ with} \\ \\ M_\lambda^+ u(-a,\xi) := u(-a,\xi)e^{-2a\frac{(\lambda+\sigma_1(\xi))}{|\xi|}}, \text{ if } 0 < \xi < 1 \\ M_\lambda^- u(a,\xi) \;\; := u(a,\xi)e^{-2a\frac{(\lambda+\sigma_1(\xi))}{|\xi|}}, \text{ if } -1 < \xi < 0 \end{cases}$$

$$\begin{cases} B_\lambda : X_p^i \longrightarrow X_p, \quad B_\lambda u := \chi_{(-1,0)}(\xi)B_\lambda^- u + \chi_{(0,1)}(\xi)B_\lambda{}^+ u \text{ with} \\ \\ B_\lambda^+ u(x,\xi) := u(-a,\xi)e^{-\frac{(\lambda+\sigma_1(\xi))|a+x|}{|\xi|}}, \text{ if } 0 < \xi < 1 \\ B_\lambda^- u(x,\xi) := u(a,\xi)e^{-\frac{(\lambda+\sigma_1(\xi))|a-x|}{|\xi|}}, \text{ if } -1 < \xi < 0 \end{cases}$$

$$\begin{cases} G_\lambda : X_p \longrightarrow X_p^o, \quad G_\lambda\varphi := (G_\lambda^+\varphi, G_\lambda^-\varphi) \text{ with} \\ G_\lambda^+\varphi(-a,\xi) := \dfrac{1}{|\xi|}\displaystyle\int_{-a}^{a} e^{-\frac{(\lambda+\sigma_1(\xi))|a-x|}{|\xi|}}\varphi(x,\xi)dx, \text{ if } 0<\xi<1 \\ G_\lambda^-\varphi(a,\xi) \;:= \dfrac{1}{|\xi|}\displaystyle\int_{-a}^{a} e^{-\frac{(\lambda+\sigma_1(\xi))|a+x|}{|\xi|}}\varphi(x,\xi)dx, \text{ if } -1<\xi<0 \end{cases}$$

and finally, we consider

$$\begin{cases} C_\lambda : X_p \longrightarrow X_p, \quad C_\lambda\varphi := \chi_{(-1,0)}(\xi)C_\lambda^-\varphi + \chi_{(0,1)}(\xi)C_\lambda^+\varphi \text{ with} \\ C_\lambda^+\varphi(x,\xi) := \dfrac{1}{|\xi|}\displaystyle\int_{-a}^{x} e^{-\frac{(\lambda+\sigma_1(\xi))|x-x'|}{|\xi|}}\varphi(x',\xi)dx', \text{ if } 0<\xi<1 \\ C_\lambda^-\varphi(x,\xi) := \dfrac{1}{|\xi|}\displaystyle\int_{x}^{a} e^{-\frac{(\lambda+\sigma_1(\xi))|x-x'|}{|\xi|}}\varphi(x',\xi)dx', \text{ if } -1<\xi<0, \end{cases}$$

where $\chi_{(-1,0)}(.)$ and $\chi_{(0,1)}(.)$ denote, respectively, the characteristic functions of the intervals $(-1,0)$ and $(0,1)$. The operators M_λ, B_λ, G_λ, and C_λ are bounded on their respective spaces. Their norms are bounded above, respectively, by $e^{-2a(\mathrm{Re}\lambda+\lambda_1^*)}$, $(p\mathrm{Re}\lambda+\lambda_1^*)^{-1/p}$, $(\mathrm{Re}\lambda+\lambda_1^*)^{-1/q}$ and $(\mathrm{Re}\lambda+\lambda_1^*)^{-1}$, where q denotes the conjugate of p. By using the above defined operators, and by recalling the fact that ψ_1 must satisfy the boundary conditions, we can write Eqs. (13.8.3) and (13.8.4) in the operators form $\psi_1^o = M_\lambda H_1\psi_1^o + G_\lambda\varphi$. From the norm estimate of M_λ, we deduce that $\|M_\lambda H_1\| < 1$ for $\mathrm{Re}\lambda > \lambda_0^1$. This gives

$$\psi_1^o = \sum_{n\geq 0}(M_\lambda H_1)^n G_\lambda\varphi. \tag{13.8.5}$$

Moreover, Eq. (13.8.2) can be written as

$$\psi_1 = B_\lambda H_1\psi_1^o + C_\lambda\varphi. \tag{13.8.6}$$

Substituting (13.8.5) into (13.8.6), we get $\psi_1 = \sum_{n\geq 0} B_\lambda H_1(M_\lambda H_1)^n G_\lambda\varphi + C_\lambda\varphi$. Therefore,

$$(\lambda - T_{H_1})^{-1} = \sum_{n\geq 0} B_\lambda H_1(M_\lambda H_1)^n G_\lambda + C_\lambda. \tag{13.8.7}$$

13.8.2 Compactness Results

Lemma 13.8.1. *If* $\frac{\kappa_{21}(x,\xi,\xi')}{|\xi'|}$ *defines a regular operator then,* $K_{21}(\lambda - T_{H_1})^{-1}$ *is weakly compact on* X_1. ◇

Proof. In view of (13.8.7), the operator $K_{21}(\lambda - T_{H_1})^{-1}$ is given by

$$K_{21}(\lambda - T_{H_1})^{-1} = \sum_{n\geq 0} K_{21} B_\lambda H_1 (M_\lambda H_1)^n G_\lambda + K_{21} C_\lambda.$$

Then, in order to prove the weak compactness of $K_{21}(\lambda - T_{H_1})^{-1}$, it is sufficient to prove the weak compactness of the operators $K_{21}B_\lambda$ and $K_{21}C_\lambda$. Let us notice that C_λ is nothing else but $(\lambda - T_1)^{-1}$, where T_1 is the streaming operator for the vacuum boundary conditions. It suffices to prove that $K_{21}B_\lambda$ is weakly compact on X_1 (the same reasoning can be applied for the operator $K_{21}C_\lambda$). Let $u \in X_1^i$. Then, we have

$$\begin{aligned} K_{21}B_\lambda u(x,\xi) &= \int_{-1}^{1} \kappa_{21}(x,\xi,\xi') B_\lambda u(x,\xi') d\xi' \\ &= \tilde{K}_{21}\tilde{B}_\lambda u(x,\xi), \end{aligned}$$

where

$$\begin{cases} \tilde{K}_{21} : X_1 \longrightarrow X_1 \\ \qquad \psi \longrightarrow \tilde{K}_{21}u(x,\xi) = \displaystyle\int_{-1}^{1} \frac{\kappa_{21}(x,\xi,\xi')}{|\xi'|} u(x,\xi')\, d\xi', \end{cases}$$

and $\tilde{B}_\lambda = |\xi'| B_\lambda$. Then, it is sufficient to establish the weak compactness of $\tilde{K}_{21}\tilde{B}_\lambda$. The fact that $\tilde{K}_{21}$ is regular and the use of Lemma 2.4.1 allow us to get the result for an operator whose kernel is $\frac{\kappa_{21}(x,\xi,\xi')}{|\xi'|} = \sum_{j=1}^{n} \alpha_j(x) f_j(\xi) g_j(\xi')$, where $\alpha_j(.) \in L_\infty(-a,a)$, $f_j(.) \in L_1(-1,1)$ and $g_j(.) \in L_\infty(-1,1)$. Therefore, we restrict ourselves to $\frac{\kappa_{21}(x,\xi,\xi')}{|\xi'|} = \alpha(x) f(\xi) g(\xi')$, where $\alpha(.) \in L_\infty(-a,a)$, $f(.) \in L_1(-1,1)$, and $g(.) \in L_\infty(-1,1)$, since the weak compactness is stable by summation. We claim that the operator $\tilde{K}_{21}\tilde{B}_\lambda$ satisfies the following estimate

$$\|\tilde{K}_{21}\tilde{B}_\lambda\| \leq 2a \|g\|_\infty \|\alpha\|_\infty \|f\|. \tag{13.8.8}$$

Indeed, let $u \in X_1^i$. Then, we have

$$\begin{aligned} \tilde{K}_{21}\tilde{B}_\lambda u(x,\xi) = \alpha(x) f(\xi) \Bigg[&\int_0^1 g(\xi') u(-a,\xi')\, e^{-\frac{(\lambda+\sigma_1(\xi'))|a+x|}{|\xi'|}} |\xi'| d\xi' \\ &+ \int_{-1}^{0} g(\xi') u(a,\xi')\, e^{-\frac{(\lambda+\sigma_1(\xi'))|a-x|}{|\xi'|}} |\xi'| d\xi' \Bigg]. \end{aligned}$$

Therefore,

$$\begin{aligned} \left|\tilde{K}_{21}\tilde{B}_\lambda u(x,\xi)\right| \leq \|g\|_\infty \|\alpha\|_\infty |f(\xi)| \Bigg[&\int_0^1 |u(-a,\xi')| e^{-\frac{(\mathrm{Re}\lambda+\lambda_1^*)|a+x|}{|\xi'|}} |\xi'| d\xi' \\ &+ \int_{-1}^{0} |u(a,\xi')| e^{-\frac{(\mathrm{Re}\lambda+\lambda_1^*)|a-x|}{|\xi'|}} |\xi'| d\xi' \Bigg]. \end{aligned}$$

Hence, for Re$\lambda > -\lambda_1^*$, we have $|\tilde{K}_{21}\tilde{B}_\lambda u(x,\xi)| \leq \|g\|_\infty \|\alpha\|_\infty |f(\xi)| \|u, X_1^i\|$. Then, the claim is proved. The inequality (13.8.8) shows that the operator $\tilde{K}_{21}\tilde{B}_\lambda$ depends continuously (in the uniform topology) on $f(.)$. Since the set of bounded functions, which vanish in the neighborhood of $\xi = 0$, is dense in $L_1(-1,1)$, $\tilde{K}_{21}\tilde{B}_\lambda$ is a limit (in the uniform topology) of integral operators with bounded kernels. The use of Theorem 2.4.5 allows us to conclude that $\tilde{K}_{21}\tilde{B}_\lambda$ is weakly compact on X_1^i. Now, the weak compactness of $K_{21}(\lambda - T_{H_1})^{-1}$ follows immediately. Q.E.D.

Lemma 13.8.2. *Let $\lambda \in \rho(T_{H_1})$ be such that $r_\sigma\left((\lambda - T_{H_1})^{-1}K_{11}\right) < 1$, where $r_\sigma(.)$ is the spectral radius.*

(*i*) *If $\frac{\kappa_{21}(x,\xi,\xi')}{|\xi'|}$ defines a regular operator, then the operator $F(\lambda) = K_{21}(\lambda - A_{11})^{-1}$ is weakly compact on X_1.*
(*ii*) *If K_{21} is regular, then the operator $F(\lambda) = K_{21}(\lambda - A_{11})^{-1}$ is compact on X_p for $1 < p < \infty$.*
(*iii*) *If the operator K_{12} is regular, then $G(\lambda) = (\lambda - A_{11})^{-1}K_{12}$ is compact on X_p for $1 < p < \infty$ and is weakly compact on X_1.* ◇

Proof. By using a similar reasoning to the proof of Lemma 13.1.1, we can show the following equality $\lim_{\text{Re}\lambda \to +\infty} \|(\lambda - T_{H_1})^{-1}\| = 0$. Then, there exists a $\lambda \in \rho(T_{H_1})$, such that $r_\sigma\left((\lambda - T_{H_1})^{-1}K_{11}\right) < 1$. For such a λ, the equation $(\lambda - T_{H_1} - K_{11})\varphi = \psi$ may be transformed into $\varphi - (\lambda - T_{H_1})^{-1}K_{11}\varphi = (\lambda - T_{H_1})^{-1}\psi$, since $\lambda \in \rho(T_{H_1})$. The fact that $r_\sigma\left((\lambda - T_{H_1})^{-1}K_{11}\right) < 1$, implies that

$$(\lambda - A_{11})^{-1} = \sum_{n\geq 0}\left[(\lambda - T_{H_1})^{-1}K_{11}\right]^n (\lambda - T_{H_1})^{-1}. \tag{13.8.9}$$

(*i*) The use of Lemma 13.8.1 implies that, for all n in $\mathbb{N}$, $K_{21}\left[(\lambda - T_{H_1})^{-1}K_{11}\right]^n (\lambda - T_{H_1})^{-1}$ is weakly compact on X_1. Now, the result follows immediately from Eq. (13.8.9) and from the fact that $\mathcal{W}(X_1)$ is a closed two-sided ideal of $\mathcal{L}(X_1)$.
(*ii*) The proof of this assertion follows immediately from both Eq. (13.8.9) and Theorem 13.4.8.
(*iii*) Eq. (13.8.9) leads to $G(\lambda) = \sum_{n\geq 0}\left[(\lambda - T_{H_1})^{-1}K_{11}\right]^n (\lambda - T_{H_1})^{-1}K_{12}$. Therefore, the hypothesis on K_{12}, together with Theorem 13.4.8, imply the compactness of $G(\lambda)$ on X_p, for $1 < p < \infty$, and also its weak compactness on X_1. Q.E.D.

13.8.3 Essential Spectra

The essential spectra of the operator T_j, with $j = 1, 2$ (T_j designates the streaming operator with vacuum boundary conditions, i.e., $H_j = 0$) were analyzed in detail in Lemma 13.4.9 and Eq. (13.4.36). In particular, it was shown that $\sigma_{ei}(T_j) = \{\lambda \in \mathbb{C}$ such that $\mathrm{Re}\lambda \leq -\lambda_j^*\}$ for $i = 1, \ldots, 6$. In view of Eq. (13.8.7), and for $\mathrm{Re}\lambda > \lambda_0^j$, with $j = 1, 2$, we have

$$(\lambda - T_{H_j})^{-1} - (\lambda - T_j)^{-1} = \sum_{n \geq 0} B_\lambda H_j (M_\lambda H_j)^n G_\lambda$$

(C_λ is nothing else but $(\lambda - T_j)^{-1}$). If the operators H_j, with $j = 1, 2$ are strictly singular on X_p, for $1 \leq p < \infty$, then $(\lambda - T_{H_j})^{-1} - (\lambda - T_j)^{-1}$ are strictly singular, too. Therefore, the use of Theorem 7.5.4 implies that

$$\sigma_{ei}(T_{H_j}) = \{\lambda \in \mathbb{C} \text{ such that } \mathrm{Re}\lambda \leq -\lambda_j^*\} \text{ for } i = 1, \ldots, 5. \tag{13.8.10}$$

The fact that $\mathbb{C} \backslash \sigma_{e5}(T_{H_j})$, with $j = 1, 2$ are connected and that $\rho(T_{H_j}) \neq \emptyset$ imply that $\sigma_{e5}(T_{H_j}) = \sigma_{e6}(T_{H_j})$. The previous equality in Eq. (13.8.10) is not optimal. An example is given in [212]. Indeed, fix $p = 2$ and suppose that the collision frequency is constant ($\sigma(.) = \sigma$). Let $\tilde{H}$ be the following boundary operator:

$$\begin{cases} \tilde{H} : X_{1,2}^o \times X_{2,2}^o \longrightarrow X_{1,2}^i \times X_{2,2}^i \\ \tilde{H} \begin{pmatrix} u_1 \\ u_2 \end{pmatrix} = \begin{pmatrix} H_{11} & 0 \\ H_{21} & 0 \end{pmatrix} \begin{pmatrix} u_1 \\ u_2 \end{pmatrix}, \end{cases}$$

where

$$\begin{cases} H_{11} : X_{1,2}^o \longrightarrow X_{1,2}^i \\ \quad u(-a, \xi) \longrightarrow u(-a, -\xi) \end{cases}$$

and

$$\begin{cases} H_{21} : X_{1,2}^o \longrightarrow X_{2,2}^i \\ H_{21} \in \mathcal{L}(X_{1,2}^o, X_{2,2}^i) \end{cases}$$

with H_{21} an arbitrary operator. Note that since H_{11} and H_{21} are not compact $\tilde{H}$ is not compact either. Our objective is to show $\sigma(T_{\tilde{H}}) = \{\lambda \in \mathbb{C}$ such that $\mathrm{Re}\lambda \leq -\sigma\}$. To this end, we shall use the following classical result (see, for instance, [89, p. 1134]). Let for $0 < \delta < \frac{1}{2}$ and $\mathrm{Re}\lambda + \sigma = \beta < 0$, $u_\delta(x, \xi) = e^{-\frac{1}{\xi}\beta(x-a)} \chi_{(\delta^2, \delta)}(\xi) \frac{1}{\delta} \frac{a+x}{a}$. It is clear that $u_\delta \in \mathcal{D}(T_{\tilde{H}})$. Let us now estimate its norm

$$\begin{aligned}
\|u_\delta\|^2 &= \int_{-a}^{a}\int_{-1}^{1} |u_\delta(x,\xi)|^2 dx d\xi \\
&= \int_{-1}^{1}\int_{-a}^{a} e^{-\frac{2}{|\xi|}\beta(x-a)} \chi_{(\delta^2,\delta)}(\xi)\frac{1}{\delta^2}\left(\frac{a+x}{a}\right)^2 dx d\xi \\
&> \frac{1}{2|\beta|}\int_{\delta^2}^{\delta}\frac{\xi}{\delta^2}(1-e^{-\frac{2a\beta}{\xi}})d\xi \\
&\geq \frac{c_1}{\delta^2}\int_{\delta^2}^{\delta}\xi d\xi \\
&= \frac{c_1}{2}(1-\delta^2) \\
&> c_2,
\end{aligned}$$

where c_1 and c_2 are independent of δ. Thus

$$\|u_\delta\| \geq c > 0. \tag{13.8.11}$$

Second

$$(\lambda - T_{\tilde{H}})u_\delta(x,\xi) = \xi e^{-\frac{\beta}{\xi}(x-a)} \chi_{(\delta^2,\delta)}(\xi)\frac{1}{\delta}\left(\frac{a+x}{a}\right)' := S,$$

$$\begin{aligned}
\|S\| &= \int_{-a}^{a}\int_{-1}^{1} \xi^2 e^{-2\frac{\beta}{\xi}(x-a)}\left(\chi_{(\delta^2,\delta)}(\xi)\frac{1}{\delta}\right)^2\left(\frac{a+x}{a}\right)'^2 dx d\xi \\
&= \frac{1}{a^2}\int_{-1}^{1}\xi^3\left(\chi_{(\delta^2,\delta)}(\xi)\frac{1}{\delta}\right)^2\frac{1}{2|\beta|}\left(\left[e^{-2\frac{\beta}{\xi}(x-a)}\right]_{-a}^{a}\right)d\xi \\
&= \frac{1}{2a^2\delta^2|\beta|}\int_{\delta^2}^{\delta}\xi^3\left(1-e^{\frac{4a}{\xi}\beta}\right)d\xi \\
&\leq \frac{1}{8a^2\delta^2|\beta|}(\delta^4-\delta^6) \\
&= \frac{\delta^2-\delta^4}{8a^2|\beta|} \to 0 \text{ as } \delta \to 0.
\end{aligned}$$

This proves that

$$\|(\lambda - T_{\tilde{H}})u_\delta\| \to 0 \text{ as } \delta \to 0. \tag{13.8.12}$$

Using Eqs. (13.8.11) and (13.8.12), we have λ belongs to $\sigma(T_{\tilde{H}})$ (see [89, p. 1134]). Now, by using the fact that the spectrum is a closed set, we get $\{\lambda \in \mathbb{C}$ such that

$\mathrm{Re}\lambda \leq -\sigma\} \subset \sigma(T_{\tilde{H}})$. On the other hand, for all $\lambda \in \mathbb{C}$ satisfying $\mathrm{Re}\lambda > -\sigma$, the resolvent equation $(\lambda - T_{\tilde{H}})\psi = \varphi$ (φ is a given function in X_2 and ψ is unknown), has a unique solution. Therefore, we obtain $\sigma(T_{\tilde{H}}) = \{\lambda \in \mathbb{C}$ such that $\mathrm{Re}\lambda \leq -\sigma\}$. Let us consider the eigenvalue problem

$$(\lambda - T_{\tilde{H}})\psi = 0, \quad (\mathrm{Re}\lambda \leq -\sigma). \tag{13.8.13}$$

The solution of Eq. (13.8.13) is formally given by $\psi(x,\xi) = k(\xi)e^{-\frac{1}{\xi}(\lambda+\sigma)x}$. Moreover, ψ must satisfy the boundary conditions which imply that Eq. (13.8.13) has only the trivial solution. So, we conclude that the point spectrum of $T_{\tilde{H}}$ is empty, $\sigma_p(T_{\tilde{H}}) = \emptyset$. Next we shall show that the residual spectrum of $T_{\tilde{H}}$ is empty. Indeed, the dual operator of $T_{\tilde{H}}$ is given by

$$\begin{cases} T^*_{\tilde{H}} : \mathcal{D}(T^*_{\tilde{H}}) \subset X_2 \longrightarrow X_2 \\ \qquad \psi \longrightarrow T^*_{\tilde{H}}\psi(x,\xi) = \xi\dfrac{\partial\psi}{\partial x}(x,\xi) - \sigma\psi(x,\xi) \\ \mathcal{D}(T^*_{\tilde{H}}) = \{\psi \in W_2 \text{ such that } \tilde{H}^*\psi^i = \psi^o\}, \end{cases}$$

where $\tilde{H}^*$ is given by

$$\begin{cases} \tilde{H}^* : X^i_{1,2} \times X^i_{2,2} \longrightarrow X^o_{1,2} \times X^o_{2,2} \\ \tilde{H}^*\begin{pmatrix} u_1 \\ u_2 \end{pmatrix} = \begin{pmatrix} H^*_{11} & H^*_{21} \\ 0 & 0 \end{pmatrix}\begin{pmatrix} u_1 \\ u_2 \end{pmatrix}, \end{cases}$$

and where H^*_{11} and H^*_{21} are, respectively, the dual operators of H_{11} and H_{21}. A similar reasoning as above shows that $\sigma_p(T^*_{\tilde{H}}) = \emptyset$. Therefore, the residual spectrum of $T_{\tilde{H}}$ is empty, $\sigma_r(T_{\tilde{H}}) = \emptyset$.

Proposition 13.8.1. *With the notation above, we have*

$$\sigma_{e5}(T_{\tilde{H}}) = \sigma_c(T_{\tilde{H}}) = \{\lambda \in \mathbb{C} \textit{ such that } \mathrm{Re}\lambda \leq -\sigma\}.$$

◇

Now, we are ready to express the essential spectra of two-group transport operators with general boundary conditions.

Theorem 13.8.1. *If the operators $H_j \in \mathcal{S}(X_p)$, with $j = 1,2$ and the operators K_{11}, K_{22}, and K_{12} are regular, and if, in addition, $\kappa_{21}(x,\xi,\xi')$ (resp. $\frac{\kappa_{21}(x,\xi,\xi')}{|\xi'|}$) defines a regular operator on X_p, for $1 < p < \infty$ (resp. on X_1), then*

$$\sigma_{ei}(A_H) = \{\lambda \in \mathbb{C} \textit{ such that } \mathrm{Re}\lambda \leq -\min(\lambda_1^*, \lambda_2^*)\}, \textit{ for } i = 1,\ldots,6.$$

◇

Proof. Let $\lambda \in \rho(T_{H_1})$, such that $r_\sigma\left((\lambda - T_{H_1})^{-1}K_{11}\right) < 1$, then $\lambda \in \rho(A_{11}) \bigcap \rho(T_{H_1})$. From Eq. (13.8.9), we have

$$(\lambda - A_{11})^{-1} - (\lambda - T_{H_1})^{-1} = \sum_{n\geq 1}\left[(\lambda - T_{H_1})^{-1}K_{11}\right]^n (\lambda - T_{H_1})^{-1}. \tag{13.8.14}$$

Since K_{11} is regular, then from Theorem 13.4.8, it follows that the operator $(\lambda - A_{11})^{-1} - (\lambda - T_{H_1})^{-1}$ is compact on X_p, for $1 < p < \infty$, and also weakly compact on X_1. The use of Eq. (13.8.14) leads to

$$\sigma_{ei}(A_{11}) = \sigma_{ei}(T_{H_1}) = \{\lambda \in \mathbb{C} \text{ such that } \mathrm{Re}\lambda \leq -\lambda_1^*\}, \text{ with } i = 1, \ldots, 6. \tag{13.8.15}$$

Let $\mu \in \rho(A_{11})$. The operator $S(\mu)$ is given by $S(\mu) = A_{22} - K_{21}G(\mu)$. By using Lemma 13.8.2, we deduce that the operator $K_{21}G(\mu)$ is compact on X_p, for $1 < p < \infty$, and also weakly compact on X_1. Then, from Theorem 7.5.4, it follows that $\sigma_{ei}(S(\mu)) = \sigma_{ei}(A_{22})$, with $i = 1, \ldots, 6$. By using a similar reasoning to the previous one, we have

$$\sigma_{ei}(S(\mu)) = \sigma_{ei}(A_{22}) = \{\lambda \in \mathbb{C} \text{ such that } \mathrm{Re}\lambda \leq -\lambda_2^*\}, \text{ with } i = 1, \ldots, 6. \tag{13.8.16}$$

By applying Theorem 10.2.2, and by using Eqs. (13.8.15) and (13.8.16), we get

$$\sigma_{ei}(A_H) = \{\lambda \in \mathbb{C} \text{ such that } \mathrm{Re}\lambda \leq -\min(\lambda_1^*, \lambda_2^*)\}, \text{ for } i = 1, \ldots, 6.$$

Q.E.D.

13.9 Elliptic Problems with λ-Dependent Boundary Conditions

For more details concerning the following elliptic problems with λ-dependent boundary conditions, the reader may refer to [50].

13.9.1 The Problem

Let Ω be an open bounded domain in $\mathbb{R}^n$ with a closure $\overline{\Omega}$ and a boundary $\partial\Omega$ of class C^∞, see [234, Section 1.7]. We refer the reader to [234, Section 2.1] for the following notions related to the theory of elliptic operators. Let us assume that the operator

$$A(x, \partial) := -\sum_{j,k=1}^{n} a_{jk}(x)\partial_j\partial_k + \sum_{j=1}^{n} a_j(x)\partial_j + a_0(x)$$

with $a_{jk} \in C(\overline{\Omega})$, $a_0, a_j \in L_\infty(\overline{\Omega})$, $j, k = 1, \ldots, n$ is properly elliptic in $\overline{\Omega}$. This means that, for any $x \in \overline{\Omega}$ and any linearly independent vectors ξ, $\eta \in \mathbb{R}^n$, the equation

$$A_0(x, \xi + t\eta) = 0 \tag{13.9.1}$$

has exactly one root t with a positive imaginary part. Here,

$$A_0(x, \xi) := \sum_{j,k=1}^{n} a_{jk}(x)\xi_j\xi_k \tag{13.9.2}$$

is the main symbol of the operator $A(x, \partial)$. The trace operator of the restriction of smooth functions in Ω to the boundary $\partial\Omega$ is denoted by γ. Let $B_j = \sum_{k=1}^{n} \gamma b_{jk}(x)\partial_k + \gamma b_{j0}(x)$, $j = 0, 1$, be two boundary operators with $b_{jk} \in C^1(\partial\Omega)$ and $b_{j0} \in C^2(\partial\Omega)$, such that B_1 is normal on $\partial\Omega$ and covers $A = A(x, \partial)$. That is, the problem $\{A, B_1\}$ is supposed to be regular elliptic in Ω. Let $m(x, y)$ be a bounded measurable function of $x \in \Omega$ and $y \in \partial\Omega$. We introduce an operator B, acting from $L_2(\partial\Omega)$ into $L_2(\Omega)$, via $(Bg)(x) := \int_{\partial\Omega} m(x, y)g(y)dy, \quad x \in \Omega$. Finally, let D be a properly elliptic operator on $\partial\Omega$ of second order with smooth coefficients. This means that, in the local coordinate system $\{y_1, y_2, \ldots, y_{n-1}\}$ of the point $y \in \partial\Omega$, the operator D is represented by

$$D(y) := -\sum_{j,k=1}^{n-1} d_{jk}(y)\partial_j\partial_k + \sum_{j=1}^{n-1} d_j(y)\partial_j + d_0(y),$$

where $\partial_j := \frac{\partial}{\partial y_j}$ and $d_{jk} \in C(\partial\Omega)$, $d_0, d_j \in L_\infty(\partial\Omega)$, $j, k = 1, \ldots, n-1$, and that a root condition which is analogous to (13.9.1) holds for D. A typical example of such a D is $-\Delta_{\partial\Omega}$, which represents the negative Laplace–Beltrami's operator on $\partial\Omega$. Now, let us consider the following boundary value of the spectral problem:

$$Au + B\gamma u = \lambda u \text{ in } \Omega \tag{13.9.3}$$

$$B_0 u + D\gamma u = \lambda B_1 u \text{ on } \partial\Omega. \tag{13.9.4}$$

The corresponding dynamic problem describes the motion of a Markovian particle which moves in Ω, according to a diffusion law, and possibly jumps at random times from $x \in \Omega$ into a set $\Gamma_0 \subset \partial\Omega$ (if $\int_{\Gamma_0} m(x, y)dy > 0$). After reaching the boundary (either by jump or diffusion), the particle can be reflected into Ω and can be absorbed in $\partial\Omega$, or also can move in $\partial\Omega$, this behavior on the boundary is governed by the terms in the boundary condition see [315]. We can easily notice that this problem (13.9.3) and (13.9.4) can be represented in the form $\mathcal{A}_0 u = \lambda u$ for the operator

$$\mathcal{A}_0 := \begin{pmatrix} A & B \\ C & D \end{pmatrix} \quad \text{in the Hilbert's space} \quad \mathcal{H} := L_2(\Omega) \oplus L_2(\partial\Omega)$$

with $C = B_0$, and that

$$\mathcal{D}(\mathcal{A}_0) := \left\{ \begin{pmatrix} u \\ g \end{pmatrix}, \text{ such that } u \in H^2(\Omega),\ g \in H^2(\partial\Omega) \text{ and } B_1 u = g \right\}.$$

The operator $\mathcal{A}_0$ takes the form of the previous sections with the following identification for the spaces and operators involved: $X = L_2(\Omega)$, $Y = L_2(\partial\Omega)$, $Z = H^{-3/2}(\partial\Omega)$, A, B, and D as above stated, with $\mathcal{D}(A) = H^2(\Omega)$ and $\mathcal{D}(D) = H^2(\partial\Omega)$, $C = B_0$, $\Gamma_X = B_1$, and Γ_Y being the natural embedding of $L_2(\partial\Omega)$ into $H^{-3/2}(\partial\Omega)$.

13.9.2 Verification of the Assumptions (J1)–(J8) of Chap. 10, Sect. 10.3

From the general theory of elliptic operators, we deduce that the operator A with a domain $H^2(\Omega)$ is closable in $X = L_2(\Omega)$, and that its closure $\overline{A}$ has a domain

$$\mathcal{D}(\overline{A}) := \{u \in X, \text{ such that } Au \in X\},$$

where, as usual, the same notation A is used for the operator in the distributional sense, see [140]. For any $u \in \mathcal{D}(\overline{A})$, $\Gamma_X u$ exists as an element of $Z = H^{-3/2}(\partial\Omega)$, and the mapping $\Gamma_X : \mathcal{D}(\overline{A}) \longrightarrow Z$ is bounded, see [140]. Obviously, the same is true for the operator C. As a result, C is closable as a mapping from $\mathcal{D}(\overline{A})$ into $Y = L_2(\partial\Omega)$. These arguments have been used in order to establish the properties $(J1)$, $(J2)$, and $(J3)$ in Chap. 10, Sect. 10.3. Since D is a uniformly elliptic operator on a compact manifold without boundary, then D is closed and also has a discrete spectrum. Hence, the property $(J6)$ of Chap. 10 Sect. 10.3 is satisfied (see [320, Section 5.1] for the case where D is the Laplacian on $\partial\Omega$). Then, the general case follows from the standard techniques. Moreover, the properties $(J7)$ and $(J8)$ are trivially satisfied here, as Γ_Y is the embedding of Y into Z noted above, and B is a bounded operator. It remains to verify $(J4)$ and $(J5)$. The first condition is a consequence of the following proposition.

Proposition 13.9.1 ([15, Theorem 2.1]). *Under the assumptions of Sect. 13.9.1, the operator A_1 in $L_2(\Omega)$, defined on the domain $\mathcal{D}(A_1) := \{u \in H^2(\Omega)$, such that $B_1 u = 0\}$ by $A_1 u := Au$, is closed and also has a discrete spectrum.* ◇

Now, let us assume that λ belongs to the resolvent set of A_1. It is known (see, e.g., [234, Section 2.7.3]) that, for any $g \in H^{1/2}(\partial\Omega)$, the problem $Au - \lambda u = 0$, $\Gamma_X u = g$ has a unique solution u which belongs to $H^2(\Omega)$. In particular, the operator K_λ, mapping $g \in H^{1/2}(\partial\Omega)$ to this solution $u \in H^2(\Omega)$, is well defined. Therefore, we can identify the subspace Z_1 of Z with $H^{1/2}(\partial\Omega)$. In order to verify $(J5)$, we show that K_λ extends to a bounded mapping $\overline{K}_\lambda$ from Z into X. In fact, an even stronger result holds.

Proposition 13.9.2 ([234, Section 2.7.3]). *With the assumptions of Sect. 13.9.1, let $\lambda \in \rho(A_1)$ and $l \geq 0$. Then, for any $g \in H^{l-3/2}(\partial\Omega)$, the problem $Aw - \lambda w = 0$, $B_1 w = g$ has a unique solution w. Moreover, w belongs to $H^l(\Omega)$ and the operator $\overline{K}_\lambda : H^{l-3/2}(\partial\Omega) \longrightarrow H^l(\Omega)$, defined by $\overline{K}_\lambda g = w$, is bounded.* ◇

For $l = 0$, the above proposition leads to the desired boundedness of $\overline{K}_\lambda$. By choosing $l = 3/2$, and recalling that the embedding of $H^{3/2}(\Omega)$ into $X = L_2(\Omega)$ is compact, we reach the following corollary.

Corollary 13.9.1. *$\overline{K}_\lambda$ is compact as an operator from $Y = L_2(\partial\Omega)$ into $X = L_2(\Omega)$.* ◇

The above properties of $\overline{K}_\lambda$ imply that, for any $y \in Y$, we have $\overline{K_\lambda \Gamma_Y} y = \overline{K}_\lambda y \in H^{3/2}(\Omega)$. In particular, we are allowed to write $\overline{K}_\lambda$ instead of $\overline{K_\lambda \Gamma_Y}$.

13.9.3 The Closure of the Operator $\mathcal{A}_0$

In order to describe the closure $\mathcal{A}$ of the operator $\mathcal{A}_0$, let us start with the following lemma.

Lemma 13.9.1. *For any $\lambda \in \rho(A_1)$, the operator $C\overline{K}_\lambda$ is bounded in $L_2(\partial\Omega)$.* ◇

Proof. Since, by hypothesis, B_1 is normal on $\partial\Omega$, then the vector field

$$b_1(x) = (b_{11}(x) \dots b_{1n}(x))$$

is never tangential on $\partial\Omega$. Hence, there exist a continuous function $\rho(x)$ as well as a vector field $b_t = (b_1 \dots b_n)$ which is tangential to $\partial\Omega$ for $x \in \partial\Omega$, such that

$$(b_{01}(x) \dots b_{0n}(x)) = \rho(x) b_1(x) + b_t(x).$$

Now, we have $Cu = \rho(x)\Gamma_X u + B_t u + b_0(x)\gamma u$, where $B_t = \sum_{k=1}^{n} \gamma b_k(x)\partial_k$, $b_0(x) = b_{00}(x) - \rho(x) b_{10}(x)$. Since b_t is tangential to $\partial\Omega$, then the operator B_t acts continuously from $H^1(\partial\Omega)$ into $L_2(\partial\Omega)$. Moreover, $\gamma\overline{K}_\lambda$ is continuous from $L_2(\partial\Omega)$ into $H^1(\partial\Omega)$ and hence, $B_t\overline{K}_\lambda = B_t\gamma\overline{K}_\lambda : L_2(\partial\Omega) \longrightarrow L_2(\partial\Omega)$ is continuous. Finally, we have

$$C\overline{K}_\lambda = \rho B_1 \overline{K}_\lambda + B_t \overline{K}_\lambda + b_0 \gamma \overline{K}_\lambda = \rho I + B_t \overline{K}_\lambda + b_0 \gamma \overline{K}_\lambda. \tag{13.9.5}$$

Q.E.D.

Corollary 13.9.2. *The operator $D+C\overline{K}_\lambda - C_\lambda B$ with a domain $\mathcal{D}(D) = H^2(\partial\Omega)$ is closed in Y and hence, coincides with $\overline{M}_\lambda$.* ◇

As a result, the operator $\mathcal{A}_0$ is closable. Now, we are ready to give an explicit description of its closure $\mathcal{A}$.

Theorem 13.9.1. *The operator $\mathcal{A}_0$ is closable in $\mathcal{H}$ and its closure $\mathcal{A}$ acts according to*

$$\mathcal{A}\begin{pmatrix} x + K_\lambda y - (A_1 - \lambda)^{-1}By \\ y \end{pmatrix} = \begin{pmatrix} Ax + \lambda K_\lambda y - \lambda(A_1 - \lambda)^{-1}By \\ Cx + \overline{M}_\lambda y \end{pmatrix}$$

on the domain

$$\mathcal{D}(\mathcal{A}) = \left\{ \begin{pmatrix} x + K_\lambda y - (A_1 - \lambda)^{-1}By \\ y \end{pmatrix}, \text{ such that } x \in \mathcal{D}(A_1) \text{ and } y \in \mathcal{D}(D) \right\}.$$

◇

13.9.4 Spectrum of the Operator $\mathcal{A}$

In this subsection, we will use the results of Chap. 10 Sect. 10.3 in order to study the spectrum of the operator $\mathcal{A}$. Recall that the operator A_1 has a discrete spectrum according to Proposition 13.9.1. If we fix $\lambda_0 \in \rho(A_1)$, then the resolvent $(A_1 - \lambda_0)^{-1}$ maps X into $H^2(\Omega)$ boundedly, and C maps $H^2(\Omega)$ into $L_2(\partial\Omega)$ compactly. Hence, C is A_1-compact. Theorem 10.3.2 implies that the essential spectra of $\mathcal{A}$ coincides with the essential spectra of $\overline{M}_{\lambda_0}$. Moreover, D has a discrete spectrum (see Sect. 13.9.2) and $\overline{M}_{\lambda_0} = D + C\overline{K}_{\lambda_0} - C_\lambda B$ represents a bounded perturbation of D. Hence, the essential spectra of $\overline{M}_{\lambda_0}$ is empty, and we have proved the following proposition

Proposition 13.9.3. *Under the assumptions of Sect. 13.9.1, the operator $\mathcal{A}$ has a discrete spectrum.* ◇

Under additional assumptions related to the operators A and D, more can be said about the distribution of the eigenvalues of $\mathcal{A}$. For example, let us assume that, for the main symbol $A_0(x,\xi)$ of A given by (13.9.2), there exists $\theta \in [0, 2\pi)$, such that

$$\arg A_0(x,\xi) \neq \theta \tag{13.9.6}$$

for all $x \in \overline{\Omega}$ and all vectors $\xi \in \mathbb{R}^n \backslash \{0\}$. Therefore, the problem $\{A - \lambda, B_1\}$ is elliptic with a parameter $\lambda = re^{i\theta}$, $r > 0$ (see [15], [16, Chapter I]). Besides, A_1 has the following property.

Proposition 13.9.4 ([15, Theorem 2.1]). *If (13.9.6) holds, then* $R_\theta := \{re^{i\theta}$ *such that* $r > 0\}$ *is a ray of minimal growth of the resolvent of* A_1, *i.e., there exist positive numbers* r_θ *and* c_θ *such that, for all* $r > r_\theta$, $\lambda = re^{i\theta} \in \rho(A_1)$, *and* $\|(A_1 - re^{i\theta})^{-1}\|_{L_2(\Omega)} \leq \frac{c_\theta}{r}$. ◇

Lemma 13.9.2. *If (13.9.6) holds, then* $C\overline{K}_\lambda$ *is uniformly bounded on the set* $R_\theta^0 := \{\lambda = re^{i\theta}$ *such that* $r \geq r_\theta\}$, *with* r_θ *as in Proposition 13.9.4.* ◇

Proof. First, let us notice that, by definition, R_θ^0 belongs to the resolvent set of A_1, so that K_λ is well defined for $\lambda \in R_\theta^0$. In view of relation (13.9.5) (recall the proof of Lemma 13.9.1), it is sufficient to prove that the mapping $K_\lambda : L_2(\partial\Omega) \longrightarrow H^{3/2}(\Omega)$ is uniformly bounded in $\lambda \in R_\theta^0$. So, let us assume that $g \in H^{l-3/2}(\partial\Omega)$ for some $l \geq 3/2$ and let $u = \overline{K}_\lambda g$ for $\lambda \in R_\theta^0$. Then, u solves the problem $Au - \lambda u = 0$, $B_1 u = g$, and [16, Theorem 4.1] implies that there exists a constant $C > 0$ such that the inequality $\|u\|_{H^l(\Omega)} + |\lambda|^l \|u\|_{L_2(\Omega)} \leq C\left(\|g\|_{H^{l-3/2}(\partial\Omega)} + |\lambda|^{l-3/2}\|g\|_{L_2(\partial\Omega)}\right)$ holds, for all $g \in H^{l-3/2}(\partial\Omega)$ and all $\lambda \in R_\theta^0$. Choosing $l = 3/2$ allows us to reach the desired conclusion. Q.E.D.

Now, let us suppose that, with the same value of $\theta \in [0, 2\pi)$ as in (13.9.6), we have

$$\arg D_0(y, \eta) \neq \theta, \tag{13.9.7}$$

for all $y \in \partial\Omega$ and all nonzero η in the tangent space $T_{\partial\Omega}(y)$ to $\partial\Omega$ at the point y. Therefore, the ray R_θ is also a ray of minimal growth for the resolvent set of D, and a suitable combination of the above results leads to the following theorem.

Theorem 13.9.2. *In addition to the assumptions at the beginning of Sect. 13.9.1, let us assume that the conditions (13.9.6) and (13.9.7) are both satisfied for the same* $\theta \in [0, 2\pi)$. *Then, all sufficiently large* $\lambda \in R_\theta$ *belong to the resolvent set of the operator* $\mathcal{A}$. ◇

Proof. In view of the Frobenius–Schur's factorization, it is sufficient to show that the operator $\overline{M}_\lambda - \lambda$ is boundedly invertible, for all sufficiently large $\lambda \in R_\theta^0$. We find that $\overline{M}_\lambda - \lambda = (D - \lambda)(I + (D - \lambda)^{-1}[C\overline{K}_\lambda - C_\lambda B])$. Since the ray R_θ is also a ray of minimal growth for the resolvent set of D, and since the operators $C\overline{K}_\lambda$ and $C_\lambda B$ are bounded in Y uniformly in $\lambda \in R_\theta^0$, according to Lemma 13.9.2 and Inequality (10.3.18), the operator $I + (D - \lambda)^{-1}[C\overline{K}_\lambda - C_\lambda B]$ is then boundedly invertible for all sufficiently large $\lambda \in \rho(D) \bigcap R_\theta^0$ and hence, the same reasoning holds for $\overline{M}_\lambda - \lambda$. Q.E.D.

13.9.5 Semigroup Generation

Now, let us suppose that both A and D generate holomorphic operator semigroups in X and Y, respectively. A sufficient (and in fact necessary) condition for this is that the main symbols $A_0(x, \xi)$ and $D_0(y, \eta)$ of A and D are sectorial, i.e., there exists a $\theta_0 \in (0, \pi/2)$ such that $|\arg A_0(x, \xi)| < \theta_0$, $|\arg D_0(y, \eta)| \leq \theta_0$ for all $x \in \Omega$, $y \in \partial\Omega$, and all nonzero $\xi \in \mathbb{R}^n$ and $\eta \in T_{\partial\Omega}(y)$, see [238, Theorem 3.1.3]. In fact, under this condition, any ray R_θ with $|\theta| \leq \pi - \theta_0$ constitutes a ray of minimal growth for the resolvents of A_1 and D and hence, A_1 and D generate holomorphic operator semigroups in X and Y, respectively. Then, $\overline{M}_\lambda$, $\lambda \in \rho(A_1)$, is also a generator of a holomorphic semigroup in Y. Applying Theorem 10.3.5 and Proposition 10.3.1 allows us to reach the following theorem.

Theorem 13.9.3. *In addition to the assumptions at the beginning of Sect. 13.9.1, let us assume that A_1 and D generate holomorphic operator semigroups in X and Y, respectively. Then, $\mathcal{A}$ is the generator of a holomorphic semigroup in $\mathcal{H}$.* ◇

This result is also deduced from the observation that, under the assumptions of Theorem 13.9.2, all R_θ with $|\theta| < \pi - \theta_0$ actually constitute some rays of minimal growth for the resolvent of $\mathcal{A}$.

13.10 Delay Differential Equations

Partial differential equations with delay have been studied using several methods. In an abstract way, and using the standard notation (see [178]), they can be written as:

$$\begin{cases} u'(t) = Cu_t + Du(t), & t \geq 0 \\ u_0 = x, \\ u(0) = y, \end{cases}$$

where

- $y \in Y$, Y being a Banach space,
- $D : \mathcal{D}(D) \subseteq Y \longrightarrow Y$ is a linear, closed, and densely defined operator which generates a strongly continuous semigroup,
- $x \in L_p([-1, 0], Y) := X$, $p \geq 1$,
- $C : W^{1,p}([-1, 0], Y) \longrightarrow Y$ is a linear and bounded operator,
- $u : [-1, \infty) \longrightarrow Y$ and $u_t : [-1, 0] \longrightarrow Y$ is defined by $u_t(\sigma) := u(t + \sigma)$.

It is well known that this problem is equivalent to an abstract Cauchy problem in $\mathcal{H} := X \times Y$ with the vector function

$$v(t) := \begin{pmatrix} u_t \\ u(t) \end{pmatrix}$$

$$\begin{cases} v'(t) = \mathcal{A}v(t), & t \geq 0, \\ v(0) = \begin{pmatrix} x \\ y \end{pmatrix}, \end{cases}$$

where

$$\mathcal{A} := \begin{pmatrix} \dfrac{d}{ds} & 0 \\ C & D \end{pmatrix}$$

is defined on the domain

$$\mathcal{D}(\mathcal{A}) := \left\{ \begin{pmatrix} x \\ y \end{pmatrix}, \text{ such that } x \in W^{1,p}([-1,0], Y),\ y \in \mathcal{D}(D) \text{ and } x(0) = y \right\}.$$

Now, it is easy to check that this operator is of the matrix form of Chap. 10, Sect. 10.3 with $Z := Z_1 := Y, \Gamma_Y := I, \Gamma_X x := x(0)$ with $\mathcal{D}(\Gamma_X) := W^{1,p}([-1,0], Y) \subset X, A = \dfrac{d}{ds}$ with $\mathcal{D}(A) = W^{1,p}([-1,0], Y)$ and $B = 0$. The operator $A_1 = \dfrac{d}{ds}$ with the domain $\mathcal{D}(A_1)$ given as follows $\mathcal{D}(A_1) = \{x \in W^{1,p}([-1,0], Y),$ such that $x(0) = 0\}$, is the generator of the left shift semigroup

$$(T(t)y)(s) := \begin{cases} y(t+s), & t+s < 0, \\ 0, & t+s \geq 0, \end{cases}$$

which is nilpotent. For $\lambda \in \mathbb{C}$, we get $N(A - \lambda) = \{e^{\lambda} y \text{ such that } y \in Y\}$. In the applications, the operator C often has such a representation

$$Cx := \int_{-1}^{0} d\eta(s) x(s), \tag{13.10.1}$$

where $\eta(.) \in BV([-1,0], \mathcal{L}(Y))$ is a given function. Then, $K_\lambda y = e^{\lambda \cdot} y$, and $CK_\lambda y = \int_{-1}^{0} e^{\lambda s} d\eta(s) y$. Now, and under these assumptions, we can verify the conditions $(J1)$–$(J8)$ of Chap. 10, Sect. 10.3. Since A is closed and $\mathcal{D}(A) = \mathcal{D}(\Gamma_X)$, the assumptions $(J1)$ and $(J2)$ are satisfied. Assumption $(J3)$ follows directly from the fact that A_1 generates a strongly continuous semigroup. Since C is of the form (13.10.1), $(J4)$ follows and $(J5)$ is a consequence of $K_\lambda y = e^{\lambda \cdot} y$. Assumption $(J6)$ was made earlier, and $(J7)$ and $(J8)$ follow trivially. Moreover, in this case, the operator $\mathcal{A}$ is already closed. For delay differential equations, the imposed conditions on the operator C_λ in Theorem 10.3.2 are satisfied if, and only if, Y is finite-dimensional (recall that, in this case, $\sigma_{ei}(D) = \emptyset$, $i = 1, \ldots, 6$). So, $\sigma_{ei}(\mathcal{A}) = \emptyset$, with $i = 1, \ldots, 6$.

13.11 A λ-Rational Sturm–Liouville's Problem

Let $p, u \in L_1(0,1)$ be real, $q \in L_2(0,1)$ be real, and let σ be a bounded nonnegative measure on $\mathbb{R}$, and $(\alpha_j, \beta_j) \in \mathbb{R}^2$, such that $\alpha_j^2 + \beta_j^2 \neq 0$, $j = 0, 1$. We will consider the spectral problem

$$-f'' + pf + \frac{qf}{u-\lambda} - \lambda f = 0 \text{ on } [0,1], \tag{13.11.1}$$

$$b_1(f) + N(\lambda) b_0(f) = 0, \quad f(1) = 0, \tag{13.11.2}$$

where $N(\lambda)$ is the Nevanlinna's function $N(\lambda) := \int_{\mathbb{R}} \frac{d\sigma(t)}{t-\lambda}$ and $b_j(f) := \alpha_j f'(0) + \beta_j f(0)$, $j = 0, 1$. These equations make sense, at least, for $\lambda \in \mathbb{C} \backslash \mathbb{S}$, where $\mathbb{S} := \overline{u([0,1])} \bigcup \operatorname{supp} \sigma$. If $q = 0$ then we can take $\mathbb{S} = \operatorname{supp} \sigma$ and also the following considerations can be simplified.

Definition 13.11.1. A point $\lambda \in \mathbb{C} \backslash \mathbb{S}$ is called an eigenvalue of the problem (13.11.1), (13.11.2), if there exists a nonzero function f, such that f' is absolutely continuous and these equations are satisfied. ◇

The problems (13.11.1) and (13.11.2) are equivalent to the following operator matrix. Set $X = L_2(0,1)$, $Y = L_2(0,1) \times \mathcal{H}$, where $\mathcal{H}$ is the Hilbert space $L_2(\sigma, \mathbb{R})$ of all functions g on $\mathbb{R}$, with $\|g\|^2 = \int_{\mathbb{R}} |g(t)|^2 d\sigma(t) < \infty$, and $Z = \mathbb{C}$. Let us define the operators A, B, C, D, Γ_X and Γ_Y as follows: $Af = -f'' + pf$ on $\mathcal{D}(A) = \{f \in W^{2,2}[0,1] \text{ such that } f(1) = 0\}$,

$$B\begin{pmatrix} h \\ g \end{pmatrix} = qh, \quad Cf = \begin{pmatrix} -f \\ b_0(f) \end{pmatrix}, \quad D\begin{pmatrix} h \\ g \end{pmatrix} = \begin{pmatrix} Uh \\ Tg \end{pmatrix},$$

$$\Gamma_X f = b_1(f), \quad \Gamma_Y \begin{pmatrix} h \\ g \end{pmatrix} = \int_{\mathbb{R}} g(t) d\sigma(t),$$

where $(Uh)(t) = u(t)h(t)$ a.e. in $[0,1]$ and $(Tg)(t) = tg(t)$ σ-a.e. on $\mathbb{R}$. The operator D is defined on all vectors $\begin{pmatrix} h \\ g \end{pmatrix}$, $h \in L_2(0,1)$, $g \in \mathcal{H}$, such that $Uh \in L_2(0,1)$ and $Tg \in \mathcal{H}$.

We will show that the assumptions $(J1)$–$(J8)$ of Chap. 10, Sect. 10.3 are satisfied. In fact, it is well known that the operator A is densely defined and closed (so $(J1)$ holds), and the functionals b_0 and b_1 are defined and continuous on X_A, which implies $(J2)$. The operator A_1 is the (self-adjoint) restriction of A by the boundary condition $b_1(f) = 0$. Hence, $(J3)$ is satisfied, and since the embedding of X_A into $L_2(0,1)$ is continuous and b_0 is continuous on X_A, C is continuous as an operator from X_A into Y and then, $(J4)$ holds. The operator K_λ is defined, at least, for nonreal λ and, for $z \in \mathbb{C}$, $K_\lambda z$ represents the solution of the boundary value problem $-f'' + pf = \lambda f$, $b_1(f) = z$, $f(1) = 0$, which depends continuously on z with respect to the norm of $L_2(0,1)$ and, then $(J5)$ holds. Obviously, D is self-adjoint in Y and hence, $(J6)$ holds. The estimate

$$\left|\int_{\mathbb{R}} g(t)d\sigma(t)\right|^2 \leq \int_{\mathbb{R}} |g(t)|^2 d\sigma(t) \int_{\mathbb{R}} d\sigma(t)$$

implies that the condition $(J7)$ is satisfied. Finally, since the operator $(A_1-\lambda)^{-1}$ acts boundedly from $L_1(0,1)$ into $L_2(0,1)$, the assumption $q \in L_2(0,1)$ implies the boundedness of the operator $(A_1-\lambda)^{-1}B$ and hence, $(J8)$ holds for a nonreal λ.

If $f \in \mathcal{D}(A)$ and $\begin{pmatrix} h \\ g \end{pmatrix} \in \mathcal{D}(D)\bigcap\mathcal{D}(B)$, then

$$\mathcal{A}_0\begin{pmatrix} f \\ h \\ g \end{pmatrix} = \begin{pmatrix} -f''+pf+qh \\ \begin{pmatrix} -f \\ b_0(f) \end{pmatrix} + \begin{pmatrix} Uh \\ Tg \end{pmatrix} \end{pmatrix}, \qquad f(1)=0,$$

and the matrix equation $\mathcal{A}_0\begin{pmatrix} f \\ h \\ g \end{pmatrix} = \lambda\begin{pmatrix} f \\ h \\ g \end{pmatrix}$ becomes

$$-f''+pf+qh=\lambda f, \quad f(1)=0, \quad -f+uh=\lambda h, \quad b_0(f)+Tg=\lambda g. \tag{13.11.3}$$

Finally, $\Gamma_X f = \Gamma_Y\begin{pmatrix} h \\ g \end{pmatrix}$ yields

$$b_1(f)=\int_{\mathbb{R}} g(t)d\sigma(t). \tag{13.11.4}$$

By solving the last two equations in (13.11.3) for h and g, we get $h=\frac{f}{u-\lambda}$, $g(t)=-\frac{b_0(f)}{t-\lambda}$, and the first two equations in (13.11.3), combined with (13.11.4), give

$$-f''+pf+q\frac{f}{u-\lambda}=\lambda f, \quad f(1)=0, \quad b_1(f)+N(\lambda)b_0(f)=0,$$

and these are exactly the already found relations (13.11.1) and (13.11.2). Now, the operator $\mathcal{A}$ can be defined as in Chap. 10, Sect. 10.3.1. Since A_1 has a compact resolvent, C is A_1-compact and $\overline{K}_\lambda$, as well as $(A_1-\lambda)^{-1}B$ are also compact operators. Therefore, the assumptions of Theorem 10.3.3 are satisfied. Since the spectrum of A_1 is discrete, then the essential spectra of $\mathcal{A}$ coincide with the essential spectra of $\overline{M}_\mu$ (and therefore of D). Hence, we get $\sigma_{ei}(\mathcal{A})=\sigma_{ei}(U)\bigcup\sigma_{ei}(T)$, $i=1,\ldots,6$.

13.12 Two-Group Radiative Transfer Equations in a Channel

13.12.1 Functional Setting

Let $X := L_1(D, dxdv)$, where $D = (0, 1) \times K$ with K being the unit sphere of $\mathbb{R}^3$, $x \in (0, 1)$ and $v = (v_1, v_2, v_3) \in K$. Let us define the following sets representing the incoming D^i and the outgoing D^0 boundary of the phase space D:

$$\begin{aligned} D^i &= D_1^i \bigcup D_2^i = \{0\} \times K^1 \bigcup \{1\} \times K^0, \\ D^0 &= D_1^0 \bigcup D_2^0 = \{0\} \times K^0 \bigcup \{1\} \times K^1, \end{aligned}$$

for $K^0 = K \bigcap \{v_3 < 0\}$ and $K^1 = K \bigcap \{v_3 > 0\}$. Moreover, we will denote the boundary spaces X^i and X^0 by the following ways:

$$\begin{aligned} X^i &:= L_1(D^i, |v_3|dv) \\ &:= L_1(D_1^i, |v_3|dv) \oplus L_1(D_2^i, |v_3|dv) \\ &:= X_1^i \oplus X_2^i \end{aligned}$$

endowed with the norm:

$$\begin{aligned} \|\psi^i, X^i\| &= \|\psi_1^i, X_1^i\| + \|\psi_2^i, X_2^i\| \\ &= \int_{K^1} |\psi(0, v)||v_3|\, dv + \int_{K^0} |\psi(1, v)||v_3|\, dv \end{aligned}$$

and,

$$\begin{aligned} X^0 &:= L_1(D^0, |v_3|dv) \\ &:= L_1(D_1^0, |v_3|dv) \oplus L_1(D_2^0, |v_3|dv) \\ &:= X_1^0 \oplus X_2^0 \end{aligned}$$

equipped with the norm:

$$\begin{aligned} \|\psi^0, X^0\| &= \|\psi_1^0, X_1^0\| + \|\psi_2^0, X_2^0\| \\ &= \int_{K^0} |\psi(0, v)||v_3|\, dv + \int_{K^1} |\psi(1, v)||v_3|\, dv. \end{aligned}$$

Let us denote by $\mathcal{W}$ the space defined by $\mathcal{W} = \left\{\psi \in X \text{ such that } v_3\frac{\partial\psi}{\partial x} \in X\right\}$. It is well known that any function $\psi \in \mathcal{W}$ possesses traces (see [89]) on the spatial boundary denoted by $\psi^0 = (\psi_1^0, \psi_2^0)^T$, and $\psi^i = (\psi_1^i, \psi_2^i)^T$ given by:

$$\begin{cases} \psi_1^i(v) = \psi(0, v), & v \in K^1 \\ \psi_2^i(v) = \psi(1, v), & v \in K^0 \\ \psi_1^0(v) = \psi(0, v), & v \in K^0 \\ \psi_2^0(v) = \psi(1, v), & v \in K^1. \end{cases}$$

We consider the following two-group transport operators $\mathcal{A} = \mathcal{T} + \mathcal{K}$ with

$$\mathcal{T}\psi := \begin{pmatrix} -v_3\frac{\partial\psi_1}{\partial x}(x,v) - \sigma_1(x,v)\psi_1(x,v) & 0 \\ 0 & -v_3\frac{\partial\psi_2}{\partial x}(x,v) - \sigma_2(x,v)\psi_2(x,v) \end{pmatrix}$$

$$= \begin{pmatrix} T_1 & 0 \\ 0 & T_2^H \end{pmatrix}\begin{pmatrix} \psi_1 \\ \psi_2 \end{pmatrix}$$

and $\mathcal{K} = \begin{pmatrix} 0 & K_{12} \\ K_{21} & K_{22} \end{pmatrix}$, where $x \in [0,1], v = (v_1, v_2, v_3) \in K$ and $K_{ij}, (i,j) \in \{(1,2),(2,1),(2,2)\}$ are bounded linear operators on X by

$$\begin{cases} K_{ij} : X \longrightarrow X \\ \qquad \psi \longrightarrow K_{ij}\,\psi(x,v) = \displaystyle\int_K \kappa_{ij}(x,v,v')\,\psi(x,v')\,dv', \end{cases}$$

and the kernels $\kappa_{ij} : (0,1) \times K \times K \longrightarrow \mathbb{R}$ are assumed to be measurable. Now, let us introduce the boundary operator H,

$$\begin{cases} H : X^0 \longrightarrow X^i \\ H\begin{pmatrix} u_1 \\ u_2 \end{pmatrix} = \begin{pmatrix} H_{11} & H_{12} \\ H_{21} & H_{22} \end{pmatrix}\begin{pmatrix} u_1 \\ u_2 \end{pmatrix} \end{cases}$$

with for $j,k \in \{1,2\}$, $H_{jk} : X_k^0 \longrightarrow X_j^i$, $H_{jk} \in \mathcal{L}(X_k^0, X_j^i)$, defined such that, on natural identification, the boundary conditions can be written as $\psi^i = H(\psi^0)$. The operators T_1 and T_2^H are defined by:

$$\begin{cases} T_1 : \mathcal{D}(T_1) \subseteq X \longrightarrow X \\ \qquad \varphi \longrightarrow T_1\varphi(x,v) = -v_3\frac{\partial\varphi}{\partial x}(x,v) - \sigma_1(x,v)\varphi(x,v) \\ \mathcal{D}(T_1) = \mathcal{W} \end{cases}$$

and

$$\begin{cases} T_2^H : \mathcal{D}(T_2^H) \subseteq X \longrightarrow X \\ \qquad\qquad \psi \longrightarrow T_2^H \psi(x,v) = -v_3 \dfrac{\partial \psi}{\partial x}(x,v) - \sigma_2(x,v)\psi(x,v) \\ \mathcal{D}(T_2^H) = \Big\{ \psi \in \mathcal{W} \text{ such that } \psi_{|D^i} := \psi^i \in X^i, \psi_{|D^0} := \psi^0 \in X^0 \text{ and } \psi^i = H(\psi^0) \Big\}, \end{cases}$$

where $\sigma_j(.,.)$, $j = 1,2$ is a positive bounded function. It is clear that the operator $\mathcal{T}$ is defined on $\mathcal{W} \times \mathcal{D}(T_2^H)$. Next, we will define $\mathcal{A}$ by

$$\mathcal{D}(\mathcal{A}) := \left\{ \begin{pmatrix} \psi_1 \\ \psi_2 \end{pmatrix} \in \mathcal{W} \times \mathcal{D}(T_2^H) \text{ such that } \psi_1^i = \psi_2^i \right\}$$

and

$$\mathcal{A} := \begin{pmatrix} A & B \\ C & D \end{pmatrix},$$

where

$$\begin{cases} A = \quad T_1 \\ B = \quad K_{12} \\ C = \quad K_{21} \\ D = T_2^H + K_{22}. \end{cases}$$

Now, it is easy to check that this operator is of the matrix form of Chap. 10, Sect. 10.3 with $Z := Z_1 := X^i$, $X = Y := L_1(D, dxdv)$,

$$\begin{cases} \Gamma_X : \mathcal{W} \longrightarrow X^i \\ \qquad \varphi \longrightarrow \Gamma_X \varphi = \varphi^i, \\ \text{with } \mathcal{D}(\Gamma_X) = \mathcal{D}(T_1) \end{cases}$$

and

$$\begin{cases} \Gamma_Y : X \longrightarrow X^i \\ \qquad \psi \longrightarrow \Gamma_Y \psi = H\psi^0. \end{cases}$$

We define the operator A_1 by:

$$\begin{cases} A_1\psi(x,v) := T_1\psi(x,v) := -v_3 \dfrac{\partial \psi}{\partial x}(x,v) - \sigma_1(x,v)\psi(x,v) \\ \mathcal{D}(A_1) = \{\psi \in \mathcal{D}(T_1) \text{ such that } \psi^i = 0\}. \end{cases}$$

Remark 13.12.1. It is well known that the operators T_1 and T_2^H are closed linear operators. Moreover, the derivative of ψ in the definition of T_1 and T_2^H is meant in distributional sense. Note that, if $\psi \in \mathcal{D}(T_1)$ [resp. $\psi \in \mathcal{D}(T_2^H)$], then it is absolutely continuous with respect to x. Hence, the restrictions of ψ to D^i and D^0 are meaningful. Note also that $\mathcal{D}(T_1)$ [resp. $\mathcal{D}(T_2^H)$] is dense in X because it contains $C_0^\infty(D^0)$. ◇

13.12.2 The Expression of the Resolvent of the Operator T_2^H

Now, we may determine the expression of the resolvent of the operator T_2^H. Let $\varphi \in X$, $\lambda \in \mathbb{C}$ and let us consider the problem

$$(\lambda - T_2^H)\psi = \varphi, \tag{13.12.1}$$

where the unknown ψ must be sought in $\mathcal{D}(T_2^H)$. Let λ_j^*, $j = 1, 2$ denote the real number defined by $\lambda_j^* := \text{ess-}\inf\{\sigma_j(x, v) \text{ such that } (x, v) \in D\}$, $j = 1, 2$, and

$$\lambda_0 := \begin{cases} -\lambda_2^* & \text{if } \|H\| \leq 1 \\ -\lambda_2^* + \log(\|H\|) & \text{if } \|H\| > 1. \end{cases}$$

Thus, for $\operatorname{Re}\lambda > -\lambda_2^*$, the solution of (13.12.1) is formally given by

$$\psi(x, v) = \psi(0, v)\, e^{-\int_0^x \frac{\sigma_2(s,v)+\lambda}{|v_3|} ds} + \frac{1}{|v_3|} \int_0^x e^{-\int_{x'}^x \frac{\sigma_2(s,v)+\lambda}{|v_3|} ds} \varphi(x', v)dx', \quad v \in K^1 \tag{13.12.2}$$

$$\psi(x, v) = \psi(1, v)e^{-\int_x^1 \frac{\sigma_2(s,v)+\lambda}{|v_3|} ds} + \frac{1}{|v_3|} \int_x^1 e^{-\int_x^{x'} \frac{\sigma_2(s,v)+\lambda}{|v_3|} ds} \varphi(x', v)dx', \quad v \in K^0 \tag{13.12.3}$$

whereas $\psi(1, v)$ and $\psi(0, v)$ are given by

$$\psi(1, v) = \psi(0, v)e^{-\int_0^1 \frac{\sigma_2(s,v)+\lambda}{|v_3|} ds} + \frac{1}{|v_3|} \int_0^1 e^{-\int_{x'}^1 \frac{\sigma_2(s,v)+\lambda}{|v_3|} ds} \varphi(x', v)dx', \quad v \in K^1 \tag{13.12.4}$$

$$\psi(0, v) = \psi(1, v)e^{-\int_0^1 \frac{\sigma_2(s,v)+\lambda}{|v_3|} ds} + \frac{1}{|v_3|} \int_0^1 e^{-\int_0^{x'} \frac{\sigma_2(s,v)+\lambda}{|v_3|} ds} \varphi(x', v)dx', \quad v \in K^0. \tag{13.12.5}$$

In order to clarify our subsequent analysis, we introduce the following bounded operators depending on the parameter λ,

$$\begin{cases} N_\lambda : X^i \longrightarrow X^0, N_\lambda u := (N_\lambda^+ u, N_\lambda^- u) & \text{with} \\ (N_\lambda^+ u)(0, v) := u(1, v)\, e^{-\int_0^1 \frac{\sigma_2(s,v)+\lambda}{|v_3|}\, ds}, & v \in K^0, \\ (N_\lambda^- u)(1, v) := u(0, v)\, e^{-\int_0^1 \frac{\sigma_2(s,v)+\lambda}{|v_3|}\, ds}, & v \in K^1, \end{cases}$$

$$\begin{cases} B_\lambda : X^i \longrightarrow X, B_\lambda u := \chi_{K^0}(v) B_\lambda^+ u + \chi_{K^1}(v) B_\lambda^- u \text{ with} \\ (B_\lambda^- u)(x, v) := u(0, v)\, e^{-\int_0^x \frac{\sigma_2(s,v)+\lambda}{|v_3|}\, ds}, & v \in K^1, \\ (B_\lambda^+ u)(x, v) := u(1, v)\, e^{-\int_x^1 \frac{\sigma_2(s,v)+\lambda}{|v_3|}\, ds}, & v \in K^0, \end{cases}$$

$$\begin{cases} G_\lambda : X \longrightarrow X^0, G_\lambda \varphi := (G_\lambda^+ \varphi, G_\lambda^- \varphi) \quad \text{with} \\ G_\lambda^- \varphi := \dfrac{1}{|v_3|} \displaystyle\int_0^1 e^{-\int_x^1 \frac{\sigma_2(s,v)+\lambda}{|v_3|}\, ds} \varphi(x, v)\, dx, & v \in K^1, \\ G_\lambda^+ \varphi := \dfrac{1}{|v_3|} \displaystyle\int_0^1 e^{-\int_0^x \frac{\sigma_2(s,v)+\lambda}{|v_3|}\, ds} \varphi(x, v)\, dx, & v \in K^0, \end{cases}$$

and,

$$\begin{cases} F_\lambda : X \longrightarrow X, F_\lambda \varphi := \chi_{K^0}(v) F_\lambda^+ \varphi + \chi_{K^1}(v) F_\lambda^- \varphi \text{ with} \\ F_\lambda^- \varphi := \dfrac{1}{|v_3|} \displaystyle\int_0^x e^{-\int_{x'}^x \frac{\sigma_2(s,v)+\lambda}{|v_3|}\, ds} \varphi(x', v)\, dx', & v \in K^1, \\ F_\lambda^+ \varphi := \dfrac{1}{|v_3|} \displaystyle\int_x^1 e^{-\int_x^{x'} \frac{\sigma_2(s,v)+\lambda}{|v_3|}\, ds} \varphi(x', v)\, dx', & v \in K^0, \end{cases}$$

where $\chi_{K^0}(.)$ and $\chi_{K^1}(.)$ denote, respectively, the characteristic functions of the sets K^0 and K^1. The operators $N_\lambda, B_\lambda, G_\lambda$, and F_λ are bounded on their respective spaces. Their norms are bounded above, respectively, by $e^{-(\mathrm{Re}\lambda+\lambda_2^*)}$, $(\lambda_2^* + \mathrm{Re}\lambda)^{-1}$, $(\lambda_2^* + \mathrm{Re}\lambda)^{-1}$ and $(\mathrm{Re}\lambda + \lambda_2^*)^{-1}$. The fact that ψ must satisfy the boundary conditions, Eqs. (13.12.4) and (13.12.5) can be written in the space X^0 in the operator form $\psi^0 = N_\lambda H \psi^0 + G_\lambda \varphi$, and $(I - N_\lambda H)\psi^0 = G_\lambda \varphi$. If $\mathrm{Re}\lambda > \lambda_0$, the solution of the last equation is reduced to the following form:

$$\psi^0 = \sum_{n \geq 0} (N_\lambda H)^n G_\lambda \varphi. \tag{13.12.6}$$

Moreover, Eqs. (13.12.2) and (13.12.3) can be rewritten as

$$\psi = B_\lambda H \psi^0 + F_\lambda \varphi. \tag{13.12.7}$$

Substituting (13.12.6) into (13.12.7), we get $\psi = \sum_{n \geq 0} B_\lambda H (N_\lambda H)^n G_\lambda \varphi + F_\lambda \varphi$.

Hence

$$(\lambda - T_2^H)^{-1} = \sum_{n\geq 0} B_\lambda H(N_\lambda H)^n G_\lambda + F_\lambda. \tag{13.12.8}$$

13.12.3 Essential Spectra

Let us observe that the operator F_λ is, nothing else but, $(\lambda - T_2^0)^{-1}$ where T_2^0 designates the operator T_2^H with a boundary condition $H = 0$. Let $\lambda \in \mathbb{C}$ such that $\mathrm{Re}\lambda > \lambda_0$ then, $\lambda \in \rho(T_2^H)\bigcap \rho(T_2^0)$. From Eq. (13.12.8), we have

$$(\lambda - T_2^H)^{-1} - (\lambda - T_2^0)^{-1} = \sum_{n\geq 0} B_\lambda H(N_\lambda H)^n G_\lambda. \tag{13.12.9}$$

It is easy to notice that

$$\sigma(T_2^0) = \sigma_c(T_2^0) = \{\lambda \in \mathbb{C} \text{ such that } \mathrm{Re}\lambda \leq -\lambda_2^*\}. \tag{13.12.10}$$

According to Remark 7.1.1, together with Eq. (13.12.10), we get the following:

$$\sigma_{ei}(T_2^0) = \{\lambda \in \mathbb{C} \text{ such that } \mathrm{Re}\lambda \leq -\lambda_2^*\}, i = 1, \ldots, 6.$$

In addition, it is well known that the spectrum of the operator A_1 is reduced to a continuous spectrum. More precisely, we have

$$\sigma_{ei}(A_1) = \sigma_c(A_1) = \{\lambda \in \mathbb{C} \text{ such that } \mathrm{Re}\lambda \leq -\lambda_1^*\}, i = 1, \ldots, 6 \tag{13.12.11}$$

Since H is weakly compact, then from Eq. (13.12.9), it follows that the operator $(\lambda - T_2^H)^{-1} - (\lambda - T_2^0)^{-1}$ is weakly compact. Therefore, by using Theorem 7.5.4 and Proposition 13.1.1, we get

$$\sigma_{ei}(T_2^H) = \{\lambda \in \mathbb{C} \text{ such that } \mathrm{Re}\lambda \leq -\lambda_2^*\}, i = 1, \ldots, 5. \tag{13.12.12}$$

Also, the fact that $\mathbb{C}\backslash\sigma_{e5}(T_2^H)$ is connected and $\rho(T_2^H) \neq \emptyset$ yields $\sigma_{e5}(T_2^H) = \sigma_{e6}(T_2^H)$.

13.12.3.1 Verification of the Assumptions $(J1)$–$(J8)$ of Chap. 10, Sect. 10.3

The conditions $(J1)$–$(J2)$ are satisfied since A is closed with $\mathcal{D}(A) = \mathcal{D}(\Gamma_X)$. From the general theory of $C_0^\infty(D^0)$, we deduce that the operator A_1 satisfies $(J3)$. The result defined by assumption $(J4)$ follows from the fact that K_{21} is bounded.

In order to verify the hypothesis $(J5)$, we will determine the expression of the solution of the equation $(\lambda - A)\psi = 0$, where the unknown ψ must be sought in $\mathcal{D}(A)$. Thus, for each $\mathrm{Re}\lambda > -\lambda_1^*$, the solution is given by

$$\psi(x,v) = \begin{cases} \psi(0,v)\, e^{-\int_0^x \frac{\sigma_1(s,v)+\lambda}{|v_3|}ds} & v \in K^1 \\ \psi(1,v) e^{-\int_x^1 \frac{\sigma_1(s,v)+\lambda}{|v_3|}ds} & v \in K^0. \end{cases}$$

Hence, the operator K_λ is defined on X^i by

$$\begin{cases} K_\lambda : X^i \longrightarrow X, K_\lambda u := \chi_{K^0} K_\lambda^+ u + \chi_{K^1} K_\lambda^- u \quad \text{with} \\ (K_\lambda^- u)(x,v) := u(0,v)\, e^{-\int_0^x \frac{\sigma_1(s,v)+\lambda}{|v_3|}ds}, \quad v \in K^1 \\ (K_\lambda^+ u)(x,v) := u(1,v) e^{-\int_x^1 \frac{\sigma_1(s,v)+\lambda}{|v_3|}ds}, \quad v \in K^0 \end{cases}$$

which is bounded by $(\mathrm{Re}\lambda + \lambda_1^*)^{-1}$. Finally, $(J6)$–$(J8)$ can be easily verified. Let us notice that the collision operator K_{ij}, $(i,j) \in \{(1,2),(2,1),(2,2)\}$ acts only on the variables v', so x may be viewed merely as a parameter in $[0,1]$. Then, we will consider K_{ij} as a function $K_{ij}(.) : x \in [0,1] \longrightarrow K_{ij}(x) \in \mathcal{L}(L_1(K,dv))$. The class of regular operators introduced in Definition 2.4.3 satisfies the following approximate property:

Lemma 13.12.1. *We assume that the collision operator K_{21} is nonnegative, with its kernel $\kappa_{21}(.,.,.)$ satisfying that $\left\{\frac{\kappa_{21}(x,.,v')}{|v_3'|} \text{ such that } (x,v') \in (0,1)\times K\right\}$ is a relatively weak compact subset of $L_1(K,dv)$. Then, for any $\lambda \in \mathbb{C}$ satisfying $\mathrm{Re}\lambda > -\lambda_1^*$, $K_{21}(\lambda - A_1)^{-1}$ is a weakly compact operator on X.* ◇

Proof. Let $\lambda \in \mathbb{C}$ be such that $\mathrm{Re}\lambda > -\lambda_1^*$. We have $K_{21}(\lambda - A_1)^{-1} = K_{21}F_\lambda$. It is sufficient to show that $K_{21}F_\lambda$ is weakly compact on X. Let $\varphi \in X$.

$$\begin{aligned}
&(K_{21}F_\lambda\varphi)(x,v) \\
&= \int_K K_{21}(x,v,v')(F_\lambda\varphi)(x,v')dv' \\
&= \int_{K^0} K_{21}(x,v,v')F_\lambda^+\varphi(x,v')dv' + \int_{K^1} K_{21}(x,v,v')F_\lambda^-\varphi(x,v')dv' \\
&= \int_{K^0} \frac{\kappa_{21}(x,v,v')}{|v_3'|}\int_x^1 e^{-\int_x^{x'} \frac{\sigma_1(s,v')+\lambda}{|v_3'|}ds}\varphi(x',v')dx'dv' \\
&\quad + \int_{K^1} \frac{\kappa_{21}(x,v,v')}{|v_3'|}\int_0^x e^{-\int_{x'}^{x} \frac{\sigma_1(s,v')+\lambda}{|v_3'|}ds}\varphi(x',v')dx'dv' \\
&= K_{21}'\tilde{F}_\lambda\,\varphi,
\end{aligned}$$

where K'_{21} and $\tilde{F}_\lambda$ denote the following bounded operators:

$$\begin{cases} K'_{21} : X \longrightarrow X \\ \qquad \varphi \longrightarrow \displaystyle\int_K \frac{\kappa_{21}(x, v, v')}{|v'_3|}\varphi(x, v')\, dv', \end{cases}$$

and

$$\begin{cases} \tilde{F}_\lambda : X \longrightarrow X \\ \psi \longrightarrow \chi_{K^0}(v') \displaystyle\int_0^x e^{-\int_{x'}^x \frac{\sigma_1(s,v')+\lambda}{|v'_3|} ds} \psi(x', v')\, dx' + \chi_{K^1}(v') \\ \displaystyle\int_x^1 e^{-\int_x^{x'} \frac{\sigma_1(s,v')+\lambda}{|v'_3|} ds} \psi(x', v')\, dx'. \end{cases}$$

We claim that $K'_{21}\tilde{F}_\lambda$ depends continuously on K'_{21}. Let $\varphi \in X$, we have

$$\begin{aligned} \|\tilde{F}_\lambda\varphi\| &\le \int_0^1 \left[\int_{K^0} dv' \int_0^x e^{-\int_{x'}^x \frac{\sigma_1(s,v')+\mathrm{Re}\lambda}{|v'_3|} ds} |\varphi(x', v')| dx'\right] dx \\ &\quad + \int_0^1 \left[\int_{K^1} dv' \int_x^1 e^{-\int_x^{x'} \frac{\sigma_1(s,v')+\mathrm{Re}\lambda}{|v'_3|} ds} |\varphi(x', v')| dx'\right] dx \\ &\le \int_0^1 \left[\int_{K^0} dv' \int_0^x e^{-\int_{x'}^x \frac{\mathrm{Re}\lambda+\lambda_1^*}{|v'_3|} ds} |\varphi(x', v')| dx'\right] dx \\ &\quad + \int_0^1 \left[\int_{K^1} dv' \int_x^1 e^{-\int_x^{x'} \frac{\mathrm{Re}\lambda+\lambda_1^*}{|v'_3|} ds} |\varphi(x', v')| dx'\right] dx \\ &\le \int_0^1 \left[\int_{K^0} dv' \int_0^x |\varphi(x', v')| dx'\right] dx + \int_0^1 \left[\int_{K^1} dv' \int_x^1 |\varphi(x', v')| dx'\right] dx \\ &\le \left[\int_{K^0} \int_0^x |\varphi(x', v')| dx' dv' + \int_{K^1} \int_x^1 |\varphi(x', v')| dx' dv'\right] \\ &\le \|\varphi\|. \end{aligned}$$

Then, $\|K'_{21}\tilde{F}_\lambda\| \le \|K'_{21}\|$. According to both Theorem 2.4.4 and Proposition 2.3.1(i), it is sufficient to prove the result when K'_{21} is dominated by a rank-one operator in $\mathcal{L}(L_1(K, dv))$. This asserts that K'_{21} has a kernel $\kappa'_{21}(v, v) = \frac{\kappa_{21}(x,v,v')}{|v'_3|} = \kappa'_1(v)\kappa'_2(v')$, $\kappa'_1(.) \in L_1(K)$, $\kappa'_2(.) \in L_\infty(K)$. Let $\mathcal{O}$ be a bounded set of X, and let $\psi \in \mathcal{O}$. We have, for all measurable subsets E of D,

$$\int_E |K'_{21}\tilde{F}_\lambda \psi(x,v)|dxdv$$

$$\leq \int_E \int_{K^0} \int_0^x |\kappa'_1(v)|\,|\kappa'_2(v')|e^{-\int_{x'}^{x} \frac{\sigma_1(s,v')+\mathrm{Re}\lambda}{|v'_3|}ds} |\varphi(x',v')|dx'dv'dxdv$$

$$+ \int_E \int_{K^1} \int_x^1 |\kappa'_1(v)|\,|\kappa'_2(v')|e^{-\int_{x}^{x'} \frac{\sigma_1(s,v')+\mathrm{Re}\lambda}{|v'_3|}ds} |\varphi(x',v')|dx'dv'dxdv$$

$$\leq \|\kappa'_2\|_\infty \int_E |\kappa'_1(v)|dxdv$$

$$\times \left[\int_{K^1} \int_x^1 |\psi(x',v')|dx'dv' + \int_{K^0} \int_0^x |\psi(x',v')|dx'dv'\right]$$

$$\leq \|\kappa'_2\|_\infty \|\psi\|_1 \int_E |\kappa'_1(v)|dxdv.$$

Since $\lim_{|E|\longrightarrow 0} \int_E |\kappa'_1(v)|dxdv = 0$, ($\kappa'_1 \subseteq L_1(K,dv)$) and, according to Theorem 2.4.5, we infer that the set $K'_{21}\tilde{F}_\lambda(\mathcal{O})$ is weakly compact. Hence the weak compactness of $K_{21}(\lambda - A_1)^{-1}$ is proved. Q.E.D.

Lemma 13.12.2. *We assume that the collision operator K_{12} is nonnegative, regular in the sense of Definition 2.4.3. Then, for $\lambda \in \mathbb{C}$ satisfying $\mathrm{Re}\lambda > -\lambda_1^*$, the operator $(\lambda - A_1)^{-1}K_{12}$ is weakly compact on X.* ◇

Proof. Let $\lambda \in \mathbb{C}$ such that $\mathrm{Re}\lambda > -\lambda_1^*$. It follows that $\|(\lambda - A_1)^{-1}\| \leq \|F_\lambda\| \leq \frac{1}{(\mathrm{Re}\lambda+\lambda_1^*)}$. Let $\varepsilon > 0$, for $\mathrm{Re}\lambda > -\lambda_1^*+\varepsilon$, we get $\|(\lambda-A_1)^{-1}K_{12}\| \leq \frac{1}{\varepsilon}\,\|K_{12}\|$. Then, $(\lambda - A_1)^{-1}K_{12}$ depends continuously on K_{12}, unfortunately on $\{\mathrm{Re}\lambda > -\lambda_1^* + \varepsilon\}$. From Theorem 2.4.4 and Proposition 2.3.1(i), it suffices to prove the result when K_{12} is dominated by a rank-one operator in $\mathcal{L}(L_1(K,dv))$. This asserts that the kernel of K_{12} can be rewritten as follows $\kappa_{12}(v,v') = \kappa_1(v)\kappa_2(v')$, $\kappa_1(.) \in L_1(K)$, $\kappa_2(.) \in L_\infty(K)$. Now, we may show that $F_\lambda K_{12}$ is weakly compact. Let $\varphi \in X$, we have

$$\begin{aligned}(F_\lambda^+ K_{12}\varphi)(x,v) &= \frac{1}{|v_3|}\int_x^1 e^{-\int_x^{x'} \frac{\sigma_1(s,v)+\lambda}{|v_3|}ds} K_{12}\varphi(x',v)dx' \\ &= \frac{1}{|v_3|}\int_x^1 \int_K e^{-\int_x^{x'} \frac{\sigma_1(s,v)+\lambda}{|v_3|}ds} \kappa_{12}(x',v,v')\varphi(x',v')dx'\,dv',\, v \in K^0 \\ &= J_\lambda U_\lambda \varphi,\end{aligned}$$

where J_λ and U_λ denote the following bounded operators

$$\begin{cases} U_\lambda : X \longrightarrow L_1((0,1), dx) \\ \qquad \varphi \longrightarrow \int_K \kappa_2(v')\varphi(x', v')\, dv', \end{cases}$$

and

$$\begin{cases} J_\lambda : L_1((0,1), dx) \longrightarrow X_1^0 \\ \qquad \varphi \longrightarrow \frac{1}{|v_3|} \int_x^1 e^{-\int_x^{x'} \frac{\sigma_1(s,v)+\lambda}{|v_3|}\, ds} \kappa_1(v)\varphi(x')\, dx'. \end{cases}$$

Now, it is sufficient to show that J_λ is weakly compact. To do so, let $\mathcal{O}$ be a bounded set of $L_1((0,1), dx)$, and let $\psi \in \mathcal{O}$. It follows that, for all measurable subsets E of K^0, we have $\int_E |J_\lambda\psi(v)||v_3|\, dv \leq \|\psi\| \int_E |\kappa_1(v)|\, dv$. Since $\lim_{|E|\to 0} \int_E |\kappa_1(v)|\, dv = 0$, $(\kappa_1(.) \in L_1(K, dv))$, where $|E|$ is the measure of E. By applying Theorem 2.4.5, we deduce that the set $J_\lambda(\mathcal{O})$ is weakly compact. Hence, the weak compactness of $F_\lambda^+ K_{12}$ is checked. A similar reasoning shows that the operator $F_\lambda^- K_{12}$ is weakly compact. Q.E.D.

We have the following theorem.

Theorem 13.12.1. *If the operator H is weakly compact, positive, and if the operators K_{12}, K_{22} are nonnegative, regular in the sense of Definition 2.4.3 and if, in addition, $\left\{\frac{\kappa_{21}(x,.,v')}{|v_3'|}\right.$ such that $(x, v') \in (0,1) \times K\left.\right\}$ is relatively weakly compact, then $\sigma_{ei}(\mathcal{A}) = \{\lambda \in \mathbb{C}$ such that $\mathrm{Re}\lambda \leq -\min(\lambda_1^*, \lambda_2^*)\}$, $i = 1, \ldots, 6$.* ◇

Proof. Let $\mu \in \rho(A_1)$. The operator M_μ is given by $M_\mu = D + K_{21}B_\mu H\psi^0 - K_{21}(\mu - A_1)^{-1}K_{12}$. By using Lemma 13.12.2 and the weak compactness of H, we conclude that $K_{21}B_\mu H\psi^0 - K_{21}(\mu - A_1)^{-1}K_{12}$ is weakly compact. Then, from Theorem 7.5.3, it follows that $\sigma_{ei}(\overline{M}_\mu) = \sigma_{ei}(D)$, for $i = 1, \ldots, 6$. For $\mu \in \rho(T_2^H)$ such that $r_\sigma((\mu - T_2^H)^{-1}K_{22}) < 1$, then $\mu \in \rho(T_2^H) \bigcap \rho(T_2^H + K_{22})$. Hence, we have

$$(\mu - D)^{-1} - (\mu - T_2^H)^{-1} = \sum_{n\geq 1} ((\mu - T_2^H)^{-1}K_{22})^n(\mu - T_2^H)^{-1}. \tag{13.12.13}$$

Since K_{22} is a nonnegative, and regular operator, implies that, for all $n \geq 1$, $((\mu - T_2^H)^{-1}K_{22})^n(\mu - T_2^H)^{-1}$ is weakly compact on X. So, $(\mu - D)^{-1} - (\mu - T_2^H)^{-1}$ is weakly compact. The use of both Eq. (13.12.13) and Theorem 7.5.4 leads to $\sigma_{ei}(D) = \sigma_{ei}(T_2^H) = \{\lambda \in \mathbb{C}$ such that $\mathrm{Re}\lambda \leq -\lambda_2^*\}$, $i = 1, \ldots, 6$. Then, from Eq. (13.12.12), we have

$$\sigma_{ei}(\overline{M}_\mu) = \{\lambda \in \mathbb{C} \text{ such that } \mathrm{Re}\lambda \leq -\lambda_2^*\}, i = 1, \ldots, 6. \tag{13.12.14}$$

Now, by applying Theorem 10.3.3, and by using Eqs. (13.12.11) and (13.12.14), we get $\sigma_{ei}(\mathcal{A}) = \{\lambda \in \mathbb{C}$ such that $\mathrm{Re}\lambda \leq -\min(\lambda_1^*, \lambda_2^*)\}$, $i = 1, \ldots, 6$. Q.E.D.

Remark 13.12.2. In this application, we have determined the essential spectra of the two-group transport operator $\mathcal{A}$ on L_1-space without knowing the essential spectra of the operator A. However, we only know the essential spectra of the restriction of the operator A on the intersection densely domain, $\mathcal{D}(A) \bigcap \mathcal{N}(\Gamma_X)$. ◇

Open question. If $K_{11} \neq 0$, can we determine the essential spectra of $\mathcal{A}$? □

Open question. Consider the multidimensional two-group neutron transport operators on L_1-spaces

$$\mathcal{A} := \begin{pmatrix} A_0 & R \\ R & A_0 \end{pmatrix},$$

where A_0 and R are the operators already defined in (13.6.6). If R is a regular collision operator on $L_1(\mathcal{D} \times V, dxdv)$ and $\mathrm{Re}\lambda > \eta$, where η is the type of the C_0-semigroup generated by $A_0 - R$, it was proved in [261] that $R(\lambda - A_0 - R)^{-1}R$ is weakly compact on $L_1(\mathcal{D} \times V, dxdv)$. However, if $\sigma(x, v) = \sigma(v)$ and if $\mathcal{D}$ is convex, then it was proved that $R(\lambda - A_0 - R)^{-1}R$ is compact on $L_1(\mathcal{D} \times V, dxdv)$. In both cases, can we determine the essential spectra of the matrix $\mathcal{A}$? □

13.13 Three-Group Transport Equation

The work presented in this section concerns the application of Theorems 11.2.2, 11.2.3 and 11.3.2 to a three-group transport operator on L_1-spaces. Let $X_1 := L_1([-a, a] \times [-1, 1], dxd\xi)$, $a > 0$, and $X = Y = Z := X_1$. We consider the boundary spaces

$$X_1^o := L_1(\{-a\} \times [-1, 0]; |\xi| d\xi) \times L_1(\{a\} \times [0, 1], |\xi| d\xi) =: X_1^o \times X_2^o$$

and

$$X_1^i := L_1(\{-a\} \times [0, 1], |\xi| d\xi) \times L_1(\{a\} \times [-1, 0], |\xi| d\xi) =: X_1^i \times X_2^i$$

respectively equipped with the norms

$$\|u^o, X_1^o\| := \left(\|u_1^o, X_1^o\| + \|u_2^o, X_2^o\|\right) = \left[\int_{-1}^{0} |u(-a, \xi)||\xi| d\xi + \int_{0}^{1} |u(a, \xi)||\xi| d\xi\right]$$

and

$$\|u^i, X_1^i\| := \left(\|u_1^i, X_1^i\| + \|u_2^i, X_2^i\|\right) = \left[\int_{0}^{1} |u(-a, \xi)||\xi| d\xi + \int_{-1}^{0} |u(a, \xi)||\xi| d\xi\right].$$

We will consider the matrix of the operator $\mathcal{L} = \mathcal{T} + \mathcal{K}$, where

$$\mathcal{T}\psi = \begin{pmatrix} -\xi\dfrac{\partial\psi_1}{\partial x} - \sigma_1(\xi)\psi_1 & 0 & 0 \\ 0 & -\xi\dfrac{\partial\psi_2}{\partial x} - \sigma_2(\xi)\psi_2 & 0 \\ 0 & 0 & -\xi\dfrac{\partial\psi_3}{\partial x} - \sigma_3(\xi)\psi_3 \end{pmatrix}$$

$$= \begin{pmatrix} T_1 & 0 & 0 \\ 0 & T_2 & 0 \\ 0 & 0 & T_H \end{pmatrix}\begin{pmatrix} \psi_1 \\ \psi_2 \\ \psi_2 \end{pmatrix}$$

and

$$\mathcal{K} = \begin{pmatrix} 0 & \mathcal{K}_{12} & \mathcal{K}_{13} \\ \mathcal{K}_{21} & \mathcal{K}_{22} & 0 \\ \mathcal{K}_{31} & \mathcal{K}_{32} & \mathcal{K}_{33} \end{pmatrix}$$

with $\mathcal{K}_{ij}$, $i, j = 1, 2, 3$ and $(i, j) \neq (1, 1)$, $(2, 3)$ are bounded linear operators defined on X_1 by

$$\begin{cases} \mathcal{K}_{ij} : X_1 \longrightarrow X_1 \\ \qquad u \longrightarrow \mathcal{K}_{ij}u(x,\xi) = \displaystyle\int_{-1}^{1} \kappa_{ij}(x,\xi,\xi')u(x,\xi')d\xi', \end{cases} \tag{13.13.1}$$

and the kernels $\kappa_{ij} : [-a, a] \times [-1, 1] \times [-1, 1] \longrightarrow \mathbb{R}$ are assumed to be measurable. The operator T_1 is defined by

$$\begin{cases} T_1 : \mathcal{D}(T_1) \subset X_1 \longrightarrow X_1 \\ \qquad \varphi_1 \longrightarrow T_1\varphi_1(x,\xi) = -\xi\dfrac{\partial\varphi_1}{\partial x}(x,\xi) - \sigma_1(\xi)\varphi_1(x,\xi) \\ \mathcal{D}(T_1) = \mathcal{W}_1, \end{cases}$$

where $\mathcal{W}_1$ is the partial Sobolev's space defined by $\mathcal{W}_1 = \Big\{\varphi \in X_1$ such that $\xi\frac{\partial\varphi}{\partial x} \in X_1\Big\}$. T_2 is the streaming operator defined by

$$\begin{cases} T_2 : \mathcal{D}(T_2) \subset X_1 \longrightarrow X_1 \\ \qquad \varphi_2 \longrightarrow T_2\varphi_2(x,\xi) = -\xi\dfrac{\partial\varphi_2}{\partial x}(x,\xi) - \sigma_2(\xi)\varphi_2(x,\xi) \\ \mathcal{D}(T_2) = \mathcal{W}_1 \end{cases}$$

and

$$\begin{cases} T_H : \mathcal{D}(T_H) \subset X_1 \longrightarrow X_1 \\ \qquad\qquad \varphi_3 \longrightarrow T_H\varphi_3(x,\xi) = -\xi \dfrac{\partial \varphi_3}{\partial x}(x,\xi) - \sigma_3(\xi)\varphi_3(x,\xi) \\ \mathcal{D}(T_H) = \{\varphi_3 \in \mathcal{W}_1 \quad \text{such that} \quad \varphi_3^i = H\varphi_3^o\}, \end{cases}$$

where $\sigma_j(.) \in L_\infty(-1,1)$, $j = 1,2,3$. It is well known that any function $\varphi \in \mathcal{W}_1$ has traces on the spacial boundary $\{-a\}\times(-1,0)$ and $\{a\}\times(1,0)$ respectively in X_1^o and X_1^i. They are denoted, respectively, by φ^o and φ^i, and represent the outgoing and the incoming fluxes ("o" for outgoing and "i" for incoming). Let λ_j^* be the real defined by $\lambda_j^* := \liminf_{|\xi|\to 0} \sigma_j(\xi)$, $j = 1,2,3$. In the following, we will define the operator $\mathcal{L}$ on the domain

$$\mathcal{D}(\mathcal{L}) = \left\{ \begin{pmatrix} \psi_1 \\ \psi_2 \\ \psi_2 \end{pmatrix}, \psi_1 \in \mathcal{W}_1, \psi_2 \in \mathcal{D}(T_2), \psi_3 \in \mathcal{D}(T_H) \text{ and } \psi_1^i = \psi_2^i = \psi_3^i \right\}.$$

We will denote by

$$\begin{cases} \Gamma_X : X_1 \longrightarrow X_1^i \\ \qquad \psi_1 \longrightarrow \psi_1^i \end{cases}$$

$$\begin{cases} \Gamma_Y : X_1 \longrightarrow X_1^i \\ \qquad \psi_2 \longrightarrow \psi_2^i \end{cases}$$

and

$$\begin{cases} \Gamma_Z : X_1 \longrightarrow X_1^i \\ \qquad \psi_3 \longrightarrow H\psi_3^o. \end{cases}$$

Let A_1 be the operator defined by

$$\begin{cases} A_1 = T_1 \\ \mathcal{D}(A_1) = \{\psi_1 \in \mathcal{W}_1 \text{ such that } \psi_1^i = 0\}. \end{cases}$$

Remark 13.13.1. It is well known that the operators T_1, T_2, and T_H are closed and linear operators. Moreover, the derivative of ψ_j in the definition of T_j, $j = 1,2$, and T_H is meant in the distributional sense. Note that, if $\psi \in \mathcal{D}(T_H)$ or $\mathcal{D}(T_j)$, then it is absolutely continuous with respect to x. Hence, the restrictions of ψ to X_1^i and X_1^o are meaningful. Let us also notice that $\mathcal{D}(T_H)$ is dense in X_1 because it contains $\mathcal{C}_0^\infty(X_1^o)$. ◇

Let us observe that, for λ such that $\mathrm{Re}\lambda > -\lambda_1^*$, we have $\lambda \in \rho(A_1)$. Indeed, the solution of $(\lambda - T_1)\psi_1 = \varphi_1$, for $\psi_1 \in \mathcal{D}(A_1)$ is formally given by:

$$\psi_1(x,\xi) = \begin{cases} \dfrac{1}{|\xi|}\displaystyle\int_{-a}^{x} e^{-\frac{(\lambda+\sigma_1(\xi))|x-x'|}{|\xi|}}\varphi_1(x',\xi)dx', & 0<\xi<1 \\ \dfrac{1}{|\xi|}\displaystyle\int_{x}^{a} e^{-\frac{(\lambda+\sigma_1(\xi))|x-x'|}{|\xi|}}\varphi_1(x',\xi)dx', & -1<\xi<0. \end{cases}$$

For any λ, such that $\mathrm{Re}\lambda > -\lambda_1^*$, the operator K_λ is chosen as follows:

$$K_\lambda u = \varphi_1, \text{ for } u \in X_1^i \text{ if, and only if, } \begin{cases} \varphi_1 \in \mathcal{D}(T_1) \text{ and } \Gamma_X\varphi_1 = u \\ (T_1 - \lambda)\varphi_1 = 0 \end{cases}$$

i.e.,

$$\begin{cases} (T_1 - \lambda)\varphi_1 = 0, \ \varphi_1 \in \mathcal{W}_1 \\ \varphi_1^i = u. \end{cases}$$

It is easy to check that K_λ is the following operator

$$\begin{cases} K_\lambda : X_1^i \longrightarrow X_1, \ K_\lambda := \chi_{[-1,0]}(\xi)K_\lambda^- u + \chi_{[0,1]}(\xi)K_\lambda^+ u \ \text{ with} \\ K_\lambda^+ u(x,\xi) := u(-a,\xi)e^{-\frac{(\lambda+\sigma_1(\xi))|a+x|}{|\xi|}}, \quad 0<\xi<1, \\ K_\lambda^- u(x,\xi) \ := u(a,\xi)e^{-\frac{(\lambda+\sigma_1(\xi))|a-x|}{|\xi|}}, \quad -1<\xi<0, \end{cases}$$

where $\chi_{[-1,0]}(.)$ and $\chi_{[0,1]}(.)$ denote respectively the characteristic functions on the intervals $[-1,0]$ and $[0,1]$. It is easy to see that K_λ is bounded and $\|K_\lambda\| \leq [(\mathrm{Re}\lambda + \lambda_1^*)]^{-1}$. The domain Y_1, defined in (11.2.2), is given by $Y_1 = \{\psi_2 \in \mathcal{W}_1$ such that $\psi_2^i \in \Gamma_X(\mathcal{W}_1)\}$. Then, $Y_1 \bigcap \mathcal{N}(\Gamma_Y) = \{\psi_2 \in \mathcal{W}_1$ such that $\psi_2^i = 0\}$. The operator J_λ can be defined as follows:

$$\begin{cases} J_\lambda : X_1^i \longrightarrow X_1 \\ \mathcal{D}(J_\lambda) = \{\psi_2^i \text{ such that } \psi_2 \in Y_1\} \end{cases}$$

$$J_\lambda u = \varphi_2, \text{ for } u \in X_1^i \text{ if, and only if, } \begin{cases} \varphi_2 \in Y_1 \text{ and } \Gamma_Y\varphi_2 = u \\ (S(\lambda) - \lambda)\varphi_2 = 0 \end{cases}$$

and the equation $(S(\lambda) - \lambda)\varphi_2 = 0$ leads to

$$(T_2 + K_{22} - \lambda)\varphi_2 + K_{21}K_\lambda\varphi_2^i - K_{21}(T_1 - \lambda)^{-1}K_{12}\varphi_2 = 0.$$

Then, $\left((T_2 + K_{22} - \lambda) - K_{21}(T_1 - \lambda)^{-1}K_{12}\right)\varphi_2 = -K_{21}K_\lambda u$. For $\lambda \in \rho(T_1)\bigcap\rho(T_2)$, such that $r_\sigma((T_2 - \lambda)^{-1}K_{22}) < 1$, we deduce that $\lambda \in$

$\rho(T_1)\bigcap\rho(T_2)\bigcap\rho(T_2+K_{22})$ and, if $r_\sigma\left((T_2+K_{22}-\lambda)^{-1}K_{21}(T_1-\lambda)^{-1}K_{12}\right)<1$, then J_λ can be given by

$$J_\lambda=-\sum_{n\geq 0}\left((T_2+K_{22}-\lambda)^{-1}K_{21}(T_1-\lambda)^{-1}K_{12}\right)^n(T_2+K_{22}-\lambda)^{-1}K_{21}K_\lambda$$

and so, J_λ is bounded. Now, the operator $S_1(\lambda)$ which is defined on $\mathcal{D}(S_1(\lambda))=\{\varphi_2\in\mathcal{W}_1$ such that $\varphi_2^i=0\}$ is given by $S_1(\lambda)=(T_2+K_{22})-K_{21}(T_1-\lambda)^{-1}K_{12}$. The operator $M(\lambda)$, already defined in Chap. 11, is weakly compact on $X_1\times X_1\times X_1$, since the operators $F_i(\lambda)$ and $\tilde{G}_i(\lambda)$, $i=1,2,3$ are weakly compact. Let us notice that the collision operators K_{ij}, $i,j=1,2,3$, $(i,j)\neq(1,1)$, $(2,3)$ defined in (13.13.1), act only on the velocity v, so x may be simply viewed as a parameter in $[-a,a]$. Then, we will consider K_{ij} as a function $K_{ij}(.):x\in[-a,a]\longrightarrow K_{ij}(x)\in\mathcal{L}\left(L_1([-1,1],d\xi)\right)$.

Theorem 13.13.1. *If the operator H is weakly compact and positive, and if the operators $K_{12}, K_{21}, K_{13}, K_{31}, K_{33}$ are nonnegative and regular, and if, in addition, $\frac{\kappa_{21}(x,\xi,\xi')}{|\xi'|}$ and $\frac{\kappa_{31}(x,\xi,\xi')}{|\xi'|}$ define two regular operators on X_1, then*

$$\sigma_{ei}(\mathcal{L})=\sigma_{e7}(\mathcal{L})=\sigma_{e8}(\mathcal{L})=\{\lambda\in\mathbb{C}\text{ such that }\operatorname{Re}\lambda\leq-\min(\lambda_1^*,\lambda_2^*,\lambda_3^*)\},\ i=1,\ldots,6.$$

◇

Open question. If $\mathcal{K}_{11}\neq 0$ or $\mathcal{K}_{23}\neq 0$, can we determine the essential spectra of $\mathcal{L}$? □

Bibliography

1. B. Abdelmoumen, Stabilité des spectres essentiels et applications à des modéles cinétiques, Thesis, University of Sfax, 2010
2. B. Abdelmoumen, H. Baklouti, Fredholm perturbations and seminorms related to upper semi-Fredholm perturbations. Filomat **27**(6), 1147–1155 (2013)
3. B. Abdelmoumen, O. Jedidi, A. Jeribi, Time asymptotic description of an abstract Cauchy problem's solution and application to transport equation. Appl. Math. **59**(1), 53–67 (2014)
4. B. Abdelmoumen, A. Jeribi, M. Mnif, Time asymptotic description of the solution to an abstract Cauchy problem and application to transport equation. Math. Z. **268**(3–4), 837–869 (2011)
5. B. Abdelmoumen, A. Jeribi, M. Mnif, Invariance of the Schechter essential spectrum under polynomially compact operators perturabation. Extracta Math. **26**(1), 61–73 (2011)
6. B. Abdelmoumen, A. Jeribi, M. Mnif, On graph measures in Banch spaces and description of essential spectra of multidimensional transport equation. Acta Math. Sci. Ser. B Engl. Ed. **32**(5), 2050–2064 (2012)
7. B. Abdelmoumen, A. Jeribi, M. Mnif, Measure of weak noncompactness, some new properties in Fredholm theory, characterization of the Schechter essential spectrum and application to transport operators. Ricerche Mat. **61**, 321–340 (2012)
8. B. Abdelmoumen, A. Dehici, A. Jeribi, M. Mnif, Some new properties of Fredholm theory, essential spectra and application to transport theory. J. Inequal. Appl. **2008**, 1–14 (2008)
9. F. Abdmouleh, Fredholm operators, essential spectra of sum of two bounded linear operators and applications to a transport operators, Thesis, University of Sfax, 2009
10. F. Abdmouleh, A. Ammar, A. Jeribi, A characterization of the pseudo-Browder essential spectra of linear operators and application to a transport equation. J. Comp. Theo. Tran. (2015). doi:10.1080/23324309.2015.1033222
11. F. Abdmouleh, A. Jeribi, Symmetric family of Fredholm operators of indices zero, stability of essential spectra and application to transport operators. J. Math. Anal. Appl. **364**, 414–423 (2010)
12. F. Abdmouleh, A. Jeribi, Gustafson, Weidman, Kato, Wolf, Schechter, Browder, Rakočević and Schmoeger essential spectra of the sum of two bounded operators and application to a transport operator. Math. Nachr. **284**(2–3), 166–176 (2011)
13. F. Abdmouleh, A. Ammar, A. Jeribi, Stability of the S-essential spectra on a Banach space. Math. Slovaca **63**(2), 299–320 (2013)
14. F. Abdmouleh, S. Charfi, A. Jeribi, On a characterization of the essential spectra of the sum and the product of two operators. J. Math. Anal. Appl. **386**(1), 83–90 (2012)

A. Jeribi, *Spectral Theory and Applications of Linear Operators and Block Operator Matrices*, DOI 10.1007/978-3-319-17566-9

15. S. Agmon, On the eigenfunctions and on the eigenvalues of general elliptic boundary value problems. Commun. Pure Appl. Math. **15**, 119–147 (1962)
16. M.S. Agranovich, M.I. Vishik, Elliptic problems with parameter and parabolic problems of a general form. Uspekhi Matem. Nauk **19**, 53–161 (1964) (Russian)
17. P. Aiena, *Fredholm and Local Spectral Theory, with Applications to Multipliers* (Kluwer Academic, Dordrecht, 2004)
18. R.R. Akhmerov, M.I. Kamenskii, A.S. Potapov, A.E. Rodkina, B.N. Sadovskii, *Measures of Noncompactness and Condensing Operators* (Birkhäuser, Basel, 1992)
19. C.D. Aliprantis, O. Burkinshaw, *Positive Operators* (Academic Press, Orlando, 1985)
20. A. Ammar, S-essential spectra, the Weyl pseudospectra of linear operators, perturbation theory of semi regular and essentially semi regular operators, Thesis, University of Sfax, 2013
21. A. Ammar, A. Jeribi, A characterization of the essential pseudo-spectra on a Banach space. Arab. J. Math. **2**(2), 139–145 (2013)
22. A. Ammar, A. Jeribi, A characterization of the essential pseudo-spectra and application to a transport equation. Extracta Math. **28**(1), 95–112 (2013)
23. A. Ammar, A. Jeribi, Measures of noncompactness and essential pseudo-spectra on Banach Space. Math. Methods Appl. Sci. **37**(3), 447–452 (2014)
24. A. Ammar, B. Boukattaya, A. Jeribi, Stability of the S-left and S-right essential spectra of a linear operator. Acta Math. Sci. **34B**(5), 1–13 (2014)
25. A. Ammar, A. Jeribi, N. Moalla, On a characterization of the essential spectra of a 3×3 operator matrix and application to three-group transport operators. Ann. Funct. Anal. **4**(2), 153–170 (2013) (electronic only)
26. F. Andreu, J. Martinez, J.M. Mazon, A spectral mapping theorem for perturb ed strongly continuous semigroup. Math. Ann. **291**, 453–462 (1991)
27. N. Angelescu, N. Marinescu, V. Protopopescu, Linear monoenergetic transport with reflecting boundary conditions. Rev. Roum. Phys. **19**, 17–26 (1974)
28. N. Angelescu, N. Marinescu, V. Protopopcu, Neutron transport with periodic boundary conditions. lkansp. Theor. Stat. Phys. **5**, 115–125 (1976)
29. C. Anné, N. Torki-Hamza, The Gauß-Bonnet operator of infinite graph, preprint, arXiv: 1301.0739 (2013)
30. P.M. Anselone, *Collectively Compact Operator Approximation Theory* (Prentice-Hall, Englewood Cliffs, 1971)
31. C. Apostol, The reduced minimum modulus. Mich. Math. J. **32**, 279–294 (1985)
32. K. Appel, W. Haken, Every planar map is four, clorable, Part I. Discharging. lllinois J. Math. **21**, 429–490 (1977)
33. K. Appel, W. Haken, Every planar map is four clorable. Part II. Reducibility. lllinois J. Math. **21**, 491–567 (1977)
34. W. Arendt, Resolvent positive operators. Proc. Lond. Math. Soc. **54**, 321–349 (1987)
35. W. Arend, R. Nagel (ed.), *One-Parameter Semigroups of Positive Operators*. Lecture Notes in Mathematics, vol. 1184 (Springer, Heidelberg, 1986)
36. Z. Artstein, Continuous dependence of solutions of operator equations, I. Trans. Am. Math. Soc. **231**(1), 143–166 (1977)
37. K. Astala, On measure of noncompactness and ideal variations in Banach spaces. Ann. Acad. Sci. Fenn. Ser. A. I. Math. Diss. **29**, (1980)
38. K. Astala, H.-O. Tylli, Seminorms related to weak compactness and to Tauberian operators. Math. Proc. Camb. Philos. Soc. **107**, 367–375 (1990). Printed in Great Britain
39. F.V. Atkinson, The normal solubility of linear equations in normed spaces. Math. Sb. (N.S) **28**(70), 3–14 (1951) (Russian)
40. F.V. Atkinson, H. Langer, R. Mennicken, A.A. Shkalikov, The essential spectrum of some matrix operators. Math. Nachr. **167**, 5–20 (1994)
41. B. Aupetit, *A Primer on Spectral Theory* (Springer, New York, 1991)
42. J.M. Ayerbe Toledano, T. Dominguez Benavides, G. López Acedo, *Measures of Noncompactness in Metric Fixed Point Theory* (Birkhäuser, Basel, 1997)

43. G. Ball, Diffusion approximation of the radiative transfer equations in a chanel. Trans. Theor. Stat. Phys. **30**(2 & 3), 269–293 (2001)
44. H. Baloudi, S. Golénia, A. Jeribi, The adjacency matrix and the discrete Laplacian acting on forms, (preprint) (2015)
45. H. Baloudi, A. Jeribi, Left-Right Fredholm and Weyl spectra of the sum of two bounded operators and application. Mediterr. J. Math. **11**, 939–953 (2014)
46. J. Banaś, Applications of measure of weak noncompactness and some classes of operators in the theory of functional equations in the Lebesgue space, in *Proceedings of the second World Congress of Nonlinear Analysis*, Part 6, Athen, 1966. Nonlinear Anal. **30**, 3283–3293 (1997)
47. J. Banaś, K. Geobel, *Measures of Noncompactness in Banach Spaces*. Lecture Notes in Pure and Applied Mathematics, vol. 60 (Marcel Dekker, New York, 1980), pp. 259–262.
48. J. Banaś, A. Martinón, On measures of weak noncompactness in Banach sequence spaces. Portugal. Math. **52**, 131–138 (1995)
49. J. Banaś, J. Rivero, On measures of weak noncompactness. Ann. Mat. Pura Appl. **151**, 213–262 (1988)
50. A. Bátkai, P. Binding, A. Dijksma, R. Hryniv, H. Langer, Spectral problems for operator matrices. Math. Nachr. **278**, 1408–1429 (2005)
51. R. Beals, V. Protopopescu, Abstract time dependent transport equations. J. Math. Anal. Appl. **121**, 370–405 (1987)
52. L.W. Beineke, Derived graphs and digraphs, in *Beitrage zur Graphentheorie*, ed. by H. Sachs, H. Voss, H. Walther (Tenbner, Leipzig, 1968), pp. 17–33
53. A. Belleni-Morante, Neutron transport in a nonuniform slab with generalized boundary conditions. J. Math. Phys. **11**, 1553–1558 (1970)
54. M. Belzad, A criterion for the planarity of a total garaph. Proc. Camb. Philos. Soc. **63**, 679–681(1967)
55. N. Ben Ali, Base de Riesz de vecteurs propres d'une famille d'opérateurs, spectres essentiels d'un opérateur matriciel et applications, Thesis, University of Sfax, 2011
56. N. Ben Ali, A. Jeribi, N. Moalla, Essential spectra of some matrix operators. Math. Nachr. **283**(9), 1245–1256 (2010)
57. A. Ben Amar, Spectral and fixed point theories and applications to problems arising in kinetic theory of gas and in growing cell populations, Thesis, University of Sfax, 2007
58. A. Ben Amar, A. Jeribi, M. Mnif, Some applications of the regularity and irreducibility on transport theory. Acta Appl. Math. **110**, 431–448 (2010)
59. A. Ben Amar, A. Jeribi, B. Krichen, Essential spectra of a 3x3 operator matrix and application to three-group transport equation. Integr. Equ. Oper. Theory **68**, 1–21 (2010)
60. A. Ben Amar, A. Jeribi, M. Mnif, Some results on Fredholm and semi-Fredholm operators. Arab. J. Math. **3**(3), 313–323 (2014)
61. M. Benharrat, A. Ammar, A. Jeribi, B. Messirdi, On the Kato, semi-regular and essentially semi-regular spectra. Funct. Anal. Approx. Comput. **6**(2), 9–22 (2014)
62. M. Berkani, A. Ouahab, Opérateur essentiellement régulier dans les espaces de Banach. Rend. Circ. Math. Palermo Serie II **46**, 131–160 (1997)
63. N. Biggs, E. Lioyd, R. Wilson, *Graph Theory* (Oxford University Press, Oxford, 1986), pp. 1736–1936
64. P. Binding, R. Hryniv, Relative boundedness and relative compactness for linear operators in Banach spaces. Proc. Am. Math. Soc. **128**, 2287–2290 (2000)
65. G. Borgioli, S. Totaro, On the spectrum of the transport operator with mixed type boundary conditions, in *Atti Congruso*, Aimeta, vol. 1 (1986), pp. 393–398
66. F.E. Browder, On the spectral theory of elliptic differential operators, I. Math. Ann. **142**, 22–130 (1961)
67. S.R. Caradus, Operators of Riesz type. Pac. J. Math. **18**, 61–71 (1966)
68. S.R. Caradus, W.E. Plaffenberger, B. Yood, *Calking Algebras and Algebras of Operators on Banach Spaces*. Lecture Notes, vol. 9 (Marcel Dekker, New York, 1974)
69. R. Carlson, Adjoint and self-adjoint differential operators on graphs. J. Differ. Equ. **6**, 1–10 (1998)

70. A.L. Cauchy, Recherche sur les polyodres premier mémoire. Journal de l'école Polytechnique **9**, 66–86 (1813)
71. A. Cayley, On the theory of the analytical forms called trees. Philos. Mag. **13**, 172–176 (1857)
72. A. Cayley, Ueber die Analytischen Figuren, welche in der Mathematic Baume genannt werden und ihre Anwendung auf die Theorie chemischer Verbindungen, Berichteder deutshen chemischen Gesellsoft **8**(2), 1056–1059 (1875)
73. W. Chaker, A. Jeribi, B. Krichen, Demicompact linear operators, essential spectrum and some perturbation results. Math. Nachr. 1–11 (2015). doi:10.1002/mana.201200007
74. S. Charfi, Spectral properties of operator matrices, perturbed linear operators, systems of evolution equations and applications, Thesis, University of Sfax, 2010
75. S. Charfi, On the time asymptotic behavior of a transport operator with diffuse reflection boundary condition. Transp. Theory Stat. Phys. **41**(7), 529–551 (2012)
76. S. Charfi, A. Jeribi, On a characterization of the essential spectra of some matrix operators and applications to two-group transport operators. Math. Z. **262**(4), 775–794 (2009)
77. S. Charfi, A. Jeribi, N. Moalla, Time asymptotic behavior of the solution of an abstract Cauchy problem given by a one-velocity transport operator with Maxwell boundary condition. Collect. Math. **64**, 97–109 (2013)
78. S. Charfi, A. Jeribi, R. Moalla, Essential spectra of operator matrices and applications. Methods Appl. Sci. **37**(4), 597–608 (2014)
79. S. Charfi, A. Jeribi, I. Walha, Essential spectra, matrix operator and applications. Acta Appl. Math. **111**(3), 319–337 (2010)
80. F.R.K. Chung, *Spectral Graph Theory*. CBMS Regional Conferance Series in Mathematics, vol. 92 (American Mathematical Society, Providence, 1997), xi, 207 pp.
81. Ph. Clement, *One-Parameter Semigroups* (North-Holland, Amsterdam, 1987)
82. Y. Colin de Verdiére, Spectres de graphes, in *Cours Spécialisés*, vol. 4 (Société Mathématique de France, Paris, 1998)
83. A. Corciovei, V. Protopopescu, On the spectrum of the linear transport oper ator with diffuse reflections. Rev. Roum. Phys. **21**, 713–719 (1976)
84. J.R. Cuthbert, On semigroups such that $U(t) - I$ is compact for some $t > 0$. Z. Wahrscheinlichkeitstheorie und Verw. Gebiete **18**, 9–16 (1971)
85. D. Cvetković, On gaps between bounded operators. Publ. Inst. Math. **72**(86), 49–54 (2002)
86. M. Damak, A. Jeribi, On the essential spectra of some matrix operators and applications. Electron. J. Differ. Equ. **11**, 1–16 (2007)
87. J. Danes, On the Istratescu measure of noncompactness. Bull. Math. Soc. R. S. Roum. **16**(64), 403–406 (1972)
88. E.B. Davies, *Spectral Theory and Differential Operators* (Cambridge University Press, Cambridge, 1996)
89. R. Dautray, J.L. Lions, *Analyse Mathématique et Calcul Numérique*, vol. 9 (Masson, Paris, 1988)
90. F.S. De Blasi, On a property of the unit sphere in a Banach spaces. Bull. Math. Soc. Sci. Math. R. S. Roum. **21(69)**, 259–262 (1977)
91. S. Degong, Some notes on the spectral properties of C_0-semigroups generated by linear transport operators. Trans. Theor. Stat. Phys. **26**(1–2), 233–242 (1997)
92. A. Dehici, K. Latrach, A. Jeribi, On a transport operator arising in growing cell populations. Spectral analysis. Adv. Math. Res. **1**, 159–185 (2002) (Nova Sci. Publ., Hauppauge)
93. A. Dehici, A. Jeribi, K. Latrach, Spectral analysis of a transport operator arising in growing cell populations. Acta Appl. Math. **92**(1), 37–62 (2006)
94. J. Diestel, *Geometry of Banach Spaces-Selected Topics*. Lecture Notes in Mathematics, vol. 485 (Springer, New York, 1975)
95. J. Diestel, *A survey of results related to Dunford-Pettis property,* in *Cont. Math.2, Amer. Math. Soc. of Conf. on Integration, Topology and Geometry in Linear Spaces* (1980), pp. 15–60
96. P. Dodds, D.H. Fremlin, Compact operators in Banach lattices. Isr. J. Math. **34**, 287–320 (1979)

97. T. Dominguez Benavides, Some properties of the set and ball measures of noncompactness and applications. J. Lond. Math. Soc. **34**(2), 120–128 (1986)
98. R. Drnovšek, Bounds for the spectral radius of positive operators. Comment. Math. Univ. Carol. **41**(3), 459–467 (2000)
99. J.J. Duderstart, W.R. Martin, *Transport Theory* (Willey, New York, 1979)
100. N. Dunford, B.J. Pettis, Linear operations on summable functions. Trans. Am. Math. Soc. **47**, 323–392 (1940)
101. N. Dunford, J.T. Schwartz, *Linear Operators, Part I. General Theory* (Interscience, New York, 1958)
102. D.E. Edmum, W.D. Evans, *Spectral Theory and Differential Operators* (Oxford Science Publications, Oxford, 1987)
103. Y. Eidelman, V. Milman, A. Tsolomitis, *Functional Analysis, Graduate*. Studies in Mathematics, vol. 66 (American Mathematical Society, Providence, 2004) (An introduction)
104. G. Emmanuele, Measure of weak non compactness and fixed point theorems. Bull. Math. Soc. Sci. Math. R.S. Roum. **25**, 353–358 (1981)
105. K.J. Engel, Positivity and stability for one-sided coupled operator matrices. Positivity **1**, 103–124 (1997)
106. K.J. Engel, R. Nagel, *One-Parameters Semigroup for Linear Evolutions Equations*. Graduate text in Mathematics (Springer, New York, 2000)
107. I.D. Evzerov, Domains of fractional powers of ordinary differential operators in L_p-spaces. Math. Zametki (Engl. Transl. in Math. Notes) **21**(4), 509–518 (1977)
108. I.D. Evzerov, P.E. Sobolevskii, Fractional powers of ordinary differential operators. Differencial'nye Uravnenija **9**, 228–240 (1973)
109. P. Exner, J. Keating, P. Kuchment, T. Sunada, A. Teplyaev, *Analysis on Graphs and Its Applications* (American Mathematical Society, Providence, 2008)
110. M. Faierman, R. Mennicken, M. Möller, The essential spectrum of a system of singular ordinary differential operators of mixed order. Part I: The general problem and an almost regular case. Math. Nachr. **208**, 101–115 (1999)
111. F. Fakhfakh, M. Mnif, Perturbation of semi-Browder operators and stability of Browder's essential defect and approximate point spectrum. J. Math. Anal. Appl. **347**(1), 235–242 (2008)
112. J.M.G. Fell, R.S. Doran, *Representations of *-Algebras Locally compact Groups, and Banach *-Algebraic Bundles*, vol. 1. *Basic Representation Theory of Groups and Algebras*. Pure Appl. Math., vol. 125 (Academic Press, Boston 1988)
113. I. Fredholm, Sur une classe d'équations fonctionelles. Acta Math. **27**, 365–390 (1903)
114. M. Garden, *Fractal Music, Hypercads, and More Mathematical Recreations from Scientific American* (W. H. Freeman and Company, San Francisco, 1992), p. 203
115. V. Georgescu, S.Golénia, Compact perturbations and stability of the essential spectrum of singular differential operators. J. Oper. Theory **59**, 115–155 (2008)
116. I. Ghoberg, S. Goldberg, M.A. Kaashoek, *Classes of Linear Operators*, vol. 1 (Birkhäuser, Basel, 1990)
117. F. Gilfeather, The structure and asymptotic behavior of polynomially compact operators. Proc. Am. Math. Soc. **25**, 127–134 (1970)
118. S.K. Godunov, V.S. Ryabenki, *Theory of Difference Schemes, an Introduction*. Translated by E. Godfredsen (North-Holland, Amsterdam; Interscience Publishers, New York; Wiley, New York, 1964)
119. I.C. Gohberg, On linear equations in Hilbert space. Dokl. Akad. Nauk SSSR (N.S.) **76**, 9–12 (1951) (Russian)
120. I.C. Gohberg, On linear equations in normed spaces. Dokl. Akad. Nauk SSSR (N.S.) **76**, 477–480 (1951) (Russian)
121. I.C. Gohberg, On linear operators depending analytically on a parameter. Dokl. Akad. Nauk SSSR (N.S.) **78**, 629–632 (1951) (Russian)
122. I.C. Gohberg, On the index of an unbounded operator. Mat. Sb. (N.S) **33**(75), 193–198 (1951) (Russian)

123. I.C. Gohberg, G. Krein, Fundamental theorems on deficiency numbers, root numbers and indices of linear operators. Am. Math. Soc. Transl. Ser. 2 **13**, 185–264 (1960)
124. I.C. Gohberg, A.S. Markus, I.A. Feldman, Normally solvable operators and ideals associated with them. Am. Math. Soc. Transl. Ser. 2, **61**, 63–84 (1967)
125. I.C. Gohberg, S. Goldberg, M.A. Kaashoek, *Classes of Linear Operators Vol. I. Operator Theory: Advances and Applications*, vol. 49 (Birkhauser, Basel, 1990)
126. S. Goldberg, *Unbounded Linear Operators* (McGraw-Hill, New York, 1966)
127. S. Goldberg, Perturbations of semi-Fredholm operators by operators converging to zero compactly. Proc. Am. Math. Soc. **45**(1), 93–98 (1974)
128. M.A. Goldman, S.N. Krackovskii, Behaviour of the space of zero elements with finite-dimensional salient on the Riesz kernel under perturbations of the operator. Dokl. Akad. Nauk SSSR **221**, 532–534 (1975); English transl., Soviet Math. Dokl. **16**, 370–373 (1975)
129. S. Golénia, Unboundedness of adjacency matrices of locally finite graphs. Lett. Math. Phys. **93**, 127–140 (2010)
130. S. Golénia, Hardy inequality and asymptotic eigenvalue distribution for discrete Laplacians. J. Funct. Anal. **266**, 2662–2688 (2014)
131. S. Golénia, C. Schumacher, The problem of deficiency indices for discrete Schrödinger operators on locally finite graphs. J. Math. Phys. **52**, 063512, 17 pp. (2011)
132. M. Gonzàlez, A. Martinon, On the generalized Sadovskii functor. Rev. Acad. Canaria Cien. **1**, 109–117 (1990)
133. M. Gonzàlez, E. Saksman, H.-O. Tylli, Representing non-weakly compact operators. Stud. Math. **113**, 265–282 (1995)
134. W.T. Gowers, A solution to the Schroeder-Bernstein problem for Banach spaces. Bull. Lond. Math. Soc. **28**, 297–304 (1996)
135. W.T. Gowers, B. Maurey, The unconditional basic sequence problem. J. Am. Math. Soc. **6**, 851–874 (1993)
136. S. Grabiner, Ascent, descent, and compact perturbations. Proc. Am. Math. Soc. **71**, 79–80 (1978)
137. B. Gramsch, D. Lay, Spectral mapping theorems for essential spectra. Math. Ann. **192**, 17–32 (1971)
138. W. Greenberg, C. Van der Mee, V. Protopopescu, *Boundary Value Problems in Abstract Kinetic Theory* (Birkhäuser, Basel, 1987)
139. G. Greiner, Spectral properties and asymptotic behavior of the linear transport equation. Math. Z. **185**, 167–177 (1984)
140. P. Grisvard, *Elliptic Problems in Nonsmooth Domains*. Monographs and Studies in Mathematics, vol. 24 (Pitman, Boston, 1985)
141. J.J. Grobler, A note on the theorems of Jentzsch-Perron and Frobenius. Indagationes Math. **49**, 381–391 (1987)
142. J.J. Grobler, Spectral theory on Banach lattice, in *Operator Theory in Function Spaces and Banach Lattice*. Oper. Theory Adv. Appl., vol. 75 (Birkhäuser, Basel, 1995), pp. 133–172
143. A. Grothendieck, Sur les applications linéaires faiblement compactes d'espaces du type $C(K)$. Can. J. Math. **5**, 129–173 (1953)
144. K. Gustafson, J. Weidmann, On the essential spectrum. J. Math. Anal. Appl. **25**, 121–127 (1969)
145. P.R. Halmos, V.S. Sunder, Bounded integral operator on L^2 spaces, in *Ergebnisses der Mathematic und ihrer Grenzgebiete*. Results in Mathematics and Related Areas, vol. 96 (Springer, Berlin, 1978)
146. Y.M. Han, S.H. Lee, W.Y. Lee, On the structure of polynomially compact opertors. Math. Z. **232**, 257–263 (1999)
147. V. Hardt, Uber ein im eigenwertparameter rationales randeigenwertproblem bei differentialgleichungssystemen zweiter ordnung, Dissertation, Regensburg, 1992
148. A. Harrabi, Pseudospectrum of a sequence of bounded operators. RAIRO Modél. Math. Anal. Numér. **32**(6), 671–680 (1998)

149. H. Heesch, Untersuchumgen zum weirfarbenproblem Mannhein. Bibliographiscles Institut (1969)
150. H.J.A.M. Heijmans, Structured populations, linear semigroups and positivity. Math. Z. **191**, 599–617 (1986)
151. H. Henriquez, Cosine operator families such that $C(t) - I$ is compact for all $t > 0$. Indian J. Pure Appl. Math. **16**, 143–152 (1985)
152. E. Hille, R.S. Phillips, *Functional Analysis and Semigroups*, vol. 31 (American Mathematical Society Colloquium Publications, Rhode Island, 1957)
153. P.D. Hislop, I.M. Segal, *Introduction to STheory with Applications to Schrodinger Operators* (Springer, New York, 1966)
154. S. Huillier, Mémoire sur la polyédrométrie. Annales de Mathématiques **3**, 169–189 (1861)
155. V.I. Istratescu, Some remarks on a class of semigroups of operators. Z. Wahrscheinlichkeitstheorie und Verw. Gebiete **26**, 241–243 (1973)
156. V.I. Istrateescu, *Introduction to Linear Operator Theory* (Mareel Dekker, New York, 1981)
157. O. Jedidi, Spectral theory of C^0-semigroups and stability of some essential spectra of linear relations on Banach spaces, Thesis, University of Sfax, 2013
158. A. Jeribi, Quelques remarques sur les opérateurs de Fredholm et application à l'équation de transport. C. R. Acad. Sci. Paris Sér. I **325**, 43–48 (1997)
159. A. Jeribi, Quelques remarques sur le spectre de Weyl et applications. C. R. Acad. Sci. Paris Sér. I **327**, 485–490 (1998)
160. A. Jeribi, Développement de certaines propriétés fines de la théorie spectrale et applications à des modèles monocinétiques et à des modèles de Reggeons, Thesis of Mathematics, University of Corsica, Frensh, 16 Janvier 1998
161. A. Jeribi, Une nouvelle caractérisation du spectre essentiel et application. C. R. Acad. Sci. Paris Sér. I **331**, 525–530 (2000)
162. A. Jeribi, A characterization of the essential spectrum and applications. Boll. dell. Unio. Mate. Ital. **8 B-5**, 805–825 (2002)
163. A. Jeribi, A characterization of the Schechter essential spectrum on Banach spaces and applications. J. Math. Anal. Appl. **271**, 343–358 (2002)
164. A. Jeribi, Some remarks on the Schechter essential spectrum and applications to transport equations. J. Math. Anal. Appl. **275**, 222–237 (2002)
165. A. Jeribi, On the Schechter essential spectrum on Banach spaces and applications. Ser. Math. Inf. **17**, 35–55 (2002)
166. A. Jeribi, Time asymptotic behavior for unbounded linear operator arising in growing cell populations. Nonlinear Anal. Real World Appl. **4**, 667–688 (2003)
167. A. Jeribi, Fredholm operators and essential spectra. Arch. Inequal. Appl. **2**(2–3), 123–140 (2004)
168. A. Jeribi, K. Latrach, Quelques remarques sur le spectre essentiel et application à l'équation de transport. C. R. Acad. Sci. Paris Sér. I **323**, 469–474 (1996)
169. A. Jeribi, K. Latrach, H. Megdiche, Time asymptotic behavior of the solution to a Cauchy problem governed by a transport operator. J. Integral Equ. Appl. **17**(2), 121–139 (2005)
170. A. Jeribi, M. Mnif, Fredholm operators, essential spectra and application to transport equation. Acta Appl. Math. **89**, 155–176 (2005)
171. A. Jeribi, N. Moalla, Fredholm operators and Riesz theory for polynomially compact operators. Acta Appl. Math. **90**(3), 227–245 (2006)
172. A. Jeribi, N. Moalla, A characterization of some subsets of Schechter's essential spectrum and application to singular transport equation. J. Math. Anal. Appl. **358**, 434–444 (2009)
173. A. Jeribi, I. Walha, Gustafson, Weidmann, Kato, Wolf, Schechter and Browder essential spectra of some matrix operator and application to two-group transport equation. Math. Nachr. **284**(1), 67–86 (2011)
174. A. Jeribi, H. Megdiche, N. Moalla, On a transport operator arising in growing cell populations II. Cauchy problem. Math. Methods Appl. Sci. **28**, 127–145 (2005)
175. A. Jeribi, N. Moalla, I. Walha, Spectra of some block operator matrices and application to transport operators. J. Math. Anal. Appl. **351**(1), 315–325 (2009)

176. A. Jeribi, N. Moalla, S. Yengui, S-essential spectra and application to an example of transport operators. Math. Methods Appl. Sci. **37**(16), 2341–2353 (2014)
177. A. Jeribi, S.A. Ould Ahmed Mahmoud, R. Sfaxi, Time asymptotic behavior for a one-velocity transport operator with Maxwell boundary condition. Acta Appl. Math. **3**, 163–179 (2007)
178. Wu. Jianhong, Theory and applications of partial functional equations. Appl. Math. Sci. **119** (1996)
179. K. Jörgens, An asymptotic expansion in the theory of neutron transport. Commun. Pure Appl. Math. **11**, 219–242 (1958)
180. K. Jörgens, *Linear Integral Operators* (Pitman Advenced Publishing Program, London, 1982)
181. P.E.T. Jorgensen, Essential self-adjointness of the graph-Laplacian. J. Math. Phys. **49**, 073510, 33 pp. (2008)
182. S.J. Joseph, chemistry and algebra. Nature **17**, 284 (1878). doi: 10.1038-017284a0
183. M.A. Kaashoek, D.C. Lay, Ascent, descent and commuting perturbations. Trans. Am. Math. Soc. **169**, 35–47 (1972)
184. H.G. Kaper, C.G. Lekkerkerker, J. Hejtmanek, *Spectral Methods in Linear Transport Theory* (Birkhauser, Basel, 1982)
185. T. Kato, Perturbation theory for nullity, deficiency and other quantities of linear operators. J. Anal. Math. **6**, 261–322 (1958)
186. T. Kato, *Perturbation Theory for Linear Oerators* (Springer, New York, 1966)
187. A. Kechris, *Classical Descriptive Set Theory* (Springer, New York, 1995)
188. M. Keller, D. Lenz, Unbounded Laplacians on graphs, Basis spectral properties and the heat equation. Math. Model. Nat. Phenom. **5**(2), 27 (2009)
189. G. Kirchhoff, Graph theory and crystal physics, in *Graph Theory and Theoretical Physics*, ed. by F. Harary, Chap. 1 (Academic Press, London, 1967), pp. 44–110
190. D. König, *Theorie der Endichen and Unendlichen Graph: Kombinatorishe Topologie der Streckenkomplexe* (Akad, Leipzig, 1936)
191. H. Konig, *Eigenvalue Distribution of Compact Operators* (Birkauser, Basel, 1986)
192. V. Kordula, V. Müller, The distance from the Apostol spectrum. Proc. Am. Math. Soc. **124**, 3055–3061 (1996)
193. M.A. Krasnoselskii, *Positive Solutions of Operator Equations* (Noordhoff, Groningen, 1964)
194. M.A. Krasnosel'skii, et al., *Integral Operators in Space of Summabie Functions* (Noordhoff, Leyden, 1976)
195. M.G. Kreĭn, M.A. Krasnoselskiĭ, Fundamental theorems on the extension of Hermitian operators and certain of their applications to the theory of orthogonal polynomials and the problem of moments. Uspehi Matem. Nauk. **2**(3(19)), 60–106 (1947)
196. M.G. Krein, M.A Krasnoselskii, Stability of index of an unbounded operator. Mat. Sb. (N.S.) **30**(92), 219–224 (1952) (Russian)
197. R. Kress, *Linear Integral Equations*. Applied Mathematical Sciences, vol. 82 (Springer, New York, 1989)
198. B. Krichen, Spectral properties, fixed point theory of block operator matrices and applications to transport equations, Thesis, University of Sfax, 2011
199. B. Krichen, Relative essential spectra involving relative demicompact unbounded linear operators. Acta Math. Sci. **34**(2), 546–556 (2014)
200. A. Kryczka, S. Prus, Measures of weak noncompactness under complex interpolation. Stud. Math. **147**, 89–102 (2000)
201. A. Kryczka, S. Prus, M. Szczepanik, Measures of weak noncompactness and real interpolation of operators. Bull. Aust. Math. Soc. **62**, 389–401 (2000)
202. P. Kuchment, Quantum graphs, an introduction and a brief survey, 'Analysis on graphs and its applications', in *Proc. Symp. Pure Math.* (American Mathematical Society, Providence, 2008), pp. 291–314
203. K. Kuratowski, Sur les espaces complets. Fund. Math. **15**, 301–309 (1930)
204. K. Kuratowski, *Topology* (Hafner, New York, 1966)
205. J-P. Labrousse, Les opérateurs quasi-Fredholm une généralisation des opérateurs semi-Fredholm. Rend. Circ. Math. Palermo **29**(2), 161–258 (1980)

206. J-P. Labrousse, Inverses généralisés d'opérateurs non bornés. Proc. Am. Math. Soc. **115**(1), 125–129 (1992)
207. V. Lakshmikantham, S. Leela, *Nonlinear Differential Equations in Abstract Spaces* (Pergamon Press, Oxford, 1981)
208. H.J. Landau, On Szegö's eingenvalue distribution theorem and non-Hermitian kernels. J. Anal. Math. **28**, 335–357 (1975)
209. E.W. Larsen, P.F. Zweifel, On the spectrum of the linear transport operator. J. Math. Phys. **15**, 1987–1997 (1974)
210. K. Latrach, Théorie spectrale d'équations cinétiques, Thèse, Université de Franche-Comte, 1992
211. K. Latrach, Compactness properties for linear transport operator with abstract boundary conditions in slab geometry. Trans. Theor. Stat. Phys. **22**, 39–64 (1993)
212. K. Latrach, Some remarks on the essential spectrum of transport operators with abstract boundary conditions. J. Math. Phys. **35**(11), 6199–6212 (1994)
213. K. Latrach, Time asymptotic behavior for linear mono-energetic transport equations with abstract boundary conditions in slab geometry. Trans. Theor. Stat. Phys. **23**, 633–670 (1994)
214. K. Latrach, Essential spectra on spaces with the Dunford-Pettis property. J. Math. Anal. Appl. **223**, 607–622 (1999)
215. K. Latrach, *Compactness properties for perturbed semigroups and application to transport equation*, preprint (2004)
216. K. Latrach, A. Dehici, Relatively strictly singular perturbations, essential spectra and application to transport operators. J. Math. Anal. Appl. **252**, 767–789 (2000)
217. K. Latrach, A. Dehici, Fredholm, semi-Fredholm perturbations and essential spectra. J. Math. Anal. Appl. **259**, 277–301 (2001)
218. K. Latrach, A. Dehici, Remarks on embeddable semigroups in groups and a generalization of some Cuthbert's results. Int. J. Math. Math. Sci. **22**, 1421–1431 (2003)
219. K. Latrach, A. Jeribi, On the essential spectrum of transport operators on L_1-spaces. J. Math. Phys. **37**(12), 6486–6494 (1996)
220. K. Latrach, A. Jeribi, Sur une équation de transport intervenant en dynamique des populations. C. R. Acad. Sci. Paris Sér. I **325**, 1087–1090 (1997)
221. K. Latrach, A. Jeribi, Some results on Fredholm operators, essential spectra, and application. J. Math. Anal. Appl. **225**, 461–485 (1998)
222. K. Latrach, B. Lods, Regularity and time asymptotic behavior of solutions to transport equations. Trans. Theor Stat. Phys. **30**, 617–639 (2001)
223. K. Latrach, H. Megdiche, A. Jeribi, Time asymptotic behavior of the solution to a Cauchy problem governed by a transport operator. J. Intergr. Equ. Appl. **17**(2), 121–140 (2005)
224. K. Latrach, J.M. Paoli, Relatively compact-like perturbations, essentilal spectra and applications. J. Aust. Math. Soc. **77**(1), 73–89 (2004)
225. K. Latrach, J.M. Paoli, Polynomially compact-like strongly continuous semigroups. Acta Appl. Math. **82**, 87–99 (2004)
226. K. Latrach, J.M. Paoli, An extension of a Phillips's theorem to Banach algebras and application to the uniform continuity of strongly continuous semigroups. J. Math. Anal. Appl. **326**, 945–959 (2007)
227. K. Latrach, J.M. Paoli, P. Simonnet, Some facts from descriptive set theory concerning essential spectra and applications. Stud. Math. **171**, 207–225 (2005)
228. K. Latrach, J.M. Paoli, P. Simonnet, A spectral characterization of the uniform continuity of strongly continuous groups. Arch. Math. **90**, 420–428 (2008)
229. K. Latrach, J.M. Paoli, M.A. Taoudi, A characterization of polynomially Riesz strongly continuous semigroups. Comment. Math. Univ. Carol. **47**(2), 275–289 (2006)
230. A. Lebow, M. Schechter, Semigroups of operators and measures of noncompactness. J. Funct. Anal. **7**, 1–26 (1971)
231. J. Lehner, M. Wing, On the spectrum of an unsymetric operator arisingin the transport theory of neutrons. Commun. Pure Appl. Math. **8**, 217–234 (1955)

232. J. Lehner, M. Wing, Solution of the linearized Boltzmann transport equation for the slab geometry. Duke Math. **23**, 125–142 (1956)
233. A.E. Lifschitz, *Magnetohydrodynamics and Spectral Theory* (Springer, Dordrecht, 1989)
234. J.-L. Lions, E. Magenes, *Non-Homogeneous Boundary Value Problems and Applications*, vol. 1 (Springer, Berlin-Heidelberg, 1972)
235. C. Lizama, Uniform continuity and compactness for resolvent families of operators. Acta Appl. Math. **38**, 131–138 (1995)
236. B. Lods, On linear kinetic equations involving unbounded cross-sections. Math. Models Methods Appl. Sci. **27**, 1049–1075 (2004)
237. H.P. Lotz, Über das Spektrum positiver Operatoren. Math. Z. **108**, 15–32 (1968)
238. A. Lunardi, *Analytic Semigroups and Optimal Regularity in Pàarabolic Problems* (Birkhauser, Basel, 1995)
239. J. Lutgen, On essential spectra of operator-matrices and their Feshbach maps. J. Math. Anal. Appl. **289**, 419–430 (2004)
240. D. Lutz, Compactness properties of operator cosine functions. C. R. Math. Rep. Acad. Sci. Can. **2**, 277–280 (1980)
241. I. Marek, Frobenius theory of positive operators : Comparison theorems and applications. SIAM J. Appl. Math. **19**, 607–628 (1970)
242. I. Marek, Fundamental decay and asymptotic behavior of positive semig roups. Czechoslov. Math. J. **30**(105), 579–590 (1980)
243. A.S. Markus, *Introduction to the Spectral Theory of Polynomial Operator Pencils* (American Mathematical Society, Providence, 1988)
244. J.E. Marsden, *Basic Complex Analysis* (W. H. Freeman and Campany, San Francisco, 1973)
245. A. Mashaghi, Inverstigation of a protein complex network. Eur. Phys. **41**(1), 113–121 (2004)
246. M. Mbekhta, A. Ouahab, Opérateur s-régulier dans un espace de Banach et théorie spectrale. Acta Sci. Math. **59**, 525–543 (1994)
247. V. Menon, On repeated interchange graphs. Am. Math. Mon. **13**, 986–989 (1966)
248. R. Mennicken, S. Naboko, C. Tretter, Essential spectrum of a system singular differential operators and the asymptotic Hain-Lüst operator. Am. Math. Soc. **130**, 1699–1710, (2001)
249. P. Meyer-Nieberg, *Banach Lattices* (Springer, New York, 1991)
250. O. Milatovic, Essential self-adjointness of magnetic Schrödinger operators on locally finite graph. Integr. Equ. Oper. Theory **71**, 13–27 (2011)
251. O. Milatovic, A sears-type self-adjointness result for discrete magnetic Schrödinger operators. J. Math. Anal. Appl. **369**, 801–809 (2012)
252. V.D. Milman, Some properties of strictly singular operators. Funct. Anal. Appl. **3**, 77–78 (1969)
253. M.M. Milovanović-Arandjelović, Measures of noncompactness on uniform spaces- the axiomatic approach, in *IMC "Filomat 2001"*, Niš (2001), pp. 221–225
254. N. Moalla, Developpement de certaines propriétés fines de la théorie spectrale et applications à l'équation de transport, Thesis, University of Sfax, 2006
255. N. Moalla, A characterization of Schechter's essential spectra by mean of measure of non-strict-singularity and application to matrix operator. Acta Math. Sci. Ser. B Engl. Ed. **32**(6), 2329–2340 (2012)
256. N. Moalla, M. Damak, A. Jeribi, Essential spectra of some matrix operators and application to two-group transport operators with general boundary conditions. J. Math. Anal. Appl. **323**(2), 1071–1090 (2006)
257. B. Mohar, W. Woess, A survey on spectra of infinite graphs. J. Bull. Lond. Math. Soc. **21**(3), 209–234 (1989)
258. M. Mokhtar-Kharroubi, Propriétés spéctrales de l'opérateur de transport dans le cas anisotrope, Thèse de Doctorat de 3ème cycle, Université Paris 6, 1983
259. M. Mokhtar-Kharroubi, Quelques applications de la positivité en théorie du transport. Ann. Fac. Sci. Toulouse. **11**, 75–99 (1990)
260. M. Mokhtar-Kharroubi, Compactness results for positive semigroups on Banach Lattices and applications. Houston J. Math. **17**(1), 25–38 (1991)

261. M. Mokhtar-Kharroubi, Time asymptotic bahaviour and compactness in Neutron Transport Theory. Eur. J. Mech. B Fluids **11**(1), 39–68 (1992)
262. M. Mokhtar-Kharroubi, *Mathematical Topics in Neutron Transport Theory. New Aspects*, vol. 46 (World Scientific, Singapore, 1997)
263. V. Müller, On the regular spectrum. J. Oper. Theory **31**, 363–80 (1994)
264. V. Müller, Spectral theory of linear operators and spectral system in Banach algebras. Oper. Theor. Adv. Appl. *139* (2003)
265. R. Nagel, Towards a "matrix theory" for unbounded operator matrices. Math. Z. **201**(1), 57–68 (1989)
266. R. Nagel, The spectrum of unbounded operator matrices with non-diagonal domain. J. Funct. Anal. **89**(2), 291–302 (1990)
267. B.Sz. Nagy, On the stability of the index of unbounded linear transformations. Acta. Math. Acad. Sic. Hungar. **3**, 49–52 (1952)
268. M.A. Naimark, *Linear Differential Operators* (Frederick Ungar, New York, 1987)
269. L.I. Nicolaescu, On the space of Fredholm operators. An. Stiint. Univ. Al. I. Cuza Iasi. Mat. (N.S.) **53**(2), 209–227 (2007)
270. J.I. Nieto, On Fredholm operators and the essential spectrum of singular integral operators. Math. Ann. **178**, 62–77 (1968)
271. R.D. Nussbaum, Positive operators and elliptic eigenvalue problems. Math. Z. **186**, 247–264 (1984)
272. W. Obershelp, *Theory of Graphs*, vol. 38 (American Mathematical Society Colloquium Publications, Providence, 1963)
273. Z. Opial, *Nonexpansive and Monotone Mappings in Banach Spaces* (Center for Dynamical Systems, Brown University, Providence, 1967), pp. 1–67
274. A. Palczewski, Spectral properties of the space nonhomogeneous linearized-Boltzmann operator. 'Ikansp. Theor. Stat. Phys. **13**, 409–430 (1984)
275. C.V. Pao, Asymptotic behavior of the solution for the time-dependent neutron transport problem. J. Integr. Equ. **1**, 31–152 (1979)
276. A. Pazy, *Semigroups of Linear Operators and Applications to Differential Equations*. Applied Mathematical Sciences, vol. 44 (Springer, New York, 1983)
277. A. Pelczynski, On strictly singular and strictly cosingular operators. I. Strictly singular and strictly cosingular operators in $C(X)$-spaces. II. Strictly singular and strictly cosingular operators in $L(\mu)$-spaces. Bull. Acad. Polon. Sci. **13**, 13–36, 37–41 (1965)
278. S. Pemmaraju, S. Skiena, Cycles, stars, and wheels, in *Computational Discrete Mathematics Combinatiorics and Graph Theory in Mathematica*, section 6.4 (Cambridge University Press, Cambridge, 2003), pp. 284–249
279. W.V. Petryshyn, Construction of fixed points of demicompact mappings in Hilbert space. J. Math. Anal. Appl. **14**, 276–284 (1966)
280. W.V. Petryshyn, Remarks on condensing and k-set-contractive mappings. J. Math. Anal. Appl. **39**, 717–741 (1972)
281. R.S. Phillips, Spectral theory for semigroups of linear operators. Trans. Am. Math. Soc. **71**, 393–415 (1951)
282. O. Post, First order approach and index theorems for discrete and metric graph. Am. Heni. Poincaré **10**, 823–866 (2009)
283. F. Rabiger, W.J. Ricker, C_0-groups and C_0-semigroups of linear operators on hereditarily indecomposable Banach spaces. Arch. Math. **66**, 60–70 (1996)
284. V. Rakoćević, On one subset of M. Schechter's essential spectrum. Mat. Vesnik **5**(18)(33)(4), 389–391 (1981)
285. V. Rakočević, Approximate point spectrum and commuting compact perturbation. Glasgow Math. J. **28**, 193–198 (1986)
286. V. Rakocevič, Generalized spectrum and commuting compact perturbations. Proc. Edinb. Math. Soc. **36**, 197–209 (1993)
287. V. Rakočević, Semi-Fredholm operators with finite ascent or descent and perturbations. Am. Math. Soc. **123**(12) (1995)

288. V. Rakočević, Semi-Browder operators and perturbations. Stud. Math. **122**(2), 131–137 (1997)
289. V. Rakočević, Measures of noncompactness and some applications. Filo-Mat. **12**(2), 87–120 (1998)
290. J.S. Raymond, Boréliens à coupes K_σ. Bull. Soc. Math. France **104**, 389–406 (1976)
291. M. Reed, *Linear graphs and Electrical Networks* (Addisson Wesky, Reading, 1961)
292. M. Reed, B. Simon, *Methods of Modern Mathematical Physics, I-IV. Analysis of Operators* (Academic Press, New York, 1978)
293. M. Ribaric, I. Vidav, *Analytic Properties of the Inverse $A(z)^{-1}$ of an Analytic Linear Operator Valued Function $A(z)$*. Arch. Ration. Mech. Anal. **32**(4), 298–310 (1969)
294. F. Riesz, Über lineare funktionalgleichungen. Acta Math. **41**, 71–98 (1918)
295. N. Robertson, D. Sanders, P. Seymour, R. Thomas, The four color theorem. J. Comb. Theory Ser. B **70**, 2–44 (1997)
296. M. Rotenberg, Transport theory for growing cell populations. J. Theor. Biol. **103**, 181–199 (1983)
297. H. Sachs, Graph derivatives. Math. Z. **76**, 385–401 (1961)
298. H.H. Schaefer, Banach lattices and positive operators. Grundlehren Math. Wiss. Bd., vol. 215 (Springer, New York, 1974)
299. M. Schechter, On the essential spectrum of an arbitrary operator. J. Math. Anal. Appl. **13**, 205–215 (1966)
300. M. Schechter, Basic theory of Fredholm operators. Anna. Scuola Norm. Sup. Pisa **21**(3), 261–280 (1967)
301. M. Schechter, *Spectra of Partial Differential Operators* (North-Holland, Amsterdam, 1971)
302. M. Schechter, *Principles of Functional Analysis*. Graduate Studies in Mathematics, vol. 36 (American Mathematical Society, Providence, 2002)
303. C. Schmoeger, Perturbation properties of some class of operators. Rend. Math. Appl. **7**, 533–541 (1994)
304. C. Schmoeger, The spectral mapping theorem for the essential approximate point spectrum. Colloq. Math. **74**(2), 167–176 (1997)
305. I. Schur, Bemerkungen Zur theorie der Beschrankten Bilinear formen mit unendhich vielen Veranderhichen, J. Reine Angew. Math. **140** 1–28 (1911)
306. G.P. Shannoa, Strictly singular and cosingular operators and topological vector spaces. Proc. R. lr. Acad. Sect. A **73**, 303–308 (1973)
307. J. Shapiro, M. Schechter, A generalized operational calculus developed from Frdholm operator theory. Trans. Am. Math. Soc. **175**, 439–667 (1973)
308. J. Shapiro, M. Snow, The Fredholm spectrum of the sum and product of two operators. Trans. Am. Math. Soc. **191**, 387–393 (1974)
309. A.A. Shkalikov, On the essential spectrum of some matrix operators. Math. Notes **58**(5–6), 1359–1362 (1995)
310. A.A. Shkalikov, C. Tretter, Spectral analysis for linear pencils $N - \lambda P$ of ordinary differential operators. Math. Nachr. **179**, 275–305 (1996)
311. Yu.L. Smul'Yan, Completely continuous perturbation of operators. Dokl. Akad. Nauk SSSR (N.S.) **101**, 35–38 (1955) (Russian)
312. D. Song, Some notes on the spectral properties of C_0-semigroups generated by linear transport operators. Trans. Theor. Stat. Phys. **26**, 233–242 (1997)
313. D. Song, On the spectrum of neutron transport equations with reflecting boundary conditions, PhD Thesis, Blacksburg, 2000
314. S. Steinberg, Meromorphic families of compact operators. Arch. Rational Mech. Anal. **31**, 372–379 (1968)
315. K. Taira, A. Favini, S. Romanelli, Feller semigroups and degenerate elliptic operators with Wentzell boundary conditions. Stud. Math. **145**, 17–53 (2001)
316. P. Takac, A spectral mapping theorem for the exponential function in linear transport theory. Trans. Theor. Stat. Phys. **14**, 655–667 (1985)
317. W.T. Tatte, *Graph Theory* (Cambridge university Press, Cambridge, 2001), p. 30

318. A.E. Taylor, Spectral theory of closed distributive operators. Acta Math. **84**, 189–224, MR 12, 717 (1951)
319. A.E. Taylor, Theorems on ascent, descent, nullity, and defect of linear operators. Math. Ann. **163**, 18–49 (1966)
320. M. Taylor, *Partial Differential Equations*. Basic Theory, vol. 1 (Springer, New York, 1996)
321. T. Toka, *Perturbations of Unbounded Fredholm Linear Operators in Banach Spaces,* Handbook on Operator Theory (Springer, 2015)
322. L.N. Trefethen, Pseudo-spectra of matrices, in *Numer. Anal. 1991* (Longman Scientific & Technical, Harlow, 1992), pp. 234–266
323. C. Tretter, Spectral issues for block operator matrices, in *Differential Equations and Mathematical Physics*, Birmingham, 1999; *AMS/IP Studies in Advanced Mathematics*, vol. 16 (American Mathematical Society, Providence, 2000), pp. 407–423
324. C. Tretter, *Spectral Theory of Block Operator Matrices and Applications* (Impe. Coll. Press, London, 2008)
325. C. Tretter, Spectral inclusion for unbounded block operator matrices. J. Funct. Anal. **11**, 3806–3829 (2009)
326. R. Van Norton, On the real spectrum of a monoenergetic neutron transport operator. Commun. Pure Appl. Math. **15**, 149–158 (1962)
327. J.M. Varah, *The Computation of Bounds for the Invariant Subspaces of a General Matrix Operator*, Stan. Univ. Comp. Sci., Dept. Tech. Report (1967)
328. I. Vidav, Existence and uniqueness of nonnegative eigenfunction of the Boltzmann operator. J. Math. Anal. Appl. **22**, 144–155 (1968)
329. I. Vidav, Spectra of perturbed semigroups with applications to transport - theory. J. Math. Anal. Appl. **30**, 264–279 (1970)
330. Ju.I. Vladimirskii, Stricty cosingular operators. Sov. Math. Dokl. **8**, 739–740 (1967)
331. J. Voigt, A perturbation theorem for the essential spectral radius of strongly continuous semigroups. Mh. Math. **90**, 153–161 (1980)
332. J. Voigt, *Functional Analytic Treatment of the Initial Boundary Value Problem for Collisionlgs Gases* (Habilitationsschrift, Munchen, 1981)
333. J. Voigt, Spectral properties of the neutron transport equation. J. Math. Anal. Appl. **106**, 140–153 (1985)
334. J. Voigt, On resolvent positive operators and positive C_0-semigroup on AL-spaces. Semigroup Forum **38**, 263–266 (1989)
335. I. Walha, Essential spectra of some operator matrices, Riesz basis and applications, Thesis, University of Sfax, 2010
336. G.F. Webb, *Theory of Nonlinear Age-Dependent Population Dynamics* (Marcel Dekker, New York, 1985)
337. A. Weber, Analysis of the physical Laplacian and the heat flow on a locally finite graphs. J. Math. Anal. **370**, 146–158 (2010)
338. L. Weis, Perturbation class of semi-Fredholm operators. Math. Z. **178**, 429–442 (1981)
339. L.W. Weis, A generalization of the Vidav-Jorgens perturbation theorem for semigroups and its application to transport theory. J. Math. Anal. Appl. **129**, 6–23 (1988)
340. T.T. West, Riesz operators in Banach spaces. Proc. Lond. Math. Soc. **16**, 131–140 (1966)
341. T.T. West, A Riesz-Schauder theorem for semi-Fredholm operators. Proc. R. Ir. Acad. Sect. A **87**, 137–146 (1987)
342. H. Weyl, Uber beschrankte quadratiche Formen, deren Differenz vollsteig ist. Rend. Circ. Mat. Palermo **27**, 373–392 (1909)
343. R.J. Whitley, Strictly singular operators and their congugates. Trans. Am. Math. Soc. **18**, 252–261 (1964)
344. M. Wing, *An Introduction to Transport Theory* (Wiley, New York, 1962)
345. R. Wojciechowski, Stochastic compactetness of graph, Ph.D. thesis, City University of New York, 72 pp., 2007
346. R. Wojciechowski, Stochatically incomplete manifolds and graphs. Progr. Probab. **64**, 163–179 (2011)

347. F. Wolf, On the invariance of the essential spectrum under a change of the boundary conditions of partial differential operators. Indag. Math. **21**, 142–147 (1959)
348. F. Wolf, On the essential spectrum of partial differential boundary problems. Commun. Pure Appl. Math. **12**, 211–228 (1959)
349. M.P.H. Wolff, Discrete approximation of unbounded operators and approximation of their spectra. J. Approx. Theory **113**, 229–244 (2001)
350. Z. Xianwen, Spectral properties of a streaming operator with diffuse reflection boundary condition. J. Math. Anal. Appl. **238**, 20–43 (1999)
351. M. Yahdi, Théorie descriptive des ensembles en géométrie des espaces de Banach, exemples, Thése de Doctorat de Mathématiques, Université Paris 6, 1998
352. S. Yengui, S-spectres essentiels, theorie de perturbation et applications à l'équation de transport, Thesis, University of Sfax, 2012
353. B. Yood, Properties of linear transformations preserved under addition of a completely continuous transformation. Duke Math. J. **18**, 599–612 (1951)
354. K. Yosida, *Functional Analysis* (Springer, Heidelberg, 1978)
355. A.C. Zaanen, *Riesz Spaces II* (North Holland, Amsterdam, 1983)
356. M. Zerner, Quelques propriétés spectrales des opérateurs positifs. J. Funct. Anal. **72**, 381–417 (1987)
357. X. Zhang, B. Liang, On the spectum of a one-velocity transport operator with Maxwell boundary condition. J. Math. Anal. Appl. **202**, 920–936 (1996)
358. S. Živković, Semi-Fredholm operators and perturbation. Publ. Inst. Math. Beo. **61**, 73–89 (1997)
359. S. Živković-Zlatanoviȉc, D.S. Djordjević, R.E. Harte, On left and right Browder operators. J. Kor. Math. Soc. **485**, 1053–1063 (2011)

Index

A. Jeribi, *Spectral Theory and Applications of Linear Operators and Block Operator Matrices*, DOI 10.1007/978-3-319-17566-9

N

O

P

Q

R

S

T

U

MIX
Papier aus verantwortungsvollen Quellen
Paper from responsible sources
FSC® C105338

If you have any concerns about our products,
you can contact us on
ProductSafety@springernature.com

In case Publisher is established outside the EU,
the EU authorized representative is:
Springer Nature Customer Service Center GmbH
Europaplatz 3, 69115 Heidelberg, Germany

Printed by Libri Plureos GmbH
in Hamburg, Germany